THE MATHEMATICAL PAPERS OF ISAAC NEWTON

VOLUME VI

1684–1691

postquam corpus pervenit ad centrum S. id non amplius redibit in orbem sed abibit in tangente. In spirali quæ secat radios omnes in dato angulo vis centripeta tendens ad spiralis principium est in ratione triplicata distantiæ reciprocè, sed in principio illo recta nulla positione determinata spiralem tangit.

57

Prob 2. Gyrat corpus in Ellipsi veterum: requiritur lex ~~[illegible]~~ vis centripetæ tendentis ad centrum Ellipseos.

Sunto CA, CB semi-axes Ellipseos. GP, DK diametri conjugatæ, PF, Qt perpendicula ad diametros, QV ordinatim applicata ad diametrum GP et QVPR parallelogrammum. His constructis erit (ex Conicis) PVG ad QV^q ut PC^q ad CD^q et QV^q ad Qt^q ut PC^q ad PF^q et conjunctis rationibus PVG ad Qt^q ut PC^q ad CD^q et PC^q ad PF^q, id est VG ad $\frac{Qt^q}{PV}$ ut PC^q ad $\frac{CD^q \times PF^q}{PC^q}$. Scribe QR pro PV et BC×CA pro CD×PF, nec non (punctis P et Q coeuntibus) 2PC pro VG et ductis extremis et medijs in se mutuo, fiet $\frac{Qt^q \times PC^q}{QR} = \frac{2BC^q \times CA^q}{PC}$. Est ergo vis centripeta reciprocè ut $\frac{2BC^q \times CA^q}{PC}$ id est (ob datum 2BC^q × CA^q) ut $\frac{1}{PC}$, hoc est directè, ut distantia PC. Q.E.I.

~~[illegible]~~

a Per Lem 4

Prob. 3. Gyrat corpus in ellipsi: requiritur lex vis centripetæ tendentis ad umbilicum Ellipseos.

Esto Ellipseos superioris umbilicus S. Agatur SP secans Ellipseos diametrum DK in E. Patet EP æqualem esse semiaxi majori AC eo, quod acta ab altero Ellipseos umbilico H linea HI ipsi EC parallela, ob æquales CS, CH æquentur ES, EI, adeo ut EP semisumma sit ipsarum PS, PI id est (ob parallelas HI, PR et angulos æquales IPR, HPZ) ipsarum PS, PH quæ conjunctim axem totum 2AC adæquant. Ad SP demittatur perpendicularis QT. Et Ellipseos latere recto principali (seu $\frac{2BC^q}{AC}$) dicto L, erit L×QR ad L×PV ut QR ad PV id est ut PE (seu AC) ad PC et L×PV ad GVP ut L ad GV et GVP ad QV^q ut CP^q ad CD^q, et QV^q ad QX^q posita ut M ad N et QX^q ad QT^q ut EP^q ad PF^q id est ut CA^q ad PF^q sive ut CD^q ad CB^q. et conjunctis his omnibus rationibus L×QR ad QT^q ut AC ad PC + L ad GV + CP^q ad CD^q + M ad N + CD^q ad CB^q, id est ut AC×L (seu 2BC^q) ad PC×GV + CP^q ad CB^q + M ad N, sive ut 2PC ad GV + M ad N. Sed punctis Q et P coeuntibus rationes 2PC ad GV et M ad N fiunt æqualitatis. Ergo et ex his composita ratio L×QR ad QT^q. Ducatur pars utraque in $\frac{SP^q}{QR}$ et fiet $L \times SP^q = \frac{SP^q \times QT^q}{QR}$. Ergo vis centripeta reciprocè est ut L×SP^q id est in ratione duplicata distantiæ SP. Q.E.I.

a Per Lem. 4.

~~Corol. [illegible]~~

Schol. Gyrant ergo Planetæ majores in ellipsibus habentibus umbilicum in centro Solis, et radijs ad solem ductis describunt areas temporibus proportionales, omnino ut supposuit Keplerus. Et harum Ellipseon latera recta sunt $\frac{Qt^q}{QR}$, ~~[illegible]~~ punctis P et Q spatio quam minimo et quasi infinite parvo distantibus.

Theorem 4

Motion in an ellipse under a diverting force to its centre or to a focus (I, §1).

THE MATHEMATICAL PAPERS OF ISAAC NEWTON

VOLUME VI

1684-1691

EDITED BY

D. T. WHITESIDE

WITH THE ASSISTANCE IN PUBLICATION OF

M. A. HOSKIN AND A. PRAG

CAMBRIDGE
AT THE UNIVERSITY PRESS
1974

CAMBRIDGE UNIVERSITY PRESS
Cambridge, New York, Melbourne, Madrid, Cape Town, Singapore, São Paulo

Cambridge University Press
The Edinburgh Building, Cambridge CB2 8RU, UK

Published in the United States of America by Cambridge University Press, New York

www.cambridge.org
Information on this title: www.cambridge.org/9780521087193

First published 1974
This digitally printed version 2008

A catalogue record for this publication is available from the British Library

Library of Congress Catalogue Card Number: 65–11203

ISBN 978-0-521-08719-3 hardback
ISBN 978-0-521-04585-8 paperback
ISBN 978-0-521-72054-0 paperback set (8 volumes)

TO I. BERNARD COHEN
OUR LEARNED GUIDE TO THE BY-WAYS
OF THE 'PRINCIPIA'

PREFACE

Even though Newton himself, half a century afterwards, preferred to look to the two 'plague years' of 1665 and 1666 as the 'prime' of his 'age for invention' when he 'minded Mathematicks & Philosophy more then at any time since',* the dozen and a half months from August 1684, when Edmond Halley first travelled to Cambridge to seek his opinion on the currently vexing question of how dynamically to determine the closed orbits of the planets round the sun, have (so it seems to us) an overriding claim to be regarded as the most deeply fruitful *annus mirabilis* of Newton's life. Over that year and a half he was encouraged to expand his immediate confident answer to Halley, that the planetary ovals are exact ellipses having the sun at a focus, into a developed treatise 'On the motion of bodies' and thereafter further to augment its scope to be the magisterial *Philosophiæ Naturalis Principia Mathematica* which made its public bow in 1687. While it manifestly transcends the purpose of the present edition systematically to review the full content of this *chef d'œuvre* of classical theoretical physics, it would be equally intolerable for us here to ignore this supreme exhibition of Newton's scientific genius. It is, of course, a truism that his expert ability to give appropriate and effective formal expression whether geometrical or (more rarely) algebraic to the various astronomical and physical problems which he therein came to confront, and the power and adequacy of his available mathematical techniques exactly or approximately to resolve them once thus precisely defined, were prime limiting factors on the success with which he was able to fulfil his central aim of reducing individual empirical phenomena to be deducible instances of a handful of guiding kinematical and dynamical 'principles of natural philosophy'. In imitation of the phrase taken up by the present Master of Newton's Cambridge college—and not wholly to depart from its context of human frailty—we might with good reason call the *Principia* the art of what was then possible. Merely to affirm as much *tout court* is in no way to effect the detailed enodation, point by point and proposition by proposition, of its arguments by which alone its mathematical substructure may be probed and analysed, and which in selected portions of its content we attempt in the present volume. Since here we run parallel to that crowded avenue of present-day scholarly research which seeks to establish and evaluate

---

* See ULC. Add. 3968.41: 85ʳ; a more connected quotation of this familiar excerpt from a draft letter of Newton's to Pierre Desmaizeaux in about the summer of 1718 has already been made on I: 152. The justice of his not everywhere documentable claims to priority of discovery in mathematics, mechanics and optics is examined more fully in D. T. Whiteside, 'Newton's Marvellous Year: 1666 and all that', *Notes and Records of the Royal Society of London*, **21**, 1966: 32–41.

its verbal niceties and physical connotations, and more narrowly because the variants in its published and publicly circulated versions are now accurately recorded in the 'variorum' edition of its text by Professors Alexandre Koyré and I. B. Cohen (which appeared just too late, unfortunately, to permit detailed reference here to its pages), our own attention is fixed squarely on the extant preliminary manuscripts, beginning with Newton's initial tract 'De motu Corporum' of autumn 1684, and culminating with the preserved portion of the refined 'De motu Corporum Liber primus' which went to make up the *Principia*'s first book and the opening propositions of its second; the printed text itself we employ only minimally as an auxiliary to fill in small gaps in their deductive sequence and, on rare occasion, to complement their content. The opportunity has also been taken to deal more fully with some individual topics—namely, the resistance of surfaces of revolution to uniform translation along their axes, the construction of parabolic cometary paths from terrestrial sightings, and the dynamical derivation of the mean annual advance of lunar apogee—which are given summary or considerably diverse discussion in the published *Principia.* And a concluding section explores a number of provisional revisions of the opening pages of its first book, which Newton had it in mind, in or shortly after 1690, to incorporate in the *edito princeps.* As ever, the reader will judge best how far our severely mathematical commentary serves to penetrate below the formal façade which Newton presents.

For permitting reproduction of the manuscripts in their care I am once more chiefly indebted to the Librarian and Syndics of the University Library, Cambridge, but also, in the case of short individual items, to their equivalents in Trinity College and Kings College, Cambridge, the Bodleian and Corpus Christi College, Oxford, and the Royal Society, London; and in one instance I owe thanks to a private owner whose *carte blanche* to publish documents in his possession we have freely taken advantage of in earlier volumes. Let me also express my appreciation of the courtesy and efficiency of the staff of what, with the demise of the old Anderson Room there, has now divided to be the Manuscript and Rare Book Rooms at the University Library. The financial support without which the several years of preliminary research encapsulated in this volume could never have been possible has continued to be furnished by the Sloan Foundation, the Leverhulme Trust and the Master and Fellows of Trinity College, Cambridge. In this present period of economic stringency their generously renewed subsidy is especially welcome. On a sad note, that grand old man of British science, Sir Harold Hartley, is no longer with us. Let me add one more tribute to his memory by stressing that the quality and strength of the present edition (such as it may be) reflect the professional advice, worldly wisdom and affectionate (if sometimes stern) encouragement which he offered unstintingly over the dozen years before his death.

To my *confrères*, Michael Hoskin and Adolf Prag, a very personal note of thanks for again helping me to guide a complex volume through the press. As before, Dr Hoskin has helped in preparing the text-figures, while the eagle eye of Mr Prag has spotted many a slip in my editorial phrasings, and to him also is due the credit for the terminal index of names.

Lastly, but never least, to the Syndics of Cambridge University Press and to the University Printer must go my collective appreciation of the efforts of all those—editors, draughtsmen, printers and other production staff—who have jointly laboured (amid the exigency and despair ensuing upon a three-day working week) to create out of a clumsily penned submitted manuscript a most elegant example of the typographer's art.

D.T.W.

Easter Day, 1974

EDITORIAL NOTE

A brief reminder that this volume narrowly follows the style and conventions of previous volumes (on which see especially pages x–xiv of the first), and that faithfulness to the spirit of the manuscript originals here reproduced is our chief concern, in regard not merely to their verbal text but also its illustrative figures. Where, however, the text is not autograph but Humphrey Newton's secretarial transcript, we have minimally standardised its accidence and punctuation to Newton's preferred norm; and in reproducing Newton's own sprawling numerical calculations we have taken some small liberties in recasting their spatial arrangement to fit the confines of the printed page and to point their sense. (Comparison of the photo-facsimiles in Plates I and III with our accompanying editorial selection and rearrangement of their detail will, we trust, allay any fear on the reader's part that we have there gone too far in deviating from the strict letter of the original text as it is preserved.) We should add a caution to the unwary reader that many of the figures once appended to, or intended to be set with, the manuscript are now there lacking, and that these are perforce here restored: every such instance is specified in a pertinent footnote. As before, we have been free in smoothing out trivial slips of Newton's pen and in rounding out the sparseness of his original where we have deemed this necessary (usually in the case of groups of computations devoid of verbal interlinking) by appropriate editorial correction and interpolation within square brackets. Within the bounds of fluency and modern idiom we have deliberately kept our English translations (facing the Latin text or, in footnotes, following it in parenthesis) closely literal, designing them to be an auxiliary tool and not an independent voice of Newton's meaning. Once more, for forwards and backwards reference within the present volume we employ the formula typified in '3, §2, Appendix 2. 3' (thereby understanding '[Section] 3, [Subsection] 2, Appendix 2. [Division] 3'); reference to preceding volumes is by the similar code 'III: 244–54' (understand '[Volume] III: [pages] 244–54'). Throughout, two thick vertical rules in the left-hand margin denote that the text alongside has been cancelled by Newton in the original. Finally, a number of non-standard mathematical symbols used by him in the manuscript are explained in footnote at the relevant places and also (where pertinent) keyed to analogous occurrences in previous volumes.

GENERAL INTRODUCTION

In our two previous volumes we took leave of Newton in the summer of 1684, hard at work in Cambridge, on a variety of mathematical—and also, indeed, chemical—projects. Edmond Halley's visit to him in August, we need scarcely repeat, abruptly changed the focus of those researches. The open challenge from Christopher Wren theoretically to define the paths of the solar planets which Halley then communicated to Newton re-kindled in him a devouring flame of interest in the general dynamical motion of bodies which had more than five years before flickered fitfully into life during his correspondence with Robert Hooke and since lain dormant. Before, however, we enter upon the mathematical detail of the several successive and increasingly bulky treatises 'De motu Corporum' which Newton drafted over the next two years and ultimately subsumed into his *Principia* in 1687 we may, as in previous volumes, momentarily turn aside to consider the man himself as he entered the early forties of his life.

While in the privacy of his Trinity rooms he laboured untiringly to compose what was to prove his scientific masterpiece, the outward pattern of Newton's day-to-day life changed but little. Except for two short absences (away in his native Lincolnshire the first time) at Easter and in mid-June of 1685(1) he passed the whole thirty months up to March 1687 in Cambridge, carrying out his minimal professorial duty of lecturing weekly—on the substance of his new findings in celestial mechanics, it would appear(2)—to the few who bothered to

(1) In his list of Newton's exits and redits from and back to Trinity College (*Correspondence of Sir Isaac Newton and Professor Cotes*, London, 1850: lxxxv) Joseph Edleston records absences from Cambridge between 27 March and 11 April and again from 11 to 20 June; on the preceding 23 February Newton had written to Francis Aston that 'now I am to goe into Lincolnshire for a Month or six weeks' (*Correspondence of Isaac Newton*, **2**, 1960: 415). No other departures occur till an exit on 25 March 1687; though no corresponding redit is found, Newton's weekly buttery bills of the period (which Edleston lists on his ensuing p. lxxxvii) establish that, except for a few days in the last week in April, he was not to return to college—from promoting the University's statutory rights before Judge Jeffreys in London, as we shall see—till late May. In 1688 he was briefly away from Cambridge, in Lincolnshire we assume, between 30 March and 25 April and again from 22 June to 17 July. After his election to be one of the University Members of Parliament in January 1689, his buttery bills confirm that he was from then on continuously out of residence, other than for a couple of days in the middle of the following October, till a redit on 4 February 1690 marked his return after the Convention Parliament was prorogued. Further brief absences in 1690 from 10 March to 12 April and from 22 June to 2 July—once more in Lincolnshire, we may guess—preluded a year's unbroken return to academic routine. A similar pattern of long stays in Cambridge, punctuated only rarely by short excursions to London or his native Lincolnshire, prevailed during his remaining five year's residence there.

(2) See I. B. Cohen, *Introduction to Newton's 'Principia'* (Cambridge, 1971): 84, 302, 307–9; and compare 1, §2: note (1) below.

come to listen to him at the university schools,[3] and fulfilling such minor college responsibilities as voting in elections, but above all keeping a paternal eye on the new Trinity library which was then being built to Wren's design.[4] Humphrey Newton, his secretary and constant attendant during the five years from spring 1684 till Newton himself went off to London in January 1689 on a prolonged stay,[5] long afterwards set down on paper for John Conduitt his none too coherent reminiscences of that time, seeking (as he put it) to 'return as perfect & as faithful [an] Account of [his] Transactions as possibly

(3) We have already (II: xix) quoted Humphrey Newton's observations to John Conduitt in 1728 that when Newton at this time 'read in ye Schools, as being Lucasianus Professor, ...so few went to hear Him, & fewer y^{t} understood him, y^{t} oftimes he did in a manner, for want of Hearers, read to ye Walls' and that he then 'usually staid about half an hour, [though] when he had no Audit^{rs} he com̄only return'd in a 4th part of that time or less' (King's College, Cambridge. Keynes MS 135; on which see note (6) following). As a comment on the accuracy of this testimony we may remark that the walking distance from Newton's rooms in the Great Court of Trinity to the University—now the 'Old'—Schools is only about one-third of a mile. Whiston's still later statement that he had in his student days heard Newton read 'one or two of [his *Principia*] lectures in the publick Schools, though I understood them not at all at that time' (*Memoirs of the Life and Writings of Mr. William Whiston... Written by himself* (London, 1749): 36) must refer to no earlier than 'the Middle of the Year 1686' when he was 'admitted of *Clare-Hall*...where I earnestly pursued my Studies, and particularly the Mathematicks, eight Hours in a Day' (*ibid.*: 19); subsequently, after his ordination at Lichfield in September 1693, Whiston returned to Clare College 'and went on with my own Studies there, particularly the Mathematicks, and the *Cartesian* Philosophy; which was alone in Vogue with us at that Time. But it was not long before I, with immense Pains, but no Assistance, set myself with the utmost Zeal to the Study of Sir *Isaac Newton*'s wonderful Discoveries in his *Philosophiæ Naturalis Principia Mathematica*.... Being indeed greatly excited thereto by a Paper of Dr. [*David*] *Gregory's* when he was Professor in *Scotland*; wherein he had given the most prodigious Commendations to that Work, ...and had already caused several of his Scholars to keep *Acts*, as we call them, upon several Branches of the *Newtonian* Philosophy; while we at *Cambridge*, poor Wretches, were ignominiously studying the fictitious Hypotheses of the *Cartesian*, which Sir *Isaac Newton* had also himself done formerly, as I have heard him say' (*ibid.*: 35–6). (While after his election to the Savilian Chair of Astronomy at Oxford in late 1691 Gregory did lecture extensively on the *Principia*—see P. D. Lawrence and A. G. Molland, 'David Gregory's Inaugural Lecture [21 April 1692] at Oxford', *Notes and Records of the Royal Society*, **25**, 1970: 143–78, especially 157–8 where the yet unpublished manuscript (Aberdeen, University Library 2206/8) of his professional lectures during 1692–7 is briefly described—, we would add that those of his earlier Edinburgh theses and lectures on mechanics and astronomy which survive are wholly elementary in character. It would be an understandable slip for Whiston, writing of an event more than half a century old, to have mistakenly put 'Scotland' for 'Oxford'.) Alternatively, his reference may have confusedly been to Davids's brother, James, who at Edinburgh in 1690 published a list of M.A. theses mostly Newtonian in theme.

(4) Edleston (*Correspondence* (note (1)): lviii, note (90)) prints an entry in the College Account Book of the building of the library: 'May 28, 1687. P^{d}...for erecting a scaffold for Mr Newton to measure the fret work of the staircase: 4^{s} 6^{d}.'

(5) 'In ye last year of K. Charles 2^{d}', he told Conduitt, 'Sr Isaac was pleas'd through ye Mediation of Dr Walker, (then Schoolmaster at Grantham) to send for me up to Cambridge, of Whom I had the Opportunity as well Hon^{r} to wait of, for about 5 years.'

does at this Time come to my Memory'.[6] These half-forgotten impressions of a man whose intellectual depth was far beyond his own meagre capacity to comprehend it do yet, for all their essential superficiality and excess of maudlin eulogy, have a freshness and immediacy undiluted by stale interpretation which will, with some slight rearrangement of their sequence into a more logical order, bear extensive quotation:

His Carriage then was very meek, sedate & comely; I cannot say, I ever saw him laugh, but once[7]....I never knew him take any Recreation or Pastime, either in Riding out to take ye Air, Walking, Bowling or any other Exercise whatever, thinking all Hours lost, y^t^ was not spent in his Studyes, to wch he kept so close...so intent, so serious upon [them], y^t^ he eat very sparingly, nay, oft times he has forgot to eat at all, so y^t^ going into his Chamber I have found his Mess untouch'd, of wch when I have reminded him, [he] would reply, Have I; & then making to ye Table, would eat a bit or two standing, for I cannot say, I ever saw Him sit at Table by himself....

I cannot say I ever saw him drink, either Wine Ale or Bear, excepting [at] Meals, & then but very sparingly. He very rarely went to Dine in ye Hall unless upon some Publick Dayes, & then, if He has not been minded, would go very carelessly, wth Shooes down at Heels, Stockins unty'd, Surplice on, & his Head scarcely comb'd....At some seldom Times when he design'd to dine in ye Hall [he] would turn to ye left hand,[8] & go out into ye street, where making a Stop, when he found his mistake, [he] would hastily turn back & then sometimes instead of going into ye Hall, would return to his Chamber again....

In his Chamber he walk'd...very much.... He very seldom sat by ye Fire..., excepting ye long frosty winter, wch made him creep to it against his will. I can't say I

(6) This account, delivered by Humphrey in two letters to Conduitt on 17 January and 14 February 1727/8 now in King's College, Cambridge (Keynes MS 135), was first printed—in an extensively modernised transcription which we do not here follow—by David Brewster in his *Memoirs of the Life, Writings and Discoveries of Sir Isaac Newton* (Edinburgh, 1855), **2**: 91–8. To this, his opening sentence, Humphrey adjoined apologetically that 'Had I had ye least Thought of gratifying after this Manner, S^r^ Is.'s Friend, I should have taken a much stricter view of his Life & Actions.'

(7) As William Stukeley had six months earlier recounted the event to Richard Mead in his letter of 15 July 1727 (now King's College, Cambridge. Keynes MS 136B) 'Twas upon occasion of asking a friend to whom he had lent Euclid to read, what progress he had made in that author, & how he liked him? he answered by desiring to know what use & benefit in life that study would be to him? upon which Sr Isaac was very merry.' For what it is worth, this put Humphrey 'in mind of ye Ephesian Phylosopher, who laugh'd only once in his Life Time, to see an Ass eating Thistles, when Plenty of Grass was by.' At any rate Newton clearly did not encourage him to play the buffoon.

(8) Having gone down the staircase from his rooms, that is, and instead of then going virtually straight ahead across the Great Court at Trinity. The often reproduced general view of Trinity College at this time which David Loggan published as one of the engravings in his *Cantabrigia Illustrata* (London, 1690) will clarify the topography; see also Brewster's *Memoirs* (note (6)), **2**: 85–6, and more generally G. N. Watson's essay on 'Trinity College in the time of Newton' (in (ed.) W. J. Greenstreet, *Isaac Newton 1642–1727*, London, 1927: 144–7).

ever saw him wear a Night-Gown, but his wearing Cloathes, that he put off at Night, at Night, do I say, yea rather towards ye Morning,[9] he put on again at his Rising. He never slept in ye Day time, y^{t} I ever perceiv'd. I believe he grudg'd y^{t} short Time he spent in eating & sleeping....In a Morning he seem'd to be as much refresh'd with his few hours Sleep, as though he had taken a whole Night's Rest. He kept neither Dog nor Cat[10] in his Chamber, wch made well for ye old Woman, his Bedmaker,...for in a Morning she has somtimes found both Dinner & Supper scarcely tasted of, w^{ch} [she] has very pleasantly & mumpingly gone away with....In Winter Time he was a Lover of Apples, and sometimes at Night would eat a smal roasted Quince....He was only once disorder'd with Pains at ye Stomach, wch confin'd Him for Days to his Bed, wch he bare with a great deal of Patience & Magnanimity, seemingly indifferent either to live or dye....He has given ye Porter many a Shilling, not for leting him at ye Gates at unseasonable Hours, for y^{t} he abhor'd, [I] never knowing him out of his Chamber at such Times....

Near y^{e} [east end of ye Chappel] was his Garden, wch was kept in Order by a Gardiner. I scarcely ever saw him do anything (as pruning &c) at it himself....in his Garden, wch was never out of Order, ...he would, at some seldom Times, take a short Walk or two, not enduring to see a Weed in it....When he has some Times taken a turn or two [he] has made a sudden Stand, turn'd himself about, run up ye Stairs[11] [&] like another A[r]chimides, with an εὕρηκα fall to write on his Desk standing, without giving himself the Leasure to draw a Chair to sit down on....

On ye left end[12] of the Garden was his Elaboratory..., where he...employ'd himself in, with a great deal of Satisfaction & Delight....[It] was well furnished with chymical Materials, as Bodyes, Receivers, Heads, Crucibles &c, wch was made very little use of, ye Crucibles excepted, in wch he fused his Metals.... His Brick Furnaces *p*[*ro*] *re natâ*, he made & alter'd himself, wthout troubling a Brick-layer....At Spring or Fall of ye Leaf...he used to imploy about 6 weeks in his Elaboratory, the Fire scarcely going out either Night or Day, he siting up one Night, as I did another, till he had

(9) Humphrey elsewhere observed more precisely that 'He very rarely went to Bed till 2 or 3 of ye Clock, sometimes not till 5 or 6, lying about 4 or 5 hours, especially at Spring or Fall of ye Leaf...'.

(10) It was not always so. In an unpublished jotting in one of his small green notebooks John Conduitt recorded at about this time that 'His cat at the University grew very fat' (King's College, Cambridge. Keynes MS 130.6^{2}: 3^{v}).

(11) Those (demolished in the early nineteenth century) which led up from his garden to a wooden verandah outside his bedroom; again see David Loggan's contemporary view of Trinity (note (8)), in whose bottom right corner it is plainly visible. Humphrey elsewhere remarks that 'His Telescope [his reflector, presumably], wch was at y^{t} Time, as near as I could guess,...near a foot long...he plac'd at ye head of ye Stairs, going down into ye Garden, buting towards ye East. What Observations he might make, I know not'.

(12) Looking eastwards from Newton's room, that is, and so to the north. David Loggan shows the 'Elaboratory' as a small shed tucked away right up against the Chapel end, only yards—and the thickness of a brick wall—away from its altar. Newton's smoking chimney must have been in clear view of all who attended the Chapel services, and his chemical experiments common gossip in the College!

finished his chymical Experiments, in ye Performances of wch he was ye most accurate, strict [&] exact....[13]

As for his private Prayers I can say nothing of them, I am apt to believe his intense Studyes depriv'd him of ye better Part....He very seldom went to y^e [college] Chappel, y^t being ye Time he chiefly took his Repose; And as for ye Afternoons, his earnest & indefatigable Studyes retain'd Him, so y^t he scarcely knew ye House of Prayers. Very frequently on Sundays he went to [Great] St Mary's Church, especially in ye fore Noons....

S^r Is. at y^t Time had no Pupils, nor any Chamber Fellow, for y^t, I presume to think, would not in ye least have been agreeable to his Studies....

This pen-picture of a quirky, retiring scholar as neurotically precise in his private manner as he was utterly careless of his outward appearance, so deeply engrossed in elaborating the intricacies of his thought and in exploring the fiery secrets of his alchemist's furnace as to be all but oblivious of the bustle of life around him, is of course both overdrawn and incomplete. Newton's occasional petulances, flashing angers and outbursts of puritanical sternness are doubtless here unconsciously suppressed in favour of his more endearing and graphic eccentricities.[14] His social and intellectual isolation at this period is certainly overstated. While Humphrey goes on to report that Newton 'always kept...to his Studyes...so close, y^t he seldom left his Chamber, unless at Term Time, when he read in y^e Schools', he also adds that 'When invited to a Treat,...[he] used to return it very handsomely, freely, & wth much Satisfaction to Himself....At [which] seldom $Entertainm^{ts}$ y^e Masters of Colledges were chiefly his Guests', and also that 'Foreigners [from outside Cambridge] He received wth a great deal of Freedom, Candour & Respect'. Of the '2 or 3 Persons' who were regular visitors to Newton's Trinity rooms Humphrey identifies 'Mr Ellis of Keys [Caius], Mr Laughton of Trinity, &

(13) Humphrey added: 'Nothing extraordinary, as I can remember, happen'd in making his Experiments, wch if there did, He was of so sedate & even Temper, y^t I could not in ye least discern it...He would sometimes, tho' very seldom, look into an old mouldy Book, wch lay in his Elaboratory, I think it was titled, *Agricola de metallis*, the transmuting of metals being his chief Design, for wch purpose Antimony was a great ingredient'. Since Newton's library copy (now Trinity NQ.11.4) of the 1621 edition of Agricola's *De Re Metallica* is unstained, this is clearly not the volume which Humphrey saw Newton using. As we have already observed (v: xiv), since Newton's record of these chemical experiments in the notebook ULC. Add. 3975 pass immediately from an entry on 'Friday May 23 [1684]' (p. 149) to one overleaf on 'Apr. 26 1686' (p. 150) he would seem to have intermitted them during the crucial twenty months from August 1684 in which he was pre-occupied in elaborating the principles of dynamical motion under a central force and applying them to explain the physical system of the world.

(14) Such mentions by Humphrey as that of 'His Behaviour mild & meek, without Anger, Peevishness or Passion, so free from y^t, that you might take him for a Stoick' are, of course, so closely tailored to his hagiographic purpose of setting down for Conduitt 'things...worthy to be inserted into ye Life of so great, so good, & so illustrious a Person as Sr Isaac' as to be meaningless.

M[r] Vigani, a Chymist, in whose Company he took much Delight and Pleasure at an evening, when they come to wait upon Him'. John Ellis, now Master of Caius College, had a couple of years earlier been tutor to Henry Wharton, the only Cambridge student we know ever to have had access to Newton's private mathematical writings;(15) John Laughton, 'y[e] Library Keeper of Trin. Coll.', in particular 'resorted much to his Chamber', Humphrey tells us; while John Francis Vigani, a private lecturer in chemistry, was an especial 'favourite' of Newton's 'till he told a loose story about a Nun & then S[r] I. left off all confidence with him', as Catherine Conduitt later recorded.(16) To this trio of Newton's close friends at Cambridge in the middle 1680's we should continue to add his Trinity colleague Humphrey Babington, whom Humphrey Newton elsewhere lists among 'others of his Acquaintance'.(17) Who the other 'Masters of Colledges' were we may only guess, but it is significant that Newton's only contact at this time with undergraduates and others of less senior status in the University came from the height and distance of a professorial podium(18) on those occasions when anyone at all turned up to hear him lecture. The visiting 'Foreigners' hospitably received by him included Halley, of course, on his several visits to Cambridge and also, by his own later testimony, the Scotsman John Craige.(19)

(15) See IV: 11, note (30) and 188–9, note (1).

(16) So her husband John initially wrote it down in one of his jotters (King's College, Cambridge. Keynes MS 130.6[2]). In his later amplification of this anecdote (Keynes MS 130.7, first printed by Brewster in his *Memoirs* (note (6)), **2**: 92–3, note 3) Conduitt describes Newton as being 'very intimate' with Vigani and taking 'great pleasure in discoursing on Chymistry' with him. The little else known of Vigani's activities in Cambridge at this time is summarised by L. J. M. Coleby in *Annals of Science*, **8**, 1952: 46–60, especially 46–8. Not till 1703 was he accorded the (honorary) title of University Professor of Chemistry.

(17) As we have seen (I: 8, note (21)) Babington was a close friend of Newton's mother and had perhaps been chiefly instrumental in securing him his sizar's place at Trinity in 1661: 'He was', William Stukeley wrote to Mead on 15 July 1727 (see note (7) above), 'own uncle to M[rs] Vincent [Miss Storey], i.e. brother to her mother M[r] Clarks wife where S[r] Is: lodgd [as a day boy at Grantham Grammar School in the 1650's], & that seems to be the reason why he went to this college. The D[r] is said to have had a particular kindness for him...'.

(18) By the additions to the statutes governing the Lucasian Professorship made under the royal seal on 18 January 1664, the holder of the Chair was (under penalty of deprivation) expressly forbidden to tutor any pupils other than fellow-commoners or to hold any teaching or administrative office, other than the Professorship and the College Fellowship tied to it, in the University. (See III: xxvii, where the Latin text of the prohibition is given.) At this period only a 'Mr Robt. Sacheverell', who is otherwise unknown to us, entered (on 16 September 1687) fellow-commoner under Newton; see Edleston, *Correspondence* (note (1)): xlv, note (16).

(19) See his *De Calculo Fluentium Libri Duo* (London, 1718); Præfatio: [b2[r]]. In the course of his few days' stay, in about the early summer of 1685, Newton showed Craige both early drafts of his *Principia* (compare Cohen's *Introduction* (note (2)): 204) and the manuscript of his 1671 tract on infinite series and fluxions (see III: 354, note (1)). We will return to discuss the significance of this visit in the next volume when we reproduce the yet unpublished original text of Newton's treatise 'De quadratura Curvarum'.

We must be disappointed that Humphrey Newton says so little of his namesake's intellectual pursuits and academic activities as they must have, in some measure at least, been known to him. His sole qualification of the *Principia* 'w^{ch}...by his Order, I copied out before it went to ye Press' as a 'stupendous Work' is as shallow and imperceptive as his more general observation that Newton's 'Thoughts were his Books, [&] tho' he had a large Study [he] seldom consulted with them' or his uncomprehending judgement on his 'chymical Experiments' that 'What his Aim might be, I was not able to penetrate into, but his Pains, his Diligence [therein]...made me think, he aim'd at something beyond ye Reach of humane Art & Industry'. In filling out this disappointingly meagre reaction of his secretary to the genius and vigour of a man then at the height of his mathematical and scientific maturity Newton's own public writings at this time add not a great deal. Over the three years from August 1684 in which he was preoccupied with composing the several drafts of his *Principia* and then helping Halley to see its final version through the press he understandably wrote few letters not directly related to that end, and of these even fewer have survived: for two crucial periods during this interval, namely from late May to mid-September 1685 and again from mid-October 1685 to late May 1686, the record of his extant correspondence is completely blank. In brief exchanges during the month from 17 December 1684 and in a reprise over the three weeks from 19 September 1685[(20)] John Flamsteed put his Greenwich observations and astronomical expertise freely at Newton's disposal in furnishing him with a variety of requested information regarding corrected sightings of the comet of 1680–1[(21)] and his computed orbital elements and periods of Jupiter and Saturn and their satellites; for his hard work (much of whose detail was incorporated in the *Principia*'s first edition) Flamsteed was rewarded with a curt acknowledgement from Newton that 'Your observations of ye Comet, being so exact...will save me a great deale of pains. I shall have no need to give you further trouble at present, but after a while I believe I may have occasion to beg your further assistance'.[(22)] A long sequence of eighteen letters exchanged with Halley between 22 May 1686 and 5 July 1687 (together with at least two more which are no longer

(20) See *The Correspondence of Isaac Newton*, **2**: 403–15/419–30.

(21) Compare v: 524–5, notes (1) and (3); and see also notes (1) and (3) on page 82 below.

(22) Newton to Flamsteed, 14 October 1685 (*Correspondence*, **2**: 430). When eleven months afterwards Newton gave 'kind enterteinm^{t}' in Cambridge to Flamsteed's 'friend Mr Philips' even though, as the latter acknowledged on 9 September 1686, 'hee has beene with me but three weekes and has onely learnt some few propositions of Euclid with his plaine trigonometry in this time' (*ibid.*: 449), Newton refused to be drawn anew on the topics of 'Cassinis new Planets about Saturn' and the possible oblateness of Jupiter's sphere with which Flamsteed had primed Philips (see *ibid.*: 448) despite Flamsteed's further attempt (*ibid.*: 449–50) to provoke from him a 'better...conceipt' regarding the cause of Jupiter's apparent ellipticity.

extant) is narrowly concerned with resolving the various problems attendant on editing and printing so complicated and technical book as the *Principia*, and with responding to Robert Hooke's claim to prior discovery of the principle of universal inverse-square gravitation and also evaluating John Wallis' independent researches in the theory of resisted projectile motion. While the hot anger of Newton's verbal outburst against Hooke on 20 June 1686—'is not this very fine? Mathematicians that find out, settle & do all the business must content themselves with being nothing but dry calculators & drudges & another that does nothing but pretend & grasp at all things must carry away all the invention...'(23)—says more about his inner passion and forcefulness than all Humphrey Newton's subsequent platitudes, and though his genuine anxiety lest Wallis pre-empt the credit for pioneering the study of resisted motion(24) is in a subtler way equally revealing of his deep-seated but long repressed desire for public recognition of his genius and originality, since pertinent quotation from these letters is made widely in editorial footnote in our main text we need not here linger to summarise their detail.(25)

Somewhat surprisingly, Halley himself is, for all the closeness of their working relationship over many months, always addressed by Newton in a tone of distant formality which belies any overt personal affection for him, and this may explain why, once the *Principia* was published, their association ceased for several years thereafter. For Edward Paget, who flits as a (now) obscure and tantalising shadow though Newton's correspondence in the middle 1680's (and whom he had earlier on 3 April 1682 commended to Flamsteed(26) as a fit

(23) *Correspondence*, **2**: 438. On Newton's reaction to Hooke's claim to 'ye duplicate proportion' of terrestrial—and indeed universal—gravitation see more generally note (46) on page 15 below, and compare note (61) on pages 20–1.

(24) Having learnt from Paget that 'Dr Wallis has sent up some things about projectiles pretty like those of mine in ye papers [*De motu Corporum*]', Newton made haste on 13 February 1686/7 to find out the truth of the matter from Halley (see *Correspondence*, **2**: 464); the latter was able to reassure him eleven days later that Wallis' result was 'much the same with yours, and he had the hint from an account I gave him of what you had demonstrated' (*ibid.*: 469). The matter is discussed in detail in note (93) on pages 64–5 below.

(25) The text of the extant correspondence between Newton and Halley during this period is printed in *Correspondence*, **2**: 431–47, 452–4, 464, 469–74, and 481–2. It is broadly treated by I. B. Cohen in his *Introduction* (note (2)): 131–42.

(26) *Correspondence*, **2**: 373. In a testimonial of the same date to the Governors of Christ's Hospital Newton described Paget, 'Maister of Arts & Fellow of Trinity-Colledge in this University', as 'ye most promising person for this end I could think of;...of a temper very sober & industrious.... He understands ye several parts of Mathematicks, Arithmetick, Geometry, Algebra, Trigonometry, Geography, Astronomy, Navigation & wch is ye surest character of a true Mathematicall Genius, learned these of his owne inclination, & by his owne industry without a Teacher...' (*ibid.*: 375). What a sad fulfilment is there implied of the high hopes of encouraging mathematical studies in the University with which Newton's Cambridge professorship had been established twenty years before! Since he already knew

candidate for the post of Master of the Mathematical School at Christ's Hospital to which he was soon after appointed), Newton evidently held a warmer esteem. Paget it was who in November 1684 'when last with us', as Newton shortly afterwards informed the newly appointed Secretary of the Royal Society,[27] unsuccessfully 'pusht forward' the 'designe of a [regular] Philosophick Meeting here'[28] and who on his return journey from Cambridge to London carried with him the fair copy of Newton's first tract 'De motu Corporum' whose text was subsequently transcribed by Halley and also entered in the Royal Society's Register;[29] thereafter, he acted more than

---

Paget well enough to add that 'his hand is very steady & accurate, as well as his fancy & apprehension, good; as may be seen by his writing & drawing wth his Pencil very well', further noting his 'long acquaintance also wth variety of Learning here', perhaps Newton had indeed given him—like Henry Wharton a year later (see IV: 11, note (30))—some minimal mathematical instruction in the privacy of his room?

(27) His old friend and colleague at Trinity, Francis Aston, who ten years before—unlike Newton—had failed to obtain a royal dispensation to remain a Fellow without taking holy orders (see II: xxiv, note (10)).

(28) Newton to Aston, 23 February 1684/5 (*Correspondence*, **2**: 415). 'I concurred with him', he went on, 'and engaged D^{r} [Henry] More to be of it, and others were spoke to partly by me, partly by Mr Charles Montague but that which chiefly dasht the buisiness was the want of persons willing to try experiments, he [Vigani?] whom we chiefly relyed on refusing to concern himself in that kind him self. And more what to add further about this buisiness I know not, but only this that I should be very ready to concurre with any persons for promoting such a designe so far as I can doe it without engaging the loss of my own time in those thinges'.

(29) Two years afterwards, in clarifying for Newton (at the height of his squabble with Hooke over prior discovery of inverse-square planetary gravitation) the crowded sequence of events during 1684, Halley recalled in his letter to Cambridge on 29 June 1686 that in 'August...when I...learnt the good news that you had brought this demonstration to perfection, ...you were pleased to promise me a copy thereof, which the November following I received with a great deal of satisfaction from Mr Paget' (*Correspondence*, **2**: 442); compare Newton's subsequent reference to 'ye papers Mr Paget first shewed you' in his later re-opening of correspondence with Halley on 13 February 1686/7 (*ibid.*: 464). At the Royal Society meeting on 10 December 1684 Newton's 'curious treatise *De Motu*..., upon Mr *Halley*'s desire, was promised to be sent to the Society to be entered upon their register. Mr *Halley* was desired to put Mr *Newton* in mind of...securing his invention to himself till he could be at leisure to publish it. Mr *Paget* was desired to join with Mr *Halley*' (Thomas Birch, *History of the Royal Society of London*, **4** (London, 1757): 347). Flamsteed, who had also (in a lost letter from Newton passed on by Paget) been 'kindly offered ye perusall of your papers' as he wrote back in acknowledgement on 27 December (*Correspondence*, **2**: 403), was in fact—perhaps, as Flamsteed guessed a week later, because the prevailing 'hard weather...prevented him as it did me from goeing to London' (*ibid.*: 410) but also because, as Newton replied on 12 January 1684/5, 'Mr Paget...has been laid up sick of an ague' (*ibid,*: 412)—not to gain a view of the 'De motu Corporum' for nearly another month when, he wrote to Newton on 27 January to thank him, 'a benifice haveing been bestowed upon me in the meane time I have not had leasure to peruse it yet' (*ibid.*: 414). The original 'papers' of Newton's communicated 'Notions about Motion', whose text had been registered at the Royal Society by mid-February following when Newton thanked Aston for entering them (*ibid.*: 415)—whether before or after they passed to Flamsteed is not clear—, thereafter disappear from the record. In 1, §1 below we

once over the next couple of years as Newton's trusted informant in London[(30)] and during a further stay in Cambridge with him in autumn 1686 he 'mended' a number of verbal slips in the proof-sheets of the *Principia*'s first book,[(31)] being duly rewarded the following summer with a gift copy of the published volume.[(32)] Here, one feels, is yet one more interesting minor acquaintance of Newton's of whom we should have liked to know more than his drunken downfall and ensuing death in exile.[(33)]

For the rest, the dedicatory epistle which Newton contributed in April 1685 to the Latin version of William Briggs' *Theory of Vision*[(34)] and the similar certificate of approval which he set a few months later to George Mabbot's *Tables for renewing and purchasing of the leases of Cathedral Churches and Colleges*[(35)]

---

reproduce them from the much corrected and overwritten draft retained by Newton, suitably filling out its minimal lacunas in line with Halley's (incompletely extant) transcript and the secretary copy in the Royal Society Register Book (see note (2) on pages 30–1 following).

(30) See note (24) above for Newton's enquiry in February 1687 of Paget as to 'how things were' in current 'differences in y[e] R. Society' (*Correspondence*, **2**: 464). Paget was also, we assume, Newton's unnamed informant in June 1686 regarding Hooke's 'great stir' at the Royal Society, 'pretending I had all from him' (see *ibid.*: 437).

(31) See Newton to Halley, 18 October 1686 (*Correspondence*, **2**: 454).

(32) Along with 'the R. Society, M[r] Boyle, ...M[r] Flamsteed and if there be any elce in town that you design to gratifie that way' as Halley wrote to Newton on 5 July 1687, having 'at length brought your Book to an end' (*Correspondence*, **2**: 481).

(33) In the late 1690's Flamsteed added at the foot of Newton's letter introducing Paget to him in April 1682 (see note (26) above) a Latin note: 'Ebrietati deinde...nimium addictus immemor officij, pueros neglexit in Flandriam transiit deposuit munus in Indiam tandem navigavit...' (*Correspondence*, **2**: 373); while H. W. Turnbull in a following footnote quotes an extract from a letter from Roger Cotes to John Smith of 11 December 1703 (Trinity College, Cambridge, MS R. 4) observing that 'Pagett...died at Isp[h]ahan in his return from y[e] great Mogul...Tis believed he had made severall valuable Observations in those parts...' (*ibid.*: 375, note (6)). Before the black sheep showed his true colour Newton maintained an active relationship with Paget, arranging to visit Flamsteed with him on a trip to London in the summer of 1691 (see Newton's exploratory letter to the latter on 10 August 1691, printed in *Correspondence*, **3**, 1961: 164) and writing to him in the late spring of 1694 regarding Paget's revised scheme of instruction for the Christ's Hospital Mathematical School (see Newton's unpublished letter to Paget of 25 May 1694, whose draft is now ULC. Add. 3965.12: 330[r], and the mass of related letters and alternative schemes of Newton's preserved in ULC. Add. 4005.16, only partially printed in *Correspondence*, **3**, 357–68 from copies in the Christ's Hospital Record Book).

(34) This dedicatory letter is reprinted in Edleston's *Correspondence* (note (1)): 272–3, and also in *The Correspondence of Isaac Newton*, **2**: 417–18. In it Newton took care to mention that Briggs' book 'in usum cedet juventuti Academicæ, & provectiores ad ulteriores in Philosophiâ progressus manuducet.'

(35) Newton's commendation states pithily: 'Methodus hujus Libri rectè se habet, numerique ut ex quibusdam ad calculum revocatis judico, satis exactè computantur. Is. Newton, Math. Prof. Luc.'. The author of these tables—directly attributed to Newton after their fifth edition (London, 1729)—was manciple (caterer) at King's College and, as far as we know, no personal friend of Newton's. (See Edleston, *Correspondence* (note (1)): lvi, note (78);

together remind us that Newton was never reluctant to lend his name (and professorial title) to promote the sale of others' books. Nor was he averse to allowing minor mathematical contributions of his own to appear in print, as the 'Directions' for constructing the volume of a paraboloid of revolution which appeared under his name in William Hunt's *Guager's Magazine* in 1687 illustrate.[36] Four letters from the retired Cambridge mathematics don Gilbert Clerke between September and November 1687 querying small points of mathematical style in the published *Principia*[37] are less memorable than Newton's extant reply to the first[38] where he displays an unusual tolerance of his aged correspondent's tiresome and near-trivial 'scruples' regarding the opening pages of his book. The stray draft, finally, of a letter of his[39] to the wayward tenant of his childhood home of Woolsthorpe in January 1688 both adds at one point the incidental information that he is now 'short-sighted' and, at a deeper level, underlines the continuity of his concern with his holdings of property in Lincolnshire.

The even tenor of Newton's existence at Cambridge, for so long undisturbed, was to begin to crack even while his *Principia* was still in press in London. The initial break came, as is well known, with the issue by James II in February 1687 of a *mandamus* to the University Vice-chancellor, Pechell, directing that a Benedictine monk, Father Francis, be admitted M.A. without taking the required oaths of supremacy and allegiance. Disturbed less by the political and religious implications of this assertion of the royal prerogative than by his

---

and also D. F. McKenzie, 'The Author of *Tables for purchasing Leases* attributed to Sir Isaac Newton', *Transactions of the Cambridge Bibliographical Society*, 3, 1960: 165–6.) A note by John Conduitt in 1729 (King's College, Cambridge. Keynes MS 130.5) adds the information that 'The leases for lives &c wch are printed in his name, a bookseller prevailed with him only to look over one page & so he writt what is there mentioned'.

(36) These 'Directions...necessary to be understood by every *Gager*...as I received [them] from the learned M. *Isaac Newton, Professor of the Mathematicks* in *Cambridge*' are conveniently reprinted from Hunt's *Magazine* (London, 1687: 272–4) in *The Correspondence of Isaac Newton*, 2: 478–80. In effect Newton obtains the segment of a paraboloid of revolution of height h and base radius r which is cut off at a distance z from the axis by a vertical parabolic section (of height $(h/r^2)(r^2-z^2)$ and semi-base $y = \sqrt{[r^2-z^2]}$) as

$$\tfrac{4}{3}h\int_r^z r^{-2}(r^2-z^2)^{\frac{3}{2}}.dz = \tfrac{4}{3}h\left[\int_r^z y.dz - \int_r^z r^{-2}yz^2.dz\right]$$
$$= \tfrac{4}{3}h\int_r^z y.dz - \tfrac{1}{6}h\int_{2r}^x v.dx,$$

where $x = 2z^2/r$ and $v = \sqrt{[x(2r-x)]}$ are readily constructed as abscissa and ordinate in the base circle. (Compare III: 31, note (11).)

(37) *Correspondence*, 2: 485–6, 488–98.

(38) *Correspondence*, 2: 487.

(39) *Correspondence*, 2: 502–4, especially, 502.

outrage that statutory law had been cynically flouted,(40) Newton was a leading advocate of defying this mandate when the question was debated in the Senate in March, and a prominent member of the University delegation whose tentative appeal to legal precedent before the Ecclesiastical Commission in April was tersely dismissed by Judge Jeffreys from his President's chair with a typically stern rebuke.(41) His reward came nearly two years later when, after the 'Glorious Revolution' which in December 1688 proclaimed William of Orange king in James' place, Newton was in January 1689 elected—not with the greatest share of the vote, we may add(42)—one of the two University members of the Convention Parliament. If not conspicuously active therein, during the next thirteen months he lived an exciting existence on the fringe of political power at Westminster and Whitehall, actively involved at first in the preparation of the new declaration of allegiance to be sworn by members of the Universities(43) and thereafter with little to do but listen on his parliamentary back-bench. We may vicariously conceive the deep impact upon him of his sudden translation from a dowdy, monotonous scholarly retreat into the glittering, sophisticated world of a metropolis in which he had never before passed more than a few fleeting days. There he met not only influential politicians but the philosopher John Locke, with whom he was to maintain contact over the next ten years both by letter and by personal visit,(44) and the

(40) If we read aright the letter which he drafted on this issue to an unknown correspondent on 19 February 1686/7 (*Correspondence*, **2**: 467–8).

(41) 'Go your way and sin no more, lest a worse thing come unto you 'as Jeffreys quoted from scripture in reprimanding the delegates on 12 May (cited from Edleston, *Correspondence*: lviii, note (90) where it is added that 'Newton does not appear at all as a speaker during the proceedings'). A full account of the Father Francis affair is given by C. H. Cooper in his *Annals of Cambridge*, **3** (London, 1845): 614–33; see also Brewster's *Memoirs* (note (6)), **2**: 104–9 and—though we need not wholly accept his individual interpretations of Newton's motives for making this unprecedented incursion into University politics—F. E. Manuel's *Portrait of Isaac Newton* (Cambridge, Massachusetts, 1968): 108–14.

(42) Newton attracted 122 votes to Sir Robert Sawyer's 125, while the third candidate in the election gained 117—a close result! (See *The Correspondence of Isaac Newton*, **3**: 8, note (1).) 'In many of the voting papers [still preserved in the University archives] his name is preceded by the words "præclarum virum", in some the adjective is "doctissimum", "integerrimum", "venerabilem", "reverendum". Pulleyn, his old tutor [see I: 10, note (26)], calls him "summum virum"' (Edleston, *Correspondence*: lix, note (93)).

(43) See the thirteen extant letters written by him to the Cambridge Vice-Chancellor, John Covel, between 12 February 1688/9 and the following 15 May (*Correspondence*, **3**: 10–23).

(44) The circumstances of Newton's first meeting with Locke are not recorded, but they were already well enough acquainted by 'Mar[ch 16]89/90' (as Brownover's secretarial inscription specifies) for Newton to send him a simplified 'Demonstration That the Planets by their gravity towards the Sun may move in Ellipses' (Locke's copy of which is now Bodleian MS. Locke.c.31: 101–4; on this see 3, §1.4: note (33) below). Of the letters which passed between them over the next decade and a half, many dealing with intricate points of biblical exegesis, eighteen are preserved (consult the indexes to *Correspondence*, **3**/**4**, *s.v.* 'Locke'), and in

eminent Dutch scientist Christiaan Huygens(45) with whom (through Oldenburg) he had briefly corresponded sixteen years before; more unexpectedly, he became close friends with the émigré Swiss mystic and mathematician Nicolas Fatio de Duillier,(46) of whom we shall hear more in the next volume. And, to show that he had arrived at last in the public world, he for the first time in his life commissioned a portrait of himself.(47)

Once implanted, the taste for political power and temporal prestige was never

the intervening periods Newton was a frequent visitor to Oates in Essex where Locke mostly lived from 1691 (till his death in October 1704) as the guest of Sir Francis Masham. In his last known letter to Locke on 15 May 1703 Newton wrote that 'I had thoughts of going to Cambridge this summer & calling at Oates in my way [from London]...' (*Correspondence*, **4**, 1967: 406).

(45) During Huygens' extended stay in London in 1689, namely. Their first meeting was at the Royal Society on 12 June when 'Mr Hugens of Zulichen being present gave an account that he himself was now about publishing a Treatise concerning the Cause of Gravity, and another about Refractions giving amongst other things the reasons of the double refracting Island Chrystall' while 'Mr Newton considering a piece of the Island Chrystall' erroneously asserted that the extraordinary ray 'suffered no refraction, when [it] came parallel to the oblique sides of the parallelepiped...' (Royal Society Journal Book, quoted from *Correspondence*, **3**: 31–2). At another meeting two months later Newton gave Huygens two short papers on motion under resistance (see *ibid.*: 25–8, 33–4).

(46) The two had already met by late January 1690 when Fatio was the intermediary through whom Huygens arranged to send Newton a presentation copy (now Trinity College, Cambridge. NQ. 16.186) of his newly published *Traité de la Lumière...Avec un Discours de la Cause de la Pesanteur* which (compare the previous note) had been the major topic of their first meeting the June before; see Huygens' letter of 7 February 1690 (N.S.) to Fatio (*Œuvres complètes de Christiaan Huygens*, **9**: 357–9), the pertinent extracts from which are given in *Correspondence*, **3**: 67. Fatio wrote to Newton himself on 24 February 1689/90 that 'I shall have I think to morrow an exemplar of Mr Hugens his book that he hath sent you' (*Correspondence*, **3**: 390) and an unpublished following letter two months later on 17 April evidently pursues an inquiry from Newton in assuring him that 'Mr Hugens is not about ye making of one of your telescopes, but onely some very large object glasses...which may, tho but short, bear a vast aperture for to discover more easily by them ye satellits already known and their Eclipses, the fixed starrs and perhaps new Planets &c' (Brotherton Library, Leeds. MS 248).

(47) The half-length study by Godfrey Kneller dated '1689', now in possession of the Earl of Portsmouth, which is from time to time to be seen on public view in London—at the Royal Academy winter exhibition of 1960–1 on 'The Age of Charles II', for example, where it was catalogued as No. 217—and is often reproduced in print as uniquely depicting Newton in the prime of his intellectual maturity. The main features of the painting—but not its subdued near-monochrome—are accurately described by Manuel in his *Portrait* (note (41): 106–7 where he writes that at the age of forty-six 'Newton...is presented with his own shoulder-length gray hair. A white shirt, open at the neck, is largely concealed by an academic gown from the right sleeve of which spidery fingers emerge. The face is angular, the sharp chin cleft, the mouth delicately shaped. A long, thin nose is elevated at the bridge. Beneath brows knit in concentration, blue, rather protuberant eyes are fixed in a gaze that is abstracted'. Why Newton commissioned his likeness from the most fashionable painter in London is not known but, without unduly stressing the psychological motives he may have had for so doing, we may treat it as a visible sign of his new-found concern with the outside world and the way in which others should look upon him. How true to reality this suggestive visualisation of his appearance was we have no precise way of telling.

to leave Newton, even though in February 1690 he was, on the dissolution of Parliament, seemingly happy enough to forego the bustle and sparkle of the capital for the dull provincial routine of the quiet Fenland university town where he had, till the last year, spent all his adult life.(48) While (as we shall see in the next volume) he continued over the next six years to make further significant advances in the fields of geometry and calculus and also attempted new systematised expositions of his earlier mathematical and scientific discoveries, he never really settled once more to the seclusion of dedicated scholarship. When in March 1696 he effectively severed his academic ties with Cambridge to pass his remaining thirty years of life in London, existing intellectually thereafter on the treasure of an achieved reputation in science, but careful to bolster it from time to time by publishing a carefully polished nugget or two from his private hoard of already mined scholarly gold, we cannot pessimistically sigh for all that he might have gone on to discover had the attractions of the London Mint not tempted him away. In the early 1690's Newton's mathematical and scientific powers passed their peak and entered on their decline, imperceptibly at first and with not a few momentary ascents to their former heights; after 1695 they began seriously to deteriorate, and we cannot believe that, had Newton continued to pursue his unhurried life of scholarship in Cambridge into his late middle age, he would have made any radically new discovery.

But in the context of the present volume this is all in the future. Here we may enjoy the full maturity of Newton's mathematical insight as he applied it in immediate preliminary—and partially in epilogue—to composing his *œuvre maîtresse*. We will delay the reader no longer from savouring its delights and illuminations in the field of central-force dynamics for himself.

(48) He may well of course, have had the ulterior motive of seeking academic preferment either within his own Cambridge college, Trinity, or elsewhere in the University. While in London the previous summer he had—evidently with some degree of support from others—tentatively allowed his name to go forward for consideration as the new Provost of King's. His appointment would, however, have required the variation of the college statutes which prescribed that the Provost should be both in holy orders and already a Fellow of King's; and when on 'Aug. 29, 1689. Before the King & Council was heard the matter of King's College about Mr Isaac Newton, why he or any other not of that foundation should be Provost, ... after the reasons shewed & argued Mr Newton was laid aside' (Alderman Newton's Diary among the Bowtell MSS at Downing College, quoted from Edleston, *Correspondence*: lix, note (96); for a fuller history of the affair see King's College. Keynes MSS 117/117^A: 'An Account of King's College's Recovery of their Right to choose their own Provost', and John Saltmarsh's discussion in (ed. J. P. C. Roach) *The Victoria County History of Cambridgeshire*, **3** (London, 1959): 397–8). A full year afterwards the rejection still rankled with Newton and when, shortly after his return to Cambridge, moves were made to obtain for him the Mastership of Charterhouse—whose emoluments of '200^lib *per an* besides a Coach (wch I reccon not) & lodgings' he spurned—he justified his refusal in the draft of a letter to Locke in mid-December 1691 with the words 'the competition is hazzardous & I am loath to sing a new song to ye tune of King's College' (*Correspondence*, **3**: 184).

ANALYTICAL TABLE OF CONTENTS

unresisted motion: rectilinear gravity-free 'inertial' motion is premised, 32. Theorem 1: Kepler's area law is deduced in generalised form to hold for an arbitrary central trajectory (and so be a universal measure of orbital time), 34. Theorem 2: the 'Huygenian' measure of the central force inducing uniform motion in a circle, 36–8. The circular case of Kepler's third law as its corollary when the force-field is inverse-square, 40. Theorem 3: the measure of the central force inducing motion in any given orbit, 40–2. Problem 1: applied to compute the law of force tending to a point in a circular orbit, 42. Problem 2: the (direct-distance) force induced by elliptical orbit round its centre, 44. Problem 3: the (inverse-square) force similarly induced towards a focus, 46–8. Theorem 4: the full third Keplerian law given *ad hoc* proof, 48–50. Construction of an 'upper' planetary orbit given the solar focus of the ellipse, a number of its focal *radii vectores* and the length of its major axis, 50. That of an inferior planet, given its maximum solar elongation, 52. Problem 4: to construct an elliptical orbit, given the speed and direction of motion at a point, 54. Similar construction of parabolic and hyperbolic orbits, 56. Its application to construct (elliptical) cometary paths from given terrestrial sightings, 58. Approximate (Cavalierian) construction of Kepler's problem, 60. Problem 5: to define the distance of rectilinear fall in given time under an inverse-square force, 62. Determination (independently of Huygens) of motion under simple gravity which is resisted instantaneously as the orbital speed. Problem 6: rectilinear motion when the gravity is zero, 64. Problem 7: when the gravity acts in line with the motion, 66–8. The two compounded to yield a logarithmic trajectory, 70. Minor complements: comparison is made with Huygens' prior investigation (1668), 72–4.

Appendix 1 (ULC Add. 3965.7: 40^r–54^r [extracts]). The augmented tract 'De motu Corporum' (December 1684?) The opening definitions are repeated without essential change, 74–5. Expanded 'laws' of motion, 76. Four lemmatical riders, 76–7. Augmented scholium to Theorem 4, 78. New, long scholium to Problem 5, 79–80.

Appendix 2 (ULC. Add. 3965.11: 163^r). Approximate computation of the (curved) path of the comet of 1680/1 by a modified rectilinear technique. The given terrestrial sightings (in the corrected form had from Flamsteed in September 1685), 81. Four ensuing calculations of orbital points, 82–5.

Appendix 3 (*Principia*, $_1$1687: 241–2/$_2$1713: 217/$_1$1687: 242–5). The ballistic curve (resistance proportional to velocity) as reworked in Book 2, Proposition IV of the published *Principia*. The logarithmic trajectory constructed, 85–7. Seven corollaries, 87–90.

§2. The revised treatise 'De motu Corporum' (winter/early spring 1684–5). [1] (ULC. Add. 3965.5: 21^r). Five preliminary definitions: of 'quality' of matter/motion and of the force 'innate in matter', 92. Of impressed and centripetal force, 94–6. [2] (ULC. Dd. 9.46: 13^r–19^r/22^r–31^r/64^r–71^r/56^r–63^r). The main surviving portion of the text. 'Laws of Motion': Axioms I and II, on uniform rectilinear motion and its change under the action of an impressed 'motive' force, 96. Law III, that 'to any action there is always a contrary, equal reaction, 98. Corollary 1: the parallelogram of forces is introduced, 98–100. Corollary 2: its employment in compounding motions exemplified, 102–4. Corollaries 3–6: the mutual interaction of bodies does not change their centre of gravity, and is independent of the uniform or uniformly accelerated motion of the surrounding space, 104–6. Scholium: Galileo (and Harriot) on the descent of 'heavy' bodies; Wren, Wallis and Huygens on collision and recoil, 106. Eleven lemmas 'on the method of first and last ratios': Lemma I, on the limit-equality of quantities, 106. Lemmas II/III: the area of a curve is the (common) limit of its inscribing/circumscribing *mixtilinea*, 108. Lemma IV: the areas of curves are as the limit-ratio of their (inscribing) *mixtilinea*, 110. Lemma V: similar figures are to one another as the squares of corresponding line-elements, 112. Lemmas VI/VII:

Lemma XV in §2 preceding, 238–40. Proposition XXI: repeats the earlier Proposition XIX and its scholium, 240–2.

Article V: 'On finding [conic] orbits, given neither focus'. Lemmas XVII–XXI, Proposition XXII and Case 1 of Proposition XXIII lightly remould the seven opening propositions of Newtons' earlier 'Solutio Problematis Veterum de Loco solido'. Lemmas XVII–XIX: the Greek 4-line locus is demonstrated to be one or other species of conic and is so constructed, 242–52. The name 'conic' is taken in its broadest sense (and includes line-pairs), 248. Lemma XX: Newton's semi-projective defining 'symptom' of a conic is stated and proved, 252–4. Lemma XXI: and used to demonstrate his 'organic' construction of a general conic by moving angles, 254–6. Proposition XXII: to describe a conic through five points, using Newton's symptom, 256–8. And by employing the organic construction, 258–60. Proposition XXIII: drawing a conic to pass through four points and touch a straight line, by means of Newton's symptom and by the organic method, 262–4. Proposition XXIV: thence to construct one through three points and touch two straight lines, 264–6. Lemma XXII: 'To change [plane] geometrical curves into others of the same class' (by compounding an affine translation and a perspectivity), 268–72. Comparison with La Hire's equivalent 'planiconic' transformation, 271. Halley's 'objection' (October 1686), 272–3. Proposition XXV: the transformation used to construct a conic through two points and to touch three straight lines, 272–4. Proposition XXVI: and through a single point to touch four lines, 276. Lemma XXIII: an old rectilinear locus is resurrected as a rider, 276–8. Lemma XXIV: anharmonic property of parallel tangents to a conic [= Apollonius, *Conics* I, 37/39], 278. Lemma XXV: generalised to hold where the tangents are no longer parallel, 278–80, In corollary, the locus of the centres of conics touching a given quadrilateral is a straight line, 280. Proposition XXVII: thereby to draw a conic to touch five straight lines, 280–2. The axes and foci of an organically constructed conic, 282–4. The 'bisecant' locus of a conic is a homothetic conic, 286. Lemma XXVI: to set a given triangle between three given straight lines, one corner on each, 286–8. Proposition XXVIII: the similar fitting of a given 'trajectory', three points of which are to lie one each on the lines, 290. Lemma XXVII: to set a quadrilateral given in species between four straight lines, each corner lying on one, 290–2. The construction holds where the quadrilateral collapses into a straight line (to be cut by the lines into portions having a given ratio to each other), 292. A variant construction of this case (employs the technique following), 294–6. Proposition XXIX: the similar fitting of a quadrilateral given in species to the given lines, 296. An ingenious variant technique of constructing the problem (using the meets of auxiliary rectilinear loci), 296–8.

Article VI: 'On finding the motions in given orbits' (*viz.* by solving Kepler's Problem). Proposition XXX: repeats the previous exact construction of the parabolic case (§2, Proposition XX), 298–300. Lemma XXVIII: 'there exists no oval whose area cut off by straight lines may generally be found by finite equations'. Newton's 'proof': its basic fallacy is pinpointed, 302–4. A similar ill-founded conjecture (already made in October 1665) regarding the oval's perimeter, 306. Contemporary reactions: a firm counter-instance is given, 306–7. 'Hence the general sector of an ellipse cannot rationally be determined' (so that Kepler's Problem in general has no exact algebraic construction), 308. Proposition XXXI: restates Wren's construction of the elliptical case by means of a (transcendental) 'stretched' cycloid, 308–10. An involved approximate geometrical construction of this case is stated (no proof given, but it is highly accurate), 310–12. An efficient iterative inversion of the equivalent equation $N = \theta - e\sin\theta$ (later simplified into standard 'Newton–Raphson' form), 314–16. A (weak) geometrical approximation in the hyperbolic case, 318. An improved (but still faulty) Wardian 'equation' of the upper-focus angle in the planetary ellipse, 318–22. Its error computed, 322.

Article VII: 'On the rectilinear ascent and descent of bodies'. Propositions XXXII–XXXV repeat (with minimal improvements) the earlier Propositions XXI–XXIV (of §2),

LIST OF PLATES

GEOMETRY AND THE DYNAMICS OF MOTION
(1684–1686/*c.* 1691)

INTRODUCTION

Much has been written in recent years(1) on the early development of Newton's kinematical and dynamical ideas, and his application of them, notably in an astronomical context,(2) to define the 'motion of bodies', either as uniform, mechanically directed rectilinear and rotary movements or, more subtly, as accelerated orbital trajectories traversed round a given centre-point under the continuous impulsion thereto of some externally impressed *vis centripeta*.(3) In preface to the extracts from his mature papers on this theme which are reproduced and, with heavy emphasis on their mathematical undertones, editorially commented upon in following pages we may accordingly be very brief, singling out only a few historical details for special mention and otherwise sketching a highly impressionistic résumé of the highlights of the long sequence of his discoveries 'De motu Corporum' which were afterwards subsumed by him into his masterly *Philosophiæ Naturalis Principia Mathematica*.

That Newton's primary indoctrination in the principles of terrestrial and celestial motion came initially through his study of the published work of Descartes—rather than, as was once thought, from reading Galileo and Kepler—is now firmly documented.(4) In late 1664, on taking up the bulky *Principia Philosophiæ* in which the Cartesian world-view had first been publicly presented

---

(1) See especially J. W. Herivel, *The Background to Newton's 'Principia'* (Oxford, 1966) and R. S. Westfall, *Force in Newton's Physics* (London, 1971). The bibliography to the latter (pp. 551–65) is a useful guide to the spate of periodical articles on the topic which have appeared over the last two decades.

(2) See D. T. Whiteside, 'Newton's Early Thoughts on Planetary Motion: A Fresh Look', *British Journal for the History of Science*, **2**, 1964: 117–37; and C. A. Wilson, 'From Kepler's Laws, So-called, to Universal Gravitation: Empirical Factors', *Archive for History of Exact Sciences*, **6**, 1970: 89–170, particularly 127–44.

(3) As Newton was to name this 'centre-seeking force' in 1684 on the analogy of the contrary *vis centrifuga* which Christiaan Huygens had earlier identified as induced in outwards 'flight' from the centre by constrained uniform motion in a circle; see 1, §1: note (4) below.

(4) Compare D. T. Whiteside, 'Before the *Principia:* the Maturing of Newton's Thoughts on Dynamical Astronomy, 1664–1684', *Journal for the History of Astronomy*, **1**, 1970: 5–19, especially 7–8, 10–12. In his 'Newton's Attribution of the First Two Laws of Motion to Galileo' (*Atti del Simposio su 'Galileo Galilei nella Storia e nella Filosofia della Scienza'*, Florence, 1967: xxiii–xliii) I. B. Cohen has laid stress (p. xl) on 'the lack of positive evidence...that Newton had ever seen Galileo's *Discorsi*'. While Newton had ready access in Cambridge—both in his college library and (from 1669 at least) on the well-stocked private shelves of his senior colleague at Trinity, Isaac Barrow, as well as in the University Library—to Kepler's principal published works, the inventory made at his death of his own books (British Museum Add. 25424, partially printed in R. de Villamil, *Newton: the Man* (London, 1931): 104–11) lists no copies of these and we know of no direct annotation by Newton of any passage in Kepler's writings.

twenty years before,[5] he had rapidly absorbed its multiplicity of secondary explanations of a variety of physical and astronomical phenomena, and in its wake in the following January he drafted a short paper on the 'reflection' of bodies,[6] departing (in elaboration of the basic 'leges Naturæ' which Descartes had set out in paragraphs 37–40 of his *Principia*'s second part) from a universal 'Ax[iom] 100. Every thing doth naturally persevere in that state in which it is unlesse it bee interrupted by some externall cause, hence... A body once moved will always keepe the same celerity, quantity and determinacon of its motion'[7] and a remodelled law of direct elastic impact ordaining that 'motion' (our modern momentum) is preserved when bodies collide.[8] Conceiving in consectary that a body which bounces along at a uniform speed inside a given circle imparts at each successive contact an infinitesimal 'pression' outwards from its centre, he was able therefrom by a simple *ad hoc* argument correctly to compute, in the limit where the inscribed polygon becomes 'ye circle it selfe', a measure of the centrifugal 'endeavour' which a body moving uniformly in a constraining circle continuously exerts upon it in given time.[9] From such an argument from first principles there is no ready extension to the case where a body is restricted to move non-uniformly in a given general curve, while—even granting the broad and far from immediately obvious assumption that no other force acts[10]—the analogous problem of defining the 'free' motion of bodies in orbits round a centre under the contrary 'pull' thereto of a comparable inwards 'endeavour' is still less tractable. Although Newton in about 1670 made the minimal application of his measure of circular centrifugal force, now derived in a simpler

(5) The *princeps* edition was published at Amsterdam in 1644, but Newton appears to have read the 1664 reprint which had just gone on sale in London.

(6) ULC. Add. 4004: 10^r–15^r/38^v/38^r, printed (with some minor inadequacies) in Herivel's *Background* (note (1)): 132–82.

(7) *ibid.*: 12^r; compare 1, §1: note (9) below.

(8) *ibid.*: 10^v/11^r: 'If two bodys...[refle]ct against one another...none of theire motion shall bee lost.... For at their occursion they presse equally uppon one another & therefore one must loose noe more motion y^n ye other doth [gain].' Compare v: 149, note (153), and see also 1, §2: note (16) below.

(9) ULC. Add. 4004: 1^r, reproduced in Herivel's *Background* (note (1)): 129–30. Where the body travels at a speed v along the perimeter of a circle of radius r, imparting at each impact a 'pression' (force-impulse) dv in infinitesimal time dt, it is readily demonstrable (compare 1, §1: note (24) below) that the *vis centrifuga* thus induced is $(dv/dt =)\ v^2/r$. Newton long afterwards repeated the gist of this primitive deduction of this Huygenian measure of circular 'endeavour from the centre' in a late addition to the scholium to Proposition IV of the first book of his published *Principia*; see 1, §2, Appendix 2.6: note (24).

(10) At a macrocosmic level, in fact, Newton continued into the 1670's to allow that the motions of the celestial bodies are modified by complex interactions between the deferent Cartesian vortices in which they are borne along; see D. T. Whiteside, 'Newton's Early Thoughts...' (note (2)): 119, note 11 and 'Before the *Principia*...' (note (5)): 10–12.

Huygenian form,[(11)] to deduce from Kepler's empirical rule that 'ye meane distances of ye primary Planets from ye Sunne are in sesquialter proportion to the periods of their revolutions in time' (as he had jotted it down in late 1664 'out of Street'[(12)]) the corollary that their 'endeavours to recede from the sun will be reciprocally as the squares of their distances from it',[(13)] and some two years afterwards, as we have seen,[(14)] was stimulated to discuss in some detail the

(11) In a short untitled paper (ULC. Add. 3958. 5: 87^r/87^v) first printed by A. R. Hall in 'Newton on the Calculation of Central Forces', *Annals of Science*, **13**, 1957: 62–71 and, slightly more fully, by H. W. Turnbull in *The Correspondence of Isaac Newton*, **1**, 1959: 297–9 (with a complete photocopy of the manuscript on the accompanying Plates V and VI). There is no evidence, contextual or circumstantial, to indicate that Newton was at this time aware of Huygens' equivalent prior (but still unpublished) determination of the measure of what in his 1673 *Horologium Oscillatorium* he referred briefly to as 'vis centrifuga'; see 1, §1: note (24) below.

(12) ULC. Add. 3966: 29^r, a straightforward annotation of an equivalent passage in Thomas Street's *Astronomia Carolina. A New Theorie of the Cœlestial Motions* (London, 1661): 39–40: 'Of the Primary Planets. The Proportion of their Orbes to the Periods of their Revolutions'. In later years Newton grew more chary of accepting this 'regula Kepleriana' as more than a loose approximation to empirical truth; compare 1, §1: note (26) following.

(13) 'in Planetis primarijs cum cubi distantiarum a sole reciproce sunt ut quadrati numeri periodorum in dato tempore: conatus a sole recedendi reciproce erunt ut quadrata distantiarum a sole' (Add. 3958.5: 87^v). The fundamental re-interpretation that these inverse-square centrifugal 'endeavours' are but the apparent effect of a centripetal force of solar gravitation was at this time manifestly still to be made. Nothing in Newton's extant early scientific papers exists to support his celebrated claim to Pierre Desmaizeaux in old age that 'the same year [1666] I began to think of gravity extending to ye orb of the Moon, & having found out how to estimate [see note (9) above] the force with w^{ch} [a] globe revolving within a sphere presses the surface of the sphere: from Keplers Rule of the periodical times of the Planets being in a sesquialterate proportion of their distances from the centers of their Orbs, I deduced that the forces wch keep the Planets in their Orbs must [be] reciprocally as the squares of their distances from the centers about wch they revolve: & thereby compared the force requisite to keep the Moon in her Orb with the force of gravity at the surface of the earth, & found them answer pretty nearly' (ULC. Add. 3968.41: 85^r, first published in preface to *A Catalogue of the Portsmouth Collection of Books and Papers written by or belonging to Sir Isaac Newton...* (Cambridge, 1888): xviii). The first documented record of Newton's making such a test of the inverse-square variation of terrestrial gravitation with distance is an off-hand reference to a recent 'computation' of his to that end which he set down about January 1685, without citing any details, in the augmented scholium to Problem 5 of his revised preliminary tract 'De motu Corporum'; see 1, §1, Appendix 1: note (12) below. Since J. C. Adams first conceived the point nearly a century ago (see note (59) following), it has become commonplace to observe that Newton could earlier have had no great confidence in the result of any numerical comparison of the moon's accelerated fall towards the earth with that of a body near the latter's surface, for, by his own testimony in June 1686 to Edmond Halley, 'before a certain demonstration'—proving, namely, that in an inverse-square force field a uniform sphere attracts an external point as if its mass were concentrated at its centre, much as in Proposition LXXI of the published *Principia*'s first book—'[which] I found ye last year [I] have suspected [ye duplicate proportion] did not reach accurately enough down so low' (*Correspondence*, **2**, 1960: 435; compare 1, §2: note (186) below).

(14) In the paper on pendular oscillation in a cycloid reproduced in III: 420–30.

isochronous 'harmonic' motion of a body vibrating to and fro in an upright cycloidal arc under the accelerating 'efficacy' (its component in the instantaneous direction of swing) of a constant 'gravity' acting vertically downwards, it is understandable that he made no real further progress towards the latter goal until in the winter of 1679/80 or shortly afterwards he attained the insight, in generalisation of Kepler's areal hypothesis for planetary motion under the specific force of solar gravitation, that 'whatsoever was the law of the [centripetal] forces wch kept the Planets in their Orbs, the areas described by a Radius drawn from them to the Sun would be proportional to the times in wch they were described'[(15)] and would therefore afford a universal measure of those orbital times.

In the early summer of 1674, meanwhile, Newton's interest was fleetingly kindled in the dynamics of projectile motion under (locally constant) terrestrial

(15) As Newton long afterwards affirmed in a stray draft (ULC. Add. 3968.9: 101^{r}) of a letter in late July to Pierre Varignon, where he briefly reviewed the *Principia*'s prehistory; a fuller version is reproduced by I. B. Cohen in his *Introduction to Newton's 'Principia'* (Cambridge, 1971): 293–4. Though he had long known Kepler's areal hypothesis in the vague form in which it had been enunciated by Christopher Wren in 1659 in John Wallis' *Tractatus Duo. Prior De Cycloide*...(Oxford, 1659): 80 and in the clearer statement of it given more recently by Nicolaus Mercator in *Philosophical Transactions*, **5**, No. 57 (for 25 March 1670): 1174–5 (compare C. A. Wilson, 'From Kepler's Laws...' (note (2)): 127–31; see also IV: 668–9, note (38) and III: 565–6, note (13) respectively), he continued to prefer to govern the motions of the planets in their elliptical orbits round the sun by a variety of equant mechanisms centred on one or other of the foci—that, for instance, which is featured in the 'Prob. De inventione Apheliorum & verorum motuum Planetarum' reproduced on V: 514–16. His several elegant refinements on the simple Boulliau–Ward hypothesis that the 'empty' (non-solar) focus is the centre of mean motion (see D. T. Whiteside, 'Newton's Early Thoughts...' (note (2)): 121–8, and compare 1, §2: note (163) below) were ultimately subsumed by him into the scholium to Proposition XXXI in the published *Principia*'s first book; compare 1, §3: notes (157) and (159). His confident assertion therein (*Principia*, $_1$1687: Book 3, Hypothesis VIII: 404) that 'Propositio est Astronomis notissima' should be tempered by the truer picture of the restricted awareness of and trust in the areal law sketched by J. L. Russell in 'Kepler's Laws of Planetary Motion: 1609–66', *British Journal for the History of Science*, **2**, 1964: 1–24. Newton had no great familiarity with the primary astronomical sources of his time, and his knowledge of current theory and practice was largely derived from inferior textbooks and secondary compendia. At his death, as H. Zeitlinger has observed in an analysis of its contents ('Newton's Library and its Discovery' (*Library of Sir Isaac Newton: Presentation...to Trinity College, Cambridge* [*on*] *30 October 1943*, Cambridge, 1944: 13–24): 21–4, especially 22–3), he owned relatively few works on astronomy of any kind, and in particular not only nothing of Kepler's (compare note (4) above) but none by Ptolemy, Copernicus, Tycho Brahe or Hevelius. Even those which he did possess manifest a noteworthy concentration of interest: whereas his copies (now Trinity College. NQ. 18.36 and NQ. 10.152 respectively) of Vincent Wing's elementary compendium *Astronomia Britannica* (London, 1669) and Nicolaus Mercator's introductory *Institutionum Astronomicarum Libri Duo* (London, 1676) are well-thumbed and, especially on their terminal flysheets, lavishly annotated by him, that (now Trinity. NQ. 16.79^{1}) of G. A. Borelli's more provocative *Theoricæ Mediceorum Planetarum ex Causis Physicis deductæ* (Florence, 1666) shows no obvious signs of use.

gravity. Two years before, James Gregory in his minute but richly stocked tract 'De Motu Penduli & Projectorum'[16] had, in extension of Galileo's simple theory of unimpeded ballistic motion in an upright parabola, imagined—in unknowing repetition of an equivalent, unpublished argument by Thomas Harriot seventy years before, as we now know[17]—that a 'uniformly retarding motion' acts to resist the speeding projectile continuously along the initial direction of fire, and correctly deduced therefrom that the ensuing trajectory is, in closer agreement with empirical reality, a tilted parabola.[18]. Gregory's stress on the necessity for allowing—in this or other ways—for the resistance of the air in establishing an exact theory of exterior ballistics was soon afterwards discountenanced by the amateur London mathematical practitioner Robert Anderson, who from repeated test-firing of guns and mortars of his own design claimed, highly implausibly,[19] to have experimentally confirmed the accuracy of the simple

(16) *Tentamina Quædam Geometrica De Motu Penduli & Projectorum* (appended to Patrick Mathers [William Sanders], *The Great and New Art of Weighing Vanity*, Glasgow, 1672): 5–9. The kernel of Gregory's preceding discussion of motion in a simple circular pendulum is given in III: 393, note (6). Whether or not we were correct in sequel thereto (in III: 425, note (15)) to suggest that Newton received a copy of this booklet from John Collins upon its publication, he had ready access to that which Isaac Barrow had in his private library at Trinity, and he had manifestly studied its content by 23 June 1673 when (see *ibid.*) he communicated one of its points to Huygens.

(17) See Harriot's worksheets on the topic preserved *passim* in British Museum. Add. 6789: 1^r–87^r, and especially 69^r where he sets down an accurate semi-analytical demonstration 'To prove the [tilted] parabola universally'. J. A. Lohne has fruitfully contrasted Harriot's approach to resisted projectile motion with Newton's later more sophisticated discussion in his examination of 'The Increasing Corruption of Newton's Diagrams' (*History of Science*, **6**, 1967: 69–89): 76–80: '4. The Ballistic Curve of 1684'.

(18) Where (in the notation of 1, §1: note (109) below) v_x ($= dx/dt$) and v_y ($= dy/dt$) are the horizontal and vertical components of speed in trajectory after time t from the firing point (0, 0), the simple Galileian theory (see 1, §2: note (35)) departs implicitly from the defining equations $dv_x/dt = 0$ and $dv_y/dt = -g$ (the downwards acceleration of terrestrial gravity, taken to be effectively constant over the whole trajectory) to posit the kinematic equations $v_x = V_x$ and $v_y = V_y - gt$, where V_x, V_y are the initial horizontal/vertical firing speeds, and thence to deduce the upright parabolic path ($x = V_x t$ and so) $y = V_y t - \frac{1}{2}gt^2 = (V_y/V_x)x - \frac{1}{2}(g/V_x^2)x^2$. In amendment of this the Harriot–Gregory supposition of a force ρ of resistance constantly decelerating the projectile's speed in the initial direction of motion requires the defining equations to be $dv_x/dt = -\rho V_x/V$ and $dv_y/dt = -\rho V_y/V - g$ where $V = \sqrt{[V_x^2 + V_y^2]}$, so producing the same trajectory as that which ensues under a single constant deceleration whose horizontal and vertical components are $\rho V_x/V$ and $\rho V_y/V + g$ respectively—namely, a tilted Apollonian parabola whose axis is parallel to the direction of that compound force of resistance to the motion.

(19) Anderson professed, for instance, to have observed—purportedly over ranges of more than ten miles!—that the horizontal distances attained by bullets of identical weight fired off with equal powder charges from the same gun at elevations equally departing on either side from 45° were the same. Nonetheless his book on gunnery rapidly achieved the status of authority; see A. R. Hall, *Ballistics in the Seventeenth Century* (Cambridge, 1952): 120–1, and E. G. R. Taylor, *The Mathematical Practitioners of Tudor & Stuart England* (Cambridge, 1954): 249.

Galileian parabolic trajectory. In thanking John Collins on 20 June 1674 for his 'kind present' of 'Mr Andersons book'[20] wherein this criticism was made public, Newton mildly commented that it 'is very ingenious, & may prove as useful if his principles be true. But I suspect one of them, namely y^t ye bullet moves in a Parabola. This would be so indeed were ye horizontal celerity of ye bullet uniform, but I should think its motion decays considerably in ye flight'.[21] Continuing Galileo's supposition that, where the 'bullet' is 'shot horizontally' (and, by implication, does not verge greatly from that initial direction in its subsequent trajectory), the distances fallen vertically in successive 'moments of time' are 'in proportion as ye square numbers 1, 4, 9, 16 &c', in lieu of likewise assuming that the corresponding 'horizontall lines...are in Arithmeticall progression' Newton now proposed an alternative 'rule...pretty nearly approach[ing] ye truth' according to which the related horizontal 'celerities' form 'a decreasing Geometricall progression'.[22] The extension in which this kinematic principle is applied to the orbital velocities is readily made once it is seen that the decelerating resistance in the instantaneous direction of motion—and in consequence its horizontal and vertical components also—is likewise geometrically decreased at successive moments, but there is no record that Newton achieved it for another decade.[23] For the moment his caution asserted itself, and he concluded with a request to Collins that 'If you should have occasion to speak of this to y^e Author, I desire you would not mention me becaus I have no mind to concern my self further about it'.

(20) *The Genuine Use and Effects of the Gunne demonstrated* (London, 1674).

(21) *Correspondence*, 1, 1959: 309; compare IV: 659.

(22) In the terms of note (18) above, on setting $V_y = 0$ there ensues ($v_y = -gt$ and so) $y = -\frac{1}{2}gt^2$, while the horizontal speed $v_x = dx/dt$ is now defined to be of the form

$$a+ab+ab^2+\ldots+ab^{t-1} = V_x(1-e^{kt})$$

where $V_x = a/(1-b)$ and $b = e^k$; at once $dv_x/dt = -kv_x$, from which $v_x = V_x - kx$ and hence $kt = \log(1-v_x/V_x) = \log(1+(k/V_x)x)$.

(23) Namely, in the concluding scholium—combining the preceding Problems 6 and 7—of his 1684 tract 'De motu Corporum' (reproduced as 1, §1 below). On suitably absorbing the constant k, the defining equations of motion in the previous note now become $dv_x/dt = -v_x$ and $dv_y/dt = -v_y \pm g$, which may severally be integrated and then combined to yield the *curva logarithmica* $y = ((V_y \mp g)/V_x)\,x \pm g \log(V_x/(V_x - x))$ as the resulting trajectory; see 1, §1: notes (98), (108) and (113). The insight that $-v_x$ and $-v_y \pm g$ are the horizontal and vertical components of the orbital acceleration $(dv/dt =) -v \pm g.dy/ds$, where s is the length of the orbital arc traversed from the firing point (0, 0) and $v = ds/dt$ is the speed attained at the point (x, y) after time t is not trivial: indeed, the tempting extension which Leibniz made in 1689 in Article VI of his 'Schediasma de Resistentia Medii, & Motu projectorum gravium in medio resistente' to resolve the comparable orbital acceleration $dv/dt = -v^2 + g.dy/ds$ into horizontal and vertical 'components' $dv_x/dt = -v_x^2$ and $dv_y/dt = -v_y^2 + g$ is invalid. (See 1, §1: note (109) below, and compare E. J. Aiton, 'Leibniz on Motion in a Resisting Medium', *Archive for History of Exact Sciences*, 9, 1972: 257–74, especially 270–3.)

Five years later his attention was abruptly re-focussed on the subject of celestial dynamics when Robert Hooke, acting formally in his new capacity as secretary to the Royal Society, but also in a commendable personal attempt to re-establish a working *rapport* with him after his increasingly acidic disputes years before over the nature and properties of light had terminated in Newton's brusque resolution to 'bid adew to [Philosophy] eternally, excepting what I do for my privat satisfaction',(24) wrote to Cambridge on 24 November 1679, both expressing the hope that 'you will please to continue your former favours to the Society by communicating what shall occur to you that is philosophicall' and adding that 'For my own part I shall take it as a great favour...particularly if you will let me know your thoughts of that [hypothesis of mine] of compounding the celestiall motions of the planetts of a direct motion by the tangent & an attractive motion towards the centrall body'.(25) Newton—freshly returned from six months in Lincolnshire tending his mother on her death-bed and thereafter, as executor, sorting out her estate(23)—was at first in no mood to begin a new controversy with his old antagonist, and in his reply four days afterwards he reminded Hooke that he had 'for some years past been endeavouring to bend

(24) As he wrote despondently to Oldenburg on 18 November 1676 (*Correspondence*, **2**, 1960: 183). Compare IV: xviii.

(25) *Correspondence*, **2**: 297. Hooke had earlier concluded his *An Attempt to Prove the Motion of the Earth by Observations*...(London, 1674) by outlining his vision of a 'System of the World' dependent 'upon three Suppositions. First, That all Cœlestial Bodies whatsoever, have an attraction or gravitating power towards their own Centers, whereby they attract not only their own parts...but...also all the other Cœlestial Bodies that are within the sphere of their activity;...second..., That all bodies whatsoever that are put into a direct and simple motion, will so continue to move forward in a streight line, till they are by some other effectual powers deflected and bent into a Motion, describing a Circle, Ellipsis, or some other more compounded Curve Line....third..., That these attractive powers are so much the more powerful in operating, by how much the nearer the body wrought upon is to their own Centers. Now what these several degrees are, I have not yet experimentally verified; but it is a notion, which if fully prosecuted as it ought to be, will mightily assist the Astronomer to reduce all the Cœlestial Motions to a certain rule which I doubt will never be done true without it' (*ibid.*: 27–8). We may readily join A. Koyré in here admiring the 'boldness and clarity of Hooke's thought and the depth of his intuition', as he commented upon it in his penetrating exegesis of 'An Unpublished Letter of Robert Hooke to Isaac Newton [9 December 1679]' (*Isis*, **43**, 1952: 312–37 [= *Newtonian Studies* (London, 1965): 221–60]): 318–19 [= 232–4].

(26) Newton's mother, Hannah (*née* Ayscough) died after a short illness in the first week of June 1679. Her will, 'Proved 11 June 1679, at Lincoln' and naming 'my son Isaac' as sole executor, directed that her legacies to her sister and her offspring by her second marriage to Barnabas Smith 'be payd by mine executor within three months after my decease'. (The probate copy, L.C.C. 1679, **2**: 406–7, is reproduced by C. W. Foster in 'Sir Isaac Newton's Family' (*Reports and Papers of the Archaeological and Architectural Societies*, **39**, 1928: 1–62): 50–3.) In his present letter to Hooke he wrote with feeling that 'I have been this last half year in Lincolnshire cumbred wth concerns amongst my relations...so y^{t} I have had no time...so much as to study or mind any thing els but Countrey affairs' (*Correspondence*, **2**: 300).

my self from Philosophy...unless it be perhaps at idle hours sometimes for a diversion', further asserting that 'I did not before ye receipt of your last letter, so much as heare (y^t^ I remember) of your Hypothes[i]s of compounding ye celestial motions of ye Planets, of a direct motion by the tang^t^ to ye curve'. In apology, however, for being 'at present unfurnished wth matter answerable to your expectations', he went on to outline for Hooke a 'fansy of my own about discovering the earth's diurnal motion' with an accompanying rough sketch of the 'spiral line' which he conceived to be traversed (in the uniformly rotating plane of a terrestrial observer) by a body released to fall freely under 'gravity... towards ye center of ye Earth', showing that 'it will not descend in ye perpendicular..., but outrunning ye parts of ye earth will shoot forward to ye east side'.(27) In his response on 9 December Hooke showed himself to be much less interested in the mere fact of this departure of the descending body from the vertical—which, if it could convincingly be detected by experiment,(28) would indeed prove the reality of the earth's rotation round its axis—than in the shape(29) which Newton gave to its onward path of free fall, and skilfully guided the discussion round once more to his preferred hypothesis of compounding such 'planetary' motions; for, he remarked, 'as to the curve Line which you seem to suppose it to des[c]end by (though that was not then at all discoursed of) vizt a kind of spirall which after sume few revolutions(30) leave it in the Center of the Earth my theory of circular motion makes me suppose it would be very differing and nothing att all akin to a spirall but rather a kind [of] Elleptueid...*AFGH* and that the body *A* would never approach neerer the Center...then *G* were it not for the Impediment of the medium as Air or the like....I could adde many other

(27) *Correspondence*, **2**: 300–1.

(28) The immense practical difficulties in so doing are well documented by A. Armitage in his historical survey of increasingly refined attempts over succeeding centuries to measure 'The Deviation of Falling Bodies' (*Annals of Science*, **5**, 1947: 342–51). The similar deflection to the south which both Hooke and Newton also agreed in their present correspondence to exist is even more elusive; compare J. A. Lohne, 'Hooke *versus* Newton. An Analysis of the Documents in the Case on Free Fall and Planetary Motion' (*Centaurus*, **7**, 1960: 6–52): 31–3.

(29) A photocopy of Newton's original penned figure in his letter of 28 November 1679 (now Trinity College, Cambridge. R.4.48^1) is reproduced on page 9 of Lohne's 'Hooke *versus* Newton'. While a small hole in the paper at the crucial place makes it no longer possible to determine whether Newton continued his manuscript spiral right to the centre, its visible remaining portion is highly approximated by the *lituus* ('shepherd's crook') whose polar defining equation is $\phi = k(\sqrt{[R^2/r^2-1]}-\cos^{-1}[r/R])$, $k = \frac{1}{18}\pi$, and which corresponds (in a stationary reference system) to the infinitely looping spiral $\phi = k\sqrt{[R^2/r^2-1]}$. Since for small r the latter is freely traversed in an approximately inverse-cube force field, it is evident that we should not lay too great an emphasis on the precise shape of Newton's roughly sketched fall curve.

(30) In Newton's manuscript figure (see previous note) only about one and a third revolutions are now visible; however, Hooke here implicitly refers the fall curve to a stationary polar reference grid in which it will indeed correspondingly make several loops round the centre; compare D. T. Whiteside, 'Newton's Early Thoughts...' (note (2)): 132, note 52.

con[s]iderations which are consonant to my Theory of Circular motions compounded by a direct motion and an attractive one to a Center'.(31) Thus put on his mettle Newton could no longer afford to treat Hooke's hypothesis lightly, and he carefully phrased his reply on 13 December: 'I agree wth you y^{t} ye body ...if its gravity be supposed uniform...will not descend in a spiral to ye very center but circulate wth an alternate ascent & descent made by it's *vis centrifuga* & gravity alternately overballancing one another.(32) Yet I imagine ye body will not describe an Ellipsoeid but rather such a figure as is represented by *AFOGHIKL* &c'.(33) This time his accompanying diagram(34) was much more

(31) *Correspondence*, **2**: 304–5. Since the original letter (whose photocopy is reproduced in Koyré's 'An Unpublished Letter of Robert Hooke...' (note (25)): 328/330) is in an amanuensis hand except for its signature, we have dared minimally to standardise its orthography.

(32) This echoes, perhaps unconsciously so, Borelli's earlier notion of equivalently compounding the radial 'impetus' of a freely orbiting solar planet from 'duo motus directi inter se contrarij, alter perpetuus, ac uniformis, quo planeta...impulsus à propria magnetica virtute sibi connaturali sese successivè admovet solari corpori, alter verò difformis, et continuè decrescens, quo planeta...expellitur à Sole vi motus circularis' (*Theoricæ Mediceorum Planetarum*... (note (15)): 77). This intuitively plausible notion of a perpetual imbalance along the rotating radius vector between a centrifugal 'force of circular motion' directed outwards from the centre and a contrary inwards force of attractive 'gravity' was again invoked by Newton seventeen months later in explaining how a solar comet 'attracted all ye time of its motion... by ye Sun's magnetism' may be conceived 'by this continuall attraction to have been made to fetch a compass about the sun..., the *vis centrifuga* at [perihelion] overpow'ring the attraction & forcing the Comet there notwithstanding the attraction, to begin to recede from y^{e} Sun' (Newton to Crompton for Flamsteed, ? April 1681; see *Correspondence*, **2**: 361). Similar ways of phrasing have often been employed in recent years to explain the motion of lunar probes near to the moon's surface, likewise making appeal to a pristine state ($\ddot{r} = 0$) of dynamic radial 'balance' at distance r from the force-centre so as therefrom to compound the instantaneously 'imbalanced' radial acceleration $\ddot{r}$ in a general orbit correctly as the difference between an outwards *vis centrifuga* c^2/r^3 induced by the orbiting body's angular revolution round that centre and a contrary *vis centripeta* ever directed thereto. We may appreciate in hindsight how exceedingly fortunate it was for Newton that Hooke pressed him so strongly to conceive of 'planetary' free fall as generated continuously by a central-force deviation from an instantaneously uniform, rectilinear path tangential to the orbit, whose radial acceleration c^2/r^3 is but the component (by implication) of the ideal *vis insita* which from moment to moment sustains it, and from which, as Newton was to demonstrate in Theorem 1 of his 'De motu Corporum' (1, §1 below) in equivalent geometrical terms, the Keplerian areal law is a direct consectary. For Leibniz, who some four years later in his 'Tentamen de Motuum Cœlestium Causis' (*Acta Eruditorum* (January 1689): 82–96) founded his variant dynamics of conic motion on an analysis of the resulting radial equation $\ddot{r} = c^2/r^3 - k^2/r^2$ in which not only the *solicitatio paracentrica* $[f(r) =]\ k^2/r^2$ but also the *conatus centrifugus* c^2/r^3 are physically real forces, the Keplerian areal law had extraneously to be introduced as an axiom independently varying the transverse angular speed of the orbiting body. (Compare E. J. Aiton, 'The Celestial Mechanics of Leibniz', *Annals of Science*, **16**, 1960: 65–82, especially 67–77; and his sequel, 'The Celestial Mechanics of Leibniz: A New Interpretation', *ibid.*, **20**, 1964: 111–23.)

(33) *Correspondence*, **2**: 307.

(34) Reproduced in photocopy (of the original in the British Museum) by J. A. Lohne on page 27 of his 'Hooke *versus* Newton' (note (28)).

carefully drawn, but, although he went on to affirm that 'Your acute Letter having put me upon considering...ye species of this curve, I might add something about its description by points *quam proximè*', it swings the orbiting body through impossibly large (equal) central angles between successive furthest departures A, H, K, ... from the centre:[35] clearly, his technique for summing the 'innumerable & infinitly little' motions of the falling body which are—'for I here consider motions according to ye method of indivisibles—'continually generated by...ye impresses of gravity in every moment of it's passage' was yet too primitive to encompass such subtleties.[36]

Inadequate or no, even such relatively unsophisticated reasonings were above Hooke's head and he could only compliantly answer on 6 January 1679/80 that 'Your Calculation of the Curve by a body attracted by an æquall power at all Distances...is right and the two auges[37] will not unite by about a third of a Revolution' before passing quickly on to object:

'But my supposition is that the Attraction always is in a duplicate proportion to the Distance from the Center Reciprocall, and Consequently that the Velocity will be in a subduplicate proportion to the Attraction and Consequently as Kepler supposes Reciprocall to the Distance.[38] And that with such an attraction the auges will unite in the

(35) Compare Lohne's 'Hooke *versus* Newton': 44–5, and D. T. Whiteside, 'Newton's Early Thoughts...' (note (2)): 134, note 55. See also 1, §2: note (127) below.

(36) See *Correspondence*, 2 307–8. We find it not unreasonable to suppose that, where the body is released at $(R, 0)$ with initial transverse speed V to attain after fall over an arc s in time t the orbital speed v $(= ds/dt)$, Newton could at this point derive the intrinsic equation (dv/dt or) $v.dv/ds = -g.dr/ds$, where g is the constant force of radial 'gravity', and then straightforwardly integrate it to deduce that $v^2 = V^2+2g(R-r)$; and thence, on assuming that r nowhere differs overmuch in magnitude from R (so that the fall path departs little from a circle), infer that the radius of curvature at the general point (r, ϕ) is very nearly

$$v^2/g = V^2/g+2(R-r).$$

The construction 'by points' (or rather by small successive circle arcs of slightly varying curvature) which ensues in an obvious manner would, to be sure, exaggerate the central angles between successive 'aphelia' in the same sense and roughly to the same degree as in Newton's manuscript figure. Nowhere in the present letter, however, is there any indication that he had as yet gained the crucial insight that Kepler's areal law, in its differential form $r^2 d\phi/dt = c$ (where, here, $c = RV$), opens the way, on eliminating the element dt of orbital time, to deriving the polar differential equation $d\phi/dr = RV/r^2 \sqrt{[V^2+2g(R-r)-R^2V^2/r^2]}$ of the fall path, without which its precise construction is impossible.

(37) The 'aphelion' points of furthest departure from the force-centre; see 1, §2: note (128) below.

(38) Observe that Hooke continues to place his faith in Kepler's approximate 'sum-distance' form of the areal law: namely, in the terms of note (36) preceding, the working rule that ($r.\,ds \propto dt$ and so) $v \propto 1/r \propto \sqrt{[f(r)]}$, where $f(r) = k/r^2$ is the inverse-square radial 'Attraction'. (Compare E. J. Aiton, 'Kepler's Second Law of Planetary Motion', *Isis*, **60**, 1969: 75–90, especially 76.) Acceptance of such a 'distance' law for regulating the orbital speed v effectively blocks the way to further theoretical advance, since from the basic intrinsic

same part of the Circle and that the neerest point of accesse to the center will be opposite to the furthest Distant. Which I conceive doth very Intelligibly and truly make out all the Appearances of the Heavens. And therefore (though in truth I agree with You that the Explicating the Curve in which a body Descending to the Center of the Earth, would circumgyrate were a Speculation of noe Use[39] yet) the finding out the proprietys of a Curve made by two such principles will be of great Concerne to Mankind...: for the composition of two such motions I conceive will make out that of the moon....But in the Celestiall Motions the Sun Earth or Centrall body are the cause of the Attraction, and though they cannot be supposed mathematicall points yet they may be Conceived as physicall and the attraction at a Considerable Distance may be computed according to the former proportion as from the very Center.'[40]

Though Newton responded neither to this[41] nor to a brief following letter by Hooke on 17 January where he reiterated that 'It now remaines to know the proprietys of a curve Line (not circular nor concentricall) made by a centrall attractive power which makes the velocitys of Descent from the tangent Line or equall straight motion at all Distances in a Duplicate proportion to the Distances Reciprocally taken',[42] the key question had been squarely and unambiguously put to him: *does* a 'centrall attractive power' which, continuously directed to the sun (set by Keplerian hypothesis at a focus of the orbit), deviates a planet from its instantaneous uniform 'direction motion by the tangent' into its elliptical path vary in action inversely as the square of the distance? The work-sheet on which Newton penned the affirmative computation which he was thus

equation (dv/dt or) $v.dv/ds = -f(r).dr/ds$, where $f(r)$ is a general attraction, there ensues the anomaly that $v^2 = V^2+2\int_R^r -f(r).dr \propto 1/r^2$, and hence that r is fixed to have one (or more) of a finite number of particular values; in Hooke's present instance of an inverse-square 'attraction' $f(r) = k/r^2$, for which $v^2 = V^2+2k(r^{-1}-R^{-1})$ it is restricted to be a root of the quadratic $(V^2-2k/R)r^2+2kr = r^2v^2$, constant.

(39) Cautious as ever, Newton had closed his preceding letter with the typical disclaimer that 'the thing being of no great moment I rather beg your pardon for having troubled you thus far wth this second scribble wherein if you meet wth any thing inept or erroneous I hope you will pardon ye former & ye latter I submit & leave to your correction' (*Correspondence*, **2**: 308). Hooke here somewhat disingenuously accepts it at its face value.

(40) *Correspondence*, **2**: 309. Hooke added as further bait that 'This Curve truly Calculated will shew the error of those many lame shifts [*sc*. equant theories of elliptical motion; compare note (15) above] made use of by astronomers to approach the true motions of the planets with their tables'—no doubt a sly dig at John Flamsteed who was then on the point of publishing his highly accurate table of lunar motions in appendix to his *Doctrine of the Sphere* (London, 1680).

(41) '[I] never answered his third [letter]', Newton affirmed to Edmond Halley on 20 June 1686 (*Correspondence*, **2**: 436).

(42) *Correspondence*, **2**: 313. He once more flatteringly adjoined that 'I doubt not but that by your excellent method you will easily find out what that Curve must be, and its proprietys, and suggest a physicall Reason of this proportion. If you have had any time to consider of this matter, a word or two of your Thoughts of it will be very gratefull to the Society...'.

led to make shortly afterwards has not survived, and he seemingly did not at the time redraft it into a more finished form,(43) but there can be no doubt that his demonstration that periodic motion in an ellipse may be maintained by an inverse-square deviating force continuously directed to a focus was founded, exactly as he always later claimed,(44) on the novel insight that the Keplerian law of the proportionality of the areas swept out by the radius vector to the times of their description held universally true, whatever the law of central 'attraction'.(45)

---

(43) J. W. Herivel has tentatively identified this with a putative antecedent version of a later autograph manuscript (ULC. Add. 3965.1: 1^r–3^v) which is a light revision of 'A Demonstration That the Planets by their gravity towards the Sun may move in Ellipses' sent to John Locke in 'Mar$\frac{89}{90}$' (or so the inscription on the verso of its secretarial transcript, now Bodleian. MS Locke. c. 31. 101^r–104^r, records); see his 'The Originals of the Two Propositions Discovered by Newton in December 1679?', *Archives Internationales d'Histoire des Sciences*, **14**, 1961: 23–33, and his revised statement of the thesis in *The Background to Newton's 'Principia'* (note (1)): 108–17. But the time at, and purpose for, which Newton composed such a prior version—one whose very existence may be established only by circumstantial extrapolation—can only, of course, be an educated guess and our own equally irrefutable surmise as to its origin (see note (52) following, and *History of Science*, **5**, 1966: 115, note 4) is somewhat different. R. S. Westfall, who initially proposed a number of further secondary arguments—none cogent—in favour of Herivel's conjecture in 'A note on Newton's demonstration of motion in ellipses' (*Archives Internationales d'Histoire des Sciences*, **22**, 1969: 111–17), has now a deal less cautiously elevated it to be basic in his reconstruction of the development of Newton's dynamic thought (see his *Force in Newton's Physics* (note (1)): 429–31, and 'Circular Motion in Seventeenth-Century Mechanics', *Isis*, **63**, 1972: 184–9, especially 188). We may add that, although Newton toyed briefly in about 1691 with the idea of summarising this slightly cumbrous variant 'Lockean' proof of inverse-square elliptical motion in a short *idem aliter* to Proposition XI of the *Principia's* first book (see 3, §1.4: note (33) below), he in fact—perhaps deterred by William Whiston's printing of the complete paper in a Latin version, 'qualem nempe eam è charta MS Ipsius Newtoni olim acceperam', in Lectures XIV/XV of his *Prælectiones Physico-Mathematicæ* (London, $_1$1710: 136–45)—never saw fit to publish it himself in any form.

(44) See note (15) above. Only six years after the event Newton affirmed to Halley, in a letter of 27 July 1686, that Hooke's 'correcting my Spiral occasioned my finding y^e Theorem by wch I afterward examined y^e Ellipsis' (*Correspondence*, **2**: 447), though he was careful immediately to adjoin: 'yet am I not beholden to him for any light into y^t business but only for ye diversion he gave me from my other studies to think on these things & for his dogmaticalnes in writing as if he had found ye motion in ye Ellipsis, wch inclined me to try it after I saw by what method it was to be done'.

(45) Namely, as a theoretical consectary of the fundamental principle that, at each instant of its orbit, the undeviated motion of the 'planetary' body is uniform and in a straight line. His proof could have differed only trivially from that which he subsequently expounded in Theorem 1 of his 'De motu Corporum' (1, §1 below). We have elsewhere ('Newton's Early Thoughts...' (note (2)): 136) hazarded the guess that his computation in the particular elliptical case (see the figure on page 47 below) deriving the infinitesimal focal deviation $RQ = \frac{1}{2}SP^{-2}.\ dt^2$ to be equal to QT^2/L, where L is the ellipse's (principal) *latus rectum*, would initially have brought home to him the necessity of working with the exact areal law in its differential form, according to which the focal sector $(PSQ) = \frac{1}{2}SP \times QT$ is proportional to the time dt of orbit over the vanishingly small arc $\widehat{PQ}$.

At long last, in direct consequence of Hooke's brief but forceful intervention (however much in later years he might scorn the notion), Newton had a sound, if still rudimentary, basis on which to begin to build the 'System of the World' which the former had earlier envisaged in his *Attempt to Prove the Motion of the Earth*.(46) Even though, diverted by other more immediately pressing chemical, astronomical and theological interests, he was not to start to convert his rough scheme into the towering edifice of his *Principia* for another four and a half years,(47) it was henceforth possible accurately to comprehend—and in principle

(46) See note (25) above. It is well known that Newton, reluctant to admit any debt to Hooke lest the latter should magnify it out of all proportion, was afterwards ungenerous in admitting the catalytic rôle which he played in his own progress to dynamical enlightenment in the winter of 1679/80. When in the late spring of 1686 Hooke began to lay public claim in London to 'the invention of ye rule of the decrease of Gravity, being reciprocally as the squares of the distances from the Center', as Halley wrote to Newton on 22 May (*Correspondence*, **2**: 431), 'though he owns the Demonstration of the Curves generated therby to be wholly your own', Newton responded that 'in one of may papers [ULC. Add. 3958.5: 87^r^/87^v^; see note (13) above] writ...above fifteen years ago the proportion of ye forces of ye Planets from ye Sun reciprocally duplicate to their distances from him is exprest & ye proportion of our gravity to ye Moon's *conatus recedendi a centro Terræ* is calculated [namely, as 1: 4000_+ where 'Luna...distat 59 vel 60 semidiametris terrestribus a terra'] thô not accurately enough.... Which shews that I had then my eye upon comparing ye forces of ye Planets arising from their circular motion & understood it: so that a while after w^n^ Mr Hook propounded ye Probleme in his *Attempt to prove ye motion of ye earth*, if I had not known ye duplicate proportion before I could not but have found it now....And I hope I shall not be urged to declare in print that I understood not ye obvious mathematical conditions of my own Hypothesis. But grant I received it afterwards from Mr Hook, yet have I as great a right to it as to ye Ellipsis. For as Kepler knew ye Orb to be not circular but oval & guest it to be Elliptical, so Mr Hook without knowing what I have found out since his letters to me, can know no more but that ye proportion was duplicate *quam proximè* at great distances from ye center, & only guest it to be so accurately [*sc.* near the earth's surface]' (*Correspondence*, **2**: 436–7). Later, when he was told 'by one who had it from another lately present at one of your [Royal Society] meetings' that Hooke continued to 'make a great stir, pretending I had all from him', Newton blew up: 'he has done nothing & yet written in such a way as if he knew & had sufficiently hinted all but what remained to be determined by ye drudgery of calculations & observations.... Now is not this very fine? Mathematicians that find out, settle & do all the business must content themselves with being nothing but dry calculators & drudges & another that does nothing but pretend & grasp at all things must carry away all the invention as well of those that were to follow him as of those that went before' (*Correspondence*, **2**: 438). Afterwards, having been suitably mollified by Halley, and confident that the world at large would recognise that 'when Hugenius [in his *Horologium Oscillatorium* (Paris, 1673): 160; see 1, §1: note (24) below] had told how to find y^e^ force in all cases of circular motion, he had told 'em how to do it in this [of solar gravitation, whose 'duplicate proportion' $v^2/r \propto r^{-2}$ is an immediate consequence of Kepler's third law, that $v^2 \propto 1/r$ for uniform speeds v of orbit in concentric circular orbits of radius r] as well' (*ibid.*), he thereafter was persuaded to insert a bare mention of Hooke's name—along with those of Wren and Halley (see note (49) following)—in scholium to the pertinent Proposition IV of the published *Principia*'s first book; see 1, §2, Appendix 2.6: note (22) below.

(47) Compare v: xiii. In astronomy, the 'great comet' which appeared in November 1680 and remained visible till the following March for long pre-empted his attention, leading him

exactly to compute—all celestial motions, and notably those of the earth and her moon and of the transient comets, within a unified dynamical framework founded on the universal action of inverse-square gravitation, one which would soon seem extendible to explain not only all observable macrocosmic reality but also the complex interplay of corpuscular forces (both central and centrifugal) which Newton conceived to act at the micro-level.

In the contingent event a gentlemanly wager was circuitously to spur Newton to undertake its preliminary construction. For 'in January [16]83/4', Halley later wrote to him from London,(48)

'I, having from the consideration of the sesquialter proportion of Kepler, concluded that the centripetall force decreased in the proportion of the squares of the distances reciprocally,(49) came one Wednesday to town, where I mett with Sr Christ. Wrenn and Mr Hook, and falling in discourse about it, Mr Hook affirmed that upon that principle all the Laws of the celestiall motions were to be demonstrated, and that he himself had done it; I declared the ill success of my attempts; and Sr Christopher to encourage the

to exchange a flurry of letters with John Flamsteed in the opening months of 1681 (*Correspondence*, **2**: 315–17, 336–72; compare v: 524, note (1)) not only regarding the determination of its path around—or rather, as Newton for long continued to believe, past—the sun, but also discussing the physical nature and properties of comets in general. (See the extended discussion of the topic given by J. A. Ruffner in his 'The Background and Early Development of Newton's Theory of Comets' (Ph.D. thesis, Indiana University, 1966): 239–321). Newton himself afterwards recalled to Halley on 14 July 1686 that 'when I had tried [the method of determining Figures] in ye Ellipsis, I threw the calculation by being upon other studies & so it rested for about 5 yeares...' (*Correspondence*, **2**: 444).

(48) In his letter of 29 June 1686 (*Correspondence*, **2**: 442), responding to Newton's reiterated request to ask Christopher Wren if he knew the inverse-square law of solar gravitation before Hooke, for 'I am almost confident by circumstances that Sr Christ...knew ye duplicate proportion wn I gave him a visit...about 9 yeares since at his Lodgings [during which he] discoursed of this Problem of Determining the Hevenly motions upon philosophicall principles ...& then Mr Hook (by [not mentioning the proportion in] his book *Cometa* written afterward [1678]) will prove ye last of us three yt knew it' (*ibid.*: 433–4, 435). In introduction to the excerpt which follows, Halley responded somewhat negatively that 'According to your desire I waited upon Sr Christopher Wren, to inquire of him if he had the first notion of the reciprocall duplicate proportion from Mr Hook. His answer was, that he himself very many years since had had his thoughts upon making out the Planets motions by a composition of a Descent towards the sun, & an impress motion; but that at length he gave over, not finding the means of doing it. Since which time Mr Hook had frequently told him that he had done it, and attempted to make it out to him, but that he never satisfied him, that his demonstrations were cogent' (*ibid.*: 441–2). This seems to refute Newton's firm conclusion a fortnight later on 14 July that 'Sr Christopher Wren's examining ye Ellipsis over against ye Focus shews yt he knew [ye duplicate proportion] many yeares ago before he left of his enquiry after ye figure by an impress motion & a descent compounded together' (*ibid.*: 445). We may guess that Wren, like Hooke, was blocked in his investigation by failure either to compute the limit-ratio $RQ/QT^2 = 1/L$ for the ellipse (see note (95)) or to apply the Keplerian areal law as a measure of orbital time.

(49) Compare notes (13) and (46) above.

Inquiry, s^d that he would give Mr Hook or me 2 months time to bring him a convincing demonstration thereof, and besides the honour, he of us that did it, should have from him a present of a book of 40^s. Mr Hook then s^d that he had it, but that he would conceale it for some time that others triing and failing, might know how to value it, when he should make it publick; however I remember Sr Christ. was little satisfied that he could do it, and tho Mr Hook then promised to show it him, I do not yet find that in that particular he has been as good as his word.(50) The August following when I did my self the honour to viset you, I then leart̄ the good news that you had brought this demonstration to perfection, and you were pleased, to promise me a copy thereof, which the November following I received with a great deal of satisfaction from M^r [Edward] Paget.'(51)

Or rather, as Newton took trouble to insist on in his reply to Halley on 14 July 1686, during that August visit 'upon your request I sought for y^t paper [containing the calculation which he had thrown by] & not finding it did it again & reduced it into ye Propositions shewed you by Mr Paget'.(52) Having thus

(50) Which is not, of course, to deny that Hooke had a deep understanding of the dynamics of circular motion; compare J. A. Lohne, 'Hooke *versus* Newton' (note (28)): 10–17. The cryptic phrase 'perfect Theory of Heavens' jotted down by Hooke on 4 January 1679/80 (see H. W. Robinson and W. Adams, *The Diary of Robert Hooke, 1672–1680* (London, 1935): 435) has puzzled recent historians; we surmise that he there made reference to the 'Curve truly Calculated' traversed by a planetary body drawn from its 'direction motion by the tangent' by an 'Attraction always...in a duplicate proportion to the Distance from the Center Reciprocall', by which alone the 'true motions of the planets' may be approached (as he wrote to Newton two days later), without himself being able more than to 'guess' that it is an ellipse.

(51) Though Halley went on to assert that 'As to the manner of Mr Hooks claiming this discovery, I fear it has been represented in worse colours than it ought; for he neither made publick application to the Society for Justice, nor pretended you had all from him', he here added his forthright opinion that 'as all this past Mr Hook was acquainted with it; and according to the philosophically ambitious temper he is of, he would, had he been master of a like demonstration, no longer conceald it, the reason he told Sr Christopher & I now ceasing. But now he sais that this is but one part of an excellent System of Nature, which he has conceived, but has not yet compleatly made out, so that he thinks not fit to publish one part without the other. But I have plainly told him, that unless he produce another differing demonstration, and let the world judge of it, neither I nor any one else can belive it' (*Correspondence*, **2**: 442).

(52) *Correspondence*, **2**: 444–5. More colourfully, as Abraham de Moivre related to John Conduitt in November 1727, when 'Dr Halley asked him for his calculation...Sr Isaac looked among his papers but could not find it, but he promised him to renew it, & then to send it him. Sr Isaac in order to make good his promise fell to work again, but he could not come to that conclusion wch he thought he had before examined with care. However he attempted a new way which tho' longer than the first, brought him again to his former conclusion, then he examined carefully what might be the reason why the calculation he had undertaken before did not prove right, & he found that having drawn an Ellipsis coursely with his own hand, he had drawn the two Axes of the Curve, instead of drawing two Diameters somewhat inclined to one another, whereby he might have fixed his imagination to any two conjugate diameters, which was requisite he should do. That being perceived, he made both his calculations agree together'. (The immediately preceding sentences of this memorandum, whose Conduitt auto-

renewed the original computation in which he applied his general 'method of determining [the centripetal forces in] Figures' to the trial case of the planetary 'Ellipsis', and thereafter elaborated it into a connected discourse on the elementary dynamical 'motion of bodies', Newton manifestly had no further need to preserve its 'paper'—if indeed it still existed (or, the cynic will add, ever was in existence)—and it is never heard of again.

The tract 'De motu Corporum'(53) in which he came to expound these augmented 'Propositions'—in fact, four Theorems and seven Problems, prefaced by three Definitions and four Hypotheses, and complemented by a number of appended explanatory 'Scholia'—opens by deducing the Keplerian areal law in general form (Theorem 1) as a consequence of the principle of rectilinear motion in an arbitrary central-force field, and gathering therefrom, in extension of the separately derived 'Huygenian' measure of the 'centre-seeking force' which sustains uniform motion in a circle (Theorem 2), a novel universal limit-formula(54) for the *vis centripeta* induced towards an arbitrary point in its plane by orbit in a given curve (Theorem 3); in sequel this is applied to calculate the central force accelerating the motion of a body revolving in a circle by 'pulling' it towards a fixed point in its perimeter (Problem 1) and then, much more importantly, that—varying as the direct first power and inverse square of the distance respectively—generated by motion in an ellipse under a deviating force to its centre and a focus (Problems 2 and 3, with Problem 5 the special case of the latter in which the ellipse shrinks into its main axis, and the ensuing reciprocating motion is continued past the force-centre into a symmetric rectilinear 'orbit' on its further side and back again(55)). The primary model of the periodic elliptical motion of a planetary mass-point under an inverse-square pull to its solar focus is further developed to yield a first proof (Theorem 4) of the exactness of Kepler's third 'law' asserting the proportionality of the square of

---

graph is now in the private possession of J. H. Schaffner, have already been quoted on IV: xx.) If this anecdote—which, we may presume, de Moivre heard from Newton's own lips in his old age, doubtless with a certain embroidery of the historical fact—is true in its essence, it is tempting to suppose (as we have earlier conjectured in *History of Science*, 5, 1966: 115, note 4) that this 'new way' was some preliminary version of the variant 'Demonstration That the Planets by their gravity towards the Sun may move in Ellipses' afterwards, in March 1690, sent to Locke (on which see note (43) above). We ignore as unfounded—not that it matters much—Conduitt's marginal query 'May?' as to the month of Halley's visit, recorded merely as 'in 1684' by de Moivre. Not even Newton's own faded recollection to Varignon some thirty-five years afterwards that it took place in 'Spring 1684' (ULC. Add. 3968.9: 101r; compare note (15) above) can controvert Halley's sharp memory in June 1686 of a crucial event then not two years old.

(53) Reproduced from the autograph original (not for the first time, but with a number of novelties in our editorial commentary) in 1, §1 below.

(54) See 1, §1: notes (19) and (30).

(55) Compare 1, §1: note (89) below.

the periodic time of revolution to the cube of the mean radius vector, and thereafter (Problem 5), given the speed and direction of motion at some point, to specify the construction of the ensuing conic curve. In a final *tour de force* Newton departed from this dominant theme of central-force orbits to annex his improved solution of the problem of terrestrial exterior ballistics,(56) justly compounding the component horizontal and vertical motions of a missile resisted in direct proportion to its speed (Problems 6 and 7) so as, in a terminal scholium, accurately to trace its logarithmic trajectory.

The rest of the story is briefly told. When he was shown the copy of Newton's 'Propositions' which Paget carried to London in November 1684, Halley not only made haste to transcribe its content(57) but 'therupon', as he reminded Newton nineteen months later,(58) 'took another Journy down to Cambridge, on purpose to conferr with you about it', finding on his arrival that the latter was already hard at work revising and enlarging its text into a full-scale treatise 'about Motion'. Although—'the examining severall things' having, in his own words 'taken a greater part of my time then I expected, and a great deale of it to no purpose'(59)—Newton's initial expectation of quickly completing the task was

(56) In extension of his tentative inferior 'rule' to the same purpose of a decade earlier; see note (22) above. The minimally corrected construction of the present logarithmic trajectory which Newton afterwards set—there with formal demonstration—as Proposition IV of the published *Principia's* second book is reproduced in 1, §1, Appendix 3 following.

(57) Halley's slightly variant autograph copy—less its Problems 6 and 7 and concluding scholium which were (see 1, §1: note (93) below) passed by him in December 1686 to John Wallis to see if they 'might not be (especially the 7th problem) better illustrated'—is now ULC. Add. 3965.7: 63r–70r; compare 1, §1: note (2).

(58) See his letter of 29 June 1686 (*Correspondence*, **2**: 422). On 10 December following Halley reported back to the Royal Society that he had 'lately seen Mr Newton at Cambridge, who had shewed him a curious treatise, *De Motu*; which was...promised to be sent to the Society to be entered upon their register' (Thomas Birch, *History of the Royal Society of London*, **4** (London, 1757): 347); the motion was then passed that Halley be 'desired to put Mr Newton in mind of his promise for the securing his invention to himself till such time as he could be at leisure to publish it' (*ibid.*). In the meantime as a preliminary safeguard over his author's rights of 'invention' the initial 'Notions about Motion' which Newton had sent up the previous month were transcribed into the Society's Register Book (**6**: 218 ff.); compare the next note.

(59) As he wrote to the Royal Society's secretary, Francis Aston, on 23 February 1684/5 (*Correspondence*, **2**: 415) in thanking him for entering his earlier 'Notions about Motion' into 'their Register' (see the previous note). Remarking in sequel that 'now I am to goe into Lincolnshire for a Month', he adjoined optimistically that 'Afterwards I intend to finish it [*sc.* the revised tract] as soon as I can conveniently' (*ibid.*). We may readily conjecture that part of Newton's largely fruitless effort in these first weeks of 1685 was spent in seeking to determine whether 'ye duplicate proportion' of the total attraction of a uniform sphere on an external point is preserved, as Hooke had asserted, near to its 'superficies'; for, he told Halley on 20 June 1686, 'before a certain demonstration [that of Proposition XL of the amplified 'De motu Corporum', later renumbered to be Proposition LXXI of the revised 'Liber primus'] I found ye last year [I] have suspected it did not reach accurately down so low' (*Correspondence*,

inevitably soon dashed, over the next two years(60) and under Halley's solicitous editorial eye and personal encouragement(61) that treatise 'De motu Corporum'

2: 435; compare note (13) above and 1, §2: note (186) below), further adding the exaggerated rhetorical flourish that 'There is so strong an objection against ye accurateness of this proportion, y^t without my Demonstrations, to wch Mr Hook is yet a stranger, it cannot be believed by a judicious Philosopher to be any where accurate' (*ibid.*: 437)—'Nam fieri posset ut proportio illa in majoribus distantiis satis obtineret, at prope superficiem Planetæ, ob inæquales particularum distantias & situs dissimiles, notabiliter erraret' as he afterwards wrote in Proposition VIII of the published *Principia*'s third book ($_1$1687: 413). Still more overdone is J. W. L. Glaisher's claim, embroidering a suggestion privately communicated to him by J. C. Adams, that Newton's recognition that a uniform sphere in an inverse-square force-field attracts an external point exactly as though its mass were concentrated at its centre is 'The great event that stands out by itself in this memorable period [1684–6], and forms a dividing point in the history of this wonderful work.... No sooner had Newton proved this superb theorem—and we know from his own words that he had no expectation of so beautiful a result till it emerged from his mathematical investigation—than all the mechanism of the universe at once lay spread before him.' (See his address at Trinity College, Cambridge on 19 April 1888 on 'The Bicentenary of Newton's *Principia*' [*The Cambridge Chronicle and University Journal*..., Friday, 20 April 1888: 7–8]: 7.)

(60) Newton's often quoted subsequent affirmation—in July 1719, in fact, in the rough draft of a letter to Varignon whose precision of detail is otherwise considerably suspect—that 'I wrote [the book] in 17 or 18 months, beginning in the end of December 1684 & sending it to ye R. Society in May 1686: excepting that about ten or twelve of the Propositions were composed before...' (ULC. Add. 3968.9: 106^v) telescopes this time-interval so as to excuse the several mistakes in the *Principia's* first edition—when he could not father these off on to 'an Emanuensis [Humphrey Newton] who understood not Mathematicks' or as 'faults...of the Press'—by 'reason of the short time in wch I wrote it' (*ibid.*: 101^r/101^v; compare note (15) above). Though Halley had written to Newton on 22 May 1686 thanking him for 'Your Incomparable treatise intituled *Philosophiæ Naturalis Principia Mathematica*' which 'was by Dr Vincent presented to the R. Society on the 28th [April] past' (*Correspondence*, 2: 431), this was only the 'Liber primus' of the work eventually published; the 'second book' which, Newton wrote to Halley on 13 February 1686/7, 'I made ready for you in Autumn...that it should come out wth ye first & be ready against ye time you might need it' (*ibid.*: 464) was not sent on to Halley till early March (*ibid.*: 472), while not till 5 April 1687 was the latter able to confirm receipt of 'the last part of your divine Treatise' which 'came to town [yester]day sennight [*viz.* 28 March]' (*ibid.*: 473).

(61) Halley's 'pains' on Newton's behalf, as the latter dubbed them on 1 March 1686/7 (*Correspondence*, 2: 471), were by no means confined, as used to be thought, to undertaking the time-consuming subeditorial labour and financial risk of seeing a finished manuscript through press; rather, as the preserved sheets of his interim *critique* (now ULC. Add. 3965.9: 94–9) of the preliminary states of the ensuing treatise reveal, he played throughout an active, if minor, rôle in shaping the verbal surface of Newton's thoughts and in bringing to his notice a number of technical imperfections. (See Cohen's *Introduction to Newton's 'Principia'* (note (15)): 122–4, 138–42 and especially 336–44; compare also 1, §2: note (1) below.) To cite perhaps the most famous instance of his benign influence on the published form of the *Principia*, when in late June 1686 Newton, stung by Hooke's claim to 'ye duplicate proportion' of terrestrial gravitation, announced his 'designe' of suppressing the revised 'Liber Tertius, De Mundi Systemate' which he was then busy preparing (see the previous note)—for 'Philosophy is such an impertinently litigious Lady that a man had as good be engaged in Law suits as have to do with

slowly evolved and broadened in scope, swelling first, in the early summer of 1685 to be a pair of related 'books' on the same enunciated theme, and thereafter (with the 'Liber primus' further enlarged and itself bifurcated, and the companion 'Liber secundus' replaced by a new and considerably more technical 'Liber tertius' expounding the application of Newton's theoretical dynamical principles to the reality of the existing physical 'system of the world') rapidly maturing to become the bulky volume on general 'mathematical principles of natural philosophy' which was finally—and, with the Royal Society demurring to give more than its *imprimatur*, ultimately at Halley's expense(62)—published to the world in the summer of 1687.

From these preliminary manuscript tracts 'on the motion of bodies', insofar as their none too complete surviving state permits, and from related revisions of their printed text which Newton in the early 1690's tentatively had it in mind to make we here, in following pages, reproduce extracts of mathematical interest and significance. There is, in prelude, no need to go deeply into their detail. In his enlarged treatise 'De motu Corporum'(63) Newton takes particular care to augment and closely explain the introductory definitions, axioms and general lemmas(64) on which its fuller discussion of dynamical motion is founded, and then in the body of the text—as we may now know it(65)—gives a much amplified

her. I found it so formerly, & now I no sooner come near her again but she gives me warning' (*Correspondence*, **2**: 437)—Halley successfully begged him 'not to let your resentments run so high as to deprive us of your third book, wherin the application of your Mathematicall doctrine...will undoubtedly render it acceptable to those that will call themselves philosophers without Mathematicks, which are by much the greater number' (*ibid.*: 443). His last service to Newton was to write a long, appreciative review of the published volume in the *Philosophical Transactions*, **16**, No. 186 [for January–March 1687]: 291–7 (reproduced in (ed.) I. B. Cohen, *Isaac Newton's Papers and Letters on Natural Philosophy* (Cambridge, 1958): 405–11). Newton, whose own usual author's attitude to time-consuming policy decisions on typographical design and production and to the boring but very necessary attendant routine of proof-reading was an impatient hurry to be the soonest done with both, was exceedingly fortunate to have so conscientious and intelligent a colleague to see his severely mathematical masterpiece through press.

(62) We may add—understanding, as would seem realistic by contemporary standards, an average cost to him of some 12*s*. per quarto sheet and that two-thirds of the total run of 250–300 copies were sold for cash rather than given away as presentations—that Halley, good entrepreneur that he was, made a reasonable (and, in the circumstances, very hard-earned) profit on the budgeted sale price of '9 shillings...bound in Calves leather [or] in Quires [*sc.* unbound sheets]...5 sh....for ready [money]' which he quoted to Newton on 5 July 1687 when he announced that 'I have at length brought your Book to an end' (*Correspondence*, **2**: 480–1).

(63) Partially reproduced in 1, §2 below.

(64) The preparatory 'Definitiones' and 'Leges' of an intermediate draft 'De motu Corporum in medijs regulariter cedentibus', herein subsumed, are reproduced in 1, §2, Appendix 1.

(65) Of its opening preliminaries only the concluding Lemmas VII–XI are in fact now

treatment of the conic motion induced under an inverse-square force to a focus, adding to its primary elliptical case (here Proposition X) a parallel discussion (in Propositions XI/XII) of its complementary hyperbolic and parabolic instances, introducing a simplified, more basic demonstration (Propositions XIII/XIV) of the exactness of Kepler's third law regulating—on the understanding that these do not perturb one another significantly—the motions of bodies in ellipses round a common focus, extending his earlier constructions of conic orbits given a focus and three points or tangents (Propositions XVII–XIX), adjoining an elegant construction (Proposition XXI) for determining the point in a given parabolic orbit attained in a given time, extending (in Propositions XXII–XXIV) his previous definition of the reciprocating rectilinear inverse-square motion of a body falling directly towards (and past) the force-centre, and, most notably of all perhaps, ingeniously proving (Proposition XL) that in a general inverse-square force-field a uniform sphere acts in total upon an external point as though its mass were concentrated at its centre.(66)

With the ensuing revised 'Liber primus', whose manuscript 'Articles' IV–X are reproduced in sequel,(67) we may be still more succinct since its text (minor last-minute changes and insertions apart) is essentially that of the corresponding Sections IV–X of the published *Principia's* first book.(68) Newton's previous constructions of conic orbits with a given focus (now yet further augmented to fill Article IV) are here complemented by a novel following section (V)—as

preserved, but we make good the deficiency by setting in introduction a rough draft of its definitions 'De motu Corporum' (ULC. Add. 3965.5: 21ʳ, reproduced in 1, §2.1 below) and then restore the intervening text by interpolating corresponding portions of the revised 'Liber primus'; see 1, §2: notes (1) and (2). Reasoned conjectures as to the content of the missing following Propositions XXV–XXXIII (uniquely restorable) and XLIV–[*c.* LXXV] are given in 1, §2: notes (174) and (188).

(66) See 1, §2: note (184), and compare note (59) above. In sequel (1, §2, Appendices 3/4) we reproduce, first, the manuscript of a parallel computation of the total inverse-square 'pull' of a laminar circle, and thereafter the printed text of the published 'Liber primus' where extension is made both (Propositions LXXVII–LXXXI) to determine the total attraction of a uniform sphere under other hypotheses of force, and again (Propositions XC/XCI) to compute the similar potency of an oblate spheroid upon a point situated externally in its axis.

(67) See 1, §3 below. The major variations from the previous 'De motu Corporum' which occur in the earlier Articles I–III are listed in 1, §2, Appendix 2 preceding.

(68) In its 1687 *editio princeps*, that is; the standard revised text effectively established in the 1713 *editio secunda, auctior et emendatior* is, as we shall have repeated occasion to notice below, in places somewhat different. With the recent appearance—too late, unfortunately, to be of extensive use to us—of the accurate and comprehensive assemblage by A. Koyré and I. B. Cohen of *Isaac Newton's 'Philosophiæ Naturalis Principia Mathematica': the third edition (1726) with variant readings* (Harvard/Cambridge, 1972) it is now easy for anyone to make comparison not only between these three principal editions, but also with the press manuscript (now Royal Society MS LXIX) from which the *Principia* was initially set in type during 1686–7.

geometrically ingenious as it is, in present context, but minimally relevant(69)—on the comparable determination of such orbits 'neutro foco dato'. Thereafter, 'Kepler's problem' is again broached (in Article VI), being given exact solution once more in the parabolic case and approximate resolution, both geometrically by means of a Wrennian prolate cycloid, and 'mechanically' by a variety of equant and iterative procedures,(70) in the far more difficult instance of a general ellipse. Next (Article VII) Newton complements his previous discussion of the periodic oscillation generated in a line by an inverse-square force to a point of it with a sketch of the isochronous 'harmonic' motion equivalently generated by a direct-distance force, before treating the general case of the linear motion induced by an arbitrary force acting along it; this leads directly on (in Article VIII) to the similar construction of the general curvilinear orbit which is traversed where the given force acts to deviate the moving body from its instantaneous direction of motion.(71) In Article IX (with one eye on his primary model of the advancing lunar orbit(72)) he analyses the rotation in the stationary elliptical path which is consequent on minimally perturbing the inverse-square 'pull' to a focus which induces orbital motion therein. And in Article X, finally, he treats—in extension of his earlier pendulum researches(73)—of isochronous oscillatory motion in a general force-field, concentrating his attention on the direct-distance case where, as he cleverly shows, the tautochrone is a hypocycloid.

For all the Grecian façade of 'Definitions', 'Axioms', 'Lemmas', 'Theorems', 'Problems' and 'Scholia' by which this rich harvest of penetrating dynamical

(69) Indeed, the main portions of its opening Lemmas XVII–XXI and Propositions XXII/XXIII are (compare 1, §3: note (23) below) repeated practically word for word from his earlier 'Solutio Problematis Veterum de Loco solido' (IV: 282–320, especially 282–302) where, in divorce from any dynamical or astronomical context, he restored the Greek analysis of the 4-line conic locus now here taken over. In May 1694, some years after the publication of the *editio princeps* of the *Principia*, he told David Gregory that, to resolve the incongruity, he intended to excise these two extraneous sections from it, setting them in a separate companion tract 'de Veterum geometria' where 'their true purpose shall be explained'; see 3, §3: note (1).

(70) Further revisions and extensions of these are recorded in 1, §3, Appendix 2, along with a numerical check upon the accuracy of his favourite 'upper-focus' equant equation.

(71) The primary instance of the inverse-square focal conic orbit, simple though it is to derive in this inverse manner (compare 2, §3: note (209) below), was not in fact so obtained by Newton, who preferred to apply his polar formula for the general central-force orbit at this time (see *ibid.*: note (213)) to the novel construction of the inverse-cube Cotesian spirals, and shortly afterwards (see 1, §3, Appendix 3.4) to determine the trajectory of a light 'ray' (corpuscle stream) through the earth's atmosphere in the hypothesis that it is, in effect, continuously deviated from its onwards rectilinear path by a central *vis refractiva*.

(72) See 1, §3: note (260).

(73) See note (14) above. In Propositions XXV–XXX of the published *Principia's* second book (reproduced in 1, § 3, Appendix 5) Newton afterwards extended his discussion of pendular oscillation in a simple upright cycloid to treat the more realistic circumstance where the motion is instantaneously resisted as some given power (or combination of powers) of the speed of swing.

insights is displayed, it will be evident that the underlying mathematical edifice is in the main built up from demonstrations, synthetically reframed as they might have been,(74) appealing in a wholly non-classical manner—and very much in the preferred style of Newton's earlier 'Geometria Curvilinea'(75)— to the limit-ratios of infinitesimal ('vanishingly small' but not 'indivisible'(76)) increments of related line-segments varying coordinately in tune with an independent parameter of 'time'.(77) In so constructing it Newton did not set an

(74) And as he was in old age at pains to urge against those who could see no obvious traces of the 'new Analysis' of infinitesimal calculus in his *Principia*; for, though by its 'help', he afterwards wrote in anonymous review of his own *Commercium Epistolicum D. Johannis Collins et Aliorum de Analysi promota* (London, 1712), he had 'found out most of the Propositions in his *Principia Philosophiæ*:... because the Ancients for making things certain admitted nothing into Geometry before it was demonstrated synthetically' he had thus recast their proofs 'that the Systeme of the Heavens might be founded upon good Geometry. And this makes it now difficult for unskilful Men to see the Analysis by which these Propositions are found out' (*Philosophical Transactions*, **29**, No. 342 [for January–February 1714/15]: 172–224, especially 206). We shall return to this point in our next volume, but may here stress that Newton's claim was very much in the current fashion: even the doughty Johann Bernoulli, arch-defender of the Leibnizian 'analysis', could affirm to Abraham de Moivre two years earlier of his own very similar treatise 'De motu corporum gravium, pendulorum, & projectilium in medijs non resistentibus & resistentibus, supposita gravitate uniformi & non uniformi atque ad quodvis datum punctum tendente...' (*Acta Eruditorum*, February/March 1713: 77–95/115–32) that 'Il n'y a dans cette piece peu ou point d'analyse, je démontre les théoremes par des syntheses, et je résous les problemes par des constructions géométriques (K. Wollenschläger, 'Der mathematische Briefwechsel zwischen Johann I Bernoulli und Abraham de Moivre', *Verhandlungen der Naturforschenden Gesellschaft in Basel*, **43**, 1933: 151–317, especially 282).

(75) See IV: 420–84.

(76) As Newton stated in scholium to the prefatory 'Lemmas on first and last ratios' of his revised 'De motu Corporum' (1, §2 below), 'in sequentibus siquando quantitates tamquam ex particulis constantes consideravero, vel si pro rectis usurpavero lineolas curvas, nolim indivisibilia sed evanescentia divisibilia, non summas et rationes partium determinatarum sed summarum & rationum limites semper intelligi', further adding afterwards (see 1, §2, Appendix 2.2) that 'Ultimæ rationes illæ quibuscum quantitates evanescunt...sunt limites ad quos quantitatum sine limite decrescentium rationes semper appropinquant,...nunquam verò transgredi [possunt]...nec tamen ideò dabuntur quantitates ultimæ.... Igitur in sequentibus, siquando facili rerum imaginationi consulens, dixero quantitates quam minimas vel evanescentes vel ultimas, cave ne intelligas quantitates magnitude determinatas sed cogita semper diminuendas sine limite.' The Newtonian infinitesimal is, it will be clear, vastly different from the similarly named 'infinitely small' entity recently given rigorous logical definition by A. Robinson in his *Non-standard Analysis* (Amsterdam, 1966).

(77) We have developed this point at length—particularly in regard to the highly interesting Proposition X of the published *Principia's* second book, numerically vitiated in the *editio princeps* by an inexact manner of approximating a tangential increment (compare 1, §1: note (109) below), whose corrected 1712 version will concern us in our final volume—in 'The Mathematical Principles underlying Newton's *Principia Mathematica*', *Journal for the History of Astronomy*, **1**, 1970: 116–38; see also V. I. Antropova, 'O geometricheskom Metode "Matematicheskikh Nachal Natural'noj Filosofii" i N'jutona', *Istoriko-matematicheskie Issledovanija*, **17**, 1966: 205–28.

impossibly high estimate on the mathematical competence and technical expertise of his reader, who is assumed to be familiar with the Euclidean *Elements* of the geometry of the straight line and circle but otherwise expected to know only the simplest Apollonian properties of conics,(78) the rest being proved *ab initio* as needed in the progress of argument or (in rare instances) justified by an appeal to the general algebraic 'quadrature of curves' which Newton had expounded at considerable length some fifteen years before in his yet unpublished 1671 tract on infinite series and fluxions(79) and of which no adequate alternative account was yet available in print. Why then the *Principia* so quickly gained its ill-deserved popular reputation of being impossibly difficult(80) is not easy to understand: certainly, though his natural terseness of style and crabbed mode of presentation was no help to its comprehension and assimilation, there is no evidence that Newton sought deliberately to be any more esoteric therein than he needed be. While the undiluted richness of their intricate mix no doubt played its part in creating the myth of the work's impenetrability, all too few of the methods there employed will individually—in divorce from the often highly ingenious manner of their dynamical application—seem novel to the student of our earlier volumes.

---

(78) Although in Proposition LXXIX—and again, equivalently, in the following Proposition LXXXIII—of the first book of his published *Principia* ($_1$1687: 204, 212) Newton makes a unique citation of the 'demonstrata *Archimedis* in Lib. de Sphæra & Cylindro', this is no specific citation of a theorem in Archimedes' work but a loose reference to a Barrovian corollary to *On the Sphere and Cylinder*, I, 42; see 1, §2, Appendix 4.1: note (8) below. Isaac Todhunter has accurately observed that 'It is not true that any large amount of familiarity with the conic sections is required [to determine the Keplerian laws of motion of the planets]; a very small fraction of the treasures accumulated by the Greek geometers would suffice...: probably a dozen pages would supply the necessities of a student who wished to master even the *Principia* of Newton' (*William Whewell: An Account of his Writings*, 1 (London, 1876): 132).

(79) See III: 32–328. On particular integrals in its comprehensive 'Catalogus Curvarum Aliquot ad Conicas Sectiones relatarum...' (*ibid.*: 244–54), notably, both Corollary 3 to Proposition XLI and Corollary 2 to Proposition XCI of the 'Liber primus' are founded; compare 1, §3: note (213) and 1, §2, Appendix 4.2: note (32) below. At some time in the middle 1680's Newton had it in mind, so as to give some justification for his several unspecified appeals to 'concessis figurarum curvilinearum quadraturis', to introduce an explicit 'Prob. Figuras curvilineas geometricè rationales...quadrare' summarising his earlier papers on such quadrature; see 1, §3, Appendix 6 where reproduction of the pertinent autograph manuscript is made. In after years, as we shall see in following volumes, he several times contemplated appending one version or another of his 1691 tract 'De quadratura Curvarum' in appendix to a revised edition of the *Principia*.

(80) A familiar early eighteenth-century anecdote, credited by John Conduitt to 'M^r^ [Martin] Folkes', has it that 'After Sr I. printed his *principia*, as he passed by the students at Cambridge said there goes the man who has writt a book that neither he nor any one else understands' (King's College, Cambridge. Keynes MS 130.5). Such popular reactions to the appearance of the *Principia*—which it is not our purpose to discuss—are well analysed by J. L. Axtell in 'Locke, Newton and the Two Cultures' (in (ed.) J. E. Yolton, *John Locke: Problems and Perspectives* (Cambridge, 1969): 169–82).

The one considerable exception to this generality is, we may now know, Newton's prior analysis of the problem of determining the curved surface of least resistance to uniform motion in the direction of its axis, the bare enunciation of whose construction he tucked away—along with a note on the similar cone frustum of minimal resistance—in an obscure scholium in the published *Principia's* second book.(81) In a fashion hinted at neither in that printed summary nor even in his later recomputation of the defining differential equation of the surface's generating curve for David Gregory in 1694(82) his original worksheet calculation reveals the depth of his mastery of the general problem of minimising (or maximising) an integral which is allowed to vary over a given range.(83) We may well appreciate that consideration for his overworked reader rather than contempt for his mathematical inadequacy guided his decision not to set the fine detail of his analytical computation in print.

Elsewhere, in seeking to fit his simplified dynamical models to the complexities of the real physical and celestial world, even Newton's comprehensive mathematical tool-kit of convergent series-expansions and refined limit-infinitesimal approximations was not always adequate to the task he set himself. While, after a number of false starts,(84) he succeeded in late 1685 in constructing the parabolic plane orbit of a solar comet from given terrestrial sightings of it—first to a rough approximation in the closing paragraphs of his preliminary 'De motu Corporum Liber secundus'(85) and then much more exactly so in the terminal Proposition

---

(81) To Proposition XXXV, namely, in the 1687 *editio princeps* (renumbered to be XXXIV in later editions). Newton's previously unpublished preliminary computation of the surface (see Plate III) and rough draft of the published scholium are reproduced in 2, §1 below, together with (in the immediately following Appendix 1) the initial version of the theorem to which the latter was attached.

(82) See 2, §1, Appendix 2.

(83) In the notation of 2, §1: note (14) below, on setting $a = dY$ and $x = dX$ to be infinitesimal increments of the perpendicular Cartesian coordinates of the general point (X, Y) of the diametral curve, Newton—without worrying overmuch about the tricky problem of existence (compare 2, §1: note (24))—starts from the necessary criterion for $\int f(Y, a/x).\, dX$, x free, to be minimal that $f(Y, a/x) + f(Y, a/(x-o))$, $x-\frac{1}{2}o$ fixed, should also be minimal to deduce by simple differentiation that

$$f_x(Y, a/x) - f_x(Y, a/(x-o)) = 0;$$

whence, on dividing through by the limit-increment o of x, there ensues the Eulerian condition $f_x(Y, a/x) =$ constant. In Newton's particular case $f \equiv \pi Ya^2/(a^2+x^2)$ is the resistance on the surface-element formed by rotating the arc-increment $ds = \sqrt{[a^2+x^2]}$ round $Y = 0$, whence the minimality criterion determines $f_x \equiv -2\pi Ya^2x/(a^2+x^2)^2 =$ constant, that is,

$$Y(dY/ds)^3\,(dX/ds) = k.$$

In sequel, from the similar instance of the cone frustum, he is able to define the bounding inequality $x \geqslant a$.

(84) See, for instance, v: 524–31 and 1, §1, Appendix 2 below.

(85) See the extracts reproduced in 2, §2.

XLI of the published *Principia's* third book[86]—and was able in consequence, partly graphically and partly by an iterative procedure, to calculate the elements of the 'great comet' of 1680 to a high degree of accuracy,[87] when shortly afterwards he attempted the more difficult feat of deducing the observable inequalities of the moon's motion as solar inverse-square perturbations of the Keplerian ellipse in which it would otherwise travel round the central body of the earth at a focus Newton more than met his match. In the *Principia's* third book,[88] it is true, he put up a brave public show of deriving such principal periodic and secular inequalities as its annual variation, latitudinal wobble and mean nodal regress reasonably well from the simplified hypothesis that the undeviated lunar orbit is a circle uniformly traversed round the earth at its centre, but the mass of his contemporary and subsequent worksheets[89] where he sought doggedly to deepen and refine those basic findings tell a different tale of repeated false starts, the myopic pursuit of dead-end trails and a near-total lack of success. Notably, in his relentless, courageous efforts[90] to compute the periodic 'hourly' motion of lunar apogee—whence its mean secular rate of advance is readily derived by a simple integration—he failed utterly to give adequate theoretical justification for his preferred formula regulating this horary speed of rotation of orbit and was reduced to fudging it in a sophisticated but logically unfounded manner.[91] We may well understand why, when long afterwards John Machin praised his lunar theory as 'all sagacity', he sadly 'smiled & said his head never ached but with his studies on the moon'.[92] A cogent explanation of the inequality was not in fact to be achieved, through the combined efforts of Euler, D'Alembert and Clairaut,[93] for more than

(86) Reproduced, along with the related preceding Lemmas VII–XI, in 2, §2, Appendix.

(87) See 2, §2, Appendix: note (26), and compare A. N. Kriloff, 'On Sir Isaac Newton's Method of Determining the Parabolic Orbit of a Comet', *Monthly Notices of the Royal Astronomical Society*, **85**, 1925: 640–56, especially 655–6.

(88) Namely, in its Propositions XXVI–XXXV; compare 2, §3: note (1) below.

(89) Now for the most part preserved in ULC. Add. 3966.

(90) Reproduced in 2, §3 below.

(91) See our remarks in 2, §3: note (26). A short scholium (*Principia*, $_1$1687: 462–3, quoted in 2, §3: note (63) below) concluded the propositions on lunar motion in the *editio princeps* by citing numerical 'computed' values for the extremes of horary motion of the apsides and mean annual advance of perigee which derive from a minor variant on Newton's formula, but this was suppressed in later editions.

(92) Or so Conduitt recorded in one of his little green pencilled notebooks (King's College. Keynes MS 130.6[3]) shortly after Newton's death. On the previous leaf he similarly jotted down that 'Dr Halley told me he pressed Sr I. to complete his Theory of the Moon saying nobody else could do it. Sr I. said it has broke my rest so often I will think of it no more', but added that Newton 'afterwards told me that when Halley had made six year's observations he would have another stroke at the moon'.

(93) See Robert Grant, *History of Physical Astronomy from the Earliest Ages to the Middle of the*

another sixty years, and then only by discarding Newton's ineffectual approach by geometrical limit-approximations in favour of more powerful analytical series methods.(94)

The concluding extracts(95) from the wide variety of revisions and assorted addenda to the published 'Liber primus' which Newton drafted for one purpose and another in the early 1690's are of only minor importance, serving merely as an epilogue to its earlier profundities. The little that is known—principally through David Gregory(96)—of their *raison d'être* (and the not very much more which may circumstantially be conjectured) is set out at length in footnote to the texts themselves, and we need not detail the ways in which their content refines and improves upon his preceding derivation of a general measure of central force by its deviation in given time from the rectilinear path of uniform inertial motion, or his consequent discussion of inverse-square conic motion round a focus. The more radical schemes of revision(97) were never implemented, and many of the more innovatory individual items found no niche in any published edition. Our interest in even the most significant of these, the fluxional measure of central force which he stated in all its generality(98) long before its rediscovery in equivalent form by Varignon, de Moivre and Johann Bernoulli,(99) must be considerably dampened when we notice that Newton twice(100) failed in his attempt to apply it correctly to a conic orbit.

But enough of editorial anticipation: let us now pass from setting the scene to

---

Nineteenth Century (London, 1852): 44–6; and especially N. I. Idel'son, 'Èakon Vsemirnogo Tjagotenija i Teorija Dvizhenija Lun'ja' [= (ed.) S. I. Vavilov, *Isaak N'juton, 1643–1727* (Moscow/Leningrad, 1943): 160–210]: 194–202.

(94) William Whewell's pronouncement (*History of the Inductive Sciences, from the Earliest to the Present Time*, London, $_1$1837: 167 [= $_2$1857: 128]) that 'No one for sixty years after the publication of the *Principia*, and, with Newton's methods, no one up to the present day, has added anything of any value to his deductions [of] all the principal lunar inequalities' is thoroughly misconceived, whatever one may think of his often quoted following rhetorical query: '...who has presented, in his beautiful geometry, or deduced from his simple principles, any of the inequalities which he left untouched? The ponderous instrument of synthesis, so effective in his hands, has never since been grasped by one who could use it for such purposes; and we gaze at it with admiring curiosity, as on some gigantic implement of war, which stands idle among the memorials of ancient days, and makes us wonder what manner of man he was who could wield as a weapon what we can hardly lift as a burden' (*ibid.*). Whether, in so dextrously wielding it, Newton fought always by the established rules of rigorous logical inference is, of course, another matter.

(95) Reproduced in 2, §§1–3 below.

(96) See especially 3, §2: notes (1), (21), (23) and (34).

(97) Those reproduced in 3, §2: and 3, §3.

(98) See 3, §2.2: note (53).

(99) Compare 3, §1: note (25).

(100) See 3, § 2.2: note (58) and 3, §2, Appendix 2.3: note (12).

examining the primary record of the texts themselves, which alone will truly reveal how deeply, broadly and ingeniously he applied his sophisticated mathematical insight and masterly technical expertise to build the structure of the work in which the fundamental principles of classical Newtonian dynamics were first expounded to the world by their 'onlie begetter'.

1

FUNDAMENTAL INVESTIGATIONS 'ON THE MOTION OF BODIES'[1]

[Autumn 1684–Winter 1685/6]

Excerpts from the originals in the University Library, Cambridge

§1. THE FIRST TRACT 'DE MOTU CORPORUM' (AUTUMN 1684).[2]

De motu corporum in gyrum.[3]

Def. 1. Vim centripetam[4] appello qua corpus impellitur vel attrahitur versus aliquod punctum quod ut centrum spectatur.

Def. 2. Et vim corporis seu corpori insitam qua id conatur perseverare in motu suo secundum lineam rectam.

---

(1) We here reproduce Newton's original autograph draft of the (now lost) tract 'De motu Corporum' which he sent to London in November 1684, and the two successive versions of its major revision in 1685 which he subsequently deposited (in an incomplete state) in Cambridge University Library, purportedly as the text of lectures delivered from the Lucasian chair during 1684–5. As stated in the preceding introduction, it is our primary concern to stress internal mathematical aspects of these documents, seeking to pinpoint their place in the broad sequence of Newton's mathematical development. To that end we relatively—and injustly—neglect to give detailed examination of the often revolutionary dynamical principles which they embody. For this we refer to the many able and penetrating discussions of the quality, novelty and sequence of Newton's developing notions of space, motion and force which we have already listed.

(2) ULC. Add. 3965.7: 55^r–$62bis^r$, first published by J. W. Herivel in *The Background to Newton's Principia. A Study of Newton's Dynamical Researches in the Years 1664–84* (Oxford, 1966): 257–74. In the much corrected and overwritten manuscript as we now have it two principal layers in its composition may be identified. The text of its initial state agrees narrowly with those of the two known contemporary transcripts of the putative fair copy (by Newton's amanuensis, Humphrey Newton?) which was sent to London by way of Edward Paget in November 1684—that, namely, which was entered in (or shortly after) early December following in the Royal Society's Register Book (**6**: 218–34, first printed by S. P. Rigaud in his *Historical Essay on the first Publication of Sir Isaac Newton's 'Principia'* (Oxford, 1838): Appendix No. 1: 1–19 under the title 'Isaaci Newtoni Propositiones De Motu') and that, made perhaps a little earlier, by Edmond Halley (whose first five propositions, afterwards returned to Newton, are now ULC. Add. 3965.7: 63^r–70^r). The only significant deficiences in Newton's original

Translation

On the motion of bodies in an orbit.[3]

Definition 1. A 'centripetal' force[4] I name that by which a body is impelled or attracted towards some point regarded as its centre.

Definition 2. And the force of—that is, innate in—a body I call that by which it endeavours to persist in its motion following a straight line.

---

autograph as against the copy subsequently registered—namely, the explicit enunciation in the latter of 'Hyp[othesis] 4' and the addition there of two opening Lemmas already employed as riders in the body of the former draft—are here made good by appropriate editorial insertions in its text (see notes (12), (13) and (15) below). The changes and revisions which were afterwards effected in this primary state of the 'De motu Corporum in gyrum' merely convert it to be identical with corresponding portions of the augmented tract 'De motu sphæricorum Corporum in fluidis' (ULC. Add. 3965.7: 40r–54r; see Appendix 1 following) which Humphrey Newton penned from it a while later. (The main innovations in this latter text were first published by W. W. R. Ball in his *An Essay on Newton's 'Principia'* (London, 1893): 51–6 in sequel to his straightforward repeat (*ibid.*: 35–51) from Rigaud of the Royal Society transcript of the original fair copy sent by Newton to London; reproduction *in toto* of its text is made by A. R. and M. B. Hall in their *Unpublished Scientific Papers of Isaac Newton. A Selection from the Portsmouth Collection in the University Library, Cambridge* (Cambridge, 1962): 243–67.) For the historical background to Newton's composition of the present piece see the preceding introduction, and compare I.B. Cohen's *Introduction to Newton's Principia* (Cambridge, 1971): Chapter III, 'Steps towards the *Principia*': 47–81, especially 54–62.

(3) Literally a closed circuit, but understand any path which is everywhere convex round some internal point. While Newton has principally in view the minimally eccentric ellipses which the orbits of the solar planets narrowly approximate, in his 'Prob. 4' below he will not for instance, exclude the open parabolas and hyperbolas which are equally possible orbits under an inverse-square force directed to a focus. The revised manuscript (see note (2)) bears the more sophisticated title 'De motu sphæricorum Corporum in fluidis' (On the motion of spherical bodies in fluids) which better defines its theme. A Newtonian 'fluid' is, of course, any uniform medium—such as the terrestrial atmosphere— which may or may not offer appreciable resistance to the passage of a body through it, while the insistence that the latter be a spherical mass may just possibly suggest that Newton had by this time already achieved the insight that in an inverse-square force-field such a body behaves as though its mass were concentrated at its centre, as he was soon rigidly to demonstrate (see §2.3: note (188)).

(4) The first occurrence of this classical *terminus technicus*, contrived as the complement of the term 'vis centrifuga (ex motu circulari) used by Christiaan Huygens to denote the 'endeavour' outwards from the centre of a body constrained to rotate uniformly in a circle, and first published in his *Horologium Oscillatorium sive De Motu Pendulorum ad Horologia aptato* (Paris, 1673): 159. When in 1719 Newton wrote to Des Maizeaux regarding Leibniz' correpondence during 1715–16 'sur l'invention des Fluxions & du Calcul Differentiel' (as Des Maizeaux headed his gathering of it in Tome II of his *Recueil de Diverses Pieces*..., Amsterdam, 11720), he observed at one point in a critique of Leibniz' celebrated 'Apostille' to his letter to Conti in December 1715 that 'Mr Hygens gave the name of *vis centrifuga* to the force by wch revol[v]ing bodies recede from the centre of their motion. Mr Newton in honour of that author retained the name & called the contrary force *vis centripeta*' (ULC. Add. 3968.28: 415v; see also A. Koyré and I. B. Cohen, 'Newton & the Leibniz–Clarke Correspondence', *Archives Internationales d'Histoire des Sciences*, **15**, 1962: 63–126, especially 122–3).

Def. 3.(5) Et resistentiam(6) quæ est medij regulariter impedientis.

Hypoth 1.(7) Resistentiam in proximis novem propositionibus nullam esse, in sequentibus esse ut corporis celeritas et medij densitas conjunctim.(8)

Hypoth 2. Corpus omne sola vi insita uniformiter secundum rectam lineam in infinitum progredi nisi aliquid extrinsecus impediat.(9)

Hyp. 3. Corpus in dato tempore viribus conjunctis eo ferri quo viribus divisis in temporibus æqualibus successivè.(10)

Hyp. 4. [Spatium quod corpus urgente quacunq; vi centripeta ipso motus initio describit esse in duplicata ratione(11) temporis.](12)

[*Lem. 1.* Quantitates differentijs suis proportionales sunt continuè proportionales. Ponatur A ad $A-B$ ut B ad $B-C$ & C ad $C-D$ &c et dividendo fiet A ad B ut B ad C et C ad D &c.](13)

---

(5) A late insertion in the manuscript, evidently added when (in afterthought?) Newton decided to append Problems 6 and 7 and their scholium on motion resisted as the instantaneous speed.

(6) Understood to be a Newtonian *vis* acting instantaneously in a direction contrary to that of the body's motion.

(7) Newton initially here made the blanket assumption that 'Corpora nec medio impediri nec alijs causis externis quo minus viribus insitæ et centripetæ exquisitè cedant' (Bodies are hindered neither by the medium nor by other external causes from yielding perfectly to their innate and to centripetal forces). The lack of reference to resisted motion in this preliminary supposition strongly supports our earlier suggestion (note (5)) that Newton's final Problems 6 and 7 below were added in afterthought.

(8) In later redraft (see Appendix 1, and compare note (3) above) Newton interjected 'et corporis moti sphærica superficies' (and the spherical surface of the moving body), but without further elaborating the addendum. In about the autumn of 1685 he returned to the topic, computing the total resistance to uniform translation of a hemispherical surface to be half that of its great-circle plane; see 2, §1, Appendix 1 below.

(9) In other words, it is supposed that 'natural' (force-free) motion of a body takes place at a uniform rate in an infinite straight line, in which state it is sustained by its (internal) *vis insita*. It is now well established that Newton arrived at this fundamental postulate of inertial rectilinearity by combining the *prima/altera leges naturæ* ('quod unaquæq; res quantum in se est, semper in eodem statu [*sc.* movendi] perseveret' and 'quod omnis motus ex se ipso sit rectus' respectively) which Descartes set down in his *Principia Philosophiæ* (Amsterdam, 1644): Pars II, §§ XXXVII/XXXIX: 51–4. In his own earliest notes on mechanics, penned by him in January 1665 in his Waste Book (ULC. Add. 4004: 10r–15r/38v, printed in Herivel's *Background* (note (2)): 132–82), Newton inserted (f. 12r) an 'Ax: 100. Every thing doth naturally persevere in yt state in wch it is unlesse it bee interrupted by some external cause, hence... A body once moved will always keep ye same celerity, quantity & determinacon of its motion'.

(10) This late addition in the manuscript's margin, given lemmatical status and formal proof in Newton's immediate revise (see Appendix 1: note (7) below) in effect enunciates the familiar 'parallelogram' rule for compounding uniform speeds, here generated 'instantaneously' by the single, simultaneous application of forces at a point: in Theorem 1 following, as a subtlety, one of these is taken to be a general *vis centripeta*, but the other the *vis insita* which, according to Hypothesis 2, sustains a given uniform motion in a given straight line.

(11) That is, the square. Below, similarly, we render 'triplicata ratio' (cube) as 'tripled ratio', and so on.

Definition 3.(5) While 'resistance'(6) is that which is the property of a regularly impeding medium.

Hypothesis 1.(7) In the ensuing nine propositions the resistance is nil; thereafter it is proportional jointly to the speed of the body and to the density of the medium.(8)

Hypothesis 2. Every body by its innate force alone proceeds uniformly into infinity following a straight line, unless it be impeded by something from without.(9)

Hypothesis 3. A body is carried in a given time by a combination of forces to the place where it is borne by the separate forces acting successively in equal times.(10)

Hypothesis 4. [The space which a body, urged by any centripetal force, describes at the very beginning of its motion is in the doubled ratio(11) of the time.](12)

[*Lemma 1.* Quantities proportional to their differences are in continued proportion. Set $A:(A-B) = B:(B-C) = C:(C-D) = \ldots$ and there will come, *dividendo*, to be $A:B = B:C = C:D = \ldots$.](13)

---

(12) Only the opening phrase 'Hyp[othesis] 4' is present—as a late marginal addition—in the manuscript. The inserted text is that of the putative fair copy later sent to London (as settled by the Royal Society and Halley transcripts; see note (2)). It will be evident that Newton presupposes that the central force acting upon a body may, over a vanishingly small length of its orbital arc, be assumed not to vary significantly in magnitude or direction, and hence that that infinitesimal arc is approximated to sufficient accuracy by a parabola whose diameter passes through the force-centre, with its deviation from the inertial tangent-line accordingly proportional to the square of the time. The point is further explored in note (19) below.

(13) This necessary lemma is likewise (compare the preceding note) here inserted from the putative fair copy, as the Royal Society and Halley transcripts (see note (2)) establish its text. The corollary that, when A is 'prima & maxima' and the 'quantitates proportionales' are 'numero infinitæ', then 'erit A–B ad A ut A ad summam omnium' (as James Gregory stated it in Propositio I of his 'N. Mercatoris Quadratura Hyperboles [*sc.* in his 1668 *Logarithmotechnia*; see II: 166] Geometrice Demonstrata' [=*Exercitationes Geometricæ* (London, 1668): 9–13, especially 9]) is all-important in Newton's application of the lemma in Problems 6 and 7 below. (Gregory's assertion that this limit-summation of a converging geometrical progression 'passim demonstratur apud Geometras' is considerably exaggerated: the result was widely used by the 'calculators' of early 14th century Oxford—Richard Swineshead and others—and was widely familiar by the early 16th century, while Archimedes in his *Quadrature of the Parabola* had given rigid proof of the particular case when the proportion factor is $\frac{1}{4}$ by a technique generalisable to instances where the factor is less than $\frac{1}{2}$, but the first completely general demonstration of Gregory's proposition appeared only in Grégoire de Saint-Vincent's *Opus Geometricum Quadraturæ Circuli et Sectionum Coni* (Antwerp, 1647): 51–177; see II: 246, note (146).) In effect Newton derives in each case the solution $\log(x_t/x_0) = -kt$ as the general solution of the fluxional equation $\dot{x}_t$ $(= dx_t/dt) = -kx_t$ by setting (in the geometrical equivalents of his hyperbolic model) $A = x_0$, $B = x_{t/n}$, $C = x_{2t/n}$, $D = x_{3t/n}$, ... together with

$$x_0/(x_0 - x_{t/n}) = x_{t/n}/(x_{t/n} - x_{2t/n}) = \ldots = x_{(n-1)t/n}/(x_{(n-1)t/n} - x_t) = 1/(kt/n),$$

where $x_0/x_{t/n} = x_{t/n}/x_{2t/n} = \ldots = x_{(n-1)t/n}/x_t = 1/(1-kt/n)$ and therefore $x_t/x_0 = (1-kt/n)^n$: accordingly, in the limit as n becomes infinitely great there results $x_t/x_0 = e^{-kt}$.

[*Lem. 2*. Parallelogramma omnia circa datam Ellipsin descripta,(14) esse inter se æqualia. Constat ex Conicis.](15)

Theorema 1. Gyrantia(16) *omnia radijs ad centrum* (17) *ductis areas temporibus proportionales describere.*

Dividatur tempus in partes æquales, et prima temporis parte describat corpus vi insita rectam *AB*. Idem secunda temporis parte si nil impediret ^a^rectà pergeret ad *c* describens lineam *Bc* æqualem ipsi *AB* adeo ut radijs *AS*, *BS*, *cS* ad centrum actis confectæ forent æquales areæ *ASB*, *BSc*. Verum ubi corpus venit ad *B* agat vis centripeta impulsu unico sed magno, faciatq corpus a recta *Bc* deflectere et pergere in recta *BC*. Ipsi *BS* parallela agatur c*C* occurrens *BC* in *C* et completa secunda temporis parte ^b^corpus reperietur in *C*. Junge *SC* et triangulum *SBC* ob parallelas *SB*, *Cc* æquale erit triangulo *SBc* atq adeo etiam triangulo *SAB*. Similí argumento si vis centripeta successivè agat in *C*, *D*, *E* &c, faciens corpus singulis temporis momentis singulas describere rectas *CD*, *DE*, *EF* &c[,] triangulum *SCD* triangulo *SBC* et *SDE* ipsi *SCD* et *SEF* ipsi *SDE* æquale erit. Æqualibus igitur temporibus æquales areæ describuntur. Sunto jam hæc triangula numero infinita et infinitè parva, sic, ut singulis temporis momentis singula respondeant triangula, agente vi centripeta sine intermissione,(18) & constabit propositio.(19)

[a] Hyp. 1.

[b] Hyp. 3.

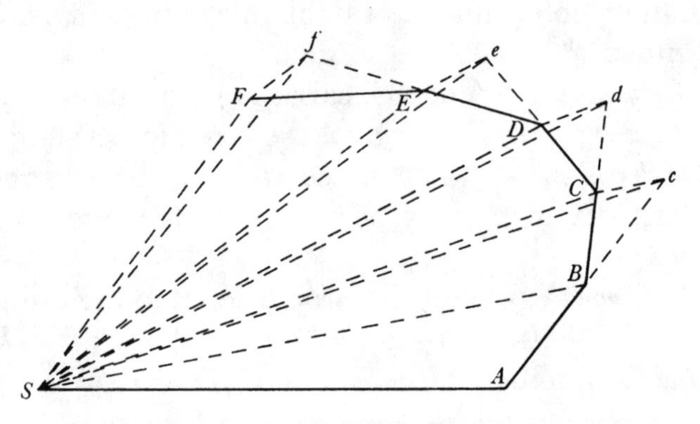

(14) Understand so as to touch it at the end-points of a pair of conjugate diameters. The phrasing of the sequel suggests that Newton is unaware that the present lemma is Apollonius, *Conics*, VII, 31 (first printed by G. A. Borelli in his *Apollonii Pergæi Conicorum Lib. V. VI. VII* (Florence, 1661): 370–1). We have already, in demolishing Whiston's absurd claim (*Memoirs*, London, $_1$1749: 39) that Newton's enunciation of this theorem, in the little modified form in which it was to appear in the *Principia*, manifested his mathematical ability to 'see almost by Intuition, even without Demonstration', remarked (see IV: 9, note (24)) that the property is all but self-evident when the ellipse is viewed as the orthogonal projection of a circle. It may be, however, that Newton by his 'Constat ex Conicis' makes oblique reference to the ingenious proof by area-dissection given by Grégoire de Saint-Vincent in his *Opus Geometricum* (note (12)): Liber IV, Propositio LXXII, 'Rectangulum sub dimidijs axibus æquale est parallelogrammo sub diametris conjugatis': 281.

(15) Like the preceding, this lemma (needed in the proof of Problems 2 and 3 below) is here inserted from the putative fair copy (as we know it from the Royal Society and Halley transcripts).

[*Lemma 2*. All parallelograms described about a given ellipse(14) are equal to one another. This is established from the *Conics*.](15)

Theorem 1. All orbiting bodies(16) describe, by radii drawn to their centre,(17) areas proportional to the times.

Let the time be divided into equal parts, and in the first part of the time let the body by its innate force describe the straight line *AB*. It would then in the second part of time, were nothing to impede it, proceed directly[a] to *c*, describing the line *Bc* equal to *AB* so as, when rays *AS*, *BS*, *cS* were drawn to the centre, to make the areas *ASB*, *BSc* equal. However, when the body comes to *B*, let the centripetal force act in one single but mighty impulse and cause the body to deflect from the straight line *Bc* and proceed in the straight line *BC*. Parallel to *BS* draw *cC* meeting *BC* in *C*, and when the second interval of time is finished the body will[b] be found at *C*. Join *SC* and the triangle *SBC* will then, because of the parallels *SB*, *Cc*, be equal to the triangle *SBc* and hence also to the triangle *SAB*. By a similar argument, if the centripetal force acts successively at *C*, *D*, *E*, ..., making the body in separate moments of time describe the separate straight lines *CD*, *DE*, *EF*, ..., the triangle *SCD* will be equal to the triangle *SBC*, *SDE* to *SCD*, *SEF* to *SDE* (and so on). In equal times, therefore, equal areas are described. Now let these triangles be infinitely small and infinite in number, such that to each individual moment of time there corresponds an individual triangle (the centripetal force acting now without interruption(18)), and the proposition will be established.(19)

[a] Hypoth. 1.

[b] Hypoth. 3.

(16) Understand 'in plano', as Newton will later make explicit in revised enunciation of this theorem (see §2, Appendix 2.3).

(17) Of force, that is, and not (necessarily) of their orbit.

(18) The Royal Society transcript here reads 'remissione' (abatement).

(19) Newton's proof of this fundamental generalisation of Kepler's areal law (contrived by the latter—more than a little shakily—to regulate the varying speeds of the solar planets in their minimally eccentric elliptical orbits and on a purely kinematic basis; compare E. J. Aiton, 'Kepler's Second Law of Planetary Motion', *Isis*, **60**, 1969: 75–90) is more subtle and considerably less cogent than it may at first appear. His unanalysed procedure of breaking down the action of a continuous central force directed instantaneously towards the centre *S* as a body covers the orbital arc $\widehat{BF}$ (it will be obvious that the initial, vanishingly small segment *AB* merely serves to define the direction of force-free motion at *B*) by splitting it into the compound of an infinity of infinitesimal force-impulses, equal one to the other and each directed to the centre *S*, applied at equal (likewise infinitesimal) intervals of time at a corresponding infinity of intervening points *C*, *D*, *E*, ... of the orbit does indeed guarantee that the focal triangles *BSC*, *CSD*, *DSE*, ... are equal in area to each other, and hence that the time in which a body under the continual bombardment of such force-impulses traverses the limit-polygonal arc *BCDE* ... is, at any point, proportional to the related focal segment (*SBCDE* ...). But he ignores whether significant error is introduced in the total action by supposing at each stage that a continuous force instantaneously directed to the centre *S* (and so varying infinitesimally in direction over the vanishingly small continuous arc in question) is adequately approxi-

Theorem. 2. Corporibus in circumferentijs circulorum uniformiter gyrantibus vires centripetas esse ut[20] *arcuum simul*[21] *descriptorum quadrata applicata ad radios circulorum.*

Corpora B, b in circumferentijs circulorum BD, bd gyrantia simul[21] describant

---

mated by an equivalent impulse of force striking instantaneously at just one point; nor does he justify his assumption that the limit-polygonal arc $BCDE \ldots F$ passes into a unique orbital arc $\widehat{BF}$ as the number of bombarding force-impulses increases to infinity. Furthermore, he will in sequel at once suppose that the direction and length of the total 'spatium superatum' (deviation from the initial rectilinear path AB) is also uniquely—and intuitively—given.

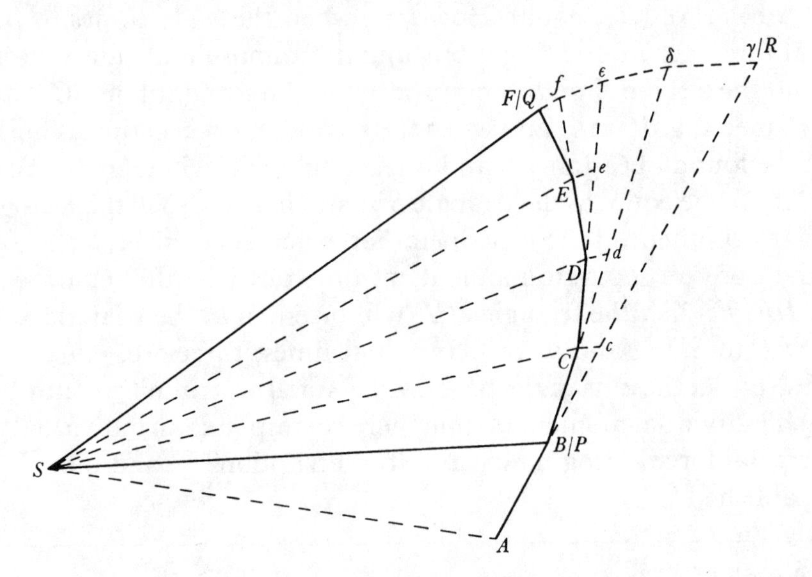

However, in terms of Newton's model, this deviation in the total time from B to F of the impulse (applied at B) which in the time of passage from B to C generates the infinitesimal segment cC is measured in length and direction by a line $\gamma\delta$, parallel to BS, where $B\gamma$ is the force-free path travelled from B in that total time; similarly, the deviation in the time from C to F of the impulse (applied at C) which in the time from C to D generates the segment dD is

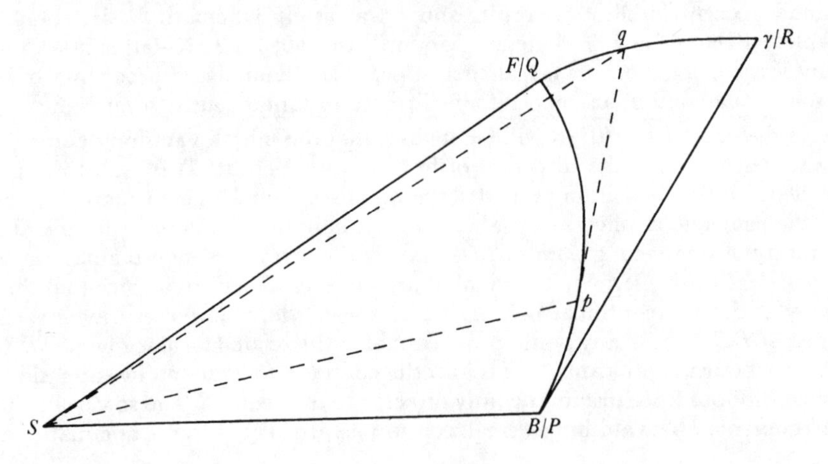

Theorem 2. Where bodies orbit uniformly in the circumferences of circles, the centripetal forces are as the squares[20] *of arcs simultaneously*[21] *described, divided by the radii of their circles.*

Let the bodies B, b orbiting in the circumferences of the circles BD, bd simultaneously[21] describe the arcs $\widehat{BD}$, $\widehat{bd}$. By their innate force alone they

---

measured in length and direction by the line $\delta\epsilon$, parallel to CS; and so on: whence the deviation effected by the totality of impulses is representable by the polygonal arc $\gamma\delta\epsilon \ldots F$, whose limiting form (as may readily be proved) is the *curvilinear* arc $\widehat{\gamma qF}$, any point q in which is such that, if pq is drawn tangent (at p) to the orbital arc $\widehat{BpF}$, then the triangle pSq is equal in area to the orbital sector (pSF). The necessary requirement for these unconsidered subtleties to be (in present context) negligible is, in fact, that the total orbital arc $\widehat{BF}$ be itself infinitesimally small, in which case the general distance Sp of the orbiting body does not vary appreciably in magnitude, and therefore the central force, f say, acting over the whole arc $\widehat{BF}$ may be considered to be constant. It follows that, if the time of orbit over that infinitesimal arc be dt, and n equal impulses of force act (at $B, C, D, E, \ldots$) at intervals of time dt/n, then each (second-order infinitesimal) deviation $cC = dD = eE = \ldots = f.(dt/n)^2$, so that

$$\gamma\delta = nf.(dt/n)^2, \quad \delta\epsilon = (n-1)f.(dt/n)^2, \quad \ldots$$

and hence the total polygonal arc $\gamma\delta\epsilon\ldots F$ is $\sum_{0 \leqslant i \leqslant n-1} (n-i)f.(dt/n)^2 = \frac{1}{2}(1+1/n)f.dt^2$. In the limit, accordingly, as $n \to \infty$ the deviation arc $\widehat{\gamma F} = \frac{1}{2}f.dt^2 \propto dt^2$ (compare note (9)), and has, moreover, a maximum slope—at F—which is that of SF, where $\widehat{BSF}$ is of infinitesimal size, so that it is adequately approximated in length and direction by any line drawn through F to $B\gamma$ whose slope does not exceed that of SF. In the sequel Newton will, as it suits him, take the deviation 'line' γF—or rather CD in Theorem 2 following and RQ thereafter—to be either the chord γF or, most often, the parallel through F to SB. Since in the dynamical contexts where the generalised areal law receives its Newtonian application the central angle $\widehat{BSF}$—or, correspondingly below, $\widehat{BSD}$ and $\widehat{PSQ}$—is invariably infinitesimal, such sophisticated complexities do not there bedevil the logical cogency of its present proof. Whether Newton himself fully appreciated these underlying subtleties and the validity of neglecting them is not clear. Certainly, neither in subsequent redraft nor in any of the editions of his *Principia* (where the theorem is rightly given pride of place as Proposition I of Book 1) did he ever insert any modification which would suggest that he later did other than continue to believe in its superficial simplicities, but its is only fair to add that none of his contemporaries and immediate successors—even Johann Bernoulli, his arch critic and an equal grand master of the infinitely small—saw fit to impugn the adequacy of Newton's demonstration. On the historical questions, finally, of how early and directly Newton became aware of Kepler's prior statement of the areal law in the case of the solar planets (and planetary satellites), and when he came to realise the fundamental rôle which its generalisation plays in the creation of a mathematical theory of central-force orbital dynamics, see the preceding introduction and compare D. T. Whiteside, 'Newton's Early Thoughts on Planetary Motion: A Fresh Look' (*British Journal for the History of Science*, **2**, 1964: 117–37): 120–2, 128–31.

(20) Newton has deleted a tentative insertion 'celeritatum sive' (of the speeds or), evidently because this generalisation to the case where the arcs $\widehat{BD}$ and $\widehat{bd}$ are described in differing times is delayed to be his prime corollary below.

(21) Or, of course, 'æqualibus temporibus' (in equal times).

arcus BD, bd. Sola vi insita describerent tangentes BC, bc his arcubus æquales.(22) Vires centripetæ sunt quæ perpetuò retrahunt corpora de tangentibus ad circumferentias, atq; adeo hæ sunt ad invicem ut spatia ipsis superata CD, cd, id est productis CD, cd ad F et f(23) ut $\frac{BS^{\text{quad}}}{CF}$ ad $\frac{bc^{\text{quad}}}{cf}$ sive ut $\frac{BD^{\text{quad}}}{\frac{1}{2}CF}$ ad $\frac{bd^{\text{quad}}}{\frac{1}{2}cf}$. Loquor de spatijs BD, bd minutissimis inq; infinitum diminuendis sic ut pro $\frac{1}{2}CF$, $\frac{1}{2}cf$ scribere liceat circulorum radios SB, sb. Quo facto constat Propositio.

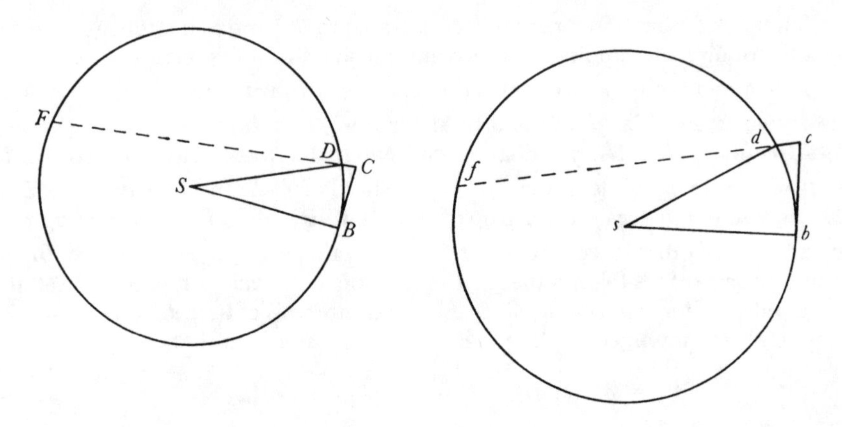

Cor 1. Hinc vires centripetæ sunt ut celeritatum(24) quadrata applicata ad radios circulorum.

Cor 2. Et reciprocè ut quadrata temporum periodicorum applicata ad radios.

Cor 3. Unde si quadrata temporum periodicorum sunt ut radij circulorum vires centripetæ sunt æquales. Et vice versa.

Cor 4. Si quadrata temporum periodicorum sunt ut quadrata radiorum vires centripetæ sunt reciprocè ut radij. Et vice versa.

---

(22) In the terms of the second diagram in note (19) above, if the orbit $\widehat{BF}$ (here $\widehat{BD}$) is a circle-arc of centre S, then the tangent pq at its general point p (extended to its meet with the deviation arc $\widehat{F\gamma}$, here named $\widehat{DC}$) is perpendicular to its corresponding radius Sp, so that, since the circle-sector (pSF) is equal in area to the triangle pSq, in every case $\widehat{pF} = pq$, and in particular $\widehat{BF}$ (that is, $\widehat{BD}$) $= B\gamma(BC)$. The infinitesimal arc $\widehat{F\gamma}(\widehat{DC})$ is evidently a circle-involute, a curve already identified in parallel circumstances by Huygens in his yet unpublished tract 'De Vi Centrifuga [ex Motu Circulari]' (*Œuvres complètes*, **16**, 1929: 255–301, especially 265/7). It is interesting that Huygens there approximates the arc $\widehat{DC}$ by its tangent at D (extended to its meet with BC), while Newton here (see next note) makes equivalent use of its chord DC.

(23) In effect Newton approximates the infinitesimal involute arcs $\widehat{CD}$, $\widehat{cd}$ by their chords CD, cd, and accurately sets the points F, f a little above the extensions of the radii BS, bs. Indeed, from his researches in 1676 (compare IV: 670, note (44)) into the approximate rectification of the arc of a central conic he was aware that, on taking $BS = DS = 1$ and $\widehat{BSD} = \phi$ (so that, where $D\alpha$ is drawn perpendicular to BS, at once $D\alpha = \sin\phi$ and $S\alpha = \cos\phi$), the near-equality $\phi/3 \gtrapprox \sin\phi/(2+\cos\phi)$ determines that CD meets BS in β such that $B\beta \lessapprox 3BS$, whence

would describe the tangent lines BC, bc equal to these arcs.(22) The centripetal forces are those which perpetually drag the bodies back from the tangents to the circumferences and hence are to each other as the distances CD, cd surmounted by them, that is, on producing CD, cd to F and f,(23) as BC^2/CF to bc^2/cf or as $\widehat{BD}^2/\frac{1}{2}CF$ to $\widehat{bd}^2/\frac{1}{2}cf$. I speak here of distances $\widehat{BD}$, $\widehat{bd}$ which are very minute and indefinitely to be diminished, so that in place of $\frac{1}{2}CF$, $\frac{1}{2}cf$ it is allowable to write the radii SB, sb of the circles. And once this is done the proposition is established.

Corollary 1. Hence the centripetal forces are as the squares of the speeds(24) divided by the radii of the circles.

Corollary 2. And reciprocally as the squares of the periodic times divided by the radii.

Corollary 3. Whence, if the squares of the periodic times are as the radii of the circles, the centripetal forces are equal. And conversely so.

Corollary 4. If the squares of the periodic times are as the squares of the radii, the centripetal forces are reciprocally as the radii. And conversely so.

$\widehat{B\beta C} \approx \frac{1}{3}\widehat{BSD}$. (This familiar inequality, publicly enunciated by Willebrord Snell in Propositio XXVIII of his *Cyclometricus. De Circuli Dimensione...ad Mechanicem accuratissima* (Leyden, 1621): 42, was later given rigid proof by Huygens in Theorema XIII of his widely read *De Circuli Magnitudine Inventa* (Leyden, 1654) [= *Œuvres complètes*, **13**, 1910: 113–81, especially 159–63].)

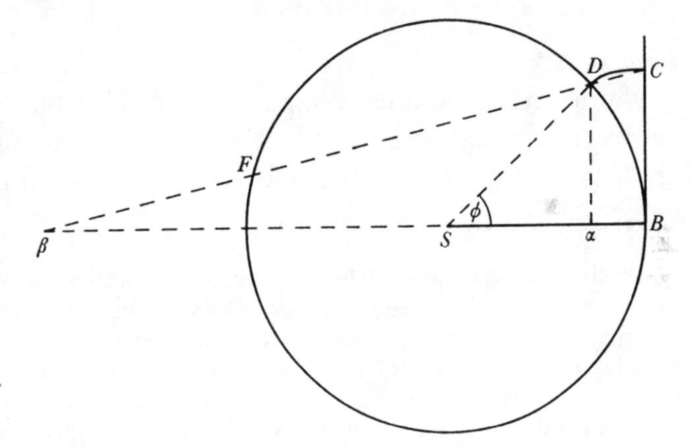

(24) That is, the length of the previous infinitesimal arcs ($\widehat{BD}$, $\widehat{bd}$) divided by the equal times in which they are uniformly traversed. This fundamental Huygenian result, here after some thought (see note (20)) delayed to be but a corollary to the preceding main theorem had initially been attained by Newton about January 1665 (ULC. Add. 4004: 1ʳ) by, much as in his present demonstration of Theorem 1, breaking down the action of a constant, continuously impressed *conatus recedendi a centro* into the aggregate of an infinity of infinitesimal impulses spaced at equal, correspondingly infinitesimal periods of time. (See J. W. Herivel, 'Newton's Discovery of the Law of Centrifugal Force', *Isis*, **51**, 1960: 546–53 and his *Background to Newton's Principia* (note (2)): 129–30; compare also D. T. Whiteside, 'Newtonian Dynamics' (*History of Science*, **5**, 1966: 104–17): 108–10.) The revised argument here offered, one making a direct appeal to the notion of a *vis centrifuga* acting continuously over an infinitesimal time-interval exactly as Huygens' earlier (but yet unpublished) demonstration in his 1659 manuscript 'De Vi Centrifuga' (note (22)), was subsequently evolved by Newton in an untitled paper of about 1670 (ULC. Add. 3958.5: 87ʳ/87ᵛ, first published by A. R. Hall in 'Newton on the Calculation of Central Forces', *Annals of Science*, **13**, 1957: 62–71, especially 64–6; see also Herivel's *Background...*: 192–8). Newton's derivation further parallels Huygens' prior investigation in there approximating the deviation arc $\widehat{DC}$ by the tangent to the involute at D, but the

Cor 5. Si quadrata temporum periodicorum sunt ut cubi radiorum vires centripetæ sunt reciprocè ut quadrata radiorum. Et vice versa.(25)

Schol. Casus Corollarij quinti obtinet in corporibus cœlestibus. Quadrata temporum periodicorum sunt ut cubi distantiarum a communi centro circum quod volvuntur. Id obtinere in Planetis majoribus circa Solem gyrantibus inqꝫ minoribus circa Jovem et Saturnum(26) jam statuunt Astronomi.

Theor. 3. Si corpus P circa centrum S gyrando, describat lineam quamvis curvam APQ, et si tangat recta PR curvam illam in puncto quovis P et ad tangentem ab alio quovis curvæ puncto Q agatur QR distantiæ SP parallela(27) *ac demittatur QT perpendicularis ad distantiam SP: dico quod*(28) *vis centripeta sit reciprocè ut solidum* $\frac{SP^{quad}\times QT^{quad}}{QR}$, *si modò solidi illius ea semper sumatur quantitas quæ ultimò fit ubi coeunt puncta P et Q.*

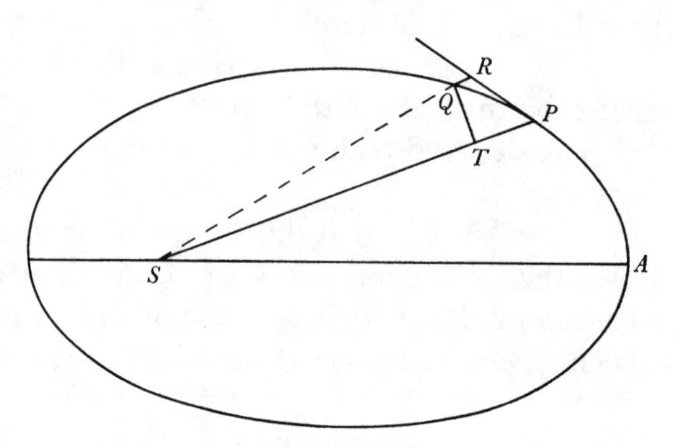

Namqꝫ in figura indefinitè parva *QRPT* lineola(29) *QR* dato tempore est ut vis centripeta et data vi ut [a]quadratum temporis atqꝫ adeo neutro dato ut vis centripeta et quadratum temporis conjunctim, id est ut vis centripeta semel et

[a]Hyp 4.

possibility that Newton in 1670 could even have been aware of the dynamical researches of his Dutch contemporary is remote. The 'De Vi Centrifuga' appeared publicly only in 1703 in Huygens' *Opera Posthuma*, though bare enunciations of its 'Theoremata' (especially 'III. Si duo mobilia æqualia in circumferentijs æqualibus ferantur, celeritate inæquali, sed utraque motu æquabili, ...erit vis centrifuga velocioris ad vim tardioris in ratione duplicata celeritatum') were appended, without prior introduction or any explanation, on pages 159–61 of his 1673 *Horologium Oscillatorium* (see note (4) above).

(25) If, where T is the time of periodic orbit in a circle of radius r at a uniform speed v ($\propto r/T$), we suppose generally that $T^2 \propto r^n$, then the central force inducing this motion is $v^2/r \propto r/T^2 \propto r^{1-n}$.

(26) This reference to the satellites of Saturn was copied by Humphrey Newton into the revised version (see note (2)), but soon afterwards, having by way of Edward Paget asked Flamsteed's opinion of 'y[e] supernumeray satellits of ♄' and been given on 27 December 1684 the dampening news that 'I can not find the 2 new ones [announced by Cassini in 1681] with a 24 foot glasse' (*Correspondence of Isaac Newton*, **2** (Cambridge, 1960): 405), Newton deleted the phrase 'et Saturnum' (and Saturn): in Book 3 'De Mundi Systemate' of the first (1687) edition of his *Principia*, likewise, Newton makes no reference to any 'planetæ circumsaturnii', but the satellites of Saturn were restored to grace—after due verification of their existence was given—in Phænomenon II of the second edition ($_2$1713: 359–60). In his further letter to Flamsteed on 30 December 1684 Newton queried whether the mean radius of orbit of Saturn itself, as 'defined' by Kepler in his *Tabulæ Rudolphinæ* (Ulm, 1627), 'is...too little for y[e] sesquialterate proportion' (*Correspondence*, **2**: 407) and failed to receive in Flamsteed's reply on 5 January following (*ibid.*: 408–9) a confident rejection of his worry that it might suffer a perturbed

Corollary 5. If the squares of the periodic times are as the cubes of the radii, the centripetal forces are reciprocally as the squares of the radii. And conversely so.[25]

Scholium. The case of the fifth corollary holds true in the heavenly bodies: the squares of their periodic times are as the cubes of their distances from the common centre round which they revolve. That it does obtain in the major planets circling round the Sun and also in the minor ones orbiting round Jupiter and Saturn[26] astronomers are agreed.

Theorem 3. If a body P in orbiting round the centre S shall describe any curved line APQ, and if the straight line PR touches that curve at any point P and to this tangent from any other point Q of the curve there be drawn QR parallel to the distance SP,[27] *and if QT be let fall perpendicular to this distance SP: I assert that*[28] *the centripetal force is reciprocally as the 'solid'* $SP^2 \times QT^2/QR$, *provided that the ultimate quantity of that solid when the points P and Q come to coincide is always taken.*

For in the indefinitely small configuration *QRPT* the[29] line-element *QR* is, given the time, as the centripetal force and, given the force, as[a] the square of the time, and hence, when neither is given, as the centripetal force and the square of the time jointly; that is, as the centripetal force taken once and the area *SQP*

[a] Hypoth. 4.

'exorbitation' of significant size when in the vicinity of Jupiter. Fifteen years before, in a manuscript annotation on 'pag 173 & 304' of Vincent Wing's *Astronomia Britannica* (London, 1669) inserted in the endpapers of his library copy of it (now Trinity College, Cambridge. NQ. 18.36) he had similarly written: 'An Jovis orbita ad hanc analogiam reduci potest haud scio, id vero suspicor sed hæ ejus tabulæ non satis bene conveniunt cum observationibus'. None the less, in his published Hypothesis V of Book 3 of his *Principia* ($_1$1687: 403) he matched his present confidence in the general validity of Kepler's third planetary 'law'—first propounded in the latter's *Harmonices Mundi Libri V* (Linz, 1619)—by asserting without reservation that '*Planetarum quinque primariorum,* &... *Terræ circa Solem tempora periodica esse in ratione sesquialtera mediocrium distantiarum à Sole.* Hæc à *Keplero* inventa ratio in confesso est apud omnes [Astronomos]'.

(27) In his manuscript figure, whether intentionally or no, Newton has in fact drawn *QR* to be more nearly in line with *SQ*. It is tempting to think that he still wishes to make *QR* closely approximate the true deviation arc, which will (see note (19)) be exactly parallel to *SP* only at its end-point *R*. The distinction will, of course, have no significance in the sequel, where only the length of *QR*—and not its infinitesimal slope to *SP*—matters.

(28) Newton here first began to write 'punctis [?*P* et *Q* coeuntibus]' (with the points [?*P* and *Q* coming to coincide]). His intention that this 'ultimate' (limit) value of the following ratio must be taken is somewhat cumbrously rephrased in the sequel.

(29) At this place in the redrafted version (see note (2)) Newton further sought to stress that the deviation is vanishingly small by inserting the adjective 'nascens' (nascent). His following 'demonstration' that, where the deviation *QR* is due to the continuous action of some central force, *f* say, directed instantaneously to the point *S* as the body in some infinitesimal time, say *dt*, traverses the infinitely small orbital arc $\widehat{PQ}$, then $QR \propto f.dt^2$ must inevitably seem superficial to modern eyes attuned to the subtleties glossed over in Newton's present assumptions; compare notes (12) and (19).

area SQP tempori proportionalis (seu duplum ejus $SP \times QT$) bis. Applicetur hujus proportionalitatis pars utraꝙ ad lineolam QR et fiet unitas ut vis centripeta et $\frac{SP^q \times QT^q}{QR}$ conjunctim, hoc est vis centripeta reciprocè ut $\frac{SP^q \times QT^q}{QR}$.(30) Q.E.D.

Corol. Hinc si detur figura quævis et in ea punctum ad quod vis centripeta dirigitur, inveniri potest lex vis centripetæ quæ corpus in figuræ illius perimetro gyrare faciet. Nimirum computandum est solidum $\frac{SP^q \times QT^q}{QR}$ huic vi reciprocè proportionale. Ejus rei dabimus exempla in problematîs sequentibus.

Prob. 1. Gyrat corpus in circumferentia circuli [:] requiritur lex vis centripetæ(31) *tendentis ad punctū aliquod in circumferentia.*

Esto circuli circumferentia $SQPA$, centrum vis centripetæ(31) S, corpus in circumferentia latum P, locus proximus in quem movebitur Q. Ad SA diametrum et SP demitte perpendicula PK QT et per Q ipsi SP parallelam age LR occurrentem circulo in L et tangenti PR in R.(32) Erit RP^q (hoc est QRL) ad QT^q ut SA^q ad SP^q. Ergo $\frac{QRL \times SP^q}{SA^q} = QT^q$. Ducantur hæ æqualia in $\frac{SP^q}{QR}$ et punctis P et Q coeuntibus scribatur SP pro RL. Sic fiet $\frac{SP^{qc}}{SA^q} = \frac{QT^q \times SP^q}{QR}$. Ergo

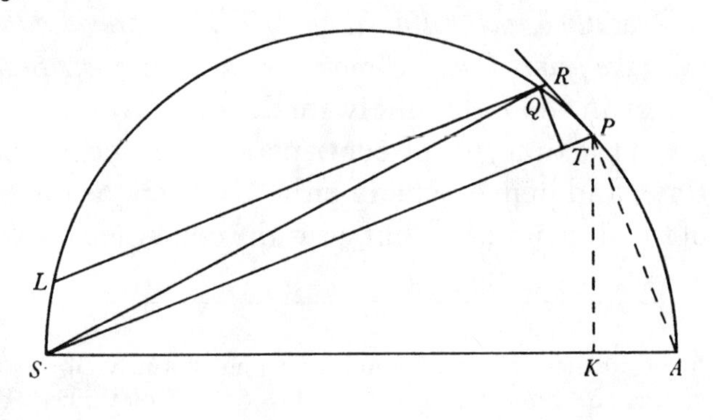

(30) More precisely, since (in the terms of the previous note) $QR = \frac{1}{2}f.dt^2$ accurately (compare note (19)), on setting $SP \times QT = c.dt$ there ensues $f = 2(c^2/SP^2).\lim_{Q\to P}(QR/QT^2)$. The full possibilities of this classical Newtonian measure of central force do not well appear in this geometrical form. If we take $S(0, 0)$ to be the origin of a system of polar coordinates in which P is the general point (r, ϕ) and its radius vector $SP = r \equiv r_\phi$ is regarded as a function of the polar angle ϕ, whose infinitesimal increment is $\widehat{PSQ} = o$, then to the infinitely near point $Q(r_{\phi+o}, \phi+o)$ corresponds the radius vector $SQ = r_{\phi+o} = r+o.dr/d\phi+\frac{1}{2}o^2.d^2r/d\phi^2+\ldots$, while the perpendicular $QT = r_{\phi+o}.\sin o = o.r+o^2.dr/d\phi+\ldots$ cuts off

$$ST = r_{\phi+o}.\cos o = r+o.dr/d\phi+\tfrac{1}{2}o^2.(d^2r/d\phi^2-r)+\ldots,$$

and therefore (since PR is tangent at P to the arc $\widehat{PQ}$)

$$QR = SP-ST+QT.(dr/r\,d\phi) = \tfrac{1}{2}o^2.(r-d^2r/d\phi^2+(2/r).(dr/d\phi)^2)+\ldots,$$

so that $\lim_{o\to\text{zero}}(QR/QT^2) = \frac{1}{2}r^{-2}(r-d^2r/d\phi^2+(2/r).(dr/d\phi)^2 = \frac{1}{2}(r^{-1}+d^2(r^{-1})/d\phi^2)$, and so the central force $f \equiv f(r)$ is $c^2r^{-2}(r^{-1}+d^2(r^{-1})/d\phi^2)$. In geometrical equivalent, Newton in his

proportional to the time (or its double, $SP \times QT$) taken twice. Divide each side of this proportionality by the line-element QR and there will come to be 1 as the centripetal force and $SP^2 \times QT^2/QR$ jointly, that is, the centripetal force will be reciprocally as $SP^2 \times QT^2/QR$.(30) As was to be proved.

Corollary. Hence if any figure be given and in it a point to which the centripetal force is directed, there can be ascertained the law of centripetal force which shall make a body orbit in the perimeter of that figure: specifically, you must compute (the quantity of) the 'solid' $SP^2 \times QT^2/QR$ reciprocally proportional to this force. Of this procedure we shall give illustrations in following problems.

Problem 1. A body orbits in the circumference of a circle: the law of centripetal force(31) *tending to some point in its circumference is required.*

Let $SQPA$ be the circle's circumference, S the centre of centripetal force,(31) P the body borne along in the circumference, Q a closely proximate position into which it shall move. To the diameter SA and to SP let fall the perpendiculars PK, QT and through Q parallel to SP draw LR meeting the circle in L and the tangent PR in R.(32) There will be RP^2 (that is, $QR \times LR$) to QT^2 as SA^2 to SP^2, and therefore $QR \times LR \times SP^2/SA^2 = QT^2$. Multiply these equals into SP^2/QR and, with the points P and Q coalescing, let SP be written in place of LR. In this

---

following Problems 1–3 will effectively compute the value of this function from the given polar defining equations

$$r^{-1} = R^{-1}\sec\phi, \quad r^{-1} = R^{-1}\sqrt{[1+(e^2/(1-e^2))\sin^2\phi]} \quad \text{and} \quad r^{-1} = R^{-1}(1+e\cos\phi)/(1-e^2)$$

to deduce respectively $f(r) \propto r^{-5}$, $f(r) \propto r$ and $f(r) \propto r^{-2}$. Though Newton himself never made such an inverse application of his present measure, there is in principle no bar to our deducing—by two integrations of the ensuing equation $f(r) = c^2r^{-2}(r^{-1}+d^2(r^{-1})/d\phi^2)$—the polar equation of the general orbit traversible in any given central-force field $f(r)$. In what was to prove historically an *experimentum crucis* of Newton's general dynamical method at the hands of Johann Bernoulli during 1710–19 (see D. T. Whiteside, 'The Mathematical Principles underlying Newton's *Principia Mathematica*', *Journal for the History of Astronomy*, **1**, 1970: 116–38, especially 125–6), it is all but immediate that the conic $r^{-1} = A+B\cos\phi$, resolving the equation $r^{-1}+d^2(r^{-1})/d\phi^2 = k/c^2$, is the most general orbit traversible in the inverse-square force-field $f(r) = k/r^2$. In Newton's own geometrical formulation, unfortunately, such a consequence is far from clearly obvious, and he himself felt forced (as we shall see in commentary upon the scholium to Propositions X–XII of §2 following) to use a less direct approach in demonstrating this inverse of his present Problem 2.

(31) Newton here—and *mutatis mutandis* widely in sequel—initially wrote 'gravitatis'. This broadening beyond the immediate model of solar and terrestrial gravitation is of considerable significance in the developing sequence of his dynamical thought. Compare also note (93).

(32) The logical passage to the next sentence was later amplified by Newton's insertion—for Humphrey Newton to copy into the revised draft—of 'et coeant TQ, PR in Z. Ob similitudinem triangulorum ZQR, ZTP, SPA [erit...]' (and let TQ, PR meet in Z. Because the triangles ZQR, ZTP, SPA are similar [there will be...]). In his accompanying figure, correspondingly, he extended TQ and PR to meet in Z.

vis centripeta(33) reciproce est ut $\frac{SP^{qc}}{SA^q}$, id est (ob datum SA^q) ut quadrato-cubus distantiæ SP.(34) Quod erat inveniendum.

Schol.(35) Cæterum in hoc casu et similibus concipiendum est quod postquam corpus pervenit ad centrum S, id non amplius redibit in orbem sed abibit in tangente.(36) In spirali quæ secat radios omnes in dato angulo(37) vis centripeta tendens ad Spiralis principium est in ratione triplicata distantiæ reciprocè, sed in principio illo recta nulla positione determinata spiralem tangit.

Prob 2. Gyrat corpus in Ellipsi veterum: requiritur lex vis centripetæ tendentis ad centrum Ellipseos.

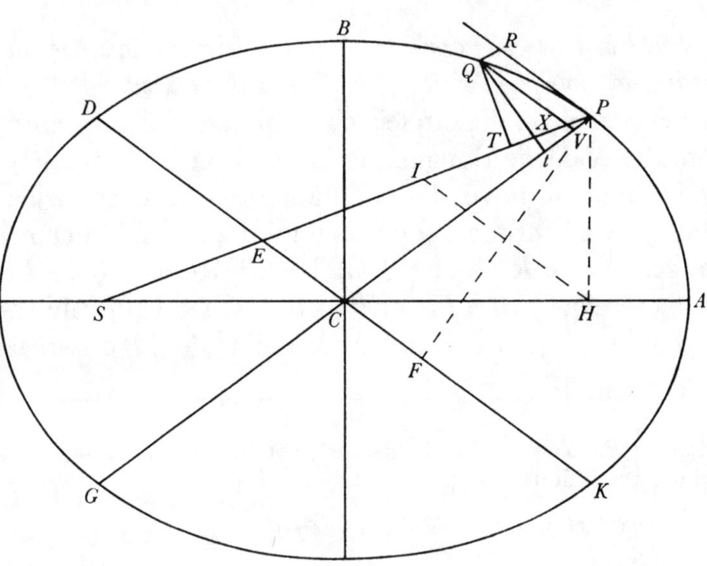

Sunto CA, CB semi-axes Ellipseos, GP, DK diametri conjugatæ, PF, Qt perpendicula ad diametros[,] QV ordinatim applicata ad diametrum GP et $QVPR$ parallelogrammum.(38) His constructis erit (ex Conicis(39)) PVG ad QV^q ut PC^q ad CD^q et QV^q ad Qt^q ut PC^q ad PF^q[,] et conjunctis rationibus PVG ad Qt^q ut PC^q ad $\frac{CD^q \times PF^q}{PC^q}$. Scribe QR pro PV et(40) $BC \times CA$ pro $CD \times PF$, nec non (punctis P et Q coeuntibus) $2PC$ pro VG[,] et ductis extremis et medijs in se mutuò fiet

(33) Newton again (compare note (31)) initially wrote 'gravitas' (gravity). We will not pinpoint further instances of this change.

(34) In effect, Newton computes $\lim_{Q \to P} (QR/QT^2) = R^2/r^3$ as $r^{-3}(ds/d\phi)^2$, where (in analytical equivalent) the polar equation $r = 2R\cos\phi$ defining $P(r, \phi)$ determines it to be on the circle whose diameter joins $S(0, 0)$ and $A(2R, 0)$. The same result follows equally from computing this limit in the equivalent form $\frac{1}{2}(r^{-1} + d^2(r^{-1})/d\phi^2)$; compare note (30).

(35) This head was probably omitted—by carelessness on its copyist's part?—from the putative fair copy subsequently sent to London (see note (2)) since it is absent in both the Royal Society and Halley transcripts.

(36) If, of course, we may uniquely define that direction: Newton at once proceeds to give a counter-instance.

(37) The logarithmic (equiangular) spiral, that is. In this curve (already familiar to him as the stereographic projection of the spherical loxodrome which meets all meridians at a constant angle; see IV: 126, note (24)) the infinitesimal elements will, for equal vanishingly

way there will come $SP^5/SA^2 = QT^2 \times SP^2/QR$. Therefore the centripetal force(33) is reciprocally as SP^5/SA^2, that is, (because SA^2 is given) as the fifth power of the distance SP.(34) As was to be found.

Scholium.(35) In this case, however, and similar instances you must conceive that after the body reaches the centre S it will no more return into its orbit but depart along the tangent.(36) In a spiral which cuts all its radii at a given angle(37) the centripetal force tending to the spiral's pole is reciprocally in the tripled ratio of the distance, but at that pole there is no straight line fixed in position which touches the spiral.

Problem 2. A body orbits in a classical ellipse: there is required the law of centripetal force tending to the ellipse's centre.

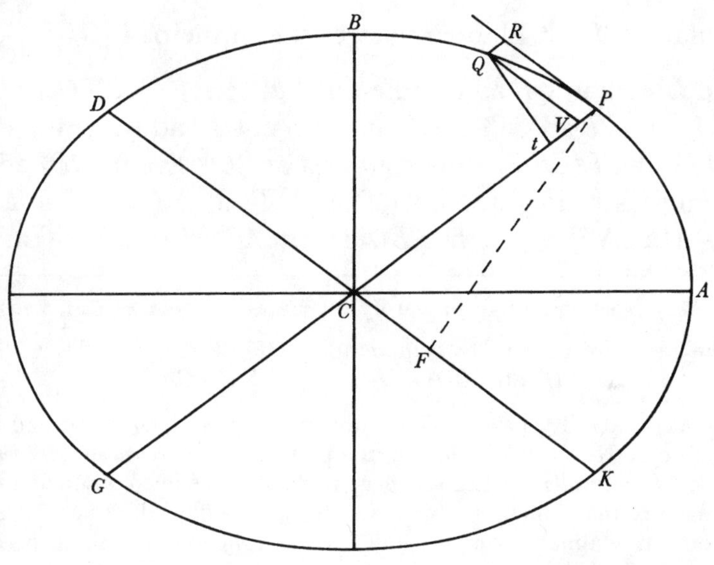

Let CA, CB be the ellipse's semi-axes, GP and DK conjugate diameters, PF and Qt perpendiculars to these diameters, QV ordinate to the diameter GP, and $QVPR$ a parallelogram.(38) With this construction there will (from the *Conics*(39)) be $PV \times VG$ to QV^2 as PC^2 to CD^2 and again QV^2 to Qt^2 as PC^2 to PF^2, and on combining these proportions $PV \times VG$ to Qt^2 as PC^2 to $CD^2 \times PF^2/PC^2$. Write QR in place of PV and(40) $BC \times CA$ in place of $CD \times PF$, and in addition (with the points P and Q coalescing) $2PC$ in place of VG, and when extremes and middles are multiplied into each other

small increments $\widehat{PSQ}$ of its polar angle, retain the same proportion both to themselves and to the radius vector SP; whence at once $\lim_{Q \to P} (QR/QT^2) \propto 1/SP$, and so the central force to the pole S which induces such a spiral orbit varies as $1/SP^3$. This example is given the status of a separate Propositio VIII in the amplified text reproduced in §2 following.

(38) However, in Newton's accompanying figure (which is economically but somewhat confusingly fashioned to illustrate both this and the following problem) QR is slightly distorted in direction to be parallel to some mean between CVP and SXP. In our English version, here and below, we split the manuscript figure into the two simpler diagrams which it combines, there accurately drawing QR parallel to CP or SP accordingly as the text directs.

(39) Understand Apollonius, *Conics* I, 11–19 or equivalent propositions in any of its more modern reformulations. The reference was inserted as an afterthought.

(40) 'Per Lem: [2]' (by Lemma 2), as Newton was to add in his revise.

$\frac{Qt^q \times PC^q}{QR} = \frac{2BC^q \times CA^q}{PC}$. Est ergo vis centripeta reciprocè ut $\frac{2BC^q \times CA^q}{PC}$, id est (ob datum $2BC^q \times CA^q$) ut $\frac{1}{PC}$, hoc est directè, ut distantia PC.(41) Q.E.I.

Prob. 3. Gyrat corpus in ellipsi: requiritur lex vis centripetæ tendentis ad umbilicum(42) *Ellipseos.*

Esto Ellipseos superioris umbilicus S. Agatur SP secans Ellipseos diametrum DK in E.(43) Patet EP æqualem esse semi-axi majori AC eò, quod actâ ab altero Ellipseos umbilico H linea HI ipsi EC parallela, ob æquales CS, CH æquentur ES, EI, adeo ut EP semisumma sit ipsarum PS, PI id est(44) ipsarum PS, PH quæ conjunctim axem totum $2AC$ adæquant.(45) Ad SP demittatur perpendicularis QT. Et Ellipseos latere recto principali $\left(\text{seu } \frac{2BC^q}{AC}\right)$ dicto L, erit $L \times QR$ ad $L \times PV$ ut QR ad PV id est ut PE (seu AC) ad PC. et $L \times PV$ ad GVP ut L ad GV. et GVP ad QV^q ut CP^q ad CD^q. et QV^q ad QX^q puta ut M ad N.(46) et QX^q ad QT^q ut EP^q ad PF^q id est ut CA^q ad PF^q sive ut CD^q ad CB^q. et conjunctis his omnibus rationibus, $L \times QR$ ad QT^q ut AC ad $PC + L$ ad $GV + CP^q$ ad $CD^q + M$ ad N(47) $+ CD^q$ ad CB^q, id est ut $AC \times L$ (seu $2BC^q$) ad $PC \times GV + CP^q$ ad CB^q

(41) Newton straightforwardly evaluates his geometrical measure $CP^{-2}.\lim_{Q\to P}(QR/Qt^2)$ of the force directed to C by compounding $QR \times VG : QV^2 = PC^2 : CD^2$ and $QV^2 : Qt^2 = PC^2 : PF^2$ to yield $\lim_{V\to P}(PC^2/VG \times CD^2 \times PF^2) = \frac{1}{2}PC/BC^2 \times CA^2 \propto PC$.

(42) Literally 'navel'. This anthropomorphic designation had been invoked twenty years before by Nicolaus Mercator in the preface of his *Hypothesis Astronomica Nova, et Consensus ejus cum Observationibus* (London, 1664) to denote—as one component of an elaborate 'humanist' image of the planetary ellipse, according to which the line of apsides is divided in the proportions of a male figure set with head at aphelion and feet at perihelion—the 'belly-button' centre of a Keplerian 'vicarious' equant circle dividing the distance between the solar focus (at the 'knees') and upper focus (at the 'breast') in divine section. With the rise to popularity of a simpler Boulliauist hypothesis of elliptical motion (see §2: note (163) below) the 'umbilic' soon came to be identified—for instance, by Claude Milliet Dechales in his *Mundus Mathematicus* (Paris, 1674)—rather with the upper focus itself, and then by extension to be applied to any focus of a conic in general: Isaac Barrow so denotes the focus of a parabola in his *Lectiones XVIII...Opticorum Phænomenωn*...(London, 1669): 28: §xii. 1, for example. Newton's anatomically freakish present innovation of permitting his planetary conic to have two 'navels' S and H was afterwards continued by him in his published *Principia*. Earlier, in his youth (see i: 32) he had been content to employ the now standard term 'focus'—introduced by Kepler in Caput iv, §4 of his *Ad Vitellionem Paralipomena*...(Frankfurt, 1604) and afterwards popularised by Mydorge and Descartes—and to this nomenclature he was, as we shall see in the next volume, to return in his private papers in the 1690's.

(43) In his revise (see note (2)) Newton later added in amplification 'et lineam QV in X et compleatur parallelogrammum $QXPR$' (and the line QV in X, and complete the parallelogram $QXPR$).

(44) Newton here subsequently inserted the parenthesis 'ob parallelas HI, PR & angulos æquales IPR, HPZ' (because of the parallels HI, PR and the equal angles $\widehat{IPR}$, $\widehat{HPZ}$), correspondingly extending the tangent RP in his preceding figure to Z.

there will come to be $Qt^2 \times PC^2/QR = 2BC^2 \times CA^2/PC$. The centripetal force is therefore reciprocally as $2BC^2 \times CA^2/PC$, that is, (because $2BC^2 \times CA^2$ is given) as $1/PC$, or in other words directly as the distance PC.[41] As was to be found.

Problem 3. A body orbits in an ellipse: there is required the law of centripetal force tending to a focus[42] *of the ellipse.*

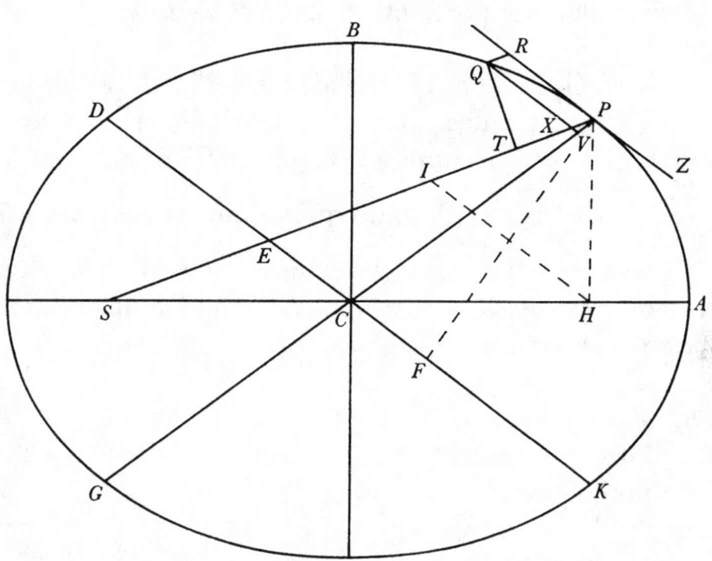

Let S be a focus of the preceding ellipse. Draw SP cutting the ellipse's diameter DK in E.[43] It is evident that EP is equal to the semi-major-axis AC, seeing that, when from the ellipse's other focus H the line HI is drawn parallel to CE, because CS and CH are equal so are ES and EI, and hence EP is the half-sum of PS and PI, that is,[44] of PS and PH which are jointly equal to the total axis $2AC$.[45] To SP let fall the perpendicular QT. Then, on calling the ellipse's principal *latus rectum* (viz. $2BC^2/AC$) L, there will be $L \times QR$ to $L \times PV$ as QR to PV, that is, as PE (or AC) to PC; and $L \times PV$ to $GV \times VP$ as L to GV; and $GV \times VP$ to QV^2 as CP^2 to CD^2; and QV^2 to QX^2 as, say, M to N;[46] and QX^2 to QT^2 as EP^2 to PF^2, that is, as CA^2 to PF^2 or as CD^2 to CB^2: and, when all these proportions are combined, $L \times QR$ is to QT^2 as

$$(AC \text{ to } PC) \times (L \text{ to } GV) \times (CP^2 \text{ to } CD^2) \times (M \text{ to } N)^{(47)} \times (CD^2 \text{ to } CB^2),$$

(45) This 'evident' property of the ellipse—and also, as Newton will soon show (in Proposition XI of §2 following), of the hyperbola—is clearly regarded by Newton as his present discovery, nor indeed have we found it listed in any preceding work on conics: of course, crucial though it is in the context of the present argument, it may (if known) have not been regarded by geometers at large as basic enough to be accorded separate status as a theorem.

(46) This separate denomination of the ratio QV^2 to QX^2 seems entirely unnecessary, particularly since in the sequel it will, in the limit as Q coincides with P, become unity. Newton realised as much when he amended his subsequent revise (see note (2)), here altering the sentence to read 'et QV^q ad QX^q punctis Q et P coeuntibus fit ratio æqualitatis' (and QV^2 to QX^2 comes, as the points Q and P coincide, to be a ratio of equality), continuing in sequel 'et QX^q seu QV^q est ad QT^q...' (and QX^2, or QV^2, is to QT^2...).

(47) These three '$+M$ ad N' ($\times(M \text{ to } N)$) were deleted by Newton in the revise in line with the emendation recorded in the previous note. The use of '+' to denote 'addition' (composition) of geometrical ratios is copied from Isaac Barrow, who introduced this some-

$+ M$ ad N,(47) sive ut $2PC$ ad $GV + M$ ad N.(47) Sed punctis Q et P coeuntibus rationes $2PC$ ad GV et M ad N fiunt æqualitatis: Ergo et ex his composita ratio $L \times QR$ ad QT^q.(48) Ducatur pars utraqȝ in $\frac{SP^q}{QR}$ et fiet $L \times SP^q = \frac{SP^q \times QT^q}{QR}$. Ergo vis centripeta reciprocè est ut $L \times SP^q$ id est in ratione duplicata distantiæ.(49) Q.E.I.

(50)*Schol.* Gyrant ergo Planetæ majores in ellipsibus habentibus umbilicum in centro solis, et radijs ad Solem ductis describunt areas temporibus proportionales, omninò ut supposuit Keplerus.(51) Et harum Ellipseon latera recta sunt $\frac{QT^q}{QR}$,(52) punctis P et Q spatio quàm minimo et quasi infinitè parvo distantibus.

Theorem. 4. Posito quod vis centripeta sit reciprocè proportionalis quadrato distantiæ a centro,(53) *quadrata temporum periodicorum in Ellipsibus sunt ut cubi transversorum axium.*

Sunto Ellipseos axis transversus AB,(54) axis alter PD, latus rectum L,(55) umbilicus alteruter S. Centro S intervallo SP describatur circulus PMD. Et eodem tempore describant corpora duo gyrantia arcum Ellipticum PQ et circulum PM, vi centripeta ad umbilicum S tendente. Ellipsin et circulum

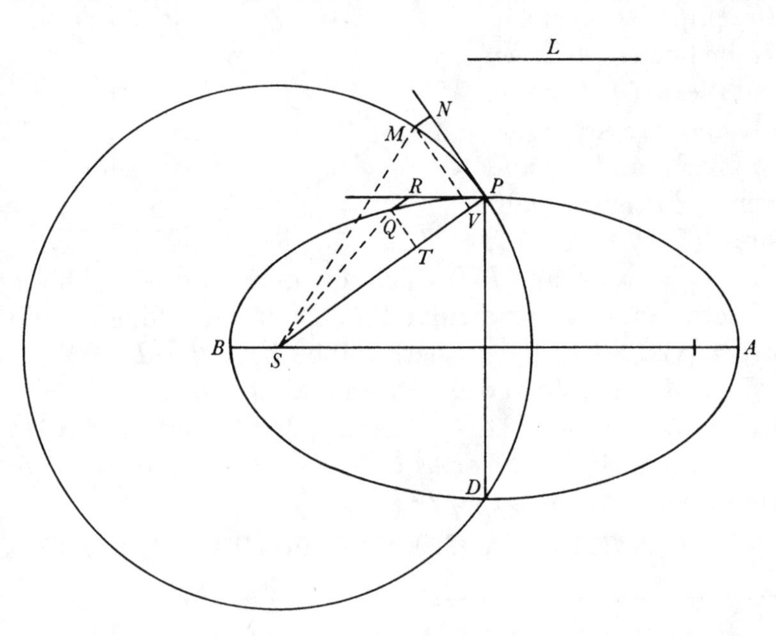

what confusing notation in his *Lectiones XXIII... In quibus Opticorum Phænomenωn Genuinæ Rationes investigantur, ac exponuntur* (London, 1669); see especially his introductory list of symbols (signature $a2^v$), where '$A.B+C.D$' is deemed to signify the 'Rationes A ad B, & C ad D compositæ'.

(48) In his revise Newton likewise recast this sentence to read '...punctis Q et P coeuntibus æquantur $2PC$ & GV: Ergo et $L \times QR$ & QT^q æquantur' (...with the points Q and P coalescing, $2PC$ and GV are equal; therefore $L \times QR$ and QT^2 also are equal).

(49) In sum, Newton here evaluates his measure $SP^{-2}. \lim_{Q \to P} (QR/QT^2)$ of the central force to S by compounding (QR or) PX: PV = (PE or) CA: PC, $PV \times VG$: $QV^2 = PC^2$: CD^2 and QX^2: QT^2 = (PE^2 or) CA^2: PF^2 to produce

$$SP^{-2}. \lim_{Q,V,X \to P} (CA^3 \times PC \times QV^2/VG \times CD^2 \times PF^2 \times QX^2) = SP^{-2}/L \propto SP^{-2},$$

where the (principal) *latus rectum* $L = 2BC^2/CA$.

that is, as

$$(AC\times L(\text{or } 2BC^2) \text{ to } PC\times GV)\times(CP^2 \text{ to } CB^2)\times(M \text{ to } N),^{(47)}$$

or as $(2PC$ to $GV)\times(M$ to $N)$.[47] But, with the points Q and P coalescing, the ratios $2PC$ to GV and M to N become ones of equality, and so also therefore does the ratio $L\times QR$ to QT^2.(48) Multiply each member by SP^2/QR and there will come to be $L\times SP^2 = SP^2\times QT^2/QR$. Therefore the centripetal force is reciprocally as $L\times SP^2$, that is, (reciprocally) in the doubled ratio of the distance.(49) As was to be found.

(50)*Scholium*. The major planets orbit, therefore, in ellipses having a focus at the centre of the Sun, and with their *radii* (*vectores*) drawn to the Sun describe areas proportional to the times, exactly as Kepler supposed.(51) And the *latera recta* of these ellipses are QT^2/QR,(52) where the distance between the points P and Q is the least possible and, as it were, infinitely small.

Theorem 4. Supposing that the centripetal force be reciprocally proportional to the square of the distance from the centre,(53) *the squares of the periodic times in ellipses are as the cubes of their transverse axes.*

Let AB be an ellipse's transverse axis,(54) PD its other axis, L its *latus rectum*,(55) S one or other of its foci. With centre S and radius SP describe the circle PMD. Then in the same time let two orbiting bodies describe (respectively) the ellipse-arc $\widehat{PQ}$ and the circle-arc $\widehat{PM}$, with the centripetal force tending to the focus S.

(50) Newton first began to enter a '*Cor.* Punctis P et Q coeu[ntibus ratio $L\times QR$ ad QT fit æqualitatis?]' (*Corollary*. With the points P and Q coming to coalesce [? the ratio of $L\times QR$ to QT^2 becomes one of equality]).

(51) The ellipticity of Mars' orbit was established by Kepler in his *Astronomia Nova* ΑΙΤΙΟΛΟΓΗΤΟΣ, *seu Physica Cœlestis, tradita commentariis De Motibus Stellæ Martis* (Prague, 1609) and of that of the other solar planets—with some remaining degree of doubt as to its exactness—in his later *Epitome Astronomiæ Copernicanæ* (Linz, 1618–21); compare C. A. Wilson's well-documented recent analysis of the former in 'Kepler's Derivation of the Elliptical Path', *Isis*, **59**, 1968: 5–25. On Kepler's formulation of the planetary areal law in his *Astronomia Nova* see note (19) above.

(52) Newton first began to write in sequel 'existentibus figuris $QTPR$ [? quam minimis]' (where the figures $QTPR$ are minimally small).

(53) Understand the centre of force (at the ellipses's focus S).

(54) Newton first began: 'Sunto Ellipseos umbilici S, H, centrum C, axis transversus PA, tangens ad verticem PR' (Let the ellipse's foci be S and H, its centre C, transverse axis PA, vertex tangent PR). Correspondingly, in his figure he originally denoted the vertex B by P, and marked the position of the second focus (as here shown) but without naming it; centred on S, furthermore, he drew two vanishingly small focal sectors SPQ and (in general position) SEG such that the force-deviations, RQ and FG respectively, from the tangents PR and EF to the orbit were equal. It is clear that the simplification embodied in rotating $SEFG$ round to coincide with $SPNM$ in the revised figure and resiting $SPRQ$ in its present position occurred to him only as he began to pen the present demonstration.

(55) That is, $L = PD^2/AB = PD^2/2SP$.

tangant PR, PN in puncto P. Ipsi PS agantur parallelæ QR, MN tangentibus occurrentes in R et N. Sint autem figuræ PQR, PMN indefinitè parvæ sic ut (per Schol. Prob. 3) fiat $L\times QR = QT^q$ et[56] $2SP\times MN = MV^q$. Ob communem a centro S distantiam SP et inde æquales vires centripetas sunt MN et QR æquales. Ergo QT^q ad MV^q est ut L ad $2SP$, et QT ad MV ut medium proportionale inter L et $2SP$ seu PD ad $2SP$. Hoc est area SPQ ad aream SPM ut area tota Ellipseos ad aream totam circuli.[57] Sed partes arearum singulis momentis genitæ sunt ut areæ SPQ et SPM atq; adeo ut areæ totæ[,] et proinde per numerum momentorum multiplicatæ simul evadent totis æquales. Revolutiones igitur eodem tempore in ellipsibus perficiuntur ac in circulis quorum diametri sunt axibus transversis Ellipseon æquales. Sed (per Cor. 5 Theor 2) quadrata temporum periodicorum in circulis sunt ut cubi diametrorum. Ergo et in Ellipsibus. Q.E.D.[58]

Schol. Hinc in Systemate cœlesti[59] ex temporibus periodicis Planetarum innotescunt proportiones transversorum axium Orbitarum. Axem unum[60] licebit assumere. Inde dabuntur cæteri. Datis autem axibus determinabuntur Orbitæ in hunc modum. Sit S locus Solis seu Ellipseos umbilicus unus[,] A, B, C, D loca Planetæ observatione inventa et Q axis transversus[61] Ellipseos. Centro A radio $Q-AS$ describatur circulus FG et erit ellipseos umbilicus alter in hujus circumferentia. Centris B, C, D, &c intervallis $Q-BS$, $Q-CS$, $Q-DS$ describantur itidem alij quotcunq; circuli & erit umbilicus ille alter in omnium circumferentijs atq; adeo in omniū intersectione communi F. Si intersectiones omnes non coincidunt, sumendum erit punctum medium[62] pro umbilico.

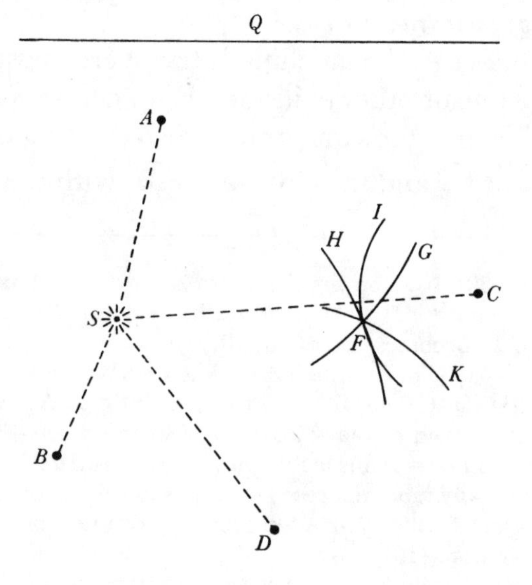

Praxis hujus commoditas est quod ad unam conclusionem eliciendam adhiberi possint et inter se expeditè comparari observationes quamplurimæ. Planetæ

(56) In the case of the circle it is immediate that the *latus rectum* is equal in length to the diameter.

(57) That is, $\frac{1}{4}\pi.AB\times PD:\pi.SP^2 = \frac{1}{2}PD:SP$, since $AB = 2SP$.

(58) In revision (see Propositions XIII/XIV of §2 following) Newton will prefer to prove this third Keplerian planetary law directly from the result $\lim_{Q\to P}(QR/QT^2) = 1/L$ without intervening appeal to the particular case of concentric-circle orbits.

(59) Understand the Keplerian *Systema Copernicanum* in which the planets traverse exact ellipses round the sun set at a common focus.

Let PR, PN be tangent to the ellipse and circle at the point P; and parallel to PS draw QR, MN meeting those tangents in R and N. Now let the figures PQR, PMN be indefinitely small, so that (by the Scholium to Problem 3) there comes to be $L \times QR = QT^2$ and [56] $2SP \times MN = MV^2$. Because of their common distance SP from the centre S and therefore equal centripetal forces producing them, MN and QR are equal. In consequence QT^2 is to MV^2 as L to $2SP$, and so QT to MV as the mean proportional between L and $2SP$, that is, PD to $2SP$; accordingly, the area (SPQ) is to the area (SPM) as the total area of the ellipse to the total area of the circle.[57] But the parts of area generated in individual moments are as the areas (SPQ) and (SPM), and hence as the total areas, and consequently when multiplied by the number of these moments they will simultaneously end up equal to the total areas. Revolutions in ellipses, therefore, are completed in the same time as those in circles whose diameters are equal to the transverse axes of the ellipses. But (by Corollary 5 of Theorem 2) the squares of the periodic times in circles are as the cubes of their diameters. And so also in ellipses, therefore. As was to be proved.[58]

Scholium. Hereby in the heavenly system[59] from the periodic times of the planets are ascertained the proportions of the transverse axes of their orbits. It will be permissible to assume one axis:[60] from that the rest will be given. Once their axes are given, however, the orbits will be determined in this manner. Let S be the position of the Sun—one focus of the ellipse, that is—, A, B, C, D positions of the planet found from observation, and Q the transverse axis[61] of the ellipse. With centre A and radius $Q-AS$ describe the circle FG and the ellipse's other focus will be in its circumference. Correspondingly, with centres $B, C, D, \ldots$ and intervals $Q-BS$, $Q-CS$, $Q-DS$, ... describe any number of other circles, and that other focus will be in all their circumferences and hence at the common intersection of all of them. If all their intersections do not coincide, you will need to take a mean point[62] for the focus. The advantage of this technique is that a large number of observations, no matter how many,

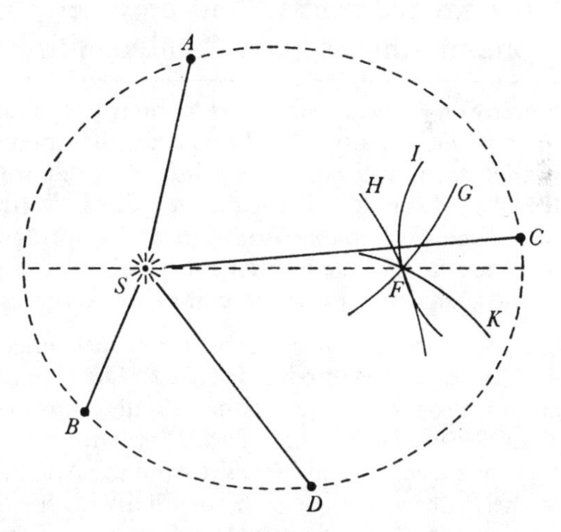

(60) In practice this will, of course, usually be the diameter of the Earth's orbit (the 'Telluris orbis magnus' as Newton names it below).

(61) More precisely, 'longitudo axis transversi' (the length of the transverse axis).

(62) Assuming that, according to some criterion, numerical 'weights' can be set on the relative accuracy with which the initial planetary positions A, B, C, D, ... are computed, and

autem loca singula A, B, C, D &c ex binis observationibus, cognito Telluris orbe magno invenire docuit Halleus.[(63)] Si orbis ille magnus nondum satis exactè determinatus habetur, ex eo propè cognito, determinabitur orbita Planetæ alicujus puta Martis propius: Deinde ex orbita Planetæ[(64)] per eandem methodum determinabitur orbita telluris adhuc propius: Tum ex orbita Telluris determinabitur orbita Planetæ multò exactiùs quam priùs: Et sic per vices donec circulorum intersectiones in umbilico orbitæ utriusꝗ exactè satis conveniunt.

Hac methodo determinare licet orbitas Telluris, Martis, Jovis et Saturni, orbitas autem Veneris et Mercurij sic. Observationibus in maxima Planetarum a Sole digressione factis, habentur orbitarum tangentes. Ad ejusmodi tangentem KL demittatur a Sole perpendiculum SL centroꝗ L et intervallo dimidij axis Ellipseos describatur circulus KM. Erit centrum Ellipseos in hujus circumferentia,[(65)] adeoꝗ descriptis hujusmodi pluribus circulis reperietur in omnium intersectione. Tum cognitis orbitarum dimensionibus, longitudines[(66)]

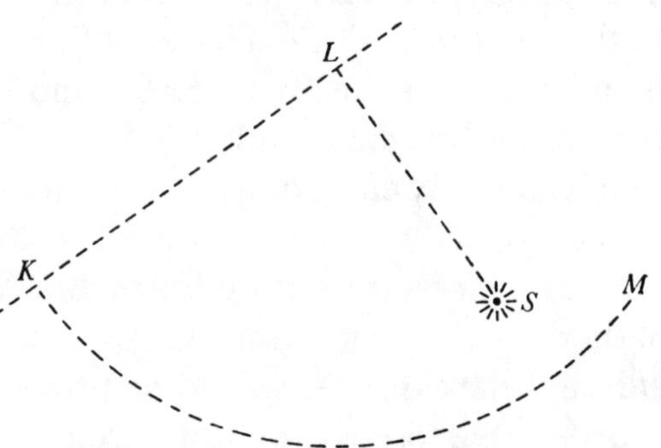

thereby on the trustworthiness of the trial positions F (say F_i, $i = 1, 2, 3, \ldots$) constructed from these taken in pairs, it will be natural to choose the mean point F as the 'centre of gravity' of the severally constructed points F_i; compare the concluding scholium (page 22) of Roger Cotes's 'Æstimatio Errorum in Mixta Mathesi' (published posthumously in his *Opera Miscellanea* [=*Harmonia Mensurarum* (Cambridge, 1722): 1–121]: 1–22). The simplest instance, in which each of the positions A, B, C, D, ... are taken to be computed with equal accuracy, would then define the mean point F as satisfying $\sum_i (F' - F'_i) = 0$, where F', F'_i are the distances of F, F_i from some arbitrary straight line—effectively the modern Gaussian least-squares test.

(63) In his 'Methodus directa & Geometrica, cujus ope investigantur Aphelia, Eccentricitates, Proportionesque orbium Planetarum primariorum...', *Philosophical Transactions of the Royal Society*, **11**, No. 128 [for 25 September 1676]: 683–6. (Halley's original English version, 'A direct Geometrical Process to find the Aphelion, Eccentricities, and Proportions of the Orbs of the Primary Planets...', enclosed with his letter to Oldenburg on the previous 11 July, is reproduced in S. P. Rigaud, *Correspondence of Scientific Men of the Seventeenth Century*, **1** (Oxford, 1841): 237–41.) Newton is seemingly unaware that the technique here cited, of fixing the position and relative sizes of planetary *radii vectores* by means of solar oppositions, is Halley's straightforward (if unacknowledged) borrowing from Kepler's *Astronomia Nova* (note (51))—a work which Newton himself almost certainly never read. The method is, of course applicable only to determining the orbits of upper planets, since Mercury and Venus can never be in opposition to the Sun (as viewed from the Earth).

(64) Newton first wrote 'Martis' (Mars), carrying over his preceding instance.

(65) For, if S' is the mirror-image of the orbital focus S in the tangent KL, the line $S'H$ drawn from S' to the second focus H will meet KL in the point P of tangency, and also (since $SL = LS'$ and $SC = CH$) be parallel to LC, where C is the required centre; at once

$$LC = \tfrac{1}{2}S'H = \tfrac{1}{2}(SP + PH) = \tfrac{1}{2}AB,$$

may be employed to elicit a single conclusion and speedily be compared one with another. How, however, the individual positions *A*, *B*, *C*, *D*, ... of a planet may be found from pairs of observations once the 'great' (annual) orbit of the Earth is known, Halley has explained.(63) If that great orbit is not yet considered to be determined with sufficient exactness, the orbit of some planet—say Mars—will be determined more closely from its close delimitation; then from the orbit of the planet(64) the orbit of the Earth will be determined still more closely by the same method; and thereafter from the orbit of the Earth the orbit of the planet will be determined much more exactly than before; and so on in turn until the intersections of the circles concur sufficiently exactly in the focus of each planet.

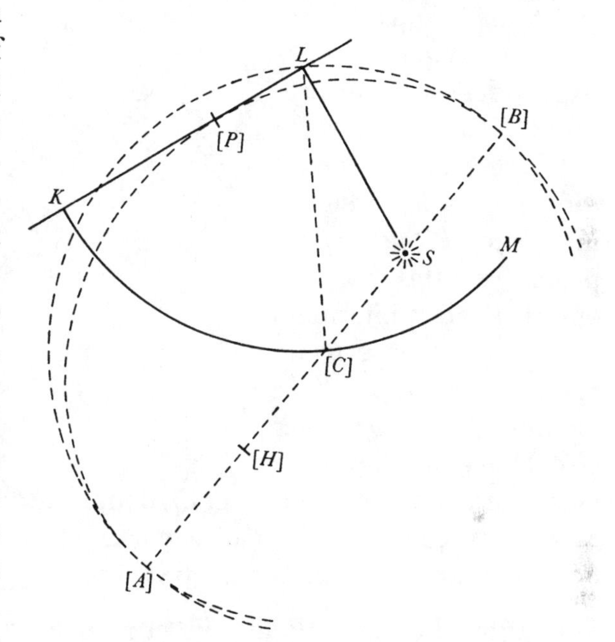

By this method we are free to determine the orbits of the Earth, Mars, Jupiter and Saturn: to ascertain the orbits of Venus and Mercury, however, do as follows. From observations made at the maximum digression of these planets from the Sun tangents to their orbits are obtained. To a tangent *KL* of this type let fall the perpendicular *SL* from the Sun, and then with centre *L* and radius half the ellipse's axis describe the circle *KM*: the centre of the ellipse will be in its circumference,(65) and hence when several circles of this sort are described it will be found at their joint intersection. Thereafter, once the dimensions of their orbits are known, the lengths(66)

where *AB* is the orbit's axis. As Newton may well here assume his reader to know, the equivalent proposition that the foot of the perpendicular from the focus of an ellipse to any tangent lies on its circumcircle is Apollonius, *Conics*, III, 49.

(66) Understand their absolute sizes (as distinct from their relative proportions, now—by the two previous techniques—assumed to be known). This use of (twin) observations of the transits of Mercury and especially Venus across the solar disc to determine, by parallax, the distance of the Sun from the Earth (and therefrom the absolute dimensions of the orbits of the solar planets) had been first publicly suggested by James Gregory in a brief scholium to Propositio 87 of the *Appendix, Subtilissimorum Astronomiæ Problematôn resolutionem exhibens* concluding his *Optica Promota* (London, 1663), where (page 130) he wrote that 'Hoc Problema pulcherrimum habet usum, sed forsan laboriosum, in observationibus Veneris, vel Mercurii particulam Solis obscurantis: ex talibus enim solis parallaxis investigari poterit.' Halley, who had witnessed

horum Planetarum exactiùs ex transitu per discum Solis determinabuntur.[67]

Prob. 4. Posito quod vis centripeta sit reciprocè proportionalis quadrato distantiæ a centro, et cognita vis illius quantitate, requiritur Ellipsis[68] *quam corpus describet de loco dato cum data celeritate secundum datam rectam emissum.*

Vis centripeta tendens ad punctum S ea sit quæ corpus π in circulo $\pi\chi$ centro S intervallo quovis $S\pi$ descripto gyrare faciat. De loco P secundum lineam PR emittatur corpus P[69] et mox inde cogente vi centripeta deflectat in Ellipsin PQ. Hanc igitur recta PR tanget in P. Tangat itidem recta $\pi\rho$ circulum in π sitꝗ PR ad $\pi\rho$ ut prima celeritas corporis emissi P ad uniformem celeritatem corporis π. Ipsis SP et $S\pi$ parallelæ agantur RQ et $\rho\chi$[,] hæc circulo in χ illa Ellipsi in Q occurrens, et a Q et χ ad SP et $S\pi$ demittantur perpendicula QT et $\pi\tau$. Est RQ ad $\rho\chi$ ut vis centripeta in P ad vim centripetam in π id est ut $S\pi^{\text{quad.}}$ ad $SP^{\text{quad.}}$, adeoꝗ datur illa ratio. Datur etiam ratio QT ad $\chi\tau$. De hac ratione duplicata auferatur ratio data QR ad $\chi\rho$[70] et manebit data ratio $\frac{QT^q}{QR}$ ad $\frac{\chi\tau^q}{\chi\rho}$, id est (per Schol. Prob. 3) ratio lateris recti Ellipseos ad diametrum circuli: Datur igitur latus rectum Ellipseos. Sit istud L. Datur præterea Ellipseos umbilicus S. Anguli RPS complementū ad duos rectos fiat angulus RPH et dabitur positione linea PH in qua umbilicus alter H locatur. Demisso ad PH perpendiculo SK et erecto semiaxe minore BC est

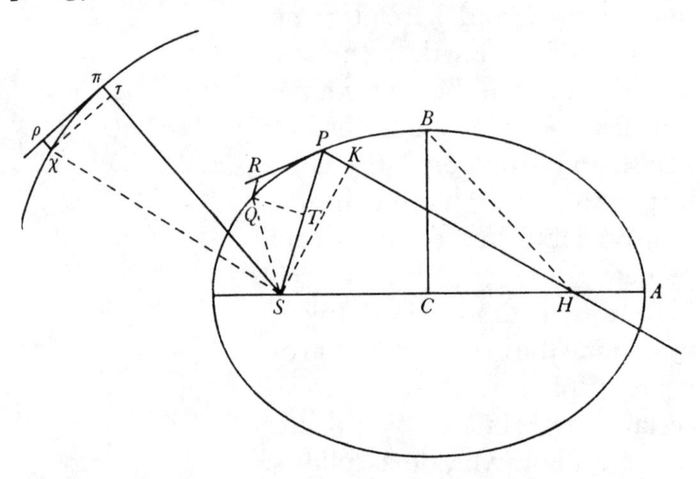

$$SP^q - 2KPH + PH^q = SH^{q\,(71)} = 4BH^q - 4BC^q = \overline{SP+PH}^{\text{quad.}} - L \times \overline{SP+PH}$$
$$= SP^q + 2SPH + PH^q - L \times \overline{SP+PH}.$$

---

a transit of Mercury at St. Helena in October 1677, and may well have directed Newton's attention to the possibilities during his visit to Cambridge in August 1684, was later (in papers published in the *Philosophical Transactions* in 1691 and 1716; see A. Armitage, *Edmond Halley* (London, 1966): 103–4) successfully to overcome the formidable practical difficulties involved in pinpointing the moment of occultation, synchronising the clocks of widely distant observers, and making adequate allowance for the diurnal motion of the Earth and inferior planet. Even so, it proved impossible to test the worth of his procedure till the time (1761 and 1769) of the next pair of Venus transits.

of these planets will more exactly be determined from their passage over the Sun's disc.[67]

Problem 4. Supposing that the centripetal force be reciprocally proportional to the square of the distance from its centre, and with the quantity of that force known, there is required the ellipse[68] *which a body shall describe when released from a given position with a given speed following a given straight line.*

Let the centripetal force tending to the point S be that which shall make the body π orbit in the circle $\pi\chi$ described with centre S and any radius $S\pi$. Let the body P be released from the position P following the line PR[69] and directly thereafter under the compulsion of the centripetal force be deflected into the ellipse $\widehat{PQ}$. This, therefore, the straight line PR will touch at P. Let the straight line $\pi\rho$ correspondingly touch the circle at π, and let PR be to $\pi\rho$ as the initial speed of the body P on release to the uniform speed of the body π. Parallel to SP and $S\pi$ draw RQ and $\rho\chi$, the latter meeting the circle in χ, the former the ellipse in Q, and from Q and χ to SP and $S\pi$ let fall the perpendiculars QT and $\chi\tau$. Now RQ is to $\rho\chi$ as the centripetal force at P to the centripetal force at π, that is, as $S\pi^2$ to SP^2, and hence that ratio is given. The ratio of QT to $\chi\tau$ is also given. From this latter ratio doubled take away the given ratio of QR to $\chi\rho$[70] and there will remain given the ratio of QT^2/QR to $\chi\tau^2/\chi\rho$, that is, (by the Scholium to Problem 3) the ratio of the ellipse's *latus rectum* to the circle's diameter; accordingly, the ellipse's *latus rectum* is given. Let that be L. There is given, furthermore, the ellipse's focus S. Make the angle $R\widehat{P}H$ the supplement of the angle $R\widehat{P}S$ and there will be given in position the line PH in which the other focus H is located. On letting fall the perpendicular SK to PH and erecting the semi-minor-axis BC, there is

$$SP^2 - 2KP \times PH + PH^2 = SH^{2\,(71)} = 4BH^2 - 4BC^2 = (SP+PH)^2 - L \times (SP+PH)$$
$$= SP^2 + 2SP \times PH + PH^2 - L \times (SP+PH).$$

(67) In his revise (see note (2)) Newton here added in sequel a further paragraph, touching on the system of solar planets as a collective whole and remarking on the 'superhuman' difficulty of reducing their mutual gravitational perturbations to a 'suitable calculus'. We reproduce it for its intrinsic interest in Appendix 1 following.

(68) Assuming, of course, that the given speed is low enough for this species of conic to be the orbit, as Newton will make clear below.

(69) A cancelled first version of the next two sentences reads in immediate sequel at this point 'ea celeritate quæ sit ad celeritatem uniformem corporis π ut recta quævis PR ad rectam quamvis $\pi\rho$' (with that speed which shall be to the uniform speed of the body π as any straight line PR to any straight line $\pi\rho$).

(70) That is, divide the given ratio $QT^2 : \chi\tau^2$ by the given ratio $QR : \chi\rho$. Newton originally continued 'et dabitur' (and there will be given...).

(71) In his later revise (see note (2)) Newton subsequently inserted '$= 4CH^q$' ($= 4CH^2$) in minimal clarification of the next equality.

Addantur utrobiꝗ $2KPH+L\times\overline{SP+PH}-SP^q-PH^q$ et fiet

$$L\times\overline{SP+PH}=2SPH+2KPH,$$

seu $SP+PH$ ad PH ut $2SP+2KP$ ad L. Unde datur umbilicus alter H. Datis autem umbilicis una cum axe transverso $SP+PH$, datur Ellipsis. Q.E.I.

Hæc ita se habent ubi figura Ellipsis est. Fieri enim potest ut corpus moveat(72) in Parabola vel Hyperbola. Nimirum si tanta est corporis celeritas ut sit latus rectum L æquale $2SP+2KP$, Figura erit Parabola umbilicum habens in puncto S et diametros omnes parallelas lineæ PH. Sin corpus majori adhuc celeritate emittitur movebitur id in Hyperbola habente umbilicum unum in puncto S alterum in puncto H sumpto ad contrarias partes puncti P et axem transversum æqualem differentiæ linearū PS et PH.(73)

Schol. Jam verò beneficio hujus Problematis soluti [Com]etarum(74) orbitas definire concessum est, et inde revolutionum tempora, et ex orbitarum magni-

(72) This was afterwards augmented to the more natural 'moveatur' in Newton's revise.

(73) For, as Newton will prove explicitly in Propositions XI and XII of §2 following, $\lim_{Q\to P}(QT^2/QR)$ may likewise be shown to be the length of the *latus rectum* L when the orbit is a hyperbola or parabola of focus S, and hence the preceding demonstration holds unchanged for the general conic. The very concreteness of Newton's geometrical argument tends to conceal certain of its general implications. If, in modern analytical equivalent, we suppose that the body P sets off in the direction PR—at an angle $\widehat{SPR}=\alpha$, say—with speed v, and is thereupon 'instantaneously' diverted towards the centre S by an inverse-square force of magnitude g at P, and if υ be the speed of the body π which rotates in a circle in the same force-field, then in the limit as the arcs $\widehat{PQ}$, $\widehat{\pi\chi}$ vanish to zero there will follow by Newton's line of reasoning (since $PR:\pi\rho=v:\upsilon$ and $\upsilon^2/S\pi=g.S\pi^{-2}/SP^{-2}$)

$$L/2S\pi=(QT^2/QR)/(\pi\rho^2/\chi\rho)=(PR^2.\sin^2\alpha/\pi\rho^2)\times(\chi\rho/QR)$$
$$=(v^2.\sin^2\alpha/\upsilon^2)\times(S\pi^{-2}/SP^{-2})=v^2.\sin^2\alpha/g.S\pi,$$

whence $L=2(v^2/g).\sin^2\alpha$. (This circuitous appeal to an auxiliary circle orbit is not, we may remark, at all necessary, since, where dt is the time of passage from P to Q, at once

$$L=\lim_{Q\to P}(QT^2/QR)=\lim_{R\to P}(PR^2.\sin^2\alpha/\tfrac{1}{2}g.dt^2)$$

with $\lim_{R\to P}(PR/dt)=\lim_{Q\to P}(\widehat{PQ}/dt)=v$.) Further, on taking $SP=R$, there ensues

$$PK=-R\cos 2\alpha$$

and so $SP+PK=2R\sin^2\alpha$, whence $(SP+PH)/PH=2(SP+KP)/L=2/(v^2/gR)$ and therefore $SP+PH=2R/(2-v^2/gR)$; accordingly, the conic's eccentricity is

$$\surd[1-L/(SP+PH)]=\surd[1-(v^2/gR)\,(2-v^2/gR)\sin^2\alpha].$$

The unexpected consequence, considerably veiled in Newton's geometrical guise, that the length $SP+PH$ of the transverse axis of the resulting conic orbit depends only on the speed of projection v and the size g of the central force at P was later to be implicitly invoked by him in criticism of a 'Platonic' hypothesis of Galileo's that, as he noted it in a letter to Richard Bentley on 17 January 1692/3, 'y^e^ motion of y^e^ planets is such as if they had all of them been created by God in some region very remote from our Systeme & let fall from thence towards y^e^ Sun, & so soon as they arrived at their several orbs their motion of falling turned aside into a

Add $2KP \times PH + L \times (SP + PH) - SP^2 - PH^2$ to each side and there will come $L \times (SP + PH) = 2SP \times PH + 2KP \times PH$, that is, $SP + PH$ to PH as $2SP + 2KP$ to L. Whence the other focus H is given. Given the foci, however, along with the transverse axis $SP + PH$, the ellipse is given. As was to be found.

This argument holds when the figure is an ellipse. It can, of course, happen that the body moves in a parabola or hyperbola. Specifically, if the speed of the body is so great that the *latus rectum* L is equal to $2SP + 2KP$, the figure will be a parabola having its focus at the point S and all its diameters parallel to the line PH. But if the body is released at a still greater speed, it will move in a hyperbola having one focus at the point S, the second at the point H taken on the opposite side of the point P, and its transverse axis equal to the difference of the lines PS and PH.(73)

Scholium. A bonus, indeed, of this problem, once it is solved, is that we are now allowed to define the orbits of comets,(74) and thereby their periods of revolution,

---

transverse one' (*Correspondence of Isaac Newton*, **3** (Cambridge, 1961): 240); rather, as he wrote again a month later on 25 February, 'there is no common place from whence all the planets being let fall & descending wth uniform & equal gravities (as Galileo supposes) would at their arrival to their several Orbs acquire their several velocities wth wch they now revolve in them. If we suppose ye gravity of all the Planets towards the Sun to be of such a quantity as it really is [*sc.* varying as the inverse-square of their distance from it] & that the motions of the Planets [in their effectively concentric-circle orbits] are turned upwards, every Planet will ascend to twice its height from ye Sun....And then by falling down again from ye places to wch they ascended they will arrive again at their several orbs wth the same velocities they had at first & wth wch they now revolve' (*ibid.*: 255). (Compare I. B. Cohen's exhaustive discussion of this 'Galileo–Plato' problem in his 'Galileo, Newton and the Divine order of the solar system' [in (ed. E. McMullin) *Galileo: Man of Science* (New York, 1967): 207–31]. Newton himself went on to remark that, if after a fall from infinity in an inverse-square field 'the gravitating power of ye Sun' doubled at the moment each planet was diverted into circle orbit, all would be well: equally, he might have supposed that by 'divine' judgement the solar field was inverse-cube at the time the planetary system was created!) The still more fundamental corollary that none but conic orbits are traversible in an inverse-square central-force field—since for every initial speed v and angle of projection α a unique conic trajectory (an ellipse, parabola or hyperbola according as $v < \surd[2gR]$, $v = \surd[2gR]$ or $v > \surd[2gR]$, namely) may correspondingly be defined, so exhausting all possibilities of motion from the given point P, distant R from the force-centre, under a central 'gravity' of intensity g—is here (not unreasonably?) taken by Newton to be self-evident. Twenty five years later, after strong criticism from Johann Bernoulli for a like 'deficiency' in thus failing to underline the obvious in the essentially unaltered equivalent Proposition XVII of Book 1 of his published *Principia* ($_1$1687: 58–9), he decided to insert in its second edition—in the closely similar context of Corollary 1 to the preceding Propositions XI–XIII (*Principia*, $_2$1713: 53)—a sentence justifying the uniqueness of inverse-square conic motion by just such an exhaustion of the possibilities of motion.

(74) In the manuscript Newton initially wrote 'Planetarum' (of planets), a careless slip copied not only into the putative fair copy—since it occurs in Halley's transcript (see note (2))—but also into the revise before Newton caught it, though he then took pains to correct both versions.

tudine, excentricitate, Aphelijs,(75) inclinationibus ad planum Eclipticæ et nodis inter se collatis cognoscere an idem Cometa ad nos sæpius redeat.(76) Nimirum ex quatuor observationibus locorum Cometæ, juxta Hypothesin quod Cometa movetur in linea recta, determinanda est ejus via rectilinea.(77) Sit ea $APBD$, sintꝗ A, P, B, D loca cometæ in via illa temporibus observationum, et S locus Solis. Ea celeritate qua Cometa uniformiter percurrit rectam AD finge ipsum emitti de locorum suorum aliquo P et vi centripeta mox correptum deflectere a recto tramite et abire in Ellipsi $Pbda$. Hæc Ellipsis determinanda est ut in superiore Problemate. In ea sunto a, P, b, d loca Cometæ temporibus observationum. Cognoscantur horum locorum e terra longitudines et latitudines. Quanto majores vel minores sunt his longitudines et latitudines observatæ tantò majores vel minores observatis sumantur longitudines et latitudines novæ.(78) Ex his novis inveniatur denuò via rectilinea cometæ et inde via Elliptica ut priùs. Et loca quatuor nova in via Elliptica prioribus erroribus aucta vel diminuta jam congruent cum observationibus exactè satis.(79) Aut si fortè errores etiamnum sensibiles manserint potest opus

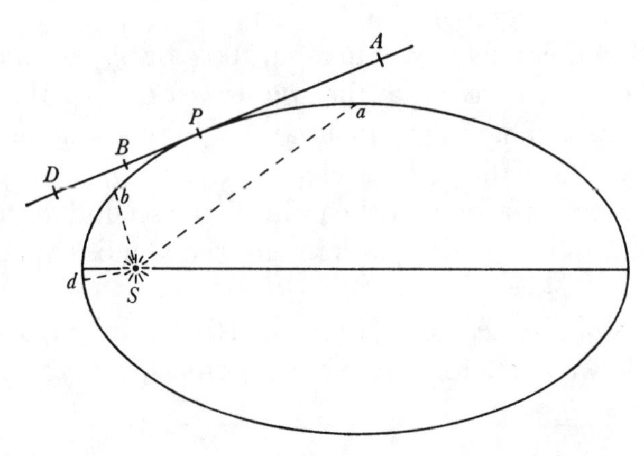

(75) Another carelessness: Newton manifestly intends 'Perihelijs' (perihelia). No solar comet known in his day is visible in the region of its aphelion from the Earth.

(76) This periodicity Halley was afterwards in his rare (but frequently reprinted) folio pamphlet, *Astronomiæ Cometicæ Synopsis* (Oxford, 1705), to verify in the celebrated instance of 'his' 1682 comet, not only deducing from his detailed computation of their elements of orbit that it was identical with those which had appeared before in 1531 and 1607—and hence, from the rough equality (75–76 years) of the intervening time-periods, probably also with still earlier ones more vaguely recorded—but accurately predicting its reappearance in late 1758 (after his death).

(77) Doubtless by the Wrennian method in 'Prob. 52' of his contemporary Lucasian lectures on algebra (see v: 298–302) or—where the Earth may be considered as motionless to sufficient accuracy—by the simpler 'Prob: 16' preceding (v: 210–12). We have also reproduced in the previous volume (v: 524–9) two versions of an abortive attempt by Newton in the autumn of 1685 to apply the former technique to locating in rough position the out-going arc of the 1680–1 comet (or, more accurately, its orthogonal projection upon the ecliptic).

(78) In his revise (see note (2)) Newton later added the clarifying phrase 'id adeo ut correctiones respondeant erroribus' (this so as to make the corrections correspond to the errors).

(79) This makeshift construction—quickly to be superseded by more viable methods (see 2, §2 below) if indeed it was ever at any time put into practice—seems more optimistic of a chance success than solidly reasoned. If T_A, T_P, T_B, T_D are the points in the Earth's solar

and then to ascertain from a comparison of their orbital magnitude, eccentricities, aphelia,(75) inclinations to the ecliptic plane and their nodes whether the same comet returns with some frequency to us.(76) To be specific, from four observations of a comet's positions we need, under the hypothesis that a comet moves in a straight line, to determine its rectilinear path.(77) Let it be *APBD*, with *A*, *P*, *B*, *D* the positions of the comet in that path at the times of observation and *S* the position of the Sun. At the speed with which it uniformly traverses the straight line *AD* imagine that the comet is released from some one of its places *P* and, snatched up immediately by the centripetal force, is deflected from its straight-line course, going off in the ellipse *Pbda*. This ellipse is to be determined as in the above problem. In it let *a*, *P*, *b*, *d* be the positions of the comet at the times of observation, and ascertain the longitudes and latitudes of these places from the Earth. As much as the observed longitudes and latitudes are greater than these take new longitudes and latitudes greater or less than the observed ones.(78) From these new ones let the comet's rectilinear path be found afresh, and therefrom the elliptical path as before. The four new positions in the elliptical path, increased or diminished by the previous errors, will now agree exactly enough with their observations.(79) Or, should perhaps the errors even

orbit (effectively a minimally eccentric circle round *S*) from which the primitive sightings $T_A A$, $T_P P$, $T_B B$ and $T_D D$ of the orbiting comet at α, P, β and δ respectively are made, Newton's hypothesis that the points A', B' and D' from and to which the comet would uniformly travel in the tangent at *P* with the speed which it there has (in the same time as it in fact, drawn towards the Sun *S*, orbits from and to α, β and δ) approximately satisfy $a\widehat{T_A}A = A\widehat{T_A}A'$, $b\widehat{T_B}B = B\widehat{T_B}B'$ and $d\widehat{T_D}D = D\widehat{T_D}D'$ is palpably false in general. Where the tangential points A, B, D; A', B', D' rigorously correspond to contemporaneous points a, b, d; a', b', d' in the

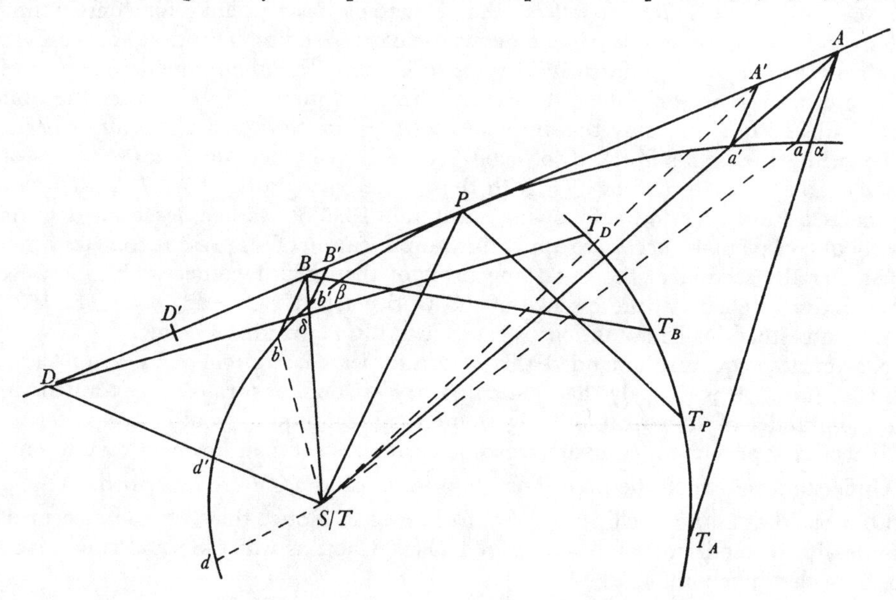

totum repeti. Et nè computa Astronomos molestè habeant suffecerit hæc omnia per praxin Geometricam(80) determinare.

Sed areas *aSP*, *PSb*, *bSd* temporibus(81) proportionales assignare difficile est. Super Ellipseos axe majore *EG* describatur semicirculus *EHG*. Sumatur angulus *ECH* tempori proportionalis. Agatur *SH* eiqȝ parallela *CK* circulo occurrens in *K*. Jungatur *HK* et circuli segmento *HKM* (per tabulam segmentorum vel secus) æquale fiat triangulum *SKN*. Ad *EG* demitte perpendiculum *NQ*, et in eo cape *PQ* ad *NQ* ut Ellipseos axis minor ad axem majorem et erit punctum *P* in Ellipsi[,] atqȝ acta recta *SP* abscindetur area Ellipseos *EPS* tempori(82) proportionalis. Namqȝ area *HSNM* triangulo *SNK* aucta et huic æquali segmento *HKM* diminuta fit triangulo *HSK* id est triangulo *HSC* æquale. Hæc æqualia adde areæ *ESH*, fient areæ(83) æquales *EHNS* et *EHC*. Cùm igitur Sector *EHC* tempori proportionalis sit et area *EPS* areæ *EHNS*, erit etiam area *EPS* tempori proportionalis.(84)

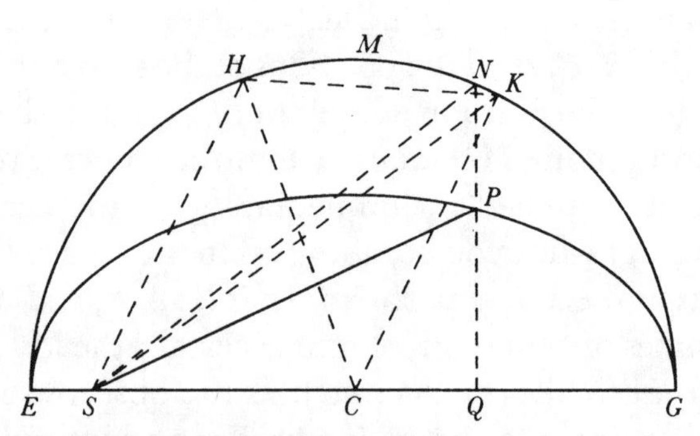

cometary orbit, then *Aa*, *Bb*, *Dd*; *A'a'*, *B'b'*, *D'd'* are the deviations due to solar gravity, and therefore (since the total cometary arc $\widehat{aPbd}$ to be determined is, by implication, small) are all very nearly parallel to the solar vector *SP*; whence to a close approximation $\widehat{aSA} = \widehat{ASA'}$, $\widehat{bSB} = \widehat{BSB'}$ and $\widehat{dSD} = \widehat{DSD'}$. It follows that Newton's basic premiss for constructing his revised longitudes (and, by simple trigonometry, the corresponding latitudes) holds only when the Earth's orbit $T_A T_P T_B T_D$ effectively shrinks to a point *T* in the immediate region of the Sun *S*; here, however, we do not need the full Wrennian method to construct the uniform tangential path *A'PB'D'* but only the simpler 'Prob: 16' of Newton's algebraic *lectiones* (see note (76)), while the points *a'*, *b'*, *d'* of orbit are derived immediately as the meets of the parallels *A'a'*, *B'b'*, *D'd'* to (*SP* or) *TP* with the respective sightings *TA*, *TB*, *TD*. Newton came to realise as much, for in the following autumn of 1685 he made at least one determined effort, without conspicuous success, to apply this simplification of his present cometary method to constructing the section of the out-going orbit of the 1680–1 comet visible between 21 December and 25 February; the essence of the worksheet (ULC. Add. 3965.11: 163^r) on which he then penned his computations is reproduced in Appendix 2 below.

(80) Newton subsequently amended this to read 'per descriptionem linearum' (by the description of lines). It is unlikely that contemporary astronomers, rigorously trained (in the main) in the practice of numerical and trigonometrical techniques, would have preferred the relative inaccuracy of such an equivalent geometrical construction, however easy to effect.

(81) Understand in which the preceding ellipse-arcs $\widehat{aP}$, $\widehat{Pb}$, $\widehat{Pd}$ are described.

(82) Of orbit, that is, in the ellipse-arc $\widehat{EP}$, from *E* to *P*. Notice that Newton measures this 'mean anomaly' in modern style from perihelion (and not, as was the usual contemporary practice, from the aphelion *G*).

now remain sensible, the whole process can be repeated. And, in case the computations prove troublesome to astronomers, it will be enough to determine all these things by a geometrical procedure.[80]

But to assign areas *aSP*, *PSb*, *bSd* proportional to the times[81] is difficult. On the major axis *EG* of an ellipse describe the semicircle *EHG*. Take the angle $\widehat{ECH}$ proportional to the time. Draw *SH* and parallel to it *CK*, meeting the circle in *K*; then join *HK* and make the triangle *SKN* equal to the circle's segment *HKM* (by means of a table of segments or otherwise). To *EG* let fall the perpendicular *NQ* and in it take *PQ* to *NQ* as the ellipse's minor axis is to its major axis, and the point *P* will then be in the ellipse, while the straight line *SP* will, when drawn, cut off an area (*EPS*) of the ellipse which is proportional to the time.[82] For the area (*HSNM*), augmented by the triangle *SNK* and diminished by the segment (*HKM*) equal to this, becomes equal to the triangle *HSK*, that is, to the triangle *HSC*. When these equals are added to the area (*ESH*) they will form equal areas (*EHNS*) and (*EHC*). Since, therefore, the sector (*EHC*) is proportional to the time, and the area (*EPS*) to the area (*EHNS*), the area (*EPS*) also will then be proportional to the time.[84]

---

(83) This was trivially changed in Newton's revise (note (2)) to read equivalently '...addita areæ *ESH*, facient areas'.

(84) Since Newton equates the segment (*HKM*) to the area of the triangle *SKN*—and not the minimally larger focal sector (*SKN*)—his construction will not be rigorously exact, though we may readily show how finely it approximates to the truth. If, in simplest analytical equivalent, we suppose the ellipse to have semi-axes $EC = CG = 1$ and eccentricity $SC = e$, then define the position *P* of the orbiting body in the arc $\widehat{EPG}$ by the eccentric angle $\widehat{ECN} = \theta$ (not shown in Newton's figure) and the time of orbit over $\widehat{EP}$ by the angle $\widehat{ECH} = T$, where (by the areal law)

$$T : \pi = (ESP) : (ESGP) = (ESN) \text{ [or } \tfrac{1}{2}(\theta - e\sin\theta)] : (ESGN) \text{ [or } \tfrac{1}{2}\pi],$$

it will be clear that Newton's construction approximately resolves the ensuing equation $\theta - e\sin\theta = T$ (given), and hence the general 'Astronomicum Problema' propounded by Kepler in 1609 in Chapter 60 of his *Astronomia Nova* (see IV: 668, note (38)) which is its geometrical model: namely, he there successively adduces to that end

$$\widehat{HCK}(=\widehat{SHC}) = \alpha = \tan^{-1}[e\sin T/(1-e\cos T)]$$
$$= e\sin T + \tfrac{1}{2}e^2\sin 2T + \tfrac{1}{3}e^3\sin 3T + \tfrac{1}{4}e^4\sin 4T + \ldots,$$

segment $(HKM) = \tfrac{1}{2}(\alpha - \sin\alpha)$ where

$$\alpha - \sin\alpha = \beta = \tfrac{1}{6}\alpha^3 + \ldots = \tfrac{1}{24}e^3(3\sin T - \sin 3T) + \tfrac{1}{16}e^4(2\sin 2T - \sin 4T)\ldots,$$

and sector $\quad (CKN) \approx (SKN)/(1 + (SC/CK).\cos\widehat{KCG}) \approx \tfrac{1}{2}\beta/(1 - e\cos\theta)$

to derive in equivalent terms the solution, correct to $O(e^5)$,

$$\theta \approx T + \alpha - \beta/(1 - e\cos T) = T + e\sin T + \tfrac{1}{2}e^2\sin 2T + \tfrac{1}{8}e^3(3\sin 3T - \sin T)$$
$$+ \tfrac{1}{6}e^4(2\sin 4T - \sin 2T) + \ldots.$$

As Newton may well have known, the simpler approximation

$$\theta \approx \widehat{ECK} = T + \sin\alpha = T + e\sin T/\sqrt{[1 - 2e\cos T + e^2]}$$
$$= T + e\sin T + \tfrac{1}{2}e^2\sin 2T + \tfrac{1}{8}e^3(3\sin 3T - \sin T) + \ldots,$$

Prob. 5. *Posito quod vis centripeta sit reciprocè proportionalis quadrato distantiæ a centro,*[85] *spatia definire quæ corpus rectà cadendo datis temporibus describit.*

Si corpus non cadit perpendiculariter describet id Ellipsin[86] puta *APB* cujus umbilicus inferior puta *S* congruet cum centro terræ.[87] Id ex jam demonstratis constat. Super Ellipseos axe majore *AB* describatur semicirculus *ADB* et per corpus decidens transeat recta *DPC* perpendicularis ad axem, actisq *DS*, *PS*, erit area *ASD* areæ *ASP* atq adeò tempori proportionalis. Manente axe *AB* minuatur perpetuò latitudo Ellipseos, et semper manebit area *ASD* tempori proportionalis. Minuatur latitudo illa in infinitum et Orbita *APB* jam coincidente cum axe *AB* et umbilico *S* cum axis termino *B*[88] descendet corpus in recta *AC* et area *ABD* evadet tempori proportionalis.[89] Definietur itaq spatium *AC* quod corpus de loco *A* perpendiculariter cadendo tempore dato describit si modò tempori proportionalis capiatur area *ABD* et a puncto *D* ad rectam *AB* demittatur perpendicularis *DC*. Q.E.F.

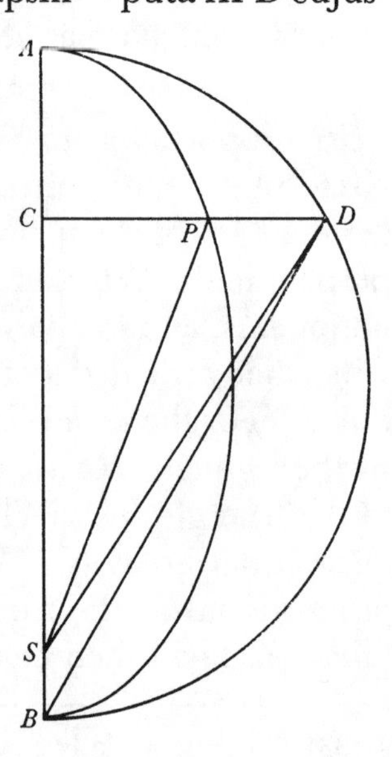

Schol.[90] Priore Problemate[91] definiuntur motus projectilium[92] in aere nostro,

correct to $O(e^4)$, had long before been derived by Bonaventura Cavalieri in an effectively identical manner (by drawing *CK* parallel to *SH*) in his *Directorium Generale Uranometricum, In quo Trigonometriæ Logarithmicæ Fundamenta, ac Regulæ demonstrantur, Astronomicæque Supputationes ad solam fere Vulgarem Additionem reducuntur* (Bologna, 1632): 152, and repeated—with full credit given to Cavalieri—in G. B. Riccioli's widely studied *Almagestum Novum, Astronomiam Veterem Novamque complectens* (Bologna, 1651): 535. Just three years before Newton contrived his present improvement—unpublished in his lifetime—Christiaan Huygens had made the Cavalierian approximation the basic regulator of the varying planetary speeds (in eccentric circle orbits) in the 'automatic' planetarium whose model he completed in 1682; see his *Œuvres complètes*, **21** (The Hague, 1944): 143–8.

(85) Newton has deleted the qualification 'terræ' (of the earth) in line with his systematic alteration throughout the present text of (solar or terrestrial) 'gravitas' into a general, unspecified 'vis centripeta' (compare note (31)). Correspondingly, the unrestricted 'corpus' *P* whose motion is here determined was initially described throughout as a 'grave' (gravitating body).

(86) Since Newton's argument requires only a vanishingly small initial speed of projection transverse to the path *AS* of rectilinear fall, there is no deficiency in his thus restricting the ensuing conic orbit.

(87) This unwanted survivor of an earlier, more restricted dynamical viewpoint—still uncancelled in both the Royal Society and Halley transcripts of the putative fair copy (see note (2))—was afterwards deleted by Newton (compare §2: note (166)) in line with his preceding enunciation (see note (85)). In making similar omission of a terrestial location for *S* we stress in our English version that it is a general 'centrum virium'.

(88) Since the eccentricity of the ellipse *APB* approaches indefinitely close to unity.

(89) Newton's procedure of maintaining the length of the orbital axis *AB* unchanged while

Problem 5. Supposing that the centripetal force be reciprocally proportional to the square of the distance from the centre,(85) *to define the distances which a body falling straight down describes in given times.*

If the body does not fall perpendicularly it will describe an ellipse,(86) suppose $\widehat{APB}$, whose lower focus, say *S*, will coincide with the centre [of force].(87) This is settled from what we have already demonstrated. On the ellipse's major axis *AB* describe the semicircle *ADB* and let the straight line *DPC* pass through the dropping body perpendicular to the axis, and when *DS*, *PS* are drawn the area (*ASD*) will be proportional to the area (*ASP*) and so also to the time. With the axis *AB* remaining fixed, perpetually diminish the width of the ellipse and the area (*ASD*) will ever remain proportional to the time. Diminish that width indefinitely and, with the orbit $\widehat{APB}$ now coming to coincide with the axis *AB* and the focus *S* with the end-point *B* of the axis,(88) the body will descend in the straight line *AC* and the area (*ABD*) will turn out to be proportional to the time.(89) In consequence, the distance *AC* described by a body falling perpendicularly from the position *A* in a given time will be defined if only the area (*ABD*) be taken proportional to the time and then from the point *D* the perpendicular *DC* be let fall to the straight line *AB*. As was to be done.

Scholium.(90) By the previous problem (91) the motions of projectiles(92) in our

continuously enlarging the initial distance *AS* of the body from the centre of force till it coincides with *AB*—and so, correspondingly, varying the gravitational constant in the force-field, k/SP^2 round *S*—seems needlessly complicated and its subtlety is far from adequately explored. The natural approach, avoiding such a complex detour, is to keep the distance *AS* unchanged but (by allowing the initial velocity of projection at *A*, normal to *AS*, to become vanishingly small) continuously to diminish the orbital diameter *AB* till *B* coincides with *S*; the 'last' ratio of the elliptical sector (*ASP*) and the infinitely narrow semi-ellipse (*ABP*) is then straightforwardly that of the segment (*ASD*) and the full semicircle, and Newton's desired result that the time of fall along the orbital arc $\widehat{AP}$, that is (in the limit) *AC*, is measured by the circle segment (*ASD*) follows immediately. Unless the falling body, when it attains the force-centre *S*, is (by its impact with it or some other means) conceived to have its direction of motion there instantaneously reversed, there will be a discontinuity in this argument from the vanishingly small approximating ellipse which Newton fails to appreciate: namely, the body will continue past *S* at a slowing speed analogously proportional to the inverse-square of its distance from it till it comes momentarily to a halt at (say) *A'* where $SA' = AS$, thereafter falling back towards *S* and past it to halt momentarily at *A* and then going on to repeat this cycle indefinitely.

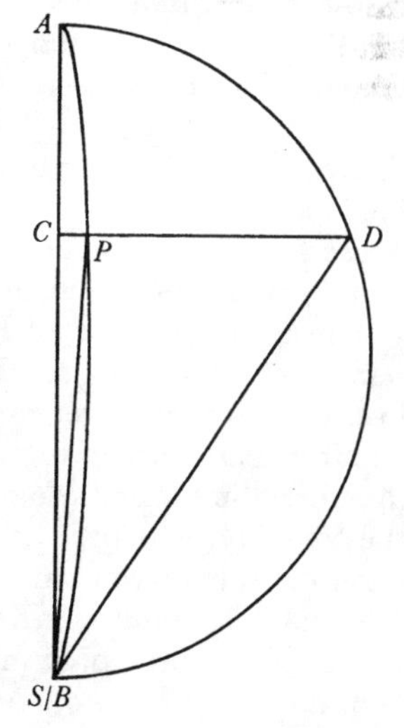

(90) In his revise (see note (2)) Newton extensively amplified this scholium, asserting that any postulated interplanetary *æther* cannot offer appreciable resistance to orbiting bodies (and so will not be a set-back to the application of the preceding Problems 3 and 4) and referring to his correspondence with Robert Hooke (in the winter of 1679–80) on the apparent deviation

hacce motus gravium perpendiculariter cadentium ex Hypothesi quod gravitas reciprocè proportionalis sit quadrato distantiæ a centro terræ quodq medium aeris nihil resistat. Nam gravitas est species una vis centripetæ.(93)

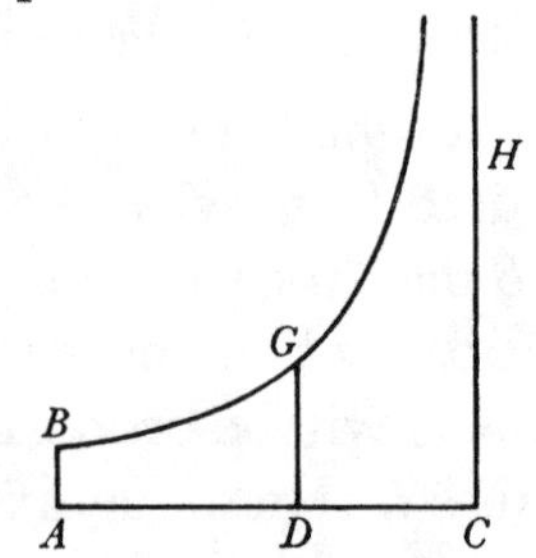

Prob. 6. Corporis sola vi insita per medium similare resistens delati motum definire.

Asymptotis rectangulis *ADC*, *CH* describatur Hyperbola secans perpendicula *AB*, *DG* in *B*, *G*. Exponatur tum corporis celeritas tum resistentia medij ipso motus initio per lineam(94) *AC*[,] elapso tempore aliquo per lineam(95) *DC*[,] et tempus exponi potest per aream *ABGD* atq spatium eo tempore descriptum per lineam *AD*. Nam celeritati proportionalis est resistentia medij et resistentiæ proportionale est decrementum celeritatis,(96) hoc est, si tempus in partes æquales

—to the south as well as east in the northern hemisphere, he would argue—of a body let fall from a height onto the rotating Earth. For its considerable intrinsic interest we reproduce its major portion in Appendix 1 below.

(91) Problem 4, that is.

(92) Newton first wrote 'gravium', carelessly anticipating his next clause. It is, of course, his wish to distinguish here between bodies impelled 'artificially' at some point and those which fall 'naturally' under terrestrial (inverse-square) gravitation alone.

(93) We have already pointed (see note (31)) to the significance of Newton's specification of 'gravity' (solar or terrestrial) as but one instance of the general central force (*vis centripeta*) whose accelerative effects he begins—for the first time—to discuss abstractly in his present treatise. In the manuscript he has, for no clear reason, cancelled an immediately preceding sentence 'Sequentibus resistentia medij similaris primùm absq gravitate dein cum gravitate consideratur' (In the following [problems] the resistance of a homogeneous medium is considered, first in divorce from gravity and then in company with it). This smooth transition to the final Problems 6 and 7 (and their scholium), so harshly broken, was repaired in Newton's revise by the insertion of an appropriate subhead underscoring the change of theme to 'the motion of bodies in resisting media'. Compare Appendix 1 below.

The extant portion of Halley's transcript of the fair copy terminates at this point. The remaining pages were sent by him to John Wallis two years after he penned them, and are now lost. In his covering letter to Wallis on 11 December 1686 he recalls that 'Mr Isaac Newton about 2 years since gave me the inclosed propositions, touching the opposition of the Medium to a direct impressed Motion, and to falling bodies, upon supposition that the opposition is as the Velocity; which tis possible is not true: however I thought any thing of his might not be unacceptable to you, and I begg your opinion thereupon, if it might not be (especially the 7th problem) somewhat better illustrated' (from the original in Trinity College, Cambridge. R.4.45.111B, first published in E. F. MacPike, *Correspondence and Papers of Edmond Halley* (London, 1932): 74–5). As newly appointed Clerk to the Royal Society, responsible among other things for restoring the health of its ailing *Transactions*, Halley was primarily concerned not to divulge the content of Newton's researches into resisted motion (the text of which, let it be said, had already (see note (2)) been transcribed into the Society's Register Book, and so was accessible to any Fellow) but to use the communicated propositions as bait to persuade out of Wallis his own 'conclusions concerning the opposition of the Medium to projects moving through it; ...not doubting but that your extraordinary talent in matters of this nature, will be able to clear up this subject which hitherto seems to have been only mencōned among

atmosphere are defined, and by the present one those of heavy bodies falling perpendicularly, in accord with the hypothesis that gravity is reciprocally proportional to the square of the distance from the earth's centre and that the medium of the air resists not at all. For gravity is one species of centripetal force.(93)

Problem 6. To define the motion of a body borne by its innate force alone through a homogeneous resisting medium.

With rectangular asymptotes *ADC*, *CH* describe a hyperbola cutting the perpendiculars *AB*, *DG* in *B* and *G*. Express both the body's speed and the resistance of the medium by the(94) line *AC* at the very start of motion and by the(95) line *DC* after some lapse of time: the time can then be expressed by the area *ABGD* and the distance described in that time by the line *AD*. For the resistance of the medium is proportional to the speed and the decrement of the speed(96) is proportional to the resistance; that is, if the time be divided into equal

---

Mathematicians, never yet fully discussed' (*ibid.*). (Like every one else in England at the time, Newton included, he was unaware that Huygens had fully solved the problem of motion under simple gravity when the 'opposition' is instantaneously proportional to the flight speed more than seventeen years earlier; see note (113) below.) After some initial hesitancy (compare A. R. Hall, *Ballistics in the Seventeenth Century* (Cambridge, 1952): 130–1) Wallis was soon afterwards coaxed into composing a short, woollily argued 'Discourse concerning the Measure of the Airs resistance to Bodies moved in it' (*Philosophical Transactions*, **16**, No. 186 [for January–March 1687]: 269–80) in which the horizontal and vertical components of the projectile path defined by $dv/dt = -rv + g.dy/ds$ are accurately rendered as the limit ($\Delta t \to 0$, where t is the base variable) of the respective difference-equations

$$v_{x+1} - v_x = -rv_x \quad \text{and} \quad v_{y+1} - v_y = -rv_y + g, \quad x, y = 0, 1, 2, 3, \ldots,$$

and the former's 'integral' $t \propto \log(v_0/v_x)$ is correctly deduced in the geometrical model of a hyperbolic area, but the corresponding solution ($v_x \to v_y - g/r$) is missed, and accordingly no rigorous justification is given for his assertion (*ibid.*: 27) that 'the line of Projects...resembles a Parabola deformed', an insight perhaps derived largely from his private view of Newton's concluding scholium. That Halley had not sought Newton's prior approval for his action is clear from his letter to Newton on 24 February 1686/7, where, in referring to 'Dr Wallis his papers', he remarked that Wallis 'had the hint from an account I gave him of what you had demonstrated' (*Correspondence of Isaac Newton*, **2**, 1960: 469). A fortnight or so before Newton had made independent enquiries, through Edward Paget, of Wallis' 'things about projectiles pretty like those of mine in ye papers Mr Paget first shewed you' and had been reassured that he would be 'consulted whether I intend to print mine', as he wrote to Halley on 13 February (*ibid.*: 464). Luckily for Halley (and Wallis), Newton had the manuscript of the second book of his *Principia* with, in its first section, improved versions of his present Problems 6 and 7 and terminal scholium (see Appendix 3 following) ready to go to press, and a potentially nasty squabble was averted by its arrival in London on 28 March (*ibid.*: 473).

(94) The clarification 'quamvis datam' (any given) is deleted. In revise (see note (2)) Newton inserted the equivalent 'datæ longitudinis' (of given length) after '...*AC*'.

(95) Later qualified as 'indefinitam' (indefinite) by Newton in his revise.

(96) Understand, as Newton at once effectively specifies, over a correspondingly infinitesimal increment of time.

dividatur, celeritates ipsarum initijs sunt differentijs suis proportionales. Decrescit ergo celeritas in [a]proportione Geometrica dum tempus crescit in Arithmetica. Sed tale est decrementum lineæ *DC* et incrementum areæ *ABGD*,(97) ut notum est. Ergo tempus per aream et celeritas per lineam illam rectè exponitur. Q.E.D. Porrò celeritati atq; adeo decremento celeritatis proportionale est incrementum spatij descripti[,] sed et decremento lineæ *DC* proportionale est incrementum lineæ *AD*. Ergo incrementum spatij per incrementum lineæ *AD*, atq; adeo spatium ipsum per lineam illam rectè exponitur. Q.E.D.(98)

[a]Lem [1]

Prob. 7. Posita uniformi vi centripeta, motum corporis in medio similari(99) rectà ascendentis ac descendentis definire.(100)

Corpore ascendente exponatur vis centripeta per datum quodvis rectangulum *BC* et resistentia medij initio ascensus per rectangulum *BD* sumptum ad contrarias partes. Asymptotis rectangulis *AC*, *CH*, per punctum *B* describatur Hyperbola secans perpendicula *DE*, *de* in *G*, *g* et corpus ascendendo tempore *DGgd* describet spatium *EGge*, tempore *DGBA* spatium ascensus totius *EGB*, tempore AB^2G^2D spatium descensus $B^{[2]}E^2G$ atq; tempore $^2D^2G^2g^2d$ spatium descensus $^2G^{[2]}E^{[2]}e^2g$: et celeritas corporis resistentiæ medij proportionalis, erit in horum temporum periodis *ABED*, *ABed*, nulla,(101) $AB^{[2]}E^2D$, $AB^{[2]}e^2d$; atq; maxima celeritas quam corpus descendendo potest acquirere erit *BC*.

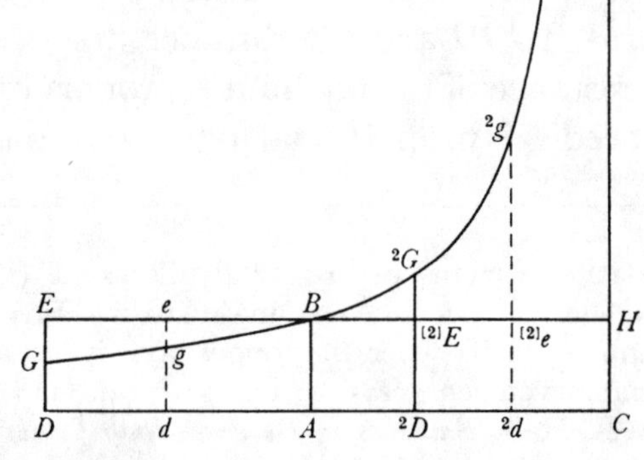

Resolvatur enim rectangulum(102) *AH* in rectangula innumera *Ak*, *Kl*, *Lm*, *Mn* &c quæ sint ut incrementa celeritatum æqualibus totidem temporibus facta[,] et erunt *Ak*, *Al*, *Am*, *An* &c ut celeritates totæ atq; adeo [a]ut resistentiæ medij in fine(103) singulorum temporum æqualium. Fiat *AC* ad *AK*, vel *ABHC* ad *ABkK* ut vis cen-

[a]Hyp [1]

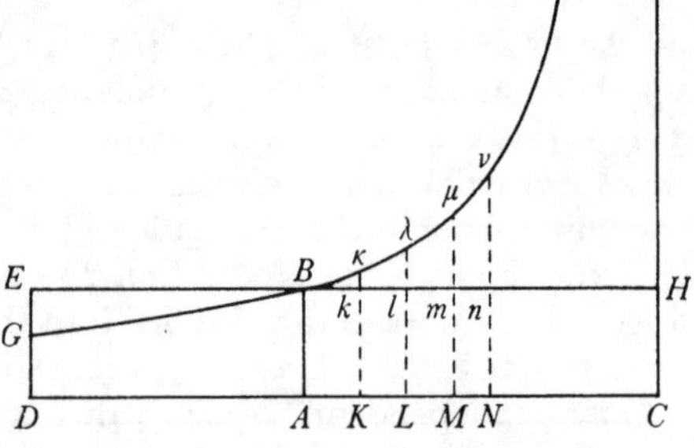

(97) Changed by Newton in his revise to read 'Sed proportione priore decrescit linea *DC* et posteriore crescit area *ABGD*' (But the line *DC* decreases in the former proportion and the area *ABGD* increases in the latter one).

(98) If (in more easily assimilable modern analytical equivalent) we suppose that the body, starting off from *A* along *ADC* with an initial speed V_x, is in time t slowed by the resistance acting over $AD = x$ down to the speed v_x at *D*, then, (with Newton) appropriately absorbing the constant of proportionality into t, we may set the resisting force $\dot{v}_x$ $(= dv_x/dt)$ to be equal to the speed $v_x = \dot{x}$ $(= dx/dt)$; accordingly, by Lemma 1 (compare note (13) above) the fluxional equation $\dot{v}_x = -v_x$ yields as its 'integral' $\log(v_x/V_x) = -t$, while its equivalent

parts, the speeds at the beginnings of these are proportional to their own differences. The speed therefore decreases in a geometrical proportion[a] while the time increases in an arithmetical one. But this is the manner of decrease of the line DC and of increase of the area $ABGD$,[97] as is known. Therefore the time is correctly expressed by the area and the speed by that line. As was to be proved. Furthermore, the increment of the space described is proportional to the speed and hence to the decrement of the speed, but so also is the increment of the line AD proportional to the decrement of the line DC. Therefore the increment of the space is correctly expressed by the increment of the line AD, and hence the space itself by that line. As was to be proved.[98]

[a] Lemma 1

Problem 7. Supposing a uniform centripetal force, to define[100] the motion of a body ascending and descending straight up and down in a homogeneous[99] medium.

Where the body ascends, let the centripetal force be represented by any arbitrary rectangle $(A)B(H)C$ and the resistance of the medium at the start of ascent by the rectangle $(A)B(E)D$ taken the opposite way. With rectangular asymptotes AC, CH through the point B describe a hyperbola cutting the perpendiculars DE, de in G and g: the body in ascent in the time $(DGgd)$ will then describe the distance $(EGge)$ and in the time $(DGBA)$ a distance of total ascent (EGB), while in the time (AB^2G^2D) it will cover the distance (B^2E^2G) in descent and in the (further) time $({}^2D^2G^2g^2d)$ a descent $({}^2G^2E^2e^2g)$; also the body's speed, proportional to the resistance of the medium, will at these points in time be $(ABED)$, $(ABed)$, nothing,[101] (AB^2E^2D) and (AB^2e^2d), while the greatest speed which the body can acquire in its descent will be BC.

For resolve the rectangle $A(B)H(C)$ into innumerable rectangles Ak, Kk, Lm, Mn, ..., which shall be as the increments of speed brought about in a corresponding number of divisions of time, and Ak, Al, Am, An, ..., will then be as the whole speeds and hence[a] as the resistances of the medium at the end[103] of each of the corresponding equal times. Make AC to AK, or $ABHC$ to $ABkK$, as the

[a] Hypoth. 1.

$\dot{v}_x = -\dot{x}$ produces straightforwardly $v_x - V_x = -x$. In the terms of Newton's hyperbolic model, therefore, it follows, on putting $AC = V_x$, that

$$DC\,(= V_x - x) = v_x \quad\text{and}\quad (ABGD) \propto \log\,(AC/DC) = \log\,(V_x/v_x) = t.$$

His proof demonstrates somewhat cumbrously that, where dt is a vanishingly small increment of the base variable t, then $d(DC) = -d(AD)$ and $d(ABGD) = GD \times d(AD) \propto -d(DC)/DC$, so that (as Newton puts it) if the area $(ABGD)$ increases arithmetically—and so its increments $d(ABGD)$ are constant—then $d(DC) \propto DC$ and therefore the increments of DC vary geometrically with their distance DC from C.

(99) Understand 'resistente' (resisting) as before.

(100) Newton first wrote 'exponere' (to represent).

(101) This replaces 'nihil' (nil).

(102) The equivalent 'parallelogrammum' was first written.

(103) In revise Newton improved his phrasing slightly to read ' et erunt nihil, *Ak*, *Al*, *Am*, *An* &c...in principio' (and nil, *Ak*, *Al*, *Am*, *An*, ... will then be...at the beginning).

tripeta ad resistentiam in fine temporis primi(104) et erunt $ABHC$, $KkHC$, $LlHC$, $[Mm]HC$ &c ut vires absolutæ quibus corpus urgetur atq; adeo ut incrementa celeritatum, id est ut rectangula Ak, Kl, Lm, Mn &c & [b]proinde in progressione geometrica. Quare si rectæ Kk, Ll, Mm, Nn [&c] productæ occurrant Hyperbolæ in κ, λ, μ, ν &c(105) erunt areæ $AB\kappa K$, $K\kappa\lambda L$, $L\lambda\mu M$, $M\mu\nu N$ &c æquales, adeoq; tum temporibus æqualibus tum viribus centripetis semper æqualibus analogæ. Subducantur rectangula Ak, Kl, Lm, Mn &c viribus absolutis analoga et relinquentur areæ $Bk\kappa$, $k\kappa\lambda l$, $l\lambda\mu m$, $m\mu\nu n$ &c resistentijs medii in fine singulorum temporum, hoc est celeritatibus atq; adeo descriptis spatijs analogæ.(106) Sumantur analogarum summæ et erunt areæ $Bk\kappa$, $Bl\lambda$, $Bm\mu$, $Bn\nu$ &c spatijs totis descriptis analogæ, nec non areæ $AB\kappa K$, $AB\lambda L$, $AB\mu M$, $AB\nu N$ &c temporibus.(107) Corpus igitur inter descendendum tempore quovis $AB\lambda L$ describit spatium $Bl\lambda$ et tempore $L\lambda[\nu]N$ spatium $\lambda ln\nu$. Q.E.D. Et similis est demonstratio motus expositi in ascensu. Q.E.D.(108)

[b] Lem. [1]

Schol. Beneficio duorum novissimorum problematum innotescunt motus projectilium in aëre nostro, ex hypothesi quod aer iste similaris sit quodq; gravitas uniformiter & secundum lineas parallelas agat. Nam si motus omnis obliquus corporis projecti distinguatur in duos, unum ascensus vel descensus, alterum progressus horizontalis: motus posterior determinabitur per Problema sextum, prior per septimum ut fit in hoc diagrammate.(109)

(104) This, correspondingly, was later changed to 'in principio temporis secundi' (at the beginning of the second time-division), while the sequel was considerably amplified to read 'deq; vi centripeta subducantur resistentiæ et manebunt $ABHC$, ...quibus corpus in principio singulorum temporum urgetur...' (from the centripetal force take away the resistances and $ABHC$, ...will then remain...by which the body is urged at the beginning of the individual times).

(105) Newton subsequently inserted a clarifying parenthesis 'ob proportionales AK ad KL ut KC ad LC hoc est ut $L\lambda$ ad $K\kappa$' (because of the proportionals AK to KL as KC to LC, that is, as $L\lambda$ to $K\kappa$).

(106) This sentence was afterwards substantially augmented to read: 'Est autem area $AB\kappa K$ ad aream $Bk\kappa$ ut $K\kappa$ ad $\frac{1}{2}k\kappa$ seu AC ad $\frac{1}{2}AK$ hoc est ut vis centripeta ad resistentiam in medio temporis primi. Et simili argumento areæ $\kappa KL\lambda$, $\lambda LM\mu$, $\mu MN\nu$ &c sunt ad areas $\kappa kl\lambda$, $\lambda lm\mu$, $\mu mn\nu$ &c ut vires centripetæ ad resistentias in medio temporis secundi tertij quarti &c. Proinde cum areæ æquales $BAK\kappa$, $\kappa KL\lambda$, $\lambda LM\mu$, $\mu MN\nu$ &c sint viribus centripetis analogæ, erunt areæ $Bk\kappa$, $\kappa kl\lambda$, $\lambda lm\mu$, $\mu mn\nu$ &c resistentijs in medio singulorum temporum, hoc est... descriptis spatijs analogæ' (Now the area ($AB\kappa K$) is to the area ($Bk\kappa$) as $K\kappa$ to $\frac{1}{2}k\kappa$ or AC to $\frac{1}{2}AK$, that is, as the centripetal force to the resistance at the middle of the first time-interval. And by a similar reasoning the areas ($\kappa KL\lambda$), ($\lambda LM\mu$), ($\mu MN\nu$),...are to the areas ($\kappa kl\lambda$), ($\lambda lm\mu$), ($\mu mn\nu$),...as the centripetal forces to the resistances at the middle of the second, third, forth,...time-intervals. Consequently, since the equal areas ($BAK\kappa$), ($\kappa KL\lambda$), ($\lambda LM\mu$), ($\mu MN\nu$),...are proportional to the centripetal forces, the areas ($Bk\kappa$), ($\kappa kl\lambda$), ($\lambda lm\mu$), ($\mu mn\nu$),...will be proportional to the resistances at the middle of the individual time-intervals, that is,...to the distances described.)

(107) In revise Newton stressed the passage to the limit implied in sequel by inserting 'Et hæ areæ ubi rectangula numero infinita et infinite parva evadunt coincidunt cum Hyperbolicis'

centripetal force to the resistance at the end of the first time-division,(104) and $ABHC$, $KkHC$, $LlHC$, $MmHC$, ... will then be as the absolute forces by which the body is urged and hence as the increments of speed, that is, as the rectangles Ak, Kl, Lm, Mn, ... and consequently[b] in geometrical progression. Wherefore, if the lines Kk, Ll, Mm, Nn, ... when produced meet the hyperbola in κ, λ, μ, ν, ...,(105) the areas $(AB\kappa K)$, $(K\kappa\lambda L)$, $(L\lambda\mu M)$, $(M\mu\nu N)$, ... will be equal and hence in proportion both to the equal times and to the ever equal centripetal forces. Take away the rectangles Ak, Kl, Lm, Mn, ... proportional to the absolute forces and there will be left areas $(Bk\kappa)$, $(k\kappa\lambda l)$, $(l\lambda\mu m)$, $(m\mu\nu n)$, ... proportional to the resistances of the medium at the end of the separate intervals of time, that is, to the speeds and hence to the distances described.(106) Take the sums of these proportionals and the areas $(Bk\kappa)$, $(Bl\lambda)$, $(Bm\mu)$, $(Bn\nu)$, ... will be in proportion to the total distances described, and also the areas $(AB\kappa K)$, $(AB\lambda L)$, $(AB\mu M)$, $(AB\nu N)$, ... to the times.(107) The body, therefore, during its descent in any time $(AB\lambda L)$ describes the distance $(Bl\lambda)$ and in the (further) time $(L\lambda\nu N)$ the distance $(\lambda ln\nu)$. As was to be proved. The demonstration for the motion represented in ascent is similar. As was to be proved.(108)

[b] Lemma 1.

Scholium. With the aid of the two most recent problems the motions of projectiles in our air are discoverable, on the hypothesis that this air is homogeneous and that gravity acts uniformly and following parallel lines. For if every oblique motion of a projected body be distinguished into two, one of ascent or descent, the other of horizontal advance, the latter motion will be determined by the sixth problem and the former by the seventh, as happens in this diagram.(109)

(And when the rectangles come to be infinite in number and infinitely small, these areas coincide with the [corresponding] hyperbolie ones).

(108) If, in analytical clarification, we suppose that the moving body starts off with speed V_y and, ever subject to a constant decelerating force g, is in time t further slowed by the resistance, acting over a distance y, down to the speed $v_y = dy/dt$, then we may set the equation of motion to be $dv_y/dt = -v_y - g$, that is, $d(v_y+g)/dt = -(v_y+g)$; whence (compare note (98)) at once $\log((v_y+g)/(V_y+g)) = -t$ and $(v_y+g)-(V_y+g) = -y-gt$. In the terms of Newton's hyperbolic model, therefore, it follows on putting $CA = g$, $AD = V_y$, $Ad = v_y$ and (for simplicity) $AB = DE = 1$ that

$$(DGgd) = CA\log(CD/cd) = g\log((V_y+g)/(v_y+g)) = gt$$

and

$$(DEed) = Dd = V_y - v_y = y+gt,$$

so that $(GEeg) = y$. Conversely, where the constant force g accelerates the motion, the corresponding equation of motion $dv_y/dt = -v_y+g$ yields

$$\log((v_y-g)/(V_y-g)) = -t \quad\text{and}\quad (v_y-g)-(V_y-g) = -y+gt:$$

hence in the geometrical model, on taking $CA = g$ and $AB = 1$ as before but now $A^2D = V_y$ and $A^2d = v_y$, there ensues

$$({}^2D^2G^2g^2d) = CA\log(C^2D/C^2d) = gt \quad\text{and}\quad ({}^2D^2E^2e^2d) = {}^2D^2d = -y+gt,$$

so that $({}^2G^2E^2e^2g) = y$.

(109) Here, if the body shot off at D along DP traverses the arc $\widehat{Dar} = s$ under simple gravity g (acting vertically downwards) to $r(x, y)$, defined by the coordinates $DR = x$ and $Rr = y$,

Ex loco quovis D ejaculetur corpus secundum lineam quamvis rectam DP, & per longitudinem DP exponatur ejusdem[110] celeritas sub initio motus. A puncto P ad lineam horizontalem DC demittatur perpendiculum PC, ut et ad DP[111] perpendiculum CI, ad quod sit DA ut est resistentia medij ipso motus initio ad vim gravitatis. Erigatur perpendiculum AB cujusvis longitudinis et completis parallelogrammis $DABE$, $CABH$, per punctum B asymptotis DC CP describatur Hyperbola secans DE in G. Capiatur linea N ad EG ut est DC ad CP et ad rectæ DC punctum quodvis R erecto perpendiculo RtT, in eo cape

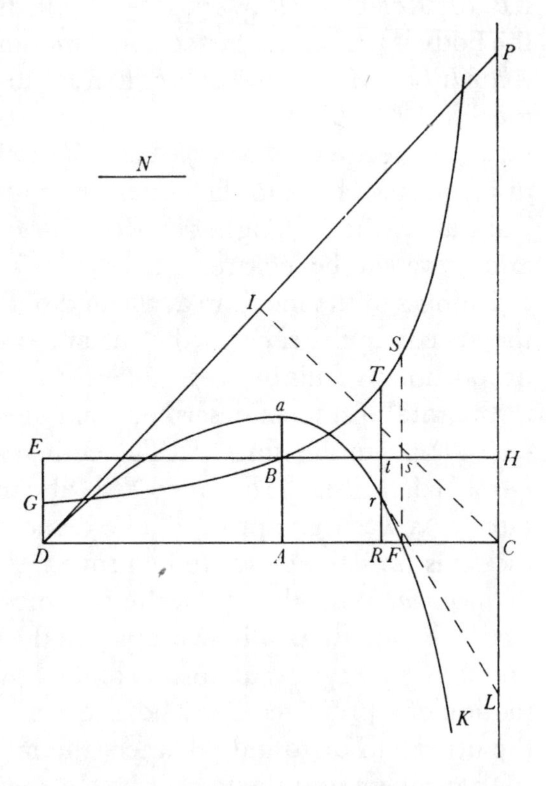

$$Rr = \frac{DRtE - DRTBG}{N}$$

et projectile tempore $DRTBG$ perveniet ad punctum r, describens curvam lineam $DarFK$ quam punctum r semper tangit, perveniens autem ad maximam altitudinem a in perpendiculo AB, deinde incidens in lineam horizontalem DC ad F ubi areæ $DFsE$, $DFSBG$ æquantur,[112] et postea semper appropinquans Asymptoton PCL. Estq; celeritas ejus in puncto quovis r ut curvæ tangens rL.[113]

---

there attaining the orbital speed $v = ds/dt$ after time t, then (on presuming, with Newton, that the constant of proportionality of resistance to instantaneous velocity is suitably absorbed) the equation of motion of the body at r is $dv/dt = -v - g \,.\, dy/ds$. Newton's present assumption that this may be split into horizontal and vertical components, respectively $dv_x/dt = -v_x$, $v_x = dx/dt$ (see note (98)) and $dv_y/dt = -v_y - g$, $v_y = dy/dt$ (see note (108)), is accurate (for, because $\widehat{DRr}$ is right, $v^2 = v_x^2 + v_y^2$ and therefore

$$dv/dt = v \,.\, dv/ds = v_x \,.\, dv_x/ds + v_y \,.\, dv_y/ds = (dx/ds) \,.\, dv_x/dt + (dy/ds) \,.\, dv_y/dt)$$

but, for all its plausibility, not a self-evident truth; it is, indeed, impermissible to split the general equation $dv/dt = -v^n - g \,.\, dy/ds$ $(n \neq 1)$ into comparable 'components' $dv_x/dt = -v_x^n$ and $dv_y/dt = -v_y^n - g$, as Leibniz in 1688 claimed might be done—'sed prolixitas hic vitanda est'—in the instance $n = 2$ (that of *resistentia respectiva*, defined by $dv/ds = -v$ when $g = 0$) in Article VI of his insufficiently pondered 'Schediasma de Resistentia Medii, & Motu projectorum gravium in medio resistente' (purportedly mapped out 'jam pro magna parte Parisiis duodecim abhinc annis' [*sc.* in 1676] but composed only shortly before its publication in *Acta Eruditorum* (January 1689): 38–47 [= (ed. C. I. Gerhardt) *Leibnizens Mathematische Schriften*, **6** (Halle, 1860): 135–43]). Newton seemingly remained ignorant of the 'Schediasma'

Let the body be hurled from any position D following any straight line DP, and let its[110] speed at the start of motion be expressed by the length DP. From the point P let fall the perpendicular PC to the horizontal DC, and also drop the perpendicular CI to DP;[111] then let DA be to CI as the resistance of the medium at the very start of motion is to the force of gravity. Erect the perpendicular AB of any length and, having completed the parallelograms $DABE$, $CABH$, through the point B and with asymptotes DC, CP describe an hyperbola cutting DE in G. Take the line N to EG as DC is to CP and, having erected the perpendicular RtT at any point R of the line DC, take in it $Rr = ((DRtE)-(DRTBG))/N$ and the projectile will in time $(DRTBG)$ reach the point r, describing the curve $DarFK$ with which the point r is ever in contact—attaining, indeed, its maximum height at a in the perpendicular AB, then falling onto the horizontal DC at F, where the areas $(DFsE)$ and $(DFSBG)$ are equal,[112] and ever afterwards approaching the asymptote PCL; and its speed at any point r will be as the curve's tangent rL.[113]

for a quarter of a century till John Keill freshly brought it to his notice in the heat of the fluxion priority squabble as one more stick with which to belabour Leibniz, but afterwards he was only too ready to criticise Leibniz' 'erroneous' attempt 'to find the Curve described in a Medium where the resistance was in a duplicate ratio of the velocity [by] compos[ing] the horizontal & perpendicular motions of the projectile' (ULC. Add. 3968.31: 457^r, extracted from a draft 'Supplement' to Desmaizeaux' editorial *Preface* introducing his *Recueil* (note (4)), **1**: i–lxxxi in 1720). Above all, in a developed critique of Leibniz' paper which he sent to John Keill in May 1714 he wrote: 'In sexto [Articulo] Propositiones sunt tantum duæ, et utraq falsa est. Corpus enim, ubi resistentia est in duplicata ratione velocitatis, non fertur motu composito ex motibus duorum Articulorum præcedentium [$dv_x/dt = -v_x^2$ and $dv_y/dt = -v_y^2+g$ respectively]' (ULC. Res. 1893.8, published in Joseph Edleston, *Correspondence of Sir Isaac Newton and Professor Cotes* (London, 1850): 309–10). Newton himself straightforwardly discussed these 'component' Leibnizian rectilinear motions in Propositions V–VII and VIII/IX respectively of Book 2 of his published *Principia* (see Appendix 2: note (17) below), but, in seeking in his following Proposition X to reduce the general equation $dv/dt = -\rho+g.\ dy/ds$ of motion under constant gravity g and resistance ρ, he was forced to concoct *ad hoc* a cumbrous, circuitous method which in a particular instance, as Johann Bernoulli brought to his attention in autumn 1712 (though he himself could not accurately trace the error—an inexact infinitesimal approximation—in Newton's general argument), led ineluctably to a numerically false result. (Compare D. T. Whiteside, 'The Mathematical Principles underlying Newton's *Principia Mathematica*' (note (30)): 128–9.) Extracts from the sequence of manuscripts in which Newton came to appreciate his mistake and to give several variant proofs of the correct result—only one of which was set, in a last minute stop-gap, in the new edition of his *Principia* ($_2$1713: 232–40)—will be reproduced in the eighth volume.

(110) Initially 'projecti' (the projectile's) was written.

(111) Accurately, this should (see note (113) below) be '...ad parallelam ipsi DP ductam per A' (to the parallel to DP drawn through A), or some equivalent which reduces the length of Newton's perpendicular CI in the ratio AC/DC.

(112) Perhaps because it merely states the obvious, this clause 'deinde...æquantur' (then ...are equal) was cancelled by Newton in his redraft.

(113) Where (as in note (109)) the body, shot off at D under simple downwards gravity g and perpetually slowed by a resistance equal to its orbital speed, traverses $\widehat{Dar}$ in time t,

Si proportio resistentiæ aeris ad vim gravitatis nondum innotescit: cognoscantur (ex observatione aliqua) anguli *ADP*, *AFr*(114) in quibus curva *DarFK* secat lineam horizontalem *DC*. Super *DF* constituatur rectangulum *DFsE* altitudinis cujusvis, ac describatur Hyperbola rectangula ea lege ut ejus una Asymptotos sit *DF*, ut areæ *DFsE*, *DFSBG* æquentur et ut *sS* sit ad *EG* sicut tangens anguli *AFr*(114) ad tangentem anguli *ADP*. Ab hujus Hyperbolæ centro *C* ad rectam *DP*(115) demitte perpendiculum *CI* ut et a puncto *B* ubi ea secat rectam *Es*, ad rectam *DC* perpendiculum *BA*, et habebitur proportio quæsita *DA* ad *CI*, quæ est resistentiæ medij ipso motus initio ad gravitatem projectilis. Quæ omnia ex prædemonstratis facilè eruuntur. Sunt et alij modi inveniendi resistentiam aeris quos lubens prætereo. Postquam autem inventa est hæc resistentia in uno casu, capienda est ea in alijs quibusvis ut corporis celeritas et superficies sphærica conjunctim, (nam projectile sphæricum esse passim suppono;) vis autem gravitatis innotescit ex pondere. Sic habebitur semper pro-

---

attaining at point r (defined by $DR = x$, $Rr = y$) the speed $ds/dt = v$ of respective horizontal and vertical components $dx/dt = v_x$ and $dy/dt = v_y$, we may integrate the corresponding equations of motion $dv_x/dt = -v_x$ and $dv_y/dt = -v_y - g$ to produce $V_x - v_x = x$, $\log(V_x/v_x) = t$ (note (98)) and $V_y - v_y = y + gt$, $\log((V_y+g)/(v_y+g)) = t$ (note (108)), where V_x, V_y are the initial values (at D) of v_x, v_y. At once $V_x/v_x = V_x/(V_x - x) = (V_y+g)/(v_y+g)$ and so

$$x/V_x = (V_y - v_y)/V_y + g),$$

whence $y = ((V_y+g)/V_x)x - g\log(V_x/(V_x - x))$ is the defining Cartesian equation of $r(x, y)$. Clearly, the projectile will reach its maximum height at $a(x, y)$ when $v_y = 0$ and hence $DA = V_xV_y/(V_y+g)$, and attain a maximum horizontal range $DC = X$ at $v_X = V_x - X = 0$ and so $DC = V_x$; whence $CP = V_y$, $DP = V$, $AC = g.V_x/(V_y+g)$ and $RC = V_x - x = v_x$, so that the tangent $rL = v_x.ds/dx = v$. Further, since $AB \times AC/EG = AC \times DC/DA = (V_x/V_y)g$, on taking (with Newton) $V_x/V_y = N/EG$, there results

$$(DRtE) = AB \times DR = (N.g/AC)\, DR = N.((V_y+g)/V_x)x$$

and $$(DRTBG) = AB \times AC.\log(DC/RC) = N.gt \propto t,$$

so that $Rr = ((DRtE) - (DRTBG))/N$. Lastly, the 'resistentia medij ipso motus initio' V is to the 'vis gravitatis' g as $V_xV_y/g.CI$, that is,—on correcting a trivial Newtonian slip—as DA/CI', where $CI' = CI.g/(V_y+g) = CI.(AC/DC)$.

Newton was to set Problems 6 and 7 and the preceding portion of the present scholium with minimal revision as the opening Section I of Book 2 of his *Principia* ($_1$1687: 236–45), there presenting in Proposition IV a geometrical demonstration of what he here merely asserts. For purposes of comparison with our modern analytical justification we reproduce the text of this geometrical revision in Appendix 3. 1/3 following. The novelty of his present solution to the problem of motion under resistance varying as the speed and a constant uni-directional diverting force should not be over-stressed: though Newton was still himself unaware of it in 1687, Christiaan Huygens had here anticipated him by nearly two decades. (Huygens made public announcement of his own near-identical researches only three years afterwards in the concluding pages of his reworked *Discours de la Cause de la Pesanteur* (Leyden, 1690): 168–80 [=*Œuvres complètes*, **21** (The Hague, 1944): 478–93], there remarking that 'J'ay vu avec plaisir ce que Mr. Newton écrit touchant les chûtes & les jets des corps pesants dans l'air, ou dans quelqu' aut[r]e milieu qui resiste au mouvement; m'estant appliqué autrefois à la mesme

If the ratio of the resistance of the air to the force of gravity is not yet ascertained, let there (from some observation) be learnt the angles $\widehat{ADP}$ and $\widehat{AFr}$(114) in which the curve *DarFK* intersects the horizontal *DC*. On *DF* form the rectangle *DFsE* of any height, then describe a rectangular hyperbola with the restrictions that *DF* be one of its asymptotes, that the areas (*DFsE*), (*DFSBG*) be equal, and that *sS* be to *EG* as the tangent of the angle $\widehat{AFr}$(114) to the tangent of $\widehat{ADP}$. From the centre *C* of this hyperbola let fall the perpendicular *CI* to the line *DP*,(115) and also from the point *B* where it cuts the line *Es* drop the perpendicular *BA* to the line *DC*, and the required ratio *DA* to *CI*—that of the resistance of the medium at the very start of motion to the gravity of the projectile—will be had. All these results are easily derived from what has previously been demonstrated. There are other methods, too, of finding out the resistance of the air, but these I readily pass over. After this resistance has been determined in one case, however, it needs to be taken in any others jointly as the body's speed and its spherical surface (for I suppose throughout that the projectile is spherical); while the force of gravity is ascertainable from its

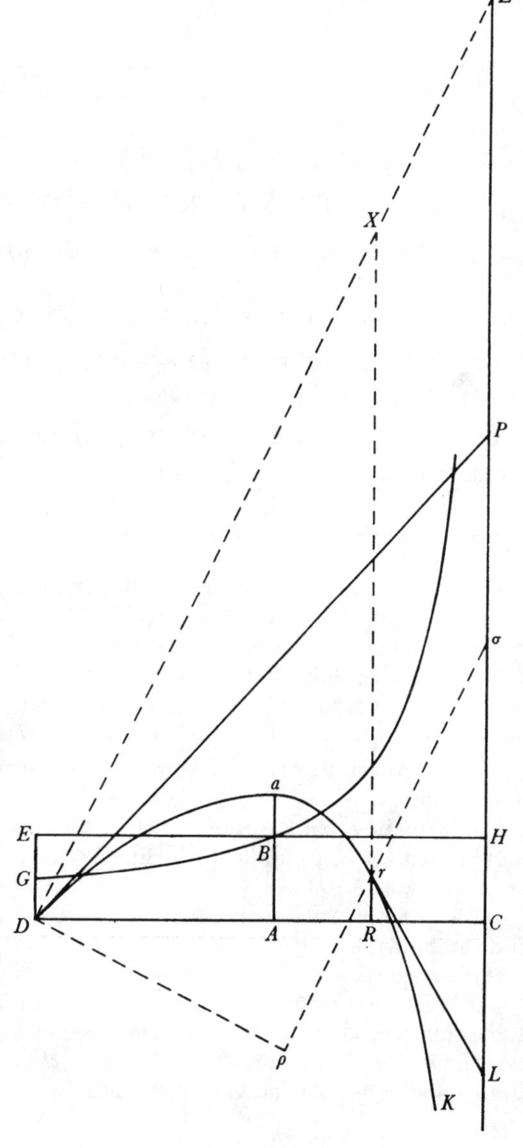

recherche....J'examinay premierement ces mouvemens, en supposant que les forces de la Resistance sont comme les Vitesses des corps, ce qui alors me paroissoit fort vraisemblable' (*ibid.*: 168–9). His original manuscript article 'De proportione gravium cadentium habita ratione resistentiæ aeris vel aquæ', precisely dated in his usual way by a 'εὕρηκα 28 Oct. 1668' [N.S.], is printed in his *Œuvres complètes*, **19**, 1937: 102–18.) In the second edition of his *Principia* ($_2$1713: 215–19) Newton amplified his earlier discussion of the logarithmic projectile-path *DarK*, inserting two new Corollaries 1 and 2 which we likewise reproduce in Appendix 3.2. In essence, if *CP* is extended to *Z* so that $PZ = g$ and *Rr* drawn to meet *DZ* in *X*, then $XR = (V_x/(V_y+g))\,x$ and so $rX = gt$. It follows at once that the body *r* moves uniformly away from *DZ* as it traverses $\widehat{DarK}$—in other words, if $D\rho$ is the perpendicular from *D* to the line $\rho r\sigma$ drawn through *r* parallel to *DZ*, then

portio resistentiæ ad gravitatem seu lineæ DA ad lineam CI. Hac proportione et angulo ADP determinatur specie figura $DarFKLP$: et capiendo longitudinem DP proportionalem celeritati projectilis in loco D determinatur eadem magnitudine sic ut altitudo Aa maximæ altitudini projectilis et longitudo DF longitudini horizontali inter ascensum et casum projectilis semper sit proportionalis, atq; adeò ex longitudine DF in agro semel mensurata semper determinet tum longitudinem illam DF tum alias omnes dimensiones figuræ $DarFK$ quam projectile describit in agro. Sed in colligendis hisce dimensionibus usurpandi sunt logarithmi pro area Hyperbolica $DRTBG$.(116)

Eadem ratione determinantur etiam motus corporum gravitate vel levitate & vi quacunq; simul et semel impressa moventium in aqua.

APPENDIX 1. THE AUGMENTED TRACT 'DE MOTU CORPORUM' (DECEMBER 1684?).(1)

Excerpts from the corrected amanuensis copy(2) in the University Library, Cambridge

De motu sphæricorum Corporum in fluidis.(3)

Def. 1. Vim centripetam appello qua corpus attrahitur vel impellitur versus punctum aliquod quod ut centrum spectatur.

Def. 2. Et vim corporis seu corpori insitam qua id conatur perseverare in motu suo secundum lineam rectam.

$D\rho\,(=(V_x/V)g.t) \propto t$. The basic subtangential property of the *logarithmica DarK*, defined by $Xr = (-g.\log(RC/DC)$ or$)\,-g.\log(ZX/ZD)$ with respect to the oblique Cartesian coordinate-lengths ZX and Xr, gives straight-forwardly $L\sigma = ZX.\,d(Xr)/d(ZX) = g = PZ$, constant, with $rL = v$ its oblique projection; whence, as a corollary, the terminal speed of the projectile as it nears the asymptote ZPC is g. (See also J. A. Lohne, 'The Increasing Corruption of Newton's Diagrams' [*History of Science*, **6**, 1967: 69–89]: 76–80.)

(114) Newton assumes, for simplicity of reference, that the small arc $\widehat{Fr}$ is effectively a straight line.

(115) Again (compare note (111)) this should be '... ad parallelam rectæ DP per A transeuntem' (to the parallel to the line DP passing through A), or some equivalent diminishing CI in the ratio AC/DC.

(116) A somewhat needless reminder to anyone who has read the preceding paragraphs with understanding, surely?

(1) As we have earlier indicated (see §1: note (2) preceding), this immediate revise is basically Humphrey Newton's secretarial transcript of the corrected state of Newton's primary autograph, amplified by an augmented set of introductory 'Definitions' (now also including five 'Laws') and 'Lemmas' and by new *scholia* to Theorem 4 and Problem 5, and further altered and lightly amended by Newton's own hand. By and large these latter changes convert

weight. In this way the ratio of the resistance to gravity—that is of the line DA to the line CI—will always be had. From this ratio and the angle $A\widehat{D}P$ the configuration $DarFKLP$ is determined in species; and by taking the length DP proportional to the speed of the projectile at the position D it is determined in magnitude: accordingly the altitude Aa is ever proportional to the maximum altitude of the projectile and the length DF to the horizontal length covered by the projectile during its rise and fall, and hence, once the length DF is measured in the field, it will always determine not merely that length DF but also all other dimensions of the figure $DarFK$ which the projectile describes in the field. But in obtaining these dimensions logarithms must be employed in place of the hyperbolic area ($DRTBG$).(116)

By the same procedure are determined also the motions of bodies moving in water under gravity or levity and any arbitrary force impressed once and instantaneously.

Def. 3. Et resistentiam quæ est medij regulariter impedientis.

Def. 4. Exponentes quantitatum sunt aliæ quævis quantitates proportionales expositis.(4)

---

the present text into corresponding portions of a still more developed tract 'De motu Corporum' (§2 following) which Humphrey Newton was likewise entrusted to pen out. Though its principal innovations are already published in more than one place (see next note) and not of narrow mathematical interest, we have thought fit to outline the broad features of this present interim revise because of the unique glimpse it affords of Newton's rapidly changing and developing ideas on general celestial and terrestrial motion in the early winter of 1684–5. This date of composition, while not explicitly supported by contemporary documentation, is narrowly delimited on the one hand by the preparation about early November 1684 of the putative fair copy of the original 'De motu Corporum' (see §1: note (2)), and on the other by the elaboration in the first months of 1685 of its several revises, beginning with the remoulding of its preliminaries (§2, Appendix 1) and then continuing with its large-scale recasting (§§2/3).

(2) ULC. Add. 3965.7: 40r–54r, first published in full—but interblended with the analytical table of contents set by Halley at the head of his transcript of the primary version (see §1: note (2))—by A. R. and M. B. Hall in their *Unpublished Scientific Papers of Isaac Newton* (Cambridge, 1962): 243–5/247–67(with English translation on 267–70/271–92 following). The principal innovations in its text were earlier recorded by W. W. R. Ball in his *Essay on Newton's Principia* (London, 1893): 51–6, and are also given in J. W. Herivel's *Background to Newton's Principia* (Oxford, 1966): 294–9 (with English renderings on 299–303).

(3) On this change in title see §1: note (3) above.

(4) A somewhat trivial innovation in the primary text. In practice, of course, Newton will employ an 'exponent' which, by suitably absorbing a constant factor of proportionality, will result in a simplest possible representing mathematical (fluxional) equation.

(5)*Lex 1.* Sola vi insita corpus uniformiter(6) in linea recta semper pergere si nil impediat.

Lex 2. Mutationem status movendi vel quiescendi proportionalem esse vi impressæ et fieri secundum lineam rectam qua vis illa imprimitur.

Lex 3. Corporum dato spatio inclusorum eosdem esse motus inter se sive spatium illud quiescat sive moveat id perpetuò et uniformiter in directum absq; motu circulari.

Lex 4. Mutuis corporum actionibus commune centrum gravitatis non mutare statum suum motus vel quietis. Constat ex Lege 3.

Lex 5. Resistentiam medij esse ut medij illius densitas et corporis moti sphærica superficies & velocitas conjunctim.

(7)*Lemma 1.* Corpus viribus conjunctis diagonalem parallelogrammi eodem tempore describere quo latera separatis.

Si corpus dato tempore vi sola *M* ferretur ab *A* ad *B* et vi sola *N* ab *A* ad *C*, compleatur parallelogrammum *ABDC* et vi utraq; feretur id eodem tempore ab *A* ad *D*. Nam quoniam vis *M* agit secundum lineam *AC* ipsi *BD* parallelam, hæc vis per Legem 2 nihil mutabit celeritatem accedendi ad lineam illam *BD* vi altera impressam. Accedet igitur corpus eodem tempore ad lineam *BD* sive vis *AC* imprimatur sive non, atq; adeò in fine illius temporis reperietur alicubi in linea illa *BD*. Eodem argumento in fine temporis ejusdem reperietur alicubi in linea *CD*, et proinde in utriusq; lineæ concursu *D* reperiri necesse est.

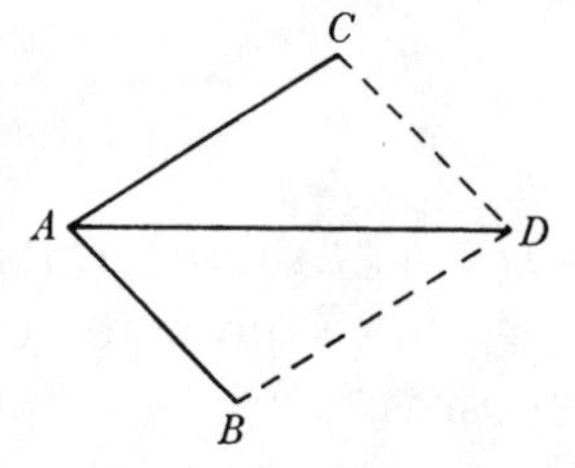

Lemma 2. Spatium quod corpus urgente quacunq; vi centripeta ipso motus initio describit, esse in duplicata ratione temporis.

Exponantur tempora per lineas *AB*, *AD* datis *Ab Ad* proportionales, et urgente vi centripeta æquali exponentur spatia descripta per areas rectilineas *ABF*, *ADH* perpendiculis *BF*, *DH* et rectâ quavis *AFH* terminatas ut exposuit Galilæus. Urgente autem vi centripeta inæquabili(8) exponantur spatia descripta

(5) In the following recasting, 'Laws' 1 and 5 replace the earlier equivalent 'Hypotheses' 2 and 1 respectively, and Law 2 explicitly justifies the basic Newtonian 'Axiom 2' of motion (as it will be renamed in §2) which is fundamental in the demonstration of Theorems 1–3 but was earlier merely assumed; Laws 3 and 4 (needed in the newly introduced final paragraph of the scholium to Theorem 4 below) have, of course, no earlier correlatives.

(6) Initially 'motu uniformi' before being changed back by Newton to his original adverb.

(7) Of these four following 'Lemmas', 1 and 2 restate—now with full demonstration (and explicit appeal to *Lex* 2 preceding)—Newton's earlier 'Hypotheses' 3 and 4 (compare §1: notes (10) and (12)), while 3 and 4 merely reiterate his earlier Lemmas 1 and 2.

(8) Lightly changed by Newton from the initially copied equivalent 'Sit autem vis centripeta inæquabilis et perinde'.

per areas ABC, ADE curva quavis ACE quam recta AFH tangit in A, comprehensas. Age rectam AE parallelis BF, bf, dh occurrentem in G, g, e, et ipsis $bf\,dh$ occurrat AFH producta in f et h. Quoniam area ABC major est area ABF minor area $ADEG$ erit area ABC ad aream $ADEG$ major quam area ABF ad aream $ADEG$ minor quam area ABG ad aream ADH[,] hoc est major quam area Abf ad aream Ade minor quam area Abg ad aream Adh. Diminuantur jam lineæ AB, AD in ratione sua data usq; dum puncta ABD coeunt et linea Ae conveniet cum tangente Ah; adeoq; ultimæ rationes Abf ad Ade et Abg ad Adh evadent eædem cum ratione Abf ad Adh. Sed hæc ratio est dupla rationis Ab ad Ad seu AB ad AD[,] ergo ratio ABC ad $ADEC$ ultimis illis intermedia jam fit dupla rationis AB ad AD id est ratio ultima evanescentium spatiorum seu prima nascentium dupla est rationis temporum.

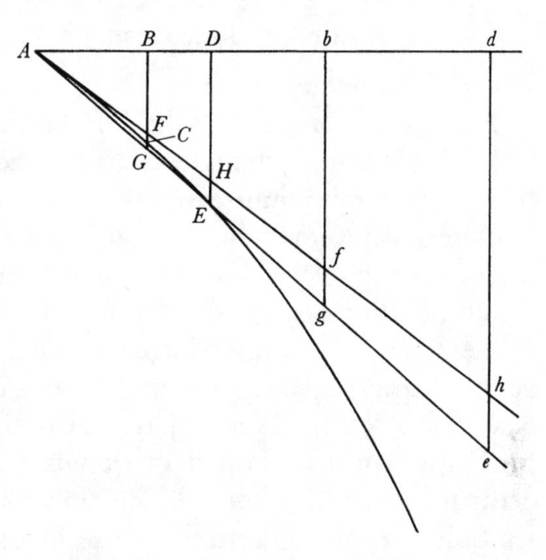

Lemma 3. Quantitates differentijs suis proportionales sunt continuè proportionales. Ponatur A ad $A-B$, ut B ad $B-C$ & C ad $C-D$ &c et dividendo fiet A ad B ut B ad C et C ad D &c.

Lemma 4. Parallelogramma omnia circa datam Ellipsin descripta, esse inter se æqualia. Constat ex Conicis.

De motu corporum in mediis non resistentibus.(9)

Theorema 1. Gyrantia omnia... describere. Dividatur tempus ...... et constabit propositio.

Theorem. 2. Corporibus in... circulorum. Corpora B, b ...... constat Propositio.

Cor. 1. .... *Cor. 2.* .... *Cor. 3.* .... *Cor. 4.* .... *Cor. 5.* ....

Schol. Casus Corollarij quinti... circa Jovem jam statuunt Astronomi.

Theor. 3. Si corpus P... puncta P et Q. Namq; in figura.... Q.E.D.

Corol. Hinc si... in problematîs sequentibus.

Prob. 1. Gyrat corpus... in circumferentia. Esto circuli ...... inveniendum.

Schol. Cæterum in hoc casu... tangit.

Prob. 2. Gyrat corpus... centrum Ellipseos. Sunto CA, CB .... Q.E.I.

(9) This inserted subhead counterbalances the parallel one '... in medijs resistentibus' which fills (compare §1: note (93)) a more necessary rôle below.

Prob. 3. Gyrat corpus...umbilicum Ellipseos. Esto Ellipsis.... Q.E.I.

Schol. Gyrant ergo Planetæ majores...sunt quantitates $\frac{QT^q}{QR}$, quæ ultimò fit ubi coeunt puncta P et Q.

Theor. 4. Posito quod vis...axium. Sunto Ellipseos...Ellipsibus. Q.E.D.

Schol. Hinc in Systemate cœlesti...conveniant. Hac methodo determinare licet...determinabuntur.

Cæterum totum cœli Planetarij spatium vel quiescit (ut vulgò creditur) vel uniformiter movetur in directum et perinde Planetarum commune centrum gravitatis (per Legem 4) vel quiescit vel una movetur. Utroq; in casu motus Planetarum inter se (per Legē 3) eodem modo se habent, et eorum commune centrum gravitatis respectu spatij totius quiescit, atq; adeo pro centro immobili Systematis totius Planetarij haberi debet. Inde verō Systema Copernicæum probatur a priori. Nam si in quovis Planetarum situ computetur commune centrum gravitatis[,] hoc vel incidet in corpus Solis vel ei semper proximum erit. Eo Solis a centro gravitatis errore fit ut vis centripeta non semper tendat ad centrum illud immobile et inde ut planetæ nec moveantur in Ellipsibus exactè neq; bis revolvant in eadem orbita. Tot sunt orbitæ Planetæ cujusq; quot revolutiones, ut fit in motu Lunæ, et pendet orbita unaquæq; ab omnium Planetarum motibus conjunctis, ut taceam eorum omnium actiones in se invicem. Tot autem motuum causas simul considerare et legibus exactis calculum commodum admittentibus motus ipsos definire superat ni fallor vim omnem humani ingenij. Omitte minutias illas et orbita simplex et inter omnes errores mediocris erit Ellipsis de qua jam egi. Si quis hanc Ellipsin ex tribus observationibus per computum trigonometricum (ut solet) determinare tentaverit, hic minus caute rem aggressus fuerit. Participabunt observationes illæ de minutijs motuum irregularium hic negligendis adeoq; Ellipsim de justa sua magnitudine et positione (quæ inter omnes errores mediocris esse debet) aliquantulum deflectere facient, atq; tot dabunt Ellipses ab invicem discrepantes quot adhibentur observationes trinæ. Conjungendæ sunt igitur et una operatione inter se conferendæ observationes quamplurimæ, quæ se mutuò contemperent et Ellipsin positione et magnitudine mediocrem exhibeant.[10]

(10) Curtis Wilson has given a lengthy analysis of this paragraph in his 'From Kepler's Laws, So-called, to Universal Gravitation: Empirical Factors' (*Archive for History of Exact Sciences*, **6**, 1970: 89–170): 161–2. We concur in his criticism that the concept of a 'Planetarum commune centrum gravitatis' (instantaneous centre of interacting planetary force) to which Newton here appeals is ill-defined and too readily supposed to lie within or closely near to the Sun, and can likewise trace no earlier Newtonian statement regarding 'eorum omnium actiones in se invicem' or dismissal of such mutual planetary interactions as merely yielding minimal periodic (and non-cumulative) divergences from the 'true' mean Keplerian, exactly elliptical orbits. (Kepler himself had been less sure, in his 1627 *Tabulæ Rudolphinæ*, that such 'physic[æ] minim[æ] intensiones et remissiones extra ordinem' were effectively negligible

Prob. 4. Posito quod...emissum. Vis centripeta tendens...Ellipsis. Q.E.I. Hæc ita se habent...*PS* et *PH*.

Schol. Jam verò...determinare. Sed areas...proportionalis.

Prob. 5. Posito quod vis...describit. Si corpus...perpendicularis *DC*. Q.E.F.

Schol: Hactenus motum corporum in medijs non resistentibus exposui; id adeo ut motus corporum cœlestium in æthere determinarem. Ætheris enim puri resistentiam quantum sentio vel nulla est vel perquam exigua....(11) Interfluit æther liberrimè nec tamen resistit sensibiliter. Cometas infra orbitam Saturni descendere jam sentiunt Astronomi saniores quotquot distantias eorum ex orbis magni parallaxi præterpropter colligere norunt: hi igitur celeritate immensa in omnes cœli nostri partes indifferenter feruntur, nec tamen vel crinem seu vaporem capiti circundatum resistentia ætheris impeditum et abreptum amittunt. Planetæ verò jam per annos millenos in motu suo perseverarunt, tantum abest ut impedimentum sentiant.

Demonstratis igitur legibus reguntur motus in cœlis. Sed et in aere nostro, si resistentia ejus non consideratur, innotescunt motus projectilium per Prob. 4. et motus gravium perpendiculariter cadentium per Prob. 5, posito nimirum quod gravitas sit reciprocè proportionalis quadrato distantiæ a centro terræ. Nam virium centripetarum species una est gravitas; et computanti mihi prodijt vis centripeta qua luna nostra detinetur in motu suo menstruo circa terram, ad vim gravitatis hic in superficie terræ, reciprocè ut quadrata distantiarum a centro terræ quamproximè.(12) Ex horologij oscillatorij motu tardiore in cacu-

over a period or the product of inaccurate observations or inadequately computed planetary elements, but such an 'extreme' view was later heavily criticised by Jeremiah Horrocks; see Curtis Wilson, 'Kepler's Derivation of the Elliptical Path' (*Isis*, **59**, 1968: 5–25): 24.)

(11) We omit several sentences digressing to consider the relative resistance of 'air' (the terrestrial atmosphere), quicksilver (mercury) and water.

(12) The first unimpeachable reference by Newton to a reasonably successful testing of the moon's orbit as traversible—solar and other deviations apart—in an inverse-square terrestrial force-field. Notice that he is still unwilling to identify the lunar *vis centripeta* with terrestrial *gravitas* (which is but one species of centripetal force) but merely states that their deviating effects are 'very nearly' the same. It is well known that Henry Pemberton (in the preface to his *A View of Sir Isaac Newton's Philosophy* (London, 1728): [a1^r/a1^v]), Abraham de Moivre (in a private 'Memorandum' he gave to John Conduitt in November 1727 [compare ULC. Add. 4007: 706^r–707^r]) and William Whiston (in his *Memoirs* (London, 1749): 36–8) agree in asserting that Newton had much earlier—Pemberton says 'when he retired from Cambridge in 1666', but this is considerably suspect—had such an 'Inclination...to try, whether the same Power did not keep the Moon in her Orbit, notwithstanding her projectile Velocity,... which makes Stones and all heavy Bodies with us fall downward, and which we call Gravity', but that when he tested this 'Postulatum' he was 'in some Degree' disappointed to find that 'the Power that restrained the Moon in her Orbit, measured by the versed Sines of that Orbit, appeared not to be quite the same that was to be expected, had it been the Power of Gravity alone, by which the Moon was there influenc'd. Upon this Disappointment, which made [him] suspect that this Power was partly that of Gravity, and partly that of *Cartesius's* Vortices,

mine montis prælti quàm in valle liquet etiam gravitatem ex aucta nostra a terræ centro distantia diminui, sed qua proportione nondum observatum est.

Cæterum projectilium motus in aere nostro referendi sunt ad immensum et revera immobile cœlorum spatium, non ad spatium mobile quod una cum terra et aere nostro convolvitur, et a rusticis ut immobile spectatur. Invenienda est Ellipsis quam projectile describit in spatio illo verè immobili et inde motus ejus in spatio mobili determinandus. Hoc pacto colligitur grave, quod de ædeficij sublimis vertice demittitur, inter cadendum deflectere aliquantulum a perpendiculo, ut et quanta sit illa deflexio et quam in partem. Et vicissim ex deflexione experimentis comprobata colligitur motus terræ. Cum ipse olim hanc deflexionem Clarissimo Hookio significarem, is experimento ter facto rem ita se habere confirmavit, deflectente semper gravi a perpendiculo versus orientem et austrum ut in latitudine nostra boreali oportuit.(13)

---

he threw aside the Paper of his Calculation, and went to other Studies' (Whiston, *Memoirs*: 36–7). (Newton's early belief in the physical existence of solar, lunar and terrestrial Cartesian vortices is otherwise documented; see D. T. Whiteside, 'Before the *Principia*: the Maturing of Newton's Thoughts on Dynamical Astronomy, 1664–1684' (*Journal of the History of Astronomy*, **1**, 1970: 5–19): 11–12.) No such 'old imperfect Calculation' can now be traced, but its present successful reworking in late 1684 was doubtless along the lines of that afterwards set down by him in Proposition IV, 'Lunam gravitare in terram, & vi gravitatis retrahi semper à motu rectilineo, & in orbe suo retineri', of Book 3 of his published *Principia* ($_1$1687: 406–7), founded on the Copernican estimate of the semi-diameter of lunar orbit being approximately sixty times the Earth's radius.

(13) On the letters relating to this topic (printed in *The Correspondence of Isaac Newton*, **2**, 1960: 297–313) which passed between Hooke and Newton during the six weeks from late November 1679 see A. Koyré, 'An Unpublished Letter of Robert Hooke to Isaac Newton' (*Isis*, **43**, 1952: 312–37, reprinted with slight emendations in his *Newtonian Essays* (London, 1965): 221–60) and especially J. A. Lohne, 'Hooke *versus* Newton. An Analysis of the Documents in the Case on Free Fall and Planetary Motion' (*Centaurus*, **7**, 1960: 6–52). Newton had originally communicated his 'fansy' that a falling body 'outrunning ye parts of ye earth will shoot forward to ye east side of the perpendicular describing in it's fall a spiral line' (*Correspondence*, **2**: 301) on 28 November, but, after Hooke pointed out in reply on 9 December that (because its plane of motion must pass through the Earth's centre) 'the fall...will not be exactly east of the perpendicular but South East and indeed more to the south then the east' (*ibid.*: 306) he came quickly to agree four days later that 'ye body in o^r latitude will fall more to ye south then east if ye height it falls from be any thing great' (*ibid.*: 307). Hooke reported on 6 January 1679/80 that he had by then 'made three tryalls of the Experiment of the falling body in Every of which the Ball fell towards the south east of the perpendicular...the Least being about a quarter of an inch' (*ibid.*: 310), adding on 17 January that 'by two tryalls since made in two several places wthin doors it succeeded also. Soe that I am now perswaded the Experiment is very certaine' (*ibid.*: 312–13). According to Lohne's computations ('Hooke *versus* Newton': 31–3), however, the deflection to the east which Hooke measured is several tens of times too much, while he queries whether—since an equal Coriolis force acts on the plumb-line which fixes the vertical—any deviation at all to the south could possibly have been observed.

DE MOTU CORPORUM IN MEDIJS RESISTENTIBUS

Prob. 6. Corporis sola vi...definire. Asymtotis rectangulis...exponitur. Q.E.D. Porrò celeritati...exponitur. Q.E.D.

Prob. 7. Posita uniformi vi...definire. Corpore ascendente...erit *BC*. Resolvatur enim...in ascensu. Q.E.D.

Schol. Beneficio...diagrammate. Ex loco quovis...tangens *rL*. Si proportio ...*DRTBG*.

Eadem ratione...moventium in aqua.

APPENDIX 2. COMPUTATION OF AN APPROXIMATION TO THE CURVED PATH OF THE COMET OF 1680–1 BY A MODIFIED RECTILINEAR TECHNIQUE.(1)

[*c*. October 1685?]

From the original worksheet(2) in the University Library, Cambridge

[1] Dec 21. [6^h. 36′. 59″. *a*.]
[Dec] 26. [5^h. 20′. 44″. *b*.]
[Dec] 30. [8^h. 10′. 26″. *C*.]
Jan 5. [6^h. 1′. 38″. *d*.]
[Jan] 13. [7^h. 8′. 55″. *E*.]
[Jan] 30. [8^h. 21′. 53″. *f*.]
Feb 25 [8^h. 41′. 9″. *g*.](3)

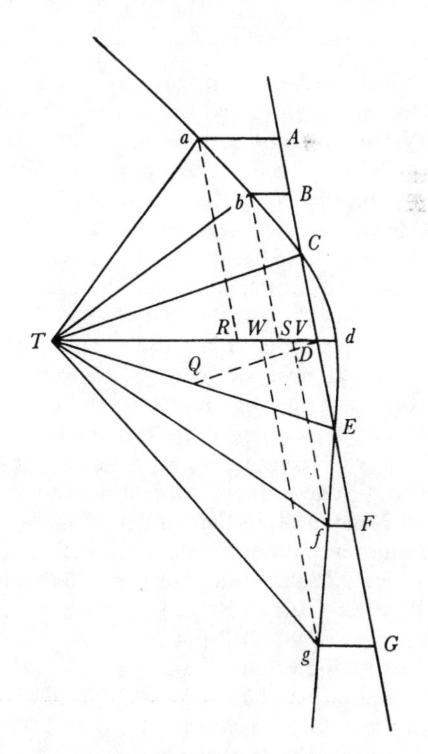

---

(1) As we have already remarked (see §1: note (79) above) this represents a considerable simplification of the method loosely outlined in the preceding 'De motu Corporum' (opening paragraph of the scholium to Problem 4) for approximately tracing the projection upon the

DTa 63^{gr}⌊6922.	*DA* 359^{h}⌊410833.	*DTb* 40⌊41777.	*DB* 240⌊68166.
DTC 21⌊201888.	*DC* 141⌊85333.	*DTE* 17⌊17333.	*DE* 193⌊12138.
DTf 34⌊507217.	*DF* 602^{h}⌊3375.	*DTg* 47⌊50333.	*DG* 1226⌊6586.

s.*Q*	38⌊37472.	[Ejus Log.]	9. 79295306.
TD	10000000		10.
s. *TDQ*	21⌊20139		9. 55828509
TQ			9. 76533203.
s. *DTQ*	17⌊17333		9. 47020925
DQ			9. 67725619.
CD) *CE*(4)			0. 3731724
DE) *CE*(4)			0. 2391815
TE	13756391.		10. 1385044.
TC	8249606.		9. 9164377.

TC+*TD*	18249606.	4.2612534.	*TE*+*TD*	23756391.	4. 3757804.
TC−*TD*	1750394.	3. 2431358.	*TE*−*TD*	3756391.	3. 5747707.
[co]tang $\frac{1}{2}CTD$	79⌊3993.	10.7277928	cotang $\frac{1}{2}DTE$	81⌊41333.	10. 8210296
Tang(5)	27⌊13414.	9. 7096752.	Tang(6)	46⌊32132.	10. 0200199.

ecliptic of the curving orbit of a comet—here, in exemplification, that of 1680–1. In line with that earlier imposed on Newton's parallel calculations at this period using the unmodified Wrennian rectilinear technique (v: 524–9), we hazard the following date of composition on the basis of his observation to Flamsteed on 19 September 1685 that 'I have not yet computed ye orbit of a comet but am now going about it...taking that of 1680 into fresh consideration' (*Correspondence of Isaac Newton*, **2**, 1960: 419), and our coupled surmise (see note (3) following) that the timed sightings from which he here works are those, corrected for atmospheric refraction, which were sent by Flamsteed with his reply to Newton's letter a week later.

(2) ULC. Add. 3965.11: 163^{r}. For brevity and clarity we have slightly compressed and trivially reordered Newton's rough calculations.

(3) As in v: 525, note (3) we follow J. A. Ruffner's eminently plausible conjecture that the cometary sightings made on these dates (from which the following differences in cometary longitude, as viewed from the earth, and in corrected 'true' times of observation are straightforwardly computed) were—except for that on 25 February, made by Newton himself at Cambridge—among those listed in a (now lost) 'tablet' included by Flamsteed with his letter to Newton of 26 September 1685, 'in which you will not wonder...to find a difference of some few minutes from y^{e} former I sent you [on 7 March 1680/1]' (*Correspondence of Isaac Newton*, **2**: 422; compare *ibid.*: 354) and which Newton afterwards published in Book 3 of his *Principia* ($_{1}$1687: 490) along with his own sighting on February 25 at an unadjusted *tempus apparens* of '8^{h}.30'' (*ibid.*: 491).

(4) Understand 'ratio *CE* ad *CD*' and 'ratio *CD* ad *DE*' respectively. Since Newton's computation of *TE* (from *TQ*) and of *TC* (from *DQ*) requires him to find only the logarithms of the analogous ratios *TE*/*TQ* and *TC*/*QD*, he does not bother to list their explicit numerical values alongside.

(5) Read 'Tang $\overline{\frac{1}{2}TCD-\frac{1}{2}TDC}$', corresponding (in this application of the familiar trigonometrical tangent-rule to resolving the triangle *CTD*, given its sides *TC*, *TD* and their included angle $C\widehat{T}D$) to 'cotang $\frac{1}{2}CTD$' [= tan $\frac{1}{2}(T\widehat{C}D+T\widehat{D}C)$] in the previous line.

$TCD = 106\lfloor 53344.$ $\qquad$ $TED = 35\lfloor 092$

$TDC = 52\lfloor 26516.$ $\qquad$ $TDE = 127\lfloor 73465.$

[sive] $TDC = 52\lfloor 26535$ ut supra.(7)

$\sin DTC$			[Ejus Log.]		9. 55828509
$\sin DCT$					9. 98166183
$DC =$	$141\lfloor 853333$				2. 1518395
$TD =$	$376^{h}\lfloor 02464.$				2. 57521624.
s: aTD	$63\lfloor 6922.$	9. 9525145	s: bTD	$40\lfloor 41777.$	9. 8118136
s: TaR	$64\lfloor 04245.$	9. 9538177	s: TbS	$87\lfloor 31688.$	9. 9995236
AD	$359\lfloor 41083.$	2. 5555911	BD	$240\lfloor 68166.$	2. 3814430
$TR =$	$360\lfloor 49041.$	2. 5568937.	$TS =$	$370\lfloor 81138.$	2. 5691530.
$Aa = RD =$	$15^{h}\lfloor 53423.$	Dec 21.	$Bb[=SD] =$	$5\lfloor 21326.$	Dec. 26.
s: fTD	$34\lfloor 507217.$	9. 7532088	s: gTD	$47\lfloor 50333.$	9. 86765406
s. TfV	$17\lfloor 758133.$	9.4842991	s: TfW	$4\lfloor 76202.$	8. 91916831
DF	$602\lfloor 3375.$	2. 7798399	DG	$1226\lfloor 6586.$	3. 08872366
$TV =$	$324\lfloor 2875.$	2. 5109302.	TW	$138\lfloor 1141.$	2. 14023791.
$Ff[=VD] =$	$51\lfloor 7371.$	Jan 30.	$Gg[=WD] =$	$137\lfloor 91054.$	Feb. 25.(8)

---

(6) Here, similarly, read

$$\text{'Tang } \tfrac{1}{2}\overline{TDE - \tfrac{1}{2}TDE}\text{'}$$

in parallel to 'cotang $\frac{1}{2}DTE$' [$= \tan \frac{1}{2}(\widehat{TDE} + \widehat{TED})$] above.

(7) The slight divergence in the last two figures is of course the additive accumulation of small inaccuracies in the seventh decimal places of the logarithmic and trigonometrical tables here employed. It will be clear that, given timed sightings *Ta, Tb, TC, Td, TE, Tf*, ... of the comet from the earth ('T[erra]', here assumed to be fixed in position), Newton's first step is to determine the direction of motion of the comet at some mean point d: this he approximates as that of the chord CE of the small surrounding (near-parabolic) arc $\widehat{CdE}$, further supposing that (because the comet's speed over this arc is effectively uniform?) the base-line Td will intersect CE at D such that $CD : DE = \text{time}_{C\to D} : \text{time}_{D\to E}$. The construction of the angle $\widehat{TDC}$ at which the line AG, parallel to the tangent at d, is inclined to TD (assigned a conventional length of 10^7 units) is then accomplished by (geometrical) Problem 16 of Newton's contemporary Lucasian lectures on algebra (see v: 210–12): namely, on drawing DQ parallel to CT, the ratios

$$TD : TQ : QD\,(= \sin \widehat{TQD} : \sin \widehat{TDQ} : \sin Q\widehat{TD}),\quad TE : TQ\,(= CE : CD)$$

and

$$TC : QD\ (= CE : DE)$$

are given, and hence the figure $TCDE$ is given in species and its elements straightforwardly calculable. As a gloss, Newton in immediate sequel converts the length of TD to the horary units in which CD and DE are expressed.

(8) Making the final blanket assumption (which we have sought tentatively to justify in §1: note (79) above) that the oblique distances $aR\,(= AD)$, $bS\,(= BD)$, $fV\,(= FD)$, $gW\,(= GD)$ from TDd in the direction of $ACEG$—taken (see note (7)) to be that of the tangent at d, or so

[2] Jan 25. [7^{h}. 58′. 42″. *f*.]
Feb 5. [7^{h}. 4′. 41″. *g*.][3]

DTf 30⌊77722.[9] *DF* 481⌊95111. *DTg* 38⌊17833. *DG* 745⌊050833.

s. *fTD*	30⌊77722.	9. 7090164	s: *gTD*	38⌊168333.	9. 7909708
s. *TfV*	21⌊48813.	9. 5638454	s. *TfW*	14⌊09702.	9. 3861968
DF	481⌊95111.	2. 6830029	*DG*	745⌊050833.	2. 8721859
TV=	345⌊010238.	2. 5378319.	*TW*	293⌊3675.	2. 4674119.
Ff=	31⌊01440.	Jan 25.	*Gg*=	82⌊65714.	Feb 5.

---

[3] Dec 29. [8^{h}. 3′. 2″. *b*.]
Jan 9. [7^{h}. 0′. 53″. *f*.][3]

[*DTb* 25⌊62361. *DB* 165⌊97666.] *DTf* 9, 90222. *DF* 96⌊9875.

	Dec 29.			Jan 9.	
s: *bTD*	25⌊62361.	9. 6359432	s: *fTD*	9⌊90222.	9. 2354458
s: *TbG*	102⌊11104.	9.9902257	s: *TfV*	42⌊36313.	9. 8285484
DB	165⌊97666.	2. 2200469	*DF*	96⌊9875.	1. 9867157
ST	375⌊25756.	2. 5743294.	*TV*	380⌊030434.	2. 5798183.
Bb	0⌊76708.		[*Ff*]	4⌊005794.[10]	

---

[4] Jan 10. [6^{h}. 6′. 10″. *f*.][3]

DTf	11,863055.	*DF*	120⌊07555.
s: *fTD*	11⌊863055.		9. 3129664
s *TfV*	40⌊40230.		9. 8116759

---

we surmise—are proportional to the corresponding differences in time between the sighting *Td* and those along *Ta*, *Tb*, *Tf*, *Tg* and hence given in ratio to *CD*, *DE* and so to *TD*, Newton by single applications of the sine-rule straightforwardly computes the ratios of the subtenses *TR*, *TS*, *TV*, *TW* and therefrom those of the cometary 'deviations' *aA* (= *RD*), *bB* (= *SD*), *fF* (= *VD*) and *gG* (= *WD*) to *TD* which fix the 'orbital' points *a*, *b*, *f*, *g* in position. Though the 'vertex' *d* is manifestly not constructable in a like manner, in his manuscript figure (here accurately reproduced) Newton roughly locates it by eye at the intersection of a 'smooth' parabolic curve drawn freehand through *a*, *b*, *C*, *E*, *f* and *g*.

(9) This should be (30° 6′ 38″ =) '30^{gr}⌊11055'. The effect of the correction will be slightly to decrease Newton's ensuing value for *Ff*.

(10) Strictly '−4⌊005794' since *TV* is greater than *TD*; whence (as it should) the point *f* will lie to the right of *ADG* in the arc $\widehat{dE}$. In this position *f* the 'comet' will, in Newton's scheme of its orbit, attain its rightmost point *d* very nearly. The final computation following reveals that a day later the 'comet' is only a little more than 2½ horary units to the right of *ADG*.

DF	$120\lfloor 07555.$	2.0794545
$[TV]$	$378\lfloor 611755.$	$2.5781940.$
$[Ff]$	$2\lfloor 587115.$(11)	

APPENDIX 3. THE BALLISTIC CURVE (RESISTANCE PROPORTIONAL TO VELOCITY) REWORKED.(1)

[spring 1685/*c.* mid-1692]

Extracts from the first and second editions of Newton's *Principia*

[1](2) PROP. IV. PROB. II.

Posito quod vis gravitatis in Medio aliquo similari uniformis sit, ac tendat perpendiculariter ad planum Horizontis; definire motum Projectilis, in eodem resistentiam velocitati proportionalem patientis.

E loco quovis D egrediatur Projectile secundum lineam quamvis rectam DP, & per longitudinem DP exponatur ejusdem velocitas sub initio motus. A puncto P ad lineam Horizontalem DC demittatur perpendiculum PC, & secetur DC in

(11) With TV again greater than TD, this should read '$-2\lfloor 587115$', whence f is (as it should be, once more) in $\widehat{dE}$ to the right of ADG. Newton's computations terminate abruptly at this point, and we have no reason to think that he ever again was tempted to apply this badly deficient method to the computation of any real cometary orbit. The would-be 'simplifying' assumption (compare §1: note (79)) that the earth be supposed to be at rest in the immediate vicinity of the sun is, even over a very short interval of time, here a crucial defect. As the rapidly varying cometary longitudes at this period indicate (see Flamsteed's tabulation in *Principia*, $_1$1687: 490), the earth in late December 1680/January 1681 was moving almost directly away from the very nearly rectilinear path of the 1680–1 comet, travelling at a little more than half its mean speed. Newton's present calculation yields a 'cometary' path which, in wide divergence from physical reality, closely approximates a parabola with its vertex near to the point d.

(1) The classically composed demonstration here reproduced (in [1]) from Book 2 of Newton's published *Principia* ($_1$1687) essentially mirrors our analytical justification (§1: note (113)) of his unproved equivalent construction of the present projectile orbit in his preceding 'De motu Corporum', and may straightforwardly be recast in its terms. To it we append from the *Principia's* second edition ($_2$1713) two opening corollaries which reduce the geometrical definition of this *logarithmica* to standard form. In [3] we reproduce the five corollaries originally added in 1687 (but here renumbered 3–7 as in the second edition) which minimally elaborate the basic construction and, in the case of the last, somewhat forlornly attempt to determine the ballistic orbit empirically by points 'ex Phænomenis quamproximè'.

(2) *Philosophiæ Naturalis Principia Mathematica* (London, $_1$1687): 241–2. The main emendations made in revise in 1713 are noticed in following footnotes.

A ut sit *DA* ad *AC* ut resistentia Medii ex motu in altitudinem sub initio orta,(3) ad vim gravitatis; vel (quod perinde est) ut sit rectangulum sub *DA* & *DP* ad rectangulum sub *AC* & *CP* ut resistentia tota sub initio motus ad vim Gravitatis. [Asymptotis *DC*, *CP*](4) [d]escribatur Hyperbola quævis *GTBS* secans erecta perpendicula *DG*, *AB* in *G* & *B*; & compleatur parallelogrammum *DGKC*, cujus latus *GK* secet *AB* in *Q*. Capiatur linea *N* in ratione ad *QB* qua *DC* sit ad *CP*; & ad rectæ *DC* punctum quodvis *R* erecto perpendiculo *RT*, quod Hyperbolæ in *T*, & rectis *GK*, *DP* in *t* & *V*(5) occurrat; in eo cape *Vr* æqualem $\frac{tGT}{N}$,(6) & Projectile tempore *DRTG* perveniet ad punctum *r*, describens curvam lineam *DraF*, quam punctum *r* semper tangit; perveniens autem ad maximam altitudinem *a* in perpendiculo *AB*, & postea semper appropinquans ad Asymptoton PLC. Estq; velocitas ejus in puncto quovis *r* ut Curvæ Tangens *rL*. Q.E.[I].

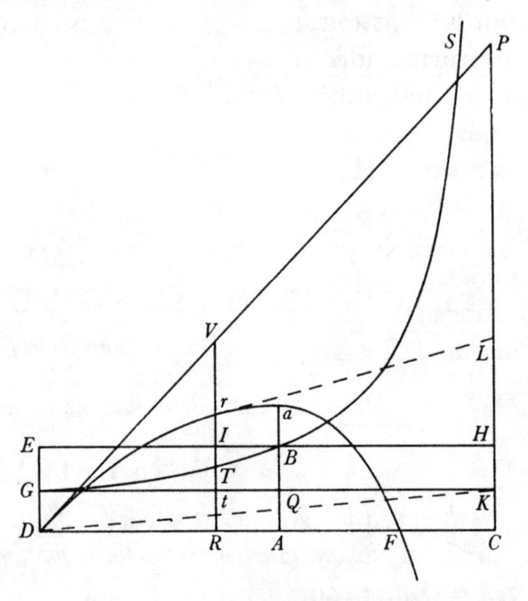

Est enim *N* ad *QB* ut *DC* ad *CP* seu *DR* ad *RV*, adeoq; *RV* æqualis $\frac{DR \times QB}{N}$, & *Rr* $\left(\text{id est } RV - Vr \text{ seu } \frac{DR \times QB - tGT}{N}\right)$ æqualis $\frac{DR \times AB - RDGT}{N}$. Exponatur jam tempus per aream *RDGT*, & (per Legum Corol. 2)(7) distinguatur motus

(3) That is, the vertical component of the initial 'resistentia medij' (V_y in the notation of §1: notes (108) and (113)). Observe that Newton now implicitly corrects his earlier slip (see §1: note (110)) in assigning the ratio of $V = (DP/CP).V_y$ to g.

(4) This very necessary phrase was later inserted by Newton himself in the second edition.

(5) In preparation for his addition in the next line, Newton in his 1713 edition expanded this to read '...rectis *EH*, *GK*, *DP* in *I*, *t* & *V*'. The intersection of *EH* and *RV* in the accompanying figure was correspondingly marked '*I*' as shown.

(6) In 1713 (compare previous note) Newton added in sequel the minimal clarification 'vel quod perinde est, cape *Rr* æqualem $\frac{GTIE}{N}$'.

(7) See §2: note (20) following. This corollary formally (in a Cartesian coordinate system, as here, where the acceleration in the tangential direction is zero) justifies Newton's splitting of the orbital acceleration $dv/dt = -v - g.dy/ds$ into horizontal and vertical components (constant in direction), but his further assumption in sequel that these are $(dv_x/dt =) -v_x$ and $(dv_y/dt =) -v_y - g$ respectively is not as immediately obvious, perhaps, as he would have it (compare §1: note (109)).

corporis in duos, unum ascensus, alterum ad latus. Et cum resistentia sit ut motus,[8] distinguetur etiam hæc in partes duas partibus motus proportionales & contrarias: ideoq; longitudo a motu ad latus descripta erit (per Prop. II. hujus)[9] ut linea DR, altitudo vero (per Prop. III. hujus)[9] ut area

$$DR \times AB - RDGT,$$

hoc est ut linea Rr. Ipso autem motus initio area $RDGT$ æqualis est rectangulo $DR \times AQ$, ideoq; linea illa Rr

$$\left(\text{seu } \frac{DR \times AB - DR \times AQ}{N}\right)$$

tunc est ad DR ut $AB - AQ$ (seu QB) ad N, id est ut CP ad DC; atq; adeo ut motus in altitudinem ad motum in longitudinem sub initio. Cum igitur Rr semper sit ut altitudo, ac DR semper ut longitudo, atq; Rr ad DR sub initio ut altitudo ad longitudinem: necesse est ut Rr semper sit ad DR ut altitudo ad longitudinem, & propterea ut corpus moveatur in linea $DraF$, quam punctum r perpetuo tangit. Q.E.D.

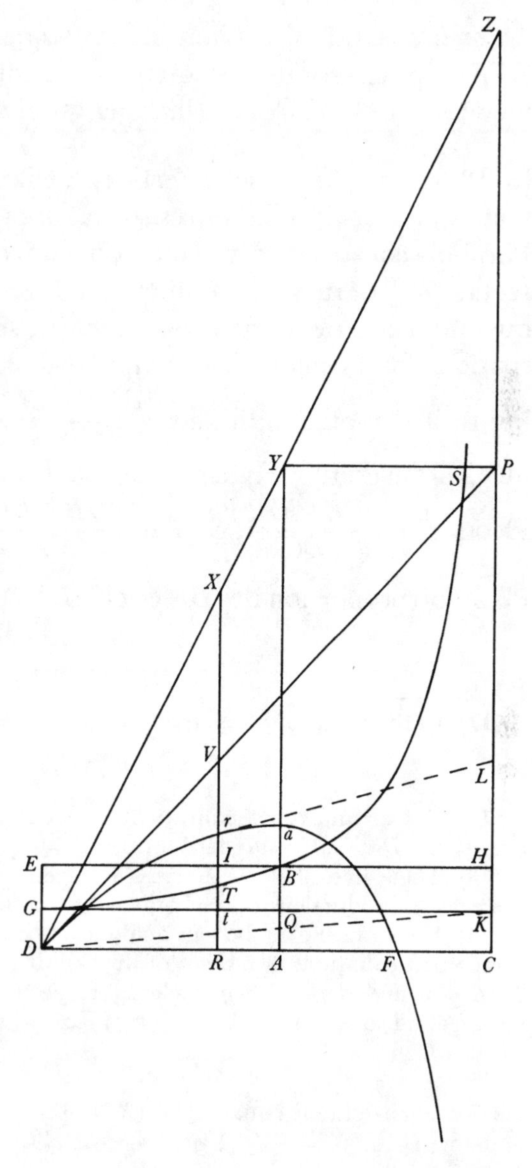

[2][10] *Corol. 1.* Est igitur Rr æqualis $\frac{DR \times AB}{N} - \frac{RDGT}{N}$, ideoque si producatur RT ad X ut sit RX æqualis $\frac{DR \times AB}{N}$, (id est, si compleatur parallelogrammum $ACPY$, jungatur DY

(8) In his working, of course, Newton suitably absorbs the constant factor of proportionality, thereby (for simplicity) equating the instantaneous resistance upon the body to its *motus* (orbital speed).

(9) These are essentially identical with Problems 6 and 7 respectively of the preceding 'De motu Corporum'; for their analytical equivalents see §1: notes (98) and (108).

(10) These two opening corollaries are additions in the revised edition of the *Principia* ($_2$1713: 217).

secans CP in Z, & producatur RT donec occurrat DY in X;) erit Xr æqualis $\frac{RDGT}{N}$, & propterea tempori proportionalis.(11)

Corol. 2. Unde si capiantur innumeræ CR vel, quod perinde est, innumeræ ZX, in progressione Geometrica; erunt totidem Xr in progressione Arithmetica.(12) Et hinc Curva $DraF$ per tabulam Logarithmorum facile delineatur.

[3](13) *Corol.* [3]. Hinc si Vertice D, Diametro DE deorsum producta, & latere recto quod sit ad $2DP$ ut resistentia tota, ipso motus initio, ad vim gravitatis, Parabola construatur: velocitas quacum corpus exire debet de loco D secundum rectam DP, ut in Medio uniformi resistente describat Curvam $DraF$, ea ipsa erit quacum exire debet de eodem loco D, secundum eandem rectam $D[P]$, ut in spatio non resistente describat Parabolam.(14) Nam Latus rectum Parabolæ hujus, ipso motus initio, est $\frac{DV^{\text{quad.}}}{Vr}$ & Vr est $\frac{tGT}{N}$ seu $\frac{DR\times Tt}{2N}$. Recta autem quæ, si duceretur, Hyperbolam GTB tangeret in G, parallela est ipsi DK,(15) ideoq; Tt est $\frac{CK\times DR}{DC}$, & N erat $\frac{QB\times DC}{CP}$. Et propterea Vr est $\frac{DR^q\times CK\times CP}{2CD^q\times Q[B]}$, id est (ob proportionales DR & DC, DV & DP) $\frac{DV^q\times CK\times CP}{2DP^q\times QB}$. & Latus rectum

(11) In the terms of §1: note (113), since $N = AB\times AC/g$, therefore

$$rX = (RDGT)/N = g.\log(DC/RC) = gt \propto t.$$

(12) For in analytical equivalent (see §1: note (113)) the defining 'symptom' of the *logarithmica* $DraF$ is, in standard form, $Xr = -g.\log(ZX/ZD)$.

(13) These five final corollaries (here renumbered as in the second edition) are lightly corrected reproductions of the equivalent Corollaries 1–5 in *Principia* ($_1$1687): 242–5.

(14) When, in the equivalent analytical terms of §1: note (113), the resistance is zero, the component equations of motion of the orbiting body $r(x, y)$ defined by $DR = x$, $Rr = y$ are $dv_x/dt = 0$ and $dv_y/dt = -g$, yielding respectively $v_x - V_x = 0$ and $v_y - V_y = -gt$, whence $x = V_x t$ and so $y = V_y t - \frac{1}{2}gt^2 = (V_y/V_x)x - \frac{1}{2}(g/V_x^2)x^2$. On setting

$$DV = (V/V_x).x = X \quad\text{and}\quad Vr = (V_y/V_x)x - y = Y,$$

the Cartesian defining equation of $r(X, Y)$ proves to be $Y = \frac{1}{2}(g/V^2)\,X^2$, a 'Galileian' parabola of diameter $Aa\,(X = VV_y/g)$ parallel to DE $(X = 0)$ and of *latus rectum* $2V^2/g = 2DP.(V/g)$. Newton's following derivation of this last result ingeniously proceeds from the assumption that the resisted *logarithmica* of the main proposition coincides with this parabola in the immediate vicinity of the firing point D (before the resistance has had opportunity to decelerate the projectile's orbital speed), but has somewhat unsatisfactorily to assume the verticality of the parabola's diameter as its initial step.

(15) For, since $CD\times DG = CA\times AB$, constant, at once $-d(DG)/d(CD) = DG$ (or CK)$/CD$. Newton assumes, of course, that in the limit as r comes to coincide with D the vanishingly small hyperbolic segment (GTt) approaches a right triangle whose hypotenuse is tangent to the arc $\widehat{GBS}$ at G.

$\frac{DV^{\text{quad.}}}{Vr}$ prodit $\frac{2DP^q \times QB}{CK \times CP}$, id est (ob proportionales QB & CK, DA & AC) $\frac{2DP^q \times DA}{AC \times CP}$, adeoq; ad $2DP$ ut $DP \times PA$ ad $PC \times AC$; hoc est ut resistentia ad gravitatem. Q.E.D.

Corol. [*4*]. Unde si corpus de loco quovis D, data cum velocitate, secundum rectam quamvis positione datam DP projiciatur, & resistentia Medii ipso motus initio detur, inveniri potest Curva $DraF$, quam corpus idem describet. Nam ex data velocitate datur latus rectum Parabolæ, ut notum est. Et sumendo $2DP$ ad latus illud rectum ut est vis Gravitatis ad vim resistentiæ, datur DP. Dein secando DC in A, ut sit $CP \times AC$ ad $DP \times DA$ in eadem illa ratione Gravitatis ad resistentiam, dabitur punctum A. Et inde datur Curva $DraF$.

Corol. [*5*]. Et contra, si datur curva $DraF$, dabitur & velocitas corporis & resistentia Medii in locis singulis r. Nam ex data ratione $CP \times AC$ ad $DP \times DA$, datur tum resistentia Medii sub initio motus, tum latus rectum Parabolæ: & inde datur etiam velocitas sub initio motus. Deinde ex longitudine tangentis rL, datur & huic proportionalis velocitas, & velocitati proportionalis resistentia in loco quovis r.

Corol. [*6*]. Cum autem longitudo $2DP$ sit ad latus rectum Parabolæ ut gravitas ad resistentiam in D; & ex aucta Velocitate augeatur resistentia in eadem ratione, at latus rectum Parabolæ augeatur in ratione illa duplicata: patet longitudinem $2DP$ augeri in ratione illa simplici, adeoq; velocitati semper proportionalem esse, neq; ex angulo CDP mutato augeri vel minui, nisi mutetur quoq; velocitas.

Corol. [*7*]. Unde liquet methodus determinandi Curvam $DraF$ ex Phænom[e]nis quamproxime, & inde colligendi resistentiam & velocitatem quacum corpus projicitur. Projiciantur corpora duo similia & æqualia eadem cum velocitate, de loco D, secundum angulos diversos CDP, [C]Dp (minuscul[æ] liter[æ] loc[o] subintellect[o]) & cognoscantur loca F, f ubi incidunt in horizontale planum DC. Tum assumpta quacunq; longitudine pro DP vel Dp, fingatur quod resistentia in D sit ad gravitatem in ratione qualibet, & exponatur ratio illa per longitudinem quamvis SM. Deinde per computationem, ex longitudine illa assumpta DP, inveniantur longitudines DF, Df, ac de ratione $\frac{Ff}{DF}$ per calculum inventa, auferatur ratio eadem per experimentum inventa, & exponatur differentia per perpendiculum MN. Idem fac iterum ac tertio, assumendo semper novam resistentiæ ad gravitatem rationem SM, & colligendo novam differentiam MN. Ducantur autem differentiæ affirmativæ ad unam partem rectæ SM, & negativæ ad alteram; & per puncta

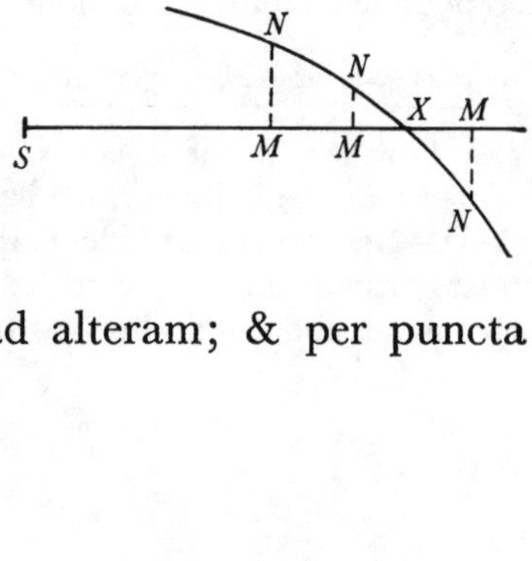

N, *N*, *N* agatur curva regularis *NNN* secans rectam *SMMM* in *X*,(16) & erit *SX* vera ratio resistentiæ ad gravitatem, quam invenire oportuit. Ex hac ratione colligenda est longitudo *DF* per calculum; & longitudo quæ sit ad assumptam longitudinem *DP* ut modo inventa longitudo *DF* ad longitudinem eandem per experimentum cognitam, erit vera longitudo *DP*. Qua inventa, habetur tum Curva Linea *DraF* quam corpus describit, tum corporis velocitas & resistentia in locis singulis.(17)

---

(16) A familiar Newtonian technique of attaining a rough approximation; compare IV: 560, note (113).

(17) As Newton well knew, the hypothesis that (in the earth's atmosphere) a projectile is instantaneously decelerated by resistance in proportion to its speed is exceedingly unrealistic, even for the low muzzle velocities obtaining in his day. It is difficult to believe that in outlining this present elaborate method for determining the elements of his ensuing *logarithmica DraF* 'ex Phænomenis quamproximè' he did not have his tongue stuck firmly—if still a little hopefully?—in this cheek. More than eighteen years before the *Principia* was published Christiaan Huygens had modestly declined to foist so mathematically elegant but physically useless a 'Theorie' upon the learned world, preferring to devote his effort to exploring the experimentally truer supposition that 'la resist[a]nce de l'air, & de l'eau, estoit comme les quarrez des vitesses', accurately divining that in 'ce veritable fondement des Resistances...la chose estoit beaucoup plus difficile, & sur tout en ce qui regarde la ligne courbe que parcourent les corps jettez obliquement' (*Discours de la Cause de la Pesanteur* (Leyden, 1690): 169; compare §1: note (113) above). Much like Huygens before him in 1669 (see his *Œuvres complètes*, **19**, 1937: 144–57; and compare A. R. Hall, *Ballistics in the Seventeenth Century* (Cambridge, 1952): 111–17, especially 116), Newton was able in the following Propositions V–IX of his *Principia* ($_1$1687: 246–60) correctly to resolve the problem of motion under a decelerating force varying instantaneously as the square of the speed in cases where the motion is confined to be in a straight line: namely, if [Propositions V–VII] $d^2x/dt^2 = dv_x/dt = -v_x^2$, then, on assuming the initial conditions $x = t = 0$ and $v_x = V_x$, from $dt/dv_x = -1/v_x^2$ there follows $t = 1/v_x - 1/V_x$, whence $dx/dt = v_x = 1/(t+1/V_x)$ and so $x = \log(V_x t+1)$; while if more generally [Propositions VIII/IX] $d^2y/dt^2 = dv_y/dt = -v_y^2 \pm g$, g constant, then with similar initial conditions $y = t = 0$ and $v_y = V_y$, from $dt/dv_y = 2dy/d(v_y^2) = -(v_y^2 \mp g)$ there ensues

$$t = \log((V_y+g^{\frac{1}{2}})/(V_y-g^{\frac{1}{2}})) - \log((v_y+g^{\frac{1}{2}})/(v_y-g^{\frac{1}{2}}))$$

or alternatively $t = \tan^{-1}(V_y/g^{\frac{1}{2}}) - \tan^{-1}(v_y/g^{\frac{1}{2}})$ according as $-g$ or $+g$ is taken, while in either case $y = \frac{1}{2}\log((V_y^2 \mp g)/(v_y^2 \mp g))$. His following Propositions XI–XIV (*Principia*, $_1$1687: 274–84) comparably—but to no real purpose other than to reveal his mastery of the geometrical quadratures (in terms of logarithmic/inverse-tangent functions) involved—depart from the compounded equations of motion

$$dv_x/dt = -v_x^2 - 2kv_x \quad (\text{or } dv_x/dx = -v_x - 2k) \quad \text{and} \quad dv_y/dt = v_y . dv_y/dy = -v_y^2 - 2kv_y \pm g.$$

Wise, however, to the fallacy of seeking vectorially to combine these resultant 'component' motions—unlike Leibniz when he four years later vainly proffered this 'solution' (see §1: note (109))—Newton was unable to approach the problem of defining the general orbit traversed, under constant 'gravity' directed vertically downwards, in a medium resisting (in the instantaneous direction of motion) as the square of the orbital speed other than by implicitly subsuming it, as a particular case not there further explored but which (he told David Gregory in May 1694) was in his 'power', in the intervening Proposition X (*Principia*, $_1$1687: 260–9) where an arbitrary resistance to motion is assumed and then related to the gravity by the equivalent of a third-order differential equation whose solution—here readily effectable parametrically—resolves the problem. (Compare D. T. Whiteside, 'The Mathematical Principles underlying Newton's *Principia Mathematica*' (§1: note (30)): 126–30 and 137, note 5.)

---

§2. THE REVISED TREATISE 'DE MOTU CORPORUM' (WINTER/EARLY SPRING? 1684–5).(1)

[1](2) De motu Corporum.

Definitiones

1. *Quantitas materiæ* est quæ oritur ex ipsius densitate et magnitudine conjunctim. Corpus duplo densius in duplo spatio quadruplum est. Hanc quantitatem per nomen corporis vel massæ designo.

2. *Quantitas motus* est quæ oritur ex velocitate et quantitate materiæ conjunctim. Motus totius est summa motuum in partibus singulis, adeoqꝫ in corpore duplò majore æquali cum velocitate duplus est et dupla cum velocitate quadruplus.

3. *Materiæ vis insita* est(3) potentia resistendi qua corpus unumquodqꝫ quantum in se est(4) perseverat in statu suo vel quiescendi vel movendi uniformiter in directum: Estqꝫ corpori suo proportionalis, neqꝫ differt quicquam ab *inertia* massæ nisi in modo conceptus nostri. Exercet verò corpus hanc vim solummodò in mutatione status sui facta per vim aliam in se impressam(5) estqꝫ exercitium

(1) In strict truth there now exist of the eighty or so leaves of this first major revise of §1 preceding no more than (in Newton's contemporary pagination) folios 9–32 [ULC. Dd. 9.46: 24^r–31^r/64^r–71^r/56^r–63^r] and 41–8 [ULC. Add. 3965.3: 7^r–14^r], comprising Lemma VII–(halfway through) Proposition XXIV and (latter portion of) Proposition XXXIV–Proposition XLIII (incomplete enunciation only). These are now reproduced in [2] and (in part) in [3] respectively. The first eight missing leaves are closely enough restored, so we would argue, by introducing (in [1] here) the text of a contemporary autograph draft of five opening Definitions, and then interpolating (at the beginning of [2]) extracts [ULC. Dd. 9.46: 13^r–19^r/22^r–23^r] from the further augmented 'De motu Corporum Liber primus' elaborating the three fundamental 'Laws of Motion' and their several corollaries widely quoted in sequel and also inserting the Lemmas I–VI 'On the Method of First and Last Ratios' which are lacking. We purposely abjure the meaningless and grossly misleading marginal division into *lectiones* which Newton afterwards set on his folios 9–32 before depositing them, in superficial fulfilment of statutory requirement, as part of his 'Octob 1684' series of Lucasian lectures. For detailed justification of such an iconoclastic act and for an extended survey of the likely content of those of the other missing folios whose propositions may be in any way identified or surmised, see I. B. Cohen's *Introduction to Newton's 'Principia'* (Cambridge, 1971): 89–91 and especially 311–19, where the present piece is designated as 'LL_α'. Of the reasons which now prompted Newton to expand the scope of his initial 'De motu Corporum' when a little while before he had been content merely to give its text a verbal polish (see §1, Appendix 1 above) we have no firm knowledge but may surmise that when, in (or shortly after) November 1684, Halley paid his second visit to Cambridge he would enthusiastically have urged Newton to continue his dynamical and astronomical researches. There is, to be sure, good evidence (see Cohen's *Introduction:* 341–2) that Halley played thereafter an active and continuous, if minor, rôle in reading through the present draft treatise, checking a number of small slips and suggesting several helpful clarifications and emendations which were accepted by Newton into his mature 'De motu Corporum Liber primus' (partially reproduced in §3 following). The

Translation

[1][2] ON THE MOTION OF BODIES.

DEFINITIONS.

1. *Quantity of matter* is what ensues from its density and magnitude jointly. A body twice as dense in twice the space is four-fold. This quantity I designate by the name of body or mass.

2. *Quantity of motion* is what ensues from the speed and quantity of matter jointly. The motion of the whole is the sum of the motions in the separate parts, and hence in a body twice as big with equal velocity is twice as great, while with twice the speed it is four-fold.

3. *Force innate in matter* is the power of resisting whereby each individual body, inasmuch as it is in it to do so,[4] perseveres in its state of resting or of moving uniformly straight on: it is, futhermore, proportional to bodily bulk, nor does it differ at all from the *inertia* of its mass other than in the manner of our conceiving it. A body in fact exerts this force only during a change of its state effected by the impress of another force upon it,[5] and the exertion is its *Resistance* and

---

date we propose for its composition is conjectural but cannot be far out, being firmly bounded by Newton's preparation of the parent tract in the autumn of 1684 (see §1: note (2)) and his initial completion of its augmented revise by early the following summer (compare §3: note (1) below).

(2) ULC. Add. 3965.5: 21^r, first published by A. R. and M. B. Hall in their *Unpublished Scientific Papers of Isaac Newton* (Cambridge, 1962): 239–41, and again by J. W. Herivel in his *The Background to Newton's Principia* (Oxford, 1966): 315–16, where however the manuscript is erroneously 'completed' by a preliminary set of *Definitiones* '6–12' and '14' on f. 25^v following (see Appendix 1: note (1) below). These introductory terminological distinctions were soon afterwards further refined and elaborated in the corresponding opening leaves (ULC. Dd. 9.46: 4^r–7^r) of Newton's revised 'De motu Corporum Liber primus', and from there—after yet more changes and emendations had been made in its text—subsumed into his printed *Principia* ($_1$1687: 1–5) along with a long following scholium (*ibid.*: 7^r–12^r = *Principia*: 5–11) discussing the differences between 'true' (absolute) and relative space, time and motion. While it is in no way our intention to dwell on the conceptual and ontological foundations of Newton's dynamical thought, we may here not irrelevantly stress his growing awareness of the verbal subtleties and epistemological difficulties involved in creating a viable mathematical theory of forces and the accelerated motions these induce in bodies moving 'naturally' with uniform speed in a straight line.

(3) Newton subsequently added and then cancelled 'inertia sive' (inertia or), elaborating his meaning in a separate sentence.

(4) That is, 'naturally...in y^t state in wch it is' as he had phrased it in 1665 (see §1: note (9)). On the Lucretian origin of this Cartesian phrase compare also I. B. Cohen, '*Quantum in se est*: Newton's concept of inertia in relation to Descartes and Lucretius', *Notes and Records of the Royal Society of London*, **19**, 1964: 131–55.

(5) Whence to all mathematical purpose this 'innate' reaction is equal to the component of the *vis impressa* acting in the direction of motion, and so not comparable to the preceding

ejus *Resistentia* et *Impetus*[,] respectu solo ab invicem distincti: *Resistentia* quatenus corpus[6] reluctatur vi impressæ, *Impetus* quatenus corpus difficulter cedendo conatur mutare statum corporis alterius. Vulgus insuper resistentiam quiescentibus & impetum moventibus[7] tribuit: sed motus et quies ut vulgo concipiuntur respectu solo distinguuntur ab invicem: neq; verè quiescunt quæ vulgò tanquam quiescentia spectantur.

4.[8] *Vis impressa* est actio in corpus exercita ad mutandum statum ejus vel quiescendi vel movendi uniformiter in directum. Consistit hæc vis in actione sola neq; post actionem permanet in corpore. Est autem diversarū originum, ut ex impetu, ex pressione, ex vi centripeta.

5. *Vis centripeta* est vel actio vel potentia quælibet qua corpus versus punctum aliquod tanquam ad centrū trahitur impellitur vel utcunq; tendit. Hujus generis est gravitas qua corpus tendit ad centrum terræ, vis magnetica qua ferrum petit centrū magnetis, et vis illa, quæcunq; sit, qua Planetæ retinentur in orbibus suis et perpetuò cohibentur ne abeant in eorum tangentibus. Est autem vis centripetæ quantitas triplex: *absoluta*, *acceleratrix* et *motrix*. *Quantitas absoluta* (quæ et *vis absoluta* dici potest) major est ad unum centrum minor ad aliud, nullo habito respectu ad distantias et magnitudines attractorum corporum; uti virtus magnetica major in uno magnete minor in alio. *Quantitas* seu *vis acceleratrix* est velocitati proportionalis[9] quam dato tempore generat; uti virtus magnetis ejusdem major in minori distantia minor in majori, vel vis gravitans major prope terram minor in regionibus superioribus. *Quantitas* seu *vis motrix* est motui proportionalis[9] quem dato tempore producit; uti pondus majus in majori corpore minus in minore.[10] Ita se habet igitur *vis motrix* ad *vim acceleratricem* ut *motus* ad *celeritatem*. Namq; oritur *quantitas motus* ex *celeritate* ducta in corpus mobile et *quantitas vis motricis* ex *vi acceleratrice* ducta in idem corpus. Unde juxta superficiem terræ ubi gravitas acceleratrix in corporibus universis eadem est, gravitas

---

'vis inertiæ'—essentially a traditional kinematic *impetus*—which causes the body 'naturally' to persevere in a state of uniform rectilinear motion (or rest) at all other times. Newton himself initially added such a clarification in a cancelled first continuation at this point: 'inq; hujus exercitio consistit Resistentia. Ea igitur non corpori quidem resistenti sed vi impressæ proportionalis est. Namq; exercitium vis insitæ in status conservationem [non] est ut vis in ipsius mutationem impressa' (and in its exertion the resistance consists. This, then, is proportional not indeed to the resisting body but to the impressed force: for the exertion of the innate force to preserve a state is not comparable to the force impressed to change it). W. A. Gabbey has percipiently explored the historical roots and ontological complexities of this confusing, but not inconsistent, dual kinematical and dynamical function of Newtonian *vis innata* in his 'Force and Inertia in Seventeenth-century Dynamics', *Studies in History and Philosophy of Science*, **2**, 1971: 1–66, especially 32–50.

(6) Newton has here deleted the tentative addition 'in conservationem motus proprij' (to conserve its own motion).

(7) 'magis' (more usually) is cancelled, along with a first final clause 'sed quies ut vulgo concipitur non est vera quies' (but rest as it is commonly conceived is not true rest).

Impetus, to be distinguished one from the other only in relative regard: it is *resistance* insofar as the body[6] struggles against the impressed force, *impetus* insofar as the body by its reluctance to give way may endeavour to change the state of a second body. Common parlance, I may add,[7] assigns resistance to things at rest and impetus to ones in motion; but motion and rest as commonly conceived are distinguished one from the other only in their relative respect, and things which are commonly regarded as at rest are not really so.

4.[8] *Impressed force* is an action exerted on a body to change its state either of resting still or moving uniformly straight on. This force consists in that action alone, and does not endure in the body after the action is over. It is diverse in its origin, arising, for instance, from impetus, pressure and centripetal force.

5. *Centripetal force* is any action or power whereby a body is drawn, impelled or in any manner tends towards some point—its centre as it were. Of this kind is the gravity whereby a body tends to the centre of the earth, the magnetic force whereby iron seeks the centre of a magnet, and the force, whatever it be, by which planets are retained in their orbits and perpetually constrained from going off in their tangents. Centripetal force is, however, three-fold in its quantity: *absolute*, *accelerative* and *motive*. Its *absolute quantity* (which can also be called the *absolute force*) is greater to one centre and less to another, without regard being paid to the distances and sizes of the bodies attracted; as the magnetic strength is greater in one magnet, less in another. Its *accelerative quantity* or *force* is proportional to the speed which it generates in a given time; as the strength of the same magnet is greater at a less distance and less at a greater, or the force of gravity greater close to the earth and less in higher regions. Its *motive force* or *quantity* is proportional to the 'motion' (momentum) which it produces in a given time; as weight effects more in a greater body and less in a smaller one.[10] Accordingly, therefore, *motive force* is to *accelerative force* as *motion* to *speed*; for the *quantity of motion* arises from the *speed* multiplied into the moving body, and the *quantity of motive force* from the *accelerative force* multiplied into the same bulk. In consequence, near the surface of the earth where accelerative gravity is the same in all bodies without exception, motive gravity—that is,

---

(8) A first version of this Definition reads: '*Vis impressa* est quæ corpus ex statu suo vel quiescendi vel movendi uniformiter in directum deturbare nititur: estꝗ multarum originum, ut ex impetu, ex pressione, ex resistentia, ex vi centripeta' (*Impressed force* is what strives to disturb a body from its state either of resting still or moving uniformly straight on: it has many origins, arising, for instance, from impetus, pressure, resistance and centripetal force).

(9) Originally 'ut velocitas' and 'ut motus' respectively.

(10) As he will go on to assert, it is henceforth Newton's fundamental postulate that in the real macroscopic world both the 'absolute' and 'motive' forces in matter are universally constant, while—within the solar system at least—the 'accelerative' force varies as the inverse-square of the distance.

motrix seu pondus est ut corpus: at si longius recedatur a terra, inq; regiones ascendatur ubi gravitas acceleratrix fit minor, pondus pariter minuetur eritq; semper ut corpus in gravitatem acceleratricem ductum. Porrò *attractiones* & *impulsus* eodem sensu *acceleratrices* & *motrices* nomino. Voces autem *attractionis impulsus* vel *propensionis* cujuscunq; in centrum indifferenter & pro se mutuò usurpo, has vires non *physicè* sed *mathematicè* tantum considerando. Unde caveat Lector ne per hujusmodi voces cogitet me speciem vel modum actionis causamve aut rationem physicam alicubi definire.(11)

[2](12) LEGES MOTUS.(13)

Lex I.

Corpus omne perseverare in statu suo quiescendi vel movendi uniformiter in directum, nisi quatenus a viribus impressis cogitur statum illum mutare.

Projectilia perseverant in motibus suis nisi quatenus a resistentia aeris retardantur et vi gravitatis impelluntur deorsum. Trochus,(14) cujus partes cohærendo perpetuo retrahunt sese a motibus rectilineis, non cessat rotari nisi quatenus ab aere retardatur. Majora autem Planetarum et Cometarum corpora motus suos et progressivos et circulares in spatijs minus resistentibus factos conservant diutius.

Lex II.

Mutationem motus proportionalem esse vi motrici impressæ et fieri secundum lineam rectam qua vis illa imprimitur.

(11) In his revised 'De motu Corporum Liber primus' (ULC. Dd. 9.46: 7r = *Principia*, 11687: 5) Newton later added 'vel centris (quæ sunt puncta mathematica) vires verè et physicè tribuere, si forte aut centra trahere, aut vires centrorum esse dixero' (or to centres (which are mathematical points) really and physically attribute forces if perchance I say either that centres attract or that forces pertain to centres). As we have already remarked in note (2), the corresponding section of the text of 'De motu Corporum Liber primus' now continues (ULC. Dd. 9.46: 7r–12r) with a long scholium defining and distinguishing absolute and relative time, space and motion. Clearly its presently restored preliminary version contained some equivalent passage based on the opening *Definitiones* 1–11 of the manuscript 'De motu corporum in medijs regulariter cedentibus' reproduced in Appendix 1 below, but it goes beyond our present purpose to make a conjecture as to its detailed content. We pass immediately on to the three classical Newtonian 'Laws of Motion' here—or rather (see next note) in the lost original text—given their first collective presentation.

(12) ULC. Dd. 9.46: 13r–19r/22r–31r/64r–71r/56r–63r. As we have earlier observed (note (1) above), ff. 13r–19r/22r–23r strictly form part of the revised 'De motu Corporum Liber primus' (later portions of which are reproduced in §3 following), but they are here interpolated to fill the place of lost preliminary sheets—up to and including page 8 in Newton's numbering—of the present text without, we hope, significantly distorting the verbal sense and content of the three 'Laws of Motion' and following Lemmas I–VI 'On the Method of First and Last Ratios' for which they here deputise. The manuscript, it may be noted, lacks the text-figures

weight—is as bodily mass; but, in receding further from the earth and ascending to regions where accelerative gravity comes to be less, weight will be correspondingly diminished and be always as the body multiplied into the accelerative gravity. *Attractions* and *impulses*, furthermore, I name *accelerative* and *motive* in the same sense. The words 'attraction', 'impulse' or any 'propensity' to a centre, however, I employ indifferently and interchangeably, considering these forces not *physically* but merely *mathematically*. The reader should hence beware lest he think that by words of this sort I anywhere define a species or mode of action, or a physical cause or reason.(11)

---

[2](12)

LAWS OF MOTION.(13)

Law I.

Every body perseveres in its state of resting or moving uniformly straight on, except inasmuch as it is compelled by impressed forces to change that state.

Projectiles persevere in their motions except insofar as they are slowed by the resistance of the air and driven downwards by the force of gravity. A hoop(14) whose parts are perpetually, by their cohesion, diverted from rectilinear motions does not cease to rotate except insofar as it is retarded by the air. The more massive bodies of planets and comets conserve for a longer period, however, the motions, both progressive and circular, which they make in less resisting spaces.

Law II.

Change in motion is proportional to the motive force impressed and takes place following the straight line along which that force is impressed.

---

which are listed by number in its margins. These we have appropriately reconstructed, *mutatis mutandis*, from their equivalents in the published *Principia*; we have, however, chosen—as Newton himself recommended to Edmond Halley on 14 July 1686 (*Correspondence of Isaac Newton*, **2**, 1960: 444), adding that he initially 'crouded y^m^ into one to save ye trouble of altering ye numbers in ye schemes you have'—to split Figure 1 into two (here numbered '1[a]' and '1[b]'), while Figure 19 is wholly our restoration on the basis of the text (omitted in redraft) which it illustrates.

(13) The revised section is headed (f. 13^r^) 'AXIOMATA/sive/LEGES MOTÛS' (AXIOMS, or LAWS OF MOTION), but Halley in his contemporary critique (ULC. Add. 3965.9: 94^r^–95^r^/97^r^/99^r^/96^r^/98^r^; compare note (1) above) of the present 'De motu Corporum' and its revise labels (f. 95^r^) his pertinent section 'De legibus Motus' in agreement with the terser subtitle—that of the preliminary draft reproduced in Appendix 1—we here employ. It will be evident that Laws I and II following are trivially expanded elaborations of 'Lex 1' and 'Lex 2' of §1, Appendix 1.

(14) Understand maybe a 'roundabout' of some sort that is placed horizontally and allowed to spin freely round its centre. Andrew Motte renders it somewhat restrictively as a 'top' in his *The Mathematical Principles of Natural Philosophy. By Sir Isaac Newton*, **1** (London, 1729): 19.

Si vis aliqua motum quemvis generet, vis dupla duplum, tripla triplum generabit, sive simul et semel,[15] sive gradatim et successive impressa fuerit. Et hic motus quoniam in eandem semper plagam cum vi generatrice determinatur, si corpus antea movebatur, motui ejus vel conspiranti additur, vel contrario subducitur, vel obliquo oblique adjicitur et cum eo secundum utriusqȝ determinationem componitur.

Lex III.[16]

Actioni contrariam semper et æqualem esse reactionem: sive, corporum duorum actiones in se mutuo semper esse æquales et in partes contrarias dirigi.

Quicquid premit vel trahit alterum tantundem ab eo premitur vel trahitur. Siquis lapidem digito premit, premitur et hujus digitus a lapide. Si equus lapidem funi allegatum trahit, retrahetur etiam et equus æqualiter in lapidem: nam funis utrinqȝ distentus eodem relaxandi se conatu urgebit equum versus lapidem, ac lapidem versus equum, tantumqȝ impediet progressum unius quantum promovet progressum alterius. Si corpus aliquod in corpus aliud impingens, motum ejus vi sua quomodocunqȝ mutaverit, idem quoqȝ vicissim in motu proprio eandem mutationem in partem contrariam vi alterius (ob æqualitatem pressionis mutuæ) subibit. His actionibus[17] æquales fiunt mutationes non velocitatum sed motuum (scilicet in corporibus non aliunde impeditis:) Mutationes enim velocitatum in contrarias itidem partes factæ, quia motus æqualiter mutantur, sunt corporibus reciprocè proportionales.[18]

Corol. 1.[19]

Corpus viribus conjunctis diagonalem parallelogrammi eodem tempore describere, quo latera separatis.

Si corpus dato tempore, vi sola M, ferretur ab A ad B et vi sola N ab A ad C; compleatur parallelogrammum $ABDC$, et vi utraqȝ feretur id eodem tempore

(15) As a single impulse, that is. This notion is basic for Newton, since he can conceive of the action of a continuous force only as the 'gradual' cumulative effect of an infinity of discrete, infinitesimal force-impulses impressed at 'successive' instants of time. It will be clear that a force f acting continuously over a time t will generate the same 'motion' (instantaneous speed) $f.t$ as an equal force-impulse applied continually over the same period at successive instants. See also 3, §1: note (8) below.

(16) In the present treatise 'De motu Corporum' as we now inadequately know it this third Newtonian law of motion is, in fact, applied only in a single instance, 'Prop. XXXVIII. Theor. XX' (ULC. Add. 3965.3: 11^r, an early version of Proposition LXIX of Book 1 of the *Principia*), which is here omitted. We have earlier (see v: 148–9, note (152)) summarised the historical context—that of the linear 'occursion' of perfectly elastic bodies wherein 'motion' (momentum) is conserved—in which this generalised Cartesian law first presented itself to Newton's attention in the middle 1660's, there pointing to a preliminary formulation of it in his Lucasian lectures on algebra, 'corpus utrumqȝ tantum reactione pat[i]tur quantum agit in

Should some force generate any motion, then twice the force will generate its double and three times it its triple, whether it be impressed once and instantaneously[15] or successively and gradually. And since this motion is determined to be ever in the same direction as the generating force, it is, if the body was beforehand in motion, either added to it when in unison, or taken from it when contrary, or adjoined obliquely to it when oblique and combined with it according to the determined direction of each.

Law III.[16]

To any action there is always a contrary, equal reaction: in other words, the actions of two bodies each upon the other are always equal and opposite in direction.

Whatever presses or draws another is pressed or drawn by it the same amount. If anyone presses a stone with his finger, that man's finger is also pressed by the stone. If a horse drags a stone tied to a rope, the horse will also be equally dragged back to the stone: for the rope, taut in each direction, will by the same effort to loosen itself pull both the horse towards the stone and the stone towards the horse, and will hinder the advance of the one as much as it assists the advance of the other. If some body, knocking against another body, changes its motion by its own force in any manner whatsoever, it will also in turn (because of the equality of their mutual pressure) undergo the same change in its own motion the opposite way by the force of the other. By these actions[17] the changes not in their velocities but in their 'motions' (momenta) will come to be equal—in bodies not otherwise impeded, of course. For the changes in speed likewise produced in opposite directions are, because the 'motions' are equally changed, reciprocally proportional to (the mass of) the bodies.[18]

Corollary 1.[19]

A body under the joint action of forces traverses the diagonal of a parallelogram in the same time as it describes the sides under their separate actions.

If the body in a given time by the action of force M alone were to be carried from A to B, and by the force N alone from A to C, complete the parallelogram $ABDC$ and by both forces it will in the same time be carried from A to D. For,

alterum' (v: 148), which is essentially identical with the initial enunciation of the present 'Lex III' (see Appendix 1: note (34) below).

(17) Understand of ideal elastic impact between bodies.

(18) A cancelled final sentence here reads: 'Hæc ita se habent in corporibus quæ non aliunde impediuntur' (These conditions obtain in bodies which are not otherwise impeded).

(19) Lemma 1 of the preliminary revise (given in §1, Appendix 1 above), transcribed word for word. The forces conjoined at a point—and their combined resultant—are, as before, taken to be measured, in both magnitude and direction, by the uniform speeds (themselves geo-

ab A ad D. Nam quoniam vis N agit secundum lineam AC ipsi BD parallelam, hæc vis nihil mutabit velocitatem accedendi ad lineam illam BD a vi altera genitam. Accedet igitur corpus eodem tempore ad lineam BD sive vis N imprimatur, sive non, atq; adeo in fine illius temporis reperietur alicubi in linea illa BD. Eodem argumento in fine temporis ejusdem reperietur alicubi in linea CD, et idcirco in utriusq; lineæ concursu D reperiri necesse est.

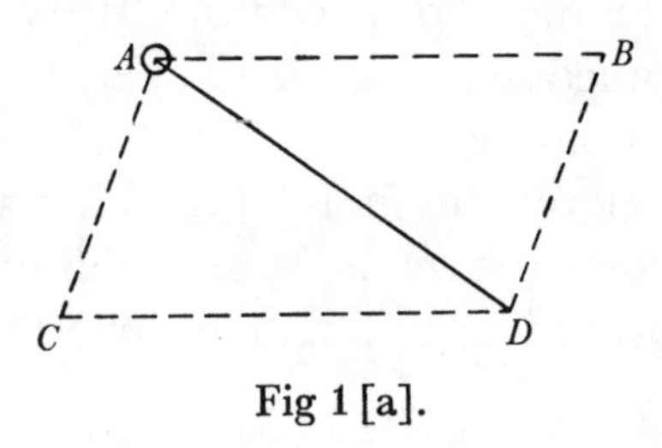

Fig 1 [a].

Corol. 2.[20]

Et hinc patet compositio vis directæ AD ex viribus quibusvis obliquis AB et BD, et vicissim resolutio vis cujusvis directæ AD in obliquas quascunq; AB et BD.[21] *Quæ quidem compositio et resolutio abunde confirmatur ex Mechanica.*

Ut si de rotæ alicujus centro O exeuntes radij inæquales OM, ON, filis MA, NP sustineant pondera A et P, et quærantur vires ponderum ad movendam rotam: per centrum O agatur recta KOL filis perpendiculariter occurrens in K et L centroq; O et intervallorum OK, OL majore OL describatur circulus occurrens filo MA in D: et actæ rectæ OD parallela sit AC et perpendicularis DC. Quoniam nihil refert utrum filorum puncta K, L, D affixa sint vel non affixa ad planum rotæ, pondera idem valebunt ac si suspenderentur a punctis K et L vel D et L. Ponderis autem A exponatur vis tota per lineam AD et hæc resolvetur in vires AC, CD, quarum AC trahendo radium OD directè a centro nihil valet ad movendam rotam; vis autem altera [CD][22] trahendo radium DO perpendiculariter idem valet ac si perpendiculariter traheret radium OL ipsi OD æqualem[,] hoc est idem atq; pondus P quod sit ad pondus A ut vis [CD][22] ad vim

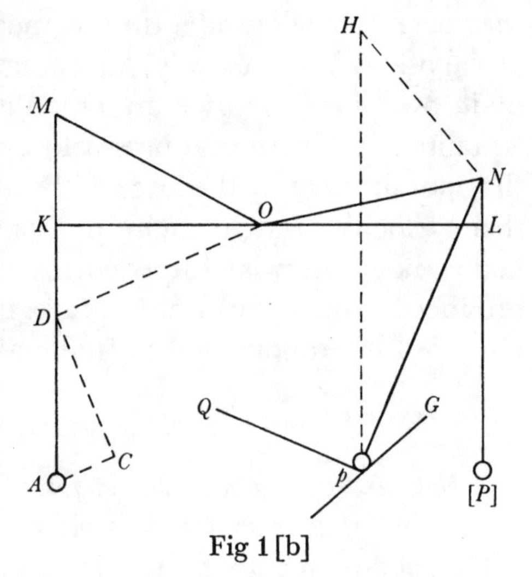

Fig 1 [b]

metrically expressed as vectorial line-lengths) which they generate from rest in a given time. As in the published *Principia*, this familiar 'parallelogram' rule for compounding such speeds is here, other than in the immediately following Corollary 2, applied uniquely to the case (Theorem I below; compare §1: note (10) preceding) where one of the generating forces is taken straightforwardly to be an arbitrary impressed *vis centripeta*, but the other is conceived to be the (ideal) *vis insita* which, according to Definition 3, is potent to generate—and instantaneously sustain—any given uniform rectilinear motion in any direction; whence the parallelogram's diagonal will denote, both in magnitude and direction, the new uniform rectilinear motion ensuing after the *vis centripeta* is impressed.

since the force N acts following the line AC parallel to BD, this force will do nothing to change the speed of approach to that line BD generated by the other force. The body will therefore approach the line BD in the same time whether the force N be impressed or not, and hence at the end of that time will be found somewhere in that line BD. By the same reasoning it will at the end of the same time be found somewhere in the line CD, and so must of necessity be found at the meet D of both lines.

Corollary 2.(20)

Obvious from this is the compounding of the direct force AD from any oblique forces AB and BD and, conversely, the resolving of any direct force AD into any oblique ones AB and BD whatsoever.(21) *The validity of this composition and resolution is, indeed, abundantly confirmed from mechanics.*

If, for instance, the unequal spokes OM, ON going out from the centre O of some wheel support weights A and P by the threads MA and NP, and the forces of the weights needed to move the wheel be sought: through the centre O draw the straight line KOL meeting the threads perpendicularly at K and L, then on centre O and with radius the greater, OL, of the intervals OK, OL describe a circle meeting the thread MA in D; to the straight line OD, when drawn, let AC be parallel and CD perpendicular. Seeing that it matters not at all whether the points K, L and D in the threads be fixed to the plane of the wheel or not, the effect of the weights will be the same as if they were suspended from the points K and L or D and L. Let the total force of the weight A, however, be represented by the line AD and this will be resolved into the forces AC and CD, of which AC by drawing the spoke DO directly from the centre is without effect to move the wheel; but the other force CD by dragging the spoke DO perpendicularly has the same effect as if it perpendicularly dragged the spoke OL equal to DO, that is, the same as the weight P should that be to the weight A as the force CD to the

(20) As an immediate corollary to the preceding argument since the instantaneous speeds generated in given time from rest by single forces may be compounded by the parallelogram rule, so too may the forces themselves which they represent both in direction and relative magnitude. Because it lends itself readily to constructing equations of accelerative motion at a point, this corollary finds frequent application both in the present 'De motu Corporum' and even more so in the published *Principia*. The three final paragraphs of the explanatory text which develops the corollary's bare enunciation were afterwards much altered by Newton, redrafted by his amanuensis Humphrey Newton (on ULC. Add. 3965.2: 15^r), further recast by Newton (on ULC. Add. 3965.3: 20^v) and then copied out by Humphrey (on ULC. Dd. 9.46: 14^v) in the final form in which it went to press in 1686 (*Principa*, $_1$1687: 15–16).

(21) Humphrey Newton first began to write in sequel 'Obliquas velocitates ...', evidently copying some half-cancelled phrase in Newton's original text—perhaps 'Obliquas velocitates [eandem compositionem et resolutionem habere]' (Oblique speeds have the same composition and resolution) in repetition of Corollary 1?

(22) The manuscript has 'DE' both times.

DA[,] id est (ob similia triangula *ADC*, *DOK*) ut *OK* ad *DO* (seu *OL*).[23] Pondera igitur *A* et *P* quæ sunt reciproce ut radij in directum positi *OK* et *OL* idem pollebunt et sic consistent in æquilibrio (quæ est proprietas notissima Libræ Vectis et Axis in Peritrochio[24])[,] sin pondus alterutrum sit majus quam in hac ratione erit vis ejus ad movendam rotam tanto major.

Quod si pondus *p* ponderi *P* æquale partim suspendatur filo *Np*, partim incumbat plano obliquo *pG*: agantur *pH*, *NH*, prior horizonti posterior plano *pG* perpendicularis, et si vis ponderis *p* deorsum[25] exponatur per lineam *pH*, resolvetur hæc in vires *pN*, *HN*. Pondus autem vi *pN* trahit filum directe et vi *HN* urget planum *pG* huic vi directè oppositum. Unde tensio fili hujus obliqui erit ad tensionem fili alterius perpendicularis *PN* ut *pN* ad *pH*.[26] Ideoq; si pondus *p* sit ad pondus *A* in[27] ratione reciproca minimarum distantiarum filorum suorum *AM*, *pN* a centro rotæ et ratione directa *pH* ad *pN* conjunctim; pondera idem valebunt ad rotam movendam, atq; adeo se mutuo sustinebunt, ut quilibet experiri potest.

Simili virium divisione[28] innotescit vis qua pondus simul urget plana duo oblique quæ et vis cunei est. Nam erecto ad lineam *pN* plano perpendiculari *pQ*, si corpus *p* planis *pQ*, *pG* utrinq; incumbat, hoc inter plana illa consistens, rationem habebit cunei inter corporis fissi facies internas: vis autem qua urget planum vel faciem *pQ*, eadem erit qua, sublato hoc plano, distenderet filum *pN*,

(23) We invert an erroneous inversion 'ut *DO* (seu *OL*) ad *OK*' (as *DO* or *OL* to *OK*). The slip passed into print in the *Principia* ($_1$1687: 14) but was caught in the concluding *Errata sensum turbantia* (*ibid.*: [511]).

(24) Three of the classical 'mechanical powers'—usually discussed in conjunction with the 'Trochlea' (pulley), 'Cuneus' (wedge) and, more rarely, the 'Cochlea' (screw). Knowledge of these in the seventeenth century derived variously from, in the main, the mechanical *addendum* to Book 8 of Pappus' *Mathematical Collection;* compare P. Ver Eecke, *Pappus d'Alexandrie: La Collection Mathématique*, **2** (Brussels, 1933): 837 and especially 873–80. As we have seen (IV: 222–4) Newton was familiar with the Manolessi edition (Bologna, 1660) of Commandino's Latin version of Pappus' work (first published at Pesaro in 1588), but he also had in his library a presentation copy (now Trinity College. NQ. 16.149.) of John Wallis' *Mechanica: sive, De Motu Tractatus Geometricus*, (London, 1669–71) [= *Opera Mathematica*, **1** (Oxford, 1695): 571–1063], where the 'Libra' (*Pars prima*, Caput III), 'Vectis' (*Pars tertia*, Caput VI) and 'Axis in Peritrochio' (*Pars tertia*, Caput VII) are treated at exhaustive length.

(25) Newton subsequently inserted a clarifying participle 'tendens' (tending).

(26) This sentence was afterwards replaced (on f. 14^v) by: 'Si filo *pN* perpendiculare esset planum aliquod *pQ* secans planum alterum *pG* in linea ad horizontem parallela; et pondus *p* his planis *pQ*, *pG* solummodo incumberet; urgeret illud hæc plana viribus *pN*, *HN* perpendiculariter, nimirum planum *pQ* vi *pN* et planum *pG* vi *HN*. Ideoq; si tollatur planum *pQ* ut pondus tendat filum, quoniam filum sustinendo pondus, jam vicem præstat plani sublati, tendetur illud eadem vi *pN* qua planum antea urgebatur. Unde tensio fili hujus obliqui erit ad tensionem fili alterius perpendicularis *PN*, ut *pN* ad *p*H' (If there were, perpendicular to the thread *pN*, some plane *pQ* cutting the other plane *pG* in a line parallel to the horizontal, and the weight *p* were to recline solely on these planes *pQ* and *pG*, it would press these planes per-

force DA, that is (because of the similar triangles ADC, DOK) as OK to DO or OL.[23] Weights A and P, then, which are reciprocally as the spokes OK and OL stationed in line will be equipollent and thus stand still in equilibrium (which is the very well known property of the balance, the lever and the windlass[24]), but if either weight be greater than it is in this ratio, its force to move the wheel will be that much greater.

If now the weight p, equal to the weight P, be partly suspended on the thread Np, partly leant on the oblique plane pG: draw pH and NH, the former perpendicular to the horizontal, the latter to the plane pG, and if the force of the weight p[25] downwards be represented by the line pH, this will be resolved into the forces pN, HN. By the force pN, however, the weight draws the thread directly, while by the force HN it presses on the plane pG directly opposed to this latter force. Hence the tension in this oblique thread will be to the tension in the other, perpendicular thread PN as pN to pH.[26] Consequently if the weight p be to the weight A in[27] the reciprocal ratio of the least distances of their respective threads pN, AM from the centre of the wheel and in the direct ratio of pH to pN jointly, the effectiveness of the weights to move the wheel will be the same and hence they will mutually support each other, as anyone can test.

By a similar splitting of forces[28] there may be ascertained the force with which a weight simultaneously presses on two oblique planes—the force of a wedge, that is. For, on erecting the plane pQ perpendicular to the line pN, if the body p leans on its either side on the planes pQ, pG, it will, as it lies still between those planes, have the relationship of a wedge between the internal faces of a cleft body: here the force with which it presses the plane or face pQ will be the same as that with which, after this plane was removed, it would stretch the thread pN,

---

pendicularly with the forces pN and HN; namely, the plane pQ with the force pN, and the plane pG with the force HN. Consequently, if the plane pQ be removed, so that the weight shall stretch the thread, because the thread, by supporting the weight, now fills the rôle of the plane taken away, it will stretch it with the same force pN as the plane was previously pressed with. Whence the tension of this oblique thread will be to the tension of the other, perpendicular thread PN as pN to pH).

(27) Newton later inserted the pedantic clarification 'ratione quæ componitur ex' (the ratio which is compounded of).

(28) Subsequently minimally altered to the equivalent 'Per similem virium divisionem' (Through a similar...), and then the present paragraph was considerably recast to read (on f. 14ᵛ): 'Pondus autem p planis illis duobus obliquis incumbens rationem habet cunei inter corporis fissi facies internas: et inde vires cunei et mallei innotescunt, utpote cum vis qua pondus p urget planum pQ sit ad vim qua idem vel gravitate sua vel ictu mallei impellitur secundum lineam pH in plano ut pN ad pH, atq...' (The weight p, however, reclining on those two oblique planes plays the part of a wedge between the interior faces of a body which is split; and thereby the forces of a wedge and its hammer may be defined, seeing that the force with which the body p presses the plane pQ shall be to the force with which it is driven, either by its gravity or by a hammer blow, along the line pH in the plane as pN to pH, and so...).

atq; adeo est ad vim qua vel pondere suo vel ictu mallei, impellitur secundum lineam Hp in plano ut pN ad pH, et ad vim qua urget planum alterum pG up pN ad NH. Sed et vis Cochleæ simili virium divisione colligitur: quippe quæ cuneus est a vecte impulsus.

Usus igitur Corollarij hujus latissimè patet,[29] et latè patendo veritatem ejusdem evincit, cùm pendeat ex jam dictis Mechanica tota. Ex hisce enim innotescunt vires machinarum (ab Authoribus diversimodè demonstratæ)[30] quæ ex rotis, tympanis, trochleis, vectibus, radijs volubilibus, nervis tensis & ponderibus directè vel obliquè ascendentibus ac descendentibus, cæterisq; potentijs Mechanicis componi solent; ut et vires nervorum a musculis contractis tensorum ad animalium ossa movenda.

Corol. 3.

Quantitas motûs quæ colligitur capiendo summam motuum factorum ad eandem partem et differentiam factorum ad contrarias non mutatur ab actione corporum inter se.[31] ......

Corol. 4.

Commune gravitatis centrum ab actionibus corporum inter se non mutat statum suum vel motus vel quietis, et propterea corporum omnium in se mutuò agentium (exclusis actionibus et impedimentis externis) commune centrum gravitatis vel quiescit vel movetur uniformiter in directum.[32] ......

Corol. 5.

Corporum dato spatio inclusorum ijdem sunt motus inter se, sive spatium illud quiescat, sive moveatur uniformiter in directum absq; motu circulari.[33] ......

---

(29) Compare note (20) above. In the published *Principia* this corollary is indeed cited more frequently than all the remaining Corollaries 1 and 3–6 together.

(30) Newton evidently refers not merely to Wallis' exhaustive technical account of such simple *machinæ* in his *Mechanica* (see note (24)) but also to more popular accounts such as John Wilkins' *Mathematical Magick: or, The Wonders that may be performed by Mechanical Geometry* (London, 1648) and equivalent continental compendia of mechanical inventions. This parenthesis was subsequently transferred into the previous sentence.

(31) An amplified assertion (compare note (16)) that the total 'motion' (momentum) is conserved in any elastic interaction between bodies. In sequel we omit (as foreign to our purpose) Newton's evident justification of this principle ('Etenim actio eiq; contraria reactio æquales sunt per Leg. 3...') and a further exemplification of it in the linear 'concourse' and 'reflection' after impact of bodies.

(32) An extended revise of 'Lex 4' of §1, Appendix 1, there justified by appeal to 'Lex 3' (= Corollary 5 following in redraft). In sequel we again omit Newton's very evident justification that 'si puncta duo progrediantur uniformi cum motu in lineis rectis et distantia eorum dividatur in ratione data, punctum dividens vel quiescet vel progredietur uniformiter in

and hence is to the force with which, either by its own weight or by a mallet blow, it is impelled following the line *Hp* in the plane as *pN* to *pH*, and so to the force with which it presses on the other plane *pG* as *pN* to *NH*. The force of a screw, moreover, is gathered from a similar splitting of forces, for it is a wedge impelled by a lever.

The use of this corollary, therefore, stretches very wide(29) and the breadth of its extent proves its truth since the whole of mechanics derives from what has just now been stated. For from it (as is demonstrated in different ways by differing authors)(30) are ascertained the forces of machines commonly made up of wheels, drums, pulleys, levers, revolvable spokes, stretched sinews and weights directly or obliquely ascending and descending, and other mechanical powers; and also the forces of the tendons to move by muscular contraction the bones of animals.

Corollary 3.

The quantity of 'motion' which is gathered by taking the sum of 'motions' made in one direction and the difference of those made the opposite way is not altered by the interaction of bodies with each other.(31) . . . . . .

Corollary 4.

The common centre of gravity of bodies does not by their interactions change its state of motion or of rest, and in consequence the common centre of gravity of all bodies acting mutually on each other is (when external actions and hindrances are excluded) either at rest or in uniform motion in a straight line.(32) . . . . . .

Corollary 5.

The motions of bodies comprised in a given space are the same with respect to each other whether that space is at rest or moves uniformly in a straight line without any circular motion.(33) . . . .

---

linea recta' (if two points advance with uniform motion in straight lines and their distance be divided in a given ratio, the dividing point will either be stationary or advance uniformly in a straight line). In proof he would here doubtless appeal to paragraphs 28 and 29 of his 1665 Waste Book notes on motion (ULC. Add. 4004: $13^v/14^r$, reproduced on IV: 270–3), though subsequently in his revise for the 'De motu Corporum Liber primus' he added (ULC. Dd. 9.46: 17^r) an explicit forward reference that 'Hoc postea in Lemmate XXI[II] demonstratur in plano et eadem ratione demonstrari potest in loco solido' (This is afterwards in Lemma XXIII demonstrated in the plane case, and can be proved in the same manner in the 'solid' locus) and so it passed into the first edition of the *Principia* ($_1$1687: 18); in the second ($_2$1713: 17) the less than lucid phrase 'in loco solido' was replaced by 'si motus illi non fiant in eodem plano' (if those motions do not take place in the same plane). We further omit a long discussion of the extension of this corollary to a system of several bodies.

(33) An unchanged repeat of 'Lex 3' of §1, Appendix 1 preceding. Newton's following justification (here omitted) appeals in an entirely obvious way to Law II above, since only an impressed accelerative force can alter the total momentum of any system of bodies.

Corol. 6.

Si corpora moveantur quomodocunç inter se et a viribus acceleratricibus æqualibus secundum lineas parallelas urgeantur; pergent omnia eodem modo moveri inter se ac si viribus illis non essent incitata.[34] ......

Schol.

Hactenus principia tradidi a Mathematicis recepta et experientia multiplici confirmata. Per Leges duas primas et Corollaria duo prima adinvenit Galilæus descensum gravium esse in duplicata ratione temporis, et motum projectilium fieri in Parabola, conspirante experientia, nisi quatenus motus illi per aeris resistentiam aliquantulum retardantur.[35] Ab ijsdem Legibus et Corollarijs pendent demonstrata de temporibus oscillantium Pendulorum,[36] suffragante horologiorum experientia quotidiana. Ex his ijsdem et Lege tertia D. Christopherus Wrennus, Johannes Wallisius S.T.D.[37] et D. Christianus Hugenius, hujus ætatis Geometrarum facile Principes, regulas congressuum et reflexionum duorum corporum seorsim adinvenerunt, et eodem fere tempore cum Societate Regia communicarunt inter se (quoad has leges) omnino conspirantes.[38] ...

... ...

[LEMMATA
DE METHODO RATIONUM PRIMARUM ET ULTIMARUM.][39]

Lemma I.

Quantitates, ut et quantitatum rationes, quæ ad æqualitatem dato[40] *tempore constanter tendunt et eo pacto propiùs ad invicem accedere possunt quàm pro data quavis differentia; fiunt ultimò æquales.*[41]

---

(34) Of this corollary, too, we omit Newton's entirely obvious justification—namely, that any linear change, accelerative or not, in a system of bodies as a whole will not alter the positions of the individual bodies relative to each other.

(35) And ignoring, of course, that bodies gravitating to the earth's centre do not fall in exactly parallel lines, as indeed Galileo also assumed in the tract 'De Motu Proiectorum' which he published in the *Giornata Quarta* of his *Discorsi e Dimostrazioni Matematiche intorno à due nuoue scienze* (Leyden, 1638): 236–82. On Thomas Harriot's priority in deriving—and indeed surpassing—this result see the preceding introduction.

(36) See III: 391–401 and compare §3: note (296) below.

(37) 'S[anctæ] T[heologiæ] D[octor]' (Doctor of Sacred Theology). We should not read too much into this bracketing of Wallis' name with those of Wren and Huygens as the master mathematicians of the age, one which is little more than a rhetorical flourish on Newton's part. He was well aware that Wren had ceased to pursue serious mathematical research more than fifteen years before (his swan-song being his 1669 paper on the line-generation of the one-sheet hyperboloid of revolution; see V: 216, note (257)), while his private opinion of Wallis' mathematical ability was distinctly cool and critical.

Corollary 6.

If bodies move in any manner whatsoever with respect to each other and are urged on in parallel directions by equal accelerative forces, they will all proceed to move in the same manner with respect to each other, as if they had not been stirred by those forces.(34) . . . . . .

Scholium.

Thus far I have delivered principles accepted by mathematicians and confirmed by a multiplicity of experiment. By means of the first two Laws and their first two corollaries Galileo discovered that the descent of heavy bodies is in the doubled ratio of the time and that the motion of projectiles takes place in a parabola, in agreement with experience except inasmuch as those motions are somewhat slowed by the air's resistance.(35) From the same Laws and corollaries depend what has been demonstrated regarding the periods of vibrating pendulums,(36) supported by our daily experience of clocks. From these same ones again, along with the third Law, Mr Christopher Wren, Dr(37) John Wallis and Mr Christiaan Huygens—easily the principal geometers of the present age—separately derived rules for the collision and recoil of two bodies, communicating them much about the same time to the Royal Society in forms exactly (in regard to these rules) in agreement with each other.(38) . . .

. . . .

LEMMAS

ON THE METHOD OF FIRST AND LAST RATIOS.(39)

Lemma I.

Quantities, and also ratios of quantities, which tend constantly in a given(40) *time to equality, and are in that manner able to approach one another to within any given difference, come ultimately to be equal.*(41)

(38) See R. Dugas, *La mécanique au XVIIe siècle* (Paris, 1954): 287–93; and compare our Summary in v: 148–9, note (152). We here omit the remainder of this long scholium (=*Principia*, $_1$1687: 20–5) where, irrelevantly to our present purpose, Newton elaborates on the 'experimentum pendulorum' which Wren performed in December 1668 before the Royal Society and 'quod etiam Clarissimus Mariottus libro integro [*Traité de la Percussion, ou du Choc des Corps* (Paris, 1673)] exponere mox dignatus est' (see Dugas' *Mécanique:* 293–8).

(39) In the revised 'De motu Corporum Liber primus' (ULC. Dd. 9.46: 22^r^), where such division of the main text into 'Articula' is first made, the corresponding subhead is 'ARTIC. I./continens/METHODUM RATIONUM PRIMARUM ET ULTIMARUM' (ARTICLE I, containing THE METHOD OF FIRST AND LAST RATIOS), to which Newton subsequently added 'cujus ope sequentia demonstrantur' (by whose help what follows is demonstrated): lacking the original text of the first six lemmas (see note (12) above), we have in default of the exact subtitle introduced an appropriately restored one. In the sequel this group of lemmas does not

Si negas[,] sit earum ultima differentia *D*. Ergo nequeunt propiùs ad æqualitatem accedere quàm pro data differentia *D*, contra hypothesin.

Lemma II.

Si in figura quavis AacE rectis Aa, AE, et curva [a]cE comprehensa, inscribantur parallelogramma(42) *quotcunq Ab, Bc, Cd &c sub basibus AB, BC, CD &c æqualibus et lateribus Bb, Cc, Dd &c figuræ lateri Aa parallelis contenta, et compleantur parallelogramma aKbl, bLcm, cMdn &c, dein horum parallelogrammorum latitudo minuatur et numerus augeatur in infinitum: dico quod ultimæ rationes quas habent ad se invicem figura inscripta AKbLcMdD, circumscripta AalbmcndoE et curvilinea AabcdE sunt rationes æqualitatis.*

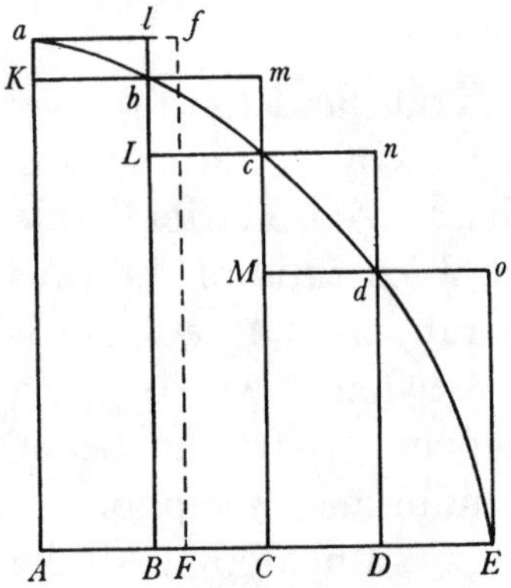

Nam figuræ inscriptæ et circumscriptæ differentia est summa parallelogrammorum

$$Kl+Lm+Mn+Do,$$

hoc est (ob æquales omnium bases) rectangulum sub unius basi *Kb* et altitudinum summa *Aa*, id est rectangulum *ABla*. Sed hoc rectangulum, eò quod latitudo ejus *AB* in infinitum minuitur, fit minus quovis dato. Ergo per Lemma I, figura inscripta et circumscripta et multò magis figura curvilinea intermedia fiunt ultimò æquales. Q.E.D.

Fig 3.

Lemma III.

Eædem rationes ultimæ sunt etiam æqualitatis, ubi parallelogrammorum latitudines AB, BC, CD &c sunt inæquales, et omnes minuuntur in infinitum.

---

in fact everywhere play the central auxiliary rôle which Newton here foresees, and they are only rarely invoked in the new propositions and lemmas which he subsequently introduced into his revised 'De motu Corporum Liber primus' (but see §3: Proposition XLIX; and compare Appendix 3: Lemma XXIX below). It would appear that his initial vision of presenting a logically tight exposition of the principles of motion under accelerative forces faded more and more when he came in detail to cast his arguments, and that he was happy after a while to lapse into the less rigorously justified mode of presentation which he largely exhibits in his published *Principia*. Whatever be the truth of the matter, these lemmas are undeniably a retrospective gloss on the arguments which they now collectively and generally justify, but in whose initial contrivance they played at best a subdued and unstated part. Nor should we unduly stress their novelty, for in seeking to formulate general lemmatical theorems to serve as a basis for the many particular geometrical quadratures, rectifications and limit-equalities he will thereafter make, Newton is solidly in a contemporary tradition comprehending the varied researches and systematisations of such men as Fermat, Blaise Pascal, Huygens, James Gregory and his lately deceased Cambridge colleague Isaac Barrow; compare D. T. Whiteside, 'Patterns of Mathematical Thought in the later Seventeenth Century' (*Archive for History of Exact Sciences*, **1**, 1961: 179–388): Chapters IX/X: 331–55. It ought not, therefore, to be greatly

If you deny it, let their ultimate difference be D. They are therefore unable to approach equality to within the given difference D, contrary to supposition.

Lemma II.

If in any figure ($AacE$) *comprehended by the straight lines* Aa, AE *and the curve* acE *there be inscribed any number of rectangles*[42] $A[K]\,b[B]$, $B[L]\,c[C]$, $C[M]\,d[D]$, ... *contained by equal bases* AB, BC, CD, ... *and sides* Bb Cc, Dd, ... *parallel to the side* Aa *of the figure, and the rectangles* $aKbl$, $bLcm$, $cMdn$, ... *be completed, and then the width of these rectangles be diminished and their number increased indefinitely: I assert that the ultimate ratios which the inscribed figure* $AKbLcMdD$, *the circumscribed one* $AalbmcndoE$ *and the curvilinear one* ($AabcdE$) *have to one another are ratios of equality.*

For the difference of the inscribed and circumscribed figures is the sum of the rectangles $Kl+Lm+Mn+Do$, that is, (because all their bases are equal) the rectangle beneath the base Kb of one of them and the sum Aa of their heights, and so is the rectangle $ABla$. But this rectangle, in consequence of its width AB being indefinitely diminished, comes to be less than any given one. Therefore, by Lemma 1, the inscribed and circumscribed figures and much more so the intervening curvilinear figure come ultimately to be equal. As was to be proved.

Lemma III.

The same ultimate ratios are also ones of equality when the widths AB, BC, CD, ... *of the rectangles are unequal and they are all indefinitely diminished.*

---

surprising that a still unpublished manuscript by James Gregory, written some fifteen years earlier and comprehending 'Geometriæ propositiones quædam generales' (ULE. M16; see (ed.) H. W. Turnbull, *James Gregory Tercentenary Memorial Volume* (London, 1939): 445), effectively duplicates Newton's present Lemmas II, IV, V, VII and IX. What is more original here is his perceptive recognition in the concluding scholium that the measure of curvature at a general point of a curve by comparison with the osculating circle—implicitly invoked in Lemmas IX and XI (and the fundamental following Theorems III and V to which they are keyed)—in fact delimits the nature of the curvilinear arcs to which his lemmas apply, even though he does not clearly explain why he cannot, in the context of central-force orbital dynamics, admit points of 'straightness' or 'infinite curving' into his scheme (on which see note (57) below).

(40) That is, finite. Newton has no desire to consider asymptotic convergence to equality over an infinite period.

(41) A general enunciation of the classical mode of proving equality by 'exhausting' all measurable difference: the demonstration which follows proceeds by a traditional *reductio ad absurdum*.

(42) For convenience in application Newton here, as universally in his later *Principia*, understands that the angle between the base and ordinates to his general (smoothly convex) curve is right. In the next paragraph, correspondingly, the 'parallelogram' $ABla$ is explicitly referred to as a 'rectangulum'. Of course, the present areal comparison—as indeed those displayed in Lemmas III and IV—will hold true, *mutatis mutandis*, where the ordinate angles are equal but oblique.

Sit enim AF æqualis latitudini maximæ et compleatur parallelogrammum $FAaf$. Hoc erit majus quam differentia figuræ inscriptæ et figuræ circumscriptæ, at latitudine sua AF in infinitum diminuta, minus fiet quam datum quodvis rectangulum.

Corol. 1. Hinc summa ultima parallelogrammorum evanescentium coincidit omni ex parte cum figura curvilinea.

Corol. 2. Et multo magis figura rectilinea(43) quæ chordis evanescentium arcuum ab, bc, cd &c comprehenditur coincidit ultimò cum figura curvilinea.

Corol. 3. Ut et figura rectilinea quæ tangentibus eorundem arcuum circumscribitur.

Corol. 4. Et propterea hæ figuræ ultimæ (quoad perimetros acE) non sunt rectilineæ, sed rectilinearum limites curvilinei.

Lemma IV.

Si in duabus figuris AacE, PprT inscribantur (ut supra) duæ parallelogrammorum series, sitq idem amborum numerus, et ubi latitudines in infinitum diminuuntur, rationes ultimæ parallelogrammorum in una figura ad parallelogramma in altera, singulorum ad singula, sint eædem, dico quod figuræ duæ AacE, PprT sunt ad invicem in eadem illa ratione.

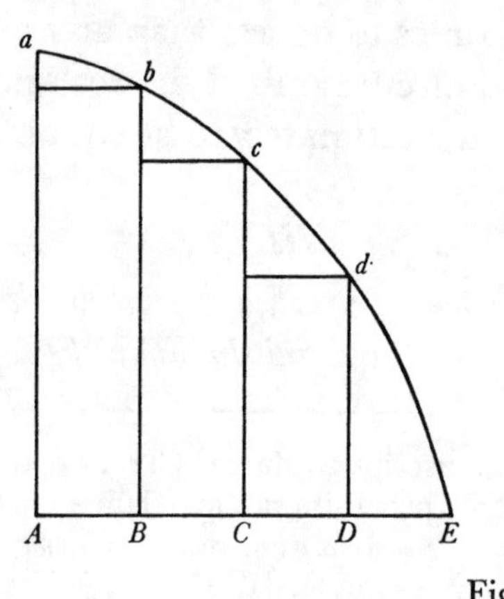

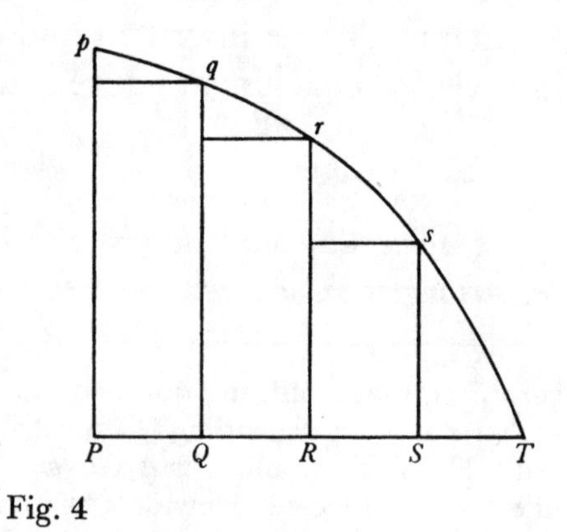

Fig. 4

Etenim ut sunt parallelogramma singula ad singula, ita (componendo) fit summa omnium ad summam omnium, et ita figura ad figuram, existente nimirum figura priore (per Lemma III) ad summam priorem et posteriore figura ad summam posteriorem in ratione æqualitatis.

Corol. Hinc si duæ cujuscunq generis quantitates in eundem partium numerum utcunq dividantur, et partes illæ ubi numerus earum augetur et magnitudo diminuitur in infinitum datam obtineant rationem ad invicem, prima ad primam secunda ad secundam cæteræq suo ordine ad cæteras: erunt tota ad invicem in eadem illa data ratione. Nam si in Lemmatis hujus figuris sumantur parallelogramma inter se ut partes, summæ partium semper erunt ut summæ parallelogrammorum: atq adeò, ubi partium et parallelogrammorum numerus augetur et magnitudo diminuitur in infinitum, in ultima ratione parallelogrammorum.(44)

For let AF be equal to the maximum width, and complete the rectangle $FAaf$. This will be greater than the difference of the inscribed figure and circumscribed figure, and yet, when its width AF is indefinitely diminished, it will come to be less than any given rectangle.

Corollary 1. Hence the ultimate sum of the vanishing rectangles coincides in its every part with the curvilinear figure.

Corollary 2. And much more so the rectilinear figure(43) which is comprised by the chords of the vanishing arcs $\widehat{ab}, \widehat{bc}, \widehat{cd}, \ldots$ coincides ultimately with the curvilinear figure.

Corollary 3. And so too the rectilinear figure which is circumscribed by the tangents of these same arcs.

Corollary 4. And in consequence these ultimate figures (as regards their perimeters $a..c..E$) are not rectilinear, but the curvilinear limits of rectilinear ones.

Lemma IV.

If in two figures AacE, PprT there be inscribed (as above) two series of rectangles, each having the same number, and, when their widths are indefinitely diminished, the ultimate ratios of the rectangles in one figure to corresponding rectangles in the other be individually the same, I assert that the two figures AacE, PprT are to one another in the same ratio.

For as the individual rectangles are to each other, so (by compounding) comes their total sum to be one to the other, and so too the figures, the former figure being of course (by Lemma III) to the former sum and the latter figure to the latter sum in a ratio of equality.

Corollary. Hence if two quantities of any kind be divided in any manner into the same number of parts and, when their number is increased and their size diminished indefinitely, those parts maintain a given ratio to one another, the first to the first, the second to the second, and the rest to the rest in their sequence, then their totals will be to one another in that same given ratio. For if the rectangles in the figures of the present Lemma be taken to each other as those parts, the sums of the parts will always be as the sums of the rectangles; and consequently, when the number of parts and rectangles is increased and their magnitude diminished indefinitely, they will be in the last ratio of the rectangles.(44)

(43) This is evidently the arithmetic mean of the preceding inscribed and circumscribed step-figures ($AKbLc \ldots D$) and ($Aalbmc \ldots oE$).

(44) Newton subsequently recast this to read '...in ultima ratione parallelogrammi ad parallelogrammum, id est (per Hypothesin) in ultima ratione partis ad partem' (...in the last ratio of rectangle to rectangle, that is, (by hypothesis) in the last ratio of part to part).

Lemma V.

Similium figurarum latera omnia quæ sibi mutuo respondent sunt proportionalia, tam curvilinea quàm rectilinea, et areæ sunt in duplicata ratione laterum.

Lemma VI.

Si arcus quilibet positione datus AB subtendatur chordâ AB et in puncto aliquo A in medio curvaturæ continuæ tangatur a recta utrinqꝫ producta AD, dein puncta A, B ad invicem accedant et coeant; dico quod angulus BAD sub chorda et tangente contentus minuetur in infinitum et ultimò evanescet.

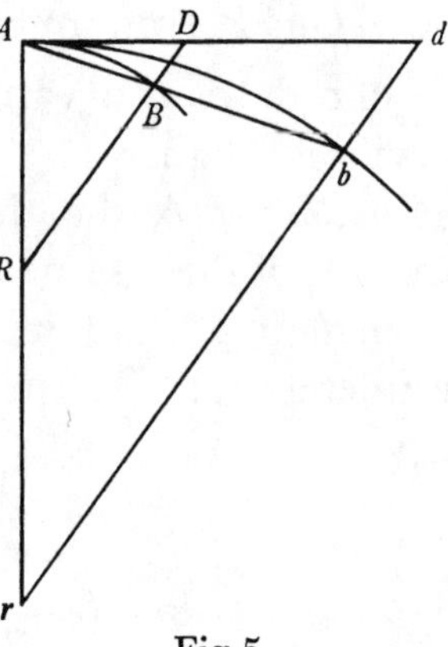

Fig 5.

Nam producatur *AB* ad *b* et *AD* ad *d*, et punctis *A*, *B* coeuntibus, nullaqꝫ adeò ipsius *Ab* parte *AB* jacente amplius intra curvam, manifestum est quod hæc recta *Ab* vel coincidet cum tangente *Ad* vel ducetur inter tangentem et curvam. Sed casus posterior est contra naturam curvaturæ, ergo prior obtinet. Q.E.D.

Lemma VII.

Iisdem positis dico quod ultima ratio arcus chordæ et tangentis(45) *ad invicem est ratio æqualitatis.*

Nam producantur *AB* & *AD* ad *b* et *d* et secanti *BD* parallela agatur *bd*. Sitqꝫ arcus *Ab* similis arcui *AB*. Et punctis *A*, [*b*] coeuntibus, angulus *dAb* per Lemma superius evanescet, adeoqꝫ lineæ *Ab*, *Ad* una cum arcu intermedio *Ab*(46) coincident et propterea æquales erunt. Unde et hisce semper proportionales rectæ *AB*, *AD*, et arcus intermedius *AB* rationem ultimam habebunt æqualitatis. Q.E.D.

Corol. 1. Unde si per *B* ducatur tangenti parallela *BF* rectam quamvis *AF* per *A* transeuntum perpetuo secans in *F*, hæc ultimò ad arcum evanescentem *AB* rationem habebit æqualitatis, eò quod completo parallelogrammo *AFBD* rationem semper habet æqualitatis ad *AD*.

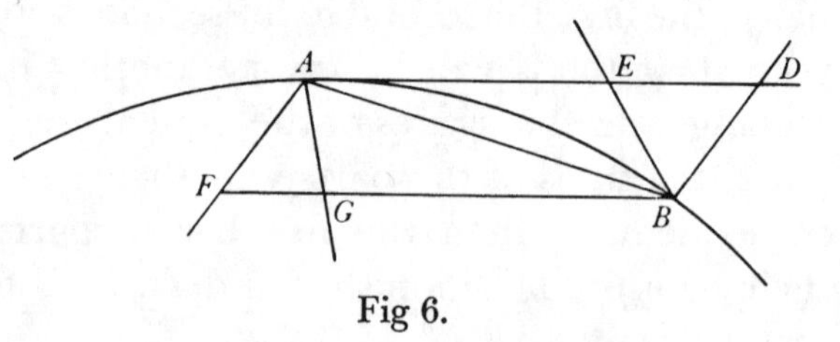

Fig 6.

Corol. 2. Et si per *B* et *A* ducantur plures rectæ *BE*, *BD*, *AF*, *AG* secantes tangentem *AD* et ipsius paral-

(45) Understand this to be the segment (*AD*) cut off in the indefinite tangent between the point of its contact (*A*) and a line (*BD*) drawn, in a fixed direction, through the common endpoint (*B*) of the preceding arc and its chord. We have noticed (IV: 485, note (5)) that Newton had earlier intended to introduce a primitive form of this present Lemma into the revised version of his 'Geometria Curvilinea'.

Lemma V.

In similar figures all mutually corresponding sides, curvilinear as well as rectilinear, are proportional and their areas are in the doubled ratio of their sides.

Lemma VI.

If any arc $\widehat{AB}$ given in position be subtended by the chord AB and at some mean point A in its continuous curvature it be touched by the straight line AD extended either way, and then the points A and B approach one another and coalesce: I assert that the angle $B\widehat{A}D$ contained by the chord and tangent will indefinitely diminish and ultimately vanish.

For produce AB to b and AD to d, and then, as the points A and B coalesce and hence with no part AB of Ab lying any longer within the curve, it is manifest that this straight line Ab will either coincide with the tangent Ad or be drawn between the tangent and the curve. But the latter event is contrary to the nature of curvature, and therefore the former case obtains. As was to be proved.

Lemma VII.

With the same suppositions I assert that the last ratios of the arc, chord and tangent(45) *to each other are ones of equality.*

For extend AB and AD to b and d and parallel to the secant BD draw bd. Let also the arc $\widehat{Ab}$ be similar to the arc $\widehat{AB}$. Then, as the points A and b coalesce, the angle $d\widehat{A}b$ will, by the previous Lemma, vanish and so the lines Ab, Ad together with the intervening arc $\widehat{Ab}$(46) will coincide and accordingly be equal. Hence also the straight lines AB, AD ever proportional to these and their intervening arc $\widehat{AB}$ will have their last ratios one of equality. As was to be proved.

Corollary 1. Hence if through B parallel to the tangent there be drawn BF perpetually cutting in F any straight line AF passing through A, this will ultimately have to the vanishing arc $\widehat{AB}$ a ratio of equality, seeing that, once the parallelogram $AFBD$ is completed, it will always have one of equality to AD.

Corollary 2. And if through B and A there be drawn several straight lines BE, BD, AF, AG intersecting the tangent AD and its parallel BF, the last ratios

(46) Newton afterwards minimally altered this to read 'rectæ *Ab*, *Ad* & arcus intermedius *Ab*' (the straight lines *Ab*, *Ad* and so the intervening arc $\widehat{Ab}$). There is a slight lacuna here since it has not been shown that $\widehat{Ab}$ must be 'intermediate' (understand in length) between Ab and Ad for some suitable direction of the ordinate bd. If, however, we conceive that the tangent to the arc at b meets Ad in δ (compare the figure in IV: 468, note (20)), then it follows by Archimedes' convexity lemmas that $Ab < \widehat{Ab} < A\delta + \delta b$; whence, on setting $\widehat{Abd}$ (as in Newton's figure) to be acute, when $\widehat{Ab}$ is small enough (and so $\widehat{b\delta d}$ sufficiently close to zero) there ensues $\widehat{\delta db} < \widehat{\delta bd}$ and so $\delta b < \delta d$, that is, $\widehat{Ab} < Ad$.

lelam *BF*, ratio ultima abscissarum omnium *AD*, *AE*, *BF*, *BG* chordæq; et arcus *AB* ad invicem erit ratio æqualitatis.

Corol. 3. Et proinde hæ omnes lineæ in omni de rationibus ultimis argumentatione pro se invicem usurpari possunt.

Lemma VIII.

Fig 5. *Si rectæ datæ AR, BR cum arcu AB, chorda AB et tangente AD triangula tria ARB, ARB, ARD constituunt, dein puncta A, B, R accedunt ad invicem: dico quod ultima forma triangulorum evanescentium est similitudinis et ultima ratio æqualitatis.*

Nam producantur *AB*, *AD*, *AR* ad *b*, *d* et *r*. Ipsi *RD* agatur parallela *rbd* et arcui *AB* similis ducatur arcus *Ab*. Coeuntibus punctis *A*, *B*, angulus *bAd* evanescet et proinde triangula tria *rAb*, *rAb*, *rAd* coincident, suntq; eo nomine similia et æqualia. Unde et hisce semper similia et proportionalia *RAB*, *RAB*, *RAD* fient ultimò sibi invicem similia & æqualia. Q.E.D.

Corol. Et hinc triangula illa in omni de rationibus ultimis argumentatione pro se invicem usurpari possunt.

Lemma IX.(47)

Si recta AE et curva AC positione datæ se mutuò secent in angulo dato A, et ad rectam illam in alio dato angulo(48) *ordinatim applicentur DB, EC curvæ occurrentes in B, C; dein puncta B, C accedant ad punctum A: dico quod areæ triangulorum ADB AEC erunt ultimò ad invicem in duplicata ratione laterum.*

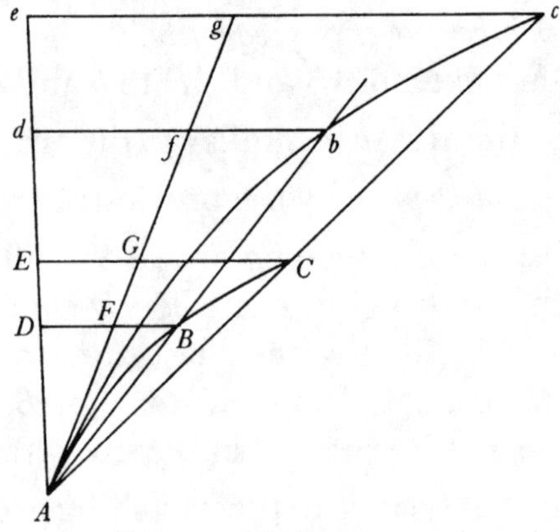

Fig 7.

Etenim in *AD* producta capiantur *Ad*, *Ae* ipsis *AD*, *AE* proportionales, et erigantur ordinatæ *db*, *ec* ordinatis *DB*, *EC* parallelæ et proportionales. Producatur *AC* ad *c*, ducatur curva *Abc* ipsi *ABC* similis, et rectâ *Ag* tangatur curva utraq; in *A* et secentur ordinatæ in *F*, *G*, *f*, *g*. Tum coeant puncta *B*, *C* cum puncto *A*, et angulo *cAg* evanescente, areæ curvilineæ *Abd*, *Ace* coincident cum rectilineis *Afd*, *Age*, adeoq; per Lemma V erunt in duplicata ratione laterum *Ad Ae*. Sed his areis proportionales semper sunt areæ *ABD*, *ACE* et his lateribus latera *AD*, *AE*. Ergo et areæ *ABD*, *ACE* sunt ultimò in duplicata ratione laterum *AD*, *AE*. Q.E.D.(49)

(47) This and Lemma X following derive in an immediate way from Lemma 2 of §1, Appendix 1 (and its unchanged redraft in Appendix 1 below).

(48) In the figure published in the *Principia* (see note (12)) this is, for simplicity (or laziness?), drawn as right.

(49) Newton fails to make it clear that, as the arc $\widehat{ABC}$ (and with it the lengths *AD*, *AE* and the parallels *DFB*, *EGC*) decreases indefinitely, the lines *Ad*, *Ae* remain unchanged in length and

of all the lines AD, AE, BF, BG and the chord AB and its arc to each other will be one of equality.

Corollary 3. And as a consequence all these lines can in any deduction regarding last ratios be mutually employed in one another's place.

Lemma VIII.

If the given straight lines AR and BR constitute with the arc $\widehat{AB}$, *its chord AB and tangent AD the three triangles* $\widehat{AB}R$, *ABR and ADR, and then the points A, B and R approach each other: I assert that the ultimate form of the triangles as they vanish is one of similarity and their last ratio one of equality.*

For extend AB, AD, AR to b, d and r. Draw rbd parallel to RD and the arc $\widehat{Ab}$ similar to arc $\widehat{AB}$. As the points A and B coalesce the angle $b\widehat{A}d$ will vanish, and in consequence the three triangles $\widehat{Ab}r$, Abr and Adr will coincide, and on that head they are similar and equal. Hence also $\widehat{AB}R$, ABR and ADR ever similar and proportional to these will come ultimately to be similar and equal to each other. As was to be proved.

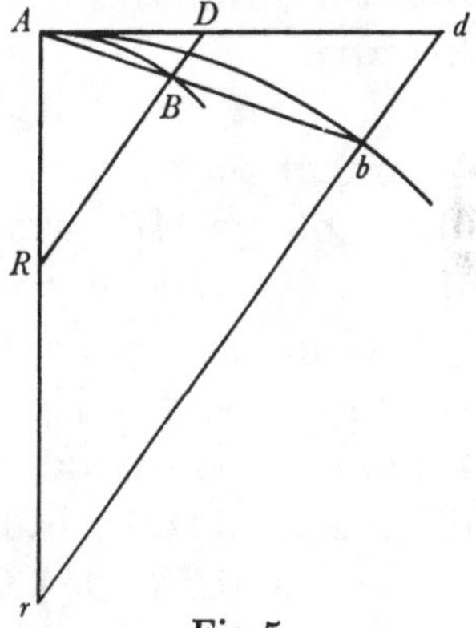

Fig 5.

Corollary. And hence those triangles in every deduction regarding last ratios can mutually be employed in one another's place.

Lemma IX.(47)

If the straight line AE and curve AC given in position mutually intersect at the given angle $\hat{A}$ *and DB, EC be applied as ordinates to that straight line at any given angle,*(48) *meeting the curve in B and C, and if the points B, C then approach the point A: I assert that the areas of the triangles ADB, AEC will be ultimately to one another in the doubled ratio of the sides.*

For in AD produced take Ad, Ae proportional to AD, AE and erect the ordinates db, ec parallel and proportional to the ordinates DB, EC. Produce AC to c, draw the curve $\widehat{Abc}$ similar to $\widehat{ABC}$, and let each curve be touched at A by the straight line Ag and the ordinates be cut by it in F, G, f, g. Then let the points B and C coalesce with the point A and, with the angle $c\widehat{A}g$ vanishing, the curvilinear areas (Abd), (Ace) will coincide with the rectilinear ones Afd, Age and hence, by Lemma V, will be in the doubled ratio of the sides Ad, Ae. But to these areas are ever proportional the areas (ABD), (ACE) and to these sides the sides AD, AE. Therefore also the areas (ABD), (ACE) are ultimately in the doubled ratio of the sides AD, AE. As was to be proved.(49)

the parallels dfb, egc fixed in position; in other words, that the configuration $Abcdefg$ is a finite 'blow-up' of $ABCDEFG$ as the latter configuration becomes vanishingly small, and hence the

Lemma X.

Spatia quæ corpus urgente quacunq; vi centripeta(50) *describit, sunt ipso motus initio in duplicata ratione temporum.*

Exponantur tempora per lineas *AD*, *AE* et velocitates genitæ per ordinatas *DB*, *EC*. Et spatia prima his velocitatibus descripta erunt ut areæ primæ *ABD*, *ACE* his ordinatis descriptæ, hoc est(51) (per Lemma IX) in duplicata ratione temporū *AD*, *AE*. Q.E.D.(52)

Lemma XI.

Subtensa evanescens anguli contactûs est ultimò in ratione duplicata subtensæ arcus contermini.

Cas. 1. Sit arcus ille *AB*, tangens ejus *AD*, subtensa anguli contactus ad tangentem perpendicularis *BD*, subtensa arcus(53) *AB*. Huic subtensæ *AB* et tangenti *AD* perpendiculares erigantur *AG*, *BG* concurrentes in *G*, dein accedant puncta *D*, *B*, *G* ad puncta *d*, *b*, *g* sitq; γ intersectio linearum *BG*, *AG* ultimò facta ubi puncta *D*, *B* accedunt usq; ad *A*. Manifestum est quod distantia $G\gamma$ minor esse potest quàm assignata quævis. Est autem (ex natura circulorū(54) per puncta *ABG*, *Abg* transeuntiū) AB^q æquale $AG\times BD$ et Ab^q æquale $Ag\times bd$, adeoq; ratio AB^q ad Ab^q componitur ex rationibus *AG* ad *Ag* & *BD* ad *bd*. Sed quoniam γG assumi potest minor longitudine quavis assignata, fieri potest ut ratio *AG* ad *Ag* minus differat a ratione æqualitatis quàm differentia quavis assignata, et proinde ut ratio AB^q ad Ab^q minùs differat a ratione *BD* ad *bd* quàm pro differentia quavis assignata. Est ergo, per Lemma I, ratio ultima AB^q ad Ab^q æqualis rationi ultimæ *BD* ad *bd*. Q.E.D.

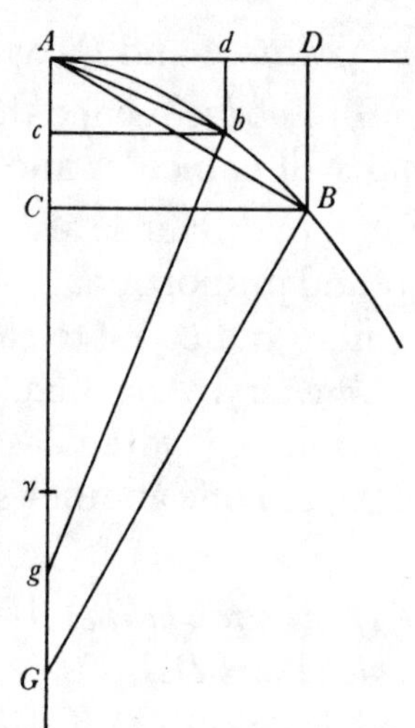

Fig 8.

Cas. 2. Inclinetur jam *BD* ad *AD* in angulo quovis dato et eadem semper erit ratio ultima *BD* ad *bd* quæ priùs adeóq; eadem(55) ac AB^q ad Ab^q. Q.E.D.

---

ratios of its elements represent the 'last' ratios of the latter's elements as they become infinitesimal. In his annotation of the equivalent Lemma in the published *Principia* Leibniz was considerably confused, writing that 'hoc idem statim dici poterat de *B.F.D.C.G.E* quod hic dicitur de assumtis prius *b.f.d.c.g.e.* Itaq; frustra assumuntur' and adding the Horatian quip (here back-firing!) that 'Quandoq; bonus dormitat Homerus'. (These Leibnizian annotations, made about the late summer of 1689 when he first read the *Principia* and was largely ignorant of its subtleties, are now published by E. A. Fellmann in his *G. W. Leibniz: Marginalia in Newtoni Principia Mathematica* (Paris, 1973).)

(50) Subsequently altered by Newton to the unrestricted qualification 'regulari' (standard).

(51) Understand 'ipso motus initio' (at the very beginning of motion), a phrase indeed which Newton here afterwards inserted.

Lemma X.

Spaces which a body describes at the urge of any centripetal[50] *force are, at the very start of motion, in the doubled ratio of the times.*

Represent the times by the lines AD, AE and the speeds generated by the ordinates DB, EC. The first spaces described with these speeds will then be as the first areas (ABD), (ACE) described by these ordinates, and so[51] (by Lemma IX) in the doubled ratio of the times AD, AE. As was to be proved.[52]

Lemma XI.

The vanishing subtense of an angle of contact is ultimately in the doubled ratio of the subtense of the bounding arc.

Case 1. Let $\widehat{AB}$ be the arc, AD its tangent, BD the subtense of the angle of contact perpendicular to the tangent, AB the subtense of the arc.[53] Perpendicular to the latter subtense AB and to the tangent AD erect BG, AG concurring in G; then let the points D, B, G approach the points d, b, g, and let γ be the intersection of the lines BG, AG ultimately occurring when the points D and B approach right up to A. It is obvious that the distance $G\gamma$ can be less than any assigned one. Now (from the nature of the circles[54] passing through the points A, B, G and A, b, g) AB^2 is equal to $AG \times BD$ and Ab^2 equal to $Ag \times bd$, and hence the ratio of AB^2 to Ab^2 is compounded of the ratios of AG to Ag and BD to bd. But since γG can be assumed less than any assigned length, it can be arranged that the ratio of AG to Ag shall differ from one of equality by less than any assigned amount, and consequently that the ratio of AB^2 to Ab^2 shall differ from the ratio of BD to bd by less than any assigned amount. The last ratio of AB^2 to Ab^2 is therefore, by Lemma I, equal to the last ratio of BD to bd. As was to be proved.

Case 2. Let BD be now inclined to AD at any given angle and the last ratio of BD to bd will ever be the same as before, and hence the same[55] as that of AB^2 to Ab^2. As was to be proved.

---

(52) Two corollaries later appended in sequel (on f. 24ᵛ) are reproduced in Appendix 2.1 below.

(53) That is, its chord.

(54) The tacit assumption here made that the small arcs $\widehat{AB}$, $\widehat{Ab}$ may adequately be approximated by corresponding arcs of their osculating circles (tangent at A and passing through B, b respectively) is taken up in the following scholium. Since its radius of curvature $A\gamma$ at A remains finite when the arc $\widehat{AbB}$ becomes vanishingly small, Newton proceeds to approximate the (now unique) osculating circle through A, b and B by the arc $\widehat{AbB}$ of a parabola of parameter $2A\gamma$.

(55) Leibniz here noted in 1689 (compare note (49) above) that 'non videtur sequi', and indeed Newton's conclusion requires further justification: in fact, since (in the limit as B, b vanish into A) BD, bd are infinitely less than AD, Ad, any change in the inclination of the

Cas. 3. Et quamvis angulus D non detur, tamen anguli D, d ad æqualitatem semper vergent & propius accedent ad invicem quàm pro differentiâ quâvis assignatâ, adeoq; ultimò æquales erunt per Lem I, et propterea lineæ BD, bd in eadem ratione ac prius. Q.E.D.

Corol. 1. Unde cùm tangentes AD, Ad, arcus AB, Ab et eorum sinus BC, bc fiunt ultimò chordis AB, Ab æquales; erunt etiam illorum quadrata ultimò ut subtensæ BD, bd.

Corol. 2. Triangula rectilinea ADB, Adb sunt ultimò in triplicata ratione laterum AD, Ad inq; sesquiplicata laterum DB, db: utpote in composita ratione laterum AD & DB, Ad & db existentia. Sic et triangula ABC, Abc sunt ultimò in triplicata ratione laterum BC, bc.

Corol 3. Et quoniam DB, db sunt ultimò parallelæ et in duplicata ratione ipsarum AD, AD; erunt areæ ultimæ curvilineæ ADB, Adb (ex natura Parabolæ) duæ tertiæ partes triangulorum rectilineorum ADB, Adb, et segmenta AB, Ab partes tertiæ eorundem triangulorum.(56) Et inde hæ areæ et hæc segmenta erunt in triplic[at]a ratione tum tangentium AD, Ad tum chordarum et arcuum AB, Ab.

Scholium.

Cæterum in his omnibus intelligimus(57) angulum contactus nec infinitè majorem esse nec infinitè minorem angulis contactuum quos circuli continent cum tangentibus suis, hoc est curvaturam ad punctum A nec infinitè parvam esse nec infinitè magnam, seu intervallum $A\gamma$ finitæ esse magnitudinis. Capi enim potest DB ut AD^3: quo in casu circulus nullus per punctum A inter tangentem AD et curvam AB duci potest, et proinde angulus contactus erit infinitè minor circularibus. Et simili argumento si fiat DB successive ut AD^4, AD^5, AD^6, AD^7 &c habebitur series angulorum contactus pergens in infinitum, quorum quilibet posterior est infinitè minor priore. Et si fiat DB successive ut AD^2, $AD^{\frac{3}{2}}$, $AD^{\frac{4}{3}}$, $AD^{\frac{5}{4}}$, $AD^{\frac{6}{5}}$, $AD^{\frac{7}{6}}$ &c habebitur alia series infinita angulorum contactus quorum primus est ejusdem generis cum circularibus, secundus infinitè major, et quilibet posterior infinitè major priore. Sed et inter duos quosvis ex his angulis potest series utrinq; in infinitum pergens angulorum intermediorum inseri quorum quilibet posterior erit infinitè major priore. Ut si inter terminos AD^2 et AD^3 inseratur series $AD^{\frac{13}{6}}$. $AD^{\frac{11}{5}}$. $AD^{\frac{9}{4}}$. $AD^{\frac{7}{3}}$. $AD^{\frac{5}{2}}$. $AD^{\frac{8}{3}}$. $AD^{\frac{11}{4}}$.

---

ordinates through B and b to ADd will not appreciably increase or diminish the lengths of AD and Ad. The osculating parabola (see previous note) will now, of course, have its principal diameter (through A) parallel to the ordinates BD, bd.

(56) We need scarcely insist that the standard property to which Newton here appeals was first proved by Archimedes in his *On the Quadrature of the Parabola*: in modern terms, since $BD = k.AD^2$, at once $(ADB) = \int_0^{AD} BD \times \sin \hat{D}.\, d(AD) = \frac{1}{3}AD \times BD \times \sin \hat{D} = \frac{2}{3}\Delta ADB$, whence $(ADB):\Delta ADB:$ segment $(BD) = 2:3:1$.

Case 3. And even though the angle $\hat{D}$ be not given, the angles $\hat{D}$, $\hat{d}$ will still always verge towards equality and approach nearer to each other than is determined by any prescribed difference and hence, by Lemma I, will ultimately be equal, and accordingly the lines BD, bd will be in the same ratio as before. As was to be proved.

Corollary 1. Whence, since the tangents AD, Ad, arcs $\widehat{AB}$, $\widehat{Ab}$ and their sines BC, bc come ultimately to be equal to the chords AB, Ab, their squares also will be ultimately as the subtenses BD, bd.

Corollary 2. The rectilinear triangles ABD, Abd are ultimately in the tripled ratio of the sides AD, Ad and in the sesquialteral one of the sides DB, db (being, namely, in the compounded ratio of the sides AD and DB, Ad and db). So also the triangles ABC, Abc are ultimately in the tripled ratio of the sides BC, bc.

Corollary 3. And, seeing that DB, db are ultimately parallel and in the doubled ratio of AD, Ad, the ultimate curvilinear areas (ADB), (Adb) will (from the nature of a parabola) be two-thirds of the rectilinear triangles ADB, Adb, and the segments (AB), (Ab) one-third of the same triangles.(56) And thence these areas and segments will be in the tripled ratio both of the tangents AD, Ad and the chords AB, Ab and their arcs.

Scholium.

We understand(57) throughout, however, that the angle of contact is neither infinitely greater nor infinitely less than the angles of contact which circles contain with their tangents, that is, that the curvature at the point A is neither infinitely small nor infinitely large—in other words that the interval $A\gamma$ is of a finite size. For DB can be taken proportional to AD^3: but in this case no circle can be drawn through the point A between the tangent AD and the curve $\widehat{AB}$, and accordingly the angle of contact will be infinitely less than circular ones. And by a similar argument, if DB be made successively proportional to AD^4, AD^5, AD^6, AD^7, ..., there will be had a sequence of angles of contact continuing to infinity, any succeeding one of which is infinitely less than the one before it. While if DB be made successively proportional to AD^2, $AD^{\frac{3}{2}}$, $AD^{\frac{4}{3}}$, $AD^{\frac{5}{4}}$, $AD^{\frac{6}{5}}$, $AD^{\frac{7}{6}}$, ..., there will be had a second infinite sequence of angles of contact, the first of which is of the same class as circular ones, the second infinitely greater than it, and any succeeding one infinitely greater than the one before. But also between any two of these angles can be inserted a sequence of intermediate angles, continuing either way to infinity, any succeeding one of which will be infinitely greater than the one before: for instance, between the terms AD^2 and AD^3 may be inserted the sequence ..., $AD^{\frac{13}{6}}$, $AD^{\frac{11}{5}}$, $AD^{\frac{9}{4}}$, $AD^{\frac{7}{3}}$, $AD^{\frac{5}{2}}$, $AD^{\frac{8}{3}}$, $AD^{\frac{11}{4}}$,

(57) For some reason Newton subsequently altered this to 'supponimus' (we suppose).

$AD^{\frac{14}{5}}$. $AD^{\frac{17}{6}}$. &c.(58) Sed et rursus inter binos quosvis angulos hujus seriei inseri potest series nova angulorum intermediorum ab invicem infinitis intervallis differentium. Neq; novit natura limitem.(59)

Quæ de curvis lineis deq; superficiebus comprehensis demonstrata sunt facile applicantur ad solidorum superficies curvas et contenta.(60) Præmisi vero hæc Lemmata ut effugerē tædium deducendi perplexas demonstrationes more veterum Geometrarum ad absurdum. Contractiores enim redduntur demonstrationes per methodum indivisibilium.(61) Sed quoniam durior est indivisibilium Hypothesis, et propterea methodus illa minus Geometrica censetur, malui demonstrationes rerum sequentiū ad ultimas quantitatum evanescentium summas & rationes primasq; nascentium[,] id est ad limites summarum & rationū deducere et propterea limitum illorum demonstrationes qua potui brevitate præmittere. His enim idem præstatur quod per methodum indivisibilium et principiis demonstratis jam tutius utemur. Proinde in sequentibus siquando quantitates tamquam ex particulis constantes consideravero, vel si pro rectis usurpavero lineolas curvas, nolim indivisibilia sed evanescentia divisibilia, non summas et rationes partium determinatarum sed summarum & rationum

(58) The indices of AD here are $\ldots, 2\frac{1}{6}, 2\frac{1}{5}, 2\frac{1}{4}, 2\frac{1}{3}, 2\frac{1}{2} = 3-\frac{1}{2}, 3-\frac{1}{3}, 3-\frac{1}{4}, 3-\frac{1}{5}, 3-\frac{1}{6}, \ldots$.

(59) This excursive paragraph on the infinite orders of infinitesimal angles of contact at a point echoes a similar aside in the 1671 tract where Newton likewise concludes (III: 166) that 'patet curvas in quibusdam punctis posse infinitè rectiores esse vel infinitè curviores quolibet circulo et tamen formam curvarum non ideo amittere'. The generalisation implied in this last sentence is not perhaps entirely true. For if, with respect to the perpendicular coordinates $AD = x$ and $DB = y$, we consider a sequence of curves $y = x^m$ $(m > 0)$ through the origin $A(0, 0)$, because their radius of curvature at a general point is $(1+m^2x^{2(m-1)})^{\frac{3}{2}}/m(m-1)\,x^{m-2}$, it follows that at A their curvature—and hence angle of contact with their tangent—is to that of a comparable curve $y = x^n$ there in the ratio $\lim_{x\to 0}\left[\frac{m(m-1)}{n(n-1)}x^{m-n}\left(\frac{1+n^2x^{2(n-1)}}{1+m^2x^{2(m-1)}}\right)^{\frac{3}{2}}\right]$ to 1, and hence infinitely greater than it for $1 \leqslant m \leqslant n$ (the range in which all of the instanced sequences lie) and for $m < n \leqslant 1$ and also for $m < 1 < n$ when $2(1-m) > n-1$, but is infinitely less in this last case when $2(1-m) < n-1$, while when $2(1-m) = n-1$ it has the finite value $-1/2m^2n$. (Whether in fact Newton would allow that the contact angles of $y = x^m$ and $y = x^n$, $m < 1 < n$, are thus geometrically comparable is doubtful: he would probably rather argue that, since by mere interchange of x and y the curve $y = x^m$, $m < 1$, is converted into $x = y^m$ and so $y = x^{m'}$, $m' = 1/m > 1$, it is significant to consider only those curves for which $m > 1$.) Much more importantly, having elaborated the point that an infinity of curves having points of zero or infinite curvature exist, Newton rather harshly leaves his reader to understand for himself why curves possessed of such points need here, in a context of central-force dynamics, to be excluded from consideration. In the case of an orbit traversed freely about a centre under the continuously varying action of a centripetal force whose instantaneous intensity may (see note (86) below) be measured by $v^2/\rho \sin\alpha$, where v is the body's speed at the pertinent point, ρ the radius of curvature of the orbit there, and α the angle between the direction of motion and the radius vector from the point to the force-centre, we may readily see why ρ should be neither zero nor infinite. It is less easy to comprehend why

$AD^{\frac{14}{5}}$, $AD^{\frac{17}{6}}$, ...[58] And yet again between any pair of angles in this sequence there can be inserted a fresh sequence of intermediate angles differing from one another by infinite intervals. Nor does this in its nature admit any bound.[59]

What has been demonstrated regarding curved lines and the areas they comprehend is easily applied to the curved surfaces and contents of solids.[60] I set these lemmas in introduction to avoid the monotony of adducing complicated proofs by *reductio ad absurdum* in the manner of the ancient geometers. For, of course, proofs are rendered more compact by the method of indivisibles.[61] Yet, because the hypothesis of indivisibles is a rather harsh one, and for this reason that method is reckoned less geometrical, I have preferred to reduce proofs of following matters to the last sums and ratios of vanishing quantities and the first ones of nascent quantities, that is, to the limits of these sums and ratios, and accordingly to set the proofs of those limits in introduction with all possible conciseness. For by these the same result is achieved as by the method of indivisibles and we may more safely employ them as principles now that they are proved. Consequently, whenever in the sequel I consider quantities as consisting of particles, or if ever in place of straight lines I use minute curved ones, I wish it always to be understood that these are not indivisibles but vanishing divisibles, not the sums and ratios of definite parts but the limits of

a curve possessed equivalently of points of 'infinite cavity' or 'straightness' may not be permitted to constrain the accelerative motion impressed by some central force: to be sure, Newton himself obeyed no such restriction when later, in Propositions XLIX–LII of his amplified 'De motu Corporum Liber primus' (§3 following), he came to discuss the isochronous motion so induced by a direct-distance force in a hypocycloid, the curvature at whose cusps is infinitely great.

(60) With appropriate changes, of course; Lemma V, for instance, must be adapted to state that the surface-areas and volumes of similar solids are respectively as the squares and cubes of their corresponding linear elements. The extension is of little more than theoretical interest: even in the published *Principia* the only solid figures which come into consideration are the simplest ones formed by the revolution of a plane figure about an axis, and for these no general lemmas on surface-area and solid content beyond those known to Archimedes are required.

(61) Understand absolutely small (but non-zero) quantities which are, in a Euclidean sense (see note (68) below) unmeasurable by any finite magnitudes, even those vanishingly infinitesimal. Whether in fact any of the many seventeenth-century geometers—Descartes, Torricelli, Fermat, Pascal, Wallis, Huygens and even Cavalieri and Leibniz—who for practical convenience appealed to 'indivisibles' of continuous magnitudes in their proofs ever conceived of these as other than the 'evanescent divisibles' to which they are here contrasted is highly dubious. Newton was not entirely unversed (see the short Archimedean tract reproduced on III: 408–18) in the complexities of constructing classical 'exhaustion' proofs of simple quadratures and rectifications with their monotonous, repetitive appeals to prolix reductions to the absurd, but like Huygens before him (compare the 1658 *cri de cœur* quoted in III: 389, note (3)) had come largely to settle for the concise display of the essence of such demonstrations which the employment of infinitesimals—coupled in his case with his theory of first and last ratios—permitted.

limites semper intelligi; et vim(62) demonstrationum ad methodum præcedentiū Lemmatum semper revocari quamvis brevitatis gratia id non semper exprimatur.

Objectio est, sed futilis admodum,(63) quod quantitatū evanescentium nulla sit ultima proportio; quippe quæ, ante quam evanuerunt, non est ultima, ubi evanuerunt nulla est. Eodem argumento dici potest(64) nullam esse corporis ad certum locum pergentis velocitatem ultimam. Hanc enim antequam corpus attingit locum non esse ultimam, ubi attigit nullam esse. Sed responsio facilis est. Per velocitatem ultimam intelligi eam qua corpus movetur neq antequam attingit locum ultimum et motus cessat, neq posteà, sed tunc cum attingit, id est illam ipsam velocitatem quacum corpus attingit locum ultimum & quâcum motus cessat.(65) Et similiter per ultimam rationem quantitatum evanescentiū intelligenda est ratio quantitatum non antequam evanescunt, non postea, sed quacum evanescunt. Pariter et ratio prima nascentium est ratio quacum nascuntur. Et summa prima et ultima est quacum esse (vel augeri et minui) incipiunt et cessant. Extat limes quem velocitas in fine motus attingere potest[,] non autem transgredi. Hæc est velocitas ultima. Et par est ratio(66) quantitatum et proportionum omnium incipientium et cessantium. Cumq hic limes sit certus et determinatus,(67) Problema est verè Geometricum eundem determinare. Geometrica verò omnia in alijs Geometricis determinandis ac demonstrandis legitimè usurpantur.(68)

Propositiones

De motu corporum in spatijs non resistentibus.(69)

Prop. I. Theorema I.(70)

Gyrantia omnia, radijs ad centrum(71) *ductis, areas temporibus proportionales describere.*

---

(62) A clarifying 'talium' (such) was afterwards here inserted.

(63) After some hesitation and minor redrafting of it, Newton later deleted this rather disdainful phrase. Far from being ineffectual, of course, the objection which follows has considerable weight and he can in sequel refute it only by specifying his contrary notion of continuity: namely, by defining the limit-value of a varying quantity at a point in its variation where it is indeterminate to be the Dedekindian meet of the continua of its values immediately before and after this 'threshold' point of discontinuity.

(64) This opening was later attenuated to read 'Sed et eodem argumento æque contendi posset' (Yet also by the same argument it might equally be contended).

(65) In our English version we have brought out the (limit) instantaneity of the referent time-interval, which is less sharply implied in Newton's Latin phrases 'tunc cum' and 'illam ipsam'.

(66) In later clarification Newton added 'limitis' (of the limit).

(67) Afterwards amended to 'definitus' (defined), probably merely to avoid repetitition with the following infinitive.

(68) In a subsequently inserted final paragraph on f. 27^v (reproduced in Appendix 2.2

such sums and ratios; and that the force of[62] proofs is always to be referred to the method of the preceding lemmas, even though for brevity's sake that may not always be expressly stated.

There is the objection to this—a somewhat futile one, however[63]—that there exists no last proportion of vanishing quantities, inasmuch as, before they vanish, there is no last one while, once they have vanished, there is none at all. By the same argument it can be asserted[64] that there is no last speed of a body proceeding to a specified position: for, before the body reaches the place, there can be no last one while, once it has reached it, there is none at all. But the answer is easy. By its last speed is understood that with which a body is moving, not before it attains its last position and its motion ceases, nor afterwards, but precisely when it reaches it—the exact speed, that is, with which the body reaches its last position and with which its motion ceases.[65] And similarly by the last ratio of vanishing quantities you must understand not the ratio of the quantities before they vanish, nor that afterwards, but that with which they vanish. Correspondingly, the first ratio of nascent quantities is the ratio with which they come into being. And their first and last sum is that with which they begin and cease to be (or be increased and diminished). There exists a limit which their speed can at the end of its motion attain, but not, however, surpass. This is their last speed. And there is a matching ratio[66] of all quantities and proportions beginning and ceasing to be. Since, again, this limit is fixed and determined,[67] it is properly a geometrical problem to determine it. And, to be sure, everything that is geometrical is legitimately employed in determining and demonstrating other geometrical things.[68]

Propositions

on the motion of bodies in non-resisting media.[69]

Proposition I. Theorem I.[70]

All orbiting bodies describe, by radii drawn to their centre,[71] *areas proportional to the times.*

following) Newton points the distinction between vanishingly small magnitudes and the finite limit-ratios which they ever more nearly approximate, urging that whenever in sequel he seems to appeal to 'minimal' or 'vanishing' or 'ultimate (last)' magnitudes these should be understood never to be absolutely 'indivisible'—and hence in Euclidean terms (*Elements*, v, Def. 5 [Barrow]; compare Appendix 2: note (5)) incommensurable with finite magnitudes—but universally to be quantities 'diminishable without limit'.

(69) In the revised 'De motu Corporum Liber primus', where (compare note (39)) a narrower division into 'Articula' is made, this subhead—repeating an equivalent in the preliminary revise (see §1, Appendix 1: note (9))—was replaced by the title 'Artic. II./continens/ Inventionem virium centripetarum' (Article II, embracing the discovery of centripetal forces), now comprehending only the first nine propositions following together with the inserted 'Prob. I' (f. 31^v) which is reproduced in Appendix 2.7. Initially, as in his pristine tract (§1 above) Newton continued to distinguish his various enunciations merely into

Dividatur tempus in partes æquales, et prima temporis parte describat corpus vi insita rectam *AB*. Idem secunda temporis parte si nil impediret [a]rectà pergeret ad *c* describens lineam *Bc* æqualem ipsi *AB*, adeo ut radijs *AS*, *BS*, *cS* ad centrum actis confectæ forent æquales areæ *ASB*, *BSc*. Verùm ubi corpus venit ad *B* agat vis centripeta impulsu unico sed magno, faciatq; corpus a recta *Bc* deflectere et pergere in recta *BC*. Ipsi *BS* parallela agatur *cC* occurrens *BC* in *C*, et completa secunda temporis parte [b]corpus reperietur in *C*. Junge *SC* & triangulum *SBC* ob parallelas *SB*, *Cc* æquale erit triangulo *SBc* atq; adeo etiam triangulo *SAB*. Simili argumento si vis centripeta successivè agat in *C*, *D*, *E* &c faciens corpus singulis temporis momentis singulas describere rectas *CD*, *DE*, *EF* &c[,] triangulum *SCD* triangulo *SBC* et *SDE* ipsi *SCD* et *SEF* ipsi *SDE* æquale erit. Æqualibus igitur temporibus æquales areæ describuntur. Sunto jam hæc triangula numero infinita et infinitè parva, sic, ut singulis temporis momentis singula respondeant triangula, agente vi centripeta sine intermissione, utq; area *SAF* sit summa ultima triangulorū evanescentium, et per Corollarium quartum Lemmatis tertij constabit propositio. Q.E.D.

[a] Lex 1.

[b] Legum Cor. 1.

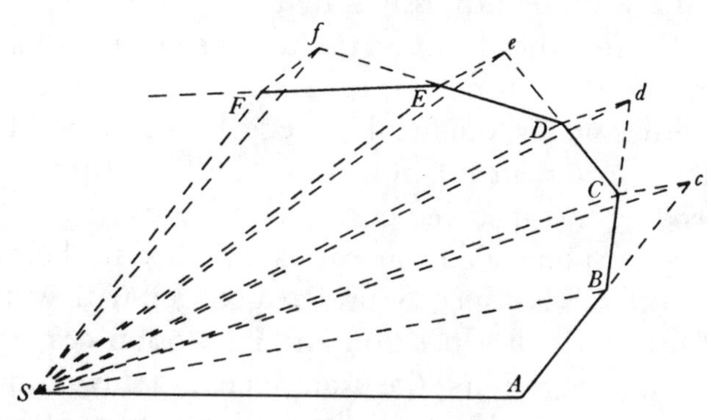

Fig. 9.

Prop. II. Theorema II.

Corpus omne quod movetur in linea aliqua curva et radio ad punctum vel immobile, vel motu rectilineo uniformiter progrediens ducto describit areas[(72)] *temporibus proportionales, urgetur vi centripeta tendente ad idem punctum.*

Cas. 1. Nam corpus omne quod movetur in linea curva, detorquetur de cursu rectilineo per vim aliquam in ipsum agentem (per Leg. 1) et vis illa qua corpus de cursu rectilineo detorquetur et cogitur triangula[(73)] *SAB*, *SBC*, *SCD* &c circa punctū immobile *S*, temporibus æqualibus æqualia describere, agit [a]in loco *B* secundum lineam parallelam ipsi *cC* hoc est secundū lineam *BS*[,] et in loco *C*

[a] per prop. 39[(74)] lib. 1 Elem. et Legem 2.

'Lemmas', 'Theorems' and 'Problems': the somewhat confusing duplicate categorisation of the two latter groups as 'Propositions' (to which the lemmas are now regarded as but auxiliary mathematical riders)—a convention which passed unchanged into the published *Principia*—was here interpolated only in afterthought.

(70) A but minimally augmented repeat of the equivalent 'Theorema 1' in §1 preceding. More considerable changes were made in its enunciation and proof in the revised 'De motu Corporum Liber primus', and to point these we reproduce the amended version in Appendix 2.3 below.

Let the time be divided into equal parts, and in the first part of the time let the body by its innate force describe the straight line *AB*. It would then in the second part of time, were nothing to impede it, proceed directly[a] to *c*, describing the line *Bc* equal to *AB* so as, when rays *AS*, *BS*, *cS* were drawn to the centre, to make the areas *ASB*, *BSc* equal. However, when the body comes to *B*, let the centripetal force act in one single but mighty impulse and cause the body to deflect from the straight line *Bc* and proceed in the straight line *BC*. Parallel to *BS* draw *cC* meeting *BC* in *C*, and when the second interval of time is completed the body will[b] be found at *C*. Join *SC* and the triangle *SBC* will then, because of the parallels *SB*, *Cc*, be equal to the triangle *SBc* and hence also to the triangle *SAB*. By a similar argument, if the centripetal force acts successively at *C*, *D*, *E*, ..., making the body in separate moments of time describe the separate straight lines *CD*, *DE*, *EF*, ..., the triangle *SCD* will be equal to the triangle *SBC*, *SDE* to *SCD*, *SEF* to *SDE* (and so on). In equal times, therefore, equal areas are described. Now let these triangles be infinitely small and infinite in number, such that to each individual moment of time there corresponds an individual triangle (the centripetal force acting now without interruption) and that the area (*SAF*) is the ultimate sum of the vanishing triangles, and the proposition will then, by Corollary 4 of Lemma III, be established. As was to be proved.

[a]Law I.

[b]Corol. 1 of the Laws.

Proposition II. Theorem II.

Every body which moves in some curved line and by a radius drawn to a point, either stationary or advancing uniformly in a rectilinear motion, shall describe areas(72) *proportional to the times is urged on by a centripetal force tending to the same point.*

Case 1. For every body which moves in a curved line is bent aside from its rectilinear path by some force acting upon it (by Law 1) and the force by which the body is bent aside from its rectilinear path and in equal times compelled to describe round a stationary point *S*(73) triangles *SAB*, *SBC*, *SCD*, ... which are equal acts[a] at the place *B* following a line parallel to *cC*, that is, along the line *BS*, and at the place *C* following a line parallel to *dD*, that is, along the line *CS*,

[a]By *Elements* I, 39(74) and Law II.

(71) To avoid any potential confusion on his reader's part, Newton afterwards in his revised 'De motu Corporum Liber primus' specified this to be the 'centrum virium' (force-centre); compare §1: note (17).

(72) Later, having discarded a tentative preliminary qualification of the preceding 'punctum' as the curve's 'centrum [*sc.* virium]' (compare previous note), Newton here equivalently inserted 'circa punctum' (round that point). The necessary further restriction, here understood, that the body's motion shall be 'in plano' was not made explicit till the *Principia's* second edition ($_2$1713: 36).

(73) A necessarily understood 'quam minima' (minimal) was later here inserted.

(74) Since the triangles *SBc*, *SBC*; *SCd*, *SCD*; ... are equal in area and have common bases *SB*, *SC*, ... For some reason, in the published *Principia* ($_1$1687: 38) appeal was here made unnecessarily to the more general 'Prop. 40 Lib. I Elem.' (*Elements* I, 40).

secundum lineam ipsi *dD* parallelam hoc est secundum lineam *CS* &c. Agit ergo semper secundum lineas tendentes ad punctum illud immobile *S*. Q.E.D.

Cas. 2. Et per Legum Corollarium quintum perinde est sive quiescat superficies,(75) in qua corpus describit figuram curvilineam, sive ipsa una cum corpore, figurâ descripta et puncto suo *S*, moveatur uniformiter in directum.

Schol.

Urgeri potest corpus vi centripetâ composita ex pluribus viribus. In hoc casu sensus Propositionis est quod vis illa quæ ex omnibus componitur, tendit ad punctum *S*. Porro si vis aliqua agat secundum lineam superficiei descriptæ perpendicularem, hæc faciet corpus deflectere a plano, sed quantitatem superficiei descriptæ nec augebit nec minuet, et proinde in compositione virium negligenda est.

Prop. III. Theor. III.

Corpus omne quod radio ad alterum corpus utcunq motum ut centrum ducto describit areas(76) *temporibus proportionales, urgetur vi composita ex vi centripeta tendente ad corpus alterum et ex vi*(77) *qua corpus alterum pro mole suo*(78) *urgetur.*

Nam per Legum Coroll. 6 si vi nova quæ æqualis et contraria sit illi qua corpus alterum urgetur, urgeatur corpus utrumq,(79) perget corpus primum describere circa corpus alterum areas easdem ac prius: vis autem qua corpus alterum urgebatur jam destruetur per vim sibi æqualem et contrariam et proinde (per Leg. 1.) corpus illud alterum vel quiescet vel movebitur uniformiter in directum, et corpus primum urgente differentia virium perget areas temporibus proportionales circa corpus alterum describere. Tendit igitur (per Theor. [II]) differentia virium ad corpus illud alterum ut centrum. Q.E.D.

(80)*Coroll. 1.* Hinc si corpus unum radio ad alterum ducto describit areas temporibus proportionales, urgetur hoc corpus nulla alia vi præter compositam illam ex vi centripeta ad corpus alterum tendente, et ex vi omni quæ agit in corpus alterum et in utrumq æqualiter (pro mole corporum) et secundum lineas parallelas agere intelligitur. Namq additio et subductio viriū in hoc Theoremate fit secundum situm linearum, ut in Legū Coroll. 1 exponitur.

Corol. 2. Et ijsdem positis si areæ sint temporibus quamproximè propor-

(75) Understand 'plana' (plane); compare note (72).

(76) As in his preceding theorem (see note (72)), Newton here too subsequently inserted 'circa punctum illud' (round that point). Once more understand that the motion takes place *in plano*.

(77) Later qualified as 'omni acceleratrice' (every accelerative). Newton hesitated whether to insert an explicit mention that this second force acts 'in eandem partem secundum lineam parallelam' (in the same direction along a parallel line) but, evidently deciding that this was obvious, finally omitted to do so.

(78) This somewhat misleading phrase was later deleted.

and so on. It acts, therefore, always along lines tending to that stationary point *S*. As was to be proved.

Case 2. And by Corollary 5 of the Laws it is exactly the same whether the[75] surface in which the body describes the curvilinear figure stays at rest or whether, along with the body, the figure described and its point *S*, it moves uniformly on in a straight line.

Scholium.

A body can be urged by a centripetal force compounded of several forces. In this case the sense of the proposition is that the force which is compounded from all tends to the point *S*. Furthermore, should some force act following a line drawn perpendicular to the surface, this will cause the body to deviate from the plane, but will neither increase nor diminish the size of the surface-area described, and is in consequence to be neglected in compounding the forces.

Proposition III. Theorem III.

Every body which by a radius drawn to another, arbitrarily moving body as centre describes areas[76] *proportional to the times is urged by a force compounded of the centripetal force tending to the other body and of the*[77] *force by which the second body is, in regard to its own mass,*[78] *urged.*

For (by Corollary 6 of the Laws) if a new force which is equal and contrary to that by which the second body is urged shall press each body,[79] the first body will proceed to describe round the second body the same areas as before; however, the force by which the second body was urged will now be annulled by the force equal and contrary to it, and in consequence (by Law I) that second body will either stay at rest or move uniformly on in a straight line, while the first body, urged by the difference of the forces, will proceed to describe areas proportional to the times round the second body. Therefore (by Theorem II) the difference of the forces tends to that second body as centre. As was to be proved.

[80]*Corollary 1*. Hence, if one body by a radius drawn to a second one describes areas proportional to the times, this body will be urged by no force except that compounded of the centripetal force tending to the other body and of every force which acts on the second body and is understood to act equally on each (in regard to their bodily mass) and following parallel lines. For the addition and subtraction of forces occurs in this Theorem along linear alignments, as is set out in Corollary 1 to the Laws.

Corollary 2. And with the same suppositions, if the areas be very nearly pro-

(79) Newton here subsequently inserted 'secundum lineas parallelas' (along parallel lines); compare note (77) above.

(80) In later revise the following corollaries were considerably emended (the first three being wholly rewritten on f. 29v): their corrected text is reproduced in Appendix 2.4 below.

tionales: vis illa communis aut æqualiter agit in corpus utrumq; quamproximè, aut agit secundum lineas quamproxime parallelas, aut perquam exigua est si cum vi centripeta ad corpus alterum tendente conferatur.

Corol. 3. Et vice versa si hæc tria contingunt, corpus radio ad alterum corpus ducto describet areas quamproxime proportionales temporibus.

Coroll. 4. At si corpus radio ad alterum corpus ducto describit areas quæ cum temporibus collatæ sunt valde inæquabiles, et corpus illud alterum vel quiescit vel movetur uniformiter in directum[,] actio vis centripetæ ad corpus illud alterū tendentis miscetur & componitur cum actionibus admodum potentibus aliarum virium. Idem obtinet ubi corpus alterum motu quocunq; movetur, si modo vis centripeta sumatur quæ restat post subductionem vis omnis agentis in corpus illud alterum.

Schol.

Quoniam æquabilis arearum descriptio index est centri cujus vi centripeta(81) corpus maximè afficitur, corpus autē vi ad hoc centrum tendente retinetur in orbita sua, et motus omnis circularis(82) rectè dicitur circa centrum illud fieri cujus vi corpus retrahitur de motu rectilineo et retinetur in orbita: quidni usurpemus in sequentibus æquabilem arearum descriptionem ut indicem centri circa quod motus omnis circularis(82) in spatijs liberis vi cæca naturæ(83) peragitur?

Prop. IV. Theor. IV.(84)

Corporibus in circumferentijs circulorum uniformiter gyrantibus, vires centripetas esse ut arcuum simul descriptorum quadrata applicata ad radios circulorum.

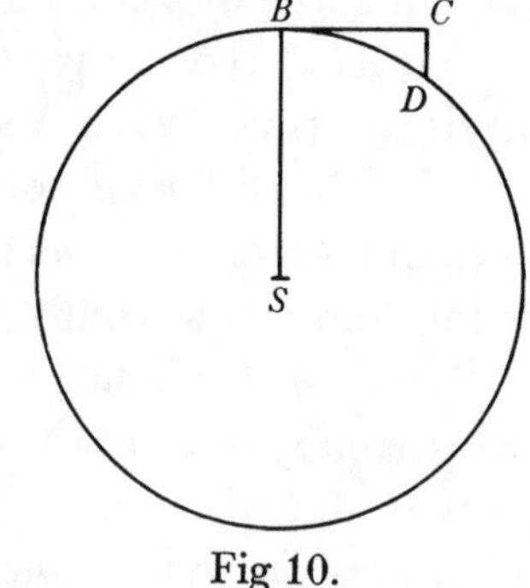

Fig 10.

Corpora B, b in circumferentijs circulorum BD, bd gyrantia, simul describant arcus BD, bd. Sola vi insita describerentur tangentes BC, bc his arcubus æquales: Vires centripetæ sunt quæ perpetuò retrahunt corpora de tangentibus ad circumferentias, atq; adeo hæ sunt ad invicem in ratione prima spatiorum nascentium CD, cd. Fiat figura $\delta\kappa b$ figuræ DCB similis, et per Lemma [V], lineola CD erit ad lineolam cd

(81) Newton afterwards recast this inadequately expressive clause to read '... centri quod vis illa respicit qua' (the centre which is the target of the force whereby).

(82) Understand a general, looping revolution.

(83) Newton later replaced this vivid phrase by the relatively colourless adverbs 'non consultò sed naturaliter' (not purposefully but naturally).

(84) A slightly emended and amplified version of Theorem 2 of the pristine 'De motu Corporum' (§1 above). For ease of argument—if not understanding of the underlying complexities (compare §1: notes (22) and (23))—Newton now assumes that the infinitesimal contemporaneous force-deviations CD, cd may (to adequate approximation, it is implied) be

portional to the times, that common force acts either very nearly equally on each body or following lines very nearly parallel, or is exceedingly slight if it be compared with the centripetal force tending to the second body.

Corollary 3. And conversely, if these three things chance to happen, the body by a radius drawn to the second body shall describe areas very nearly proportional to the times.

Corollary 4. But if the body by a radius drawn to the second body describes areas which, when compared with the times, are markedly unequal while that second body either remains at rest or moves uniformly on in a straight line, the action of the centripetal force tending to that second body is mixed and compounded with the exceedingly powerful actions of other forces. The same circumstance obtains when the second body moves with any motion whatever if but the centripetal force is assumed which remains after subtracting every force acting on that second body.

Scholium.

Seeing that equable description of areas is an indicator of the centre by whose centripetal force(81) a body is most affected, while the body is by the force tending to this centre kept in its orbit, and every circular motion(82) is correctly stated to take place round that centre by whose force the body is drawn aside from its rectilinear motion and kept in orbit: why should we not in the sequel employ the equable description of areas as our criterion for the centre round which all circular motion(82) is by the blind force of nature(83) performed in free space?

Proposition IV. Theorem IV.(84)

Where bodies orbit uniformly in the circumferences of circles, their centripetal forces are as the squares of the arcs simultaneously described, divided by the radii of their circles.

Let the bodies B, b orbiting in the circumferences of the circles BD, bd simultaneously describe the arcs $\widehat{BD}$, $\widehat{bd}$. By their innate force alone they would describe the tangent lines BC, bc equal to these arcs: the centripetal forces are those which perpetually drag the bodies back from the tangents to the circumferences, and hence are to one another in the first ratio of the nascent intervals

taken to be parallel to the corresponding circle radii BS, bs and there is accordingly no longer any need to consider their second meets F, f with the circles, thereafter arguing that, since $\widehat{BD}$, $\widehat{bd}$ are vanishingly small, the lengths $CF = \widehat{BD}^2/CD$ and $cf = \widehat{bd}^2/cd$ may be replaced by the respective diameters $2BS$ and $2bs$. In further revise in his 'De motu Corporum Liber primus' (see Appendix 2.5 below), Newton considerably rephrased the theorem's main text and added two extra corollaries, numbered 3 and 7, the latter of which (see note (86)) adumbrates the extension of this Huygenian measure of *vis centripeta* sustaining motion in a circle about the force-centre to embrace that of a general impressed force acting continuously over a given infinitesimal arc of known curvature towards a given point.

ut arcus BD ad arcum $b\delta$, necnon per Lemma XI(85) lineola nascens $\delta\kappa$ ad lineolam nascentem dc ut $b\delta^{\text{quad}}$ ad bd^{quad}, et ex æquo lineola nascens DC ad lineolam nascentem dc ut $BD \times b\delta$ ad bd^{quad}. Sunt ergo vires centripetæ ut $BD \times b\delta$ ad bd^{quad}, hoc est, ut

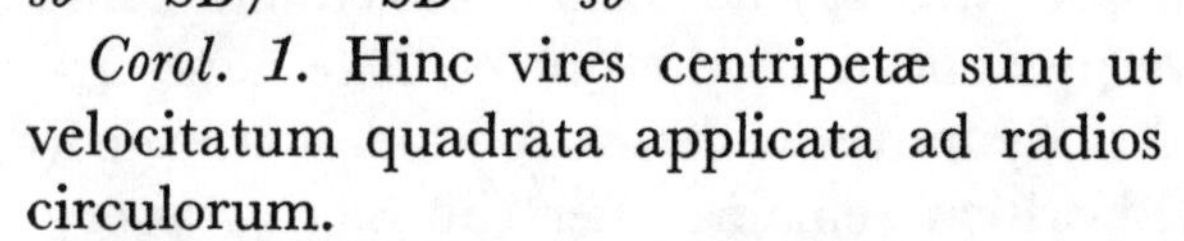

$\frac{BD \times b\delta}{sb}$ ad $\frac{bd^q}{sb}$ adeoq; $\left(\text{ob æquales rationes } \frac{b\delta}{sb} \text{ et } \frac{BD}{SB}\right)$ ut $\frac{BD^q}{SB}$ ad $\frac{bd^q}{sb}$. Q.E.D.

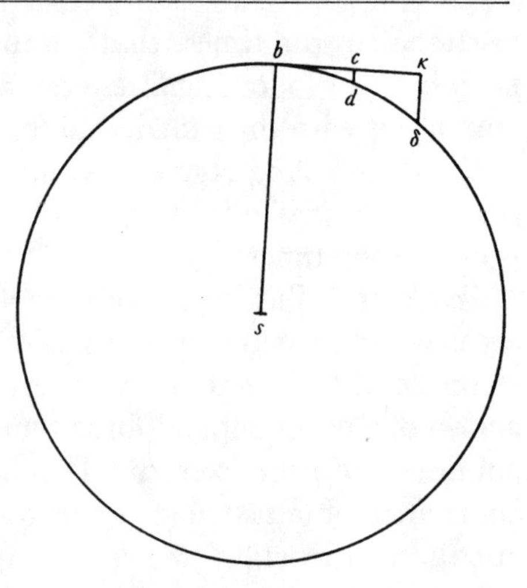

Fig 11.

Corol. 1. Hinc vires centripetæ sunt ut velocitatum quadrata applicata ad radios circulorum.

Corol. 2. Et reciprocè ut quadrata temporum periodicorum applicata ad radios. Vel (ut cum Geometris loquar) hæ vires sunt in ratione compositâ ex duplicata ratione velocitatum directe et ratione radiorum reciprocè: nec non in ratione compositâ ex ratione radiorum directè et ratione duplicata temporum periodicorum reciproce.

Corol. 3. Unde si quadrata temporum periodicorum sunt ut radij circulorum[,] vires centripetæ sunt æquales. Et vice versa.

Corol. 4. Si quadrata temporum periodicorum sunt ut quadrata radiorum, vires centripetæ sunt reciprocè ut radij: et vice versa.

Corol. 5. Si quadrata temporum periodicorum sunt ut cubi radiorum[,] vires centripetæ sunt reciprocè ut quadrata radiorum: et vice versa.(86)

Schol.

Casus Corollarij quinti obtinet in corporibus cœlestibus. Illorum tempora periodica sunt in sequiplicata ratione distantiarum a centris(87) et propterea

(85) Since in the limit as they become vanishingly small the arcs $\widehat{BD}$, $\widehat{b\delta}$ come to differ by less than any finite quantity from their corresponding 'tangents' BC, bc.

(86) Exactly as before (compare §1: note (25)), if the square of the time T of periodic orbit in a circle of radius r at a uniform speed v ($\propto r/T$) be proportional to r^n, then the impressed central force inducing the circular motion is measured by $v^2/r \propto r^{1-n}$. In his revised 'De motu Corporum Liber primus' (= *Principia*, ${}_1$1687: 42) Newton appended a final corollary (on f. 31r, reproduced in Appendix 2.5 below) asserting in generalisation that 'Eadem omnia de temporibus, velocitatibus et viribus quibus corpora similes Figurarum quarumcunq; similium centraq; similiter posita habentium partes describunt, consequuntur ex Demonstratione præcedentium ad hosce casus applicata' (Identical assertions regarding the times, speeds and forces in, at and by which bodies describe similar portions of any similar [orbital] figures having their centres similarly placed ensue in every case from applying thereto the proof of those preceding). For, of course, we may—assuming (compare note (59)) that it is always

CD, cd. Make the figure $\delta\kappa b$ similar to the figure DKB and then, by Lemma V,[85] the line-element CD will be to the element cd as the arc $\widehat{BD}$ to the arc $b\delta$, while again, by Lemma XI, the nascent line-element $\delta\kappa$ will be to the nascent element dc as $b\delta^2$ to bd^2, and so *ex æquo* the nascent line-element DC will be to the nascent line dc as $BD \times b\delta$ to bd^2. The centripetal forces are therefore as $BD \times b\delta$ to bd^2, that is, as $BD \times b\delta/sb$ to bd^2/sb, and consequently (because the ratios $b\delta/sb$ and BD/SB are equal) as BD^2/SB to bd^2/sb. As was to be proved.

Corollary 1. Hence the centripetal forces are as the squares of the speeds divided by the radii of the circles.

Corollary 2. And reciprocally as the squares of the periodic times divided by the radii.

Or (to talk in geometrical jargon) these forces are in a ratio compounded of the doubled ratio of the speeds directly and the ratio of the radii reciprocally; and, again, in a ratio compounded of the ratio of the radii directly and the doubled ratio of the periodic times reciprocally.

Corollary 3. Whence if the squares of the periodic times are as the radii of the circles, the centripetal forces are equal. And conversely so.

Corollary 4. If the squares of the periodic times are as the squares of the radii, the centripetal forces are reciprocally as the radii. And conversely so.

Corollary 5. If the squares of the periodic times are as the cubes of the radii the centripetal forces are reciprocally as the squares of the radii. And conversely so.[86]

Scholium.

The case of Corollary 5 holds true in the heavenly bodies. The periodic times of these are in the sesquialteral ratio of the distances from their centres,[87] and

possible to do so—replace the orbital arc in the immediate vicinity of any given point in it by the corresponding arc of the circle of curvature which there osculates it, and thereafter apply the present theorem to measure the component normal to the orbit of the centripetal force which is there instantaneously directed to any point in the orbital plane. In precise terms, if the orbital speed at the point be v and the radius of curvature there be ρ, then v^2/ρ will measure the component $f.\sin\alpha$ normal to the orbit of a deviating centripetal force f acting towards some point at an angle α to the instantaneous (tangential) direction of motion in the orbit at the point, whence the force f is measured by $v^2/\rho.\sin\alpha$. No use of this corollary is made—explicitly so at least (but see note (124) below)—either in the present 'De motu Corporum' or in the enlarged 'Liber primus' of the *Principia* in which it was afterwards (in 1687) published; in the early 1690's, however, when he began seriously to revise this *editio princeps* for a future reprinting, Newton introduced a consectary to his following Proposition VI in which (see 3, §1: note (25)) this highly convenient formula is elegantly derived as an offshoot of the basic measure of central force there expounded.

(87) Almost at once Newton changed this to read 'Planetarum tempora periodica sunt in sesquiplicata ratione distantiarum ab orbium suorum centro seu nodo communi' (The periodical times of the planets are in the sequialteral ratio of their distances from the common centre or node of their orbits). We have seen (§1: note (26) above) why Newton was at this

quæ spectant ad vim centripetam decrescentem in duplicata ratione distantiarū a centris decrevi fusius in sequentibus exponere.(88)

Prop. V. Theor. V.(89)

Si corpus P circa centrum S gyrando, describat lineam quamvis curvam APQ, et si tangat recta PR curvam illam in puncto quovis P et ad tangentem ab alio quovis curvæ puncto Q agatur QR distantiæ SP parallela ac demittatur QT perpendicularis ad distantiam SP: dico quod vis centripetasit reciprocè ut solidum $\frac{SP^{\text{quad}}\times QT^{\text{quad}}}{QR}$, *si modò solidi illius ea semper sumatur quantitas quæ ultimò fit ubi coeunt puncta P et Q.*

Namq; in figura indefinitè parva *QRPT* lineola nascens *QR* dato tempore est ut [a]vis centripeta, et data vi ut [b]quadratum temporis, atq; adeo neutro dato ut vis centripeta et quadratum temporis conjunctim,(90) id est ut vis centripeta semel et area *SPQ* tempori proportionalis (vel duplum ejus $SP\times QT$) bis. Applicetur hujus proportionalitatis pars utraq; ad lineolam *QR* et fiet unitas ut vis centripeta et $\frac{SP^q\times QT^q}{QR}$ conjunctim, hoc est vis centripeta reciprocè ut $\frac{SP^q\times QT^q}{QR}$. Q.E.D.

[a]Leg. [II].
[b]Lem. [X].

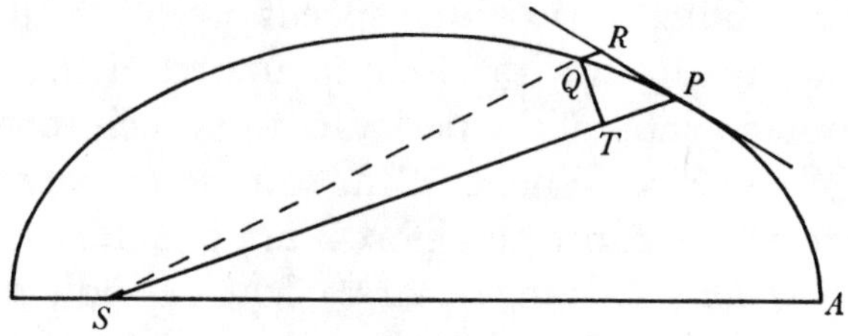

Fig 12.

Corol. Hinc si detur figura quævis, et in ea punctum ad quod vis centripeta dirigitur, inveniri potest lex vis centripetæ quæ corpus in figuræ illius perimetro

time (early 1685) reluctant to comprehend the totality of the known planetary satellites within Kepler's third law, for in December 1684 Flamsteed had been unable to confirm observationally the very existence of the two new 'satellits of [Saturn]' whose discovery Cassini had recently announced.

(88) In mid-1686, even as the sheet (G) of the *Principia* which was to print the present text was being typeset, Newton acceded to a diplomatic request by Halley that he should 'make some mention' of Hooke's 'pretensions upon the invention of ye rule of the decrease of Gravity' by enclosing with his letter of 14 July 1686 (*Correspondence*, **2**: 444–5) a much augmented replacement for this scholium: therein he not only gives appropriate credit to Wren and Halley as well as to Hooke for independently applying Huygens' measure of circular *vis centrifuga* (as announced in his 1673 *Horologium*) to Kepler's third planetary law to deduce the inverse-square decrease of solar gravity with distance, but, much more interestingly, appends a summary of his own independent derivation in 'some old papers' (of twenty years before) of the basic Huygenian measure. We return to this latter point in more detail in Appendix 2.6, where we reproduce the text of the augmented scholium. Having come subsequently to realise in Proposition XV below, by an argument which is manifestly not restricted to the conic case

I have in consequence resolved to elaborate more fully in the sequel what regards centripetal force decreasing (inversely) in the doubled ratio of the distances from the centres.(88)

Proposition V. Theorem V.(89)

If a body P in orbiting round the centre S shall describe any curved line APQ, and if the straight line PR touches that curve at any point P and to this tangent from any other point Q of the curve there be drawn QR parallel to the distance SP, and if QT be let fall perpendicular to this distance SP: I assert that the centripetal force is reciprocally as the 'solid' $SP^2 \times QT^2/QR$, *provided that the ultimate quantity of that solid when the points P and Q come to coincide is always taken.*

For in the indefinitely small configuration *QRPT* the nascent line-element *QR* is, given the time, as[a] the centripetal force and, given the force, as[b] the square of the time, and hence, when neither is given, as the centripetal force and the square of the time jointly;(90) that is, as the centripetal force taken once and, proportional to the time, the area *SPQ* or rather its double $SP \times QT$ taken twice. Divide each side of this proportionality by the line-element *QR* and there will come to be 1 as the centripetal force and $SP^2 \times QT^2/QR$ jointly, that is, the centripetal force will be reciprocally as $SP^2 \times QT^2/QR$. As was to be proved.

[a]Law II.
[b]Lemma X.

Corollary. Hence if any figure be given and in it a point to which the centripetal force is directed, there can be ascertained the law of centripetal force which shall make a body orbit in the perimeter of that figure: specifically, you must

---

there exemplified, that the perpendicular let fall from the force-centre to the tangent at any point of a given orbit is inversely proportional to the instantaneous speed at that point, Newton in revision (on f. 31[v]) here inserted a new 'Prop. V. Prob. I' in which he somewhat artificially applies the property to construct the orbit's force-centre, given only the speeds at three points of the orbit: this we reproduce in Appendix 2.7 below.

(89) Theorem 3 of the preceding 'De motu Corporum' (§1 above) transcribed word for word.

(90) Leibniz in 1689 could not here see why the square—rather than the simple quantity—of the time is needed, and in attempted correction of Newton's argument in his copy (see note (49)) altered the text to read '...lineola nascens *QR*, eodem dato momento est ut velocitas [!] centripeta...& data velocitate, ut tempus', subsequently qualifying the assertion that *QR* is 'data vi, ut quadratum temporis' as an 'error' which he then proceeded suitably to amend in sequel. We may well wonder if anyone—even Huygens—could begin to appreciate the underlying subtleties of the *Principia's* dynamical basis on first reading it, when Leibniz failed so miserably. In later revise Newton considerably changed the conclusion of his verbal demonstration to read in sequel 'adeoq vis centripeta ut lineola *QR* directè et quadratum temporis inversè. Est autem tempus ut area *SPQ* ejusve dupla $SP \times QT$, id est ut *SP* et *QT* conjunctim, adeoq vis centripeta ut *QR* directè atq $SP^{\text{quad.}} \times QT^{\text{quad.}}$ inversè, id est ut $\frac{SP^{\text{quad}} \times QT^{\text{quad}}}{QR}$ inversè. Q.E.D.' (and hence the centripetal force is as the line-element *QR* directly and the square of the time inversely. But the time is as the area (*SPQ*) or its double $SP \times QT$, that is, as *SP* and *QT* jointly, and hence the centripetal force is as *QR* directly and $SP^2 \times QT^2$ inversely, that is, as $SP^2 \times QT^2/QR$ inversely. As was to be proved).

gyrare faciet. Nimirum computandum est solidum $\frac{SP^q \times QT^q}{QR}$ huic vi reciprocè proportionale. Ejus rei dabimus exempla in problematîs sequentibus.(91)

Prop. VI. Prob. I.(92)

Gyrat corpus in circumferentia circuli, requiritur lex vis centripetæ tendentis ad punctum aliquod in circumferentia.

Esto circuli circumferentia $SQPA$, centrum vis centripetæ S, corpus in circumferentia latum P, locus proximus in quem movebitur Q. Ad SA diametrum et SP(93) demitte perpendicula PK, QT et per Q ipsi SP parallelam age LR occurrentem circulo in L et tangenti PR in R, et coeant TQ, PR in Z. Ob similitudinem triangulorum ZQR, ZTP, SPA erit RP^q (hoc est QRL) ad QT^q ut SA^q ad SP^q

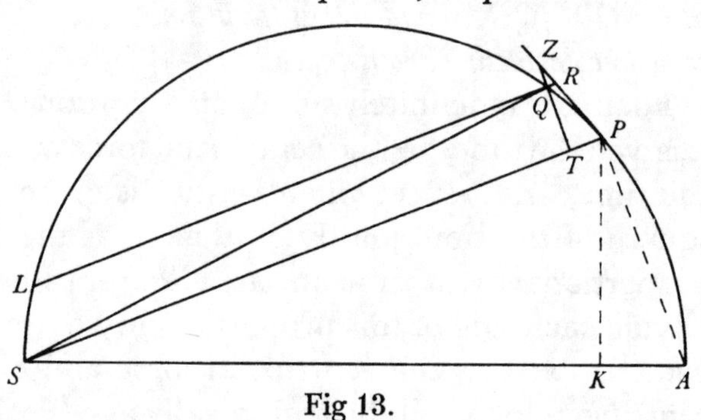

Fig 13.

Ergo $\frac{QRL \times SP^q}{SA^q} = QT^q$. Ducantur hæc æqualia in $\frac{SP^q}{QR}$ et punctis P et Q coeuntibus scribatur SP pro RL. Sic fiet $\frac{SP^{qc}}{SA^q} = \frac{QT^q \times SP^q}{QR}$. Ergo [a]vis centripeta reciprocè est ut $\frac{SP^{qc}}{SA^q}$, id est (ob datum SA^q) ut quadrato-cubus distantiæ SP. Quod erat inveniendum.(94)

[a]Cor. Theor. [V].

Prop. VII. Prob. II.(95)

Movetur corpus in circulo PQA: ad hunc effectum requiritur lex vis centripetæ tendentis ad punctum(96) *adeo longinquum ut lineæ omnes PS, RS ad id ductæ pro parallelis habeantur.*

(91) Much as before (in §1) Newton applies this fundamental measure $SP^{-2}.\lim_{Q\to P}(QR/QT^2)$ to compute the 'law' of centripetal force inducing motion in a circle (Propositions VI/VII: centre of force on the circle/at infinity) or conic (Propositions IX/X–XII: centre of force at the conic's centre/focus) or logarithmic spiral (Proposition VIII: centre of force at the pole). In each case a parallel analytical computation may (compare §1: note (30)) be effected by taking $r^{-2}(r^{-1}+d^2(r^{-1})/d\phi^2)$ as the equivalent measure of the central force acting at (r, ϕ) towards the origin (0, 0) of the polar coordinates in whose terms the orbit is equivalently defined.

(92) A straightforward repeat of Problem 1 of the pristine 'De motu Corporum' (§1 preceding), minimally improved by explicit demonstration of the proportion $RP:QT = SA:SP$.

(93) Newton later added a balancing 'rectam' (the straight line).

(94) For the equivalent analytical computation see §1: note (34). In his preliminary revisions of the first edition of the *Principia* (where the theorem appears as 'Prop. VII, Prob. II'

compute the solid $SP^2 \times QT^2/QR$ reciprocally proportional to this force. Of this procedure we shall give illustrations in following problems.(91)

Proposition VI. Problem I.(92)

A body orbits in the circumference of a circle: the law of centripetal force tending to some point in its circumference is required.

Let $SQPA$ be the circle's circumference, S the centre of centripetal force, P the body borne along in the circumference, Q a closely proximate position into which it shall move. To the diameter SA and to(93) SP let fall the perpendiculars PK, QT and through Q parallel to SP draw LR meeting the circle in L and the tangent PR in R, and let TQ, PR unite in Z. Because of the similarity of the triangles ZQR, ZTP and SPA there will be RP^2 (that is $QR \times LR$) to QT^2 as SA^2 to SP^2, and therefore $QR \times LR \times SP^2/SA^2 = QT^2$. Multiply these equals into SP^2/QR and, with the points P and Q coalescing, let SP be written in place of LR. In this way there will come $SP^5/SA^2 = QT^2 \times SP^2/QR$. Therefore[a] the centripetal force is reciprocally as SP^5/SA^2, that is, (because SA^2 is given) as the fifth power of the distance SP. As was to be found.(94)

[a] Corollary to Theorem V.

Proposition VII. Problem II.(95)

A body moves in the circle PQA: to this effect there is required the law of centripetal force tending to a point(96) *so far distant that all lines PS, RS drawn to it may be considered as parallel.*

---

of the first book, allowing—see note (88) above—for an inserted 'Prop. V. Prob. I'), Newton came in the early 1690's (see 3, §1.3 below) to treat of the extension in which the force-centre S is in general position in the plane of the circle, employing a minimally variant form of his present elegant geometrical demonstration. If in equivalent polar terms the circle $P(r, \phi)$ has centre $(a, 0)$ and radius R, and so is defined by the equation $r = a\cos\phi + \sqrt{[R^2 - a^2\sin^2\phi]}$, it follows that $d^2(r^{-1})/d\phi^2 = -1/r + R^2/(r - a\cos\phi)^3$ and therefore the centripetal force $r^{-2}(r^{-1} + d^2(r^{-1})/d\phi^2)$ acting at P towards $S(0, 0)$ is $R^2/r^2(r - a\cos\phi)^3$: the present problem is the particular case in which $a = R$ and the circle's defining equation becomes $r = 2R\cos\phi$.

(95) While without exact parallel in the earlier tract 'De motu Corporum', this is but an extremal instance of the generalisation of the preceding 'Problema': namely (in the analytical terms of the previous note) that in which a, and hence r, becomes indefinitely large and ϕ vanishingly small, so that, if in this limiting case we set

$$a\sin\phi = CM = x \quad \text{and} \quad r - a\cos\phi = MP = y,$$

the (now Cartesian) defining equation of the semicircle $\widehat{AP}$ assumes the form $y = \sqrt{[R^2 - x^2]}$ and the central force at (Px, y) acting along PM is proportional to $1/y^3$, on taking the infinitesimal ratio R^2/r^2 to be determinate (compare note (98) following).

(96) That is, S in the present standardised nomenclature. Following a convention which Newton himself uses elsewhere not infrequently, we have in our restoration of his original figure inserted a corresponding 'infinite' point midway between the indefinite extensions of PM and QN. Halley's remark that 'in fig[ura] deest littera S' (ULC. Add. 3965.9: 97r)—one of a miscellany of contemporary editorial criticisms of the present manuscript—confirms that

A circuli centro C agatur semidiameter CA parallelas istas perpendiculariter secans in M et N, et jungatur CP. Ob similia triangula CPM et TPZ, [a]vel TPQ(97) est CP^q ad PM^q ut PQ^q vel[b] PR^q ad QT^q et ex natura circuli rectangulum $QR \times \overline{RN+QN}$ æquale est PR quadrato. Coeuntibus autem punctis P, Q fit $RN+QN$ æqualis $2PM$. Ergo est CP^q ad PM^q ut $QR \times 2PM$ ad QT^q, adeoq;

[a]Lem VIII

[b][Lem VII]

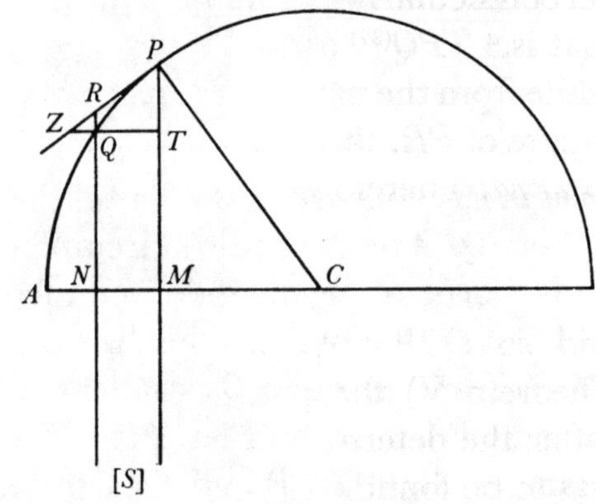

Fig 14.

$$\frac{QT^q}{QR} = \frac{2PM^{cub}}{CP^q},$$

et $$\frac{QT^q \times SP^q}{QR} = \frac{2PM^{cub} \times SP^q}{CP^q}.$$

Est ergo (per Corol. Theor. V) vis centripeta reciprocè ut $\frac{2PM^{cub} \times SP^q}{CP^q}$[,] hoc est $\left(\text{neglecta ratione determinata } \frac{2SP^q}{CP^q}\right)$(98) reciprocè ut PM^{cub}. Q.E.I.

Et simili argumento corpus movebitur in Ellipsi(99) vi centripeta quæ sit reciprocè ut cubus ordinatæ ad centrum maximè longinquum tendentis.

Prop. VIII. Prob. III.(100)

Gyrat corpus in spirali PQS secante radios omnes SP, SQ &c in dato angulo: requiritur lex vis centripetæ tendentis ad centrum spiralis.

Detur angulus indefinitè parvus PSQ, et ob datos omnes angulos dabitur specie figura $SPQRT$. Ergo datur ratio $\frac{QT}{QR}$ estq; $\frac{QT^q}{QR}$ ut QT[,] hoc est ut SP. Mutetur jam(101) angulus PSQ et recta QR angulum contactus QPR subtendens mutabitur (per Lemma XI) in duplicata ratione ipsius PR vel QT. Ergo manebit $\frac{QT^q}{QR}$ eadem quæ prius, hoc est ut SP. Quare

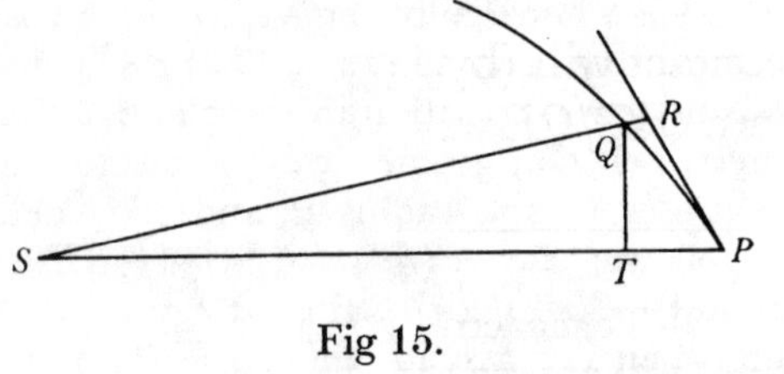

Fig 15.

Newton made no indication at all in his draft figure of the infinite force-centre: the diagram as finally printed (*Principia*, ₁1687: 46) repairs the omission by attaching an 'S' at the foot of each of the extensions of PM and QN, but this appeal to a variant convention may be Halley's rather than Newton's.

(97) This limit-equivalence will not hold good when P comes to coincide with A (or the point diametrally opposite to it), but in compensation—and as a mark of the discontinuity before it passes into the *vis centrifuga* under which the lower semicircular arc (not shown by Newton) is similarly traversed—the ensuing centripetal force comes there to be indefinitely large.

From the circle's centre C draw the radius CA intersecting those parallels perpendicularly in M and N, and join CP. Because the triangles CPM and TPZ, that is,[a] TPQ[97] are similar, there is CP^2 to PM^2 as PQ^2, that is,[b] PR^2 to QT^2, while from the nature of the circle the rectangle $QR \times (RN+QN)$ is equal to the square of PR. But as the points P, Q come to coalesce $RN+QN$ becomes equal to $2PM$. Therefore CP^2 is to PM^2 as $QR \times 2PM$ to QT^2, and hence

[a]Lemma VIII.
[b]Lemma VII.

$$QT^2/QR = 2PM^3/CP^2$$

and so $QT^2 \times SP^2/QR = 2PM^3 \times SP^2/CP^2$. Therefore (by the Corollary to Theorem V) the centripetal force is reciprocally as $2PM^3 \times SP^2/CP^2$, that is, (after the determinate ratio $2SP^2/CP^2$ is neglected)[98] reciprocally as PM^3. As was to be found.

And by a similar argument a body will move in an ellipse[99] under a centripetal force which shall be reciprocally as the cube of its ordinate when directed to a centre exceedingly far distant.

Proposition VIII. Problem III.[100]

A body orbits in a spiral PQS which intersects all its radii SP, SQ, ... at a given angle: there is required the law of centripetal force tending to the spiral's centre.

Let there be given the indefinitely small angle $\widehat{PSQ}$ and then, because all its angles are given, the configuration $SPQRT$ will be given in species. The ratio QT/QR is therefore given and so QT^2/QR is as QT, that is, as SP. Now let the angle $\widehat{PSQ}$ change[101] and the straight line QR subtending the angle $\widehat{QPR}$ of contact will (by Lemma XI) change in the doubled ratio of PR or QT. Therefore QT^2/QR will stay the same as before, that is, proportional to SP. In conse-

(98) To modern eyes this omission of a ratio which, for all that it is 'determined', is infinitely great will appear curious, but we are to understand an absolute constant for the *vis centripeta* which is likewise infinitely great in compensation.

(99) Newton afterwards added 'vel etiam in Hyperbola vel Parabola' (or, also, in a hyperbola or parabola)—to no purpose since, even though it is repeated *verbatim* in all editions of the published *Principia*, this present attempted generalisation is false! In correction (see 3, §2: note (46) below), where CE, drawn parallel to the tangent PR, meets PS in E, the centripetal force will be directly as PE^3; only in the case of a circle, where $\widehat{ECP}$ is ever a right angle and hence the triangles PMC, PCE are similar (so that $PE = CP^2/PM \propto 1/PM$), can there be

$$PE^3 \propto 1/PM^3.$$

(100) This elaborates with explicit demonstration what Newton had in the scholium to Problem 1 of his earlier 'De motu Corporum' (compare §1: note (37)) previously been content merely to notice.

(101) Newton later inserted 'utcunq;' (in any manner whatever).

[a]Cor. Theor. V. $\frac{QT^q \times SP^q}{QR}$ est ut SP^{cub}, id est, [a]vis centripeta reciprocè ut cubus distantiæ *SP*. Q.E.I.[(102)]

Lemma XII.

Parallelogramma omnia circa datam Ellipsin[(103)] *descripta esse inter se æqualia. Idem intellige de Parallelogrammis in Hyperbola descriptis circum diametros figuræ.*

Prop. IX. Prob. IV.[(104)]

Gyrat corpus in Ellipsi veterum: requiritur lex vis centripetæ tendentis ad centrum Ellipseos.

Sunto *CA*, *CB* semi-axes Ellipseos, *GP*, *DK* diametri conjugatæ, *PF*, *Qt* perpendicula ad diametros, *QV* ordinatim applicata ad diametrum *GP*, et *QVPR* parallelogrammum. His constructis erit[(105)] (ex Conicis) *PVG* ad QV^q ut PC^q ad CD^q et[(106)] QV^q ad Qt^q ut PC^q ad PF^q et conjunctis rationibus *PVG* ad Qt^q ut PC^q ad CD^q et PC^q ad PF^q, id est *VG* ad $\frac{Qt^q}{PV}$ ut PC^q ad $\frac{CD^q \times PF^q}{PC^q}$. [a]Lem. XII. Scribe *QR* pro *PV* et[a] $BC \times CA$ pro $CD \times PF$, nec non (punctis *P* et *Q* coeuntibus) $2PC$ pro $VG_{[,]}$ et ductis extremis & medijs in se mutuo fiet

$$\frac{Qt^q \times PC^q}{QR} = \frac{2BC^q \times CA^q}{PC}.$$

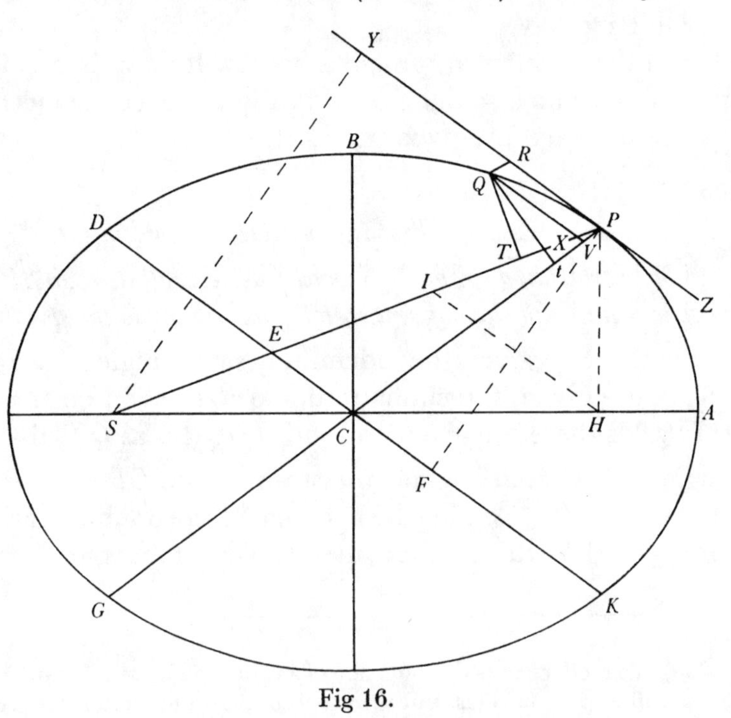

Fig 16.

(102) In equivalent analytical terms (compare note (91)) we may analogously set *S* to be the origin of the polar system in which, on making $SP = r$ and $\widehat{SPR} = \alpha$, the logarithmic spiral $P(r, \phi)$: $\log(r/R) = -\phi \cot \alpha$ is defined by the differential condition $dr/rd\phi = -\cot\alpha$, whence $d^2(r^{-1})/d\phi^2 = \cot^2\alpha/r$ and so the centripetal force at *P* acting towards *S* is measured by $r^{-2}(r^{-1}+d^2(r^{-1})/d\phi^2) = \text{cosec}^2\alpha/r^3$. As Newton soon came to know, the logarithmic spiral is only one of an infinity of possible inverse-cube orbits, and in Corollary 3 to Proposition XLI of his subsequent 'De motu Corporum Liber primus' (§3 following) he will construct the two most general species of such orbits.

(103) As in Lemma 2 of his earlier 'De motu Corporum' (see §1: note (14)) Newton here intends 'circa diametros conjugatas datæ Ellipsis' (around the conjugate diameters of a given ellipse). A parallel following justification that 'Constat utrumq; ex Conicis' (Both are established from the *Conics*) was afterwards inserted at the paragraph's end.

quence $QT^2 \times SP^2/QR$ is as SP^3, that is,[a] the centripetal force is reciprocally as the cube of the distance *SP*. As was to be found.(102)

[a]Corollary to Theorem V.

Lemma XII.

All parallelograms described around a given ellipse(103) *are equal to one another. Understand the same for parallelograms described in a hyperbola round diameters of the figure.*

Proposition IX. Problem IV.(104)

A body orbits in a classical ellipse: there is required the law of centripetal force tending to the ellipse's centre.

Let *CA*, *CB* be the ellipse's semi-axes, *GP* and *DK* conjugate diameters, *PF* and *Qt* perpendiculars to these diameters, *QV* ordinate to the diameter *GP*, and *QVPR* a parallelogram. With this construction(105) there will (from the *Conics*) be $PV \times VG$ to QV^2 as PC^2 to CD^2, and(106) QV^2 to Qt^2 as PC^2 to PF^2, and on combining these proportions $PV \times VG$ to Qt^2 as PC^2 to CD^2 and PC^2 to PF^2, that is, *VG* is to Qt^2/PV as PC^2 to $CD^2 \times PF^2/PC^2$. Write *QR* in place of *PV* and[a] $BC \times CA$ in place of $CD \times PF$, and again (with the points *P* and *Q* coalescing) $2PC$ in place of *VG*, and when extremes and middles are multiplied into each other there will come to be $Qt^2 \times PC^2/QR = 2BC^2 \times CA^2/PC$.

[a]Lemma XII.

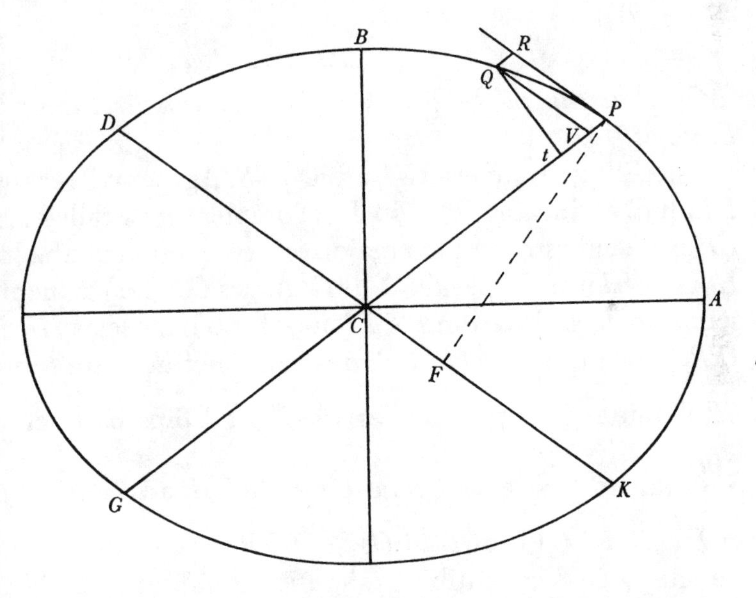

(104) An unaltered transcription of Problem 2 of the earlier 'De motu Corporum' (§1 preceding).

(105) In revise Newton changed the text here to read more simply '...et si compleatur parallelogrammum *QVPR*, erit' (and if the parallelogram *QVPR* be completed, then). As with that jointly illustrating Problems 2 and 3 of the earlier tract (see §1: note (38)), his accompanying diagram is made to do double duty both here and in Proposition X following, whence *QR* is slightly distorted in direction to be parallel to a mean between *CP* and *SP*. For ease of understanding, in our English version we again introduce suitably simplified separate figures for each case, drawing *QR* accurately parallel to either *CP* or *SP*, as is pertinent, but in both omitting the perpendicular *SY* (which is needed only in Proposition XV below).

(106) The justification 'ob similia triangula *QVt PCF*' (because of the similarity of the triangles *QVt*, *PCF*) was afterwards inserted.

[b]Cor. Theor. V.

Est ergo [b]vis centripeta reciprocè ut $\frac{2BC^q \times CA^q}{PC}$, id est (ob datum $2BC^q \times CA^q$) ut $\frac{1}{PC}$, hoc est directè ut distantia PC. Q.E.I.[(107)]

Schol.

Si Ellipsis, centro in infinitum abeunte, vertatur in Parabolam, corpus movebitur in hac Parabola, et vis ad centrum infinite distans jam tendens, evadet æquabilis. Hoc est Theorema Galilæi.[(108)] Et si conisectio Parabolica, inclinatione plani ad conum sectum mutata, vertatur in Hyperbolam, movebitur corpus in hujus perimetro vi centripeta in centrifugam[(109)] versa.

Prop. X. Prob. V.[(110)]

Gyrat corpus in Ellipsi: requiritur lex vis centripetæ tendentis ad umbilicum Ellipseos.

Fig 16.

Esto Ellipseos superioris umbilicus S. Agatur SP secans Ellipseos diametrum DK in E, et lineam[(111)] QV in X et compleatur parallelogrammum $QXPR$. Patet EP æqualem esse semi-axi majori AV eò quod actâ ab altero Ellipseos umbilico H lineâ HI ipsi EC parallelâ (ob æquales CS, CH) æquentur ES, EI, adeo ut EP semisumma sit ipsarum PS, PI, id est (ob parallelas HI, PR & angulos æquales IPR, HPZ) ipsarum PS, PH, quæ conjunctim axem totum $2AC$ adæquant. Ad SP demittatur perpendicularis QT, et Ellipseos latere recto principali $\left(\text{seu } \frac{2BC^q}{AC}\right)$ dicto L, erit $L \times QR$ ad $L \times PV$ ut QR ad PV id est ut PE (seu AC) ad PC, et $L \times PV$ ad GVP ut L ad GV, et GVP ad QV^q ut CP^q ad CD^q, et QV^q ad QX^q punctis Q et P coeuntibus [a]est ratio æqualitatis, et QX^q seu QV^q est ad QT^q ut EP^q ad PF^q id est ut CA^q ad PF^q sive [b]ut CD^q ad CB^q. Et conjunctis his omnibus rationibus, $L \times QR$ fit ad QT^q ut AC ad $PC + L$ ad $GV + CP^q$ ad

[a]Lem. VIII. [b]Lem. XII.

(107) Two corollaries which Newton added in sequel in his revised 'De motu Corporum Liber primus' are reproduced in Appendix 2.8 below.

(108) In his 1638 *Discorsi e Dimostrazioni Matematiche*; see note (35).

(109) Still varying, of course, directly as the distance from the centre.

(110) A trivially rephrased repeat of Problem 3 of the earlier 'De motu Corporum', retaining its argument—and Barrovian logical connectives '+' (see §1: note (47))—unaltered.

(111) Afterwards emended to 'ordinatim applicatam' (ordinate).

The centripetal force is therefore[b] reciprocally as $2BC^2 \times CA^2/PC$, that is, (because $2BC^2 \times CA^2$ is given) as $1/PC$, or in other words directly as the distance PC. As was to be found.(107)

[b] Corollary to Theorem V.

Scholium.

Should the ellipse, with its centre going off to infinity, turn into a parabola, the body will move in this parabola and the force, now tending to an infinitely distant centre, will prove to be constant. This is Galileo's theorem.(108) And should the parabolic cone-section, by altering the slope of the plane to the cone which it cuts, turn into a hyperbola, the body will move in the latter's perimeter with the centripetal force now changed into a centrifugal one.(109)

Proposition X. Problem V.(110)

A body orbits in an ellipse: there is required the law of centripetal force tending to a focus of the ellipse.

Let S be a focus of the previous ellipse. Draw SP cutting the ellipse's diameter DK in E and the line(111) QV in X, and complete the parallelogram $QXPR$. It is evident that EP is equal to the semi-major-axis AC, seeing that, when from the ellipse's other focus H the line HI is drawn parallel to CE, (because CS and CH are equal) ES and EI are equal, and hence EP is the half-sum of PS and PI, that is, (because HI, PR are parallel and the angles $I\widehat{P}R$, $H\widehat{P}Z$ equal) of PS and PH,

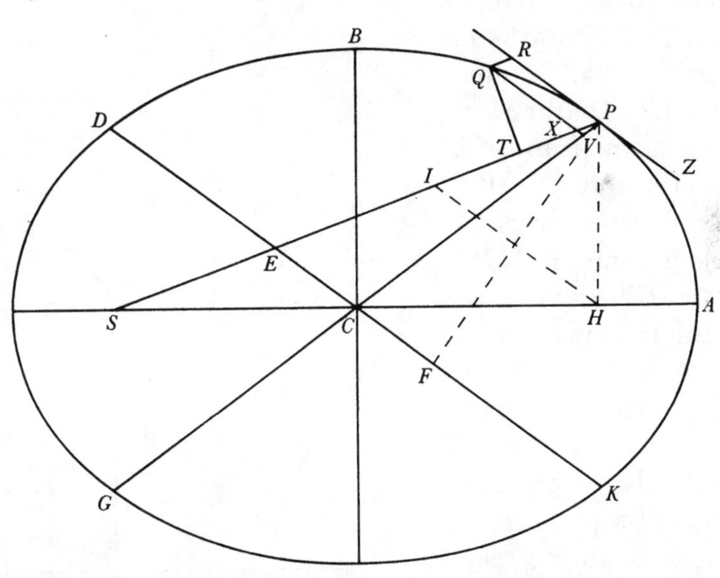

which are jointly equal to the total axis $2AC$. To SP let fall the perpendicular QT, and then, on calling the ellipse's principal *latus rectum* (viz. $2BC^2/AC$) L, there will be $L \times QR$ to $L \times PV$ as QR to PV, that is, as PE (or AC) to PC; and $L \times PV$ to $GV \times VP$ as L to GV; and $GV \times VP$ to QV^2 as CP^2 to CD^2; and QV^2 to QX^2 is, with the points Q and P coalescing,[a] a ratio of equality; and QX^2 (or QV^2) is to QT^2 as EP^2 to PF^2, that is, as CA^2 to PF^2 or[b] as CD^2 to CB^2: and, when all these ratios are combined, $L \times QR$ comes to be to QT^2 as

[a] Lemma VIII.
[b] Lemma XII.

$$(AC \text{ to } PC) \times (L \text{ to } GV) \times (CP^2 \text{ to } CD^2) \times (CD^2 \text{ to } CB^2),)$$

CD^q+CD^q ad CB^q, id est ut $AC\times L$ (seu $2CB^q$) ad $PC\times GV+CP^q$ ad CB^q, sive ut $2PC$ ad GV. Sed punctis Q et $[P]$ coeuntibus æquantur $2PC$ & GV. Ergo et[112] $L\times QR$ & QT^q æquantur. Ducatur pars utraq; in $\frac{SP^q}{QR}$ et fiet $L\times SP^q=\frac{SP^q\times QT^q}{QR}$.

*c*Cor. Theor. V.

Ergo *c*vis centripeta reciprocè est ut $L\times SP^q$. id est reciproce in ratione duplicata distantiæ SP. Q.E.I.

Eâdem brevitate quâ traduximus Problema quartum ad Parabolam, et Hyperbolam, liceret idem hìc facere. Verũ ob dignitatem Problematis & usum ejus in sequentibus, non pigebit casus cæteros[113] demonstratione confirmare.

Prop. XI. Prob. VI.[114]

Movetur corpus in Hyperbola, requiritur lex vis centripetæ tendentis ad umbilicum figuræ.

Sunto $CA_{[,]}$ CB semiaxes Hyperbolæ, GP, DK diametri conjugatæ; PF, Qt[115] perpendicula ad diametros, & QV ordinatim applicata ad diametrum GP. Agatur SP secans diametrum DK in E, et lineam[111] QV in X et compleatur parallelogrammum $QRPX$. Patet EP æqualem esse semi-axi transverso AC eò, quod acta ab altero [Hyperbolæ] umbilico H linea HI ipsi EC parallela, ob æquales CS, CH æquentur ES, EI, adeo ut EP semidifferentia sit ipsarum PS, PI, id est (ob parallelas HI, PR et angulos æquales IPR, HPZ) ipsarũ $P[S]$, PH quæ conjunctim[116] axem totum $2AC$ adæquant. Ad SP demittatur perpendicularis QT. Et

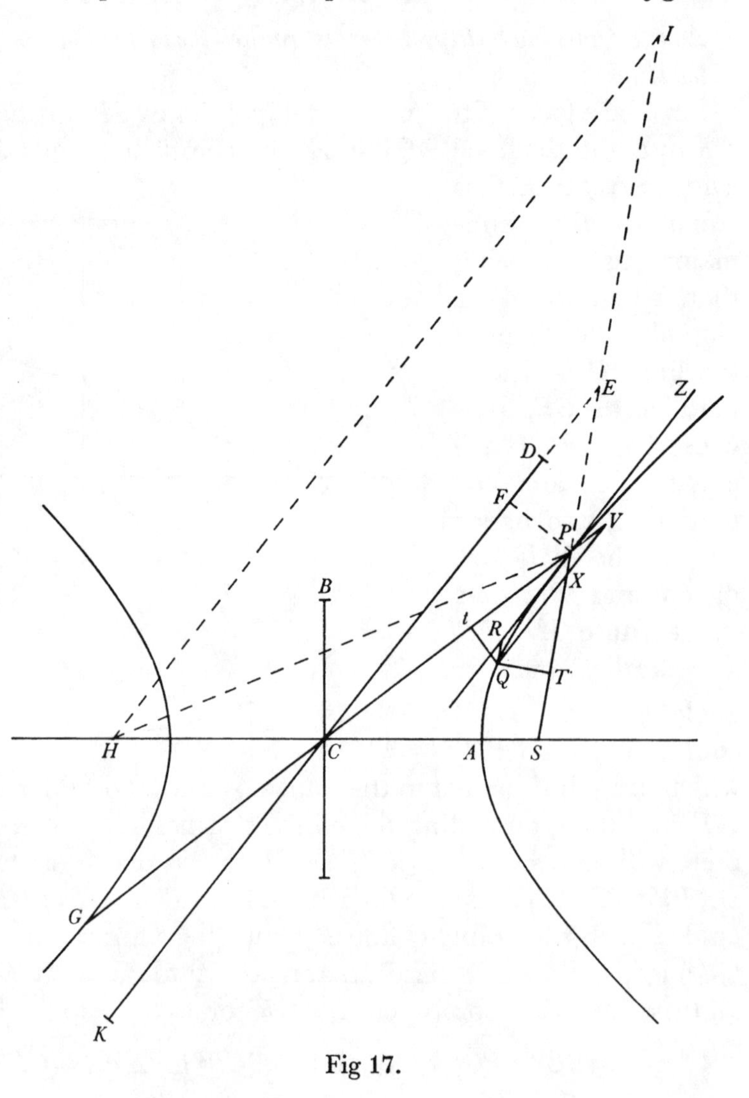

Fig 17.

that is, as ($AC \times L$ (or $2CB^2$) to $PC \times GV$) $\times$ (CP^2 to CB^2), or as $2PC$ to GV. But, with the points Q and P coalescing, $2PC$ and GV are equal. Therefore also[112] $L \times QR$ and QT^2 are equal. Multiply each member by SP^2/QR and there will come to be $L \times SP^2 = SP^2 \times QT^2/QR$. Therefore[c] the centripetal force is reciprocally as $L \times SP^2$, that is, reciprocally in the doubled ratio of the distance SP. As was to be found.

[c] Corollary to Theorem V

With the same conciseness with which we transposed Problem IV to treat of the parabola and hyperbola, we would be at liberty to do the same here. But, because of the merit of the problem and its usefulness in the sequel, it will be no imposition to corroborate the remaining cases[113] by (individual) proof.

Proposition XI. Problem VI.[114]

A body moves in a hyperbola: there is required the law of centripetal force tending to a focus of the figure.

Let CA, CB be the hyperbola's semi-axes, GP and DK conjugate diameters, PF and Qt[115] perpendiculars to the diameters, and QV ordinate to the diameter GP. Draw SP cutting the diameter DK in E and the line QV, in X, and complete the parallelogram $QRPX$. It is evident that EP is equal to the semi-major-axis AC, seeing that, when from the hyperbola's focus H the line HI is drawn parallel to CE, because CS and CH so also are ES and EI, and hence EP is the half-difference of PS and PI, that is, (because HI, PR are parallel and the angles $\widehat{IPR}$, $\widehat{HPZ}$ equal) of PS and PH, [whose difference][116] equals the total axis $2AC$. To SP let fall the perpendicular QT. Then, on calling the hyperbola's principal

(112) Newton subsequently added 'his proportionalia' (the proportionals to these) in both instances, likewise beginning the following sentence with the restyled opening 'Ducantur hæc æqualia in...' (Multiply these equals by...).

(113) As Newton will afterwards make implicit in Corollary 1 to Proposition XII below, the two ensuing cases of impressed motion in a hyperbola/parabola round a force-centre at a focus—together with the limiting case of such motion in a straight line through the force-centre which he considers afterwards in Propositions XXI–XXVI (compare note (125))—exhaust the possibilities of orbit in an inverse-square force-field. What use he here foresees for the hyperbolic orbit is difficult to envisage: no planet or then known solar comet has this shape of curve.

(114) This parallel theorem, here newly contrived, closely—and on occasion too slavishly—follows the pattern of Proposition X preceding.

(115) This thoughtless re-introduction of Qt from Proposition X is devoid of any purpose (since a hyperbola cannot—unlike the previous ellipse—be traversed under a centripetal force to its centre) but lingered on into print (*Principia*, $_1$1687: 52; $_2$1713: 50), to be suppressed only in the ensuing *editio ultima* ($_3$1726: 56), with a corresponding deletion in the accompanying figure. In the Latin text below we have corrected a similarly thoughtless repeat of 'Ellipseos'.

(116) Newton intends the converse 'disjunctim' and so we render it in our English version, which is an exact translation of the recast clause 'quarum differentia...adæquat' in the published *Principia* ($_1$1687: 51).

Hyperbolæ latere recto principali $\left(\text{seu}\,\frac{2BC^q}{AC}\right)$ dicto L, erit $L \times QR$ ad $L \times PV$ ut QR ad PV id est ut PE (seu AC)ad PC[,] et $L \times PV$ ad GVP ut L ad GV[,] et GVP ad QV^q ut CP^q ad CD^q, et QV^q ad QX^q punctis Q et P coeuntibus [a]fit ratio æqualitatis,(117) et QX^q seu QV^q est ad QT^q ut EP^q ad PF^q id est ut CA^q ad PF^q sive [b]ut CD^q ad CB^q[,] et conjunctis his omnibus rationibus $L \times QR$ fit ad QT^q ut AC ad $PC+L$ ad $GV+CP^q$ ad CD^q+CD^q ad CB^q[,] id est ut $AC \times L$ (seu $2BC^q$) ad $PC \times GV+CP^q$ ad CB^q sive ut $2PC$ ad GV. Sed punctis Q et [P] coeuntibus æquantur $2PC$ et GV. Ergo(112) $L \times QR$ & QT^q æquantur. Ducatur pars utraq; in $\frac{SP^q}{QR}$ et fiet $L \times SP^q = \frac{SP^q \times QT^q}{QR}$. Ergo [c]vis centripeta reciprocè est ut $L \times SP^q$ id est in ratione duplicata distantiæ SP. Q.E.I.(118)

[a]Lem. VIII.
[b]Lem XII.
[c]Cor. Theor. V.

Lemma XIII.

Quadratum perpendiculi quod ab umbilico Parabolæ ad tangentem ejus demittitur, est ad quadratum intervalli inter umbilicum et punctum contactus ut latus rectum principale ad latus rectum quod pertinet ad diametrum transeuntem per punctum contactus.

Sit AQP Parabola, S umbilicus ejus, A vertex principalis, P punctum contactus, PM tangens diametro principali occurrens in M et SN linea perpendicularis ab

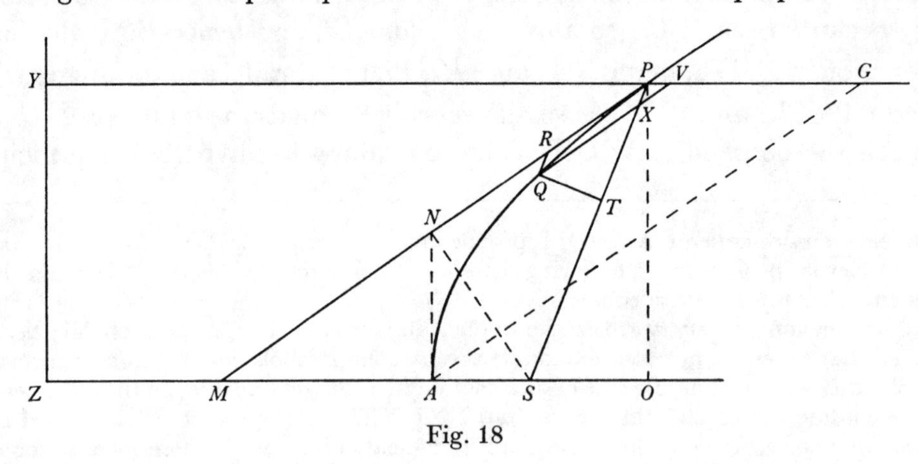

Fig. 18

umbilico in tangentem. Produc SA ad Z ut sit AZ latus rectū principale. Huic AZ erige perpendicularem ZY, cui occurrat PY ipsi AZ parallela, et erit hæc PY latus rectum pertinens ad diametrum transeuntem per punctum contactus P. Id ex Conicis patet.(119) Dico igitur quod sit PS^q ad SN^q ut PY ad AZ. Nam completis parallelogrammis $AGPM$, $ZYPO$, ob æqualia triangula PSN, MSN et

(117) 'puta 1 ad 1' (say, as 1 to 1) is deleted in sequel.

(118) In revise Newton squashed in a following afterthought that 'Eodem modo demon-

latus rectum (viz. $2BC^2/AC$) L, there will be $L \times QR$ to $L \times PV$ as QR to PV, that is, as PE (or AC) to PC; and $L \times PV$ to $GV \times VP$ as L to GV; and $GV \times VP$ to QV^2 as CP^2 to CD^2; while QV^2 to QX^2, with the points Q and P coalescing, becomes[a] a ratio of equality;(117) and QX^2 or QV^2 is to QT^2 as EP^2 to PF^2, that is, as CA^2 to PF^2 or[b] as CD^2 to CB^2: and, when all these ratios are combined, $L \times QR$ comes to be to QT^2 as (AC to PC) $\times$ (L to GV) $\times$ (CP^2 to CD^2) $\times$ (CD^2 to CB^2), that is, as ($AC \times L$ (or $2BC^2$) to $PC \times GV$) $\times$ (CP^2 to CB^2), or as $2PC$ to GV. But, with the points Q and P coalescing, $2PC$ and GV are equal. Therefore(112) $L \times QR$ and QT^2 are equal. Multiply each member by SP^2/QR and there will come to be $L \times SP^2 = SP^2 \times QT^2/QR$. Therefore[c] the centripetal force is reciprocally as $L \times SP^2$, that is, in the doubled ratio of the distance SP. As was to be found.(118)

[a]Lemma VIII.

[b]Lemma XII.

[c]Corollary to Theorem V.

Lemma XIII.

The square of the perpendicular let fall from the focus of a parabola to its tangent is to the square of the distance intervening between the focus and the point of contact as the principal latus rectum to the latus rectum pertaining to the diameter passing through the point of contact.

Let AQP be the parabola, S its focus, A the principal vertex, P the point of contact, PM the tangent meeting the main diameter in M, and SN the line perpendicular from the focus onto the tangent. Extend SA to Z so that AZ is the principal *latus rectum*. To this AZ erect the perpendicular, met by PY parallel to AZ, and this line PY will then be the *latus rectum* pertaining to the diameter passing through the point P of contact. This is evident from the *Conics*.(119) I say, therefore, that PS^2 shall be to SN^2 as PY to AZ. For on completing the parallelogram $AGPM$ and rectangle $ZYPO$, because the triangles PSN, MSN are equal

stratur quod corpus hac vi centripeta in centrifugam versa movebitur in hyperbola conjugata' (In the same manner it is proved that, with the present centripetal force converted to be centrifugal, a body will move in a 'conjugate' [*sc.* opposite] hyperbola). Such a 'conjugate' hyperbolic branch as that through G is evidently the only conic arc possessing (as required) a focus S not contained within its convexity. At Pemberton's latter suggestion the confusing 'conjugata' was changed to be 'opposita' (*Principia*, ${}_3$1726: 58).

(119) Not so at all! For, since by the ensuing argument $YP/ZA = PS^2/SN^2 = PS/AS$ where (by definition) $ZA = 4AS$, at once $YP = 4PS = 4MS = ZA + 4AO \neq ZO$ and so the asserted construction is false. Fortunately, what is to be proved requires only that the *latus rectum* YP (whose existence is established in Apollonius, *Conics* I, 49) be in proportion to ZA as PS^2 to SN^2. When Newton caught his error, he initially modified his demonstration (see Appendix 2.9 below) to depart from the unproven assertion that 'ex natura Parabolæ' $4PS$ will be the *latus rectum* pertaining to the diameter through P. At a subsequent stage, perhaps aware that this auxiliary theorem is not to be found in Apollonius' *Conics* (where, indeed, the focus of a parabola is not even defined), he set the latter as an initial Lemma XIII—justifying it with a vague 'Patet ex Conicis'—and demonstrated the present one in its train as a new Lemma XIV, making appropriate adjustments in the proof of the following proposition; the finished revise is reproduced in Appendix 2.10.

similia *MSN*, *MPO*,(120) est PS^q ad SN^q ut PM^q seu AG^q ad PO^q, hoc est ut rectangulum *YPG* ad rectangulum *ZAO*, id est, ob æquales *PG*, *AM*, *AO*, ut *PY* ad *AZ*. Q.E.D.

Prop. XII. Prob. VII.(121)

Movetur corpus in perimetro Parabolæ, requiritur lex vis centripetæ tendentis ad umbilicum hujus figuræ.

Fig. 18 Maneat constructio Lemmatis, sitꝗ *P* corpus in perimetro Parabolæ; et a loco *Q* in quem corpus proxime movetur, age ipsi *SP* parallelam *QR* et perpendicularem *QT*, nec non *QV* tangenti parallelam et occurrentem tum diametro *YPG* in *V*, tum distantiæ *SP* in *X*. Jam ob similia triangula *PXV*, *MSP* et æqualia unius latera *SM*, *SP*, æqualia sunt alterius latera *PX* seu *QR*, et *PV*. Sed ex Conicis,(122) quadratum ordinatæ *QV* æquale est rectangulo sub latere recto et segmento diametri *PV*, et punctis *P* et *Q* coeuntibus ratio *QV* ad *QX* fit æqualitatis, Ergo QX^q æquale est rectangulo $YP\times QR$. Est autem QX^q ad QT^q ut PS^q ad SN^q[,] hoc est (per Lem. XIII) ut *YP* ad *ZA* id est ut $YP\times QR$ ad $ZA\times QR$, et proinde QT^q et $ZA\times QR$ æquantur. Ducantur hæc æqualia in $\frac{SP^q}{QR}$ et fiet $\frac{SP^q\times QT^q}{QR}$ æquale $SP^q\times ZA$. Quare per Cor. Theor. [V] vis centripeta est reciprocè ut $SP^q\times ZA$[,] id est, ob datum *ZA*, [reciprocè] in duplicata ratione distantiæ *SP*. Q.E.I.

Corol. 1. Ex tribus novissimis Propositionibus consequens est quod si corpus quodvis *P* secundùm lineam quamvis rectam *PR* quacunꝗ cum velocitate, exeat de loco *P*, et vi centripeta, quæ sit reciprocè proportionalis quadrato distantiæ a centro, simul corripiatur,(123) movebitur hoc corpus in aliquâ Sectionum Conicarum.(124)

(120) Justification of this not immediately obvious premiss follows straightforwardly from the familiar properties of a parabola that the tangency of *MP* at *P* determines that $MA = AO$ (Apollonius, *Conics* I, 35) and $MS = SP$ (a corollary to the parabola's focus-directrix definition cited by Pappus in his *Mathematical Collection* VII, 238): both are standard theorems in seventeenth-century geometrical texts.

(121) As we have mentioned (note (119)) Newton came in revise to modify the detailed demonstration of this proposition somewhat. The mature version (that finally published in the *Principia* in 1687) is, for comparison's sake, reproduced in Appendix 2.10 below.

(122) Apollonius, *Conics* I, 49 is the *locus classicus* where demonstration of this basic property is given (though Archimedes cites it as Theorem 3 of his *Quadrature of the Parabola*, omitting its proof as being already standard in the 'Elements of Conics'—presumably by Euclid and Aristæus—extant in his day, but now lost).

(123) Afterwards altered to be 'agitetur' (...disturbed).

(124) As it stands, this is a none too happy or—because no mention of instantaneous speed is made in the preceding Propositions X–XII to which it is appended—too narrowly logical a corollary. All that Newton has there done is to demonstrate that in the case of the three (non-degenerate) species of conic the limit-value of QT^2/QR is equal to its *latus rectum*, that is, in the

and MSN, MPO are similar,(120) there is PS^2 to SN^2 as PM^2 (or AG^2) to PO^2, that is, as the rectangle $YP \times PG$ to the rectangle $ZA \times AO$ and accordingly, because PG, MA and AO are equal, as PY to AZ. As was to be proved.

Proposition XII. Problem VII.(121)

A body moves in the perimeter of a parabola: there is required the law of centripetal force tending to the focus of this figure.

Maintaining the construction of the lemma, let P be the body in the parabola's perimeter, and then, from the position Q to which the body next moves, parallel to SP draw QR and perpendicular to it QT and also QV parallel to the tangent and meeting both the diameter YPG in V and the distance SP in X. Now, because the triangles PXV, SMP are similar and the sides SM, SP of one are equal, the sides PX (or QR) and PV of the other are equal. But from the *Conics*(122) the square of the ordinate QV is equal to the rectangle contained by its *latus rectum* and the segment PV of its diameter, while, with the points P and Q coalescing, the ratio of QV to QX becomes one of equality. Therefore QX^2 is equal to the rectangle $YP \times QR$. However, QX^2 is to QT^2 as PS^2 to SN^2, that is, (by Lemma XIII) as YP to ZA and so as $YP \times QR$ to $ZA \times QR$, and in consequence QT^2 and $ZA \times QR$ are equal. Multiply these equals by SP^2/QR and there will come to be $SP^2 \times QT^2/QR$ equal to $SP^2 \times ZA$. Whence by the Corollary to Theorem V the centripetal force is reciprocally as $SP^2 \times ZA$, that is, because ZA is given, reciprocally in the doubled ratio of the distance SP. As was to be found.

Corollary 1. It is a consequence of the three most recent propositions that, should any body P depart from the place P following any straight line PR and with any velocity whatsoever, and if it be instantaneously snatched up(123) by a centripetal force which is reciprocally proportional to the square of the distance from its centre, this body shall move in some one of the conic sections.(124)

---

parabola its parameter, and in a central conic $a(1-e^2)$, where a is the main axis and e its eccentricity. While it is clear that variations in the instantaneous speed at P (and so in the length of PR) and in its direction (and therefore in the magnitude of $\widehat{SPR}$) will correspondingly alter $QT = PR.\sin \widehat{SPR}$, and that an increase in the magnitude of the centripetal force acting at P towards S will analogously augment the deviation QR, it is nowhere specified how such changes will alter the conic's axis and eccentricity, that is, define the size and shape of the conic orbit which, passing through P and with a focus at S, has its new *latus rectum* equal in length to $\lim_{Q,R \to P} (QT^2/QR)$. An adequate characterisation of this sort might (compare §1: note (73)) be made by an anticipatory appeal to Problem 4 of the earlier 'De motu Corporum', here delayed to be Proposition XVI following, and in later years Newton did indeed assert that 'the first Corollary...being very obvious, I...contented myself with adding ...Proposition whereby it is proved that a body in going from any place with any velocity will in all cases describe a conic Section: wch is that very Corollary' (ULC. Add. 3968.9: 101^v^, part of a draft preface to an abortive intended edition of the *Principia* in about late 1714). Likewise, the hidden assumption here made that no curve other than a conic may, in an inverse-square

Corol. 2. Et si velocitas, quâcum corpus exit de loco suo *P*, ea sit, quâ lineola *PR* in minima aliqua temporis particula describi possit, et vis centripeta potis sit eodem tempore corpus idem movere per spatiũ *QR*: movebitur hoc corpus in Conica aliqua sectione cujus latus rectum est quantitas illa $\frac{QT^q}{QR}$ quæ ultimò fit ubi lineolæ *PR*, *QR* in infinitum diminuuntur. Circulum in his Corollarijs refero ad Ellipsin, et casum excipio ubi corpus rectà descendit ad centrum.(125)

Schol.(126)

Si vis centripeta ageret in omnibus distantijs æqualiter, corpus autem hac vi urgente describeret curvam *ABCGE* et in *A* longissimè distaret a centro *S*, perveniret idem corpus ad minimam a centro distantiam in *C* ubi angulus *ASC*

force-field centred on *S*, satisfy all possibilities of motion at *P* wants—for all its manifest plausibility—an explicit, rigorous justification, and Newton was later fairly criticised by Johann Bernoulli for merely (in the unchanged statement of it published in his *Principia*, $_1$1687: 55) presupposing its truth 'sans le démontrer'. (See the 'Extrait' from Bernoulli's letter to Jakob Hermann of 7 October 1710 (N.S.) which was read out to the Paris Académie des Sciences at its session on 13 December following and subsequently printed in its *Mémoires de Mathématique & de Physique pour [l'Année. MDCC. X]* (Paris, [$_1$1713 →] $_2$1732): 521–32. His added comment (*ibid.*) that the furnishing of such a justification would demand 'bien plus d'adresse' than mere compounding of direct, geometrical arguments permits is more challengeable, since it is here easy to show that Propositions X–XII do exhaust all the possibilities of motion under an inverse-square central force. His own preferred analytical proof of the uniqueness of the conic orbit is itself less than wholly original, for in effect it but works through the particular inverse-square case of the general solution of the 'inverse problem' of central forces presented in Proposition XLI of Newton's 'De motu Corporum Liber primus' (§3 following) and afterwards published in the *Principia* ($_1$1687): 127–9; compare D. T. Whiteside, 'The Mathematical Principles underlying Newton's *Principia Mathematica*', *Journal for the History of Astronomy*, **1**, 1970: 116–38, especially 123–6.) As it happens, Newton had himself noticed the *non-sequitur* in his corollary the year before, just as the second edition of the *Principia* was going to press, and to repair the lapse in it on 11 October 1709 requested its editor Roger Cotes (see Joseph Edleston, *Correspondence of Sir Isaac Newton and Professor Cotes* (London, 1850): 5) to implement its bare statement with the added sentences 'Nam datis umbilico et puncto contactus & positione tangentis, describi potest Sectio conica quæ curvaturam datam ad punctum illud habebit. Datur autem curvatura ex data vi centripeta: et Orbes duo se mutuo tangentes eadem vi describi non possunt' (For, given a focus, the point of contact and the position of the tangent, there can be described the conic which shall have a given curvature at that point. The curvature, however, is given from the given centripetal force: and two [such] orbits which are tangent one to the other cannot be described by the same force). This abrupt *addendum* passed without further elaboration into print in 1713 (*Principia*, $_2$1713: 53) and was not afterwards altered, Newton leaving his reader to supply the necessary supporting detail. Given that Newton knew the polar equation of a general conic referred to a focus—that of the ellipse he had derived in a stray astronomical fragment of c. 1665 entered on ULC. Add. 4004: 1191^r (compare D. T. Whiteside, 'Newton's Early Thoughts on Planetary Motion: A Fresh Look', *British Journal for the History of Science*, **2**, 1964: 117–37, especially 123, note 23), while those for the hyperbola and parabola are entirely analogous—and that he could (see III: 170) construct at an arbitrary point (y, x) of a general

Corollary 2. And if the speed with which the body departs from its place P be that with which the line-element PR might be described in some minimal particle of time, and the centripetal force have the power to move the same body in the same time through the interval QR: this body will then move in some conic whose *latus rectum* is the quantity QT^2/QR which ultimately results when the elements PR and QR are infinitely diminished. The circle in these corollaries I list with the ellipse, and I exclude the case where the body descends straight downwards to the centre.(125)

Scholium.(126)

Should the centripetal force act uniformly at all distances, and were a body under the pressure of this force to describe the curve $ABCGE$, being farthest distant from the centre S at A, the same body would attain its least distance from

polar curve the radius of curvature $(1+z^2)^{\frac{3}{2}}/y^{-1}(1+z^2-r)$, in which (in anachronous fluxional terminology) $z = \dot{y}/y\dot{x}$ and $r = \dot{z}/\dot{x}$, it will be evident he was thereby able to compute the elements of the conic $y^{-1} = A+B\cos(x+\epsilon)$, A, B and ϵ constant, which passes through the point $P(R, 0)$ where the instantaneous orbital speed is v in a direction at an angle α to the radius vector SP along which the impressed force acting centrally to the focus $S(0, 0)$ is of magnitude g. For at $y = R$ the conic's slope $z = \sqrt{[B^2y^2-(Ay-1)^2]}$ is $\cot\alpha$ and its radius of curvature $OP = (1+z^2)^{\frac{3}{2}}/A$, that is, $1/A\sin^3\alpha$, is $v^2/g\sin\alpha$ (since $v^2/OP = g\sin\alpha$, the component of g acting perpendicularly to the conic at P; see note (86)); hence $A = g/v^2\sin^2\alpha$ and therefore the conic is $y^{-1} = A(1-e\cos(x+\epsilon))$ of eccentricity $e = B/A = \sqrt{[1-(v^2/gR)(2-v^2/gR)\sin^2\alpha]}$ and main axis $2/A(1-e^2) = 2R/(2-v^2/gR)$, exactly as in §1: note (73). (That the radius of curvature at P is $(\mathrm{cosec}\,\alpha)^3/A$ follows directly from the Corollary to 'Exempl: 1' on III: 158 on taking $1/A = \frac{1}{2}a = DP.\sin\alpha$.) Accordingly, since for every v, g, R and α a unique corresponding conic orbit through P may be defined, from Newton's premiss that no two distinct orbits may be drawn to satisfy these initial conditions of motion it follows that only conic inverse-square orbits are possible.

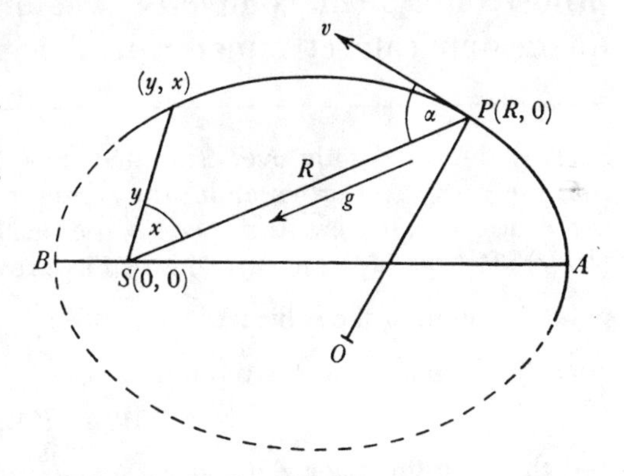

(125) As we have remarked above (see note (113)), this limit case of the inverse-square conic orbit is treated at considerable length in the ensuing Propositions XXI ff.

(126) This scholium, none too accurate in its numerical detail (see notes (127) and (129)), was later cancelled by Newton and finds no place in the revised 'De motu Corporum Liber' except as an implied corollary to Proposition XLV (see § 3: note (251)), where its loose extrapolations from the few central-force orbits hitherto constructed by him are subsumed under a generalised analytical argument in the case where the orbit does not significantly depart from a circle. The manuscript lacks an accompanying figure: that here reproduced is (compare note (12) above) our restoration in line with the text (which was first published without diagram or comment by J. W. Herivel in his *The Background to Newton's Principia* (Oxford, 1966): 325).

est 110 graduum circiter,(127) deinde ad Augem(128) seu maximam a centro distantiam in D ubi angulus CSD est æqualis angulo ASC, postea ad minimam a centro distantiam in E ubi angulus DSE est æqualis angulo CSD[,] et sic [in] infinitum. Quod si vis centripeta reciproce proportionalis esset distantiæ a centro, corpus de loco maximæ sui a centro distantiæ A descenderet ad locum minimæ a centro distantiæ, puta ad G, ubi angulus ASG est quasi 136 vel 140 graduum,(129) dein hoc angulo repetito ascenderet rursus ad maximam a centro distantiam et sic per vices in infinitum. Et universaliter, si vis centripeta decresceret in minori quam duplicata ratione distantiæ a centro corpus ad Augem prius rediret quam compleret circulum, (130) sin vis illa decresceret in majore quam duplicata et minore quam triplicata ratione corpus prius compleret

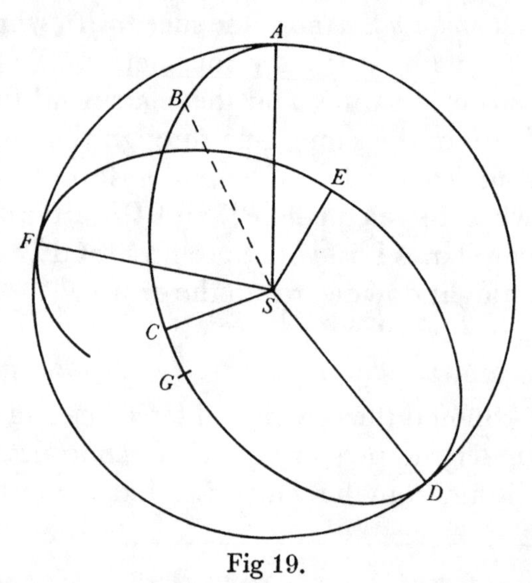

Fig 19.

(127) This is slightly over-estimated. For, if in modern analytical equivalent we suppose that the body, as it orbits under the constant centripetal force g instantaneously directed to the centre $S(0, 0)$, has the speed V perpendicular to the radius $SA = R$, then its speed $(RV/r)\sqrt{[1+(dr/rd\phi)^2]}$ at $B(r, \phi)$ is readily proved to be $\sqrt{[V^2+2g(R-r)]}$, whence the defining polar equation of the orbit is $\widehat{ASB} = \phi = \int_R^r \frac{RV}{r^2\sqrt{[V^2(1-R^2/r^2)+2g(R-r)]}}.dr$; when, consequently, the body is at the minimum distance $SC = \rho(\neq R)$ from S, the speed

$$V = \sqrt{[2g\rho^2/(R+\rho)]}$$

and therefore the angle $\widehat{ASC}$ through which the body has passed in traversing the arc $\widehat{AC}$ is

$$\int_R^\rho \frac{R\rho}{r\sqrt{[(R-r)(r-\rho)(r(R+\rho)+R\rho)]}}.dr = \int_{-\frac{1}{2}\pi}^{\frac{1}{2}\pi} \sqrt{\frac{1+\lambda\sin\theta}{3+\lambda\sin\theta}}.d\theta \text{ radians,}$$

that is, $180/\sqrt{[3+\frac{1}{2}\lambda^2+\frac{1}{2}\lambda^4]}$ degrees very nearly, on therein setting $\lambda = (R-\rho)/(R+\rho)$ and $\sin\theta = (r(R+\rho)-2R\rho)/r(R-\rho)$. Lacking his autograph figure (see previous note) and hence precise knowledge of the ratio of $SA = R$ and $SC = \rho$ to which it relates, we cannot narrowly assess the accuracy of Newton's present assertion that the angle $\widehat{SAC}$ is 'about' 110°, but it is readily seen that for $R > \rho \geqslant \frac{1}{2}R$ the true value is more nearly 103° and that in every case it is less than $180/\sqrt{3} \approx 103.9°$ (seeing that this theoretical maximum is attained only when $|\lambda|$ becomes vanishingly small and the orbit is a circle, traversed uniformly at the speed $\sqrt{[gR]}$, in which the points $A, C, D, E, F, \ldots$ of greatest and least distance from S are not defined). Five years before when, on 13 December 1679, Newton had disingenuously introduced this Borellian orbit—there conceived to be generated by 'inumerable & infinitly little ...impresses of [uniform] gravity in every moment of it's passage' (*Correspondence of Isaac Newton*, 2: 308)—in his correspondence with Robert Hooke regarding his earlier notion (*ibid.*: 301) of the 'spiral' path of fall downwards of a body released at the earth's surface, he

the centre at C where the angle $A\widehat{S}C$ is about 110°,(127) and next its auge(128) or greatest distance from the centre at D where the angle $C\widehat{S}D$ is equal to the angle $A\widehat{S}C$, thereafter its least distance from the centre at E where the angle $D\widehat{S}E$ is equal to the angle $C\widehat{S}D$, and so on indefinitely. But were the centripetal force reciprocally proportional to the distance from the centre, the body would descend from the position of its maximum distance from the centre at A to the position of its least distance from the centre at G, say, where the angle $A\widehat{S}G$ is something like 136° or 140°,(129) and would then on repeating this angle rise again to a maximum distance from the centre, and so on turn by turn indefinitely. And universally, if the centripetal force were to decrease in less than the doubled ratio of the distance from the centre, the body would return to its auge before it could complete a circle,(130) while if that force were to decrease in a greater than doubled but less than tripled ratio, the body would complete a

---

had roughly sketched a figure in which $\rho \approx \frac{3}{8}R$, but yet the corresponding central angle is considerably more than 115°. (See J. Pelseneer, 'Une lettre inédite de Newton', *Isis*, **12**, 1929: 237–54; J. A. Lohne, 'Hooke *versus* Newton', *Centaurus*, **7**, 1960: 6–52, especially 23–30, 43–5; and D. T. Whiteside, 'Newton's Early Thoughts on Planetary Motion' (note (124)): 131–5. A photocopy of the manuscript figure—none too accurately reproduced in a number of recent secondary studies—is given on page 27 of Lohne's article.) While he here, doubtless guided by a more careful and sophisticated analysis on the lines of that we have given above, certainly goes some way towards refining the crudity of this more primitive 'description by points quam proximè' as he sketched it for Hooke in 1679, it will be clear that Newton has still not achieved a perfect understanding of its subtleties.

(128) The 'aphelion' points from which the orbital speed, there at its slowest, begins to increase. As John Harris defined it a little afterwards, 'AUGE...is that Point of the Orbit, wherein a Planet being, is furthest distant from the Central Body about which it revolves, and is then slowest in Motion, insomuch that from this Point the distance of a Planet is [in the standard seventeenth-century convention] reckoned, to find thereby the Inequality of its Motion' (*Lexicon Technicum. Or, An Universal English Dictionary of Arts and Sciences* (London, 1704): signature K1^v). The etymological derivation is from the Arabic 'awj' (summit).

(129) Since, in the limiting case of the circle where SG comes to be equal to SA, the angle $A\widehat{S}G$ attains a maximum value of $180/\sqrt{2} \approx 127.3°$, this is well out. If (in the terms of note (127) with the instantaneous speed at $B(r, \phi)$ now $\sqrt{[V^2+2g(\log R-\log r)]}$, or equivalently so) Newton here went to the trouble of deriving the general polar equation of the inverse-distance orbit, we may picture his dismay when faced with an equivalent to the integral

$$\int_R^\rho \frac{R\rho}{r\sqrt{[-\rho^2(R^2-r^2)+r^2(R^2-\rho^2)(\log R-\log r)/(\log R-\log \rho)]}}.dr$$

which ensues for $A\widehat{S}G$ on substituting $V^2 = 2g\rho^2(\log R-\log\rho)/(R^2-\rho^2)$ in it. His conjecture in lieu that $A\widehat{S}G$ is 'virtually' 140° evidently comes from intercalating in the sequence 180°, *c.* 110° and 90° of corresponding angles where the central force acting at the general point B varies as SB^n, $n = -2$, 0 and 1 respectively, the value for $n = -1$: wisely, he does not extrapolate this sequence $(5n^2-25n+110)°$ to embrace the inverse-cube case $(n = -3)$!

(130) Understand a single revolution round the centre S.

circulum(130) quam rediret ad Augem. At si vis eadem decresceret in triplicata vel plusquam triplicata ratione distantiæ a centro, et corpus inciperet moveri in curva quæ in principio motus secaret radium AS perpendiculariter, hoc si semel inciperet descendere, pergeret semper descendere usq; ad centrum, [ac] si semel inciperet ascendere abiret in infinitum.(131)

Prop. XIII. Theorema VI.(132)

Si corpora plura circa commune centrum volvantur(133) *et vis centripeta decrescat in duplicata ratione distantiarum a centro; dico quod Latera recta orbitarum*(133) *sunt in duplicata ratione arearum quas corpora radijs ad centrum ductis eodem tempore describunt.*

Nam per Cor. 2. Prob. [VII] latus rectum L est æquale quantitati $\frac{QT^q}{QR}$ quæ ultimò fit ubi coeunt puncta P et Q. Sed linea minima QR dato tempore est ut vis centripeta generans, hoc est [a]reciprocè ut SP^q. Ergo $\frac{QT^q}{QR}$ est ut $QT^q \times SP^q$, hoc est, latus rectum L in duplicata ratione areæ $QT \times SP$.(134) Q.E.D.

[a]Hypoth.

Corol. Hinc Ellipseos area tota, eiq; proportionale rectangulum sub axibus, est in ratione composita ex dimidiata ratione lateris recti & integra ratione temporis periodici.

Prop. XIV. Theorema VII.(135)

Iisdem positis dico quod tempora periodica in Ellipsibus sunt in ratione sesquiplicata transversorum axium.

Namq; axis minor est medius proportionalis inter axem majorem(136) et latus

(131) If the central force acting at $B(r, \phi)$ is taken to be $g(r/R)^n$, then, much as before (see notes (127) and (129)), on setting $V^2 = 2g\rho^2(R^{n+1}-r^{n+1})/(n+1)\, R^n(R^2-\rho^2)$ in the ensuing polar equation of orbit we may determine that $(n \neq -1)$ the size of $\widehat{ASG}$ is

$$\int_R^\rho \frac{R\rho}{r\sqrt{[-\rho^2(R^2-r^2)+r^2(R^2-\rho^2)\,(R^{n+1}-r^{n+1})/(R^{n+1}-\rho^{n+1})]}}\cdot dr.$$

In the inverse-cube case $(n = -3)$ the denominator of the integrand is zero, and hence $\widehat{ASG} = \infty$, while for $n < -3$ a more involved argument will establish the same result. Newton's own reasoning here is almost certainly more qualitative: his assertion is manifestly true in the logarithmic spiral (Proposition VIII preceding), which is seemingly the only inverse-cube central orbit he yet knows, and, because (when $r < R$) $g(r/R)^n > g(r/R)^{-3}$, *à fortiori* where the central force varies inversely as a higher power of the distance than the cube it is stronger and 'hugs' the orbiting body still more closely to itself. Such intuitive notions of the continuous alteration in angular distance between successive orbital points of maximum/minimum distance from the force-centre as the power-index of the force is conceived to vary are not unprecedented. In his reply to Newton's letter of 13 December 1679 communicating his sketch of the constant-force orbit (see note (127) above) Robert Hooke wrote back on the following 6 January that 'Your Calculation of the Curve [described] by a body attracted by an æquall power at all Distances from the center...is right and the two auges will not unite by about a third of a Revolution. But my supposition is that the Attraction always is in a duplicate proportion to the Distance from the Center Reciprocall...And that with such an

circle(130) before it could return to its auge. If, however, the same force were to decrease in the tripled or more than tripled ratio of the distance from the centre, and the body should begin to move in a curve which at the start of motion intersected the radius *AS* at right angles, this, were it once to begin to descend, would continue to descend right to the centre, while if it were once to begin to ascend, it would go off to infinity.(131)

Proposition XIII. Theorem VI.(132)

Should several bodies revolve round a common centre and the centripetal force decrease in the doubled ratio of the distances from the centre: I say that the latera recta of orbits are in the doubled ratio of the areas which bodies by rays drawn to the centre describe in the same time.

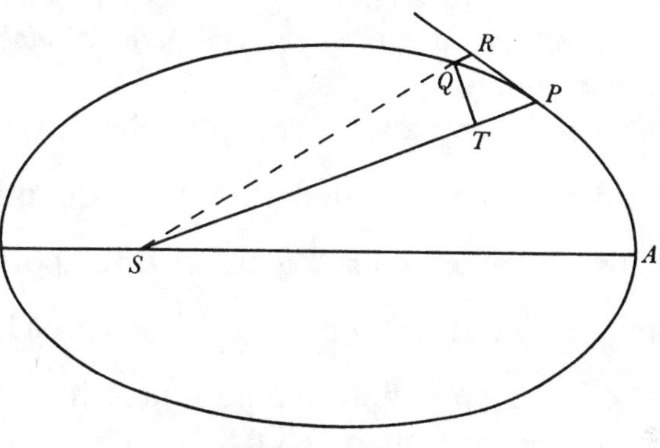

For by Corollary 2 to Problem VII the *latus rectum L* is equal to the quantity QT^2/QR which ultimately results when the points *P* and *Q* coalesce. But the very small line *QR* is in a given time as the generating centripetal force, that is,[a] reciprocally as SP^2. Therefore QT^2/QR is as $QT^2 \times SP^2$, or in other words the *latus rectum L* is in the doubled ratio of the area $QT \times SP$.(134) As was to be proved.

[a] By hypothesis.

Corollary. Hence the total areas of the ellipses, and the rectangles contained by the axes proportional to them, are in the ratio compounded of the halved ratio of the *latera recta* and the whole ratio of the periodic times.

Proposition XIV. Theorem VII.(135)

With the same suppositions I assert that the periodic times in ellipses are in the sesquialteral ratio of the transverse axes.

For the minor axis is a mean proportional between the major(136) axis and the

attraction the auges will unite in the same part of the Circle...' (*Correspondence of Isaac Newton*, **2**: 309).

(132) A minor rider to the new proof of Kepler's third law of planetary motion which is given in Proposition XIV following.

(133) Newton subsequently replaced these with the respective equivalents 'revolvantur' and 'orbium'.

(134) Strictly, this should be '$\frac{1}{2}QT \times SP$' of course.

(135) A recast of Theorem 4 of the earlier 'De motu Corporum', in which (compare §1: note (58)) a new and simpler proof is furnished of Kepler's third law of planetary motion.

(136) Newton later added in parenthesis 'quem transversum appello' (which I call 'transverse').

rectum, atq; adeo rectangulum sub axibus est in ratione composita ex dimidiata ratione lateris recti et sesquiplicata ratione axis transversi. Sed hoc rectangulum per Corollarium Theorematis sexti est in ratione composita ex dimidiata ratione lateris recti et integra ratione periodici temporis. Dematur utrobiq; dimidiata ratio lateris recti et manebit sesquiplicata ratio axis transversi æqualis rationi periodici temporis. Q.E.D.

Corol. Sunt igitur tempora periodica in Ellipsibus eadem ac in circulis quorum diametri æquantur majoribus axibus Ellipseôn.

Prop. XV. Theorema VIII.

Iisdem positis et actis ad corpora lineis rectis quæ ibidē tangant orbitas, demissisq; ab umbilico communi ad has tangentes lineis perpendicularibus: dico quod velocitates corporū sunt in ratione composita ex ratione perpendiculorum inversè et dimidiata ratione laterum rectorum directè.

Fig. 16.(137) Ab umbilico S ad tangentem PR demitte perpendiculū SY et velocitas corporis P erit reciprocè in dimidiata ratione quantitatis $\frac{SY^q}{L}$. Nam velocitas illa est ut arcus quàm minimus PQ in data temporis particula descriptus, [a]hoc est ut tangens PR, id est(138) ut $\frac{SP\times QT}{SY}$ sive ut SY reciprocè et $SP\times QT$ directè, estq; $SP\times QT$ ut area dato tempore descripta, id est per Theor. VI in dimidiata ratione lateris recti.(139) Q.E.D.

[a]Lem VII.

Corol. 1. Latera recta sunt in ratione composita ex duplicata ratione perpendiculorum et duplicata ratione velocitatum.

Corol. 2. Velocitates corporum in maximis et minimis ab umbilico communi distantijs sunt in ratione composita ex ratione distantiarum inversè et dimidiata ratione laterū rectorum directè. Nam perpendicula jam sunt ipsæ distantiæ.

Corol. 3. Ideoq; velocitas in Conica sectione in minima ab umbilico distantia est ad velocitatem in circulo in eadem a centro distantia in dimidiata ratione lateris recti ad distantiam illam duplicatam.

(137) Instead of repeating this figure—and its weight of (here) extraneous detail—from Proposition IX above, we insert in our English version the simplified diagram which Henry Pemberton introduced at this point in the *editio ultima* of the *Principia* (London, $_3$1726: 61).

(138) Newton later inserted the justification 'ob proportionales PR ad QT & SP ad SY' (because PR is in proportion to QT as SP to SY). More generally, of course, since (by Theorem I) $SP\times QT$ is proportional to the time of passage from P to Q in any arbitrary central-force orbit, there likewise the perpendicular $SY = \lim_{Q,T\to P} (SP\times QT/\widehat{PQ}) = SP.\sin\widehat{SPR}$ is inversely proportional to the instantaneous orbital speed at P.

(139) In polar analytical equivalent, where a central force of magnitude g instantaneously directed to $S(0, 0)$ varies at $P(R, 0)$ the uniform 'inertial' speed v in the tangential path PR inclined at $\widehat{SPR} = \alpha$ to the radius vector $SP = R$, we may readily deduce (see §1: note (73),

latus rectum, and therefore the rectangle contained by the axes is in the ratio compounded of the halved ratio of the *latus rectum* and the sesquialteral ratio of the transverse axis. But by the Corollary to Theorem VI this rectangle is in the halved ratio of the *latus rectum* and the whole ratio of the periodic time. Take away from each portion the halved ratio of the *latus rectum*, and there will be left the sesquialteral ratio of the transverse axis equal to the ratio of the periodic time. As was to be proved.

Corollary. Periodic times in ellipses are therefore the same as those in circles whose diameters are equal to the major axes of the ellipses.

Proposition XV. Theorem VIII.

With the same suppositions, and on drawing straight lines through the bodies which touch their orbits at these places and letting fall perpendiculars from the common focus to these tangents: I state that the speeds of the bodies are in a ratio compounded of the ratio of the perpendiculars inversely and the halved ratios of the latera recta directly.

From the focus S let fall SY perpendicular to the tangent PR, and the speed of the body P will then be reciprocally in the halved ratio of the quantity SY^2/L. For that speed is as the extremely minute arc $\widehat{PQ}$ described in a given particle of time, that is,[a] as the tangent PR or as $SP\times QT/SY$; in other words,[(138)] as SY reciprocally and $SP\times QT$ directly, with $SP\times QT$ as the area described in given time, that is, (by Theorem VI) in the halved ratio of the *latus rectum*.[(139)] As was to be proved.

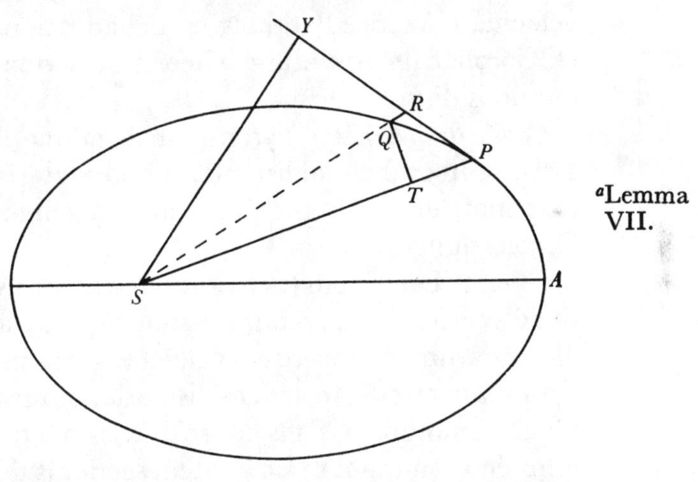

[a] Lemma VII.

Corollary 1. The *latera recta* are in a ratio compounded of the doubled ratio of the perpendiculars and the doubled ratio of the speeds.

Corollary 2. The speeds of bodies at their greatest and least distances from the common focus are in a ratio compounded of the ratio of the distances inversely and the halved ratio of the *latera recta* directly. For the perpendiculars are now those very distances.

Corollary 3. Consequently, the speed in a conic at the least distance from the focus is to the speed in a circle at the same distance from the centre in the halved ratio of the *latus rectum* to double that distance.

---

or note (124) above) that the ensuing conic orbit has *latus rectum*

$$L = 2(1-e^2)R/(2-v^2/gR) = 2(v^2/g)\sin^2\alpha,$$

whence $v^2 = \frac{1}{2}gL/\sin^2\alpha = \frac{1}{2}gR^2.L/SY^2$.

Corol. 4. Corporum in Ellipsibus gyrantium velocitates in mediocribus distantijs ab umbilico communi sunt eædem quæ corporum gyrantium in circulis ad easdem distantias, hoc est[a] reciprocè in dimidiata ratione distantiarum. Nam perpendicula jam sunt semi-axes minores, et hi sunt ut mediæ proportionales inter distantias & latera recta. Componatur hæc ratio inversè cum dimidiata ratione laterum rectorum directè et fiet ratio dimidiata distantiarum inversè.

[a] Cor. [5] Theor. IV.

Corol. 5. In eadem vel æqualibus figuris(140) velocitas corporis est reciprocè ut perpendiculum demissum ab umbilico ad tangentem. Idem obtinet in figuris inæqualibus quarum æqualia sunt latera recta.

Corol. 6. In Parabola velocitas est reciprocè in dimidiata ratione distantiæ;(141) in Ellipsi minor est, in Hyperbola major quàm in hac ratione. Nam (per Lemma XIII) perpendiculum demissum ab umbilico ad tangentem Parabolæ est in dimidiata ratione distantiæ.

Coroll. 7. In Parabola velocitas ubiq; est ad velocitatē corporis gyrantis in circulo ad eandem distantiam, in dimidiata ratione numeri binarij ad unitatem; in Ellipsi minor est, in Hyperbola major. Nam per Corollarium secundum velocitas in vertice Parabolæ est in hac ratione, et per Corollarium sextum hujus et Corollarium quintum Theorematis quarti, servatur eadem proportio in omnibus distantijs.(142)

Corol. 8. Velocitas gyrantis in Sectione quavis Conica est ad velocitatem gyrantis in circulo in distantia dimidij lateris recti sectionis, ut distantia illa ad perpendiculum ab umbilico in tangentem sectionis demissum. Patet per Corollarium quintum.

Corol. 9. Unde cum (per Corol. 5 Theor. IV) velocitas gyrantis in hoc circulo sit ad velocitatem gyrantis in circulo quovis alio, reciprocè in dimidiata ratione distantiarum; fiet ex æquo, velocitas gyrantis in Conica sectione ad velocitatem gyrantis in circulo in eadem distantia, ut media proportionalis inter distantiam illam communem, et semissē lateris recti sectionis, ad perpendiculum ab umbilico communi in tangentem sectionis demissum.(143)

Prop. XVI. Prob. VIII.(144)

Posito quod vis centripeta sit reciproce proportionalis quadrato distantiæ a centro et cognita vis illius quantitate;(145) *requiritur linea*(146) *quam corpus describet, de loco dato cum data velocitate secundum datam rectam emissum.*

(140) Continue to understand that these are conics, of course.

(141) For added clarity Newton afterwards inserted 'corporis ab umbilico figuræ' (of the body from the focus of the figure). More precisely, since (in the analytical terms of note (139) preceding) the eccentricity of the general conic orbit is $e = \sqrt{[1-(v^2/gR)(2-v^2/gR)\sin^2\alpha]}$, when this is a parabola ($e = 1$) the orbital speed $v = \sqrt{[2gR]}$. Likewise, accordingly as v is greater or less than $\sqrt{[2gR]}$—that is, as $e > 1$ or $e < 1$—the conic will be a hyperbola or ellipse (and a circle when $v = \sqrt{[gR]}$ and $\alpha = \frac{1}{2}\pi$).

Corollary 4. The speeds of bodies orbiting in ellipses are at their mean distances from the common focus the same as those of bodies orbiting in circles at the same distances, namely,[a] reciprocally in the halved ratio of the distances. For the perpendiculars are now (equal to) the semi-minor-axes, and these are the mean proportionals between the distances and the *latera recta*. Compound this ratio inversely with the halved ratio of the *latera recta* directly and it will become the halved ratio of the distances inversely.

[a] Corollary 5 to Theorem IV

Corollary 5. In the same figure, or equal ones,(140) the speed of a body is reciprocally as the perpendicular let fall from the focus to its tangent. The same holds in unequal figures whose *latera recta* are equal.

Corollary 6. In a parabola the speed is reciprocally in the halved ratio of the distance;(141) in the ellipse it is less than in this ratio, in the hyperbola greater. For (by Lemma XIII) the perpendicular let fall from the focus to the tangent of a parabola is in the halved ratio of the distance.

Corollary 7. In a parabola the speed is everywhere to the speed of a body orbiting in a circle at the same distance in the halved ratio of the number 2 to unity; in the ellipse it is less, in the hyperbola greater. For by Corollary 2 the speed at the parabola's vertex is in this ratio, and by the present Corollary 6 and Corollary 5 of Theorem IV the same proportion is preserved at all distances.(142)

Corollary 8. The orbital speed in any conic is to the orbital speed in a circle at a distance of half the section's *latus rectum* as that distance to the perpendicular let fall from the focus to the section's tangent. This is evident by Corollary 5.

Corollary 9. Whence (by Corollary 5 to Theorem IV) the speed of orbit in this circle shall be to the speed of orbit in any other circle reciprocally in the halved ratio of the distances; and *ex æquo* the orbital speed in a conic will come to be to the speed of orbit in a circle at the same distance as a mean proportional between that common distance and half the section's *latus rectum* to the perpendicular let fall from the common focus to the section's tangent.(143)

Proposition XVI. Problem VIII.(144)

Supposing that the centripetal force be reciprocally proportional to the square of the distance from its centre, and with the quantity of that force known,(145) *there is required the line*(146) *which a body shall describe when released from a given position with a given speed following a given straight line.*

---

(142) This complement to the preceding Corollary 6 is an immediate deduction from the analytical equivalents given in the preceding note.

(143) In the analytical terms of notes (139) and (141) these two final corollaries assert the equivalent proportions $v : \sqrt{[g.\frac{1}{2}L]} = R : R\sin\alpha$ and $v : \sqrt{[gR]} = \sqrt{[R.\frac{1}{2}L]} : R\sin\alpha$.

(144) Problem 4 of the earlier 'De motu Corporum' (§1), now remodelled in its dynamical portion so as to depend on Theorem VIII preceding.

(145) Newton afterwards in his revised 'De motu Corporum Liber primus' (f. 45r) altered

Vis centripeta tendens ad punctum S ea sit quæ corpus π in orbita quavis data(147) $\pi\chi$ gyrare faciat et cognoscatur hujus velocitas in loco π. De loco P secundum lineam PR emittatur corpus P cum data velocitate et mox inde cogente vi centripeta deflectat in Conisectionem PQ. Hanc igitur recta PR tanget in P. Tangat itidem recta aliqua $\pi\rho$ orbitā $\pi\chi$ in π, et si ab S ad has tangentes demitti intelligantur perpendicula, erit per Cor. 1. Theor. VIII latus rectum Conisectionis ad latus rectum orbitæ datæ in ratione composita ex duplicata ratione perpendiculorum, et duplicata ratione velocitatū, atqȝ adeo datur. Sit istud L. Datur præterea Conisectionis umbilicus S. Anguli RPS complementum ad duos rectos fiat angulus RPH et dabitur positione linea PH in qua umbilicus alter H locatur. Demisso ad PH perpendiculo SK et erecto semiaxe conjugato BC, est[a]

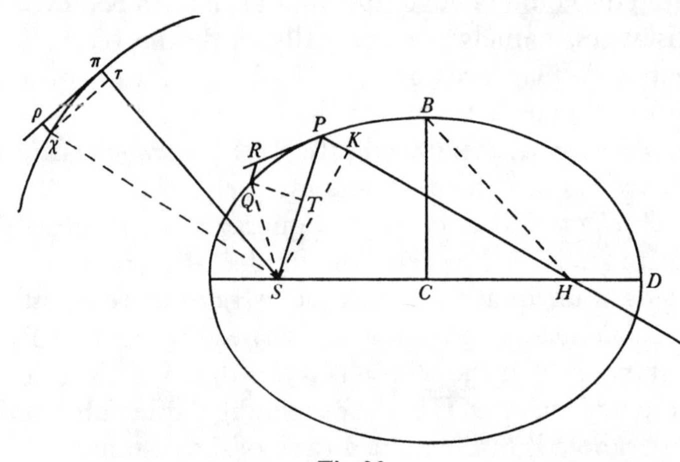

Fig 20.

[a]Prop 13. lib. 2 Elem.

$$SP^q - 2KPH + PH^q = SH^q = 4CH^q = 4BH^q - 4BC^q = \overline{SP+PH}^{\text{quad.}} - L \times \overline{SP+PH}$$
$$= SP^q + 2SPH + PH^q - L \times \overline{SP+PH}.$$

Addantur utrobiqȝ $2KPH + L \times \overline{SP+PH} - SP^q - PH^q$ et fiet

$$L \times \overline{SP+PH} = 2SPH + 2KPH,$$

seu $SP+PH$ ad PH ut $2SP+2KP$ ad L. Unde datur PH longitudine et positione.(148) Nimirum si ea sit corporis in P velocitas, ut latus rectum L minus fuerit quam $2SP+2KP$, jacebit PH ad eandem partem tangentis PR cum linea PS, adeoqȝ figura erit Ellipsis, et ex datis umbilicis S, H et axe principali $SP+PH$ dabitur. Sin tanta sit corporis velocitas ut latus rectum L æquale fuerit $2SP+2KP$ longitudo PH infinita erit et proinde figura erit Parabola axem

---

this to read '...*et quod vis illius quantitas absoluta sit cognita*' (*and that the absolute quantity of that force be known*).

(146) A conic, namely, with a focus at the force-centre. In his earlier 'De motu Corporum' Newton had (see §1: note (68)) referred more restrictively to an 'Ellipsis'.

(147) Again understand a conic with a focus at the force-centre S. This specious generalisation from Newton's earlier choice of a circle (π) of centre S and arbitrary radius merely serves to obscure the reason for making such comparison, and indeed effectively vitiates its very

Let the centripetal force tending to the point S be that which shall make the body π orbit in any given path[147] $\pi\chi$ and let the latter's speed at the position π be ascertained. Let the body P be released with given speed from the position P following the line PR and directly thereafter under the compulsion of the centripetal force be deflected into the conic $\widehat{PQ}$. This, therefore, the straight line PR will touch at P. Let some straight line $\pi\rho$ correspondingly touch the orbit $\pi\chi$ at π, and, if perpendiculars are understood to be let fall from S to these tangents, the conic's *latus rectum* will then, by Corollary 1 to Theorem VIII, be to the *latus rectum* of the given orbit in a ratio compounded of the doubled ratio of the perpendiculars and the doubled ratio of the speeds, and hence is given. Let it be L. There is given, moreover, the conic's focus S. Make the angle $R\widehat{P}H$ the supplement of the angle $R\widehat{P}S$, and there will be given in position the line PH in which the other focus H is located. On letting fall the perpendicular SK to PH and erecting the conjugate semi-axis BC, there *is*[a]

[a] *Elements*, II, 13.

$$SP^2 - 2KP \times PH + PH^2 = SH^2 = 4CH^2$$
$$= 4BH^2 - 4BC^2 = (SP+PH)^2 - L \times (SP+PH)$$
$$= SP^2 + 2SP \times PH + PH^2 - L \times (SP+PH).$$

Add $2KP \times PH + L \times (SP+PH) - SP^2 - PH^2$ to each side and there will come $L \times (SP+PH) = 2SP \times PH + 2KP \times PH$, that is, $SP+PH$ to PH as $2SP+2KP$ to L. Whence PH is given in length and position.[148] Specifically, if the speed of the body at P be such that the *latus rectum* L proves to be less than $2SP+2KP$, then PH will lie on the same side of the tangent PR as the line PS, and hence the figure will be an ellipse, given in consequence of its foci S, H and principal axis $SP+PH$ being given. But should the speed of the body be so great that the *latus rectum* L shall be equal to $2SP+2KP$, the length of PH will be infinite and as a

---

purpose. For if, in the now well-established analytical equivalent (see note (139)) we suppose that v is the comparable orbital speed at π in the present conic, and λ its *latus rectum*, with $S\pi = R'$, $S\widehat{\pi\rho} = \alpha'$ and γ the quantity of force directed at π towards S, then (because the force varies as the inverse-square of the distance from S) $g:\gamma = R^{-2}:R'^{-2}$ and therefore (by Corollary 1 of Theorem VIII)

$$L:\lambda = v^2 \times (R\sin\alpha)^2 : v^2 \times (R'\sin\alpha')^2 = (v^2/g)\sin^2\alpha : (v^2/\gamma)\sin^2\alpha',$$

whence $L = (\lambda\gamma/v^2\sin^2\alpha').(v^2/g)\sin^2\alpha$. The point of choosing (π) to be a circle of centre S is that in this simplest comparison conic $\alpha' = \frac{1}{2}\pi$ and $v^2/\gamma = \rho = \frac{1}{2}\lambda$, from which

$$L = 2(v^2/g)\sin^2\alpha$$

(compare note (139) above).

(148) Exactly as in §1: note (73), it likewise follows, in the comparable analytical equivalent in which v is the orbital speed at P, $SP = R$, $\widehat{SPR} = \alpha$ and g the quantity of the force acting at P, that $PK = -R\cos 2\alpha$ and so $SP+PK = 2R\sin^2\alpha$, whence, since $L = 2(v^2/g)\sin^2\alpha$ and therefore $(SP+PH)/PH = 2/(v^2/gR)$, the conic's main axis is $SP+PH = 2R/(2-v^2/gR)$.

habens SH parallelum lineæ PK, et inde dabitur. Quod si corpus majori adhuc velocitate de loco suo P emittitur, capienda erit longitudo PH ad alteram partem tangentis, adeoq; tangente inter umbilicos pergente Figura erit Hyperbola axem habens principalem æqualem differentiæ linearum SP & PH, et inde dabitur. Q.E.I.

Corol 1. Hinc in omni Conisectione ex dato vertice principali D latere recto L et umbilico S datur umbilicus alter H capiendo DH ad DS ut est latus rectum ad differentiam inter latus rectum et $4DS$. Nam proportio $SP+PH$ ad PH ut $2SP+2KP$ ad L, in casu hujus corollarij fit $DS+DH$ ad DH ut $4DS$ ad L, et divisim DS ad DH ut $4DS-L$ ad L.

Corol. 2. Unde si datur corporis velocitas in vertice principali D, invenietur Orbita expeditè, capiendo latus rectum ejus ad duplam distantiam DS, in duplicata ratione velocitatis hujus datæ ad velocitatem corporis in circulo ad distantiam DS gyrantis (per Corol. 3. Theor. VIII,) dein DH ad DS ut latus rectum ad differentiam inter latus rectum et $4DS$.(149)

Prop. XVII. Prob. IX.(150)

Datis axibus transversis et umbilico describere Trajectorias Ellipticas et Hyperbolicas quæ transibunt per puncta data, et rectas positione datas contingent.

Sit S communis umbilicus figurarum, AB axis principalis orbitæ cujusvis, P punctum per quod debet transire, et TR recta quam debet tangere. Centro P intervallo $AB-SP$ si orbita sit Elliptica vel $AB+SP$ si ea sit Hyperbolica, describatur circulus HG. In hoc circulo locabitur umbilicus alter. Ad tangentem TR demittatur perpendicularis ST et producatur ea ad V ut sit TV æqualis ST. Centro V intervallo AB(151) describatur circulus FH. In hoc circulo locabitur etiam umbilicus ille alter. Hac methodo, sive dentur duo puncta P, p, sive duæ tangentes TR, tr sive punctum P et tangens

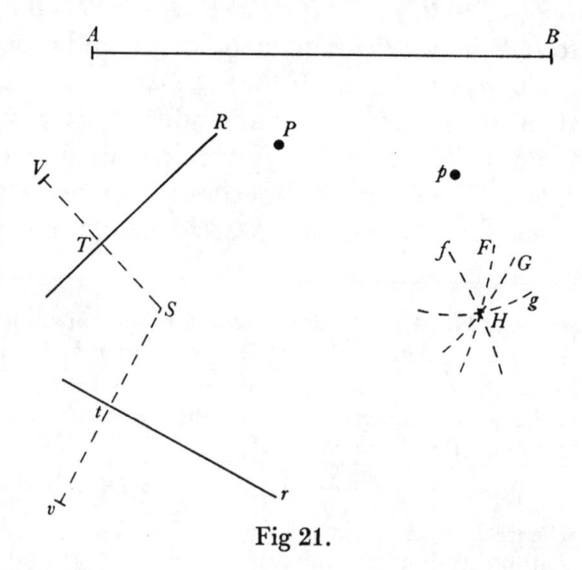

Fig 21.

(149) Newton subsequently appended (on f. 56v) two further corollaries outlining the evident way in which the present proposition may be used to compute the ensuing conic orbit when the motion is disturbed by the impulse, singly discrete or—by repeated blows at infinitesimal moments of time apart—continuous, of an external force. These are reproduced in Appendix 2.11 below.

(150) An elaboration of the opening paragraph of the scholium to Theorem 4 in the earlier

result the figure will be a parabola having its axis SH parallel to the line PK, and be given from this. While if the body is released from its position P with a still greater speed, the length PH will need to be taken on the other side of the tangent, and so, with its tangent proceeding between its foci, the figure will be a hyperbola having its principal axis equal to the difference of the lines SP and PH, and will be given thereby. As was to be found.

Corollary 1. Hence in every conic, given the principal vertex D, the *latus rectum* L and focus S, the other focus H is given therefrom by taking DH to DS as the *latus rectum* to the difference between the *latus rectum* and $4DS$. For the proportion $SP+PH$ to PH as $2SP+2KP$ to L becomes, in the case of the present corollary, $DS+DH$ to DH as $4DS$ to L, and *dividendo* DS to DH as $4DS-L$ to L.

Corollary 2. Whence if there is given the speed of the body at the principal vertex D, the orbit will swiftly be ascertained by taking its *latus rectum* to twice the distance DS in the doubled ratio of this given speed to the speed of a body revolving in a circle at the distance DS (by Corollary 3 to Theorem VIII), and thereafter DH to DS as the *latus rectum* to the difference between the *latus rectum* and $4DS$.(149)

Proposition XVII. Problem IX.(150)

Given their transverse axes and a focus, to describe elliptical and hyperbolical trajectories which shall pass through given points and touch straight lines given in position.

Let S be the common focus of the figures, AB the principal axis of any orbit, P a point through which it ought to pass, and TR a straight line which it ought to touch. With centre P and radius $AB-SP$ if the orbit be elliptical, or $AB+SP$ if it be hyperbolical, describe the circle HG. The second focus will be located in this circle. Let fall the perpendicular ST to the tangent TR and extend it to V so that TV is equal to ST, then with centre V and radius AB(151) describe the circle FH. The second focus will be located in this circle also. By this method, whether there be given two points P and p, two tangents TR and tr, or a point P and a

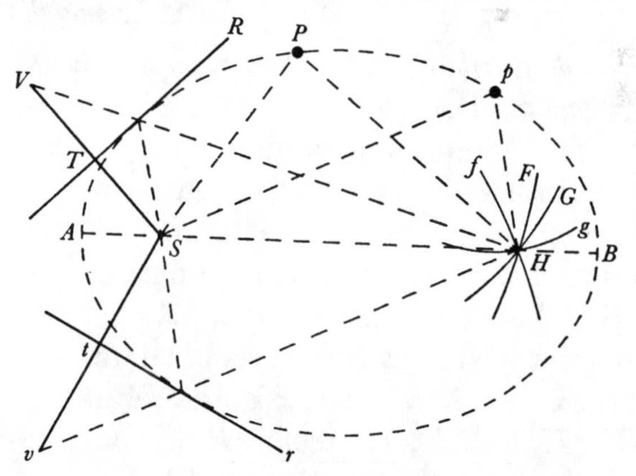

'De motu Corporum', now expanded to embrace the cases where the given conic shall not only pass through given points but touch given lines.

(151) Newton quickly cancelled an overhasty sequel 'si Orbita sit Elliptica vel $AB-SV$ si ea sit Hyperbolica' (if the orbit be elliptical, or $AB-SV$ if it be hyperbolic). In either species of central conic, where π is its point of contact with the tangent RT, the focal rays $S\pi$ and πH will (as Newton knew already from his first youthful reading of Schooten's *Exercitationes*

TR describendi sunt circuli duo et in eorum intersectione communi *H* reperietur umbilicus quæsitus. Datis autem umbilicis et axe principali datur Trajectoria. Q.E.I.

Prop. XVIII. Prob. X.(152)

Circa datum umbilicū Trajectoriam Parabolicam describere quæ transibit per puncta data et rectas positione datas continget.

Sit *S* umbilicus, *P* punctum et *TR* tangens trajectoriæ describendæ. Centro *P*, intervallo *PS* describe circulum *FG*. Ab umbilico ad tangentem demitte perpendicularem *ST*, et produc eam ad *V*, ut sit *TV* æqualis *ST*. Eodem modo describendus est alter circulus *fg* si datur alterum punctum *p*; vel inveniendum alterum punctum *v* si datur altera tangens *tr*: dein ducenda recta *IF* quæ tangat duos circulos *FG*, *fg* si dantur duo puncta *P*, *p*, vel transeat per duo puncta *V*, *v* si dantur duæ tangentes *TR*, *tr*, vel tangat circulum *FG* et transeat per punctum *V* si datur punctum *P* et tangens *TR*. Ad *FI* demitte perpendicularem *SI*, eamq̧ biseca in *K* et erit *K* vertex principalis et *SK* axis Parabolæ. Q.E.I.

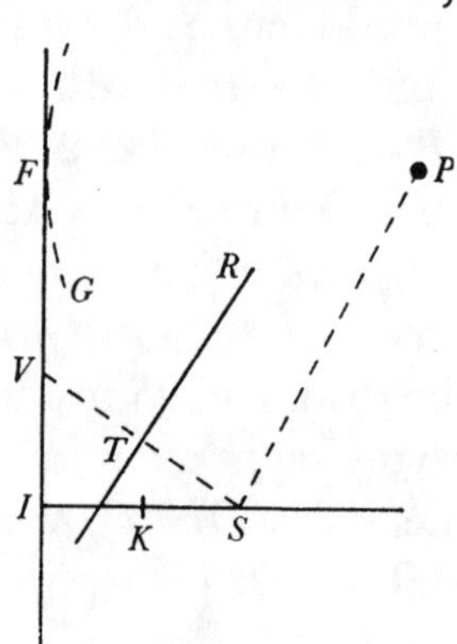

Fig 22.

Demonstrationes hujus et præcedentis ut nimis obvias non adjungo.(153)

Lemma XV.

A datis tribus punctis ad quartum non datum, inflectere tres rectas quarum differentiæ vel dantur vel nullæ sunt.

Cas. 1. Sunto puncta illa data *A*, *B*, *C*, et punctum quartū *Z* quod invenire oportet. Ob datam differentiam linearum *AZ*, *BZ* locabitur punctum *Z* in Hyperbola cujus umbilici sunt *A* et *B* et axis transversus differentia illa data. Sit axis ille *MN*. Cape *PM* ad *MA* ut est *MN* ad *AB*[,] et erecto *PR* perpendiculari ad *AB* demissoq̧ *ZR* perpendiculari ad *PR* erit ex natura hujus Hyperbolæ *ZR* ad *AZ* ut est *MN* ad *AB*. Simili discursu punctum *Z* locabitur in alia Hyperbola cujus umbilici sunt *A*, *C* et axis transversus differentia

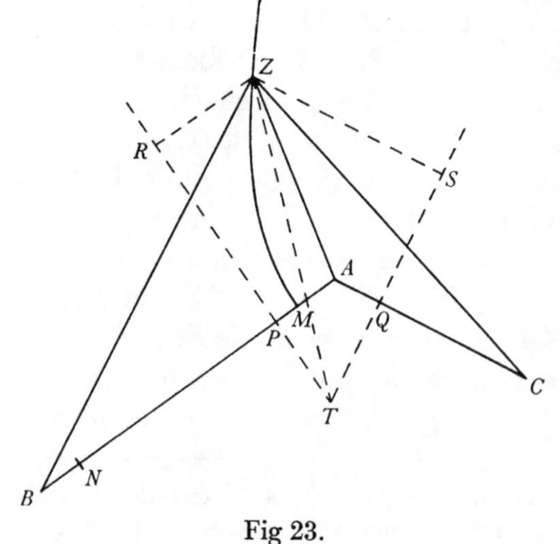

Fig 23.

Mathematicæ; see I: 32, 36) be equally inclined thereto, and so $H\pi$ will pass through V; accordingly $S\pi = \pi V$ and therefore $VH = \pm S\pi + \pi H = \pm SP + PH = AB$ in either case.

tangent TR, two circles are to be described and the required focus will then be found to be at their common intersection. Given the foci, however, and the principal axis, the trajectory is given. As was to be found.

Proposition XVIII. Problem X.(152)

To describe round a given focus a parabolic trajectory which shall pass through given points and touch straight lines given in position.

Let S be the focus, P a point and TR a tangent of the trajectory to be described. With centre P and radius PS describe the circle FG. Let fall the perpendicular ST from the focus to the tangent and extend it to V so that TV is equal to ST. There needs, if a second point p is given, to be described a second circle fg; or, if a second tangent tr is given, a second point v needs to be found: then you must describe a straight line IF which, if two points P and p are given, shall touch the two circles FG, fg; or, if two tangents TR and tr are given, shall pass through the two points V and v; or, if a point P and tangent TR is given, shall touch the circle FG and pass through the point V. To FI let fall the perpendicular SI and bisect it in K, and K will then be the principal vertex of the parabola and SK its axis. As was to be found.

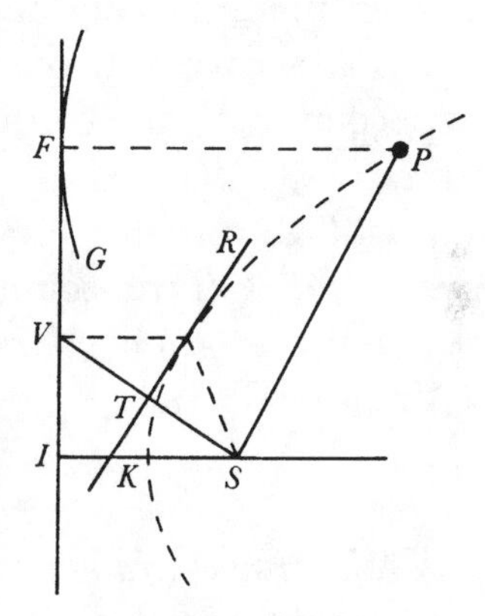

Proofs of this and the preceding proposition I do not, as being overly obvious, adjoin.(153)

Lemma XV.

From three given points to incline to a fourth one, not given, three straight lines whose differences are either given or nil.

Case 1. Let A, B and C be the given points and Z the fourth point which it is necessary to find. Because of the given difference of the lines AZ and BZ, the point Z will be located in a hyperbola whose foci are A and B and transverse axis is that given difference. Let that axis be MN. Take PM to MA as MN to AB and, on erecting PR perpendicular to AB and dropping ZR perpendicular to PR, there will from the nature of this hyperbola be ZR to AZ as MN to AB. In a like way the point Z will be located in another hyperbola whose foci are A and C

Doubtless prompted by this initial blunder, in his revised 'De motu Corporum Liber primus' Newton introduced a separate preceding Lemma XV to demonstrate the point.

(152) The limiting case of the preceding problem in which the second focus H passes to infinity and so the bifocal defining property of the central conic becomes the classical focus-directrix 'symptom' of the parabola cited by Pappus in his *Mathematical Collection* VII, 238 and noted by Newton as an undergraduate from Schooten's *Exercitationes* (see I: 35).

(153) Newton afterwards changed his mind, adding short demonstrations to the equivalent Propositions XVIII and XIX of his 'De motu Corporum Liber primus' (§3 following).

inter AZ et CZ duciq potest QS ipsi AC perpendicularis[,] ad quam si ab Hyperbolæ hujus puncto quovis Z demittatur normalis ZS, hæc fuerit ad AZ ut est differentia inter AZ et CZ ad AC. Dantur ergo rationes ipsarum ZR et ZS ad AZ et proinde datur earundem ZR, ZS ratio ad invicem, adeoq rectis RP, SQ concurrentibus in T, locabitur punctum Z in recta TZ positione data. Eadem methodo per Hyperbolam tertiam cujus umbilici sunt B et C et axis transversus differentia rectarum BZ, CZ, inveniri potest alia recta in qua punctum Z locatur. Habitis autem duobus locis rectilineis, habetur punctum quæsitum Z in earum intersectione. Q.E.I.

Cas 2. Si duæ ex tribus lineis, puta AZ et BZ, æquantur, punctum Z locabitur in perpendiculo bisecante distantiam AB, et locus alius rectilineus invenietur ut supra. Q.E.I.

Cas 3. Si omnes tres æquantur, locabitur punctum Z in centro circuli per puncta A, B, C transeuntis. Q.E.I.

Solvitur etiam hoc Lemma problematicum per librum Tactionum Apollonij a Vieta restitutum.(154)

Prop. XIX. Prob. XI.(155)

Trajectoriam circa datum umbilicum describere quæ transibit per puncta data et rectas positione datas continget.

Detur umbilicus S, punctum P, et tangens TR, et inveniendus sit umbilicus alter H. Ad tangentem demitte perpendiculum ST et produc idem ad V ut sit

(154) For if it be required to draw a circle which shall (externally) touch circles constructed on centres A, B, C with respective radii a, b and c, at once $AZ-BZ = a-b$, given, and $AZ-CZ = a-c$, given, exactly as Newton's lemma demands. Apollonius' two books *On Tangencies* [*sc.* of circles] are lost, and our modern knowledge of their content derives uniquely from the detailed description given by Pappus in the preamble to Book VII of his *Mathematical Collection* (rendered into French by P. Ver Eecke in his *Pappus d'Alexandrie: La Collection Mathématique* (Brussels, 1933): 483–5). Soon after the appearance of Commandino's Latin *editio princeps* (Pesaro, 1588) of Pappus' book, François Viète challenged the Flemish geometer Adriaen van Roomen to restore Apollonius' solution of the 3-circles tangency problem, but the latter in his published *Problema Apolloniacum quo datis tribus circulis, quæritur quartus eos contingens, antea a...Francisco Vieta...omnibus mathematicis...ad construendum propositum, jam vero per Belgam...constructum* (Würzburg, 1596) could only produce the straightforward 'solid' solution by intersecting hyperbolas which Newton here ingeniously adapts, locating the desired point Z (the required circle-centre in the Apollonian model) as the common union of rectilinear loci through the meets of their directrices. In his own restoration, printed four years after under the counter-title *Apollonius Gallus. Seu, Exsuscitata Apollonii Pergæi ΠΕΡῚ 'ΕΠΑΦΩ͂Ν Geometria* (Paris, 1600) [=(ed. Frans van Schooten) *Opera Mathematica* (Leyden, 1646): 324–38], Viète employed only 'plane' (straight-edge/circle) methods, but could attain the construction of the general problem only by way of a hierarchy of successive particular cases. Newton himself had, as we have seen (V: 262, note (332)), earlier made use of one of these reductions—that (Problema VII) of the tangency of two circles and a line to

and transverse axis is the difference between AZ and CZ, and QS can be drawn perpendicular to AC, while, if the normal ZS be let fall to it from any point Z of the hyperbola, this will be to AZ as the difference between AZ and CZ to AC. Consequently, the ratios of ZR and ZS to AZ are given, and therefore the ratio of ZR and ZS to each other is given; hence, where RP and SQ are concurrent in T, the point Z will be located in the straight line TZ given in position. By the same method, with the aid of a third hyperbola whose foci are B and C and transverse axis is the difference of the straight lines BZ and CZ, there can be found another straight line in which the point Z is located. Once these two rectilinear loci are had, however, the required point Z is obtained at their intersection. As was to be found.

Case 2. If two of the three lines, AZ and BZ say, are equal, the point Z will be located in the perpendicular bisector of the distance AB, and the other rectilinear locus will be found as above. As was to be found.

Case 3. If all three are equal, the point Z will be located at the centre of the circle passing through the points A, B and C. As was to be found.

This lemmatical problem is solved also by means of Apollonius' book *On Tangencies*, restored by Viète.(154)

Proposition XIX. Problem XI.(155)

To describe round a given focus a trajectory which shall pass through given points and be tangent to straight lines given in position.

Let there be given the focus S, point P and tangent TR, and let H be the second focus which is to be found. To the tangent let fall the perpendicular ST and

---

the equivalent lower-level tangency of a circle, line and point—in composing Problem 40 of his Lucasian lectures on algebra. We may add that none of these solutions closely imitates Apollonius' original construction as Robert Simson plausibly restored it on the basis of one of Pappus' pertinent lemmas (*Mathematical Collection* VII, 117) in 1734—'Feb. 9....mane, post horam 1mam antemeridiem' as the eureka in his original manuscript records. (See his posthumous *Opera Quædam Reliqua* (Glasgow, 1776): Appendix: 20–3; and compare William Trail, *Account of the Life and Writings of Robert Simson* (London, 1812): 57. The more recent equivalent restorations by H. G. Zeuthen, *Die Lehre von den Kegelschnitten im Altertum* (Copenhagen, 1886): 381–3 and T. L. Heath, *A History of Greek Mathematics*, **2** (Oxford, 1921): 184–5 are inessentially variant generalisations of Simson's basic insight.) The Apollonian problem has received numerous modern solutions, both synthetic—notably that, using sophisticated notions of inversion and antihomothety, by J. D. Gergonne (1817, further improved by J. Fouché in 1892)—and especially analytical—by Euler, Lambert, Gauss and others; on these see Ver Eecke's *Pappus*: lxvi–lxix and É. Callandreau, *Célèbres Problêmes mathématiques* (Paris, 1949): 219–26.

(155) The generalisation of Proposition XVII in which the length of the main axis is no longer given, but an additional given tangent to the locus or point on it compensates for the deficiency.

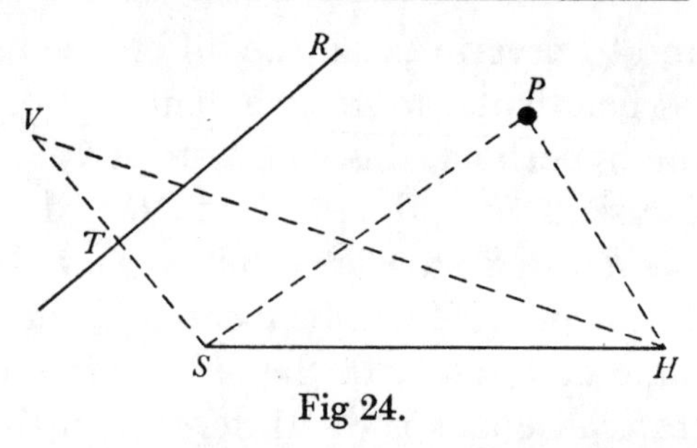

Fig 24.

TV æquale *ST*, et erit *VH* æqualis axi transverso. Junge *SP*, *HP* et erit *SP* differentia inter *HP* et axem transversum. Hoc modo si dentur plures tangentes *TR* vel plura puncta *P* devenietur semper ad lineas totidem *VH*, vel *PH* a datis punctis *V* vel *P* ad umbilicum *H* ductas quæ vel æquantur axibus vel datis longitudinibus *SP* differunt ab ijsdem, atqȝ adeo quæ vel æquantur sibi invicem vel datas habent differentias, & inde per Lemma superius datur umbilicus alter *H*. Habitis autem umbilicis unà cum axis longitudine *VH*, vel si trajectoria Ellipsis est $PH+SP$, sin Hyperbola $PH-SP$[,] habetur Trajectoria. Q.E.I.

Scholium.

Casus ubi dantur tria puncta sic solvitur expeditiùs. Dentur puncta *B*, *C*, *D*. Junctas *BC*, *CD* produc ad *E*, *F* ut sit *EB* ad *EC* ut *SB* ad *SC* et *FC* ad *FD* ut *SC* ad *SD*. Ad *EF* ductam & productam demitte normales *SG*, *BH*, inqȝ *GS* producta cape *GA* ad *AS* et *Ga* ad *aS* ut est *HB* ad *BS*[,] et erit *A* vertex et *Aa* axis transversus trajectoriæ quæsitæ: quæ perinde ut *GA* minor æqualis vel major(156) fuerit quam *AS*, erit Ellipsis Parabola vel Hyperbola; puncto *a* in primo casu cadente ad eandem partem lineæ *GK* cum puncto *A*, in secundo casu abeunte in infinitum, in tertio cadente ad contrariam partem lineæ *GK*. Nam si demittantur ad *GF* perpendicula *CI*, *DK* erit *IC* ad *HB* ut *EC* ad *EB* hoc est ut *SC* ad *SB*[,] et vicissim *IC* ad *SC* ut *HB* ad *SB* seu *GA* ad *SA*. Et simili argumento probabitur esse *KD* ad *SD* in eadem ratione. Jacent ergo puncta *B*, *C*, *D* in conisectione circa umbilicum *S* descripta, ea lege ut rectæ omnes ab umbilico *S* ad singula sectionis puncta ductæ sint ad perpendicula punctis ijsdem ad rectam *GK* demissa in data illa ratione.(157)

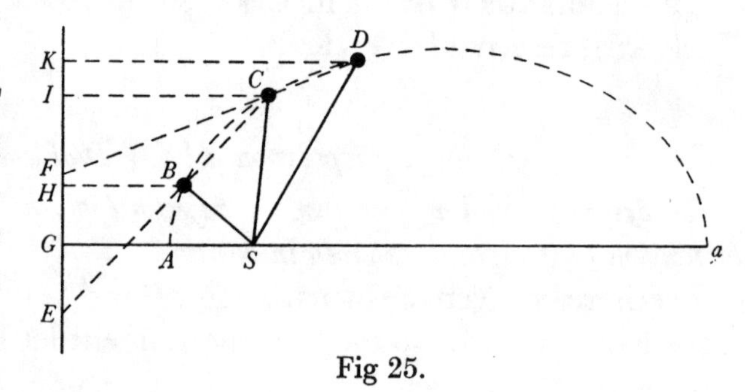

Fig 25.

(156) Read in reverse order 'major, æqualis vel minor' (greater than, equal to, or less than): the slip endured through the ensuing 'De motu Corporum Liber primus' (§3) into the equivalent scholium of the published *Principia* ($_1$1687: 69) to be corrected in its revise ($_2$1713: 65). Since *EGK* will, by the construction, be a directrix of the conic, it follows that the common value of the equal ratios $GA/AS = Ga/aS = HB/BS = IC/CS = KD/DS$ is not the eccentricity but its reciprocal.

extend it to V so that TV be equal to ST, and VH will then be equal to the transverse axis. Join SP and HP, and SP will be the difference between HP and the transverse axis. In this way, should there be given more tangents TR or more points P, we will always arrive at an equal number of lines VH or PH, drawn from the given points V or P to the focus, which either are equal to the axes or differ from them by given lengths SP, and which hence either equal one another or have given differences; and therefrom by means of the previous lemma the second focus H is given. Once, however, the foci along with the length of the axis VH—or of $SP+PH$ if the trajectory is an ellipse, but of $PH-SP$ if an hyperbola—are had, the trajectory is obtained. As was to be found.

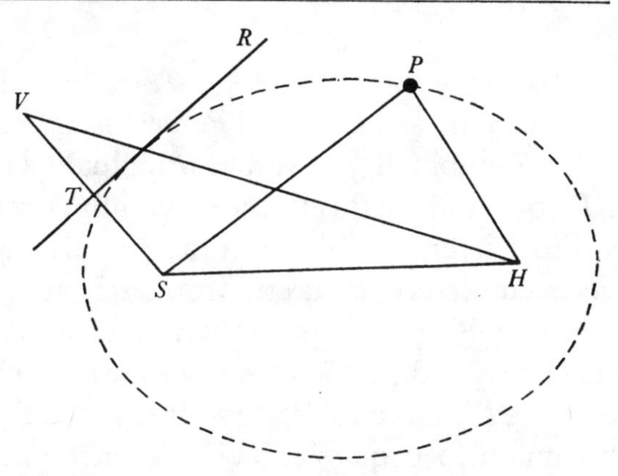

Scholium.

The case where three points are given is solved more speedily in this fashion. Let there be given the points B, C, D. Join BC, CD and extend them to E, F so that EB be to EC as SB to SC, and FC to FD as SC to SD. Draw and extend EF and to it let fall the normals SG, BH, then in GS produced take GA to AS and Ga to aS as HB is to BS, and A will be the vertex and Aa the transverse axis of the trajectory required. According as GA is less than, equal to or greater than(156) AS this will be an ellipse, parabola or hyperbola, with the point a in the first case falling on the same side of the line GK as the point A, in the second case going off to infinity, and in the third falling on the opposite side of the line GK. For, if perpendiculars CI, DK be let fall to GF, there will be IC to HB as EC to EB, that is, as SC to SB, and so *alternando* IC to SC as HB to SB or GA to SA. And by a similar argument it will be proved that KD is to SD in the same ratio. The points B, C and D lie, therefore, in a conic described round the focus S with the restriction that all straight lines drawn from the focus S to the individual points of the section shall be to the perpendiculars let fall from the same points to the straight line GK in that given ratio.(157)

(157) We have already pointed out (v: 516, note (32)) that, unbeknown to Newton, his present employment of the focus-directrix property to construct a conic 'trajectory', given a focus and three points of the orbit, had effectively been anticipated by James Gregory eleven years before. While in the construction which Gregory enunciated—without proof—in his letter to Colin Campbell on 30 April 1674 (first published in *Archæologia Scotica*, 3, 1831: 280–2; the original is now lost) no explicit mention of it is made, close examination will reveal that he

Prop. XX. Prob. XII.

Corporis in data trajectoria Parabolica moventis, invenire locum ad tempus assignatum.

Sit S umbilicus[,] A vertex principalis Parabolæ, et $4AS \times M$ area Parabolica APS quæ radio SP vel post excessum corporis de vertice descripta fuit vel ante appulsū ejus ad verticem describenda est. Innotescit area illa ex tempore(158) ipsi proportionali. Biseca AS in G, erige perpendiculū GH æquale 3M, & circulus centro H intervallo HS descriptus secabit Parabolam in loco quæsito P.(159) Nam demissa ad axem perpendiculari PO, est

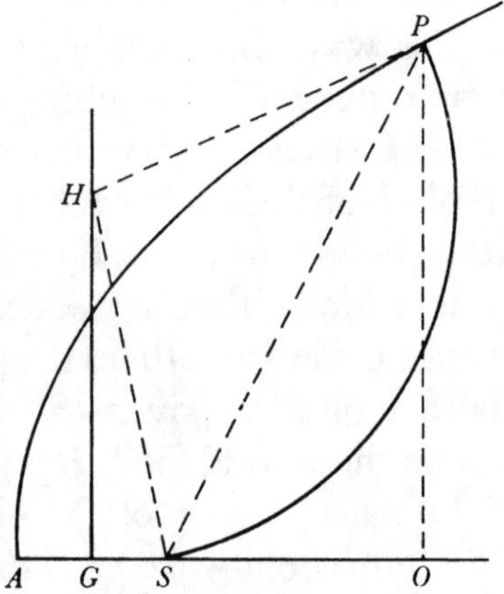

$$HG^q + GS^q (= HS^q = HP^q = GO^q + \overline{HG - PO}^q)$$
$$= GO^q + HG^q - 2HG \times PO + PO^q.$$

Fig. 26.

Et deleto utrinꝗ HG^q fiet

$$GS^q = GO^q - 2HG \times PO + PO^q,$$

seu $2HG \times PO (= GO^q + PO^q - GS^q =^{(160)} AO^q - 2GAO + PO^q) = AO^q + \frac{3}{4}PO^q$.

Pro AO^q scribe $AO \times \dfrac{PO^q}{4AS}$, et applicatis terminis omnibus ad $3PO$ ductisꝗ in $2AS$ fiet

$$\tfrac{4}{3}HG \times AS (= \tfrac{1}{6}AO \times PO + \tfrac{1}{2}AS \times PO = \frac{AO + 3AS}{6} PO = \frac{4AO - 3SO}{6} PO$$
$$= \text{areæ } APO - SPO) = \text{areæ } APS.$$

Sed $\frac{4}{3}HG \times AS$ est $4AS \times M$[,] ergo area APS æqualis est $4AS \times M$. Q.E.D.

Schol.

Problema novissimum in Ellipsi et Hyperbola constructionem Geometricam(161) non admittit, conficitur verò quamproxime in Ellipsi ut

assumes the focus-directrix property throughout and that his primary goal is to determine the meet (G in Newton's present figure) of the directrix with the conic's axis, whence, since $(SD - SC)/(DK - CI)$ measures the eccentricity, he straightforwardly deduces that

$$GA/Aa = ((DK - CI)^2 - (SD - SC)^2)/2(SD - SC)(DK - CI).$$

(158) Of orbit over the arc $\widehat{AP}$, that is.

(159) The preliminary analysis by which Newton arrived at this construction—doubtless more revealing to modern eyes than the somewhat contrived geometrical synthesis which follows—has not survived, but we may readily restore it by taking A to be the origin of the perpendicular Cartesian coordinate system in which $AO = x$, $OP = y$ defines the general point P, and setting $AS = p$, $AG = r$ and $GH = s$. At once, the circle $x^2 + y^2 = 2rx + 2sy$ of

Proposition XX. Problem XII.

Where a body moves in a given parabolic trajectory, to find its position at an assigned time.

Let S be the focus and A the principal vertex of the parabola, and $4AS \times M$ the parabolic area (APS) which either has been described by the radius (vector) SP after the body's departure from the vertex or is yet to be described by it before its arrival there. That area is ascertained from the time [158] proportional to it. Bisect AS in G, erect the perpendicular GH equal to $3M$, and the circle described on centre H and with radius HS will intersect the parabola in the required place P.[159] For, on letting fall the perpendicular PO to the axis, there is

$$HG^2+GS^2$$
$$= (HS^2 = HP^2 = GO^2+(HG-PO)^2 =)\, GO^2+HG^2-2HG\times PO+PO^2,$$

and with HG^2 deleted from each side there will come to be

$$GS^2 = GO^2-2HG\times PO+PO^2,$$

that is,

$$2HG\times PO(= GO^2+PO^2-GS^2 =^{(160)} AO^2-2AG\times AO+PO^2) = AO^2+\tfrac{3}{4}PO^2.$$

In place of AO^2 write $AO\times PO^2/4AS$ and, when all the terms are divided by $3PO$ and multiplied by $2AS$, there will prove to be

$$\tfrac{4}{3}HG\times AS(=\tfrac{1}{6}AO\times PO+\tfrac{1}{2}AS\times PO = \tfrac{1}{6}(AO+3AS)\times PO=\tfrac{1}{6}(4AO-3SO)\times PO$$
$$= \text{area } (APO)-(SPO)) = \text{area } (APS).$$

But $\tfrac{4}{3}HG\times AS$ is $4AS\times M$, and therefore the area (APS) is equal to $4AS\times M$. As was to be proved.

Scholium.

The most recent problem does not allow of a geometrical construction[161] in the ellipse and hyperbola; it is, however, very closely accomplished in the

---

centre $H(r, s)$ and through $A(0, 0)$ meets the parabola $y^2 = 4px$ in the point $P(x, y)$ such that $y = x(x+4p-2r)/2s$, while from equating

$$(ASP) = (AOP)-\triangle SOP = (\tfrac{2}{3}xy-\tfrac{1}{2}(x-p)\,y \text{ or})\ \tfrac{1}{6}(x+3p)y$$

to $4pM$ (where M is Newton's measure of the time of orbit over $\widehat{AP}$) there ensues

$$(x+3p)/6M = 4p/y = y/x,$$

whence $3M(x+4p-2r) \equiv s(x+3p)$ and therefore $r = \tfrac{1}{2}p$, $s = 3M$.

(160) Since $GO = AO-AG$ and $GS = AG$.

(161) In the Cartesian sense of being effectable by the meets of 'geometrical' (algebraic) curves. Where the general point (r, ϕ) of a given central conic is defined by the polar equation $r = (1-e^2)/(1\pm e\cos\phi)$ referred to a focus as origin (and so, in contemporary astronomical parlance, r is its *radius vector* and ϕ the 'true anomaly'), its general focal sector is

$$\tfrac{1}{2}\surd[1-e^2]\ (\theta\mp e\sin\theta) = \tfrac{1}{2}\surd[1-e^2]\ T,$$

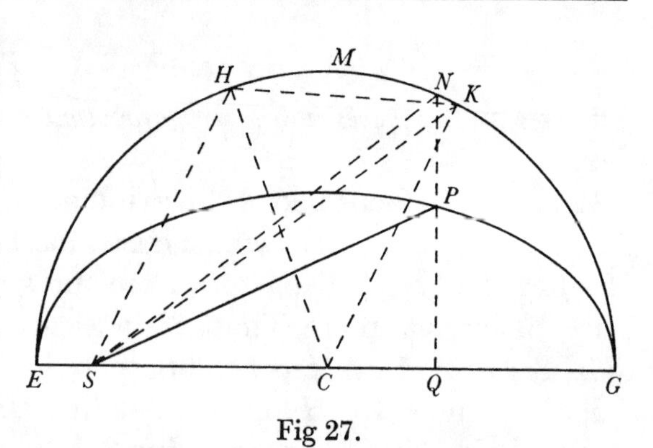

Fig 27.

sequitur.(162) Super Ellipseos axe majore *EG* describatur semicirculus *EHG*. Sumatur angulus *ECH* tempori proportionalis. Agatur *SH* eiq; parallela *CK* circulo occurrens in *K*. Jungatur *HK* et circuli segmento *HKM* (per tabulam segmentorum vel secus) æquale fiat triangulum *SKN*. Ad *EG* demittatur perpendiculum *NQ*, et in eo capiatur *PQ* ad *NQ* ut est Ellipseos axis minor ad axem majorem et erit punctum *P* in Ellipsi atq; acta recta *SP* abscindet aream Ellipseos *EPS* tempori proportionalem. Namq; area *HSNM* triangulo *SNK* aucta et huic æquali segmento *HKM* diminuta fit triangulo *HSK*, id est triangulo *HSC* æquale. Hæc æqualia addita areæ *ESH*, facient areas æquales *EHNS* et *EHC*. Cùm igitur sector *EHC* tempori proportionalis sit et area *EPS* areæ *EHNS*, erit etiam area *EPS* tempori proportionalis.

Insistendo vestigijs eorum quæ Viri celeberrimi Dr Sethus Wardus nunc Episcopus Sarum mihi plurimum colendus et Ismael Bullialdus adinvenerunt,(163) idem sic porrò conficimus. Existentibus *S*, *H* umbilicis et *AC*, *CB*, *CQ* semiaxibus Ellipseos, junge *SQ* et quære angulum *CQR* qui sit ad angulum rectum ut est umbilicorum distantia *SH* ad perimetrum circuli descripti diametro *AC*. Hoc invento, cape angulum *BHK* proportionalem tempori, angulumq; *BHL* cujus tangens sit ad tangentē anguli *BHK* ut est Ellipseos axis major ad axem minorem, et angulum *LHP* qui sit ad

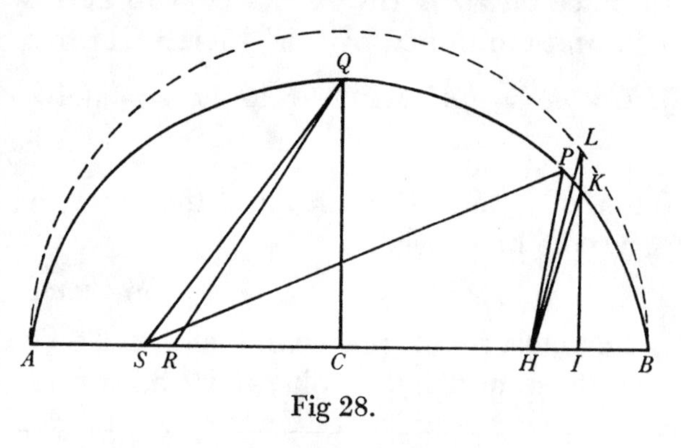

Fig 28.

on introducing the 'eccentric anomaly' $\theta = \cos^{-1}(r\cos\phi \pm e)$ related to the 'mean anomaly' T measuring the time of orbit by the Keplerian equation $T = \theta \mp e\sin\theta$ (see §1: note (84)). In these terms Newton's present statement is equivalent to asserting that this Keplerian equation cannot be inverted to yield θ as an algebraic function of T (for, of course, ϕ is derivable from θ by a simple geometrical—that is, finitely algebraic—construction), so implying that $\sin\theta$ is not in general 'geometrically' constructible from θ, and this he very ingeniously attempts—and fails (see §3: note (121))—to demonstrate in Lemma XXVIII of his revised 'De motu Corporum Liber primus'.

(162) The sequel is lifted word for word from the latter half of the scholium to Problem 4

ellipse as follows.(162) On the ellipse's major axis EG describe the semicircle EHG. Take the angle $\widehat{ECH}$ proportional to the time. Draw SH and parallel to it CK, meeting the circle in K; then join HK and make the triangle SKN equal to the circle's segment HKM (by means of a table of segments or otherwise). To EG let fall the perpendicular NQ and in it take PQ to NQ as the ellipse's minor axis is to its major axis, and the point P will then be in the ellipse, while the straight line SP will, when drawn, cut off an area (EPS) of the ellipse which is proportional to the time. For the area ($HSNM$), augmented by the triangle SNK and diminished by the segment (HKM) equal to this, becomes equal to the triangle HSK, that is, to the triangle HSC. When these equals are added to the area (ESH) they will form equal areas ($EHNS$) and (EHC). Since, therefore, the sector (EHC) is proportional to the time, and the area (EPS) to the area ($EHNS$), the area (EPS) also will be proportional to the time.

By treading in the footsteps of procedures devised by the very celebrated Dr Seth Ward—now Bishop of Salisbury, a man to whom I am exceedingly devoted—and Ismael Boulliau(163) we achieve the same end in this way moreover. Where S, H continue to be the ellipse's foci and AC, CB, CQ its semi-axes, join SQ and seek the angle $\widehat{CQR}$ which shall be to a right angle as the foci's distance SH to the perimeter of a circle described with diameter AC. Once this is ascertained, take the angle $\widehat{BHK}$ proportional to the time, and the angle $\widehat{BHL}$ whose tangent shall be to the tangent of the angle $\widehat{BHK}$ as the ellipse's major axis to its minor axis, and also the angle $\widehat{LHP}$ which shall be to the angle $\widehat{SQR}$ as

---

in the earlier 'De motu Corporum'. We there showed (§1: note (84)) that Newton's construction is equivalent to inverting the equation $\theta - e\sin\theta = T$ to yield, correct to $O(e^5)$,

$$\theta \approx T + e\sin T + \tfrac{1}{2}e^2\sin 2T + \tfrac{1}{8}e^3(3\sin T - \sin T) + \tfrac{1}{6}e^4(2\sin 4T - \sin 2T).$$

(163) In the terms of Newton's following figure, Seth Ward had deduced the simple equant hypothesis that the angle $\widehat{BHP}$ at the 'empty' focus H is the mean anomaly, measuring the time of passage from aphelion B to the general point P over the ellipse-arc $\widehat{BP}$, as an immediate corollary—there unstated—to a superficially more complicated model of planetary motion expounded by Boulliau in his *Astronomia Philolaïca* (Paris, 1645): 25–36, according to which the elliptical orbit is conceived to be described in an oblique cone such that its focus H lies in the line joining the apex to the centre of the base circle, and the planet's motion in it postulated to be uniform round that axis. (Compare C. A. Wilson, 'From Kepler's Laws, So-called, to Universal Gravitation: Empirical Factors' (*Archive for History of Exact Sciences*, **6**, 1970: 89–170): 111–17.) Therefrom, concisely in his *In Ismaelis Bullialdi Astronomiæ Philolaicæ Fundamenta, Inquisitio Brevis* (Oxford, 1653) and more elaborately in the first book of his *Astronomia Geometrica: ubi Methodus proponitur qua Primariorum Planetarum Astronomia... possit Geometricè absolvi* (London, 1656), Ward developed an elegant mathematical theory of 'Astronomia Elliptico-Copernicana', adjoining in Book 2 of the latter work a parallel theory of 'Astronomia Circularis' in which the ellipse is replaced by an approximating eccentric circle and its 'empty' focus becomes a corresponding Ptolemaic 'bissextile' *punctum æquans* set symmetrically

angulum SQR ut est quadratum sinûs anguli BHL ad quadratum radij.(164) Jaceat HP inter HL et HA occurrens Ellipsi in $P_{[,]}$ et acta SP, abscindet areã ASP tempori proportionalem quamproximè.

Hactenus speculati sumus motum corporum in lineis curvis. Fieri autem potest ut descendant vel ascendant secundu[m] lineas rectas. Quæ ad casus istos spectant, jam pergimus exponere.

Prop. XXI. Prob. XIII.(165)

Posito quod vis centripeta sit reciproce proportionalis quadrato distantiæ a centro, spatia definire quæ corpus recta cadendo datis temporibus describit.

Cas. 1. Si corpus non cadit perpendiculariter describet id sectionem aliquam conicam cujus umbilicus inferior congruet cum centro.(166) Id ex modò demonstratis constat. Sit sectio illa Conica $ARPB$ et umbilicus inferior S. Et primò si Figura illa Ellipsis est, super hujus axe majore AB describatur semicirculus ABD et per corpus decidens transeat recta DPC perpendicularis ad axem, actisq; DS, PS, erit area ASD areæ ASP atq; adeo etiam tempori proportionalis. Manente axe AB minuatur perpetuò latitudo Ellipseos, et semper manebit area ASD

opposite the sun in line through the centre. The corrected Wardian hypothesis in which, on taking $B\widehat{H}K$ to be the mean motion, the planet is decreed to be at the intersection of HL with the elliptical orbit was published by Boulliau in his riposte the next year, *Ismaelis Bullialdi Astronomiæ Philolaicæ Fundamenta clarius explicata, & asserta, Adversus Clarissimi Viri Sethi Wardi... impugnationem* (Paris, 1657): Caput III: 14–17 (compare Wilson, 'From Kepler's Laws...': 119–20) and shown by him to agree with a number of Tycho Brahe's near-acronychal observations of Mars to within some 5′ of arc, the threshold of accuracy of Tycho's naked-eye techniques. Newton's present further attempted improvement (examined in detail in the next note) is a theoretical deduction from the Keplerian measure of orbital time by the focal sector (BSP), and has no practical basis or indeed—inasmuch as it ignores the distorting effect of mutual planetary perturbations which becomes pronounced at this higher-order level of accuracy—even significance. The unusual warmth with which he here refers to Ward very possibly mirrors an intimacy between the two which is not otherwise documented. As we have seen (III: xvii), it was Ward who in December 1671 sponsored Newton in his election to the fellowship of the Royal Society.

(164) Accurately, this should read '...ut est quadruplum cubi sinûs anguli BHL ad cubum radij' (as four times the cube of the sine of the angle $B\widehat{H}L$ to the cube of its radius). While the worksheet on which Newton entered his preliminary analysis of this erroneous construction has not, here as so often elsewhere, survived, we may restore it with confidence on the basis of his later observation of *c.* 1692 in an equivalent context that, since (or so it is implied) the focal radii SP and HP are equally inclined to the ellipse at P and hence

$$SP.d(\widehat{BSP}) = HP.d(\widehat{BHP}),$$

'si area [$2(BSP)$] fluit uniformiter sitq; ejus fluxio $1_{[,]}$ fluxio anguli circa [S] erit [$1/SP^2$] et fluxio anguli circa [H] erit [$1/SP \times HP$]' (ULC. Add. 3965.18: 725 *bis*r). In modern analytical

the square of the sine of the angle $B\widehat{H}L$ to the square of its radius.(164) Let *HP* lie between *HL* and *HA*, meeting the ellipse in *P*, and then *SP* will, when drawn, cut off an area (*ASP*) very nearly proportional to the time.

Thus far we have considered the motion of bodies in curved lines. It can, however, happen that they descend or ascend following straight lines. What regards those cases we now proceed to reveal.

Proposition XXI. Problem XIII.(165)

Supposing that the centripetal force be reciprocally proportional to the square of the distance from its centre, to define the places which a body falling straight down describes in given times.

Case 1. If the body does not fall perpendicularly, it will describe some conic whose lower focus will coincide with the centre.(166) That is settled from what has already been demonstrated. Let the conic be *ARPB* and its lower focus *S*. Then, first, if the figure is an ellipse, on its major axis *AB* describe the semicircle *ADB* and let the straight line *DPC* pass through the dropping body perpendicular to the axis, and when *DS*, *PS* are drawn the area (*ASD*) will be proportional to the area (*ASP*) and hence also to the time. With the axis *AB* remaining fixed, perpetually diminish the width of the ellipse and the area (*ASD*) will ever

---

terms, on setting $AC = CB = 1$, $SC = CH = e$ and $B\widehat{H}P = \psi$, and taking T to be the corresponding mean anomaly (proportional to the time of orbit over $\widehat{BP}$ such that $T = \pi$ when $B\widehat{S}P = \pi$), it follows that $HP = (1-e^2)/(1+e\cos\psi)$ with $SP = 2-HP$, whence

$$T = (1/\sqrt{[1-e^2]})\int_{\psi=0}^{\psi=\psi} SP^2.d(BSP) = (1/\sqrt{[1-e^2]})\int_0^{\psi} SP\times HP.d\psi,$$

that is,

$$T = (1-e^2)^{\frac{1}{2}}\int_0^{\psi}(1+2e\cos\psi+e^2)/(1+e\cos\psi)^2.d\psi = \psi-\tfrac{1}{4}e^2\sin 2\psi-\tfrac{2}{3}e^3\sin^3\psi\ldots$$

and therefore $\psi = T+\frac{1}{4}e^2\sin 2T+\frac{2}{3}e^3\sin^3 T+O(e^4)$. Newton's procedure constructs $B\widehat{H}K = T$ and then $\tan B\widehat{H}L/\tan T = 1/\sqrt{[1-e^2]}$, so yielding

$$B\widehat{H}L = T+\tfrac{1}{4}e^2\sin 2T+O(e^4);$$

and thereafter makes $C\widehat{Q}R = \frac{1}{2}\pi.(2e/\pi) = e$, whence $R\widehat{Q}S = \sin^{-1}e-e$, thus determining $L\widehat{H}P$ to be $\frac{1}{6}e^3\sin^2 T+O(e^5)$ in considerable departure from its true value

$$(\psi-B\widehat{H}L =)\ \tfrac{2}{3}e^3\sin^3 T+O(e^4).$$

Compare §3: note (157) below.

(165) An amplification of Problem 5 of the earlier 'De motu Corporum', here subsumed to be Case 1 following.

(166) Again understand 'virium' (or force).

tempori proportionalis. Minuatur latitudo illa in infinitum et orbe *APB* jam coincidente cum axe *AB* et umbilico *S* cum axis termino *B* descendet corpus in recta *AC* et area *ABD* evadet tempori proportionalis. Definietur itaq; spatium *AC* quod corpus de loco *A* perpendiculariter cadendo tempore dato describit si modò tempori proportionalis capiatur area *ABD* et a puncto *D* ad rectam *AB* demittatur perpendicularis *DC*. Q.E.F.

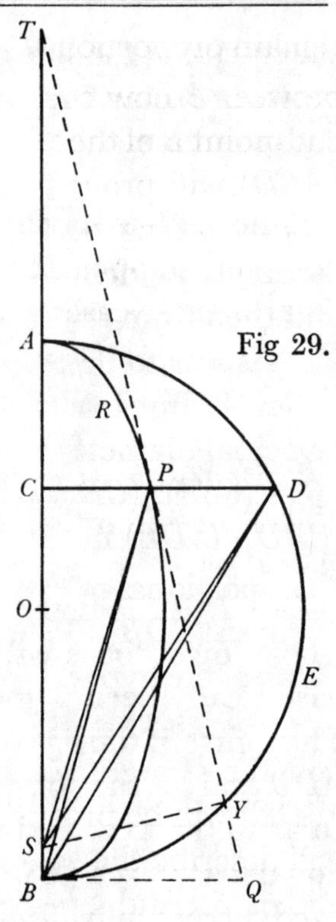

Fig 29.

Cas. 2. Sin figura superior *RPB* Hyperbola est, describatur super eadem diametro principali *AB* Hyperbola rectangula *BD*, et quoniam areæ *CSP*, *CBP*, *SPB* sunt ad areas *CSD*, *CBD*, *SDB* singulæ ad singulas in data ratione altitudinum *CP*, *CD*, et area *SPB* proportionalis est tempori quo corpus *P* movebitur per arcum *PB*, erit etiã area *SDB* eidem tempori proportionalis. Minuatur latus rectum Hyperbolæ *RBP* in infinitum manente latere transverso et coibit arcus *PB* cum recta *CB* & umbilicus *S* cum vertice *B* et recta *SD* cum recta *BD*. Proinde area *BDE* proportionalis erit tempori quo corpus *C* recto descensu describit lineam *CB*. Q.E.I.

Cas. 3. Et simili argumento si figura *RPB* parabola est(167) et eodem vertice principali *B* describatur alia Parabola *BED* quæ semper maneat data interea dum parabola prior in cujus perimetro corpus *P* movetur, diminuto et in nihilũ redacto ejus latere recto, conveniat cum linea *CB*, fiet segmentum Parabolicum *BDE* proportionale tempori quo corpus illud *P* vel *C* descendet ad centrum *B*. Q.E.I.(168)

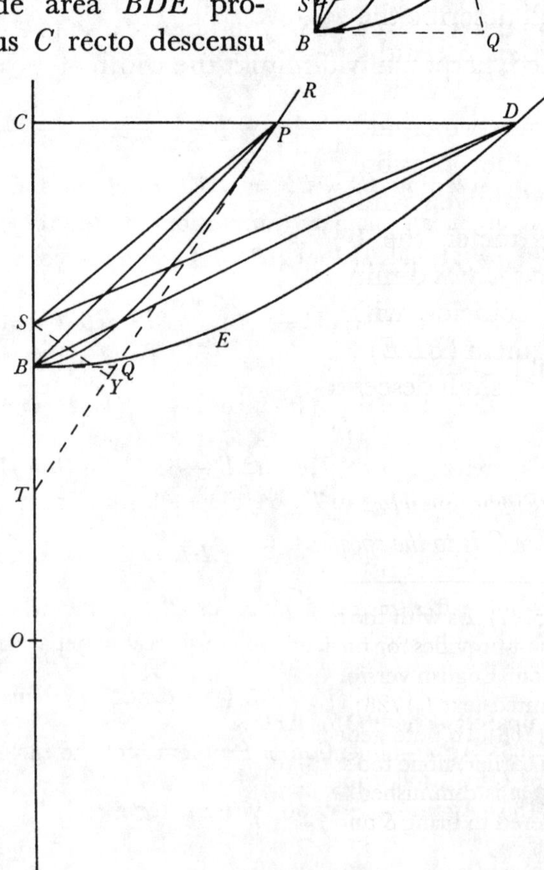

Fig 30.

Prop. XXII. Theor. IX.

Positis jam inventis, dico quod corporis cadentis velocitas in loco quovis C est ad velocitatem corporis centro B intervallo BC circulum describentis, in dimidiata ratione quam CA distantia

remain proportional to the time. Diminish that width indefinitely and, with the orbit $\widehat{APB}$ now coming to coincide with the axis *AB* and the focus *S* with the end-point *B* of the axis, the body will descend in the straight line *AC* and the area (*ABD*) will prove to be proportional to the time. The space *AC* described by a body falling perpendicularly from the position *A* in a given time will accordingly be defined if only the area (*ABD*) be taken proportional to the time and then from the point *D* the perpendicular *DC* be let fall to the straight line *AB*. As was to be done.

Case 2. But if the previous figure *RPB* is a hyperbola, describe on the same principal diameter *AB* the rectangular hyperbola *BD*, and then, seeing that the areas (*CSP*), (*CBP*), (*SPB*) are individually to the corresponding areas (*CSD*), (*CBD*), (*SDB*) in the given ratio of the altitudes *CP*, *CD*, while the area (*SPB*) is proportional to the time in which the body *P* shall move through the arc $\widehat{PB}$, the area (*SDB*) also will be proportional to this same time. Diminish the *latus rectum* of the hyperbola *RBP* indefinitely, its main diameter remaining fixed, and its arc $\widehat{PB}$ will coincide with the straight line *CB*, its focus *S* with the vertex *B*, and the straight line *SD* with the straight line *BD*. Consequently, the area (*BDE*) will be proportional to the time in which the body *C* in straight descent will describe the line *CB*. As was to be found.

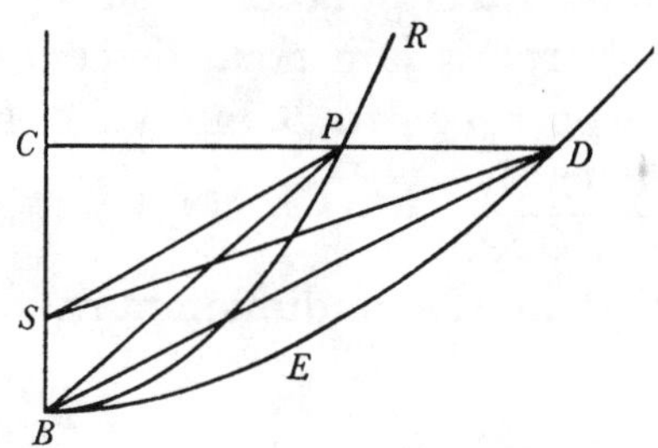

Case 3. And by a similar argument, if the figure *RPB* is a parabola(167) and with the same principal vertex *B* there be described another parabola *BED* which shall ever remain given while the first parabola (in whose perimeter the body moves) comes, with its *latus rectum* diminished and reduced to nothing, to coincide with the line *CB*, the parabolic segment (*BDE*) will prove to be proportional to the time in which that body *P*, or *C*, shall descend to the centre *B*. As was to be found.(168)

Proposition XXII. Theorem IX.

Supposing what has just now been found, I assert that the speed of the falling body at any place C is to the speed of a body describing a circle of centre B and radius BC in the halved

(167) As with the corresponding 'Cas. 3' in the published *Principia* ($_1$1687: 116), the manuscript provides for no separate figure to illustrate this parabolic instance. For clarity, we insert in our English version that which Pemberton introduced at this point in his *editio ultima* of the printed text ($_3$1726: 113).

(168) In here generalising it to comprehend the hyperbola and parabola, Newton retains his earlier, none too satisfactory mode of proof (see §1: note (89)) in which, as the conic's *latus rectum* is diminished, the position of the force-centre *S*—rather than that of the vertex *B*—is altered to bring *S* and *B* into ultimate coincidence. The modern reader will perhaps be better

corporis a circuli vel Hyperbolæ vertice ulteriore A habet ad figuræ semidiametrum $\frac{1}{2}AB$.

Fig 29 & 30. Namq; ob proportionales CD, CP(169) communis est utriusq; figuræ RPB, DEB diameter AB. Bisecetur hæc in O et rectâ PT tangatur figura RPB in P et secetur diameter (si opus est producta) in T, sitq; SY(170) ad hanc rectam & BQ ad hanc diametrum perpendicularis, atq; figuræ RPB latus rectum ponatur L. Constat per Cor. 9. Theor. VIII quod corporis in linea RPB circa centrum S moventis velocitas in loco quovis P sit ad velocitatem corporis intervallo SP circulum circa centrum idem describentis in dimidiata ratione rectanguli $\frac{1}{2}L\times SP$ ad SY quadratum. Est autem ex Conicis ACB ad CP^q ut $2AO$ ad L adeoq; $\dfrac{2CP^q\times AO}{ACB}=L$. Ergo velocitates illæ sunt in dimidiata ratione $\dfrac{CP^q\times AO\times SP}{ACB}$ ad SY^q. Porrò ex Conicis est CO ad BO ut BO ad TO et divisim ut CB ad BT. Unde componendo fit $CO\pm BO$ ad BO ut CT ad BT id est AC ad AO ut CP ad BQ indeq;

$$\frac{CP^q\times AO\times SP}{ACB}=\frac{BQ^q\times AC\times SP}{AO\times BC}.$$

Minuatur jam in infinitum figuræ RPB latitudo CP sic ut punctum P coeat cum puncto C et punctum S cum puncto B et linea SP cum BC lineaq; SY cum BQ, et corporis jam rectà descendentis in linea CB, velocitas fiet ad velocitatem corporis centro B intervallo BC circulum describentis, in dimidiata ratione $\dfrac{BQ^q\times AC\times SP}{AO\times BC}$ ad SY^q, hoc est (neglectis æqualitatis rationibus SP ad BC et BQ^q ad SY^q) in dimidiata ratione AC ad AO. Q.E.D.(171)

Prop. XXIII. Theor. X.

Si figura BED Parabola est, dico quod corporis cadentis velocitas in loco quovis C æqualis est velocitati qua corpus centro B dimidio intervalli sui BC circulum uniformiter describere potest.

---

persuaded by the equivalent analytical argument in which, on setting $AS=\pm R$ (accordingly as A is above S or—in the hyperbolic case—'unattainably' positioned below it) and $CS=r$, and taking $v=-dr/dt$ to be the body's instantaneous speed at C after a fall in time t from rest at A, where the force (directed to centre S) has magnitude g, over a distance $AC=R-r$ which in the hyperbolic case is 'more than infinite', then the equation of motion is

$$d^2r/dt^2=v.dv/dr=-gR^2/r^2,$$

whence $v^2=2gR(R-r)/r$, and so the time of fall from C to S (or B coincident with it) is

$$\int_0^r\sqrt{[r/2gR(R-r)]}.dr=\sqrt{[1/2gR]}.(\tfrac{1}{2}R\cos^{-1}(1-2r/R)-\sqrt{[r(R-r)]}),$$

that is, $\sqrt{[8/gR^3]}.(BDE)$ in the case of a central conic. The limiting parabolic case in which A is at infinity and therefore $R=\infty$ affords no difficulty on setting $gR^2=\gamma$, finite, and hence $v^2=2\gamma/r$.

ratio which the distance CA of the body from the circle's or hyperbola's farther vertex A has to the semi-diameter $\frac{1}{2}AB$ of the figure.

For, because of the proportionals CD, CP,(169) the diameter AB is common to either figure DEB, RPB. Bisect this in O, let the figure RPB be touched by the straight line PT at P and the diameter (extended if need be) cut by it in T, then let SY(170) be perpendicular to this straight line and BQ to this diameter, and set the *latus rectum* of the figure RPB to be L. It is established by Corollary 9 to Theorem VIII that the speed at any place P of the body moving in the line RPB round the centre S shall be to the speed of a body describing a circle with the radius SP round the same centre in the halved ratio of the rectangle $\frac{1}{2}L \times SP$ to the square of SY. However, from the *Conics*, $AC \times BC$ is to CP^2 as $2AO$ to L, and hence $L = 2CP^2 \times AO/AC \times BC$. Accordingly, those speeds are in the halved ratio of $CP^2 \times AO \times SP/AC \times BC$ to SY^2. From the *Conics*, moreover, CO is to BO as BO to TO, and so *dividendo* as CB to BT. Whence by compounding there comes to be $CO \pm BO$ to BO as CT to BT, that is, AC to AO as CP to BQ, and thence $CP^2 \times AO \times SP/AC \times BC = BQ^2 \times AC \times SP/AO \times BC$. Now let the width CP of the figure RPB be indefinitely diminished, so that the point P comes to coalesce with the point C, the point S with point B, the line SP with BC and the line SY with BQ, and the speed of the body now descending straight downwards in the line CB will come to be to the speed of a body describing a circle on centre B with radius BC in the halved ratio of $BQ^2 \times AC \times SP/AO \times BC$ to SY^2, that is, (with the ratios of equality SP to BC and BQ^2 to SY^2 neglected) in the halved ratio of AC to AO. As was to be proved.(171)

Figures 29 and 30

Proposition XXIII. Theorem X.

If the figure BED is a parabola, I assert that the speed of the falling body at any place C is equal to the speed with which a body can uniformly describe a circle of centre B at half its interval BC.

---

(169) These are to one another, of course, as the *latera recta* of the affinely correspondent conics (D) and (P).

(170) In either species of central conic the locus of Y is (by Apollonius, *Conics*, III, 49) the eccentric circle of diameter AB. While Newton correctly positions SY in his illustration of the elliptical case in the published *Principia* ($_1$1687: 115)—here reproduced as Figure 29—, in the companion hyperbolic diagram (*ibid.*: 116, repeated without correction in *Principia*, $_2$1713: 106) it is drawn to meet the tangent at P in an impossible site between P and Q (a little above the latter): our Figure 30 is here emended in line with the properly traced figure incorporated by Pemberton in the third edition (*Principia*, $_3$1726: 114).

(171) In continuation of note (168) we may equivalently argue that, since the body's speed at C is $v = \sqrt{[2gR(R-r)/r]}$, while the speed, v' say, of uniform rotation in a circle at distance $CS = r$ from the force-centre is $\sqrt{[(gR^2/r^2).r]} = \sqrt{[gR^2/r]}$, their ratio is at once

$$v/v' = \sqrt{[(R-r)/\tfrac{1}{2}R]} = \sqrt{[AC/\tfrac{1}{2}AB]}.$$

In effect, Newton's following theorem merely states the corollary that $v = v'\sqrt{2}$ when $R = \infty$

Nam corporis Parabolam *RPB* circa centrum *S* describentis velocitas in loco quovis [*P*] ᵃæqualis est velocitati corporis dimidio intervalli *SP* circulum circa idem *S* uniformiter describentis. Minuatur Parabolæ latitudo *CP* in infinitum eo ut arcus Parabolicus [*B*]*P*(172) cum recta *CB*, centrū *S* sum vertice *B*, et intervallum *SP* cum intervallo *C*[*B*](172) coincidat, et constabit Propositio. Q.E.D.

ᵃCor. 7. Theor. VIII.

Prop. XXIV. Theor. XI.

Iisdem positis dico quod area figuræ DES radio SD descripta, æqualis sit areæ quam corpus radio dimidium lateris recti figuræ DES æquante, circa centrum S uniformiter gyrando, eodem tempore describere potest.

Fig. 31 et 32.(173)

Nam concipe corpus *C* quàm minima temporis particula lineolam *Cc* inter cadendum describere, et interea corpus aliud *K*, uniformiter in circulo *OKk* circa centrum *S* gyrando, arcum *Kk* [describere.](174) ... ...

---

[3](175) ... ...

Schol.

His Propositionibus(176) manuducimur ad speculationem proportionis(177) inter vires centripetas et corpora centralia ad quæ vires illæ dirigi solent. Rationi enim consentaneum est vires quæ ad corpora diriguntur pendere ab eorundem corporum natura et quantitate, ut fit in Magneticis,(178) atqȝ adeò ex viribus

---

(172) The manuscript (f. 63ʳ) reads '*CP*' in each case: the slip passed into the published *Prinicipia* ($_1$1687: 118) but was caught in the concluding *errata* (*ibid.*: signature Ooo 4ʳ).

(173) These are the complementary figures illustrating the elliptical/hyperbolic cases which we reproduce with the revision of the present proposition in §3 following (see next note).

(174) The manuscript page terminates at this point and the next eight leaves of the original, numbered by Newton as 33–40 (see note (1) above) are lost. It will, however, be patently clear that the missing portion of the present demonstration can have been only minimally variant from that of the analogous Proposition XXXV of the revised 'De motu Corporum Liber primus' which we reproduce in § 3. From the further eight leaves of Newton's original which are still extant—namely his pages 41–8 (ULC. Add. 3965.3: 7ʳ–14ʳ), comprehending all but the first few lines of the demonstration of Proposition XXXIV, Propositions XXXV–XLII and the uncompleted enunciation of Proposition XLIII—we extract for their mathematical interest the excerpts given in [3] following. It would seem possible confidently to identify the intervening Propositions XXV–XXXIII as first versions of Propositions XXXVI, XXXVII and LVII–LXIII of the published *Principia's* first book (compare I. B. Cohen's *Introduction* (note (1)): 314–17).

(175) ULC. Add. 3965.3: 11ʳ–14ʳ, embracing (see previous note) Newton's pages 45–8. In this extract he sets out for the first time his familiar *ad hoc* techniques for computing the total attraction, in an inverse-square force-field, of a uniform spherical shell—and hence of a sphere whose density varies only as the distance from its centre—upon both an internal and an external point.

For the speed of the body describing the parabola *RPB* round the centre *S* is[a] at any place *P* equal to the speed of a body uniformly describing a circle round the same *S* at half the interval *SP*. Now let the width *CP* of the parabola be indefinitely diminished with the result that the parabolic arc $\widehat{BP}$ comes to coincide with the straight line *BC*, the centre *S* with the vertex *B*, and the interval *SP* with the interval *BC*, and the proposition will be manifest. As was to be proved.

[a] Corollary 7 to Theorem VIII.

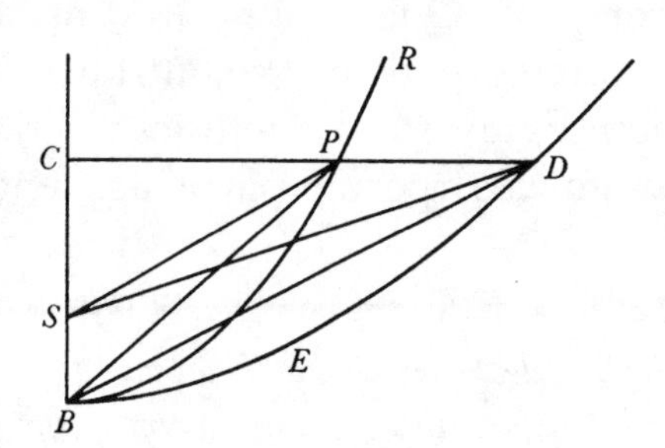

Proposition XXIV. Theorem XI.

With the same suppositions, I assert that the area of the figure DES described by the radius SD shall be equal to the area which a body can, by orbiting uniformly round the centre S at a radius equalling half the latus rectum of the figure DES, describe in the same time.

For imagine that the body *C* during its fall describes in a minutely small particle of time the line-element *Cc*, and meanwhile that another body *K*, by orbiting uniformly in the circle *OKk* round the centre *S*, [describes] the arc $\widehat{Kk}$.(174) . . . . .

Figures 31 and 32.(173)

---

[3](175) . . . .

Scholium.

We are guided by these propositions(176) to a consideration of the proportion(177) between centripetal forces and the central bodies to which those forces are usually directed. For it accords with reason that forces which are directed to bodies should depend on the nature and qualitity of those bodies—as happens in the case of magnetic ones—(178) and hence be compounded of innumerable smaller

(176) Namely XXVII–XXXVIII, corresponding to the—in part considerably—revised Propositions LVII—LXIV/LXVI–LXIX of Book I of the published *Principia*. The manuscript of XXXIV–XXXVIII alone survives (see note (174)).

(177) Newton afterwards altered this to read 'ad analogiam' (to the analogy).

(178) The remainder of the scholium was subsequently deleted and in its stead Newton wrote: 'Et quoties hujusmodi casus incidunt[,] æstimandæ erunt corporum attractiones assignando singulis eorum particulis vires proprias et colligendo summas virium. Videamus igitur quibus viribus corpora sphærica ex particulis attractivis constantia debeant in se mutuo agere & quales motus inde consequantur' (And as often as cases of this sort chance to occur, we shall need to estimate the attractions of bodies by assigning to their individual particles each its own force and gathering the sums of these forces. Let us then see, where spherical

minoribus innumeris ad singula corporum particulas attractivas tendentibus componi. Quare cum vires, ut ad unicum tantum punctum mathematicū in centro corporis cujusq; attrahentis constitutum tendentes, hactenus speculati sumus: e re erit conditiones etiam et leges virium ex viribus minoribus ad singulas corporum particulas tendentibus compositarum jam paucis exponere.

Prop. XXXIX. Theor. XXI.

Si ad sphæricæ superficiei puncta singula tendunt vires æquales centripetæ decrescentes in duplicata ratione distantiarum a punctis[,] *dico quod corpusculum intra superficiem constitutum his viribus nullam in partem attrahitur.*

Sit *HIKL* superficies illa sphærica et *P* corpusculū intus constitutum. Per *P* agantur ad hanc superficiem lineæ duæ *HK*, *IL* arcus quam minimos *HI*, *KL* intercipientes. Et ob[a] similia triangula *HPI*, *LPK*, arcus illi erunt distantijs *HP*, *LP* proportionales, et superficiei sphæricæ particulæ quævis ad *HI* et *KL* rectis per punctum *P* transeuntibus undiq; terminatæ, erunt in duplicata illa ratione. Ergo vires harum particularum in corpus *P* exercitæ sunt inter se æquales. Sunt enim ut particulæ directè et quadrata distantiarum inversè, et hæ duæ rationes componunt rationem æqualitatis. Attractiones igitur in contrarias partes æqualiter factæ se mutuò destruunt. Et simili argumento attractiones omnes per totam sphæricam superficiē contrarijs attractionibus destruuntur. Proinde corpus *P* nullā in partem his attractionibus impellitur. Q.E.D.[179]

[a] Cor. 3. Lem. VII.

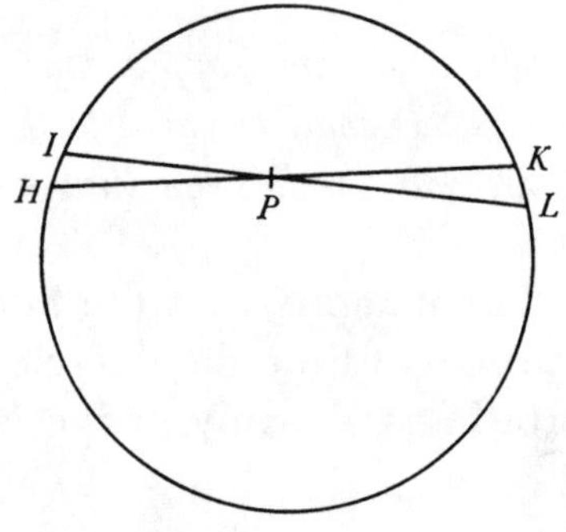

Fig 41.

Prop. XL. Theor. XXII.

Iisdem positis dico quod corpusculum extra sphæricam superficiem constitutum attrahitur versus centrum sphæræ vi reciprocè proportionali quadrato distantiæ a[180] *centro.*

---

bodies are composed of attractive particles, with what forces these ought mutually to act upon one another and what motions ensue therefrom). He afterwards further amplified the equivalent scholium in the published *Principia* ($_1$1687: 191–2), there noting that 'Vocem attractionis hic generaliter usurpo pro corporum conatu quocunq; accedendi ad invicem.... Eodem sensu generali usurpo vocem impulsus, non species virium & qualitates physicas, sed quantitates & proportiones Mathematicas in hoc Tractatu expendens: ut in Definitionibus explicui. In Mathesi investigandæ sunt virium quantitates & rationes illæ, quæ ex conditionibus quibuscunq; positis consequentur: deinde ubi in Physicam descenditur, conferendæ sunt hæ rationes cum Phænomenis, ut innotescat quænam virium conditiones singulis corporum attractivorum generibus competant. Et tum demum de virium speciebus, causis & rationibus physicis tutius disputare licebit' (I here employ the phrase 'attraction' generally for any endeavour whatever of bodies to approach each other.... In the same general sense I use the expression 'impulse', considering in this tract not the species and physical qualities of forces but, as I explained in

forces tending to the individual attractive particles of bodies. Accordingly, since we have so far regarded forces as tending but to a single mathematical point set at the centre of each attracting body, it will now be apposite to set out in brief the circumstances and also the laws of forces composed of lesser forces tending to the individual particles of bodies.

Proposition XXXIX. Theorem XXI.

If to the individual points of a spherical surface there tend equal centripetal forces decreasing in the doubled ratio of the distances from the points, I assert that a corpuscle set within the surface is nowhere attracted by these forces.

Let *HIKL* be the spherical surface and *P* a corpuscle set within it. Through *P* draw to this surface two lines *HK*, *IL* intercepting the two minutely small arcs $\widehat{HI}$, $\widehat{KL}$ in it. Then, because[a] the triangles *HPI*, *LPK* are similar, those arcs will be proportional to the distances *HP*, *LP*, and any particles of the spherical surface terminated on their either side at *HI* and *KL* by straight lines passing through the point *P* will be in that ratio doubled. Consequently, the forces of these particles exerted on the body *P* are equal to each other: for they are as the particles directly and the squares of the distances inversely, and these two ratios together make up a ratio of equality. Therefore their attractions, occurring equally in opposite directions, mutually destroy one another. And by a similar argument all attractions throughout the spherical surface are destroyed by contrary attractions. In consequence, the body *P* is nowhere impelled by these attractions. As was to be proved.(179)

[a] Corollary 3 to Lemma VII.

Proposition XL. Theorem XXII.

With the same suppositions I state that a corpuscle set outside a spherical surface is attracted towards the centre of the sphere by a force reciprocally proportional to the square of its distance from the(180) *centre.*

---

the *Definitions*, their quantities and mathematical proportions. We are in mathematics to find out the quantities and ratios of forces which shall ensue from any conditions presupposed; thereafter, when descent to physics is made, we must compare these ratios with observed appearances so as to ascertain what particular conditions of force relate to individual kinds of attractive bodies; and then we shall at last be free to argue in some safety over the species, causes and physical proportions of forces).

(179) This not inelegant *ad hoc* demonstration, that upon an internal point in an inverse-square force-field a spherical shell exerts an attraction which in every direction balances itself and hence is totally without effect, passed unchanged into Proposition LXX of Book 1 of the published *Principia* ($_1$1687: 192–3), where in sequel it is generalised, in Corollary 3 to Proposition XCI (*ibid.*: 221–2, reproduced as the terminal problem of Appendix 4 below), to comprehend the analogous case of an ellipsoidal *superficies*.

(180) Evidently to stress that this is not to be confused with a centre of force, Newton subsequently trivially amplified this to read 'ab eodem' (from the same).

Sint *AHKB*, *ahkb* æquales duæ superficies sphæricæ centris *S*, *s*[,] diametris *AB*, *ab* descriptæ, et *P*, *p* corpuscula sita extrinsecus in diametris illis productis. Agantur a corpusculis lineæ *PHK*, *PIL*, *phk*, *pil* auferentes de circulis maximis *AHB*, *ahb* æquales arcus quam minimos *HK*, *hk* & *IL*, *il*. Et ad eas demittantur

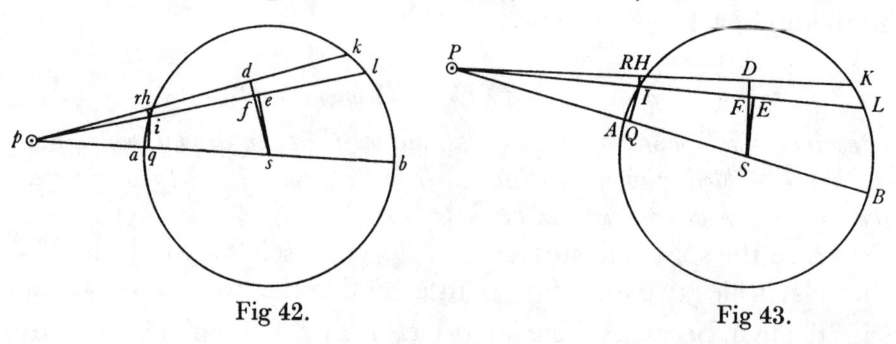

Fig 42. Fig 43.

perpendicula *SD*, *sd*, *SE*, *se*, *IR*, *ir*, quorū *SD*, *sd* secent *PL*, *pl* in *F* et *f*. Demittantur etiam ad diametros perpendicula *IQ*, *iq*, et ob æquales *DS*, *ds* et *ES*, *es*, et angulos evanescentes *DPE*, *dpe* lineæ *PE*, *PF* et *pe*, *pf* et lineolæ *DF*, *df* pro æqualibus habeantur: quippe quarum ratio ultima, angulis illis *DPE*, *dpe* simul evanescentibus, est æqualitatis. His ita constitutis, erit *PI* ad *PF* ut *RI* ad *DF*, et *pf* ad *pi* ut *DF* vel *df* ad *ri*, et ex æquo $PI \times pf$ ad $PF \times pi$ ut *RI* ad *ri*, hoc est[a] ut arcus *IH* ad arcum *ih*. Rursus *PI* ad *PS* ut *IQ* ad *SE* et *ps* ad *pi* ut *SE* vel *se* ad *iq*[,] et ex æquo $PI \times ps$ ad $PS \times pi$ ut *IQ* ad *iq*. Et conjunctis rationibus $PI^q \times pf \times ps$ ad $pi^{\text{quad.}} \times PF \times PS$ ut $IH \times IQ$ ad $ih \times iq$, hoc est ut superficies circularis(181) quam arcus *IH* convolutione semicirculi *AKB* circa diametrum *AB* describet ad superficiem circularem(181) quam arcus *ih* convolutione semicirculi *akb* circa diametrum *ab* describet. Et vires quibus hæ superficies attrahunt corpuscula *P* et *p* secundum lineas obliquas ad se tendentes sunt (per Hypothesin) ut ipsæ superficies applicatæ ad quadrata distantiarum suarum a corporibus[,] hoc est ut $pf \times ps$ ad $PF \times PS$.(182) Suntq; hæ vires ad vires quæ ex ipsis per Legum Corol. 2 componuntur et(183) secundum lineas *PS*, *ps* ad centra tendunt, ut *PI* ad *PQ* et *pi* ad *pq* id est ut *PS* ad *PF* et *ps* ad *pf*. Unde ex æquo fit attractio corpusculi *P* versus *S* ad attractionem corpusculi *p* versus *s*, ut $\dfrac{PF \times pf, ps}{PS}$ ad $\dfrac{pf \times PF \times PS}{ps}$, hoc est, ut $ps^{\text{quad.}}$ ad $PS^{\text{quad.}}$. Et simili argumento vires quibus superficies convolutione arcuum *KL*, *kl* descriptæ trahunt corpuscula, erunt ut $ps^{\text{quad.}}$ ad $PS^{\text{quad.}}$, inq; eadem ratione erunt vires superficierum omnium circularium in quas utraq; superficies sphærica, capiendo semper $sd = SD$ et $se = SE$, distingui potest. Et per compositionem vires totarum superficierum sphæricarum in corpuscula exercitæ erunt in eadem ratione. Q.E.D.(184)

[a] Cor. 3 Lem. VII.

(181) Literally 'circular surface', that is, the surface (of revolution) traced out by the circle-arc round the axis *AB*.

Let $AHKB$, $ahkb$ be two equal spherical surfaces described with centres S, s and diameters AB, ab, and P, p corpuscles stationed externally in those diameters produced. From the corpuscles draw lines PHK, PIL, phk, pil cutting off from the great circles AHB, ahb equal minimally differing arcs $\overset{\frown}{HK}$, $\overset{\frown}{hk}$ and $\overset{\frown}{IL}$, $\overset{\frown}{il}$. Then to these let fall perpendiculars SD, sd, SE, se, IR and ir, of which SD, sd shall intersect PL, pl in F and f. Let fall also the perpendiculars IQ, iq to the diameters, and, because DS, ds and ES, es are equals and the angles $D\widehat{P}E$, $d\widehat{p}e$ vanishing, the lines PE, PF and pe, pf and the line-elements DF, df may be regarded as equal (seeing that their ultimate ratio as those angles $D\widehat{P}E$, $d\widehat{p}e$ come simultaneously to vanish, is one of equality). With this set-up there will be PI to PF as RI to DF, and pf to pi as DF or df to ri, and so *ex æquo* $PI \times pf$ to $PF \times pi$ as RI to ri, that is,[a] as the arc $\overset{\frown}{IH}$ to the arc $\overset{\frown}{ih}$. Again, PI is to PS as IQ to SE, and ps to pi as SE or se to iq, and *ex æquo* $PI \times ps$ to $PS \times pi$ as IQ to iq. And so, on conjoining these ratios, $PI^2 \times pf \times ps$ is to $pi^2 \times PF \times PS$ as $\overset{\frown}{IH} \times IQ$ to $\overset{\frown}{ih} \times iq$, that is, as the spherical zone(181) which the arc $\overset{\frown}{IH}$ shall describe by the revolution of the semicircle $\overset{\frown}{AKB}$ round the diameter AB to the spherical zone(181) which the arc ih shall describe by the revolution of the semicircle $\overset{\frown}{akb}$ round the diameter ab. And the forces with which these surfaces attract the corpuscles P and p along lines tending obliquely to them are (by hypothesis) as those surfaces themselves divided by the squares of their distances from the bodies, that is, as $pf \times ps$ to $PF \times PS$.(182) And these forces are to the forces which are, by Corollary 2 of the Laws, compounded from them and(183) tend to the centres along the lines PS and ps, as PI to PQ and pi to pq, that is, as PS to PF and ps to pf. Whence *ex æquo* the attraction on the corpuscle P towards S comes to be to the attraction on the corpuscle p towards s as $PF \times pf \times ps/PS$ to $pf \times PF \times PS/ps$, that is, as ps^2 to PS^2. And by a similar argument the forces by which the surfaces described by the revolution of the arcs KL, kl attract the corpuscles will be as ps^2 to PS^2, and in this same ratio will be the forces of all 'circular' surfaces into which each of the spherical surfaces can, by taking always $sd = SD$ and $se = SE$, be distinguished. And, by compounding, the forces of the total spherical surfaces exerted on the corpuscles will be in the same ratio. As was to be proved.(184)

[a] Corollary 3 to Lemma VII.

(182) Namely, on multiplying the previous ratio by that of $1/PI^2$ to $1/pi^2$.

(183) Afterwards altered by Newton to the less clumsy opening 'Suntq; hæ vires ad ipsarum partes obliquas quæ (facta per Legum Corollarium 2 resolutione virium)' (And these forces are to their oblique parts which, on resolving the forces by means of Corollary 2 of the Laws...).

(184) While Newton's prior analysis of this opaque and overlong synthetic demonstration has seemingly not been preserved, we may readily restore it on the pattern of the parallel (and considerably more difficult) computation of the total attraction of an annular ring which he came subsequently, about the autumn of 1692, to make in the course of his unpublished researches into the distorting effect of the sun upon the moon's terrestrial orbit, there—to a

Prop. XLI. Theor. XXIII

Si ad sphæræ cujusvis puncta singula tendant vires æquales centripetæ decrescentes in duplicata ratione distantiarum a punctis, ac datur ratio diametri sphæræ ad distantiam corpusculi a centro ejus; dico quod vis illa qua corpusculum attrahitur proportionalis erit semidiametro sphæræ.

... ...(185)

Prop. XLII. Theor. XXIV.

Si ad sphæræ alicujus datæ puncta singula tendant æquales vires centripetæ decrescentes in duplicata ratione distantiarum a punctis: dico quod corpusculum intra sphæram constitutum attrahitur vi proportionali distantiæ suæ ab ipsius centro.

In sphæra *ACBD* centro *S* descripta locetur corpusculum *P*, et centro eodem *S* intervallo *SP* concipe sphæram interiorem *PEQF* describi. Manifestum est per Theor. XXI quod sphæricæ superficies concentricæ ex quibus sphærarum differentia *AEBF* componitur, attractionibus per attractiones contrarias destructis nil agunt in corpus *P*. Restat sola attractio sphæræ interioris *PEQF*. Et per Theor. XXIII hæc est ut distantia *PS*. Q.E.D.(186)

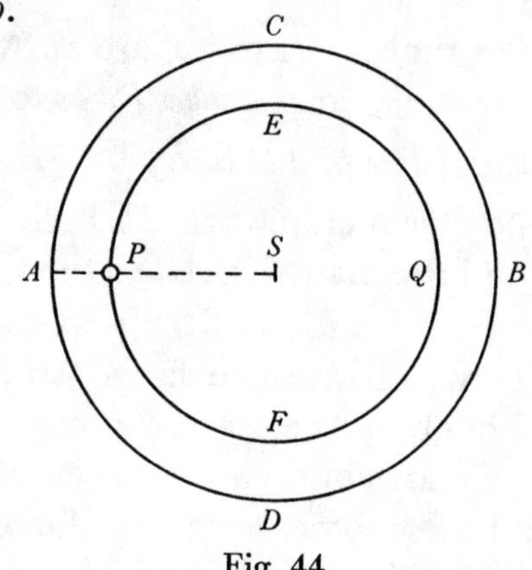

Fig. 44.

first approximation—assumed to be an excentric circle. (Newton's calculations are reproduced in Appendix 3, along with the two lemmas in which their results were incorporated.) Here, to restrict our attention initially to the laminar spherical shell of centre *S* and radius *AS* (Figure 43), because in the limit as the angle $\widehat{KPL}$ becomes vanishingly small there ensues

$$\widehat{IH}: IR = IS \text{ (or } AS): IE \quad \text{and} \quad IR : PI = FD : PF,$$

while (because the triangles *PIQ*, *PSE* are similar) $IQ: PI = SE: PS$, we may straightforwardly determine that the attraction upon the external point *P* of the spherical zone (of surface-area $2\pi.\widehat{IH}\times IQ$) formed by revolving $\widehat{IH}$ round the axis *PS* is proportional to $2\pi.\widehat{IH}\times IQ/PI^2 = 2\pi.AS\times SE\times FD/IE\times PF\times PS$; and that, since $IE = EL$, this is equal to the corresponding attraction of the zone which results from rotating the complementary infinitesimal arc $\widehat{KL}$ round *PS*: whence the component of their joint attraction acting along *PS* (at an inclination $\widehat{FPS} = \cos^{-1}[PF/PS]$ to *PIL*) is $4\pi.(AS\times SE/IE\times PS^2)\times FD$, and accordingly, because in the limit as $\widehat{KPL}$ vanishes (and so the decremented *SD* comes to coincide with *SE*) we may set $FD = SD-SE = -d(SE)$, it follows that the total attraction by the spherical shell on *P* in the direction *PS* is $4\pi.AS/PS^2.\int_{AS}^{0}(SE/IE).d(SE)$. At once, since $SE^2+IE^2 = SI^2 = AS^2$, constant, and so $SE.d(SE)+IE.d(IE) = 0$, this integral equals $4\pi.AS/PS^2.\int_0^{AS} d(IE) = 4\pi.AS^2/PS^2$: which, as Newton requires, is the attraction on *P* of the

Proposition XLI, Theorem XXIII.

If to the individual points of any sphere there tend equal centripetal forces decreasing in the doubled ratio of the distances from the points, and there is given the ratio of the diameter of the sphere to the distance of a corpuscle from its centre, I assert that the force by which the corpuscle is attracted will be proportional to the sphere's radius.

... ...(185)

Proposition XLII, Theorem XXIV.

If to the individual points of any given sphere there tend equal centripetal forces decreasing in the doubled ratio of the distances from the points, I assert that a corpuscle set within the sphere is attracted by a force proportional to its distance from the latter's centre.

In the sphere *ACBD* described with centre *S* let there be placed a corpuscle *P*, and then with the same centre *S* and radius *SP* conceive the inner sphere *PEQF* to be described. It is manifest by Theorem XXI that the concentric spherical surfaces of which the difference (*AEBF*) of the spheres is made up have, with their attractions destroyed by contrary attractions, no (combined) effect on the body *P*. There remains the attraction of the inner sphere *PEQF* alone, and by Theorem XXIII this is as the distance *PS*. As was to be proved.(186)

'mass' $4\pi . AS^2$ of the laminar *superficies sphærica* concentrated at its centre *S* distant *PS* from it. In his present synthesis, however, Newton rejects so neat and natural a conclusion—whether by this simple integral transformation (in line with those employed by him in his preceding argument) or by an equivalent direct evaluation of $\int_0^{SE} (SE/\sqrt{[AS^2-SE^2]}).d(SE)$ as $\sqrt{[AS^2-SE^2]} = IE$—in favour of a less sophisticated (but much more turgidly involved) comparison with the attraction of an equal spherical surface of centre *s* and radius $as = AS$ on a similar point *p* (Figure 42), where the lines *pl*, *pk* cut off from its central circular section arcs $\widehat{il} = \widehat{IL}$, $\widehat{hk} = \widehat{HK}$ whose semi-chords $ie = IE$, $rd = RD$ are at distances $se = SE$, $sd = SD$ from *s*; whence in the limit as $\widehat{KPL}$ and therefore $(\widehat{IL}-\widehat{HK} = \widehat{il}-\widehat{hk}$ and so) $\widehat{kpl}$ becomes vanishingly small there ensues $fd(= sd-se = SD-SE) = FD$, that is, $d(se) = d(SD)$, and consequently the ratio of the total attractions on *P* and *p* of the respective spherical surfaces of centres *S* and *s* is that of their corresponding infinitesimal elements, namely

$$4\pi.AS\times SE\times FD/IE\times PS^2 : 4\pi.as\times se\times fd/ie\times ps^2 = 1/PS^2 : 1/ps^2.$$

(185) We omit Newton's somewhat vaguely expressed verbal demonstration, and also two following corollaries: the text of these passed without any change into that of the equivalent Proposition LXXII of Book 1 of the *Principia* ($_1$1687: 195–6). Its sole function in sequel, in the following Proposition XLII, is to support his assertion that in spheres of uniform density (and whose mass therefore varies as the cube of their radius) the total attraction on a surface point (which by the previous proposition varies inversely as the square of the radius) is effectively proportional to the radius.

(186) For, since the gravitational mass is assumed everywhere to be uniform, the total attractive power of the inner sphere is jointly as its volume $\frac{4}{3}\pi . PS^3$ and as its average intensity, which—by making a concealed anticipatory appeal to the result demonstrated in Proposition

Schol.

Superficies ex quibus solida componuntur, hic non sunt pure mathematicæ[(187)] sed orbes adeo tenues ut eorum crassitudo instar nihili sit, nimirum orbes evanescentes ex quibus sphæra ultimò constat ubi orbium illorum numerus augetur et crassitudo minuitur in infinitum juxta methodum sub initio in Lemmatîs generalibus expositam. Similiter per puncta ex quibus lineæ superficies et solida componi dicuntur, intelligendæ sunt particulæ æquales magnitudinis contemnendæ.

Prop XLIII. Theor. XXV.

Iisdem positis, dico quod corpusculum extra sphæram constitutum attrahitur vi reciprocè proportionali quadrato [distantiæ suæ ab ipsius centro.][(188)]

XLIII following —is inversely as PS^2, and hence the total attraction varies directly as *PS*. In the geophysical model in which the sphere *ACBD* represents the earth (taken, both for simplicitly and in default of any exact contemporary knowledge of its internal constitution, to be uniformly dense) the theorem has an obvious relevance to the letters exchanged five years before between Newton and Robert Hooke on the ideal path of free fall—were such motion possible—below the earth's surface. Hooke had in fact written to him on 6 January 1679/80 (*Correspondence of Isaac Newton*, **2**: 309) that in his preceding letter of 9 December, where he had propounded 'a kind [of] Elleptueid' (*ibid.*: 305) as his preferred fall curve, 'my supposition is that the Attraction always is in a duplicate proportion to the Distance from the Center Reciprocall... not that I beleive there really is such an attraction to the very Center of the Earth, but on the Contrary I rather Conceive that the more the body approaches the Center, the lesse will it be Urged by the attraction—possibly somewhat like the Gravitation on a pendulum or a body moved in a Concave sphere [*sc.* in a hypocycloid; compare §3: note (298)] where the power Continually Decreases the neerer the body inclines to a horizontall motion, which it hath...in the lowest point'. Newton was conveniently to forget Hooke's codicil when on 20 June 1686, in a retrospect of 'w[t] past between us in our Letters so far as I...remember', he disingenuously complained to Halley that 'what he told me of y[e] duplicate proportion was erroneous, namely that it reacht down from hence to y[e] center of y[e] earth', adding that he himself 'never extended y[e] duplicate proportion lower then to y[e] superficies of y[e] earth & before a certain demonstration [that of Proposition XL above] I found y[e] last year have suspected it did not reach accurately enough down so low' (*Correspondence*, **2**: 435–6). In modern hindsight we may add that Newton's deduction from the postulate of uniform gravitational mass that the effective internal attraction of a sphere varies approximately as the distance from the centre does not mirror geophysical reality: because of the earth's relatively heavy core, terrestrial gravity continues in fact slightly to increase during the first twelve hundred or so miles of descent below the earth's surface before levelling off and thereafter decreasing—somewhat 'bumpily'—to zero as the centre is neared. (See K. E. Bullen, 'The Earth and Mathematics', *Mathematical Gazette*, **54**, 1970: 352–3; and compare F. Tisserand, *Traité de Mécanique Céleste*, **2** (Paris, 1891): 244.)

(187) That is, of no thickness at all—or at least (compare note (68) above) of one which is so 'indivisibly' small as to have no common Archimedean measure by any finite magnitude.

(188) The manuscript here (on f. 14[r]) breaks off with the enunciation not quite complete.

Scholium.

The surfaces of which solids are composed are here not purely mathematical ones[(187)] but shells so slim that their thickness is tantamount to nothing: specifically, the vanishingly thin shells of which a sphere consists when the number of those shells increases and their thickness diminishes indefinitely in accord with the method developed near the beginning in the general lemmas. Similarly, by points of which lines, surfaces and solids are said to be composed you must understand equal particles of spurnable size.

Proposition XLIII, Theorem XXV.

With the same suppositions, I assert that a corpuscle set outside a sphere is attracted by a force reciprocally proportional to the square [of its distance from the latter's centre.][(188)]

It will be plain that Newton's ensuing demonstration can have been only trivially variant from that of the equivalent Proposition LXXIV in Book 1 of the published *Principia* ($_1$1687: 197), namely '...distinguatur Sphæra in superficies Sphæricas innumeras concentricas, & attractiones corpusculi a singulis superficiebus oriundæ erunt reciproce proportionales quadrato distantiæ corpusculi a centro, per Theor. [XXII]. Et componendo, fiet summa attractionum, hoc est attractio Sphæræ totius, in eadem ratione' (let the sphere be separated into innumerable concentric spherical surfaces, and the attractions of the corpuscle arising from the individual surfaces will, by Theorem XXII, be reciprocally proportional to the square of the corpuscle's distance from the centre; and, by compounding, the sum of the attractions—that is, the attraction of the total sphere—will prove to be in the same ratio). As we have already partially remarked (in note (186)), some fourteen months afterwards in June 1686 Newton affirmed to Halley that, until he had 'ye last year' contrived a demonstration of the preceding Proposition XL on which the present theorem is squarely founded, he had 'suspected...ye duplicate proportion [of the inverse variation of terrestrial gravitation]...did not reach accurately enough down...to ye superficies of ye earth, & therefore in ye doctrine of projectiles never used it nor considered [*sc.* thereby] ye motions of ye heavens' (*Correspondence*, **2**: 435). Plainly, if in such a field Newton doubted whether, corresponding to an arbitrary point on its surface, the earth's effective centre of attraction does not coincide—approximately at least—with the mathematical centre of its geosphere, then he could attain no precise confirmation of the inverse-square variation of terrestrial gravitation simply by comparing the square of the ratio of the mean radius of lunar orbit to the earth's semidiameter with the ratio of the acceleration of 'gravity' at the earth's surface to the readily computable central 'endeavour' presumed instantaneously to deviate the circling moon towards the earth. While in the same paragraph of a paper of about 1670 (ULC. Add. 3958.3: 87^r/87^v, reproduced in photocopy in *The Correspondence of Isaac Newton*, **1**, 1959: 297–8) Newton could both assert that 'Luna...distat 59 vel 60 semidiametris terrestribus a terra' and calculate—on the basis of an erroneous value of 3500 miles for the earth's radius—that the terrestrial 'vis gravitatis est 4000 vicibus major conatu lunæ recedendi a centro terræ, et amplius', there is, in confirmation of his later remark to Halley, nothing in any of his extant early mechanical or astronomical manuscripts to suggest that before beginning his correspondence with Hooke in late 1679 he ever connected the two. (Compare D. T. Whiteside, 'Before the *Principia*: The Maturing of Newton's Thoughts on

APPENDIX 1. THE PRELIMINARY 'DEFINITIONS' AND 'LAWS' OF THE RECAST TRACT 'DE MOTU CORPORUM'.(1)

From the corrected amanuensis copy(2) in the University Library, Cambridge

De motu corporum in medijs regulariter cedentibus.

Definitiones.

Def. 1. Tempus absolutum est quod sua natura absq; relatione ad aliud quodvis æquabiliter fluit. Tale est cujus æquationem investigant Astronomi, alio nomine dictum Duratio.

Def. 2. Tempus relative spectatum est quod respectu fluxionis seu transitus rei alicujus sensibilis consideratur ut æquabile. Tale est tempus dierum mensium et aliarum periodorum cœlestium apud vulgus.(3)

Def. 3. Spatium absolutum(4) est quod sua natura absq; relatione ad aliud quodvis semper manet immobile. Ut partium temporis ordo immutabilis est sic etiam partium spatij. Moveantur hæ de locis suis et movebuntur de seipsis. Nam tempora et spatia sunt sui ipsorum et rerum omnium loca. In tempore quoad ordinē successionis, in spatio quoad ordinem situs locantur universa. De

---

Dynamical Astronomy, 1664–1684', *Journal for the History of Astronomy*, 1, 1970: 5–19, especially 11–13.) There is, we might add, nothing in Newton's 1686 observation—which he nowhere repeated—to lend support to J. C. Adams' widely accepted conjecture (first announced by J. W. L. Glaisher in a Trinity College address on 'The Bi-centenary of Newton's Principia', given on the afternoon of 19 April 1888 and published on the following day in *The Cambridge Chronicle and University Journal*...: 7–8) that lack of the present theorem led him at any time before 1685 to lay aside his computations for any such test of the moon's normal acceleration against the earth's gravity. Nor should we suppose that its proof held any great difficulties for him when he came to demonstrate it: as we showed above (in note (184)) a straightforward analysis in a manner analogous to that which Newton afterwards used to evaluate the attractive potential of a laminar ring (see Appendix 3) leads easily and naturally to an exact algebraic integral which may at once be resolved by a simple change of variable, while his mathematical abilities were subsequently little more strained in dealing with the generalised problem—where the law of attraction is arbitrary—whose solution he elegantly effected in Propositions LXXIX–LXXXI of his ensuing *Principia* ($_1$1687: 204–11; reproduced, for their intrinsic interest and as a complement to the present particular solution, in Appendix 4).

What exactly Newton comprehended in the remaining portion of his original manuscript (now, it would appear, irretrievably lost) we do not know other than that it extended at least as far as a 'Prop. LXXII' (so referred to on f. 44r of the companion 'De motu Corporum Liber secundus, now ULC. Add. 3990) whose context readily allows it to be identified as a preliminary version of Proposition XXII of Book 2 of the later *Principia* ($_1$1687: 298–9; compare I. B. Cohen's *Introduction*...(note (1)): 317): this evidently there formed part of a complementary set of 'Propositiones De motu Corporum in spatijs resistentibus' (compare note (69) above) which would have opened with revised versions of Problems 6 and 7 of the initial

illorum essentia est ut sint loca et loca primaria moveri absurdum est.[5] Porrò vi illata moveatur una pars spatij et vi tanta ad omnes in infinitum partes applicata movebitur totum, quod rursus absurdum est.

Def. 4. Spatium relativum est quod respectu rei alicujus sensibilis consideratur ut immobile: uti spatium aeris nostri respectu terræ. Distinguuntur autem hæc spatia ab invicem ipso facto per descensum gravium quæ in spatio absoluto rectà petunt centrum[,] in relativo absolute gyrante deflectunt ad latus.

Def. 5. Corpora in sensus omnium incurrunt[,] ut res [6]mobiles quæ se mutuo penetrare nequeunt.

[7][*Def.*] *6.* Densitas corporis est quantitas seu copia materiæ collata cum quantitate occupati spatij.

[*Def.*] *7.* Per pondus intelligo quantitatem seu copiam materiæ movendæ abstracta gravitationis consideratione quoties de gravitantibus non agitur. Quippe pondus gravitantium[8] proportionale est quantitati materiæ, et analoga per se invicem exponere et designare licet. Analogia verò sic colligebitur.

---

tract 'De motu' (§1) and an amplification of their following scholium much as in Proposition IV of Book 2 of the *Principia* ($_1$1687: 241–5, reproduced in Appendix 3 to §1). To what theorems in the published *Principia* the remainder of the original intervening Propositions XLIV–LXXI might have corresponded must remain highly conjectural—*faute de mieux*, we would tentatively urge the plausibility of roughly identifying these with LXXV–LXXVIII, LXXXV–XC and XCII/XCIII of Book 1, and V–IX, XI–XIV and XIX–XXI of Book 2.

(1) This revised and much amplified opening to Newton's earlier augmented tract 'De motu sphæricorum Corporum in fluidis' (see §1, Appendix 1 above) is here reproduced to reveal how its Definitions and Laws came rapidly to evolve into the mature 'Definitiones' and 'Leges Motûs' of the revised 'De motu Corporum' (§2 preceding). As ever, it is not our purpose to discuss in any detail the increasingly sophisticated conceptual basis for his technical dynamical arguments which Newton here sets down, and (other than trivially to establish the text) we restrict our commentary to a few points of especial interest.

(2) ULC. Add. 3965.5^a: 25^r–26^r/23^r–24^r, first printed—with several errors of transcription and some confusion over textual sequence—by J. W. Herivel in his *The Background to Newton's Principia* (Oxford, 1966): 304–8/316–17; compare also I. B. Cohen's *Introduction to Newton's Principia* (Cambridge, 1971): 93–5. Like its parent text (§1, Appendix 1) the manuscript was initially drafted in Humphrey Newton's careful secretarial script but Newton himself has afterwards extensively corrected and augmented it in his own hand.

(3) This initially affirmed: 'Tempus relativum est quod ex rei alterius fluxione seu transitu mensuratur. Tale est tempus dierum mensium et aliarum periodorum cœlestium, quod propterea cum hoc mundo cœpisse creditur'. In preliminary revise Newton altered the last clause to read 'ex hypothesi quod hæ periodi sunt æquabiles' and then to 'quas vulgus ut æquabiles considerat'.

(4) Originally—and less dogmatically—'absolute dictum' was written.

(5) This concluding phrase replaces a less forceful '...et loca moveri nequeunt'.

(6) 'extensæ et' is deleted.

(7) The two following Definitions replace a preliminary 'Def. 6. Centrum corporis cujusq; est quod vulgo dicitur centrum gravitatis et axis corporis est linea quævis recta per centrum transiens' which was afterwards partially subsumed into Definition 11 below.

(8) 'copia materiæ' was initially repeated here.

Pendulis æqualibus numerentur oscillationes corporum duorum ejusdem ponderis et copia materiæ in utroq; erit reciprocè ut numerus oscillationum eodem tempore factarum. Experimentis autem in auro, argento, plumbo, vitro, arena, sale communi, aqua, ligno, tritico diligenter factis(9) incidi semper in eundem oscillationum numerum.(10)

Def. 8. Locus corporis est pars spatij in quo corpus existit, estq; pro genere spatij vel absolutus vel relativus.

Def. 9. Quies corporis est perseverantia ejus in eodem loco, estq; vel absoluta vel relativa pro genere loci.

Def. 10. Motus corporis est translatio ejus de loco in locum, estq; itidem vel absolutus vel relativus pro genere loci. Distinguitur autem ipso facto motus absolutus a relativo in gyrantibus per conatum recedendi a centro, quippe qui ex gyratione nudè relativa nullus est,(11) in relative quiescentibus permagnus esse potest, ut in corporibus cœlestibus quæ ex mente Cartesianorum quiescunt et conantur tamen a sole recedere. Conatus ille certus semper et determinatus arguit certam aliquam et determinatam esse motus realis quantitatem in singulis corporibus, a relationibus quæ innumeræ sunt totidemq; motus relativos constituunt minimè pendentem. Porro motum et quietem absolute dictos non pendere a situ et relatione corporum ad invicem manifestum est ex eo quod hæ nunquam mutantur nisi vi in ipsum corpus motum vel quiescens impressa, tali autē vi semper mutantur; at relativæ mutari possunt vi solummodo impressa in altera corpora ad quæ fit relatio et non mutari vi impressa in utraq; sic ut situs relativus conservetur.

---

(9) Originally 'institutis', followed in sequel by 'reperi'.

(10) A cancelled final sentence reads: 'Ob hanc analogiam & defectu vocis commodioris expono et designo quantitatem materiæ per pondus, etiam in corporibus quorum gravitatio non consideratur.'

(11) The remainder of this sentence and the whole of the next is a retrospective addition in the margin.

(12) 'Celeritas motus est quantitas momentanea' was initially written.

(13) Originally 'quantitate' in a first version of this Definition (on f. 26r, there numbered '11' and augmented with two sentences, 'Æstimatur autem quantitas corporis ex copia materiæ corporeæ quæ gravitati suæ proportionalis esse solet. Pendulis æqualibus numerentur oscillationes corporum duorum ejusdem ponderis, et copia materiæ in utroq; erit reciprocè ut numerus oscillationum eodem tempore factarum', which were subsequently subsumed into Definition 7).

(14) Newton has omitted to change the original numbering '12' in the manuscript: we increase this by unity to allow for the replacement of the original 'Def. 6' by the two Definitions 6/7 above (see note (7)).

(15) That is, 'innate, inherent and essential' as Newton later rendered this phrase for Richard Bentley on 25 February 1692/3 (*Correspondence*, **3**, 1961: 254; compare Cohen's *Introduction* (note (2)): 68). The manuscript here originally opened with 'Vis corporis seu corpori insita et innata'.

Def. 11. Velocitas est quantitas[12] translationis quoad longitudinem itineris certo tempore confecti. Iter verò est quod corporis puncto medio describitur a Geometris dicto centro gravitatis. Loquor de motu progressivo.

Def. 12. Quantitas motus est quæ oritur ex velocitate et pondere[13] corporis translati conjunctim. Motus additione corporis alterius tanto cum motu fit duplus et duplicata velocitate quadruplus.

Def. 1[3].[14] Corporis vis insita innata et essentialis[15] est potentia qua id perseverat[16] in statu suo quiescendi vel movendi uniformiter in linea recta, estqꝫ corporis quantitati proportionalis. Exercetur verò proportionaliter mutationi status[17] et quatenus exercetur dici potest corporis vis exercita conatus et reluctatio. Hujus una species est vis centrifuga gyrantium.[18]

Def. 1[4].[19] Vis motus seu corpori ex motu suo adventitia est qua corpus quantitatem totam sui motus conservare conatur. Ea vulgo dicitur impetus estqꝫ motui proportionalis, et pro genere motus vel absoluta est vel relativa. Ad absolutam referenda est vis centrifuga gyrantium.[20]

Def. 15. Vis corpori illata et impressa[21] est qua corpus urgetur mutare statum suum movendi vel quiescendi: estqꝫ diversarum specierum, sicut pulsus seu pressio percutientis, pressio continua, vis centripeta, resistentia medij[22] &c.

Def. 16. Vim centripetam appello qua corpus impellitur vel attrahitur versus punctum aliquod quod ut centrum spectatur. Hujus generis est gravitas tendens

(16) In an exact repeat of the Cartesian phrase in 'Def. 2' of the parent text (§1, Appendix 1) 'conatur perseverare' was first written.

(17) Originally 'allatæ'.

(18) Inasmuch, apparently, as such a *vis centrifuga* maintains—or is maintained by—a uniform 'motus circularis' in the rotating body. Such a force cannot, however, in any sense be conceived to conserve rectilinear motion, and it is distinctly illogical of Newton to include it here, having (see note (20)) cancelled the following Definition which initially introduced it as a (unique?) species of a Cartesian *vis motus* which he speedily came to reject. The original 'Def. 12' (on f. 26r) initially comprehended only the first sentence, and was given a preliminary recasting (on f. 25v) which reads: 'Corporis vis exercita est qua id conatur conservare status sui movendi vel quiescendi partem illam quam singulis momentis amittit, estqꝫ status illius mutationi seu parti singulis momentis amissæ proportionalis, nec improprie reluctatio vel resistentia corporis dicitur. Huic referenda est vis centrifuga gyrantium.'

(19) Much as before (see note (14)) the manuscript reads '13' without allowance, in revise, for the additional Definition 7.

(20) Newton would seem here to introduce this generalised notion of a speed-preserving 'vis motus' only to afford the uniform circular motion which induces such a *vis centrifuga* a special status in his dynamical scheme. In cancelling this Definition he was not yet prepared to subsume such a 'motus uniformis circularis'—as he will call it in *Lex* 1 below—into a general theory of constrained motion induced by a central force, but preferred illogically to attach it to the preceding Definition 13 of uniform motion in a straight line which a body's *vis insita* maintains (see note (18)).

(21) Originally termed 'Vis impulsus' simply.

(22) 'aut corporis cujusvis alterius' is deleted.

ad centrum terræ, vis magnetica tendens ad centrum magnetis et vis cœlestis[23] cohibens Planetas ne abeant in tangentibus orbitarum.

Def. 17. Per medij resistentiam in sequentibus intelligo vim medij regulariter impedientis. Sunt et aliæ vires ex corporum elasticitate, mollitie, tenacitate &c pendentes quas hic non considero.[24]

Def. 18. Exponentes temporum spatiorum motuum celeritatū et virium sunt quantitates quævis proportionales exponendis.[25]

Hæc omnia fusius explicare visum est ut Lector præjud[i]cijs quibusdam vulgaribus liberatus[26] et distinctis principiorum Mechanicorum conceptibus imbutus accederet ad sequentia. Quantitates autem absolutas et relativas ab invicem sedulò distinguere necesse fuit[27] eò quod phænomena omnia pendeant ab absolutis, vulgus autem qui cogitationes a sensibus abstrahere nesciunt semper loquuntur de relativis, usq; adeo ut absurdum foret vel sapientibus vel etiam Prophetis apud hos aliter loqui. Unde et sacræ literæ[28] et scripta Theologorum de relativis semper intelligenda sunt, et crasso laboraret præjudicio qui inde de rerum naturalium motibus philosophicis disputationes moveret.[29]

LEGES MOTUS.[30]

Lex 1. Vi insita corpus omne perseverare in statu suo quiescendi vel movendi uniformiter in linea recta nisi quatenus viribus impressis[31] cogitur statum illum mutare. Motus autem uniformis hic est duplex, progressivus secundum lineam

(23) Notice that Newton no longer identifies this with a 'gravitas solaris' but embraces within it the gravitational attractions to all other members of the celestial world (perhaps even including the fixed stars?).

(24) We omit a following cancelled 'Def. 16. Momenta quantitatum sunt ipsarum principia generantia vel alterantia fluxu continuo: ut tempus præsens præteriti et futuri, motus præsens præteriti, vis centripeta aut alia quævis momentanea impetus, punctum lineæ, linea superficiei, superficies solidi et angulus contactus anguli rectilinei.' In Lemma II of Book 2 of his published *Principia* ($_1$1687: 250–3, reproduced on IV: 521–5) Newton was similarly to assert that 'quantitates ut indeterminatas & instabiles, & quasi motu fluxuve perpetuo crescentes vel decrescentes hic considero, & eorum incrementa vel decrementa momentanea sub nomine momentorum intelligo' (*ibid.*: 251).

(25) This trivially expands 'Def. 4' of the parent text (§1, Appendix 1 preceding).

(26) This adjectival phrase replaces an earlier 'claris' attached (in parallel with 'distinctis') to 'conceptibus' following.

(27) Originally 'acriter distinguere coactus sum'.

(28) 'holy writ'.

(29) Newton first went on to add, and then cancelled, a not very revealing astronomical parallel: 'Perinde est ac si quis Lunam (in Gen[ere] 1) magnitudine non apparente sed absoluta inter duo maxima lumina [*sc.* together with the Sun] numerari contenderet.'

(30) An amplified version of the five laws of the parent text (§1, Appendix 1), augmented with a new 'Lex 3' in which Newton for the first time enunciates his now familiar third law of motion in a dynamical context (see note (34) below).

(31) 'et impedientibus' is deleted.

rectam quam corpus centro suo æquabiliter lato describit & circularis(32) circa axem suum quemvis qui vel quiescit vel motu uniformi latus semper manet positionibus suis prioribus parallelus.

Lex 2. Mutationem motus proportionalem esse vi impressæ et fieri secundum lineam rectam quâ vis illa imprimitur.

Hisce duabus Legibus jam receptissimis Galilæus invenit(33) projectilia gravitate uniformiter et secundum lineas parallelas agente in medio non resistente lineas Parabolicas describere. Et suffragatur experientia nisi quatenus motus projectilium resistentia aeris aliquantulum retardatur.

Lex 3. Corpus omne tantum pati reactione quantum agit in alterum.(34) Quicquid premit vel trahit alterum, ab eo tantum premitur vel trahitur. Si vesica aere plena premit vel ferit alteram sibi consimilem cedet utraꝙ æqualiter introrsum. Si corpus impingens in alterum vi sua mutat motum alterius et ipsius motus (ob æqualitatem pressionis mutuæ) vi alterius tantum mutabitur. Si magnes trahit ferrum ipse vicissim tantum trahitur, et sic in alijs. Constat verò hæc Lex per Def. 1[3] et 14 in quantum vis corporis ad status sui conservationem exercita sit eadem cum vi in corpus alterum ad illius statum mutandum impressa, et vi priori proportionalis sit mutatio status prioris[,] posteriori ea posterioris.

Lex 4. Corporum dato spatio inclusorum eosdem esse motus inter se sive spatium illud absolutè quiescat sive moveat id perpetuò et uniformiter in directum absꝙ motu circulari. E.g. Motus rerum in navi perinde se habent sive navis quiescat sive moveat ea uniformiter in directum.

Lex 5. Mutuis corporum actionibus commune centrum gravitatis non mutare statum suum motus vel quietis. Hæc Lex et duæ superiores se mutuò probant.

Lex 6. Resistentiam medij esse ut medij illius densitas et sphærici corporis moti superficies et velocitas conjunctim. Hanc legem exactam esse non affirmo. Sufficit quod sit vero proxima. Corpora verò sphærica esse suppono in sequentibus, ne opus sit circumstantias diversarum figurarum considerare.(35)

(32) Newton again (compare note (18) above) sets uniform rotary motion on a par with 'inertial' motion in a straight line. This privileged status is abandoned in the revised 'De motu Corporum' (§2 preceding).

(33) In his *Discorsi e Dimostrazioni Matematiche...* (Leyden, 1638): 236 ff; see §2: note (35).

(34) This is effectively the enunciation of his celebrated third law of motion earlier given by Newton in the 12th arithmetical problem of his Lucasian lectures on algebra (v: 148); it was straightaway to be recast into its more familiar Newtonian form in §2 preceding.

(35) This amplification of 'Lex 5' of the parent text (§1, Appendix 1) evidently became superfluous as soon as Newton began to consider resistances varying as some arbitrary function of the speed—as we have conjectured that he did in Propositions LXI–LXVIII (corresponding to Propositions V–IX/XI–XIV of Book 2 of the *Principia*) of the lost portion of the revised 'De motu Corporum' (see §2: note (188)).

LEMMATA.

Lem. 1. Corpus viribus conjunctis diagonalem parallelogrammi eodem tempore describere quo latera separatis.

Si corpus dato tempore...reperiri necesse est.

Lem. 2. Spatium quod corpus urgente quacunqꝫ vi centripeta ipso motus initio describit esse in duplicata ratione temporis.

Exponantur tempora per lineas...per areas rectilineas *ABF ADH* perpendiculis(36)

APPENDIX 2. LATER ADDITIONS TO THE REVISED 'DE MOTU CORPORUM'.(1)

[mid-1685/July 1686]

From originals in the University Library, Cambridge and the Royal Society, London

[1](2) [*Lemma X.*]

Coroll. 1. Et hinc facilè colligitur quod corporum similes similium figurarum partes temporibus proportionalibus describentium errores(3) qui viribus æqualibus in partibus istis ad corpora similiter applicatis generantur et mensurantur a locis figurarum ad quæ corpora temporibus ijsdem proportionalibus absqꝫ viribus istis in fine temporum pervenirent sunt ut quadrata temporum in quibus generantur quamproximè.

Coroll. 2. Errores autem qui viribus proportionalibus similiter applicatis generantur sunt ut vires et quadrata temporum conjunctim.

(36) The manuscript, as we now have it, ends (at the foot of f. 24^r) in mid-sentence, its continuation seemingly lost. It will, however, be clear that the present Lemma 2—as the preceding 'Lem. 1'—is a word-for-word repeat of its equivalent in the parent text (§1, Appendix 1). Newton doubtless went on similarly to copy the ensuing Lemmas 3 and 4, but it would appear impossible to say what further additions (if any) were made in sequel.

(1) With the exception of [6]—added in July 1686 to the press copy after it had passed into the printer's hands (see note (21) below)—these minor complements and additions to the text of the Lemmas and Propositions I–XVI in §2 preceding convert it into Articles I–III of the revised 'De motu Corporum Liber primus' (effectively Sections I–III of *Principia*, $_1$1687: 26–60) whose sequel we reproduce in §3 following. The suggested date of middle (say May/June) 1685 for their composition will accordingly not be far out.

(2) ULC. Dd. 9. 46: 24^v; compare §2: note (52). This trivial extension of Lemma X is nowhere applied in the later text of the 'De motu Corporum' (or the ensuing *Principia*).

(3) Understand instantaneous deviations from the tangential path.

[2][4] *Scholium* [ad Lemmata].

... ...

Contendi etiam potest quod si dentur ultimæ quantitatum evanescentium rationes, dabuntur et ultimæ magnitudines; et sic quantitas omnis constabit ex indivisibilibus, contra quam Euclides de incommensurabilibus in libro decimo Elementorum demonstravit.[5] Verum hæc objectio falsæ innititur hypothesi. Ultimæ rationes illæ quibuscum quantitates evanescunt, revera[6] non sunt rationes quantitatum ultimarum sed limites ad quos quantitatum sine limite decrescentium rationes semper appropinquant, et quos propius assequi possunt quam[7] data quavis differentia, nunquam verò transgredi, neq; prius quam quantitates diminuuntur in infinitum. Res clarius intelligetur in infinite magnis. Si quantitates duæ quarum data est differentia augeantur in infinitum, dabitur[8] harum ultima ratio, nimirum ratio æqualitatis, nec tamen ideò dabuntur quantitates ultimæ seu maximæ quarum ista sit ratio. Igitur in sequentibus, siquando facili rerum imaginationi consulens, dixero quantitates quam minimas vel evanescentes vel ultimas, cave ne intelligas quantitates magnitudine determinatas sed cogita semper diminuendas sine limite.[9]

---

[3][10] *Prop. I. Theorema I.*

Areas quas corpora in gyros ad immobile centrum virium ductis describunt, et in planis immobilibus consistere et esse temporibus proportionales.

Dividatur tempus in partes æquales, et prima temporis parte describat corpus vi insita rectam AB. Idem secunda temporis parte si nil impediret rectà pergeret

(4) ULC. Dd. 9. 46: 27v; compare §2: note (68).

(5) More accurately 'postulavit': Newton first wrote 'demonstratam reliquit'. The demand made by Euclid (in *Elements* x, 1) is an implicit appeal—clarified in Isaac Barrow's text (*Euclidis Elementorum Libri XV. breviter demonstrati*, Cambridge, 1655), as Newton well knew, by an immediately preceding interpolated *Postulatum* (p. 192) asserting that 'quamlibet magnitudinem toties posse multiplicari, donec quamlibet magnitudinem ejusdem generis excedat'—to *Elements* v, Definition 5 [Barrow]: 'Rationem habere inter se magnitudines dicuntur; quæ possunt multiplicatæ se mutuò superare' (*ibid.*: 90). In his 1671 tract, as we have remarked (III: 166, note (316)), Newton had earlier introduced this Eudoxian axiom in an analogous context to classify angles of contact into *genera* having the same (infinitesimal) measure; compare §2: note (59).

(6) Newton initially wrote 'si rigidè loquemur'.

(7) In the printed *Principia* ($_1$1687: 36) a clarifying 'pro' is here inserted.

(8) The equivalent 'erit æqualis dato' was first written.

(9) Let any modern mathematical logician tempted to argue that the Newtonian infinitesimal is a non-Archimedean quantity beware!

(10) ULC. Dd. 9. 46: 27v–28v; compare §2: note (70). We may leave the points of improvement in this reworking of Newton's opening proposition to be noticed by the reader: none involve any fundamental change in the underlying argument.

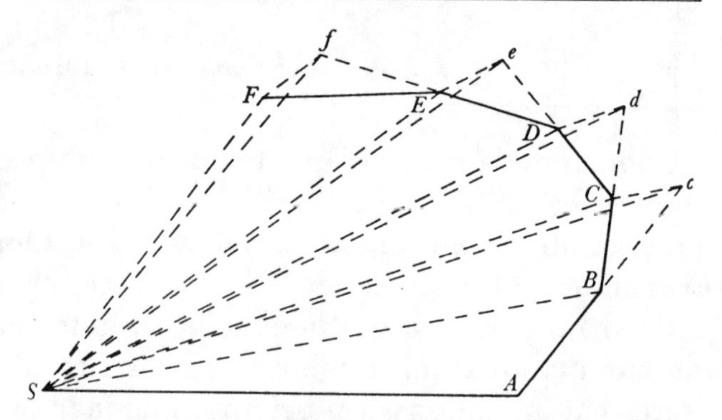

ad *c* describens lineam *Bc* æqualem ipsi *AB*, adeo ut radijs *AS*, *BS*, *cS* ad centrum actis confectæ forent æquales areæ *ASB*, *BSc*. Verùm ubi corpus venit ad *B* agat vis centripeta impulsu unico sed magno, faciatq corpus a recta *Bc* deflectere et pergere in recta *BC*. Ipsi *BS* parallela agatur *cC* occurrens *BC* in *C*, et completa secunda temporis parte corpus reperietur in *C* in eodem plano cum triangulo *ASB*. Junge *SC* & triangulum *SBC* ob parallelas *SB*, *Cc* æquale erit triangulo *SBc* atq adeo etiam triangulo *SAB*. Simili argumento si vis centripeta successivè agat in *C*, *D*, *E* &c faciens ut corpus singulis temporis particulis singulas describat rectas *CD*, *DE*, *EF* &c, jacebunt hæ in eodem plano et triangulum *SCD* triangulo *SBC* et *SDE* ipsi *SCD* et *SEF* ipsi *SDE* æquale erit. Æqualibus igitur temporibus æquales areæ in plano immoto describuntur[:] et componendo, sunt arearum summæ quævis *SADS*, *SAFS* inter se ut sunt tempora descriptionum. Augeatur jam numerus et minuatur latitudo triangulorum in infinitum et eorum ultima perimeter *ADF*, per Corollarium quartum Lemmatis tertij, erit linea curva, adeoq vis centripeta qua corpus de tangente hujus curvæ perpetuò retrahitur, aget indesinenter, areæ verò quævis descriptæ *SADS*, *SAFS* temporibus descriptionum semper proportionales, erunt ijsdem temporibus in hoc casu proportionales. Q.E.D.

Coroll. 1. In medijs non resistentibus si areæ non sunt temporibus proportionales, vires non tendunt ad concursum radiorum.

Coroll. 2. In medijs omnibus si arearum descriptio acceleratur vires non tendunt ad concursum radiorum, sed inde declinant in consequentia.

---

[4][11] *Prop. III. Theor. III.*

... ...

Corol. 1. Hinc si corpus unum radio ad alterum ducto describit areas temporibus proportionales; atq de vi tota (sive simplici sive ex viribus pluribus juxta[12] Legum Corollarium secundum composita) qua corpus prius urgetur, sub-

(11) ULC. Dd. 9. 46: 29^v; compare §2: note (80). These revised corollaries express Newton's intentions more clearly and concisely without adding anything new.

(12) Originally 'secundum'.

ducatur (per idem Legum Corollarium) vis tota acceleratrix qua corpus alterum(13) urgetur: vis omnis reliqua qua corpus prius urgetur tendet ad corpus alterum ut centrum.

Corol. 2. Et si areæ illæ sunt temporibus quamproximè proportionales, vis reliqua tendet ad corpus alterum quamproximè.

Corol. 3. Et vice versa si vis reliqua tendit quamproximè ad corpus alterum, erunt areæ illæ temporibus quamproximè proportionales.

Corol. 4. Si corpus radio ad alterum corpus ducto describit areas quæ cum temporibus collatæ sunt valde inæquabiles, et corpus illud alterum vel quiescit vel movetur uniformiter in directum[,] actio vis(14) centripetæ ad corpus illud alterū tendentis vel nulla est vel miscetur & componitur cum actionibus admodum potentibus aliarum virium visq; tota, ex omnibus si plures sunt vires, composita, ad aliud (sive immobile sive mobile) centrum dirigitur circum quod æquabilis est arearum descriptio. Idem obtinet ubi corpus alterum motu quocunq; movetur, si modo vis centripeta sumatur quæ restat post subductionem vis totius agentis in corpus illud alterum.

---

[5](15) *Prop. IV. Theor. IV.*

Corporum quæ diversos circulos æquabili motu describunt,(16) *vires centripetas ad centra eorundem circulorum tendere, et esse inter se ut arcuum simul descriptorum quadrata applicata ad circulorum radios.*

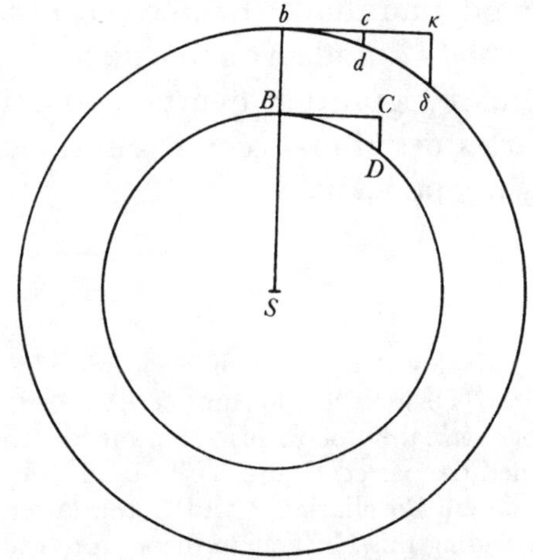

Corpora *B*, *b* in circumferentijs circulorum *BD*, *bd* gyrantia, simul describant arcus *BD*, *bd*. Quoniam sola vi insita describerent tangentes *BC*, *bc* his arcubus æquales[,] manifestum est quod(17) vires centripetæ sunt quæ perpetuò retrahunt corpora de tangentibus ad circumferentias circulorum, atq; adeo hæ sunt ad invicem in ratione prima spatiorum nascentium *CD*, *cd*[:] tendunt vero ad centra circulorum per Theor. [II], propterea quod areæ radijs descriptæ ponuntur temporibus proportionales. Fiat figura $\delta\kappa b$ *DCB* similis, et per Lemma [V], lineola *CD* erit ad lineolam [$\kappa\delta$] ut arcus *BD* ad arcum

---

(13) Newton initially added 'pro mole sua' and then converted this to 'pro pondere suo' before deleting the phrase.

(14) 'totius' is deleted in sequel.

(15) ULC. Dd. 9. 46: 30v/31r; compare § 2: note (84). Several of the verbal recastings in

$b\delta$, necnon per Lemma XI lineola nascens $\delta\kappa$ ad lineolam nascentem dc ut $b\delta^{\text{quad}}$ ad bd^{quad}, et ex æquo lineola nascens DC ad lineolam nascentem dc ut $BD\times b\delta$ ad bd^{quad}. Sunt ergo vires centripetæ ut $BD\times b\delta$ ad bd^{quad}, seu quod perinde est, ut $\frac{BD\times b\delta}{Sb}$ ad $\frac{bd^q}{Sb}$ adeoq; $\left(\text{ob æquales rationes } \frac{b\delta}{Sb} \text{ et } \frac{BD}{SB}\right)$ ut $\frac{BD^q}{SB}$ ad $\frac{bd^q}{Sb}$. Q.E.D.

Corol. 1. Hinc vires centripetæ sunt ut velocitatum quadrata applicata ad radios circulorum.

Corol. 2. Et reciprocè ut quadrata temporum periodicorum applicata ad radios, ita sunt hæ vires inter se.(18) Id est (ut cum Geometris loquar) hæ vires sunt in ratione compositâ ex duplicata ratione velocitatum directe et ratione simplici radiorum inverse: nec non in ratione compositâ ex ratione simplici radiorum directè et ratione duplicata temporum periodicorum inversè.

Corol. 3. Unde si tempora periodica æquantur,(19) erunt tum vires centripetæ tū velocitates ut radij, et vice versa.

Corol. 4. Si quadrata temporum periodicorum sunt ut radij circulorum[,] vires centripetæ sunt æquales et velocitates in dimidiata ratione radiorum. Et vice versa.

Corol. 5. Si quadrata temporum periodicorum sunt ut quadrata radiorum, vires centripetæ sunt reciprocè ut radij et velocitates æquales: Et vice versa.

Corol. 6. Si quadrata temporum periodicorum sunt ut cubi radiorum[,] vires centripetæ sunt reciprocè ut quadrata radiorum, velocitates autem in radiorum dimidiata ratione: Et vice versa.

Corol. 7. Eadem omnia, de temporibus velocitatibus et viribus quibus corpora similes Figurarum quarumcunq; similium centraq; similiter posita habentium partes describunt, consequuntur ex Demonstratione præcedentium ad hosce casus applicata.(20)

---

this ameliorated version of Proposition IV were suggested by Halley in his manuscript critique of the 'De motu Corporum' (ULC. Add. 3965.9: 92r–99r, especially 97r) though Newton did not invariably follow his recommendations to the letter. Newton himself has elegantly combined the paired Figures 10/11 of his original text into a single composite and has also added two new Corollaries, 3 and 7, the latter of which (see note (20) below) points the way to extending this Huygenian theorem to relate the times, speeds and central forces in arbitrary similar orbits.

(16) Newton initially changed this opening phrase of his enunciation to be 'Corporibus circulos æquabiliter describentibus': the present further clarification was suggested by Halley (see previous note).

(17) In his critique (see note (15)) Halley had urged merely that Newton set his original sentence 'Sola vi insita describerent tangentes *BC*, *bc* his arcubus æquales' in parentheses.

(18) A trivial variant on Halley's suggested rephrasing '...ita sunt vires centripetæ inter se'.

(19) Newton first wrote 'sunt æquales'.

[6][21] *Scholium* [ad Prop. IV].

Casus Corollarij sexti obtinet in corporibus cœlestibus (ut seorsim colligerunt Wrennus, Halleus et Hookius[22]) et propterea quæ spectant ad vim centripetam decrescentem in duplicata ratione distantiarum a centris decrevi fusius in sequentibus exponere.

Porro præcedentis demonstrationis beneficio colligitur etiam proportio vis centripetæ ad vim quamlibet notam, qualis est ea gravitatis. Nam cum vis illa, quo tempore corpus percurrit arcum *BC*, impellat ipsum per spatium *CD*, quod ipso motus initio æquale est quadrato arcus illius *BD* ad circuli diametrum applicato; et corpus omne vi eadem in eandem semper plagam continuata, describat spatia in duplicata ratione temporum: Vis illa, quo tempore corpus revolvens arcum quemvis datum describit, efficiet ut corpus idem rectà progrediens describat spatium quadrato arcus illius ad circuli diametrum applicato æquale; adeoq est ad vim gravitatis ut spatium illud ad spatium quod grave

(20) If v be the instantaneous speed at a point in a given figure where the curvature is ρ, then (compare §2: note (86)) v^2/ρ will measure the component $f.\sin\alpha$ perpendicular to the direction of motion of a force f there continuously diverting the orbiting body at an angle α towards some centre distant, say, R away. If in a similar figure at a point similarly placed at a distance R' away from the force-centre the corresponding speed, curvature and deviating force (acting at an equal angle α) be v', ρ' and f', then $\rho:\rho' = R:R'$ and hence

$$f:f' = (v^2/\rho\sin\alpha : v'^2/\rho'\sin\alpha \text{ or}) \; v^2/R : v'^2/R',$$

exactly as in the particular instance of the circle.

(21) Royal Society MS LXIX: 38. Newton 'inclosed' this new Scholium to Proposition IV in his letter to Halley of 14 July 1686 in order 'to compose y^e present dispute...between Mr Hook & me' (*Correspondence* 2: 445)—that, namely, over priority of discovery of 'ye rule of the [inverse-square] decrease of Gravity' (see the preceding general introduction and compare §2: note (88)). The variant demonstration of the preceding theorem thereto appended he had, he went on, 'met with...in turning over some old papers' (probably his 1665 Waste Book notes on motion; see note (24) below), and he now repeated it in support of his assertion earlier in the same letter that he had gathered 'y^e duplicate proportion' thereby 'from Keplers Theorem [third planetary law] about 20 yeares ago'.

(22) Halley tactfully changed the sequence in the published *Principia* ($_1$1687: 42) to be 'Wrennus, Hookius & Halleus'. In his letter to Halley on 27 May 1686 Newton mentioned that 'about 9 years since, Sir Christopher Wren upon a visit...I gave him at his Lodgings, discoursd of this Problem of Determining the Hevenly motions upon philosophicall principles' (*Correspondence*, 2: 433–4) and, having been told by Halley a month later on 29 June that he himself had 'from the consideration of the sesquialter proportion of Kepler, concluded that the centripetall force decreased in the proportion of the squares of the distances reciprocally' some time in (or before?) 'January 83/4' (*ibid.*: 442), in his reply on 14 July affirmed that 'Sr Christopher Wren's examining ye Ellipsis over against ye Focus shews y^t he knew it many yeares ago before he left of his enquiry after ye figure by an imprest motion & a descent compounded together' (*ibid.*: 445). Hooke never claimed to have deduced the inverse-square decrease of solar gravitation with distance from Kepler's third law till after his correspondence with Newton in the early winter of 1679/80.

cadendo eodem tempore describit. Et hujusmodi Propositionibus Hugenius in eximio suo Tractatu de Horologio oscillatorio vim gravitatis cum revolventium viribus centrifugis contulit.(23)

Demonstrari etiam possunt præcedentia in hunc modum.(24) In circulo quovis describi intelligatur Polygonum laterum quotcunq;. Et si corpus in Polygoni lateribus data cum velocitate movendo ad ejus angulos singulos a circulo reflectatur, vis(25) qua singulis reflexionibus impingit in circulum erit ut ejus velocitas, adeoq; summa virium in dato tempore erit ut velocitas illa et numerus reflexionum conjunctim, hoc est (si Polygonum detur specie) ut longitudo dato illo tempore descripta et longitudo eadem applicata ad Radium circuli, id est ut quadratum longitudinis illius applicatum ad Radium; adeoq; si Polygonum lateribus infinite diminutis coincidat cum circulo, ut quadratum arcus dato tempore descripti applicatum ad radium. Hæc est vis(26) qua corpus urget circulum, et huic æqualis est vis contraria qua circulus continuo repellit corpus centrum versus.

---

[7](27) *Prop V. Prob. I.*

Data quibuscunq; in locis velocitate, qua corpus figuram datam viribus ad commune aliquod centrum tendentibus describit, centrum illud invenire.(28)

---

(23) Included in the list of the propositions of his still unpublished 'De Vi Centrifuga' which (compare §1: note (24)) Christiaan Huygens appended to his *Horologium Oscillatorium sive De Motu Pendulorum ad Horologia aptato Demonstrationes geometricæ* (Paris, 1673): 159–161 was an exact parallel to Newton's present deduction, namely: 'V. Si mobile in circumferentia circuli feratur ea celeritate, quam acquirit cadendo ex altitudine, quæ sit quartæ parti diametri æqualis; habebit vim centrifugam suæ gravitati æqualem...' (*ibid.*: 160). Manifestly, if a body revolves uniformly with a speed $v = \sqrt{[gR]}$ in a circle of radius R, the *vis centrifuga* v^2/R thereby generated will be equal to the *gravitas* g accelerating a second body from rest to attain, after travelling a distance $\frac{1}{2}R$ (in time $\sqrt{[R/g]} = v/g$), the same instantaneous speed.

(24) It was first pointed out by J. W. Herivel ('Newton's Discovery of the Law of Centrifugal Force', *Isis* **51**, 1960: 546–53, especially 552–3) that the following variant derivation of the Huygenian measure of circular *vis centrifuga*—one which Newton himself told Halley he had 'met with...in turning over some old papers' (see note (21))—is effectively that adumbrated by him in about January 1665 on the opening page of his Waste Book (ULC. Add. 4004: 1r, reproduced in Herivel's *Background to Newton's Principia* (Oxford, 1966): 129–30). The equivalent argument of the preceding main Proposition 4, which makes direct appeal to the deviating action of a continuous force, was attained by Newton (almost certainly in ignorance of Huygens' then unpublished prior discovery of it) only some half dozen years later; see §1: note (24).

(25) Understand, more precisely, the impulse of speed which such an accelerative force generates in given (infinitesimal) time.

(26) That is, the 'vis centrifuga' which constraining the body to be in a circle engenders, as Newton was to make clear in the second edition of his *Principia* ($_2$1713: 40) at this point. In

Figuram descriptam tangant rectæ tres PT, TQV, VR in punctis totidem P, Q, R, concurrentes in T et V. Ad tangentes erigantur perpendicula PA, QB, RC velocitatibus corporis in punctis illis P, Q, R quibus eriguntur reciprocè proportionalia; id est ita ut sit PA ad QB ut velocitas in Q ad velocitatem in P[,] et QB ad RC ut velocitas in R ad velocitatem in Q. Per perpendiculorum terminos(29) A, B, C ad angulos rectos ducantur AD, DBE, EC concurrentia in D et E: et actæ TD, VE concurrent in centro quæsito S.

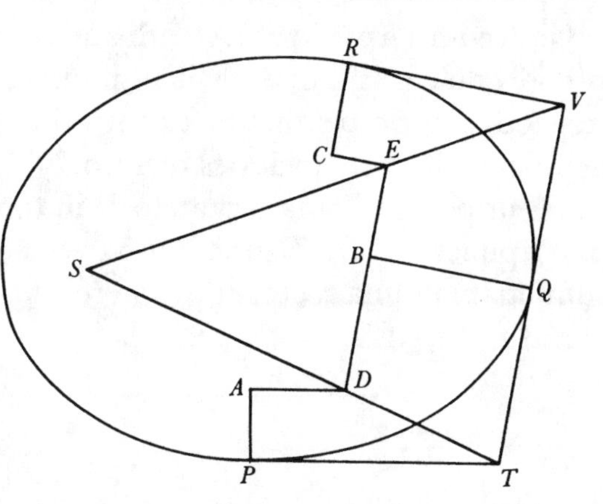

clarification of the somewhat over-brief preceding argument we may suppose that the body moves uniformly along the perimeter $ABCDE\ldots$ of a regular polygon given in species such that at the vertices $B, C, D, \ldots$ it bounces back, without change in speed, from the circumscribed circle of centre S. If the circle's radius is R and in time t the body, covering a total distance s, traverses n of the equal polygon sides, then, where these sides $BC = CD = DE = \ldots = s/n$ represent the uniform speed $v(= s/t)$ of the body as it moves in the polygon's perimeter, the inwards impulses $(f.t/n)$ of the force of 'reflection' f generated in time t/n by the constraining circle at each occursion $B, C, D, \ldots$ will correspondingly be denoted by the deviations

$$cC = dD = eE = \ldots (= f.t^2/n^2)$$

ensuing in the same time-interval between the 'unreflected' paths Bc, Cd, De, ... and the equal ones BC, CD, DE, ... taken after impact; whence, because of the similarity of the equal isosceles triangles BcC CdD, DeE, ... to the congruent central triangles SBC, SCD, SDE, ... given in species, Newton's assertion that (the impulse generated in that time by) the central *vis* f is as the speed v readily follows, the given factor of proportionality being $cC/BC = (BC/SC$ or) $s/nR = k$, say. In the whole time t, therefore, the *summa virium* (the total impulse $f.t$) will be $kv.n = v.s/R = s^2/Rt \propto s^2/R$ 'dato tempore'. On considering the limiting case in which the number n of sides becomes infinite and the polygon's perimeter comes to coincide with its circumcircle—and, as a further subtlety (compare §1: note (19)), presuming that the total distance s and time t themselves become infinitesimal—we elegantly derive the Huygenian result (Corollary 1 of the main Proposition IV preceding) that the *vis centrifuga* f with which a body moving at a uniform speed v in a circle arc $\widehat{BCDE\ldots}$ of radius R is urged outwards from the centre S is equal to v^2/R.

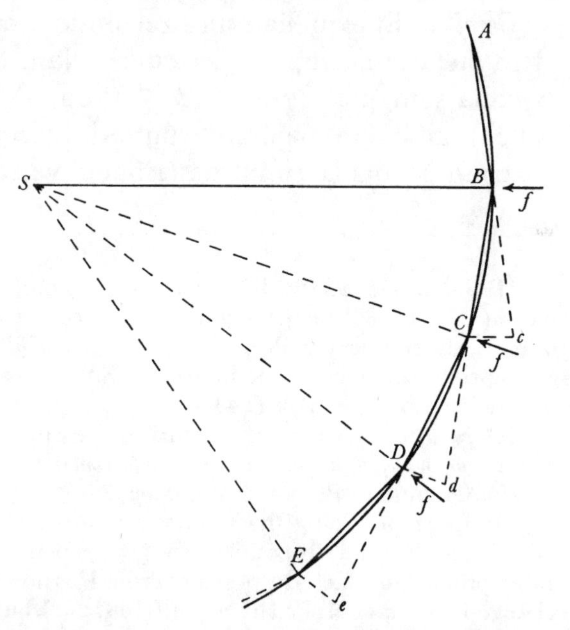

Nam cum corpus in P et Q radijs ad centrum ductis areas describat temporibus proportionales, sintꝗ areæ illæ simul descriptæ ut velocitates in P et Q ductæ respectivè in perpendicula a centro in tangentes PT, QT demissa: erunt perpendicula illa ut velocitates reciprocè,(30) adeo ut perpendicula AP, BQ directè, id est ut perpendicula a puncto D in tangentes demissa. Unde facile colligitur quod puncta S, D, T sunt in una recta.(31) Eodem argumento puncta S, E, V sunt etiam in una recta, et propterea centrum S in concursu rectarum TD, VE versatur. Q.E.D.

---

[8](32) *Prop. X. Prob. V.*

... ...

Coroll. [*1*]. Unde vicissim si vis sit ut distantia, movebitur corpus in Ellipsi centrum habente in centro virium aut fortè in circulo in quem Ellipsis migrare potest.(33)

[*Corol. 2.* Et æqualia erunt revolutionum in Figuris universis circa centrum idem factarum tempora periodica. Nam tempora illa in Ellipsibus similibus æqualia sunt per Corol. 3 & 7 Prop. IV: in Ellipsibus autem communem habentibus axem majorem, sunt ad invicem ut Ellipseôn areæ totæ directe & arearum particulæ simul descriptæ inverse; id est ut axes minores directe &

---

(27) ULC. Dd. 9. 46: 31^v; compare §2: note (88). With the insertion—somewhat out of place (see note (30) below)—of this minor constructional problem, one of wholly theoretical interest, the subsequent Propositions V–XVI of §2 preceding become, after renumbering and appropriate revision, Propositions XI–XVII of the enlarged 'De motu Corporum Liber primus' (= *Principia*, $_1$1687: 44–60).

(28) Newton first enunciated this in the equivalent form '*Datur ubiꝗ velocitas quâ corpus datam figuram viribus ad commune centrum tendentibus describit, & requiritur illud centrum.*'

(29) Originally 'Per puncta' *tout court.*

(30) In §2 preceding the demonstration of this basic property of a general central-force orbit is not given till Proposition XV (= Proposition XVI of the revised 'De motu Corporum Liber primus' in which the present 'Prob. I' is now set): the inconsistency was carried without change into the published *Principia* ($_1$1687: 44) but amended in revise by introducing a new '*Corol. 1.* Velocitas corporis in centrum immobile attracti est in spatiis non resistentibus reciproce ut perpendiculum a centro illo in Orbis tangentem rectilineam demissum....' (*Principia*, $_2$1713: 35; compare 3, §1.2 below) and then altering the present proof to read: 'Nam perpendicula a centro S in tangentes PT, QT demissa (per Corol. 1. Prop. I.) sunt reciproce ut velocitates corporis in punctis P & V; adeoꝗ per constructionem ut perpendicula AP, BQ directe...' (*ibid.*: 41; compare 3, §2.1).

(31) For the ratio of the perpendicular distances of S from the tangents PT and QT is as that of the speeds at P and Q, and hence as the perpendicular distances of D (namely, by parallels, AP and BQ respectively) from the same tangents.

(32) Newton's autograph of the first of these inserted corollaries to the old 'Prop. IX' (compare §2: note (107)) is entered at the top of f. 48^r in the preliminary 'De motu Corporum'

corporum velocitates in verticibus principalibus(34) inverse, hoc est ut axes illi directe & ordinatim applicatæ ad axes alteros inverse, & propterea (ob æqualitatem rationum directarum & inversarum) in ratione æqualitatis.]

---

[9](35) *Lemma XIII.*

Quadratum perpendiculi quod ab umbilico Parabolæ ad tangentem ejus demittitur æquale est rectangulo sub quarta parte lateris recti principalis et distantia inter umbilicum et punctum contactus.

Sit AQP Parabola, S umbilicus ejus, PM tangens diametro principali occurrens in M, et SN linea perpendicularis ab umbilico in tangentem[,] et ex natura Parabolæ erit AS pars quarta lateris recti principalis et PS pars quarta lateris recti pertinentis ad diametrum quæ transit per punctum contactus P. Dico igitur quod sit SN^q æquale $AS \times SP$.(36)

(ULC. Dd. 9. 46). To this we here append a complementary 'Corol. 2' added by Newton in the published *Principia* ($_1$1687: 49). Both prepare the way into his subsequent discussion of simple harmonic motion in Proposition XXXVIII of the revised 'De motu Corporum Liber primus' (see §3: note (186)).

(33) Newton first concluded with the equivalent conditional phrase '...nisi ubi Ellipsis in circulum migrat'. In the analytical terms of §1: note (30) this corollary effectively asserts that the differential equation $c^2r^{-2}(r^{-1}+d^2(r^{-1})/d\phi^2) \propto r$, $c = r^2 d\phi/dt$ constant—that is,

$$r^{-1}+d^2(r^{-1})/d\phi^2 = (k/c^2)r^3$$

on introducing a suitable factor k of proportionality—has the unique solution

$$r^{-1} = \sqrt{[A-B\cos 2(\phi+\epsilon)]},\quad A = \sqrt{[B^2+k/c^2]} > B,$$

which is the polar equation, referred to its centre as origin, of an ellipse having semi-axes $1/\sqrt{[A-B]}$ and $1/\sqrt{[A+B]}$ in length.

(34) Since (to continue the previous note) the total area of the previous ellipse is

$$\pi/\sqrt{[A^2-B^2]},$$

the 'Keplerian' equation defining the time t of orbit over its general arc is

$$r^2 d\phi/dt = 1/\sqrt{[A^2-B^2]},$$

whence the orbital speed at the end of the principal axis $R = 1/\sqrt{[A-B]}$ is measured by $1/R\sqrt{[A^2-B^2]} = 1/\sqrt{[A+B]}$, the minor semi-axis.

(35) A first attempt, written in over the original text on ULC. Dd. 9. 46: 68^r, at a valid proof of Lemma XIII in the preceding 'De motu Corporum'; compare §2: note (119).

(36) Newton breaks off, evidently realising that he assumes without any justification the fundamental parabolic property that the *latus rectum* pertaining to the diameter through P is $4PS$: this he at once (in [10] following) proceeds to introduce as a separate lemma, patent 'ex Conicis' (see note (38) below).

[10](37) *Lemma XIII.*

Latus rectum Parabolæ ad verticem quemvis pertinens, est quadruplum distantiæ verticis illius ab umbilico figuræ.

Patet ex Conicis.(38)

Lemma XIV.

Perpendiculum quod ab umbilico Parabolæ ad tangentem ejus demittitur, medium est proportionale inter distantias umbilici a puncto contactus et a vertice principali figuræ.

Sit enim *APQ* Parabola, *S* umbilicus ejus, *A* vertex principalis, *P* punctum contactus, *PO* ordinatim applicata ad diametrum principalem, *PM* tangens

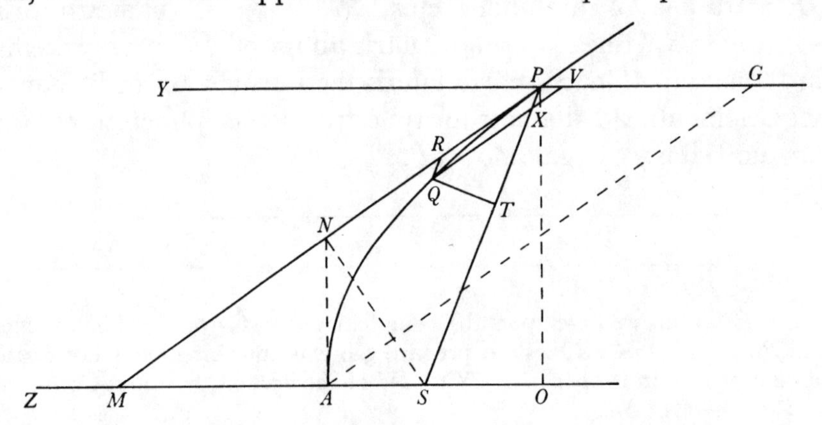

diametro principali occurrens in *M*, & *SN* linea perpendicularis ab umbilico in tangentem.(39) Jungatur *AN*, et ob æquales *MS* et *SP*, *MN* et *NP*, *MA* et *AO* parallelæ erunt rectæ *AN* et *OP*, et inde triangulum *SAN* rectangulum erit ad *A* et simile triangulis æqualibus *SMN*, *SPN*. Ergo *PS* est ad *SN* ut *SN* ad *SA*. Q.E.D.

(37) ULC. Dd. 9. 46: 67v. To this immediate revise of [9] preceding, whose argument is now split into a general rider (Lemma XIII) and the particular property of the parabola (Lemma XIV) which Newton thereby demonstrates, we append (from ff. 68r/69r) the corrected text of the ensuing Proposition (XII, now renumbered) XIII where the latter is employed.

(38) While (compare §2: note (119)) the existence of the *latus rectum* pertaining to the general vertex *P* is established in Apollonius, *Conics* I, 49, the present definition of its length is seemingly Newton's independent discovery. Formal proof that $AG^2 = PG \times 4PS$ derives straightforwardly from the parabola's defining 'symptom' $PO^2 = AO \times 4AS$ by way of the equalities $AG = MP$ and $PG = (MA =)\ AO$ and the proportion (dependent on the similarity of the triangles *MPO*, *PSN* and *NSA*)

$$AG^2/PG : PO^2/AO = MP^2 : PO^2 = PS^2 : SN^2 = PS : AS.$$

(39) Notice that the now superseded ordinate *AG* and diametral extensions *YP* and *ZM* linger superfluously on in Newton's figure, repeated without change in the first two editions of the *Principia* ($_1$1687: 53/54 = $_2$1713: 51/52) to be finally deleted by Henry Pemberton in the dual figures of his *editio ultima* ($_3$1726: 58, 59).

Corol. 1. PS^q est ad SN^q ut PS ad SA.

Corol. 2. Et ob datam SA, est SN^q ut PS.

Corol. 3. Et concursus tangentis cujusvis PM cum recta SN quæ ab umbilico in ipsam perpendicularis est,(40) incidit in rectam AN quæ Parabolam tangit in vertice principali.

Prop. XIII. Prob. VIII.

Moveatur corpus in perimetro Parabolæ, requiritur lex vis centripetæ tendentis ad umbilicum hujus figuræ.

Maneat constructio Lemmatis, sitꝗ P corpus in perimetro Parabolæ; et a loco Q in quem corpus proxime movetur, age ipsi SP parallelam QR et perpendicularem QT, nec non QV tangenti parallelam et occurrentem tum diametro YPG in V, tum distantiæ SP in X. Jam ob similia triangula PXV, MSP et æqualia unius latera SM, SP, æqualia sunt alterius latera PX seu QR, et PV. Sed ex Conicis, quadratum ordinatæ QV æquale est rectangulo sub latere recto et segmento diametri $PV_{[,]}$ id est (per Lemma XIII) rectangulo $4PS \times PV$, seu $4PS \times QR$, et punctis P et Q coeuntibus ratio QV ad QX ([per] Lem. VIII) fit æqualitatis. Ergo QX^q, eo in casu, æquale est rectangulo $4PS \times QR$. Est autem (ob æquales angulos QXT, MPS, PMO) QX^q ad QT^q ut PS^q ad SN^q, hoc est ([per] Cor. 1, Lem. XIV) ut PS ad AS id est ut $4PS \times QR$ ad $4AS \times QR$, et inde ([per] Prop. 9. lib. v. Elem) QT^q et $4AS \times QR$ æquantur. Ducantur hæc æqualia in $\frac{SP^q}{QR}$ et fiet $\frac{SP^q \times QT^q}{QR}$ æquale $SP^q \times 4AS$: et propterea (per Cor. Theor. V) vis centripeta est reciprocè ut $SP^q \times 4AS$ id est, ob datam $4AS$, reciprocè in duplicata ratione distantiæ SP. Q.E.I.

---

[11](41) *Prop. XVII. Prob. IX.*

... ...

Corol. 3. Hinc etiam si corpus moveatur in sectione quacunꝗ Conica et ex orbe suo impulsu quocunꝗ exturbetur, cognosci potest orbita(42) in quo postea cursum suum peraget. Nam componendo proprium corporis motum cum motu

(40) Newton first wrote less complicatedly 'Et SN ab umbilico in tangentem quamlibet PM perpendicularis'. Evidently SN bisects PM, and hence (since $MA = AO$) AN is parallel to the perpendicular ordinate OP and so tangent at A.

(41) ULC. Dd. 9. 46: 56ᵛ. These two further corollaries to the equivalent Proposition XVI in the preceding 'De motu Corporum' (see §2: note (149)) sketch its evident application to computing the orbit ensuing when motion in a Keplerian conic is disturbed by a single or repeated external impulse.

(42) As in the main proposition (compare §2: note (147)) understand this to be a conic with a focus at the force-centre.

illo quem impulsus solus generaret habebitur motus quocum corpus de dato impulsûs loco secundum rectam positione datam exibit.

Corol: 4. Et si corpus illud vi aliqua extrinsecus impressa continuò perturbetur, innotescet cursus quamproximè, colligendo mutationes quas vis illa in punctis quibusdam inducit, & ex seriei analogia mutationes continuas in locis intermedijs æstimando.

APPENDIX 3. THE GRAVITATIONAL ATTRACTION OF A THIN CIRCULAR RING UPON AN EXTERNAL AND AN INTERNAL POINT.(1)

[late 1692?]

From the original worksheets in the University Library, Cambridge

[1](2) [Pone] $Ee+Ff=o$. [erit](3) $PD.PE::\frac{1}{2}o.\frac{o,PE}{2PD}=Ee$.

[necnon] $Ff=\frac{o,PF}{2PD}$.(4)

[unde] $\frac{o}{2PD,PE}=$ vi Ee.

[et] $\frac{o}{2PD,PF}=$ vi Ff.

[Ergo] $\frac{\overline{PE+PF},o}{2PD,PE,PF}=\frac{o}{PG^2}$

$=\frac{o}{APB}=$ vi $Ee+Ff$ secundum PD. [unde]

$\frac{o,PD}{APB,PC}=$ vi $Ee+Ff$

secundum PH.

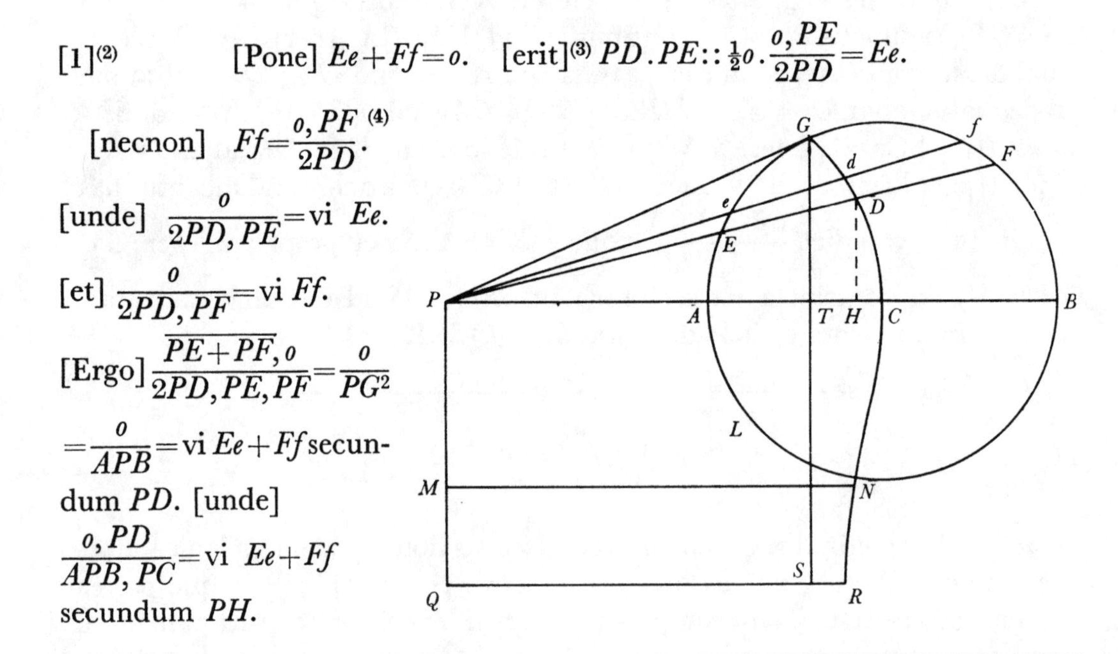

(1) This determination—and approximate evaluation—of the total attraction of a laminar ring upon a point outside its circumference is here reproduced to exemplify Newton's general approach to calculation of the gravitational potential of a given figure, and thereby to support the restoration given in §2: note (184) of his prior analysis of the parallel case of a spherical surface whose unilluminating and over-contrived synthesis he presented in Proposition XL of the preceding 'De motu Corporum'. The context in which these calculations find their application is that of lunar theory, where (see note (18) below) Newton seeks the effective gravitational centre of the moon's orbit, assumed to be a perfect circle, round the earth under the distorting action of the sun, taken to be so far away that its change in apparent angular

[adeoq; faciendo] $PM = \dfrac{\text{arc}\, AE + \text{arc}\, BF}{2}$. $MN = PD$.

[ut et] $PQ = \dfrac{\text{arc}\, AG + [\text{arc}]\, BG}{2}$. $QR = PG$.

[erit area] $PCNM$ attractio arcuum $AE + BF$(5) [necnon area tota] [4]$PQRC$ attractio annuli totius

$$= [4\ \text{in}]\ \frac{PC+QR}{2} PQ + \frac{\sqrt{PC^2 - \frac{PC^2 - QR^2}{2}} - \frac{PC - QR}{2}}{2} PQ$$

$$= \frac{PC+QR}{4} + \tfrac{1}{4}\sqrt{2PC^2 + 2QR^2}\ \text{in}\ [4]\ PQ^{(6)}$$

$$= PC + QR + \sqrt{2PC^2 + 2QR^2}\ \text{in circumf: } AGFBL.$$

---

position (as seen from the earth) may be assumed to be negligible during the interval of a lunar month. The date of composition assigned is suggested by the quality of Newton's handwriting in the autograph originals, but there is strong documentary evidence that the jumbled mass of contemporary lunar researches which are now preserved in ULC. Add. 3966 were pursued for the most part in the autumn/winter of 1692–3.

(2) ULC. Add. 3966.2: 16 *bis*r. Newton computes the total attraction of the circle AGB upon the external point P by summing the components in the direction PC of the joint pull of the infinitesimal arcs $\widehat{Ee}$ and $\widehat{Ff}$ cut off between immediately neighbouring chords EF and ef each through P.

(3) Understand that $\widehat{CDG}$ is the portion within the given circle of the semicircle on diameter PC, whence (since CD is perpendicular to the chord EF) D is the mid-point of EF for all lines PEF drawn through P to meet the circle.

(4) Seeing that the infinitesimal arcs $\widehat{Ee}$ and $\widehat{Ff}$ are equally inclined to the chord EF.

(5) Read 'semiarcuum $\dfrac{AE+BF}{2}$': we have made appropriate correction in the sequel. It will be clear that Newton has here silently multiplied by a factor $PA \times PB \times PC = PG^2 \times PC$. In equivalent algebraic terms, on taking $AC = CB = CE = r$, $PC = s$ and $CD = x$, whence $PM = \frac{1}{2}(\widehat{AE} + \widehat{BF}) = AC \times \widehat{CED} = r\sin^{-1}(x/r)$, $MN = PD = \sqrt{[s^2 - x^2]}$ and $PG = \sqrt{[s^2 - r^2]}$, the area $(PCNM)$ will be

$$\int_0^x \sqrt{[s^2 - x^2]}\,.\,d(r\sin^{-1}(x/r)) = \int_0^x r\sqrt{[(s^2 - x^2)/(r^2 - x^2)]}\,.\,dx,\ s > r.$$

(6) In terms of the previous note, on setting $\widehat{CED} = \theta$ (and so $x = r\sin\theta$) the area $(PQRC)$ is

$$\int_0^r r\sqrt{[(s^2 - x^2)/(r^2 - x^2)]}\,.\,dx = r\int_0^{\frac{1}{2}\pi} \sqrt{[s^2 - r^2\sin^2\theta]}\,.\,d\theta$$
$$= rs\int_0^{\frac{1}{2}\pi} (1 - \tfrac{1}{2}(r^2/s^2)\sin^2\theta - \tfrac{1}{8}(r^4/s^4)\sin^4\theta \ldots)\,.\,d\theta$$
$$= \tfrac{1}{2}\pi rs(1 - \tfrac{1}{4}r^2/s^2 - \tfrac{3}{64}r^4/s^4 + O(r^6/s^6)).$$

As a first approximation

$$\tfrac{1}{2}(PC + QR)\,.\,PQ = \tfrac{1}{2}(s + \sqrt{[s^2 - r^2]})\,.\,\tfrac{1}{2}\pi r = \tfrac{1}{2}\pi rs(1 - \tfrac{1}{4}r^2/s^2 - \tfrac{1}{16}r^4/s^4 \ldots)$$

Et hæc attractio est ad attractionem materiæ annuli totius in centro C locatæ ut area $PCRQ$ ad rectangulum $P[T]\,SQ_{[,]}$(7) hoc est ut $PC+QR+\sqrt{2PC^2+2QR^2}$ ad $4P[T]$, (existente PC ad PG ut PG ad $P[T]$). id est ut

$$PC+PG+\sqrt{2PC^2+2PG^2} \text{ ad } 4P[T]=\frac{4PG^2}{PC}.$$

---

[2](8)

Dato $[Mm+Ll=o]$ datur $\dfrac{MS,o}{ML}=Mm.\ \dfrac{[SL],o}{[ML]}=Ll.$(9) [unde vis $Mm-Ll$ secundum ST æqualis est]

$$\frac{[o],Sh}{SM^2[,ML]}-\frac{o,SH}{SL^2[,ML]}$$

$$[=]\frac{SL^2,Sh-SM^2,SH}{GS^4[,ML]}[o]^{(10)}$$

$$[=]\frac{SL,SH,SM-SM^2,SH}{GS^4[,ML]}[o]$$

$$=\frac{SH,SM,SK}{GS^4,ML}[o]$$

$$=\frac{SK^2}{GS^2,ML,SR}[o].^{(11)}$$

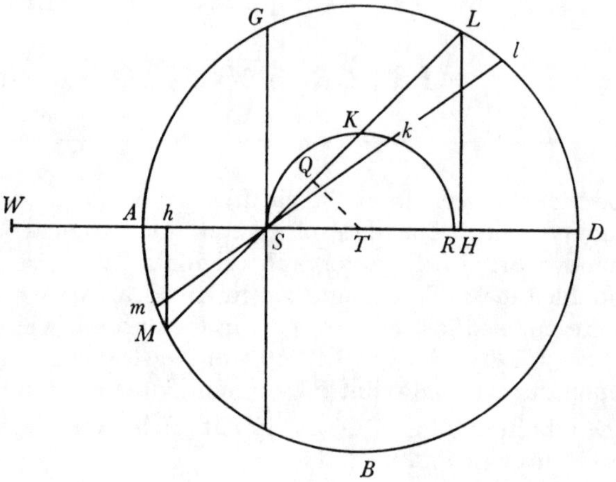

---

and

$$\sqrt{[PC^2-\tfrac{1}{2}(PC^2-QR^2)]}\,.PQ = \sqrt{[s^2-\tfrac{1}{2}r^2]}\,.\tfrac{1}{2}\pi r = \tfrac{1}{2}\pi rs(1-\tfrac{1}{4}r^2/s^2-\tfrac{1}{32}r^4/s^4\ldots),$$

differing by $\tfrac{1}{2}\pi rs(-\tfrac{1}{32}r^4/s^4\ldots)$, whence to $O(r^6/s^6)$

$$(PQRC) = \tfrac{1}{2}(\tfrac{1}{2}(PC+QR)+\sqrt{[PC^2-\tfrac{1}{2}(PC^2-QR^2)]})\,.PQ$$
$$= \tfrac{1}{4}(PC+QR+\sqrt{[2(PC^2+QR^2)]}).\ PQ$$

as Newton finds.

(7) That is, $PQ\times PT=\tfrac{1}{2}\widehat{AGD}\times PG^2/PC$, which on dividing by the previous multiplying factor $PG^2\times PC$ (see note (5)), represents the attraction $\tfrac{1}{4}\widehat{AGBL}/PC^2$ of one-fourth of the given circle at the distance PC.

(8) ULC. Add. 3966.2: 14r. Newton computes the analogous gravitational attraction of the circle $\widehat{AGDB}$ upon an interior point S by summing the opposite pulls of infinitesimal arcs $\widehat{Ll}$ and $\widehat{Mm}$ cut off by immediately neighbouring chords through S. Understand that $\widehat{SKR}$ is a concentric semicircle through S.

(9) Because (compare note (4)) the infinitesimal arcs $\widehat{Ll}$ and $\widehat{Mm}$ are equally inclined to the chord LM.

(10) Where LH, Mh and $GS=\sqrt{[LS\times SM]}$ are perpendicular to ASD, namely.

(11) For $GS^2=SM\times SL$, while (because the triangles SHL and SKR are similar)

$$SH/SL = SK/SR.$$

[Unde] Attractio pūcti S ab annulo ad attractionem(12) materiæ ejusdem in G [est] ut $\frac{SR}{AD}+\frac{SR}{ML}$ ad $\frac{4AD}{AD}$, id est ut

$$\frac{1}{AD}+\frac{1}{ML}\text{ ad }\frac{4}{SR}::\frac{1}{AT}+\frac{1}{MQ}\cdot\frac{4}{ST}\text{[ubi] }MQ=\sqrt{AT^2-\tfrac{1}{2}ST^2}.^{(13)}$$

[Cape igitur] SW ad SG in subdupl. rat.(14) $\frac{4}{SR}$ ad $\frac{1}{AD}+\frac{1}{ML}$ seu $\frac{2}{ST}$ ad $\frac{1}{AD}+\frac{1}{ML}$, seu $2AD$ ad $ST+\frac{[2]\,AD,\,ST}{\sqrt{AD^2-2ST^2}}$.

---

[3](15) *Lemma 4.*

Vis qua perimeter circuli uniformis $AGDB$ centro T descripti attrahitur a puncto S extra circulum illum sito æqualis est vi qua materia eadem in puncto V consistens attraheretur ab eodem S si modo triangula STG, SGR rectangula sint & sumatur SV ad ST in subduplicata ratione $4SR$ ad

$ST+SG$
$+2\sqrt{ST^q-\tfrac{1}{2}AT^q}$(16).

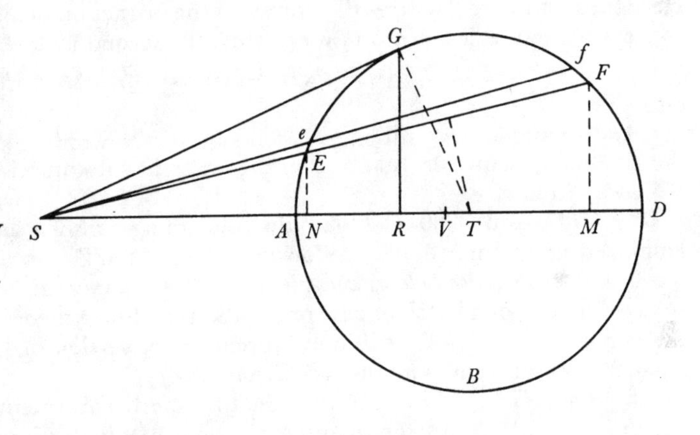

Corol. Si punctum S valde longinquum est, erit $VT=\tfrac{3}{8}RT$.(17)

---

(12) 'Gravitas annuli ad grav[itatem]' was first written.

(13) Understand that LM now bisects the right angle $D\widehat{S}G$. Much as in notes (5) and (6) preceding, on taking $AT = TD = r$, $ST = s$ and $QT = x$ it follows that $ML = 2\sqrt{[r^2-x^2]}$, $SK = 2\sqrt{[s^2-x^2]}$, $SR = 2s$, $GS = \sqrt{[r^2-s^2]}$ and, since $\widehat{AM}+\widehat{DL} = AT\times 2L\widehat{S}T = r.2\sin^{-1}(x/s)$, the total attraction of the annular ring on S is

$$2\int_0^s 2r(s^2-x^2)/(r^2-s^2)\,s\sqrt{[r^2-x^2]}\,.\,d(\sin^{-1}(x/s)) = (4r/(r^2-s^2)\,s)\int_0^s\sqrt{[(s^2-x^2)/(r^2-x^2)]}\,.\,dx,\ s<r;$$

that is, on setting $L\widehat{S}T = \theta$ and hence $x = s\sin\theta$,

$$(4s/(r^2-s^2))\int_0^{\frac{1}{2}\pi}\cos^2\theta/\sqrt{[1-(s^2/r^2)\sin^2\theta]}\,.\,d\theta$$
$$= (4s/(r^2-s^2))\int_0^{\frac{1}{2}\pi}(1+\tfrac{1}{2}(s^2/r^2)\sin^2\theta+\tfrac{3}{8}(s^4/r^4)\sin^4\theta\ldots)\cos^2\theta\,.\,d\theta$$
$$= (2\pi r/(r^2-s^2))\times\tfrac{1}{2}(s/r+\tfrac{1}{8}s^3/r^3+\tfrac{3}{64}s^5/r^5\ldots).$$

Lemma 5.

Vis qua perimeter uniformis $AGDB$, centro T descripti attrahitur a dato puncto S intra circulum illum sito æqualis est vi qua materia eadem in puncto W consistens attraheretur si modo sumatur $S[W]$ ad SG in subduplicata ratione $\frac{2}{ST}$ ad $\frac{1}{AD}+\frac{1}{MN}$ existentibus scilicet angulis DSN, NSG semirectis.(18)

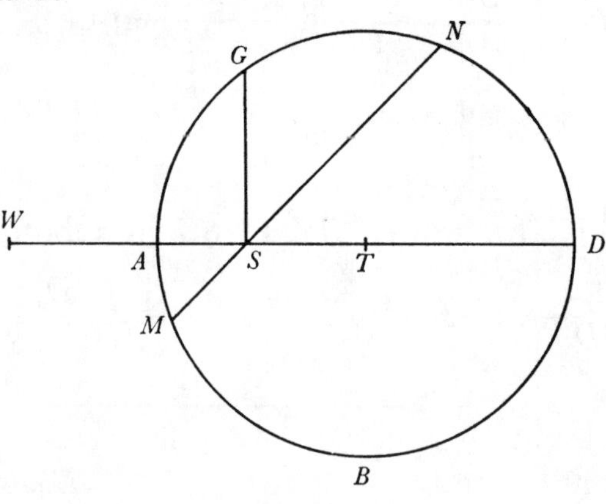

Here the first factor $2\pi r/(r^2-s^2)$ represents the attraction on S of the ring when its 'matter' is concentrated at a distance GS away, while the second is, to $O(s^7/r^7)$,

$$\tfrac{1}{4}(s/r+s/\sqrt{[r^2-\tfrac{1}{2}s^2]}) = \tfrac{1}{4}(SR/AD+SR/ML)$$

on fixing $\widehat{LST} = \frac{1}{4}\pi$.

(14) 'in subduplicata ratione'. It follows from Newton's construction that the attraction of the ring on S is equal to that of the ring when its substance is concentrated at a point distant SW away from it.

(15) ULC. Add. 3966.4: 32r/32v the concluding pair of a set of five lemmas (on ff. 32r–33r) employed in an immediately following '*Prop. XXXIV. Invenire excentricitatem Orbis Lunæ qua Terra et Orbis ille circa Solem revolventes in æquilibrio permanebit*' (*ibid.*: 33r/33v) intended to be inserted, as one of a batch of new propositions on lunar theory, in the *Principia's* third book in revision of its first (1687) edition. A preliminary version of Lemma 4 exists on ULC. Add. 3966.2: 16r (where it is numbered 'Lemma 3').

(16) Since $ST^2-AT^2 = SG^2$ this result is exactly equivalent to that computed in [1]; compare note (6) above. In the manuscript about 6–8 lines of space are left for the proof to be inserted.

(17) For, when ST is large in comparison with AT, because $AT^2/ST = RT$ it follows that $SG = \sqrt{[ST^2-AT^2]} = ST-\frac{1}{2}RT\ldots$ and $\sqrt{[ST^2-\frac{1}{4}AT^2]} = ST-\frac{1}{4}RT\ldots$, whence

$$SV/ST \approx \sqrt{[4(ST-RT)/(4ST-RT)]} \approx 1-\tfrac{3}{8}RT/ST$$

and so $VT \approx \frac{3}{8}RT$. In preliminary draft (see note (15)) Newton wrote equivalently 'capiatur $[S]V = [S]G+\frac{1}{8}R[T]$'.

(18) Again a space of some dozen lines is left in the manuscript for later insertion of the demonstration of this result, which is taken over without essential change from [2] preceding. In the immediately following 'Prop. XXXIV' (see note (15) above) Newton applies Lemma 4 to find the effective centre of attraction of the moon on the sun during the period of a month, making the simplifying postulates that 'Sit S Sol, $ABDG$ orbis Lunæ & $[T]$ centrum ejus et vice Lunæ unius fingamus Lunulas innumeras per totam Orbis circumferentiam spargi', whence 'Si hæ Lunulæ ad æquales ab invicem distantias uniformi cum motu circa Terram in centro $[T]$ locatam revolvantur, gravitas annuli Lunularum in Solem eadem erit ac si tota annuli materia in puncto quodam $[V]$ consisteret quod invenitur per Lemma IV capiendo $R[T]$ ad Orbis semidiametrum $G[T]$ ut est semidiameter ista ad Solis distantiam $S[T]$, dein $[VT]$ ad $R[T]$ ut 3 ad 8. Gravitas autem Terræ in Solem eadem erit ac si tota ejus materia in Orbis centro $[T]$ consisteret' (f. 33r).

APPENDIX 4. THE 'PULL' OF A SPHERE UPON AN EXTERNAL POINT, AND OF A SPHEROID ON A POINT IN ITS AXIS.(1)

[summer? 1685/August 1686]

Extracts from the first edition of Newton's *Principia*

[1](2) *Prop. LXXVII. Theor. XXXVII.*

Si ad singula sphærarum puncta tendant vires centripetæ proportionales distantiis punctorum a corporibus attractis: dico quod vis composita, qua sphæræ duæ se mutuo trahent, est ut distantia inter centra sphærarum.

Cas. 1. Sit *ACBD* sphæra, *S* centrum ejus, *P* corpusculum attractum, *PASB* axis sphæræ per centrum corpusculi transiens, *EF*, *ef* plana duo quibus sphæra secatur, huic axi perpendicularia, & hinc inde æqualiter distantia a centro sphæræ; *G*[,] *g* intersectiones planorum & axis, & *H* punctum quodvis in plano *EF*. Puncti *H* vis centripeta in corpusculum *P* secundum lineam *PH* exercita est ut distantia *PH*, & (per Legum Corol. 2) secundum lineam *PG*, seu versus centrum *S*, ut longitudo *PG*. Igitur punctorum omnium in plano *EF*, hoc est plani totius vis, qua corpusculum *P* trahitur versus centrum *S*, est ut numerus

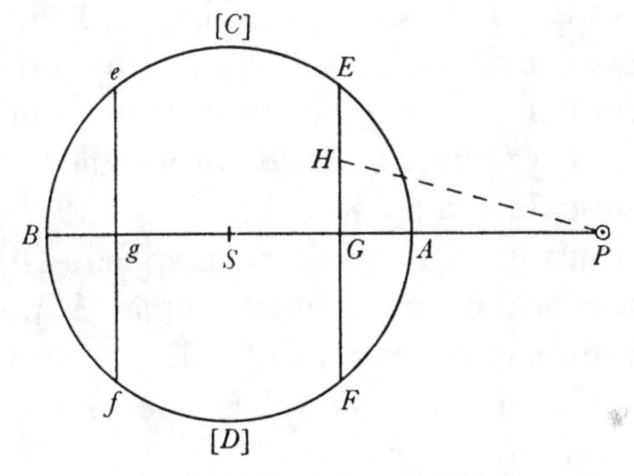

(1) Of the lemmas and propositions here reproduced the first was probably included (perhaps as 'Prop. XLVI'; see §2: note (188)) in the missing final portion of the revised 'De motu Corporum' as an analogous direct-distance complement to the inverse-square case of spherical attraction considered in the earlier Proposition XL, while the remainder—all but the concluding Corollaries 2 and 3 to Proposition XCI, which (see note (30) below) Newton appended in August 1686 as a late afterthought—doubtless came into being in the early summer of 1685 in the lost concluding pages of his revised 'De motu Corporum Liber primus' (whose extant text terminates in the middle of Proposition LIV; see §3 following). Except for the equivalent passages in Humphrey Newton's trivially variant secretarial copy, now Royal Society MS LXIX, from which the *Philosophiæ Naturalis Principia Mathematica* was printed during 1686–7, no preliminary drafts of these propositions—or rough calculations pertinent thereto—appear to have survived, and we accordingly restrict our editorial attention to noticing minor emendations and rephrasings of the text incorporated in later editions and to elucidating historical points and technical niceties as these arise. We have, we should add, silently raised all final indices '*q*' (or 'quad.') and 'cub' in mathematical square and cube powers to superscript position, following Newton's invariable custom in his autograph manuscripts: it is only, of course, for typographical convenience that these are lowered onto the line in the printed version.

(2) *Principia*, $_1$1687: 200–2/203–11.

punctorum ductus in distantiam PG:(3) id est ut contentum sub plano ipso EF & distantia illa PG. Et similiter vis plani ef, qua corpusculum P trahitur versus centrum S, est ut planum illud ductum in distantiam suam Pg; sive ut huic æquale planum EF ductum in distantiam illam Pg; & summa virium plani utriusq; ut planum EF ductum in summam distantiarum $PG+Pg$, id est, ut planum illud ductum in duplam centri & corpusculi distantiam PS, hoc est ut duplum planum EF ductum in distantiam PS, vel ut summa æqualium planorum $EF+ef$ ducta in distantiam eandem. Et simili argumento, vires omnium planorum in sphæra tota, hinc inde a centro sphæræ distantium, sunt ut summa planorum ducta in distantiam PS, hoc est, ut sphæra tota ducta in distantiam centri sui S a corpusculo P. Q.E.D.(4)

Cas. 2. Trahat jam corpusculum P sphæram $ACBD$. Et eodem argumento probabitur quod vis, qua sphæra illa trahitur, est ut distantia PS. Q.E.D.

Cas. 3. Componatur jam sphæra altera ex corpusculis innumeris P; & quoniam vis, qua corpusculum unumquodq; trahitur, est ut distantia corpusculi a centro sphæræ primæ ducta in sphæram eandem, atq; adeo eadem est ac si prodiret tota de corpusculo unico in centro sphæræ; vis tota qua corpuscula omnia in sphæra secunda trahuntur, hoc est, qua sphæra illa tota trahitur, eadem erit ac si sphæra illa traheretur vi prodeunte de corpusculo unico in centro sphæræ primæ, & propterea proportionalis est distantiæ inter centra sphærarum. Q.E.D.

Cas. 4. Trahant sphæræ se mutuo, & vis geminata proportionem priorem servabit. Q.E.D.

Cas. 5. Locetur jam corpusculum p intra sphæram $ACBD$, & quoniam vis plani ef in corpusculum est ut contentum sub plano illo & distantia pG; & vis contraria plani EF ut contentum sub plano illo & distantia pG; erit vis ex utraq; composita ut differentia contentorum, hoc est, ut summa æqualium planorum ducta in semissem differentiæ distantiarum, id est, ut summa illa ducta in pS, distantiam corpusculi a centro sphæræ. Et simili argumento attractio planorum omnium EF, ef in sphæra tota, hoc est attractio

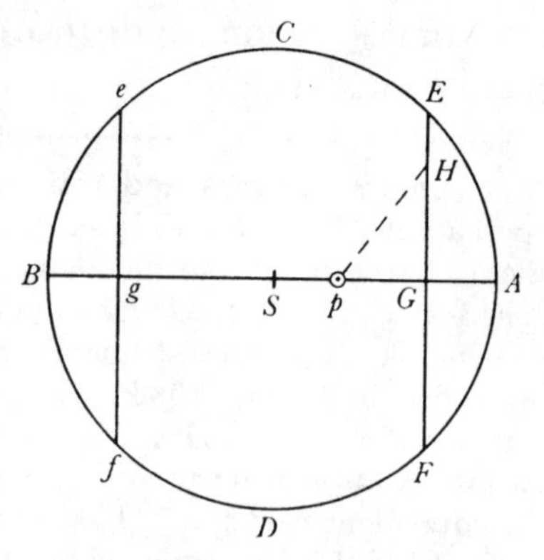

(3) In the third edition (*Principia*, ₃1726: 197) the sequence of this phrase was altered to read 'ut distantia PG multiplicata per numerum punctorum', the text continuing in sequel with the minor amelioration 'id est, ut solidum quod continetur sub...'.

(4) Since the attraction on P of the diametrically opposite unit-particles E and f (or e and F) is $PG+Pg = 2PS$, this result holds true for any non-uniformly dense sphere in which pairs of particles set symmetrically round the centre S have equal gravitational mass.

sphæræ totius, est[5] ut summa planorum omnium, seu sphæra tota, ducta in *pS* distantiam corpusculi a centro sphæræ. Q.E.D.

Cas. 6. Et si ex corpusculis innumeris *p* componatur sphæra nova intra sphæram priorem *ACBD* sita, probabitur ut prius, quod attractio, sive simplex sphæræ unius in alteram, sive mutua utriusqꝫ in se invicem, erit ut distantia centrorum *pS*. Q.E.D.

... ...

Scholium.

Attractionum Casus duos insigniores jam dedi expositos;[6] nimirum ubi vires centripetæ decrescunt in duplicata distantiarum ratione, vel crescunt in distantiarum ratione simplici; efficientes in utroqꝫ casu ut corpora gyrentur in Conicis sectionibus, & componentes corporum sphæricorum vires centripetas eadem lege in recessu a centro decrescentes vel crescentes cum seipsis. Quod est notatu dignum. Casus cæteros, qui conclusiones minus elegantes exhibent, sigillatim percurrere longum esset: Malim cunctos methodo generali simul comprehendere ac determinare, ut sequitur.

Lemma XXIX.

Si describantur centro S circulus quilibet AEB, &[7] *centro P circuli duo EF, ef, secantes priorem in E, e, lineamqꝫ PS in F, f; & ad PS demittantur perpendicula ED, ed: dico quod si distantia arcuum EF, ef in infinitum minui intelligatur, ratio ultima lineæ evanescentis Dd ad lineam evanescentem Ff ea sit, quæ lineæ PE ad lineam PS.*

Nam si linea *Pe* secet arcum *EF* in *q*; & recta *Ee*, quæ cum arcu evanescente *Ee* coincidit, producta occurrat rectæ *PS* in *T*; & ab *S* demittatur in *PE* normalis *SG*: ob similia triangula *EDT*, *edT*, *EDS*, erit *Dd* ad *Ee* ut *DT* ad *ET* seu *DE* ad *ES*, & ob triangula *Eqe*, *ESG* (per Lem. VIII. & Corol. 3 Lem VII.) similia, erit *Ee* ad *qe* seu *Ff*, ut *ES* ad *EG*, & ex æquo *Dd* ad *Ff* ut *DE* ad *SG*; hoc est (ob similia triangula *PDE*, *PGS*) ut *PE* ad *PS*. Q.E.D.

---

(5) A clarifying 'conjunctim' was inserted at this point in the third edition (*Principia*, $_3$1726: 198).

(6) In the earlier Propositions LXXI and LXXVII, namely, the former of which (*Principia*, $_1$1687: 193–5) is a lightly revised repeat of Proposition XL in the preceding 'De motu Corporum' (compare §2: note (184)).

(7) We here omit an editorial direction 'Vide Fig. Prop. sequentis', preferring to advance the figure in question (*Principia*, $_1$1687: 204) to be with the present lemma. In compensation, in Propositions LXXIX/LXXX and LXXXI we introduce in illustration two appropriately simplified versions of it which were there analogously inserted in the third edition (*Principia*, $_3$1726: 201/202 and 203/205/206).

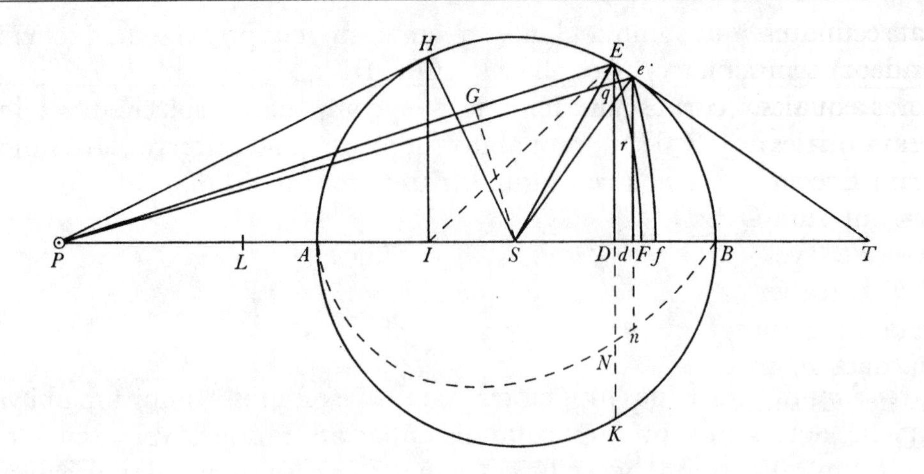

Prop. LXXIX. Theor. XXXIX.

Si superficies ob latitudinem infinite diminutam jamjam evanescens EFfe, convolutione sui circa axem PS, describat solidum sphæricum concavo-convexum, ad cujus particulas singulas æquales tendant æquales vires centripetæ: dico quod vis, qua solidum illud trahit corpusculum situm in P, est in ratione composita ex ratione solidi $DE^q \times Ff$ *& ratione vis qua particula data in loco Ff traheret idem corpusculum.*

Nam si primo consideremus vim superficiei sphæricæ *FE*, quæ convolutione arcus *FE* generatur, & [a] linea *de* ubivis secatur in *r*; erit superficiei pars annularis, convolutione arcus *rE* genita, ut lineola *Dd*, manente sphæræ radio *PE*, (uti demonstravit *Archimedes* in Lib. de Sphæra & Cylindro.)(8) Et hujus vis secundum lineas *PE* vel *Pr* undiq; in superficie conica sitas exercita, ut hæc ipsa superficiei pars annularis; hoc est, ut lineola *Dd*, vel quod perinde est, ut rectangulum sub dato sphæræ radio *PE* & lineola illa

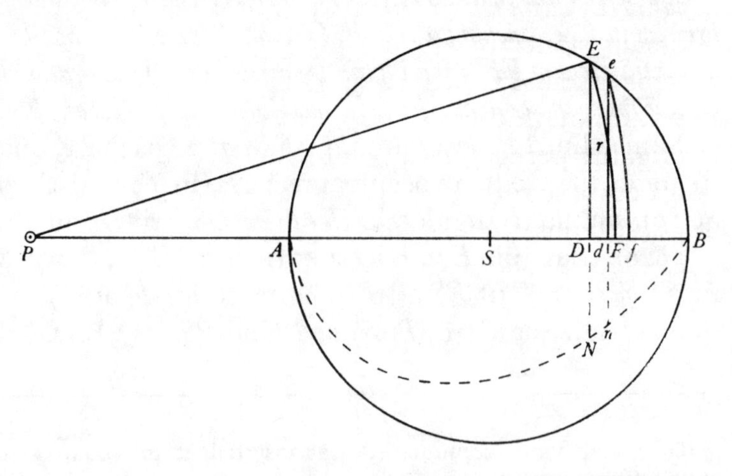

(8) More precisely, the surface of the spherical zone formed by rotating the circle arc $\widehat{Er}$ (of radius $PE = PF$) round AB is $2\pi . PE \times Dd$. Though this is an immediate corollary to Archimedes' *Sphere and Cylinder* I, 42 (which demonstrates that the surface of the spherical segment formed by similarly rotating the whole arc $\widehat{EF}$ round AB is equal in area to a circle of radius $EF = \sqrt{[2PF \times DF]}$), the theorem as such—earlier cited by Newton as 'Prop. 8' of his

Dd: at secundum lineam *PS* ad centrum *S* tendentem minor in ratione *PD* ad *PE*, adeoq; ut $PD \times Dd$. Dividi jam intelligatur linea *DF* in particulas innumeras æquales, quæ singulæ nominentur *Dd*; & superficies *FE* dividetur in totidem æquales annulos, quorum vires erunt ut summa omnium $PD \times Dd$, hoc est, cum lineolæ omnes *Dd* sibi invicem æquentur, adeoq; pro datis haberi possint, ut summa omnium *PD* ducta in *Dd*, id est, ut $\frac{1}{2}PF^q - \frac{1}{2}PD^q$ sive $\frac{1}{2}PE^q - \frac{1}{2}PD^q$ vel $\frac{1}{2}DE^q$ ductum in *Dd*; hoc est, si negligatur data $\frac{1}{2}Dd$, ut DE^{quad}. Ducatur jam superficies *FE* in altitudinem *Ff*, & fiet solidi *EFfe* vis exercita in corpusculum *P* ut $DE^q \times Ff$: puta si detur vis quam particula aliqua data *Ff* in distantia *PF* exercet in corpusculum *P*. At si vis illa non detur, fiet vis solidi *EFfe* ut solidum $DE^q \times Ff$ & vis illa non data conjunctim. Q.E.D.

Prop. LXXX. Theor. XL.

Si ad sphæræ alicujus AEB, centro S descriptæ, particulas singulas æquales tendant æquales vires centripetæ, & ad sphæræ axem AB, in quo corpusculum aliquod P locatur, erigantur de punctis singulis D perpendicula DE, sphæræ occurrentia in E, & in ipsis capiantur longitudines DN, quæ sint ut quantitas $\frac{DE^q \times PS}{PE}$ *& vis quam sphæræ particula sita in axe ad distantiam PE exercet in corpusculum P conjunctim: dico quod vis tota, qua corpusculum P trahitur versus sphæram, est ut area comprehensa sub axe sphæræ AB & linea curva ANB, quam punctum N perpetuo tangit.*

Etenim stantibus quæ in Lemmate & Theoremate novissimo constructa sunt, concipe axem sphæræ *AB* dividi in particulas innumeras æquales *Dd*, & sphæram totam dividi in totidem laminas sphæricas concavo-convexas *EFfe*; & erigatur perpendiculum *dn*. Per Theorema superius, vis qua lamina *EFfe* trahit corpusculum *P* est ut $DE^q \times Ff$ & vis particulæ unius ad distantiam *PE* vel *PF* exercita conjunctim. Est autem per Lemma novissimum, *Dd* ad *Ff* ut *PE* ad *PS*, & inde *Ff* æqualis $\frac{PS \times Dd}{PE}$, & $DE^q \times Ff$ æquale *Dd* in $\frac{DE^q \times PS}{PE}$, & propterea vis laminæ *EFfe* est ut *Dd* in $\frac{DE^q \times PS}{PE}$ & vis particulæ ad distantiam *PF* exercita conjunctim, hoc est (ex Hypothesi) ut $DN \times Dd$, seu area evanescens

untitled paper on cylindrical mensuration on III: 418—is not in Archimedes' extant works. In default, Newton would here probably appeal to Corollary 4 ('Cujusvis portionis [sphæræ] superficies...æquatur curvæ superficiei cylindri...habentis eandem altitudinem...et diametrum...æqualem sphæræ diametro') of the third lecture of Isaac Barrow's 'Lectiones IV. In quibus Theoremata & Problemata Archimedis *De Sphærâ & Cylindro*, Methodo Analyticâ eruuntur', recently appended to his posthumously published *Lectiones Mathematicæ XXIII; In quibus Principia Matheseôs exponuntur* (London, 1683): $_2$341–88.

DNnd. Sunt igitur laminarum omnium vires in corpus *P* exercitæ, ut areæ omnes *DNnd*, hoc est sphæræ vis tota ut area tota *ABNA*. Q.E.D.[(9)]

Corol. 1. Hinc si vis centripeta ad particulas singulas tendens, eadem semper maneat in omnibus distantiis, & fiat *DN* ut $\frac{DE^q \times PS}{PE}$: erit vis tota qua corpusculum a sphæra attrahitur ut area *ABNA*.

Corol. 2. Si particularum vis centripeta sit reciproce ut distantia corpusculi a se attracti, & fiat *DN* ut $\frac{DE^q \times PS}{PE^q}$: erit vis qua corpusculum *P* a sphæra tota attrahitur ut area *ABNA*.

Corol. 3. Si particularum vis centripeta sit reciproce ut cubus distantiæ corpusculi a se attracti, & fiat *DN* ut $\frac{DE^q \times PS}{PE^{qq}}$: erit vis qua corpusculum a tota sphæra attrahitur ut area *ABNA*.

Corol. 4. Et universaliter si vis centripeta ad singulas sphæræ particulas tendens ponatur esse reciproce ut quantitas *V*, fiat autem *DN* ut $\frac{DE^q \times PS}{PE \times V}$; erit vis qua corpusculum a sphæra tota attrahitur ut area *ABNA*.

Prop. LXXXI. Prob. XLI.

Stantibus jam positis, mensuranda est area ABNA.

A puncto *P* ducatur recta *PH* sphæram tangens in *H*, & ad axem *PAB* demissa normali *HI*, bisecetur *PI* in *L*, & erit (per Prop. 12, Lib. 2 *Elem.*) PE^q æquale

(9) In modernised résumé (compare V. I. Antropova, 'O geometricheskom metode "Matematicheskikh Nachal Natural' noi Filosofii" in N'juton', *Istoriko-Matematicheskie Issledovanija*, **17**, 1966: 205–28, especially 223–8) of the essence of the preceding propositions, on taking the attraction of a unit-particle at a distance n to be $f(n)$, the component in the direction *PS* of the attraction exerted on *P* by the spherical zone formed by rotating the arc $\widehat{Er}$ (of centre *P*) round *AB* is (see note (8))

$$2\pi . PE \times Dd \times f(PE) \times PD/PE = 2\pi . f(PE) \times PD \times d(PD),$$

whence the similar component of the attraction on *P* of the whole arc $\widehat{EF}$ comes to be

$$2\pi . f(PE) \int_{PD=PD}^{PD=PE} PD . d(PD) = \pi . f(PE) \times (PE^2 - PD^2) = \pi . f(PE) \times DE^2.$$

Accordingly the component of attraction of the whole sphere along *PS* is

$$\pi \int_{PF=PA}^{PF=PB} f(PE) \times DE^2 . d(PF) = \pi \int_{SD=-AS}^{SD=SB} f(PE) \times DE^2 \times PS/PE . d(SD)$$

since (by Lemma XXIX) $d(PF)/d(SD)$ (or Ff/Dd) $= PS/PE$. (If we set $AS = SB = r$, $PS = s$ and $SD = x$, so that $DE = \sqrt{[r^2 - x^2]}$ and PE(or $\sqrt{[(PS+SD)^2 + DE^2]}$) $= \sqrt{[r^2+s^2+2sx]}$, this assumes the form $\pi \int_{-r}^{r} s(r^2-x^2) \times f(\sqrt{[r^2+s^2+2sx]}) / \sqrt{[r^2+s^2+2sx]} . dx$ more palatable to modern taste.) In sequel Newton will denote the inverse *vis centripeta* $1/f(PE)$ by *V*.

$PS^q + SE^q + 2PSD$. Est autem SE^q seu SH^q (ob similitudinem triangulorum SPH, SHI) æquale rectangulo PSI. Ergo PE^q æquale est contento sub PS & $PS + SI + 2SD$, hoc est, sub PS & $2LS + 2SD$, id est, sub PS & $2LD$. Porro DE^{quad} æquale est $SE^q - SD^q$ seu

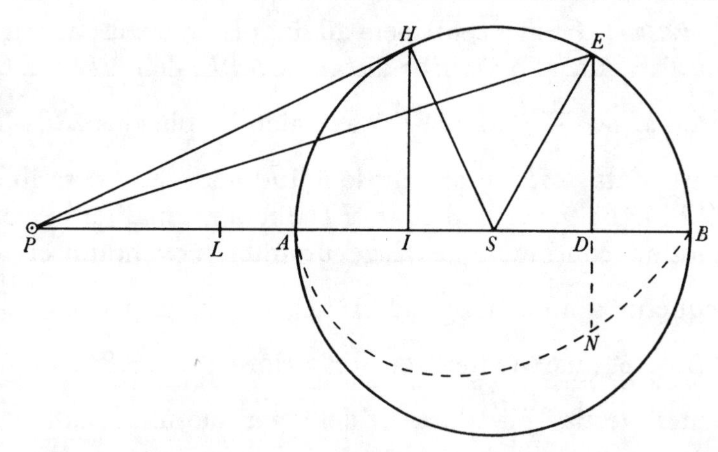

$$SE^q [-] LS^q + 2SLD - LD^q,$$

id est [2] $SLD - LD^q - ALB$. Nam $LS^q - SE^q$ seu $LS^q - SA^q$ (per Prop. 6 Lib. 2. *Elem.*) æquatur rectangulo ALB. Scribatur itaq; $2SLD - LD^q - ALB$ pro DE^q & quantitas $\frac{DE^q \times PS}{PE \times V}$, quæ secundum Corollarium quartum Propositionis præcedentis est ut longitudo ordinatim applicatæ DN, resolvet sese in tres partes $\frac{2SLD \times PS}{PE \times V} - \frac{LD^q \times PS}{PE \times V} - \frac{ALB \times PS}{PE \times V}$: ubi si pro V scribatur ratio inversa vis centripetæ, & pro PE medium proportionale inter PS & $2LD$, tres illæ partes evadent ordinatim applicatæ linearum totidem curvarum, quarum areæ per Methodos vulgatas innotescunt. Q.E.F.[10]

(10) The effect of Newton's introduction of the new origin L (and related variable LD) defined by $LS = \frac{1}{2}(PS + SE^2/PS)$ is to simplify the representation of $PE = \sqrt{[2PS]} \times LD^{\frac{1}{2}}$ at the expense of rendering $DE^2 = SE^2 - (LD - LS)^2 = -(LS^2 - SE^2) + 2LS \times LD - LD^2$ in a slightly more complicated manner. It follows that, where the *vis centripeta* $f(PE)$ is $k.PE^n$, the total attraction of the sphere acting along PS is

$$\pi \int_{LD=LA}^{LD=LB} k.PS(-LA \times LB + 2LS \times LD - LD^2) \times (2PS \times LD)^{\frac{1}{2}(n-1)}.d(LD);$$

that is, on setting $AS = SB = r$, $PS = s$, $LS = \frac{1}{2}(s + r^2/s) = R$ and $LD = y$,

$$k\pi s \int_{R-r}^{R+r} (-(R^2 - r^2) + 2Ry - y^2)\,(2sy)^{\frac{1}{2}(n-1)}.dy,$$

where $R \pm r = \frac{1}{2}(s \pm r)^2/s$. It would now be natural foı Newton to repeat the results of his preceding Propositions LXXI and LXXVII by taking $n = 1$ and -2, thence deducing the spherical attraction to be respectively

$$k\pi s \int_{R-r}^{R+r} (-(R^2 - r^2) + 2Ry - y^2).dy = k\pi s.\Big[-(R^2 - r^2)y + Ry^2 - \tfrac{1}{3}y^3\Big]_{y=R-r}^{y=R+r} = \tfrac{4}{3}\pi r^3.ks$$

and, where $\sigma = (2sy)^{\frac{1}{2}}$,

$$k\pi s \int_{R-r}^{R+r} (-(R^2 - r^2) + 2Ry - y^2)\,(2sy)^{-\frac{3}{2}}.dy$$

$$= k\pi s^{-2}.\Big[(R^2 - r^2)s^2/\sigma + Rs\sigma - \tfrac{1}{12}\sigma^3\Big]_{\sigma=s-r}^{\sigma=s+r} = \tfrac{4}{3}\pi r^3.ks^{-2},$$

Exempl. 1. Si vis centripeta ad singulas sphæræ particulas tendens sit reciproce ut distantia, pro V scribe distantiam PE, dein $2PS\times LD$ pro PE^q, & fiet DN ut $SL-\frac{1}{2}LD-\frac{ALB}{2LD}$. Pone DN æqualem duplo ejus $2SL-LD-\frac{ALB}{LD}$: & ordinatæ pars data $2SL$ ducta in longitudinem AB describet aream rectangulam $2SL\times AB$; & pars indefinita LD ducta normaliter in eandem longitudinem per motum continuum, ea lege ut inter movendum crescendo vel decrescendo æquetur semper longitudini LD, describet aream $\frac{LB^q-LA^q}{2}$, id est, aream $SL\times AB$, quæ subducta de area priore $2SL\times AB$ relinquit aream $SL\times AB$. Pars autem tertia $\frac{ALB}{LD}$ ducta itidem per motum localem normaliter in eandem longitudinem, describet aream Hyperbolicam; quæ subducta de area $SL\times AB$ relinquet aream quæsitam $ABNA$. Unde talis emergit Problematis constructio. Ad puncta L, A, B erige perpendicula Ll, Aa, Bb, quorum Aa ipsi LB, & Bb ipsi LA æquetur. Asymptotis Ll, LB, per puncta a, b describatur Hyperbola ab. Et acta chorda ba claudet aream aba areæ quæsitæ $ABNA$ æqualem.(11)

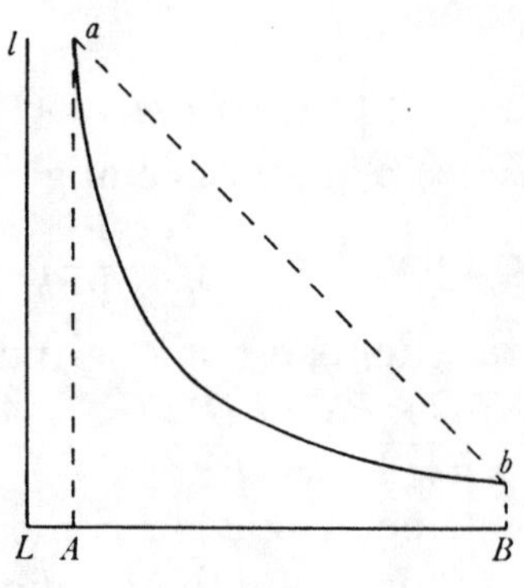

Exempl. 2. Si vis centripeta ad singulas sphæræ particulas tendens sit reciproce ut cubus distantiæ, vel (quod perinde est) ut cubus ille applicatus ad planum quodvis

so that in each case the sphere pulls on an external point as if its mass were concentrated at its centre. (This can happen for no other power of the *vis centripeta*, since in general the sphere attracts the point A on its surface—at which $s = R = r$—with a pull along AS of magnitude

$$k\pi r\int_0^{2r}(2ry-y^2)\,(2ry)^{\frac{1}{2}(n-1)}.dy = \frac{2^{n+4}}{(n+3)\,(n+5)}\,k\pi r^{n+3}$$

which is equal to $\frac{4}{3}\pi r^3.kr^n$ only when $3.2^{n+2} = (n+3)\,(n+5)$; compare Colin Maclaurin, *A Treatise of Fluxions* (Edinburgh, 1742): Book II, Chapter V: 723. Of the three solutions $n = 1$, -2 and *c.* $-5\frac{1}{8}$ possible the last is manifestly not valid when $s > r$.) But in his three following examples Newton prefers to consider the fresh, if physically unrewarding, instances $n = -1$, -3 and -4.

(11) In the analytical terms of the preceding note the attraction of the sphere is here

$$\pi s\int_{R-r}^{R+r}(-(R^2-r^2)+2Ry-y^2)/2sy.dy = \tfrac{1}{2}\pi(-(R^2-r^2)\log((R+r)/(R-r))+2Rr)$$
$$= \pi s^{-2}(-\tfrac{1}{4}(s^2-r^2)^2\log((s+r)/(s-r))+\tfrac{1}{2}rs(s^2+r^2)),$$

that is, very nearly $\frac{4}{3}\pi r^3.(1-\frac{1}{20}r^2/s^2)\,s^{-1}$ where r/s is small. Newton elegantly constructs the solution as proportional (by a factor $\frac{1}{2}\pi$) to the area $\int_{R-r}^{R+r}(z_1-z_2).dy$ cut off between the straight line $z_1 = 2R-y$ and the rectangular hyperbola $z_2 = (R^2-r^2)/y$. This particular case had been broached by Descartes in the long geostatic 'Examen de la question sçavoir si un

datum; scribe $\frac{PE^{cub}}{2AS^q}$ pro V, dein $2PS \times LD$ pro PE^q; & fiet DN ut $\frac{SL \times AS^q}{PS \times LD} - \frac{AS^q}{2PS} - \frac{ALB \times AS^q}{2PS \times LD^q}$, id est (ob continue proportionales PS, AS, SI) ut $\frac{LSI}{LD} - \frac{1}{2}SI - \frac{ALB \times SI}{2LD^q}$. Si ducantur hujus partes tres in longitudinem AB, prima $\frac{LSI}{LD}$ generabit aream Hyperbolicam; secunda $\frac{1}{2}SI$ aream $\frac{1}{2}AB \times SI$; tertia $\frac{ALB \times SI}{2LD^q}$ aream $\frac{ALB \times SI}{2LA} - \frac{ALB \times SI}{2LB}$, id est $\frac{1}{2}AB \times SI$. De prima subducatur summa secundæ ac tertiæ, & manebit area quæsita $ABNA$. Unde talis emergit Problematis constructio. Ad puncta L, A, S, B erige perpendicula Ll, Aa, Ss, Bb, quorum Ss ipsi SI æquetur, perq; punctum s Asymptotis Ll, LB describatur Hyperbola asb occurrens perpendiculis Aa, Bb in a & b; & rectangulum $2ASI$ subductum de area Hyperbolica $AasbB$ relinquet aream quæsitam $ABNA$.(12)

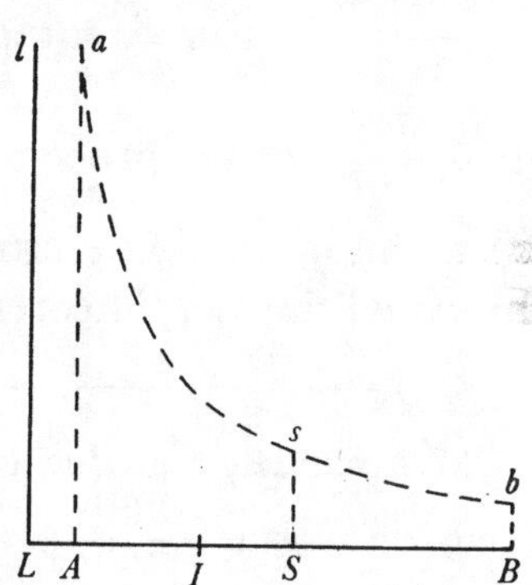

Exempl. 3. Si vis centripeta ad singulas sphæræ particulas tendens decrescit in quadruplicata ratione

corps pese plus...estant proche du centre de la Terre qu'en estant éloigné' which was enclosed with his letter of 13 July 1638 (N.S.) to Mersenne and first published by Clerselier in his *Lettres de Mr Descartes*, **1** (Paris, 1657): 327–46 [= (ed.) C. de Waard and B. Rochot, *Correspondance du P. Marin Mersenne*, **7** (Paris, 1962): 346–68], whose Latin edition (London, 1668) Newton knew well. The exact solution was, of course, beyond Descartes' mathematical capabilities, but even his qualitative argument which confidently concludes that 'clairement ...le centre de gravité de toute cete sphere n'est pas au...centre de sa figure, mais quelque peu plus bas, en la ligne droite qui tend de ce centre...vers celuy de la terre' (*ibid.* 346 [= 368]) is vitiated by his failure to consider the components acting along the central line of the 'pesanteurs' of individual particles of the sphere. It follows from above, in fact, that in an inverse-distance field a sphere attracts an external point as if its mass $\frac{4}{3}\pi r^3$ were concentrated at a distance $s(1+\frac{1}{20}r^2/s^2...)$ greater than that, s, of its centre from the point. Descartes' parting shot that his conclusion will appear 'veritablement fort paradoxe, lorsqu'on n'en considere pas la raison; mais en la considerant, on peut voir que c'est vne verité mathematique tres assurée' badly misfires.

(12) In the algebraic equivalent of note (10) the attraction of the sphere along PS is here ($k = 2r^2$, $n = -3$) given by

$$2\pi r^2 s \int_{R-r}^{R+r} (-(R^2-r^2)+2Ry-y^2)\,(2sy)^{-2}.dy = \pi(R\log((R+r)/(R-r))-2r)\,r^2/s$$
$$= \pi r^2((1+r^2/s^2)\log((s+r)/(s-r))-2r/s),$$

or $\frac{4}{3}\pi r^3.(1+\frac{3}{5}r^2/s^2)\,2r^2/s^3$ very nearly, where as before $LS = R$, $AS = SB = r$ and $PS = s$, so that $IS = Ss = r^2/s$. Newton constructs the former value as proportional (by a factor of π, it is understood) to $\int_{LA}^{LB} z.dy - 2AS \times IS$ on drawing the rectangular hyperbola $z = LS \times IS/y$ through s.

distantiæ a particulis, scribe $\frac{PE^4}{2AS^3}$ pro V, dein $\sqrt{2PS \times LD}$ pro PE, & fiet DN ut $\frac{SL \times SI^{\frac{3}{2}}}{\sqrt{2} \times LD^{\frac{3}{2}}} - \frac{SI^{\frac{3}{2}}}{2\sqrt{2} \times LD^{\frac{1}{2}}} - \frac{ALB \times SI^{\frac{3}{2}}}{2\sqrt{2} \times LD^{\frac{5}{2}}}$. Cujus tres partes ductæ in longitudinem AB producunt Areas totidem, viz.

$$\frac{\sqrt{2} \times SL \times SI^{\frac{3}{2}}}{LA^{\frac{1}{2}}} - \frac{\sqrt{2} \times SL \times SI^{\frac{3}{2}}}{LB^{\frac{1}{2}}}, \quad \frac{LB^{\frac{1}{2}} \times SI^{\frac{3}{2}} - LA^{\frac{1}{2}} \times SI^{\frac{3}{2}}}{\sqrt{2}} \;\&\; \frac{ALB \times SI^{\frac{3}{2}}}{3\sqrt{2} \times LA^{\frac{3}{2}}} - \frac{ALB \times SI^{\frac{3}{2}}}{3\sqrt{2} \times LB^{\frac{3}{2}}}.$$

Et hæ post debitam reductionem, subductis posterioribus de priori, evadunt $\frac{[4]SI^{cub}}{3LI}$(13). Igitur vis tota, qua corpusculum P in sphæræ centrum trahitur, est ut $\frac{SI^{cub}}{PI}$, id est reciproce ut $PS^{cub} \times PI$.(14) Q.E.I.

Eadem methodo determinari potest attractio corpusculi siti intra sphæram, sed expeditius per Theorema sequens.

---

(13) Accurately reproducing a numerical slip in the press copy (see note (1) above), the printed text reads $\frac{\text{'}8SI^{cub}\text{'}}{3LI}$ (*Principia*, $_1$1687: 209). In later revise this abrupt sentence was expanded in the second edition into the more helpful amplification 'Cujus tres partes ductæ in longitudinem AB, producunt areas totidem, *viz.* $\frac{2SI^q \times SL}{\sqrt{2SI}}$ in $\frac{1}{\sqrt{LA}} - \frac{1}{\sqrt{LB}}$; $\frac{SI^q}{\sqrt{2SI}}$ in $\sqrt{LB} - \sqrt{LA}$; & $\frac{SI^q \times ALB}{3\sqrt{2SI}}$ in $\frac{1}{\sqrt{LA^{cub}}} - \frac{1}{\sqrt{LB^{cub}}}$. Et hæ post debitam reductionem fiunt $\frac{2SI^q \times SL}{LI}$, SI^q, & $SI^q + \frac{2SI^{cub}}{3LI}$. Hæ vero, sub[du]ctis posterioribus de priore, evadunt $\frac{4SI^{cub}}{3LI}$' (*Principia*, $_2$1713: 188). The reader is here denied the hint, earlier inserted by Newton in his preliminary draft of this passage (ULC. Add. 3965.17: 643r), that he should effect the reduction 'substituendo $\frac{AI^q}{2SI}$ & $\frac{BI^q}{2SI}$ pro AL et BL', whence $LB^{\frac{1}{2}} - LA^{\frac{1}{2}} = (2IS)^{\frac{1}{2}}$ and therefore, since

$$LA^{\frac{1}{2}} \times LB^{\frac{1}{2}} = (AI \times IB \text{ or}) \; HI^2/2IS = LI, \quad \text{also} \quad 1/LA^{\frac{1}{2}} - 1/LB^{\frac{1}{2}} = (2IS)^{\frac{1}{2}}/LI,$$

while

$$LA \times LB(1/LA^{\frac{3}{2}} - 1/LB^{\frac{3}{2}}) = (LB^{\frac{3}{2}} - LA^{\frac{3}{2}})LI = (3AS^2 + IS^2)/LI \times (2IS)^{\frac{1}{2}},$$

where $AS^2 = IS^2 + 2LI \times IS$.

(14) Seeing that $PS \times IS = HS^2$, the square of the given sphere's radius. On taking $AS = SB = r$, $PS = s$, $LS = R = (s^2 + r^2)/2s$ and $LD = y$ as before (see note (10)) in this instance ($k = 2r^3$, $n = -4$) the total attraction of the sphere along PS will be

$$\begin{aligned} 2\pi r^3 s \int_{R-r}^{R+r} (-(R^2 - r^2) + 2Ry - y^2)\,(2sy)^{-\frac{5}{2}}.dy \\ &= \pi(r^2/2s)^{\frac{3}{2}}.\left[\tfrac{2}{3}(R^2 - r^2)\,y^{-\frac{3}{2}} - 4Ry^{-\frac{1}{2}} - 2y^{\frac{1}{2}}\right]_{y=R-r}^{y=R+r} \\ &= \tfrac{8}{3}\pi(r^2/2s)^{\frac{3}{2}}\,(\sqrt{[R+r]} - \sqrt{[R-r]})\,(R/\sqrt{[R^2 - r^2]} - 1) \\ &= \tfrac{8}{3}\pi r^6/s^2(s^2 - r^2) = \tfrac{4}{3}\pi r^3.(1 + r^2/s^2 \ldots)\,2r^3/s^4, \end{aligned}$$

since $R \pm r = (s \pm r)^2/2s$.

Prop. LXXXII. Theor. XLI.

In sphæra centro S intervallo SA descripta, si capiantur SI, SA, SP continue proportionales: dico quod corpusculi intra sphæram in loco quovis I attractio est ad attractionem ipsius extra sphæram in loco P, in ratione composita ex dimidiata ratione distantiarum a centro IS, PS & dimidiata ratione virium centripetarum, in locis illis P & I ad centra tendentium.

Ut si vires centripetæ particularum sphæræ sint reciproce ut distantiæ corpusculi a se attracti; vis, qua corpusculum situm in I trahitur a sphæra tota, erit ad vim qua trahitur in P, in ratione composita ex dimidiata ratione distantiæ SI ad distantiam SP & ratione dimidiata vis centripetæ in loco I, a particula aliqua in centro oriundæ, ad vim centripetam in loco P ab eadem in centro oriundam, id est, ratione dimidiata distantiarum SI, SP ad invicem reciproce. Hæ duæ rationes dimidiatæ componunt rationem æqualitatis, & propterea attractiones in I & P a sphæra tota factæ æquantur. Simili computo, si vires particularum sphæræ sunt reciproce in duplicata ratione distantiarum, colligetur quod attractio in I sit ad attractionem in P, ut distantia SP ad sphæræ semidiametrum SA:(15) si vires illæ sunt reciproce in triplicata ratione distantiarum, attractiones in I & P erunt ad invicem ut SP^{quad} ad SA^{quad}; si in quadruplicata, ut SP^{cub} ad SA^{cub}. Unde cum attractio in P, in hoc ultimo casu, inventa fuit reciproce ut $PS^{\text{cub}} \times PI$, attractio in I erit reciproce ut $SA^{\text{cub}} \times PI$, id est (ob datum SA^{cub}) reciproce ut PI. Et similis est progressus in infinitum. Theorema vero sic demonstratur.

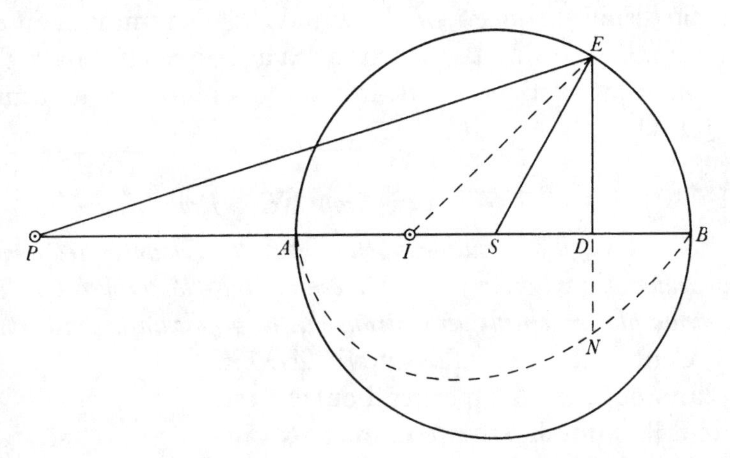

Stantibus jam ante constructis, & existente corpore in loco quovis P, ordinatim applicata DN inventa fuit ut $\frac{DE^q \times PS}{PE \times V}$. Ergo si agatur IE, ordinata

(15) Whence, since by Proposition XL of §2 preceding (which is effectively Proposition LXXI of the first book of the published *Principia*) the component along PS of the attraction of the sphere on P is proportional to the mass of the sphere ($\frac{4}{3}\pi . AS^3$) multiplied by the inverse-square, PS^{-2}, of the distance of its centre from it, it again follows that the comparable attraction of the sphere on I will be as its mass multiplied by $1/PS \times AS = IS/AS^3 \propto IS$, as Newton has already demonstrated from first principles in the previous Proposition XXXIX (which passed unchanged into Proposition LXX of the *Principia*'s first book).

illa ad alium quemvis locum I mutatis mutandis evadet ut $\frac{DE^q \times IS}{IE \times V}$. Pone vires centripetas e sphæræ puncto quovis E manantes esse ad invicem in distantiis IE, PE ut PE^n ad IE^n, (ubi numerus n designet indicem potestatum PE & IE) & ordinatæ illæ fient ut $\frac{DE^q \times PS}{PE \times PE^n}$ & $\frac{DE^q \times IS}{IE \times IE^n}$, quarum ratio ad invicem est ut $PS \times IE \times IE^n$ ad $IS \times PE \times PE^n$. Quoniam ob similia triangula SPE, SEI(16) fit IE ad PE ut IS ad SE vel SA, pro ratione IE ad PE scribe rationem IS ad SA & ordinatarum ratio evadet $PS \times IE^n$ ad $SA \times PE^n$. Sed PS ad SA dimidiata est ratio distantiarum PS, SI; & IE^n ad PE^n(17) dimidiata est ratio virium in distantiis PS, IS. Ergo ordinatæ, & propterea areæ quas ordinatæ describunt hisꝗ proportionales attractiones, sunt in ratione composita ex dimidiatis illis rationibus. Q.E.D.

---

[2](18) *Prop. XC. Prob. XLIV.*

Si ad singula circuli cujuscunꝗ puncta tendant vires centripetæ decrescentes(19) *in quacunꝗ distantiarum ratione: invenire vim qua corpusculum attrahitur ubivis in recta quæ ad planum circuli per centrum ejus perpendicularis consistit.*

Centro A intervallo quovis AD, in plano cui recta AP perpendicularis est, describi intelligatur circulus; & invenienda sit vis qua corpus(20) quodvis P in eundem attrahitur. A circuli puncto quovis E ad corpus(20) attractum P agatur recta PE: In recta PA capiatur PF ipsi PE æqualis, & erigatur normalis FK, quæ sit ut vis qua punctum E trahit corpusculum P. Sitꝗ IKL curva linea quam punctum K perpetuo tangit. Occurrat eadem circuli plano in L. In PA capiatur PH æqualis PD, & erigatur perpendiculum HI curvæ prædictæ occurrens in I; & erit corpusculi P attractio in circulum ut area $AHIL$ ducta in altitudinem AP. Q.E.I.

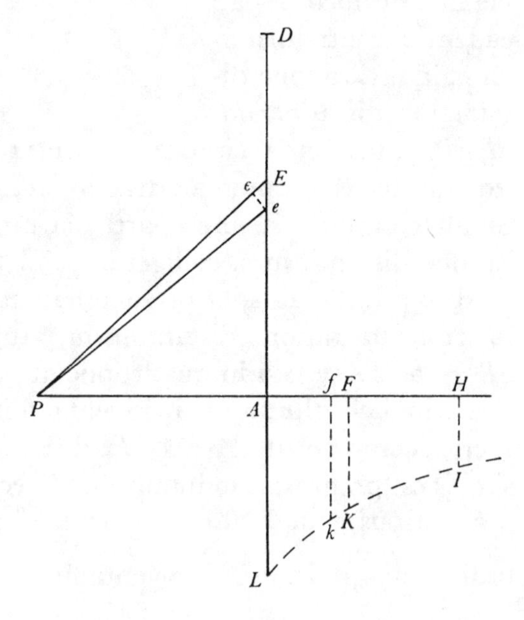

Etenim in AE capiatur linea quam minima Ee. Jungatur Pe, & in [PE,] PA

(16) This was minimally amplified in the third edition to read 'ob continue proportionales *SI*, *SE*, *SP* similia sunt triangula *SPE*, *SEI* et inde' (*Principia*, ₃1726: 208).

(17) A clarifying parenthesis 'ob proportionales *IE* ad *PE* et *IS* ad *SA*' was here inserted in the *Principia's* third edition (*ibid.*).

(18) *Principia*, ₁1687: 218–22, given modern paraphrase in A. N. Kriloff, 'On Sir Isaac

capiatur $[P\epsilon,]$ Pf ipsi Pe æqualis. Et quoniam vis, qua annuli(21) punctum quodvis E trahit ad se corpus P, ponitur esse ut FK, & inde vis qua punctum illud trahit corpus P versus A, ut $\frac{AP\times FK}{PE}$, & vis qua annulus totus trahit corpus P versus A, ut annulus & $\frac{AP\times FK}{PE}$ conjunctim; annulus autem iste est ut rectangulum sub radio AE & latitudine Ee, & hoc rectangulum (ob proportionales PE & AE, Ee & $[\epsilon]E$) æquatur rectangulo $PE\times[\epsilon]E$ seu $PE\times Ff$; erit vis qua annulus iste trahit corpus P versus A ut $PE\times Ff$ & $\frac{AP\times FK}{PE}$ conjunctim, id est, ut contentum $Ff\times AP\times FK$, sive ut area $FKkf$ ducta in AP. Et propterea summa virium, quibus annuli omnes in circulo qui centro A & intervallo AD describitur, trahunt corpus P versus A, est ut area tota $AHIKL$ ducta in AP. Q.E.D.(22)

Corol. 1. Hinc si vires punctorum decrescunt in duplicata distantiarum ratione, hoc est, si sit FK ut $\frac{1}{PF\,\text{quad.}}$, atq; adeo area $AHIKL$(23) ut $\frac{1}{PA}[-]\frac{1}{PH}$; erit attractio corpusculi P in circulum ut $1-\frac{PA}{PH}$, id est, ut $\frac{AH}{PH}$.

Corol. 2. Et universaliter, si vires punctorum ad distantias D(24) sint reciproce

Newton's Formula for the Attraction of a Spheroid...', *Monthly Notices of the Royal Astronomical Society*, **85**, 1925: 571–5.

(19) Later amended (*Principia*, $_2$1713: 196 [+$_3$1726: 214]) to read more precisely 'vires æquales [crescentes vel] decrescentes'.

(20) Changed for consistency's sake to 'corpusculum' in the second edition (*Principia*, $_2$1713: 196).

(21) Understand 'centro A intervallo AE in plano prædicto descripti' as Pemberton specified it in his *editio ultima* (*Principia*, $_3$1726: 215).

(22) In sum, where $f(s)$ is the *vis centripeta* of a unit 'corpuscle' at distance s, the component along PA of the total attraction on P of the laminar circle formed by rotating AD round the axis PA is

$$2\pi\int_0^{AD} f(PE)\times(PA/PE)\times AE.d(AE) = 2\pi\int_{PA}^{PD} PA\times f(PE).d(PE)$$

since ($AE^2 = PE^2-PA^2$ and so) $d(AE)/d(PE) = eE/\epsilon E = PE/AE$; that is, on taking $PD = PH$, $PE = PF$ and $f(PE) = FK$ the attraction along PA is $2\pi\times PA\int_{PA}^{PH} FK.d(PF) = 2\pi\times PA.(AHIL)$.

(23) Here, on setting $f(PE) = k.PE^{-2}$ to be the *vis centripeta*, it follows that

$$(AHIL) = \int_{PA}^{PH} k.PF^{-2}.d(PF),$$

whence the attraction along PA of the laminar circle will be $2k\pi.\Big[-PA\times PF^{-1}\Big]_{PF=PA}^{PF=PH}$.

(24) That is, $PE = PF$. In this more general case, where the *vis centripeta* is $k.PE^{-n}$, the area $(AHIL) = \int_{PA}^{PH} k.PF^{-n}.d(PF)$ and hence the attraction along PA of the laminar circle will be $2\pi\times PA.(AHIL) = 2(k\pi/(n-1)).\Big[-PA\times PF^{-(n-1)}\Big]_{PF=PA}^{PF=PH}$.

ut distantiarum dignitas quælibet D^n, hoc est, si sit FK ut $\frac{1}{D^n}$, adeoq; area $AHIKL$ ut $\frac{1}{PA^{n-1}}$ [−] $\frac{1}{PH^{n-1}}$; erit attractio corpusculi P in circulum ut $\frac{1}{PA^{n-[2]}} - \frac{PA}{PH^{n-1}}$.

Corol. 3. Et si diameter circuli augeatur in infinitum, & numerus n sit unitate major; attractio corpusculi P in planum totum infinitum erit reciproce ut PA^{n-2}, propterea quod terminus alter $\frac{PA}{PH^{n-1}}$ evanescet.

Prop. XCI. Prob. XLV.

Invenire attractionem corpusculi siti in axe solidi,(25) *ad cujus puncta singula tendunt vires centripetæ*(26) *in quacunq; distantiarum ratione decrescentes.*

In solidum $ADEF^{[\prime]}G$(27) trahatur corpusculum P situm in ejus axe AB. Circulo quolibet RFS ad hunc axem perpendiculari secetur hoc solidum, & in ejus [semi]diametro FS, in plano aliquo $PALKB$ per axem transeunte, capiatur (per Prop. XC.) longitudo FK vi qua corpusculum P in circulum illum attrahitur proportionalis. Tangat autem punctum K curvam lineam LKI planis extimorum circulorum AL & BI occurrentem in [L] & [I]; & erit attractio corpusculi P in solidum ut area $LABI$. Q.E.D.(28)

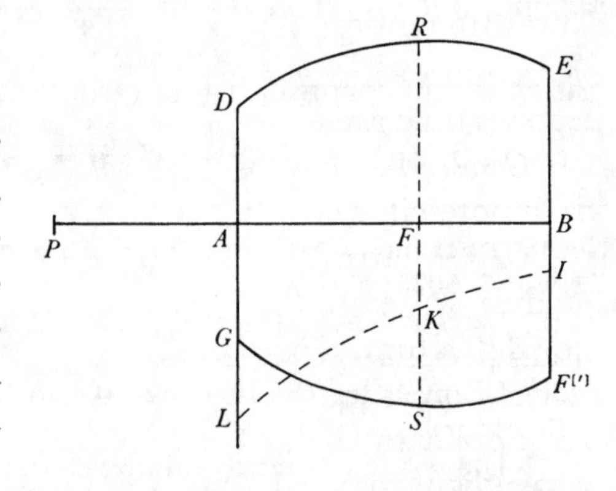

(25) A very necessary 'rotundi' was subsequently inserted (*Principia*, $_2$1713: 215).

(26) Again qualified more precisely as 'vires æquales centripetæ' in the second edition (*ibid.*).

(27) The general figure accompanying is here interpolated from the third edition (*Principia*, $_3$1726: 216); previously ($_1$1687: 220; $_2$1713: 197) the reader had to make his own imaginative leap from the simpler diagram illustrating Corollary 1 (where $\widehat{DRE}$ and $\widehat{GSF'}$ are straight lines parallel to AFB). To avoid confusion, Newton's two points F are distinguished by accenting that which marks the mirror-image of E in AB.

(28) More precisely, where the *vis centripeta* at distance s is ks^{-n} and hence (compare note (24)) on taking $FK = PF^{-(n-2)} - PF \times PR^{-(n-1)}$ the component along PF of the attraction on P of the laminar circle formed by rotating FR round AB is $2(k\pi/(n-1)).FK$, the similar component of the attraction on P of the 'rotund solid' formed by rotating the area

$$(ADEB) = \int_{PA}^{PB} FR.d(PF)$$

round AB will be $2(k\pi/(n-1))\int_{PA}^{PB} FK.d(PF) = 2(k\pi/(n-1)).(ALIB)$. In particular, where the law of attraction is the inverse-square ($n = 2$), the total pull on P is

$$2k\pi\int_{PA}^{PB}(1-PF/PR).d(PF).$$

Corol. 1. Unde si solidum Cylindrus sit, parallelogrammo $ADEB$ circa axem AB revoluto descriptus, & vires centripetæ in singula ejus puncta tendentes sint reciproce ut quadrata distantiarum a punctis: erit attractio corpusculi P in hunc Cylindrum ut $BA-PE+PD$. Nam ordinatim applicata (per Corol. 1. Prop. XC) erit ut $1-\frac{PF}{PR}$. Hujus pars 1 ducta in longitudinem AB describit aream $1\times AB$; & pars altera $\frac{PF}{PR}$ ducta in longitudinem PB describit aream 1 in $\overline{PE-PD}$ (id quod ex curvæ LKI quadratura facile ostendi potest:) & similiter pars eadem ducta in longitudinem PA describit aream 1 in $PD-AD$, ductaꝗ in ipsarum PB, PA differentiam AB describit arearum differentiam 1 in $\overline{PE-PD}$. De contento primo $1\times AB$ auferatur contentum postremum 1 in $\overline{PE-PD}$, & restabit area $LABI$ æqualis 1 in $\overline{AB-PE+PD}$.(29) Ergo vis huic areæ proportionalis est ut $AB-PE+PD$.

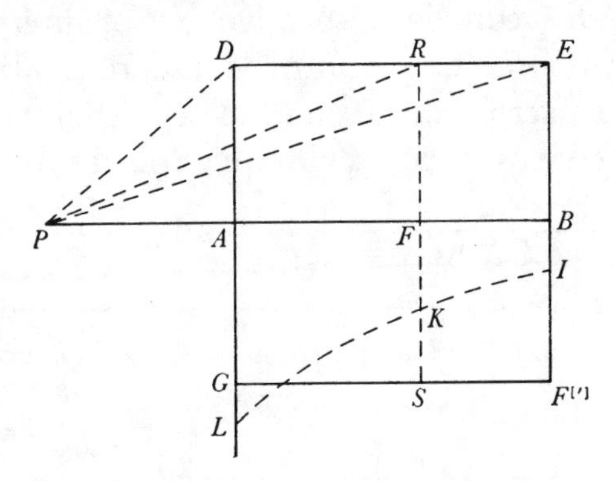

(30)*Corol. 2.* Hinc etiam vis innotescit qua Sphærois $AGBCD$(31) attrahit corpus quodvis P, exterius in axe suo AB situm. Sit $NKRM$ Sectio Conica cujus ordinatim applicata ER, ipsi PE perpendicularis, æquetur semper longitudini PD, quæ ducitur ad punctum illud D in quo applicata ista sphæroidem secat. A sphæroidis

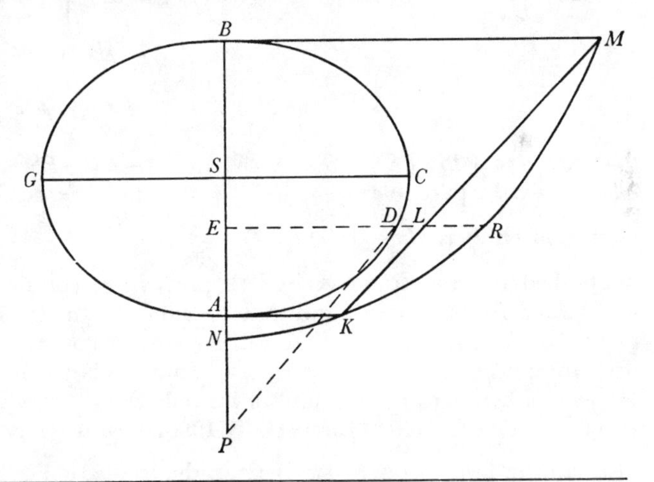

(29) Since the cylinder has a fixed radius, and therefore $PR^2-PF^2 = FR^2$ is constant, whence $PF.d(PF)/d(PR) = PR$, at once

$$\int_{PA}^{PB}(1-PF/PR).d(PF) = \Big[PF-PR\Big]_{PF=PA}^{PF=PB}.$$

(30) These two final corollaries, whose text is added out of sequence in the *Principia's* press copy (Royal Society MS LXIX: 212), were evidently sent by Newton to his editor in a lost letter of 20 August 1686 (see his *Correspondence*, **2**, 1960: 445, note (5)), for when he again wrote to Halley on 13 February 1686/7 he added that 'I hope you received a letter, wth two Corollaries I sent you in Autumn' (*ibid.*: 464).

(31) Understand that this spheroidal volume is formed by the rotation of the area of the semi-ellipse ($ABCD$) round either one of its axes AB.

verticibus A, B ad ejus axem AB erigantur perpendicula AK, BM ipsis AP, BP æqualia respective, & propterea sectioni Conicæ occurrentia in K & M; & jungantur KM auferens ab eadem segmentum $KMRK$. Sit autem sphæroidis centrum S & semidiameter maxima SC: & vis qua sphærois trahit corpus P erit ad vim qua Sphæra diametro AB descripta trahit idem corpus, ut $\dfrac{AS\times CS^q - PS\times KMRK}{PS^q+CS^q-AS^q}$ ad $\dfrac{AS^{\text{cub.}}}{3PS^{\text{quad.}}}$.(32)

(32) For (compare note (28)) the spheroid's attraction along PS is

$$2k\pi\int_{PA}^{PB}(1-PE/PD).d(PE)$$

where $DE^2 = (CS^2/AS^2)\times AE\times EB$ and so $PD = \sqrt{[PE^2+(CS^2/AS^2)(AS^2-ES^2)]}$; that is, $2k\pi\int_{s-r}^{s+r}(1-x/v).dx$ on taking $CS = q$, $AS = SB = r$, $PS = s$, $PE = x$ and hence

$$PD = ER = v = \sqrt{[x^2+(q^2/r^2)(r^2-(s-x)^2)]}.$$

By setting $z^\eta = x$, $e = -q^2(s^2-r^2)/r^2$, $f = 2q^2s/r^2$ and $g = -(q^2-r^2)/r^2$ in species 1 and 2 of the 'Ordo octavus' of conic integrals tabulated by Newton in his 1671 tract (III: 252) it follows that the spheroid's attraction is

$$2k\pi.\left[x-\left(-2(fx+2e)+4f\int v.dx\right)\Big/(f^2-4eg)\right]_{x=s-r}^{x=s+r}$$
$$= 4k\pi\left(r(q^2+2s^2)-s\int_{s-r}^{s+r}v.dx\right)\Big/(q^2-r^2+s^2)$$
$$= 4k\pi\left(rq^2-s\int_{s-r}^{s+r}(v-x).dx\right)\Big/(q^2-r^2+s^2),$$

that is, $4k\pi.(AS\times CS^2-PS\times(KRM))/(CS^2-AS^2+PS^2)$ on making $EL = PE$, whence $LR = v-x$. In particular, when $q = r$ and the spheroid is a sphere of radius $CS = AS$, the attraction on P is $4k\pi s^{-2}\left(r(r^2+2s^2)-s\int_{s-r}^{s+r}\sqrt{[-(s^2-r^2)+2sx]}.dx\right) = \frac{4}{3}\pi r^3.ks^{-2}$ (as already established in Proposition XL of §2 preceding; compare also note (10) above); that is, $4k\pi.\frac{1}{3}AS^3/PS^2$. Bearing in mind that Newton was (in 1685) unable to refer his reader to any printed equivalent of his 1671 table of the areas of curves 'comparable with [central] conics'—itself to appear only much later as a 'Tabula Curvarum simpliciorum quæ cum Ellipsi & Hyperbola compar[ar]i possunt' in his published *Tractatus de Quadratura Curvarum* (= *Opticks* (London, $_1$1704): $_2$163–211, especially 199–204)—we may readily comprehend his reluctance in default to justify from first principles the evaluation of $\int x/v.dx$, $v = \sqrt{[e+fx+gx^2]}$, which underlies the present elegant geometrical statement. By the early readers of the *Principia*, however, its analytical subtleties went unappreciated. In about 1690 David Gregory tried to reproduce Newton's analysis but, on noticing that v was the square root of a trinomial in x, erroneously concluded that the integrand x/v 'could not be squared by...Newton's method' (ULE. Gregory C60, quoted in *The Correspondence of Isaac Newton*, **3**, 1961: 389, note (23)). Even a quarter of a century afterwards it was with manifest pride in his achievement that Roger Cotes announced to Newton on 18 August 1709 that 'Some days ago I was examining the 2d Cor: of Prop 91 Lib I and found it to be true by ye Quadratures of ye 1st & 2d Curves of ye 8th Form of ye second Table in yr Treatise *De Quadrat.*' (Joseph Edleston, *Correspondence of Sir Isaac Newton and Professor Cotes* (London, 1850): 3–4).

The purpose for which Newton contrived this corollary was, in Proposition XIX of the *Principia's* third book, to compare the gravitational pull of the earth (conceived to be an oblate

Corol. 3. Quod si corpusculum intra sphæroidem in data quavis ejusdem diametro collocetur, attractio erit ut ipsius distantia a centro. Id quod facilius colligetur hoc argumento. Sit *AGOF* sphærois attrahens, *S* centrum ejus & *P* corpus attractum. Per corpus illud *P* agantur tum semidiameter *SPA*, tum rectæ duæ quævis *DE*, *FG* sphæroidi hinc inde occurrentes in *D* & *E*, *F* & *G*: sintqȝ *PCM*, *HLN* superficies sphæroidum duarum interiorum, exteriori similium & concentricarum, quarum prior transeat per corpus *P* & secet rectas *DE* & *FG* in *B* & *C*, posterior secet easdem rectas in *H*, *I* & *K*, *L*. Habeant autem sphæroides omnes axem communem, & erunt rectarum partes hinc inde interceptæ *DP* & *BE*, *FP* & *CG*, *DH* & *IE*, *FK* & *LG* sibi mutuo æquales, propterea quod rectæ *DE*, *PB* & *HI* bisecantur in eodem puncto, ut & rectæ *FG*, *PC* & *KL*. Concipe jam *DPF*, *EPG* designare Conos oppositos angulis verticalibus *DPF*, *EPG* infinite parvis descriptos, & lineas etiam *DH*, *EI* infinite parvas esse; & conorum particulæ sphæroidum superficiebus abscissæ *DHKF*, *GLIE* ob æqualitatem linearum *DH*, *EI* erunt ad invicem ut quadrata distantiarum suarum a corpusculo *P*, & propterea corpusculum illud æqualiter trahent. Et pari ratione, si superficiebus sphæroidum innumerarum similium

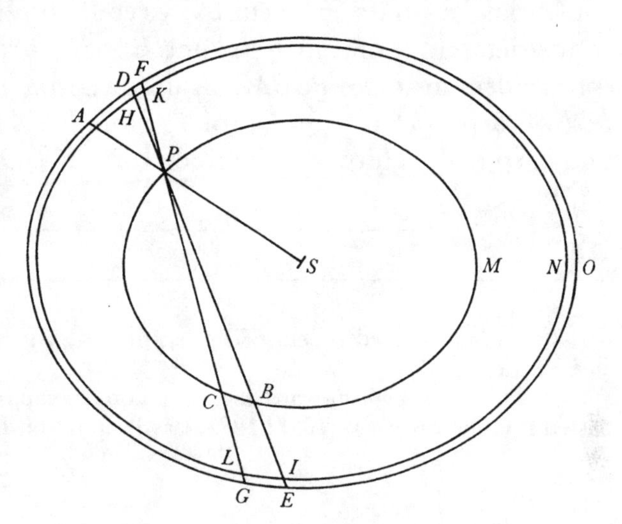

spheroid of uniform density) on a body situated at one of its poles with the corresponding attraction of a sphere (of the same density) which has the same polar axis. In this case the spheroids' equatorial diameter *CG* will not be greatly larger than its axis *AB*, while *P* may be considered to coincide with *A*, so that we can set $q = r(1+\epsilon)$, ϵ small, and $r = s$; accordingly the component along *AS* of the spheroid's (inverse-square) attraction on *A* is

$$4k\pi r\left(1+2(1+\epsilon)^{-2}-(1+\epsilon)^{-1}\int_0^{2r}(2x/r)^{\frac{1}{2}}\,(1-\epsilon(1+\tfrac{1}{2}\epsilon)(1+\epsilon)^{-2}x/r)^{\frac{1}{2}}/r\,.\,dx\right),$$

that is, $4k\pi r(\frac{1}{3}+\frac{4}{15}\epsilon\ldots) = \frac{4}{3}\pi r^3.(1+\frac{4}{5}\epsilon\ldots)kr^{-2}$ on expanding the integrand in powers oı ϵ and then integrating term by term. In revise, about 1692, Newton made a few inconclusive calculations to this effect (see ULC. Add. 3965.8: 80ʳ), employing ratios $q/r = 1{\cdot}005$ and also $1{\cdot}01$ (as in *Principia*, ₁1687: 423). It was not till some time after his death that successful approaches were made—notably by Clairaut and Laplace—to the much more difficult generalised problem of computing the attraction of a spheroid on a point not in its axis; see Isaac Todhunter's *History of the Mathematical Theories of Attraction and Figure of the Earth from Newton to Laplace* (Cambridge, 1873).

concentricarum & axem communem habentium dividantur spatia *DPF*, *EGCB* in particulas, hæ omnes utrinq; æqualiter trahent corpus *P* in partes contrarias. Æquales igitur sunt vires coni *DPF* & segmenti conici *EGCB*, & per contrarietatem se mutuo destruunt. Et par est ratio virium materiæ omnis extra sphæroidem intimam *PCBM*. Trahitur igitur corpus *P* a sola sphæroide intima *PCBM*, & propterea (per Corol. 3. Prop. LXXII.(33)) attractio ejus est ad vim, qua corpus *A* trahitur a sphæroide tota *AGOD*, ut distantia *PS* ad distantiam *AS*.(34) Q.E.I.

---

(33) The unchanged repeat of Proposition XLI in the preceding 'De motu Corporum' (see §2: note (185)).

(34) This is a straightforward generalisation of Proposition XXXIX of §2 preceding, which passed into the *Principia* ($_1$1687: 192–3) as Proposition LXX of the first book.

§3. ARTICLES IV–X OF THE AUGMENTED 'DE MOTU CORPORUM LIBER PRIMUS' (EARLY SUMMER 1685/WINTER 1685–6).(1)

(1) Reproduced from Humphrey Newton's secretarial copy (ULC, Dd. 9. 46: 48^v–55^r/40^r–47^r/88^r–103^r/32^r–39^r/80^r–87^r/72^r–79^v) which was afterwards corrected and further altered by Newton himself and then—once its amended text had been transcribed into the *Principia's* press copy (now Royal Society MS LXIX) and its own useful life was at an end—deposited by him in Cambridge University Library, purportedly as the 'exemplar nitidè descriptum' required by professorial statute (see III: xviii, xxii) of Lucasian *lectiones* delivered publicly in the autumn terms commencing in 'Octob. 1684' and 'Octob. 1685'. As thus preserved, the manuscript terminates abruptly in mid-sentence at the end of f. 79^v, halfway through the proof of Propositon LIV, and the sheets on which its continuation was originally penned (beginning '103' in Newton's own numbering of the leaves) have vanished without trace, no doubt discarded by their author when their immediate purpose was fulfilled and they were not required for deposit as 'lectures'. Of the remainder of this 'De motu Corporum Liber primus', indeed, only the latter part (now ULC. Add. 3970.3: 428 *bis*^{r}/3970.9: 615^r–617^r) of its concluding(?) 'ARTIC. XIV', beginning *in medio* halfway through the demonstration of Proposition XCVI and ending with the scholium to Proposition XCVIII, appears to have survived: the mathematically significant portion of this tailpiece (a preliminary version of *Principia*, $_1$1687: 230, l. 22–235) has already been reproduced (on III: 549–53) as an appendix to the geometrical passages of the 'Optica' (ULC. Dd. 9. 67; see III: 474–512) to which it narrowly pertains. The text (ULC. Dd. 9. 46: 4^r–31^r/64^r–71^r/48^r/49^r) of the Definitions, Axioms and Articles I–III differs only minimally from the equivalent portion of the preceding 'De motu Corporum' over whose manuscript it is (see §2: note (1)) in part written and the principal additions to which are given in §2, Appendix 2: it is effectively identical with pages 1–60 of the first book of the published *Principia* ($_1$1687) into which it was soon after subsumed, and so we safely here omit it. Our justification for printing the following Articles IV–X (effectively *Principia*, $_1$1687: 61–159) is not merely that their manuscript shows several interesting variations from the equivalent Sections IV–X of the published book, but that they embrace a number of purely mathematical propositions which are not otherwise found in Newton's extant papers. The date of early summer 1685 which we loosely assign to the main portion of the 'Liber primus' is founded on his statement to Halley on 20 June 1686 that the following 'second [book] was finished last summer being short & only wants transcribing' (*Correspondence of Isaac Newton*, **2**, 1960: 437), to which he straightaway added that 'In Autumn last [1685] I spent two months in calculations [on comets] to no purpose for want of a good method, wch made me afterwards return to ye first Book & enlarge it wth divers Propositions some relating to Comets others to other things found out last Winter' (*ibid.*). The manuscript bears no trace of any additions relating to the improved theory of comets in which their orbits are taken to be parabolic (round the Sun at their common focus) and it may be that Newton delayed these in afterthought to be in Book 3 (compare 2, §2, Appendix below); among the other minor improvements to the 'Liber primus' which he records that he effected in the winter of 1685–6 we may probably point to Corollary 3 to Proposition XLI, which is certainly a late replacement (on f. 37^v) for the erroneous Corollary 7 to the following Proposition XLIV. We again (compare §2: note (1)) omit the highly suspect and self-inconsistent marginal chronology which Newton imposed on the manuscript when he deposited it as his 'lectures'; see also I. B. Cohen, *Introduction to Newton's 'Principia'* (Cambridge, 1971): 87–9, 318–19, where the present text is given the coding 'LL_β'.

ARTIC. IV.

DE INVENTIONE ORBIUM ELLIPTICORUM, PARABOLICORUM ET HYPERBOLICORUM EX UMBILICO DATO.(2)

Lemma XV.(3)

Si ab Ellipseos vel Hyperbolæ cujusvis umbilicis duobus S, H ad punctum quodvis tertium V inflectantur rectæ duæ SV, HV quarum una HV æqualis sit axi transverso figuræ, altera SV a perpendiculo TR in se demisso bisecetur in T; perpendiculum illud TR Sectionem Conicam alicubi tanget: et contra, si tangit, erit VH æqualis axi figuræ.

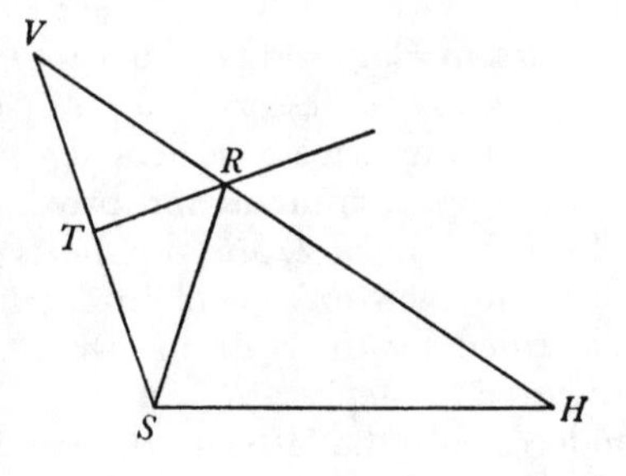

Secet enim *VH* Sectionem conicam in *R*, et jungatur *SR*. Ob æquales rectas *TS*, *TV*, æquales erunt anguli *TRS*, *TRV*. Bisecat ergo *RT* angulum *VRS* et propterea figuram tangit: et contra. Q.E.D.

Prop. XVIII. Prob. X(4)

Datis umbilico et axibus transversis describere Trajectorias Ellipticas et Hyperbolicas quæ transibunt per puncta data, et rectas positione datas contingent.

Sit *S* communis umbilicus figurarum, *AB* longitudo axis transversi Trajectoriæ cujusvis, *P* punctum per quod Trajectoria debet transire, et *TR* recta quam debet tangere. Centro *P* intervallo $AB-SP$ si orbita sit Ellipsis,

---

(2) Originally 'ARTIC. IV/continens/INVENTIONEM ORBIUM...EX CONDITIONIBUS DATIS' (ARTICLE IV, containing THE FINDING OF...ORBITS FROM GIVEN CONDITIONS). This division into 'Articula'—later changed by Halley to be 'Sectiones', a variant obligingly offered by Newton in his letter of 20 June 1686 (*Correspondence*, 2, 1960: 437)—was introduced only after much of the manuscript had been transcribed by Humphrey Newton; for, whereas the titles of this Article and V–IX following are late insertions in Newton's own hand (either crowded in between lines or, as here, marked for incorporation on the facing verso), that of Article X (on f. 72r) has been entered by Humphrey in the ordinary way with normal spacing. The only changes in this Article from Propositions XVII–XIX and Lemma XV of §2 preceding which it otherwise exactly repeats is that an opening lemma is introduced as a rider to the proofs now added of the first two Propositions XVIII and XIX—whose demonstration had earlier (see §2: note (153)) been reckoned to be 'overly obvious'—and the introduction of a new Proposition XX in which the various component cases of the problem of constructing a conic o given eccentricity to pass through given points and touch given straight lines are separately examined.

(3) This late autograph addition in the manuscript was perhaps (compare note (1)) inserted in the winter of 1685–6. While the property demonstrated in this lemma is immediately derivable from Apollonius, *Conics* III, 48 and 51/52 taken together, it is not itself classical, though Newton was well aware that Seth Ward—and after him, for instance, Thomas Street

Translation

Article IV.

Given a focus, the finding of elliptical, parabolic and hyperbolical orbits therefrom.(2)

Lemma XV.(3)

If from the two foci S, H of any ellipse or hyperbola to any third point V there be inclined two straight lines SV and HV, one of which, HV, shall be equal to the transverse axis of the figure, while the other, SV, shall by the perpendicular TR let fall to it be bisected in T, then that perpendicular TR will touch the conic at some point; and, conversely, if it so touches, VH will then be equal to the figure's axis.

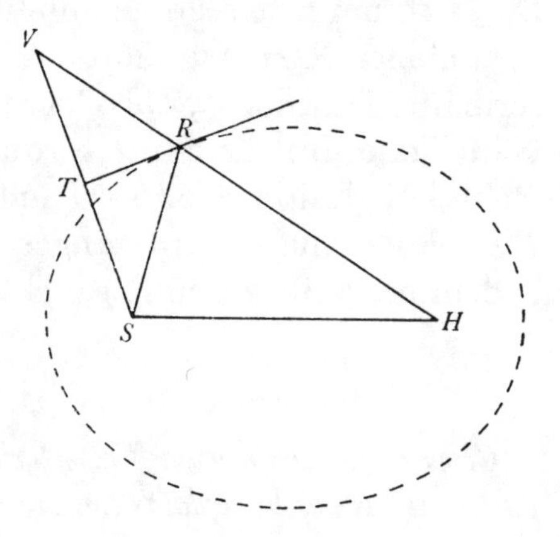

For let *VH* intersect the conic at *R*, and join *SR*. Because the straight lines *TS* and *TV* are equal, the angles $T\widehat{R}S$, $T\widehat{R}V$ will be equal. Consequently *RT* bisects the angle $V\widehat{R}S$ and therefore touches the figure; and conversely so. As was to be proved.

Proposition XVIII, Problem X.(4)

Given a focus and their transverse axes, to describe elliptical and hyperbolical trajectories which shall pass through given points and touch straight lines given in position.

Let *S* be the common focus of the figures, *AB* the length of the transverse axis of any trajectory, *P* a point through which the trajectory ought to pass, and *TR* a straight line which it ought to touch. With centre *P* and radius $AB-SP$ if

and Nicolaus Mercator—had made frequent use of its elliptical case in the 1650's in his analyses of elliptical planetary motion under the simple Boulliau hypothesis of uniform motion round the non-solar focus. (Compare §2: note (163). In introducing the property in Caput VI of his *In Ismaelis Bullialdi Astronomiæ Philolaicæ Fundamenta, Inquisitio Brevis* (Oxford, 1653): 26 Ward had assumed it to be manifest 'ex...ellipsi'.) It is perhaps not without significance that Newton initially stumbled in stating the complementary hyperbolic property in the first version of his following problem (see §2: note (151)).

(4) Changed from 'Prop. XVII. Prob. IX' when the new 'Prop. V. Prob. I' reproduced in §2, Appendix 2.7 preceding was inserted in the manuscript of the 'Liber primus' (on f. 31v). The addition was made by Newton only after Humphrey Newton had transcribed Proposition (XXXVI →) XXXVII below, thus necessitating (compare §2: note (88)) that all intervening numberings of Propositions—and, where pertinent, of Problems—be increased thereafter by unity.

vel $AB+SP$ si ea sit Hyperbola, describatur circulus HG.(5) Ad tangentem TR demittatur perpendiculum ST, et producatur ea ad V ut sit TV æqualis ST, centroq; V & intervallo AB describatur circulus FH.(5) Hac methodo, sive dentur duo puncta P, p, sive duæ tangentes TR, tr sive punctum P et tangens TR; describendi sunt circuli duo. Sit H eorum intersectio communis et umbilicis S, H, axe illo dato describatur Trajectoria. Dico factum. Nam Trajectoria descripta, eò quod $PH+SP$ in Ellipsi, et $PH-SP$ in Hyperbola æquatur axi, transibit per punctum P et per Lemma superius tanget rectam TR. Et eodem argumento vel transibit eadem per puncta duo P, p, vel tanget rectas duas TR, tr. Q.E.F.

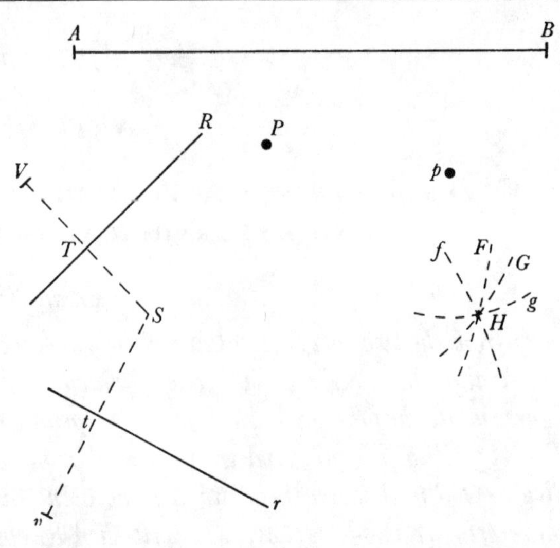

Prop. XIX. Prob. XI.(6)

Circa datum umbilicum Trajectoriam Parabolicam describere quæ transibit per puncta data et rectas positione datas continget.

Sit S umbilicus, P punctum et TR tangens trajectoriæ describendæ. Centro P, intervallo PS describe circulum FG. Ab umbilico ad tangentem demitte perpendicularem ST, et produc eam ad V, ut sit TV æqualis ST. Eodem modo describendus est alter circulus fg si datur alterum punctum p; vel inveniendum alterum punctum v si datur altera tangens tr; dein ducenda recta IF quæ tangat duos circulos FG, fg si dantur duo puncta P, p, vel transeat per duo puncta V, v si dantur duæ tangentes TR, tr vel tangat circulum FG et transeat per punctum V si datur punctum P et tangens TR. Ad FI demitte perpendicularem SI, eamq; biseca in K, et axe SK, vertice principali K describatur Parabola. Nam Parabola ob æquales SK, IK et SP, FP transibit per punctum P et (per Lemmatis XIV Corol. 3) ob æquales ST, TV et angulum rectum STR, tanget rectam TR. Q.E.F.(7)

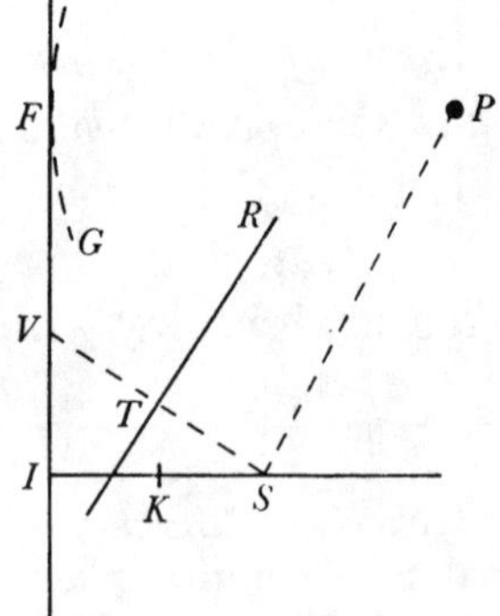

Prop. XX. Prob. XII.

Circa datum umbilicum Trajectoriam quamvis specie datam(8) *describere quæ per data puncta transibit et rectas tanget positione datas.*

the orbit be an ellipse, or $AB+SP$ if it be a hyperbola, describe the circle *HG*.(5) Let fall the perpendicular *ST* to the tangent *TR* and extend it to *V* so that *TV* is equal to *ST*, then with centre *V* and radius *AB* describe the circle *FH*.(5) By this method, whether there be given two points *P* and *p*, two tangents *TR* and *tr*, or a point *P* and a tangent *TR*, two circles are to be described. Let *H* be their common intersection and then with foci *S*, *H* and the given axis describe a trajectory. I say it is done. For, because $PH+SP$ in the ellipse ($PH-SP$ in the hyperbola) is equal to the axis, the trajectory described will pass through the point *P* and, by the previous lemma, touch the straight line *TR*. And by the same argument it will pass through two points *P* and *p*, or touch two straight lines *TR* and *tr*. As was to be done.

Proposition XIX, Problem XI.(6)

To describe round a given focus a parabolic trajectory which shall pass through given points and touch straight lines given in position.

Let *S* be the focus, *P* a point and *TR* a tangent of the trajectory to be described. With centre *P* and radius *PS* describe the circle *FG*. Let fall the perpendicular *ST* from the focus to the tangent and extend it to *V* so that *TV* is equal to *ST*. In the same way there needs, if a second point *p* is given, to be described a second circle *fg*; or, if a second tangent *tr* is given, while a second point *v* needs to be found: then you must describe a straight line *IF* which, if two points *P* and *p* are given, shall touch the two circles *FG*, *fg*; or, if two tangents *TR* and *tr* are given, shall pass through the two points *V* and *v*; or, if a point *P* and tangent *TR* are given, shall touch the circle *FG* and pass through the point *V*. To *FI* let fall the perpendicular *SI* and bisect it in *K*, and then with axis *SK* and principal vertex *K* describe the parabola. For because of the equals *SK*, *IK* and *SP*, *FP* the parabola wili pass through the point *P* and (by Corollary 3 of Lemma XIV), because *ST* and *TV* are equal and the angle $S\widehat{T}R$ right, it will touch the straight line *TR*. As was to be done.(7)

Proposition XX, Problem XII.

To describe round a given focus any trajectory given in species(8) *which shall pass through given points and touch straight lines given in position.*

(5) Before Newton attached the demonstration of this construction in the amended sequel, Humphrey Newton here first transcribed from Proposition XVII of §2 the subsequently cancelled sentences 'In hoc circulo locabitur umbilicus alter' (The second focus will be located in this circle) and 'In hoc circulo locabitur etiam umbilicus alter' (The second focus will be located in this circle also) respectively.

(6) Except for the newly added concluding demonstration, a virtually unaltered repeat of Proposition XVIII of §2 preceding.

(7) Since two lines *IF* may be drawn from *I* to touch the circle of centre *P*, two solutions are possible.

Cas. 1. Dato umbilico S describenda sit Trajectoria ABC per puncta duo B, C. Quoniam trajectoria datur specie,(8) dabitur ratio axis transversi ad distantiam umbilicorum. In ea ratione cape KB ad BS et LC ad CS. Centris B, C intervallis BK CL describe circulos duos et ad rectam quæ tangat eosdem(9) in K et L demitte perpendiculum SG, idemq; seca in A et a ita ut sit SA ad AG et Sa ad aG ut est SB ad BK, et axe Aa, verticibus A, a describatur Trajectoria. Dico factum. Sit enim H umbilicus alter figuræ descriptæ et cum sit SA ad AG ut Sa ad aG erit divisim $Sa-SA$ seu SH ad $aG-AG$ seu Aa(10) in eadem ratione adeoq; in ratione quam habet axis transversus figuræ describendæ ad distantiam umbilicorum ejus, et propterea figura descripta est ejusdem speciei cum describenda. Cumq; sint KB ad BS et LC ad CS in eadem ratione transibit hæc Figura per puncta B, C, ut ex Conicis(11) manifestum est.

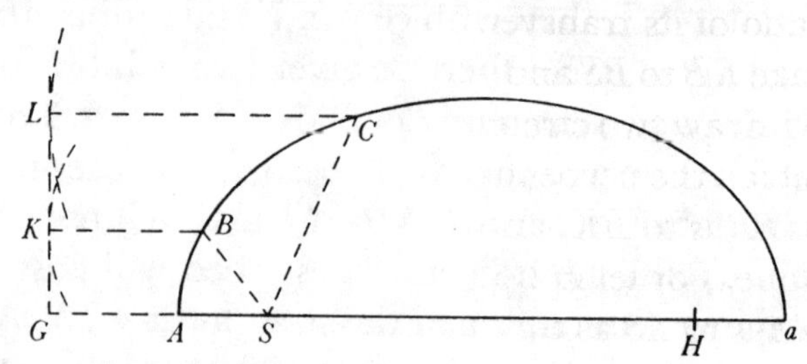

Cas. 2. Dato umbilico S describenda sit Trajectoria quæ rectas duas TR, tr alicubi contingat. Ab umbilico in tangentes demitte perpendicula ST, St et produc eadem ad V, v ut sint TV, tv æquales TS, $t[S]$. Biseca Vv in O et erige perpendiculum infinitum OH, rectamq; VS infinite productam seca in K et k ita ut sit VK ad KS et Vk ad kS ut est Trajectoriæ describendæ axis transversus ad umbilicorum distantiam. Super diametro Kk describatur circulus secans rectam OH in H; et umbilicis S, H, axe transverso ipsam VH æquante, describatur Trajectoria. Dico factum.(12) Nam biseca Kk in X, et junge HX, HS, HV, Hv. Quoniam est VK ad KS ut Vk ad kS et compositè ut $VK+Vk$ ad $KS+kS$,

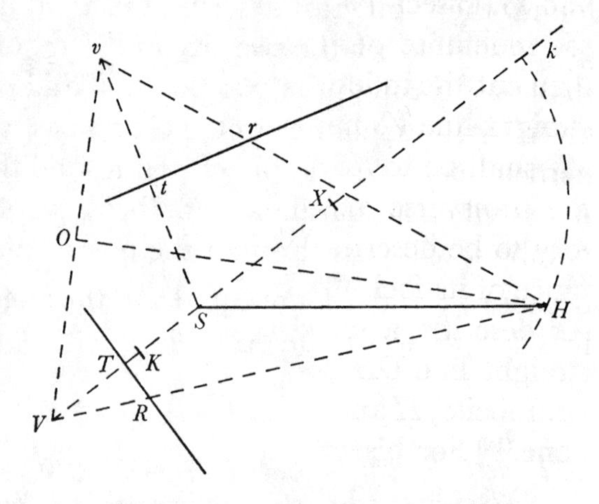

(8) That is, of given eccentricity. Since the ovalness of a solar planet or comet was determinable in contemporary astronomical theory only by first constructing its elliptical or parabolic orbit, Newton could have had no practical purpose in here introducing this mathematically interesting problem.

(9) Newton's replacement for an initial 'utrosq;' (each).

(10) To be consistent with the sequel this should read in inverse sequence 'cum sit GA ad AS ut Ga ad Sa erit divisim $Ga-GA$ seu Aa ad $Sa-AS$ seu SH' and so we translate it. The slip was caught only in Pemberton's *editio ultima* of the ensuing *Principia* ($_3$1726: 68).

Case 1. Given the focus S, it is required to describe a trajectory ABC through the two given points B and C. Seeing that the trajectory is given in species,(8) the ratio of its transverse axis to the distance of the foci will be given. In this ratio take KB to BS and LC to CS. With centres B and C and (respective) radii BK and CL draw two circles, and to the straight line which shall touch them(9) at K and L let fall the perpendicular SG; cut this in A and a so that AS be to GA and Sa to Ga as SB is to BK, and with axis Aa and vertices A, a draw a trajectory. I say it is done. For let H be the second focus of the figure described and there will, since GA is to AS as Ga to Sa, then *dividendo* be $Ga-GA$ or Aa to $Sa-AS$ or SH in the same ratio and hence in the ratio had by the transverse axis of the figure required to be drawn to the distance of its foci; accordingly, the figure described is of the same species as that which needs to be. And since KB is to BS and LC to CS in the same ratio, this figure will, as is evident from the *Conics*,(11) pass through the points B and C.

Case 2. Given the focus S, it is required to describe a trajectory which shall touch the two straight lines TR and tr at some point. Let fall the perpendiculars ST, St from the focus onto the tangents and extend them to V and v so that TV and tv are equal to TS and tS. Bisect Vv in O and erect the indefinite perpendicular OH, then cut the indefinitely extended straight line VS in K and k so that VK shall be to KS and Vk to kS as the transverse axis of the trajectory to be described is to the distance of its foci. On the diameter Kk describe a circle cutting the straight line OH in H; and then

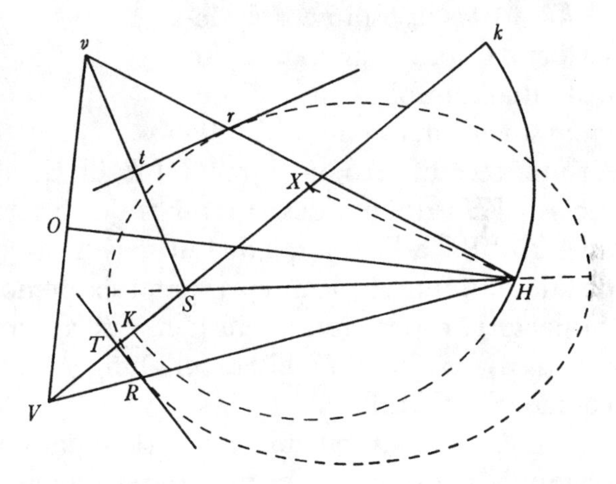

with foci S, H and transverse axis equal to VH, describe a trajectory. I say it is done.(12) For bisect Kk in X and join HX, HS, HV, Hv. Because VK is to KS as Vk to kS, and so *componendo* as $VK+Vk$ to $KS+kS$ and dividendo as $Vk-VK$

(11) Again (compare §2: note (152)) understand Pappus, *Mathematical Collection* VII, 238 or any more modern enunciation of the focus-directrix defining property of a conic to which Newton here appeals.

(12) The conditions $VH(=SR\pm RH=Sr\pm rH)=vH$ and VH (or $SR\pm RH$)$:SH=1:e$, where e is the conic's given eccentricity, respectively determine H to be both on the perpendicular bisector of Vv and on the Apollonian circle of diameter Kk: the two resulting meets H yield two positions for the conic's second focus and hence two solutions (which coincide or are not real according as OH is tangent to the circle or fails to intersect it). Newton proceeds to justify his construction of these auxiliary loci *ab initio*.

divisimq; ut $Vk-VK$ ad $kS-KS$ id est ut $2VX$ ad $2KX$ et $2KX$ ad $2SX$ adeoq; ut VX ad HX et HX ad SX, similia erunt triangula VXH, HXS, et propterea VH erit ad SH ut VX ad XH, adeoq; ut VK ad KS. Habet igitur Trajectoriæ descriptæ axis transversus VH eam rationem ad ipsius umbilicorum distantiam SH quam habet Trajectoriæ describendæ axis transversus ad ipsius umbilicorum distantiam et propterea ejusdem est speciei. Insuper cum VH, vH æquentur axi transverso et VS, vS a rectis TR, tr perpendiculariter bisecentur, liquet ex Lemmate XV rectas illas Trajectoriam descriptam tangere. Q.E.F.

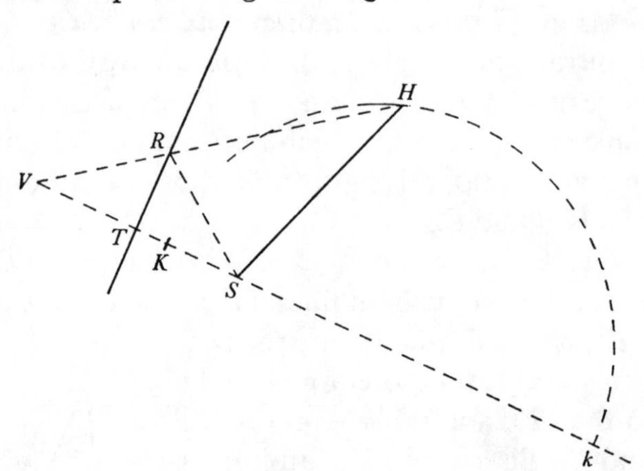

Cas. 3. Dato umbilico S describenda sit Trajectoria quæ rectam TR tanget in puncto dato R. In rectam TR demitte perpendicularem ST et produc eandem ad V ut sit TV æqualis ST. Junge VR et rectam VS infinite productam seca in K et k, ita ut sit VK ad SK et Vk ad Sk ut Ellipseos(13) describendæ axis transversus ad distantiam umbilicorum; circuloq; super diametro Kk descripto, secetur producta recta VR in H, et umbilicis S, H, axe transverso rectam VH æquante describatur trajectoria. Dico factum.(14) Namq; VH esse ad SH ut VK ad SK atq; adeo ut axis transversus Trajectoriæ describendæ ad distantiam umbilicorum ejus, patet ex demonstratis in Casu 2do, et propterea Trajectoriam descriptam ejusdem esse speciei cum describenda: rectam verò TR qua angulus VRS bisecatur, tangere trajectoriam in puncto R patet ex conicis.(15) Q.E.F.

Cas. 4. Circa umbilicum S describenda jam sit Trajectoria APa quæ tangat rectam TR transeatq; per punctum quodvis P extra tangentem datum, quæq;

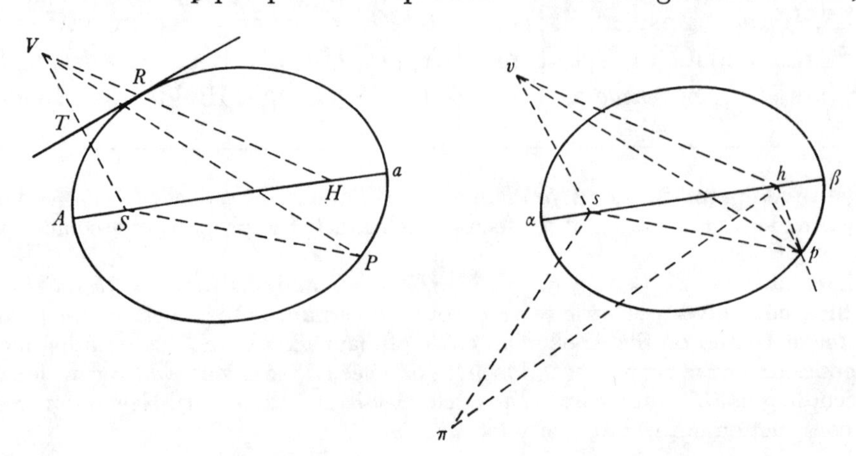

to $kS-KS$, that is, as $2VX$ to $2KX$ and $2KX$ to $2SX$ and hence as VX to HX and HX to SX, the triangles VXH, HXS will be similar and accordingly VH will be to SH as VX to XH, and hence as VK to KS. The transverse axis VH of the trajectory described has therefore the ratio to the distance SH of its foci which the transverse axis of the trajectory requiring to be drawn has to the distance of its foci, and it is accordingly of the same species. Furthermore, since VH and vH are equal to the transverse axis, while VS, vS are perpendicularly bisected by the straight lines TR, tr, it is patent from Lemma XV that those straight lines touch the trajectory described. As was to be done.

Case 3. Given the focus S, it is required to describe a trajectory which shall touch the straight line TR at the given point R. Let fall the perpendicular ST onto the straight line TR and extend it to V so that TV be equal to ST. Join VR and cut the indefinitely prolonged straight line VS in K and k so that VK be to SK and Vk to Sk as the transverse axis of the ellipse(13) to be described to the distance of its foci; then, after a circle is described on the diameter Kk, let

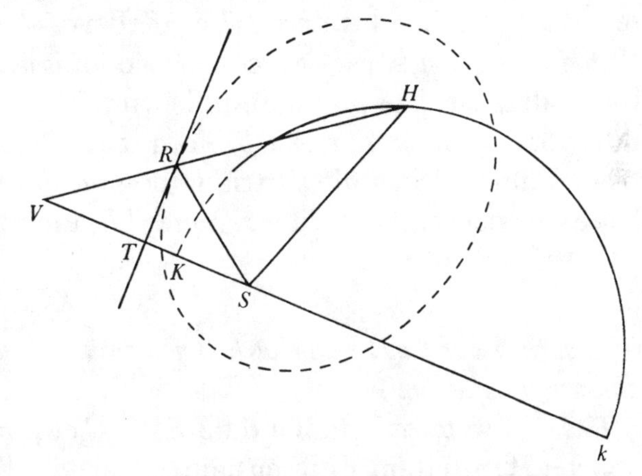

it intersect the extended straight line VR in H, and with foci S, H and transverse axis equal to the straight line VH draw a trajectory. I say it is done.(14) For that VH is to SH as VK to SK, and hence as the transverse axis of the trajectory requiring to be drawn to the distance of its foci, is evident from what is proved in Case 2, and it is accordingly clear that the trajectory described is of the same species as that needing to be drawn; but that the straight line TR by which the angle $V\widehat{R}S$ is bisected touches the trajectory at the point R is evident from the *Conics*.(15) As was to be done.

Case 4. Let it now be required to describe round the focus S a trajectory APa which shall touch the straight line TR and pass through any point P given outside the tangent and shall be similar to the figure $\alpha p\beta$ described with trans-

(13) Read 'Ellipseos vel Hyperbolæ' (of the ellipse or hyperbola). The oversight is not adjusted in any edition of the printed *Principia*.

(14) Here the second focus H is determined as the meet of the straight line VR and the Apollonian circle on diameter Kk; there will again be two possible solutions (coincident or imaginary according as VR is tangent to the circle or does not intersect it in real points).

(15) Apollonius, *Conics* III, 48. Newton himself learnt this property from reading Schooten's *Exercitationes* as an undergraduate (see I: 32, 38).

similis sit figuræ $\alpha p\beta$ axe transverso $\alpha\beta$ et umbilicis s, h descriptæ.(16) In tangentem TR demitte perpendiculum ST et produc idem ad V ut sit TV æqualis ST. Angulis autem VSP, SVP fac angulos $hs\pi$, $sh\pi$ æquales, centroq; π et intervallo quod sit ad $\alpha\beta$ ut SP ad VS describe circulum secantem figuram $\alpha p\beta$ in p. Junge sp, et age SH quæ sit ad sh ut est SP ad sp quæq; angulum PSH angulo psh et angulum VSH angulo $ps\pi$ æquales constituat. Deniq; umbilicis S, H axe distantiam VH æquante describatur Sectio conica. Dico factum.(17) Nam si agatur sv quæ sit ad sp ut est sh ad $s\pi$, quæq; constituat angulum vsp angulo $hs\pi$ et angulum vsh angulo $ps\pi$ æquales, triangula svh, $sp\pi$ erunt similia, et propterea vh erit ad $p\pi$ ut est sh ad $s\pi$, id est (ob similia triangula VSP, $hs\pi$) ut est VS ad SP seu $\alpha\beta$ ad $p\pi$. Æquantur ergo vh et $\alpha\beta$. Porrò ob similia triangula VSH, vsh, est VH ad SH ut vh ad sh, id est, axis conicæ sectionis jam descriptæ ad illius umbilicorum intervallum ut axis $\alpha\beta$ ad umbilicorum intervallum sh, et propterea figura jam descripta similis est figuræ $\alpha p\beta$. Transit autem hæc figura per punctum P eò, quod triangulum PSH simile sit triangulo psh; et quia VH æquatur ipsius axi et VS bisecatur perpendiculariter a recta TR, tangit eadem rectam TR. Q.E.F.

Lemma XVI.

A datis tribus punctis ad quartum non datum inflectere tres rectas quarum differentiæ vel dantur vel nullæ sunt.

Cas. 1. Sunto puncta illa data A, B, C et punctum quartum Z quod invenire oportet. Ob datam differentiam linearum AZ, BZ, locabitur punctum Z in Hyperbola cujus umbilici sunt A et B et axis transversus differentia illa data. Sit axis ille MN. Cape PM ad MA ut est MN ad AB, et erecto PR perpendiculari ad AB, demissoq; ZR perpendiculari ad PR, erit ex natura hujus Hyperbolæ ZR ad AZ ut est MN ad AB. Simili discursu punctum Z locabitur in alia Hyperbola cujus umbilici sunt A, C et axis transversus differentia inter AZ et CZ, duciq; potest QS ipsi AC perpendicularis[,] ad quam si ab Hyperbolæ hujus puncto quovis Z demittatur normalis ZS, hæc fuerit ad AZ ut est differentia inter AZ et

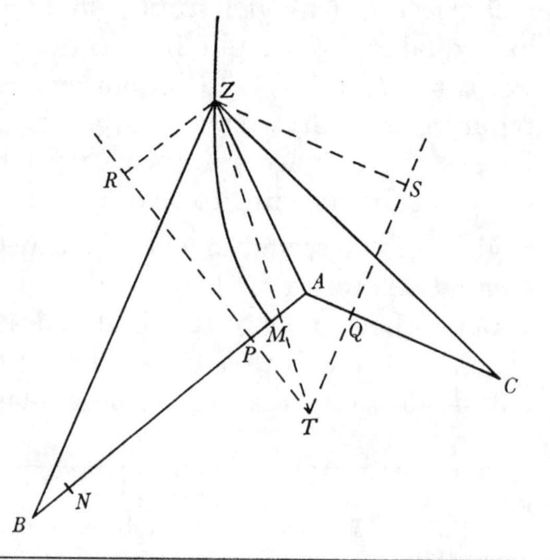

(16) It would be more natural, of course to fix the species of the conic APa by letting the value of its eccentricity SH/Aa be given, but Newton has need of the similar auxiliary conic $\alpha p\beta$ in his ensuing construction.

(17) The conditions in this case determine the second focus H of the conic APB to be the meet of the loci—an Apollonian circle and a central conic of foci V, P respectively defined by

verse axis $\alpha\beta$ and foci s, h.(16) Let fall the perpendicular ST onto the tangent TR and extend it to V so that TV be equal to ST. Now make the angles $h\widehat{s}\pi$, $s\widehat{h}\pi$ equal to the angles $V\widehat{S}P$, $S\widehat{V}P$, and with centre π and a radius which shall be to $\alpha\beta$ as SP to VS describe a circle interesecting the figure $\alpha p\beta$ in p. Join sp, and draw SH which shall be to sh as SP is to sp and constitute the angle $P\widehat{S}H$ equal to $p\widehat{s}h$ and the angle $V\widehat{S}H$ equal to $p\widehat{s}\pi$. Finally, with foci S, H and axis equal to the distance VH describe a conic. I say it is done.(17) For, if there be drawn sv which shall be to sp as sh is to $s\pi$ and constitute the angle $v\widehat{s}p$ equal to $h\widehat{s}\pi$ and the angle $v\widehat{s}h$ equal to $p\widehat{s}\pi$, the triangles svh, $sp\pi$ will be similar and accordingly vh will be to $p\pi$ as sh is to $s\pi$, that is, (because of the similar triangles VSP, $hs\pi$) as VS to SP or $\alpha\beta$ to $p\pi$. Therefore vh and $\alpha\beta$ are equal. Moreover, (because of the similar triangles VSH, vsh) there is VH to SH as vh to sh, that is, the axis of the conic just now described to the distance of its foci as the axis $\alpha\beta$ to the distance sh of the foci, and accordingly the figure now described is similar to the figure $\alpha p\beta$. This figure passes, however, through the point P by reason of the triangle PSH being similar to the triangle psh; and, because VH is equal to its axis and VS is bisected perpendicularly by the straight line TR, it touches the straight line TR. As was to be done.

Lemma XVI.

From three given points to incline to a fourth one, not given, three straight lines whose distances are either given or nil.

Case 1. Let A, B and C be the given points and Z the fourth point which it is necessary to find. Because of the given difference of the lines AZ and BZ, the point Z will be located in a hyperbola whose foci are A and B and transverse axis is that given difference. Take PM to MA as MN is to AB and, on erecting PR perpendicular to AB and letting fall ZR perpendicular to PR, there will from the nature of this hyperbola be ZR to AZ as MN is to AB. By a similar disquisition the point Z will be located in another hyperbola whose foci are A and C and transverse axis is the difference between AZ and CZ, and QS can be drawn perpendicular to AC, while, if the normal ZS be let fall to it from any point Z of the hyperbola, this will be to AZ as the difference between AZ and CZ to AC.

$SH/VH = e$, constant, and $VH \mp PH = SP$, constant; whence there will be two pairs of possible positions for H, each of which may, as before, coincide or be imaginary. In the given similar conic $\alpha p\beta$ the parallel problem is to fit the triangle vsp, similar to the given triangle VSP, round the focus s so that p be on the conic while the distance of v from the second focus h be equal to the axis $\alpha\beta$: because, constructing the triangle $hs\pi$ similar to triangle VSP, and so to vsp, the triangles vsh and $ps\pi$ come also to be similar, it ensues that as the triangle vsp rotates round s, maintaining the distance $vh = \alpha\beta$ unaltered, the locus of p is a circle of centre π and radius $\pi p = (vh \times s\pi/hs$ or) $\alpha\beta \times SP/VS$, which therefore intersects the conic $\alpha p\beta$ in two corresponding pairs of points p. Newton's following synthesis somewhat obscures this relatively straightforward preliminary analysis.

CZ ad AC. Dantur ergo rationes ipsarum ZR et ZS ad AZ et idcirco datur earundem ZR, ZS ratio ad invicem, adeoq; rectis RP, SQ concurrentibus in T, locabitur punctum Z in recta TZ positione data. Eadem methodo per Hyperbolam tertiam, cujus umbilici sunt B et C et axis transversus differentia rectarum BZ, CZ, inveniri potest alia recta in qua punctum Z locatur. Habitis autem duobus locis rectilineis, habetur punctum quæsitum Z in earum intersectione. Q.E.I.[18]

Cas 2. Si duæ ex tribus lineis, puta AZ et BZ æquantur, punctum Z locabitur in perpendiculo bisecante distantiam AB, et locus alius rectilineus invenietur ut supra. Q.E.I.

Cas. 3. Si omnes tres æquantur locabitur punctum Z in centro circuli per puncta A, B, C transeuntis. Q.E.I.

Solvitur etiam hoc Lemma problematicum per librum Tactionum Apollonij a Vieta restitutum.[19]

Prop. XXI. Prob. XIII.[20]

Trajectoriam circa datum umbilicum describere quæ transibit per puncta data et rectas positione datas continget.

Detur umbilicus S, punctum P, et tangens TR, et inveniendus sit umbilicus alter H. Ad tangentem demitte perpendiculum ST et produc idem ad V ut sit TV æqualis ST, et erit VH æqualis axi transverso. Junge SP, HP et erit SP differentia inter HP et axem transversum. Hoc modo si dentur plures tangentes TR vel plura puncta P devenietur semper ad lineas totidem VH, vel PH, a datis punctis V vel P ad umbilicum H ductas, quæ vel æquantur axibus, vel datis longitudinibus SP differunt ab ijsdem; atq; adeo quæ vel æquantur sibi invicem vel datas habent differentias; et inde per Lemma superius datur umbilicus ille alter H. Habitis autem umbilicis unà cum axis longitudine (quæ vel est VH, vel si trajectoria Ellipsis est $PH+SP$, sin Hyperbola $PH-SP$) habetur Trajectoria. Q.E.I.

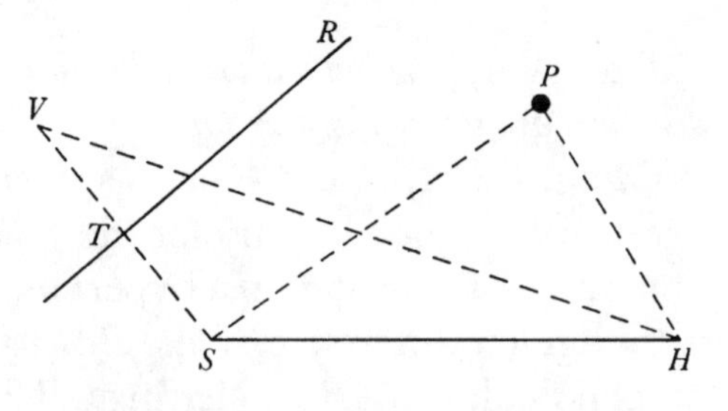

Scholium.

Casus ubi dantur tria puncta sic solvitur expeditiùs. Dentur puncta B, C, D. Junctas BC, CD produc ad E, F ut sit EB ad EC ut SB ad SC et FC ad FD ut SC ad SD. Ad EF ductam et productam demitte normales SG, BH, inq; GS infinite pro-

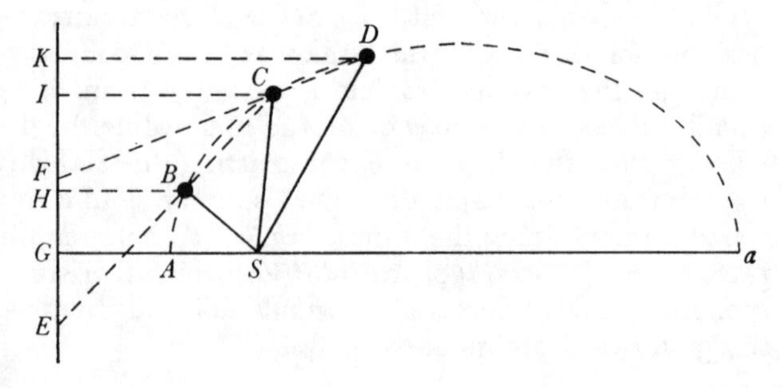

Consequently, the ratios of ZR and ZS to AZ are given, and therefore the ratio of ZR and ZS to each other is given; hence, where RP and SQ are concurrent in T, the point Z will be located in the straight line TZ given in position. By the same method, with the aid of a third hyperbola whose foci are B and C and transverse axis is the difference of the straight lines BZ and CZ, there can be found another straight line in which the point Z is located. Once these two rectilinear loci are had, however, the required point Z is obtained at their intersection. As was to be found.(18)

Case 2. If two of the three lines, AZ and BZ say, are equal, the point Z will be located in the perpendicular bisector of the distance AB, and the other rectilinear locus will be found as above. As was to be found.

Case 3. If all three are equal, the point Z will be located at the centre of the circle passing through the points A, B and C. As was to be found.

This lemmatical problem is solved also by means of Apollonius' book *On Tangencies*, restored by Viète.(19)

Proposition XXI, Problem XIII.(20)

To describe round a given focus a trajectory which shall pass through given points and be tangent to straight lines given in position.

Let there be given the focus S, point P and tangent TR, and let H be the second focus which is to be found. To the tangent let fall the perpendicular ST and extend it to V so that TV be equal to ST, and VH will then be equal to the transverse axis. Join SP and HP, and SP will be the difference between HP and the transverse axis. In this way, should there be given more tangents TR or more points P, we will always arrive at an equal number of lines VH or PH, drawn from the given points V or P to the focus, which either are equal to the axes or differ from them by given lengths SP, and which hence either equal one another or have given differences; and from these by means of the previous lemma the second focus H is given. Once, however, the foci along with the length of the axis (that is, either VH, or $PH+SP$ if the trajectory is an ellipse, but $PH-SP$ if an hyperbola) are had, the trajectory is obtained. As was to be found.

Scholium.

The case where three points are given is solved more speedily in this fashion. Let there be given the points B, C, D. Join BC, CD and extend them to E, F so that EB be to EC as SB to SC, and FC to FD as SC to SD. Draw and extend EF

(18) An exact repeat of the equivalent Lemma XV of §2.

(19) See §2: note (154).

(20) This, along with its following scholium, is an unchanged repeat of Proposition XIX of §2.

ducta cape *GA* ad *AS* et *Ga* ad *aS* ut est *HB* ad *BS*[,] et erit *A* vertex [e]t *Aa* axis transversus trajectoriæ quæsitæ: quæ, perinde ut *GA* minor æqualis vel major(21) fuerit quam *AS*, erit Ellipsis Parabola vel Hyperbola; puncto *a* in primo casu cadente ad eandem partem lineæ *GK* cum puncto *A*, in secundo casu abeuntc in infinitum, in tertio cadente ad contrariam partem lineæ *GK*. Nam si demittantur ad *GF* perpendicula *CI*, *DK*, erit *IC* ad *HB* ut *EC* ad *EB*, hoc est, ut *SC* ad *SB*, et vicissim *IC* ad *SC* ut *HB* ad *SB* seu *GA* ad *SA*. Et simili argumento probabitur esse *K*[*D*] ad *SD* in eadem ratione. Jacent ergo puncta *B*, *C*, *D* in conisectione circa umbilicum *S* ita descripta ut rectæ omnes ab umbilico *S* ad singula sectionis puncta ductæ sint ad perpendicula a punctis ijsdem ad rectam *GK* demissa in data illa ratione.(22)

ARTIC. V.

INVENTIO ORBIUM UBI UMBILICUS NEUTER DATUR.(23)

Lemma XVII.(24)

Si a datæ conicæ sectionis puncto quovis P, ad Trapezij alicujus ABCD in Conica illa sectione inscripti, latera quatuor infinite producta AB, CD, AC, DB, totidem rectæ PQ, PR, PS, PT in datis angulis ducantur, singulæ ad singula: rectangulum ductarum ad

(21) Again (compare §2: note (156)) read in inverse sequence 'major æqualis vel minor' (greater than, equal to or less than). Due adjustment was made in the second edition of the *Principia* ($_2$1713: 65).

(22) In the printed *Principia* ($_1$1687: 69) Newton adjoined a further short paragraph announcing that 'Methodo haud multum dissimili hujus problematis solutionem tradit Clarissimus Geometra *De la Hire*, Conicorum suorum Lib. VIII. Prop. XXV'; see Philippe de La Hire, *Sectiones Conicæ in Novem Libros Distributæ, In quibus quicquid hactenus observatione dignum cùm a veteribus, tùm a recentioribus Geometris traditum est, novis contractisque demonstrationibus explicatur; Multis etiam & exquisitis Propositionibus recèns inventis illustratur* (Paris, 1685): Liber Octavus, Propositio XXV: 191–2, where a parallel construction founded on the focus-directrix property is given of this '*Problema.* Sectionis Conicæ datis tribus punctis *BCD*, & focorum altero...; determinare axem positione, & magnitudine'. We may readily accept Edleston's suggestion (*Correspondence of Sir Isaac Newton and Professor Cotes*, London, 1850: xcvi, note †) that Newton's postscript to his letter to Edmond Halley on 18 October 1686 communicating his 'thanks for your note of De la Hire' (*Correspondence*, **2**, 1960: 454) is proof enough that the initial impulse to incorporate this courteous acknowledgement of La Hire's independent solution to the problem came from Halley, evidently in a lost enclosure to his own preceding letter of 14 October (*ibid.*: 452–3): there is nothing to suggest that Newton himself ever read La Hire's prolix treatise, which is not listed in the catalogue of his library at his death (British Museum MS Add. 25424, printed—not very accurately—by R. de Villamil in his *Newton: the man* (London, 1931): 62–103). It is clear that neither Halley nor Newton were aware, at this time anyway, of James Gregory's still earlier equivalent construction (which was published only in 1831; see §2: note (157)).

In the manuscript Newton has deleted two immediately following lemmas (on ff. $53^r/54^r$) which resolve in a lengthy classical manner the problem of drawing a straight line through

and to it let fall the normals *SG*, *BH*, then in *GS* indefinitely produced take *GA* to *AS* and *Ga* to *aS* as *HB* is to *BS*, and *A* will be the vertex and *Aa* the transverse axis of the trajectory required. According as *GA* is less than, equal to or greater than(21) *AS* this will be an ellipse, parabola or hyperbola, with the point *a* in the first case falling on the same side of the line *GK* as the point *A*, in the second case going off to infinity, and in the third falling on the opposing side of the line *GK*. For, if perpendiculars *CI*, *DK* be let fall to *GF*, there will be *IC* to *HB* as *EC* to *EB*, that is, as *SC* to *SB*, and so *alternando IC* to *SC* as *HB* to *SB* or *GA* to *SA*. And by a similar argument it will be proved that *KD* is to *SD* in the same ratio. The points *B*, *C* and *D* lie, therefore, in a conic described round the focus *S* such that all straight lines drawn from the focus *S* to the individual points of the section shall be to the perpendiculars let fall from the same points to the straight line *GK* in that given ratio.(22)

Article V.

The finding of orbits when neither focus is given.(23)

Lemma XVII.(24)

If from any point P of a given conic to the four infinitely extended sides AB, CD, AC, BD of some quadrilateral inscribed in that conic an equal number of straight lines PQ, PR, PS, PT are drawn at given angles, each to a separate one: the product $PQ \times PR$ *of those*

---

a given point to cut three lines given in position so that the intercepts shall be in a predetermined ratio: these we reproduce below in Appendix 1. The problem constructed—a generalisation of that solved algebraically by Newton in 'Prob: 16' of his earlier Lucasian lectures (see v: 210–12)—has an evident application in the computation of cometary paths in the simplified Wrennian hypothesis (see *ibid.*: 210, note (251)) that these are uniformly traversed straight lines, and the lemmas were cancelled probably when Newton chose to incorporate its 'Cas. 1' in the initial version (ULC. Add. 3990) of the complementary 'De motu Corporum Liber secundus'; see Appendix 1: notes (1) and (7).

(23) Evidently instructed to incorporate in sequel the opening of the unfinished tract whose existing text is reproduced in iv: 282–320, Humphrey Newton here initially transcribed its title 'Solutio Problematis Veterum/de Loco solido' (Solution of the Ancients' problem of the solid locus) and likewise proceeded to head the following theorem with an unnumbered '*Prop.* ', but soon—doubtless so ordered by Newton—cancelled the title and altered the latter to be a '*Lemma*' which was subsequently given a number by Newton himself. (The manuscript text of the similarly copied Lemmas XVIII–XXI is free of such incautious fidelity to its original.) The revised sub-heading we reproduce is a still later interlineation by Newton himself (see note (2)). With the addition of the ameliorations therein made at this time (see iv: 285, note (3)) and some inessential further recasting, the following Lemmas XVII-XXI, Proposition XXII and 'Cas. 1' of Proposition XXIII effectively repeat corresponding portions of Newton's earlier Propositions 1–7 (iv: 282–302). The succeeding 'Cas. 2' of Proposition XXIII, Proposition XXIV and—with the way prepared by the general projective 'transmutation' introduced in Lemma XXII, the now familiar rectilinear locus inserted in Lemma XXIII

opposita duo latera $PQ \times PR$, erit ad rectangulum ductarum ad alia duo latera opposita $PS \times PT$ in data ratione.

Cas. 1. Ponamus imprimis lineas ad opposita latera ductas, parallelas esse alterutri reliquorum laterum, puta *PQ* ct *PR* lateri *AC* et *PS* ac *PT* lateri *AB*. Sintq; insuper latera duo ex oppositis, puta *AC* & *BD* parallela. Et recta quæ bisecat parallela illa latera erit una ex diametris Conicæ sectionis, et bisecabit etiam *RQ*. Sit *O* punctum in quo *RQ* bisecatur, et erit *PO* ordinatim applicata ad diametrum illam. Produc *PO* ad *K* ut sit *OK* æqualis *PO*, et erit *OK* ordinatim applicata ad contrarias partes diametri.

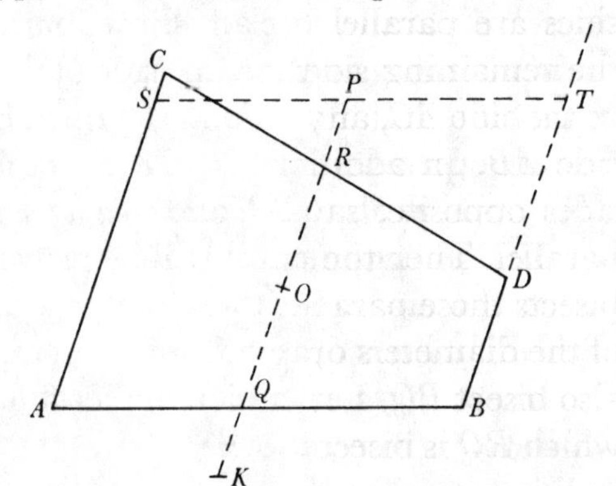

Cum igitur puncta *A*, *B*, *P* et *K* sint ad conicam sectionem, et *PK* secet *AB* in dato angulo, erit (per Prop. 17 et 18 lib. 3 Apollonij(25)) rectangulum *PQK* ad rectangulum *AQB* in data ratione. Sed *QK* et *PR* æquales sunt, utpote æqualium *OK*, *OP* et *OQ*, *OR* differentiæ, et inde etiam rectangula *PQK* et $PQ \times PR$ æqualia sunt, atq; adeo rectangulum $PQ \times PR$ est ad rectangulum *AQB*, hoc est ad rectangulum $PS \times PT$ in data ratione. Q.E.D.

Cas. 2. Ponamus jam trapezij latera opposita *AC* et *BD* non esse parallela. Age *Bd* parallelam *AC* et occurrentem tum rectæ *ST* in *t*, tum Conicæ sectioni in *d*. Junge *Cd* secantem *PQ* in *r*, et ipsi *PQ* parallelam age *DM* secantem *Cd* in *M* et *AB* in *N*. Jam ob similia triangula *BTt*, *DBN* est *Bt* seu *PQ* ad *Tt* ut *DN* ad *NB*. Sic et *Rr* est ad *AQ* seu *PS* ut *DM* ad *AN*. Ergo ducendo antecedentes in antecedentes et consequentes in consequentes, ut rectangulum *PQ* in *Rr* est ad rect-

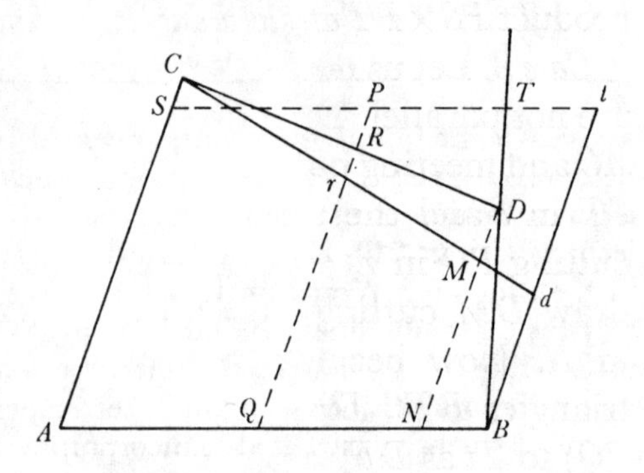

and the Apollonian conic-envelopes demonstrated in Lemmas XXIV and XXV—Propositions XXV–XXVII, which complete the possibilities of constructing conic 'trajectories' to pass through given points and to touch given lines, are here wholly new. Equally novel are the concluding Propositions XXVIII and XXIX which construct conics of given size and species (and, as a particular case, a straight line) to lie proportionally along given 'sighting' lines. It will be obvious that Newton's overriding interest in exhaustively exploring the geometrical subtleties of the problems whose ingenious solutions are here set down with elegance and economy far transcends the immediate, practical aim of effectively and accurately constructing planetary and cometary orbits which is their professed *raison d'être*. Divorced as they are from the central dynamical themes expounded in the surrounding Articles of this mature treatise 'De motu Corporum', these theorems—and to some smaller degree those in the preceding

drawn to two opposite sides will be to the product $PS \times PT$ of those drawn to the other two sides in a given ratio.

Case 1. Let us suppose in the first instance that the lines drawn to opposite sides are parallel to one or other of the remaining sides, say PQ and PR to the side AC, and PS and PT to the side AB. In addition, let two of the sides opposite, say AC and BD, be parallel. Then the straight line which bisects those parallel sides will be one of the diameters of the conic and will also bisect RQ. Let O be the point in which RQ is bisected, and PO will be ordinate to that diameter. Produce PO to K so that OK is equal to PO, and OK will be the ordinate on the opposite side of the diameter. Since, then, the points A, B, P and K are on the conic and PK intersects AB at a given angle, therefore (by Apollonius' [*Conics*](25) III, 17 and 18) the product $PQ \times QK$ will be to the product $AQ \times QB$ in a given ratio. But QK and PR are equal, being namely the differences of the equals OK, OP and OQ, OR, and thence also the products $PQ \times QK$ and $PQ \times PR$ are equal; consequently the product $PQ \times PR$ is to the product $AQ \times QB$, that is, to the product $PS \times PT$ in a given ratio. As was to be proved.

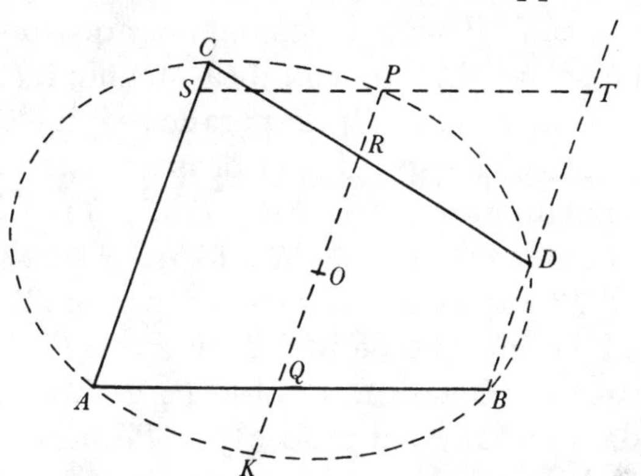

Case 2. Let us now suppose that the quadrilateral's opposite sides AC and BD are not parallel. Draw Bd parallel to AC and meeting both the straight line ST in t and the conic in d. Join Cd cutting PQ in r, and parallel to PQ draw DM cutting Cd in M and AB in N. Now because of the similar triangles BTt, DBN there is Bt (or PQ) to Tt as DN to NB. So also Rr is to AQ (or PS) as DM to AN. Therefore, on multiplying the prior and posterior members of the ratios into one another, as the product $PQ \times Rr$ is to the product $Tt \times PS$, so is the product

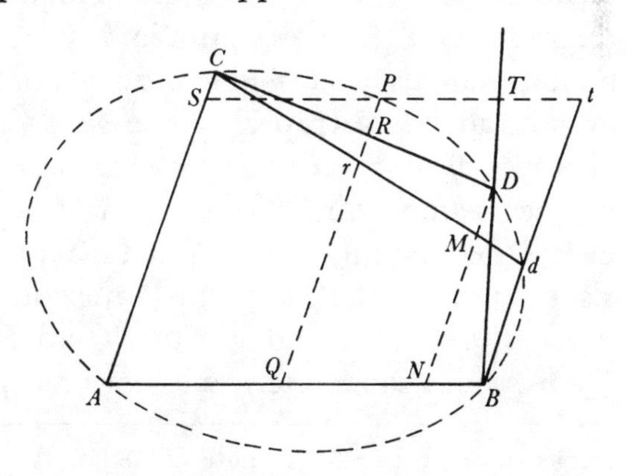

Article IV also—serve, for all their dazzling display of pure expertise, largely to unbalance and interrupt the continuity of its main development. Though Newton permitted them to pass unchanged and unabridged into the corresponding Sections IV and V of the first book of the printed *Principia* ($_1$1687: 70–103) and made no recorded effort to remove them from its later published editions, he came to appreciate the essential incongruity of their present location and in the radical restructuring of the *Principia* which he began about 1692 but never fully

angulum Tt in PS, ita rectangulum NDM est ad rectangulum ANB et (per Cas. 1) ita rectangulum QPr est ad rectangulum SPt, ac divisim ita rectangulum QPR est ad rectangulum $PS \times PT$. Q.E.D.

Cas. 3. Ponamus deniꝗ lineas quatuor PQ, PR, PS, PT non esse parallelas lateribus AC, AB, sed ad ea utcunꝗ inclinatas. Earum vice age Pq, Pr parallelas ipsi AC, et Ps, Pt parallelas ipsi AB, et propter datos angulos triangulorum PQq, PRr, PSs, PTt dabuntur rationes PQ ad Pq, PR ad Pr, PS ad Ps et PT ad Pt atꝗ adeo rationes compositæ PQ in PR ad Pq in Pr et PS in PT ad Ps in Pt. Sed per superius demonstrata ratio Pq in Pr ad Ps in Pt data est: Ergo et ratio PQ in PR ad PS in PT. Q.E.D.

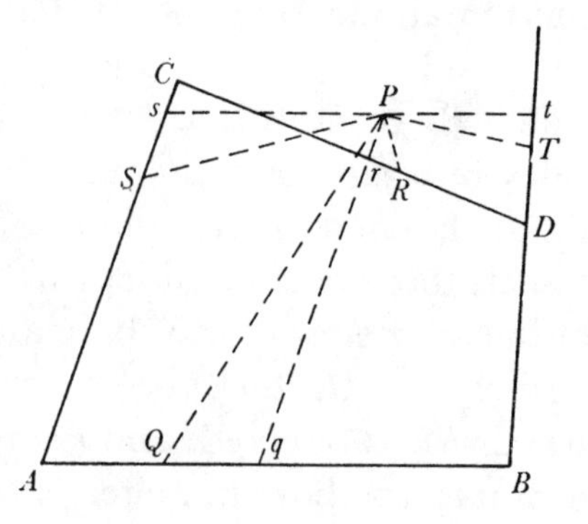

Lemma XVIII.[26]

Iisdem positis,[27] *si rectangulum ductarum ad opposita duo latera Trapezij* $PQ \times PR$ *sit ad rectangulum ductarum ad reliqua duo latera* $PS \times PT$ *in data ratione: punctum P a quo lineæ ducuntur tanget Conicam sectionem circa Trapezium descriptam.*

Per puncta A, B, C, D et aliquod infinitorum punctorum P, puta p concipe Conicam sectionem describi: dico punctum P hanc semper tangere. Si negas, junge AP secantem hanc Conicam sectionem alibi quam in P si fieri potest, puta in π. Ergo si ab his punctis p et π ducantur in datis angulis ad latera trapezij rectæ pq, pr, ps, pt et $\pi\chi$, $\pi\rho$, $\pi\sigma$, $\pi\tau$, erit ut $\pi\chi \times \pi\rho$ ad $\pi\sigma \times \pi\tau$ ita (per Lemma XVII[28]) $pq \times pr$ ad $ps \times pt$, et ita (per hypoth) $PQ \times PR$ ad $PS \times PT$. Est et, propter similitudinem Trapeziorum $\pi\chi A\sigma$, $PQAS$, ut $\pi\chi$ ad $\pi\sigma$ ita PQ ad PS. Quare applicando terminos prioris propo[r]tionis[29] ad terminos corre-

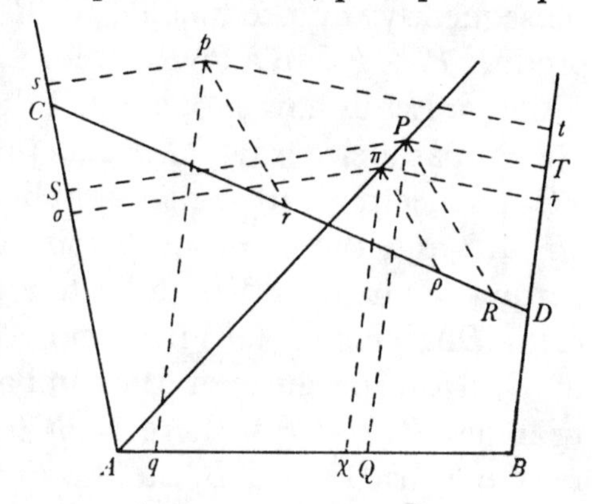

carried through (see 3, §2: note (1) below) he planned to transfer both groups of theorems *en bloc* into a separate appended 'Tractatus Geometricus' where they would occupy a more suitable place.

(24) A straightforward repeat of Proposition 1 of the earlier 'Solutio' (IV: 282–6).

(25) We have already noticed (IV: 285, note (4)) that a clarifying 'Conic.' inserted by Newton at this point in the 'Solutio' was carelessly overlooked by Humphrey Newton in making his transcription: the omission was afterwards rectified by adding an equivalent 'Conicorum' in the second edition of the *Principia* ($_2$1713: 66). In sequel Humphrey less explicably inverted the text he copied to read 'rectangulum AQB ad rectangulum PQK', but Newton has made appropriate correction in the manuscript.

$DN \times DM$ to the product $AN \times NB$ and so (by Case 1) is the product $PQ \times Pr$ to the product $PS \times Pt$, and *dividendo* so is the product $PQ \times PR$ to the product $PS \times PT$. As was to be proved.

Case 3. Let us suppose, finally, that the four lines PQ, PR, PS and PT are not parallel to the sides AC, AB but inclined to one another in any manner. In their stead draw Pq, Pr parallel to AC and Ps, Pt parallel to AB, and then, on account of the angles of the triangles PQq, PRr, PSs and PTt being given, the ratios of PQ to Pq, PR to Pr, PS to Ps and PT to Pt will be given, and consequently the compound ratios $PQ \times PR$ to $Pq \times Pr$ and $PS \times PT$ to $Ps \times Pt$ also. But, by what has previously been demonstrated, the ratio of $Pq \times Pr$ to $Ps \times Pt$ is given: so also, therefore, is that of $PQ \times PR$ to $PS \times PT$. As was to be proved.

Lemma XVIII.(26)

With the same suppositions,(27) *if the product $PQ \times PR$ of lines drawn to two opposite sides of a quadrilateral be to the product $PS \times PT$ of those drawn to the remaining two sides in a given ratio: the point P from which the lines are drawn will lie in a conic described round the quadrilateral.*

Through the points A, B, C, D and some one, p say, of the infinity of points P conceive a conic to be described: I assert that the point P lies always in this. Should you deny it, join AP cutting this conic elsewhere, if possible, than in P, say in π. In consequence, if from these points p and π the straight lines pq, pr, ps, pt and $\pi\chi$, $\pi\rho$, $\pi\sigma$, $\pi\tau$ be drawn at given angles to the sides of the quadrilateral, there will be as $\pi\chi \times \pi\rho$ to $\pi\sigma \times \pi\tau$ so (by Lemma XVII(28)) $pq \times pr$ to $ps \times pt$, and so (by hypothesis) $PQ \times PR$ to $PS \times PT$. Further, because of the similarity of the quadrilaterals $\pi\chi A\sigma$ and $PQAS$, as $\pi\chi$ to $\pi\sigma$ so PQ to PS. In consequence, by dividing the terms of the previous proportion by the corresponding terms of the

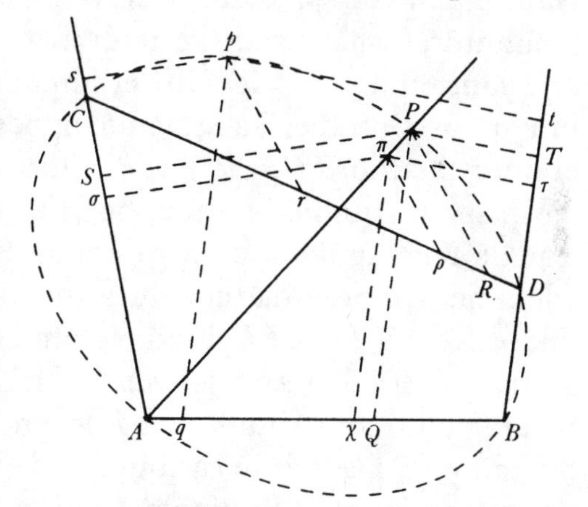

(26) A minimally altered repeat of Proposition 2 of the 'Solutio' (IV: 286–8); the following Corollary and Scholium are, however, new.

(27) Newton—or Humphrey Newton, at least—here returns to his initial opening, abandoning (as we remarked in IV: 287, note (9)) his earlier revised preference for '*Et contra*' (*Conversely*).

(28) Following his original (see IV: 288) too faithfully, Humphrey Newton here initially transcribed 'per Prop. 1' (by Proposition 1).

(29) Humphrey Newton carelessly mis-transcribed this as 'propositionis', a nonsensical variant which passed unnoticed into the *Principia's* first edition ($_1$1687: 72) to be caught in the second ($_2$1713: 68).

spondentes hujus, erit $\pi\rho$ ad $\pi\tau$ ut PR ad PT. Ergo Trapezia æquiangula $D\rho\pi\tau$, $DRPT$ similia sunt & eorum diagonales $D\pi$, DP propterea coincidunt. Incidit itaq; π in intersectionem rectarum AP, DP adeoq; coincidit cum puncto P. Quare punctum P, ubicunq; sumatur, incidit in assignatam Conicam sectionem. Q.E.D.

Corol. Hinc si rectæ tres PQ, PR, PS a puncto communi P ad alias totidem positione datas rectas AB, CD, AC singulæ ad singulas in datis angulis ducantur, sitq; rectangulum sub duabus ductis $PQ \times PR$ ad quadratum terti[æ] $PS^{qu.}$ in data ratione: punctum P a quibus rectæ ducuntur locabitur in sectione Conica quæ tangit lineas AB, CD in A et C et contra.(30) Nam coeat linea BD cum linea AC manente positione trium AB, CD, AC; dein coeat etiam linea PT cum linea PS: et rectangulum $PS \times PT$ evadet $PS^{quad.}$ rectæq; AB, CD quæ Curvam in punctis A et B, C et D secabant, jam Curvam in punctis illis coeuntibus non amplius secare possunt sed tantùm tangent.

Scholium.

Nomen Conicæ sectionis in hoc Lemmate latè sumitur, ita ut sectio tam rectilinea per verticem Coni transiens quam circularis basi parallela includatur. Nam si punctum p incidit in rectam quâ quævis ex punctis quatuor A, B, C, D junguntur, Conica sectio vertetur in geminas rectas(31) quarum una est recta illa in quam punctum p incidit, et altera recta qua alia duo ex punctis quatuor junguntur. Si trapezij anguli duo oppositi simul sumpti æquentur duobus rectis et lineæ quatuor PQ, PR, PS, PT ducantur ad latera ejus vel perpendiculariter vel in angulis quibusvis æqualibus, sitq; rectangulum sub duabus ductis $PQ \times PR$ æquale rectangulo sub alijs duabus $PS \times PT$, sectio Conica evadet circulus. Idem fiet si lineæ quatuor ducantur in angulis quibusvis et rectangulum sub duabus ductis $PQ \times PR$ sit ad rectangulum sub alijs duabus $PS \times PT$ ut rectangulum sub sinubus angulorum S, T in quibus duæ ultimæ PS, PT ducuntur ad rectangulum sub sinubus angulorum Q, R in quibus duæ primæ PQ, PR ducuntur.(32) Cæteris in casibus locus puncti P erit aliqua trium figurarum quæ vulgo nominantur Sectiones Conicæ. Vice autem Trapezij $ABCD$ substitui potest quadrilaterum cujus latera duo opposita se mutuò ad instar diagonalium decussant. Sed et e punctis quatuor A, B, C, D possunt unum vel duo abire in infinitum eoq; pacto latera figuræ quæ ad puncta illa convergunt, evadere parallela: quo in casu sectio conica transibit per cætera puncta, et in plagas parallelarum abibit in infinitum.

(30) The classical 3-line locus, here presented as the particular case of the 4-line locus in which two of the base lines coincide. As we have seen (IV: 276, note (7)), Apollonius derived this locus (essentially *Conics* III, 54–6) probably as a prelude to identifying the more general 4-line locus as a conic.

(31) A line-pair in modern terminology.

present one, there will be $\pi\rho$ to $\pi\tau$ as PR to PT. Therefore the equiangular quadrilaterals $D\rho\pi\tau$, $DRPT$ are similar and their diagonals $D\pi$, DP accordingly coincide. So π falls at the intersection of the straight lines AP, DP and hence coincides with the point P. In consequence the point P, wherever it be taken, falls in the allotted conic. As was to be proved.

Corollary. From this, if three straight lines PQ, PR, PS be drawn at given angles to an equal number of others AB, CD, AC given in position, each to a separate one, and the product $PQ \times PR$ of two of those drawn be to the square PS^2 of the third in a given ratio: the point P from which the straight lines are drawn will be in a conic which touches the lines AB, CD at A and C; and conversely so.(30) For let the line BD coincide with the line AC, the position of the three lines AB, CD and AC staying the same; then let the line PT also coincide with the line PS: and the product $PS \times PT$ will come to be PS^2, while the straight lines AB, CD which intersect the curve in the points A and B, C and D can now, with those points coinciding, no longer cut the curve in them but merely touch it.

Scholium.

The appellation 'conic' is taken in this lemma in a broad sense, and shall accordingly include both a rectilinear section passing through the vertex of the cone and a circular one parallel to the base. For, if the point p chances to be in a straight line joining any (pair) of the four points A, B, C and D, the conic will turn into twin straight lines,(31) one of which is the line in which the point p falls, the second that joining the other two of the four points. If two opposite angles of the quadrilateral, taken together, be equal to two right angles and the four lines PQ, PR, PS, PT be drawn to its sides either perpendicularly or at any equal angles, while the product $PQ \times PR$ of two of the drawn lines be equal to the product $PS \times PT$ of the other two, then the conic will prove to be a circle. The same will happen if the four lines be drawn at any arbitrary angles and the product $PQ \times PR$ of two of the lines drawn be to the product $PS \times PT$ of the other two as the product of the sines of the angles $\hat{S}$, $\hat{T}$ at which the two last PS, PT are drawn to the product of the sines of the angles $\hat{Q}$, $\hat{R}$ at which the first two PQ, PR are drawn.(32) In the rest of the cases the locus of the point P will be some one of the three figures which are generally called conic sections. Instead of the quadrilateral $ABCD$, however, there can be substituted one whose two opposite sides cross one another in the fashion of diagonals. Also, again, of the four points A, B, C, D one or two can go off to infinity and in that manner the sides of the figure which converge on those points prove to be parallel: in this case the conic will pass through the remaining points and pass off to infinity in the direction of the parallels.

(32) Compare IV: 236, note (22). In explicit proof, if from the point P perpendiculars

$$PQ = PA.\sin P\widehat{A}B,\quad PR = PD.\sin P\widehat{D}C,\quad PS = PA.\sin P\widehat{A}C$$

Lemma XIX.[33]

Invenire punctum P a quo si rectæ quatuor PQ, PR, PS, PT, ad alias totidem positione datas rectas AB, CD, AC, BD singulæ ad singulas in datis angulis ducantur, rectangulum sub duabus ductis $PQ\times PR$ sit ad rectangulum sub alijs duabus $PS\times PT$ in data ratione.

Lineæ *AB*, *CD* ad quas rectæ duæ *PQ*, *PR* unum rectangulorum continentes ducuntur, conveniant cum alijs duabus positione datis lineis in punctis *A*, *B*, *C*, *D*. Ab eorum aliquo *A* age rectam quamlibet *AH* in qua velis punctum *P* reperiri. Secet ea lineas oppositas *BD*, *CD*, nimirum *BD* in *H* et *CD* in *I*, et ob datos omnes angulos figuræ dabuntur rationes *PQ* ad *PA* et *PA* ad *PS*, adeoq; ratio *PQ* ad *PS*. Auferendo hanc a data ratione $PQ\times PR$ ad $PS\times PT$ dabitur ratio *PR* ad *PT* et addendo datas rationes *PI* ad *PR* et *PT* ad *PH* dabitur ratio *PI* ad *PH* atq; adeò punctum *P*. Q.E.I.

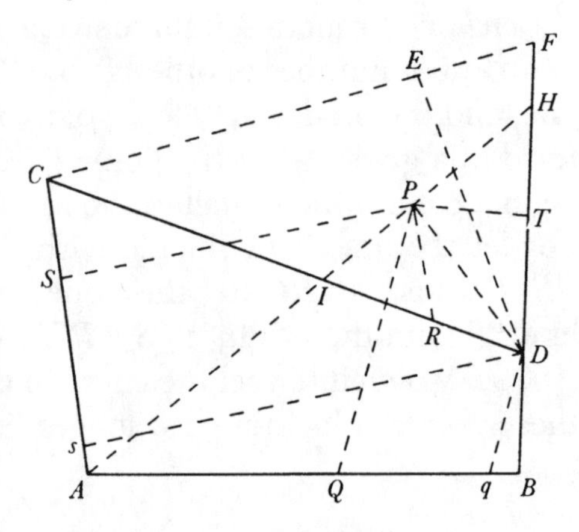

Corol. 1. Hinc etiam ad Loci punctorum infinitorum *P*, punctum quodvis *D* tangens duci potest. Nam chorda *PD* ubi puncta *P* ac *D* conveniunt, hoc est, ubi *AH* ducitur per punctum *D*, tangens evadit. Quo in casu, ultima ratio evanescentium *IP* et *PH* invenietur ut supra.[34] Ipsi igitur *AD* duc parallelam *CF* occurrentem *BD* in *F* et in ea ultima ratione sectam in *E*, et *DE* tangens erit, propterea quod *CF* et evanescens *IH* parallelæ sunt et in *E* et *P* similiter sectæ.

---

and $PT = PD.\sin\widehat{PDB}$ are let fall to the sides *AB*, *DC*, *CA* and *BD* respectively of the cyclic quadrilateral *ABCD*, then it follows that

$PQ\times PR/PS\times PT$

$= \sin\widehat{PAB}\times\sin\widehat{PDC}/\sin\widehat{PAC}\times\sin\widehat{PDB}$:

hence, if *P* itself is on the circle *ABCD*, so that $\widehat{PAB}$ = (exterior angle) $\widehat{PDB}$ and $\widehat{PDC}=\widehat{PAC}$, at once $PQ\times PR/PS\times PT=1$; conversely, if $PQ\times PR = PS\times PT$, then $\sin\widehat{PAB}\times\sin\widehat{PDC} = \sin\widehat{PAC}\times\sin\widehat{PDB}$ and so, since $\widehat{PAB}+\widehat{PAC} = \widehat{BAC}=\widehat{CDT}$ = (exterior angle)$\widehat{PDB}+\widehat{PDC}$ and therefore $\cos(\widehat{PAB}-\widehat{PDC}) = \cos(\widehat{PDB}-\widehat{PAC})$, also $\cos(\widehat{PAB}+\widehat{PDC}) = \cos(\widehat{PDB}+\widehat{PAC})$, whence $\widehat{PAB}$ = (exterior) $\widehat{PDB}$ and $\widehat{PDC}=\widehat{PAC}$, or *P* is on the circle circumscribing *ABDC*.

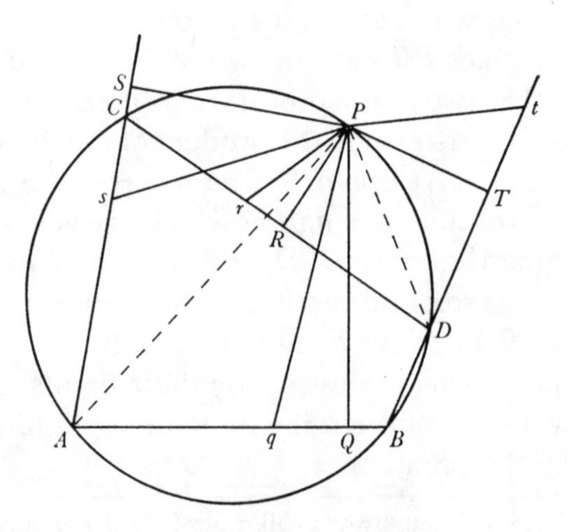

Lemma XIX.(33)

To find a point P such that, if four straight lines PQ, PR, PS, PT be drawn from it at given angles to an equal number of other straight lines AB, CD, AC, BD given in position, one to each separately, the product $PQ \times PR$ of two of those drawn shall be to the product $PS \times PT$ of the other two in a given ratio.

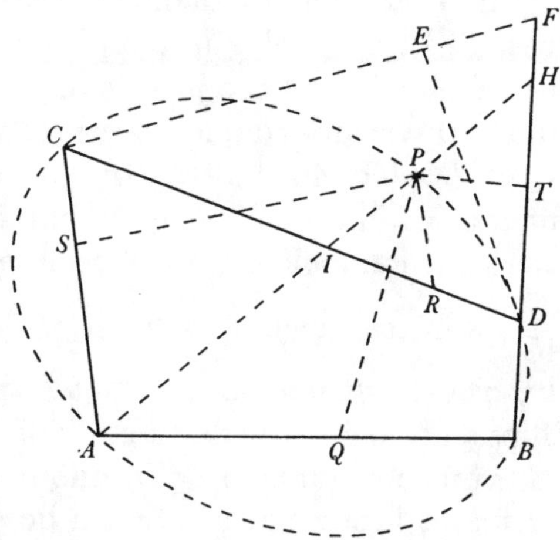

Let the lines *AB*, *CD* to which the two straight lines *PQ*, *PR* containing one of the products are drawn meet the other two lines given in position in the points *A*, *B*, *C*, *D*. From some one, *A*, of these draw any straight line *AH* in which you wish the point *P* to be found. Let it cut the opposite lines *BD*, *CD*—namely, *BD* in *H* and *CD* in *I*—and then, because all angles in the figure are given, the ratios of *PQ* to *PA* and of *PA* to *PS* will be given, and hence the ratio of *PQ* to *PS*. On eliminating this from the given ratio $PQ \times PR$ to $PS \times PT$, there will be given the ratio of *PR* to *PT*, and by adjoining the given ratios *PI* to *PR* and *PT* to *PH* the ratio of *PI* to *PH* and hence the point *P* will be given. As was to be found.

Corollary 1. Hence also at any point *D* of the locus of the infinity of points *P* a tangent can be drawn. For when the points *P* and *D* are coincident, that is, when *AH* is drawn through the point *D*, the chord *PD* comes to be tangent. In this case the ultimate ratio of the vanishing lines *IP* and *PH* will be ascertained as above.(34) Parallel to *AD*, therefore, draw *CF* meeting *BD* in *F* and cut at *E* in that ultimate ratio, and *DE* will be tangent for the reason that *CF* and the vanishing line *IH* are parallel and similarly cut at *E* and *P*.

Where, more generally, *Pq*, *Pr*, *Ps* and *Pt* are let fall at given angles $\hat{q}$, $\hat{r}$, $\hat{s}$ and $\hat{t}$ to *AB*, *CD*, *CA* and *BD*, and $PQ \times PR = PS \times PT$, it is immediate that

$$Pq \times Pr/Ps \times Pt = (PS/Ps) \times (PT/Pt)/(PQ/Pq) \times (PR/Pr) = \sin\hat{s} \times \sin\hat{t}/\sin\hat{q} \times \sin\hat{r}.$$

(33) This lemma and its following Corollary 1 are a minimal revise of Proposition 3 and its 'Corol.' on IV: 288–90; likewise, Corollary 2 below repeats Proposition 4 on IV: 290 with only trivial alteration.

(34) In the accompanying figure (here copied from *Principia*, $_1$1687: 74) observe that Newton has inserted in his original diagram (see IV: 288) the 'ultimate' positions *Dq* and *Ds* of *PQ* and *PS* as *P* comes to coincide with *D*, evidently intending at some stage explicitly to define the direction of the line *DE* by means of $\lim_{P\to D} (PR/PT) = k.Ds/Dq$ where $PQ \times PR/PS \times PT = k$, constant.

These intrinsically superfluous added lines endured into the *Principia's* second edition ($_2$1713: 69/70) to be finally suppressed by Pemberton in the third ($_3$1726:77).

Corol. 2. Hinc etiam Locus punctorum omnium P definiri potest. Per quodvis punctorum A, B, C, D, puta A, duc Loci tangentem AE et per aliud quodvis B duc tangenti parallelam BF occurrentem Loco in F. Invenietur autem punctum F per Lemma superius. Biseca BF in G, et acta AG diameter erit ad quam BG et FG ordinatim applicantur. Hæc AG occurrat Loco in H, et erit AH latus transversum ad quod latus rectum est ut BG^q ad AGH. Si AG nullibi occurrit Loco, linea AH existente infinita, Locus erit Parabola et latus rectum ejus $\frac{BG^q}{AG}$. Sin ea alicubi occurrit, Locus

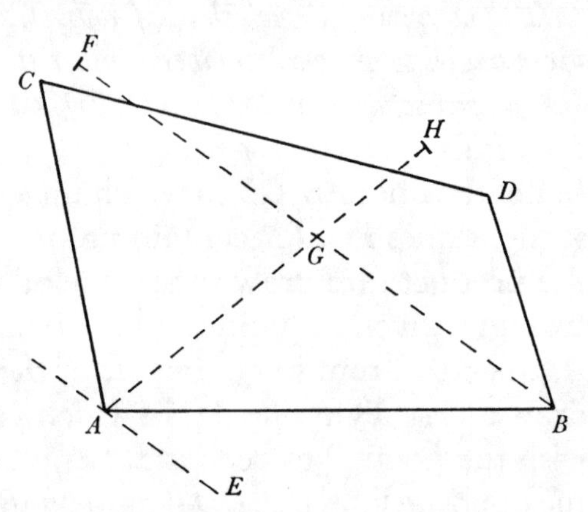

Hyperbola erit ubi puncta A et H sita sunt ad easdem partes ipsius G, et Ellipsis ubi G intermedium est, nisi forte angulus AGB rectus sit et insuper BG^q æquale rectangulo AGH, quo in casu circulus habebitur.

Atq; ita Problematis Veterum de quatuor lineis ab Euclide incœpti et ab Apollonio continuati non calculus sed compositio Geometrica qualem Veteres quærebant, in hoc Corollario exhibetur.(35)

Lemma XX.(36)

Si Parallelogrammum quodvis ASPQ angulis duobus oppositis A et P tangit sectionem quamvis conicam in punctis A et P, et lateribus unius angulorum illorum infinitè productis AQ, AS occurrit eidem sectioni conicæ in B et C; a punctis autem occursuum B et C ad quintum quodvis sectionis conicæ punctum D agantur rectæ BD, CD occurrentes alteris duobus infinitè productis parallelogrammi lateribus PS, PQ in T et R: erunt semper abscissæ laterum partes PR et PT ad invicem in data ratione. Et contra, si partes illæ abscissæ sunt ad invicem in data ratione, punctū D tanget sectionem Conicam per ſuncta quatuor A, B, P, C transeuntem.

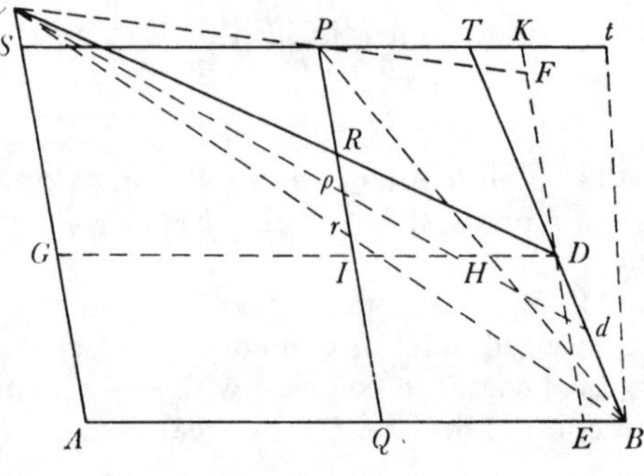

Cas. 1. Jungantur BP, CP et a puncto D agantur rectæ duæ DG, DE quarum prior DG ipsi AB parallela sit et occurrat PB, PQ, CA in [H], I, G; altera DE parallela sit ipsi AC et occurrat PC, PS, AB, in F, K, E. Et erit (per Lem. XVII) rectangulum $DE \times DF$ ad rectangulum $DG \times DH$ in ratione data. Sed est PQ ad DE seu IQ ut PB ad HB, adeoq; ut PT

(35) This paragraph is a late insertion by Newton himself in the manuscript. Clearly, having suppressed his original title of the 'Solutio' (see note (23) above) he now feels that he must

Corollary 2. Hence also the locus of the totality of points P can be defined. Through any of the points A, B, C or D, say A, draw the tangent AE of the locus and through any other point B draw BF parallel to the tangent, meeting the locus in F. The point F, however, will be found by means of the previous lemma. Bisect BF in G and when AG is drawn it will be a diameter to which BG and FG are ordinate. Let this line AG meet the locus in H and AH will be the transverse diameter whose *latus rectum* is to it as BG^2 to $AG \times GH$. If AG nowhere meets the locus, with the line AH proving to be infinite, the locus will be a parabola and BG^2/AG its *latus rectum*. But if it does meet it somewhere, the locus will be a hyperbola when the points A and H are situated on the same side of G and an ellipse when G intervenes between them, unless perchance the angle $A\widehat{G}B$ be right and in addition BG^2 is equal to the rectangle $AG \times GH$, in which case a circle will be obtained.

And in this way of the Ancients' problem of four lines, begun by Euclid and continued by Apollonius, we furnish in this corollary not an (analytical) computation but a geometrical synthesis of the sort the Ancients required.(35)

Lemma XX.(36)

If any parallelogram ASPQ at two opposite corners A and P touches any conic in the points A and P and, with the sides AQ, AS of one of those corners indefinitely extended, meets the same conic in B and C; and if, again, from the meeting-points B and C from any fifth point D on the conic there be drawn two straight lines BD, CD meeting the parallelogram's other two indefinitely extended sides PS, PQ in T and R: the portions PR and PT of the sides cut off will be to one another in a given ratio. And conversely, if those portions cut off are to one another in a given ratio, the point D will lie in a conic passing through the four points A, B, P and C.

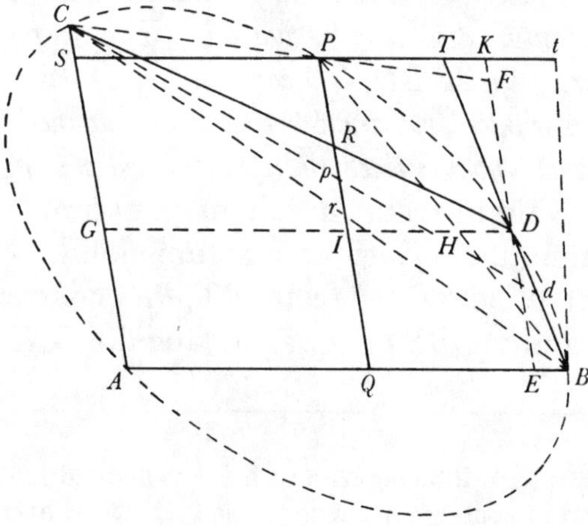

Case 1. Join BP, CP and from the point D draw two straight lines DG, DE, the first of which, DG, shall be parallel to AB and meet PB, PQ, CA in H, I and G, while the other, DE, shall be parallel to AC and meet PC, PS, AB in F, K and E. Then (by Lemma XVII) the product $DE \times DF$ will be to the product $DG \times DH$ in a given ratio. But PQ is to DE (or IQ) as PB to HB, and hence as PT to DH, and so *alternando* PQ to PT as DE to

otherwise call to the reader's attention the classical end to which the preceding lemmas were initially contrived. As we before observed (IV: 291, note (17)), the 'calculus' which he here spurns is Descartes' algebraic reduction of the 4-line locus (in the first two books of his

ad *DH*, et vicissim *PQ* ad *PT* ut *DE* ad *DH*. Est et *PR* ad *DF* ut *RC* ad *DC*, adeoq; ut *IG* vel *PS* ad *DG*, et vicissim *PR* ad *PS* ut *DF* ad *DG*, et conjunctis rationibus fit rectangulum $PQ \times PR$ ad rectangulum $PS \times PT$ ut rectangulum $DE \times DF$ ad rectangulum $DG \times DH$ atq; adeo in data ratione. Sed dantur *PQ* et *PS* et proptcrca ratio *PR* ad *PT* datur. Q.E.D.

Cas. 2. Quod si *PR* et *PT* ponantur in data ratione ad invicem, tunc simili ratiocinio regrediendo sequetur esse rectangulum $DE \times DF$ ad rectangulum $DG \times DH$ in ratione data, adeoq; punctum *D* (per Lemma XVIII) contingere Conicam sectionem transeuntem per puncta *A*, *B*, *P*, *C*. Q.E.D.[37]

Corol. 1. Hinc si agatur *BC* secans *PQ* in *r*, et in *PT* capiatur *Pt* in ratione ad *Pr* quam habet *PT* ad *PR*, erit *Bt* Tangens Conicæ sectionis ad punctum *B*. Nam concipe punctum *D* coire cum puncto *B* ita ut, chorda *BD* evanescente, *BT* Tangens evadat; et *CD* ac *BT* coincident cum *CB* et *Bt*.

Corol. 2. Et vice versa si *Bt* sit Tangens, et ad quodvis Conicæ sectionis punctum *D* conveniant *BD*, *CD*: erit *PR* ad *PT* ut *Pr* ad *Pt*. Et contra, si sit *PR* ad *PT* ut *Pr* ad *Pt* convenient *BD*, *CD* ad Conicæ sectionis punctum aliquod *D*.[38]

Corol. 3. Conica sectio non secat Conicam sectionem in punctis pluribus quam quatuor. Nam, si fieri potest, transeant duæ Conicæ sectiones per quinq; puncta *A*, *B*, *C*, *D*, *P*, easq; secet recta *BD* in punctis *D*, *d*, & ipsam *PQ* secet recta *Cd* in ρ. Ergo *PR* est ad *PT* ut $P\rho$ ad *PT*, hoc est *PR* et $P\rho$ sibi invicem æquantur, contra Hypothesin.

Lemma XXI.[39]

Si rectæ duæ mobiles & infinitæ BM, CM per data puncta B, C ceu polos ductæ, concursu suo M describant tertiam positione datam rectam MN, et aliæ infinitæ rectæ BD, CD cum prioribus duabus ad puncta illa data B, C datos angulos MBD, MCD efficientes ducantur; dico quod hæ duæ BD, CD concursu suo D describent sectionem Conicam. Et vice versa si rectæ BD, CD concursu suo D describant sectionem Conicam per puncta B, C, A transeuntem, et harum[40] *concursus tunc incidit in ejus punctum aliquod A, cùm alteræ duæ coincidunt cum linea BC; punctum M continget rectam positione datam.*

Nam in recta *MN* detur punctum *N*, et ubi punctum mobile *M* incidit in immotum *N*, incidat punctum mobile *D* in immotum *P*. Junge *CN*, *BN*, *CP*, *BP*, et a puncto *P* age rectas *PT*, *PR* occurrentes ipsis *BD*, *CD* in *T* et *R*, et facientes angulum *BPT* æqualem angulo *BNM* et angulum *CPR* æqualem angulo

Geometrie), showing it to be a curve defined by a general Cartesian equation of second degree and in consequence a conic; see IV: 220–3 and compare the opening paragraph of Newton's preliminary discussion of the 'Veterum Loca solida restituta' (IV: 274–82, especially 274–6).

(36) A trivially modified repeat of Proposition 5 and its corollaries on IV: 290–4.

(37) On the prehistory of this Newtonian defining 'symptom' of a general conic see IV: 292–3, note (19).

DH. Further, PR is to DF as RC to DC, and hence as (IG or) PS to DG, and so *alternando* PR to PS as DF to DG. And when the ratios are combined the product $PQ \times PR$ comes to be to the product $PS \times PT$ as the product $DE \times DF$ to the product $DG \times DH$, and hence in a given ratio. But PQ and PS are given, and accordingly the ratio of PR to PT is given. As was to be proved.

Case 2. But should PR and PT be supposed to be in a given ratio to one another, by a similar reverse argument it will follow that the product $DE \times DF$ is to the product $DG \times DH$ in a given ratio, and hence (by Lemma XVIII) that the point D lies in a conic passing through the points A, B, P and C. As was to be proved.(37)

Corollary 1. Consequently, if BC be drawn meeting PQ in r, and in PT there be taken PT in the ratio to Pr which PT has to PR, then Bt will be tangent to the conic at the point B. For imagine the point D to coalesce with the point B with the result that, as the chord BD vanishes, BT comes to be tangent, and CD and BT will then coincide with CB and Bt.

Corollary 2. And *vice versa*, if Bt be tangent and BD, CD meet at any point D on the conic, then will PR be to PT as Pr to Pt. Conversely, if PR be to PT as Pr to Pt, then BD and CD will meet at some point D on the conic.(38)

Corollary 3. A conic does not intersect a conic in more than four points. For, if it is possible, let two conics pass through the five points A, B, C, D and P, and let the straight line BD cut these in the points D, d with ρ the intersection of the straight lines CD and PQ. Then PR is to PT as $P\rho$ to PT, that is, PR and $P\rho$ are equal to each other, contrary to hypothesis.

Lemma XXI.(39)

If the two mobile, infinite straight lines BM, CM drawn through the given points B, C, as poles should at their meet M describe a third straight line MN given in position, and if two other infinite straight lines BD, CD be drawn, making given angles $M\widehat{B}D$, $M\widehat{C}D$ at those given points B, C with the first two: I assert that these two BD, CD will by their meet D describe a conic. Conversely, if the straight lines BD, CD by their meet D describe a conic passing through the points B, C, A, and the meet of these(40) *falls at some point A of it just when the other two coincide with the line BC, then the point M will lie in a straight line given in position.*

For in the straight line MN let the point N be given, and when the mobile point M falls at the stationary one N, let the mobile point D fall at the stationary one P. Join CN, BN, CP, BP and from the point P draw the straight lines PT, PR meeting BD, CD in T and R and making the angle $B\widehat{P}T$ equal to $B\widehat{N}M$ and

(38) This inserted converse is a late addition to the manuscript in Newton's own hand.
(39) A light remodelling of the main text of Proposition 7 of the 'Solutio' (see IV: 298–300).
(40) In sequel 'utrinq; productarum' (produced either way) is deleted.

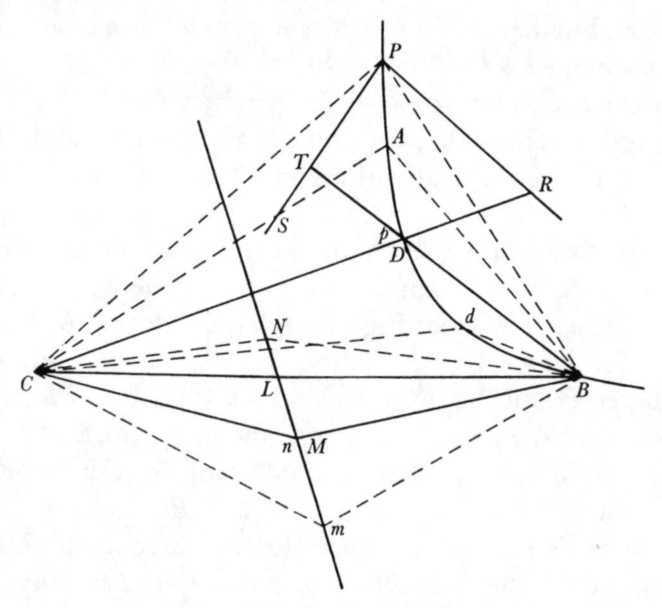

CNM.(41) Cùm ergo ex Hypothesi æquales sint anguli *MBD*, *NBP*, ut et anguli [*M*]*CD*, *NCP*: aufer communes *NBD* et [*N*]*C*[*D*] et restabunt æquales *NBM* & *PBT*, *NCM* & *PCR*: adeoq triangula *NBM*, *PBT* similia sunt, ut et triangula *NCM*, *PCR*. Quare *PT* est ad *NM* ut *PB* ad *NB*, et *PR* ad *NM* ut *PC* ad *NC*. Ergo *PT* et *PR* datam habent rationem ad *NM* proindeq datam rationem inter se, atq adeo per Lemma XX punctum [*D*] (perpetuus rectarum mobilium *BT* & *CR* concursus) contingit sectionem Conicam.(42) Q.E.D.

Et contra, si punctum *D* contingit sectionem conicam transeuntem per puncta *B*, *C*, *A* et ubi rectæ *BM*, *CM* coincidunt cum recta *BC* punctum illud *D* incidit in aliquod sectionis punctum *A*; ubi verò punctum *D* incidit successivè in alia duo quævis sectionis puncta *p*, *P*, punctum mobile *M* incidit successive in puncta immobilia *n*, *N*: per eadem *n*, *N* agatur recta *nN* et hæc erit locus perpetuus puncti illius mobilis *M*. Nam si fieri potest, versetur punctum *M* in linea aliqua curva.(43) Tanget ergo punctum *D* sectionem Conicam per puncta quinq *C*, *p*, *P*, *B*, *A* transeuntem ubi punctum *M* perpetuò tangit lineam curvam.(44) Sed et ex jam demonstratis tanget etiam punctum *D* sectionem Conicam per eadem quinq puncta *C*, *p*, *P*, *B*, *A* transeuntem ubi punctum *M* perpetuò tangit lineam rectam. Ergo duæ sectiones conicæ transibunt per eadem quinq puncta contra Corol. 3 Lem. XX. Igitur punctum *M* versari in linea curva(44) absurdum est. Q.E.D.

Prop. XXII. Prob. XIV.(45)

Trajectoriam per data quinq puncta describere.

Dentur puncta quinq *A*, *B*, *C*, *D*, *P*. Ab eorum aliquo *A* ad alia duo quævis *B*, *C* quæ poli nominentur age rectas *AB*, *AC* hisq parallelas *TPS*, *PRQ* per

(41) This ensures, of course, that *R* and *T* pass simultaneously to infinity and hence (see IV: 299, note (32)) that the ratio *PR*/*PT* is constant, as Newton requires.

(42) A hyperbola, in fact, as Newton draws it in his accompanying figure. We have added its second branch (through the second pole *C*) in the simplified equivalent diagram inserted in the English version.

the angle $\widehat{CPR}$ equal to $\widehat{CNM}$.[41] Since, therefore, by hypothesis the angle $\widehat{MBD}$, $\widehat{NBP}$ and also the angles $\widehat{MCD}$, $\widehat{NCP}$ are equal, take away the common ones $\widehat{NBD}$ and $\widehat{NCD}$ and there will remain the equals $\widehat{NBM}$, $\widehat{PBT}$ and $\widehat{NCM}$, $\widehat{PCR}$: hence the triangles NBM, PBT are similar, and so also the triangles NCM, PCR. In consequence PT is to NM as PB to NB, and PR to NM as PC to NC.

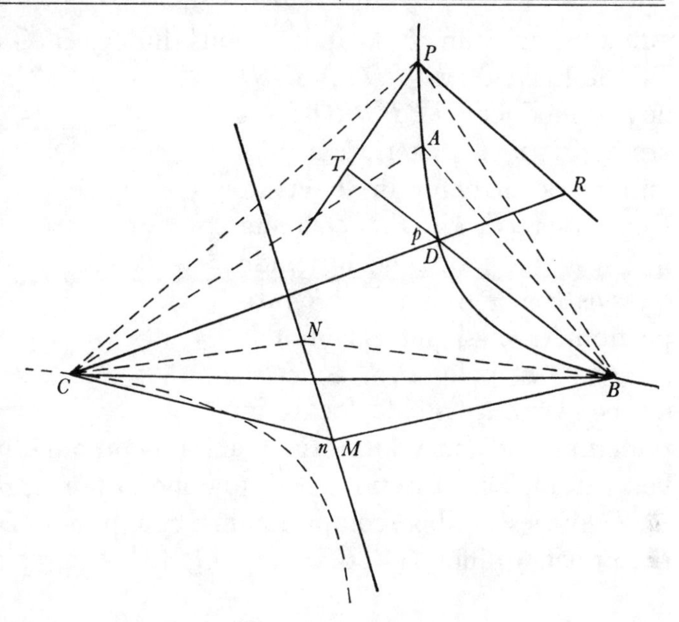

Therefore PT and PR have a given ratio to NM and accordingly a given ratio to one another, and hence by Lemma XX the point D (the perpetual meet of the mobile straight lines BT and CR) lies in a conic.[42] As was to be proved.

Conversely, if the point D lies in a conic passing through the points B, C, A and, when the straight lines BM, CM coincide with the line BC, that point D falls on some point A of the conic; while, when the point D falls successively on any two other points, p, P of the section, the mobile point M falls successively on the stationary points n, N: through these same n, N draw a straight line nN and this will be the locus for ever of the mobile point M. For, if possible, let the point M move about in some curved line.[43] The point D will therefore lie in some conic passing through the five points C, p, P, B, A when the point M lies perpetually in some curve.[44] But again, from what is just now proved, the point D will also lie in a conic passing through the same five points C, p, P, B, A when the point M lies perpetually in a straight line. Consequently two conics will pass through the same five points, contrary to Corollary 3 of Lemma XX. That the point M moves about in a curve[44] is therefore absurd. As was to be proved.

Proposition XXII, Problem XIV.[45]

To describe a trajectory through five given points.

Let there be given the five points A, B, C, D and P. From some one of them, A, to any two others B and C—which shall be named poles—draw the straight lines

(43) Originally 'exorbitet punctum M et inveniatur idem alicubi extra lineam $n[N]$' (let the point M wander away and be found in some place outside the line nN).

(44) Here correspondingly (compare the previous note) the manuscript initially read

punctum quartum P. Deinde a polis duobus B, C age per punctum quintum D infinitas duas BDT, CRD novissime ductis TPS, PRQ (priorem priori et posteriorem posteriori) occurrentes in T et R. Deniq; de rectis PT, PR actâ rectâ $t[r]$ ipsi TR parallelâ, abscinde quasvis Pt, Pr ipsis PT, PR proportionales et si per earum terminos t, r et polos B, C actæ Bt, Cr concurrant in d, locabitur punctum illud d in Trajectoria quæsita. Nam punctum illud d (per Lem. [XX]) versatur in Conica sectione per puncta quatuor A, B, P, C transeunte; et lineis, Rr Tt evanescentibus, coit punctum d cum puncto D. Transit ergo sectio conica per puncta quinq; A, B, C, D, P. Q.E.F.

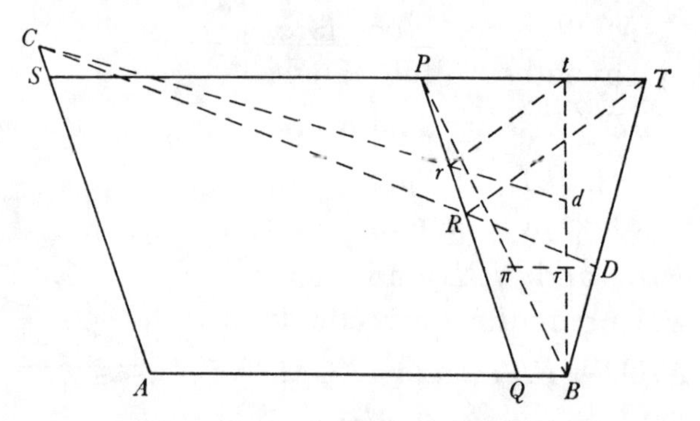

Idem aliter.

E punctis datis junge tria quævis A, B, C et circum duo eorū B, C ceu polos, rotando angulos magnitudine datos ABC, ACB,(46) applicentur crura BA, CA primò ad punctum D, deinde ad punctum P, et notentur puncta M, N in quibus altera crura BL, CL casu utroq; se decussant. Agatur(47) recta infinita MN & rotentur anguli illi mobiles circum polos suos B, C ea lege ut crurum BL, CL vel BM, CM intersectio communis quæ jam sit m perpetuò versetur in recta illa MN: et reliquorum crurum BA, CA vel BD, CD intersectio quæ jam sit d trajectoriam quæsitam $PADdB$ delineabit. Nam punctum d per Lem. XXI continget sectionem conicam per puncta B, C transeuntem et ubi punctum m accedit ad puncta L, M, N, punc-

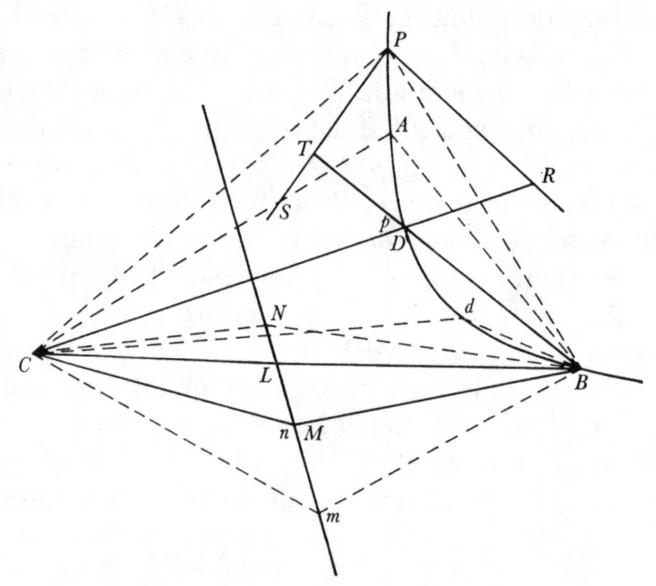

'...non tangit rectam nN' (does not lie in the straight line nN) and '...inveniri extra lineam nN' (is found outside the line nN) respectively.

(45) A reworking of Proposition 6 of the earlier 'Solutio' (see IV: 296–8), to which Newton

AB, *AC* and, parallel to these, *TPS*, *PRQ* through a fourth point *P*. Next, from the two poles *B* and *C* through the fifth point *D* draw two unterminated lines *BDT* and *CRD* meeting those, *TPS* and *PRQ*, most recently drawn (the former the former and the latter the latter) in *T* and *R*. Finally, by drawing the straight line *tr* parallel to *TR*, cut off from the straight lines *PT*, *PR* any *Pt*, *Pr* proportional to *PT*, *PR*, and then, if the lines *Bt*, *Cr* drawn through their end-points *t*, *r* and the poles *B*, *C* concur in *d*, that point *d* will be located in the required trajectory. For the point *d* (by Lemma XX) moves round in a conic passing through the four points *A*, *B*, *P* and *C*; and, with the lines *Rr* and *Tt* vanishing, the point *d* coincides with the point *D*. The conic consequently passes through the five points *A*, *B*, *C*, *D* and *P*. As was to be done.

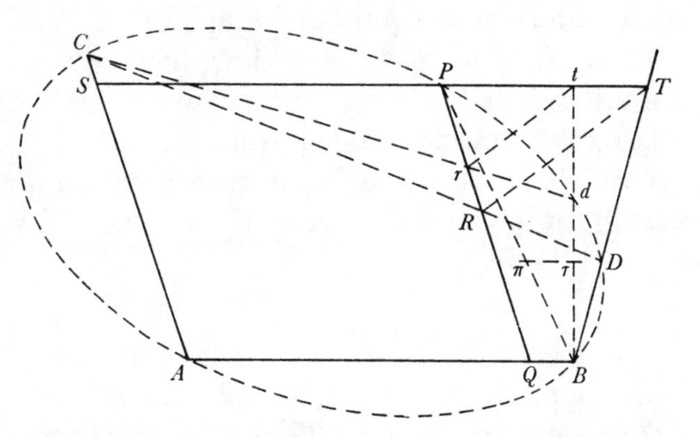

The same done another way.

Of the given points join any three *A*, *B*, *C* and, round two of them *B*, *C* as poles rotating angles $\widehat{ABC}$, $\widehat{ACB}$ given in magnitude,(46) apply the legs *BA* and *CA* first to the point *D*, then to the point *P*, and mark the points *M* and *N* at which the other legs *BL* and *CL* cross in each case. Draw(47) the indefinite straight line *MN* and rotate the mobile angles round their poles *B*, *C* under the restriction that their common intersection—let it now be *m*—move perpetually about in that straight line *MN*: the intersection—let it now be *d*—of the legs *BA* and *CA* or *BD* and *CD* will then trace out the required trajectory *PADdB*. For the point *d* will, by Lemma XXI, lie in a conic passing through the points *B* and *C*, while when the point *m* attains the points *L*, *M*, *N*, the point *d* will (by construction)

adds an *Idem aliter* wherein the organic description expounded in the preceding Lemma XXI is applied—as it was originally designed to be (see II: 118–20, and compare IV: 327–8)—straightforwardly likewise to construct a conic 'trajectory' passing through five given points. To emphasise the arbitrariness of the chronology which Newton later sought to impose upon the manuscript when (see note (1)) he afterwards deposited it as the 'polished' text of his contemporary Lucasian lectures, in the margin alongside he subsequently claimed that he here began 'Lect. 1' of his 'Octob. 1685' series, having delivered the above lemmas on which this and the two following propositions depend in the previous autumn term.

(46) This determines that the pair of lines *BC* and *MN* together correspond to the described conic *PADBC*.

(47) This is written in over 'Eruatur' (Draw out).

tum d (per constructionem) accedet ad puncta A, D, P. Describetur itaq; sectio conica transiens per puncta quinq; A, B, C, D, P. Q.E.F.

Corol. 1. Hinc rectæ expeditè duci possunt quæ trajectoriam in punctis quibusvis datis B, C tangent. In casu utrovis(48) accedat punctum d ad punctum C et recta Cd evadet tangens quæsita.

Corol. 2. Unde etiam Trajectoriarū centra diametri et latera recta inveniri possunt ut in Corollario secundo Lemmatis XX.

Schol.

Constructio in casu priore evadet paulo simplicior jungendo BP, et in ea, si opus est producta, capiendo(49) $B\pi$ ad BP ut est PR ad PT et per π agendo rectam infinitam $\pi\tau$ ipsi SPT parallelam; inq; ea capiendo semper $\pi\tau$ æqualem Pr, et agendo rectas $B\tau$, Cr concurrentes in d. Nam cum sint Pr ad Pt, PR ad PT, πB ad PB, & $\pi\tau$ ad Pt in eadem ratione[,] erunt $\pi\tau$ et Pr semper æquales. Hac methodo puncta trajectoriæ inveniuntur expeditissimè, nisi mavis curvam ut in casu secundo(50) describere Mechanicè. (51)Nam curvarum descriptio per motum ad Mechanicam pertinet. Rectam et circulum describere Geometria non docet sed postulat[,] id est postulat Tyronem antequam is incipit esse Geometra descriptiones eorum didicisse. Et quamvis Principia scientiarum debeant esse simplicia, neq; cuiquam conceditur Geometriæ limen attingere qui non prius didicit postulata;(52) et propterea Geometria nihil omnino postulat nisi quod sit in omni Mechanica simplicissimum, ipsa tamen sphæram Conum Cylindrum et paritate rationis figuras universas etiam spirales Quadratrices et similes descriptionibus non postulatis definit et vi definitionum considerat, oblatasq;(53) mensurat et utitur(54) in constructione problematum et non oblatas definitivè determinat in usum peritiorum Artificum qui figuras illas describere didicere.

Prop. XXIII. Prob. XV.(55)

Trajectoriam describere quæ per data quatuor puncta transibit et rectam continget positione datam.

---

(48) To be strict, the sequel relates only to the latter case, but Newton understands that a parallel conclusion holds when, as d approaches B, Bd comes to be tangent there.

(49) Newton originally employed a turn of phrase which is not necessarily accurate, writing 'eamq; producendo ad π ut sit' (and producing it to π so that there is).

(50) This replaces a slightly vaguer 'modo secundo' (by the second method).

(51) The remaining sentences of this scholium—which strongly echo a similar passage in his contemporary algebraic lectures (see v: 470–4)—were subsequently cancelled in the manuscript and do not appear in the published *Principia*. Perhaps Newton considered them in retrospect to be a little too mathematically didactic for his future scientific reader?

(52) We are reminded of the saying μηδεὶς ἀγεωμέτρητος εἰσίτω μοῦ τὴν στέγην which

attain the points A, D, P. There will accordingly be described a conic passing through the five points A, B, C, D, P. As was to be done.

Corollary 1. Hence straight lines can readily be drawn which shall touch a trajectory in any given points B and C. In either case[48] let the point d approach the point C and the straight line Cd will come to be the tangent required.

Corollary 2. Whence, too, the centres, diameters and *latera recta* of trajectories can be found as in Corollary 2 of Lemma XX.

Scholium.

The construction in the former case will prove a little simpler by joining BP and in it, produced if need be, taking[49] $B\pi$ to BP as PR is to PT and through π drawing an indefinite straight line $\pi\tau$ parallel to SPT; and then ever taking in it $\pi\tau$ equal to Pr and drawing the straight lines $B\tau$, Cr concurrent in d. For, since Pr is to Pt, PR to PT, πB to PB, and $\pi\tau$ to Pt in the same ratio, $\pi\tau$ and Pr will then ever be equal. By this method points on the trajectory are most readily found, unless you prefer, as in the second case,[50] to describe the curve mechanically. [51]For the description of curves by motion is the province of mechanics. Geometry does not teach how to describe the straight line and circle, but postulates them as drawn; that is, it postulates that before ever a beginner starts to be a geometer he shall have learned their descriptions. Though the principles of a science ought to be simple, not to anyone is it granted to gain entrance to geometry without first he learn its postulates;[52] and for that reason geometry makes no mechanical postulate of any kind unless it be the very simplest in all that field. But the sphere, cone, cylinder and, by a parity of reasoning, figures generally—spirals, quadratrixes and the like too—it defines by description, not postulate, and examines them by dint of definitions, measuring them when presented[53] and using [them] in the construction of problems, and, when those figures are not so offered, determining them in a precise manner to the profit of the more expert practitioners who have learnt how to describe them.

Proposition XXIII, Problem XV.[55]

To describe a trajectory which shall pass through four given points and be tangent to a straight line given in position.

Plato is traditionally (see Tzetzes, *Chiliad* 8, 972) said to have had inscribed over the entrance to his Academy at Athens. On Newton's notion of geometrical 'simplicity' see v: 422–6.

(53) As curves already described or otherwise defined. Compare v: 472.

(54) To be pedantically accurate, this should be 'adhibet' or some equivalent verb which may take the accusative 'oblatas' as its object.

(55) 'Cas. 1' of this construction amplifies Corollary 1 to Proposition 7 of the earlier 'Solutio' (iv: 300); the following *Idem aliter* and 'Cas. 2' are newly introduced.

Cas. 1. Dentur tangens *HB*, punctum contactus *B*, et alia tria puncta *C, D, P*. Junge *BC* et agendo *PS* parallelam *BH* et *PQ* parallelam *BC* comple parallelogrammum *BSPQ*. Age *BD* secantem *SP* in *T*, et *CD* secantem *PQ* in *R*. Deniq; agendo quamvis *tr* ipsi *TR* parallelam, de *PQ*, *PS* abscinde *Pr*, *Pt* ipsis *PR*, *PT* proportionales, et actarum *Cr*, *Bt* concursus *d* incidet semper in Trajectoriam describendam.(56)

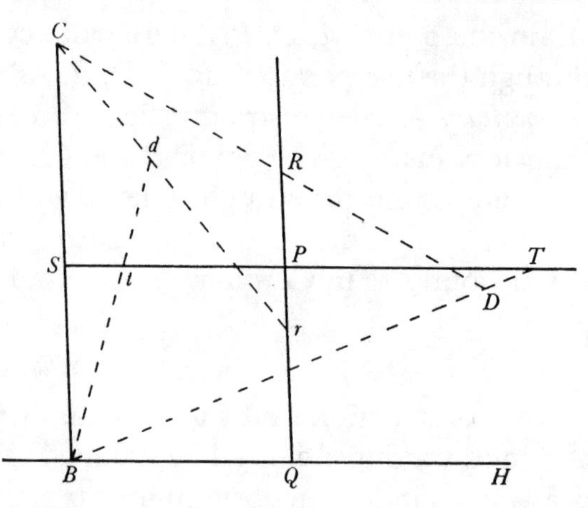

Idem aliter.

Revolvatur tum angulus magnitudine datus *CBH* circa polum *B* tum radius quilibet rectilineus(57) et utrinq; productus *DC* circa polum *C*. Notentur puncta *M*, *N* in quibus anguli crus *BC* secat radium illum ubi crus alterum *BH* concurrit cum eodem radio in punctis *D* et *P*. Deinde ad actam infinitam *MN* concurrant perpetuò radius ille(58) et anguli crus *BC* et cruris alterius *BH* concursus cum radio delineabit Trajectoriam quæsitam.

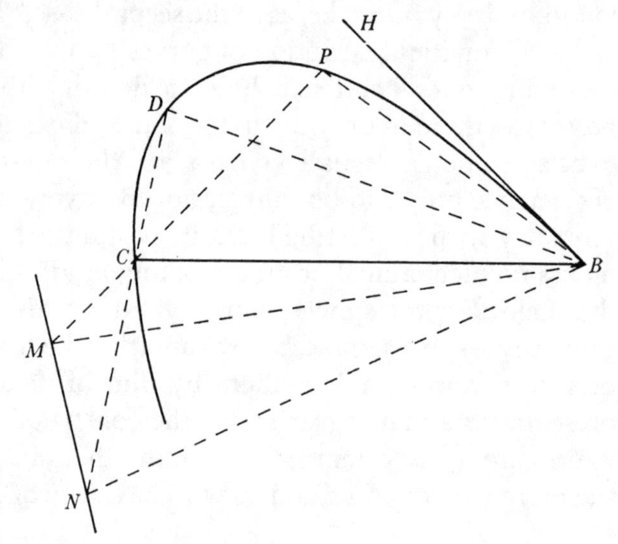

Nam si in constructionibus Problematis superioris accedat punctum *A* ad punctum *B*, lineæ *CA* et *CB* coincident, et linea *AB* in ultimo suo situ fiet tangens *BH*; atq; adeo constructiones ibi positæ evadent eædem cum constructionibus hic descriptis. Delineabit igitur cruris *BH* concursus cum radio sectionem conicam per puncta *C, D, P* transeuntem et rectam *BH* tangentem in puncto *B*. Q.E.F.

Cas. 2. Dentur puncta quatuor *B, C, D, P* extra tangentem *HI* sita. Junge bina *BD*, *CP* concurrentia in *G* tangentiq; occurrentia in *H* et *I*. Secetur tangens in *A* ita ut sit *HA* ad *AI* ut est rectangulum sub media proportionali inter *BH* et *HD* et media proportionali inter *CG* et *GP* ad rectangulum sub media proportionali

(56) For, when the points *R* and *T* pass simultaneously to infinity in the lines *PQ* and *PS*, the point *D* of the conic locus through *B*, *C* and *P* comes to coincide with *B*, and hence *BT*—

Case 1. Let there be given the tangent HB, the point B of contact, and the three other points C, D and P. Join BC and, by drawing PS parallel to BH and PQ parallel to BC, complete the parallelogram $BSPQ$. Draw BD cutting SP in T, and CD cutting PQ in R. Finally, by drawing any arbitrary line tr parallel to TR, cut off from PQ, PS lines Pr, Pt proportional to PR, PT and the meet of Cr, Bt when these are drawn will fall always in the trajectory to be described.(56)

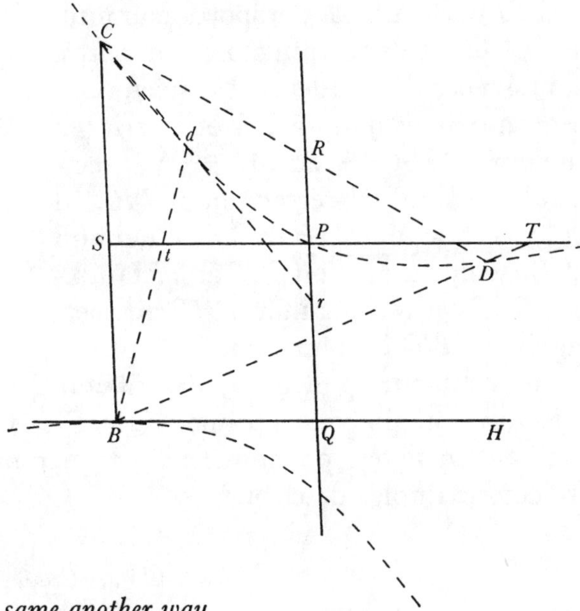

The same another way.

Revolve both the angle $\widehat{CBH}$ given in magnitude round the pole B and also any rectilinear(57) radius DC, extended either way, round the pole C. Mark the points M, N in which the leg BC of the angle intersects that radius when its other leg BH meets the same radius in the points D and P. Then, once the indefinite line MN is drawn, let that radius(58) and the leg BC of the angle meet perpetually at it, and the meet of the other leg BH with the radius will trace the required trajectory.

For, if in the constructions of the previous problem the point A shall come to concur with the point B, the lines CA and CB will coincide and the line AB in its last position will become the tangent BH; and accordingly the constructions there set down will prove to be the same as the ones here described. The meet of the leg BH with the radius will therefore trace a conic passing through the points C, D, P and touching the straight line BH at the point B. As was to be done.

Case 2. Let the four points B, C, D, P be given in situation outside the tangent HI. Join the pairs B, D and C, P (by lines) concurrent at G and meeting the tangent in H and I. Let the tangent be cut in A such that HA shall be to AI as the rectangle contained by the mean proportional between BH and DH and that between CG and PG is to the rectangle contained by the mean proportional

that is, BH—meets it there in two coincident points. Newton defers his own equivalent justification of the construction to the next paragraph but one.

(57) Originally 'rectus' (straight).

(58) Newton afterwards here inserted the clarification 'CP vel CD' (CP or CD) in his published *Principia* ($_1$1687: 82).

inter PI et IC et media proportionali inter DG et GB, et erit A punctum contactus. Nam si rectæ PI parallela HX trajectoriam secet in punctis quibusvis X et Y: erit (ex Conicis(59)) $HA^{\text{quad.}}$ ad $AI^{\text{quad.}}$ ut rectangulum XHY ad rectangulum PIC, id est, ut rectangulum XHY ad rectangulum BHD (seu rectangulum CGP ad rectangulum DGB) et rectangulum BHD ad rectangulum PIC conjunctim.(60) Invento autem contactus puncto A, describetur trajectoria ut in casu primo. Q.E.F. Capi autem potest punctum A vel inter puncta H & I vel extra; et perinde Trajectoria duplex describi.(61)

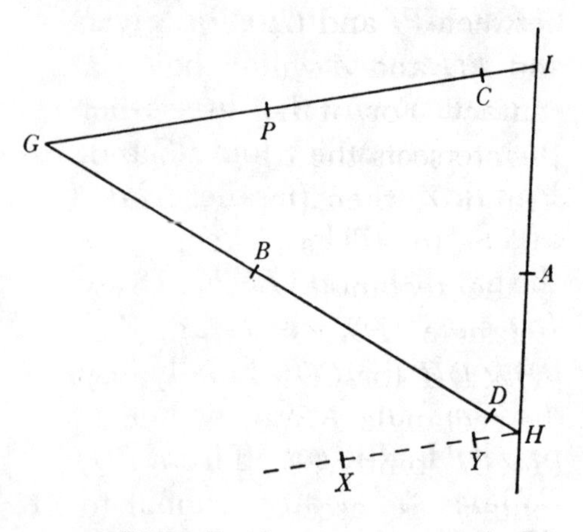

Prop. XXIV. Prob. XVI.(62)

Trajectoriam describere quæ transibit per data tria puncta et rectas duas positione datas continget.

Dentur tangentes HI, KL et puncta B, C, D. Age BD tangentibus occurrentem in punctis H, K; et CD tangentibus occurrentem in punctis I, L. Actas ita seca in R et S ut sit HR ad KR ut est media proportionalis(63) inter BH et HD ad mediam proportionalē(63) inter BK et KD, et IS ad LS ut est media proportionalis(63) inter CI et ID ad mediam proportionalem(63) inter CL et LD. Age RS secantem tangentes in A et P et erunt A et P puncta contactus. Nam si(64) per punctorum H, I, K, L quodvis I agatur recta IY tangenti KL parallela & occurrens curvæ in X et Y et in ea sumatur IZ media proportionalis(63) inter IX et IY: erit ex Conicis(59) rectangulum XIY (seu IZ^{quad}) ad LP^{quad} ut

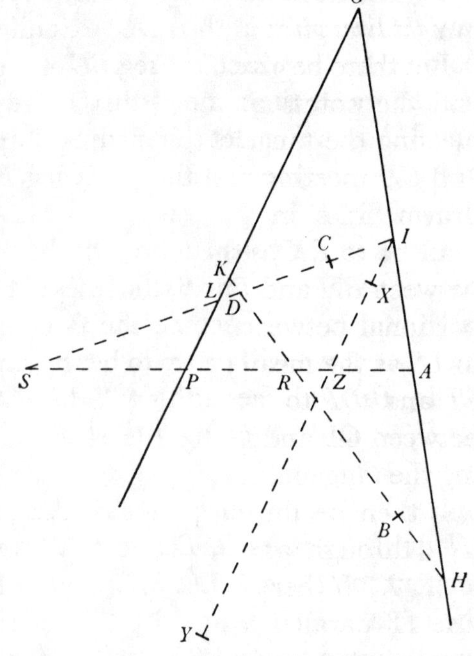

(59) Apollonius, *Conics* III, 17/18 as so often.

(60) Whence $HA/AI = \sqrt{[(CG\times PG/DG\times BG)\times(BH\times DH/PI\times CI)]}$, which is fixed. It may happen that the parallel through H to GI does not meet the conic in real points, but—from a sophisticated, non-classical viewpoint—the conjugate complex magnitudes HX and HY continue to have a real product, and Newton's deductions therefrom are unaffected. Without

between PI and CI and that between DG and BG and A will then be the point of contact. For, if HX parallel to the line PI intersects the trajectory in any points X and Y, then (from the *Conics*(59)) HA^2 will be to AI^2 as the rectangle $HX \times HY$ to the rectangle $PI \times CI$, that is, as the rectangle $HX \times HY$ to the rectangle $BH \times DH$ (or $CG \times PG$ to $DG \times BG$) and the rectangle $BH \times DH$ to the rectangle $PI \times CI$ jointly.(60) Once the point A of contact is ascertained, however, the trajectory will be described as in the first case. As was to be done. But the point A can be taken either between the points H and I or outside them; and a double trajectory may correspondingly be described.(61)

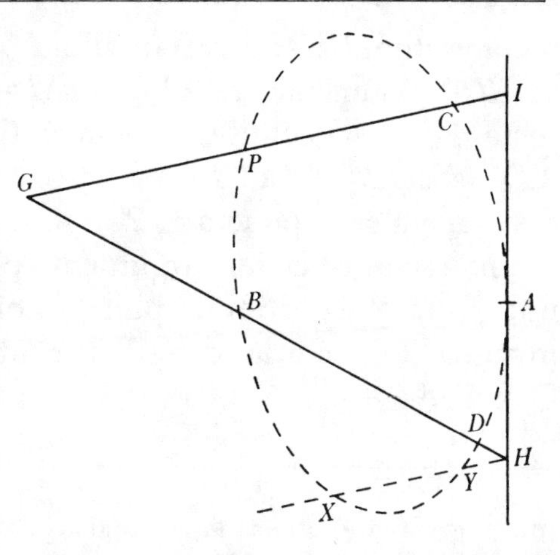

Proposition XXIV, Problem XVI.(62)

To describe a trajectory which shall pass through three given points and touch two straight lines given in position.

Let there be given the tangents HI, KL and the points B, C and D. Draw BD meeting the tangents in the points H, K and CD meeting them in I, L. Cut these drawn lines in R and S so that HR shall be to RK as the mean proportional between BH and DH is to the mean proportional between BK and DK, and IS to LS as the mean proportional between CI and DI to the mean proportional between CL and DL. Draw RS intersecting the tangents in A and P, and A and P will then be the points of contact. For, if(64) through any one, I, of the points H, I, K, L there be drawn the straight line IY parallel to the tangent KL and meeting the curve in X and Y, and in it be taken IZ the mean proportional between IX and IY: then from the *Conics* the rectangle $IX \times IY$ (or IZ^2) will be

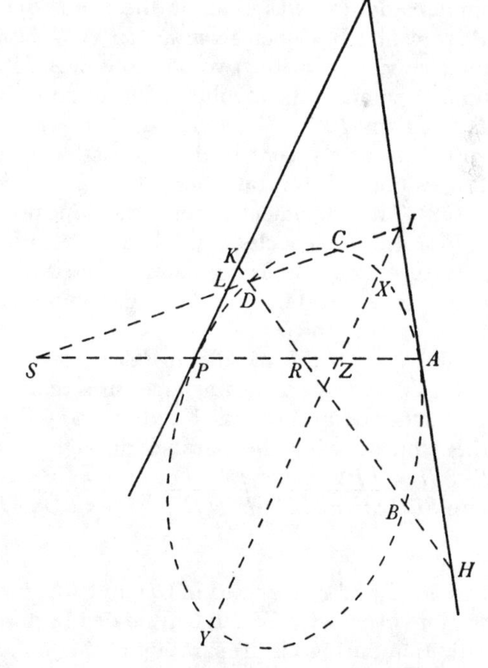

passing out of the real Euclidean plane we may readily construct a roundabout justification of the present proportion by taking some point h in AH near enough to A so that the parallels hg

rectangulum *CID* ad rectangulum *CLD*[,] id est (per constructionem) ut SI^{quad} ad SL^{quad}, atq; adeo *IZ* ad *LP* ut *SI* ad *SL*. Jacent ergo(65) puncta *S*, *P*, *Z* in una recta. Porrò tangentibus concurrentibus in [*G*] erit (ex Conicis(59)) rectangulum *XIY* (seu IZ^{quad}) ad IA^{quad} ut GP^{quad} ad GA^{quad}, adeoq; *IZ* ad *IA* ut *GP* ad *GA*. Jacent ergo puncta *P*, *Z* et *A* in una recta, adeo[q;] puncta *S*, *P* et *A* sunt in una recta. Et eodem argumento probabitur quod puncta *R*, *P* et *A* sunt in una recta. Jacent igitur puncta contactus *A* et *P* in recta *SR*. Hisce autem inventis Trajectoria describetur ut in casu primo problematis superioris. Q.E.F.(66)

(meeting *GI* in *g*) and *hx* to *HG* and *IG* do intersect the conic in pairs of real points *b*, *d* and *x*, *y* respectively: when at once

$$HA^2 : hA^2 = BH \times DH : bh \times dh \quad \text{and} \quad hA^2 = AI^2 = xh \times yh : PI \times CI$$

with $bh \times dh : xh \times yh = DG \times BG : CG \times PG$. This 'imperfection' in Newton's present proof—and the similar one in his demonstration of the following proposition—were brought to his attention by Henry Pemberton when he came to revise the corresponding passages in the printed *Principia* for his third edition in 1726, but Newton refused to make more than a bare public reference to the difficulty in the ensuing *editio ultima*; see note (66) below.

(61) This final sentence is a late insertion in the manuscript in Newton's own hand: a similar addition was made in the press copy (Royal Society MS LXIX: 33r, reproduced as Plate 6 in I.B. Cohen's *Introduction to Newton's 'Principia'*, Cambridge, 1971). From a more modern viewpoint the two positions of *A* defined by the constancy of the ratio AH/AI are the double points of the involution cut out on *HI* by the family of conics through the given points *B*, *C*, *D* and *P*.

(62) A newly contrived construction of a problem analogous to that of the preceding Proposition XXIII, on whose 'Cas. 2' its resolution is modelled.

(63) The equivalent phrase 'medium proportionale' is replaced in each case.

(64) In the concluding 'Errata Sensum turbantia' of his published *Principia* ($_1$1687: signature Ooo 4r) Newton later directed his reader here to insert in emendation '*A* & *P* sint puncta contactuum ubivis in tangentibus sita, &' (*A* and *P* be the points of contact situated in the tangents, and).

(65) Since *LP* is parallel to *IZ*.

(66) Newton's preceding analysis is readily restored. For, where the conic through *B*, *C* and *D* touches *GI* and *GK* in *A* and *P*, and *CD* meets *GI*, *GK* and *AP* in *I*, *L* and *S* respectively, then, on drawing the parallel through *I* to *GK* to meet the conic in *X* and *Y*, at once $SL : SI = LP : IZ$ and $IX \times IY : IA^2 = GP^2 : GA^2 = IZ^2 : IA^2$, so that $IX \times IY = IZ^2$ and therefore $SL^2 : SI^2 = LP^2 : IX \times IY = LC \times LD : IC \times ID$, while furthermore

$$HR^2 : KR^2 = HB \times HD : KB \times KD;$$

whence *S* and *R* are fixed in *CD* and *BD*, and in sequel *A* and *P* are determined as the intersections of *SR* with *GI* and *GK*. As Pemberton again (see note (61) above) brought to Newton's attention during his preparation of the *editio ultima* of the *Principia* in the early 1720's, there is some difficulty not allowed for in the present argument if the parallel through *I* to *GK* does not in fact meet the conic in real points *X* and *Y*. We may, implicitly passing to the complex plane, continue to allow that the product $IX \times IY$ is real even when the component (conjugate) magnitudes *IX* and *IY* are no longer so. Alternatively, we may derive a slightly more devious classical demonstration of Newton's construction by introducing auxiliary parallels *ixy* and *icd*

to LP^2 as the rectangle $CI \times DI$ to the rectangle $CL \times DL$, that is, (by construction) as IS^2 to LS^2, and hence IZ is to LP as IS to LS. The points S, P, Z lie therefore[(65)] in a single straight line. Moreover, with the tangents concurrent in G, there will (from the *Conics*) be the rectangle $IX \times XY$ (or IZ^2) to IA^2 as GP^2 to GA^2, and hence IZ to IA as GP to GA. The points P, Z and A lie therefore in a single straight line, and hence the points S, P and A are in one straight line. And by the same argument it will be proved that the points R, P and A are in one straight line. The points A and P of contact lie therefore in the straight line SR. Once these are found, however, the trajectory will be described as in the first case of the previous problem. As was to be done.[(66)]

---

through some other point i in GI, which meet the conic in respective points x, y and c, d which are all real (which may always be done by taking i sufficiently near to A) and then, where ixy intersects AP in z, suitably eliminating between the proportions

$$LP^2 : LC \times LD = (ix \times iy \text{ or}) \; iz^2 : ic \times id,$$

$$IA^2 : IC \times ID = iA^2 : ic \times id$$

and

$$IZ^2 : IA^2 = iz^2 : iA^2$$

to produce $SL^2 : SI^2 = (LP^2 : IZ^2 \text{ or}) \; LC \times LD : IC \times ID$ in this case also. Newton's immediate response was to insert a new following paragraph in the text of the *Principia* ($_3$1726: 87) stating that 'In hac propositione, & casu secundo propositionis superioris constructiones eædem sunt, sive recta XY trajectoriam secet in X & Y, sive non secet; eæque non pendent ab hac sectione. Sed demonstratis constructionibus ubi recta illa trajectoriam secat, innotescunt constructiones, ubi non secat; iisque ultra demonstrandis brevitatis gratia non immoror' (In this proposition, and Case 2 of the preceding one, the constructions are the same whether the line XY cuts the trajectory in X and Y or it does not: they are not dependent on this intersection. But, once constructions when the line cuts the trajectory are proved, the constructions when it does not are made known; and in further demonstrating them I do not, for brevity's sake, linger). In early 1726, when the *editio ultima* was on the point of being printed off, he refused to consider Pemberton's last-minute proposal in a letter of 9 February (ULC. Add. 3986. 23) that the second half of this additional paragraph be changed to read '...Si vero non secet demonstrationes perfici possunt sumendo in tangente AI punctum tam prope puncto contactus ut duæ rectæ duci possint trajectoriæ occurrentes; quarum altera in hac propositione parallela sit rectæ DC, altera tangenti PL; et in casu secundo propositionis præcedentis harum rectarum altera parallela sit rectæ BD, et altera rectæ PC', even though Pemberton persuasively urged that the correction 'will stand nearly in the same room...and will perhaps answer your purpose somewhat more fully' (*ibid.*; compare Cohen's *Introduction* (note (61)): 272–3).

Since each of the points R and S—equivalently defined by the cross-ratio equalities $(HKBR) = (HKRD)$ and $(LICS) = (LISD)$ which are an immediate consequence of their being double points in the involutions $(H, K; B, D)$ and $(L, I; C, D)$ cut out in HK and LI by the family of conics (one through B, C and D) which are tangent at A and P to GI and GK—is determined in a dual position (one within and the other outside the intervals HK and IL) by the constancy of the ratios

$$HR/KR = \sqrt{[HB \times HD/KB \times KD]} \quad \text{and} \quad IS/LS = \sqrt{[IC \times ID/LC \times LD]},$$

Newton might well have added, much as in the final sentence he appended to the previous Proposition XXIII, that there exist two real solutions to the problem (the other pair of combinations of the positions of R and S being impossible).

Lemma XXII.(67)

Figuras in alias ejusdem generis figuras mutare.

Transmutanda sit figura quævis *HGI*. Ducantur pro lubitu rectæ duæ parallelæ *AO*, *BL* tertiam quamvis positione datam *AB* secantes in *A* et *B*, et a figuræ puncto quovis *G*, ad rectam *AB* ducatur *GD*, ipsi *OA* parallela. Deinde a puncto aliquo *O* in linea *OA* dato ad punctum *D* ducatur recta *OD* ipsi *BL* occurrens in *d* et a puncto occursus erigatur recta *gd*, datum quemvis angulum cum recta *BL* continens atq; eam habens rationem ad *Od* quam habet *GD* ad *OD*, et erit *g* punctum in figura nova *hgi* puncto *G* respondens. Eadem ratione puncta singula figuræ primæ dabunt puncta totidem figuræ novæ. Concipe igitur punctum *G* motu continuo percurrere puncta omnia figuræ primæ, et punctum *g* motu itidem continuo percurret puncta omnia figuræ novæ et eandem describet. Distinctionis gratia nominemus *DG* ordinatam primam, *dg* ordinatam novam; *BD* abscissam primam, *Bd* abscissam novam; *O* polum, *OD* radium abscindentem,(68) *OA* radium ordinatum primum et *Oa* (quo parallelogrammum *OABa* completur) radium ordinatum novum.

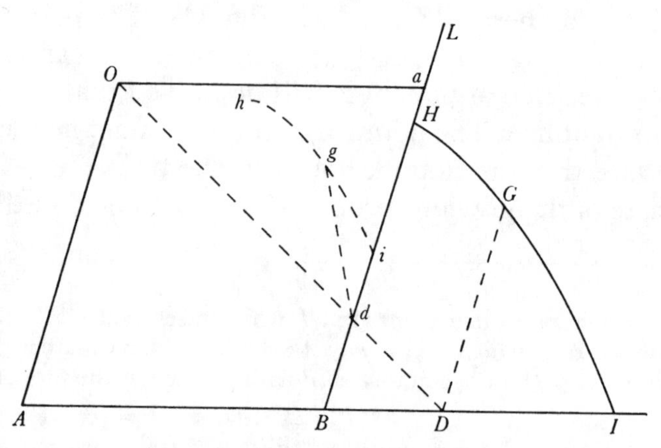

Dico jam quod si punctum *G* tangit rectam lineam positione datam, punctum *g* tanget etiam lineam rectam positione datam. Si punctum *G* tangit Conicam sectionem, punctum *g* tanget etiam conicam sectionem. Conicis sectionibus hic circulum annumero. Porrò si punctum *G* tangit lineam tertij ordinis analytici,(69) punctum *g* tanget lineam itidem tertij ordinis, et sic de curvis lineis superiorum ordinum. Lineæ duæ erunt ejusdem semper ordinis Analytici(69) quas puncta *G*, *g* tangunt. Etenim ut est *ad* ad *OA* ita sunt *Od* ad *OD*, *dg* ad *DG* et *AB* ad *AD*; adeoq; *AD* æqualis est $\frac{OA \times AB}{ad}$ et *DG* æqualis est $\frac{OA \times dg}{ad}$. Jam si punctum [*G*] tangit rectam lineam atq; adeò in æquatione quavis qua relatio inter abscissam *AD* et ordinatam *DG* habetur, indeterminatæ illæ *AD* et *DG* ad unicam tantum dimensionem ascendunt, scribendo in hac æquatione $\frac{OA \times AB}{ad}$ pro *AD* &

(67) A general, degree-preserving, 1, 1 continuous transformation of plane curves—effectively (see note (70)) the product of a simple affine transformation and a plane perspectivity which had been published by La Hire in a rare French work on conics a decade

Lemma XXII.[67]

To change figures into other figures of the same class.

Let it be required to transform any figure HGI. Draw at pleasure two parallel straight lines AO, BL intersecting any third one AB, given in position, in A and B, and from any point G of the figure to the straight line AB draw GD parallel to OA. Then from some point O given in the line OA draw to the point D the straight line OD meeting BL in d, and from the meeting-point erect a straight line dg containing any given angle with the straight line BL and having to Od the ratio which GD bears to OD: g will then be the point in the new figure hgi corresponding to the point G. By the same procedure individual points of the first figure will yield an equal number of points in the new figure. Conceive therefore, that the point G travels in a continuous movement through all points of the first figure, and the point g will travel with a similarly continuous motion through all the points of the new figure, describing it. For distinction's sake let us name DG the prime ordinate, dg the new ordinate; BD the prime abscissa, Bd the new abscissa; O the pole, OD the abscinding[68] radius, OA the prime ordinate radius and Oa (by which the parallelogram $OABa$ is completed) the new ordinate radius.

I now assert that, if the point G traces a straight line given in position, the point g will also trace a straight line given in position; if the point G traces a conic, then the point g will also trace a conic (with conics I here count the circle); while if the point G traces a line of third analytic order,[69] the point g will trace a line likewise of third order; and so with curves of higher orders: the two lines traced by the points G and g will be ever of the same analytic order.[69] For, indeed, as ad is to OA, so are Od to OD, dg to DG and AB to AD; and consequently AD is equal to $OA \times AB/Oa$ and DG equal to $OA \times dg/ad$. Now, if the point G traces a straight line, and hence in any equation by which the relationship between the abscissa AD and ordinate DG is exhibited those indeterminates AD and DG rise but to a single dimension, on writing in this equation $OA \times AB/Oa$ in place of AD and $OA \times dg/ad$ in place of DG there will be produced a new

earlier, whether or not Newton himself knew it—whose depth and power far transcends the application here made of it in sequel to convert two intersecting straight lines into a pair of parallels. The lemma was, as we shall see in the next volume, to fill a more basic rôle when Newton employed it to demonstrate that all cubic curves may be derived as the optical 'shadows' of the five (projectively distinct) 'divergent' cubic parabolas.

(68) Manifestly so dubbed because OD 'cuts off' the related *abscissa nova ad* from the (indefinitely extensible) base line aB, departing from the fixed point a.

(69) Understand the algebraic order of a curve as determined by the degree of the defining equation which relates its oblique Cartesian coordinates (here BD and DG, or ad and dg). Since, as we have already remarked (see IV: 341, note (22)), Descartes in his *Geometrie* himself preferred to grade curves by paired degree, we may justly term this more natural present order 'Newtonian'.

$\frac{OA \times dg}{ad}$ pro DG, producetur æquatio nova in qua abscissa nova ad et ordinata nova dg ad unicam tantum dimensionem ascendent atq; adeo quæ designat lineam rectam. Sin AD et GD (vel earum alterutra) ascendebant ad duas dimensiones in æquatione prima[,] ascendent itidem ad et dg ad duas in æquatione secunda. Et sic de tribus vel pluribus dimensionibus. Indeterminatæ ad, dg in æquatione secunda et AD, DG in prima ascendent semper ad eundem dimensionum numerum et propterea lineæ quas puncta G, g tangunt sunt ejusdem ordinis Analytici.

Dico præterea quod si recta aliqua tangat lineam curvam in figura prima; hæc recta translata tanget lineam curvam in figura nova: et contra. Nam si Curvæ puncta duo quævis accedunt ad invicem et coeunt in figura prima: puncta eadem translata coibunt in figura nova, atq; adeo rectæ quibus hæc puncta junguntur simul evadent curvarum tangentes in figura utraq;.(70) Componi possent harum assertionum Demonstrationes more magis Geometrico. Sed brevitati consulo.

---

(70) Where the parallel through the general point D to dg meets Og in G', it will be evident that $DG': dg = OD: Od = OA: ad = DG: dg$ and therefore $DG' = DG$. Accordingly, Newton's 'transmutation' $G \rightarrow g$ is the product of the affine projection $G \rightarrow G'$ (and in particular $H \rightarrow h$)—that is, a perspectivity in which the optical centre is at infinity in the direction of the parallels GG' and Hh—under which the curve $\widehat{IGH}$ passes into $\widehat{IG'h}$, and of the perspectivity $G' \rightarrow g$ from optical centre O (preserving h in position) by which the general point g of the 'shadow' curve $\widehat{igh}$ is the point in OG' such that $Og: OG' = Od: OD = AB: AD$

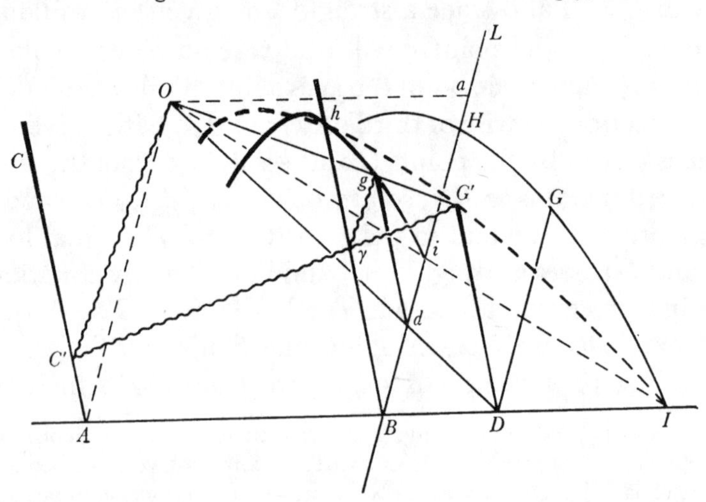

for all corresponding positions of the 'abscinded' points D and d in AB and aB. Since a perspectivity is a 1, 1 continuous correspondence preserving convexity and tangency, and because any curve (G) of n-th order—defined by some Cartesian equation

$$C_n(AD, DG) \equiv C_n(AD, DG') = 0$$

say—evidently passes into a transmuted curve (g) defined by the equation

$$C_n(OA \times AB/ad, OA \times dg/ad) \equiv C'_n(ad, dg) = 0$$

of an equal order on multiplying through by ad^n, Newton's preceding assertions ensue as

equation in which the new abscissa *ad* and new ordinate *dg* will rise but to a single dimension, and which shall hence denote a straight line. But were *AD* and *DG* (or one or other of them) rising to two dimensions in the first equation, then *ad* and *dg* will rise likewise to two in the second equation. And so for three or more dimensions: the indeterminates *ad*, *dg* in the second equation and *AD*, *DG* in the first will ever rise to the same number of dimensions, and accordingly the lines which the points *G* and *g* trace are of the same analytic order.

I further assert that, should some straight line touch a curve in the first figure, after transformation this straight line will touch the curve in the new figure; and conversely so. For, if any two points of a curve approach one another and coalesce in the first figure, the same points will after transformation coalesce in the second figure, and hence the straight lines joining these points will come simultaneously to be tangent in each figure.(70) Proofs of these assertions might be composed in a more geometrical style, but it is my intention to be brief.

---

immediate consequences. While the generality of this transformation is unprecedented, it is tempting to conjecture that he may have been stimulated to contrive it by reading Philippe de La Hire's *Nouvelle Methode en Geometrie pour les Sections des Superficies Coniques et Cylindriques: Qui ont pour bases des Cercles, ou des Paraboles, des Elipses, & des Hyperboles* (Paris, 1673): to effect his conic transform with respect to 'pôle' O, La Hire would in Newton's figure construct a 'directrice' AC through A parallel to the 'formatrice' Bh, and then determine the point g corresponding to G' by taking an arbitrary point C' in AC, joining $C'G'$ to meet Bh in γ and then drawing γg parallel to $C'O$ to intersect OG' in the required point. (Compare D. T. Whiteside, 'Patterns of Mathematical Thought in the later Seventeenth Century' [*Archive for History of Exact Sciences*, **1**, 1961: 179–388]: 286–8. Alternatively, as Michel Chasles contrives it in Note 19 to his *Aperçu Historique sur l'Origine et le Développement des Méthodes en Géométrie* (Paris, $_3$1889): 347–8, we might (somewhat more cumbrously) likewise project $G \rightarrow g'$, the position of g after it is rotated clockwise round B through an angle equal to $G'\widehat{D}G$, by constructing as 'pôle' the point O' which ensues after rotating O round A through an equal angle, and then considering OA and aB to be 'directrice' and 'formatrice' respectively.) A copy of La Hire's rare tract was certainly acquired soon after its publication by Cambridge University Library (where it is now press-marked M.10.76) and, even if Newton's attention was not arrested by (Collins'?) review of this 'New Optical or Projective Method' in *Philosophical Transactions* **11**, No. 129 (for 20 March 1676): 745–6, he was later to be made aware of its existence by Robert Hooke, who wrote to him on 24 November 1679 that 'Mr Collins shewd me a booke he Received from Paris of De la Hire, conteining first a new method of ye Conick sections and secondly a treatise *De locis solidis*. I have not perused the book, but Mr Collins commends it' (*Correspondence of Isaac Newton*, **2**, 1960: 298): much as in La Hire's appended *Les Plani-coniques* to which Hooke also here refers, we may regard our figure as the orthogonal representation of a three-dimensional configuration in which G (in the plane $AOaI$ of the paper) is affinely projected into G' (in a plane $AChI$ inclined through AI to it) and thence optically, in line with O, into g (in a second plane Bha inclined through Ba). But whether Newton did study La Hire's *Nouvelle Methode* or not, we may agree with J. L. Coolidge's judgement that his contrivance of the present generalised projective transformation 'was a real contribution to geometry, seldom mentioned in appreciations of his work' (*A History of the Conic Sections and Quadric Surfaces* (Oxford, 1945): 47; compare also H. W. Turnbull, *The Mathematical Discoveries of Newton* (London, 1945): 55–6).

Igitur si figura rectilinea in aliam transmutanda est[,] sufficit rectarum intersectiones transferre et per easdem in figura nova lineas rectas ducere. Sin curvilineam transmutare oportet, transferenda sunt puncta tangentes et aliæ rectæ quarum ope Curva linea definitur. Inservit autem hoc Lemma solutioni difficiliorum problematum, transmutando figuras propositas in simpliciores. Nam rectæ quævis convergentes transmutantur in parallelas adhibendo pro radio ordinato primo *AO* lineam quamvis rectam quæ per concursum convergentium transit: id adeò quia concursus ille hoc pacto abit in infinitum, lineæ autem parallelæ sunt quæ ad punctum infinite distans tendunt.(71) Postquam autem Problema solvitur in figura nova, si per inversas operationes transmutetur hæc figura in figuram primam, habebitur Solutio quæsita.

Utile est etiam hoc Lemma in solutione Solidorum(72) problematum. Nam quoties duæ sectiones conicæ obvenerint quarum intersectione problema solvi potest, transmutare licet unam earum in circulum. Recta item et sectio conica in constructione planorũ(72) problematũ vertuntur in rectam et circulum.

Prop. XXV. Prob. XVII.

Trajectoriam describere quæ per data duo puncta transibit et rectas tres continget positione datas.

Per concursum tangentium quarumvis duarum cum se invicẽ et concursum tangentis tertiæ cum recta illa quæ per puncta duo data transit age rectam infinitam, eaqȝ adhibita pro radio ordinato primo, transmutetur figura per

(71) When in autumn 1686 Edmond Halley came to edit the present lemma for publication in the *Principia* ($_1$1687: 85–7), he was at a loss to understand this corollary and wrote to Newton on 14 October that 'In your Transmutation of Figures according to the 22th Lemma which you use in the 2 following Problems, to me it seems that the manner of Transmuting a Trapezium into a Parallelogram needs some farther Explication; I have printed it as you sent it, but I pray you please a little farther to describe by an example the manner of doing it, for I am not perfect Master of it, a short hint will suffice' (*Correspondence*, **2**: 452). Newton responded four days later that Halley should 'conceive ye curve *HGI* to be produced both ways till it meet & intersect it self any where in ye *radius ordinatus primus AO*: & when ye point *G* moving up & down in ye curve *HI* arrives at yt intersection point, I say ye point *g* moving in like manner up & down in ye curve *hi* will become infinitely distant. For ye point *G* falling upon ye line *OA*, ye point *D* will fall upon ye point *A* & ye line *OD* upon ye line *OA* & so becoming parallel to *AB* their intersection point *d* will become infinitely distant & consequently ye line *dg* will become infinitely distant & so will its point *g*. Q.E.D. So then if any two lines of ye primary figure *HGID* intersect in ye *radius ordinatus primus AO* their intersection in ye new figure *hgid* shall become infinitely distant & therefore if the two intersecting lines be right ones they shall become parallel. For right lines which tend to a point infinitely distant do not intersect one another & diverge but are parallel. Therefore if in ye primary figure there be any Trapezium whose opposite sides converge to points in ye *radius ordinatus primus OA* those sides in ye new figure shal become parallel & so ye trapezium be converted into a parallelogram' (*ibid.*: 454).

Consequently, if one rectilinear figure is to be transmuted into another, it is enough to transfer the intersections of its straight lines and then draw straight lines through these in the new figure. But should it be necessary to transmute a curvilinear one, you will need to transfer its points, tangents and the other straight lines by whose aid the curve is defined. This lemma, I may add, serves to solve more difficult problems by transforming the figures proposed into simpler ones. For any converging straight lines are transmuted into parallels by employing as prime ordinate radius AO any straight line passing through the meet of the converging ones: this, indeed, because that meet goes off by this means to infinity, while parallel lines are ones which tend towards an infinitely distant point.(71) After the problem is solved in the new figure, however, if this figure be transmuted by a reverse procedure into the first, the required solution will be obtained.

This lemma is useful also in solving solid(72) problems. For every time there occur two conics by whose intersection a problem can be solved, we are at liberty to transmute one of them into a circle. In constructing plane(72) problems, likewise, a straight line and a conic may be turned into a straight line and a circle.

Proposition XXV, Problem XVII.

To describe a trajectory which shall pass through two given points and touch three straight lines given in position.

Through the meet of any two of the tangents with one another and the meet of the third tangent with the straight line which passes through the two given points draw an indefinite straight line, and with this employed as prime ordinate

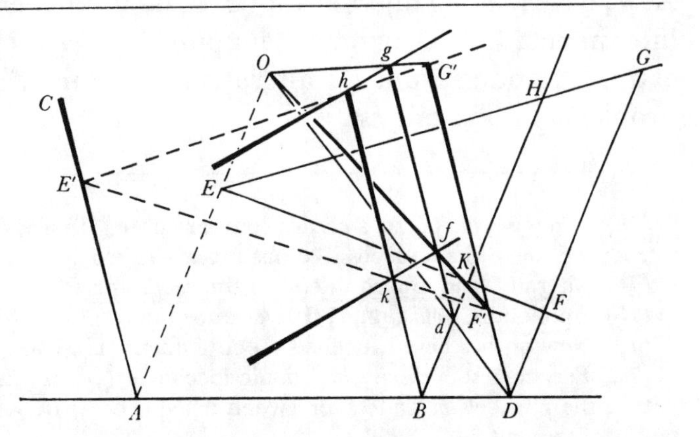

More simply, if, in the extended figure of the previous note, the straight lines GH, FK intersect in E on the 'prime ordinate radius' AO, they pass (where $Bh = BH$ and $Bk = BK$) affinely into the straight lines $G'h$, $F'k$ meeting in E' on the La Hire 'directrice' AC, and thence, by projection from the optical centre O, into the parallels hg, fk meeting at infinity in the direction OE'. Whether because Halley took it that Newton's (English) explanation was a private elucidation not intended for publication in the Latin *Principia*, or because—with a weather eye on the brevity counselled by Newton a few lines earlier—he considered it to be already too late to make any lengthy addition to a text even then being set in type, no such amplification appeared in the *editio princeps* or any subsequent edition.

(72) Constructible, that is, by the intersections of a pair of 'solid' (conic) or 'plane' (rectilinear/circle) loci respectively; see v: 423, note (617).

Lemma superius in figuram novam.(73) In hac figura tangentes illæ duæ evadent parallelæ et tangens tertia fiet parallela rectæ per puncta duo transeunti. Sunto *hi*, *kl* tangentes duæ parallelæ, *ik* tangens tertia, et [*h*]*l* recta huic parallela transiens per puncta illa *a*, *b*, per quæ Conica sectio in hac figura nova transire debet, et parallelogrammum *hikl* complens. Secentur rectæ *hi*, *ik*, *kl* in *c*, *d* et *e* ita ut sit *hc* ad latus quadratum rectanguli(74) *ahl*, *ic* ad *id* et *ke* ad *kd* ita ut est summa rectarum *hi*, *kl* ad summam trium linearum quarum prima est recta *ik*, et alteræ duæ sunt latera quadrata rectangulorum(74) *ahb* et *alb*. Et erunt *c*, *d*, *e* puncta contactus. Etenim ex Conicis(75) sunt *hc* quadratum ad rectangulum *ahb*, et *ic* quadratum ad *id* quadratum, et *ke* quadratum ad *kd* quadratum, et *el* quadratum ad *alb* rectangulum in eadem ratione[,] et propterea *hc* ad latus quadratum ipsius *ahb*, *ic* ad *id*, *ke* ad *kd* et *el* ad latus quadratum ipsius *alb* sunt in dimidiata illa ratione et compositè in data ratione omnium antecedentium $hi+kl$ ad omnes consequentes, quæ sunt(76) lat. quad. rectanguli ahb + recta ik + lat. quad. rectanguli *alb*. Habentur igitur ex data illa ratione puncta contactus *c*, *d*, *e* in figura nova. Per inversas operationes Lemmatis novissimi transferantur hæc puncta in figuram primam et ibi[,] per casum priorem Problematis XIV, describetur Trajectoria. Q.E.F. Cæterum perinde ut puncta *a*, *b* jacent vel inter puncta *h*, *l*, vel extra, debent puncta *c*, *d*, *e* vel inter puncta *h*, *i*, *k*, *l* capi vel extra.(77) Si punctorum *a*, *b* alterutrum cadit inter puncta *h*, *l* et alterum extra, Problema impossibile est.(78)

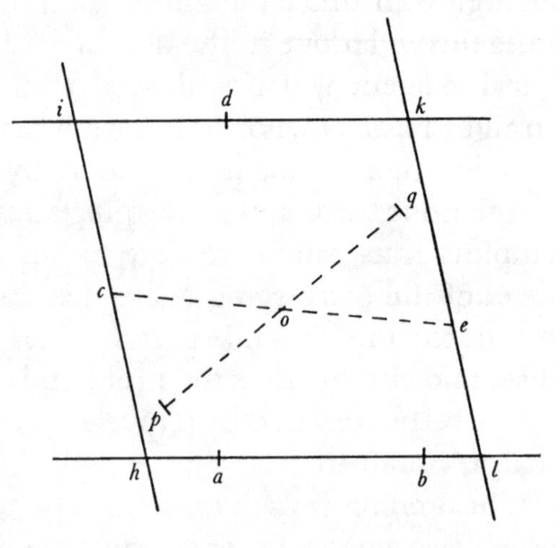

(73) Likewise, of course, a conic; compare note (70) above. In the English version we have introduced one of its four possible instances—an ellipse—in our accompanying figure.

(74) Literally 'the square side(s) of the rectangle(s)'.

(75) Apollonius, *Conics* III, 17/18 yet once more.

(76) Newton has here cancelled an equivalent 'nimirum' (to wit).

(77) For, correspondingly, the conic locus will (when it is an ellipse) be contained wholly within the parallels *hi* and *kl*, or (when a hyperbola) lie wholly outside them except at the points of tangency.

(78) Manifestly so by the previous note. There is of course no necessity to project the general problem into the present simplified case in order to construct it. Since it follows, much as in Proposition XXIV, that the families of conics drawn through *a*, *b* to touch *hi* and *ik*, *ik* and *kl*, and *hi* and *kl* respectively have their points of contact (*c* and *d*, *d* and *e*, *c* and *e*) in line with the double points—say α, β and γ—of the involutions $(h, \infty; a, b)$, $(l, \infty; a, b)$ and $(h, l; a, b)$, and because, where any transversal through γ cuts *ik* and *kl* in δ and ϵ, the meet of $\alpha\delta$ and $\beta\epsilon$

radius transmute the figure by means of the preceding Lemma into a new figure.(73) In this figure the two former tangents will prove to be parallel and the third tangent will come to be parallel to the straight line passing through the two points. Let *hi*, *kl* be the two parallel tangents, *ik* the third tangent and *hl* the straight line, parallel to this, passing through the points *a*, *b* through which the conic in this new figure ought to pass, and completing the parallelogram *hikl*. Let the straight lines *hi*, *ik*, *kl* be cut in *c*, *d*, *e* such that *hc* to the square root of the product(74) $ha\times hl$, *ic* to *id*, and *ke* to *dk* be (each) as the sum of the straight lines *hi* and *kl* to the sum of three lines, the first of which is *ik* and the other two are the square roots of the products(74) $ha\times hb$ and $la\times lb$. And then *c*, *d*, *e* will be the points of contact. For indeed, from the *Conics*(75) the square ch^2 to the product $ha\times hb$, ic^2 to id^2, ke^2 to dk^2, and el^2 to the product $la\times lb$ are in the same ratio, and accordingly *ch* to the square root of $ha\times hb$, *ic* to *id*, *ke* to *dk* and *el* to the square root of $la\times lb$ are in that ratio halved, and so by compounding in the given ratio of all the prior elements, $hi+kl$, to all the posterior ones, namely,(76)

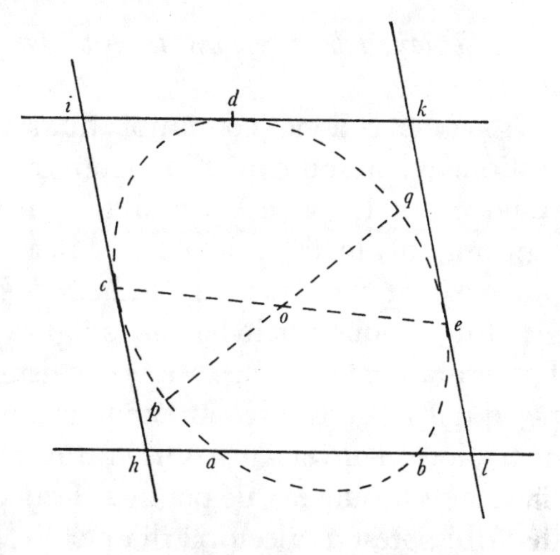

$$\sqrt{[ha\times hb]}+ik+\sqrt{[la\times lb]}.$$

From that given ratio, therefore, are had the points of contact *c*, *d*, *e* in the new figure. By reversing the procedure of the most recent lemma transfer these points to the first figure, and there, by Case 1 of Problem XIV, will the trajectory be described. As was to be done. Accordingly, however, as the points *a* and *b* lie either between the points *h* and *l* or outside them, so ought the points *c*, *d* and *e* to be taken either between the points *h*, *i*, *k*, *l* or outside them.(77) If one of the points *a*, *b* falls between the points *h* and *l* and the other outside, the problem is impossible.(78)

---

lies on a fixed straight line through *k* (a Pappus–Desargues locus; see IV: 239, note (28)), this line will intersect *hi* in the required point *c* of contact. The equivalent construction of the original problem when a reverse Newtonian 'transmutation' (preserving both rectilinearity and, because it is projective, involution) is made is effectively identical with that given by Robert Simson in 1734. (See the letter of Francis Hutcheson written to G. Smith on 28 February 1734/5 which was subsequently printed in the *Bibliothèque Raisonnée des Ouvrages des Savans de l'Europe. Avril, Mai et Juin 1735* (Amsterdam, 1735): 476–83, especially 480–1.) As befits this dual of Proposition XXIV there are two conic 'trajectories' constructible to satisfy the conditions of the problem.

Prop. XXVI. Prob. XVIII.

Trajectoriam describere quæ transibit per punctū datum et rectas quatuor positione datas continget.

Ab intersectione communi duarum quarumlibet tangentium ad intersectionem communem reliquarum duarum agatur recta infinita, et eâdem pro radio ordinato primo adhibitâ, transmutetur figura (per Lemma XXII) in figuram novam,(79) & tangentes binæ quæ ad radium ordinatum concurrebant jam evadent parallelæ. Sunto illæ *hi* & *kl*, *ik* et *hl* continentes parallelogrammum *hikl*. Sitq; *p* punctum in hac nova figura puncto in figura prima dato respondens. Per figuræ centrum *o* agatur *pq* et existente *oq* æquali *op* erit *q* punctum alterum per quod sectio Conica in hac figura nova transire debet. Per Lemmatis XXII operationem inversam transferatur hoc punctum in figuram primam et ibi habebuntur puncta duo per quæ Trajectoria describenda est.(80) Per eadem verò describi potest Trajectoria illa per Prob. XVI. Q.E.F.

Lemma XXIII.(81)

Si rectæ duæ positione datæ $AC_{[,]}$ $B[D]$ *ad data puncta A, B terminentur datamq; habeant rationem ad invicem, rectaq; CD qua puncta indeterminata C, D junguntur secetur [in K] in ratione data, dico quod punctum K locabitur in recta positione data.*

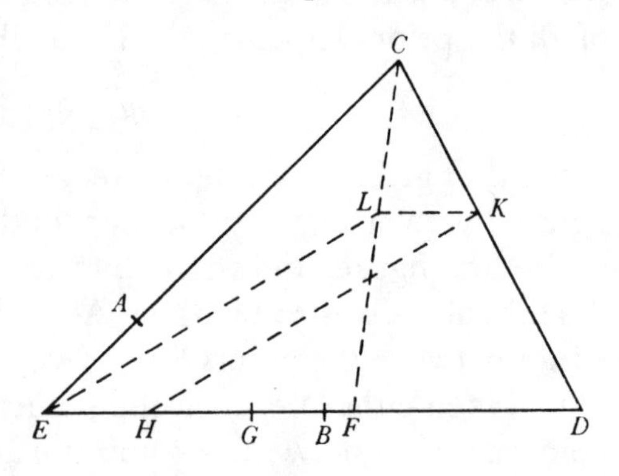

Concurrant enim rectæ *AC*, *BD* in *E* et in *BE* capiatur *BG* ad *AE* ut est *BD* ad *AC* sitq; *FD* æqualis *E*[*G*] et erit *EC* ad $GD_{[,]}$ hoc est ad *EF* ut *AC* ad *BD* adeoq; in ratione $\text{data}_{[,]}$ et propterea dabitur specie triangulum *EFC*. Secetur *CF* in *L* in ratione *CK* ad *CD* et dabitur etiam specie triangulum *EFL*, proindeq; punctum *L* locatur in recta *EL* positione data. Junge *LK* et ob datam *FD* et datam rationem

(79) Newton understands the particular case of the preceding figure in which *a* and *b* (now no longer given) coincide. For variety's sake we have introduced an appropriate diagram illustrating the alternative situation where the given point *p* lies outside the parallelogram *hikl* and the resulting locus is hyperbolic.

(80) Originally 'transire debet' (ought to pass). There will again (see note (78)) be two conic 'trajectories' constructible to satisfy this dual of Proposition XXIII preceding.

(81) This, inserted in afterthought on f. 88v to pave the way into Corollary 3 to Lemma XXV following, was subsequently (see §2; note (32)) invoked retrospectively to justify Corollary 4 to the introductory 'Leges Motûs'. We have previously remarked that Newton had derived this rectilinear locus twenty years before in its equally valid three-dimensional form

Proposition XXVI, Problem XVIII.

To describe a trajectory which shall pass through a given point and touch four straight lines given in position.

From the common intersection of any two of the tangents to the common intersection of the remaining two draw an indefinite straight line and, with this employed as prime ordinate radius, transmute the figure (by means of Lemma XXII) into a new figure:(79) the pairs of tangents before concurrent at the radius will now prove to be parallel. Let these be *hi* and *kl*, *ik* and *hl*, containing the parallelogram *hikl*. And let *p* be the point in this new figure corresponding to the point given in the first figure. Through the centre *o* of the figure draw *pq* and, where *oq* is equal to *po*, *q* will be a second point through which the conic in this new figure ought to pass. By the reverse procedure of Lemma XXII transfer this point into the first figure and in it there will be had two points through which the trajectory is to be described.(80) Through these, however, that trajectory can be described by Problem XVI. As was to be done.

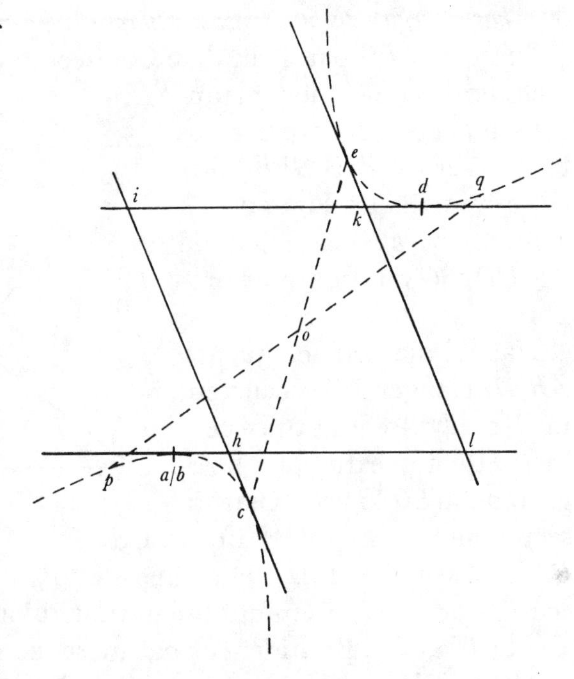

Lemma XXIII.(81)

If two straight lines AC, BD given in position and terminated at given points A, B have a given ratio one to the other, and the straight line CD by which the indeterminate points C, D are joined be cut [at K] in a given ratio, I assert that the point K will be located in a straight line given in position.

For let the straight lines *AC*, *BD* concur at *E*, and in *BE* take *BG* to *AE* as *BD* is to *AC*. Let *FD* be equal to *EG*, and *EC* will then be to *GD*, that is, to *EF*, as *AC* to *BD* and hence in a given ratio, and in consequence the triangle *EFC* will be given in species. Let *CF* be cut at *L* in the ratio of *CK* to *CD* and the triangle *EFL* will also be given in species, and accordingly the point *L* is located in the straight line *EL* given in position. Join *LK* and then, because *FD* and the ratio

(IV: 270–3) and later, about 1680, introduced it into a long list of 'Loca plana' in his Waste Book (IV: 240).

LK ad *FD* dabitur *LK*. Huic æqualis capiatur *EH* et erit *ELKH* parallelogrammum. Locatur igitur punctum *K* in parallelogrammi latere positione dato *HK*. Q.E.D.(82)

Lemma XXIV.(83)

Si rectæ tres tangant quamcunꝙ Coni sectionem quarum duæ parallelæ sint ac dentur positione, dico quod sectionis semidiameter hisce duabus parallela sit media proportionalis inter harum segmenta punctis contactuum et tangenti tertiæ interjecta.

Sunto *AF*, *BG* parallelæ duæ Conisectionem *ADB* tangentes in *A* et *B*; & *EF* recta tertia conisectionem tangens in *I* et occurrens prioribus tangentibus in *F* [et *G*];(84) sitꝙ *CD* semidiameter Figuræ tangentibus parallela: dico quod *AF*, *CD*, *BG* sunt continuè proportionales.

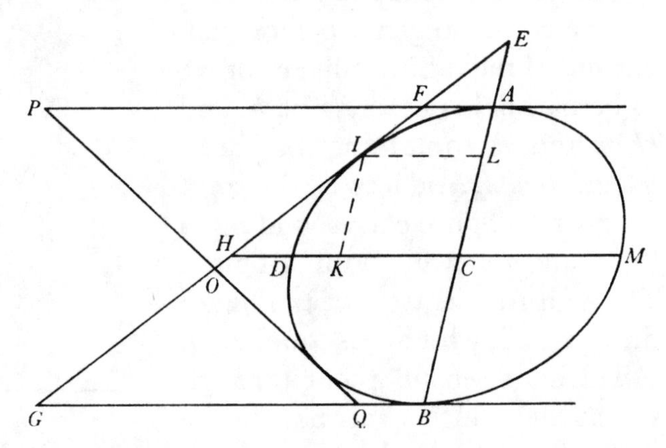

Nam si diametri conjugatæ *AB*, *DM* tangenti *FG* occurrent in *E* [et *H*](84) seꝙ mutuò secent in *C* et compleatur parallelogrammum *IKCL*, erit ex natura sectionum conicarum(85) *EC* ad *CA*, ita *CA* ad *LC* et ita divisim $EC-CA$ ad $CA-CL$ seu *EA* ad *AL*[,] et compositè *EA* ad $EA+AL$ seu *EL* ut *EC* ad $EC+CA$ seu *EB*, adeoꝙ (ob similitudinem triangulorum *EAF*, *ELI*, *ECH*, *EBG*) *AF* ad *LI* ut *CH* ad *BG*. Est itidem ex natura sectionum conicarum(85) *LI* seu *CK* ad *CD* ut *CD* ad *CH* atꝙ adeò ex æquo perturbate *AF* ad *CD* ut *CD* ad *BG*. Q.E.D.

Corol. 1. Hinc si tangentes duæ *FG*, *PQ* tangentibus parallelis *AF*, *BG* occurrant in *F*, *G*, *P* et *Q* seꝙ mutuò secent in *O*, erit (ex æquo perturbate) *AF* ad *BQ* ut *AP* ad *BG* et divisim ut *FP* ad *G*[*Q*], atꝙ adeo ut *FO* ad *OG*.

Corol. 2. Unde etiam rectæ duæ *PG*, *FQ* per puncta *P* et *G*, *F* et *Q* ductæ concurrent ad rectam *ACB* per centrum figuræ et puncta contactuum *A*, *B* transeuntem.

Lemma XXV.(86)

Si parallelogrammi latera quatuor infinite producta tangant sectionem quamcunꝙ Conicam et abscindantur ad tangentem quamvis quintam, sumantur autem abscissæ terminatæ ad angulos oppositos parallelogrammi: dico quod abscissa unius lateris sit ad

(82) The demonstration follows immediately from Newton's foreknowledge of the rectilinearity of the locus: for, when *G* coincides with *E* (and hence *D* with *F* and *K* with *L*, where $CK:CD = CL:CP$), the locus of *L* as EC/FE remains constant and so *CF* moves parallel to itself is evidently a straight line through *E*; and hence in general, since $LK = FD\times CK/CD$ is

of LK to FD are given, LK will be given. Equal to this take EH and $ELKH$ will be a parallelogram. The point K is located, therefore, in the parallelogram's side HK given in position. As was to be proved.(82)

Lemma XXIV.(83)

If three straight lines, two of which are parallel and given in position, touch any conic section, I state that the section's semidiameter parallel to these two is a mean proportional between the segments of these intercepted between the points of contact and the third tangent.

Let AF and BG be two parallels tangent to the conic ADB at A and B, and EF a third straight line tangent to the conic at I and meeting the previous tangents in F [and G];(84) and let CD be the semidiameter of the figure parallel to those tangents: I assert that AF, CD and BG are in continued proportion.

For if the conjugate diameters AB, DM meet the tangent FG in E [and H],(84) and mutually intersect in C, and if the parallelogram $IKCL$ be completed, then from the nature of conics(85) EC will be to AC as AC to LC, and so *dividendo* as $EC-AC$ to $AC-LC$ or EA to AL, and, by compounding, EA to $(EA+AL$ or) EL as EC to $(EC+AC$ or) EB, and hence (because of the similarity of the triangles EAF, ELI, ECH and EBG) AF to LI as CH to BG. Likewise from the nature of conics(85) there is LI (or CK) to CD as CD to CH, and hence *ex æquo perturbate* AF to CD as CD to BG. As was to be proved.

Corollary 1. From this, if two tangents FG, PQ meet the parallel tangents AF, BG in F, G, P and Q, and mutually intersect in O, then *ex æquo perturbate* AF will be to BQ as AP to BG, and *dividendo* as FP to GQ, and consequently as FO to OG.

Corollary 2. Whence also the two straight lines PG, FQ will, when drawn through P and G, F and Q, concur at the straight line ACB passing through the figure's centre and the points A, B of contact.

Lemma XXV.(86)

If a parallelogram's four indefinitely extended sides touch any conic and be intercepted at any fifth tangent, and the intercepts terminated at opposite corners of the parallelogram be taken up: I assert that the intercept of one side shall be to that side as the portion of the

both equal to $EH = EG\times CK/CD$ and parallel to it, the locus of K as $AC/BD = EC/GD$ remains constant is a parallel straight line through H.

(83) Newton was, we presume, not unaware that this lemmatical theorem is, along with its demonstration, essentially identical with Apollonius, *Conics* III, 42, but he here doubtless prefers to give explicit justification for the two important corollaries which he in sequel deduces from it.

(84) These trivial omissions were thus made good in the published *Principia* ($_1$1687: 90, 91).

(85) By Apollonius, *Conics* I, 37/39; compare II: 92, note (21).

(86) Newton generalises the preceding, Apollonian theorem, so effectively attaining in his following Corollary 1 (see note (87)) the anharmonic property of tangents to a conic.

latus illud ut pars lateris contermini inter punctum contactus et latus tertium ad abscissam lateris hujus contermini.

Tangant parallelogrammi *MIKL* latera quatuor *ML*, *IK*, *KL*, *MI* sectionem conicam in *A*, *B*, *C*, *D*, et secet tangens quinta *FQ* hæc latera in *F*, *Q*, *H* et *E*; dico quod sit *ME* ad *MI* ut *BK* ad *KQ*, et *KH* ad *KL* ut *AM* ad *MF*. Nam per Corollarium [1] Lemmatis superioris est *ME* ad *EI* ut *AM* seu *BK* ad *BQ* et componendo *ME* ad *MI* ut *BK* ad *KQ*. Q.E.D. Item *KH* ad *HL* ut *BK* seu *AM* ad *AF*, et dividendo *KH* ad *KL* ut *AM* ad *MF*. Q.E.D.

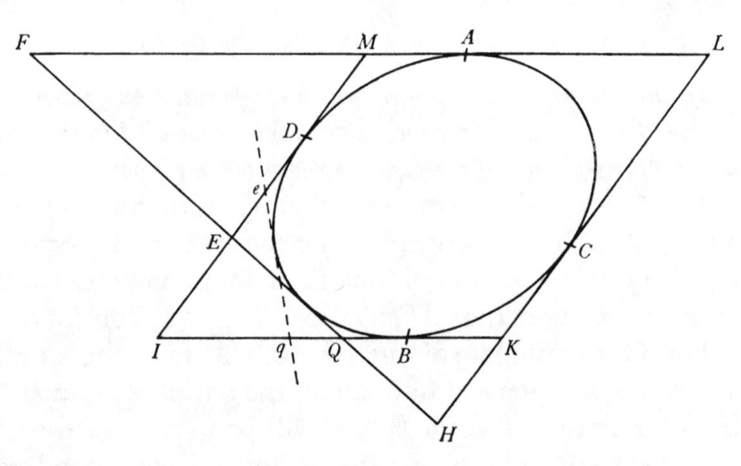

Corol. 1. Hinc si parallelogrammum *IKLM* datur, dabitur rectangulum $KQ \times ME$, ut et huic æquale rectangulum $KH \times MF$. Æquantur enim rectangula illa ob similitudinem triangularum *KQH*, *MFE*.(87)

Corol. 2. Et si sexta ducatur tangens *eq* tangentibus *KI*, *MI* occurrens in *e* et *q*[,] rectangulum $KQ \times ME$ æquabitur rectangulo $Kq \times Me$ eritꝗ *KQ* ad *Me* ut *Kq* ad *ME* et divisim ut *Qq* ad *Ee*.

Corol. 3. Unde etiam si *Eq*, *eQ* jungantur et bisecentur et recta per puncta bisectionum agatur, transibit hæc per centrum Sectionis Conicæ. Nam cum sit *Qq* ad *Ee* ut *KQ* ad *Me*, transibit eadem recta per medium omnium *Eq*, *eQ*, *MK* per Lem. XXIII et medium rectæ *MK* est centrum Sectionis.(88)

Prop. XXVII. Prob. XIX.

Trajectoriam describere quæ rectas quinꝗ positione datas continget.

Dentur positione tangentes *ABG*, *BCF*, *GCD*, *FDE*, *EA*. Figuræ quadrilateræ sub quatuor quibusvis contentæ *ABFE* diagonales *AF*, *BE* biseca et (per Cor. 3

(87) From a more general viewpoint, as the line *EQ* envelopes the conic it determines a 1, 1 correspondence $E \leftrightarrow Q$ between pairs of points in the tangents *MI* and *KI* such that the ranges (*E*) and (*Q*) are anharmonic (equi-cross): for, since in this correspondence $M \leftrightarrow \infty_{IK}$, $D \leftrightarrow I$, $I \leftrightarrow B$ and $\infty_{MI} \leftrightarrow K$, it follows that $ME \times KQ = MD \times KI = MI \times KB$, so that

$$MD/ME = KQ/KI \quad \text{and} \quad MI/ME = KQ/KB$$

or, on expressing these in cross-ratio form,

$$(M\,\infty_{MI}\,DE) = (\infty_{IK}\,KIQ) \quad \text{and} \quad (M\,\infty_{MI}\,IE) = (\infty_{IK}\,KBQ)$$

adjoining side between the point of contact and the third side is to the intercept of this adjoining side.

Let the four sides *ML, IK, KL, MI* of the parallelogram *MIKL* touch a conic at *A, B, C, D*, and let a fifth tangent *FQ* intersect these sides in *F, Q, H* and *E*: I assert that *ME* shall be to *MI* as *BK* to *KQ*, and *KH* to *KL* as *AM* to *MF*. For, by Corollary 1 of the previous Lemma, *ME* is to *EI* as *AM* or *BK* to *BQ*, and *componendo* *ME* to *MI* as *BK* to *KQ*. As was to be proved. Similarly *KH* is to *HL* as *BK* or *AM* to *AF*, and *dividendo* *KH* to *KL* as *AM* to *MF*. As was to be proved.

Corollary 1. Hence, if the parallelogram *IKLM* is given, the rectangle $KQ \times ME$ will be given, as also the rectangle $KH \times MF$ equal to it.(87) Those rectangles are equal, of course, because the triangles *KQH, MFE* are similar.

Corollary 2. And if a sixth tangent *eq* be drawn, meeting the tangents *KI* and *MI* in *e* and *q*, the rectangle $KQ \times ME$ will be equal to the rectangle $Kq \times Me$, and so *KQ* will be to *Me* as *Kq* to *ME*, and *dividendo* as *Qq* to *Ee*.

Corollary 3. Whence also, if *Eq, eQ* be joined and bisected and a straight line be drawn through the points of bisection, this will pass through the centre of the conic. For, since *Qq* is to *Ee* as *KQ* to *Me*, this same straight line will pass through the mid-point of all *Eq, eQ, MK* by Lemma XXIII, while the mid-point of the line *MK* is the section's centre.(88)

Proposition XXVII, Problem XIX.

To describe a trajectory which shall touch five straight lines given in position.

Let there be given in position the tangents *ABG, BCF, GCD, FDE* and *EA*. In the quadrilateral *ABFE* contained by any four bisect the diagonals *AF, BE* and

---

respectively. When *K* and *M*, and hence the conic's centre, lie at infinity, there ensues $(DEI\,\infty_{MI}) = (IQB\,\infty_{IK})$ and so $DI/DE = IB/IQ$; conversely, therefore, where points *E* (in *MI*) and *Q* (in *IK*) are taken such that the ratio DE/IQ is constant, the line *EQ* envelopes a parabola: a then wholly novel result which was independently derived by Christiaan Huygens by a simple algebraic argument in the case where $\widehat{MIK}$ is right (see his *Œuvres complètes*, **22**, 1950: 320) and afterwards expounded in general form (and with appropriate Apollonian proof) by Edmond Halley in an obscure concluding 'Scholion' of his edition of Apollonius' tract *On cutting off a ratio* (*Apollonii Pergæi de Sectione Rationis Libri Duo Ex Arabico MS^to^ Latine Versi. Accedunt Ejusdem de Sectione Spatii Libri Duo Restituti* (Oxford, 1706): 135–7), but first brought to general attention by J. H. Lambert in his widely read *Insigniores Orbitæ Cometarum Proprietates* (Augsburg, 1761): Section I, Lemmas IV–VI: 5–8. The hyperbolic locus which ensues when the point *M* coincides with *I* was similarly made by Halley the basis of his companion restoration of Apollonius' *On cutting off a space* (*Apollonii...de Sectione Rationis*...: 163–8). As we observed of the last in IV: 223, note (21), none of these later geometers show any appreciation of Newton's present generalisation.

(88) Conversely, of course, the centres of all conics which touch the quadrilateral *EQqe* lie on the line bisecting its diagonals *EQ* and *eq* (whence the mid-points of the three diagonals of a complete quadrilateral are collinear).

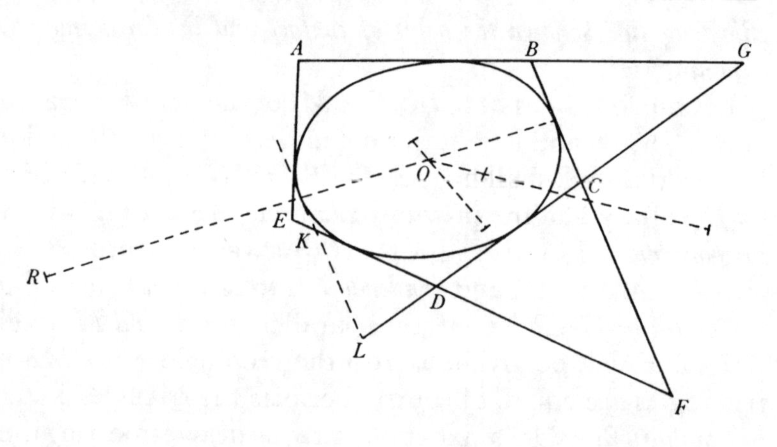

Lem. XXV) recta per puncta bisectionum acta transibit per centrum Trajectoriæ. Rursus figuræ quadrilateræ *BGDF* sub alijs quibusvis quatuor tangentibus contentæ diagonales (ut ita dicam) *BD*, *GF* biseca et recta per puncta bisectionum acta transibit per centrum sectionis. Dabitur ergo centrum in concursu bisecantium. Sit illud *O*. Tangenti cuivis *BC* parallelam age *KL* ad eam distantiam[89] ut centrum *O* in medio inter parallelas locetur, et acta *KL* tanget trajectoriam describendam. Secet hæc tangentes alias quasvis duas *CD*, *FDE* in *L* et *K*. Per tangentium quarumvis non parallelarum *CL*, *FK* cum parallelis *CF*, *KL* concursus *C* et *K*, *F* et *L* age *CK FL* concurrentes in *R* et recta *OR* ducta et producta secabit tangentes parallelas *CF*, *KL* in punctis contactuum. Patet hoc per Corol. 2. Lem. XXV. Eadem methodo invenire licet alia contactuum puncta, et tum demum per Casum 1. Prob. XIV. Trajectoriam describere. Q.E.F.[90]

Scholium.

Problemata, ubi dantur Trajectoriarum vel centra vel Asymptoti, includuntur in præcedentibus. Nam datis punctis et tangentibus una cum centro dantur alia totidem puncta aliæq; tangentes a centro ex altera ejus parte æqualiter distantes. Asymptotos autem pro tangente habenda est, et ejus terminus infinitè distans (si ita loqui fas sit) pro puncto contactus. Concipe tangentis cujusvis punctum contactus abire in infinitum et tangens vertetur in Asymptoton atq; constructiones Problematis XIV et Casus primi Problematis XV vertentur in constructiones Problematum ubi Asymptoti dantur.

Postquam Trajectoria descripta est invenire licet axes et umbilicos ejus hac methodo. In constructione Lemmatis XXI[91] fac ut angulorum mobilium *PBN*, *PCN* crura *BP*, *CP* quorum concursu trajectoria describebatur sint sibi invicem parallela eumq; servantia situm revolvantur circa polos suos *B*, *C*. Interea verò describant altera angulorum illorum crura *CN*, *BN* concursu suo [*K*] vel

(89) Originally 'ea lege' (with the restriction).

(90) For all its ingenuity Newton's construction is neither elegant nor, because of its several unwieldy auxiliary lines, easy to effect. With a knowledge of C. J. Brianchon's polar

then (by Corollary 3 of Lemma XXV) a straight line drawn through the points of bisection will pass through the centre of the trajectory. Again, in the quadrilateral *BGDF* contained by any other four tangents bisect the diagonals—if I may call them so—*BD*, *GF* and a straight line drawn through the points of bisection will pass through the section's centre. The centre will therefore be given at the meet of the bisectors. Let that be *O*. Parallel to any tangent *BC* draw *KL* at such a distance away[89] that the centre *O* shall be located midway between the parallels, and the drawn *KL* will touch the trajectory to be described. Let this intersect any other two tangents *CD*, *FDE* in *L* and *K*. Through the meets *C* and *K*, *F* and *L* of any non-parallel tangents *CL*, *FK* with the parallel ones *CF*, *KL* draw *CK* and *FL* concurring at *R*, and then the straight line *OR* will, when drawn and produced, intersect the parallel tangents *CF*, *KL* in points of contact. This is evident by Corollary 2 to Lemma XXV. By the same method we are free to find further points of contact, and then finally, by Case 1 of Problem XIV, to describe the trajectory. As was to be done.[90]

Scholium.

Problems in which either the centres or asymptotes of trajectories are given are included in the preceding ones. For, given points and tangents together with the centre, there are given an equal number of further points and tangents equidistant from the centre on its other side. An asymptote, however, is to be considered as a tangent, and its infinitely distant end-point (if it is right to speak of it so) as the point of contact. Imagine that the point of contact in any tangent goes off to infinity and the tangent will be turned into an asymptote, and the constructions of Problem XIV and Case 1 of Problem XV into constructions of problems in which asymptotes are given.

After a trajectory has been described, we are at liberty to find out its axes and foci by this method. In the construction of Lemma XXI[91] make the legs *BP*, *CP* of the mobile angles $\widehat{PBN}$, $\widehat{PCN}$ by whose meet the trajectory was described parallel to one another, and then, keeping that position, revolve round their poles *B*, *C*. During this, however, let the other legs *CN*, *BN* of those angles

dual of the Pascalian *hexagrammum mysticum* (first published in his 'Mémoire sur les Surfaces courbes du second Degré', *Journal de l'École Polytechnique*, **6**, 1806: 297–311) we may in hindsight directly determine the conic's point of contact with the tangent *BC* as the latter's intersection with the line joining *E* to the meet of *AC* and *BD*. The simplicity and economy of such a solution would have strongly appealed to Newton's puritanist aesthetic sense.

(91) In a parenthesis in the manuscript at this point Newton refers his reader back to the 'fig. 39' there illustrating this lemma. In compensation for here omitting this textual pointer to it we incorporate the essence of that diagram in the figure illustrating more fully the hyperbolic case of the present addendum which we give in our English version. The printed *Principia* ($_1$1687: 94) emends the present phrase to read 'In constructione & Figura Lemmatis XXI' (In the construction and figure of Lemma XXI).

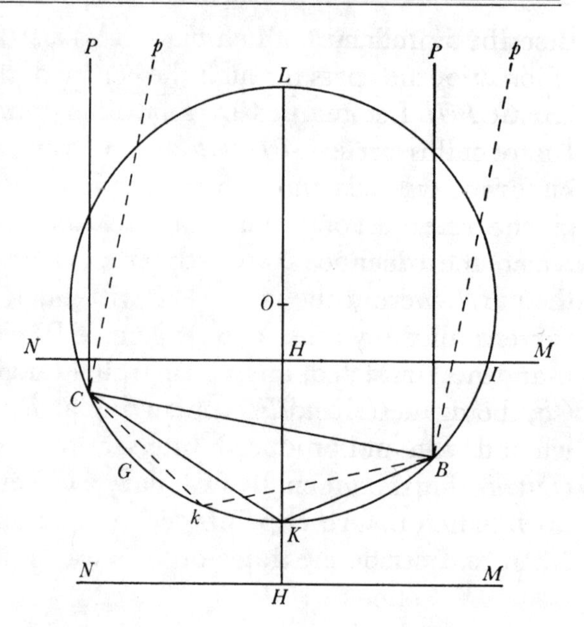

[k][(92)] circulum $IBKGC$.[(93)] Sit circuli hujus centrum O. Ab hoc centro ad Regulam MN ad quam altera illa crura CN, BN interea concurrebant dum trajectoria describe[bat]ur, demitte normalem OH circulo occurrentem in K et L. Et ubi crura illa altera CB, BN concurrunt ad punctum istud K, quod Regulæ[(94)] propius est, crura prima CP, BP parallela erunt axi majori et perpendicularia minori[,] et contrarium eveniet si crura eadem concurrunt ad punctum remotius L.[(95)] Unde si detur Trajectoriæ centrum, dabuntur axes. Hisce autem datis, umbilici sunt in promptu.

Axium verò quadrata sunt ad invicem ut KH ad LH,[(96)] et inde facile est Trajectoriam specie datam per data quatuor puncta describere. Nam si duo ex punctis datis constituantur poli C, B, tertium dabit angulos mobiles PCK, PBK. Tum ob datam specie Trajectoriam dabitur ratio OH ad OK centroq; O et

(92) Newton's original version of this autograph insertion in the manuscript was 'concursu suo qui in Figura nominatur K' (by their meet, which in the [present] figure is named K), but subsequently, in a confusing further reference back to his earlier diagram (see the previous note), he altered this to 'concursu suo N vel n'. Our textual emendation follows Newton's further change of mind, back to his initial intention, in the ensuing *Principia* ($_1$1687: 94).

(93) For, since the included angle $\widehat{BKC} = \widehat{BIC}$ (or its supplement) is equal to the sum of the given angles $\widehat{IBP}$ and $\widehat{ICP}$—the positions (in the figure of the preceding Lemma XXI, here incorporated in its essence into our expanded variant diagram) of the rotating angles $\widehat{NBP}$ and $\widehat{NCP}$ when P (in the directions marked by us as P' and P'') is at infinity—, it is itself given in magnitude.

(94) Namely, MN.

(95) Since, where the legs BN, CN of the rotating angles $\widehat{NBP}$, $\widehat{NCP}$ pass through the two intersections I (marked as I', I'' in our English figure) of the *regula* MN with the circle, their other legs are through the described conic's infinite point P (similarly differentiated as P', P'' in our illustration of the variant hyperbolic case where these are real) and hence parallel to its asymptotes, it follows that, when the legs BN, CN pass through the (diametrally opposite) points K and L, equidistant from I' and I'' on the circle, the describing legs BP, CP likewise meet at infinity in a pair of points which are in the conic's principal axes bisecting the angle between the asymptotes. Clearly, the angle between the asymptote direction corresponding to the intersection of the legs BN, CN at I' and the axis direction corresponding to their meet at K is $\widehat{I'BK} = \widehat{I'LK}$, and hence, when (as drawn) K is nearer than L to MN, the describing legs BP, CP corresponding to K are parallel to the conic's major axis, as Newton says.

describe by their meet K or k(92) the circle $IBKGC$.(93) Let the centre of this circle be O. From this centre to the ruler MN at which the other legs CN, BN were concurrent all the while that the trajectory was being described let fall the normal OH meeting the circle in K and L. And, when the other legs CN, BN concur at that point K which is nearer the ruler,(94) the first legs CP, BP will be parallel to the major axis and perpendicular to the minor one, and the converse will happen if the same legs concur at the farther point L.(95) Hence, if the centre of the trajectory be given, the axes will be given. Once, however, these are given, the foci are to hand.

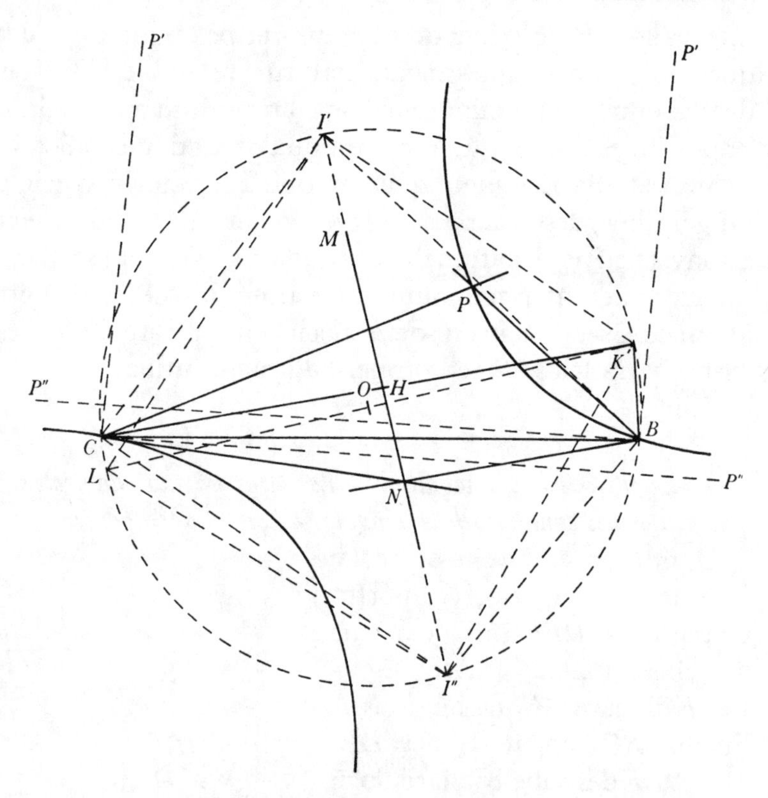

The squares of the axes, to be sure, are to one another as KH to LH,(96) and it is easy therefrom to describe a trajectory given in species through four given points. For, if two of the given points form the poles B and C, a third will yield the mobile angles $P\widehat{B}K$, $P\widehat{C}K$; then because the trajectory is given in species, the ratio of OH to OK will be given and so, on describing a circle with centre O and

(96) For, as an immediate sequel to the preceding note, in the hyperbolic case where the conic's points at infinity are real the major and minor axes are to one another as

$$I'L = \sqrt{[LH \times LK]}$$

to $I'K = \sqrt{[HK \times LK]}$, that is, as $\sqrt{LH}$ to $\sqrt{HK}$. Though Henry Pemberton would doubtless here formulate a parallel objection to that he made in 1726 regarding Proposition XXIV when IXY there no longer meets the conic (see note (66)), Newton's assertion that this proportion continues to hold in the elliptical case where the *regula* MHN no longer intersects the circle in real points I follows by continuity, but again a more roundabout elementary proof may be given without passing into the complex plane. The limiting parabolic case is, of course, that in which the *regula* MN is tangent to the circle and hence the points I coincide with one or other of K or L. See also Colin Maclaurin, *Geometria Organica: sive Descriptio Linearum Curvarum Universalis* (London, 1720): Pars Prima, Sectio I, Propositio II. 'Problema. Determinare Curvarum Assymptotos atque Species': 3–5.

intervallo OH describendo circulum et per punctum quartum agendo rectam quæ circulum illum tangat, dabitur regula MN[97] cujus ope Trajectoria describetur. Unde etiam vicissim Trapezium specie datum (si casus quidam impossibiles excipiantur) in data quavis sectione conica inscribi potest.

Sunt ct alia Lemmata quorum ope Trajectoriæ specie datæ, datis punctis et tangentibus, describi possunt. Ejus generis est quod, si recta linea per punctum quodvis positione datum ducatur quæ datam Coni sectionem in punctis duobus intersecet, et intersectionum intervallum bisecetur, punctum bisectionis tanget aliam coni sectionem ejusdem speciei cum priore, atq; axes habentem prioribus axibus parallelas.[98] Sed propero ad magis utilia.

Lemma XXVI.

Trianguli specie et magnitudine dati tres angulos ad rectas totidem positione datas, quæ non omnes sunt parallelæ, singulos ad singulas[99] *ponere.*

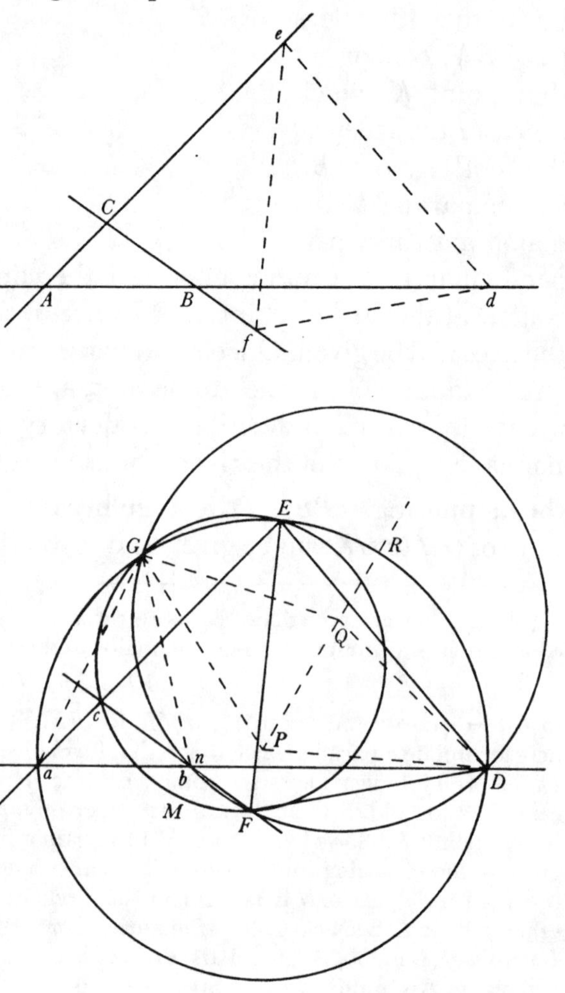

Dantur positione tres rectæ infinitæ AB, AC, BC, et oportet triangulum DEF ita locare ut angulus ejus D lineam AB, angulus E lineam AC, et angulus F lineam BC tangat. Super DE, DF et EF describe tria circulorū segmenta[100] DRE, DGF, EMF quæ capiant angulos angulis BAC, ABC, ACB æquales respectivè. Describantur autem hæc segmenta ad eas partes linearum DE, DF, EF, ut literæ $DKED$ eodem ordine cum literis $BACB$, literæ $DLFD$ eodem cum literis $ABCA$ et literæ $EMFE$ eodem cum literis $ACBA$ in orbem redeant: deinde compleantur hæc segmenta in circulos.[101] Secent circuli duo priores se mutuo in G, sintq; centra eorum P et Q. Junctis GP, PQ, cape Ga ad AB ut est GP ad PQ et centro G intervallo Ga describe circulum qui secet circulum primum GDE in a. Jungatur tum aD secans circulum secundum DGF in b,

radius OH and drawing through the fourth point a straight line to touch that circle, there will be given the ruler MN(97) with whose aid the trajectory shall be described. Whence also, in turn, a quadrilateral given in species can (if certain impossible cases be excepted) be inscribed in any given conic.

There are other lemmas, too, with whose aid trajectories given in species, and with given points and tangents, can be described. Of the sort is this: if a straight line be drawn through any point given in position to intersect a given conic section in two points, and the distance between the intersections be bisected, the point of bisection will lie on another conic of the same species as the first and having its axes parallel to the former's.(98) But I hasten on to more useful ones.

Lemma XXVI.

To set the three corners of a triangle given in species and size on the same number of straight lines (not all parallel) given in position, each on its separate one.(99)

The three indefinite straight lines AB, AC, BC are given in position, and it is required so to place the triangle DEF that its corner D touches the line AB, its corner E the line AC, and its corner F the line BC. On DE, DF and EF describe three circle segments(100) DRE, DGF, EMF which shall contain angles equal to the angles $B\widehat{A}C$, $\widehat{ABC}$ and $\widehat{ACB}$ respectively—but describe these segments on the sides of the lines DE, DF, EF which yield the letters DKE in the same closed cyclic order as the letters BAC, the letters DLF in the same one as the letters ABC, and the letters EMF in the same one as the letters ACB: then complete these segments into circles.(101) Let the first two circles mutually intersect in G, and let their centres be P and Q. On joining GP, PQ, take Ga to AB as GP is to PQ, and with centre G, radius Ga draw a circle to intersect the first circle DGE in a. Then join aD, cutting the second circle DGF in b,

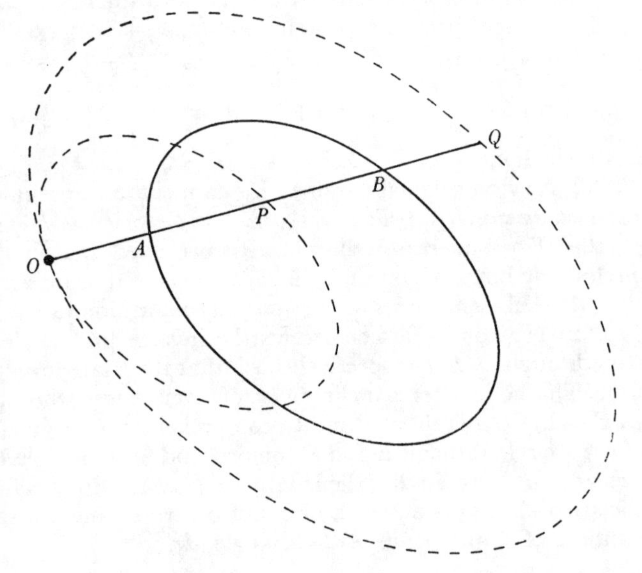

(97) More precisely, two points on MN, which will then determine its position.

(98) For, given a general conic AB and some point O outside it, the locus (Q) defined for all transversals $OABQ$ through O by $OA = BQ$ is manifestly a similar concentric conic which instantaneously shares a common conjugate diameter through the mid-point P of the chords AB and OQ; whence, since universally $OP = \frac{1}{2}OQ$, the locus (P) is by homothety a similar, similarly positioned conic. Analytically, where

tum aE secans circulum tertium $GE[F]$ in c. Et compleatur figura $ABCdef$ similis et æqualis figuræ $abcDEF$. Dico factum.(102)

Agatur enim Fc ipsi aD occurrens in n. Jungantur aG, bG, PD, QD et producatur PQ ad R. Ex constructione est angulus EaD æqualis angulo CAB et angulus EcF æqualis angulo ACB, adeoq; triangulum anc triangulo ABC æquiangulum. Ergo angulus anc seu FnD angulo ABC adeoq; angulo FbD æqualis est, et propterea punctum n incidit in punctum b. Porrò angulus GPQ æqualis est angulo ad circumferentiam GaD et angulus GQR qui dimidius est complementi(103) anguli ad centrum GQD æqualis est angulo ad circumferentiam GbD, adeoq; eorum complementa PQG abG æquantur, suntq; [i]deò triangula GPQ Gab similia, et Ga est ad ab ut GP ad PQ[,] id est (ex constructione) ut Ga ad AB. Æquantur itaq; ab et AB et propterea triangula abc, ABC quæ modo similia esse probavimus sunt etiam æqualia. Unde cum tangant insuper trianguli DEF anguli D, E, F trianguli abc latera ab, ac, bc respectivè: compleri potest figura $ABCdef$ figuræ $abcDEF$ similis et æqualis atq; eam complendo solvetur Problema. Q.E.F.

Corol. Hinc recta duci potest cujus partes longitudine datæ rectis tribus positione datis interjacebunt. Concipe(104) Triangulum DEF puncto D ad latus EF accedente, et lateribus DE, DF in directum positis mutari in lineam rectam cujus pars data DE, rectis positione datis AB, AC et pars data DF rectis positione datis AB, BC interponi debet et applicando constructionem præcedentem ad hunc casum solvetur Problema.

O is the origin of a system of Cartesian coordinates in which the given conic is defined by the general second-degree equation $ax^2+2hxy+by^2+2gx+2fy+c = 0$ and is through $A(x_1, y_1)$ and $B(x_2, y_2)$, the locus of $P(X, Y)$ satisfies $Y/X = y/x = m$, say, where

$$(a+2hm+bm^2)x^2+2(g+fm)x+c = 0 \quad \text{and so} \quad X = \tfrac{1}{2}(x_1+x_2) = -(g+fm)/(a+2hm+bm^2),$$

from which $aX^2+2hXY+bY^2+gX+fY = 0$.

(99) Newton originally framed his enunciation to require the triangle to be placed '...*ad rectas totidem specie et magnitudine*[!] *datas*' (*on the same number of straight lines given in species and size*).

(100) For some reason this phrase is preferred to an initial equivalent 'tres circulos' (three circles). By both, of course, he means 'tres circulorum arcus' (three circle arcs).

(101) This sentence is a late autograph addition in the manuscript (f. 66v).

(102) Newton in fact constructs the inverse problem of fitting straight lines ab, ac, bc to pass through D, E, F respectively such that the triangle abc they form is determined in magnitude. This he reduces to inclining ab, of given length, through D so as to lie between the circles DaE and DbG, of centres P and Q, comprehending the given angles $\hat{A}$ and $\hat{B}$ (whence the locus of c is a circle through E and F comprehending the angle $\hat{C} = \pi-\hat{A}-\hat{B}$): since the triangles Gab and GPO are similar, the length $Ga = ab\times GP/PQ$ is fixed and the point a is immediately constructible by means of a circle of centre G and that radius. There are manifestly two positions of a, and hence of the triangle abc.

and also aE cutting the third circle GEF in c. And complete the figure $ABCdef$ similar and congruent to the figure $abcDEF$. I say it is done.[102]

For draw Fc meeting aD in n. Join aG, bG, PD, QD and extend PQ to R. By construction angle $\widehat{EaD}$ is equal to angle $\widehat{CAB}$ and $\widehat{EcF}$ to $\widehat{ACB}$, and therefore the triangle anc is equiangular with the triangle ABC. In consequence the angle $\widehat{anc}$ or $\widehat{FnD}$ is equal to the angle $\widehat{ABC}$, and hence to $\widehat{FbD}$, and as a result the point n coincides with the point b. Moreover, the angle $\widehat{GPQ}$ is equal to the angle $\widehat{GaD}$ at the circumference, and the angle $\widehat{GQR}$, which is half the complement[103] of the angle $\widehat{GQD}$ at the centre, is equal to the angle $\widehat{GbD}$ at the circumference, and hence their supplements $\widehat{PQG}$, $\widehat{abG}$ are equal; the triangles GPQ, Gab are consequently similar, and so Ga is to ab as GP to PQ, that is, (by construction) as Ga to AB. Therefore ab and AB are equal, and the triangles abc, ABC which we just now proved to be similar are accordingly also congruent. Whence, since in addition the corners D, E, F of the triangle DEF touch the sides ab, ac, bc respectively of the triangle abc, the figure $ABCdef$ similar and congruent to the figure $abcDEF$ can be completed, and by completing it the problem will be solved. As was to be done.

Corollary. Hence a straight line can be drawn whose sections given in length shall lie between three straight lines given in position. Imagine[104] that the triangle DEF, as its point D joins the side EF, and so with its sides DE, DF set out in a straight line, is changed into a straight line whose given section DE must be placed between the straight lines AB, AC given in position, and given section DF between the straight lines AB, BC given in position; and then by applying the preceding construction to this case the problem will be solved.

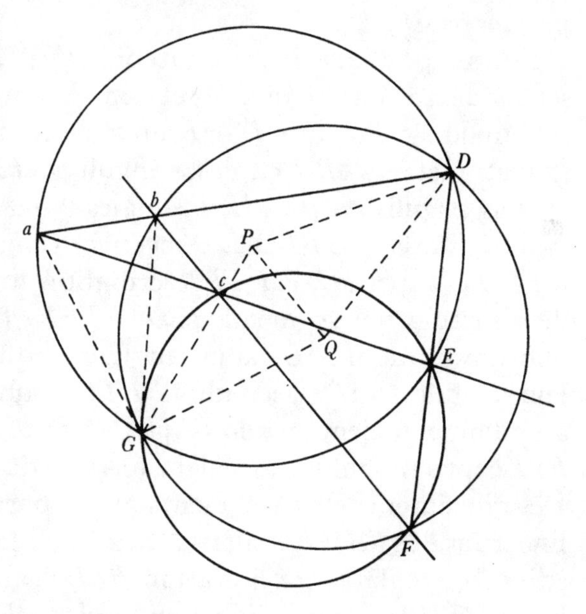

(103) Understand 'ad quatuor rectos' (to four right angles).

(104) Imagination is converted into visual reality in the appropriately remoulded figure we introduce into the English version.

Prop. XXVIII. Prob. XX.

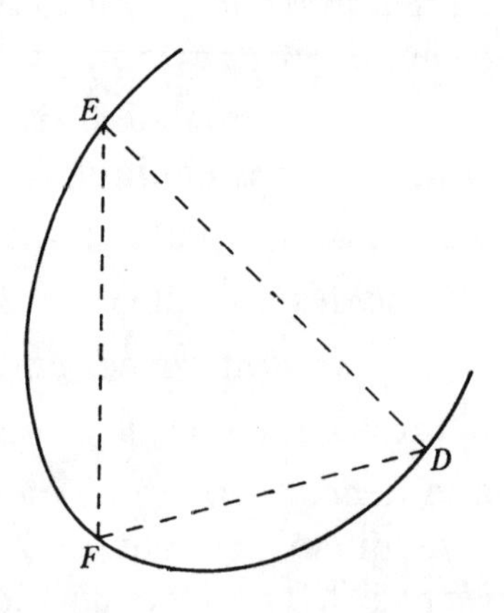

Trajectoriam specie et magnitudine datam describere cujus partes datæ(105) *rectis tribus positione datis interjacebunt.*

Describenda sit Trajectoria quæ sit similis et æqualis lineæ curvæ *DEF* quæq; a rectis tribus *AB*, *AC*, *BC* positione datis, in partes datis hujus partibus *DE* et *EF* similes et æquales secabitur. Age rectas *DE*, *EF*, *DF* et trianguli hujus *DEF* pone angulos *D*, *E*, *F* ad rectas illas positione datas (per Lem. XXVI). Dein circa Triangulum describe Trajectoriam curvæ *DEF* similem et æqualem. Q.E.F.

Lemma XXVII.

Trapezium specie datum describere cujus anguli ad rectas quatuor positione datas (quæ neq; omnes parallelæ sunt neq; ad commune punctum convergunt)(106) *singuli ad singulas consistent.*

Dentur positione rectæ quatuor *ABC*, *AD*, *BD*, *CE* quarum prima secet secundam in *A*, tertiam in *B*, et quartam in *C*: et describendum sit Trapezium *fghi* quod sit Trapezio *FGHI* simile et cujus angulus *f* angulo dato *F* æqualis tangat rectam *ABC* cæteriq; anguli *g*, *h*, *i* cæteris angulis datis *G*, *H*, *I* æquales tangant cæteras lineas *AD*, *BD*, *CE* respectivè. Jungatur *FH* et super *FG*, *FH*, *FI* describantur totidem circulorum segmenta *FSG*, *FTH*, *FVI*, quorum primum *FSG* capiat angulum æqualem angulo *BAD*, secundum *FTH* capiat angulum æqualem angulo *CB*[*D*] ac tertium *FVI* capiat angulum æqualem angulo *ACE*. Describi autem debent segmenta ad eas partes linearum *FG*, *FH*, *FI*, ut literarum *FSGF* idem sit ordo circularis qui literarum *BADB*, utq; literæ *FTHF* eodem ordine cum literis *CBEC*, et literæ *FVIF* eodem cum literis *ACEA* in orbem redeant. Compleantur segmenta in circulos, sitq; *P* centrum circuli primi *FSG*, et *Q* centrum secundi *FTH*. Jungatur et utrinq; producatur *BQ* et in ea capiatur *QR* in ea ratione ad *PQ* quam habet *BC* ad *AB*. Capiatur

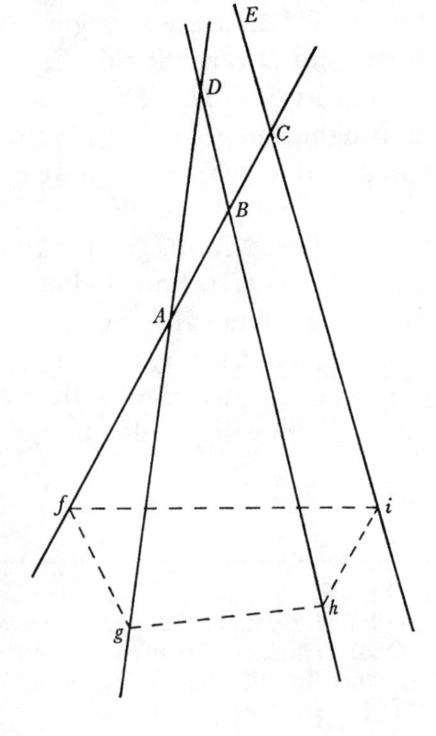

Proposition XXVIII, Problem XX.

To describe a trajectory, given in species and magnitude, whose given(105) *parts shall lie between three straight lines given in position.*

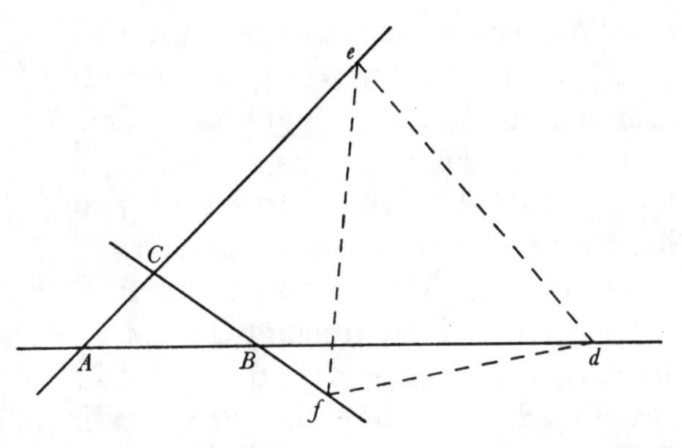

Let there need to be described a trajectory which shall be similar and congruent to the curve *DEF*, and be cut by the three straight lines *AB*, *AC*, *BC* given in position into parts similar and congruent to the given parts *DE* and *EF* of this. Draw the straight lines *DE*, *EF*, *DF* and (by Lemma XXVI) place the corners *D*, *E*, *F* of this triangle *DEF* onto those straight lines given in position. Then about the triangle describe the trajectory similar and congruent to the curve *DEF*. As was to be done.

Lemma XXVII.

To draw a quadrilateral given in species whose corners shall be positioned on four straight lines given in position (not all parallel or converging to a common point),(106) *each on their individual one.*

Let there be given in position the four straight lines *ABC*, *AD*, *BD*, *CE*, the first of which shall cut the second in *A*, the third in *B* and the fourth in *C*; and let there need to be described the quadrilateral *fghi*, similar to the quadrilateral *FGHI*, such that its angle *f* (equal to the given angle *F*) shall touch the straight line *ABC* and its other angles *g*, *h*, *i* (equal to the remaining given angles *G*, *H*, *I*) the remaining lines *AD*, *BD*, *CE* respectively. Join *FH* and on *FG*, *FH*, *FI* describe as many circle segments *FSG*, *FTH*, *EVI*, of which the first *FSG* shall contain an angle equal to $B\widehat{A}D$, the second $F\widehat{T}H$ an angle equal to $C\widehat{B}D$, and the third *FVI* one equal to $A\widehat{C}E$. The segments ought, however, to be described on the sides of the lines *FG*, *FH*, *FI* making the letters *FSG* have the same cyclical order as the letters *BAD*, the letters *FTH* the same order as the letters *CBE*, and the letters *FVI* the same rotary order as the letters *ACE*. Complete the segments into circles, and let *P* be the centre of the first circle *FSG* and *Q* the centre of the second, *FTH*. Join *BQ* and extend it either way, and in it take *QR* to *PQ* in the ratio which *BC* has to *AB*—but on the side of point *Q* which makes the letters

(105) 'imperatæ' (ordained) was initially written.

(106) This parenthesis is a late insertion by Newton in the manuscript.

autem QR ad eas partes puncti Q ut literarum P, Q, R idem sit ordo circularis atq; literarum ABC centroq; R et intervallo RF describatur circulus quartus FNc secans circulum tertium FVI in c. Jungatur Fc secans circulum primū in a et secundum in b. Agantur aG, bH et figuræ $abcFGHI$ similis constituatur figura $ABCfghi$: eritq; trapezium $fghi$ illud ipsum quod constituere oportuit.(107)

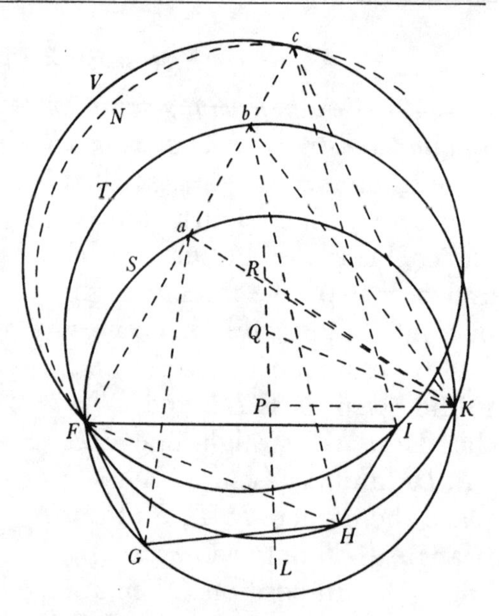

Secent enim circuli duo primi FSI, FGH se mutuò in K. Jungantur PK, QK, RK, aK, bK, cK et producatur QP ad L. Anguli ad circumferentias FaK, FbK, FcK sunt semisses angulorum FPK, FQK, FRK adeoq; angulorum illorum dimidijs LPK, LQK, LRK æquales. Est ergo figura $PQRK$ figuræ $abcK$ æquiangula et similis[,] et propterea ab est ad bc ut PQ ad QR id est ut AB ad BC. Angulis insuper FaG, FbH, FcI æquantur fAg, fBh, fCi per constructionem. Ergo figuræ $abcFGHI$ figura similis $ABCfghi$ compleri potest. Quo facto trapezium $fghi$ constituetur simile trapezio $FGHI$ et angulis suis f, g, h, i tanget rectas AB, AD, BD, CE. Q.E.F.

Corol. Hinc recta duci potest cujus partes rectis quatuor positione datis dato ordine interjectæ datam habebunt proportionem a[d] invicem. Augeantur anguli FGH, GHI usq; eo ut rectæ FG, GH, HI in directum jaceant et in hoc casu construendo Problema ducetur recta $fghi$ cujus partes fg, gh, hi rectis quatuor positione datis AB et AD, AD et BD et BD, CE interjectæ, erunt ad invicem ut lineæ FG, GH, HI eundemq; servabunt ordinem inter se.(108) Idem verò sic fit expeditius.

---

(107) Much as in his preceding Lemma XXVI, Newton again resolves the inverse problem of inclining straight lines through the corners F, G, H, I of the given quadrilateral so that, where a, b, c are on a transversal through F such that $ab : bc = AB : BC$, the angles $F\widehat{a}G$, $F\widehat{b}H$ and $F\widehat{c}I$ shall be equal to the given angles $f\widehat{A}g$, $f\widehat{B}h$ and $f\widehat{C}i$ respectively: clearly, the latter condition requires a, b, c to lie on circles FaG, FbH and $FVcI$ comprehending the given angles $\hat{A}$, $\hat{B}$ and $\hat{C}$ respectively, while the former also thereby determines c to be on a second circle FNK, of centre R and through the meet (other than F) of the circles (a) and (b), which is constructed, where P and Q are the (collinear) centres of the two latter circles, by setting $PQ : QR = AB : BC$ (since the configurations $Kabc$ and $KPQR$ are readily seen to be similar; compare IV: 240, note (31)). As an unstated corollary (which will afterwards, however, implicitly be employed in the scholium to Proposition XXIX; see note (113) below), when the circles $FVcI$ and $FNcK$ coincide—that is, when (c) is the circle through F, I and K—an infinite

P, Q, R have the same cyclical order as that of the letters A, B, C; and then with centre R and radius RF describe a fourth circle FNc intersecting the third circle FVI in c. Join Fc, cutting the first circle in a and the second in b. Draw aG, bH and let the figure $ABCfghi$ be formed similar to the figure $abcFGHI$: the quadrilateral $fghi$ will then be the very one it was required to construct.(107)

For let the first two circles FSI, FGH mutually intersect in K. Join PK, QK, RK, aK, bK, cK and extend QP to L. The angles $\widehat{FaK}$, $\widehat{FbK}$, $\widehat{FcK}$ at the circumferences are halves of the angles $\widehat{FPK}$, $\widehat{FQK}$, $\widehat{FRK}$ and hence equal to the halves $\widehat{LPK}$, $\widehat{LQK}$, $\widehat{LRK}$ of those angles. In consequence, the figure $PQRK$ is equiangular, and so similar, to the figure $abcK$ and accordingly ab is to bc as PQ to QR, that is, as AB to BC. To the angles $\widehat{FaG}$, $\widehat{FbH}$, $\widehat{FcI}$, moreover, are (by construction) equal the angles $\widehat{fAg}$, $\widehat{fBh}$, $\widehat{fCi}$. In consequence, the figure $ABCfghi$ similar to the figure $abcFGHI$ can be completed. Once this is effected, the quadrilateral $fhgi$ will be formed similar to the quadrilateral $FGHI$ and at its corners f, g, h, i touching the straight lines AB, AD, BD, CE. As was to be done.

Corollary. Hence a straight line can be drawn whose sections, intercepted in a given order by four straight lines given in position, shall have a given ratio to one another. Let the angles $\widehat{FGH}$, $\widehat{GHI}$ increase so far as to make the straight lines FG, GH, HI lie straight out, and by constructing the problem in this case a straight line $fghi$ shall be drawn whose sections fg, gh, hi intercepted by four straight lines given in position (by AB and AD, AD and BD, and BD and CE) will be to one another as the lines FG, GH, HI and preserve the same order among themselves.(108) But the same is done more speedily in this manner.

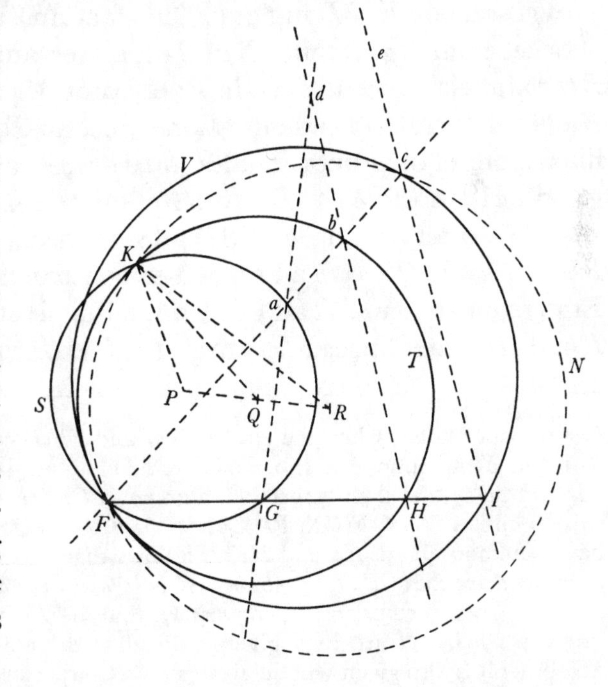

number of quadrilaterals $fghi$ similar to $FGHI$ can be constructed to lie within the given lines; whence conversely, if $fghi$ is placed so that f lies on AB, g on AD and h on BD, it follows that the remaining vertex will lie always on a unique straight line, whose intersection with CE will fix it in the position here required.

(108) A figure illustrating this particular case is introduced in our English version. As before

Producantur AB ad K et BD ad L ut sit BK ad AB ut HI ad GH et DL ad BD ut GI ad FG et jungatur KL occurrens rectæ CE in i. (109)Producatur iL ad M ut sit LM ad iL ut GH ad HI et agatur tum MQ ipsi Lh parallela rectæq; AD occurrens in g[,] tum gi secans AB, BD in f et h. Dico factum.(110)

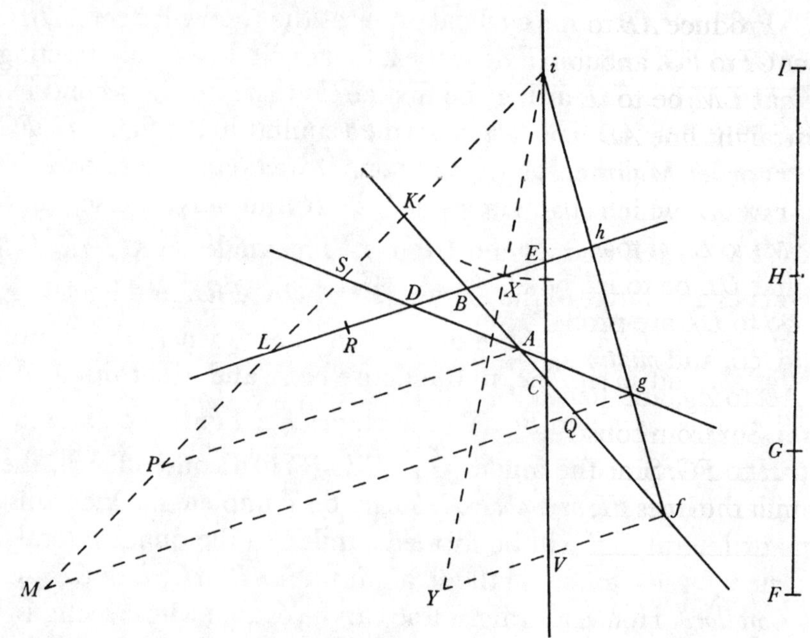

Secet enim Mg rectam AB in Q et AD rectam KL in S et agatur AP quæ sit ipsi BD parallela et occurrat iL in P; et erunt Mg ad Lh (Mi ad Li, gi ad hi, AK ad BK) et AP ad BL in eadem ratione. Secetur DL in R ut sit DL ad RL in eadem illa ratione et ob proportionales gS ad gM, AS ad AP et DS ad DL erit ex æquo ut gS ad Lh ita AS ad BL et DS ad RL et mixtim $BL-RL$ ad $Lh-BL$ ut $AS-DS$ ad $gS-AS$. Id est BR ad Bh ut AD ad Ag, adeoq; ut BD ad gQ. Et vicissim BR ad BD ut Bh ad gQ seu fh ad fg. Sed ex constructione est BR ad BD ut FH ad FG. Ergo fh est ad fg ut FH ad FG. Cum igitur sit etiam ig ad ih ut Mi ad Li id est ut IG ad IH, patet lineas FI, fi in g et h, G et H similiter sectas esse. Q.E.F.

(see previous note), when the circles $FVcI$ and $FNcK$ coincide, and so (c) is the circle through F, I and K, an infinite number of lines FI can be placed within the given lines AB, AD, BD and CE so that their respective intercepts FG, GH, HI are in a given proportion: since the trios of circles $FGaK$, $FHbK$, $FIcK$ and $HbKF$, $HdKG$, HeK (where e is the meet of bd and ce) have common chords FK and HK, it follows that necessary and sufficient conditions for this to happen are that $GH:HI=ab:bc$ and $FG:GI=bd:de$ (compare, *mutatis mutandis*, v: 301–2, note (405)). It further ensues, conversely, that, if FI is set within the lines ab, ad, bc so that its intercepts FG, GH are in a given ratio, then the point I in it such that the ratio of FI to FG (or GH) is also given will lie always on a unique straight line given in position.

(109) The manuscript originally continued: 'Producatur iE ad M ut sit EM ad iE ut GI ad FG et agatur MF ipsi BD parallela ipsiq; AB occurrens in f. Et jungatur fi secans AD et BD in g et h. Dico factum. Nam concurrentibus iL, fM in N sunt Nf ad Lh, Ni ad Li, Mi ad Ei, fi ad hi, BL ad [D]L in eadem ratione' (Produce iE to M so that EM be to iE as GI to FG, and draw MF parallel to BD and meeting AB in f. Then join fi cutting AD and BD in g and h. I say it is done. For, where iL and fM meet in N, the ratios Nf to Lh, Ni to Li, Mi to Ei, fi to hi and BL to DL are then the same).

Produce AB to K and BD to L so that BK be to AB as HI to GH, and DL to BD as GI to FG, and join KL meeting the straight line CE in i. (109)Produce iL to M so that LM be to iL as GH to HI, and draw both MQ parallel to Lh meeting the straight line AD in g, and then gi cutting AB, BD in f and h. I say it is done.(110)

For let Mg intersect the straight line AB in Q, and AD the line KL in S, and draw AP which shall be parallel to BD and meet iL in P; then will Mg to LH (Mi to Li, gi to hi, AK to BK) and AP to BL be in the same ratio. Cut DL in R so that DL be to RL in that same ratio and then, because gS to gM, AS to AP and DS to DL are proportional, there will *ex æquo* be as gS to Lh, so AS to BL and DS to RL, and *mixtim* $BL-RL$ to $Lh-BL$ as $AS-DS$ to $gS-AS$; that is, BR to Bh as AD to Ag, and hence as BD to gQ, and *alternando* BR to BD as Bh to gQ, or fh to fg. But from construction BR is to BD as FH to FG, and consequently fh is to fg as FH to FG. Since also, then, ig is to ih as Mi to Li, that is, as IG to IH, it is evident that the lines FI, fi are divided similarly at G and H, g and h. As was to be done.

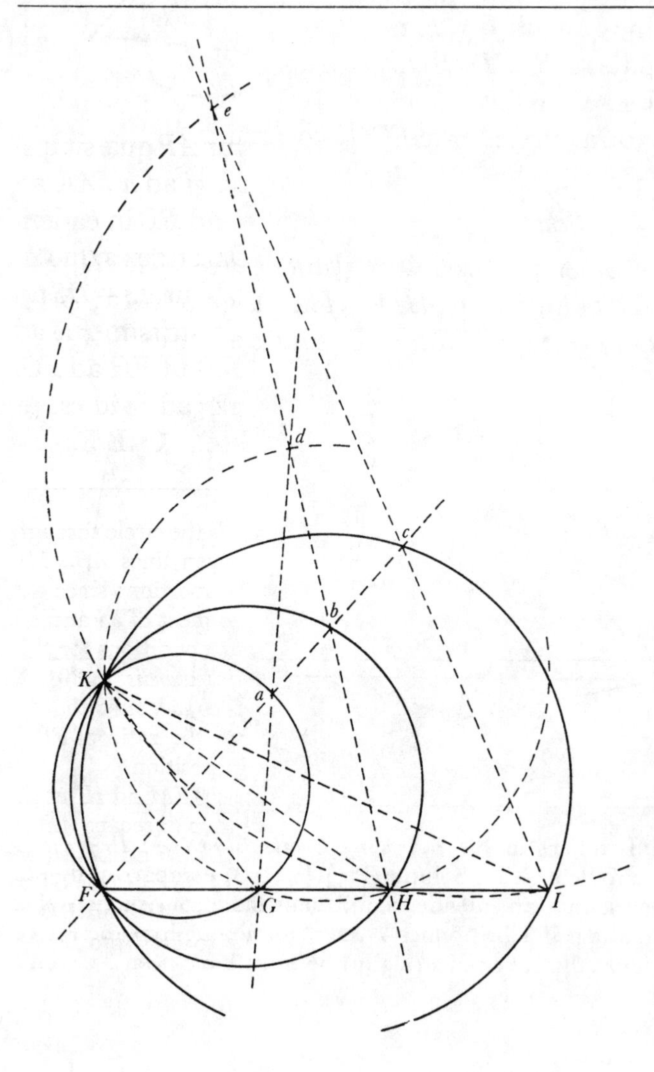

(110) All is straightforward once i is constructed in correct position on CE. To do so, let the transversal fi move over the given lines AC, AD and BD so that its intersections f, g and h continue to satisfy $fg:gh = FG:GH$, constant, and then, where in all such positions there remains

$$fg:fi = FG:FI,$$

constant, it follows from note (109) that its point i lies on a straight line KSL determined in position, accordingly as f coincides with C, A or B, by an appropriate pair of the ratios $CA:AB:BK = Ar:rD:DS = BD:DR:RL = FG:GH:HI$, on taking r in AD such that $Ar:AD = FG:FH$. Newton chooses $AB:BK = GH:HI$ and $BD:DL = FG:GI$, and then obscures the basic simplicity of the construction with a following synthetic demonstration in which the auxiliary line KL is conjured up like a rabbit from a hat, totally without explanation. In more sophisticated terms, equivalently, KL is the second tangent from i to the parabola which touches the four given lines CE, AC, AD, BD (the two latter in r and R), and

In constructione Corollarij[111] hujus postquam ducitur *SK* secans *CE* in *i*, producere licet *iE* ad *V* ut sit *EV* ad *iE* ut *FH* ad *HI* et agere *Vf* parallelam ipsi *BD*. Eodem recidit si centro *i* intervallo *IH* describatur circulus secans *BD* in *X*, producatur *iX* ad *Y* ut sit *iY* æqualis *IF*, et agatur *Yf* ipsi *BD* parallela.[112]

Prop. XXIX. Prob. XXI.

Trajectoriam specie datam describere quæ a rectis quatuor positione datis in partes secabitur ordine specie et proportione datas.

Describenda sit Trajectoria *fghi* quæ similis sit lineæ curvæ *FGHI* et cujus partes *fg*, *gh*, *hi* illius partibus *FG*, *GH*, *HI* similes et proportionales rectis *AB* et *AD*, *AD* et *BD*, *BD* et *CE* positione datis prima primis[,] secunda secundis[,] tertia tertijs interjaceant. Actis rectis *FG*, *GH*, *HI*, *FI* describatur Trapezium *fghi* quod sit Trapezio *FGHI* simile et cujus anguli *f*, *g*, *h*, *i* tangant rectas illas positione datas *AB*, *AD*, *BD*, *CE* singuli singulas dicto ordine. Dein circa hoc Trapezium describatur Trajectoria curvæ lineæ *FGHI* consimilis.

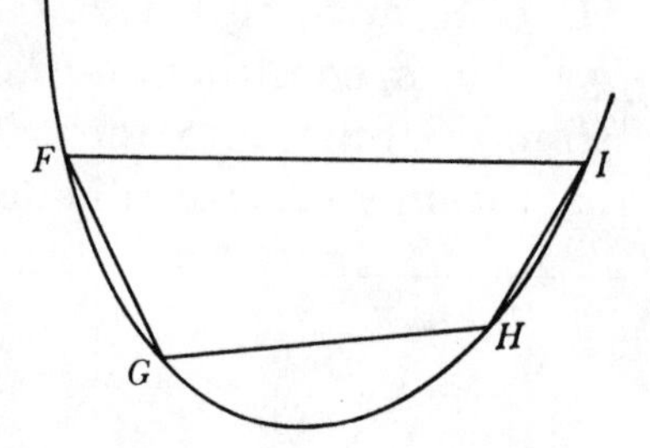

Scholium

Construi etiam potest hoc Problema ut sequitur. Junctis *FG*, *GH*, *HI*, *FI*, produc *GF* ad *V*, jungeq; *FH*, *IG*, et angulis *FGH*, *VFH* fac angulos *CAK*, *DAL* æquales. Concurrant *AK*, *AL* cum recta *BD* in *K* et *L* et inde agantur *KM*, *LN*

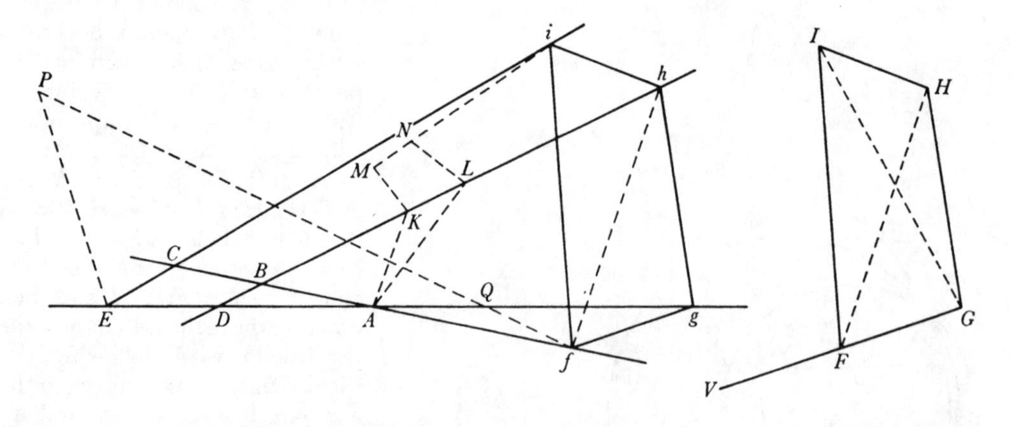

was later so constructed by J. H. Lambert in his *Insigniores Orbitæ Cometarum Proprietates* (Augsburg, 1761): Lemma XVIII. Problema VII, Solutio [Prior]: 21–3; compare also note (87) above. (We need scarcely remark that an intended equivalent later construction of a comet's orbit from four terrestrial sightings in Christopher Wren's simplifying hypothesis (see v: 302–3, note (405)) that it is a uniformly traversed straight line must have been Newton's

In the construction of this corollary(111) we are, after SK is drawn cutting CE in i, at liberty to extend iE to V so that EV be to iE as FH to HI and then draw Vf parallel to BD. It comes to the same if with centre i and radius IH a circle be described intersecting BD in X, then iX be produced to Y so that iY is equal to IF, and Yf be drawn parallel to BD.(112)

Proposition XXIX, Problem XXI.

To describe a trajectory given in species which shall be cut by four straight lines given in position into sections given in species and proportion.

Let it be required to describe a trajectory $fghi$ which is to be similar to the curve $FGHI$ and whose sections fg, gh, hi similar and proportional to its parts FG, GH, HI shall lie between the given straight lines AB and AD, AD and BD, BD and CE (the first between the first pair, the second between the second and the third between the third). On drawing the straight lines FG, GH, HI, describe a quadrilateral $fghi$ which shall be similar to the quadrilateral $FGHI$ and whose corners f, g, h, i shall touch those straight lines AB, AD, BD, CE given in position, each their separate ones in the order stated. Then about this quadrilateral describe the trajectory exactly similar to the curve $FGHI$.

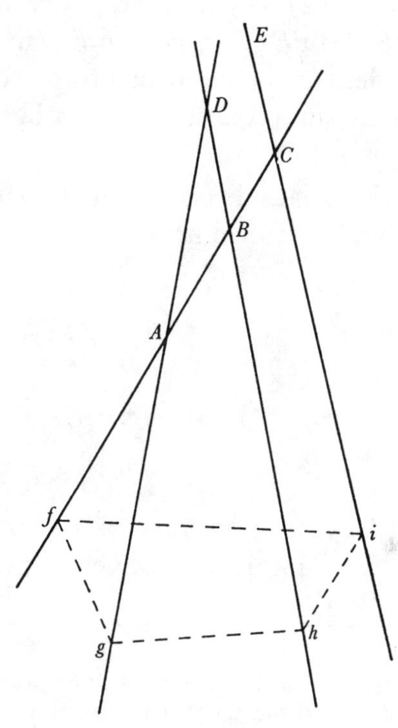

Scholium.

This problem can also be constructed as follows. On joining FG, GH, HI and FI, extend GF to V, join FH, IG and make the angles $C\widehat{A}K$, $D\widehat{A}L$ equal to the angles $F\widehat{G}H$, $V\widehat{F}H$. Let AK, AL meet the straight line BD in K and L, and from

underlying justification for inserting this present group of geometrical lemmas and related problems into his 'De motu Corporum Liber primus'; compare note (22) above and especially note (112) following.) The problem has an indeterminate solution when KL coincides with CE, in which case $AB:BC = GH:HI$ and $BD:DE = FG:GI$ exactly as before (see note (108)).

(111) A less accurate 'Lemmatis' (Lemma) is replaced in the manuscript.

(112) Newton further hinted at the parabolic *raison d'être* for the present construction by adding in sequel in the second edition of his *Principia* ($_2$1713: 95) that 'Problematis hujus solutiones alias *Wrennus* & *Wallisius* olim excogitarunt' (Of this problem other solutions were some while past contrived by Wren and Wallis). See note (110) and especially v: 210–11, note (251).

quarum KM constituat angulum AKM æqualem angulo GHI sitq; ad AK ut est HI ad GH, et LN constituat angulum ALN æqualem angulo FHI sitq; ad AL ut HI ad FH. Ducantur autem AK, KM, AL, LN ad eas partes linearum AD, AK, AL ut literæ $CAKMC$, ALK, $DALND$ eodem ordine cum literis $FGHIF$ in orbem redeant. Et acta MN occurrat rectæ CE in i. Fac angulum iEP æqualem angulo IGF sitq; PE ad Ei ut FG ad GI et per P agatur QPf quæ cum recta AED contineat angulum PQE æqualem angulo FIG rectæq; AB occurrat in f et jungatur fi. Agantur autem PE et PQ ad eas partes linearum CE, PE ut literarum $PEiP$ et $PEQP$ idem sit ordo circularis qui literarum $FGHIF$, et si super linea fi eodem quoq; literarum ordine constituatur trapezium $fghi$ trapezio $FGHI$ simile et circumscribatur Trajectoria specie data solvetur Problema.(113)

Hactenus de orbibus inveniendis. Superest(114) ut motus corporum in orbibus inventis determinemus.

Artic. VI.

De inventione motuum in Orbibus datis.(115)

Prop. XXX. Prob. XXII.(116)

Corporis in data trajectoria Parabolica moventis, invenire locum ad tempus assignatum. Sit S umbilicus & A vertex principalis Parabolæ, et $4AS \times M$ area Parabolica

(113) Newton some years afterwards justified this mode of construction in an annotation inserted at this point (page 103, line 21) in his library copy of the *Principia's* first edition (now Trinity College, Cambridge. NQ. 16.200): 'Nam si figura $fghi$ figuræ $FGHI$ semper similis existens ita moveatur et interea magnitudine augeatur vel minuatur, ut ejus puncta f, g, h rectas tres AB, AD, BD semper tangant: ejus punctum quartum i tanget locū rectilineum iM [see note (107) above]; et ubi punctum f incidit in punctum A, rectæ fh et hi incident in rectas AL et LN; ubi verò punctum g incidit in punctum A rectæ gh, hi incident in rectas AK KM. Et inde dantur puncta M, N, i ut supra. Et simili methodo triangulum ifh simile triangulo IFH locabitur ad punctum i' (For if the figure $fhgi$, continuing ever similar to the figure $FGHI$, shall, all the while increasing or diminishing in scale, so move that its points f, g, h ever lie on the three straight lines AB, AD and BD, then its fourth point i shall lie on the rectilinear locus iM; and when the point f coincides with the point A, the straight lines fg and hi will coincide with the lines AL and LN; but when the point g coincides with the point A, the straight lines gh and hi will coincide with the lines AK and KM. And thereby the points M, N and i are given as above. And by a similar method the triangle ifh, similar to the triangle IFH, will be positioned at the point i). More directly, if we erect on Af a quadrilateral $fA\alpha\beta$ similar to $fghi$, K and L will be the meets with BD of αA and the parallel through A to αh, while M will be the meet of βA with the parallel to $\alpha\beta$ through K, and N that of Mi with the parallel to βi through A; whence, because the triangle AKM is similar to $A\alpha\beta$ and so to ghi, and the triangle ALN is readily shown to be similar to fhi, it follows that $AK: KM = gh: hi = GH: HI$ and

these draw KM, LN, of which KM shall form the angle $\widehat{AKM}$ equal to $\widehat{GHI}$ and be to AK as HI is to GH, while LN shall form the angle $\widehat{ALN}$ equal to $\widehat{FHI}$ and be to AL as HI to FH—but draw AK, KM, AL, LN on the sides of the lines AD, AK, AL which shall yield the letters $CAKM$, ALK and $DALN$ to be in the same cyclical order as the letters $FGHI$. Then, on drawing MN, let it meet the straight line CE in i. Make angle $\widehat{iEP}$ equal to $\widehat{IGF}$, then let PE be to Ei as FG to GI and through P draw PQf which shall contain with the straight line AED an angle $\widehat{PQE}$ equal to $\widehat{FIG}$; let this meet the straight line AB in f and join fi—but draw PE and PQ on the sides of the lines CE, PE which make the letters PEi and PEQ in the same circular order as that of the letters $FGHI$. Then, if on the line fi, also in the same order of letters, there be constructed a quadrilateral $fghi$ similar to the quadrilateral $FGHI$ and round it be circumscribed the trajectory given in species, the problem will be solved.(113)

So much for finding orbits. It remains(114) for us to determine the motions of bodies in orbits (already) found.

Article VI

On finding the motions in given orbits.(115)

Proposition XXX, Problem XXII.(116)

Where a body moves in a given parabolic trajectory, to find its position at an assigned time.

Let S be the focus and A the principal vertex of the parabola, and $4AS \times M$ the

---

$AL: LN = fh: hi = FH: HI$, both given. (Here, since the triangles αfh and βfi are similar $\widehat{LAN}$ ($=\widehat{h\gamma i}$ where αh and βi meet in γ) is equal to $\widehat{hfi}$, and also

$$AL: \alpha h = AK: \alpha K$$
$$= AM: \beta M = AN: \beta i$$

or

$$AL: AN = \alpha h: \beta i = fh: fi.)$$

Once the point i is constructed, the rest is straightforward: for, on erecting the triangle iEP similar to igf, the triangles ifP, igE prove also to be similar and therefore, since they share a common vertex i, are inclined to one another at the given angle $\widehat{PQE} = \widehat{fig}$ ($= \widehat{FIG}$); from which f is immediately determined as the meet of PQ with AB.

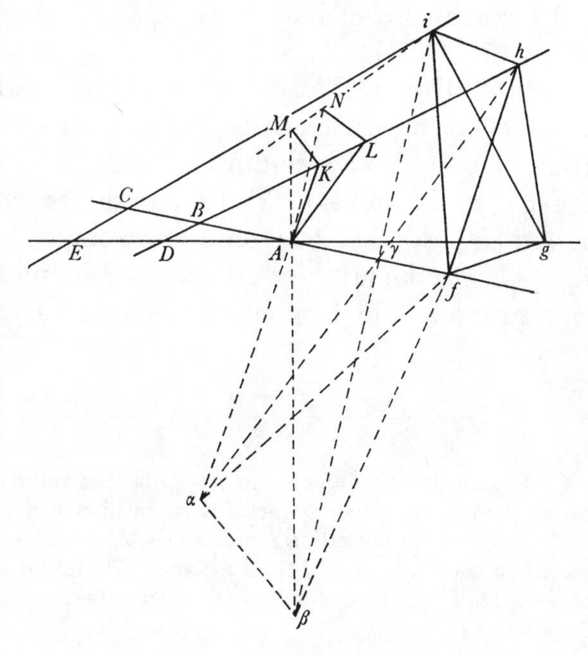

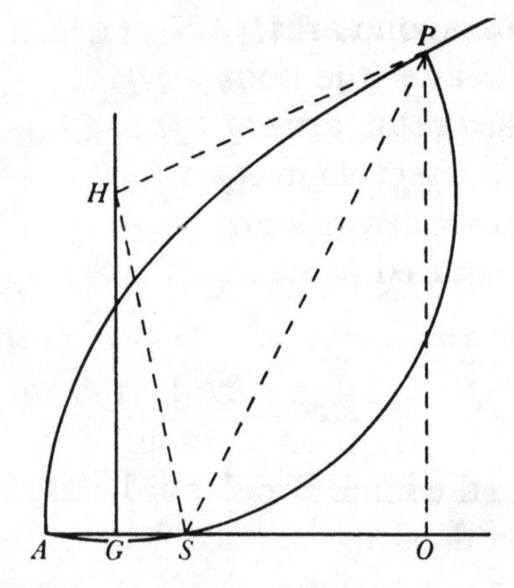

APS quæ radio SP vel post excessum corporis de vertice descripta fuit vel ante appulsum ejus ad verticem describenda est. Innotescit area illa ex tempore ipsi proportionali. Biseca AS in G, erigeq3 perpendiculum GH æqualc $3M$; ct circulus ccntro H intervallo HS descriptus secabit Parabolam in loco quæsito P. Nam demissa ad axem perpendiculari PO, est

$$HG^q + GS^q (= HS^q = HP^q = GO^q + \overline{HG - PO^q}) \\ = GO^q + HG^q - 2HG \times PO + PO^q.$$

Et deleto utrinq3 HG^q fiet

$$GS^q = GO^q - 2HG \times PO + PO^q,$$

seu

$$2HG \times PO (= GO^q + PO^q - GS^q = AO^q - 2GAO + PO^q) = AO^q + \tfrac{3}{4} PO^q.$$

Pro AO^q scribe $AO \times \dfrac{PO^q}{4AS}$, et applicatis terminis omnibus ad $3PO$ ductisq3 in $2AS$ fiet

$$\tfrac{4}{3} HG \times AS \left(= \tfrac{1}{6} AO \times PO + \tfrac{1}{2} AS \times PO = \frac{AO + 3AS}{6} PO = \frac{4AO - 3SO}{6} PO \right. \\ \left. = \text{areæ } APO - SPO \right) = \text{areæ } APS.$$

Sed GH erat $3M$ et inde $\tfrac{4}{3} HG \times AS$ est $4AS \times M$. Ergo area APS æqualis est $4AS \times M$. Q.E.D.

Corol. 1. Hinc GH est ad AS ut tempus quo corpus descripsit arcum AP ad tempus quo corpus descripsit arcum inter verticem A et perpendiculum(117) ad axem ab umbilico S erectum.

Corol. 2. Et circulo ASP per corpus movens perpetuò transeunte velocitas puncti [H] est ad velocitatem quam corpus habuit in vertice A ut 3 ad 8;(118) adeoq3 in ea etiam ratione est linea GH ad lineam rectam quam corpus tempore motus sui ab A ad P ea cum velocitate quam habuit in vertice A, describere possit.

---

(114) In order to complete the preceding discussion of motion in conic 'trajectories', that is, before passing on to more general dynamical considerations.

(115) As in the preceding Articles III–V, continue to understand these to be one or other species of conic traversed round a focus. The initial subhead, following a standard original pattern, read '... CONTINENS INVENTIONEM ...' (CONTAINING THE FINDING).

parabolic area (APS) which either has been described by the radius (vector) SP after the body's departure from the vertex or is yet to be described by it before its arrival there. That area is ascertained from the time proportional to it. Bisect AS in G, and erect the perpendicular GH equal to $3M$; and then the circle described on centre H and with radius HS will intersect the parabola in the required place P. For, on letting fall the perpendicular PO to the axis, there is

$$HG^2+GS^2 = (HS^2 = HP^2 = GO^2+(HG-PO)^2 =)\, GO^2+HG^2-2HG\times PO+PO^2,$$

and with HG^2 deleted from each side there will come to be

$$GS^2 = GO^2-2HG\times PO+PO^2,$$

that is,

$$2HG\times PO(= GO^2+PO^2-GS^2 = AO^2-2AG\times AO+PO^2) = AO^2+\tfrac{3}{4}PO^2.$$

In place of AO^2 write $AO\times PO^2/4AS$ and, when all the terms are divided by $3PO$ and multiplied by $2AS$, there will prove to be

$$\tfrac{4}{3}HG\times AS(=\tfrac{1}{6}AO\times PO+\tfrac{1}{2}AS\times PO = \tfrac{1}{6}(AO+3AS)\times PO$$
$$=\tfrac{1}{6}(4AO-3SO)\times PO = \text{area}\,(APO)-(SPO)) = \text{area}\,(APS).$$

But HG was $3M$, and thence $\tfrac{4}{3}HG\times AS$ is $4AS\times M$. Therefore the area (APS) is equal to $4AS\times M$. As was to be proved.

Corollary 1. Hence GH is to AS as the time in which the body has described the arc $\widehat{AP}$ to the time in which the body described the arc between the vertex A and the perpendicular(117) raised to the axis from the focus S.

Corollary 2. And, with the circle ASP passing perpetually through the moving body, the speed of the point H is to the speed which the body had at the vertex A as 3 to 8;(118) and therefore in that ratio also is the line GH to the straight line which the body might, in the time of its movement from A to P with the speed which it had in the vertex A, describe.

---

(116) To a minimally augmented repeat of Proposition XX of the preceding 'De motu Corporum' (§2) Newton here appends three minor corollaries.

(117) Since the general perpendicular through any point O in the abscissa is the parabola's ordinate $OP = \sqrt{[4AS\times AO]}$, this—$SP'$ say—is equal to $2AS$, so that $(AP'S) = \tfrac{4}{3}AS^2$. In general, therefore, GH (or $3M = \tfrac{3}{4}(APS)/AS$) $: AS = (APS):(AP'S)$, as Newton effectively here states.

(118) For the speed of H as it moves outward from G is inversely proportional to

$$(APS)/GH = \tfrac{4}{3}AS,$$

while the orbital speed at A is likewise inversely proportional to $\lim_{A\to P}[(APS)/\widehat{AP}] = \tfrac{1}{2}AS$.

Corol. 3. Hinc etiam vice versa inveniri potest tempus quo corpus descripsit arcum quemvis assignatum *AP*. Junge *AP* et ad medium ejus punctum erige perpendiculum rectæ *GH* occurrens in *H*.

Lemma XXVIII.

Nulla extat Figura Ovalis cujus area rectis pro lubitu abscissa possit per æquationes numero terminorum ac dimensionum finitas generaliter inveniri.(119)

Intra Ovalem detur punctum quodvis circa quod ceu polum revolvatur perpetuò linea recta, et interea in recta illa exeat punctum mobile de polo, pergatq semper cum velocitate quæ sit ut rectæ illius intra Ovalem longitudo.(120) Hoc motu punctum illud describet spiralem gyris infinitis. Jam si area Ovalis per finitam æquationem inveniri potest, invenietur etiam per eandem æquationem distantia puncti a polo quæ huic areæ proportionalis est, adeoq(121) omnia spiralis puncta per æquationem finitam inveniri possunt: et propterea rectæ cujusvis positione datæ intersectio cum spirali inveniri etiam potest per æquationem finitam. Atqui recta omnis infinite producta spiralem secat in punctis numero infinitis et æquatio qua intersectio aliqua duarum linearum invenitur exhibet earum intersectiones omnes radicibus totidem, adeoq ascendit ad tot dimensiones quot sunt intersectiones. Quoniam circuli se mutuo secant in punctis duobus[,] intersectio una non invenitur nisi per æquationem duarum dimensi[on]um qua intersectio altera etiam inveniatur. Quoniam

(119) As originally enunciated in the manuscript, this first asserted yet more universally that '*Ovalium areæ rectis quibusvis assignatis abscissæ non possunt per æquationes numero terminorum ac dimensionum finitas inveniri*' (*In ovals, areas cut off by any straight lines assigned as you will cannot be found by means of equations finite in the number of their terms and dimensions*): Newton's subsequent emendation is strictly necessary since—even if we forget its central *non-sequitur* (on which see note (121))—his following argument cannot preclude the possibility that there may exist *particular* rationally definable sectors of an 'oval' which are algebraically quadrable. As we observed in I: 545, note (2), the present lemma is founded on a parallel youthful remark by Newton (ULC. Add. 3958.2: 34r, reproduced on I: 545) that 'The length of no Elliptical line whatever...can be found. For if so the spirall lines made by them wou[l]d bee geometricall'; the gist of this is here re-introduced in the second paragraph below (see note (127)).

(120) Read 'quadratum' (square), as David Gregory somewhat confusingly recorded nine years afterwards on a 'Charta...Continens adnotata Phys: et Math: cum Neutono... Cantabrigiæ. 4 Maij. 1694' (Gregory C33 [now in the Royal Society, London], reproduced in *The Correspondence of Isaac Newton*, **3**, 1961: 311–15) in his third paragraph: 'In prop: [!] ubi probatur non posse rectam ex centro spiralis exeuntem indefinite exprimi absque serie, loco vocis *longitudo* scribendum *quadratum*' (*ibid.*: 311). Having passed unchecked into the ensuing *Principia* ($_1$1687: 105), the slip was duly corrected in the second edition ($_2$1713: 98). Leibniz independently noticed the error in autumn 1689, when he inserted in the margin of his annotated copy of the *editio princeps* (see §2: note (49)) the equivalent terminal phrase 'ductæ in sinum anguli circulationis cujus radius sit dicta longitudo si circulatio sit uniformis et sint velocitates ut longitudines quadratæ'.

(121) A subtly fallacious inference which irreparably vitiates Newton's further argument. The spiral measures not only the simple area of the corresponding sector of the oval, but also

Corollary 3. Hence also, conversely, there can be found the time in which the body described any assigned arc $\widehat{AP}$. Join AP and at its mid-point erect a perpendicular meeting the straight line GH in H.

Lemma XXVIII.

There exists no oval figure whose area cut off at will by straight lines might generally be found by means of equations finite in the number of their terms and dimensions.(119)

Within the oval let there be given any point round which as pole a straight line shall perpetually revolve, and all the while in that line a mobile point shall go out from the pole, proceeding ever with a speed which is as the length(120) of that straight line within the oval. By this motion that point will describe a spiral with an infinity of gyrations. Now if the area of the oval can be found by means of a finite equation, there will also be found by the same equation the point's distance from the pole which is proportional to this area, and hence(121) every point of the spiral can be found by means of a finite equation; and in consequence the intersection of any straight line given in position with the spiral can also be found by means of a finite equation. And yet every infinitely extended straight line cuts a spiral in an infinite number of points, while an equation by which some intersection of two (curve) lines is found exhibits all their intersections by an equal number of its roots, and hence rises to as many dimensions as there are intersections. Because circles cut one another mutually in two points, one intersection may not be found except by an equation of two dimensions whereby the other in ersection shall also be obtained. Because of two conics there can be four

---

that area together with the total area of the oval taken once, twice, three times, ... indefinitely; whence, whether or not the (single) area of the oval's sector is algebraically quadrable, the spiral must in all cases have an infinite number of intersections with the rotating straight line and therefore cannot be an algebraic curve (definable, in Newtonian terms, by a 'geometrical' equation in any system of Cartesian coordinates; compare v: 470, note (689)). In the simplest instance, where the oval is a circle of centre O and the straight line OP as it rotates round that centre from an initial position OA generates, by its continual intersection B with the circle, a related Archimedean spiral (P_i) defined by $OP_i \propto AB + 2i\pi \times OA, i = 0, 1, 2, \ldots$ (or, backwards, $i = -1, -2, \ldots$), the infinite number of points P_i corresponding to the simple circle sector $(AOB) = \frac{1}{2}OA^2 \times \widehat{AOP}$ merely reflect the periodicity of the angle $\widehat{AOP_i} = \widehat{AOP} + 2i\pi$; compare D. T. Whiteside, 'Patterns of Mathematical Thought in the later Seventeenth Century' (*Archive for History of Exact Sciences*, **1**, 1961: 179–388): 203–5.

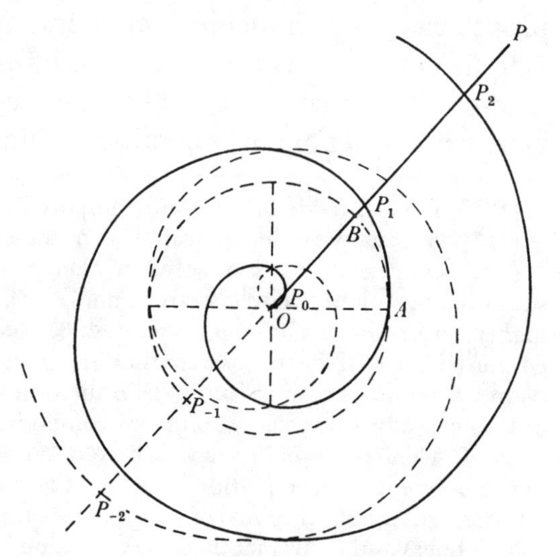

duarum sectionum conicarum quatuor esse possunt intersectiones, non potest aliqua earum generaliter inveniri nisi per æquationem quatuor dimensionum qua omnes simul inveniantur. Nam si intersectiones illæ seorsim quærantur, quoniam eadem est omnium lex et conditio idem erit calculus in casu unoquoq; et propterea eadem semper conclusio, quæ igitur debet omnes intersectiones simul complecti et indifferenter exhibere. Unde etiam intersectiones Sectionum conicarum et curvarum tertiæ potestatis,(122) eò quod sex esse possunt, simul prodeunt per æquationes sex dimensionum et intersectiones duarum curvarum tertiæ potestatis,(122) quia novem esse possunt, simul prodeunt per æquationes dimensionum novem. Id nisi necessariò fieret, reducere liceret(123) problemata omnia solida ad plana et plusquam solida ad solida.(124) Eadem de causa intersectiones binæ rectarum et sectionum conicarum prodeunt semper per æquationes duarum dimensionum, ternæ rectarum et curvarum tertiæ potestatis(122) per æquationes trium, quaternæ rectarum et curvarum quartæ potestatis(122) per æquationes dimensionum quatuor[,] et sic in infinitum. Ergo intersectiones numero infinitæ rectarum et spiralium, propterea quod omnium eadem est lex et idem calculus, requirunt æquationes numero dimensionum et radicum infinitas quibus omnes possunt simul exhiberi. Si a polo in rectam illam secantem demittatur perpendiculum, et perpendiculū una cum secante revolvatur circa polum: intersectiones spiralis transibunt in se mutuò, quæq; prima erat seu polo proxima, post unam revolutionem secunda erit[,] post duas tertia[,] et sic deinceps: nec interea mutabitur æquatio nisi pro mutata magnitudine quantitatum per quas positio secantis determinatur.(125) Unde cum quantitates illæ post singulas revolutiones redeunt ad magnitudes primas, æquatio redibit ad formam primam, adeoq; una eademq; exhibebit intersectiones omnes, et propterea radices habebit numero infinitas quibus omnes exhiberi possunt.(125) Nequit ergo intersectio rectæ et spiralis per æquationem finitam generaliter inveniri, et idcirco nulla extat Ovalis cujus area rectis imperatis abscissa possit per talem æquationem generaliter exhiberi.(126)

(122) That is, Cartesian degree; compare note (69) above.

(123) Originally 'reducerentur' (...possible to reduce).

(124) On the distinction between 'plane', 'solid' and 'sursolid' problems—ones constructible by 'plane' (circle/straight line), 'solid' (conic) and 'more than solid' (cubic and higher) loci respectively—again see v: 423, note (617). At this point in the *Principia's* second edition ($_2$1713:99) Newton inserted the disclaimer 'Loquor hic de Curvis potestate irreducibilibus. Nam si æquatio per quam Curva definitur, ad inferiorem potestatem reduci possit: Curva non erit unica, sed ex duabus vel pluribus composita, quarum intersectiones per calculos diversos seorsim inveniri possunt' (I speak here of curves irreducible in power [degree]. For if it be possible to reduce the equation by which the curve is defined to a lower power, the curve will be not singular but composed of two or more others whose intersections can be found by separate calculations) and then smoothed the transition into the following sentence with the remoulded opening 'Ad eundem modum intersectiones binæ...' (after the same manner the pairs of intersections...).

intersections, some one of these cannot generally be ascertained except by an equation of four dimensions whereby all shall simultaneously be ascertained: for should those intersections be sought separately, because the law and circumstance of all is the same, the computation will be the same in each individual case and there will accordingly ever be the same conclusion, which must therefore simultaneously embrace all the intersections, exhibiting them indifferently. Whence also the intersections of conics and curves of third power,[122] since they can be six in number, are simultaneously yielded by equations of six dimensions; and those of two curves of third power,[122] since they can be nine, result simultaneously by means of equations of nine dimensions. Unless that were to be so of necessity, it would be permissible to reduce[123] all solid problems to plane ones, and those greater than solid to solid ones.[124] For the same reason the pairs of intersections of straight lines and conics are always forthcoming by equations of two dimensions, the trios of those of straight lines and curves of third power[122] by equations of three, the quartets of those of straight lines and curves of fourth power[122] by equations of four dimensions, and so on indefinitely. In consequence, the intersections—infinite in number—of straight lines and spirals, seeing that for all there is the same law and same computation, require equations infinite in the number of their dimensions and roots whereby all can be simultaneously exhibited. If a perpendicular be let fall from the pole onto the intersecting straight line, and the perpendicular along with the intersecting line revolve round the pole, then the spiral's intersections will pass mutually one into another, and the one which was first (or nearest to the pole) shall after one revolution be the second, after two the third, and so on in turn; nor will the equation alter in the meantime, except in regard to the altering size of the quantities by which the position of the intersecting line is determined.[125] Since, therefore, those quantities return after each individual revolution to their initial sizes, the equation will return to its original form, and hence one and the same equation will exhibit all the intersections and shall accordingly have an infinity of roots whereby all can be displayed. It is consequently impossible for the intersection of a straight line with the spiral to be generally found by means of a finite equation, and on that account there exists no oval whose area cut off by appointed straight lines might generally be exhibited by means of such an equation.[126]

(125) An ingenious argument: in more general terms, if some equation $f(x) = 0$ is unchanged when its roots are each augmented by some quantity, say a, then $f(x) \equiv f(x+ia) = 0$ for all integers i and hence the equation must have an infinite number of equally spaced roots $X+ia$ where X satisfies $f(X) = 0$. In the present instance (see the figure accompanying note (121)) $P_iP_{i+1} = a$ for all integral i, $-\infty < i < \infty$, and so, where $OP_0 = X$, at once $OP_i = X+ia$.

(126) Though Newton's attempted justification of it is (see note (121)) irreparably flawed, the truth of the present lemma is not thereby gainsaid. Upon its appearance, unchanged from

Eodem argumento, si intervallum poli et puncti quo spiralis describitur capiatur Ovalis perimetro abscissæ proportionale, probari potest quod longitudo perimetri nequit per finitam æquationem generaliter exhiberi.(127)

---

its present form, in the printed *Principia* ($_1$1687: 105–7) the initial reaction of Continental mathematicians was mixed. Jakob Bernoulli accepted its truth unhesitatingly and incorporated Newton's 'insight' into his 1691 paper on the Archimedean spiral ('Specimen Calculi Differentialis in dimensione Parabolæ helicoidis...aliisque', *Acta Eruditorum* (January 1691): 13–23 [= *Opera*, **1** (Geneva, 1744): 431–42]) to 'prove' that the general quadrature of the circle is impossible: 'Obiter noto, hinc etiam ostendi posse, quadraturam circuli indefinitam, & in genere rectificationem ullius curvæ Geometricæ in se redeuntis impossibilem esse. Hæc enim si possibilis esset, dari posset relatio inter curvam & applicatam vel abscissam, cumque & harum relatio tum inter se, tum ad tangentem data ponatur, data quoque foret ipsius curvæ ad tangentem ratio; quare...prodiret alia æquatio certi & definiti gradus, cujus radices, quarum nunquam plures esse possunt, quam æquatio dimensiones habet, determinarent omnia curvæ nostræ suprema puncta, sed hoc fieri nequit, quoniam spiralis ista, si continuetur, infinitis gyris circa radium...circumvolvitur, in quibus singulis aliquod punctum supremum existit, quorumque adeo punctorum numerus infinitus est' (*ibid.*: 21). Eight months afterwards, on the other hand, Leibniz was unpersuaded and in a general critique 'De Solutionibus Problematis Catenarii...aliisque a Dn. Jac. Bernoullio propositis' (*Acta Eruditorum* (September 1691): 435–8 [=(ed. C. I. Gerhardt) *Leibnizens Mathematische Schriften*, **5** (Halle, 1858): 255–8]) remarked that 'Hæreo...circa id, quod...dictum est...nullius curvæ Geometricæ in se redeuntis rectificationem (generalem) esse possibilem. Scio alium virum Clarissimum [*sc.* Newton] simili argumento probare instituisse, nullius areæ curvæ Geometricæ in se redeuntis quadraturam indefinitam esse possibilem; visum tamen est Dn. *Hugenio* non minus quam mihi, rem non esse consectam. Et ni fallor, dantur instantiæ, quibus tamen hujusmodi argumenta applicari possunt' (*ibid.*: 437). Earlier, in a letter to Huygens on 30 February (N.S.) discussing the general quadrature of the lemniscate $a^2x^2 = a^2y^2-y^4$ (which Fatio de Duillier had previously told Huygens was impossible), Leibniz readily deduced that the area is

$$\left[\int_y^a x.dy = \right] \tfrac{1}{3}(a^2-y^2)^{\frac{3}{2}}/a$$

and then went on: 'Puisque la [courbe] retourne en elle meme, en forme de 8, on en peut juger que le theoreme de M[r] Newton p. 105, qui pretend, qu'il n'y a point de courbe recourrante (de la Geometrie ordinaire) indefiniment quadrable, ne scauroit subsister, et qu'il y a quelque faute dans sa demonstration. Mais je ne l'en estime pas moins' (C. I. Gerhardt, *Der Briefwechsel von Gottfried Wilhelm Leibniz mit Mathematikern*, **1** (Berlin, 1899): 639–41, especially 640 [=*Œuvres complètes de Christiaan Huygens*, **10** (The Hague, 1905): 49–52, especially 51])—to which he could not resist adding the Horatian *mot* 'Opere in longo fas est obrepere somnum'. To which Huygens, not to be outdone, riposted on 26 March following: 'Mr Fatio ne peut pas bien soutenir la Prop. de Mr Newton pag 105, surtout quand pour son Ovale indeterminée, je luy marque deux portions egales de parabole [say, $cx^2 = \pm d^2(y\mp c)$, $-d \leqslant x \leqslant d$] qui aient la mesme base [$2d$]' (*Briefwechsel*, **1**: 643 [= *Œuvres*, **10**: 57]). However, since Leibniz's Bernoullian lemniscate has a double loop, crossing itself at the origin, while Huygens' double parabola has (for $|x| > d$) two pairs of conjugate infinite branches—to which Newton was later (see note (127)) to take explicit exception—neither are true closed 'ovals' in the simple sense of the term.

By and large, mathematicians in the eighteenth century tended to accept the truth of Newton's lemma—Edward Waring in Theorem X of Chapter I of his *Proprietates Algebraicarum*

By the same reasoning, if the distance between the pole and the point by which the spiral is traced out be taken proportional to the oval's perimeter so cut off, it may be proved that the length of the perimeter cannot generally be exhibited by means of a finite equation.(127)

Curvarum (Cambridge, 1772): 46 even concocted an equally spurious algebraic 'proof' of the equivalent proposition that 'Nulla datur algebraica curva, quæ habet ovalem sese in dato puncto haud intersecantem, quæ generaliter quadrari potest'—while those geometers in the nineteenth who concerned themselves with proving the existence of transcendental functions developed surer analytical methods for approaching such problems and the subtleties of the *Principia's* mode of demonstration were left to the historians to unravel as they were able. Henry Brougham and E. J. Routh in their *Analytical View of Sir Isaac Newton's Principia* (London, 1855): 73 merely repeat, in 'disproof' of the lemma, the trivial generalisation of Leibniz's attempted counter-example which is represented by the class of lemniscates $y = nx^{n-1}(a^n - x^n)^{1/m}$ (all of which cross from one of their loops to the other at the origin), while W. W. Rouse Ball in his *Mathematical Recreations and Essays* (London, $_6$1914): 294–5 was finally led to remark that 'Newton proved that in any closed oval an arbitrary sector bounded by the curve and two radii cannot be expressed in terms of the co-ordinates of the extremities of the arc by [an equation of] a finite number of algebraical terms. The argument is considered and difficult to follow: ... but on... careful reflection I think that the conclusion is valid without restriction'. Others more recent have been equally impercipient. It is not, in fact, difficult to provide an unchallengeable counter-example to Newton's assertion by combining the line-quadruple defined (in a standard system of perpendicular Cartesian coordinates) by $|y| = 1 - |x|$ with the Leibnizian lemniscate $y^2 = x^2(1-x^2)$ (the locus (B') in our figure) to produce the curve

$$y^2 = (1 - \sqrt{x^2} + \sqrt{[x^2(1-x^2)]})^2.$$

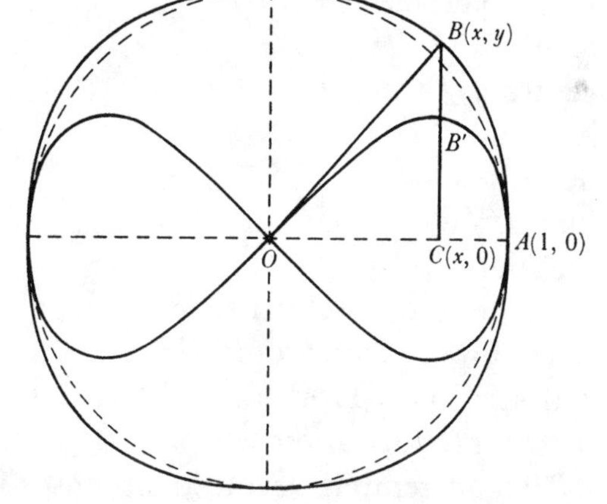

This symmetrical, slightly 'lumpy' and wholly convex oval circumscribes the unit circle

$$y^2 = 1 - x^2,$$

touching it in the four quadrant points, while, as may easily be shown, its general sector (AOB) has an area $\frac{1}{2}(1-x) + \frac{1}{6}(2+x^2)\sqrt{[1-x^2]}$, $x, y > 0$; whence in particular the area of the whole oval is $3\frac{1}{3}$.

(127) We have already (note (119)) pointed out that this repeats his earlier, youthful assertion to the same effect on I: 545. In the *Principia's* second edition ($_2$1713: 99) Newton here inserted a further sentence emphasising that 'De Ovalibus autem hic loquor quæ non tanguntur a figuris conjugatis in infinitum pergentibus' (The ovals of which I here speak, however, are not ones in contact with conjugate figures proceeding to infinity). He may have had particularly in mind the Cartesian folium $x^3 + y^3 = axy$ which had been a favourite curve of his twenty years before (see I: 184–5, 234, 288–9) and whose generally quadrable loop encloses a total area $\frac{1}{12}a^2$ (as Huygens and Johann Bernoulli independently found in 1691–2; compare J. E. Hofmann, 'Altes und Neues von der Quadratur des Descartesschen Blattes', *Centaurus*, **3**, 1954: 279–95, especially 280). It is considerably less likely that Huygens had already contrived

Corol.(128) Hinc area Ellipseos quæ radio ab umbilico ad corpus mobile ducto describitur non prodit ex dato tempore per æquationem finitam, & propterea per descriptionem curvarum Geometricè rationalium determinari nequit. Curvas Geometricè rationales appello quarum puncta omnia per longitudines æquationibus(129) definitas[,] id est per longitudinum rationes complicatas determinari possunt; cæteras (ut Spirales, Quadratrices, Trochoides) Geometricè irrationales. Nam longitudines quæ sunt vel non sunt ut numerus ad numerum(130) (quemadmodum in decimo Elementorum) sunt Arithmetice rationales vel irrationales. Aream igitur Ellipseos tempori proportionalem abscindo per Curvam Geometricè irrationalem ut sequitur.

Prop. XXXI. Prob. XXIII.

Corporis in data Trajectoria Elliptica moventis invenire locum ad tempus assignatum.

Ellipseos *APB* sit *A* vertex principalis, *S* umbilicus, *O* centrum, sitq *P* corporis locus inveniendus. Produc *OA* ad *G* ut sit *OG* ad *OA* ut *OA* ad *OS*. Erige perpendiculum *GH* centroq *O* et intervallo *OG* describe circulum *EFG*, et quoniam Problema per curvas Geometrice rationales(131) non solvitur, super regula *GH* ceu fundo progrediatur rota *GEF* revolvendo circa axem suum et interea puncto suo *A* describendo Trochoidem(132) *ALI*. Quo facto cape *GK* in ratione ad rotæ perimetrum *GEFG* ut est tempus quo corpus progrediendo ab *A* descripsit arcum *AP*(133) ad tempus revolutionis unius in Ellipsi. Erigatur perpendiculum *KL*

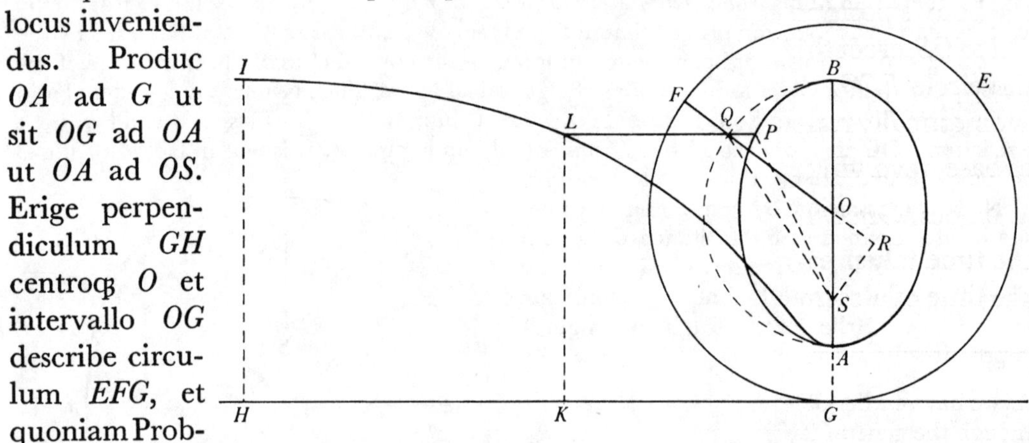

the double parabola with which (see note (126)) he later sought to refute Newton's lemma when the two met personally at London in June 1689 and conversed, among other things, on the *Principia* (see *The Correspondence of Isaac Newton*, 3, 1961: 31–2, notes (1) and (2)).

(128) This initially contained only a preliminary version of the first sentence following, namely 'Hinc area Ellipseos radio ab umbilico ad corpus mobile ducto descriptæ ex dato tempore non prodit per æquationem finitam, neq adeò per descriptionem curvarum Analyticarum determinatur graphicè' (Hence the area of an ellipse described by a radius drawn from a focus to a moving body is not forthcoming from the time given by means of a finite equation, nor consequently is it determined graphically by the tracing of analytic [algebraic] curves).

Corollary. [128]Hence the area of an ellipse which is described by a radius drawn from a focus to a moving body does not ensue from the time given by means of a finite equation, and accordingly cannot be determined by the drawing of geometrically rational curves. I call 'geometrically rational' curves whose every point can be determined by aid of lengths defined by[129] equations, that is, by involved ratios of lengths; the rest (such as spirals, quadratrixes, cycloids) I name 'geometrically irrational'. For lengths which are or are not as an integer to an integer[130] (in the manner of *Elements*, x) are arithmetically rational or irrational. I therefore cut off an area of an ellipse proportional to the time by means of a geometrically irrational curve as follows.

Proposition XXXI, Problem XXIII.

Where a body moves in a given elliptical trajectory, to find its position at an assigned time.

In the ellipse ABP let A be a main vertex, S a focus, O the centre, and let P be the body's position needing to be found. Extend OA to G so that OG be to OA as OA to OS, erect the perpendicular GH, and with centre O and radius OG describe the circle EFG; then, seeing that the problem is not to be solved by means of geometrically rational curves,[131] move the wheel GEF forward along the rule GH as base, revolving round its axis and tracing out the cycloid[132] $\widehat{ALI}$ at its point A as it goes. Once this is done, take GK in ratio to the wheel's perimeter $\widehat{GEFG}$ as the time taken by the body, proceeding from A, to describe the arc $\widehat{AP}$[133] is to the time of a complete orbit in the ellipse. Erect the perpendicular KL meeting

As we have earlier observed (§2: note (161)), this is equivalent to asserting—correctly so, even though the general lemma by which he tried to justify this corollary is invalid (see note (121) above)—that the Keplerian equation $T = \theta \mp e\sin\theta$ cannot be inverted to yield θ explicitly as an algebraic function of T. On this basis Newton proceeds, confidently and without qualms, to present Wren's 'mechanical' construction of θ from given T in his following Proposition XXXI, and thereafter, in the ensuing general scholium to the present article, to outline several ways of effecting the inversion approximately.

(129) Understand 'geometricas' (algebraic).

(130) That is, commensurable one with the other, as Newton clarifies it in his following reference to Euclid, *Elements* x, Definition 1 'Commensurabiles magnitudines dicuntur, quas eadem mensura metitur' (Isaac Barrow, *Euclidis Elementorum Libri XV. breviter demonstrati* (Cambridge, 1655): 190). In sequel 'quemadmodum' replaces an initial, over-exact 'ut' (just as in): the emendation was evidently made when Newton checked his citation and found that it was Barrow (and not Euclid himself) who exemplified commensurability in the ratio of 'a number to a number'.

(131) See note (128); this superfluous phrase was afterwards omitted in the published *Principia* ($_1$1687: 108).

(132) A 'stretched' (prolate) cycloid, that is.

(133) Originally '...quo corpus progressum fuit ab A ad P' (taken by the body in its advance from A to P).

occurrens Trochoidi in L et acta LP ipsi KG parallela occurret Ellipsi in corporis loco quæsito P.(134)

Nam centro O intervallo OA describatur semicirculus AQB et arcui AQ occurrat LP producta in Q, junganturꝗ SQ, OQ. Arcui EFG occurrat OQ in F et in eandem OQ demittatur perpendiculum SR. Area APS est ut area AQS, id est ut differentia arearum $OQA - OQS$ seu $\frac{1}{2}OQ$ in $AQ - SR$; hoc est ut differentia inter arcum AQ et rectam SR quæ est ad sinum arcus illius ut OS ad OA, adeoꝗ ut(135) GK differentia inter arcum GF et sinum arcus AQ. Q.E.D.

Scholium.

Cæterum ob difficultatem describendi hanc curvam præstat constructiones vero proximas in praxi Mechanica adhibere. Ellipseos vel Hyperbolæ(136) cujusvis APB sit AB axis major, O centrum, S umbilicus, OD semiaxis minor, et AK dimidium lateris recti. Seca AS in G ut sit AG ad AS ut BO ad BS et quære longitudinem L quæ sit ad GK(137) ut est $AO^{\text{quadr.}}$ ad rectangulum $AS \times OD$. Biseca OG in C, centroꝗ C et intervallo CG describe semicirculū GFO. Deniꝗ cape angulum GCF ad angulos quatuor rectos ut est tempus datum quo corpus descripsit arcum quæsitum AP ad tempus periodicum

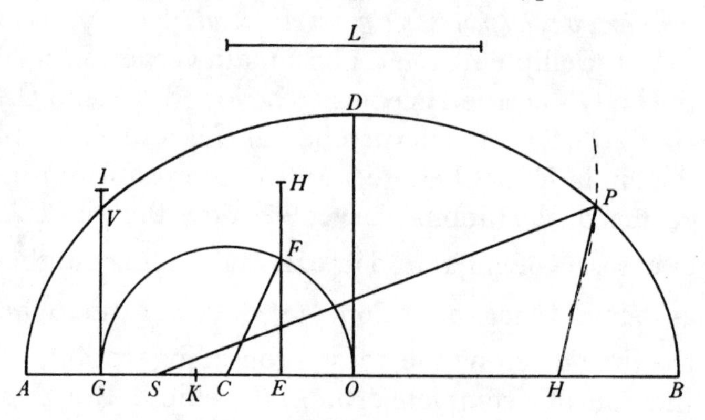

(134) This is essentially the solution contrived by Christopher Wren in 1658 and published the next year by John Wallis in his *Tractatus Duo. Prior, De Cycloide et corporibus inde genitis. Posterior, Epistolaris*... (Oxford, 1659): $_{1}80/_{2}73$–4 [= *Johannis Wallis... Opera Mathematica*, **1**, 1695: 540–1] under the subhead 'De problemate Kepleriano per Cycloidem solvendo' (compare IV: 668–9. note (38)). Like Newton after him, Wren was careful to add that 'Veruntamen, cum Cyclois sit Linea Mechanica$_{[,]}$ non Geometrica; non vere solv[i]tur problem[a], sed Mechanice perfic[i]tur' (*ibid.*: $_{2}74$). In Newton's present figure (which is but a relettered mirror-image of Wren's in all its essentials) it will, on setting the eccentric angle $\widehat{AOQ} = \theta$ and elliptical eccentricity $OS/OA = e$ (whence $OG/OA = 1/e$), be clear that in consequence $GK = (\widehat{FG}$ or) $OG \times \widehat{AOQ} - OA \times \sin \widehat{AOQ} = OG \times T$, where the mean anomaly $T = \theta - e \sin\theta$ measures the time of orbit in the ellipse from perihelion A to the general point P. While we should beware of overstressing the originality of Wren's insight that the perpendicular coordinates $GK = OG(\theta - e\sin\theta)$ and $KL = OG(1 - e\cos\theta)$ determine the locus (L) to be the 'protracted' cycloid whose general arc he had a little before equated to that of an auxiliary ellipse (see *ibid.*: $_{1}78$–9), the elegance and simplicity with which Wren's curve geometrically represents the variation of $GK/OG = \theta - e\sin\theta$ is not to be denied and we may appreciate Newton's reasons for here incorporating it as a visually stimulating prelude to

the cycloid in L, and on drawing LP parallel to KG it will meet the ellipse in the required position P of the body.(134)

For with centre O and radius OA draw the semicircle AQB and let LP (extended) meet the arc $\widehat{AQ}$ in Q, then join SQ, OQ. Let OQ meet the arc $\widehat{EFG}$ in F, and to the same line OQ let fall the perpendicular SR. The area (APS) is as the area (AQS), that is, as the difference of the areas $(AQO)-(SQO)$, or

$$\tfrac{1}{2}OQ(\widehat{AQ}-SR);$$

in other words, as the difference between the arc $\widehat{AQ}$ and the line SR which is to that arc's sine as OS to OA; and hence as(135) GK, the difference between the arc $\widehat{GF}$ and the sine of the arc AQ. As was to be proved.

Scholium.

Because of the difficulty of describing this curve, however, it is better in working practice to employ constructions approximating to the truth. In any ellipse or hyperbola(136) APB let AB be the major axis, O the centre, S a focus, OD the minor semi-axis, and AK half the *latus rectum*. Cut AS in G so that AG be to AS as BO to BS, and ascertain a length L which shall be to GK(137) as AO^2 to the rectangle $AS\times OD$. Bisect OG in C, and with centre C and radius CG describe the semicircle GFO. Lastly, take the angle $G\widehat{C}F$ to four right angles as the given time taken by the body to describe the required arc $\widehat{AP}$ is to the periodic time—

attacking, in the scholium which follows, the much more difficult but considerably more rewarding problem of accurately effecting the approximate solution of the Keplerian equation which is there embodied.

(135) In a preliminary revise of this portion of the manuscript, Newton amended his argument to read (ULC. Add. 3965.4: 17r; compare I. B. Cohen's *Introduction* (note (1)): 126): '...ut differentia inter sectorem OQA et triangulum OQS, sive ut differentia rectangulorum $\frac{1}{2}OQ\times AQ$ et $\frac{1}{2}OQ\times SR$, hoc est ob datam $\frac{1}{2}OQ$ ut differentia inter arcum AQ et rectam SR, adeoq; (ob æqualitatem rationum SR ad sinum arcûs AQ, OS ad OA, OA ad OG, AQ ad GF, et divisim $AQ-SR$ ad $GF-$sin. arc. AQ) ut' (... as the difference between the sector (OQA) and the triangle (OQS), or as the difference of the products $\frac{1}{2}OQ\times\widehat{AQ}$ and $\frac{1}{2}OQ\times SR$, that is, because $\frac{1}{2}OQ$ is given, as the difference between the arc $\widehat{AQ}$ and the straight line SR, and hence (because of the equality of the ratios SR to $\sin\widehat{AQ}$, OS to OA, OA to OG, $\widehat{AQ}$ to GF, and so *dividendo* $\widehat{AQ}-SR$ to $GF-\sin\widehat{AQ}$) as). This passed without further structural alteration into the published *Principia* ($_1$1687: 108).

(136) While the following construction is dovetailed to the ellipse, it will manifestly hold *mutatis mutandis* for the hyperbola, where the related 'Keplerian' equation is (in modern terminology) $e\sinh\theta-\theta = T$, $e>1$. In October of the next year, however, Newton—doubtless to avoid confusion—instructed Halley to delete the present reference to this alternative conic species (see note (138)), though an equivalent phrase is retained in sequel (see note (143)).

(137) This should be '$\frac{1}{2}GK$' as Newton came to see in October 1686 when urged by Halley to check the detail of the present 'Geometricall effection'; see note (138) following.

seu revolutionis unius in Ellipsi. Ad AO demitte normalem FE eamq; produc versus F ad usq; H ut sit EH ad Longitudinem L ut anguli illius sinus EF ad radium CF, centroq; H et intervallo AH descriptus circulus secabit Ellipsin in corporis loco quæsito P quamproximè.(138)

Si(139) Ellipscos latus transversum(140) multo majus sit quàm latus rectum et motus corporis prope verticem Ellipseos desideretur, (qui casus in Theoria Cometarum incidit)(141) educere licet e puncto G rectam GI axi AB perpendicularem et in ea ratione ad GK quam habet area $AVPS$ ad rectangulum $AK \times AS$, dein centro I et intervallo AI circulum describere. Hic enim secabit Ellipsin in corporis loco quæsito P quamproximè.(142) Et eadem constructione (mutatis mutandis) conficitur Problema in Hyperbola.(143)

---

(138) This paragraph was in due course, like the rest of the 'Liber primus', transcribed *verbatim* by Humphrey Newton into the fair copy (now Royal Society MS LXIX: the rejected preliminary sheet ULC. Add. 3965.4: 17^r is likewise here without variant) which was in March 1686 sent to London for publication as the first book of the *Principia*. When Halley came to edit the passage the following autumn, he found that he could neither restore Newton's derivation of the construction nor make it work, and so on 14 October sent a letter to Cambridge requesting 'to be informed concerning your Geometricall effection of the problem XXIII, as much as relates to the 63 figure, for upon triall (there being no demonstration annexed) there seems to be some mistake committed, wherfore I entreat you would please to send me, revised by your self, those few lines that relate therto, and if it be not too much trouble, be prevailed upon to subjoyn something of the Demonstration' (*Correspondence of Isaac Newton*, 2, 1961: 452). In response, after a careful check of the present 'effection' in which he both detected the small numerical slip (see note (137)) in his assigned value for L and amended his accompanying 'Fig. 63' (set on a now lost separate sheet in the manuscript, but here restored on the basis of the revised diagram on *Principia*, $_1$1687: 109) to have only one point H, Newton wrote back four days later that 'In y^e^ Scholium...the words *vel Hyperbolæ* [see note (136)] are to be struck out, &...the words *quæ sit ad GK* should be *quæ sit ad* $\frac{1}{2}GK$. I send you inclosed y^e^ beginning of this Scholium wth y^e^ 63^d^ figure as I would have them printed. I thank you heartily for giving me notice that it was amiss' (*ibid.*: 453).

Neither in his preliminary recomputations (ULC. Add. 3965.9: 96^v, reproduced in Appendix 2.1 below) nor in the much augmented ensuing revised text (Royal Society MS LXIX: 89, reproduced in Appendix 2.2) which, in Humphrey Newton's transcript, he thereafter sent Halley to be published in the *Principia* ($_1$1687: 109–11) does Newton explain in detail how he came to contrive this clever geometrical approximation, but it is manifestly founded on the parabolic construction of Proposition XXX to which—as Newton afterwards emphasised—it reduces in the limiting case where the focus S is indefinitely close to the vertex A. In the generalised problem he again seeks to determine the point P of the orbit as the latter's (second) meet with a circle, of centre H, which passes through A, but with the Keplerian equation which relates the given mean anomaly T to P's eccentric angle θ or

$$\cos^{-1}(SP.\cos \widehat{BSP}/OA-e), \quad e = OS/OA,$$

now $\theta - e\sin\theta = T$. On setting $OE = \lambda$ and $EH = \mu$ to be the coordinates of H, and also (for convenience) putting $OA = 1$, since $\cos\theta$ and $\sqrt{[1-e^2]}\sin\theta$ are the corresponding coordinates of P and because (by construction) $AH = HP$, it follows, on squaring, that

$$(1-\lambda)^2+\mu^2 = (\cos\theta-\lambda)^2+(\sqrt{[1-e^2]}\sin\theta-\mu)^2$$

or $\lambda(1-\cos\theta)-\mu\sqrt{[1-e^2]}\sin\theta-\frac{1}{2}e^2\sin^2\theta = 0$; whence to a first approximation $\lambda \approx r(1+\cos\theta)$

of one revolution, that is—in the ellipse. To AO let fall the normal FE and extend it in F's direction up to H so that EH be to the length L as the sine EF of that angle to its radius CF; then with centre H and radius AH describe a circle and it will intersect the ellipse in the required position P of the body very nearly.(138)

If(139) the ellipse's main axis be much greater than its *latus rectum* and were the body's motion near to a vertex of the ellipse desired (this is the case met with in the theory of comets),(141) you are at liberty to draw out from the point G a straight line GI perpendicular to the axis AB and in the ratio to GK had by the area $(AVPS)$ to the rectangle $AK \times AS$, and then on centre I and with radius AI to describe a circle: for this will intersect the ellipse in the required position P of the body very nearly.(142) And the problem is, *mutatis mutandis*, accomplished by the same construction in the hyperbola.(143)

---

and $\mu \approx r\sin\theta$, from which, since $\theta = T+O(e)$, it is a natural step to set $\lambda = r(1+\cos T)$ exactly, thus determining $\mu = (1+O(e))\, r\sin T$, and the perpendicular EH is thereby fixed on taking $OC = CG = r$, $G\widehat{CF} = T$ and $EH/EF(= \mu/r\sin T) = s$, constant to $O(e)$. The values which Newton assigns to r and s result naturally from his further requirement that the construction should reduce to that of Proposition XXX when the ellipse passes into a parabola (or S comes to coincide with A) and P is located in the immediate vicinity of A. For, since then $\lim_{S\to A} (AG/AS) = \lim_{e\to 1} [(1-2r)/(1-e)] = \frac{1}{2}$, but also $\lim_{S\to O} (AG/AS) = \lim_{e\to 0} [(1-2r)/(1-e)] = 1$, it is simplest to set $AG/AS = 1/(1+e) = BO/BS$ and so $r = e/(1+e)$; and again, because in general $(ASP):(GCF) = (ASBP):(GCOF) = AO \times OD : GC^2$, while in the limit as S and P coincide with A (when the normal to the vanishingly small arc $\widehat{AP}$ through H, equidistant from A and P, both bisects $\widehat{AP}$ and also passes through K, the centre of curvature of the ellipse at A) there comes to be $(ASP) = \frac{1}{2}AS \times (\widehat{AP}$ or$)\ 2EH \times AK/GK$ and $(GCF) = \frac{1}{2}GC \times (\widehat{GF}$ or$)\ EF$, it follows that $AS \times EH \times AK/GK : \frac{1}{2}GC \times EF = AO \times OD : GC^2$ and hence

$$\lim_{E\to A} (EH/EF) = s = \tfrac{1}{2}GK \times (AO^2/AS \times OD)/GC$$

on replacing AK by its equal OD^2/AO, half the *latus rectum* of the ellipse. It would seem likely (compare Appendix 2: note (11)) that Newton's omission of the factor $\frac{1}{2}$ in his equivalent statement of this in the form of a double proportion is the result of carelessly setting AK equal to the whole *latus rectum*.

(139) In a hesitant afterthought Newton converted this to be 'Siquando' (Whenever) and then deleted the change.

(140) Newton emphasised that this is the main *latus transversum* by here inserting 'principale' in his preliminary revise (ULC. Add. 3965.4: 17r) but the insertion was omitted in the printed *Principia* ($_1$1687: 111). We appropriately convey the nuance in our English version.

(141) This parenthesis, here somewhat out of place, is a late addition in Newton's hand in the manuscript.

(142) In this limiting case where the ellipse is effectively indistinguishable from a parabola of focus S and vertex curvature-radius $AK = 2AS$, it will be evident that GI is (very nearly) equal to GH in Proposition XXX, whence, since then $AG = \frac{1}{2}AS$ and so $GK = \frac{3}{2}AS$, it follows that $GI = (\frac{3}{4}(APS)/AS$ or$)\ GK \times (APS)/AK \times AS$, as Newton states.

(143) Unlike the previous reference to this alternative species of central conic (see note (136)), this passed unaltered into the printed *Principia* ($_1$1687: 111), where Newton further added in explanation for Halley that 'Hæ autem constructiones demonstrantur ut supra, & si

Si quando locus ille P accuratiùs determinandus sit[,] inveniatur tum angulus quidam B qui sit ad angulum graduum 57|29578 quem arcus radio æqualis subtendit ut est umbilicorum distantia SH ad Ellipseos diametrum AB,(144) tum etiam longitudo quædam L quæ sit ad Radium in eadem ratione inversè.(145) Quibus semel inventis Problema deinceps confit per sequentem Analysin. Per constructionem superiorem (vel utcunq conjecturam faciendo) cognoscatur corporis locus P quamproximè. Demissaq ad axem Ellipseos ordinata(146) PR, ex proportione diametrorum Ellipseos dabitur circuli circumscripti ordinatim applicata RQ, quæ sinus est anguli ACQ, existente AC radio.(147) Sufficit angulum illum rudi calculo in numeris proximis invenire. Cognoscatur etiam angulus tempori proportionalis, id est qui sit ad quatuor rectos ut est tempus quo corpus descripsit arcum AP ad tempus revolutionis unius in Ellipsi. Sit iste(146) N, tum capiatur angulus D ad angulum B ut est sinus iste anguli ACQ ad Radium, et angulus E ad angulum $N-ACQ+D$ ut est longitudo L ad longitudinem eandem L cosinu anguli $ACQ+\frac{1}{2}D$(148) diminutam ubi angulus ille recto minor est, auctam ubi major. Postea capiatur angulus F ad angulum B ut est sinus anguli $ACQ+E$ ad Radium, et angulus G ad angulum $N-ACQ-E+F$ ut est longitudo

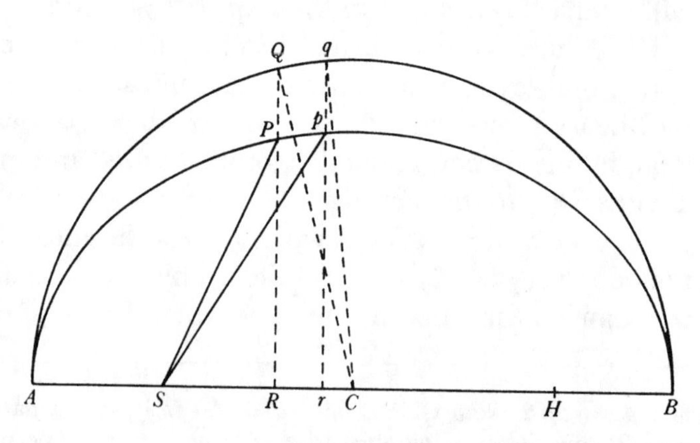

Figura (vertice ulteriore B in infinitum abeunte) vertatur in Parabolam, migrant in accuratam illam constructionem Problematis XXII' (These constructions, however, are demonstrated as above, and if the figure—with the further vertex B going away to infinity—should turn into a parabola, they pass over into the accurate construction of Problem XXII).

Though in the much augmented version of the two preceding paragraphs which he sent to London in October 1686 (see Appendix 2.2) he somewhat refined its accuracy—and indeed still further improved and simplified it in a later fragment of about 1692 (ULC. Add. 3965.17: 637r, reproduced in Appendix 2.3) which he never published—Newton fell ultimately out of love with this ingenious but impractical geometrical mode of construction, and suppressed it from the later editions of his *Principia* in favour of the computationally superior analytical methods, making use of iterations and expansions into series, which follow. To bridge the awkward ensuing gap Newton amended the opening line of the next paragraph to read (*Principia*, $_2$1713: 101 = $_3$1726: 109): 'Cæterum, cum difficilis sit hujus Curvæ [sc. the prolate cycloid of Proposition XXXI] descriptio, præstat solutionem vero proximam adhibere. Inveniatur tum angulus quidam B...' (But since the description of this curve is difficult, it is better to employ a solution approximating to the truth. Let there be found both an angle B...).

(144) That is, we are to ascertain an angle B whose radian measure is the ellipse's eccentricity.

Should ever that position P need to be determined more accurately, let there be found both an angle B which shall be to the angle 57·295 78° subtended by an arc equal to the radius as the distance SH of its foci are to the ellipse's diameter AB,(144) and also a length L which shall be to the radius in the inverse of the same ratio.(145) Once these are found, the problem is thereafter accomplished by the following analysis. By the preceding construction (or making any sort of guess at it) ascertain the body's position P very nearly. Then, on letting fall PR ordinate to the axis of the ellipse, from the proportion between the diameters of the ellipse there will be given the circumscribed circle's ordinate RQ, which is the sine of the angle $\widehat{ACQ}$, AC being the radius.(147) It is enough to find that angle approximately by a rough numerical calculation. Let there also be found out the angle proportional to the time—one, that is, which shall be to four right angles as the time taken by the body to describe the arc $\widehat{AP}$ to the time of one orbit in the ellipse. Let that (angle) be N, then take an angle D (which is) to angle B as that sine of $\widehat{ACQ}$ is to its radius, and the angle E to the angle $N-\widehat{ACQ}+D$ as the length L to the same length L diminished by the cosine of the angle $\widehat{ACQ}+\frac{1}{2}D$(148) when that angle is less than a right one, but increased by it when it is greater. Thereafter, take the angle F to the angle B as the sine of the angle $\widehat{ACQ}+E$ is to its radius, and the angle G to the angle $N-\widehat{ACQ}-E+F$ as the length

(145) This ratio was originally spelled out in the manuscript 'ut Ellipseos diameter AB ad umbilicorum distantiam' (as the ellipse's diameter AB to the distance of its foci).

(146) Amended to be 'ordinatim applicata' and 'angulus iste' respectively in Newton's preliminary revise (ULC. Add. 3965.4: 18r), whence these verbal ameliorations passed unaltered into the published *Principia* ($_1$1687: 111, 112).

(147) Notice that the ellipse's centre is now marked 'C'. In the final paragraph of his present scholium (where the preceding figure is assumed) Newton will revert to 'O'.

(148) In his copy (now Bodleian. 4° Z. 23. Art.) of the *editio princeps* Fatio de Duillier some five years later declared in the margin at this point (*Principia*, $_1$1687: 112, line 8) that 'Hic latet error, quemadmodum Cl. Newtonus post maturam examinationem mihi confessus est. Vel quantitas $\frac{1}{2}D$ prorsus delenda est, vel ejus loco $\frac{1}{2}E$ reponendum, quod quidem erit accuratius. Sic paulo inferius loco $\frac{1}{2}F$ scribendum $\frac{1}{2}G$, et loco $\frac{1}{2}H$ scribendum $\frac{1}{2}I$, nisi ambas quantitates $\frac{1}{2}F$ et $\frac{1}{2}H$ delere visum fuerit, quod jam tuto facias. Hæc si adhibeatur correctio congruet operatio præscripta cum approximatione in quam olim incidi; nisi quod calculationi consulens quantitatibus $\frac{1}{2}E$, $\frac{1}{2}G$, $\frac{1}{2}I$ non utebar. . . . Ut autem inveniatur E duplici proportione utendum est; altera in qua ope cosinus anguli ACQ inveniatur E proxime; altera verò in qua ope cosinus anguli $ACQ+\frac{1}{2}E$, qui jam proxime cognitus [est], inveniatur E accuratius' (compare S. P. Rigaud, *Historical Essay on the First Publication of Sir Isaac Newton's 'Principia'* (Oxford, 1838): 92). Appreciating the soundness of this criticism (on which see note (151) below), Newton initially proposed to incorporate Fatio's emendations in a revised determination of the 'corporis locus in usus omnes Astronomicos abunde satis correctus' (see ULC. Add. 3967.1: 5v) though in his published *editio secunda* (*Principia*, $_2$1713: 102; unchanged on $_3$1726: 110) he ultimately adopted Fatio's simpler proposal for mending his earlier mistake, there merely omitting the faulty error terms '$+\frac{1}{2}D$', '$+\frac{1}{2}F$' and '$+\frac{1}{2}H$' to leave no trace of his more sophisticated original scheme of iteration.

L ad longitudinem eandem cosinu anguli $ACQ+E+\frac{1}{2}F$(149) diminutam ubi angulus ille recto minor est[,] auctam ubi major. Tertia vice capiatur angulus H ad angulum B ut est sinus anguli $ACQ+E+G$ ad radium et angulus I ad angulum $N-ACQ-E-G+H$ ut est longitudo L ad eandem longitudinem cosinu anguli $ACQ+E+G+\frac{1}{2}H$(149) diminutam ubi angulus ille recto minor est, auctam ubi major. Et sic pergere licet in infinitum. Deniꝗ capiatur angulus ACq æqualis angulo $ACQ+E+G+I$ &c et ex cosinu ejus Cr et ordinata pr quæ est ad sinum qr ut Ellipseos axis minor ad axem majorem, habebitur corporis locus correctus p. Siquando angulus $N-ACQ+D$ negativus est debet signum $+$ ipsius E ubiꝗ mutari in $-$ et signum $-$ in $+$. Idem intelligendum est de signis ipsorum G et I ubi anguli $N-ACQ-E+F$ et $N-ACQ-E-G+H$ negativè prodeunt. Convergit autem series infinita $ACQ+E+G+I$ [&c] quam celerrimè, adeò ut vix unquam opus fuerit ultra progredi quam ad terminum secundum E. Et fundatur calculus in hoc Theoremate quod area APS sit ut differentia inter arcum AQ et rectam(150) ab umbilico S in Radiū CQ perpendiculariter demissam.(151)

(149) As Fatio observed in his (1690?) annotation (see the previous note), these expressions should, in line with his preceding correction of '$ACQ+\frac{1}{2}D$' into '$ACQ+\frac{1}{2}E$', be amended to read '$ACQ+E+\frac{1}{2}G$' and '$ACQ+E+G+\frac{1}{2}I$' respectively.

(150) Of length $SC.\sin\widehat{ACQ}$, that is. The ellipse's focal sector (ASP) is, of course, in given proportion to the corresponding sector $(ASQ) = (ACQ)-\triangle SCQ = \frac{1}{2}AC.(\widehat{AQ}-SC.\sin\widehat{ACQ})$ of the eccentric circle.

(151) On taking the trigonometrical radius to be R, the elliptical eccentricity

$$SC/AC = B = R/L = e$$

and θ to be the eccentric angle fixing the required position in orbit at a given time measured by (T or) N, both measured from perihelion A, then it is once more Newton's problem approximately to invert the Keplerian equation $N = \theta - e\sin\theta$ which ensues. Suppose that θ_i is a near solution and set $\theta = \theta_i+\epsilon$, ϵ small, so that

$$N = \theta_i+\epsilon-e(\sin\theta_i\cos\epsilon+\cos\theta_i\sin\epsilon) = \theta_i+\epsilon-e(\sin\theta_i.[1-\tfrac{1}{2}\epsilon^2...]+\cos\theta_i.[\epsilon-...])$$

and hence $$N-\theta_i+e\sin\theta_i = \epsilon(1-e[\cos\theta_i-\tfrac{1}{2}\epsilon\sin\theta_i...]) \approx \epsilon(1-e\cos(\theta_i+\tfrac{1}{2}\epsilon)),$$

or $\epsilon \approx \epsilon_i = \dfrac{N-\theta_i+e\sin\theta_i}{1-e\cos(\theta_i+\frac{1}{2}\epsilon_i')}$ where $\epsilon_i' = \dfrac{N-\theta_i+e\sin\theta_i}{1-e\cos\theta_i} \approx \epsilon_i$. As corrected by Fatio (see notes (148) and (149)) Newton's present scheme computes successively, given $\widehat{ACQ} = \theta_1$, the sequence (rapidly convergent for $|\theta-\theta_1|$ small enough) $\theta_{i+1} = \theta_i+\epsilon_i$, $i = 1, 2, 3, ...$: namely, $E = \epsilon_1$, $G = \epsilon_2$, $I = \epsilon_3$, ... (where $D = e\sin\theta_1$, $F = e\sin\theta_2$, $H = e\sin\theta_3$, ...) and so, on taking enough terms to make their difference negligible,

$$\theta = \theta_i = (\theta_1+\epsilon_1+\epsilon_2+\epsilon_3+... \text{ or}) \ \widehat{ACQ}+E+G+I....$$

In contrast, the simplified scheme introduced into later editions of the *Principia* computes, again given $\widehat{ACQ} = \theta_1 = \theta_1'$, the slightly less quickly converging sequence $\theta_{i+1}' = \theta_i'+\epsilon_i'$, $i = 1, 2, 3, ...$: namely, $E = \epsilon_1'$, $G = \epsilon_2'$, $I = \epsilon_3'$, ... (where $D = e\sin\theta_1$, but now $F = e\sin\theta_2'$, $H = e\sin\theta_3'$, ...), so yielding to adequate approximation

$$\theta = \theta_i' = (\theta_1+\epsilon_1'+\epsilon_2'+\epsilon_3'+... \text{ or}) \ \widehat{ACQ}+E+G+I...$$

L to the same length diminished by the cosine of the angle $\widehat{ACQ}+E+\frac{1}{2}F$[149] when that angle is less than a right one, but increased by it when it is greater. At a third stage, take the angle H to the angle B as the sine of the angle $\widehat{ACQ}+E+G$ to its radius, and the angle I to the angle $N-\widehat{ACQ}-E-G+H$ as the length L is to the same length diminished by the cosine of the angle $\widehat{ACQ}+E+G+\frac{1}{2}H$[149] when that angle is less than a right one, but increased by it when it is greater. And so you are free to proceed indefinitely. Finally, take the angle $\widehat{ACq}$ equal to the angle $\widehat{ACQ}+E+G+I\ldots$, and from its cosine Cr and ordinate pr (this is to its sine qr as the ellipse's minor axis to its major axis) there will be obtained the body's corrected position p. If the angle $N-\widehat{ACQ}+D$ chances to be negative, the $+$ sign of E must everywhere be changed to a $-$ and the $-$ sign to a $+$. The same is to be understood for the signs of G and I when the angles $N-\widehat{ACQ}-E+F$ and $N-\widehat{ACQ}-E-G+H$ turn out to be negative. The infinite sequence $\widehat{ACQ}+E+G+I\ldots$ converges very rapidly, however, and as a result there will scarcely ever be need to proceed farther than to the second term E. And the computation is based on this theorem, that the area (APS) is as the difference between the arc $\widehat{AQ}$ and the straight line[150] let fall from the focus S perpendicularly onto the radius CQ.[151]

---

once more. Since, in practical terms, any single iteration $\epsilon_i \rightarrow \epsilon_{i+1}$ is roughly equivalent to a double iteration $\epsilon'_{2i-1} \rightarrow \epsilon'_{2i} \rightarrow \epsilon'_{2i+1}$ in the much simpler, computationally less tricky revised scheme, Newton had good reason for expounding only the latter mode of calculation in his re-edited *Principia* text. While he here makes no mention of this relationship—and what readers would have comprehended the allusion had he done so?—it could hardly have escaped his notice that the simple error term ϵ'_i is $-f(\theta_i)/f'(\theta_i)$ where

$$f(\theta) \equiv \theta - e\sin\theta - N = 0.$$

As we remarked in the fourth volume, Newton had sketched an equivalent application of this 'Newton–Raphson' iterative procedure to the speedy calculation of general n-th roots (where $f(x) \equiv x^n - A = 0$) for the benefit of a London table-computer in July 1675; see IV: 664, and compare 665, note (24). The difficult question of establishing an effective criterion for the convergence of the error term ϵ'_i was not successfully answered for more than a century afterwards until it was independently considered by J. R. Mourraille in his *Traité de la résolution des équations en général* (Paris, 1798) and by Joseph Fourier in researches in the summer of 1804 posthumously published in the second book of his *Analyse des équations déterminées* (Paris, 1831); see F. Cajori, 'Fourier's improvement of the Newton–Raphson method of approximation anticipated by Mourraille', *Bibliotheca Mathematica*, $_3$**11**, 1910–11: 132–7, and I. Grattan-Guinness, *Joseph Fourier, 1768–1830* (Cambridge, Massachusetts, 1972): 82, note 6 and 482–3. Newton's more sophisticated iteration was little appreciated in his lifetime, though it is of course subsumed in the general 'Methodus Nova Accurata & facilis inveniendi Radices Æquationum quarumcunque generaliter, sine prævia Reductione' published by Edmond Halley in 1694 (*Philosophical Transactions* **18**, No. 210: 136–48; see V: 12, note (50)) as a transcendental extension there not allowed for; after his death it had to await 'rediscovery' by Antonio Cagnoli in his *Trigonometrica piana e sferica*—simultaneously rendered into French (by Chompré) as his *Traité de Trigonométrie rectiligne et sphérique*—(Paris, 1786):

Non dissimili calculo conficitur Problema in Hyperbola. Sit ejus centrum C[,] vertex A, umbilicus S, Asymptotos [C]K. Cognoscatur quantitas areæ APS tempori proportionalis: sit ea A, et fiat conjectura de positione rectæ SP quæ aream illam abscindat quamproximè. Jungatur CP,(152) et ab A et P ad Asymptoton agantur AI, PK Asymptoto alteri parallelæ et per tabulam logarithmorum dabitur area $AIKP$, eiq; æqualis(153) area CPA quæ subducta de triangulo CPS relinquet aream APS. Applicando arearum A et APS semidifferentiam $\frac{APS-A}{2}$ vel $\frac{A-APS}{2}$ ad lineam SN quæ ab umbilico S in tangentem PT perpendicularis est orietur longitudo PQ.(154) Capiatur autem PQ inter A et P si area APS major sit area A, secus ad puncti P contrarias partes: et punctum Q erit locus corporis accuratiùs inventus. Et computatione repetita invenietur idem accuratius in perpetuum.(155)

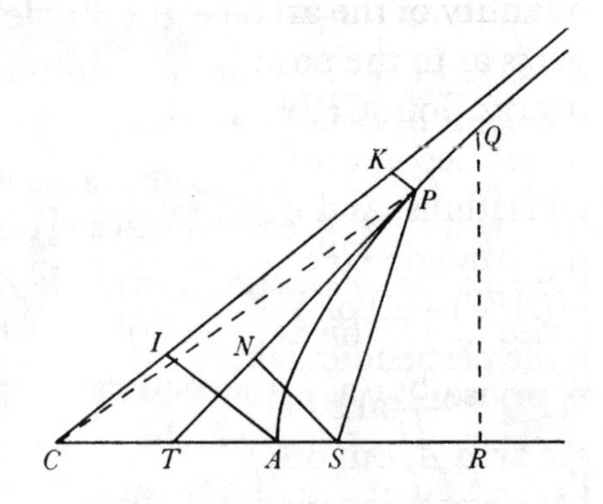

Atq; his calculis Problema generaliter confit Analyticè. Verum usibus Astronomicis accommodatior est calculus particularis qui sequitur. Existentibus(156) AO, OB, OD semiaxibus Ellipseos et L ipsius latere recto,(157) quære tum

377–8, and Newton's priority was restored only a century afterwards by John Couch Adams in a short note 'On Newton's Solution of Kepler's Problem' (*Monthly Notices of the Royal Astronomical Society*, **43**, No. 2 [8 December 1882]: 43–7, especially 44–5).

(152) Through an oversight the points C and P are not in fact joined in the corresponding figure in the *Principia* ($_1$1687: 113) but we repair the omission in our reproduction of it—there, for accuracy's sake retaining the perpendicular QR which is not used in the text (but still kept in *Principia*, $_2$1713: 103, to be at length deleted by Pemberton in $_3$1726: 111).

(153) For, because $CK \times KP = CI \times IA$, the triangles CKP and CIA are equal in area.

(154) Understand 'quamproximè' (very nearly), of course, since

$$(PSQ) \approx \tfrac{1}{2}\widehat{PQ} \times SN \approx \tfrac{1}{2}PQ \times SN.$$

That PQ (rather than its near-equal $\widehat{PQ}$) is here intended was clarified in the *Principia's* second edition ($_2$1713: 103) by (Cotes'?) interpolation of 'chordæ' (of the chord) before 'PQ'. In his *editio ultima* ($_3$1726: 111) Pemberton unnecessarily inserted the chord PQ in his figure, thus suggesting to the unwary reader that it is the continuation of the tangent TP.

(155) A somewhat fumbling half-geometrical, half-computational construction reminiscent of his earlier Cavalierian construction of the Keplerian problem in the ellipse (see §1: note (84) and §2: note (162)), here replaced by superior methods. Had he thought of so doing, it would have been easy for Newton here to contrive an efficient iteration exactly paralleling that presented by him in the previous paragraph for the elliptical case: namely, on taking the hyperbola's eccentricity $CS/AS = e\,(>1)$ and, as before, θ to be the eccentric angle fixing the required position in orbit (from 'perihelion' A) in a given time measured by the mean anomaly T (Newton's N in the preceding paragraph), there results $T = e\sinh\theta - \theta$; whence, if θ_i be a near solution, it follows—in the simplified 'Newton–Raphson' scheme—that $\theta_{i+1} = \theta_i - (T + \theta_i - e\sinh\theta_i)/(1 - e\cosh\theta_i)$ is, for $|\theta - \theta_i|$ small enough, a considerably better

The problem is accomplished in a hyperbola by a not dissimilar computation. Let C be its centre, A a vertex, S a focus, CK an asymptote. Ascertain the quantity of the area (APS) proportional to the time: let that be A, and make a guess as to the position of the straight line SP which shall cut off that area very nearly. Join CP,(152) and from A and P to the asymptote draw AI, PK parallel to the second asymptote; the area $(AIKP)$ will then be given by a table of logarithms, and equal to this(153) the area (CPA), which when subtracted from the triangle CPS will leave the area (APS). On dividing the semidifference, $\frac{1}{2}((APS)-A)$ or $\frac{1}{2}(A-(APS))$, of the areas A and (APS) by the line SN which is the perpendicular from the focus to the tangent PT there will ensue the length of PQ(154)—take PQ between A and P, however, if the area (APS) be greater than the area A, but otherwise on the opposite side of P—and the point Q will be the body's position more accurately found. And by repeating the computation it will be found ever more accurately still.(155)

These computational methods afford a general analytical solution of the problem. But the particular calculus which follows is better suited to astronomical use. Where(156) AO, OB, OD are semi-axes of the ellipse and L is its *latus rectum*,(157) seek both the angle Y whose tangent shall be to the radius as the

---

one. While it is true that Newton had no explicit notation for the hyperbolic functions $QC/AC = \cosh\theta$ and $PQ/AC = \sqrt{[e^2-1]}.\sinh\theta$, where

$$\theta = (ACP)/\tfrac{1}{2}\sqrt{[e^2-1]}.AC^2 = \log(QC/AC+\sqrt{[QC^2/AC^2-1]}),$$

it will be clear from his construction of the general inverse-cube orbit in Corollary 3 to Proposition XLI following (see also Appendix 4 below) that he was familiar with their properties in the geometrical model (as here) of the hyperbola.

(156) Newton has here deleted '*S*, *H* umbilicis et' (*S*, *H* are the foci and), evidently regarding it as by now sufficient to refer below to '...umbilicorum distantia *SH*' (the distance *SH* between the foci). As his marginal reference in the manuscript to 'Fig 63' shows, he here understands the diagram of the scholium's first paragraph (but with L now denoting the ellipse's *latus rectum* $2OD^2/AO$): for convenience—and following Pemberton's lead in the third edition of the ensuing *Principia* ($_3$1726: 111/112)—we introduce into our reproduction of the Latin text an appropriately simplified figure, in which we have inserted also the mean focal radius SD directed to be drawn in a preliminary version of the sequel (see next note).

(157) Evidently referring to an equivalent figure in which, as in the two previous paragraphs, the ellipse's centre was marked C and the other end-point of the minor semi-axis was denoted by Q, Newton first continued: 'junge SQ et quære angulos X, Y, Z quorum primus X sit ad duos rectos ut est umbilicorum distantia SH ad perimetrum circuli descripti diametro AB, secundus Y sit ad differentiam angulorum X et CQS ut est Ellipseos latus transversum AB ad istius latus rectum $\frac{4QC^{\text{quad.}}}{AB}$, tertius Z subtendatur arcu qui sit ad Radium ut est differentia inter Ellipseos semidiametros AC et CQ ad ipsius diametrum minorem [2]CQ. His angulis semel inventis, locus corporis sic deinceps determinabitur. Cape angulum L proportionalem tempori et angulum M ad angulum Z ut est sinus anguli $2L$ ad radium atq; angulum N ad angulum Y ut est quadratum sinus anguli L ad quadratum radij vel (quod perinde est) ut sinus versus anguli $2L$ ad radium duplicatum. Angulorum vel summæ $L+N+M$ si angulus L recto

angulum Y cujus tangens sit ad Radium ut est semiaxium differentia $AO-AD$ ad eorum summam $AO+OD$, tum angulum Z cujus tangens sit ad Radium ut est rectangulum sub umbilicorum distantia SH et semiaxium differentia $AO-OD$ ad triplum rectangulum sub $O[D]$ semiaxe minore et $OA-\frac{1}{4}L$ differentia inter semiaxem majorem et quartam partem lateris recti. His angulis

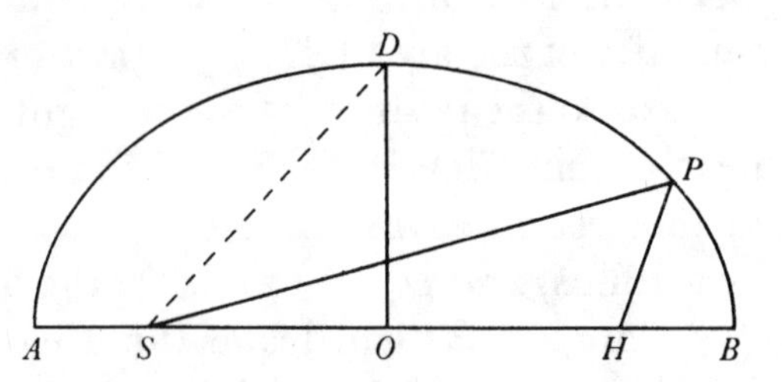

semel inventis, locus corporis sic deinceps determinabitur. Cape angulum T proportionalem tempori quo arcus BP descriptus est, seu motui medio (ut loquuntur) æqualem[,] et angulum V (primam medij motus æquationem) ad angulum Y (æquationem maximam primam) ut est sinus anguli T duplicati ad radium[,] atq; angulum X (æquationem secundam) ad angulum Z (æquationem maximam secundam) ut est sinus versus anguli T duplicati ad radium duplicatum vel (quod eodem recidit) ut est quadratum sinus anguli T ad quadratum Radij.(158) Angulorum T, V, X vel summæ $T+X+V$ si angulus T recto minor est vel differentiæ $T+X-V$ si is recto major est rectisq; duobus minor, æqualem cape angulum BHP (motum medium æquatum:) et si HP occurrat Ellipsi in P, acta SP abscindet aream BSP tempori proportionalem quamproximè.(159) Hæc praxis satis expedita videtur, propterea quod angulorum perexiguorum V et X (in minutis secundis, si placet, positorum) figuras duas tresve primas invenire sufficit.(160) Invento autem angulo motus medij æquati BHP, angulus veri

---

minor est vel differentiæ $L+N-M$ si is recto major est æqualem cape angulum BHP. Occurrat HP Ellipsi in P et acta SP abscindet aream BSP tempori proportionalem quamproximè' (join $S[D]$ and seek the angles X, Y and Z, the first of which, X, shall be to two right angles as the distance SH of the foci is to the circumference of a circle described on diameter AB, the second, Y, shall be to the difference of the angles X and $[\widehat{ODS}]$ as the ellipse's main axis AB is to its *atus rectum* $4[OD^2]/AB$, and the third, Z, shall be subtended by an arc which shall be to its radius as the difference between the ellipse's semi-diameters $A[O]$ and $[OD]$ is to its lesser diameter $2[OD]$. Once these angles are ascertained, the position of the body will thereafter be determined in this manner. Take angle L proportional to the time, and then angle M to angle Z as the sine of the angle $2L$ is to the radius, and angle N to the angle Y as the square of the sine of the angle L is to the square of the radius or (what amounts to the same) as the versine of the angle $2L$ to the radius. Equal to the sum $L+M+N$ of the angles if L is less than a right angle, or to their difference $L-M+N$ if it is greater than a right angle, take the angle $\widehat{BHP}$. Let HP meet the ellipse in P and then, when SP is drawn, it will cut off the area (BSP) proportional to the time very nearly). Where the ellipse's eccentricity (SC/AC or) SO/AO is taken to be e, Newton sets $X = \pi.(2e/2\pi) = e$, $Y = N/\sin^2 L = (\sin^{-1}e-e)/(1-e^2) = \frac{1}{6}e^3\ldots$ and $Z = M/\sin 2L = (1-\sqrt{[1-e^2]})/2\sqrt{[1-e^2]} = \frac{1}{4}e^2+\frac{3}{16}e^4\ldots$, whence $\widehat{BHP}$ (here, like the mean anomaly L, measured from aphelion) is, to $O(e^4)$, approximated as

$$L \pm M+N = L+\tfrac{1}{4}e^2\sin 2L+\tfrac{1}{6}e^3\sin^2 L,$$

much as in the earlier 'improvement' of Ismael Boulliau's 1657 'equation' which he presented

difference $AO-OD$ of the semi-axes is to their sum $AO+OD$, and the angle Z whose tangent shall be to the radius as the rectangle contained by the distance SH of the foci and the difference $AO-OD$ of the semi-axes to triple the rectangle contained by the semi-minor-axis OD and the difference $OA-\frac{1}{4}L$ between the semi-major-axis and one-fourth the *latus rectum*. Once these angles are ascertained the position of the body will thereafter be determined in this manner. Take the angle T proportional to the time in which the arc $\widehat{BP}$ is described—equal, that is, to the mean motion (as it is called)—, the angle V (the first equation of the mean motion) to the angle Y (the maximum first equation) as the sine of double the angle T to the radius, and the angle X (the second equation) to the angle Z (the maximum second equation) as the versed sine of double the angle T to double the radius, or (what amounts to the same) as the square of the sine of the angle T to the square of the radius.(158) Equal to the sum $T+V+X$ of the angles T, V and X if T is less than a right angle, or to their difference $T-V+X$ if it is greater than a right angle and less than two, take the angle $B\widehat{H}P$ (the equated mean motion); and then, if HP should meet the ellipse in P, SP when drawn will cut off the area (BSP) proportional to the time very nearly.(159) This technique seems expedient enough in view of the fact that of the exceedingly small angles V and X (reckoned in seconds, please) it suffices to find the first two or three (significant) figures.(160) Once, however, the angle

in the scholium to Proposition XX of the preceding 'De motu Corporum' (see §2: note (164)). His present variant determination of Y clearly reflects some prior—and inaccurate—attempt to improve upon his previous coefficient $\sin^{-1}e-e$ which it seems purposeless here to restore; that of Z is likewise no real advance over the simpler equivalent $(AO-OD)/AB = \frac{1}{2}\sqrt{[1-e^2]}$. Newton's replacement of his initial version by the reworked text reproduced above indicates that, on checking through his (now lost?) calculations, he came quickly to see that the term $\frac{1}{6}e^3$ which predominates in the expansion of Y is only one-fourth of true; but even in his recast construction he still—perhaps by mere oversight—retains the erroneous present proportion $N/Y = \sin^2 L$; see note (159).

(158) This should (see the next note) read '...ut est cubus sinus anguli T ad cubum Radij' (as the cube of the sine of the angle T to the cube of the radius) and it was so corrected in the *Principia*'s second edition ($_2$1713: 104).

(159) On again setting the ellipse's eccentricity SO/AO to be e, Newton's construction makes

$\tan Y = (1-\sqrt{[1-e^2]})/(1+\sqrt{[1-e^2]})$ and $\tan Z = 2e(1-\sqrt{[1-e^2]})/3\sqrt{[1-e^2]}\,(1-\frac{1}{2}(1-e^2))$, so determining $Y = \pm V/\sin 2T = \frac{1}{4}e^2+\frac{1}{8}e^4\ldots$ and $Z = N/\sin^2 T = \frac{2}{3}e^3-\frac{1}{6}e^5\ldots$, and thereby approximating $B\widehat{H}P = \psi$ by $T\pm V+X = T+\frac{1}{4}e^2\sin 2T+\frac{2}{3}e^3\sin^2 T\ldots$. Accurately (see §2: note (164)), $\psi = T+\frac{1}{4}e^2\sin 2T+\frac{2}{3}e^3\sin^3 T\ldots$. Newton made correct adjustment of the power of $\sin T$ in his stated value for X when, some seven years later, he came to revise the corresponding portion of his *Principia* ($_1$1687: 113–14) for its second edition (delayed for another twenty years); see Appendix 2.4 below.

(160) Newton's present approximation errs by about

$$\tfrac{2}{3}e^3\sin^2 T-\tfrac{2}{3}e^3\sin^3 T = \tfrac{2}{3}e^3\sin^2 T(1-\sin T) \leqslant \tfrac{8}{81}e^3;$$

for a Martian orbit $(e \approx \frac{1}{11})$, therefore, the maximum error is only some $\frac{1}{4}'$ of arc—wholly

motus(161) *HSP* et distantia *SP* in promptu sunt per methodum notissimam Dris Sethi Wardi Episcopi Salisburiensis mihi plurimùm colendi.(162)

Hactenus de motu corporum in lineis curvis. Fieri autem potest ut mobile rectà descendat vel rectà ascendat, et quæ ad istiusmodi motus spectant, pergo jam exponere.

Artic. VII.

De corporum ascensu ac descensu rectilineo.(163)

Prop. XXXII. Prob. XXIV.

Posito quod vis centripeta sit reciproce proportionalis quadrato distantiæ locorum a centro, spatia definire quæ corpus recta cadendo datis temporibus describit.

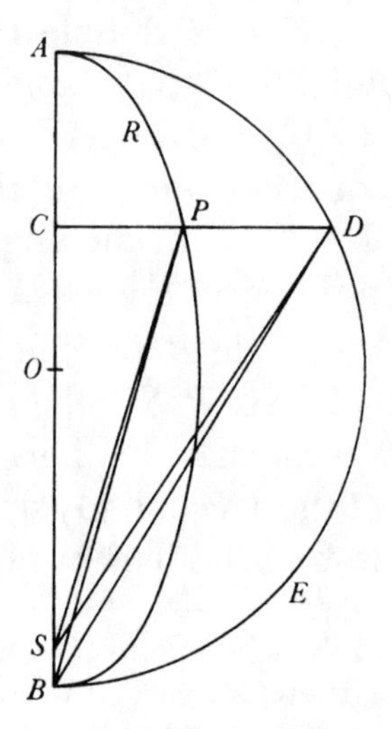

Cas. 1. Si corpus non cadit perpendiculariter[,] describet id sectionem aliquam Conicam cujus umbilicus inferior congruet cum centro.(164) Id ex Propositionibus XI, XII, XIII et earum Corollarijs(165) constat. Sit sectio illa Conica *ARPB* et umbilicus inferior *S*. Et primò si Figura illa Ellipsis est, super hujus axe majore *AB* describatur semicirculus *ADB* et per corpus decidens transeat recta *DPC*, perpendicularis ad axem, actisq; *DS*, *PS* erit area *ASD* areæ *ASP* atq; adeo etiam tempori proportionalis.

insignificant when compared with the deviations from ideal elliptical orbit (through external perturbation) of the solar planets. As far as we have been able to discover, Newton's surviving papers contain no hint that he ever made astronomical application even of the less refined 'Boulliau' equation $\psi \approx T+\frac{1}{4}e^2\sin 2T$ to relate position in Keplerian orbit to the corresponding mean anomaly T: on the rare occasions when he attempted such practical computation at all (notably on ULC. Add. 3965.11: 157r, where—probably as an adjunct to the accurate transposition of terrestrial sightings of a comet into equivalent solar ones—he was concerned to treat the earth's orbit round the sun as an ellipse of eccentricity $e = 0{\cdot}01672$) he seems to have found it enough to employ the 'Ptolemaic' approximation $\psi \approx T$. In about 1692, however, Newton did devise—and, with not a little trouble, contrive to pass—a theoretical test of the accuracy of his revised 'equation' of the upper-focus angle by constructing a semi-ellipse of eccentricity $e = \frac{1}{10}$ and thereby determining the trio of values of ψ corresponding to $T = \theta+e\sin\theta$, $\theta = \frac{1}{4}\pi, \frac{1}{2}\pi, \frac{3}{4}\pi$, and then in each case calculating the error in approximating ψ by $T+Y\sin 2T+Z\sin^3 T$, $Y = \frac{1}{4}e^2\ldots$, $Z = \frac{2}{3}e^3\ldots$. (The essential content of this check is reproduced in Appendix 2.5.)

(161) Originally 'ad centrum virium' (at the force-centre).

(162) See §2: note (163). As Seth Ward showed in his *In Ismaelis Bullialdi Astronomiæ Philolaicæ Fundamenta, Inquisitio Brevis* (Oxford, 1653): 26–8: 'Methodus Calculi primæ Inæqualitatis sive, ex dato angulo Anomaliæ mediæ, inveniendi Angulum ad solem' (compare note (3) above), and equivalently in his *Astronomia Geometrica* (London, 1656): Liber I, Caput V. 'Calculus locorum Planetæ alicujus in orbita sua, pro oculo in Sole constituto': 9–10, if (in the present figure) we extend *SP* to *K* so that $PK = PH$, and so SK (or $SP+PH$) $= AB$,

$\widehat{BHP}$ of equated mean motion is found, the angle $\widehat{HSP}$ of true motion(161) and the distance SP are readily had by the very well known method of Dr Seth Ward, Bishop of Salisbury, whom I most highly cherish.(162)

So much for the motion of bodies in curves. It may happen, however, that a body goes straight down or rises straight up. What regards motion of this type I now proceed to reveal.

Article VII.

On the rectilinear ascent and descent of bodies.(163)

Proposition XXXII, Problem XXIV.

Supposing that the centripetal force be reciprocally proportional to the square of the distance of places from its centre, to define the spaces which a body falling straight down describes in given times.

Case 1. If the body does not drop perpendicularly, it will describe some conic whose lower focus will coincide with the centre.(164) That is a settled consequence of Propositions XI, XII, XIII and their corollaries.(165) Let the conic be $ARPB$ and its lower focus S. Then first, if the figure is an ellipse, on its major axis AB describe the semicircle ADB and let the straight line DPC pass through the falling body perpendicular to the axis, and when DS, PS are drawn the area (ASD) will be proportional to the area (ASP) and hence also to the time. With

---

then, where $\widehat{BHP} = \psi$ and the 'true' anomaly $\widehat{BSP} = \phi$, it follows that

$$\widehat{PHK} = \widehat{SKH} = \tfrac{1}{2}\widehat{SPH} = \tfrac{1}{2}(\psi-\phi)$$

and so $\widehat{BHK} = \frac{1}{2}(\psi+\phi)$, whence $SH/(AB$ or) $SK = e = \sin\frac{1}{2}(\psi-\phi)/\sin\frac{1}{2}(\psi+\phi)$ and consequently (see Seth Ward's *Idea Trigonometriæ Demonstratæ* (Oxford, 1654): 6, and compare IV: 118, note (12)) $\tan\frac{1}{2}\psi/\tan\frac{1}{2}\phi = (1+e)/(1-e)$; it is then entirely straightforward to compute $SP = SH.\sin\psi/\sin(\psi-\phi)$ where, in explicit terms, $\phi = 2\tan^{-1}[((1-e)/(1+e)).\tan\frac{1}{2}\psi]$.

(163) Yet again (see note (1)) this is a late insertion in the manuscript in Newton's own hand. The account of such rectilinear motion under an inverse-square accelerating force in Propositions XXXII–XXXVII repeats that of Propositions XXI–XXIV and—very probably at least (see §2: note (174))—the untraceable following Propositions XXV and XXVI of the preceding tract 'De motu Corporum'; but the not dissimilar discussion in Proposition XXXVIII (as a preparatory to Proposition LII below) of the isochronous 'simple harmonic' motion induced in a line by a direct-distance force, and the concluding semi-analytical treatment in Proposition XXXIX (as a prelude to the construction of general central-force orbits in Propositions XL/XLI immediately after) of the rectilinear motion induced by an arbitrary accelerating force are without earlier parallel.

(164) Once more understand 'virium' (of force); compare §2: note (166).

(165) So changed in afterthought by Newton from the vaguer initial back-reference 'ex modò demonstratis' (of what has now been proved) which Humphrey Newton faithfully transcribed from the preceding 'De motu Corporum'.

Manente axe AB minuatur perpetuò latitudo Ellipseos, et semper manebit area ASD tempori proportionalis. Minuatur latitudo illa in infinitum et orbe APB jam coincidente cum axe AB et umbilico S cum axis termino B descendet corpus in recta AC et area ABD evadet tempori proportionalis. Dabitur itaq; spatiū AC quod corpus de loco A perpendiculariter cadendo tempore dato describit si modò tempori proportionalis capiatur area ABD et a puncto D ad rectam AB demittatur perpendicularis DC. Q.E.I.

Cas. 2. Sin figura superior RPB Hyperbola est, describatur ad eandem diametrum principalem AB Hyperbola rectangula BD, et quoniam areæ CSP $CBfP$, $SPfB$ sunt ad areas CSD, $CBED$, $SDEB$ singulæ ad singulas in data ratione altitudinum CP, CD, et area $SPfB$ proportionalis est tempori quo corpus P movebitur per arcum PB, erit etiam area $SDEB$ eidem tempori proportionalis. Minuatur latus rectum Hyperbolæ RBP in infinitum manente latere transverso et coibit arcus PB cum recta CB et umbilicus S cum vertice B et recta SD cum recta BD. Proinde area BDE proportionalis erit tempori quo corpus C recto descensu describit lineam CB. Q.E.I.

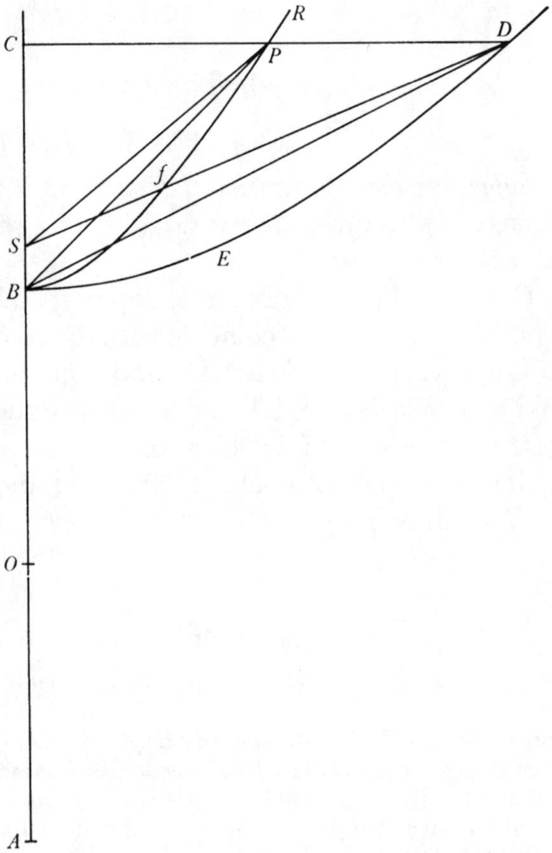

Cas. 3. Et simili argumento si figura RPB Parabola est(166) et eodem vertice principali B describatur alia Parabola BED quæ semper maneat data interea dum parabola prior, in cujus perimetro corpus P movetur, diminuto et in nihilum redacto ejus latere recto, conveniat cum linea CB, fiet segmentum Parabolicum BDE proportionale tempori quo corpus illud P vel C descendet ad centrum B. Q.E.I.(167)

(166) In default of any provision in the manuscript for a separate diagram illustrating this parabolic case, we again (compare §2: note (167)) insert Pemberton's 1726 figure in our English version.

(167) A minimally altered repeat of Proposition XXI of the preceding 'De motu Corporum'.

the axis AB remaining fixed, perpetually diminish the width of the ellipse and the area (ASD) will ever remain proportional to the time. Diminish that width indefinitely and, with the orbit $\widehat{APB}$ now coming to coincide with the axis AB and the focus S with the end-point B of the axis, the body will descend in the straight line AC and the area (ABD) will prove to be proportional to the time. The space AC described by a body falling perpendicularly from the position A in a given time will accordingly be yielded if only the area (ABD) be taken proportional to the time and then from the point D the perpendicular DC be let fall to the straight line AB. As was to be found.

Case 2. But if the previous figure RPB is a hyperbola, describe on the same principal diameter AB the rectangular hyperbola BD, and then, seeing that the areas (CSP), $(CBfP)$, $(SPfB)$ are individually to the corresponding areas (CSD), $(CBED)$, $(SDEB)$ in the given ratio of the altitudes CP, CD, while the area $(SPfB)$ is proportional to the time in which the body P shall move through the arc $\widehat{PB}$, the area $(SDEB)$ also will be proportional to the same time. Diminish the *latus rectum* of the hyperbola RBP indefinitely, its main diameter remaining fixed, and its arc $\widehat{PB}$ will coincide with the straight line CB, its focus S with the vertex B, and the straight line SD with the straight line BD. Consequently, the area (BDE) will be proportional to the time in which the body C in straight descent will describe the line CB. As was to be found.

Case 3. And by a similar argument, if the figure RPB is a parabola(166) and with the same principal vertex B there be described another parabola BED which shall ever remain given while the first parabola (in whose perimeter the body moves) comes, with its *latus rectum* diminished and reduced to nothing, to coincide with the line CB, the parabolic segment (BDE) will prove to be proportional to the time in which that body P, or C, shall descend to the centre B. As was to be found.(167)

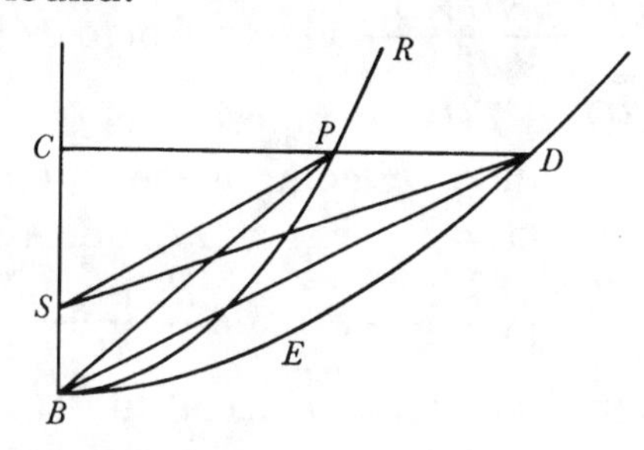

As before (see §2: note (168)), on setting $AS = \pm R$ and $CS = r$ and taking $v = -dr/dt$ to be the body's speed at C after a fall over $AC = \pm R - r$ from rest at A where the force (to centre S) has magnitude g, we may—retracing the path taken by Newton himself in Proposition XXXIX below—straightforwardly deduce from the equation of motion $d^2r/dt^2 = -gR^2/r^2$ that $v^2 = 2g(R^2/r - R)$, whence the time of fall from C to S is

$$\int_0^r \sqrt{[r/2gR(R-r)]}\,.\,dr = \sqrt{[8/gR^3]}\,.\,(BDE)$$

since $d(BDE) = \frac{1}{2}SY \times \widehat{Dd} = (\frac{1}{2}SY \times (DT/CT).Cc$ or) $\frac{1}{4}R\sqrt{[r/(R-r)]}.dr$, where (as in Proposition XXXIII) SY is the perpendicular let fall from S to the tangent DT at D.

Prop. XXXIII. Theor. IX.(168)

Positis jam inventis, dico quod corporis cadentis velocitas in loco quovis C est ad velocitatem corporis centro B intervallo BC circulum describentis, in dimidiata ratione quam CA distantia corporis a circuli vel Hyperbolæ vertice ulteriore A habet ad figuræ semidiametrum principalem $\frac{1}{2}AB$.

Namq ob proportionales CD, CP, linea AB communis est utriusq figuræ RPB, DEB diameter. Bisecetur eadem in O et agatur recta PT quæ tangatur figura RPB in P atq etiam secet communem illam diametrum AB (si opus est producta) in T, sitq SY(169) ad hanc rectam et BQ ad hanc diametrū perpendicularis, atq figuræ RPB latus rectum ponatur L. Constat per Cor. 9 Theor. VIII, quod corporis in linea RPB circa centrum S moventis velocitas in loco quovis P sit ad velocitatem corporis intervallo SP circa idem centrum circulum describentis in dimidiata ratione rectanguli $\frac{1}{2}L\times SP$ ad SY quadratum. Est autem ex Conicis ACB ad CP^q ut $2AO$ ad L adeoq $\frac{2CP^q\times AO}{ACB}$ æquale L. Ergo velocitates illæ sunt ad invicem in dimidiata ratione $\frac{CP^q\times AO\times SP}{ACB}$ ad SY^q. Porro ex Conicis est CO ad BO ut BO ad TO et compositè vel divisim ut CB ad BT. Unde dividendo vel componendo fit $BO-\text{vel}+CO$ ad BO ut CT ad BT[,] id est AC ad AO ut CP ad BQ indeq $\frac{CP^q\times AO\times SP}{ACB}$ æquale $\frac{BQ^q\times AC\times SP}{AO\times BC}$. Minuatur jam in infinitum figuræ RPB latitudo CP sic ut punctum P coeat cum puncto C punctumq S cum puncto B et linea SP cum linea BC lineaq SY cum linea BQ, et corporis jam recta descendentis in linea CB velocitas fiet ad velocitatem corporis centro B intervallo BC circulum describentis, in dimidiata ratione $\frac{BQ^q\times AC\times SP}{AO\times BC}$ ad SY^q, hoc est (neglectis æqualitatis

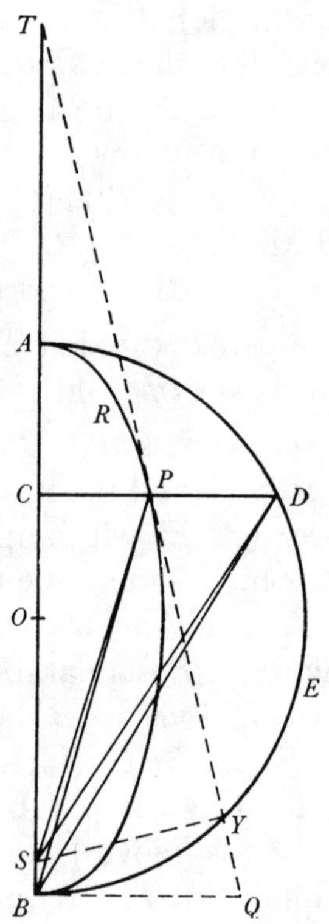

(168) Apart from the newly added corollary, this is transcribed without change from Proposition XXII in §2.

(169) In the figure (set in our English version) which illustrates the hyperbolic case we have again (see §2: note (170)) rectified Newton's error in the first two editions of the *Principia* of drawing SY to meet the tangent PT in a point above Q.

Proposition XXXIII, Theorem IX.(168)

Supposing what has just now been found, I assert that the speed of the falling body at any place C is to the speed of a body describing a circle of centre B and radius BC in the halved ratio which the distance CA of the body from the circle's or hyperbola's farther centre A has to the figure's principal semi-diameter $\frac{1}{2}AB$.

For, because of the proportionals CD, CP, the line AB is the common diameter of either figure RPB, DEB. Bisect it in O and draw the straight line PT to touch the figure RPB and also to intersect that common diameter AB (extended if need be) in T; then let SY(169) be perpendicular to this line and BQ to this diameter, and set the *latus rectum* of the figure RPB to be L. It is established by Corollary 9 to Theorem VIII that the speed at any place P of the body moving in the line RPB round the centre S shall be to the speed of a body describing a circle with the radius SP round the same centre in the halved ratio of the rectangle $\frac{1}{2}L \times SP$ to SY squared. However, from the *Conics*, $AC \times BC$ is to CP^2 as $2AO$ to L, and hence $L = 2CP^2 \times AO/AC \times BC$. Accordingly, those speeds are to one another in the halved ratio of $CP^2 \times AO \times SP/AC \times BC$ to SY^2. From the *Conics*, moreover, CO is to BO as BO to TO, and so, by compounding or dividing, as CB to BT. Whence, *dividendo* or *componendo*, there comes to be $BO \mp CO$ to BO as CT to BT, that is, AC to AO as CP to BQ, and thence

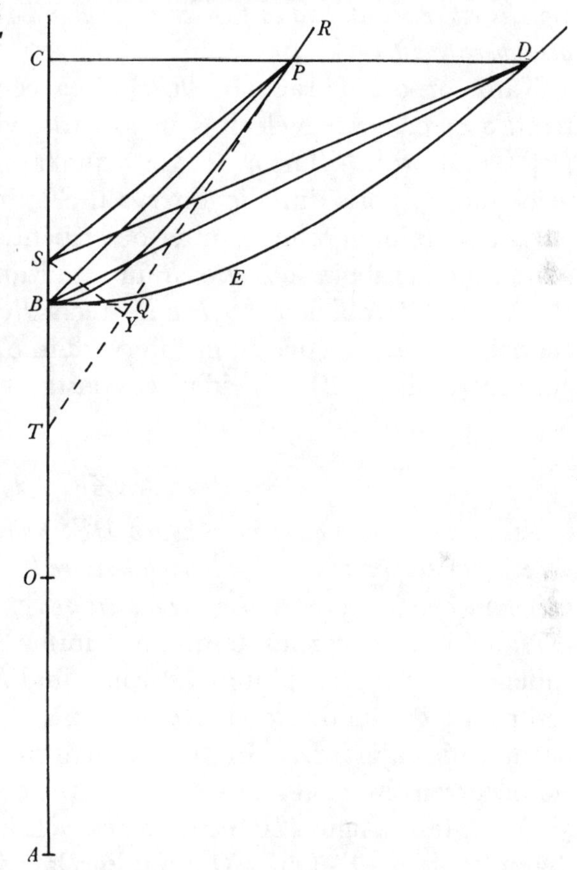

$$CP^2 \times AO \times SP/AC \times BC = BQ^2 \times AC \times SP/AO \times BC.$$

Now let the width CP of the figure RPB be indefinitely diminished, so that the point P comes to coalesce with the point C, the point S with point B, the line SP with the line BC and the line SY with the line BQ, and the speed of the body now descending straight downwards in the line CB will come to be to the speed of a body describing a circle on centre B with radius BC in the halved ratio of $BQ^2 \times AC \times SP/AO \times BC$ to SY^2, that is, (with the ratios of equality SP to

rationibus SP ad BC et BQ^q ad SY^q) in dimidiata ratione AC ad AO.(170) Q.E.D.

Corol. Punctis B et S coeuntibus, fit TC ad ST ut AC ad AO.(171)

Prop. XXXIV. Theor. X.(172)

Si figura BED Parabola est, dico quod corporis cadentis velocitas in loco quovis C æqualis est velocitati qua corpus centro B dimidio intervalli sui BC circulum uniformiter describere potest.

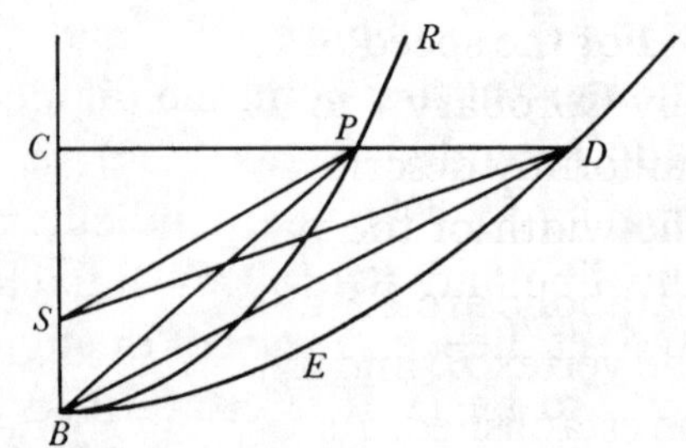

Nam corporis Parabolam RPB circa centrum S describentis velocitas in loco quovis [P] (per Corol. 7 Theor. VIII) æqualis est velocitati corporis dimidio intervalli SP circulum circa idem S uniformiter describentis. Minuatur Parabolæ latitudo CP in infinitum eò ut arcus Parabolicus [B]P cum recta BC, centrum S cum vertice B, et intervallum SP cum intervallo C[B] coincidat, et constabit Propositio. Q.E.D.

Prop. XXXV. Theor. XI.

Iisdem positis dico quod area figuræ DES radio indefinito SD descripta, æqualis sit areæ quam corpus radio dimidium lateris recti figura DES æquante, circa centrum S uniformiter gyrando, eodem tempore describere potest.

Nam concipe corpus C quam minima temporis particula lineol[a]m Cc cadendo describere, et interea corpus aliud K, uniformiter in circulo OKk circa centrum S gyrando, arcum Kk describere.(173) Erigantur perpendicula CD, cd occurrentia figuræ DES in D, d. Jungantur [SD,] Sd, SK, Sk. Ducatur Dd axi AS occurrens in T, et ad eam demittatur perpendiculum SY.

Cas. 1. Jam si figura DES circulus est vel Hyperbola[,] bisecetur ejus transversa diameter AS in O, et erit SO dimidium lateris recti. Et quoniam est TC ad TD ut Cc ad Dd, et TD ad ST ut CD ad SY, erit ex æquo TC ad ST ut $CD \times Cc$ ad

(170) In sequel to note (167), equivalently, we may as before (see §2: note (171)) argue that, because the body's speed at C is $v = \sqrt{[2gR(R-r)/r]}$, whereas the comparable speed of uniform rotation in a circle at a distance $CS = r$ from the force-centre in S is
$$v' = \sqrt{[(gR^2/r^2).r]} = \sqrt{[gR^2/r]},$$
their ratio is, in the case of a central conic, $v/v' = \sqrt{[(R-r)/\frac{1}{2}R]}$, which in the limiting parabolic case ($R = \infty$)—that of Proposition XXXIV following—becomes
$$v = v'\sqrt{2} = \sqrt{[gR^2/\tfrac{1}{2}r]},$$
the speed of rotation in a circle at a distance $\frac{1}{2}r$ from S.

BC and BQ^2 to SY^2 neglected) in the halved ratio of AC to AO.(170) As was to be proved.

Corollary. With the points B and S coincident, CT comes to be to ST as AC to AO.(171)

Proposition XXXIV, Theorem X.(172)

If the figure BED is a parabola, I assert that the speed of the falling body at any place C is equal to the speed with which a body can uniformly describe a circle of centre B with half its interval BC.

For the speed of the body describing the parabola RPB round the centre S is (by Corollary 7 to Theorem VIII) at any place P equal to the speed of a body uniformly describing a circle round the same S at half the interval SP. Now let the width of the parabola be indefinitely diminished with the result that the parabolic arc $\widehat{BP}$ comes to coincide with the straight line CB, the centre S with the vertex B, and the interval SP with the interval BC, and the proposition will be established. As was to be proved.

Proposition XXXV, Theorem XI.

With the same suppositions, I assert that the area of the figure DES described by the indefinite radius SD shall be equal to the area which a body can, by orbiting uniformly round the centre S at a radius equalling half the latus rectum of the figure DES, describe in the same time.

For imagine that the body C during its fall describes in a minutely small particle of time the line-element Cc, and meanwhile that another body K, by orbiting uniformly in the circle OKk round the centre S, describes the arc $\widehat{Kk}$.(173) Erect the perpendiculars CD, cd meeting the figure DES in D and d. Join SD, Sd, SK, Sk. Draw Dd meeting the axis AS in T, and to it let fall the perpendicular SY.

Case 1. If now the figure DES is a circle or a hyperbola, bisect its transverse diameter AS in O and then SO will be half the *latus rectum*. And, because TC is to TD as Cc to $\widehat{Dd}$, and TD to ST as CD to SY, there will *ex æquo* be TC to ST as

(171) For, as Newton has proved above, before S comes to coincide with B there is $CT: BT = AC: AO$.

(172) An essentially unaltered repeat of Proposition XXIII of the preceding 'De motu Corporum'. As before (see §2: note (172)), we have amended three small slips in the text where points in the figure are mis-named.

(173) As it is known to us (see §2: note (174)), the central text of the preceding 'De motu Corporum' terminates at this point, midway through its corresponding Proposition XXIV.

$SY \times Dd$. Sed per Corol. Prop. XXXIII est(174) TC ad ST ut AC ad AO, puta si in coitu punctorum D, d capiantur linearum rationes ultimæ. Ergo AC est ad AO[,] id est ad SK ut $CD \times Cc$ ad $SY \times Dd$. Porrò corporis descendentis velocitas in C est ad velocitatem corporis circulum intervallo SC circa centrum S describentis in dimidiata ratione AC ad AO vel SK (per Theor. IX.) et hæc velocitas ad velocitatem corporis describentis circulum OKk in dimidiata ratione SK ad SC per Cor. 6 Theor. IV, et ex æquo velocitas prima ad ultimam, hoc est lineola Cc ad arcum Kk in dimidiata ratione AC ad SC id est in ratione AC ad CD. Quare est $CD \times Cc$ æquale $AC \times Kk$, et propterea AC ad SK ut $AC \times Kk$ ad $SY \times Dd$, indeq; $SK \times Kk$ æquale $SY \times Dd$, et $\frac{1}{2}SK \times Kk$ æquale $\frac{1}{2}SY \times Dd$, id est area KSk æqualis areæ SDd. Singulis igitur temporis particulis generantur arearū duarum particulæ KSk, DSd, quæ si magnitudo earum minuatur et numerus augeatur in infinitum, rationem obtinent(175) æqualitatis et propterea per Corollarium Lemmatis IV areæ totæ simul genitæ sunt semper æquales. Q.E.D.(176)

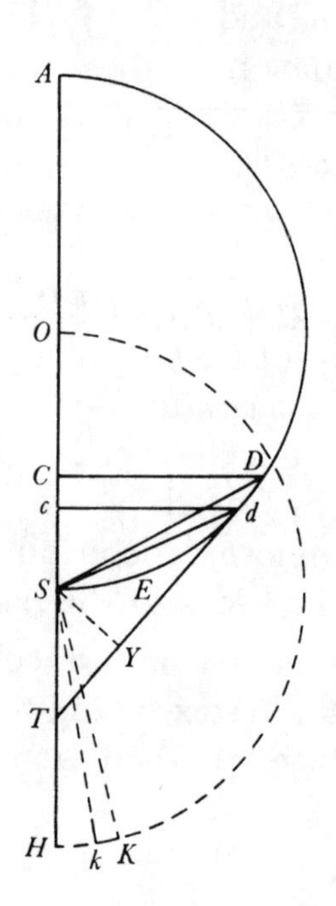

Cas. 2. Quod si figura DES Parabola sit, invenietur ut supra $CD \times Cc$ esse ad $SY \times Dd$ ut TC ad ST hoc est ut 2 ad 1, adeoq; $\frac{1}{4}CD \times Cc$ æqualem esse $\frac{1}{2}SY \times Dd$. Sed corporis cadentis velocitas in C æqualis est velocitati quo circulus intervallo $\frac{1}{2}SC$ uniformiter describi possit (per Theor. X) et hæc velocitas ad velocitatem qua circulus radio SK describi possit, hoc est lineola Cc ad arcum Kk, est in dimidiata ratione SK ad $\frac{1}{2}SC$ id est in ratione SK ad $\frac{1}{2}CD$(177)

---

(174) Originally—and somwhat less precisely—the manuscript read 'Sed supra erat' (But above there was).

(175) Understand 'ad invicem' (to one another).

(176) Further to continue note (167), where again

$$AS = \pm R, \quad CS = r \quad \text{and} \quad v = dr/dt = \sqrt{[2gR(R-r)/r]}$$

is the body's speed at C after a fall in time t from rest at A where the force has magnitude g, it follows that the areal velocity

$$d(BDE)/dt = (\tfrac{1}{4}R\sqrt{[r/(R-r)]}.v \text{ or}) \tfrac{1}{4}R\sqrt{[2gR]} = d(HSK)/dt,$$

since $\sqrt{[2gR]}$ is the speed of uniform rotation in a circle at a distance $SK = \frac{1}{2}R$ from the force-centre S.

(177) Since by construction the parabola's *latus rectum* CD^2/SC is equal to $2SK$.

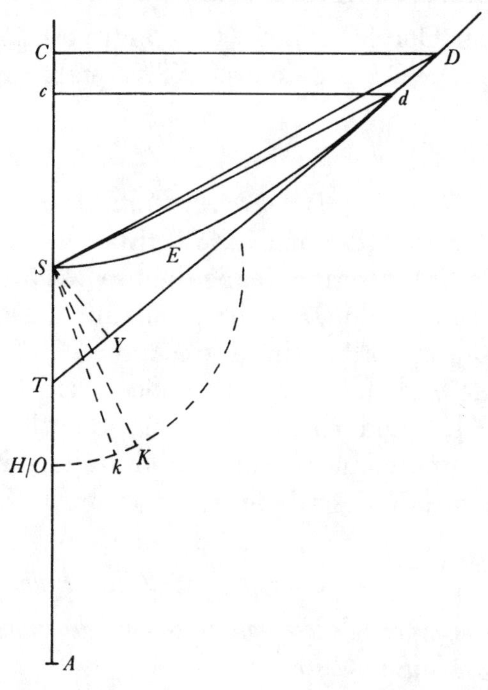

$CD \times Cc$ to $SY \times \widehat{Dd}$. But by the Corollary to Proposition XXXIII there is[174] TC to ST as AC to AO, if, namely, in the coalescing of the points D and d last ratios of lines be taken. Therefore AC is to AO, that is, to SK, as $CD \times Cc$ to $SY \times Dd$. Furthermore, the speed of the descending body at C is to the speed of a body describing a circle round the centre S with radius SC in (by Theorem IX) the halved ratio of AC to AO or SK, and this latter speed to the speed of a body describing the circle OKk in (by Corollary 6 to Theorem IV) the halved ratio of SK to SC, and so *ex æquo* the first velocity to the last one, that is, the line-element Cc to the arc $\widehat{Kk}$, is in the halved ratio of AC to SC, that is, in the ratio of AC to CD. In consequence $CD \times Cc$ is equal to $AC \times \widehat{Kk}$, and accordingly AC is to SK as $AC \times \widehat{Kk}$ to $SY \times \widehat{Dd}$, and thence $SK \times \widehat{Kk}$ is equal to $SY \times \widehat{Dd}$ and so $\frac{1}{2}SK \times \widehat{Kk}$ equal to $\frac{1}{2}SY \times \widehat{Dd}$, that is, the area (KSk) is equal to the area (DSd). In individual particles of time, therefore, there are generated particles (KSk), (DSd) of the two areas which, if their size be diminished and their number increased indefinitely, preserve a ratio of equality;[175] accordingly, by the Corollary to Lemma IV, the total areas simultaneously generated are ever equal. As was to be proved.[176]

Case 2. But if the figure DES be a parabola, it will be found as above that $CD \times Cc$ is to $SY \times \widehat{Dd}$ as TC to ST, that is, as 2 to 1, and hence $\frac{1}{4}CD \times Cc$ is equal to $\frac{1}{2}SY \times \widehat{Dd}$. But the speed of the falling body at C is (by Theorem X) equal to the speed with which a circle might uniformly be described at the interval $\frac{1}{2}SC$, and this latter speed to the speed with which a circle of radius SK might be described, that is, the line-element Cc to the arc $\widehat{Kk}$, is (by Corollary 6 of Theorem IV) in the halved ratio of SK to $\frac{1}{2}SC$, that is, in the ratio of SK to $\frac{1}{2}CD$.[177] Consequently $\frac{1}{2}SK \times \widehat{Kk}$ is equal to $\frac{1}{4}CD \times Cc$, and

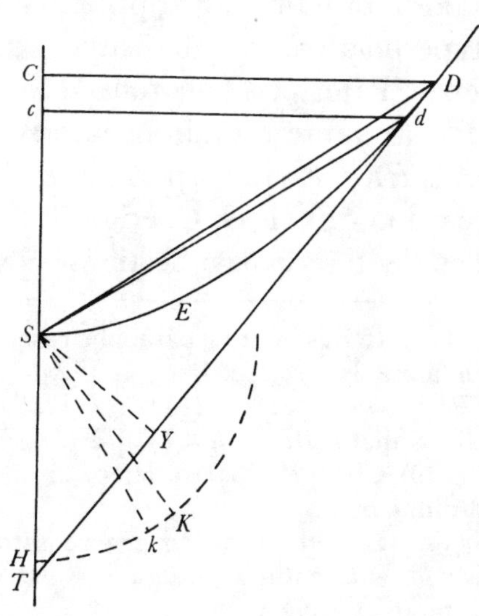

per Corol. 6 Theor. IV. Quare est $\frac{1}{2}SK \times Kk$ æquale $\frac{1}{4}CD \times Cc$, adeoq; æquale $\frac{1}{2}SY \times Dd$, hoc est area KSk æqualis areæ SDd ut supra. Q.E.D.(178)

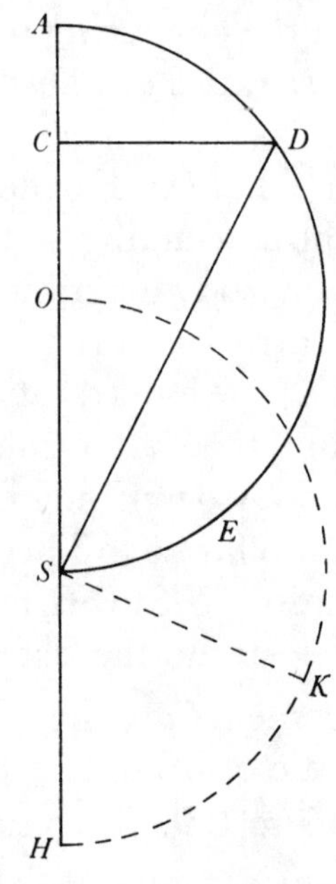

Prop. XXXVI. Prob. XXV.

Corporis de loco dato A cadentis determinare tempora descensus.

Super diametro AS (distantia corporis a centro sub initio) describe semicirculum ADS, ut et huic æqualem semicirculū OKH circa centrum S. De corporis loco quovis C erige ordinatim applicatam(179) CD. Junge SD, et areæ ASD æqualem constitue Sectorem OSK. Patet per Theor. XI, quod corpus cadendo describet spatium AC eodē tempore quo corpus aliud uniformiter circa centrum S gyrando describere potest arcum OK. Q.E.F.(180)

Prop. XXXVII. Prob. XXVI.

Corporis de loco dato sursum vel deorsum projecti definire tempora ascensus vel descensus.

Exeat corpus de loco dato G secundum lineam ASG cum velocitate quacunq;. In duplicata ratione hujus velocitatis ad uniformem in circulo velocitatem quâ corpus ad intervallum datum SG circa centrum S revolvi(181) posset, cape GA ad $\frac{1}{2}AS$. Si ratio illa est numeri binarij ad unitatem, punctum A cadet ad infinitam distantiam[,] quo in casu Parabola vertice S, axe SC, latere quovis recto describenda est. Patet hoc per Theorema X. Sin ratio illa minor vel major est quam 2 ad 1, priore casu circulus, posteriore Hyperbola rectangula super diametro SA describi debet. Patet per Theorema IX. Tum centro S intervallo æquante dimidium lateris recti describatur circulus HKk. Et ad corporis ascendentis vel descendentis loca duo quævis G, C, erigantur perpendicula GI, CD occurrentia Conicæ Sectioni(182) vel circulo in I ac D.

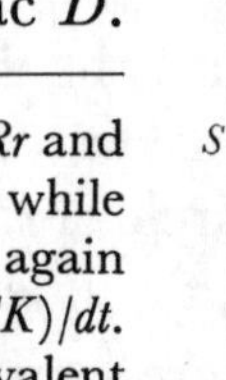

(178) In this limiting parabolic case, where now $CD^2 = Rr$ and the falling body's speed at C is $\surd[2gR^2/r]$ (see note (170)), while $d(BDE) = \frac{1}{2}SY \times Dd = (\frac{1}{2}SY \times (DT/2ST).Cc$ or) $\frac{1}{4}CD.Cc$, it again follows that $d(BDE)/dt = (\frac{1}{4}\surd[Rr].v$ or) $\frac{1}{4}R\surd[2gR] = d(HSK)/dt$.

(179) Changed, as so often, from an original equivalent 'ordinatam'.

(180) Like its following generalisaton, this construction derives in an entirely straightforward way from the preceding Proposition XXXV.

hence to $\frac{1}{2}SY \times \widehat{Dd}$, that is, the area (KSk) is equal to the area (DSd) as above. As was to be proved.(178)

Proposition XXXVI, Problem XXV.

Where a body falls from a given place A, to determine its times of descent.

On the diameter AS (the body's distance from the centre at the start) describe the semicircle ADS, and also round the centre S the semicircle OKH equal to it. From any place C of the body erect the ordinate CD. Join SD, and equal to the area (ASD) construct the sector (OSK). It is evident by Theorem XI that the body as it falls describes the space AC in the same time as another body is, by orbiting uniformly round the centre S, able to describe the arc $\widehat{OK}$. As was to be done.(180)

Proposition XXXVII, Problem XXVI.

Where a body is projected upwards or downwards from a given place, to define its times of ascent or descent.

Let the body depart from the given place G along the line ASG with any speed whatever. In the doubled ratio of this speed to the uniform speed in the circle in which a body might revolve(181) at the given interval SG round the centre S take GA to $\frac{1}{2}AS$. If that ratio is as the number two to unity, the point A will fall an infinite distance away, and in this case a parabola is to be described with vertex S, axis SC and any *latus rectum:* this is evident by Theorem X. But if that ratio is less or greater than 2 to 1, in the former case a circle and in the latter a rectangular hyperbola must be described on the diameter SA: this is evident by Theorem IX. Then with centre S and radius equalling half the *latus rectum* describe the circle HKk, and at any two positions G, C of the ascending or descending body erect the perpendiculars GI, CD meeting the conic(182) or circle in I and D. Thereafter, on joining SI and SD, make the

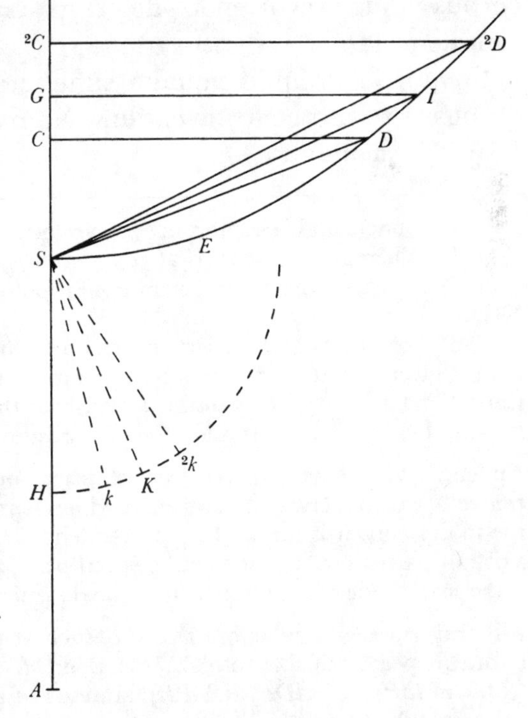

(181) A less specific 'moveri' (move) was first written.

(182) The preceding parabola or rectangular hyperbola, that is.

Dein junctis SI, SD fiant segmentis $SEIS$, $SEDS$ sectores HSK, $[H]Sk$ æquales; et per Theorema XI, corpus G describet spatium GC eodem tempore quo corpus K describere potest arcum Kk. Q.E.F.

Prop. XXXVIII. Theor. XII.

Posito quod vis centripeta proportionalis sit altitudini seu distantiæ locorum a centro, dico quod cadentium tempora velocitates et spatia descripta sunt arcubus(183) *arcuumq sinubus versis et sinubus rectis respectivè proportionales.*

Cadat corpus de loco quovis A(184) secundum rectam AS et centro virium S intervallo AS describatur circuli quadrans AE sitq CD sinus rectus arcus cujusvis AD et corpus A tempore AD cadendo describet spatium AC, inq loco C acquisierit velocitatem CD. Demonstratur eodem modo ex Propositione X quo Propositio XXXII ex Propositione XI(185) demonstrata fuit.(186) Q.E.O.

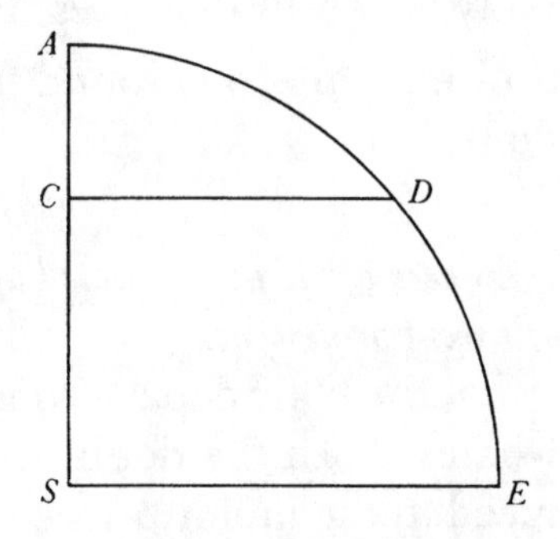

Corol. 1. Hinc æqualia sunt tempora quibus corpus unum de loco A cadendo provenit ad centrum S et corpus aliud revolvendo describit arcum quadrantalem ADE.

C[orol]. 2. Proinde æqualia sunt tempora omnia quibus corpora de locis quibusvis ad usq centrum cadunt. Nam tempora omnia periodica (per Corol. 3 Prop. IV) æquantur.

---

(183) Understand 'circulorum' (of circles).

(184) Where, of course, it is at rest.

(185) And also the complementary Propositions XII and XIII.

(186) For, if a body P departs from A at right angles to AS with a given speed, and thereafter is continuously urged towards S by a force varying instantaneously as the distance SP, it will, by Corollaries 1 and 2 to Proposition X (see §2, Appendix 2.8 above), traverse an elliptical orbit $\widehat{APQB}$ of centre S with a period of revolution which is independent of the initial outwards speed at A and the length AB of its main axis. In particular, if that starting speed be $\sqrt{[g.AS]}$ where g is the magnitude of the force-field at the distance AS, the body will traverse the circular orbit $\widehat{ADEB}$ such that, when CPD is drawn perpendicular to ACS, the point D in it is, since $(SAD):(SAP) = (SADB):(SAPB)$, attained in the same time as point P is reached in the ellipse; whence the sector $(SAD) = \frac{1}{2}AS \times \widehat{AD}$ measures the time of orbit over the corresponding elliptical arc $\widehat{AP}$. In the limiting case, therefore, when the initial speed outwards at A is zero, and so the ellipse $\widehat{APB}$ shrinks into coincidence with its diameter ACB, the time of fall

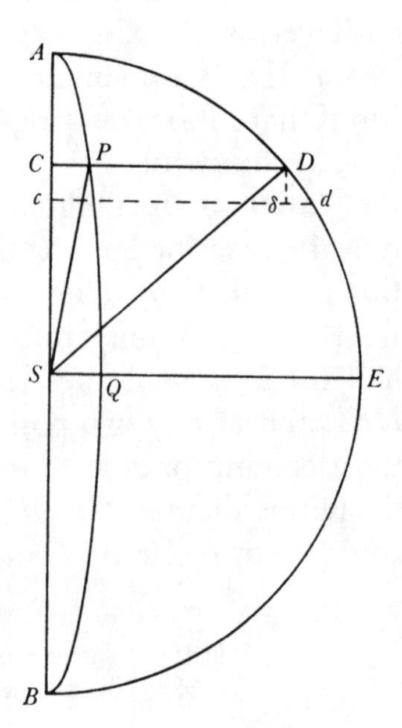

sectors (HSK), (HSk) equal to the segments (SEI) and (SED); and then, by Theorem XI, the body G will describe the space GC in the same time as the body K is able to describe the arc $\widehat{Kk}$. As was to be done.

Proposition XXXVIII, Theorem XII.

Supposing that the centripetal force be proportional to the height or distance of their places from the centre, I assert that of things as they fall the times, speeds and spaces described are respectively proportional to arcs(183) *and to the versed sines and right sines of those arcs.*

Let a body fall from any place A(184) along the straight line AS, and with (force-)centre S and radius AS describe the circle quadrant AE; let CD be the right sine of any arc $\widehat{AD}$, and then the body A falling in time $\widehat{AD}$ will describe the space AC and at the place C have acquired the speed CD. This is proved from Proposition X in the same way as Proposition XXXII was demonstrated from Proposition XI.(185) As was to be shown.(186)

Corollary 1. Hence equal are the times in which one body by falling from A attains the centre S and another body by revolving round describes the quadrantal arc $\widehat{ADE}$.

Corollary 2. And, in consequence, equal are all times in which bodies fall from any places whatever right to the centre. For (by Corollary 3 to Proposition IV) all the periodic times are equal.

---

from A to C is measured by $\widehat{AD}$, and consequently, where cd is an indefinitely close parallel to CD and $D\delta$ is let fall perpendicular to it, the speed of fall at C is proportional to $\lim_{c\to C} [Cc \text{ (or } D\delta)/Dd] = CD/SD \propto CD$ since the radius SD is fixed.

To anticipate minimally, the analogous derivation from Newton's following Proposition XXXIX is more straightforward. If we again set $AS = R$ and $CS = r$, and take $v = -dr/dt$ to be the speed at C after a fall over $AC = R-r$ in time t from rest at A, the equation of motion is $d^2r/dt^2 = v.dv/dr = -kr$, $k = g/R$, whence $v^2 = k(R^2-r^2)$ and accordingly

$$t = (1/k^{\frac{1}{2}}).\int_R^r -1/\sqrt{[R^2-r^2]}.dr = (1/k^{\frac{1}{2}})\cos^{-1}(r/R) = (1/k^{\frac{1}{2}}R).\widehat{AD};$$

inversely, therefore, $r = R\cos(k^{\frac{1}{2}}t)$, thus determining $v = -k^{\frac{1}{2}}R\sin(k^{\frac{1}{2}}t) = -k^{\frac{1}{2}}.CD$. Since k is an absolute constant of the force-field, the period $2\pi k^{-\frac{1}{2}} = 2\pi\sqrt{[R/g]}$ of this 'simple' harmonic motion (from A through S to B and back again to A) is obviously independent of the particular point from which the body begins its fall. Newton will afterwards apply his present theorem both to show, in Proposition LII—in extension of his proof from first principles some dozen years earlier (III: 420–30, especially 422) of the isochronism of the pendular vibrations induced by constant downwards gravity in a vertical cycloid—that the period of the swing correspondingly generated in a hypocycloidal pendulum under a force varying as the distance which is instantaneously directed to the centre of its deferent circle is likewise invariant; and, in Proposition LIII immediately following, to frame as a general criterion for the motion of a point (r, ϕ) oscillating in an arbitrary central-force field $f(r)$ to be isochronous the necessary

[187]*Prop. XXXIX. Prob. XXVII.*

Posita cujuscunq generis vi centripeta et concessis figurarū curvilinearum quadraturis, requiritur corporis recta ascendentis vel descendentis tum velocitas in locis singulis, tum tempus quo corpus ad locum quemvis perveniet: et contra.

De loco quovis A[188] in recta $ADEC$ cadat corpus E, deq loco ejus E erigatur semper perpendicularis EG vi centripetæ in loco illo ad centrum C tendenti proportionalis. Sitq BFG linea curva quam punctum G perpetuò tangit. Coincidat autem EG ipso motus initio cum perpendiculari AB et erit corporis velocitas in loco quovis E ut areæ curvilineæ $ABGE$ latus quadratum. Q.E.I. In EG capiatur EM areæ $ABGE$ reciproce proportionalis, sitq ALM[189] curva linea quam punctum [M] perpetuò tangit, et erit tempus quo corpus cadendo describit lineam AE ut area curvilinea $ALME$. Q.E.I.

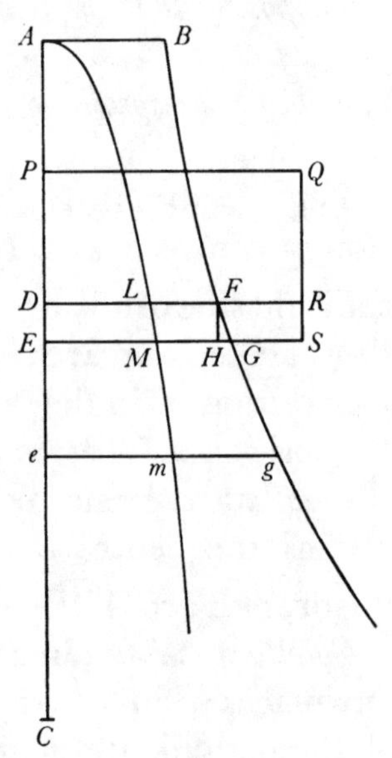

Etenim in recta AE capiatur linea quàm minima DE datæ longitudinis, sitq DLF locus lineæ EMG ubi corpus versabatur in D; et si ea sit vis centripeta ut areæ $ABGE$ latus quadratum sit ut descendentis velocitas, erit area ipsa in duplicata ratione velocitatis, id est si pro velocitatibus in D et E scribatur V et $V+I$[,] erit area $ABFD$ ut V^2 et area $ABGE$ ut $V^2+2VI+I^2$, et divisim area $DFGE$ ut $2VI+I^2$, adeoq $\frac{DFGE}{DE}$, id est, si primæ quantitatum nascentium rationes sumantur, longitudo DF ut quantitas $\frac{2I\times V}{DE}$, adeoq etiam ut quantitatis hujus dimidium $\frac{I\times V}{DE}$. Est autem tempus quo corpus cadendo describit

and sufficient requirement that the arc-length $\int_R^r \sqrt{[1+(r\,.\,d\phi/dr)^2]}\,.\,dr$ of the curve in which it is constrained to travel shall be identically equal to $\sqrt{\left[-2\int_R^r f(r)\,.\,dr\right]}$.

(187) Immediately above this (at the top of f. 34r) Newton first squashed in and then quickly deleted an unqualified subhead 'ARTIC. VIII'. Below, correspondingly, Articles VIII and IX were briefly renumbered 'IX' and 'X'. While the present semi-analytical generalised problem is more nearly related to Newton's following Propositions XL and XLI—of which, indeed, it is a particular instance—than to his previous treatment of rectilinear motion under inverse-square and direct-distance forces as the limiting case (where the initial speed is nil) of one or other of the species of conic orbit thereby generally traversed, to divorce it from both by setting it incongruously in a separate section of its own is patently absurd.

(188) As in Proposition XXXII, this point where the body is at rest need not be finite nor even real (when Newton would doubtless again choose to set it at a 'more than infinite distance' away below the force-centre C).

[187]*Proposition XXXIX, Problem XXVII.*

Supposing a centripetal force of any kind whatsoever, and granted the quadratures of curvilinear figures, there is required of a body ascending or descending straight up or down both its speed at individual places and the time in which a body shall arrive at any place; and the converse.

Let the body E fall from any place A[188] in the straight line $ADEC$, and from its place E ever erect a perpendicular EG proportional to the centripetal force at that place tending to the centre C; and let BFG be the curve in which the point G perpetually lies, on taking EG at the very start of motion to coincide with the perpendicular line AB: the body's speed at any place E will then be as the square root of the curvilinear area ($ABGE$). As was to be found. In EG take EM reciprocally proportional to the [square root of the] area ($ABGE$), and let [V]LM[189] be the curve in which the point M perpetually lies: the time in which the body by falling describes the line AE will then be as the curvilinear area (A[TV]ME). As was to be found.

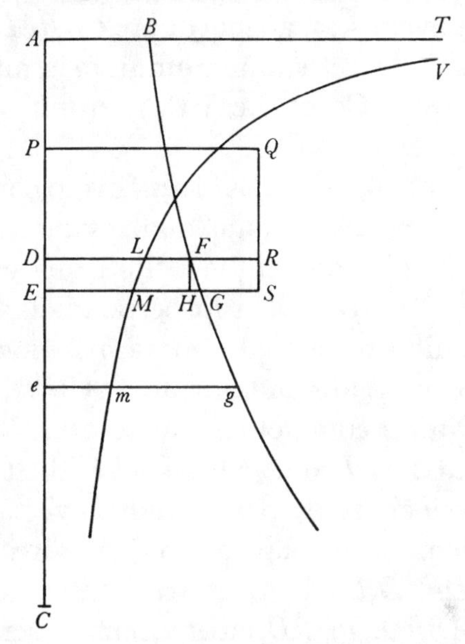

For, to be sure, in the line AE take the minimally small line (-element) DE of given length, and let DLF be the position of the line EMG when the body was situated at D. Then, if the centripetal force be such that the square root of the area ($ABGE$) is as the speed of the descending body, that area will be in the doubled ratio of the speed; in other words, if V and $V+I$ be written for the speeds at D and E, the area ($ABFD$) will be as V^2, and the area ($ABGE$) as $V^2+2VI+I^2$, and *dividendo* the area ($DEGE$) will be as $2VI+I^2$: hence ($DFGE$)/DE, that is, if first ratios of nascent quantities be now assumed, the length DF, will be as the quantity $2I\times V/DE$, and so also as half that quantity, $I\times V/DE$. However, the time in which the body by falling describes the line-element DE is as that

(189) Since EM is inversely proportional to the area ($ABGE$) and hence to the falling body's speed at E, when E coincides with A (where the body is at rest) EM must be infinite in length, so that the curve ML must tend asymptotically to the line AB. Though (see S. P. Rigaud's *Historical Essay* (note (148)): 93) Fatio de Duillier noticed the error in his copy of the published *Principia* ($_1$1687: 122–3) into which the present text passed without alteration, the slip continued unchecked into the second edition in 1713, to be finally caught by Henry Pemberton in his *editio ultima*: Pemberton's revised figure (*Principia*, $_3$1726: 120) is here inserted in the English version along with appropriate minor adjustments of the text.

lineolam DE ut lineola illa directè et velocitas V inversè, estq; vis ut velocitatis incrementum I directè et tempus inversè, adeoq; si primæ nascentium rationes sumantur[,] ut $\frac{I \times V}{DE}$ hoc est ut longitudo DF. Ergo vis ipsi DF(190) proportionalis facit corpus ea cum velocitate descendere quæ sit ut areæ $ABGE$ latus quadratum. Q.E.D.(191)

Porrò cum tempus quo quælibet longitudinis datæ lineola DE describatur, sit ut velocitas, adeoq; ut areæ $ABFD$ latus quadratum inversè, erit tempus ut area $DLME$, et summa omnium temporum ut summa omnium arearum, hoc est ([per] Corol. Lem IV) tempus totum quo linea AE describitur ut area tota AME. Q.E.D.(192)

Corol. 1. Si P sit locus de quo corpus cadere debet, ut, urgente aliqua uniformi vi centripeta nota (qualis vulgo supponitur gravitas,) velocitatem acquirat in loco D æqualem velocitati quã corpus aliud vi quacunq; cadens acquisivit in eodem loco D, et in perpendiculari DF capiatur DR quæ sit ad DF ut vis illa uniformis ad vim alteram in loco D, et compleatur rectangulũ $PDRQ$, eiq; æqualis abscindatur area $ABFD$, erit A locus de quo corpus alterum cecidit. Namq; completo rectangulo $EDRS$, cum sit area $ABFD$ ad aream $DFGE$ ut VV ad $2V \times I$, adeoq; ut $\frac{1}{2}V$ ad I, id est, ut semissis velocitatis totius ad incrementum velocitatis corporis uniformi vi cadentis: sintq; incrementa illa (ob æqualitatem temporum nascentium,) ut vires generatrices, id est ut ordinatim applicatæ DF, DR, adeoq; ut areæ nascentes $DFGE$, $DRSE$: erunt (ex æquo) areæ totæ $ABFD$, $PQRD$ ad invicem ut semissis totarum velocitatum, et propterea (ob æqualitatem velocitatum) æquantur.(193)

Corol. 2. Unde si corpus quodlibet de loco quocunq; D data cum velocitate vel sursum vel deorsum projiciatur, et detur lex vis centripetæ, invenietur velocitas

---

(190) That is, EG since Newton assumes the abscissa's increment DE to be vanishingly small. In the concluding *errata* of the published *Principia* ($_1$1687) he directed that a clarifying 'vel EG' (or EG) should here be inserted to the same purpose.

(191) To resume Newton's argument in a slightly more modern form, if the body under the urge of a given force directed to C falls in time t from rest at A to D, there attaining the speed $V = DE/dt$ where $DE = -d(CD)$ is the 'instantaneous' decrement of CD in time dt, then, on setting DF to be the force $dV/dt = f(CD)$, say, acting on the body at D, it follows that $(ABFD) = \int_{CA}^{CD} -f(CD).d(CD)$ is equal to $\frac{1}{2}V^2$: this Newton demonstrates from first principles, 'differentiating' to derive the equals

$$d(ABFD)/d(CD) = (DEGF)/DE = \lim_{E \to D} [DF + O(DE)] = DF = I/dt$$

and
$$d(\tfrac{1}{2}V^2)/d(CD) = \lim_{I \to 0} [\tfrac{1}{2}((V+I)^2 - V^2)/DE] = V \times I/DE,$$

where $I = dV$ is the increment of the speed V over DE.

element directly and the speed V inversely, while the force is as the increment I of the speed directly and the time inversely, and hence, if first ratios of the nascents be assumed, as $I\times V/DE$, that is, as the length DF. Therefore a force proportional to DF(190) makes the body descend with a speed which shall be as the square root of the area $(ABGE)$. As was to be proved.(191)

Furthermore, since the time in which any line-element DE of given length is described is as the speed—and hence as the square root of the area $(ABFD)$—inversely, the time will be as the area $(DLME)$, and the sum of all the times as the sum of all the areas; that is, (by the Corollary to Lemma IV) the total time in which the line AE is described is as the total area $(A[TV]ME)$. As was to be proved.(192)

Corollary 1. If P be the place from which a body ought to fall so as, under the urge of some known uniform centripetal force (such as gravity is commonly supposed to be), to acquire a speed at the place D equal to the velocity which another body, falling under any force whatever, has acquired at the same place D, in the perpendicular DF take DR to be to DF as the former uniform force to the second force at the place D, complete the rectangle $PDRQ$ and equal to it cut off the area $(ABFD)$: A will then be the place from which the latter body fell. For since, on completing the rectangle $EDRS$ the area $(ABFD)$ is to the area $(DFGE)$ as V^2 to $2V\times I$, and so as $\frac{1}{2}V$ to I—that is, as half the total speed to the increment in speed of the body falling under the uniform force, let those increments (because of the equality of the nascent times) be as the generating forces, that is, as the ordinates DF, DR, and hence as the nascent areas $(DFGE)$, $(DRSE)$: the total areas $(ABFD)$, $(PQRD)$ will then *ex æquo* be to one another as half the total speeds, and accordingly (because of the equality of the speeds) they are equal.(193)

Corollary 2. Whence if any body whatever be projected, upwards or downwards, with a given speed from any place D, and the law of centripetal force be

(192) To continue the previous note, on making $DL = 1/V$, that is,

$$DL = 1\Big/\sqrt{\left[2\int_{CA}^{CD} -f(CD)\,.\,d(CD)\right]},\quad \text{there ensues}\quad t = \int_{CA}^{CD} 1/V.d(CD) = (A[TV]LD);$$

whence, as Newton states it, $t+dt = (A[TV]ME)$.

(193) The effect of this substitution of $(PQRD) = \frac{1}{2}V^2$, where V is the body's speed at D, for the equal area $(ABFD) = \int_{CA}^{CD} -f(r)\,.\,dr$, where $f(r)$ is the central force acting towards C at a distance r is to free Newton's construction from its unnecessary dependence upon the point A of nil motion (which, as we have noticed, may inconveniently either be at infinity or not even real). Where $v = \sqrt{\left[V^2+2\int_{CD}^{Ce} -f(r)\,.\,dr\right]}$ is the speed at e, it is then immediate for Newton in his two remaining corollaries to set $v = k\sqrt{[(PQRD)+(DFge)]}$, $k = \sqrt{2}$, and thence, on making $em = 1/k\sqrt{[(PQRD)+(DFge)]}$, to measure the time $\int_{CD}^{Ce} 1/v\,.\,dr$ over De by its equal, $(DLme)$.

ejus in alio quovis loco *e* erigendo ordinatam *eg*, et capiendo velocitatem illam ad velocitatem in loco *D* ut est latus quadratum rectanguli *PQRD* area curvilinea *DFge* vel aucti si locus *e* est loco *D* inferior, vel diminuti si is superior est, ad latus quadratum rectanguli solius *PQRD*, id est ut $\sqrt{PQRD + \text{vel} - DFge}$ ad $\sqrt{PQRD}$.

Corol. 3. Tempus quoq; innotescet erigendo ordinatam *em* reciproce proportionalem lateri quadrato ex *PQRD*+vel−*DFge*, et capiendo tempus quo corpus descripsit lineam *De* ad tempus quo corpus alterum vi uniformi cecidit a *P* et cadendo pervenit ad *D*, ut area curvilinea *DLme* ad rectangulum $2PD \times DL$. Namq; tempus quo corpus vi uniformi(194) descendens descripsit lineam *PD* est ad tempus quo corpus idem descripsit lineam *PE* in dimidiata ratione *PD* ad *PE*, id est (lineola *DE* jamjam nascente) in ratione *PD* ad $PD + \frac{1}{2}DE$ seu $2PD$ ad $2PD + DE$, et divisim ad tempus quo corpus idem descripsit lineolam *DE* ut $2PD$ ad *DE*, adeoq; ut rectangulum $2PD \times DL$ ad aream *DLME*, estq; tempus quo corpus utrumq; descripsit lineolam *DE* ad tempus quo corpus alterum inæquabili motu descripsit lineam *De* ut area *DLME* ad aream *DLme*, et ex æquo tempus primum ad tempus ultimum ut rectangulū $2PD \times DL$ ad aream *DLme*.(195)

ARTIC. VIII.(196)

CONTINENS

INVENTIONEM ORBIUM IN QUIBUS CORPORA VIRIBUS QUIBUSCUNQ; CENTRIPETIS AGITATA REVOLVENTUR.

Prop. XL. Theor. XIII.

Si corpus cogente vi quacunq; centripeta moveatur utcunq; & corpus aliud recta ascendat vel descendat, sintq; eorum velocitates in aliquo æqualium altitudinum casu æquales, velocitates eorum in omnibus æqualibus altitudinibus erunt æquales.

Descendat corpus aliquod ab *A*(197) per *D*, *E* ad centrum *C*, et moveatur corpus aliud a *V* in linea curva *VIKk*. Centro *C* intervallis quibusvis describantur circuli *DI*, *EK* rectæ *AC* in *D* et *E*, curvæq; *VIK* in *I* et *K* occurrentes. Jungatur *IC* occurrens ipsi *KE* in *N*; et in *IK* demittatur perpendiculum *NT*, sitq; circum-

(194) Thus amended by Newton from an inexact earlier phrase 'uniformi motu' (with uniform motion). In the sequel, however, a parallel 'inæquabili motu' (with inequable motion) is inconsistently retained.

(195) To bring out the full power of this general resolution of the problem of accelerated rectilinear motion, Newton might well have attached a worked example of its application to some particular law of force $(-d^2r/dt^2 = -v.dv/dr =) f(r)$. In notes (167), (170) and (186) above we have outlined the variant demonstrations of Propositions XXXII/XXXIII and XXXVIII which straightforwardly ensue on setting $f(r)$ equal to k/r^2 and kr respectively.

given, its speed at any other place e will be found by erecting the ordinate eg, and taking that speed to the speed at the place D as the square root of the rectangle $(PQRD)$ increased (if the place e is lower than the place D) or diminished (if it is higher) by the curvilinear area $(DFge)$ to the square root of the rectangle $(PQRD)$ alone, that is, as $\sqrt{[(PQRD) \pm (DFge)]}$ to $\sqrt{[(PQRD)]}$.

Corollary 3. The time, also, will be ascertained by erecting the ordinate em proportional to the square root of $(PQRD) \pm (DFge)$, and taking the time in which the body described the line De to the time in which the other body fell under the uniform force from P, and in falling arrived at D, as the curvilinear area $(DLme)$ to the rectangle $2PD \times DL$. For the time in which the body descending under a uniform force(194) described the line PD is to the time in which the same body described the line PE in the halved ratio of PD to PE, that is, (with the line-element DE now just nascent) in the ratio of PD to $PD + \frac{1}{2}DE$ or $2PD$ to $2PD + DE$, and so *dividendo* to the time in which the same body described the line-element DE as $2PD$ to DE, and hence as the rectangle $2PD \times DL$ to the area $(DLME)$; while the time taken by both bodies to describe the line-element DE is to the time in which the second body in its non-uniform motion described the line De as the area $(DLME)$ to the area $(DLme)$; and so *ex æquo* the first time is to the last one as the rectangle $2PD \times DL$ to the area $(DLme)$.(195)

Article VIII.(196)

EMBRACING

The finding of orbits in which bodies urged upon by any centripetal forces whatever revolve.

Proposition XL, Theorem XIII.

If a body pressed by any centripetal force whatever should move in any manner while another body ascends or descends straight up or down, and their speeds be in some one case at an equal height equal, then their speeds will at all equal heights be equal.

Let some body descend from A(197) through D and E to the centre C while another body moves from V in the curve $VIKk$. With centre C and any radii describe the circle arcs DI, EK meeting the straight line AC in D and E and the curve VIK in I and K. Join IC meeting KE in N, and onto $\widehat{IK}$ let fall the perpendi-

(196) Tentatively altered to 'IX' and then changed back again to 'VIII'; compare note (187).

(197) Understand once more (see note (188) above) that this point of initial rest may be at infinity or even not real. Following Pemberton's lead in his *editio ultima* of the *Principia* we have for convenience introduced into our English version an appropriately simplified version of Newton's accompanying figure: the excess detail relates to Proposition XLI following, which it jointly illustrates.

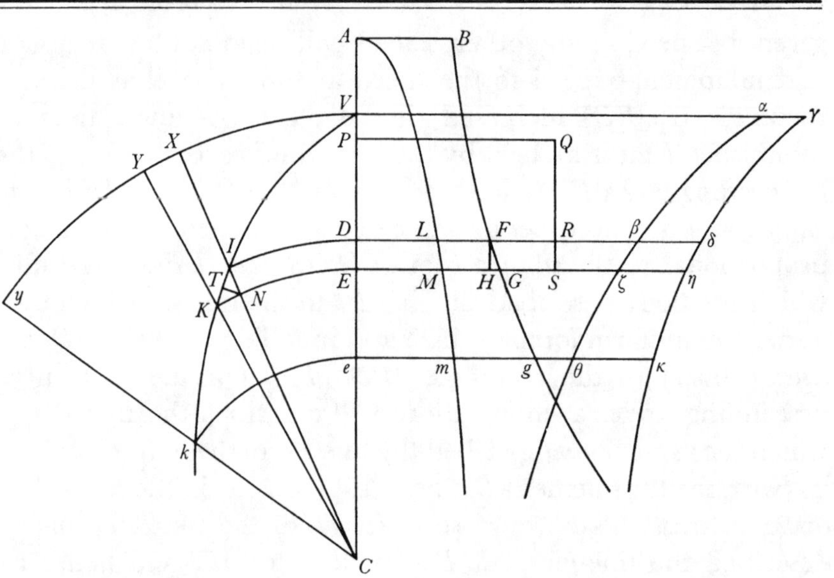

ferentiarum circulorum intervallum *DE* vel *IN* quàm minimum, et habeant corpora in *D* et *I* velocitates æquales. Quoniam distantiæ *CD*, *CI* æquantur, erunt vires centripetæ in *D* et *I* æquales. Exponantur hæ vires per æquales lineolas *DE*, *IN*, et si vis una *IN*, per Legum Corol. 2 resolvatur in duas *NT* et *IT*, vis *NT* agendo secundum lineam *NT* corporis cursui *ITK* perpendicularem, nil mutabit velocitatem corporis in cursu illo sed retrahet solummodo corpus de cursu rectilineo, facietꝗ ipsum de Orbis tangente perpetuo deflectere, inꝗ via curvilinea *ITKk* progredi. In hoc effectu producendo vis illa tota consumetur. Vis autem altera *IT* secundū corporis cursum agendo, tota accelerabit illud ac dato tempore quamminimo accelerationem generabit sibi ipsi proportionalem. Proinde corporum in *D* et *I* accelerationes æqualibus temporibus factæ (si sumantur linearum nascentium *DE*, *IN*, *IK*, *IT*, *NT* rationes primæ) sunt ut lineæ *DE*, *IT*: temporibus autem inæqualibus ut lineæ illæ et tempora conjunctim. Tempora ob æqualitatem velocitatum sunt ut viæ descriptæ *DE* et *IK*, adeoꝗ accelerationes in cursu corporum per lineas *DE* et *IK* sunt ut *DE* et *IT*, *DE* et *IK* conjunctim, id est ut $DE^{\text{quad.}}$ et $IT \times IK$ rectangulum. Sed rectangulum $IT \times IK$ æquale est IN^{quadrato}, hoc est æquale DE^{quadrato}, et propterea accelerationes in transitu corporum a *D* et *I* ad *E* et *K* æquales generantur. Æquales igitur sunt corporum velocitates, in *E* et *K* et eodem argumento semper reperientur æquales in subsequentibus æqualibus distantijs. Q.E.D.(198) Sed et eodem argumento corpora æquivelocia et æqualiter a centro distantia, in ascensu ad æquales distantias æqualiter retardabuntur. Q.E.D.

Corol. 1. Hinc si corpus vel funipendulum(199) oscilletur, vel impedimento

(198) In equivalent analytical terms, if, under the continuous urge of a force whose magnitude at the distance $CI = CD = r$ from its centre C is $f(r)$, we set t to be the time of orbit over the arc $\widehat{VI} = s$ from the point V distant $CV = R$ from C, and t' the corresponding time of direct fall over $VD = R-r$, it follows, on taking the speed $v = ds/dt$ of orbit at I to be equal to the speed $-dr/dt'$ of fall at D, that the ratio dt/dt' of the times in which the related

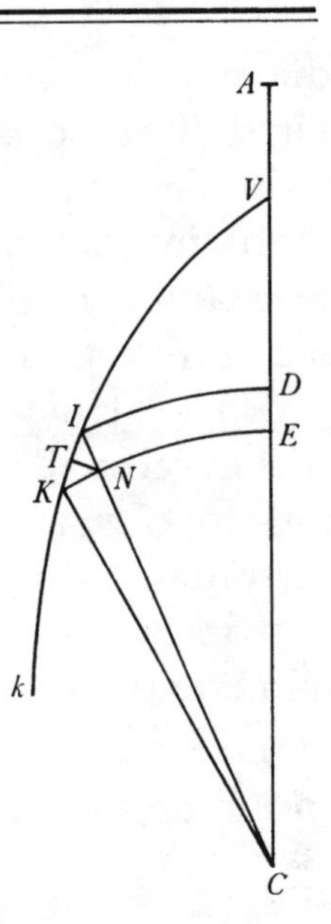

cular NT; then let the distance DE or IN between the circle circumferences be minimally small, and let the bodies have equal speeds at D and I. Since the distances CD, CI are equal, the centripetal forces at D and I will be equal. Represent these forces by the equal line-elements DE, IN, and then, if one force IN be, by Corollary 2 of the Laws, resolved into the components TN and IT, the force TN, acting along the line TN perpendicular to the course ITK of the body, will do nothing to alter the speed of the body in that course but merely draw the body aside from its rectilinear run, making it perpetually deviate from the orbit's tangent and proceed in the curvilinear path $ITKk$; and in producing this result that force will be wholly spent. The second force IT, however, acting along the body's course, will wholly accelerate it and in a given minimally small time generate an acceleration proportional to itself. As a consequence, the accelerations of the bodies effected in equal times at D and I are (if first ratios of the nascent lines DE, IN, IK, IT, TN be assumed) as the lines DE and IT; but in unequal times are jointly as those lines and the times. Because of the equality of the speeds the times are as the paths DE and IK described, and hence the accelerations in the run of the bodies along the lines DE and IK are jointly as DE and IT, and DE and IK, that is, as DE^2 and the rectangle $IT \times IK$. But the rectangle $IT \times IK$ is equal to IN^2, that is, to DE^2, and accordingly the accelerations generated in the passage of the bodies from D and I to E and K are equal. And equal, therefore, are the speeds of the bodies at E and K, and by the same argument will always be found equal at subsequent equal distances. As was to be proved.(198) Also by the same argument bodies which are equally swift and at equal distances from the centre will in their rise to equal distances equally be slowed. As was to be proved.

Corollary 1. Hence, should a body swing as a cord-pendulum(199) or be com-

increments $\widehat{IK} = ds$ (of orbit $\widehat{VI}$) and $DE = -dr$ (of fall VD) are traversed is equal to the ratio $-ds/dr$ of those increments, and therefore the 'instantaneous' increase $dv = f(r) \times (-dr/ds) \,.\, dt$ in orbital speed is equal to the corresponding increment $f(r) \,.\, dt'$ in the speed of direct fall; whence the speed in orbit at K is equal to that at E in the fall line VC, and so likewise at all ensuing pairs of points k and e at equal distances $Ck = Ce$ from the force-centre. Where V is the orbital speed at V, by Corollary 2 to Proposition XXXIX—or, more directly, by integrating the equation $dv/dt = v \,.\, dv/ds = -f(r) \,.\, dr/ds$ of motion—the speed at I is at once

$$v = \sqrt{\left[V^2 + 2\int_R^r -f(r) \,.\, dr\right]}.$$

(199) The manuscript originally had 'pendulum' *tout court*: this was in turn converted by Newton to be 'filo pendens' (hanging by a thread)—a variant trivially emended to pendens' a filo' in *Principia*, $_3$1726: 124.

quovis politissimo et perfectè lubrico(200) cogatur in linea curva moveri, et corpus aliud recta ascendat vel descendat sintq velocitates eorum in eadem quacunq altitudine æquales: erunt velocitates eorum in alijs quibuscunq æqualibus altitudinibus æquales. Namq impedimento vasis absolute lubrici(200) idem præstatur quo vi transversa NT. Corpus eo non retardatur, non acceleratur, sed tantū cogitur de cursu rectilineo discedere.

Corol. 2. Hinc etiam si quantitas P sit maxima a centro distantia,(201) ad quam corpus vel oscillans vel in trajectoria quacunq revolvens, deq quovis trajectoriæ puncto, eâ quam ibi habet velocitate sursum projectum, ascendere possit; sitq quantitas A distantia(201) corporis a centro in alio quovis orbis puncto, et vis centripeta semper sit ut ipsius A dignitas quælibet A^{n-1} cujus index $n-1$ est numerus quilibet n unitate diminutus; velocitas corporis in omni altitudine A erit ut $\sqrt{nP^n - nA^n}$,(202) atq adeo datur. Namq velocitas ascendentis ac descendentis (per Prop. XXXIX) est in hac ipsa ratione.(203)

Prop. XLI. Prob. XXVIII.

Posita cujuscunq generis vi centripeta et concessis figurarum curvilinearum quadraturis, requiruntur tum trajectoriæ in quibus corpora movebuntur, tum tempora motuum in Trajectorijs inventis.

Tendat vis quælibet ad centrum C et invenienda sit trajectoria $VITKk$. Detur circulus VXY centro C intervallo quovis CV descriptus, centroq eodem describantur alij quivis circuli ID, KE trajectoriam secantes in I et K rectamq CV in D et E. Age tum rectam CKY occurrentem

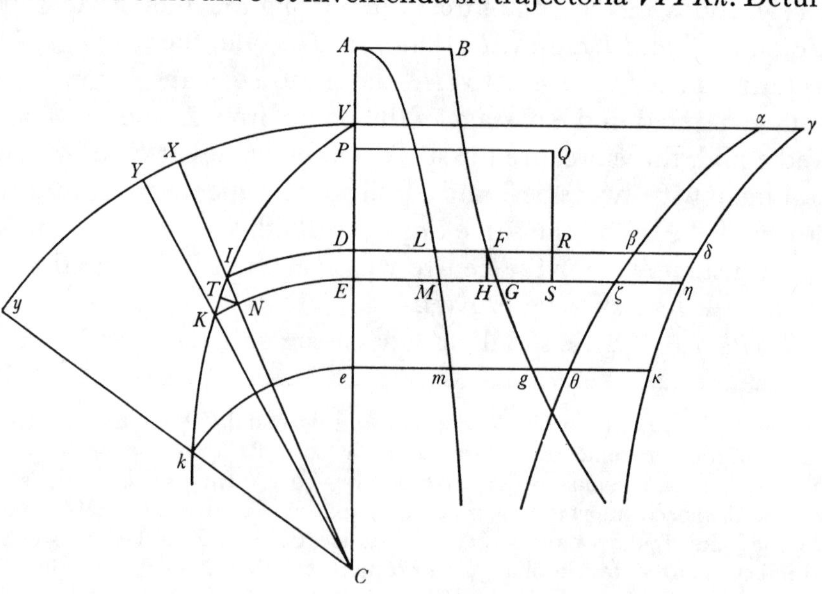

(200) That is, frictionless. The respective equivalent adjectives 'glabro' and 'glabri' (smooth-shaved) were first written.

(201) So changed by Newton from 'altitudo' (height) although a parallel 'altitudine' remains unreplaced in the sequel.

pelled by any completely smooth and perfectly slippery(200) constraint to move in a curve, while another body shall rise or descend straight up or down, if their speeds at any identical height be equal, then their speeds at any other equal heights will be equal. For the same end is achieved by the constraint of an absolutely slippery(200) bowl as by the transverse force TN. A body is neither slowed nor accelerated by it, but merely compelled to depart from its rectilinear course.

Corollary 2. Hence also if the quantity P be the greatest distance(201) from the centre to which a body, either oscillating or revolving in an arbitrary trajectory, can, when shot upwards from any point of the trajectory with the speed which it has there, ascend, let the quantity A be the distance(201) of the body from the centre in any other point of orbit, and the centripetal force there ever as any power A^{n-1} of A (where the index $n-1$ is any number n diminished by unity), and then the speed at every height A will be as $\sqrt{[(P^n-A^n)/n]}$ and so given. For the speed of ascent and descent is (by Proposition XXXIX) in this very ratio.(203)

Proposition XLI, Problem XXVIII.

Supposing a centripetal force of any kind, and granted the quadratures of curvilinear figures, there are required both the trajectories in which bodies move, and the times of motion in the trajectories so found.

Let any force tend to the centre C, and let it be required to find the trajectory $VITKk$. Let there be given the circle-arc VXY described with centre C and any radius CV, and with the same centre describe any other circle-arcs DI, EK cutting the trajectory in I and K and the straight line CV in D and E. Then draw the straight line CKY meeting the circle-arc VXY in Y. Let, however,

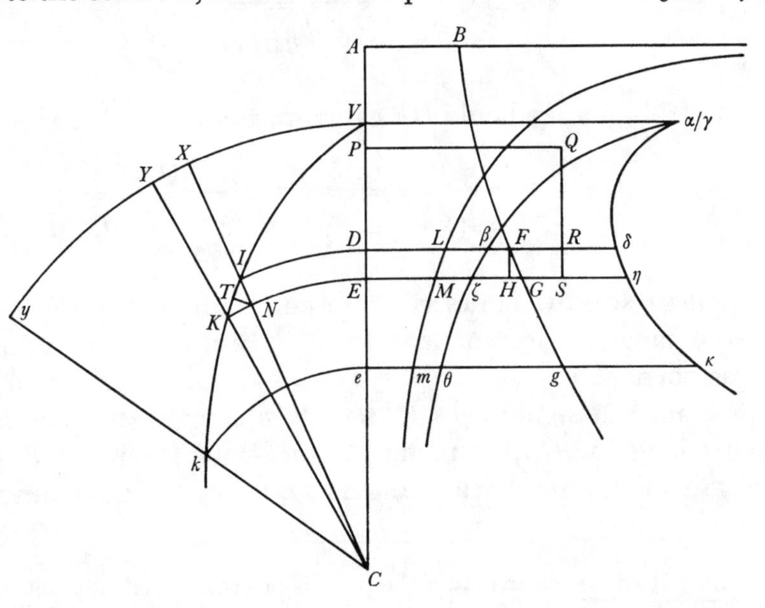

(202) Newton intends '$\sqrt{\frac{P^n}{n}-\frac{A^n}{n}}$' and we have so 'translated' it in our English version. The slip passed unnoticed into the *Principia*'s *editio princeps* ($_1$1687: 127) but was acceptably mended in the second edition ($_2$1713: 116) by shortening the root to be '$\sqrt{P^n-A^n}$' simply.

circulo VXY in Y. Sint autem puncta I et K sibi invicem vicinissima, et pergat corpus ab V per I, T, K ad k sitꝗ A altitudo de qua(204) corpus aliud cadere debet ut in loco D velocitatem acquirat æqualem velocitati corporis prioris in I: et stantibus quæ in Propositione XXXIX, quoniam lineola IK dato tempore quàm minimo descripta, est ut velocitas atꝗ adeo ut latus quadratum areæ $ABFD$, et triangulum ICK tempori proportionale datur, adeoꝗ KN est reciprocè ut altitudo IC, id est, si detur quantitas aliqua Q et altitudo IC nominetur A, ut $\frac{Q}{A}$:(205) ponamus eam esse magnitudinem ipsius Q ut sit $\sqrt{ABFD}$ in aliquo casu ad $\frac{Q}{A}$ ut IK ad KN, et erit semper $\sqrt{ABFD}$ ad $\frac{Q}{A}$ ut IK ad KN, et $ABFD$ ad $\frac{QQ}{AA}$ ut $IK^{\text{quad.}}$ ad $KN^{\text{quad.}}$ et divisim $ABFD-\frac{QQ}{AA}$ ad $\frac{QQ}{AA}$ ut $IN^{\text{quad.}}$ ad $KN^{\text{quad.}}$ adeoꝗ $\sqrt{ABFD-\frac{QQ}{AA}}$ ad $\frac{Q}{A}$ ut IN ad KN[,] et propterea $A\times KN$ æquale $\frac{Q\times IN}{\sqrt{ABFD-\frac{QQ}{AA}}}$. Unde cum $YX\times XC$ sit ad $A\times KN$(206) in duplicata ratione YC ad KC, erit $YX\times XC$ æquale

$$\frac{Q\times IN\times CX^{\text{quad.}}}{AA\sqrt{ABFD-\frac{QQ}{AA}}}.$$

Igitur si in perpendiculo DR capiantur semper $D\beta$, $D\delta$ ipsis

$$\frac{Q}{2\sqrt{ABFD-\frac{QQ}{AA}}} \quad \text{et} \quad \frac{Q\times CX^{\text{quad.}}}{2AA\sqrt{ABFD-\frac{QQ}{AA}}}$$

æquales respectivè et describantur curvæ lineæ $\alpha\beta$, $\gamma\delta$(207) quas puncta β, δ perpetuò tangent, deꝗ puncto V ad lineam AC erigatur perpendiculum $V\alpha\gamma$ abscindens areas curvilineas $VD\beta\alpha$, $VD\delta\gamma$, et erigantur ordinatæ $E\zeta$, $E\eta$: quoniam rectangulū $D\beta\times IN$ seu $D\beta\zeta E$ æquale est dimidio rectanguli $A\times KN$ seu triangulo ICK, et rectangulum $D\delta\times IN$ seu $D\delta\eta E$ æquale est dimidio rectanguli YX in CX seu triangulo XCY, hoc est, quoniam arearum $VD\beta\alpha$, VIC

---

(203) It will be clear that, if the body is projected vertically upwards at V till it comes to rest (at A), then $CA = P$ and hence (see note (198)) the orbital speed at any nearer distance $CI = A$ is

$$\sqrt{\left[2\int_P^A -f(r).dr\right]} \quad \text{where } f(r) \propto r^{n-1}.$$

(204) Read 'punctum de quo', as it was to be amended—with the unnecessary further insertion of an apposed 'locus ille' (that place)—in the *Principia's* second edition, and as we

the points I and K be exceedingly close to one another, and a body proceed from V through I, T and K to k, with A the [point] from which another body must fall so as to acquire at the place D a speed equal to that of the previous body at I. Then with things standing as in Proposition XXXIX, because the line-element IK described in a given minimal time is as the speed, and so as the square root of the area $(ABFD)$, while the triangle ICK is given proportional to the time, and therefore KN is reciprocally as the height IC, that is, given some quantity Q and on naming the height IC to be A, as Q/A:(205) let us suppose that Q's size is such that $\sqrt{[(ABFD)]}$ shall in some case be to Q/A as IK to KN, and there will always be $\sqrt{[(ABFD)]}$ to Q/A as IK to KN, and $(ABFD)$ to Q^2/A^2 as IK^2 to KN^2, and *dividendo* $(ABFD)$—Q^2/A^2 to Q^2/A^2 as IN^2 to KN^2, and so $\sqrt{[(ABFD)-Q^2/A^2]}$ to Q/A as IN to KN, and consequently $A\times KN$ equal to

$$Q\times IN/\sqrt{[(ABFD)-Q^2/A^2]}.$$

Whence, since $YX\times XC$ is to $A\times KN$(206) in the doubled ratio of YC to KC, there will be $YX\times XC$ equal to $Q\times IN\times CX^2/A^2\sqrt{[(ABFD)-Q^2/A^2]}$. Therefore if in the perpendicular DR there be taken $D\beta$, $D\delta$ ever equal respectively to

$$\frac{Q}{2\sqrt{[(ABFD)-Q^2/A^2]}} \quad \text{and} \quad \frac{Q\times CX^2}{2A^2\sqrt{[(ABFD)-Q^2/A^2]}},$$

and the curves $\alpha\beta$, $\gamma\delta$(207) which the points β, δ perpetually trace be described, from the point V to the line AC erect the perpendicular $V\alpha/\gamma$ cutting off the curvilinear areas $(VD\beta\alpha)$, $(VD\delta\gamma)$, and also the ordinates $E\zeta$, $E\eta$: then because the rectangle $D\beta\times IN$, or $(D\beta\zeta E)$, is equal to half the rectangle $A\times KN$, or the triangle ICK, and the rectangle $D\delta\times IN$, or $(D\delta\eta E)$, is equal to half the rectangle $YX\times CX$, or the triangle XCY—in other words, because the areas $(VD\beta\alpha)$,

render it in our English version. Newton evidently here momentarily confuses the point A in his figure with the 'height' $(CI=)\ A$ which he defines below.

(205) No doubt largely to ease the work of the typesetter, Edmond Halley here inserted an ugly parallel following phrase 'quam nominemus Z' (which let us name Z) in the *Principia's* press copy (Royal Society MS LXIX), replacing all occurrences of $\frac{`Q'}{A}$ and $\frac{`QQ'}{AA}$ in sequel by 'Z' and 'ZZ' correspondingly. The 'improvement' passed into all published editions.

(206) That is, $KC\times KN$ since the points I and K are taken to be indefinitely close and therefore $IN = IC-KC$ is vanishingly small.

(207) Since when D coincides with V the 'height' $(CD = CI =)\ A$ is equal to $(CD =)\ CX$, the points α and γ in which the curves (β) and (δ) meet the perpendicular to CA at V will not be distinct, as shown in Newton's figure (here reproduced from *Principia*, $_1$1687: 128, with the denotations a, b, d, c, z, x, v and w—there introduced by Halley once again to simplify the printer's task, no doubt—replaced by their Greek originals $\alpha, \beta, \gamma, \delta, \zeta, \eta, \theta$ and κ): the venial error endured into Cotes' second edition ($_2$1713: 116), to be silently corrected by Pemberton in his *editio ultima*. We make appropriate adjustment in our 'English' figure, where we have also (compare note (189)) accurately drawn the 'inverse-speed' curve mML asymptotically to approach AB.

æquales semper sunt nascentes particulæ $D\beta\zeta E$, ICK et arearum $VD\delta\gamma$, $VC[X]$ æquales semper sunt nascentes particulæ $DE\eta\delta$, XCY; erit area genita $VD\beta\alpha$ æqualis areæ genitæ VIC adeoq; tempori proportionalis, et area genita $VD\delta\gamma$ æqualis sectori genito VCX. Dato igitur tempore quovis ex quo corpus discessit de loco $V_{[,]}$ dabitur area ipsi proportionalis $VD\beta\alpha$ et inde dabitur corporis altitudo CD vel CI et area $VD\delta\gamma$ eiq; æqualis sector $VC[X]$, ejusq; angulus VCI.(208) Datis autem angulo VCI & altitudine CI datur locus I in quo corpus completo illo tempore reperietur. Q.E.I.(209)

---

(208) If, much as in note (198), we again suppose t to be the time of orbit over the arc $\widehat{VI} = s$ from the point V distant $CV = R$ from the centre C in a force-field whose accelerative pull at the distance $CI = A$ is $f(A)$, by Proposition XL the orbital speed $v = \widehat{IK}/dt = ds/dt$ is $\sqrt{\left[V^2+2\int_R^A -f(r).dr\right]}$; accordingly, if we put Newton's Keplerian constant Q equal to

$$2(ICK)/dt = CI\times KN/dt = A^2.d(\widehat{VCI})/dt,$$

it then follows that

$$(ABFD) = v^2 = (KN/dt)^2+(NI/dt)^2 = (Q/A)^2+(dA/dt)^2,$$

whence

$$D\beta = \tfrac{1}{2}Q/\sqrt{[v^2-Q^2/A^2]} = \tfrac{1}{2}Q.dt/dA \quad\text{and}\quad D\delta = \tfrac{1}{2}QR^2/A^2\sqrt{[v^2-Q^2/A^2]} = \tfrac{1}{2}R^2.d(\widehat{VCI})/dA,$$

so that

$$(V\alpha\beta D) = \int_R^A D\beta.d(CD) = \tfrac{1}{2}Qt \quad\text{and}\quad (V\gamma\delta D) = \int_R^A D\delta.d(CD) = (VCX).$$

(209) With but minimal recasting Newton's construction determines the elements of orbit in a known force-field $f(r)$ varying as the distance r, given the speed and direction of motion at some point. For, if $I(r, \phi)$ is attained in time t from $V(R, 0)$ where the initial motion is directed with speed V at an angle α to the radius vector $CV = R$ along which the force has an accelerative pull to the centre $C(0, 0)$ of magnitude g, the orbital speed at I is

$$v = \sqrt{\left[V^2+2\int_R^r -f(r).dr\right]}$$

and thence

$$t = \int_R^r 1/\sqrt{[v^2-Q^2/r^2]}.dr \quad\text{and}\quad \widehat{VCI} = \phi = \int_R^r Q/r^2\sqrt{[v^2-Q^2/r^2]}.dr,$$

where the Keplerian constant Q $(= r^2.d\phi/dt) = RV\sin\alpha$. To establish the uniqueness of the general conic as a possible orbit in an inverse-square central force-field, it is then enough to make $f(r) = gR^2/r^2$, so deriving $v = \sqrt{[V^2+2gR^2(1/r-1/R)]}$ and thence

$$\phi = \int_R^r RV\sin\alpha/r^2\sqrt{[V^2-2gR+2gR^2r^{-1}-(R^2V^2\sin^2\alpha)\,r^{-2}]}.dr$$

$$= \Big[\cos^{-1}((1-(V^2/gr)\sin^2\alpha)/e)\Big]_{r=R}^{r=r}, \quad e = \sqrt{[1-(V^2/gR)\,(2-V^2/gR)\sin^2\alpha]},$$

or $$R/r = (1-e\cos(\phi+\epsilon))/(V^2/gR)\sin^2\alpha, \quad \epsilon = \cos^{-1}((1-(V^2/gR)\sin^2\alpha)/e):$$

which is the defining polar equation, referred to a focus as origin, of a conic of eccentricity e and main axis $2(V^2/g)\sin^2\alpha/(1-e^2) = 2R/(2-V^2/gR)$. Even though the central quadrature therein is contained in the 'Catalogus posterior' of integrals in Problem 9 of his 1671 tract (on setting $z = r$, $\eta = -1$, $d = RV\sin\alpha$, $e = V^2-2gR$, $f = 2gR^2$ and $g = -R^2V^2\sin^2\alpha$ in the first species of its 'Ordo octavus' on III: 252), Newton forbore to repeat a result which he had

(VCI) ever have the equal nascent particles ($D\beta\zeta E$), (ICK) and the areas ($VD\delta\gamma$), (VCX) ever the equal nascent particles ($D\delta\eta E$), (XCY)—, the generated area ($VD\beta\alpha$) will be equal to the generated area (VCI), and hence proportional to the time, and the generated area ($VD\delta\gamma$) equal to the generated sector (VCX). Given any time, therefore, since the body departed from the place V, the area ($VD\beta\alpha$) proportional to it will be given, and therefrom the body's height CD or CI, and also the area ($VD\delta\gamma$) and the sector (VCX) equal to it, and the latter's angle $\widehat{VCI}$.(208) Given the angle $\widehat{VCI}$ and height CI, however, there is given the place I at which the body will, when the time is finished, be located. As was to be found.(209)

---

earlier demonstrated to his satisfaction in equivalent ways (see §1: note (73) and §2: note (174)), preferring in his following Corollary 3—a late insertion in the main manuscript—to determine the orbital curve in the comparable (if unrealistic) instance of an inverse-cube force-field. Here, much as before, on making $f(r) = gR^3/r^3$ there ensues

$$v = \sqrt{[V^2+gR^3(1/r^2-1/R^2)]}$$

and in consequence

$$\phi = \int_R^r RV\sin\alpha/r^2\sqrt{[V^2-gR-R^2(V^2\sin^2\alpha-gR)\,r^{-2}]}\,.\,dr = (\lambda/\mu)\left[\cos^{-1}(\mu R/r)\right]_{R=r}^{r=r}$$

where $$\lambda^2 = \mu^2+1/(V^2/gR-1) = (V^2/gR)\sin^2\alpha/(V^2/gR-1),$$

or $R/r = \cos((\mu/\lambda)\phi+\cos^{-1}\mu)/\mu$, the defining equation of a general Cotesian spiral, which Newton reduces to be $R/r = \cos k\phi$, $k = \sqrt{[1-gR/V^2]}$, by the restrictive initial condition $\alpha = \frac{1}{2}\pi$ (see note (213) below).

Newton's contemporaries were very slow to appreciate the depth and power of this construction—'granted the quadrature of [the pertinent] curves'—of the general orbit traversed in a given central force-field, or to acknowledge the ease with which it might be applied to the particular cases of the inverse-square and inverse-cube orbits; and when, more than a decade later, Pierre Varignon, Jakob Hermann and Johann Bernoulli groped their way to equivalent representations of the preceding Propositions XXXIX–XLI in terms of Leibnizian calculus (see E. J. Aiton, 'The Inverse Problem of Central Forces', *Annals of Science*, **20**, 1965: 81–99), they regarded their hard-won analytical reformulations not as mere 'translations' of Newton's geometrically couched theorems but as considerable qualitative improvements upon them. Though he went furthest in justifying such a claim to mathematical—as distinct from notational—originality, Bernoulli was also the least generous in recognising the magnitude of Newton's prior achievement. In the 'Extrait' of his letter to Hermann of 7 October 1710 (N.S.) outlining his preferred solution of the 'Problême inverse des Forces centripetes' which he subsequently published in the *Memoires de Mathématique & de Physique...de l'Academie Royale des Sciences. Année M. DCC. X* (Paris, [$_1$1713→] $_2$1732): 521–33, he grudgingly noted of his preparatory 'Lemme'—formally equivalent to the above Proposition XL—that 'La démonstration...se trouve dans le Livre de M. Newton *De Princ. Math. Nat.* [$_1$1687] pag. 125. Mais elle y est trop embarrassée' (*ibid.* 524), and similarly remarked of his dependent 'Problême. Les quadratures étant supposées, & la loi des forces centripetes..., trouver la trajectoire qu'elles doivent faire décrire au mobile'—an even more exact parallel to the present Proposition XLI—that 'ainsi la construction geometrique s'en peut aisément déduire...& même plus commodément que M. Newton ne l'a trouvée dans la *pag.* 127. &c de ses *Princ. Math.*' (*ibid.*: 526), while afterwards, having given accurate proof that the inverse-square orbit is in

Corol. 1. Hinc maximæ minimæq; corporum altitudines, id est Apsides Trajectoriarum expeditè inveniri possunt. Incidunt enim Apsides in loca illa ubi Trajectoria VIK perpendicularis est ad lineam AC per centrum ductam, id est ubi IK et NK(210) æquantur, adeoq; ubi area $ABFD$ æqualis est $\frac{QQ}{AA}$.(211)

all conditions of initial motion a conic 'ainsi que M. Newton l'a supposée pag. 55. Corol. 1 [to Propositions XI–XIII] sans le démontrer', he further emphasised Newton's 'failure' to demonstrate this converse in a concluding 'Remarque' (*ibid.*: 532–3), adding that 'Pour voir encore la nécessité de la démonstration que je viens de donner de cette inverse, il n'y a qu'à considerer que de ce qu'un corps pour se mouvoir sur une Spirale logarithmique, requiert des forces centrales en raison réciproque des centres de ses distances au foyer ou centre de cette Courbe; ce n'est pas une conséquence qu'avec de telles forces il décrivît toûjours une telle Courbe'. In later years Newton spent a great deal of time and effort in seeking to refute this far from wholly deserved criticism of his inadequate grasp of the niceties of general orbital dynamics in the *Principia*, mostly—for reasons of diplomacy rather than cowardice—conducting his counter-attack through the willing pen of his enthusiastic adjutant John Keill, who first brought Bernoulli's 'Extrait' to Newton's notice in late 1713. It would take a long monograph to examine in detail the thrust and counter-parry with which Keill and Bernoulli sought to out-foil each other in a long series of published papers and tracts, and to show how closely Newton fashioned and guided Keill's rapier. (See especially Keill's 'Observationes...de inverso Problemate Virium Centripetarum...', *Philosophical Transactions*, **29** No. 340 [July–September 1714]: 91–111, answered anonymously by Bernoulli in his 'Epistola pro eminente Mathematico, Dn. Johanne Bernoullio, contra quendam ex Anglia antagonistam scripta', *Acta Eruditorum* (July 1716): 296–315; and Keill's 'Défense du Chevalier Newton...', *Journal Literaire*, **8**, 1716: 418–33, to which Bernoulli responded under the pen of his own supporter Johann Heinrich Kruse in *Acta Eruditorum* (October 1718): 454–66, and was met by Keill's open 'Lettre...à M. Jean Bernoulli' in the *Journal Literaire*, **10**, 1719: 261–87 and a second, Latin *Epistola ad Virum Clarissimum Joannem Bernoulli* printed as a pamphlet at London in 1720. A riposte, 'Keillius Heauton Timoroumenos', which Bernoulli was preparing at the time of Keill's death in August 1721 is preserved in Basel University Library. A wealth of Newton's autograph English drafts and extensive corrections to de Moivre's French renderings of Keill's propagandist pieces exists in ULC. Add. 3968.23, Add. 3985.14 and especially Res. 1893.5.) Beneath such an outpouring of emotional and increasingly savage and embittered prose the essential justice of Newton's claim to have had complete mastery of the inverse problem long before Bernoulli was rapidly all but lost to sight, and even Jakob Hermann's wise summing-up (in his *Phoronomia, sive de Viribus et Motibus Corporum Solidorum et Fluidorum Libri Duo* (Amsterdam, 1716): 73) that 'Hoc problema primùm solutionem accepit à Cel. Newtono Prop. 41. Lib. [I] *Princ. Phil. Nat. Math.* & postea à Perspicacissimo Geometra Joh. Bernoulli gemino modo' has been largely ignored by posterity down to the present day.

We may add that, in the 'Cartesian' hypothesis—propounded by Newton in an unpublished paper 'Of Refraction & ye velocity of light according to ye density of bodies' (reproduced in Appendix 3.1) and thereafter in Section XIV of Book 1 of the *Principia* (see Appendix 3.2) and, in a simpler form, in the preliminary version of his *Opticks* a few years later (see Appendix 3.3) —according to which an 'emitted' light-'corpuscle' is refracted at an interface between media of different density as though it were instantaneously attracted normally at the interface towards the denser medium by an accelerative *vis refractiva*, the angle of inclination

$$\theta = \phi + \tan^{-1}(r.\,d\phi/dr)$$

of the tangent at I to the initial radius vector CV finds an important application in Newton's later theory of the atmospheric refraction of stellar light (never discussed in the printed

Corollary 1. Hence are the greatest and least heights of bodies, that is, the apsides of trajectories, speedily discoverable. For the apsides fall at the places where the trajectory $\widehat{VIK}$ is perpendicular to the line AC drawn through the centre, that is, where IK and NK(210) are equal, and so where the area $(ABFD)$ is equal to Q^2/A^2.(211)

Principia, but developed by him in unpublished manuscript—notably ULC. Add. 3967.3—and introduced fitfully into his correspondence with Flamsteed in the middle 1690's; compare Appendix 3.4). At once, on substituting

$$\phi = \int_R^r Qr^{-2}/\sqrt{[v^2 - Q^2/r^2]}.dr, \quad v = \sqrt{\left[V^2 + 2\int_R^r -f(r).dr\right]},$$

there ensues $d\theta/dr = Qr^{-2}/\sqrt{[v^2 - Q^2/r^2]} + d(\sin^{-1}(Qr^{-1}/v))/dr = -Qr^{-1}f(r)/v^2\sqrt{[v^2 - Q^2/r^2]}$ (compare Pierre-Simon de Laplace, *Traité de Mécanique Céleste*, **4** (Paris, An XIII [= 1804]): Livre x, Chapitre Premier. 'Des réfractions astronomiques': 231–76, especially 231–3: 'Équation différentielle du mouvement de la lumière'). Equivalently, as Brook Taylor showed in his *Methodus Incrementorum Directa & Inversa* (London, 1715): Pars II, Propositio XXVII: 108–10, if $CO = r^2.d\phi/ds = Q/v$ is let fall perpendicularly to the tangent IZ at I which forms the angle $\widehat{AZI} = \theta$ with CV, and Co, meeting IZ in ω, is similarly let fall normally to the tangent Kz at K (distant, as before, $\widehat{IK} = ds$ from I), then

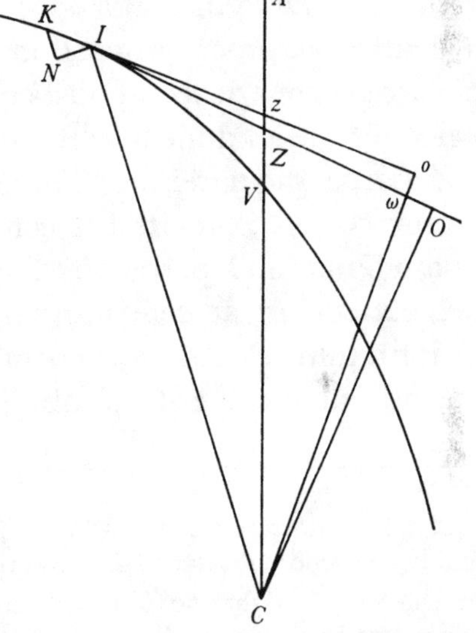

$$d\theta = \widehat{AzK} - \widehat{AZI} = \widehat{OIo} = o\omega/(I\omega \text{ or}) \; IO$$

and so

$$d\theta/dr = -(d(CO)/dr)/IO$$
$$= -(d(Q/v)/dr)/\sqrt{[r^2 - Q^2/v^2]}.$$

It follows that $d\theta/r.d\phi = -f(r)/v^2$ and hence

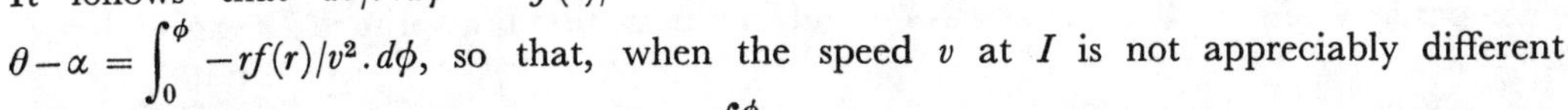

$\theta - \alpha = \int_0^\phi -rf(r)/v^2.d\phi$, so that, when the speed v at I is not appreciably different from that (V) at V, there is $\theta - \alpha \approx V^{-2}\int_0^\phi -rf(r).d\phi$. The particular case in which $f(r)$ is constant was to be, as he affirmed in a following letter a month later, the 'foundation' of the '*Tabula Refractionum*'—'computed by applying a certain Theorem to your Observations' whose 'demonstration is too intricate [here] to be set down'—which Newton sent to John Flamsteed on 17 November 1694 (*Correspondence of Isaac Newton*, **4**, 1967: 46–9; see also 61), while the physically more realistic case in which, in effect, he set $f(r) \propto r^{-2}e^{k(r^{-1}-R^{-1})}$ was the basis of a comparable '*Tabula refractionum siderum ad altitudines apparentes*' calculated by him soon afterwards, though it was not to appear in print till 1721 when, with his permission, Edmond Halley appended it to 'some Remarks' of his 'on the Allowances to be made in Astronomical Observations for the Refraction of the Air' (*Philosophical Transactions*, **31**, No. 368: 169–72). See our commentary to Appendix 3.4 below, where we reproduce a central extract from Newton's letter to Flamsteed of 20 December 1694 in which he expounds his resulting construction of the former hypothesis.

(210) In preliminary revise on the sheet rejected from the ensuing press transcript (see note

Corol. 2. Sed et angulus *KIN* in quo Trajectoria alibi secat lineam illam *IC*, ex data corporis altitudine *IC* expeditè invenitur, capiendo sinum ejus ad Radium ut *KN* ad *IK*, id est ut $\frac{Q}{A}$ ad latus quadratum areæ *ABFD*.

(212)*Corol.* 3. Si centro *C* et vertice principali *V* describatur sectio quælibet Conica *VRS*, et a quovis ejus puncto *R* agatur Tangens *RT* occurrens axi infinitè producto *CV* in puncto *T*; dein junctâ *CR* ducatur recta *CP* quæ æqualis est abscissæ *CT* angulumq; *VCP* sectori *VCR* proportionalem constituat; tendat autem ad centrum *C* vis centripeta cubo distantiæ locorum a centro reciprocè proportionalis, et exeat corpus de loco *V* justâ cum velocitate secundum lineam rectæ *CV* perpendicularem:(213) progredietur corpus illud in Trajectoria quam punctum *P* perpetuò tangit; adeoq; si Conica sectio *CVRS* Hyperbola sit, descendet idem ad centrum, sin ea Ellipsis sit, ascendet illud perpetuò et abibit in infinitum. Et contra si corpus quacunq; cum velocitate exeat de loco *V*, et perinde ut incœperit vel obliquè descendere ad centrum vel ab eo obliquè

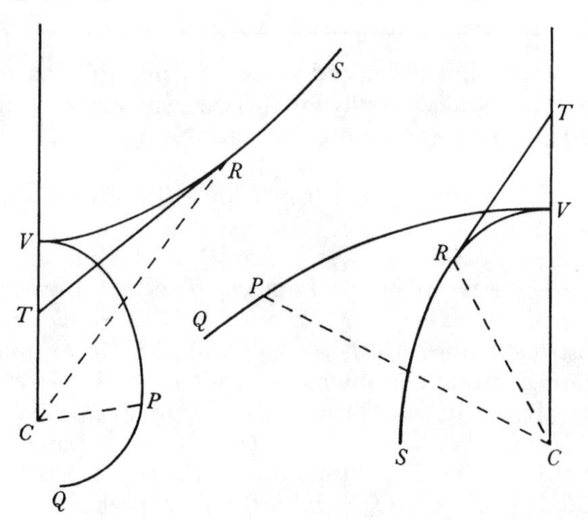

(212)) Newton rephrased this to be '...in puncta illa in quibus recta *IC* per centrum ducta incidit perpendiculariter in Trajectoriam *VIK*: id quod fit ubi rectæ *IK* et *NK*' (at the points in which the straight line *CI* drawn through the centre meets the trajectory $\widehat{VIK}$ at right angles: which occurs where the lines *IK* and *NK*); and so it passed into the printed *Principia* ($_1$1687: 129).

(211) That is, where (in the analytical terminology of note (209)) $v^2 = V^2+2\int_R^r -f(r).dr$ is equal to Q^2/r^2, and hence $Q = rv$. In general, as Newton goes on to state, the radius *CI* meets the orbit $\widehat{VI}$ at an angle $\sin^{-1}(r.d\phi/ds) = \sin^{-1}(Q/rv)$.

(212) The following Corollary was inserted at a late stage on a facing verso in the manuscript (f. 38^v), evidently to replace the erroneous 'Corol. 7' to Proposition XLIV following. Its incorporation required the substitution of a related sheet (now ULC. Add. 3965.4:19^r/20^r) in the *Principia*'s press transcript by a correspondingly augmented revise; compare I. B. Cohen's *Introduction* (note (1)): 124, 127.

(213) This supposition that the initial motion is at right angles to the radius vector—a natural carry-over from the terms of Corollary 6 to Proposition XLIV which was (see the previous note) evidently in the forefront of his mind in drafting the present corollary—is unnecessarily restrictive, since not all inverse-cube central-force orbits have apses. Specifically, whereas the polar orbit $R/r = \cos((\mu/\lambda)\phi+\cos^{-1}\mu)/\mu$, $\lambda^2 = \mu^2+1/(V^2/gR-1) = (V^2/gR)\sin^2\alpha/(V^2/gR-1)$, which (see note (209)) ensues when the angle $\alpha = \sin^{-1}(Q/RV)$ is general divides into five

Corollary 2. So, too, is the angle $K\widehat{I}N$ at which the trajectory anywhere cuts the line IC speedily found from the height IC of the body given by taking its sine to the radius as KN to IK, that is, as Q/A to the square root of the area $(ABFD)$.

(212)*Corollary 3.* If with centre C and principal vertex V there be described any conic VRS you please, from any point R of it draw the tangent RT, meeting the axis CV (indefinitely extended) in the point T, and then, joining CR, draw the straight line CP equal to the abscissa CT and forming an angle $V\widehat{C}P$ proportional to the sector (VCR); further, let there tend to the centre C a central force reciprocally proportional to the cube of the places' distance therefrom, and a body set off at a suitable speed from the place V along a line perpendicular to the straight line CV:(213) that body will proceed in the trajectory which the point P perpetually traces, and hence, if the conic $CVRS$ be a hyperbola, it will descend to the centre, while, if this be an ellipse, it will perpetually ascend and pass away to infinity. Conversely, if a body should set off at any speed whatever from the place V, and, correspondingly as it begins either obliquely to descend to the

---

principal species according as $0 < V^2/gR < 1$, $V^2 = gR$, $1 < V^2/gR < \text{cosec}^2\alpha$, $V^2 = gR\,\text{cosec}^2\alpha$ and $\text{cosec}^2\alpha < V^2/gR < \infty$, Newton's assumption that $\alpha = \frac{1}{2}\pi$ (and so $Q = RV$) leads him here to construct only the first and last of these in the simplified canonical forms $r/R = \text{sech}\,k\phi$ and $r/R = \sec k\phi$ respectively (with the second and fourth species reducing trivially to be a circle of radius $R = \sqrt{[V^2/g]}$ concentric with the force-centre C, while the third is here not real). Of course, as he was later at pains to point out in the face of criticism from Johann Bernoulli (see note (209) and compare note (215) below), Newton had previously shown in Proposition VIII of the preceding 'De motu Corporum'—renumbered IX when it was subsumed unchanged into the present 'De motu Corporum Liber primus'—that the general logarithmic spiral ($V^2 = gR$, $\alpha \neq \frac{1}{2}\pi$) $\log(R/r) = \phi\cot\alpha$ is also a possible inverse-cube orbit. The two remaining general species—namely, the hyperbolic cosecant spiral

$$(1 < V^2/gR < \text{cosec}^2\alpha)\ r/R = \text{cosech}\,k\phi$$

and the 'reciprocal' spiral ($V^2 = gR\,\text{cosec}^2\alpha$) $r/R = 1/(\phi\cot\alpha+1)$—were first identified by Roger Cotes about late October 1709 when he came to check the accuracy of Newton's present corollary (as reproduced in *Principia*, ${}_1$1687: 130), although he did not publish his enumeration of the component species of the inverse-cube orbit till some four years afterwards in his 'Logometria' (*Philosophical Transactions*, **29**, No. 338 [January–March 1714]: 5–45 [= (ed. Robert Smith) *Harmonia Mensurarum, sive Analysis & Synthesis per Rationum & Angulorum Mensuras* (Cambridge, 1722): Pars prima: 1–41]): 34–9 [=30–5]. In the interim, having earlier (in August 1710) privately communicated to Leibniz—characteristically 'disguised' as the quartic Cartesian curve $y = (x^2-2R^2)/\sqrt{[4R^2-x^2]}$ (where $x = r\sin\phi$, $y = r\cos\phi$)—the particular instance $k = \frac{1}{2}$ of the 'hyperbola spiralis' $r/R = \sec k\phi$ (see *Got. Gul. Leibnitij et Johan. Bernoullij Commercium Philosophicum et Mathematicum*, **2** (Lausanne/Geneva, 1745): 230–1 [=(ed. C. I. Gerhardt) *Leibnizens Mathematische Schriften*, **3** (Halle, 1856): 854]), Johann Bernoulli had independently drawn attention to the species of the reciprocal spiral, his 'Spiralis Archimedea inversa', $r = R'/\phi'$ (where $R' = R\tan\alpha$, $\phi' = \phi+\tan\alpha$) at the end of a long article 'De Motu Corporum Gravium, Pendulorum, & Projectilium...Demonstrationes Geometricæ' (*Acta Eruditorum* (February/March 1713): 77–95/115–32 [=*Opera*, **1** (Lausanne/Geneva, 1742): 514–58]): 128 [=552].

ascendere, figura *CVRS* vel Hyperbola sit vel Ellipsis: inveniri potest Trajectoria augendo vel minuendo angulum *VCP* in data aliqua ratione. Sed et vi centripeta in centrifugam versa ascendet corpus obliquè in Trajectoria *VPQ* quæ invenitur capiendo angulum *VCP* sectori Elliptico *CVRC* proportionalem et longitudinem *CP* longitudini *CT* æqualem ut supra. Consequuntur hæc omnia ex Propositione præcedente per Curvæ cujusdam(214) quadraturam cujus inventionem ut satis facilem brevitatis gratia missam facio.(215)

---

(214) Namely $y = d/z\sqrt{[e+fz^2]}$, where $d = RV/\sqrt{[V^2-gR]}$, $e = -R^2$ and $f = 1$, since on setting $\alpha = \frac{1}{2}\pi$ in the polar equation derived for the general inverse-cube orbit in note (109) the point $P(r, \phi)$ of the present restricted trajectory is defined by

$$\phi = \int_R^r RV/z^2\sqrt{[(V^2-gR)\ (1-R^2z^{-2})]}.dz.$$

The quadrature of this curve is at once given as the particular instance $\eta = 2$ of Ordo 7.1 of the 1671 tract's 'Catalogus Curvarum aliquot ad Conicas Sectiones relatarum' (see III: 244–54, especially 250), where by substituting ($z^{-\frac{1}{2}n}$ or) $z^{-1} = x$ there is obtained

$$\int_R^r y.dz = (2d/f).\left|\tfrac{1}{2}vx - \int v.dx\right|_{x=1/R}^{x=1/r},\ v = \sqrt{[f+ex^2]}.$$

Little differently, Newton here effectively employs the homogeneous substitution $z = R^2/x$ to derive

$$\phi = (1/\sqrt{[1-gR/V^2]})\int_R^{R^2/r} -1/\sqrt{[R^2-x^2]}.dx = (2/R^2k).\left|\tfrac{1}{2}vx - \int v.dx\right|_{x=R}^{x=R^2/r},$$

where $k^2 = \pm(1-gR/V^2)$ and $v^2 = \pm(R^2-x^2)$ accordingly as $V^2 > gR$ or $V^2 < gR$. To construct the two ensuing species of orbit—in modern terminology

$$\phi = \cos^{-1}(R/r)/\sqrt{[1-gR/V^2]} \quad \text{and} \quad \phi = \cosh^{-1}(R/r)/\sqrt{[gR/V^2-1]}$$

respectively—Newton straightforwardly introduces the auxiliary central conics *VRS* whose general point $R(x, y)$ is defined, with respect to origin C and abscissa CVT, by the Cartesian equation $y^2 = \pm\lambda^2(R^2-x^2)$, thereby determining the orbital radius $CP = r$ to be equal to their subtangent $CT = CV^2/x$ and the polar angle $\widehat{VCP} = \phi$ to be proportional to the sector $(VCR) = \left|\tfrac{1}{2}\lambda vx - \int \lambda v.dx\right|_{x=CV}^{x=CV^2/r}$. His original computations to this end seem not to have survived, but their remembered essence is presumably repeated in a recalculation of the analytical basis of this present construction of the (would-be) general inverse-cube orbit which he undertook at David Gregory's request in May 1694. Of this latter, retrospective analysis both Newton's initial jotted notes and the ensuing extended explanation which Gregory took away with him—wherein the restriction $\alpha = \frac{1}{2}\pi$ is introduced obliquely (and probably unintendedly) at a terminal stage—are reproduced in Appendix 4 below (see especially its note (13)).

(215) After Johann Bernoulli came in 1713 (see note (209) above) none too gently to point out that the logarithmic spiral ($V = \sqrt{[gR]}$, $\alpha \neq \frac{1}{2}\pi$) introduced by him in his earlier Proposition IX [=Proposition VIII of the preceding 'De motu Corporum'; compare §2: note (102)] is not embraced in either of the present categories of inverse-cube orbit

$$(V \gtrless \sqrt{[gR]},\ \alpha = \tfrac{1}{2}\pi),$$

Newton gave a good deal of thought to how he might remedy this deficiency without overstressing its relative significance. Having earlier penned a first version at the pertinent point in his interleaved personal copy of the *Principia's* second edition (ULC. Adv. b. 39.2: facing

centre or obliquely to rise away from it, the figure $CVRS$ be either an hyperbola or an ellipse, then the trajectory can be ascertained by increasing or diminishing the angle $\widehat{VCP}$ in some given ratio. However, with the centripetal force changed into a centrifugal one the body will ascend obliquely in a trajectory VPQ which is found by taking the angle $\widehat{VCP}$ proportional to the elliptical sector (CVR) and the length CP equal to the length CT, as above. All these things are consequences of the preceding proposition by way of the quadrature of a certain curve,(214) the finding of which, as being easy enough, I for brevity's sake pass over.(215)

---

page 119), in a further wad of corrections and additions to Cotes' *editio secunda* (ULC. Add. 3965.13: 497v; a much abbreviated further revise is found on f. 515v following) he afterwards instructed that a further 'sectio' (paragraph) here be added: 'Ostendimus in Prop. IX Lib. I, quod corpus vi centripeta quæ cubo distantiarum reciproce proportionalis est, moveri potest in Spirali quæ radios omnes in angulo dato secat. Corpus autem hacce virium Lege pro varia ejus velocitate movebitur in varijs Curvis. Si velocitas ea sit quacum corpus in Circulo circa centrum virium ad distantiam suam revolvi potest, movebitur idem in hac Spirali. Si velocitas minor est, movebitur corpus in Spirali VPQ quæ spiris infinitis descendet ad centrum sed non ascendet in infinitum. Si velocitas major est, movebitur corpus in Curva VPQ quæ Asymptoton habet et juxta eandem crure Hyperbolico in infinitum abit' (We have shown in Proposition IX, Book 1 that a body can, under a centripetal force which is reciprocally proportional to the cube of the distance, move in a spiral which intersects all its radii at a given angle. If the speed be that with which the body can at its distance revolve in a circle round the force-centre, it will move in this spiral. If the speed is less, the body will move in the spiral VPQ which shall descend in infinite coils to the centre, but not ascending to infinity. If the speed is greater, the body will move in the curve VPQ possessed of an asymptote, departing in line with it to infinity in a hyperbolic branch). While this adds the information that in the spiral

$$r = R\,\text{sech}\,k\phi \quad (V < \sqrt{[gR]}, \alpha = \tfrac{1}{2}\pi)$$

the radius vector r decreases indefinitely to zero as the polar angle ϕ increases to infinity, and that the complementary spiral $r = R \sec k\phi$ $(V > \sqrt{[gR]}, \alpha = \frac{1}{2}\pi)$ has a (dual) asymptote parallel to the central line defined by $\phi = \pm\frac{1}{2}\pi/k$, the full force of his self-imposed restriction (now $V \neq \sqrt{[gR]}$) to an initial angle $\alpha = \frac{1}{2}\pi$ still goes unrecognised by Newton. We may perhaps conjecture that he ultimately suppressed this intended further paragraph when—perhaps after reading (or in 1722 re-reading) the full enumeration of the types of inverse-cube orbit published by Cotes in his 1714 'Logometria' (see note (213))—he afterwards realised that the two species of the hyperbolic cosecant and reciprocal spirals for which

$$1 < V/\sqrt{[gR]} \leqslant \operatorname{cosec}\alpha, \quad \alpha \neq \tfrac{1}{2}\pi,$$

were still to be allowed for.

In about late 1714, transcribing an initial rough draft on ULC. Add. 3965.13: 374r without essential change, Newton tentatively incorporated into his interleaved copy of the *Principia*'s second edition a further '*Corol. 4.* Concessis Curvilinearum Quadraturis inveniri possunt Lineæ Curvæ in quibus corpora descendendo vel æquabiliter accedent ad Horizontem vel alias assignatas motus leges observabunt' (*Corollary 4.* Granted the quadrature of curvilinear figures, there can be ascertained the curves in which bodies shall in descending either uniformly approach the horizontal or observe other assigned laws of motion). Manifestly, the sole effect of constraining a central-force orbit other than by requiring it to be traversed in free fall to the force-centre is, in the terms of note (209) above, to replace the Keplerian restriction $r^2 . d\phi/dt = Q$, given, on the preceding Proposition XLI by a substitute deriving from the variant constraint

Prop. XLII. Prob. XXIX.

Datâ lege vis centripetæ, requiritur motus corporis de loco dato data cum velocitate secundum datam rectam egredientis.(216)

Stantibus quæ in tribus Propositionibus præcedentibus: exeat corpus de loco I secundum lineolam IT ea cum velocitate quam corpus aliud vi aliqua uniformi centripeta de loco P cadendo acquirere posset in D: sitꝗ hæc vis uniformis ad vim qua corpus primum urgetur in I ut DR ad DF. Pergat autem corpus versus k, centroꝗ C et intervallo Ck describatur circulus ke occurrens rectæ PD in e et erigantur curvarum $ALMm$, $BFGg$, $\alpha\beta\zeta\theta$, $\gamma\delta\eta\kappa$ ordinatim applicatæ em, eg, $e\theta$, $e\kappa$. Ex dato rectangulo $PDRQ$ dataꝗ lege vis centripetæ qua corpus primum agitatur, dantur curvæ lineæ $BFGg$, $ALMm$ per constructionem Problematis XXVII et ejus Corol. 1. Deinde ex dato angulo CIT datur proportio nascentium IK, KN et inde per constructionem Prob XXVIII datur quantitas Q, una cum curvis lineis $\alpha\beta\zeta\theta$, $\gamma\delta\eta\kappa$: adeoꝗ completo tempore quovis $D\beta\theta e$ datur tum corporis altitudo Ce vel Ck, tum area $D\delta\kappa e$, eiꝗ æqualis sector XCy, angulusꝗ XCy et locus k in quo corpus tunc versabitur. Q.E.I.(217)

Supponimus autem in his Propositionibus vim centripetam in recessu quidem

set upon the orbit. As his terminology reveals, Newton in his cited instance proposes a generalisation of the problem of the 'isochrone' which Leibniz, during the course of a wider dispute with him on the conservation of 'motion' (momentum), put to the Abbé Catelan in the summer of 1687 in a public 'Réponse' (see the *Nouvelles de la Republique des Lettres* (September 1687): §III: 952–6, and compare Émile Ravier, *Bibliographie des Œuvres de Leibniz* (Paris, 1937): 54, note [98]): namely, 'Invenire lineam isochronam, in qua grave descendat uniformiter, sive æqualibus temporibus æqualiter accedat ad horizontem, atque adeo sine acceleratione, & æquali semper velocitate deorsum feratur', as Leibniz restated it—with his own solution of a semicubical parabola in the limiting case where the force-centre is at infinity and the 'gravity' is constant—in his subsequent article 'De Linea Isochrona, in qua grave sine acceleratione cadit, & de controversia cum Dn. Abbate D.C.', *Acta Eruditorum* (April 1689): 195–8, especially 196. (A second solution by Huygens was published in the *Nouvelles* in October 1687, and a third by Jakob Bernoulli in the *Acta* in May 1690. To Leibniz' generalised problem 'invenire lineam, in qua descendens grave recedat uniformiter a puncto dato, vel ad ipsum accedat' (*Acta* (April 1689): 198) solutions were afterwards returned by Jakob and Johann Bernoulli and by Leibniz himself in the *Acta* in June, October and August 1694 respectively). In present terms, if we set the radial component $-dr/dt = Q'$ of the orbital speed

$$v = \sqrt{\left[V^2+2\int_r^R -f(r)\,.\,dr\right]}$$

to be constant, at once $t = (R-r)/Q'$ is the time of orbit from $V(R, 0)$ to $I(r, \phi)$ in the constrained orbit defined by $\phi = \int_R^r -\sqrt{[v^2-Q'^2]}/Q'r\,.\,dr$. The complementary restriction of setting the component of speed $r\,.\,d\phi/dt = Q''$ instantaneously perpendicular to the radius vector to be constant similarly yields $t = \int_R^r 1/\sqrt{[v^2-Q''^2]}\,.\,dr$ as the time of orbit through the polar angle $\phi = \int_R^r Q''/r\sqrt{[v^2-Q''^2]}\,.\,dr$ in the constraining curve thereby defined.

Proposition XLII, Problem XXIX.

Given the law of centripetal force, there is required the motion of a body setting out[216] *from a given place with a given speed along a given straight line.*

With the content of the three preceding propositions continuing to hold, let a body go off from the place I along the line-element IT with the speed which another in falling under some uniform centripetal force from the place P might acquire at D; and let this uniform force be to the force with which the first body is urged at I as DR to DF. Let the body proceed, however, to k, and then with

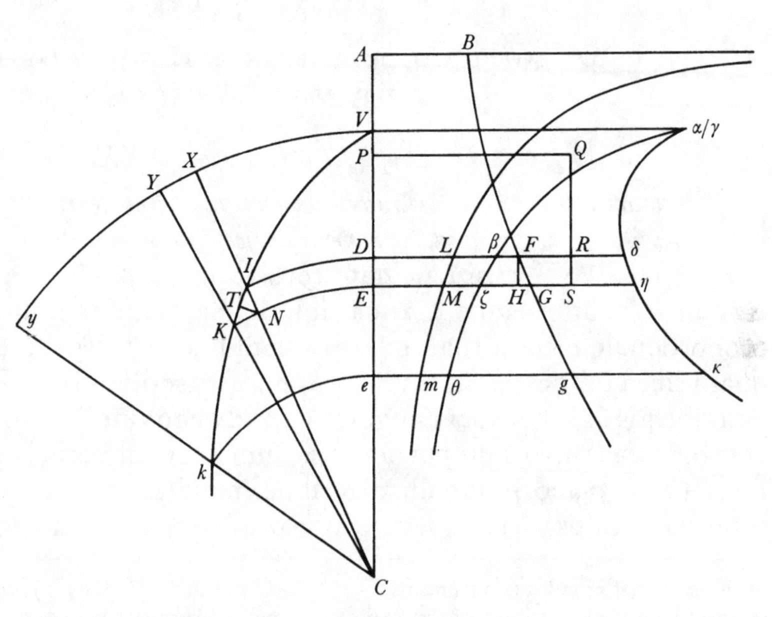

centre C and radius Ck describe the circle-arc ke, meeting the straight line PD in e, and to the curves LMm, $BFGg$, $\alpha\beta\zeta\theta$ and $\gamma\delta\eta\kappa$ erect the ordinates em, eg, $e\theta$, $e\kappa$. From the given rectangle $PDRQ$ and the given law of centripetal force by which the first body is urged on there are, by the construction of Problem XXVII and its Corollary 1, given the curves $BFGg$ and LMm. Next, from the given angle $\widehat{CIT}$ is given the ratio of the nascent elements IK, KN and thence, by the construction of Problem XXVIII, there is given the quantity Q together with the curves $\alpha\beta\zeta\theta$ and $\gamma\delta\eta\kappa$. Hence, after any period of time $(D\beta\theta e)$ is completed, there is given both the body's height, Ce or Ck, and the area $(D\delta\kappa e)$ and the sector (XCy) equal to it, and so the angle $\widehat{XCy}$ and the place k in which the body shall then be situated. As was to be found.[217]

It is, I say, our supposition in these propositions that the centripetal force

(216) 'emissi' (sent off) was first written. In the printed *Principia* ($_1$1687: 131) Newton further altered this to be 'egressi' (started out).

(217) As Newton observes, this all follows straightforwardly enough from Propositions XL and XLI preceding. In principle, we may add, the determination of the ensuing motion is no more difficult when the orbiting body is conceived to be retarded by some assignable resistance ρ acting instantaneously contrary to the direction of motion, for then, in the terms of note

a centro variari secundum legem quamcunꝗ quā quis imaginari potest, in æqualibus autem a centro distantijs esse undiꝗ eandem. Atꝗ hactenus motus corporum in orbibus immobilibus consideravimus. Superest ut de motu eorum in orbibus qui circa centrum virium revolvuntur adjiciamus pauca.

Artic. [IX](218)

De motu corporum in Orbibus mobilibus, deq3 motu Apsidum.

Prop. XLIII. Prob. XXX.

Efficiendum est ut corpus in trajectoria quacunꝗ circa centrū virium revolvente perinde moveri possit, atꝗ corpus aliud in eadem Trajectoria quiescente.

In orbe *VPK*(219) positione dato revolvatur corpus *P* pergendo a *V* versus *K*. A centro *C* agatur semper *Cp* quæ sit ipsi *CP* æqualis angulumꝗ *VCp* angulo *VCP* proportionalem constituat, et area quam linea *Cp* describit erit ad aream *VCP* quam linea *CP* describit ut velocitas lineæ describentis *Cp* ad velocitatem lineæ describentis *CP*[,] hoc est ut angulus *VCp* ad angulum *VCP* adeoꝗ in data ratione, et propterea tempori proportionalis. Cum area tempori proportionalis sit quam linea *Cp* in plano immobili describit[,] manifestum est quod corpus cogente justæ quantitatis vi centripeta revolvi possit una cum puncto *p* in curva illa linea

(198) above, the orbital acceleration is (d^2s/dt^2 or) $v.dv/ds = -f(r).dr/ds+\rho$, where ρ will in general be a function both of the radial distance r (from the centre of the force-field $f(r)$) and of the orbital speed $v = ds/dt$. In his published *Principia* Newton invokes this sophisticated notion of a resisted 'circularis motus' round a finite force-centre only in Section IV of Book 2, where he examines under what law of resistance it is possible, in a general force-field, to traverse the logarithmic 'Spiralis quæ secet radios omnes...in æqualibus angulis'—namely, $\log(R/r) = \phi\cot\alpha$ where the given angle is ($\cos^{-1}(dr/ds) =$) α. Since, as he demonstrates from first principles in his introductory Lemma III (*Principia*, ${}_1$1687: 283), the radius of curvature at the spiral's general point (r, ϕ) is $r\operatorname{cosec}\alpha$, it is immediate that the force $f(r)$ to its pole $(0, 0)$ is v^2/r; whence at once ($v.dv/dr$ or) $\frac{1}{2}d(v^2)/dr = -r^{-1}v^2+\rho\sec\alpha$. In further simplification Newton presumes that $\rho = \delta v^2$, where the 'density' δ at (r, ϕ) varies only as the radial distance r from the pole; accordingly, $\frac{1}{2}v^{-2}.d(v^2)/dr = -r^{-1}+\delta\sec\alpha$. Where $\delta \propto r^{-1}$, he was thereafter in his Proposition XV (*ibid.*: 284–5) able to deduce by a prolix geometrical limit argument that the ensuing equation of motion $v^{-2}.d(v^2)/dr \propto r^{-1}$ is satisfied by the supposition $v^2 \propto r^{-1}$, from which $f(r) = r^{-1}v^2 \propto r^{-2} \propto \delta^2$. But, where more generally $\delta \propto r^{-n}$, he mistakenly asserted in his following Proposition XVI (*ibid.*: 288–9) that $f(r) \propto r^{-(n+1)}$. The error was first noticed by Roger Cotes in the early spring of 1710 when he came to revise the *Principia* for its second edition. In his long letter to Newton on 15 April of that year (ULC. Add. 3983.1; the minimally variant copy retained by Cotes, now in Trinity College, Cambridge. R. 16. 38, was published by Edleston in his *Correspondence of Sir Isaac Newton and Professor Cotes* (London, 1850): 8–12)—otherwise listing a number of minor slips in the text of the preceding Proposition XV—Cotes wrote that 'Prop. 16. must be altered; for by my [accurate!] reckoning, if y^e^ centripetal force [$f(r)$] be as [$r^{-(n+1)}$], the velocity [v] will be as [$r^{-\frac{1}{2}n}$], the Resistance [ρ] as [$(1-\frac{1}{2}n)\cos\alpha.r^{-(n+1)}$], & consequently the Density as [$(1-\frac{1}{2}n)\cos\alpha.r^{-1}$]'. And so it passed into the *Principia's editio secunda* (${}_2$1713: 258), along with Cotes' suggested (second) 'Corollary. Si vis centripeta sit ut [r^{-3}], erit $1-\frac{1}{2}n = 0$, adeoꝗ Resistentia & Densitas Medij nulla erit, ut in

may vary in retreat from its centre following any law whatever imaginable, but that at equal distances from the centre it is everywhere the same. And so far we have considered the motions of bodies in stationary orbits. It remains for us to add a few things regarding their motion in orbits which revolve round the centre of force.

ARTICLE IX.(218)

ON THE MOTION OF BODIES IN MOBILE ORBITS, AND THE MOTION OF THE APSIDES

Proposition XLIII, Problem XXX.

It is required to bring about that a body may move in any trajectory revolving round the centre of forces exactly like another body in the same trajectory at rest.

Let a body P revolve in the given orbit VPK,(219) proceeding from V towards K. From the centre C ever draw Cp equal to CP and forming the angle $\widehat{VCp}$ proportional to $\widehat{VCP}$, and the area which the line Cp describes will be to the area (VCP) which the line CP describes as the speed of the describing line Cp to the speed of the describing line CP, that is, as the angle $\widehat{VCp}$ to $\widehat{VCP}$, and hence in a given ratio and accordingly proportional to the time. Since the area which the line Cp describes in the stationary plane is proportional to the time, it is manifest that a body might, under the pressure of a centripetal force of just the right quantity, revolve along with the point p in the curve which the same point p

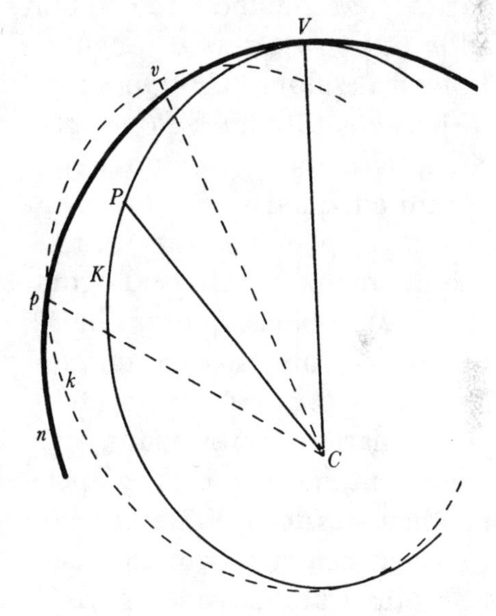

Propositione Nona Libri primi'. Johann Bernoulli independently reached a similar conclusion only a few months afterwards, remarking in a letter of 10 January 1711 (N.S.) (to Varignon?) whose central portion was subsequently printed in the *Mémoires de Mathématique & de Physique ...de l'Academie Royale des Sciences. Année M. DCCXI* (Paris, [$_1$1714→] $_2$1730): 47–54 that, since $v^2 \propto r^{-n}$ is the general solution of the equation of motion $\frac{1}{2}v^{-2}.d(v^2)/dr \propto r^{-1}$, 'la *Prop. 15*.... de M. Newton, souffre une plus grande généralité que ne porte son énoncé', while correspondingly 'sa *Prop. 16*....n'est vraye que dans les cas de $n = 1$ qui est celuy de la *Prop. 15.* qui précéde celle-ci. Car...si l'on suppose avec M. Newton [$\delta = r^{-n}$] & [$\rho = k\delta v^2$], on trouvera suivant mon Analyse, qu'il faudroit [$f(r) \propto r^{-3}e^{2k\sec\alpha . r^{1-n}/(1-n)}$], & non pas [$f(r) \propto r^{-(n+1)}$] comme M. Newton le dit, pour faire ici décrire au mobile une spirale Logarithmique' (*ibid.*: 51–2).

(218) Originally 'IX' and then (compare note (187) above) converted to be 'X' before finally, by a slip of Newton's pen, being changed 'back' to be 'VIII'.

(219) The margin of the manuscript at this point (f. 39r) fails to cite an accompanying

quam punctum idem p ratione jam exposita describit in plano immobili. Fiat angulus VCv angulo PCp et linea Cv lineæ CV atq; figura vCp figuræ VCP æqualis, et corpus in p semper existens movebitur in perimetro figuræ revolventis vCp eodemq; tempore describet arcum ejus vp quo corpus aliud P arcum ipsi similem et æqualem VP in figura quiescente VPK describere potest. Quæratur igitur, per Corollarium Propositionis VI,(220) vis centripeta qua corpus revolvi possit in curva illa linea quam punctum p describit in plano immobili, et solvetur Problema. Q.E.F.(221)

Prop. XLIV. Theor. XIV.

Differentia virium quibus corpus in orbe quiescente et corpus aliud in eodem orbe revolvente æqualiter moveri possunt est in triplicata ratione communis altitudinis inversè.

Partibus orbis quiescentis VP, PK sunto similes et æquales orbis revolventis partes vp, pk. A puncto k in rectam pC demitte perpendiculum kr idemq; produc ad m ut sit mr ad kr ut angulus VCp ad angulum VCP. Quoniam corporum altitudines PC et pC, KC et kC semper æquantur, manifestum est quod si corporum in locis P et p existentium distinguantur motus singuli (per Legum Corol. 2) in binos, quorum hi(222) versus centrum, sive secundum lineas PC, pC, alteri(222) priori perpendiculariter transversum, sive secundum lineas ipsis PC pC perpendiculares determinantur: motus versus centrum erunt æquales, et motus transversus corporis p erit ad motum transversum corporis P ut motus angularis lineæ pC ad motum angularem lineæ PC, id est ut angulus VCp ad angulum VCP. Igitur eodem tempore quo corpus P motu suo utroq; pervenit ad punctum K, corpus p æquali in centrum motu æqualiter movebitur a [p] versus C adeoq; completo illo tempore reperietur alicubi in linea mkr quæ per punctum k in

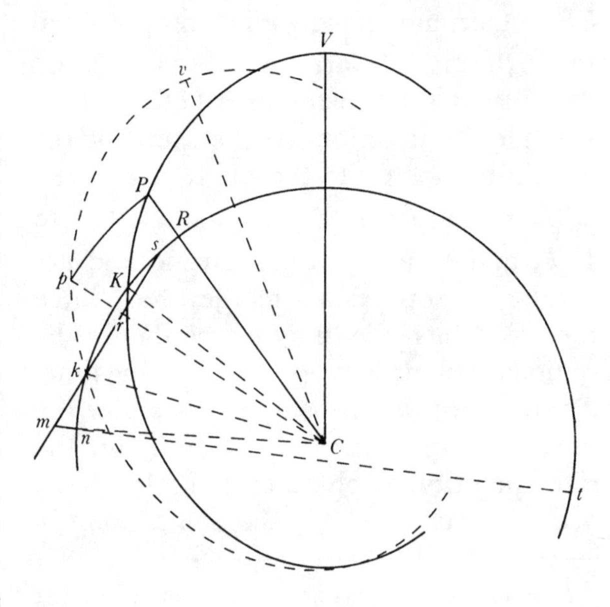

figure, but in the immediately following revise (ULC. Add. 3965.4: 20r) Newton refers to his 'Fig 76' which illustrates the added complexities introduced in Proposition XLIV following, and in the printed *Principia* ($_1$1687: 132) this more complicated configuration was here correspondingly inserted alongside. For simplicity (and to avoid any possible initial confusion) we follow Cotes' lead in his revised *editio secunda* ($_2$1713: 121) and for the moment refer the

describes, in the manner just now disclosed, in the stationary plane. Make the angle $\widehat{VCv}$ equal to $\widehat{PCp}$, the line Cv to CV, and the figure vCp to the figure VCP, and the body, being ever at p, will move in the perimeter of the rotating figure vCp, describing its arc $\widehat{vp}$ in the same time as the other body P can, in the figure VPK at rest, describe the arc $\widehat{VP}$ similar and equal to it. Ascertain therefore, by the Corollary to Proposition VI(220) the centripetal force whereby a body shall be able to revolve in the curve which the point p describes in the stationary plane, and the problem will be solved. As was to be done.(221)

Proposition XLIV, Theorem XIV.

The difference between the forces whereby bodies, one in an orbit at rest, and another in an identical orbit in revolution, are able to move equally is inversely in the tripled ratio of their common height.

Let the parts vp, pk of the revolving orbit be equal and similar to the parts VP, PK of the orbit at rest. From the point k onto the straight line pC drop the perpendicular kr and extend it to m so that mr be to kr as the angle $\widehat{VCp}$ to $\widehat{VCP}$. Because the heights PC and pC, KC and kC of the bodies are ever equal, it is obvious that, if the individual motions of the bodies at their contemporaneous positions P and p be (by Corollary 2 of the Laws) separated into two components, one of these towards the centre, that is, along the lines PC, pC, and the other determined to be perpendicularly cross-wise to the first, that is, along lines perpendicular to PC, pC, then the motions towards the centre will be equal and the transverse motion of the body p will be to the transverse motion of the body P as the angular motion of the line pC to the angular motion of the line PC, that is, as $\widehat{VCp}$ to $\widehat{VCP}$. Consequently, in the same time as the body P by its joint motion arrives at the point K, the body p will by an equal motion to the centre move equally from p towards C and hence at the finish of that time will be located somewhere in the line mkr which is the perpendicular through k to the

---

reader to the simplified version of Newton's figure (in which we have added the compound orbit $\widehat{Vpn}$) reproduced in the English version.

(220) That is, Proposition V of the previous 'De motu Corporum' (§2 preceding).

(221) It follows that, by suitably varying the *vis centripeta* directed to C under which the 'stationary' curve $\widehat{VPK}$ (of general point $P(r, \phi)$ where, as ever, $CP = r$ and $\widehat{VCP} = \phi$) is traversed, the orbiting body can be constrained to move in the 'revolving' orbit $\widehat{Vpn}$ of general point $p(r, k\phi)$, k a given constant. Newton proceeds at once to show that this central 'disturbing' force instantaneously directed through C must in all cases vary as the inverse-cube of the distance.

(222) Originally 'in binos, unum' and 'alterum' respectively, and so we render Newton's intended sense in our English version. His preferred apposition of 'hi' and 'alteri' is harsh and perhaps not immediately comprehensible.

lineam pC perpendicularis est et motu transverso acquiret distantiam a linea pC quæ sit ad distantiam quam corpus alterum acquirit a linea PC ut est hujus motus transversus ad motum transversum alterius. Quare cum kr æqualis sit distantiæ quam corpus alterum acquirit a linea [P]C, sitqꝫ mr ad kr ut angulus VCp ad angulum VCP, hoc est ut motus transversus corporis p ad motum transversum corporis P, manifestum est quod corpus p completo illo tempore reperietur in loco m. Hæc ita se habebunt ubi corpora P et p æqualiter secundum lineas pC et PC moventur adeoqꝫ æqualibus viribus secundum lineas illas urgentur. Capiatur autem angulus pCn ad angulum pCk ut est angulus VCp ad angulum VCP, sitqꝫ nC æqualis kC[,] et corpus p completo illo tempore revera reperietur in n adeoqꝫ vi majore urgetur si modo angulus mCp angulo kCp major est, id est si orbis vpk movetur in consequentia(223) et minore si orbis regreditur, estqꝫ virium differentia ut locorum intervallum mn per quod corpus illud p ipsius actione dato illo temporis spatio transferri debet. Centro C intervallo Cn vel Ck describi intellig[a]tur circulus secans lineas mr mn productas in s et t(224) et erit rectangulum $mn \times mt$ æquale rectangulo $mk \times ms$, adeoqꝫ mn æquale $\frac{mk \times ms}{mt}$. Cum autem triangula pCk, pCn dentur magnitudine,(225) sunt kr et mr earumqꝫ differentia mk et summa ms reciproce ut altitudo pC adeoqꝫ rectangulum $mk \times ms$ est reciproce ut quadratum altitudinis pC. Est [e]t mt directè ut $\frac{1}{2}mt$ id est ut altitudo pC. Hæ sunt primæ rationes linearum nascentium et hinc fit $\frac{mk \times ms}{mt}$ id est lineola nascens mn eiqꝫ proportionalis virium differentia reciproce ut cubus altitudinis pC. Q.E.D.(226)

Corol. 1. Hinc differentia virium in locis P et p vel K et k est ad vim qua corpus motu circulari revolvi potest ab r ad k eodem tempore quo corpus P in orbe immobili describit arcum PK,(227) ut $mk \times ms$ ad rk quadratum, hoc est si capi-

(223) In the direction of orbital motion, that is. Newton adapts and generalises a *terminus technicus* of contemporary descriptive astronomy: 'The Astronomers say, a Planet is in *Antecedence*, when it appears to move contrary to the usual Course or Order of the Signs of the *Zodiack*, as when it moves from *Taurus* towards *Aries*, &c, but if it go from *Aries* to *Taurus*, and thence to *Gemini*, &c, they say it goes *in Consequentia* or *in Consequence*' (John Harris, *Lexicon Technicum: Or, An Universal English Dictionary of Arts and Sciences*, **1** (London, 1704): Article 'ANTECEDENCE IN, or *in Antecedentia*': signature F4^v).

(224) We follow the corrected diagram of Pemberton's *editio ultima* of the *Principia* ($_3$1726: 131) in not making the deviation chord mn pass through the force-centre C, as it does in the two preceding editions ($_1$1687: 132; $_2$1713: 122). Compare §1: note (23).

(225) Since Newton understands that the time of passage over the corresponding arc $\widehat{PK}$ of the stationary orbit is given.

(226) In equivalent analytical terms, if the body moves in time t from $V(R, 0)$ to $P(r, \phi)$ over the arc $\widehat{VP} = s$ of the stationary orbit, so that it is instantaneously at P deviated towards $C(0, 0)$ by the central force $f(r) = -v.dv/dr = -\frac{1}{2}d(v^2)/dr$, where

$$v^2 = (ds/dt)^2 = (Q^2/r^2)((dr/rd\phi)^2+1), \quad Q = r^2 d\phi/dt,$$

line pC, while by its transverse motion it will gain a distance from the line pC which shall be to the distance which the other body gains from the line PC as its transverse motion is to the transverse motion of the other body. Therefore, since kr is equal to the distance which the other body gains from the line PC, and mr is to kr as $V\widehat{C}p$ to $V\widehat{C}P$, that is, as the transverse motion of the body p to the transverse motion of the body P, it is manifest that the body p will at the finish of the time be found at the place m. This holds true when the bodies P and p move equally along the lines pC and PC, and hence are urged by equal forces along those lines. But take the angle $p\widehat{C}n$ to $p\widehat{C}k$ as the angle $V\widehat{C}p$ to $V\widehat{C}P$, and let nC be equal to kC, and then the body p will at the end of the time be found in fact at n, and is therefore urged on by a greater force provided the angle $m\widehat{C}p$ is greater than the angle $k\widehat{C}p$, that is, if the orbit vpk advances,[223] and by a lesser one if that orbit regresses, and the difference of the forces is as the intervening distance mn over which the body p ought by its action to be borne in that given space of time. Understand that with centre C and radius Cn or Ck a circle-arc is described, cutting the lines mr and mn produced in s and t,[224] and the rectangle $mn \times mt$ will be equal to the rectangle $mk \times ms$, and therefore $mn = mk \times ms/mt$. Since, however, the triangles pCk, pCn are given in magnitude,[225] kr and mr and so their difference mk and sum ms are reciprocally as the height pC, and hence the rectangle $mk \times ms$ is reciprocally as the square of the height pC. Also, mt is directly as $\frac{1}{2}mt$, that is, as the altitude pC. These are first ratios of nascent lines, and hence there comes $mk \times ms/mt$, that is, the nascent line-element mn and the difference of the forces proportional to it, to be reciprocally as the cube of the height pC. As was to be proved.[226]

Corollary 1. Hence the difference of the forces at the places P and p or K and k is to the force by which a body can revolve in a circular motion from r to k in the same time as the body P in the stationary orbit describes the arc $\widehat{PK}$[227] as

---

then the related 'revolving orbit' $\widehat{Vp}$ in which $V\widehat{C}p/V\widehat{C}P = k$ and so $Q' = r^2d(k\phi)/dt = kQ$ is traversed under a central force to C of magnitude $f'(r) = -\frac{1}{2}d(v'^2)/dr$ where

$$v'^2 = (Q'^2/r^2)((dr/rd(k\phi))^2+1) = v^2+(Q'^2-Q^2)/r^2,$$

whence at once the 'disturbing' central force which makes the stationary orbit $\widehat{VPK}$ rotate is $f'(r)-f(r) = -\frac{1}{2}d(v'^2-v^2)/dr = (Q'^2-Q^2)/r^3$.

(227) That is, as $mn\,(= mk \times ms/2mC)$ to $qr\,(= kr^2/2qC)$, where q is the meet of Cp with the circle arc $\widehat{Kk}$ and so, in the limit as k, m and therefore q come to coincide with p, $mC = qC$. This lacuna was filled in the *Principia's* second edition ($_2$1713: 123–4) by interpolating in sequel 'ut lineola nascens mn ad sinum versum arcus nascentis [rk], id est ut $\frac{mk \times ms}{mt}$ ad $\frac{rk^q}{2kC}$ vel' (as the nascent line-element mn to the versine of the nascent arc $\widehat{rk}$, that is, as $mk \times ms/mt$ to $rk^2/2kC$, or).

antur datæ quantitates F, G in ea ratione ad invicem quam habet angulus VCP ad angulum VCp, ut G^2-F^2 ad F^2.(228) Et propterea si centro C intervallo quovis CP vel Cp describatur sector circularis æqualis areæ toti VPC quam corpus P tempore quovis in orbe immobili revolvens radio ad centrum ducto descripsit: differentia virium quibus corpus P in orbe immobili et corpus p in orbe mobili revolvuntur, erit ad vim centripetam qua corpus aliquod radio ad centrum ducto sectorem illum eodem tempore quo descripta sit area VPC uniformiter describere potuisset, ut G^2-F^2 ad F^2. Namqꝫ sector ille et area pCk sunt ad invicem ut tempora quibus describuntur.

Corol. 2. Si orbis VPK Ellipsis sit umbilicum habens C et Apsidem summam(229) V; eiqꝫ similis et æqualis ponatur Ellipsis mobilis vpk, ita ut sit semper $p[C]$ æqualis PC et angulus VCp sit ad angulum VCP in data ratione G ad F; pro altitudine autem PC vel $p[C]$ scribatur A et pro Ellipseos latere recto ponatur $2R$: erit vis qua corpus in Ellipsi mobili revolvi potest, ut $\frac{F^2}{A^2}+\frac{RG^2-RF^2}{A^3}$;(230) et contra. Exponatur enim vis qua corpus revolvatur in immota Ellipsi per quantitatem $\frac{F^2}{A^2}$ et vis in V erit $\frac{F^2}{CV^{\text{quad.}}}$. Vis autem qua corpus in circulo ad distantiam CV ea cum velocitate revolvi posset quam corpus in Ellipsi revolvens habet in V, est ad vim qua corpus in Ellipsi revolvens urgetur in Apside V ut dimidium lateris recti Ellipseos ad circuli semidiametrum CV,(231) adeoqꝫ valet $\frac{RF^2}{CV^{\text{cub}}}$, et vis quæ sit ad hanc ut G^2-F^2 ad F^2, valet $\frac{RG^2-RF^2}{CV^{\text{cub}}}$, estqꝫ hæc vis (per hujus Corol. 1) differentia virium quibus corpus P in Ellipsi immota VPK et corpus p in Ellipsi mobili vpk(232) revolvuntur. Unde cum (per hanc Prop.) differentia illa in alia quavis altitudine A sit ad seipsam in altitudine CV ut $\frac{1}{A^{\text{cub}}}$ ad $\frac{1}{CV^{\text{cub}}}$, eadem differentia in omni altitudine A valebit $\frac{RG^2-RF^2}{A^3}$. Igitur ad vim $\frac{F^2}{A^2}$ qua corpus

(228) Notice the numeral indices. In the published *Principia* ($_1$1687: 135) these were, doubtless for puristic editorial reasons of consistency, converted to be '$Gq.-Fq.$' and '$Fq.$' in line with previous 'house style'; in the second edition ($_2$1713: 124) these bastard variants were ameliorated to be '$GG-FF$' and 'FF'. Similar changes were made in the printed editions at equivalent points in the following text.

(229) The 'aphelion' point, that is. In his preceding 'De motu Corporum' Newton called it 'auge'; see §2: note (128).

(230) Here, of course, the indices '2' and '3' govern the quantities A, F and G singly. Below, in contrast, the verbal equivalents 'quad.' and 'cub' are understood to indicate the 'square' and 'cube' powers of the line CV.

(231) By Corollary 3 to Proposition XV of the preceding 'De motu Corporum' [= Proposition XVI of the present 'De motu Corporum Liber primus'].

$mk \times ms$ to rk squared, that is, if the given quantities F and G be taken to one another in the ratio had by the angle $\widehat{VCP}$ to $\widehat{VCp}$, as G^2-F^2 to F^2. In consequence, if with centre C and any radius CP or Cp a circle-sector be described equal to the total area (VCP) which the body P revolving in the stationary orbit has described in any time by a radius drawn to the centre, then the difference of the forces whereby the body P revolves in the stationary orbit and the body p in the mobile orbit will be to the centripetal force whereby some body would, by a radius drawn to the centre, have been able uniformly to describe that sector in the same time as the area (VCP) was described, in the ratio of G^2-F^2 to F^2. For that sector and the area (pCk) are to one another as the times in which they are described.

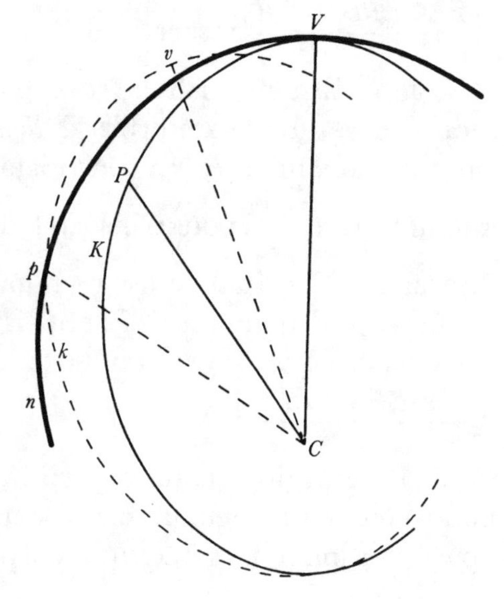

Corollary 2. If the orbit VPK be an ellipse having C for a focus and its top apsis[229] V, let the mobile ellipse vpk be supposed similar and equal to it such that always pC shall be equal to PC and the angle $\widehat{VCp}$ to $\widehat{VCP}$ shall be in the given ratio G to F, while in place of the height PC or pC there be written A and for the ellipse's *latus rectum* there be put $2R$, then the force whereby a body can revolve in the mobile ellipse will be as

$$F^2/A^2+R(G^2-F^2)/A^3;$$

and conversely so. For denote the force whereby a body may revolve in the stationary ellipse by the quantity F^2/A^2 and the force at V will be F^2/CV^2. But the force whereby a body might revolve in a circle at the distance CV with the same speed as the body revolving in the ellipse has at V is to the force whereby the body revolving in the ellipse is urged on at the apsis V as half the ellipse's *latus rectum* to the circle's semidiameter CV,[231] and hence its value is RF^2/CV^3, and so the value of the force which is to be to this as G^2-F^2 to F^2 is $R(G^2-F^2)/CV^3$; and this force is (by the present Corollary 1) the difference of the forces whereby the body P revolves in the stationary ellipse VPK and the body p in the mobile ellipse vpk.[232] Whence, since (by the present Proposition) that difference at any other height A is to its quantity at the height CV as $1/A^3$ to $1/CV^3$, the value of this same difference at every height A will be $R(G^2-F^2)/A^3$. Consequently, to the force

(232) Originally 'in loco puncti p' (in the locus of the point p).

revolvi potest in Ellipsi immobili VPK, addatur excessus $\frac{RG^2-RF^2}{A^3}$ et componetur vis tota $\frac{F^2}{A^2}+\frac{RG^2-RF^2}{A^3}$ qua corpus in Ellipsi mobili vpk ijsdem temporibus revolvi possit.(233)

Corol. 3. Ad eundem modum colligetur quod si orbis immobilis VPK Ellipsis sit centrum habens in virium centro C, eiq; similis, æqualis et concentrica ponatur Ellipsis mobilis vpk, sitq; $2R$ Ellipseos hujus latus rectum et $2T$ latus transversum atq; angulus VCp semper sit ad angulum VCP ut G ad F: vires quibus corpora in Ellipsi immobili et mobili temporibus æqualibus revolvi possunt erunt ut $\frac{F^2A}{T^3}$, et $\frac{F^2A}{T^3}+\frac{RG^2-RF^2}{A^3}$ respectivè.(234)

Corol. 4. Et universaliter, si corporis altitudo maxima CV nominetur T, et radius curvaturæ quam orbis VPK habet in V[,] id est radius circuli æqualiter curvi nominetur R, et vis centripeta qua corpus in Trajectoria quacunq; immobili VPK revolvi potest, in loco V dicatur $\frac{F^2}{T^2}V$(235) atq; alijs in locis P indefinite dicatur X,(236) altitudine CP nominata A; et capiatur G ad F in data ratione anguli VCp ad angulum VCP: erit vis centripeta qua corpus idem eosdem motus in eadem trajectoria vpk circulariter mota temporibus ijsdem peragere potest ut $X+\frac{VRG^2-VRF^2}{A^3}$.(237)

Corol. 5. Dato igitur motu corporis in orbe quocunq; immobili, augeri vel minui potest ejus motus angularis circa centrum virium in ratione data et inde inveniri novi orbes immobiles in quibus corpora novis viribus centripetis gyrentur.

---

(233) Alternatively, in direct corollary to note (226), since the central force under which the stationary ellipse traced by $P(r, \phi)$ is traversed round its focus $C(0, 0)$ is, by Proposition X of the preceding 'De motu Corporum' [= Proposition XI of the present 'Liber primus'], $f(r) = (Q^2/R)r^2$ where $Q = r^2 d\phi/dt$ and $2R$ is the ellipse's *latus rectum*, the force under which the revolving orbit traced by $p(r, k\phi)$, $k = \widehat{VCp}/\widehat{VCP} = Q'/Q = G/F$, is traversed is

$$f'(r) = (Q^2/R)/r^2+(Q'^2-Q^2)/r^3 \propto F^2/r^2+R(G^2-F^2)/r^3.$$

(234) Here, much as in the previous note, since the central force under which the stationary ellipse traced by $P(r, \phi)$ is traversed round its centre $C(0, 0)$ is, by Proposition IX of the preceding 'De motu Corporum' [= Proposition X of the present 'Liber primus'],

$$f(r) = (Q^2/T^2U^2)r$$

where $Q = r^2 d\phi/dt$ and $2T$, $2U(= 2\sqrt{[RT]}$, $2R$ the *latus rectum*) are the ellipse's major and minor axes, the force under which the revolving orbit traced by $p(r, k\phi)$, $k = Q'/Q = G/F$, is traversed is $f'(r) = (Q^2/RT^3)r+(Q'^2-Q^2)/r^3 \propto (F^2/T^3)r+R(G^2-F^2)/r^3$.

(235) Read '$\frac{F^2}{T^2}$' simply. Newton assumes, of course, that a general stationary curve VPK

F^2/A^2 by which a body can revolve in the stationary ellipse VPK add the excess $R(G^2-F^2)/A^3$ and the compound will be the total force $F^2/A^2+R(G^2-F^2)/A^3$ whereby a body might revolve in the mobile ellipse vpk in identical times.(233)

Corollary 3. In much the same way it will be gathered that, if the stationary orbit VPK be an ellipse having its centre at the force-centre C, and the mobile ellipse vpk be supposed similar, equal and concentric to it, while the *latus rectum* of this ellipse be $2R$ and its main axis $2T$, and the angle $V\widehat{C}p$ ever be to $V\widehat{C}P$ as G to F, then the forces whereby bodies can revolve in equal times in the stationary ellipse and the mobile one will be as F^2A/T^3 and

$$F^2A/T^3+R(G^2-F^2)/A^3$$

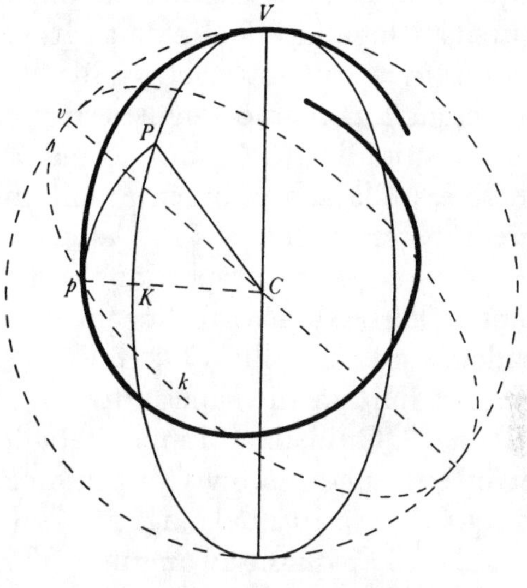

respectively.(234)

Corollary 4. And universally, if the maximum height CV be named T and the radius of curvature which the orbit VPK has at V—the radius of a circle equally curved, that is—be named R, and the centripetal force whereby a body can revolve in any stationary trajectory VPK whatever at the place V be called $(F^2/T^2)V$(235) and at any other places P be indefinitely called X,(236) with the height CP named A take G to F in the given ratio of the angle $V\widehat{C}p$ to $V\widehat{C}P$, and the centripetal force whereby the same body can accomplish the same motions in the same times in the same trajectory vpk now rotating circle-wise will be as $X+VR(G^2-F^2)/A^3$.(237)

Corollary 5. Given, then, the motion of a body in any stationary orbit, its angular motion round the centre of force can be increased or diminished in a given ratio and thereby new stationary orbits can be found in which bodies shall gyrate under new centripetal forces.

---

may be approximated in the immediate vicinity of the apsis V by a conic of focus C and having the same radius of curvature R (half its *latus rectum*) there; whence the central force at V is $(Q^2/R)/T^2 = F^2/T^2$. The less cynical may perhaps be surprised that this subtle slip is perpetuated without correction in all editions of the published *Principia*.

(236) Understand that $X \equiv (F^2/T^2).f(A/T)$. In Corollaries 2 and 3 above there is

$$X \equiv (F^2/T^2).(A/T)^{-2} \quad \text{and} \quad X \equiv (F^2/T^2).(A/T)$$

respectively.

(237) In line with note (235) this should be simply '$X+\dfrac{RG^2-RF^2}{A^3}$'.

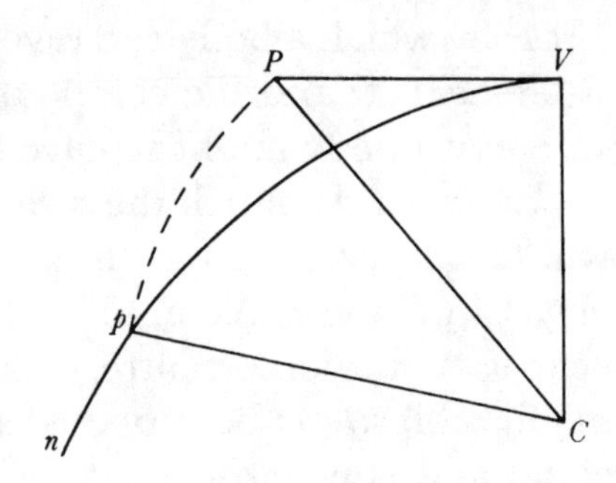

Corol. 6. Igitur(238) si ad rectam CV positione datam erigatur perpendiculum VP(239) longitudinis indeterminatæ, jungaturq; CP et ipsi æqualis agatur Cp constituens angulum VCp qui sit ad angulum VCP in data ratione: vis qua corpus gyrari potest in curva illa $Vp[n]$ quam punctum p perpetuò tangit, erit reciprocè ut cubus altitudinis Cp. Nam corpus P per vim inertiæ, nulla alia vi urgente uniformiter progredi potest in recta VP. Addatur vis in centrum C cubo altitudinis CP vel Cp reciprocè proportionalis et (per jam demonstrata) detorquebitur motus ille rectilineus in lineam curvam $Vp[n]$. Est autem hæc curva $Vp[n]$ eadem cum curva illa VPQ in Corol. 3 Prop. XLI inventa in qua ibi diximus corpora hujusmodi viribus attracta, obliquè ascendere.(240)

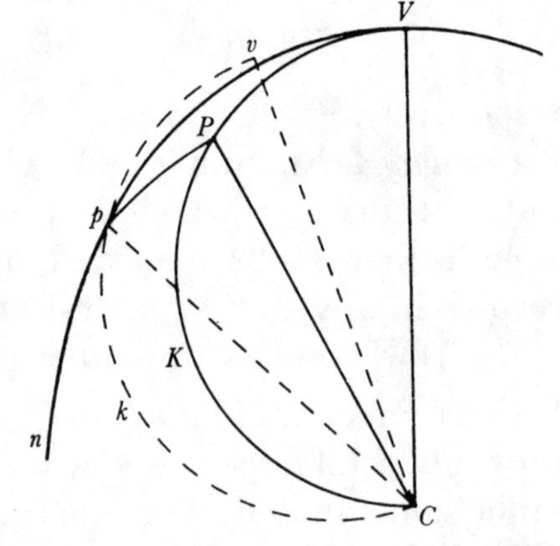

Corol. 7. Quinetiam si super data diametro VC describatur semicirculus VPC et in eo capiatur ubivis punctum P, jungaturq; CP et ipsi æqualis agatur Cp continens angulum VCp qui sit ad angulum VCP in data ratione: vis qua corpus gyrari potest in curva $Vp[n]$ quam punctum p perpetuò tangit, erit reciprocè ut cubus altitudinis CP vel Cp. Nam per Prop. [VII] vis qua corpus revolvi potest in semicirculi perimetro VPC, est reciprocè ut cubus(241) altitudinis illius CP et vis nova qua motus corporis de semicirculo in curvam lineam $Vp[n]$ transferri potest est etiam reciprocè ut ejusdem altitudinis cubus[,] et vis ex vi utraq; per additionem vel subductionem conflata (hoc est vis tota qua corpus in curva $Vp[n]$ gyrari potest) est reciprocè ut ille idem altitudinis cubus.(242)

Prop. XLV. Prob. XXXI.

Orbium qui sunt circulis maxime finitimi requiruntur motus Apsidum.

Problema solvitur Arithmeticè faciendo ut orbis quem corpus in Ellipsi mobili

(238) Originally 'Et particulariter' (And in particular).

(239) In the manuscript Newton has deleted a superfluous intervening phrase 'et in eo capiatur VP' (and in it be taken VP).

(240) This last sentence is a late insertion by Newton evidently made when (see note (212)) he added Corollary 3 to his above Proposition XLI as a replacement for the following erroneous 'Corol. 7'. If, in the analytical terms of note (226), we suppose that $P(r, \phi)$ traces the stationary line $r = T\sec\phi$ from the apsis $V(T, 0)$, it is readily shown that the radial acceleration at P to

Corollary 6. In consequence,(238) if to the straight line CV given in position there be erected the perpendicular VP(239) of indeterminate length, join CP and equal to it draw Cp forming an angle $\widehat{VCp}$ which shall be to the angle $\widehat{VCP}$ in a given ratio: the force whereby a body can orbit in the curve Vpn which the point p perpetually traces will be reciprocally as the cube of the height Cp. For the body P by its own force of inertia can, without any other force urging it on, proceed uniformly in the straight line VP. Add the force to the centre C reciprocally proportional to the cube of the height CP or Cp, and (by what has just now been proved) that rectilinear motion will be distorted into the curve Vpn. This curve Vpn is, however, the same as the curve VPQ found in Corollary 3 to Proposition XLI, wherein, we there asserted, bodies attracted by this class of force obliquely ascend.(240)

Corollary 7. Indeed, if on the given diameter VC a semicircle VPC be described and in it anywhere there be taken the point P, join CP and equal to it draw Cp containing an angle $\widehat{VCp}$ which shall be to $\widehat{VCP}$ in a given ratio: the force whereby a body can orbit in the curve Vpn perpetually traced out by the point p will then be reciprocally as the cube of the height CP or Cp. For by Proposition VII the force by which a body can revolve in the semicircle's perimeter VPC is reciprocally as the cube(241) of that height CP, and so the new force whereby the motion of the body can be transferred from the semicircle to the curve Vpn is also reciprocally as the cube of the same height, and so the force which is the conflation of both by addition or subtraction (that is, the total force whereby a body can orbit in the curve Vpn) is reciprocally as the same cube of the altitude.(242)

Proposition XLV, Problem XXXI.

In orbits which are exceedingly close to circles the motions of the apsides are required.

The problem is solved arithmetically by making the orbit described by the

---

the centre $C(0, 0)$ is $f(r) = d^2r/dt^2 = Q^2/r^3$, $Q = r^2d\phi/dt$; whence the 'revolving' orbit—the secant spiral $r = T\sec k\phi$—traced by $p(r, k\phi)$, $k = Q'/Q$, is traversed under the central force $f'(r) = Q^2/r^3 + (Q'^2 - Q^2)/r^3 = Q'^2/r^3$, $Q' = r^2d(k\phi)/dt$.

(241) Read 'quadrato-cubus' (fifth power)! Newton momentarily seems to confuse Proposition VII [= Proposition VI of the previous 'De motu Corporum'] with the following Proposition VIII [= Proposition VII]. According to the former (compare §2: note (94)) the central force at $P(r, \phi)$ directed to $C(0, 0)$ in the circular orbit $r = T\cos\phi$ from the apsis $V(T, 0)$ to C is $f(r) = 2Q^2T^2/r^5$, $Q = r^2d\phi/dt$.

(242) On correcting Newton's earlier slip (see the previous note), the central force to C in the 'revolving' orbit traced by $p(r, k\phi)$ is in fact $f'(r) = 2Q^2T^2/r^5 + (Q'^2 - Q^2)/r^3$, $Q'/Q = k$ as before, so that the whole corollary is justly cancelled. We have inserted a suitable diagram in the Latin text to fill the place of the unnumbered 'Fig.' which Newton scheduled in the margin of the manuscript alongside to illustrate the verbal text.

ut in Propositionis superioris Corol. 2 vel 3(243) revolvens describit in plano immobili accedat ad formam orbis cujus Apsides requiruntur, et quærendo Apsides orbis quem corpus illud in plano immobili describit. Orbes autem eandem acquirant formam si vires centripetæ quibus describuntur inter se collatæ, in æqualibus(244) altitudinibus reddantur proportionales.(244) Sit punctum V Apsis summa et scribantur T pro altitudine maxima CV, A pro altitudine quavis alia CP vel Cp, et X pro altitudinum differentia $CV-CP$[,] et vis qua corpus in Ellipsi circa umbilicū ejus C (ut in Corollario 2) revolvente movetur, quæꝗ in Corollario 2(245) erat ut $\frac{F^2}{A^2}+\frac{RG^2-RF^2}{A^3}$ id est ut $\frac{F^2A+RG^2-RF^2}{A^3}$, scribendo $T-X$ pro A, erit ut $\frac{RG^2-RF^2+TF^2-F^2X}{A^3}$. Reducenda similiter est vis alia quævis centripeta ad fractionem cujus denominator sit A^3, & numeratores facta homologorum terminorum collatione statuendi sunt analogi.(246) Res exemplis patebit.

Exempl. 1. Ponamus vim centripetam uniformem esse adeoꝗ ut $\frac{A^3}{A^3}$ sive (scribendo $T-X$ pro A in numeratore) ut $\frac{T^3-3T^2X+3TX^2-X^3}{A^3}$; et collatis numeratorum terminis correspondentibus, nimirum datis cum datis et non datis(247) cum non datis, fiet $RG^2-RF^2+TF^2$ ad T^3 ut

$$-F^2X \text{ ad } \overline{-3T^2+3TX-X^2} \text{ in } X,$$

sive ut $-F^2$ ad $-3T^2+3TX-X^2$. Jam cum orbis ponatur circulo quam maximè finitimus coeat orbis cum circulo et ob factas R, T æquales atꝗ X in infinitum diminutam, rationes ultimæ erunt RG^2 ad T^3 ut $-F^2$ ad $-3T^2$ seu G^2 ad T^2 ut F^2 ad $3T^2$ et vicissim G^2 ad F^2 ut T^2 ad $3T^2$ id est ut 1 ad 3, adeoꝗ G ad F hoc est angulus VCp ad angulum VCP ut 1 ad $\sqrt{3}$. Ergo cum corpus in Ellipsi immobili ab Apside summa ad Apsidem imam descendendo conficiat angulum VCP (ut ita dicam) graduum 180, corpus aliud in Ellipsi mobili atꝗ adeo in orbe immobili de quo agimus descendendo ab Apside summa ad Apsidem imam descendendo conficiet angulum VCp graduum $\frac{180}{\sqrt{3}}$: id adeo ob

(243) In these cases, where the stationary orbit VPK is an ellipse of focus C and one of centre C respectively, it follows that when these are indefinitely near to being circular (with their eccentricities become vanishingly small) they effectively coincide, thus generating similar 'revolving' curves Vpn.

(244) 'paribus' and 'pares' (equivalent) were first written.

(245) Since the difference $T-A(=X)$ is taken as indefinitely small, an identical derivation could be made in sequel from Corollary 3 to Proposition XLIV, whereby the 'vis in Ellipsi circa centrum ejus C' is found to be $F^2A/T^3+R(G^2-F^2)/A^3$; compare note (243).

(246) Originally 'ad eandem formam reducendi sunt' (are...to be reduced to the same form).

body revolving in a mobile ellipse, as in Corollary 2 or 3[243] of the previous proposition, approach in the stationary plane the shape of the orbit whose apsides are required, and seeking the apsides of the orbit which that body describes in the stationary plane. Orbits acquire the same shape, however, if the centripetal forces whereby they are described are, when compared one with another, made proportional[244] at equal[244] heights. Let the point V be the top apsis, and write T for the maximum height CV, A for any other height CP or Cp, and X for the difference $CV-CP$ of these heights: the force whereby a body moves in an ellipse revolving (as in Corollary 2) round its focus C, and which in Corollary 2[245] was as $F^2/A^2+R(G^2-F^2)/A^3$, that is, as $(F^2A+RG^2-RF^2)/A^3$, will, on writing $T-X$ in A's place, then be as

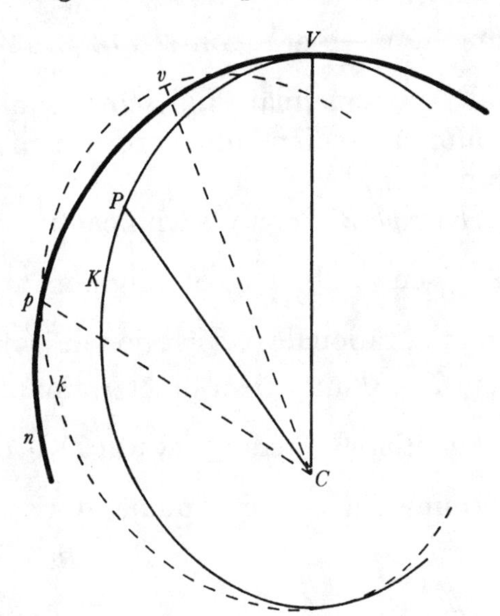

$$(RG^2-RF^2+TF^2-F^2X)/A^3.$$

Any other centripetal force is similarly to be reduced to a fraction whose denominator shall be A^3, and then the numerators are, on making comparison between corresponding terms, to be set in proportion.[246] The procedure will be clear from examples.

Example 1. Let us suppose that the centripetal force is uniform and therefore as A^3/A^3, that is, (on writing $T-X$ in A's place in the numerator) as

$$(T^3-3T^2X+3TX^2-X^3)/A^3;$$

and when homologous terms in the numerators are compared—namely, given ones with given ones, and ones not given[247] with ones not given—there will come to be $RG^2-RF^2+TF^2$ to T^3 as $-F^2X$ to $(-3T^2+3TX-X^2)X$, or as $-F^2$ to $-3T^3+3TX-X^2$. Now, since the orbit is supposed to be indefinitely close to a circle, let the orbit coincide with a circle and then, because R and T become equal and X is indefinitely diminished, the last ratios will be RG^2 to T^3 as $-F^2$ to $-3T^2$, that is, G^2 to T^2 as F^2 to $3T^2$ and *alternando* G^2 to F^2 as T^2 to $3T^2$, that is, as 1 to 3; hence G to F, that is, the angle $\widehat{VCp}$ to the angle $\widehat{VCP}$ will be as 1 to $\sqrt{3}$. Therefore, since a body in descending in the stationary ellipse from the top apsis to the bottom one completes (if you allow the expression) an angle $\widehat{VCP}$ of 180°, the other body in descending in the mobile ellipse, and hence in the stationary orbit which is our concern, from the top apsis to the bottom one will complete an angle $\widehat{VCp}$ of $180°/\sqrt{3}$: this, of course, because of the similarity

(247) Those multiplied by X, that is.

similitudinem orbis hujus quem corpus agente uniformi vi centripeta describit et orbis illius quem corpus in Ellipsi revolvente gyros peragens describit in plano quiescente. Per superiorem terminorum collationem similes redduntur hi orbes, non universaliter sed tunc cum ad formam circularem quam maximè appropinquant. Corpus igitur uniformi cum vi centripeta in orbe propemodum circulari revolvens, inter Apsidem summam et Apsidem imam conficiet semper angulum $\frac{180}{\surd 3}$ graduum seu $103^{\text{gr}}.55'$ ad centrum, perveniens ab Apside summa ad Apsidem imam ubi semel confecit hunc angulum, et inde ad Apsidem summam rediens ubi iterum confecit eundem angulum, et sic deinceps in infinitum.(248)

Exempl. 2. Ponamus vim centripetam esse ut altitudinis A dignitas quælibet A^{n-3} seu $\frac{A^n}{A^3}$ ubi $n-3$ et n significant dignitatum indices quascunq; integras vel fractas, rationales vel irrationales, affirmativas vel negativas. Numerator ille A^n seu $\overline{T-X}^n$ in seriem indeterminatam per methodum nostram serierum convergentium(249) reducta evadit $T^n - nXT^{n-1} + \frac{nn-n}{2}X^2T^{n-2}$ &c[,] et collatis hujus terminis cum terminis numeratoris alterius $RG^2-RF^2+TF^2-F^2X$ fit

$$RG^2-RF^2+TF^2$$

ad T^n ut $-F^2$ ad $-nT^{n-1}+\frac{nn-n}{2}XT^{n-2}$ &c[,] et sumendo rationes ultimas ubi orbes ad formam circularem accedunt RG^2 ad T^n ut $-F^2$ ad $-nT^{n-1}$ seu(250) G^2 ad T^{n-1} ut F^2 ad nT^{n-1} et vicissim G^2 ad F^2 ut T^{n-1} ad nT^{n-1} id est ut 1 ad n adeoq; G ad F id est angulus VCp ad angulum VCP ut 1 ad $\surd n$. Quare cum

(248) This value of ($\pi/\surd 3$ radians $=$) $180°/\surd 3$ for the apsidal angle ($\widehat{VC\pi} = \pi\widehat{CW}$ in the accompanying figure)

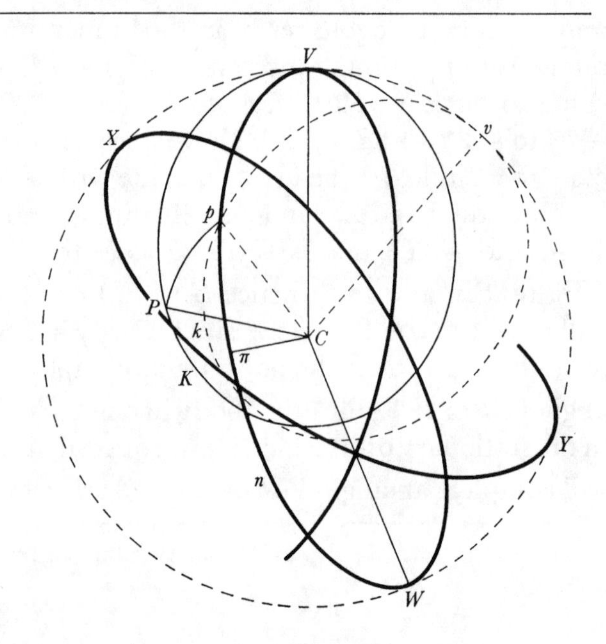

of the latter orbit described by a body under the action of a uniform centripetal force to the former orbit described by a body executing revolutions in an ellipse rotating in the stationary plane. By the preceding comparison of terms these orbits are rendered similar, not universally but on the occasion when they approximate exceedingly closely to the shape of a circle. A body, therefore, revolving with a uniform centripetal force in an orbit very nearly circular will, between its top apsis and its (next) bottom one always complete an angle of $180°/\sqrt{3}$, that is, $103° 55'$, at the centre, arriving from the top apsis at the bottom one when it has completed this angle once, and thence returning to the top apsis after it has completed the same angle a second time, and so on in turn *ad infinitum*.(248)

Example 2. Let us suppose that the centripetal force is as any power A^{n-3}, that is, A^n/A^3, of the height A, where $n-3$ and n denote any power-indices integral or fractional, rational or irrational, positive or negative. Once the numerator $A^n = (T-X)^n$ is reduced to an indeterminate series by our method of converging series(249) it proves to be $T^n - nXT^{n-1} + \frac{1}{2}(n^2-n)X^2T^{n-2}$..., and when the terms of this are compared with the other numerator's terms

$$RG^2 - RF^2 + TF^2 - F^2X$$

there comes to be $RG^2 - RF^2 + TF^2$ to T^n as $-F^2$ to $-nT^{n-1} + \frac{1}{2}(n^2-n)XT^{n-2}$... and so, on taking the last ratios when the orbits approach a circular form, RG^2 to T^n as $-F^2$ to $-nT^{n-1}$ or(250) G^2 to T^{n-1} as F^2 to nT^{n-1}, and *alternando* G^2 to F^2 as T^{n-1} to nT^{n-1}, that is, as 1 to n; hence G to F, that is, the angle $\widehat{VCp}$ to the

---

is in fact the maximum obtainable in the constant-force orbit, as we have shown in §2: note (127): in cases where the orbit does not approximate to a circle, it will vary between the limits 90° and $180°/\sqrt{3}$. Some five years earlier in December 1679 when Newton had sketched this Borellian curve for Hooke he had (see *ibid.*) made the apsidal angle to be very nearly 120°, while in the scholium following Proposition XII of his immediately preceding 'De motu Corporum' he still conjectured its value to be 'about 110°'. Of course, where the constant-force orbit departs considerably from being circular, it will no longer be accurately modelled by Newton's present 'revolving' orbit Vpn, which is (in the analytical terms of notes (226) and (233) above) traversed under the general central force

$$(3Q'^2/R)/r^2 - 2Q'^2/r^3,$$

where $Cp = r$ and R is the *latus rectum* of the stationary ellipse VPK, while $Q' = Q/\sqrt{3}$ is the Keplerian constant $r^2d(\phi/\sqrt{3})/dt$ for the 'revolving' orbit traced by $p(r, \phi/\sqrt{3})$.

(249) By simple binomial expansion, that is, with the ratio X/T understood to be vanishingly small.

(250) Since, with the 'revolving' orbit indefinitely approaching a circle of centre C through V, the stationary (focal) ellipse also comes to coincide with the same circle, and hence its *latus rectum* $2R$ becomes indistinguishable in length from its main diameter $2T$.

angulus VCP in descensu corporis ab Apside summa ad Apsidem imam in Ellipsi confectus sit graduum 180[,] conficietur angulus VCp in descensu corporis ab Apside summa ad Apsidem imam in orbe propemodum circulari quem corpus quodvis vi centripeta dignitati A^{n-3} proportionali describit æqualis angulo graduum $\frac{180}{\sqrt{n}}$[,] et hoc angulo repetito corpus redibit ab Apside ima ad Apsidem summam et sic deinceps in infinitum. Ut si vis centripeta sit ut distantia corporis a centro[,] id est ut A seu $\frac{A^4}{A^3}$[,] erit n æqualis 4 et $\sqrt{n}$ æqualis 2, adeoq angulus inter Apsidem summam et Apsidem imam æqualis $\frac{180^{gr}}{2}$ seu 90^{gr}. Completa igitur quarta parte revolutionis unius corpus perveniet ad Apsidem imam et completa alia quarta parte ad Apsidem summam et sic deinceps per vices in infinitum. Id quod etiam ex Propositione X manifestum est. Nam corpus urgente hac vi centripeta revolvetur in Ellipsi immobili cujus centrum est in centro virium. Quod si vis centripeta sit reciprocè ut distantia, id est directè ut $\frac{1}{A}$ seu $\frac{A^2}{A^3}$[,] erit n æqualis 2, adeoq inter Apsidem summam et Apsidem imam angulus erit graduum $\frac{180}{\sqrt{2}}$, seu $127^{gr}.17'$, et propterea corpus tali vi revolvens perpetua anguli hujus repetitione, vicibus alternis ab Apside summa ad imam et ab ima ad summam perveniet in æternum. Porrò si vis centripeta sit reciproce ut latus quadratoquadratum undecimæ dignitatis Altitudinis, id est reciprocè ut $A^{\frac{11}{4}}$ adeoq directe ut $\frac{1}{A^{\frac{11}{4}}}$ seu $\frac{A^{\frac{1}{4}}}{A^3}$, erit n æqualis $\frac{1}{4}$ et $\frac{180^{grad.}}{\sqrt{n}}$ æqualis $360^{gr.}$, et propterea corpus de Apside summa discedens et subinde perpetuo descendens perveniet ad Apsidem imam ubi complevit revolutionem integram, dein perpetuo ascensu complendo aliam revolutionem integram redibit ad Apsidem summam et sic per vices in infinitum.(251)

Exempl. 3. Assumentes m et n pro quibusvis indicibus dignitatum altitudinis, et b, c pro numeris quibusvis datis ponamus vim centripetam esse ut $\frac{bA^m+cA^n}{A^3}$, id est ut $\frac{b \text{ in } \overline{T-X}^m + c \text{ in } \overline{T-X}^n}{A^3}$ seu (per methodum nostram serierum convergentium)(249) ut

$$\frac{bT^m - mbXT^{m-1} + \frac{mm-m}{2}bX^2T^{m-2}\&c + cT^n - nXT^{n-1} + \frac{nn-n}{2}cX^2T^{n-2}\&c}{A^3},$$

(251) Compare Newton's earlier scholium to Proposition XII in the preceding 'De motu

angle $\widehat{VCP}$ will be as 1 to $\surd n$. Consequently, since the angle $\widehat{VCP}$ completed in a body's descent from the top apsis in the ellipse to the bottom one is 180°, the angle $\widehat{VCp}$ completed in the descent of a body from the top apsis to the (next) bottom one in the very nearly circular orbit described by any body under a centripetal force proportional to the power A^{n-3} will be equal to $180°/\surd n$; and with this angle repeated the body will return from the bottom apsis to the top one, and so on in turn *ad infinitum*. If, for instance, the centripetal force be as the distance of the body from the centre, that is, as $A = A^4/A^3$, then will n be equal to 4 and $\surd n$ equal to 2, and hence the angle between the top apsis and the bottom apsis will equal $180°/2 = 90°$. Having completed, therefore, a quarter of one revolution the body will arrive at the bottom apsis, and with another quarter completed it will reach the top apsis, and so on indefinitely turn by turn: as is also obvious from Proposition X, for a body under the urge of this centripetal force will revolve in a stationary ellipse whose centre is at the centre of force. But should the centripetal force be reciprocally as the distance, that is, directly as $1/A = A^2/A^3$, then n will equal 2, and hence the angle between the top apsis and the bottom one will be $180°/\surd 2 = 127° 17'$; accordingly, a body revolving with such a force will, by perpetual repetition of this angle, arrive alternatively from the top apsis at the bottom one, and then from the bottom apsis at the top one, for ever. Furthermore, if the centripetal force be reciprocally as the fourth root of the eleventh power of the height, that is, reciprocally as $A^{\frac{11}{4}}$ and hence directly as $1/A^{\frac{11}{4}} = A^{\frac{1}{4}}/A^3$, n will equal $\frac{1}{4}$ and so $180°/\surd n = 360°$; accordingly, a body departing from the top apsis and perpetually descending thereafter will arrive at the bottom apsis when it has completed a whole revolution, and then on completing another whole revolution in perpetual ascent will arrive at the top apsis, and so on by turns indefinitely.(251)

Example 3. Assuming m and n to be any indices of powers of the height, and b and c to be any given numbers, let us set the centripetal force to be as $\dfrac{bA^m + cA^n}{A^3} = \dfrac{b(T-X)^m + c(T-X)^n}{A^3}$, that is, (by our method of converging series)(249) as

$$\frac{b(T^m - mXT^{m-1} + \frac{1}{2}(m^2-m)X^2T^{m-2}\ldots) + c(T^n - nXT^{n-1} + \frac{1}{2}(n^2-n)X^2T^{n-2}\ldots)}{A^3},$$

Corporum', here (see §2: note (126)) suppressed because of its numerical inaccuracies. Again, of course, where the orbit traversed in the force-field, centred on C, varying as the $(n-3)$-th power of the distance differs appreciably from a circle, it will no longer be adequately approximated by the Newtonian elliptical 'revolving' orbit traced by $p(r, \phi/\surd n)$ under a central attractive pull to C of magnitude $(nQ'^2/R)/r^2 - (n-1)Q'^2/r^3$, $Q' = r^2 d(\phi/\surd n)/dt = Q/\surd n$.

et collatis numeratorum terminis fiet $RG^2 - RF^2 + TF^2$ ad $bT^m + cT^n$ ut $-F^2$ ad

$$-mbT^{m-1} - ncT^{n-1} + \frac{mm-m}{2}XT^{m-2} + \frac{nn-n}{2}XT^{n-2} \text{ \&c}$$

et sumendo rationes ultimas quæ prodeunt ubi orbes ad formam circularem accedunt fit G^2 ad $bT^{m-1} + cT^{n-1}$ ut F^2 ad $mbT^{m-1} + ncT^{n-1}$ et vicissim G^2 ad F^2 ut $bT^{m-1} + cT^{n-1}$ ad $mbT^{m-1} + ncT^{n-1}$. Quæ proportio exponendo altitudinem maximam CV seu T arithmeticè per unitatem fit G^2 ad F^2 ut $b+c$ ad $mb+nc$ adeoq; ut 1 ad $\frac{mb+nc}{b+c}$. Unde est G ad F id est angulus VCp ad angulum VCP ut 1 ad $\sqrt{\frac{mb+nc}{b+c}}$. Et propterea cum angulus VCP inter Apsidem summam et Apsidem imam in Ellipsi immobili sit 180^{gr}[,] erit angulus VCp inter easdem Apsides in orbe quem corpus vi centripeta quantitati $\frac{bA^m + cA^n}{A^3}$ proportionali describit, æqualis angulo graduum $180\sqrt{\frac{b+c}{mb+nc}}$. Et eodem argumento si vis centripeta sit ut $\frac{bA^m - cA^n}{A^3}$, angulus inter Apsides invenietur $180\sqrt{\frac{b-c}{mb-nc}}$ graduum. Nec secus resolvetur Problema in casibus difficilioribus. Quantitas cui vis centripeta proportionalis est resolvi semper debet in series convergentes denominatorem habentes A^3. Dein pars data numeratoris ad ipsius partem non datam et pars data numeratoris hujus $RG^2 - RF^2 + TF^2 - F^2X$ ad partem non datam(252) in eadem ratione ponendæ sunt; et quantitates superfluas delendo, scribendoq; unitatem pro T obtinebitur proportio G ad F.

Corol. 1. Hinc si vis centripeta sit ut aliqua altitudinis dignitas, inveniri potest dignitas illa ex motu Apsidum; et contra. Nimirum si motus totus angularis quo corpus redit ad Apsidem eandem sit ad motum angularem revolutionis unius seu graduum 360 ut numerus aliquis m ad numerum aliquem n, et altitudo nominetur A: erit vis ut altitudinis dignitas illa $A^{\frac{nn}{mm}-3}$ cujus index est $\frac{nn}{mm} - 3$. Id quod per Exempla secunda manifestum est. Unde liquet vim illam in majore quam triplicata altitudinis ratione decrescere non posse. Corpus tali vi(253) revolvens deq; Apside discedens, si cœperit descendere, nunquam perveniet ad Apsidem imam seu altitudinem minimam sed descendet usq; ad centrum[,] describens curvam illam lineam de qua egimus in Corol. 7. Prop. XLIV.(254) Sin c[œ]perit illud de Apside discedens vel minimum ascendere, ascendet in infinitum, neq; unquam perveniet ad Apsidem summam. Describet enim cur-

(252) Namely, as ($RG^2 - RF^2 + TF^2 \approx$) RG^2 to $-F^2X$.

(253) Varying as the inverse-cube of the 'height' above the force-centre, that is.

and when terms in the numerators are compared there will come to be $RG^2 - RF^2 + TF^2$ to $bT^m + cT^n$ as $-F^2$ to

$$-(mbT^{m-1} + ncT^{n-1}) + (\tfrac{1}{2}(m^2 - m)\ T^{m-2} + \tfrac{1}{2}(n^2 - n)\ T^{n-2})\ X\ldots;$$

then, on taking the last ratios which ensue when the orbits approach a circular form, G^2 comes to be to $bT^{m-1} + cT^{n-1}$ as F^2 to $mbT^{m-1} + ncT^{n-1}$, and *alternando* G^2 to F^2 as $bT^{m-1} + cT^{n-1}$ to $mbT^{m-1} + ncT^{n-1}$. On expressing the maximum height CV, that is T, arithmetically by unity this proportion becomes G^2 to F^2 as $b + c$ to $mb + nc$, and therefore as 1 to $(mb + nc)/(b + c)$. Whence G to F, that is, $V\widehat{C}p$ to $V\widehat{C}P$, is as 1 to $\sqrt{[(mb + nc)/(b + c)]}$. And in consequence, since the angle $V\widehat{C}P$ between the top apsis in the stationary ellipse and the bottom one is 180°, the angle $V\widehat{C}p$ between the same apsides in the orbit described by a body under a centripetal force proportional to the quantity $(bA^m + cA^n)/A^3$ will be equal to $180\sqrt{[(b + c)/(mb + nc)]}°$. And by the same argument, if the centripetal force be as $(bA^m - cA^n)/A^3$, the angle between the apsides will be found to be $180\sqrt{[(b - c)/(mb - nc)]}°$. The problem will be resolved no differently in more difficult cases. The quantity to which the centripetal force is proportional ought always to be resolved into a converging series having A^3 as its denominator; then the given part of the numerator to its part not given is to be set in the same ratio as the given part of this numerator $RG^2 - RF^2 + TF^2 - F^2X$ to its part not given;(252) and on deleting superfluous quantities (and writing unity in place of T) the proportion of G to F will be obtained.

Corollary 1. Hence if a centripetal force be as some power of the height, that power can be ascertained from the motion of the apsides; and conversely so. Specifically, if the total angular motion by which a body returns to the same apsis be to the angular motion of one revolution, that is, 360°, as some number m to some number n, and the height be named A, then the force will be as the height's power $A^{n^2/m^2 - 3}$, the index of which is $n^2/m^2 - 3$. This is obvious by Example 2. Whence it is clear that the force cannot decrease in a greater than tripled ratio of the height. If a body revolving with such a force(253) begins, departing from an apsis, to descend, it will never reach a bottom apsis or minimum height, but will descend right to the centre, describing the curve we discussed in Corollary 7 of Proposition XLIV.(254) But if, departing from the apsis, it begins the least bit to ascend, then it will ascend indefinitely, never ever reaching a top apsis: for it will describe the curve which was discussed in

---

(254) With this erroneous corollary rightly deleted, reference needs to be made to 'Corol. 3. Prop. XLI' which (see note (212)) replaces it, and so it was amended in the published *Principia* ($_1$1687: 142).

vam illam lineam de qua actum est in Corol. 3 Prop. XLI(255) [et] Corol. 6. Prop. XLIV. Sic et ubi vis in recessu a centro decrescit in majori quam triplicata ratione altitudinis, corpus de Apside discedens, perinde ut cœperit descendere vel ascendere, vel descendet ad centrum usq; vel ascendet in infinitum. At si vis in recessu a centro vel decrescat in minori quam triplicata ratione altitudinis, vel crescat in altitudinis ratione quacunq;, corpus nunquam descendet ad centrum usq; sed ad Apsidem imam aliquando perveniet: et contra, si corpus de Apside ad Apsidem alternis vicibus descendens et ascendens nunquam appellat ad centrum, vis in recessu a centro aut augebitur aut in minore quam triplicata altitudinis ratione decrescet, et quo citius corpus de Apside ad Apsidem redierit, eo longius ratio virium recedet a ratione illa triplicata. Ut si corpus revolutionibus 8 vel 4 vel 2 vel $1\frac{1}{2}$ de Apside summa ad Apsidem summam alterno descensu et ascensu redierit, hoc est si fuerit m ad n ut 8 vel 4 vel 2 vel $1\frac{1}{2}$ ad 1, adeoq; $\frac{nn}{mm}-3$ valeat $\frac{1}{64}-3$ vel $\frac{1}{16}-3$ [vel $\frac{1}{4}-3$] vel $\frac{4}{9}-3$[,] erit vis ut $A^{\frac{1}{64}-3}$ vel $A^{\frac{1}{16}-3}$ vel $A^{\frac{1}{4}-3}$ vel $A^{\frac{4}{9}-3}$ id est reciproce ut $A^{3-\frac{1}{64}}$ vel $A^{3-\frac{1}{16}}$ vel $A^{3-\frac{1}{4}}$ vel $A^{3-\frac{4}{9}}$. Si corpus singulis revolutionibus redierit ad Apsidem eandem immotam, erit m ad n ut 1 ad 1 adeoq; $A^{\frac{nn}{mm}-3}$ æquale A^{-2} seu $\frac{1}{A^2}$, et propterea decrementum virium in ratione duplicata altitudinis ut in præcedentibus(256) demonstratum est. Si corpus partibus revolutionis unius vel tribus quartis vel duabus tertijs vel una tertia vel una quarta ad Apsidem eandem redierit, erit m ad n ut $\frac{3}{4}$ vel $\frac{2}{3}$ vel $\frac{1}{3}$ vel $\frac{1}{4}$ ad 1 adeoq; $A^{\frac{nn}{mm}-3}$ æquale $A^{\frac{16}{9}-3}$ vel $A^{\frac{9}{4}-3}$ vel A^{9-3} vel A^{16-3} et propterea vis aut reciproce ut $A^{\frac{11}{9}}$ vel $A^{\frac{3}{4}}$ aut directe ut A^6 vel A^{13}. Deniq; si corpus pergendo ab Apside summa ad Apsidem summam confecerit revolutionem integram et præterea gradus tres,(257) adeoq; Apsis illa singulis corporis revolutionibus conf[e]cerit in consequentia gradus tres, erit m ad n ut 363gr ad 360gr, adeoq; $A^{\frac{nn}{mm}-3}$ æquale $A^{-\frac{265707}{131769}}$ et propterea vis centripeta reciprocè ut $A^{\frac{265707}{131769}}$ seu $A^{2\frac{4}{243}}$(258). Decrescit igitur vis centripeta in ratione paulo majore quam duplicata sed quæ vicibus $60\frac{3}{4}$ propiùs ad duplicatam quam ad triplicatam accedit.

Corol. 2. Hinc etiam si corpus vi centripeta quæ sit reciprocè ut quadratum altitudinis revolvatur in Ellipsi umbilicum habente in centro virium, et huic vi

(255) A late insertion in the manuscript which in the published *Principia* was, in line with the amended reference in the previous sentence (see note (254)), adjusted to be 'in eodem Corol.' (in the same corollary).

(256) Namely, Propositions X–XII of the previous 'De motu Corporum' [= Propositions XI–XIII of the present 'Liber primus']. 'Disturbing' a given inverse-square force-field by a secondary inverse-square force instantaneously directed through the force-centre merely, of

Corollary 3 of Proposition XLI(255) and Corollary 6 of Proposition XLIV. So also, when the force in retreat from the centre decreases in a greater than tripled ratio of the height, a body departing from an apsis will, correspondingly as it begins to descend or ascend, either descend right to the centre or ascend indefinitely. If, however, the force in recess from the centre should either decrease in a less than tripled ratio of the height or increase in any ratio of it whatever, the body will never descend right to the centre but will at some point reach a bottom apsis; and conversely, if a body descending and ascending in alternate stages should never call at the centre, then the force in recess from the centre will either increase or alternatively decrease in a less than tripled ratio of the height, and the more swiftly a body returns from apsis to apsis, the further will the force-ratio recede from that tripled one. If, for instance, a body shall return from top apsis to top apsis by alternate descent and ascent in 8 or 4 or 2 or $1\frac{1}{2}$ revolutions, that is, if m to n be as 8 or 4 or 2 or $1\frac{1}{2}$ to 1, and hence the value of n^2/m^2-3 be $\frac{1}{64}-3$ or $\frac{1}{16}-3$ or $\frac{1}{4}-3$ or $\frac{4}{9}-3$, then the force will be as $A^{\frac{1}{64}-3}$ or $A^{\frac{1}{16}-3}$ or $A^{\frac{1}{4}-3}$ or $A^{\frac{4}{9}-3}$, that is, reciprocally as $A^{3-\frac{1}{64}}$ or $A^{3-\frac{1}{16}}$ or $A^{3-\frac{1}{4}}$ or $A^{3-\frac{4}{9}}$. Should the body return to the same stationary apsis each single revolution, m will be to n as 1 to 1, and so A^{n^2/m^2-3} will be equal to A^{-2}, that is, $1/A^2$, and in consequence the decrease in force is in the doubled ratio of the height, as has been demonstrated in preceding propositions.(256) If the body should return to the same apsis in three-quarters or two-thirds or one-third or one-quarter of a single revolution, then m will be to n as $\frac{3}{4}$ or $\frac{2}{3}$ or $\frac{1}{3}$ or $\frac{1}{4}$ to 1, and hence A^{n^2/m^2-3} will be equal to $A^{\frac{16}{9}-3}$ or $A^{\frac{9}{4}-3}$ or A^{9-3} or A^{16-3}, and consequently the force is either reciprocally as $A^{\frac{11}{9}}$ or $A^{\frac{3}{4}}$ or directly as A^6 or A^{13}. If, finally, the body in proceeding from top apsis to top apsis completes a whole revolution and a further three degrees,(257) and hence that apse at each individual revolution of the body achieves an advance of three degrees, then m will be to n as 363° to 360°, and so A^{n^2/m^2-3} equal to $A^{-265707/131769}$, and accordingly the centripetal force will be reciprocally as $A^{265707/131769}$, that is, $A^{2\frac{4}{243}}$.(258) The centripetal force therefore decreases in a ratio slightly greater than doubled, but one which is $60\frac{3}{4}$ times nearer a doubled than a tripled.

Corollary 2. Hence also, if a body should revolve in an ellipse having a focus at the force-centre under a centripetal force reciprocally as the square of the height, and there be added to or taken away from this centripetal force any

course, converts it to be a variant inverse-square force-field in which the only unconstrained (and unresisted) orbits possible are stationary conics having a focus at the force-centre.

(257) Though Newton does not here press the connection, it is not coincidental that this is almost exactly the angular advance of the moon's apsidal line in a lunar month; compare note (260) below.

(258) Understand 'quam proximè' (very nearly). In fact $\frac{265707}{131769} = 2\frac{241}{14641}$ where

$$241/14641 = 4/243\tfrac{1}{241}.$$

centripetæ addatur vel auferatur vis alia quævis extranea: cognosci potest (per Exempla tertia) motus Apsidum qui ex vi illa extranea orietur, et contra. Et si vis qua corpus revolvitur in Ellipsi sit ut $\frac{1}{A^2}$ et vis extranea ablata ut cA adeoq vis reliqua ut $\frac{A-cA^4}{A^3}$, erit (in Exemplis tertijs) A æqualis 1 et n æqualis 4 adeoq angulus revolutionis inter Apsides æqualis angulo graduum $180\sqrt{\frac{1-c}{1-4c}}$. Ponatur vim illam extraneam esse 357⌊45 vicibus minorem quam vis altera qua corpus revolvitur in Ellipsi, id est c esse $\frac{100}{35745}$ et $180\sqrt{\frac{1-c}{1-4c}}$ evadet $180\sqrt{\frac{35645}{35345}}$ seu 180⌊7602[,] id est 180gr. 45′. 37″.(259) Igitur corpus de Apside summa discedens motu angulari 180gr. 45′. 37″ perveniet ad Apsidem imam et hoc motu duplicato ad Apsidem summam redibit: adeoq Apsis summa singulis revolutionibus progrediendo conficiet 1gr. 31′. 14″.(260)

Hactenus de motu corporum in orbibus quorum plana per centrum virium transeunt. Superest ut motus etiam determinemus in planis excentricis.(261) Nam Scriptores qui motum gravium tractant considerare solent ascensus et descensus ponderum tam obliquos in planis quibuscunq datis quam perpendiculares et pari jure motus corporum viribus quibuscunq centra petentium et planis excentricis (261) innitentium hic considerandus venit. Plana autem supponimus esse politissima et absolute lubrica ne corpora retardent. Quinimò in his

(259) Corrected to be '180, 7623, id est 180gr. 45m. 44s' (180.7623[°], that is, 180° 45′ 44″) in the *Principia's* second edition ($_2$1713: 131).

(260) Likewise changed in 1713 to be '1gr. 31m. 28sec' (1° 31′ 28″). Without explicitly saying so, Newton here sets the ratio of the disturbing force

$$R(G^2-F^2)/r^3 \approx (G^2-F^2)/T^2, \quad G/F = \sqrt{[(1-4c)/(1-c)]},$$

directed through C to the central force $F^2/r^2 \approx F^2/T^2$ maintaining motion in the near-circular stationary ellipse VPK of main axis insignificantly different from $2CV = 2T$ to be that of the component directed through the Earth's centre of the mean solar perturbation upon the Moon's orbit to the inverse-square gravitational pull of the Earth upon the Moon which, in the absence of disturbance by the Sun (and other external bodies), would cause the Moon to orbit in a perfect, stationary ellipse round the Earth. By Propositions XXV and XXVI of Book 3 of the published *Principia* ($_1$1687: 434–7) the latter ratio is approximately $\frac{3}{2}(T_M/T_E)^2$, where $T_M \approx 27$ days, 7 hours, 43 minutes and $T_E \approx 365$ days, 6 hours, 9 minutes are the periodic times of orbit of the Moon round the Earth and of the Earth round the Sun respectively; whence it follows (compare *ibid.*: 435) that, since $G^2/F^2-1 = 3c/(1-c)$,

$$c \approx \tfrac{1}{2}(T_M/T_E)^2 \approx \tfrac{1}{2}(1/178.725) = 1/357.45.$$

Since no allowance is thereby made for the (roughly equal) mean transverse component of solar perturbation (instantaneously at right angles to the lunar vector), it could have come as

other extraneous force, then there can (by Example 3) be found out the motion of the apsides ensuing from that foreign force; and conversely so. And if the force whereby the body revolves in the ellipse be as $1/A^2$ and the extraneous force taken away be as cA, and therefore the remaining force as $(A-cA^4)/A^3$, there will then (in Example 3) be A equal to 1 and n equal to 4, and hence the angle of revolution between the apsides will be equal to $180\sqrt{[(1-c)/(1-4c)]}°$. Set the extraneous force to be 357·45 times less than the other force whereby the body revolves in the ellipse, that is, c to be 100/35745, and

$$180\sqrt{[(1-c)/(1-4c)]}$$

will prove to be $180\sqrt{\frac{35645}{35345}}$ or 180·7602, that is, 180° 45′ 37″.(259) Consequently, a body departing from the top apsis will arrive at the bottom apsis by an angular motion of 180° 45′ 37″, and by double this motion will return to the top apsis: in consequence, the top apsis will achieve in each individual revolution an advance of 1° 31′ 14″.(260)

So much for the motion of bodies in orbits whose planes pass through the centre of force. It remains also to determine motions in planes which are off-centre.(261) For writers who treat of the motion of heavy bodies usually consider the ascents and descents of weighty things obliquely in any given planes whatever as well as perpendicular ones, and with equal right the motion of bodies seeking centres under any forces whatever and inclined in eccentric planes(261) here comes into consideration. We suppose, however, that the planes are highly polished and perfectly slippery so as not to retard the bodies. To be sure, in these

no surprise to Newton that the rate of angular progression of the Moon's apsidal line here computed, some 1½° per lunar month, is only about half of that observed in practice—that 'Apsis lunæ est duplo velocior circiter' (The [advance of] lunar apsis is about twice as swift), as he added in conclusion at this point in the *Principia*'s third edition ($_3$1726: 141). This deficiency he had immediately sought to remedy, indeed, in terms of a refined Horrocksian model in which the Moon is conceived to move in an ellipse which not only advances angularly round the Earth set at a focus but also is permitted simultaneously to vary its eccentricity in time with the apparent motion of the Sun round the Earth: by considering both the radial and transverse solar perturbations inducing these respective secular and periodic motions, he was then able—with some 'cooking' of basic constants in the model—to derive a respectably accurate value of some 38° 52′ for the mean annual advance of lunar apsis. (See 2, §3 below, where we reproduce the extant worksheets of this more sophisticated but partially fudged computation.)

(261) So altered by Newton in the manuscript from 'in alijs superficiebus datis' (in other given surfaces) and 'superficiebus obliquis' (on oblique surfaces) respectively—correctly so, since a force-field whose centre lies outside a general, asymmetric surface will not, as Newton requires in his two following Propositions XLVI and XLVII, induce accelerated motion within the surface to some unique point in it.

demonstrationibus vice planorum(262) quibus corpora incumbunt quasq; tangunt incumbendo, usurpamus plana(262) his parallela in quibus centra corporum moventur, et orbitas movendo describunt. Et eadem lege motus corporum in superficiebus curvis peractos subinde determinamus.

Artic. X.

De motibus corporum in superficiebus(263) datis deq; motibus reciprocis funipendulorum.

Prop. XLVI. Prob. XXXII.

Posita cujuscunq; generis vi centripeta datoq; tum virium centro tum plano quocunq; in quo corpus revolvitur et concessis figurarum curvilinearum quadraturis: requiritur motus corporis de loco dato data cum velocitate secundum rectam in plano illo datam egressi.

Sit S centrum virium, SC distantia minima centri hujus a plano data, P corpus de loco P secundum rectam PZ egrediens, Q corpus idem in trajectoria sua revolvens, et PQR Trajectoria illa in plano dato descripta quam invenire oportet. Jungantur CQ, QS et si in QS capiatur SV proportionalis vi centripetæ qua corpus trahitur versus centrum S, et agatur VT quæ sit parallela CQ et occurrat SC in T: vis SV resolvetur (per Legum Corol. 2) in vires ST, TV, quarum ST trahendo corpus secundum lineam plano perpendicularem, nil mutat motum ejus in hoc plano. Vis autem altera TV agendo secundum positionem plani trahit corpus directè versus punctum C in plano datum, adeoq; facit illud in hoc plano perinde moveri ac si vis ST tolleretur, et corpus vi sola TV revolveretur circa centrum C in spatio libero. Data autem vi centripeta TV qua corpus Q(264) in spatio libero circa centrū datum [C] revolvitur, datur per Prop. XLII tum Trajectoria PQR quam corpus describit, tum locus Q in quo corpus ad datum quodvis tempus versabitur, tum deniq; velocitas corporis in loco illo Q; et contra. Q.E.I.(265)

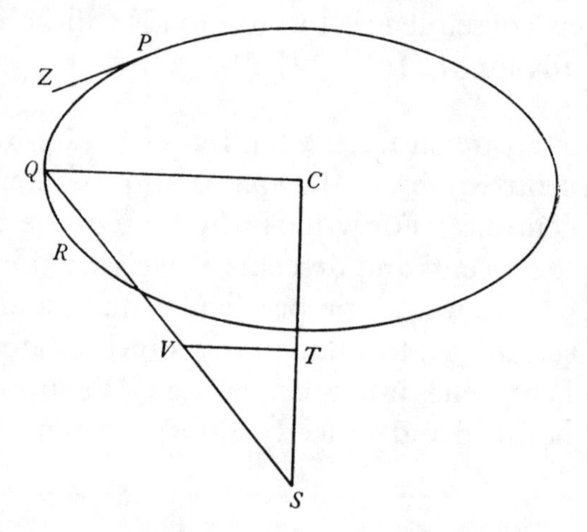

(262) Originally 'superficierum' and 'superficies' (surfaces) respectively; compare the previous note.

(263) Understand (see note (261)) 'in planis excentricis et lineis' (in off-centre planes and lines), as Newton first altered an unacceptable original manuscript head 'De motu corporum in superficiebus obliquis' (on the motion of bodies in oblique surfaces) here to read.

(264) We would expect 'P', the point at which the initial direction of motion is defined.

(265) Analytically, where the perpendicular distance from the 'excentric' force-centre S to

demonstrations we may, instead of the planes(262) in which the bodies recline and which in doing so they touch, employ planes(262) parallel to these in which the bodies' centres move and by their motion describe orbits. And by the same principle we thereafter determine the motions performed in curved surfaces by bodies.

Article X

On the motions of bodies in given surfaces,(263) and the reciprocating motions of cord-pendulums.

Proposition XLVI, Problem XXXII.

Supposing a centripetal force of any kind whatever and given both the force-centre and any plane in which a body revolves, and granting the quadratures of curvilinear figures, there is required the motion of a body setting out from a given place with given speed and along a straight line given in the plane.

Let S be the centre of force, SC the minimum distance of this centre from the given plane, P a body setting out from the place P along the straight line PZ, Q the same body revolving in its trajectory, and PQR the trajectory described in the given plane which it is required to find. Join CQ, QS and, if in QS there be taken SV proportional to the centripetal force whereby the body is drawn towards the centre S, and VT be drawn parallel to CQ and meeting SC in T, then the force SV will (by Corollary 2 of the Laws) be resolved into the forces ST and TV. Of these, ST by drawing the body along a line perpendicular to the plane changes its motion in this plane not at all; but the other force TV by acting along the lie of the plane draws the body directly towards the point C given in the plane, and hence makes it move in this plane exactly as though the force ST were removed and the body were by the force TV alone to revolve round the centre C in free space. Given, however, the centripetal force TV whereby the body Q(264) revolves in free space round the given centre C, there is given by Proposition XLII both the trajectory PQR which the body describes, and the place Q at which the body will be positioned at any given time, and lastly the speed of the body at the place Q; and conversely so. As was to be found.(265)

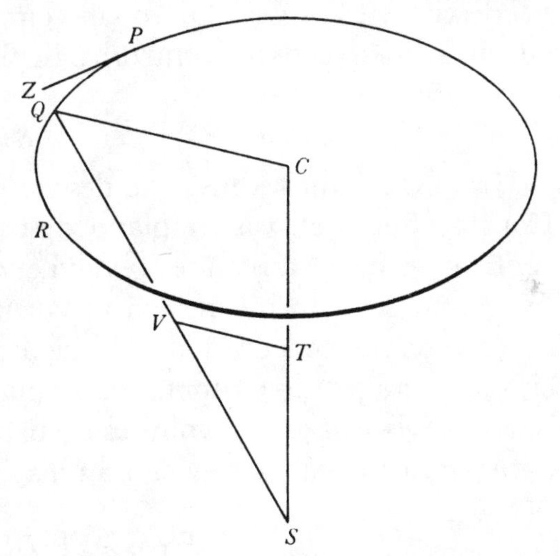

the given plane is $SC = h$, and $f(r)$ is the force directed to S at the distance $SQ = r$, this will induce within the plane at Q an apparent force instantaneously inclined through C and of magnitude $f(r).\sqrt{[r^2-h^2]}/r$, that is, $f(\sqrt{[s^2+h^2]}).s/\sqrt{[s^2+h^2]}$ on setting $CQ = \sqrt{[r^2-h^2]} = s$.

Prop. XLVII. Theor. XV.

Posito quod vis centripeta proportionalis sit distantiæ corporis a centro: corpora omnia in planis quibuscunq revolventia describent Ellipses et revolutiones temporibus æqualibus peragent; quæq moventur in lineis rectis ultro citroq discurrendo, singulas eundi et redeundi periodos ijsdem temporibus absolvent.

Nam stantibus quæ in superiore Propositione, vis SV qua corpus Q in plano quovis PQR revolvens trahitur versus centrum S est ut distantia SQ, atq adeo ob proportionales SV et SQ, TV et CQ vis TV qua corpus trahitur versus punctum C in orbis plano datum, est ut distantia CQ. Vires igitur quibus corpora in plano PQR versantia trahuntur versus punctum C sunt pro ratione distantiarum æquales viribus quibus corpora undequaq trahuntur versus centrum S, et propterea corpora movebuntur ijsdem temporibus in ijsdem figuris in plano quovis PQR circa punctum C atq in spatijs liberis circa centrum S, adeoq (per Corol. 2 Prop. X, & Cor. 2 Prop. XXXVIII) temporibus semper æqualibus vel describent Ellipses in plano illo circa centrum C vel periodos movendi ultro citroq in lineis rectis per centrum C in plano illo ductis complebunt. Q.E.D.[266]

Schol.

His affines sunt ascensus ac descensus corporum in superficiebus[267] curvis. Concipe lineas curvas in plano describi, dein circa axes quosvis datos per centrum virium transeuntes revolvi et ea revolutione superficies curvas describere, tum corpora ita moveri ut eorum centra in his superficiebus perpetuò reperiantur. Si corpora illa obliquè ascendendo et descendendo currant ultro citroq[,] peragentur eorum motus in planis per axem transeuntibus, atq adeo in lineis curvis quarum revolutione curvæ illæ superficies genitæ sunt.[268] Istis igitur in casibus sufficit motum in his lineis curvis considerare.[269]

Prop. XLVIII. Theor. XVI.

Si rota globo extrinsecus ad angulos rectos insistat et more rotarum[270] *revolvendo progrediatur in circulo maximo: longitudo itineris curvilinei quod punctum quodvis in rotæ perimetro datum, ex quo globum tetigit confecit, erit ad duplicatum sinum versum arcus* [271] *dimidij qui globum ex eo tempore inter eundum tetigit ut summa diametrorum globi et rotæ ad semidiametrum globi.*[272]

---

(266) In this simplest instance of the preceding Proposition XLVI, the external force-field centred on S is (in the analytical terms of note (265)) $f(r) \propto SQ = r$, whence the force induced towards C in the given plane is evidently proportional to $CQ = s$.

(267) 'quibusdam' (certain) is deleted in the manuscript.

(268) Originally 'fuerunt' (had been).

(269) Newton has cancelled 'ut fit in exemplis sequentibus' (as is done in the following instances).

Proposition XLVII, Theorem XV.

Supposing that the centripetal force be proportional to the distance of a body from the centre, all bodies revolving in any planes whatever shall describe ellipses and accomplish their revolutions in equal times; while those moving in straight lines with a to-and-fro run shall complete their individual periods of going and returning in the same times.

For, with the circumstances of the previous proposition standing, the force SV whereby the body Q revolving in any plane PQR is drawn towards the centre S is as the distance SQ, and hence, because of the proportionality of SV and SQ, TV and CQ, the force TV whereby the body is drawn towards the point C given in the plane of orbit is as the distance CQ. The forces, therefore, by which bodies positioned in the plane PQR are drawn towards the point C are, in proportion to the distances, equal to the forces by which bodies whencesoever are drawn towards the centre S, and accordingly bodies will in the same times move in the same figures in any plane PQR round the point C as in free space they would do round the centre S; hence (by Corollary 2 of Proposition X and Corollary 2 of Proposition XXXVIII) in times which are ever equal they will either describe ellipses in that plane around the centre C or complete periodic to-and-fro motions in straight lines drawn through the centre C in that plane. As was to be proved.(266)

Scholium.

Related to these motions are the ascents and descents of bodies in (267) curved surfaces. Imagine curves to be described in a plane and then rotated round any given axes passing through the centres of force, by that revolution describing curved surfaces, and then that bodies move such that their centres be perpetually located in these surfaces. If those bodies should in oblique ascent or descent run back and forth, their motions will be performed in planes passing through the axis, and hence in the plane curves by whose rotation those curved surfaces were(268) generated. In those cases, therefore, it is enough to examine the motion in these curves.(269)

Proposition XLVIII (XLIX), Theorem XVI (XVII).

If a wheel should stand erect on the outside (inside) of a (hollow) globe and, rolling as wheels do,(270) move forward in a great circle, the length of the curvilinear track completed by any given point in the wheel's rim will be to double the versed sine of half the arc(271) brought en route during that time into contact with the globe as the sum (difference) of the diameters of the globe and wheel to the radius of the globe.(272)

(270) That is, uniformly rotating round its centre as its rim advances (without slipping) over the sphere's surface.

(271) Of the wheel's circumference.

(272) It has seemed advantageous to us to combine these two parallel enunciations in our

Prop. XLIX. Theor. XVII.

Si rota globo concavo ad rectos angulos intrinsecus insistat et revolvendo[270] *progrediatur in circulo maximo: longitudo itineris curvilinei quod punctum quodvis in rotæ perimetro datum, ex quo globum tetigit confecit, erit ad duplicatum sinum versum arcus*[271] *dimidij qui globum toto hoc tempore inter eundum tetigit ut differentia diametrorum globi et rotæ ad semidiametrum globi.*[272]

Sit *ABD* globus, *C* centrum ejus, *BPV* rota ei insistens[,] *E* centrum rotæ, *B* punctum contactus et *P* punctum datum in perimetro rotæ. Concipe hanc

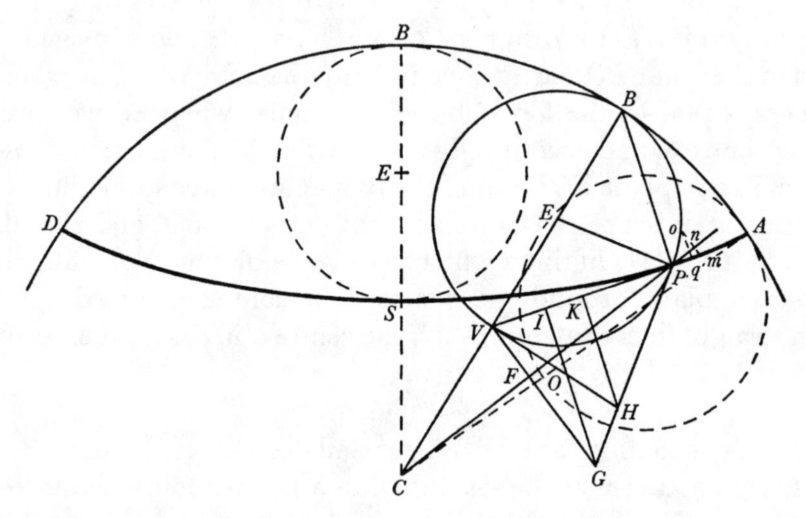

Rotam pergere in circulo maximo *ABD* ab *A* per *B* versus *D*[,] et inter eundum ita revolvi ut arcus *AB*, *PB* sibi invicem semper æquentur, atq; punctum illud *P* in perimetro Rotæ datum interea describere viam curvilineam *AP*: sit autem *AP* via tota curvilinea descripta ex quo rota globum tetigit in *A*, et erit viæ hujus longitudo *AP* ad duplum sinum versum arcus $\frac{1}{2}PB$, ut $2CE$ ad *CB*. Nam recta *CE*, si opus est producta[,] occurrat Rotæ in *V* junganturq; *CP*, *BP*, *EP*, *VP*, et in *CP* productam demittatur normalis *VF*. Tangant *PH*, *VH* circulum in *P* et *V* concurrentes in *H*, secetq; *PH* ipsam *VF* in *G* et ad *VP* demittantur normales *GI*, *HK*, centro item *C* et intervallo quovis describatur circulus *nom* secans rectam *CP* in *n*, rotæ perimetrum in *o* et viam curvilineam *AP* in *m*, centroq; *V* et intervallo *Vo* describatur circulus secans *VP* productam in *q*.

Quoniam rota eundo semper revolvitur circa punctum contactus *B*[,][273] mani-

English version, especially since Newton himself proceeds to give them a common demonstration. In contrast, we have dared for clarity's sake—and following Henry Pemberton's lead in his *editio ultima* of the *Principia* ($_3$1726: 146)—to split Newton's unwieldy and confusing original compound figure (*Principia*, $_1$1687: 148; $_2$1713: 135) into two, setting its (in context) more important hypocycloidal component (Proposition XLIX) with the Latin text and trans-

Let ABD be the globe, C its centre, BPV the wheel upright on it, E the wheel's centre, B its point of contact, and P the given point in the wheel's rim. Imagine that this wheel proceeds in the great circle ABD from A on through B towards D,

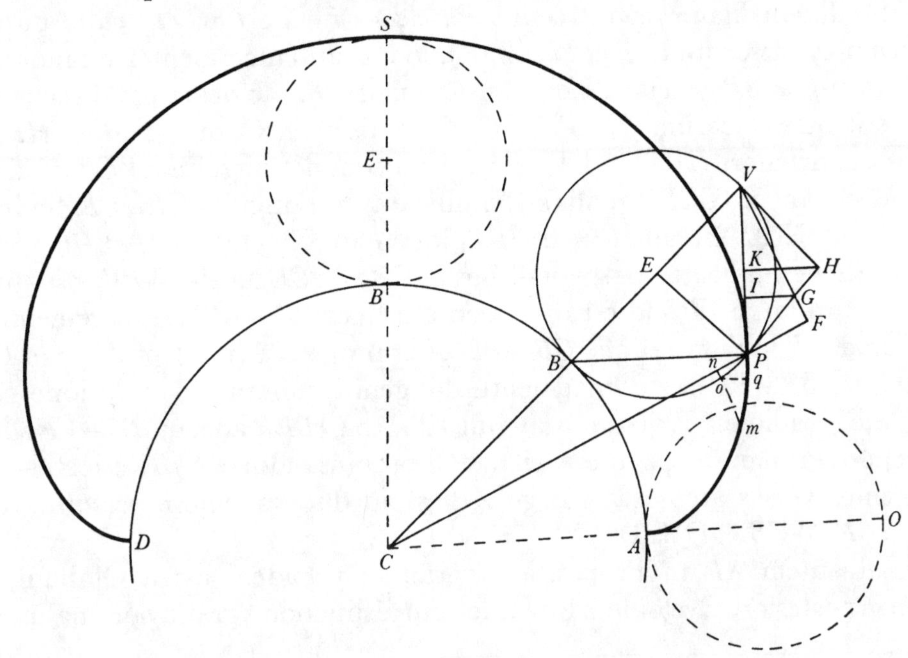

and as it goes so rotates that the arcs $\widehat{AB}$ and $\widehat{PB}$ be ever equal to one another, while the point P given in the wheel's rim meanwhile describes the curvilinear path $\widehat{AP}$; let, however, $\widehat{AP}$ be the total curvilinear path described since the wheel was in contact with the globe at A, and then the length of this arc $\widehat{AP}$ will be to twice the versed sine of half the arc $\widehat{PB}$ as $2CE$ to CB. For let the straight line CE, produced if need be, meet the wheel in V and join CP, BP, EP, VP and then onto CP extended let fall the normal VF. Let the tangents PH, VH to the circle at P and V concur at H, let PH intersect VF in G, and to VP let fall the normals GI, HK; further, with centre C and any radius describe a circle-arc $\widehat{nom}$ cutting the straight line CP in n, the wheel's rim in o and the curvilinear path $\widehat{AP}$ in m, and then with centre V and radius Vo describe a circle cutting VP produced in q.

Since the wheel as it goes on ever revolves round the point B of contact(273) it is

ferring its complementary epicycloidal part (Proposition XLVIII) to be with the facing English translation.

(273) This is untrue since, as Newton at once goes on to show in his following Proposition L, the instantaneous centre of rotation of the cycloid at P is not the 'rolling centre' B of its generating 'wheel' BPV, but the corresponding point in the cycloid's evolute (a similar cycloid coincident with $\widehat{APSD}$ only at its end-points A and D where the cycloid's curvature is

festum est quod recta BP perpendicularis est ad lineam illam curvam AP quam rotæ punctum P describit, atqꝫ adeo quod recta VP tangit hanc curvam in puncto P. Circuli *nom* radius sensim auctus(274) æquetur tandem distantiæ CP et ob similitudinem figuræ evanescentis *Pnomq* et figuræ *PFGVI*[,] ratio ultima lineolarum evanescentium Pm, Pn, Po, Pq, id est ratio incrementorum momentaneorum curvæ AP[,] rectæ CP et arcus circularis BP ac decrementi rectæ VP eadem erit quæ linearum PV, PF, PG, PI respectivè. Cùm angulus VHP ob angulos quadrilateri $HVEP$ ad V et P rectos complet angulum VEP ad duos rectos adeoqꝫ angulo CEP æqualis est, similia erunt triangula VHG, CEP et inde fiet ut EP ad CE ita HG ad HV seu HP et ita KI ad KP: et divisim ut CB ad CE ita PI ad KP et duplicatis consequentibus ut CB ad $2CE$ ita PI ad PV. Est igitur decrementum lineæ VP[,] id est incrementum lineæ $BV-VP$ ad incrementum lineæ curvæ AP in data ratione CB ad $2CE$ et propterea (per Corol. Lem. IV) longitudines $BV-VP$ et AP incrementis illis genitæ sunt in eadem ratione. Sed existente BV radio est VP cosinus anguli VBP seu $\frac{1}{2}BEP$ adeoqꝫ $BV-VP$ sinus versus ejusdem anguli, et propterea in hac Rota cujus radius est $\frac{1}{2}BV$ erit $BV-VP$ duplus sinus versus arcus $\frac{1}{2}BP$. Ergo AP est ad duplum sinum versum arcus $\frac{1}{2}BP$ ut $2CE$ ad CB. Q.E.D.(275)

Lineam autem AP in Propositione priore Cycloidem extra Globum,(276) alteram in posteriore Cycloidem intra globum distinctionis gratia nominabimus.

---

infinite). While, however, the normal go at the point q in the cycloid indefinitely near to P does not pass through B, it is inclined to PB at an angle which is vanishingly small, so that, as Newton requires, PB is indeed accurately perpendicular to the cycloidal arc $\widehat{AP}$ at P. (Compare our similar remark on I: 379, note (11) regarding Descartes' analogous 'justification' of his construction of the normal to a 'stretched' linear cycloid in his letter to Mersenne of 23 August 1638 (N.S.).)

(274) Understand 'vel diminutus' (or diminished) as Newton so clarified the point in the second edition of his *Principia* ($_2$1713: 136).

(275) In gist, where Pm, Pn, Pp and Pq are contemporaneous limit-increments of $\widehat{SP}$, CP, PB and VP, Newton on the chord VP constructs the configuration *PVFGI* similar (but directed in the contrary sense) to the differential configuration *Pmnoq*, so determining that

$$(Pm/Pq \text{ or}) \; d(\widehat{SP})/d(VP) = PV(\text{or } 2PK)/PI(\text{or } PK+KI) = 2VH/(VH+HG),$$

that is, $2CE/(CE+EP)$ or $2CE/CB$ seeing that the triangles VHG and CEP are similar; whence at once $\widehat{SP} = (2CE/CB).VP$ and therefore

$$\widehat{AP} = (CE/CB)\;(VB-VP) = (2CE/CB).VB(1-\cos\tfrac{1}{2}\widehat{BEP}).$$

In the considerably more fearsome modern analytical equivalent wherein, on setting $CE = R$, $EP(=EB=EV) = \rho$, $CP = r$, $\widehat{SP} = s$, $S\widehat{C}P = \phi$, $S\widehat{C}E = \theta$ and $V\widehat{E}P = \psi$, there ensues $r^2 = R^2+\rho^2-2R\rho\cos\psi$, $(R+\rho)\,\theta = \rho\psi$ and $r\sin(\phi-\theta) = \rho\sin\psi$, and hence, by eliminating θ and ψ, the polar equation of the cycloid $P(r, \phi)$ referred to origin C will prove to be

$$\phi = \frac{\rho}{R+\rho}\cos^{-1}\left[\frac{R^2+\rho^2-r^2}{2R\rho}\right]+\sin^{-1}\left[\frac{\rho}{r}\sqrt{\left(1-\frac{(R^2+\rho^2-r^2)^2}{4R^2\rho^2}\right)}\right],$$

since therefrom we deduce $$\frac{r\,d\phi}{dr} = \frac{R-\rho}{R+\rho}\sqrt{\left(\frac{(R+\rho)^2-r^2}{r^2-(R-\rho)^2}\right)}$$

obvious that the straight line BP is perpendicular to the curve $\widehat{AP}$ described by the point P on the wheel, and hence that the straight line VP touches this curve at its point P. When the radius of the circle $\widehat{nom}$ is gradually increased[(274)] it will at last come to be equal to the distance CP, and then, because of the similarity of the vanishing figure $Pnomq$ and of the figure $PFGVI$, the ultimate ratio of the vanishing lines Pm, Pn, Po, Pq, that is, of the instantaneous increments of the curve $\widehat{AP}$, the straight line CP, the circle-arc $\widehat{BP}$ and decrement of the straight line VP, will be that of the lines PV, PF, PG, PI respectively. Since, because the angles of the quadrilateral $HVEP$ at V and P are right, the angle $V\widehat{H}P$ is the supplement of the angle $V\widehat{E}P$ and hence equal to $C\widehat{E}P$, the triangles VHG, CEP will be similar and there will thence come to be as EP to CE so HG to VH or HP and so IK to PK, and *dividendo* as CB to CE so PI to PK, and, on doubling the latter members, as CB to $2CE$ so is PI to PV. Therefore the decrement of the line VP, that is, the increment of the line $BV-VP$, is to the increment of the curve $\widehat{AP}$ in the given ratio of CB to $2CE$, and accordingly (by the Corollary to Lemma IV) the lengths $BV-VP$ and $\widehat{AP}$ generated by those increments are in the same ratio. But, with BV as radius, VP is the cosine of the angle $V\widehat{B}P$, that is, $\frac{1}{2}B\widehat{E}P$, and hence $BV-VP$ is the versed sine of the same angle, and consequently in the present wheel, having $\frac{1}{2}BV$ for radius, $BV-VP$ will be twice the versed sine of the arc $\frac{1}{2}\widehat{BP}$. As a result $\widehat{AP}$ is to twice the versed sine of the arc $\frac{1}{2}\widehat{BP}$ as $2CE$ to CB. As was to be proved.[(275)]

For distinction's sake, however, we shall name the curve $\widehat{AP}$ in the former Proposition a 'cycloid outside the globe',[(276)] and that other in the latter one a 'cycloid inside the globe'.

---

it follows that

$$s = (2\sqrt{[R\rho]}/(R+\rho))\int_{R-\rho}^{r} r/\sqrt{[r^2-(R-\rho)^2]}\,.dr = (2\sqrt{[R\rho]}/(R+\rho))\sqrt{[r^2-(R-\rho)^2]},$$

that is, $(4R\rho/(R+\rho))\sin\frac{1}{2}\psi$. Newton's rectification generalises that given by Roberval of the simple cycloid more than forty years earlier in his tract 'De Longitudine Trochoidis' (see III: 308, note (704)), still unpublished at this time.

(276) The standard modern designation 'epicycloid' for this curve was coined—on the analogy of the 'epicycle' BPV which in ancient astronomical theory generates it by wheeling round on the circular 'deferent' traced by its centre E—only nine years afterwards by Philippe de La Hire in his *Traité des Epicycloïdes et de leurs Usages* (Paris, 1694), while the term 'hypocycloid' for the complementary interior curve was to be devised (by Leonhard Euler, it would appear) only long after Newton's death, though as an unspecified 'cycloid' its definition and main properties were known to such early eighteenth-century geometers as Johann Bernoulli and Clairaut. (Compare F. Gomes Teixeira, *Traité des courbes spéciales remarquables*, **2** (Coimbra, 1909): 155 ff; and G. Loria, *Spezielle Algebraische und Transzendente Ebene Kurven. Theorie und Geschichte*, **2** (Leipzig/Berlin, 1911): 92 ff.) We may count it as fortunate that Newton's ungainly present nomenclature did not prevail with posterity.

Corol. 1. Hinc si describatur Cyclois integra *AS*[*D*] et bisecetur ea in *S*, erit longitudo partis *PS* ad longitudinem *VP* (quæ duplus est sinus anguli *VBP* existente *EB* radio) ut 2*CE* ad *CB*, atq; adeo in ratione data.(277)

Corol. 2. Et longitudo semiperimetri Cycloidis *AS* æquabitur lineæ rectæ quæ est ad Rotæ diametrum *AO* ut 2*CE* ad *CB*.

Corol. 3. Ideoq; longitudo illa est ut rectangulum(278) *BEC*, si modo globi detur semidiameter.(279)

Prop. L. Prob. XXXIII.

Facere ut corpus pendulum oscilletur in Cycloide data.

Intra globum *QVS* centro *C* descriptum detur Cyclois *QRS* bisecta in *R* et punctis suis extremis *Q* et *S* superficiei globi hinc inde occurrens. Agatur *CR* bisecans arcum *QS* in *O* et producatur ea ad *A* ut sit *CA* ad *CO* ut *CO* ad *CR*. Centro *C* intervallo *CA* describatur globus exterior *ABD* et intra hunc globum a rota cujus diameter sit *AO* describantur duæ semicycloides *AQ*, *AS* quæ globum interiorem tangant in *Q* et *S* et globo exteriori occurrant in *A*. A puncto illo *A* filo *APT* longitudinem *AR* æquante pendeat corpus *T*, et ita intra semicycloides *AQ*, *AS* oscilletur ut quoties pendulum digreditur a perpendiculo *AR*, filum parte sui superiore *AP* applicetur ad semicycloidem illam *APS* versus quam peragitur motus et circum eam ceu obstaculum flectatur et parte reliqua *PT* cui semicyclois nondum objicitur protendatur in lineam rectam, et pondus *T* oscillabitur in Cycloide(280) data *QRS*.

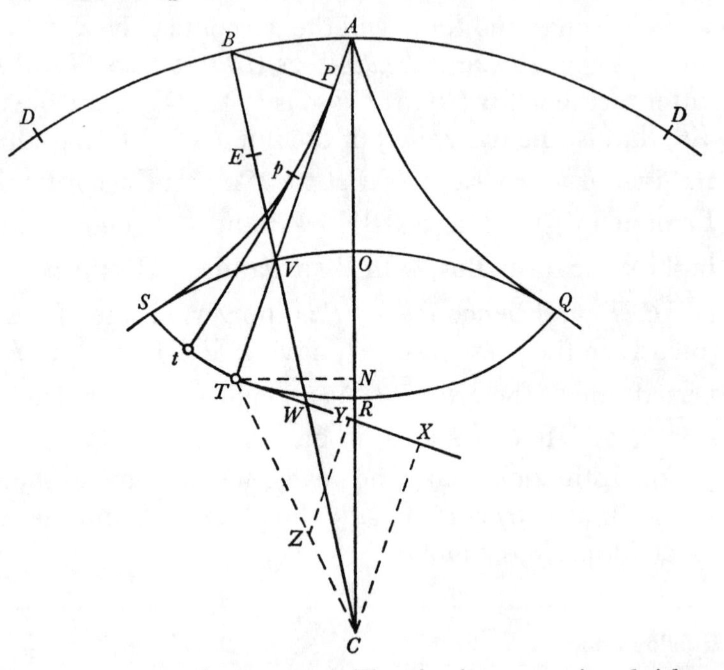

(277) For a direct analytical derivation of this subsidiary result see note (275).

(278) This replaces an erroneous initial 'latus quadratum rectanguli' (the square root of the rectangle).

(279) Namely $CB(=4BE\times EC/\widehat{AS})$. This necessary terminal stipulation is a late addition by Newton in the manuscript.

(280) A replacement in afterthought for 'Trochoide', Newton's preferred designation in his earlier 1671 tract (see III: 160) and 'Arithmetica Universalis' (see V: 426).

Corollary 1. Hence, if a complete cycloid $\widehat{ASD}$ be described and then bisected in S, the length of the portion $\widehat{SP}$ will be to the length of VP (this is, twice the sine of the angle $V\hat{B}P$ where EB is the radius) as $2CE$ to CB, and hence in a given ratio.[277]

Corollary 2. And the length of the cycloid's half-perimeter $\widehat{SA}$ will be equal to a straight line which is to the wheel's diameter OA as $2CE$ to CB.

Corollary 3. In consequence, that length is as the rectangle[278] $BE \times CE$ provided the globe's radius[279] be given.

Proposition L, Problem XXXIII.

To make a pendulating body oscillate in a given cycloid.

Within the globe QVS described with centre C let there be given the cycloid $\widehat{QRS}$ bisected at R and meeting the surface of the globe at its end-points Q and S on either side. Draw CR bisecting the arc $\widehat{QS}$ in O and extend it to A so that CA be to CO as CO to CR. With centre C and radius CA describe an exterior globe ABD, and inside this globe by help of a wheel whose diameter shall be AO describe the two half-cycloids $\widehat{AQ}$, $\widehat{AS}$ to touch the interior globe in Q and S, and to meet the exterior globe at A. From the point A let a body T hang by the thread $\widehat{APT}$ equal in length to AR, and so swing between the half-cycloids $\widehat{AQ}$, $\widehat{AS}$ that every time the pendulum departs from the perpendicular AR the thread shall in its upper part $\widehat{AP}$ be folded along the half-cycloid $\widehat{APS}$ in whose direction the motion is performed, bending about it as though round an obstacle, while in its remaining portion PT not yet obstructed by the half-cycloid it is stretched taut into a straight line, and the weight T shall then vibrate in the given cycloid $\widehat{QRS}$.

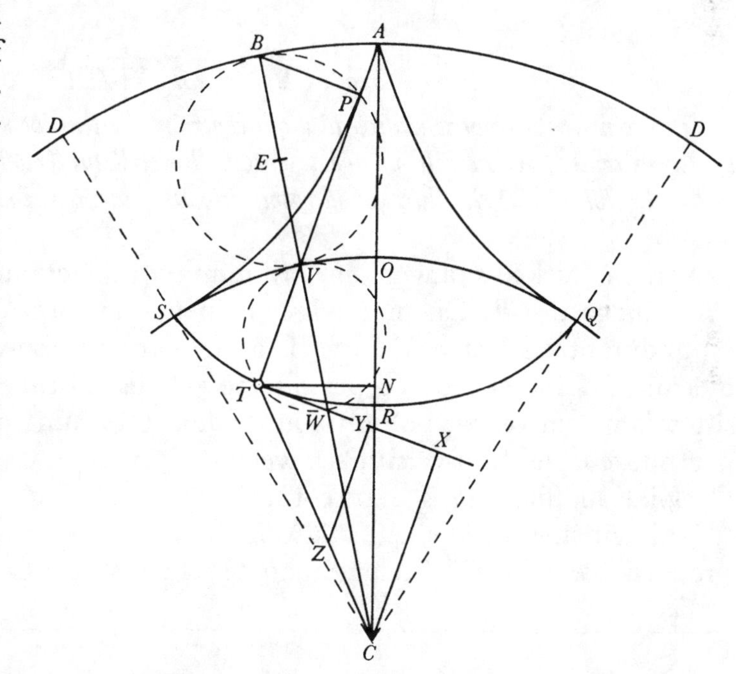

Secet enim filum *PT* tum Cycloidem *QRS* in *T*, tum circulum (281) *QOS* in *V*, agaturq; *CV* occurrens circulo *ABD* in *B*[,] et ad fili partem rectam *PT* e punctis extremis *P* ac *T*, erigantur perpendicula *PB*, *TW* occurrentia rectæ *CV* in *B* et *W*. Patet enim ex genesi Cycloidis quod perpendicula illa *PB*, *TW* abscindent de *CV* longitudines *VB*, *VW* rotarum diametris *OA*, *OR* æquales, atq; adeo quod punctum *B* incidet in circulum *ABD*. Est igitur *TP* ad *VP* duplum sinum anguli *VBP* (existente $\frac{1}{2}BV$ radio) ut *BW* ad *BV* seu $AO+OR$ ad *AO*[,] id est (cùm sint *CA* ad *CO*, *CO* ad *CR* et divisim *AO* ad *OR* proportionales) ut $CA+CO$ seu $2CE$ ad *CA*. Proinde per Corol. 1 Prop. XLIX longitudo *PT* æquatur Cycloidis arcui *PS*, et filum totum *APT* æquatur Cycloidis arcui dimidio *APS*[,] hoc est (per Corol. 2 Prop. XLIX) longitudini *AR*. Et propterea vicissim si filum manet semper æquale longitudini *AR* movebitur punctum *T* in Cycloide *QRS*. Q.E.D.(282)

Corol. Filum *AR* æquatur Cycloidis arcui dimidio *APS*.

Prop. LI. Theor. XVIII.

Si vis centripeta tendens undiq; ad globi centrum C sit in locis singulis ut distantia loci cujusq; a centro, et hac sola vi agente corpus T oscilletur (modo jam descripto) in perimetro Cycloidis QRS: dico quod oscillationum utcunq; inæqualium æqualia erunt tempora.

Nam in Cycloidis tangentem *TW* infinite productam cadat perpendiculum *CX* et jungatur *CT*. Quoniam vis centripeta qua corpus *T* impellitur versus *C* est ut distantia *CT* et vis *CT* (per Legum Corol. 2) resolvitur in partes *CX*, *TX* quarum *CX* impellendo corpus directè a *P* distendit filum *PT* et per ejus resistentiam tota cessat, nullum alium edens effectum: pars autem altera *TX* urgendo corpus transversim seu versus *X*, directè accelerat motum ejus in Cycloide, manifestum est quod corporis acceleratio huic vi acceleratrici proportionalis, sit singulis momentis ut longitudo *TX*, id est, ob datas *CV*, *WV* ijsq; proportionales *TX*, *TW* ut longitudo *TW*, hoc est (per Corol. 1. Prop. XLIX) ut

(281) Altered to 'Occurrat enim filum...tum Cycloidi...tum circulo' (For let the thread ...meet both the cycloid...and the circle-arc) in the published *Principia* ($_1$1687: 150) as a minimal amelioration.

(282) This elegant generalisation to the hypocycloid of Huygens' 1659 discovery that the evolute of the primary cycloid is a congruent cycloid (see III: 163–4, note (308)) follows readily from Proposition XLIX since, in gist,

$$\widehat{SP}/VP = 2CE\,(\text{or } CV+CB)/CB = (WV+VB)/VB = TP/VP$$

and so $\widehat{SP} = TP$. Though Newton—here concerned only with gravitating pendulums—neglects to say so, it will be clear that, by Proposition XLVIII, the epicycloid likewise has a similar epicycloid for its evolute.

For let the thread PT cut both the cycloid $Q\widehat{R}S$ in T and the circle-arc(281) $Q\widehat{O}S$ in V, then draw CV meeting the circle-arc $A\widehat{B}D$ in B and to the straight part PT of the thread erect from its end-points P and T perpendiculars PB, TW meeting the straight line CV in B and W. It is, of course, evident from a cycloid's generation that those perpendiculars PB, TW shall cut off from CV lengths VB, VW equal to the diameters OA, OR of the wheels, and hence that the point B shall fall on the circle ABD. Consequently, TP is to VP—twice the sine of $V\widehat{B}P$ (where $\frac{1}{2}BV$ is the radius)—as BW to BV or $AO+OR$ to AO, that is, (since the ratios CA to CO, CO to CR and *dividendo* AO to OR are identical) as ($CA+CO$ or) $2CE$ to CA. As a result, by Corollary 1 of Proposition XLIX the length PT is equal to the cycloid's arc $\widehat{PS}$, and the whole thread $\widehat{AP}T$ to the cycloid's half-arc $A\widehat{P}S$, that is, (by Corollary 2 of Proposition XLIX) to the length AR. And therefore should, conversely, the thread ever remain equal to the length AR, then the point T will move in the cycloid $Q\widehat{R}S$. As was to be proved.(282)

Corollary. The thread AR is equal to the cycloid's half-arc $A\widehat{P}S$.

Proposition LI, Theorem XVIII.

If a centripetal force tending on all sides to a globe's centre C be at each individual place as its distance from the centre, and by the sole action of this force the body T should vibrate (in the manner just now described) in the perimeter of the cycloid QRS, then I assert that, howsoever unequal the oscillations be, their times will be equal.

For to the cycloid's tangent TW, indefinitely extended, let fall the perpendicular CX, and join CT. Seeing that the centripetal force whereby the body T is impelled towards C is as the distance CT, and because this force CT is (by Corollary 2 of the Laws) resolved into parts CX and TX, of which CX by driving the body directly away from P stretches the thread PT and is wholly stopped by its resistance, producing no other effect, while the other part TX by urging the body on crosswise, that is, towards X, directly accelerates its motion in the cycloid, it is hence manifest that the body's acceleration, proportional to this accelerative force, shall be at each individual instant as the length TX, that is, (since CV, WV and so TX, TW proportional to these are given) as the length TW, or (by Corollary 1 of Pro-

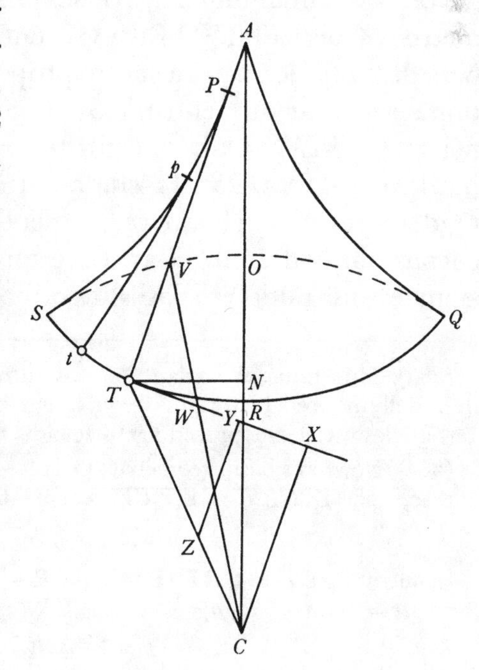

longitudo arcus Cycloidis *TR*. Pendulis igitur duobus *APT*, *Apt* de perpendiculo *AR* inæqualiter deductis et simul demissis, accelerationes eorum semper erunt ut arcus describendi *TR*, *tR*. Sunt autem partes sub initio descriptæ ut accelerationes[,] hoc est ut totæ sub initio describendæ[,] et propterea partes quæ manent describendæ et accelerationes subsequentes his partibus proportionales sunt etiam ut totæ, et sic deinceps. Sunt igitur accelerationes atq velocitates genitæ et partes his velocitatibus descriptæ partesq describendæ, semper ut totæ; et propterea partes describendæ datam servantes rationem ad invicem simul evanescent, id est corpora duo oscillantia simul pervenient ad perpendiculum *AR*. Cumq vicissim ascensus pendulorum de loco infimo *R* per eosdem arcus Trochoidales motu retrogr[a]do facti retardentur in locis singulis a viribus ijsdem a quibus descensus accelerabantur, patet velocitates ascensuum ac descensuum per eosdem arcus factorum æquales esse atq adeo temporibus æqualibus fieri, et propterea cum Cycloidis partes duæ *RS* et *RQ* ad utrumq perpendiculi latus(283) jacentes sint similes et æquales, pendula duo oscillationes suas tam totas quam dimidias ijsdem temporibus semper peragent. Q.E.D.(284)

Prop. LII. Prob. XXXIV.

Definire et velocitates pendulorum in locis singulis et tempora quibus tum oscillationes totæ tum singulæ oscillationum partes peraguntur.

[*Cas. 1.*] Centro quovis *G* intervallo *GH* Cycloidis arcum *RS* æquante describe semicirculum *HKMG* semidiametro *GK* bisectum.(285) Et si vis centripeta distantijs locorum a centro proportionalis tendat ad centrum *G*, sitq in perimetro *HIK* æqualis vi centripetæ in perimetro globi *QOS* ad ipsius centrum tendente, et eodem tempore quo pendulum *T* dimittitur a loco supremo *S* cadat corpus aliquod *L* ab *H* ad *G*: quoniam vires quibus corpora urgentur sunt æquales sub initio et spatijs describendis *TR*, *GL* semper proportionales, atq

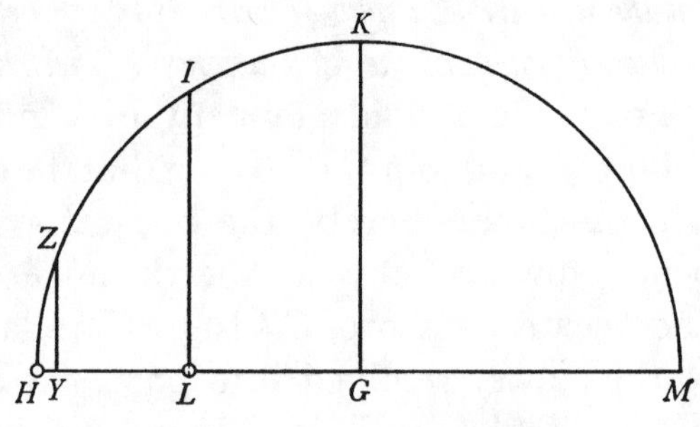

(283) This unusual variant on 'ab utroq...latere' was, in fact, born when Newton minimally altered his original equivalent phrase 'ad utramq...partem' to avoid a harsh verbal clash with the preceding 'Cycloidis partes duæ'.

(284) Newton's basic geometrical result—that, in explicit terms, $\widehat{RT}/WT = (CO+CR)/CO$ and so, since $WT/XT = WV/CV = (CO-CR)/CO$, therefore

$$\widehat{RT}/CT = (XT/CT)\,(CO^2-CR^2)/CO^2$$

—is, on setting $CT = r$, $\widehat{RT} = s$, $CO = R+\rho$ and $CR = R-\rho$, readily seen to be the derivative form $s/r = (dr/ds)((R+\rho)^2-(R-\rho)^2)/(R+\rho)^2$ of the rectification

$$s^2 = (4R\rho/(R+\rho)^2)(r^2-(R-\rho)^2)$$

position XLIX) as the length of the cycloid's arc $\widehat{TR}$. When, therefore, the two pendulums $\widehat{APT}$, $\widehat{Apt}$ are drawn back unequally from the vertical AR and simultaneously released, their accelerations will ever be as the arcs $\widehat{TR}$, $\widehat{tr}$ to be described. But the portions described at the start are as the accelerations, that is, as the wholes to be described at the start, and thereby the portions remaining to be described and the ensuing accelerations proportional to these parts are also as the wholes, and so on successively. Therefore the accelerations, and hence the speeds generated and the parts described at these speeds and the parts to be described with them, are ever as the wholes; and accordingly the parts to be described, preserving a given ratio to one another, shall simultaneously vanish, that is, the two oscillating bodies will simultaneously reach the vertical AR. Since, conversely, the ascents of pendulums made from the lowest place R along the same cycloidal arcs with a backwards motion are slowed in their individual places by the same forces as those by which their descents were accelerated, it is evident that the speeds of ascents and descents made along the same arcs are equal and hence made in equal times, and consequently, since the cycloid's two sections $\widehat{RS}$ and $\widehat{RQ}$ lying on either side of the vertical are similar and equal, that the two pendulums will perform their swings, both wholes and halves, ever in the same times. As was to be proved.(284)

Proposition LII, Problem XXXIV.

To define both the speeds of the pendulums at individual places and the times in which not only complete swings but individual parts of swings are performed.

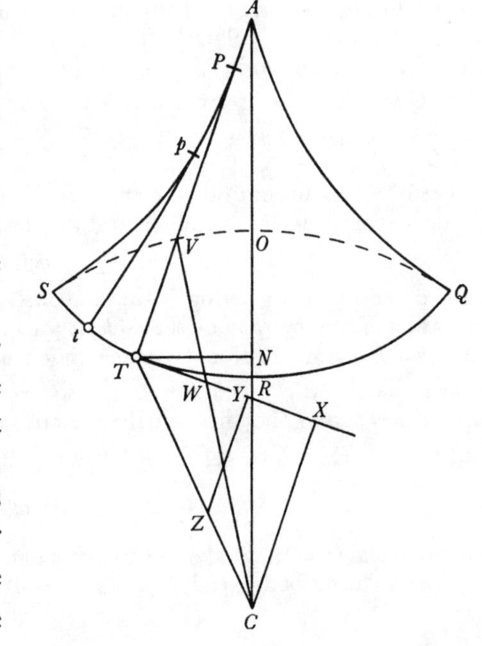

[*Case 1.*] With any centre G and radius GH equalling the cycloid's arc RS describe the semicircle $HKMG$ bisected by the radius GK.(285) Then, if a centripetal force proportional to the distances of places should tend to the centre G, let it at the circumference $\widehat{HIK}$ be equal to the centripetal force tending at the globe's perimeter $\widehat{QOS}$, and in the same time as the pendulum T is released from its highest place S let some body L fall from H to G; and, because the forces whereby the bodies are urged are equal at the start and ever proportional to the spaces $\widehat{TR}$, LG to be described, and hence, if $\widehat{TR}$ and LG be

adeo, si æquantur TR et LG, æquales in locis T et L, patet corpora illa describere spatia ST, HL æqualia sub initio, adeoq; subinde pergere æqualiter urgeri et æqualia spatia describere. Quare (per Prop. XXXVIII) tempus quo corpus describit arcum ST est ad tempus oscillationis unius ut arcus HI (tempus quo corpus H perveniet ad L) ad semicirculum HKM (tempus quo corpus H perveniet ad M) et velocitas corporis penduli in loco T est ad velocitatem ipsius in loco infimo R (hoc est velocitas corporis H in loco L ad velocitatem ejus in loco G seu incrementum momentaneum lineæ HL ad incrementum momentaneum lineæ HG, arcubus HI, HK æquabili fluxu crescentibus) ut ordinatim applicata LI ad radium GK, sive ut $\sqrt{SR^q - TR^q}$ ad SR. Unde cum in oscillationibus æqualibus describantur æqualibus temporibus arcus totis oscillationum arcubus proportionales[,] habentur ex datis temporibus et velocitates et arcus descripti in oscillationibus universis. Quæ erant primo invenienda.(286)

---

deduced in note (275). Whence, if the force $f(r)$ directed to C at T be directly proportional to $CT = r$, it is immediate that the component $f(r).dr/ds \propto r.dr/ds$ of the force acting at T instantaneously in the direction TWX of the constrained cycloidal motion is similarly proportional to $\widehat{RT} = s$, so that (as Newton here sketches and will elaborate in his following Proposition LII) the oscillatory motion thereby induced in the hypocycloid $\widehat{QRS}$ is isochronous. To reverse the argument, demonstrating that in the direct-distance force-field $f(r) \propto r$ the only possible *curva isochrona* (satisfying $f(r).dr/ds \propto s$) is a hypocycloid defined by $s^2 \propto r^2 - k^2$, k constant, is entirely straightforward, as Newton doubtless realises.

Evidently deciding in afterthought that the remark was perhaps not entirely transparent to his reader, Newton inserted at this point in the second edition of his *Principia* ($_2$1713: 140) a following '*Corol.* Vis qua corpus T in loco quovis T acceleratur vel retardatur in Cycloide, est ad totum corporis ejusdem pondus in loco altissimo S vel Q, ut Cycloidis arcus TR ad ejusdem arcum SR vel QR' (*Corollary*. The force whereby the body T is at any place T accelerated or slowed in the cycloid is to the total weight of the same body at the highest point S or Q as the cycloid's arc $\widehat{TR}$ to its arc $\widehat{SR}$ or $\widehat{QR}$).

(285) Originally just '...describe circuli quadrantem HKG' (describe the circle-quadrant HKG).

(286) Since (to continue the analytical terms of note (284)) on taking the arc $\widehat{RT}(=GL) = s$ to be traversed in time t the orbital acceleration d^2s/dt^2 is equal to the component

$$f(r).dr/ds \propto r.dr/ds \propto s$$

of the central force acting instantaneously along the constraining cycloidal arc at T, the equation of motion may be set as dv/dt (or $v.dv/ds$) $= -ks$, where $v = ds/dt$ is the orbital speed at T and k is some constant; whence, much as in Proposition XXXVIII (see note (186) above), $v.dv/ds + ks = \frac{1}{2}d(v^2 + ks^2)/ds = 0$ and so $v = ds/dt = k^{\frac{1}{2}}.\sqrt{[S^2 - s^2]} = k^{\frac{1}{2}}.\sqrt{[\widehat{RS}^2 - \widehat{RT}^2]} = k^{\frac{1}{2}}.LI$ where $v = 0$ at S (not necessarily the cusp of the hypocycloid which Newton here takes it to be) distant $\widehat{RS}(= GH = GI) = S$ from R in the cycloid, thus determining

$$t = (1/k^{\frac{1}{2}}).\int_S^s 1/\sqrt{[S^2 - s^2]}.ds = (1/k^{\frac{1}{2}}).\sin^{-1}(s/S) = (1/k^{\frac{1}{2}}S).\widehat{KI}.$$

In particular, the time taken to traverse the full half-arc $\widehat{SR}$ from rest at S is $(1/k^{\frac{1}{2}}S).\widehat{HK}$, with the pendulating body attaining at $R(s = 0)$ its maximum orbital speed

$$k^{\frac{1}{2}}S = k^{\frac{1}{2}}.KG, \quad k = g/S = g(R+\rho)/4R\rho.$$

equal, are equal at the places T and L, it is clear that those bodies describe equal spaces $\widehat{ST}$, HL at the start, and so proceed thereafter to be equally urged on and to describe equal spaces. In consequence (by Proposition XXXVIII) the time in which the body describes the arc $\widehat{ST}$ is to the time of one swing as the arc $\widehat{HI}$ (the time in which the body H will reach L) to the semicircle $\widehat{HKM}$ (the time in which the body will reach M), and the speed of the pendulating body at the place T is to its speed at the lowest place R (that is, the speed of the body H at the place L to its speed at the place G, or the instantaneous increment of the line HL to the instantaneous increment of the line HG where the arcs $\widehat{HI}$, $\widehat{HK}$ increase at a uniform flow) as the ordinate LI to the radius GK, that is, as $\sqrt{[SR^2 - TR^2]}$ to SR. Whence, since in equal oscillations there are in equal times described arcs proportional to the total arcs of swing, from the times given there are obtained both the speeds and the arcs described in entire swings. As was first to be found.(286)

Newton's three following paragraphs long-windedly elaborate—but only trivially extend—this primary result, and were in the *Principia*'s second edition ($_2$1713: 141–2) reduced pithily to be but one: 'Oscillentur jam Funipendula corpora in Cycloidibus diversis intra Globos diversos, quorum diversæ sunt etiam Vires absolutæ, descriptis: &, si Vis absoluta Globi cujusvis QOS dicatur V, Vis acceleratrix qua Pendulũ urgetur in circumferentia hujus Globi, ubi incipit directe versus centrum ejus moveri, erit ut distantia Corporis penduli a centro illo & Vis absoluta Globi conjunctim, hoc est, ut $CO \times V$. Itaque lineola HY, quæ sit ut hæc Vis acceleratrix $CO \times V$, describetur dato tempore; &, si erigatur normalis YZ circumferentiæ occurrens in Z, arcus nascens HZ denotabit datum illud tempus. Est autem arcus hic nascens HZ in subduplicata ratione rectanguli GHY, adeoque ut $\sqrt{GH \times CO \times V}$. Unde Tempus oscillationis integræ in Cycloide QRS (cum sit ut semiperipheria HKM, quæ oscillationem illam integram denotat, directe utque arcus HZ, qui datum tempus similiter denotat, inverse) fiet ut GH directe & $\sqrt{GH \times CO \times V}$ inverse, hoc est, ob æquales GH & SR, ut $\sqrt{\dfrac{SR}{CO \times V}}$, sive (per Corol. Prop. L) ut $\sqrt{\dfrac{RA}{AC \times V}}$. Itaque Oscillationes in Globis & Cycloidibus omnibus, quibuscunque cum Viribus absolutis factæ, sunt in ratione quæ componitur ex subduplicata ratione longitudinis Fili directe, & subduplicata ratione distantiæ inter punctum suspensionis & centrum Globi inverse, & subduplicata ratione Vis absolutæ Globi etiam inverse. Q.E.I.' (Now let bodies pendulating on cords swing in different cycloids described within different globes, wherein the absolute forces are also different, and then, if the absolute force of any globe QOS be called V, the accelerative force by which the pendulum is urged in the circumference of this globe when it begins to move directly towards the centre of it will be jointly as the distance of the pendulating body from that centre and as the absolute force of the globe, that is, as $CO \times V$. Consequently, the line HY which shall be as this accelerative force $CO \times V$ will be described in given time, and so, if there be erected the normal YZ meeting the circumference in Z, the nascent arc $\widehat{HZ}$ will denote that given time. But this nascent arc $\widehat{HZ}$ is in the halved ratio of the rectangle $HG \times HY$, and therefore as $\sqrt{(HG \times CO \times V)}$. Hence the time of a complete swing in the cycloid arc $\widehat{QRS}$—since it is directly as the half-circumference $\widehat{HKM}$ denoting that complete swing and inversely as the arc $\widehat{HZ}$ similarly denoting the given time—will prove to be directly as HG and inversely as $\sqrt{(HG \times CO \times V)}$, that is, because HG and $\widehat{SR}$

[*Cas. 2.*] Oscillentur jam funipendula duo corpora in Cycloidibus inæqualibus et earum semiarcubus æquales capiantur rectæ *GH*, *gh*, centrisꝗ *G*, *g* et intervallis *GH*, *gh* describantur semicirculi *HZKM*, *hzkm*. In eorum diametris *HM*, *hm* capiantur lineolæ æquales *HY*, *hy*, [et] erigantur normaliter[287] *YZ*, *yz* circumferentijs occurrentes in *Z* et *z*. Quoniam corpora pendula sub initio motûs versantur in circumferentia globi *QOS*, adeoꝗ a viribus æqualibus urgentur in centrum, incipiuntꝗ directè versus centrum moveri, spatia simul confecta æqualia erunt sub initio. Urgeantur igitur corpora *H*, *h* a viribus ijsdem in *H* et *h*, sintꝗ *HY*, *hy* spatia æqualia ipso motûs initio descripta, et arcus *HZ*, *hz* denotabunt æqualia tempora. Horum arcuum nascentium ratio prima duplicata est eadem quæ rectangulorum *GHY*, *ghy*,[288] id est eadem quæ linearum *GH*, *gh*, adeoꝗ arcus capti in dimidiata ratione semidiametrorum denotant æqualia tempora. Est ergo tempus totum in circulo *HKM*, oscillationi in una Cycloide respondens, ad tempus totum in circulo *hkm* oscillationi in altera Cycloide respondens, ut semiperiferia *HKM* ad medium proportionale inter hanc semiperiferiam et semiperiferiam circuli alterius *hkm*, id est in dimidiata ratione diametri *HM* ad diametrum *hm*, hoc est in dimidiata ratione perimetri Cycloidis primæ ad perimetrum Cycloidis alterius, adeoꝗ tempus illud in Cycloide quavis est (per Corol. 3 Prop XLIX) ut latus quadratum rectanguli *BEC* contenti sub semidiametro rotæ qua Cyclois descripta fuit et differentia inter semidiametrum illam et semidiametrum globi. Q.E.I. Est et idem tempus (per Corol. Prop. L) in dimidiata ratione longitudinis fili *AR*. Q.E.I.

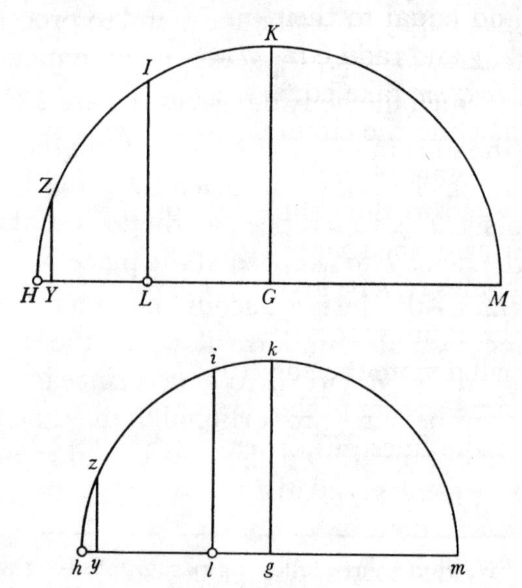

are equal, as $\sqrt{(\widehat{SR}/CO \times V)}$ or, by the Corollary to Proposition L, as $\sqrt{(AR/AC \times V)}$. In consequence, swings in all globes and cycloids, whatever be the absolute forces with which they are made, are in a ratio compounded of the halved ratio of the cord-length directly, the halved ratio of the distance between the point of suspension and the globe's centre inversely, and the halved ratio of the absolute force of the globe also inversely. As was to be found). It will be clear that, since Newton's *vis absoluta* V determines the central force to C at the distance $CT = r$ to be $f(r) = -Vr$, at the distance $CO = CS = R+\rho$ where, in the terms of note (186), it is equal to the acceleration g towards R in the hypocycloid $\widehat{STR}$ there will be $CO \times V = g$, so that here (as in the more discursive argument in the original text which follows above) the constant period $(2\pi k^{-\frac{1}{2}})$ of the 'harmonic' oscillatory motion defined by $d^2s/dt^2 = -ks$ is correctly assigned to be proportional to $\sqrt{(\widehat{SR}/g)} = \sqrt{(S/g)} = \sqrt{(1/k)}$.

[*Case 2.*] Now let two bodies pendulating on cords swing in unequal cycloids, and equal to their half-arcs take the straight lines *GH*, *gh*, then with centres *G*, *g* and radii *GH*, *gh* describe the semicircles *HZKM*, *hzkm*. In their diameters *HM*, *hm* take equal line-elements *HY*, *hy* and erect *YZ*, *yz* at right angles,(287) meeting the circumferences in *Z* and *z*. Because the pendulating bodies are at the start of motion located in the globe's circumference $\widehat{QOS}$, and are hence urged to the centre by equal forces and begin to move directly towards the centre, the spaces they simultaneously cover will at the start be equal. Let the bodies *H*, *h* be urged, therefore, by identical forces at *H* and *h*, and let *HY*, *hy* be the equal spaces described at the very start of motion, and then the arcs $\widehat{HZ}$, $\widehat{hz}$ will denote the equal times. Of these nascent arcs the doubled first ratio is the same as that of the rectangles $HG \times HY$ and $hg \times hy$,(288) that is, the same as that of the lines *HG*, *hg*, and therefore arcs taken in the halved ratio of the radii represent equal times. Consequently, the total time in the circle-arc $\widehat{HKM}$ (corresponding to an oscillation in one cyloid) is to the total time in the circle-arc $\widehat{hkm}$ (corresponding to a swing in the other cyloid) as the semi-circumference $\widehat{HKM}$ to the mean proportional between this semi-circumference and the semi-circumference of the other circle $\widehat{hkm}$, that is, in the halved ratio of the diameter *HM* to the diameter *hm*, or in the halved ratio of the perimeter of the first cycloid to the perimeter of the second cycloid; and hence the time in any cycloid is (by Corollary 3 of Proposition XLIX) as the square root of the rectangle $BE \times EC$ contained by the radius of the wheel whereby the cycloid was described and the difference between that radius and the radius of the globe. As was to be found. The same time is also (by the Corollary to Proposition L) in the halved ratio of the length of the thread *AR*. As was to be found.

(287) Originally 'ordinatæ' (ordinate).

(288) Since in the limit as they become vanishingly small

$$\widehat{HZ} = \sqrt{(HY \times HM)} = \sqrt{(HY \times 2HG)} \quad \text{and} \quad \widehat{hz} = \sqrt{(hy \times 2hg)}.$$

(289) In the published *Principia* (₁1687: 155) the paragraph concludes: 'Cum igitur Oscillationum tempora in globo dato sint in dimidiata ratione longitudinis *AR*, atq; adeo (ob datum *AC*) in dimidiata ratione numeri $\frac{AR}{AC}$, id est in ratione integra numeri $\sqrt{\frac{AR}{AC}}$; & hic numerus $\sqrt{\frac{AR}{AC}}$ servata ratione *AR* ad *AC* (ut fit in Cycloidibus similibus) idem semper maneat, & propterea in globis diversis, ubi Cycloides sunt similes, sit ut tempus: manifestum est quod Oscillationum tempora in alio quovis globo dato, atq; adeo in globis omnibus concentricis sunt ut numerus $\sqrt{\frac{AR}{AC}}$, id est, in ratione composita ex dimidiata ratione longitudinis fili *AR* directe & dimidiata ratione semidiametri globi *AC* inverse. Q.E.I.' (Since, then, the times of oscillation in the given globe are in the halved ratio of the length *AR*, and hence, because *AC* is given, in the halved ratio of the number *AR*/*AC*, that is, in the whole ratio of the number $\sqrt{(AR/AC)}$;

Porrò si in globis concentricis describantur similes Cycloides; quoniam earum perimetri sunt ut semidiametri globorum et vires in analogis perimetrorum locis sunt ut distantiæ locorum a communi globorum centro, hoc est ut globorum semidiametri, atq; adeo ut Cycloidum perimetri et perimetrorū partes similes, æqualia erunt tempora quibus perimetrorum partes similes oscillationibus similibus describuntur, et propterea oscillationes omnes erunt isochronæ. (289)Atqui latus illud quadratum cui tempus proportionale est reducitur in globis inæqualibus ad æqualitatem applicando ipsum ad globi cujusq; semidiametrum et propterea tempora oscillationum in globis quibuscunq; (qui æqualibus viribus absolutis attractivis pollent(290)) sunt ut latera illa directè et semidiametri globorum inversè[,] id est ut rectangulorum RAC quæ sub pendulorum longitudinibus AR, et centrorum a quibus pendent, centriq; globorum distantijs AC continentur latera quadrata directè et distantiæ illæ AC inversè, sive ut numerus $\sqrt{\frac{AR}{AC}}$. Q.E.I.

Deniq; si vires absolutæ diversorum globorum ponantur inæquales, accelerationes temporibus æqualibus factæ erunt ut vires. Unde si tempora capiantur in dimidiata ratione virium inversè, velocitates erunt in eadem dimidiata ratione directè et propterea spatia erunt æqualia quæ his temporibus describuntur. Ergo oscillationes in globis et Cycloidibus omnibus, quibuscunq; cum viribus absolutis factæ[,] sunt in ratione quæ componitur ex dimidiata ratione longitudinis penduli directè et dimidiata ratione distantiæ inter centrum penduli et centrum globi inversè et dimidiata ratione vis absolutæ etiam inversè, id est, si vis illa dicatur V, in ratione numeri $\sqrt{\frac{AR}{AC\times V}}$. Q.E.I.(291)

Corol. 1. Hinc etiam oscillantium cadentium et revolventium corporum tempora possunt inter se conferri. Nam si rotæ qua Cyclois intra Globum describitur diameter constituatur æqualis semidiametro globi, Cyclois evadet linea recta per centrum globi transiens,(292) et oscillatio jam erit descensus et

---

while this number $\sqrt{(AR/AC)}$ with the ratio of AR to AC preserved (as occurs in similar cycloids) remains ever the same, and consequently in different globes within which the cycloids are similar is as the time: it is manifest that the times of oscillation in any other given globe, and hence in all concentric globes, are as the number $\sqrt{(AR/AC)}$, that is, in the ratio compounded of the halved ratio of the cord-length AR directly and the halved ratio of the globe's radius AC inversely. As was to be found).

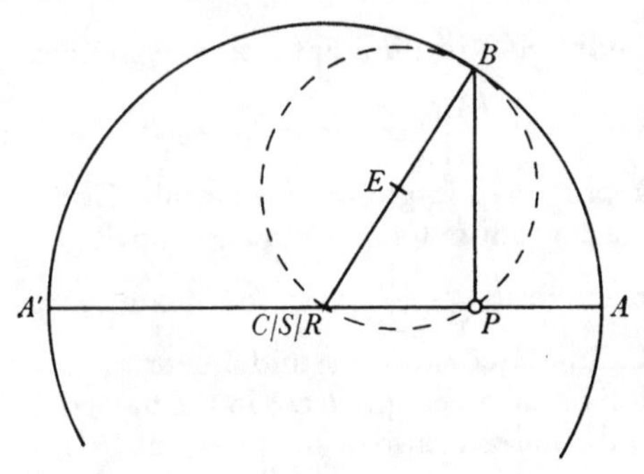

(290) Originally 'eadem vi absoluta pollentibus' (powered with the same absolute force).

(291) See our preceding observation in note (286).

(292) For, by construction, the circle-arcs

$$\widehat{BA} = BC.\widehat{BCA} \text{ and } \widehat{BP} = BC.\widehat{BCP}$$

are equal so that $\widehat{BCA} = \widehat{BCP}$ and hence P is in-

Furthermore, if in concentric globes similar cycloids be described, because their perimeters are as the radii of the globes and the forces at analogous places in the perimeters are as the distances of the places from the globes' common centre, that is, as the radii of the globes and hence as the cycloids' perimeters and similar portions of their perimeters, the times in which similar portions of the perimeters are described in similar oscillations will be equal, and accordingly all the swings will be isochronous. (289)The square root, however, to which the time is proportional is reduced in unequal globes to equality by dividing it by the radius of each globe; accordingly, the times of oscillation in any globes whatever (which have equi-potent absolute attractive forces(290)) are as those roots directly and the radii of the globes inversely, that is, as the square roots of the rectangles $AR \times AC$ contained by the lengths AR of the pendulums and the distances AC between the centres from which they hang and the centres of the globes directly and those distances AC inversely, or as the number $\sqrt{[AR/AC]}$. As was to be found.

Finally, if the absolute forces of different globes be supposed unequal, the accelerations achieved in equal times will be as the forces. Whence, if the times be taken in the halved ratio of the forces inversely, the speeds will be in the same halved ratio directly, and accordingly the spaces described in these times will be equal. In consequence, the oscillations in all globes and cycloids, whatsoever be the absolute forces with which they are accomplished, are in a ratio compounded of the halved ratio of the pendulum's length directly, the halved ratio of the distance between the centre of the pendulum and the centre of the globe inversely, and the halved ratio of the absolute force also inversely—in other terms, if that force be called V, in the ratio of the number $\sqrt{[AR/AC \times V]}$. As was to be found.(291)

Corollary 1. Hence also the times of oscillating, falling and revolving bodies can be compared one with another. For, if the diameter of the wheel by which a cycloid within a globe is described be set equal to the globe's radius, the cycloid will prove to be a straight line passing through the globe's centre,(292) and

stantaneously the orthogonal projection of B upon the rectilinear 'arc' AS. This instance is, of course, the canonical case of simple harmonic motion propounded by Newton in Proposition XXXVIII above, in which the body B rotating uniformly in the circle ABA' determines, at any point in time defined by the angle $\widehat{ASB}$, the position of the corresponding point P moving in the diameter ACA' such that instantaneously its acceleration to S is proportional to the distance SP. Shorn of its dynamical interpretation, the basic kinematical model occurs widely in late medieval Arabic and early modern European astronomical theory as a mechanism for uniformly and periodically varying the lengths of primary lunar and planetary *radii vectores*, and indeed has recently, in honour of its putative inventor Naṣīr al-Dīn al-Ṭūsī, come widely to be named a 'Ṭūsī couple'. (See J. L. E. Dreyer, *A History of the Planetary Systems from Thales to Kepler* (Cambridge, 1906): 268–9; and especially E. S. Kennedy, 'Late Medieval Planetary Theory', *Isis* **57**, 1966: 365–78, particularly 368–70 where the model is described

subsequens ascensus in hac recta. Unde datur tum tempus descensus de loco quovis ad centrum, tum tempus huic æquale quo corpus uniformiter circa centrum globi ad distantiam quamvis revolvendo arcum quadrantalem describit. Est enim hoc tempus (per casum secundum) ad tempus semi-oscillationis in Trochoide quavis $A[P]S$ ut $\frac{1}{2}BC$ ad $\sqrt{BEC}$.(293)

Corol. 2. Hinc etiam consectantur quæ D. C. Wrennus et D. C. Hugenius de Cycloide vulgari adinvenerunt. Nam si globi diameter augeatur in infinitum, mutabitur ejus superficies sphærica in plan[u]m, visq; centripeta aget uniformiter secundum lineas huic plano perpendiculares et Cyclois nostra abibit in Cycloidem vulgi. Isto autem in casu longitudo arcus Cycloidis inter planum illud et punctum describens æqualis evadet quadruplicato sinui verso dimidij arcus Rotæ inter idem planum et punctum describens, ut invenit D. C. Wrennus,(294) et pendulum inter duas ejusmodi Cycloides in simili et æquali Cycloide in temporibus æqualibus oscillabitur[,] ut demonstravit Hugenius.(295) Sed et descensus gravium tempore oscillationis unius is erit quam Hugenius indicavit.(296)

Aptantur autem Propositiones a nobis demonstratæ ad veram constitutionem Terræ, quatenus Rotæ eundo in ejus circulis maximis describunt motu

under the head 'al-Ṭūṣī's rolling device'.) Nicolaus Copernicus—who during his early stay in Italy may well (though this is not yet proven) have met with manuscripts recording the lunar and planetary models, structurally identical with his own, which had been propounded two centuries earlier by the Damascus astronomer Ibn al-Shāṭir—elaborates the kinematical model, with full proof that it generates an oscillating rectilinear motion, in his *De Revolutionibus Orbium Cœlestium, Libri VI* (Nuremberg, 1543): Liber III, Caput IIII. 'Quomodo motus reciprocus siue librationis ex circularibus constet': $67^r/67^v$, applying it thereafter in both his Moon and Mercury theories. But Newton, who was ill-read in pre-Keplerian astronomy, would have been unlikely in 1685 to have known any of these antecedents.

(293) Altered minimally in the *Principia*'s second edition ($_2$1713: 142) to conclude 'in Cycloide quavis QRS ut 1 ad $\sqrt{\frac{AR}{AC}}$' (in any cycloid $\widehat{QRS}$ as 1 to $\sqrt{[AR/AC]}$) in line with the preceding revised text reproduced in note (286). Evidently, when the hypocycloid $\widehat{QRS}$ becomes a straight line passing through the globe's centre C, the points R and C coincide and the vertex point A (the centre of curvature of $\widehat{QRS}$ at R) passes to infinity, so that the ratio AR/AC tends to unity in the limit.

(294) In his 1658 "Εὐθυσμὸς Curvæ lineæ Cycloidis primariæ', inserted the next year by John Wallis into his *Tractatus Duo. Prior, De Cycloide*... (Oxford, 1659): 72–80; compare III: 163, note (308).

(295) In early December 1659 initially, though he published the result—in a much polished, classically recast form—only in his *Horologium Oscillatorium, sive De Motu Pendulorum ad Horologium aptato Demonstrationes Geometricæ* (Paris, 1673): Part II, Propositions XVI–XXVI: 43–58; compare III: 392–5.

(296) Some twelve years earlier Newton had himself composed a short paper (reproduced on III: 420–30) wherein he examines in some detail the harmonic motion engendered in an upright primary cycloid by constant, downwards directed 'gravity'. While stressing its several

the oscillation will be now a descent and thereafter an ascent in this line. Whence there is given both the time of descent from any place to the centre, and the time equal to this in which a body by revolving uniformly round the centre of the globe at any distance describes a quadrantal arc. For (by Case 2) this time is to the time of a half-swing in any cycloid $\widehat{APS}$ as $\frac{1}{2}BC$ to $\sqrt{[BE\times EC]}$.(293)

Corollary 2. Hence also follow what Mr C. Wren and Mr C. Huygens have discovered regarding the common cycloid. For if the globe's diameter be indefinitely increased, its spherical surface will be changed into a plane, the centripetal force will act uniformly along lines perpendicular to this plane, and our cycloid will pass into a common one. In that case, however, the arc-length of the cycloid betwee nthat plane and the describing point will turn out to be four times the versed sine of half the wheel's arc between the same plane and describing point, as Mr C. Wren found out,(294) while a pendulum beating between two cycloids of this type will oscillate in equal times in a congruent cycloid, as Huygens has demonstrated.(295) Also, the descent of heavy bodies in the time of one swing will be that which Huygens has pointed out.(296)

The propositions we have demonstrated, however, conform to the true constitution of the Earth inasmuch as wheels, by going in great circles round it,

---

parallels with Huygens' kinematic proof, we earlier (see III: 391 and 422–3, note (14)) insisted on the originality of its underlying dynamical basis—one which Newton but lightly adapts in his present generalised demonstration that hypocycloidal pendular motion is isochronous in a direct-distance force-field round the centre of its parent 'globe'. In this earlier special case, the centre C lies at infinity and the generating circle of diameter OR traces the arc $\widehat{SR}=S=2OR$ of a 'common' cycloid by rolling along underneath the line SOQ (now straight): where the constant downwards gravity is g, it then follows much as before (see note (286)) that the equation of motion defining the time t of passage over the arc $\widehat{TR}=s$ is again $dv/dt=-(g/S)\,s$ on taking $v=ds/dt$ to be the instantaneous speed at T; whence $\frac{1}{2}d(v^2+(g/S)\,s^2)/ds=0$, so determining $v=(g/S)^{\frac{1}{2}}\sqrt{[S^2-s^2]}$ and thereafter $s=S\sin(g/S)^{\frac{1}{2}}t$ with the period of swing $2\pi(S/g)^{\frac{1}{2}}$, as Huygens showed by his variant Galileian argument.

It is no more difficult to formulate—though not, of course, to resolve—the equation $dv/dt=-(g/S)\,s+\rho$ of motion in the same cycloid (and *mutatis mutandis* in the hypocycloidal tautochrone in a general direct-distance field) under some decelerating force of resistance ρ instantaneously acting directly contrary to the onwards motion at T. This extension was in fact afterwards explored by Newton in Propositions XXV–XXX of Book 2 of his published *Principia* ($_1$1687: 305–15), where he successively set ρ to be constant, then proportional to the speed v and its square, and finally some combination of low powers of v. Since these (only partly successful) attempts at exact solution in these particular instances and at approximate reduction in the general case illustrate perhaps better than anything in his more narrowly mathematical papers the severe limitations of Newton's geometricised techniques for resolving differential—or, more accurately, infinitesimal—equations other than by converting their 'roots' into (as he hoped) converging infinite series, we have thought fit to reproduce substantial extracts from their text in Appendix 5.

clavorum[297] Cycloides extra globum,[298] et pendula inferius in fodinis et cavernis Terræ in Cycloidibus intra globos oscillari debent ut oscillationes omnes evadant isochronæ. Nam gravitas (ut in libro tertio[299] docebitur) decrescit a superficie Terræ sursum quidem in duplicata ratione distantiarum a centro ejus, deorsum verò in ratione simplici.

Prop. LIII. Prob. XXXV.

Concessis figurarum curvilinearum quadraturis, invenire vires quibus corpora in datis curvis lineis oscillationes semper isochronas peragent.

Oscilletur corpus T in curva quavis linea $STRQ$ cujus axis sit $[A]R$ transiens [per][300] virium centrum C. Agatur TX quæ curvam illam in corporis loco quovis T contingat, inqꝫ hac Tangente TX capiatur TY æqualis arcui TR. Nam longitudo arcus illius ex figurarum quadraturis per methodos vulgares[301] innotescit. De puncto Y educatur recta YZ Tangenti perpendicularis. Agatur CT perpendiculari illi occurrens in Z, et erit vis centripeta proportionalis rectæ TZ. Q.E.I.[302]

Nam si vis qua corpus trahitur de T versus C, exponatur per rectam TZ captam ipsi proportionalem, resolvetur hæc in vires TY, YZ quarum YZ

(297) Those set in their outside rims only: 'nails' placed elsewhere within wheels rolling over the Earth's surface (here presumed to be exactly spherical) will trace epitrochoids.

(298) That is, epicycloids (see note (276) above): in a cancelled following phrase in the manuscript these were originally qualified as 'quarum longitudines determinavimus in Propositione XLVIII' (whose lengths we determined in Proposition XLVIII). The example is near-trivial in realistic dynamical terms since, as Newton surely realised, a direct-distance *vis centrifuga* is needed to induce motion of constant period in such curves. The following superficially more plausible instance of isochronous pendular motion in '(hypo)cycloids within globes' below the Earth's surface—which, as we have already remarked (§2: note (186)), was suggested to him by Robert Hooke in his letter of 6 January 1679/80—is in fact scarcely more in tune with geophysical reality since, on account of the Earth's relatively heavy core, the 'pull' of terrestrial gravity continues slightly to increase over the first thousand or so miles of descent down from the surface towards its centre (see *ibid.*).

(299) Newton looks ahead to Propositions III and IX of Book 3 of the published *Principia* ($_1$1687: 405–6 and 416 respectively). According to the former, '*Vim qua Luna retinetur in Orbe suo respicere terram, & esse reciprocè ut quadratum distantiaæ locorum ab ipsius centro*.... Patet enim, per Corol. 1. Prop. XLV. Lib. I. quod si distantia Lunæ à centro Terræ dicatur D, vis à qua motus talis oriatur, sit reciprocè ut $D^{2\frac{4}{243}}$...hoc est in ratione distantiæ paulo majore quam duplicata inverse, sed quæ vicibus $60\frac{3}{4}$ propius ad duplicatam quam ad triplicatam accedit. Tantillus autem accessus meritò contemnendus est. Oritur verò ab actione Solis... & propterea hic negligendus est'. By the latter, '*Gravitatem pergendo à superficiebus Planetarum deorsum decrescere n ratione distantiarum à centro quam proximè*. Si materia Planetæ quoad densitatem uniformis esset, obtineret hæc Propositio accuratè per Prop. LXXIII. Lib. I [= Proposition XLII of the preliminary tract 'De motu Corporum' reproduced in §2 above]. Error igitur tantus est, quantus ab inæquabili densitate oriri possit'.

(300) The corner of the manuscript page (f. 79^v) on which this word was penned has since been torn away.

describe by the motion of their nails[297] cycloids outside the globe,[298] while pendulums farther down in mines and caverns in the Earth must oscillate in cycloids within globes so as to have all their vibrations isochronous. For gravity (as will be explained in the third book[299]) decreases upwards from the Earth's surface, indeed, in a doubled ratio of the distances from its centre, but down below it in their simple (direct) ratio.

Proposition LIII, Problem XXXV.

Granted the quadratures of curvilinear figures, to find the forces whereby bodies in given curves shall perform oscillations which are ever isochronous.

Let the body T oscillate in any curve $\widehat{STRQ}$, whose axis shall be AR passing through the force-centre C. Draw TX to touch that curve at any place T of the body, and in this tangent TX take TY equal to the arc $\widehat{TR}$. (The length of that arc is, of course, ascertained from the quadratures of figures by ordinary methods.[301]) Out from the point Y draw the straight line YZ perpendicular to the tangent, and join CT meeting that perpendicular in Z, and the centripetal force will be proportional to the line TZ. As was to be found.[302]

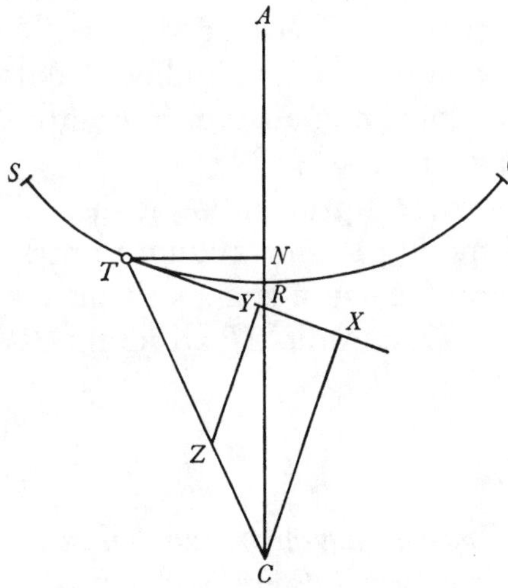

For, if the force whereby the body is drawn from T towards C be represented by the line TZ taken proportional to it, this will be resolved into the forces TY

(301) Understand the several methods of algebraic quadrature encapsulated by Newton in Problem 9 of his 1671 tract (III: 210–92, especially 236–54) and the related paper on 'The Quadrature of all curves whose æquations consist of but three terms' (III: 374–85, particularly 380–2). At about this time, we may add, he had it in mind to insert a version of the latter paper in sequel in his present 'Liber primus', there elaborating for his untutored reader the significance of his several appeals—not only here but, as we have already seen, in Proposition XLI above and also in Proposition LIV following—to be 'conceded the quadrature of curvilinear figures'. The extant preliminary drafts and revised version of this abortive, unnumbered 'Prop. Prob.' are reproduced in Appendix 6 below.

(302) In effect, where $f(r)$ is the force acting at T towards the centre C, distant $CT = r$ away, to induce an oscillatory motion in the given constraining curve SRQ, Newton states the general condition for isochronism that the component $f(r).dr/ds$ of the force in the instantaneous direction of motion at T shall ever be proportional to its arc-distance $\widehat{TR} = s$ from some fixed point R in the constraining curve. If the end-point S of the arc $\widehat{SR} = S$—that at which the motion comes to rest instantaneously—is distant $CS = R$ from the force-centre, it follows much as before that $s = \sqrt{[S^2 - v^2]}$, where v is the speed of oscillation at T.

trahendo corpus secundum longitudinem fili PT motum ejus nil mutat, vis autem altera TY motum ejus in curva $STRQ$ directè accelerat vel directè retardat. Proinde cum hæc sit ut via describenda TR, accelerationes corporis vel retardationes in oscillationum duarum (majoris et minoris) partibus proportionalibus describendis, erunt semper ut partes illæ, et propterea facient ut partes illæ simul describantur. Corpora autem quæ partes totis semper proportionales simul describunt, simul describent totas. Q.E.D.

Corol. 1. Hinc si corpus T filo rectilineo AT a centro A pendens, describat arcum circularem $STRQ$, et interea urgeatur secundum lineas parallelas deorsum a vi aliqua quæ sit ad vim uniformem gravitatis ut arcus TR ad ejus sinum TN: æqualia erunt oscillationum singularum tempora. Etenim ob parallelas TZ, AR, similia erunt triangula ANT, TYZ; et propterea TZ erit ad AT ut TY ad TN, hoc, est, si gravitatis vis uniformis exponatur per longitudinem datam AT, vis TZ qua oscillationes evadent isochronæ erit ad vim gravitatis AT ut arcus TR ipsi TY æqualis ad arcûs illius sinum TN.(303)

Corol. 2. Igitur in horologijs oscillatorijs,(304) si vires a Machina in pendulum ad motum conservandum impressæ, ita cum vi gravitatis componi possint ut vistota deorsum semper sit ut linea quæ oritur applicando rectangulum sub arcu TR et radio AR ad sinum TN, oscillationes omnes erunt isochronæ.

Prop. LIV. Prob. XXXVI.

Concessis figurarum curvilinearum quadraturis, invenire tempora quibus corpora vi qualibet centripeta in lineis quibuscunq; curvis in plano per centrum virium transeunte descriptis descendent et ascendent.

Descendat corpus de loco quovis S(305) per lineam quamvis curvam $STtR$ in plano per virium centrum C transeunte datam. Jungatur CS et dividatur eadem in partes innumeras æquales, sitq; Dd partium illarum aliqua. Centro C, intervallo CD, Cd describantur circuli DT, dt lineæ curvæ $STtR$ occurrentes in T et t.

---

(303) Whence at once (compare Newton's consectary 5 on III: 426), since $\widehat{TR} \approx TN$ for small angles $R\widehat{A}T$, gentle circular swings under (effectively constant) terrestrial gravity are virtually isochronous.

(304) That is, timepieces regulated by the periodic swing of a simple pendulum. This *terminus technicus* is, of course, taken over by Newton from the title (for which see note (295)) given by Huygens to his 1673 treatise on the theory and construction of such clocks.

(305) Where it is assumed to be at rest; compare notes (188) and (197) above.

and YZ, of which TY by drawing the body along the length of the thread PT does nothing to change its motion, but the other force TY directly accelerates or directly retards its motion in the curve $\widehat{STRQ}$. In consequence, since the latter is as the path $\widehat{TR}$ to be described, the body's accelerations or retardations in describing proportional parts of two swings—a greater one and a less—will ever be as those parts, and accordingly will cause those parts to be simultaneously described. Bodies, however, which simultaneously describe parts ever proportional to their wholes will simultaneously describe the wholes. As was to be proved.

Corollary 1. Hence, should a body T hanging from the centre A by a straight thread AT describe the circle-arc $\widehat{STRQ}$, and meanwhile be urged downwards along parallel lines by some force which shall be to the uniform force of gravity as the arc $\widehat{TR}$ to its sine TN, then the times of individual swings will be equal. For, because TZ and AR are parallel, the triangles ANT, TYZ will be similar, and accordingly TZ will be to AT as TY to TN, that is, if the uniform force of gravity be represented by the given length AT, the force TZ whereby the swings shall prove to be isochronous will be to the force AT of gravity as the arc $\widehat{TR}$ (equal to TY) to the sine TN of its arc.[303]

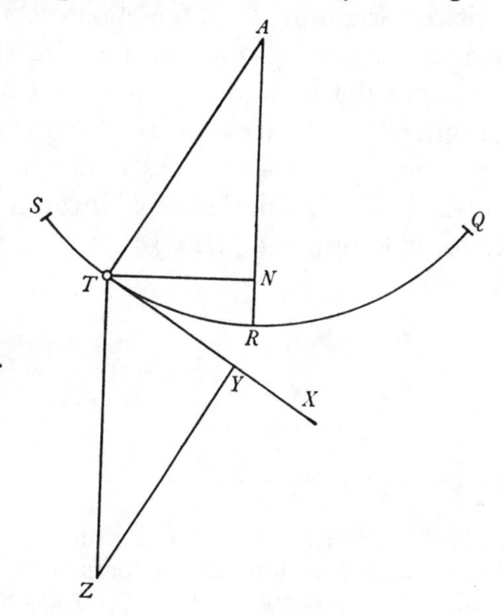

Corollary 2. In oscillatory clocks,[304] therefore, if the forces impressed by a machine on the pendulum to conserve its motion be so compoundable with the force of gravity that the total force downwards shall be ever as the line which ensues from dividing the rectangle beneath the arc $\widehat{TR}$ and the radius AR by the sine TN, all swings will be isochronous.

Proposition LIV, Problem XXXVI.

Granted the quadratures of curvilinear figures, to find the times in which bodies shall ascend and descend under any centripetal force whatever in any curves whatever described in a plane passing through the centre of force.

Let a body descend from any place S[305] by way of any curve $\widehat{STrR}$ in a plane passing through the force-centre C. Join CS and divide it into innumeral equal parts, and let Dd be some one of those parts. With centre C and radii CD, Cd describe circle-arcs $\widehat{DT}$, $\widehat{dt}$ meeting the curve $\widehat{STtR}$ in T and t. Then, given both

Et ex data tum lege vis centripetæ tum altitudine CS de qua corpus cecidit, dabitur velocitas corporis in alia quavis altitudine CT per (306)[Prop. XXXIX.(307) Tempus autem quo corpus describit lineolam Tt est ut lineolæ hujus longitudo (id est ut secans anguli tTC) directè, et velocitas inversè. Tempori huic proportionalis sit ordinatim applicata DN ad rectam CS per punctum D perpendicularis, et ob datum Dd erit rectangulum $Dd \times DN$, hoc est area $DNnd$, eidem tempori proportionale. Ergo si SNn(308) sit curva illa linea quam punctum N perpetuò tangit,(309) erit area $SNDS$(308) proportionalis tempori quo corpus descendendo descripsit lineam ST, proindeq; ex inventa illa area dabitur tempus. Q.E.I.]

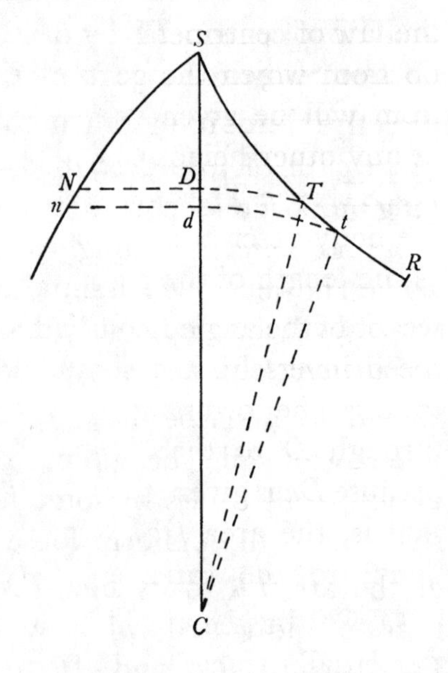

. . . .(310)

(306) The manuscript as it now survives here (at the bottom of f. 79v) terminates, though in its pristine state it manifestly continued without pause on following sheets which have now disappeared, probably (see note (1)) discarded by Newton himself when he deposited its preceding folios in Cambridge University Library as the purported written record of his Lucasian lectures during 1684–5. For completeness' sake we in sequel restore the remainder of the present proposition from the version—evidently only minimally revised—afterwards printed in the *Principia* ($_1$1687: 159).

(307) More accurately, 'XL'. Much as before, if the general point $T(r, \phi)$ of the constraining curve $\widehat{STR}$ is defined by some given polar equation referred to an origin $C(0, 0)$ at the centre of the force-field whose intensity at distance $CT = r$ is $f(r)$, that equation will yield by differentiation the ratio ds/dr of the limit-increment Tt of its arc-length $\widehat{TR} = s$ to that, Dd, of the radius vector $CD = CT$; hence, since the motion is (see note (305)) assumed to be instantaneously at rest at S, on taking $CS = R$ and setting the time of passage over $\widehat{TR}$ to be t it follows from Proposition XL that the speed $v = ds/dt$ of motion in the curve at T is $\sqrt{\left[2\int_R^r -f(r).dr\right]}$, and consequently the time $\int_R^r (ds/dr)/v.dr$ will be measured by

$$(SDN) = \int_{CS}^{CD} DN.d(CD)$$

where $DN \propto \sec \hat{T}/v$.

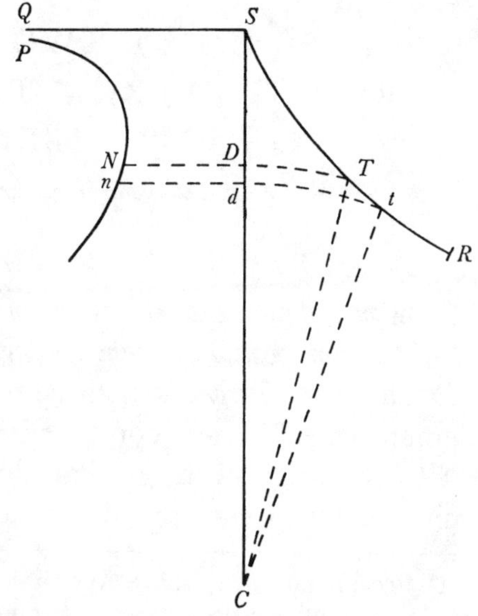

the law of centripetal force and the height CS from which the body has fallen, therefrom will be given the speed of the body at any other height CT by(306) Proposition XXXIX(307). The time, however, in which the body describes the curve-element Tt is as the length of the element (that is, as the secant of the angle $t\widehat{TC}$) directly and as the speed inversely. Let there be proportional to this the ordinate DN, perpendicular through D to the straight line CS, and, because Dd is given, the rectangle $Dd \times DN$, that is, the area $(DNnd)$, will be proportional to the same time. Therefore, if $[\widehat{P]Nn}$(308) be the curve which the point N perpetually traces,(309) the area $([P]\ NDS[Q])$(308) will be proportional to the time in which the body in descending has described the curve $\widehat{ST}$; consequently, once that area is ascertained, from it will be given the time. As was to be found.

. . . .(310)

---

(308) Forgetting that the motion in the arc $\widehat{STR}$ is assumed to come to rest at S, so that there (in the terms of the preceding note) $v = 0$ and hence $DN \propto (ds/dr)/v$ is infinite, Newton in his accompanying figure (here reproduced in the Latin text from *Principia*, $_1$1687: 159 = $_2$1713: 145) carelessly drew the locus of the end-points (N) to pass through S. The slip was caught by Henry Pemberton in his *editio ultima* ($_3$1726: 156), from whence we take the corrected figure set in the English version and the minimal textual amendments there introduced.

(309) Somewhat unnecessarily to emphasise his mending of the imperfection in Newton's original accompanying figure (see the previous note) Pemberton here inserted in the *Principia*'s third edition ($_3$1726:157) a following phrase 'ejusq; asymptotos sit recta SQ rectæ CS perpendiculariter insistens' (and its asymptote be the straight line SQ standing perpendicular to the line CS).

(310) Section X of the 'Liber primus' concludes in the published *Principia* ($_1$1687: 159–61) with two further related propositions analogously determining the planar oscillatory motion induced by a given force-field 'in any [symmetrical] curved surface whose axis passes through the centre of force'. We see no need here to reproduce these minimal generalisations of Proposition LIV.

APPENDIX 1. TWO GEOMETRICAL LEMMAS REJECTED FROM THE 'DE MOTU CORPORUM LIBER PRIMUS'.(1)

From the original transcript(2) in the University Library, Cambridge

Lemma [α].

In angulo dato Parallelogrammum magnitudine datum constituere quod angulo suo opposito rectam positione datam continget.

Docuit Euclides constructionem hujus Problematis in Prop. 28 et 29 libri sexti Elementorum.(3) Utere constructione vel Euclidea vel ea quam subjungimus.

Sit CAD angulus in quo Parallelogrammum constituendum est et CB recta positione data secans anguli latera in B et C.(4) Sitq; $CADE$ parallelogrammum

(1) See §3: note (22) above. We there suggested that Newton originally intended to apply the construction developed in the second lemma (to which the first is merely a rider) in the same astronomical context as that of Problem 16 of his earlier Lucasian lectures (see v: 210–12), where algebraic solution—mirroring the geometrical synthesis given in its own equivalent 'Cas. 1'—is afforded of the requirement 'Cometæ in linea recta...uniformiter progredientis positionem cursûs ex tribus observationibus determinare'; and that he cancelled these lemmas in favour of his ensuing Corollary to the following Lemma XXVII, whereby (compare §3: note (110)) the more advanced problem of determining a uniformly traversed rectilinear cometary path from four timed, non-concurrent terrestrial sightings is resolved. See also note (7) below.

(2) Dd. 9. 46: 53r/54r, there entered by Humphrey Newton from a lost original worksheet. The manuscript lacks the figures which once accompanied it and those reproduced are our restorations on the basis of the text. For convenience of reference we have listed Newton's two lemmas as [α] and [β] respectively.

(3) Isaac Barrow, *Euclidis Elementorum Libri XV. breviter demonstrati* (Cambridge, 1655): 134, 135. In these equivalent constructions Euclid effectively—in terms of Newton's present figure—first determines the mid-points, L and M say, of AC and BC, and then on the diagonal MC erects the parallelogram $CLMN$: on supposing that the parallelograms $ADEC$ and $AHIK$ are equal in area, it readily follows that the similar parallelogram on diagonal MI is equal to the combined areas of the parallelograms $ADEC$ and $CLMN$, from which at once

$$FH = \sqrt{[\tfrac{1}{4}BA^2+BA\times AD]}.$$

It is now usual to suppose that Euclid designed his propositions so as geometrically to solve the general quadratic equation by completing the square of the terms involving the unknown. Whatever be the historical truth of the matter, Newton's present construction equivalently determines the root $AH = \sqrt{[BA(\tfrac{1}{4}BA+AD)]}-\tfrac{1}{2}BA$ of the quadratic

$$BA\times AD = (AH\times BH \text{ or}) \; AH^2+BA\times AH.$$

(4) Assuming that the given line BI passes through the corner C of the given parallelogram $ADEC$: if it does not, we shall need to construct a parallelogram $AD'E'C'$ equal in area to it within the same angle $\widehat{CAD}$ (or, equivalently, the point D' in AD such that $AD' = AD\times AC/AC'$) and then, where F and

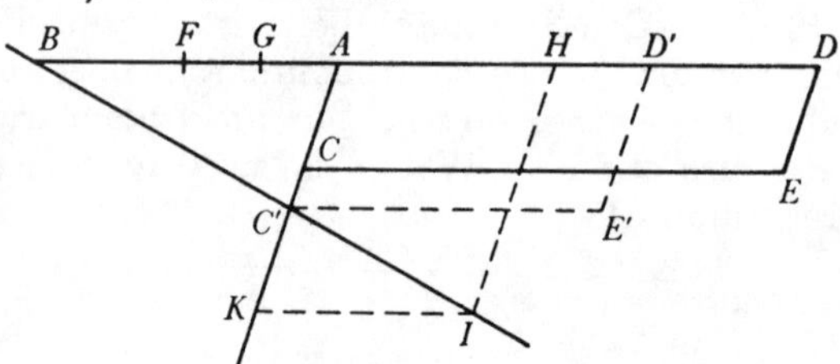

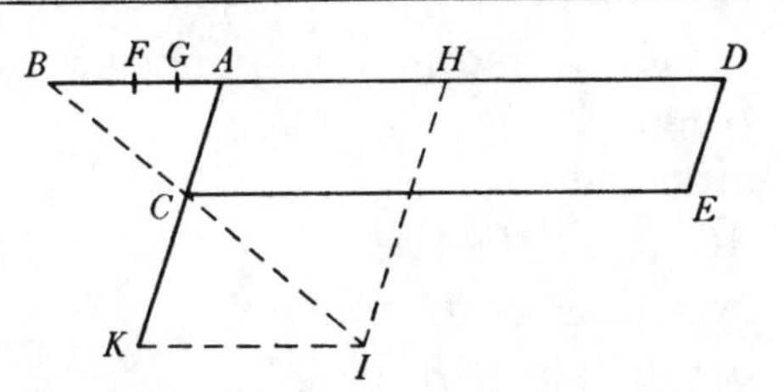

in angulo illo super AC constitutum cui parallelogrammum constituendum æquari debet. Biseca AB in F et AF in G et inter AB ac GD cape FH medium proportionale. Age HI occurrentem BC in I et comple parallelogrammum $AHIK$. Dico factum.(5)

Namq rectangulum DG in AB æquatur FH^{quadrato} et rectangulum AG in AB æquatur FG^{quadrato} indeq rectangulum DAB (id est $DG-AG$ in AB seu $FH^{\text{quad.}}-FA^{\text{quad}}$) æquatur rectangulo AHB, estq DA ad AH ut HB ad AB adeoq ut HI [ad] AC; et propterea parallelogramma $ADEC$, $AHIK$ sibi invicem æquantur. Q.E.D.

Lemma [β].

Rectam lineam per datum punctum ducere cujus partes rectis tribus positione datis interjectæ datam habebunt rationem ad invicem.(6)

Dentur positione tres rectæ AB, AC, BC et per datum punctum ducenda sit recta quarta DF cujus partes DE, EF prioribus rectis interjectæ sint ad invicem in ratione data M ad N.

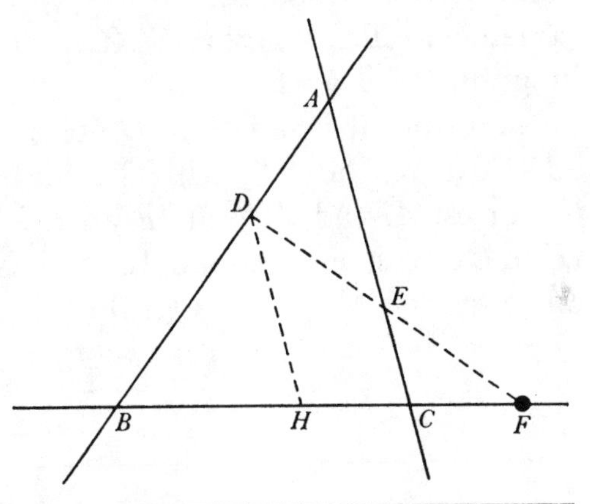

Cas. 1.(7) Primo jaceat punctum datum in aliqua rectarum positione datarum[,] puta in BC sitq illud F. In hac recta BC capiatur CH quæ sit ad CF in ratione illa data et rectæ CA agatur parallela HD occurrens rectæ tertiæ [AB] in D ac jungatur DF. Erit DE ad EF ut est CH ad CF. Q.E.F.

G remain the mid-points of BA and FA respectively (or $FA = \frac{1}{2}BA$, $GA = \frac{1}{4}BA$), take $FH = \sqrt{[BA \times GD']}$. In sequel it is this extended lemma to which Newton will, in fact, appeal (see note (8) below).

(5) For, to recreate the analysis which Newton proceeds synthetically to remould, since by supposition $AC \times AD = AH \times HI$, at once $AH/AD = (AC/HI \text{ or}) \, BA/BH$ and therefore $BA \times AD = AH \times BH = AH^2 + BA \times AH$, whence $FH^2 = (\frac{1}{2}BA + AH)^2 = \frac{1}{4}BA^2 + BA \times AD$, that is, $BA(GA + AD)$. It will be evident that two positions of H (symmetrically placed around F) are thereby determined, and hence two points: the intersections, namely, of the given line BI with the hyperbolic locus (I) defined by $AH \times HI = AD \times DE$, constant.

(6) Originally (compare Lemma XXVI in §3 preceding) this enunciated the more general requirement 'Triangulum [*sc.* specie datum] constituere cujus anguli contingant rectas positione datas [et] latera duo cum rectis [faciant datos angulos]'.

(7) In confirmation of our suggestion (in note (1)) that it was probably Newton's plan to apply this present lemma to construct cometary paths from given terrestrial sightings in the simplified hypothesis that they are uniformly traversed straight lines, this first case—essentially, as we have already noticed, the geometrical reformulation of just such a problem in his earlier Cambridge lectures (see v: 210–12)—was afterwards re-couched in its initial astronomical

Cas. 2. Jaceat jam punctum extra tres lineas positione data sitꝗ illud G. Junge AG et in ea cape hinc inde AH et GP ad AG ut est N ad M. Perꝗ puncta P et H age PX parallelam AB et HX parallelam AC ipsisꝗ PX, AB, BC occurrentem in X S et Q. Ad hanc agatur GK parallela AB et completo parallelogrammo $XKGL$ describatur (per Lemma superius[(8)]) in angulo KXL parallelogrammum $XOFT$ quod sit æquale parallelogrammo $XKGL$ et angulo suo F tangat rectam BCQ. Et jungatur GF. Dico factum.[(9)]

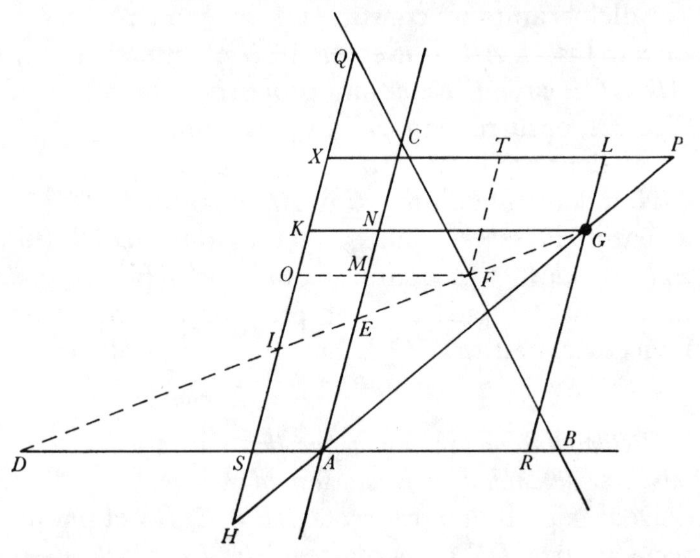

Nam cum sit IO ad IK ut OF and KG atꝗ adeo ut KX ad $XO_{[,]}$ et divisim IO ad OK ut KX ad OK, erit IO æqualis KX adeoꝗ æqualis SH et propterea ad GR ut est AH ad AG seu IE ad EG. Sed ob similia triangula IOF, GRD est IF ad GD in eadem ratione. Ergo divisim $F[E]$ est ad ED in eadem ratione, hoc est in ratione N ad M. Q.E.F.

---

context in Lemma 2 of the complementary 'De motu Corporum Liber secundus' (ULC. Add. 3990: 53r); see 2, §2: note (20) below.

(8) More exactly by the extension of Lemma [α] sketched in note (4) above.

(9) Restoration of the analysis of this case is considerably more revealing than Newton's elegant following synthetic demonstration. Where the variable point F is fixed by the oblique geometrical coordinate lengths AM (measured along AC) and MF (parallel to AB), the rotating line GF meets the given lines AB and AC in points D and E such that

$$(RG-AE)/AR = (RG-AM)/(AR-MF)$$

and therefore

$$(AR\times AM-RG\times MF)/(AR-MF) = AE = AM/(1+N/M),$$

since $AE\colon EM = DE\colon EF = M\colon N$; it follows that

$$AM\times MF+(N/M)\,AR\times AM-(1+N/M)\,RG\times MF = 0$$

and so $(MF+(N/M)\,AR)\,((1+N/M)\,RG-AM) = (1+N/M)\,AR\times(N/M)\,RG$, the defining equation of an Apollonian hyperbola referred to asymptotes $MF = -(N/M)\,AR$ and $AM = (1+N/M)\,RG$, that is, XH and XP on taking $SA = (N/M)\,AR$ and $GL = (N/M)\,RG$, whence (as Newton constructs these) $HA = GP = (N/M)\,AG$. The further restriction that F shall lie in BC is met by applying Lemma [α] to construct the (two) intersections of BC with the hyperbolic locus (F) now defined by $OF\times FT = KG\times GL$, constant.

APPENDIX 2. ADDITIONS AND CORRECTIONS TO THE SCHOLIUM TO PROPOSITION 31 OF THE 'DE MOTU CORPORUM LIBER PRIMUS'.(1)

From originals in the University Library and Trinity College, Cambridge and the Royal Society, London

[1](2)

Nam si Ellipsis, punctis *O* et *B* in infinitum abeuntibus, vertatur in Parabolam, constructio hic allata evadet eadem cum constructione Propositionis [XXX] adeoq͡ eo in casu accurata est.(3) Alijs in casibus ubi dimidium temporis periodici completur[,] corpus semper reperietur in *B* et completo tempore toto periodico redibit ad *A* ut oportet. Et ubi punctum *P* discedere incipit a puncto *A*(4) vel proxime

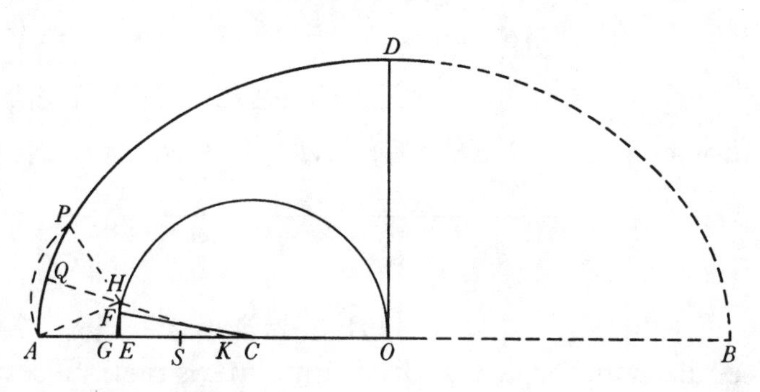

acces[s]erit ad punctum *B*, ratio prima sectoris nascentis *ASP* ad sectorem nascentem *GCF* et ultima sectoris evanescentis *PSB* ad sect[orem] evanescentem *FCB* eadem est ac semiellipseos *ABD* ad circulum, ut calculum tentanti facilè constabit:(5) et propterea Regula satis accurata est prope terminos Ellipseos *A* et *B*. Error maximus est in Quadrantibus temporis, seu e regione umbilici ulterioris *H*.(6)

(1) Specifically, of this miscellany of preliminary recomputations, minor revisions and intended amplifications of the methods there presented for approximately determining the position of a point in elliptical orbit at a given time the first three items pertain to the opening geometrical 'effection' by constructing a suitably defined auxiliary circle, the two last to the concluding refinement of the simple Boulliau–Ward theory which sets the 'upper' (non-solar) focus in a Keplerian ellipse to be an equant point.

(2) ULC. Add. 3965.9: 96^v, roughly drafted by Newton on the back of a sheet of Halley's contemporary editorial critique of the 'De motu Corporum' (*ibid.*: 94^r–99^v; see I. B. Cohen, *Introduction to Newton's 'Principia'* (Cambridge, 1971): 122–4 and especially 336–44). It will be evident (see §3: note (138)) that this was penned some time between Newton's receipt of Halley's letter to him of 14 October 1686 and the dispatch of his reply four days later.

(3) Newton first continued: 'Sin æquatis inter se semidiametris *AO* et *OD* ellipsis vertatur in circulum, puncta *G*, *S*, *K*, *O* coincident'.

(4) Originally 'ubi punctum *P* quam minime abest a puncto [*A*]'.

(5) See §3: note (138). Newton finds the calculation not at all easy, in fact, in his following recomputation and is forced to abandon an initial algebraic start when it becomes overcomplicated and leads him astray; compare note (9).

(6) Understand only in the basic mode of construction outlined in the scholium's original version (reproduced in §3): in [3] below, in fact, Newton was himself later to contrive a

[Pone] $DS = AO = q$. $AS = t$. $DO = v$. $AK = r$. [erit] $q - t = OS = \sqrt{qq - vv}$. $\frac{tq}{2q-t} = AG$. $rq = vv$.(7) $L = \frac{2qr - rt - qt}{2q - t} \times \frac{qq}{tv}$.(8) [Pone etiam] $GF = o$. [prodit]

$$\text{sector } GCF = \frac{qq - qt}{2q - t} \text{ in } \tfrac{1}{2}o. \text{ [et] } \frac{o \times \frac{2qr - rt - qt}{2q - t}}{\frac{qq - qt}{2q - t}} = \frac{2qr - rt - qt}{qq - qt} \text{ in } o = EH.^{(9)}$$

[Est] $\frac{1}{2}AP, AS.\frac{1}{2}FG, GC :: AO, OD.GC^q$. [seu] $FG, AO, OD = AP, AS, GC$. [Nam quia] $EF = FG$. $EH = HG$. [fit] $AP.HG = HE :: 2AK.GK$. [ut et] $HE.EF :: L = \frac{GK, AO^{q\,(10)}}{AS, OD}.CG :: GK, AO^q.CG, AS, OD$. [adeoq$\mathfrak{z}$]

$$AP.EF :: 2AK, AO^q.CG, AS, OD.$$

[hoc est] $AP, AS.EF, CG :: 2AK, AO^q.OD, CG^q$. [ubi] $2AK, AO = OD^q$.(11)

---

[2](12)

Ellipseos cujusvis *APB* sit *AB* axis major, *O* centrum *S* umbilicus, *OD* semiaxis minor, & *AK* dimidium lateris recti. Secetur *AS* in *G*, ut sit *AG* ad *AS*

---

refinement of the method which effectively eliminates this defect. To avoid confusion over the two points *H* which Newton here employs, we have purposely omitted to mark the ellipse's second focus in the accompanying figure (whose right-hand part is our completion of Newton's rough sketch, started by him too close to the outside edge of the manuscript page to be there fully drawn).

(7) Since the radius of curvature *AK* at *A* is half the *latus rectum* $2OD^2/AO$.

(8) That is, $L = GK \times AO^2/AS \times DO$. Newton has not yet identified his numerical slip (see §3: note (137)).

(9) With this erroneous implied equation of $EH/(GF$ or) EF to GK/GC Newton abandons this unrewarding algebraical mode of reduction for a more forthcoming direct geometrical approach.

(10) Once more the omitted factor '$\frac{1}{2}$' escapes his notice.

(11) Which should of course be '$AK, AO = OD^q$'. Here, we presume, Newton caught his numerical slip in the value he had earlier assigned for *L* (see §3: note (137)) and broke off to draft the revised and amplified account of the construction which follows in [2]. Certainly, the remainder of the manuscript page is blank.

(12) Royal Society MS LXIX: 89, the revised and amplified 'beginning' of the scholium (transcribed by Humphrey Newton with a few small verbal corrections inserted by Newton himself) which was sent on to Halley in London on 18 October 1686 (see §3: note (138)) and afterwards published—with only trivial differences—in the *Principia* ($_1$1687: 109–11). The manuscript lacks its accompanying figure: that here restored is founded on the one redrawn for the *editio princeps* (*ibid.*: 109), but we have simplified it by omitting a confusing second position of the perpendicular *EFN* and a corresponding circle arc through a second point *P* on the ellipse, and by deleting a focal radius *HP* (which serves only to determine the 'upper' focus angle $B\widehat{HP}$ employed in the last paragraph of the scholium and so here has no relevance).

ut BO ad BS: & quæratur longitudo L, quæ sit ad $\frac{1}{2}GK$ ut est $AO^{\text{quad.}}$ ad rectangulum $AS \times OD$. Bisecetur OG in C, centroq; C & intervallo CG describatur semicirculus GFO. Deniq; capiatur angulus GCF in ea ratione ad angulos quatuor rectos, quam habet tempus datum, quo corpus descripsit arcum quæsitum AP, ad tempus periodicum seu revolutionis unius in Ellipsi: Ad AO demittatur normalis FE, & producatur eadem versus F ad usq; N, ut sit EN ad longitudinem L, ut anguli illius sinus EF ad radium CF; centroq; N & intervallo AN descriptus circulus secabit Ellipsin in corporis loco quæsito P quam proxime.(13)

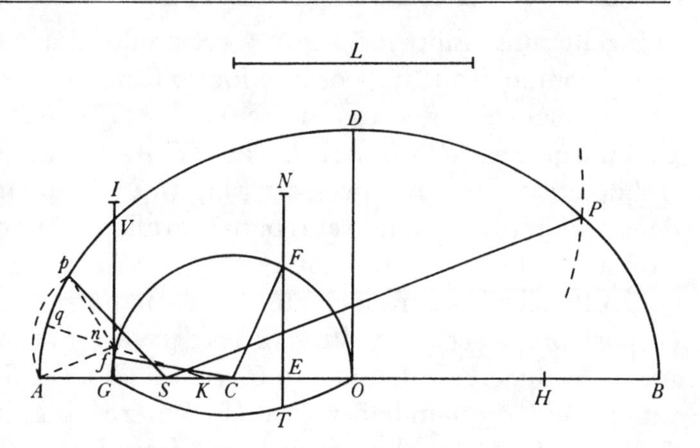

Nam completo dimidio temporis periodici, corpus P semper reperietur in Apside summa B, & completo altero temporis dimidio, redibit ad Apsidem imam, ut oportet. Ubi vero proxime abest ab Apsidibus, ratio prima nascentium sectorum ASP, GCF, & ratio ultima evanescentium BSP & OCF, eadem est rationi Ellipseos totius ad circulum totum. Nam punctis P, F & N incidentibus in loca p, f & n axi AB quam proximis, ob æquales An, pn, recta nq, quæ ad arcum Ap perpendicularis est, adeoq; concurrit cum axe in puncto K, bisecat arcum Ap. Proinde est $\frac{1}{2}Ap$ ad Gn ut AK ad GK, & Ap ad Gn ut $2AK$ ad GK. Est & Gn ad Gf ut EN ad EF, seu L ad CF, id est, ut $\dfrac{GK \times AO^q}{2AS \times OD}$ ad CF, seu $GK \times AO^q$ ad $2AS \times OD \times CF$, & ex æquo Ap ad Gf ut $2AK$ ad $GK + GK \times AO^q$ ad

$$2AS \times OD \times CF,$$

id est ut $AK \times AO^q$ ad $AS \times OD \times CF$. Proinde $Ap \times \frac{1}{2}AS$ est ad $Gf \times \frac{1}{2}GC$ ut $AO \times OD \times AS$ ad $AS \times CF \times GC$, seu $AO \times OD$ ad CG^q. id est, sector nascens $AS[p]$ ad sectorem nascentem GCf ut $AO \times OD$ ad CG^q. & propterea ut area Ellipseos totius ad aream circuli totius. Q.E.D. Argumento prolixiore(14) probari potest analogia ultima in sectoribus evanescentibus BSP, OCF: ideoq; locus puncti P prope Apsides satis accurate inventus est. In quadraturis error quasi

(13) Except for the crucial emendation of 'GK' into '$\frac{1}{2}GK$' (see §3: note (137)) this repeats the scholium's original opening paragraph with only minor verbal ameliorations. In sequel Newton proceeds to elaborate a synthetic justification of the value assigned for L; we restored the equivalent preceding analysis in §3: note (138).

(14) But, of course, on the same pattern.

quingentesimæ Ellipseos totius(15) vel paulo major obvenire solet: qui tamen propemodum evanescet per ulteriorem Constructionem sequentem.

Per puncta G, O duc arcum circularem GTO justæ magnitudinis; dein produc EF hinc inde ad T & N ut sit EN ad FT ut $\frac{1}{2}L$(16) ad CF; centroꝗ N & intervallo AN describe circulum qui secet Ellipsin in P, ut supra.(17) Arcus autem GTO determinabitur quærendo ejus punctum aliquod T; quod constructionem in illo casu accuratam reddet.

Si Ellipseos latus transversum multo majus sit quam latus rectum, & motus corporis prope verticem Ellipseos desideretur, (qui casus in Theoria Cometarum incidit,) educere licet e puncto G rectam GI axi AB perpendicularem, & in ea ratione ad GK quam habet area $AVPS$ ad rectangulum $AK \times AS$; dein centro I & intervallo AI circulum describere. Hic enim secabit Ellipsin in corporis loco quæsito P quamproxime.(18) Et eadem constructione (mutatis mutandis) conficitur Problema in Hyperbola. Hæ autem constructiones demonstrantur ut

---

(15) That is, $\frac{1}{250}\pi$ radians or about $\frac{3}{4}°$ of arc. This grossly overstates the theoretical error of the construction and, since there is no indication that Newton ever applied the construction in empirical practice, we must suppose that he has made an error in his calculation. For if, much as in §3: note (138), we set $OA = AB = 1$, $OS = e$ and take θ to be the eccentric angle defining P and T the corresponding mean anomaly (measured from perihelion A in both cases), then by Newton's construction $OE = (e/(1+e))\,(1+\cos T)$ and also $AN = NP$ so that, where $EN = \mu$, there is $\sqrt{[(1-OE)^2+\mu^2]} = \sqrt{[(\cos\theta+OE)^2+(\sqrt{[1-e^2]}\sin\theta-\mu)^2]}$ and after reduction $\mu\sqrt{[1-e^2]}\sin\theta = r(1+\cos T)\,(1-\cos\theta)-\frac{1}{2}e^2\sin^2\theta$, where $r = e/(1+e)$ and $\theta - e\sin\theta = T$, so that, on inverting, $\theta = T+e\sin T+\frac{1}{2}e^2\sin 2T+e^3\sin^2 T(1-\frac{3}{2}\sin^2 T)+\ldots$; whence, by eliminating θ, we may with some difficulty derive $\mu = (e-\frac{1}{2}e^2+e^3-e^4(\frac{3}{4}-\frac{1}{12}\sin^2 T))\sin T$ to $O(e^5)$. Newton, however, sets $EN = L\sin \widehat{GCF} = L\sin T$, where

$$L = \tfrac{1}{2}GK \times AO^2/AS \times OD = (e+\tfrac{1}{2}e^2)/(1+e)\,\sqrt{[1-e^2]} = e-\tfrac{1}{2}e^2+e^3-\tfrac{3}{4}e^4\ldots,$$

with an error $\epsilon = \mu - L\sin T \approx \frac{1}{12}e^4\sin^3 T$. If to this error ϵ in allocating the length of EN there corresponds an error η in the eccentric angle θ determining the position of P, then

$$(\mu-\epsilon)\sqrt{[1-e^2]}\sin(\theta-\eta) = r(1+\cos T)(1-\cos(\theta-\eta))-\tfrac{1}{2}e^2\sin(\theta-\eta);$$

and thence, when e is small (and so $r = e/(1+e) \approx e$), there ensues a divergence

$$\alpha \approx \epsilon\sin\theta/(e(1+\cos T)\sin\theta-\mu\cos\theta).$$

Accordingly, at quadratures ($\theta \approx T \approx \frac{1}{2}\pi$) the error $\alpha \approx \epsilon/e \approx \frac{1}{12}e^3$: for a Martian orbit ($e \approx \frac{1}{11}$) this is only about $\frac{1}{15000}$ radians or some $\frac{1}{4}'$ of arc.

(16) In a reversal of Newton's earlier omission of a factor '$\frac{1}{2}$', this should be 'ut L' simply!

(17) On allowing for the slight numerical slip in his statement of this adjustment, Newton now seeks to augment the line FE by a small length ET so that the perpendicular EN shall be accurately equal to $L \times (FE+ET)/CF$, or (in the terms of note (15) above) that

$$\mu = L(\sin T+ET/e(1+e)^{-1}),$$

whence $ET = e(1+e)^{-1}\,(\mu-L\sin T)/L \approx \frac{1}{12}e^4\sin^3 T \approx \frac{1}{12}e.(GE\times EO)^{\frac{3}{2}}$, since

$$GE = e(1+e)^{-1}\,(1-\cos T) \quad\text{and}\quad EO = e(1+e)^{-1}\,(1+\cos T).$$

His concluding approximation of the near-parabolic arc $\widehat{GTO}$ by that of a circle through the end-points G, O and some suitably determined intermediate point T will not be far out.

(18) See §3: note (142).

supra, & Si Figura (vertice ulteriore B in infinitum abeunte) vertatur in Parabolam, migrant in accuratam illam constructionem Problematis XXII.

---

[3][19]

[Ellipseos cujusvis APB sit AB axis major, O centrum, S umbilicus, OD semiaxis minor, & AK dimidium lateris recti. Secetur AS in G, ut sit AG ad AS ut BO ad BS.][20] Bisecetur OG in C et erigatur perpendiculū $C[L]$. Ab umbilico S agatur recta $S[M]$ quæ secet Ellipsin in $[M]$ & bisecet ejus aream $AB[M]D$. Jungatur $A[M]$, bisecetur eadem in V et erigatur perpendiculum $V[L]$ rectæ $C[L]$ occurrens in $[L]$, et axe minore CO semiaxe majore $C[L]$ describatur Ellipsis $[OL]$. His ita constructis capiatur angulus GCF in ea ratione ad angulos quatuor rectos[21] [quam habet tempus datum quo corpus descripsit arcum quæsitum AP ad tempus periodicum seu revolutionis unius in Ellipsi ADB. Ad AO demittatur normalis FE, & producatur versus F ad usq; N ubi occurrat Ellipsi GLO; centroq; N et intervallo AN descriptus circulus secabit Ellipsin in corporis loco quæsito P quamproximè.][22]

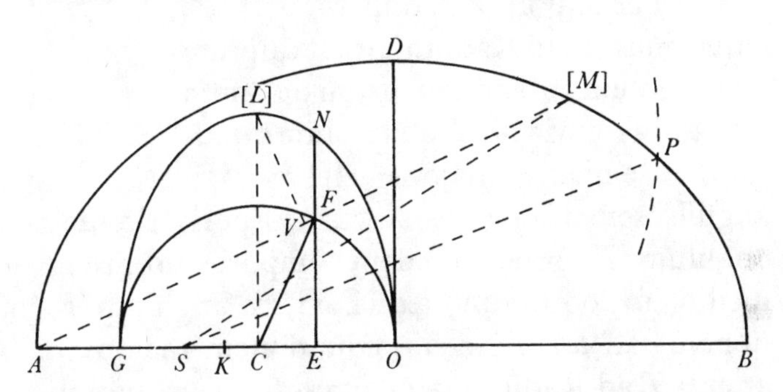

---

(19) ULC. Add. 3965.17: 637r, a rough jotting made (by its handwriting) some time after 1690 on a page which contains an equally fragmentary addition to some dynamical portion of the *Principia*. We have given the piece an appropriate opening and conclusion, and also, to render its naming of points consistent with that of the *Principia*'s figure ($_1$1687: 109, reproduced with certain deletions in [2] preceding), have replaced Newton's 'E' and 'N' by 'L' and 'M' respectively throughout.

(20) Much as in the equivalent opening to [2], we here specify what Newton understands to be given.

(21) The manuscript text here breaks off in mid-sentence. In sequel, we provide an appropriate termination.

(22) It will be clear that L is the position of N when P is at M, that is, (in the terms of note (15) above) when $\widehat{GCF} = T$ is $\frac{1}{2}\pi$; whence, to $O(e^5)$, $CL = e-\frac{1}{2}e^2+e^3-\frac{2}{3}e^4$ and so $EN = CL.\sin T$ deviates from true by $\frac{1}{12}e^4\sin T\cos^2 T \leqslant \frac{1}{18\sqrt{3}}e^4$, less than half the maximum error, about $\frac{1}{12}e^4$, of Newton's previous refinement in [2]. The chief limitation on the present construction, however, will be accurately to construct the focal radius SM which bisects the semi-ellipse.

[4][23]

Existentibus AO, OB, OD semiaxibus Ellipseos, & L ipsius latere recto, ac D differentia inter semiaxem minorem OD & lateris recti semissem $\frac{1}{2}L$, quære tum angulum Y cujus sinus sit ad Radium ut est rectangulum sub differentia illa D & semisumma axium $AO+OD$ ad quadratum axis majoris AB, tum angulum Z, cujus sinus sit ad Radium ut est duplum rectangulum sub umbilicorum distantia SH & differentia illa D ad triplum quadratum semiaxis majoris AO.[24] His angulis semel inventis, locus corporis sic deinceps determinabitur. Sume angulum T proportionalem tempori quo arcus BP descriptus est, seu motui medio (ut loquuntur) æqualem, & angulum V (primam medij motus æquationem) ad angulum Y (æquationem maximam primam) ut est sinus dupli anguli T ad Radium, atq; angulum X (æquationem secundam) ad angulum Z (æquationem maximam secundam) ut est cubus sinus anguli T ad cubum Radii. Angulorum T, V, X vel summæ $T+X+V$ si angulus T recto minor est vel differentiæ $T+X-V$ si is recto major est rectisq; duobus minor, æqualem cape angulum BHP (motum medium æquatum;)[25] & si HP occurrat Ellipsi

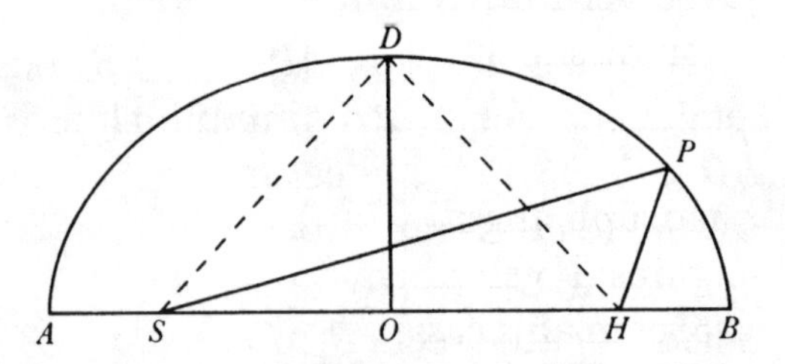

(23) This augmented revision of the scholium's last paragraph is entered, as a correction and extension of the first edition's text (*Principia*, $_1$1687: 113–14, essentially that reproduced on pages 318–22 above), in the margins of Newton's annotated library copy (now Trinity College, Cambridge. NQ. 16.200).

(24) In his initial draft of this sentence (on ULC. Add. 3965.18: 670r) Newton wrote: 'Existentibus AO, OB, OD semiaxibus Ellipseos, quære tum angulum Y cujus sinus sit ad Radium (vel qui sit ad 57|29578 gradus) ut est rectangulum sub semidifferentia Axis minoris $2OD$ et Lateris recti[,] et axium semisumma $AO+OD$ ad quadratum axis majoris AB; tum angulum Z qui sit ad maximam æquationem centri [$S\widehat{D}H$] ut est tertia pars differentiæ axium $\left(\frac{2AO-2OD}{3}\right)$ ad axis minoris semissem OD'. Where e is the ellipse's eccentricity this is equal to setting $\sin Y\,[=Y+O(e^6)] = \frac{1}{2}(2\sqrt{[1-e^2]}-2(1-e^2))\,(1+\sqrt{[1-e^2]})/4 = \frac{1}{4}e^2-\frac{1}{8}e^4\ldots$ and $Z = 2\sin^{-1}e.\frac{1}{3}(2-2\sqrt{[1-e^2]})/\sqrt{[1-e^2]} = \frac{2}{3}e^3+\frac{11}{18}e^5\ldots$. Little differently, his present modification postulates that $\sin Y = (\sqrt{[1-e^2]}-(1-e^2))\,(1+\sqrt{[1-e^2]})/4 = \frac{1}{4}e^2-\frac{1}{8}e^4\ldots$ (as before) and $\sin Z\,[=Z+O(e^7)] = 4e(\sqrt{[1-e^2]}-(1-e^2))/3 = \frac{2}{3}e^3-\frac{1}{6}e^5\ldots$.

(25) Since $\pm V/\sin 2T = Y = \frac{1}{4}e^2-\frac{1}{8}e^4\ldots$ and $X/\sin^3 T = Z = \frac{2}{3}e^3\ldots$ (see the previous note), Newton's revised approximation $T\pm V+X$ of $B\widehat{H}P = \psi$ is, to $O(e^5)$,

$$T+(1-e^2)^{\frac{1}{2}}\,(\tfrac{1}{4}e^2\sin 2T+\tfrac{2}{3}e^3\sin^3 T).$$

Accurately (as we showed in §2: note (164))

$$T = (1-e^2)^{\frac{1}{2}}\int_0^{\psi}(1+2e\cos\psi+e^2)/(1+e\cos\psi)^2.d\psi$$
$$= \psi-(1-e^2)^{\frac{1}{2}}\,(\tfrac{1}{4}e^2\sin 2\psi+\tfrac{2}{3}e^3\sin^3\psi+\tfrac{3}{32}e^4\sin 4\psi\ldots)$$

in P, acta SP abscindet aream BSP tempori proportionalem quamproxime. Hæc praxis satis expedita videtur, propterea quod angulorum perexiguorum V & X (in minutis secundis, si placet, positorum) figuras duas tresve primas invenire sufficit. Sed et satis accurata est ad Theoriam Planetarum. Nam in Orbe vel Martis ipsius cujus æquatio [centri] maxima est(26) graduum decem error vix superabit minutum unum secundum.(27) Reddi tamen potest plusquam vigecuplo accuratior capiendo angulum S ad angulum $\frac{3D, Y}{8AO}$(28) ut sinus quadrupli anguli T ad Radium, et angulo BHP addendo hunc angulum S ubi angulus T vel minor est gradibus 45, vel major gradibus 90 et minor gradibus 135, et in alijs casibus ipsum subducendo: Qua ratione angulus ille BHP evadet vel $T+X+V+S$ vel $T+X+V-S$ vel $T+X-V+S$ vel $T+X-V-S$, perinde ut angulus T vel minor est 45^{gr} vel major 45^{gr} et minor $90^{gr.}$ vel major 90^{gr} et

---

and therefore $\psi = T+(1-e^2)^{\frac{1}{2}}\,(\frac{1}{4}e^2\sin 2T+\frac{2}{3}e^3\sin^3 T+\frac{5}{32}e^4\sin 4T+O(e^5))$, whence the error in Newton's approximate 'equation' of $\widehat{BHP}$ is very nearly $\frac{5}{32}e^4\sin 4T$: this he will attempt—not entirely successfully—to allow for in sequel. (The correct evaluation of the upper-focus angle ψ in terms of mean anomaly T to the order of the fifth power of the eccentricity e was achieved only after Newton's death by Thomas Simpson in his *Miscellaneous Tracts on Some curious, and very interesting Subjects in Mechanics, Physical-Astronomy, and Speculative Mathematics* (London, 1757): 46–57: 'A very exact Method for finding the Place of a Planet in its orbit, from a Correction of Ward's hypothesis, by means of One, or more Equations applied to th motion about the upper focus'. As Simpson remarks (*ibid.*: 53–4) an earlier attempt by John Machin in his tract on 'The Laws of the Moon's Motion according to Gravity' [appended to the second volume of Andrew Motte's *The Mathematical Principles of Natural Philosophy. By Sir Isaac Newton. Translated into English* (London, 1729)]: 54–8 to compute the error in Newton's published rule incorrectly propounded, in present terms, that

$$\psi = T+(1-e^2)^{\frac{1}{2}}\,(\tfrac{1}{4}e^2\sin 2T+\tfrac{2}{3}e^3\sin^3 T+e^4\sin 3T)$$

to $O(e^5)$. Machin had earlier (*ibid.*: 40–54; compare his similar autograph manuscript, 'A Correction of the Hypothesis of Bullialdus for the motions of the Planets, by the quadrature of certain portions of the Ellipsis', now in ULC. Res. a. 1893.7) ingeniously constructed—in elaboration of Boulliau's 1657 amended 'equation' (see §2: note (163)) which incorporates $V \approx \frac{1}{4}e^2\sin 2T$ alone—an elliptical equant curve, centred on the 'planetary' ellipse's upper focus, which accurately embodies also $X \approx \frac{2}{3}e^3\sin^3 T$.)

(26) Originally 'in Orbibus Martis et Mercurij quarum æquationes maximæ sunt'. Newton, upon reflection, wisely decided to steer clear of the sun's nearest neighbour, a planet which is not only considerably eccentric but, still more intransigently, will deviate widely from any ideal, stationary elliptical orbit into which one may attempt to fit its 'wanderings'.

(27) Since Mars' eccentricity e is about $\frac{1}{11}$, its 'maximum equation of centre'

$$\widehat{SDH} = 2\sin^{-1}e$$

is slightly more than 10°, while the greatest error (at $T \approx \frac{1}{8}\pi, \frac{3}{8}\pi, \frac{5}{8}\pi, \frac{7}{8}\pi, \ldots$) in Newton's preceding rule is about $(\frac{5}{32}e^4 \approx)$ $\frac{1}{90000}$ radians, or c. $\frac{1}{25}'$.

(28) Read '$\frac{5D, Y}{4AO}$' preferably; see note (29) following.

minor 135^{gr} vel major 135^{gr} respectivè.(29) Invento autem angulo motus medij æquati BHP, angulus veri motus HSP & distantia SP in promptu sunt per Wardi methodum notissimam.(30)

---

[5](31)

$AC = CD = CB = 100000.$

$FC = 10000.$

$Cd = \surd 9900000000 = 99498\llcorner 7437.$(32)

D[ifferentia Dd] $= 501\llcorner 2563.$

$\triangle FdC = 497493718\frac{1}{2}.$

$\triangle FDC = 500000000 = 5^{gr}\llcorner 729578.$(33)

Angle $FDC = 5^{[gr]}\llcorner 710580973.$(34)

Angle $FdC = 5^{[gr]}\llcorner 73917114.$(34)

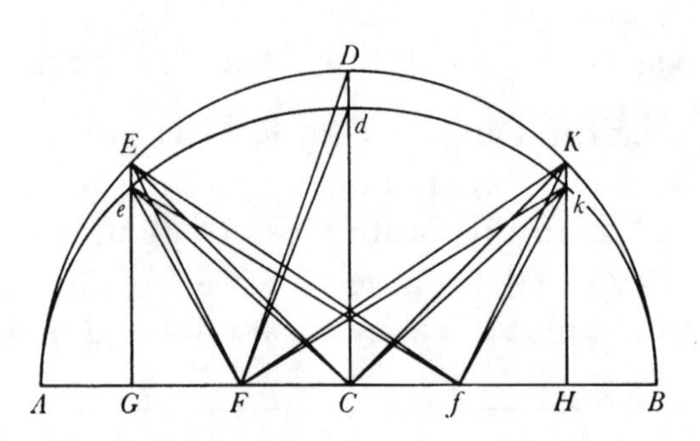

$EG = KH = 70710\llcorner 68[2].$(35)

$\triangle CEF = [\triangle]CKf = 3535534[1] = 4^{gr}\llcorner 05142355.$(33)

---

(29) Since the error in Newton's preceding approximation is (in the terms of note (24)) very nearly $\frac{5}{32}e^4\sin 4T$, while his new correction term $\pm S = \frac{3}{8}(\surd[1-e^2]-(1-e^2))Y\sin 4T$ adds only $\frac{3}{64}e^4\sin 4T + O(e^5)$, there remains an error of some $\frac{7}{64}e^4\sin 4T$. It will be evident that this will be all but eliminated by retaining the approximation $\psi \approx T \pm V + X \pm S$ but now proceeding 'capiendo angulum S ad angulum $\frac{5}{4}D \times Y/AO$ ut sinus quadrupli anguli T ad radium'.

(30) See §3: note (162). When Newton twenty years afterwards passed this revised rule for 'equating' the upper-focus angle into the second edition of his *Principia* ($_2$1713: 103–4), he there omitted the sentences 'Reddi tamen potest plusquam vigecuplo accuratior ... ... vel major 135^{gr} respectivè' in which he here sets out his 'twenty-fold more accurate' approximation by the error term S, doubtless having come in the interim to lose faith in its practical effectiveness, if not to suspect its theoretical precision.

(31) ULC. Add. 3965.18: 706r; a rough, unfinished preliminary version exists on f. 672r before. Like the preceding piece (to whose partial draft on f. 670r it is akin) this was, if we date its handwriting accurately, composed in about 1692–3. In it, Newton attempts to calculate the numerical exactness with which the upper-focus angle ψ is approximated by the previous 'equation' $T + Y\sin 2T + Z\sin^3 T$, $Y = \frac{1}{4}e^2 \ldots$, $Z = \frac{3}{8}e^3 \ldots$, by describing a semi-ellipse of a (near-Martian) eccentricity $e = \frac{1}{10}$ and therein determining the three pairs of accurate values of ψ and $T = \theta + e\sin\theta$ which ensue on setting the eccentric angle θ equal in turn to $\frac{1}{4}\pi$, $\frac{1}{2}\pi$ and $\frac{3}{4}\pi$ radians.

(32) That is, $10^5\surd[1-e^2]$ where $e = FC/AC = \frac{1}{10}$. Notice that the foci are now denoted as F and f (and not, as universally in the *Principia*, by S and H): this variant convention is found in several of Newton's astronomical worksheets written at this period (ULC. Add. 3965.11: 157r, for example, and Add. 3965.1: 2r ff.).

(33) Understand of mean anomaly, 180° of which measure the area $\frac{1}{2}\pi . 10^{10}$ of the semicircle (ADB).

(34) The value in degrees of $\tan^{-1}e$ and $\sin^{-1}e$ radians respectively.

(35) Namely, $10^5\sin\theta$ where $\theta = \widehat{BCK} = 45°$ and $\theta = \widehat{BCE} = 135°$.

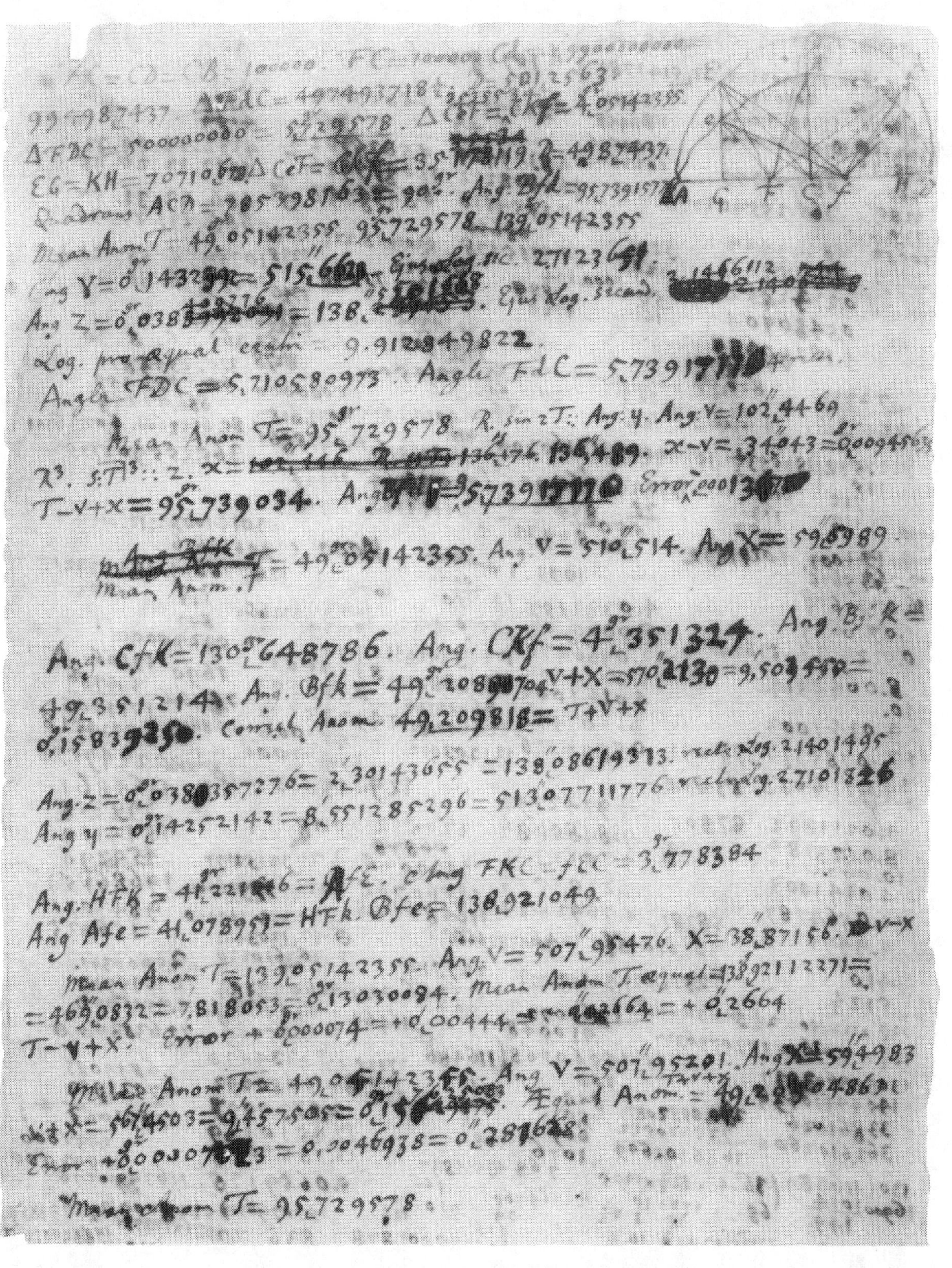

Plate I. Errors at quadratures and octants in a planetary ellipse whose upper focus is an equant (I, §3, Appendix 2.5).

$\triangle CeF = \triangle Ckf =$ 35178119. [$\frac{1}{2}$ Lat rect = 99000.] D[36] = 498⌊7437.
Quadrans ACD = 785398163[4] = 90gr.[33] Ang Bfd = 95[gr]⌊73917114.
Mean Anom. T = 49gr⌊05142355. 95gr⌊729578. 139gr⌊05142355.
Ang Y = 0gr⌊1432392[37] = 515″⌊6611. Ejus Log[arithmus] sec[andus] 2.7123651.
Ang Z = 0gr⌊038409776[38] = 138″⌊551938. Ejus Log. secand. 2.1416112.
Log. pro æquat. centri = 9.912849822.[39]

Mean Anom. T = 95gr⌊729578. R . sin $2T$ :: Ang : Y . Ang : V = 102″⌊4479.
R^3 . s: $T|^3$ :: Z . X = 136″⌊489. $X-V$ = 34″⌊043 = 0gr⌊00945635.
$T-V+X$ = 95gr⌊739034. Ang Bfd = 95[gr]⌊7391711 [recte].
Error − [0gr⌊]000137[= 0′⌊00822 = 0″⌊4932].

Mean Anom T = 49gr⌊05142355. Ang V = 507″⌊95201. Ang X = 59″⌊4983.
Ang. CfK[40] = 130gr⌊648786. Ang. CKf = 4gr⌊351214.
Ang. BfK = 49[gr]⌊351214. Ang. Bfk[41] = 49gr⌊2089704.
$V+X$ = 567″⌊4503 = 9′⌊457505 = 0gr⌊157625083.
Æquat[a] Anom[alia] $T+V+X$ = 49[gr]⌊20904863.
Error[42] + 0gr⌊00007823 = 0′⌊0046938 = 0″⌊281628.

Mean Anom. T = 139[gr]⌊05142355. Ang. V = 507″⌊95476.
[Ang] X = 38″⌊87156.
Ang. HFK = 41gr⌊221616 = [Ang] AfE. Ang. FKC = [Ang] fEC = 3gr⌊778394.
Ang Afe = 41[gr]⌊078951 = [Ang] HFk. [Ang] Bfe[43] = 138[gr]⌊921049.
$V-X$ = 469″⌊0832 = 7′⌊818053 = 0gr⌊13030084.

---

(36) That is, Cd − '$\frac{1}{2}$ Lat rect' = $10^5 . (\sqrt{[1-e^2]} - (1-e^2))$, $e = \frac{1}{10}$, as in the preceding general rule.

(37) While we would here expect Newton to evaluate ($\sin Y \approx$) Y as

$$D \times (AC+Cd)/AB^2 = \tfrac{1}{4}e^2 - \tfrac{1}{8}e^4 \ldots$$

(as he indeed does below; see note (45)), this is—a slight error in the terminal place apart—simply $\frac{1}{4}e^2$ radians. In his draft on f. 672r he initially computed 'max Y = 513″⌊0864'.

(38) Probably intended to be $\frac{3}{8}e^3$ radians = '0gr⌊03819718', in contrast to the 'corrected' value $2D \times SH/3AO^2 = \frac{3}{8}e^3 - \frac{1}{6}e^5 \ldots$ computed below (see note (46)). In his draft on f. 672r Newton originally calculated there to be 'max Z = 137″⌊165472'.

(39) This is $\log_{10}(\frac{9}{11} . 10^{10}) = 10 + \log_{10}(Bf/BF)$.

(40) That is, $180° - (\widehat{BfK}$ or) $\tan^{-1}(HK/fH)$ and so $90° + \tan^{-1}(1 - \frac{1}{10}\sqrt{2})$.

(41) That is, $\tan^{-1}(Hk/fH) = \tan^{-1}(\sqrt{[1-e^2]}/(1 - \frac{1}{10}\sqrt{2}))$.

(42) From $\widehat{Bfk}$, namely.

(43) That is, $180° - (\widehat{Afe}$ or) $\tan^{-1}(Ge/fG)$ and so $180° - \tan^{-1}(\sqrt{[1-e^2]}/(1 + \frac{1}{10}\sqrt{2}))$.

Mean Anom. T æquat[a] $= 138^{\mathrm{g[r]}}\,{}_{\llcorner}92112271 = T - V + X$.
Error(44) $+\, 0^{\mathrm{gr}}\,{}_{\llcorner}000074 = +0'\,{}_{\llcorner}00444 = +0''\,{}_{\llcorner}2664$.

Ang. $Y = 0^{\mathrm{gr}}\,{}_{\llcorner}14252142 = 8'\,{}_{\llcorner}551285296 = 513''\,{}^{\ulcorner}07711776$ recte.(45)
Log 2.7101826.

Ang. $Z = 0^{\mathrm{gr}}\,{}_{\llcorner}038357276 = 2'\,{}_{\llcorner}30143655 = 138''\,{}_{\llcorner}08619313$ recte.(46)
Log 2.1401495.(47)

APPENDIX 3. THE 'GRAVITATIONAL' THEORY OF OPTICAL REFRACTION.(1)

[1685/*c.* 1690/1694]

From autographs in private possession, the University Library, Cambridge and Corpus Christi College, Oxford, and from Newton's *Principia* ($_1$1687)

[1](2) *Of Refraction & y^e velocity of light according to y^e density of bodies.*

Theorem 1. The attraction of y^e Ray is proportional to y^e variation of density.(3)

---

(44) From $\widehat{Bfe}$ above.

(45) That is, $\frac{1}{4}e^2 - \frac{1}{8}e^4 \ldots$ radians; compare note (37) above.

(46) Some slight numerical inaccuracy apart, this is $\frac{2}{3}e^3 - \frac{1}{6}e^5 \ldots$ radians; compare note (38) above.

(47) We may remark that Newton's computation of $\psi - (T + Y\sin 2T + Z\sin^3 T)$ for values of $T = \theta + e\sin\theta$ defined by $\theta = \frac{1}{4}\pi$, $\frac{1}{2}\pi$ and $\frac{3}{4}\pi$ is not well suited to delimiting the full order of magnitude of the term $\frac{5}{32}e^4\sin 4T \approx \frac{5}{32}e^4\sin 4\theta + O(e^5)$ which there mostly predominates. More accurately, since by further refining the expansion given in note (25) there ensues at length $\psi = T + (1-e^2)^{\frac{1}{2}}\,(\frac{1}{4}e^2\sin 2T + \frac{2}{3}e^3\sin^3 T + \frac{5}{32}e^4\sin 4T + e^5(\frac{8}{3}\sin^3 T - \frac{37}{15}\sin^5 T)\ldots)$, the error in 'equating' ψ by $T + (\frac{1}{4}e^2 - \frac{1}{8}e^4\ldots)\sin 2T + (\frac{2}{3}e^3 - \frac{1}{6}e^5\ldots)\sin^3 T$ is very nearly $\frac{5}{32}e^4\sin 4T + \frac{5}{2}e^5\sin^3 T\cos^2 T$. (Compare K. Stumpff, 'Über eine Eigenschaft des zweiten Brennpunkts der Keplerschen Bahnellipse und ihre Verwendung in der Ephemeridenrechnung', *Astronomische Nachrichten*, **273**, 1942–3: 179–88.) This difference all but vanishes when $T \approx \frac{1}{2}\pi$, and is only about $\frac{5}{11}e^5$ radians $\approx 1''$ of arc when $T \approx \frac{1}{4}\pi$, but is some four times larger when $T \approx \frac{1}{8}\pi$ or $\frac{3}{8}\pi$. For all their failure to convince with the precision which Newton clearly thought he had thereby achieved, his present calculations do, however, amply stress the capability of the rounded-off 'equation' $\psi \approx T + \frac{1}{4}e^2\sin 2T + \frac{2}{3}e^3\sin^3 T$ to compute positions in elliptical orbit to well within the ceiling, say $2'$ of arc, imposed by the optical deficiencies of contemporary techniques of astronomical observation.

(1) We scarcely need to remark that Newton had been familiar since his late undergraduate days at Cambridge with the simple Cartesian 'emission' hypothesis of optical refraction—first presented by Descartes in the 'Discours Second' of his *Dioptrique* [= *Discours de la Methode*, (Leyden, 1637): first *Appendice*]: 13–23, and read by Newton in its Latin version (Amsterdam $_1$1644; see I: 549, note (2))—according to which a light 'ray' is conceived to be the observed path of a fast-moving stream of light corpuscles, each one of which, when it arrives at a given

Corol. 1. Therefore if a medium begin at BV & its density increase from thence uniformly so as in all places to be proportional to y^e^ distance from y^t^ line; & if a ray be projected from V towards B: it shall describe a parabola VCD whose ordinates AC, BD are as y^e^ squares of y^e^ lines VA, VB.(4)

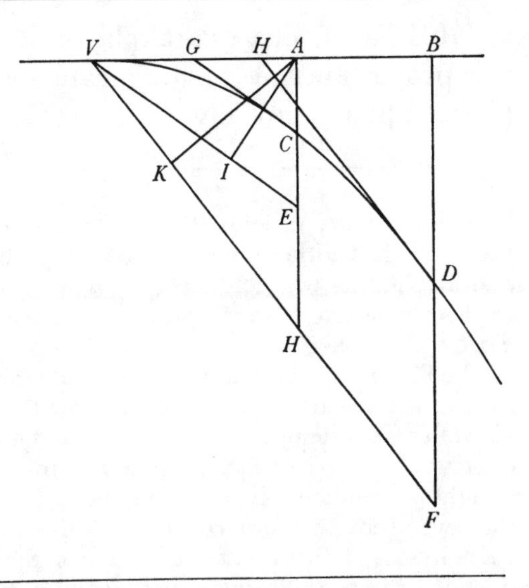

Corol. 2. And the velocity of y^e^ ray in any point C or D shall be as y^e^ tangents GC, HD applied to half y^e^ abscissæ $AG = \frac{1}{2}AV$ & $BH = \frac{1}{2}BV$;(5) that is, as $\frac{\sqrt{AC^q + \frac{1}{4}AV^q}}{\frac{1}{2}AV}$, $\frac{\sqrt{BD^q + \frac{1}{4}BV^q}}{\frac{1}{2}BV}$, that is, (if

interface between two media of differing density, is violently propelled by a 'blow' delivered normally at the interface into the denser medium (much, to use Descartes' preferred image, as a tennis ball struck sharply by a racquet) at a changed speed and in an altered direction. Since the components of motion parallel to the interface are unaffected by the refractive impulse, the Harriot–Snell sine-law is an immediate corollary: specifically, where v_i and v_r are the speeds of the corpuscles before incidence (at an angle i to the normal) and after refraction (at an angle r), the transverse components $v_i . \sin i = v_r . \sin r$ of motion are preserved, whence at once $\sin i/\sin r = v_r/v_i$, constant. Building upon a similarly computed table of the refractions of 'the Extreamely Heterogenous [red/blue] Rays' (ULC. Add. 4000: 33^v^; reproduced, from John Collins' somewhat inaccurate contemporary copy, in *The Correspondence of Isaac Newton*, **1**, 1959: 103, note (6)) set down in a later portion of the early essay 'Of Refractions' in which (see I: 559) he first noted the essence of the Cartesian theory, Newton came some half dozen years later in his Lucasian '*Lectiones opticæ*' to generalise it into a model of colour dispersion (see III: 466–8) in which, it would appear, individual corpuscles of light are supposed to travel at speeds varying as their colour, but are alike impelled into the denser medium upon refraction at an interface. Though he afterwards came to replace this by a slightly different non-kinematical one which more finely mirrored physical reality (see III: 468, note (37)), it was in early 1685 natural for Newton, with his attention solidly directed to the mathematical exploration of motion under the deviating action of an accelerative *vis centripeta*, further to generalise the simple Cartesian model by supposing the stream of light-corpuscles composing a ray to be 'instantaneously' refracted at an infinite succession of infinitesimally separated contiguous interfaces, so determining the ray path to be the orbit of the corpuscles through a medium of continuously varying density, at each point of which there is defined both in magnitude and direction a short-range Newtonian force of refractive attraction—the *vis refractiva Medij* as he had earlier denoted it in the first version of his Lucasian '*Lectiones opticæ*' (ULC. Add. 4002: 74). In this 'planetary' theory of light—though Newton was ever careful not to specify how his *vis refractiva* operated, merely requiring it to be 'attractioni...similior', somewhat akin to an attractive force (see III: 550, note (1))—the sine-law of refraction is elegantly subsumed into the generalised areal law, since again transverse components of motion are unaccelerated by the action of such a posited force of refraction. Both because they develop a non-gravitational application of the abstract theory of central dynamical force expounded at length in the preceding versions of his treatise 'De motu Corporum', and because the intrinsic sophistication of this elegant, if (see note (32) below) ultimately abortive,

ye latus Rectũ of ye Parabola be R) as $\sqrt{4RR+AC^q}$, $\sqrt{4RR+BD^q}$.(6) And in this proportion are the [co]secants of Refraction VE, VH directly, & ye sines $[V]I$, $[V]K$ reciprocally.(7)

'emission' theory of light still goes widely unappreciated, we here reproduce the principal texts—the first hitherto unprinted—in which Newton elaborated its theoretical consequences and sought, in a way emulated by many during the next century, to apply it computationally to the tabulation of angular error in observation due to the refraction of stellar light in the earth's atmosphere.

(2) This rough, unfinished autograph sheet (now in private possession), composed some time in the middle 1680s if our assessment of its handwriting be accurate, perhaps represents Newton's first attempt to frame a general dynamical theory of optical refraction, supposing the existence at every interface of a determinable *vis refractiva* which 'attracts' an emitted 'corpuscle' of light perpendicularly towards the denser medium. Throughout, where Newton intends (in the seventeenth-century convention) the trigonometrical ratio of 'radius' to sine, he has written 'secant' in error: not to overstress the slip, we convert this to '[co]secant' in our reproduction of his text. A similar idiosyncracy obtrudes into the opening sentence of his scholium to Proposition XCVI of Book 1 of the *Principia* ($_1$1687: 231; compare III: 549, note (1)) where, in a related context (see [2] following), he observed that 'Harum attractionum haud multum dissimiles sunt Lucis reflexiones & refractiones, factæ secundum datam Secantium [!] rationem, ut invenit *Snellius*, & per consequens secundum datam Sinuum rationem, ut exposuit *Cartesius*.'

(3) That is, the *vis refractiva* which impels the speeding light-corpuscles normally to the interface between two media of respective uniform density ρ_1, ρ_2 into the denser is by implication defined to be measured by their difference in density $\rho_2-\rho_1$ (where $\rho_2 > \rho_1$). As Newton immediately goes on to suppose, where a medium has a continuously varying density $\rho \equiv \rho(r)$, r some parameter, the *vis refractiva* will at each individual point be measured by the density gradient $d\rho/dr$ and be directed instantaneously along the line of increasing density. Like Thomas Harriot and others before him, Newton was well aware of the existence of such substances as olive oil and spirit of turpentine which are less dense than water, but which—in apparent contradiction of the present principle—have a greater 'power' to refract the passage of light through them. Preferring to preserve the principle (and the mathematical theory based upon it) that 'The refracting power of two contiguous bodies is the difference of their refracting powers in vacuo...proportional to their specific gravities' (as he propounded it in the two opening propositions of his projected 'Fourth book [of Opticks] concerning the nature of light & ye power of bodies to refract & reflect it'—now ULC. Add. 3970.3: 337r ff—in about 1692), he conceived that since 'sulphurous bodies cæteris paribus are most strongly refractive... therefore tis probable yt ye refracting power lies in ye sulphur & is proportional not to ye specific weight or density of ye whole body but to that of ye sulphur alone' (*ibid.*). Correspondingly, in Book 2, Part III, Proposition 10 of his mature 'Opticks' (ULC. Add. 3970.3: 17r–284r [= *Opticks: Or, A Treatise of the Reflexions, Refractions and Colours of Light* (London, $_1$1704): $_1$1–144/$_2$1–137]), where he asserted axiomatically that 'the forces of...bodies to reflect and refract light are very nearly proportional to ye densities of ye same bodies', he was careful to append a late insertion in the manuscript 'excepting that unctuous & sulphureous bodies refract more then others of the same density' (*ibid.*: 191r [= *Opticks*: $_2$70–1]), thence concluding that, since 'they partake more or less of sulphureous oyly particles, ...it seems rational to attribute the refractive power of all bodies chiefly if not wholy to the sulphureous parts wth wch they abound' (*ibid.*: 194r [= *Opticks*: $_2$76]).

(4) Evidently, since the density ρ of the medium varies as the distance r below its 'beginning' line VB, its (downward) accelerative *vis refractiva*—proportional to the density gradient $d\rho/dr$—will here be constant; whence the vertical fall AC of the 'ray' from its initially horizontal path

Corol. 3. The[re]fore if $VE = P$, $VH^{(8)} = Q$ y^e^ [co]secants of the greatest Refraction made into any two mediums be given in respect of y^e^ Radius $AV = A$: take $BF^q.AH^q::(BV^q.AV^q::BD.AC::)\,BF.AE$, that is, take $BF.AH::AH.AE$ & y^e^ density of y^e^ mediums (wch are(9) as BD to AC) shal be as BF to AE, that is, as AH^q to AE^q.

Corol. 4. And therefore y^e^ difference of the densities of y^e^ media is (as $AH^q - AE^q$[,] that is) as the difference of y^e^ squares of y^e^ [co]secants of the greatest refraction $(VH^q - VE^q)$.

Corol. 5. And the difference of y^e^ density of y^e^ medium of incidence & y^e^ medium of refraction is as y^e^ difference of y^e^ squares of y^e^ [co]secant of incidence & y^e^ [co]secant of refraction, that is reciprocally as y^e^ difference of y^e^ squares of their sines.

The Theoreme in y^e^ beginning is thus proved.(10)

---

[2](11) *Prop. XCIV. Theor. XLVIII.*

Si media duo similaria, spatio planis parallelis utrinq; terminato, distinguantur ab invicem, & corpus in transitu per hoc spatium attrahatur vel impellatur perpendiculariter

---

will be proportional to the time of passage over $\widehat{VC}$, and hence to that of the horizontal motion VA which measures the time.

(5) For VA, VB are proportional to the time over $\widehat{VC}$, $\widehat{VD}$, and hence the speeds at C and D are proportional to $d(\widehat{VC})/d(VA) = GC/GA$ and $d(\widehat{VD})/d(VB) = HD/HB$, where GC and HD are tangents to the parabola at C and D.

(6) Read '$\sqrt{4AC+R}$, $\sqrt{4BD+R}$' since $R = AV^2/AC = BV^2/BD$ *ex natura parabolæ*.

(7) Because, where v_C and v_D are the orbital speeds at C and D, the (unaccelerated) horizontal components of motion $v_C.\sin\widehat{ACG}$ and $v_D.\sin\widehat{BDH}$ are equal, and hence

$$v_C : v_D = \sin(\widehat{BDH}\text{ or})\ \widehat{AHV} : \sin(\widehat{ACG}\text{ or})\ \widehat{AEV} = VE : VH = VK : VI$$

where AI, AK are let fall perpendicularly to VE, VH.

(8) Understand (of the two possibilities in Newton's figure with its confusing two points H) the segment cut off by AE from VF.

(9) According to Newton's hypothesis that the density increases uniformly (from zero) with increasing distance below VB.

(10) The manuscript here breaks off. Exactly how Newton proposed to 'prove'—or, at least, justify—his 'Theorem 1' is not clear, for it is in effect an axiomatic principle demonstrable (if at all) by empirical test of the quantity of refraction at interfaces between a suitable variety and range of optical media. It may be that he would argue for this property of his postulated *vis refractiva* (as for those of other similar short-range forces) by analogic appeal to macroscopic gravitation: in a 'Conclusion' (ULC. Add. 3970.3: 337^v^–338^v^) to his manuscript 'Opticks' which he afterwards suppressed from the final printer's copy, having asserted as his primary hypothesis that 'The particles of bodies have certain spheres of activity w^th^in wch they attract or shun one another', he continued (*ibid.*: 338^r^/338^v^): 'As all the great motions in the world

versus medium alterutrum, neq ulla alia vi agitetur vel impediatur; sit autem attractio, in æqualibus ab utroq plano distantijs ad eandem ipsius partem captis, ubiq eadem: dico quod sinus incidentiæ in planum alterutrum erit ad sinum emergentiæ ex plano altero in ratione data.

Cas. 1. Sunto *Aa*, *Bb* plana duo parallela. Incidat corpus in planum prius *Aa* secund[u]m lineam *GH*, ac toto suo per spatium intermedium transitu attrahatur vel impellatur versus medium incidentiæ, eaq actione describat lineam curvam *HI*, & emergat secundum lineam *IK*. Ad planum emergentiæ *Bb* erigatur perpendiculum *IM*, occurrens tum lineæ incidentiæ *GH* productæ in *M*, tum plano incidentiæ *Aa* in *R*; & linea emergentiæ *KI* producta occurrat *HM* in *L*. Centro *L* intervallo *LI* describatur circulus, secans tam *HM* in *P* & *Q*, quam *MI* productam in *N*; & primo si attractio vel impulsus ponatur uniformis, erit (ex demonstratis *Galilæi*)[12] curva *HI* Parabola, cujus hæc est proprietas, ut rectangulum sub dato[13] latere recto & linea *IM* æquale sit *HM* quadrato; sed & linea *HM* bisecabitur in *L*. Unde si ad *MI* demittatur perpendiculum *LO*,

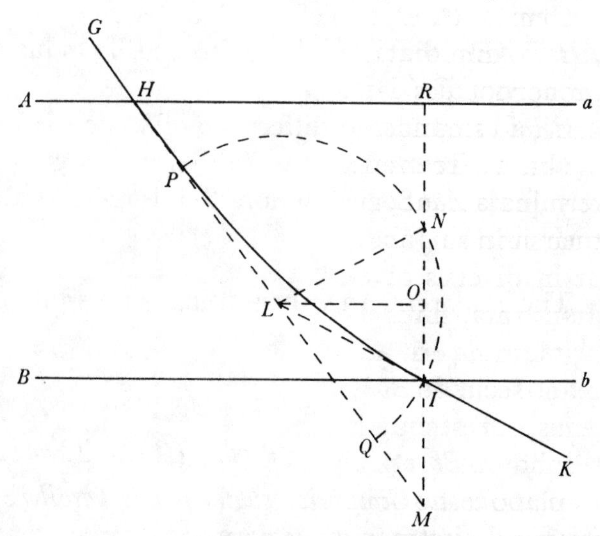

depend upon a certain kind of force (wch in this ear[th] we call gravity) whereby great bodies attract one another at great distances: so all the little motions in the world depend upon certain kinds of forces whereby minute bodies attract or dispell one another at little distances....The truth of this Hypothesis I assert not because I cannot prove it, but I think it very probable because a great part of the phænomena of nature do easily flow from it, wch seem otherways inexplicable: such as are chymical solutions, præcipitations, philtrations, detonizations, volatizations, fixations, rarefactions, condensations, unions, separations, fermentations, ...the reflexion and refraction of light, ...'.

(11) Reproduced from the printed *Philosophiæ Naturalis Principia Mathematica* (London, ₁1687): 227–30. These two propositions from Section XIV of its first book are only minimally revised by Newton in Humphrey Newton's press transcript (Royal Society MS LXIX) and their text was reissued in 1713 and 1726 without alteration.

(12) See §2: note (35).

(13) There is a subtlety here: while for each individual parabolic arc $\widehat{HI}$ the *latus rectum* HM^2/IM will be constant as *IM* moves parallel to itself to intersect the tangent *HM*, for the totality of parabolas traversible under the uniform (upwards) accelerative 'attraction or impulse' which Newton here presumes, it will be a variable parameter. His assumption in the sequel—probably thoughtless rather than intended, we would say—that this *latus rectum* shall in all cases be constant is a concealed restriction on the generality of his present model of the action of a constant *vis refractiva*; see the next note.

æquales erunt MO, OR; & additis æqualibus IO, ON, fient totæ æquales MN, IR. Proinde cum IR detur, datur etiam MN, estq; rectangulum NMI ad rectangulum sub latere recto & IM, hoc est ad HM^q, in data ratione. Sed rectangulum NMI æquale est rectangulo PMQ, id est, differentiæ quadratorum ML^q & PL^q seu LI^q; & HM^q datam rationem habet ad sui ipsius quartam partem LM^q: ergo datur ratio ML^q-LI^q ad ML^q, & divisim, ratio LI^q ad ML^q, & ratio dimidiata LI ad ML. Sed in omni triangulo LMI sinus angulorum sunt proportionales lateribus oppositis. Ergo datur ratio sinus anguli incidentiæ LMR ad sinum anguli emergentiæ LIR. Q.E.D.(14)

Cas. 2. Transeat jam corpus successive per spatia plura parallelis planis terminata $AabB$, $BbcC$ &c & agitetur vi quæ sit in singulis separatim uniformis, at in diversis diversa; & per jam demonstrata, sinus incidentiæ in planum primum Aa erit ad sinus emergentiæ ex plano secundo Bb in data ratione; & hic sinus, qui est sinus incidentiæ in planum secundum Bb, erit ad sinum emergentiæ ex plano tertio Cc in data ratione; & hic sinus ad sinum emergentiæ ex plano quarto Dd in data ratione; & sic in infinitum: & ex æquo sinus incidentiæ in planum primum ad sinum emergentiæ ex plano ultimo in data ratione.

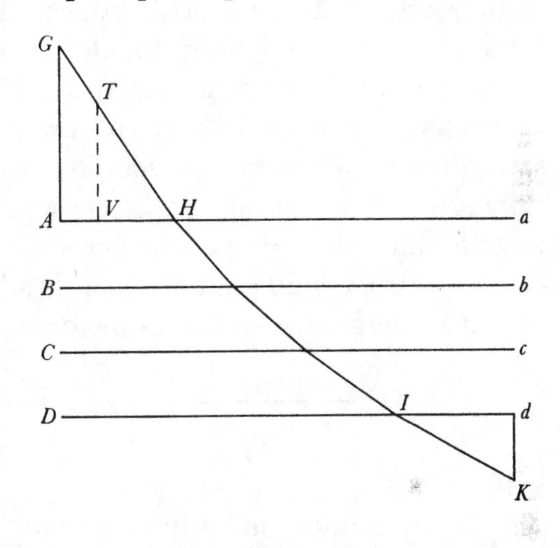

Minua[n]tur jam planorum intervalla & augeatur numerus in infinitum, eo ut attractionis vel impulsus actio secundum legem quamcunq; assignatam continua

(14) For all the ingenuity of its geometrical derivation, this result promises more than it in fact proves. If, in appropriate analytical equivalent, we suppose that under a constant *vis attractiva* g (directed vertically upwards) a body, initially shot off horizontally, attains the point I after rising through a vertical distance x and thereafter follows the parabolic path $\widehat{IH}$ to H, rising through a further vertical distance $IR = a$, then on taking the parabola's principal *latus rectum* to be R (whence its initial horizontal speed is $\sqrt{[2g.\frac{1}{4}R]} = \sqrt{[\frac{1}{2}gR]}$) it follows that the speeds attained by the body at I and H are respectively $v_I = \sqrt{[2g(x+\frac{1}{4}R)]} = \sqrt{[\frac{1}{2}g(R'-4a)]}$ and $v_H = \sqrt{[\frac{1}{2}gR']}$, where, by Lemma XIII of the preceding 'De motu Corporum Liber primus' (see §2, Appendix 2.10: note (38)), $R' = 4\sqrt{[(x+a-\frac{1}{4}R)^2+R(x+a)]} = 4(x+a)+R$ is the parabola's *latus rectum* pertaining to the point H. Newton's hidden assumption (see previous note) that R' is given determines straightforwardly that the ratio

$$\sin\hat{H}/\sin\hat{I}\,(=v_I/v_H) = \sqrt{[(R'-4a)/R']}$$

is also given, but its arbitrariness will be evident. If the magnitude g of the *vis attractiva* also be given, the speeds v_H, v_I at H and I will be given in absolute magnitude, and only the instantaneous directions of motion at the ends of the parabolic arc $\widehat{HI}$ will vary.

reddatur; & ratio sinus incidentiæ in planum primum ad sinum emergentiæ ex plano ultimo, semper data existens, etiamnum dabitur. Q.E.D.[(15)]

Prop. XCV. Theor. XLIX.

Iisdem positis, dico quod velocitas corporis ante incidentiam est ad ejus velocitatem post emergentiam ut sinus emergentiæ ad sinum incidentiæ.

Capiantur *AH*, *Id* æquales, & erigantur perpendicula *AG*, *dK* occurrentia lineis incidentiæ & emergentiæ *GH*, *IK* in *G* & *K*. In *GH* capiatur *TH* æqualis *IK*, & ad planum *Aa* demittatur normaliter *TV*. Et per Legum Corol. 2[(16)] distinguatur motus corporis in duos, unum planis *Aa*, *Bb*, *Cc* &c perpendicularem, alterum ijsdem parallelum. Vis attractionis vel impulsus agendo secundum lineas perpendiculares nil mutat motum secundum parallelas, & propterea corpus hoc motu conficiet æqualibus temporibus æqualia illa secundum parallelas intervalla quæ sunt inter lineam *AG* & punctum *H* interq; punctum *I* & lineam *dK*; hoc est, æqualibus temporibus describet lineas *GH*, *IK*. Proinde velocitas ante incidentiam est ad velocitatem post emergentiam ut *GH* ad *IK* vel *TH*, id est, ut *AH* vel *Id* ad *VH*, hoc est (respectu radii *TH* vel *IK*) ut sinus emergentiæ ad sinum incidentiæ. Q.E.D.[(17)]

---

[3][(18)]

If any motion or body or moving thing whatsoever be incident wth any velocity on any broad & thin space terminated on both sides by two parallel planes & in its passage through that space be urged[(19)] perpendicularly towards ye further plane by any force wch at given distances from ye planes is of a given

---

(15) The straightforward extension to the case where the upwards directed *vis* is no longer uniform.

(16) See §2: note (20).

(17) Newton's proof minimally recasts the standard Cartesian deduction of the law of sines as inversely proportional to the corresponding orbital speeds (see note (1) above) in the case of the extended emission model where the density now varies, but remains uniform over each infinitesimally thin horizontal layer of a continuous medium.

(18) ULC. Add. 3970.3: 341^r^/341^v^, a roughly penned and considerably rewritten autograph draft which by its handwriting and verbal style was composed in about 1692. In accord with Newton's directions in the manuscript we here combine the initial version and following revise, but reproduce significantly variant original phrasings in footnotes. The text was somewhat further altered in further revision—when it was keyed to a relettered diagram—but not essentially remodelled in its structure, and was then inserted by Newton in the press copy of his 'Opticks' (ULC. Add. 3970.3: 17^r^–284^r^; see especially 59^r^/60^r^ [= *Opticks* (London, $_1$1704): $_1$57–8]).

(19) Originally '...be accelerated by any given force urging it'.

quantity:[20] the perpendicular velocity of that motion body or thing at its emerging out of that space shall be always equal to y^e square root of y^e summe of the square of y^e perpendicular velocity of that motion body or thing at its incidence on that space & of the square of the perpendicular velocity which that motion body or thing would have at its emergence if at its incidence its perpendicular velocity was infinitely little.

And the same Proposition holds true of any motion body or thing perpendicularly retarded in its passage through y^t space if instead of ye summ of ye two squares you take their difference. The Demonstration[21] Mathematicians will easily find out & therefore I shall not trouble the Reader wth it.

Suppose now that a ray coming most obliquely in ye line AC be refracted at C by ye plane AB into ye line CQ & it be required to find ye line CE into wch any other ray [D]C shall be refracted:[22] Let AC, DK be ye sines of incidence & QM, EN ye sines of refraction & let ye equal motions of ye incident rays be represented by y^e equal lines AC & DC & the motion of the Ray DC be distinguished into two motions DK & KC, one of wch DK is parallel & y^e other KC perpendicular to y^e refracting su[r]face. In like manner let y^e motions of the emerging rays be distinguished into two whereof the perpendicular ones are $\frac{AC}{MQ}CM$ & $\frac{DK}{NE}CN$: And if y^e force of the refracting plane begin to act upon the rays at a certain distance from y^t plane on one side & end at a certain distance from it on the

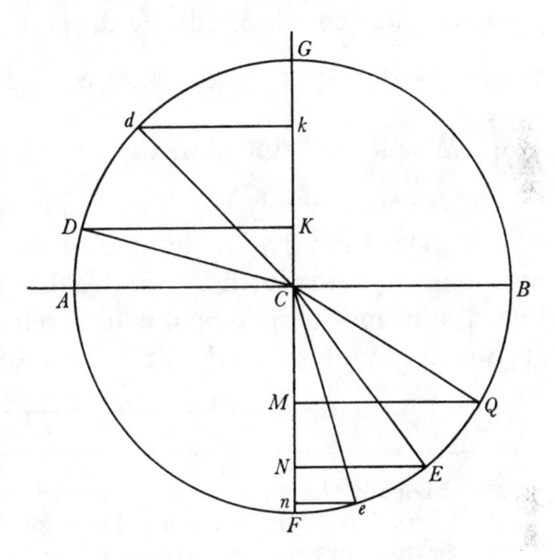

(20) Newton has deleted in sequel the amplification 'or constantly greater at some places then at others'.

(21) Since the present theorem is merely the particular case of Proposition XL of the preceding 'De motu Corporum Liber primus' (see §3: note (198)) in which the force-centre is at infinity and the *vis refractiva* is here understood to be applied as one single impulse, its proof follows straightforwardly therefrom *mutatis mutandis*: namely, if the initial speed along the incident line DC in the following figure be V, and the impulse of refractive force applied at C (perpendicular to the interface AB) generates 'instantaneously' a downwards speed W along CF, then the speed along the refracted ray CE will be $v = \sqrt{[V^2+W^2]}$. Since the horizontal velocity is unchanged by the refraction, the vertical speeds V', v' of motion before and after passage through the interface will likewise be related by $v' = \sqrt{[V'^2+W^2]}$, of course.

(22) Newton initially wrote '...be required to know into what line the ray DC be refracted: I describe wth y^e center C & any radius a circle cutting y^e incident rays in D, d & y^e refracted ones in E, e &'.

other side & in all places between those two limits act upon y^e ray perpendicularly to y^t refracting plane, & ye actions upon ye ray[s] at equal distances from ye plane be equal[,] at unequal ones unequal(23) according to any rate whatever: the motion of y^e ray wch is parallel to y^e refracting surface will suffer no alteration by that force(24) & the motion wch is perpendicular to it will be altered according to y^e foregoing rule:(25) that is[,] if for y^e perpendicular velocity of y^e emerging Ray CQ you write $\frac{AC}{MQ}CM$, then the perpendicular velocity of any other emerging ray CE wch is $\frac{DK}{NE}CN$ will be equal to $\sqrt{CK^q+\frac{AC^q}{MQ^q}CM^q}$. & by squaring these equals & adding the equals DK^q & CK^q-AC^q(26) & dividing the summs by the equals CN^q+NE^q & CM^q+MQ^q you will have $\frac{DK^q}{NE^q}$ equal to $\frac{AC^q}{MQ^q}$. Whence the sine of incidence DK is to the sine of refr[action] NE as AC to MQ & consequently in a given ratio. And this Demonstration being general w^{th}out determining what light is or by what kind of force it is refracted or assuming any thing further then this that y^e refracting body or surface acts upon the rays in lines perpendicular to it self, I take it to be a very convincing (27) argument of ye exact truth of this Proposition.(28)

---

(23) Originally '& ye actions be greater where ye distance is lesse'.

(24) A little less precisely, Newton first wrote '...will remain y^e same in all cases'.

(25) Newton's initial demonstration originally concluded: 'that is[,] if V be put for such a velocity that the perpendicular velocity of y^e emerging ray CE wch conceive here to be represented by $\frac{DK}{EN}CN$ be equal to $\sqrt{CK^q+V^q}$, then the perpendicular velocity of any other emerging ray Ce wch is $\frac{dk}{en}Cn$ will be equal to $\sqrt{Ck^q+V^q}$. From y^e squares of y^e last equals subduct y^e squares of y^e first equals & there will remain $\frac{dk^q}{en^q}Cn^q-\frac{DK^q}{EN^q}CN^q$ equal to Ck^q-CK^q, that is to DK^q-dk^q. From these equals subduct their negative parts & you will have $\frac{dk^q}{en^q}Cn^q+dk^q$ equal to $\frac{DK^q}{EN^q}CN^q+DK^q$ or $\frac{dk^q\times Cn^q+dk^q\times en^q}{en^q}$ equal to $\frac{DK^q\times[C]N^q+DK^q\times EN^q}{EN^q}$. Divide these equals by the equals Cn^q+en^q & CN^q+EN^q & you will have $\frac{dk^q}{en^q}$ equal to $\frac{DK^q}{EN^q}$[,] whence $\frac{dk}{en}$ is equal to $\frac{DK}{EN}$ & consequently dk is to en as DK to EN, that is the sines of incidence are proportional to y^e sines of refraction'.

(26) Read 'AC^q-CK^q'.

(27) Originally just 'very good'!

(28) In this over-complicated restatement of the simple Cartesian emission model Newton thoroughly obscures the basic point that the sine-law of refraction—here produced much like

Plate II. Derivation of the optical law of sines in the modified Cartesian hypothesis of an impressed 'vis refractiva' (I, §3, Appendix 3.3).

[4][29]

Let AKL represent the globe of the earth, & suppose this globe is covered wth an Atmosphere of Air whose density decreases uniformly from y^e^ earth upwards to the top[30] wch is here represented by the circle $M[O]N$. And let SO be a ray of light falling on y^e^ top of this Atmosphere at O & in its passage from thence through y^e^ Atmosphere to the spectator at A, continually refracted & bent in y^e^ curve line OBA.[31] From any point of this curve line B to y^e^ center of the earth draw the right line BC cutting the surface of the earth in D & take

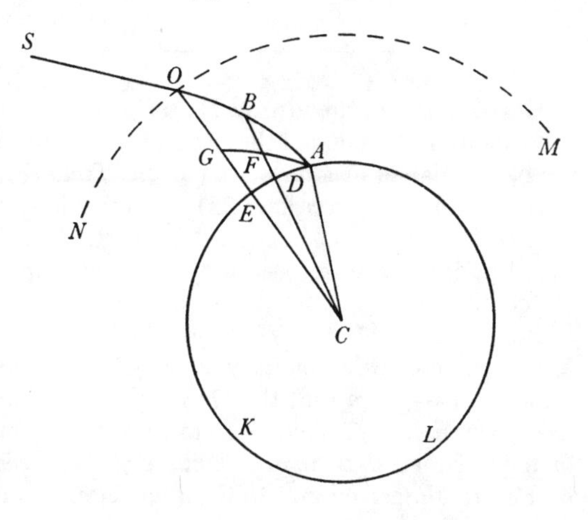

a *deus ex machina*—derives from the invariance of the horizontal component of the speed of the light 'ray' in its passage through the optical interface at which it receives its downwards impulse of refractive force, and his introduction of an undemonstrated rule for the increase in 'orbital' speed thereby engendered serves only further to confuse it. One might almost be forgiven for thinking that Newton was deliberately aiming to mystify the mathematically unsophisticated reader of his 'Opticks', but the clumsiness and opaqueness of his exposition is here doubtless unintended.

(29) Extracted from Newton's letter to John Flamsteed of 20 December 1694 (Corpus Christi College, Oxford. MS 361.65, first published by Francis Baily in *An Account of the Revd John Flamsteed, ...compiled from his own manuscripts and other authentic documents, never before published* (London, 1835): 145–6, and transcribed more exactly in *The Correspondence of Isaac Newton*, **4** 1967: 61–2). The topic of how to correct visual observations (from earth) of the heavenly bodies for the varying refractive deviation suffered by light-rays in traversing the terrestrial atmosphere at different angles and differing times of the year had arisen in their correspondence—frequent at this period—some two months before. In particular Flamsteed had sent Newton on 29 October a 'Synopsis Refractionum ab observatis Veneris a Sole distantiis deductarum cum veris distantiis puncti Solis observati a vertice ad quamlibet deductionem' (reproduced in *Correspondence*, **4**: 40–1) which he had derived from his own sightings of Venus during the previous February–April, and when Newton returned it on the following 17 November he enclosed a *Tabula Refractionum* ([Baily, *Flamsteed:* 141 =] *Correspondence*, **4**: 49) which, he observed in his covering letter, 'I have computed by applying a certain Theorem to your Observations' but did not otherwise explain. When Flamsteed again wrote on 19 December with the gentle rebuke that 'Your concealeing the foundations on which you computed [your Table] from me has caused me to bestow some time & paines on y^t^ subject. I have [*sc.* to no avail!] examined all the tables I have by me of Refractions to see on what fundamentalls they were built' (*Correspondence*, **4**: 57), Newton replied by communicating the mode of construction here described, apologetically remarking in introduction that 'The foundation of the Table of refractions I concealed not as a secret but omitted through ye hast I was in when I wrote my last letter. But since you desire it I will now set it down', and further adding that 'This Theorem is Geometrically demonstrable but the demonstration is too

CF a mean proportional between *CB* & *CD* & let *AFG* be y[e] locus of the point *F*, that is the curve line in wch ye point *F* will be allways found: & if this curve line *AFG* cut the right line *OC* in *G*; the whole refraction of ye ray in passing from *O*

---

intricate to be set down in a Letter' (*ibid.*: 61). Armed with a knowledge of P. S. Laplace's published construction of the 'Équation différentielle du mouvement de la lumière' according to the Newtonian emission theory (see his *Traité de Mécanique Céleste*, **4** (Paris, An XIII [= 1804]): Livre x, Chapitre Premier: 231–76, especially 231–4), the French physicist and historian of science J. B. Biot in a brilliant 'ANALYSE des Tables de réfraction construites par Newton, avec l'indication des procédés numériques par lesquels il a pu les calculer' (*Journal des Savants* (Décembre 1836): 735–54) appended to a series of essays in extended review of Baily's *Flamsteed* (*ibid.* (Mars/Avril/Novembre 1836): 156–66/205–22/641–58) not only plumbed the basis—still concealed by its author—of Newton's present mode of construction, but even more remarkably—given that the Newtonian worksheets which substantiate the surmise were then (see I: xxviii–xxix) locked away in Lord Portsmouth's archives, their very existence unknown—further divined that a subsequent *Tabula Refractionum* of Newton's, privately communicated by him to Flamsteed and to Edmond Halley but eventually published (by the latter) without any explanation of its structure, was similarly founded on the more realistic hypothesis of atmospheric density enunciated in Book 2, Proposition XXII of the *Principia*. (See note (31) below.) We are heavily indebted to his pioneering lead.

(30) Where understand that it is nil, so that, beyond, the ray path *SO* is a straight line.

(31) Since the density of the atmosphere is, in Newton's hypothesis uniform at all equal distances from the earth's centre, so that the interface at any point *B* is instantaneously perpendicular to the radius vector *CB*, it follows that the ray path $\widehat{OBA}$ is traced under the action of an infinite succession of short-range 'refractive powers' which are all directed through *C* and may accordingly be treated as a single, continuous central *vis refractiva*; whence the dynamical elements of the 'orbit' $\widehat{OBA}$ thereby induced may straightforwardly be defined by Propositions XL and XLI of the preceding 'De motu Corporum Liber primus'. Much as in §3: note (209), therefore, let *C* be the origin of a system of polar coordinates in which $B(r, \phi)$ is a general point of the ray path traced through the earth's atmosphere (of height $EO = h$) from its outer limit $O(R+h, \Phi)$ down to $A(R, 0)$ at the earth's surface; further, let the atmospheric densities at *B* and *A* be ρ and d, and set (as initial conditions) V to be the speed of the corpuscle-stream at *A*, where the ray path they trace is inclined at a (directly observable) angle α to the radius *CA*. By Newton's present hypothesis there ensues $\rho = (d/h)(R+h-r)$, so determining the *vis refractiva* at *B* to be the (constant) density gradient $d\rho/dr = -d/h$: at once, the instantaneous orbital speed there is $v = \sqrt{[V^2-2(d/h)(R-r)]}$; whence, on introducing the 'Keplerian' constant $Q = RV\sin\alpha$, the polar equation of the ray path is $\phi = \int_R^r Q/r^2\sqrt{[v^2-Q^2/r^2]}\,.\,dr$ and, where the tangent at *B* is inclined at an angle θ to the radius *CA*, the angular refractive deviation over the arc $\widehat{BA}$ is $\theta-\alpha = \int_0^\phi r(d/h)/v^2\,.\,d\phi \approx (d/hV^2)\int_0^\phi r\,.\,d\phi$. In sequel Newton elegantly (if unnecessarily) constructs the auxiliary curve $\widehat{AFG}$ meeting *CB* in *F* such that $CF = \sqrt{[CD\times CB]} = \sqrt{[Rr]}$, thereby determining the angular refraction over $\widehat{BA}$ to be (nearly) proportional to its corresponding sector $(ACF) = \frac{1}{2}\int_0^\phi Rr\,.\,d\phi$, and consequently 'the whole refraction $[\Phi-\alpha]$ of y[e] ray in passing from *O* to *A* will be proportional to the area

$$AFGC\left[= \tfrac{1}{2}R\int_0^\Phi r\,.\,d\phi\right]'.$$

to A will be proportional to the area $AFGC$ & the refractions in passing through any part of that line OB or BA will be proportional to the corresponding part of the area $GFCG$ or $FACF$.(32)

Of course, as he was well aware, merely geometrically to define the construction of the problem in this way is, till its underlying integrals are resolved, only halfway to attaining the explicit quantitative solution required of it in given numerical instances. Here in fact the polar equation of the ray path $\widehat{OBA}$ —closely akin, it will be evident, to the Boulliauist 'figure' of free fall under constant terrestrial gravity from a point above the earth's surface which Newton communicated to Robert Hooke on 13 December 1679 (see §2: note (127))—is not evaluable in exact algebraic terms, but, where the (effective) atmospheric height EO is very small—as in physical reality it is—in comparison with the length $CE = CA$ of the earth's radius, the orbital arc $\widehat{OBA}$ will be very nearly parabolic and may be so approximated: say, on extending CB and CO to meet the tangent at A to the path in λ and μ, by supposing that $A\lambda^2/B\lambda = A\mu^2/O\mu$, constant, for all *radii vectores* CB in the central angle $\widehat{ACO}$. Once the ray path is approximately thus determined, it will then be straightforward—though 'ye calculation is intricate', Newton added in his following letter to Flamsteed on 15 January 1694/5 (*Correspondence*, **4**: 67)—to evaluate the areas $(ADB) \approx R\int_0^\phi (R-r).d\phi$ corresponding to a suitable range of values of the curvilinear base $AD = R\phi$ by the numerical quadrature method (here effectively reducing to Simpson's rule) outlined by him in 'the 5t Lemma of my Third Book of *Principia Math.* [$_1$1687: 482–3; see IV: 73, note (6)]' and then appropriately intercalate intermediate areas by the preceding general interpolation formula (*Principia*, $_1$1687: 481–2; see IV: 70–3).

(32) Though Newton went on confidently enough to affirm to Flamsteed that 'as my Table of refractions computed from this Theoreme agrees much better with your Observations then the vulgar ones [listed by Flamsteed in his previous letter of 10 December 1694; see *Correspondence*, **4**: 57], so I beleive you will allow ye Theorem it self to be a better foundation then the vulgar [Keplerian] ones of a single refraction on ye top of an uniform Atmosphere' (*Correspondence*, **4**: 61), he remained unsatisfied with its basic premiss of a constant density gradient, which, he announced to Flamsteed on 15 January 1694/5, 'makes ye refracting power [*vis refractiva*] of ye Atmosphere as great at ye top as at ye bottom. This has put me upon thinking on a new Theoreme & I think I have found one but intend to consider it a little further' (*ibid.*: 67). Newton amplified his remark a month later, informing Flamsteed on 16 February that he had 'been ever since I wrote to you last, upon making a new Table of refractions...supposing ye Atmosphere to be of such a constitution as is described in the 22th Proposition of my second book (wch certainly is the truth)' (*ibid.*: 86). His added observation that 'Tis a very intricate & laborious piece of work. Yet something I have done towards it' is amply confirmed by his extant calculations (ULC. Add. 3967.3: [in correct sequence] 22v/29r, 23v/28r, 24v/27r, 25v/26r) in this more realistic hypothesis of atmospheric air being a 'fluid' whose 'density shall be proportional to the compression, and its parts drawn downwards by gravity reciprocally proportional to the squares of their distances from the centre' as he stated it in Proposition XXII of Book 2 of his *Principia* ($_1$1687: 298). Recast in the terms of note (30) preceding, this determines the 'differentia densitatum' $d\rho$ at any point to be proportional to the 'pressio' $r^{-2}\rho.dr$ there (see *ibid.*: 298–9), whence $\int_d^\rho \rho^{-1}.d\rho = \int_R^r -kr^{-2}.dr$ and so, on integrating, $\log(\rho/d) = k(r^{-1}-R^{-1})$; accordingly, the ray path (B) is defined by

$$\phi = \int_R^r Q/r^2 \sqrt{[v^2 - Q^2/r^2]}.dr,$$

where again $Q = RV\sin\alpha$ but now the instantaneous speed at $B(r, \phi)$ is $v = \sqrt{[V^2 - 2(d-\rho)]}$, $\rho = de^{k(r^{-1}-R^{-1})}$, while the refractive deviation $\theta - \alpha$ over $\widehat{BA}$ is proportional to

$$\int_0^\phi -r(d\rho/dr)/v^2.d\phi \approx kV^{-2}\int_0^\phi r^{-1}\rho.d\phi.$$

Since the atmospheric density is now conceived continuously to decrease outwards from the earth's centre C (till at infinity it attains the value $de^{-k/R}$), it is clear that when the ray path $\widehat{ABO}$ is extended backwards it will asympotically approach the true line of incidence of stellar light; where CB meets the orbital tangent at A, Newton in his computations accordingly approximated the ray path as the polar hyperbola defined, on taking $A\lambda = x$, by

$$B\lambda = lx^2/(mx+n),$$

l, m and n suitable constants, and then further refined the ensuing approximation to $CB = r$ and related refraction angle θ by aid of the geometrical 'fluxional' equivalent of the recursive adequality $r \approx R\sin\alpha.\left(\int_0^\phi \sin(\theta-\alpha)/\sin^3(\alpha-\phi).d\phi - \operatorname{cosec}(\alpha-\phi)\right)$, evaluating the integrals so arising by the technique of numerical quadrature sketched by him in Lemma V of the *Principia's* third book (see note (31)). Successively adjusting the constants of his model to yield a 'tota refractio horizontalis' (when $\alpha = 90°$) at A, from incidence at infinity, of various sizes between 33′ 20″ and 39′ 5″, Newton therefrom computed the corresponding total angular refraction for altitudes of incidence at A of 3° and 12°, and so (since there is zero refraction at $\alpha = 0°$) was able to interpolate a comprehensive tabulation of refractions (ULC. Add. 3967.3: 19^r) for altitudes of incidence at A of 0°(15″) 1°(30″) 10°(1°) 90°, whose copy he passed on to Flamsteed on 15 March 1694/5 with the commendation that it 'is exact to a second minute for all altitudes above 10^degr. & ... in ye altitudes between 3 & 10 degrees ye greatest error cannot be above 2 or 3 seconds' (*Correspondence*, **4**: 94) and which was afterwards published by Edmond Halley, as a *Tabula Refractionum Siderum ad Altitudines apparentes* 'such as I long since received it from its Great Author', in appendix to 'Some Remarks on the Allowances to be made in Astronomical Observations for the Refraction of the Air', *Philosophical Transactions*, **31**, No. 368 [May–August 1721]: 169–72. (Compare Biot's 1836 '*Analyse*...' (note (29)): 748–53, but notice that the 'petite différence de 8″' between Newton's 'tota refractio horizontalis' 33′ 45″ and Biot's computed 'réfraction horizontale' 33′ 36.7″ which, he conjectured, 'peut provenir soit de ce qu'il n'aurait pas serré ses quadratures autant qu'il aurait pu le faire avec plus de peine, soit de ce que la suppression des fractions de secondes dans sa table ne nous permettrait pas d'en retrouver les constantes avec une complète rigueur' ('Analyse...': 752) is misconceived: Newton himself assumes the 'tota refractio horizontalis' to be an empirical constant, here given *à priori*. The chief limitation upon the accuracy of his revised *Tabula* is probably due to his reliance on the 'fluxional' adequality

$$d(R\phi)/d(R\sin\alpha/\sin(\alpha-\phi)-r) \approx \sin^3(\alpha-\phi)/\sin\alpha.\sin(\theta-\alpha)$$

as an accurate basis for refining the primitive hyperbolic approximation to the ray path when ϕ is no longer very small. But of this Biot in 1836 could have had no knowledge.) Halley himself gave no explanation of Newton's *Tabula* other than to qualify it ('Some Remarks...': 169) as 'the first accurate Table [of] the Refractive Power of the Air...: The *Curve* which a Beam of *Light* describes, as it approaches the *Earth*, being one of the most perplext and intricate that can well be proposed, as Dr. *Brook Taylor* in the last Proposition of his *Methodus Incrementorum* [*Directa & Inversa* (London, 1715)] has made it evident'. This last citation points to one more instance in which Newton, by not publishing a discovery at the time he made it, had to endure seeing it published by a younger, independent 'inventor' of it. With repeated appeal to the pertinent theorems in 'Lib. I' and 'Lib. II. *Princip. Math.*' Taylor in his *Methodus* in fact came exactly to retrace Newton's way of resolving the problem of atmospheric refraction: specifically—on inessentially recasting his argument into present terms—in his Proposition

APPENDIX 4. RECOMPUTATION OF THE INVERSE-CUBE ORBIT (MAY 1694).[1]

From autograph originals in the University Library, Cambridge and the Royal Society, London

[1][2] [Pone] $C[D]=x$.

$$[DF]=\frac{a^4}{x^3}.^{(3)}$$

[prodit] area [infinita]

$$=-\frac{2a^4}{x^2}.^{(4)}$$

[adeoq3 posito] $AC=c$. [erit]

$$\sqrt{\frac{2a^4}{x^2}-\frac{2a^4}{c^2}}=\sqrt{ABFD}$$

$=$veloc[itati] in $[I]=KI$.[5]

$\left[\text{Insuper erit}\frac{Q}{x}=\right]Z=KN$.[6]

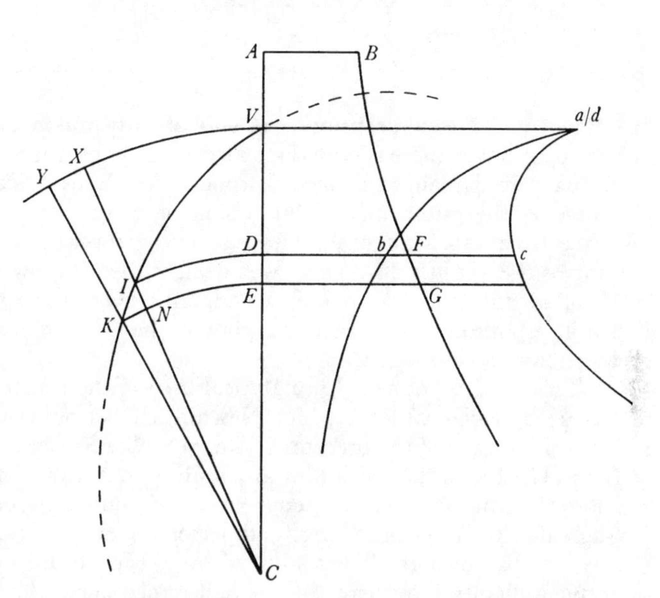

XXVI (*Methodus*: 103–4) he introduced the Newtonian defining condition that the element of density $d\rho$ be proportional to the 'incrementum pressionis' $r^{-2}\rho.dr$, so that

$$\log(\rho/d) \propto r^{-1}-R^{-1};$$

then, having in his Lemma XIII (*ibid.*: 106–7) deduced that $v^2 = V^2+2(\rho-d)$, in his terminal Proposition XXVII (*ibid.*: 108–18) he both derived Newton's measure $\int_0^\phi -r^{-1}\rho/v^2.d\phi$ of the refractive deviation (by the geometrical limit-increment argument we summarised at the end of §3: note (209) above) and, finally, expanded this integral as a 'Taylor' power-series. To what extent he was in all this aware of Newton's unpublished earlier computations to a parallel end is difficult to determine; while the two were in close contact from at least early 1712 onwards, Taylor in his *Methodus* implicitly claims originality for his central Proposition XXVII and we have no reason to believe that Newton had done more than suggest to him the way of applying his *Principia* theorems so as to develop a sophisticated dynamical theory of atmospheric refraction—indeed, Newton may well have permitted Halley to publish his own similarly based *Tabula Refractionum Siderum* as a mild, belated independent claim to priority.

Whatever the truth of this, the world at large was given no other hint by Newton of his extended researches into the topic in the middle 1690's, and his printed *Tabula* was soon all but forgotten. When, twenty-five years after Newton's death, Leonhard Euler came to apply his own variant formulation of the emission theory to the problem of atmospheric refraction, he was wholly unaware that he was but treading in his English predecessor's steps. (See Euler's letter to Tobias Mayer of 18 March 1752 (N.S.), recently rendered into English by E. G. Forbes in *The Euler–Mayer Correspondence* (*1751–1755*) (London, 1971): 50–5, where refinement is made of a primitive Newtonian approximation of the ray path as a generalised hyperbolic locus.) By the time that Biot came, nearly a century afterwards, to give due

$$[\text{unde}]\ IN = \sqrt{ABFD - ZZ} = \sqrt{\frac{2a^4}{xx} - \frac{2a^4}{cc} - \frac{Q^2}{xx}}.\quad [\text{Ergo.}]$$

$$\text{Tempus}^{(7)} = \square \frac{Q}{2\sqrt{\dfrac{2a^4 - Q^2}{xx} - \dfrac{2a^4}{cc}}}.\ [\text{sive}] = \square \frac{Qx}{2\sqrt{2a^4 - Q^2 - \dfrac{2a^4xx}{cc}}}.$$

prominence to Newton's priority it was already much too late to do more than give him historical homage. Increasingly discarded by physicists in the 1830's and 1840's because of its inadequacy to explain such other properties of light as colour dispersion and interference phenomena, the grand Newtonian vision of a 'gravitating' light-corpuscle deviated at an interface by a short-range accelerative *vis refractiva*, directed into the denser medium and varying as the density difference, was dealt its death-blow in 1850 when, in contradiction of its corollary that the corpuscular speed must ever be greater in the denser medium, J. L. Foucault established by direct experiment that the velocity of light in air is considerably greater than its velocity in water.

(1) This recalculation of the analytical basis of the construction of the two principal species of inverse-cube central-force which Newton had set out, without explanation, in Corollary 3 to Proposition XLI of the preceding 'De motu Corporum Liber primus' was made at the time of David Gregory's stay with him at Cambridge in early May 1694, doubtless in response to the latter's orally communicated request for enlightenment regarding its demonstration. Though after the lapse of so long (and personally eventful) an intervening period Newton here clearly retains no more than a blurred impression of his original computation, finding considerable difficulty in resurrecting its half-remembered detail for Gregory, his reconstruction will—for all its verbal ellipses and numerical deficiencies—adequately serve to show how he must initially have applied his general theory of central-force orbits to this particular, mathematically simplest case.

(2) ULC. Add. 3960.13: 223[r]. That this fragmentary draft was composed little before the dated revise reproduced in [2] is confirmed by Newton's jotting on the same sheet that he has 'Seen Bartram's [rent] acquittance for y[e] year [16]92...as also widdow Burtons acquittances who owes only for half a years rent ending at Lady day [25 March] last past'. We have, the better to bring out its logical sequence, sparingly fleshed out the skeleton manuscript text with editorial interpolations, and have also adjoined a simplified version of the (corrected) figure which Newton here understands.

(3) Carelessly referring not to the ordinate DF but to the indefinitely near EG, Newton here wrote '$CE = x.EG = \dfrac{a^4}{x^3}$'.

(4) Newton thoughtlessly appeals yet again (compare §3: note (202)) to the erroneous quadrature theorem $\int x^{n-1}.dx = nx^n$. The sole effect of the slip is to quadruple the measure DF of the centripetal force as enunciated in the previous line.

(5) Understanding that the increment of 'tempus' taken to traverse this infinitesimal arc is unity. More generally, if t be the time of orbit over $\widehat{VI}$, then of course the instantaneous *velocitas* at I is $\widehat{IK}/dt$.

(6) Again (see previous note) this should more accurately be KN/dt. As before,

$$Q(= CI \times KN/dt)$$

is Newton's Keplerian constant, while $Z = Q/CI$ is Halley's editorial insertion in the printed *Principia* (see §3: note (205)).

(7) Of orbit over $\widehat{VI}$, that is.

[hoc est] $$=\frac{Q}{-\frac{4a^4}{cc}}\sqrt{2a^4-Q^2-\frac{2a^4}{cc}xx}=\text{sectori } VC[\Gamma].$$

[et] Sector $$VCX=[\Box]\frac{Q,\ VC,\ VC}{2x\sqrt{2a^4-Q^2-\frac{2a^4}{cc}xx}}=\text{ang[ulo] } VC[\Gamma].^{(8)}$$

---

[2][9] [Pone] $CP=CT$(Fig. p. 130)[10]$=CD=CI$ (p. 128) $=x$. [Est] DF reciproce ut cubus altitudinis (per Hypoth. & Prop. XXXIX) vel $=\frac{a^4}{x^3}$ assumpto utcunꝗ a. Sit $AC=c$ et erit area $ADFB=\frac{2a^4}{x^2}-\frac{2a^4}{cc}$[11] et propterea velocitas corporis cadentis in D vel revolventis in I (p. 128) vel P (pag 130) id est longitudo IK (pag. 128) est ut $\sqrt{\frac{2a^4}{x^2}-\frac{2a^4}{c^2}}$ per Prop. XXXIX.

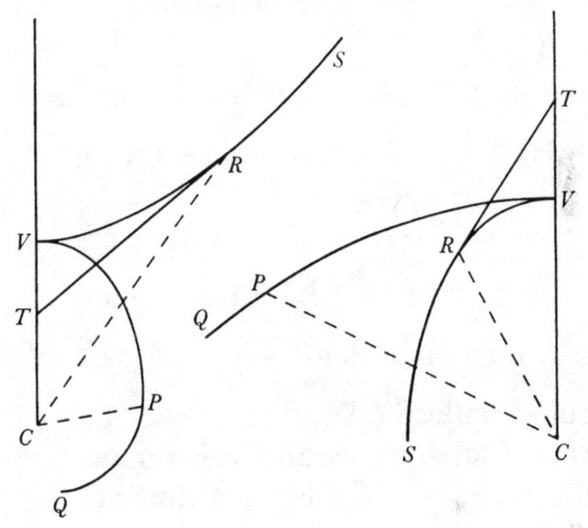

Lege præscripta pag 128, assumatur Q, et sit $\frac{Q}{x}=Z$ (pag 128) et erit[12]

$$Db=\frac{Q}{[2]\sqrt{ABFD-ZZ}}\text{(p. 129) hoc est}=\frac{Qx}{2\sqrt{2a^4-Q^2-\frac{2a^4xx}{cc}}}.$$

---

(8) Understand 'ducto in datam $\frac{1}{2}VC$, VC'. Newton here breaks off without performing the second 'squaring', evidently to pen the fuller following text which he gave to Gregory.

(9) Royal Society. Gregory MS Volume: 163, first published in *The Correspondence of Isaac Newton*, **3**, 1961: 348–9. To the original autograph David Gregory has added the title (keyed to his 'Index Chartarum in M.S. *C* in folio') '[53] Ad Corol: 3. prop XLI. [*Philos*:] Lib. 1. pag. 130. propriâ Newtoni manu 8 Maij 1694'.

(10) This figure we here repeat for convenience of reference. The essence of that (on the *Principia's* page 128) cited in sequel is conveyed in our previous simplified diagram.

(11) The erroneous quadrature $\int a^4/x^3.dx=-2a^4/x^2$ is silently carried over from [1]; see note (4) above.

(12) As in the published *Principia* (see §3: note (207)) the points denoted in sequel by a, b, c and d are those named α, β, δ and γ respectively in Proposition XLI of the preceding 'De motu Corporum Liber primus'.

Et quadrando curvam *ab* cujus hæc est ordinata, prodit area

$$VabD = \frac{ccQ}{4a^4}\sqrt{2a^4 - Q^2 - \frac{2a^4xx}{cc}} \pm \text{dat.} = \text{sectori } VCI \text{ (p. 128) seu } VCP \text{ (p. 130).}$$

Igitur ex data corporis revolventis altitudine *CP* seu *x*, datur Orbis quæsiti sector *VCP* tempori proportionalis, adeoq; datur tempus quo corpus ad altitudinem illam pervenit.

Porro ex ijsdem præmissis et assumptis erit(12)

$$Dc = \frac{Q \times CX^{\text{quad.}}}{2AA\sqrt{ABFD - ZZ}} \text{ (pag. 129) hoc est} = \frac{Q \times [ee]}{2x\sqrt{2a^4 - Q^2 - \frac{2a^4xx}{cc}}}$$

[posito $CX = e$]. Et quadrando curvam *dc* cujus hæc est ordinata prodit area *VdcD* hujusmodi. Sit $\frac{ee}{x} = z = CR$, & $\sqrt{-\frac{2a^4}{cc} + \frac{2a^4 - Q^2}{e^4} \times z^2} = \text{ordinatæ } RS$, et area quæsita *VdcD* erit = sectori *CVS* ducto in datam $\left[\frac{Qcc}{2a^4}\right]$.(13) Quare cum angulus *VCX* (pag 128) seu *VCP* (pag. 130) sit ut hæc area, erit angulus ille ut sector *VCS*. Et cum altitudo *CI* seu *x* sit $\frac{ee}{z}$ et *CT* (ad tangentem *ST* terminata) sit $\frac{CV^2}{CR}$ seu $\frac{CV^2}{z}$[,] si *e* ponatur $= CV$ erit *CI* seu $x = CT$.

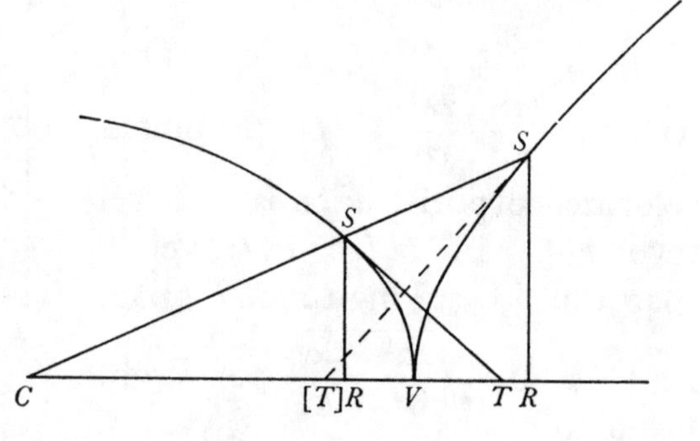

(13) The manuscript reads $\dfrac{\text{'}Q \times CX^q \times \frac{2a^4 - Q^2}{ee} \times cc\text{'}}{a^4}$, that is, $Q(2a^4 - Q^2)\,c^2/a^4$, in error. Newton's substitution $x = e^2/z$ in fact reduces $(VdcD) = \int_e^x Qe^2/2x\,\sqrt{[2a^4 - Q^2 - (2a^4/c^2)\,x^2]}\,.dx$ to be $\int_{e^2/x}^{e} Q/2y\,.dz = (Qc^2/2a^4)\,.\left|\tfrac{1}{2}yz - \int y\,.dz\right|_{z=e^2/x}^{z=e}$ where $y = \sqrt{[((2a^4 - Q^2)/e^4)\,z^2 - 2a^4/c^2]}$, while his further assertion that $(VdcD)$ is proportional to $(CVS) = \left|\tfrac{1}{2}yz - \int y\,.dz\right|_{y=y}^{y=0}$ tacitly restricts the ensuing orbit *VPQ* by the initial condition that $y = 0$ when $z = e$, or

$$(2a^4 - Q^2)/e^2 = 2a^4/c^2:$$

which is equivalent to requiring that the orbital speed $\sqrt{[2a^4/e^2 - 2a^4/c^2]}$ at *V* (when $x = e$) be Q/e, and hence—willy-nilly—that the direction of motion at *V* be perpendicular to *CV*. In the terms of §3: note (214), to which Newton's present solution transposes on making $2a^4 = gR^3$ and $(CA =)\ c = R/\sqrt{[1 - V^2/gR]}$, it will be clear that the point *A* of zero speed in *CV* is finite or

APPENDIX 5. RESISTED OSCILLATION IN A CYCLOIDAL PENDULUM.(1)

[EARLY SUMMER 1685]

Extracted from the second book of Newton's *Principia*(2)

SECT. VI.

De Motu & resistentia Corporum Funependulorum.

. . . .

Prop. XXV. Theor. XIX.

Corpora Funependula quæ in Medio quovis resistuntur in ratione momentorum temporis quæq; in ejusdem gravitatis specificæ Medio non resistente moventur, oscillationes in Cycloide eodem tempore peragunt & arcuum partes proportionales simul describunt.

---

not real according as V is less or greater than $\sqrt{[gR]}$, while when $V = \sqrt{[gR]}$—and the restricted Newtonian orbit is a circle of centre C and radius $CV = R$ (see §3: note (213))—it is at infinity.

(1) As we remarked in §3: note (296), by suitably augmenting the basic dynamical equation governing the isochronous oscillatory motion which ensues under gravity in an upright cycloidal pendulum—discussed at length by Newton in the generalised case of hypocycloidal motion in a direct-distance force-field in Propositions LI and LII of the preceding 'De motu Corporum Liber primus', and then briefly instanced in the primary cycloid in Corollary 2 to the latter Proposition—it is straightforwardly possible to define the analogous cycloidal motion subject to a known resistance, varying as the speed and density of the surrounding medium, which acts instantaneously contrary to the direction of motion. Specifically, where (in terms of the following diagram) a body falls under constant, downwards gravity g in the given primary cycloid ZCA of vertex Z and base-point C, on setting t to be the time of its fall from rest at B (distant $\widehat{CB} = S$ from C) to d (distant $\widehat{Cd} = s$), there attaining under the action of some known resistance ρ the instantaneous speed $v = -ds/dt$, it readily follows that, because the component of gravity directed along the cycloid at d is $g.s/\widehat{CZ}$ (compare III: 422, note (14)), the ensuing equation of motion is then $-d^2s/dt^2$ (or dv/dt) $= ks-\rho$, where $k = g/\widehat{CZ}$, with the initial condition that $v = 0$ when $s = S$. In Book 2, Section VI of his published *Principia* Newton somewhat clumsily—and, in the case of his Proposition XXVI, erroneously—contrived geometrically to construct the solution of this basic equation on the assumption that the density ('gravitas specifica') of the surrounding medium is everywhere uniform and consequently that the resistance ρ varies uniquely with the speed v, taking it either to be constant (Proposition XXV) or to be proportional to v or v^2 (Propositions XXVI–XXIX) or, generally, to vary as some combination of these (Proposition XXX). Though no preliminary drafts or worksheets of these complements to his preceding analysis of unresisted cycloidal motion have seemingly survived, we have thought fit here to reproduce the essence of the printed text of these attempted solutions of the 'damped' harmonic equation $d^2s/dt^2 = v.dv/ds = -ks+lv^n$, $n = 0, 1, 2$, not only for their considerable intrinsic interest but, even more, for the unique illumination they shed on Newton's mature power to formulate, and the limitations of his geometrical expertise accurately and adequately to substantiate, viable exact solutions to 'infinitesimal' equations not in immediately quadrable form.

(2) *Philosophiæ Naturalis Principia Mathematica* (London, $_1$1687): 305–15. Humphrey Newton's press transcript (Royal Society. MS LXIX) is here but trivially variant from the *editio princeps*,

Sit *AB* Cycloidis arcus, quem corpus *D* tempore quovis in Medio non resistente oscillando describit. Bisecetur idem in *C*, ita ut *C* sit infimum ejus punctum, & erit vis acceleratrix qua corpus urgetur in loco quovis *D* vel *d* vel *E* ut longitudo arcus *CD* vel *Cd* vel *CE*. Exponatur vis illa per eundem arcum, & cum resistentia sit ut momentum temporis adeoq; detur, exponatur eadem per datam arcus Cycloidis partem *CO*, & sumatur arcus *Od* in ratione ad arcum *CD* quam habet arcus *OB* ad arcum *CB*: & vis qua corpus in *d* urgetur in Medio resistente, cum sit excessus vis *Cd* supra resistentiam *CO*, exponetur per arcum *Od*, adeoq; erit ad vim qua corpus *D* urgetur in Medio non resistente in loco *D* ut arcus *Od* ad arcum *CD*; & propterea etiam in loco *B* ut arcus *OB* ad arcum *CB*. Proinde si corpora duo *D*, *d* exeant de loco *B*, & his viribus urgeantur: cum vires sub initio sint ut arcus *CB* & *OB*, erunt velocitates primæ & arcus primo descripti in eadem ratione. Sunto arcus illi *BD* & *Bd*, & arcus reliqui *CD*, *Od* proportionales manebunt in eadem ratione ac sub initio, & propterea corpora pergent arcus in eadem ratione simul describere. Igitur vires & velocitates & arcus reliqui *CD*, *Od* semper erunt ut arcus toti *CD*, *OB*, & propterea arcus illi reliqui simul describentur. Quare corpora duo *D*, *d* simul pervenient ad loca *C* & *O*, alterum quidem in Medio non resistente ad locum *C*, & alterum in Medio resistente ad locum *O*. Cum autem velocitates in *C* & *O* sint ut arcus *CB* & *OB*, erunt arcus quos corpora ulterius pergendo simul describunt in eadem ratione. Sunto illi *CE* & *Oe*. Vis qua corpus *D* in Medio non resistente retardatur in *E* est ut *CE*, & vis qua corpus *d* in Medio resistente retardatur in *e* est ut summa vis *Ce* & resistentiæ *CO*, id est ut *Oe*; ideoq; vires quibus corpora retardantur sunt ut arcubus *CE*, *Oe* proportionales arcus *CB*, *OB*, proindeq; velocitates in data illa ratione retardatæ manent in eadem illa data ratione. Velocitates igitur & arcus ijsdem descripti semper sunt ad invicem in data illa ratione arcuum *CB* & *OB*, & propterea si sumantur arcus toti *AB*, *aB* in eadem ratione, corpora *D*, *d* simul describent hos arcus & in locis *A* &

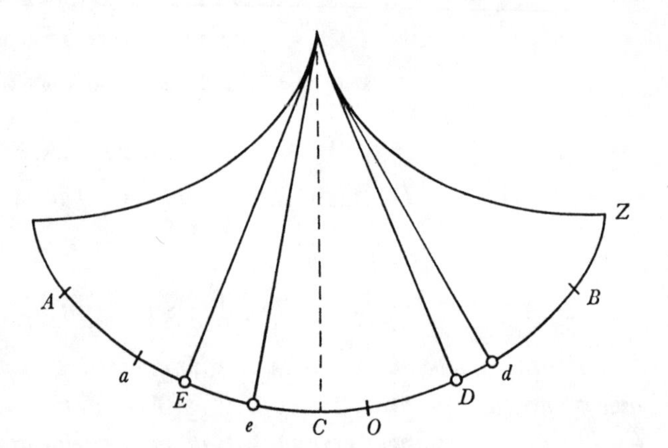

and only minor alterations—for the most part mere corrections of misprints or inessential verbal ameliorations—were introduced in its 1713 and 1726 revisions. The date of composition advanced is founded on Newton's remark to Halley on 20 June 1686 that 'the second [book] was finished last summer' (*Correspondence*, 2: 437).

a motum omnem simul amittent. Isochronæ sunt igitur oscillationes totæ, & arcubus totis *BA*, *B*[*a*] proportionales sunt arcuum partes quælibet *BD*, *Bd* vel *BE*, *Be* quæ simul describuntur. Q.E.D.[3]

Corol. Igitur motus velocissimus in Medio resistente non incidit in punctum infimum *C*, sed reperitur in puncto illo *O* quo arcus totus descriptus *aB* bisecatur. Et corpus subinde pergendo ad *a* ijsdem gradibus retardatur quibus antea accelerabatur in descensu suo a *B* ad *O*.

Prop. XXVI. Theor. XX.

Corporum Funependulorum quæ resistuntur in ratione velocitatum oscillationes in Cycloide sunt isochronæ.

Nam si corpora duo a centris suspensionum æqualiter distantia oscillando describant arcus inæquales, & velocitates in arcuum partibus correspondentibus sint ad invicem ut arcus toti: resistentiæ velocitatibus proportionales erunt etiam ad invicem ut ijdem arcus. Proinde si viribus motricibus a gravitate oriundis quæ sint ut ijdem arcus [au]ferantur vel addantur hæ resistentiæ, erunt differentiæ vel summæ ad invicem in eadem arcuum ratione; cumq velocitatum incrementa vel decrementa sint ut hæ differentiæ vel summæ, velocitates semper erunt ut arcus toti: Igitur velocitates, si sint in aliquo casu ut arcus toti, manebunt semper in eadem ratione. Sed in principio motus ubi corpora incipiunt descendere & arcus illos describere, vires, cum sint arcubus proportionales, generabunt velocitates arcubus proportionales. Ergo velocitates semper erunt ut arcus toti describendi, & propterea arcus illi simul describentur. Q.E.D.[4]

(3) On setting $\rho = l$, constant, in note (1), the equation of motion in the cycloidal arc BCA (where $\widehat{BC} = \widehat{CA} = S$) is $d^2s/dt^2 = -ks+l$, where t is the time of fall from rest at B down to d, distant $\widehat{Cd} = s$ from the base-point C. It follows, on taking $\widehat{CO} = -l/k$ to measure the magnitude of the resistance and putting $\widehat{Od} = s-l/k = \sigma$, that the equation of motion becomes $d^2\sigma/dt^2 = -k\sigma$, identical with that of unresisted motion in the cycloid from rest at a point distant $\widehat{OB} = S-l/k$ from the base-point C; in consequence, where D, d; C, O; E, e; and A, a (at which the two motions are again instantaneously at rest) are pairs of points attained in equal times from rest at B in unresisted and resisted fall in the cycloid, the ratio $\widehat{BD}/\widehat{Bd} = \widehat{BC}/\widehat{BO} = \widehat{BE}/\widehat{Be} = \widehat{BA}/\widehat{Ba}$ is that of the respective speeds at these points (whence, as Newton correctly asserts in his following Corollary, the greatest speed in the resisted motion is at O, the mid-point of $\widehat{Ba}$). In either case the half-period $\pi k^{-\frac{1}{2}}$ of motion (over $\widehat{BA}$ and $\widehat{Ba}$ respectively) will manifestly be independent of the point B from which it begins and hence all oscillations will be isochronous (though, of course, in its return swing from a the resisted motion will come to rest not at B but at some intermediate point distant $\widehat{CB}-4\widehat{CO}$ from C in the cycloid).

(4) Newton's verbally plausible suggestion that, where (to continue the terms of note (1)) $\rho = lv$, the ensuing equation of motion $d^2s/dt^2 = v.dv/ds = -ks+\rho$ has the solution $v \propto s$ is deeply misconceived. While it is true that, if $k = \frac{1}{4}l^2$, the equation has the particular solution ($s = S(\frac{1}{2}lt+1)e^{-\frac{1}{2}lt}$ and so) $v = \frac{1}{2}l(s-Se^{-\frac{1}{2}lt})$ and is then—implicitly in the presumption that

Prop. XXVII. Theor. XXI.

Si corpora Funependula resistuntur in duplicata ratione velocitatum, differentiæ inter tempora oscillationum in Medio resistente ac tempora oscillationum in ejusdem gravitatis specificæ Medio non resistente erunt arcubus oscillando descriptis proportionales, quam proxime.

... ...(5)

Prop. XXVIII. Theor. XXII.

Si corpus Funependulum in Cycloide oscillans resistitur in ratione momentorum temporis, erit ejus resistentia ad vim gravitatis ut excessus arcus descensu toto descripti supra arcum ascensu subsequente descriptum ad penduli longitudinem duplicatam.

$S = 0$!—satisfied by setting $v = \frac{1}{2}ls$, this has the unfortunate corollary, as James Stirling brought to Newton's notice in a letter from Venice on 7 August 1719, that 'the Pendulum in the lowest point [C, at which $\widehat{Cd} = s$ is zero] ha[s] no velocity [v], and consequently c[an] perform but one half oscillation and then rest' (King's College, Cambridge. Keynes MS 104, first printed by David Brewster in his *Memoirs of the Life, Writings and Discoveries of Sir Isaac Newton*, **2** (Edinburgh, 1855): 516). Stirling did not indicate at what point this 'half oscillation' might possibly commence, but went on to report: 'Mr Nicholas Bernoulli, [who] proposed to me, to enquire into the curve which defines the resistances of a Pendulum, when the resistance is proportionall to the velocity...had found that before, as also one Count Ricata [Jacopo Riccati presumably], which I understood, after I communicated to Bernoulli, what occurred to me. Then he asked me, how in that hypothesis of resistance a Pendulum could be said to oscillat, since it only fell to the lowest point of the Cycloid and then rested. So I conjecture that his uncle [Johann] sets him on to see what he can pick out of your writings that may anyways be cavill'd against'. In a short paper 'De oscillationibus penduli in medio quod resistit in ratione simplici velocitatis' composed about this time (though not published to the world at large till more than twenty years afterwards in his *Opera Omnia, tam antea sparsim edita, quam hactenus inedita*, **4** (Lausanne/Geneva, 1742): 374–7) Johann Bernoulli had in fact elegantly employed the substitution $v = zs$ to derive the partial solution

$$\tfrac{1}{2}\log(v^2 - lvs + ks^2) + \frac{\frac{1}{2}l}{\sqrt{[k - \frac{1}{4}l^2]}}\tan^{-1}\frac{v - \frac{1}{2}ls}{s\sqrt{[k - \frac{1}{4}l^2]}} = \tfrac{1}{2}\log(kS^2) - \tan^{-1}\frac{\frac{1}{2}l}{\sqrt{[k - \frac{1}{4}l^2]}}$$

and so was well aware of the complexities involved in analysing the cycloidal motion ensuing in this law of resistance, though he seems not to have attained the general resolution $(k \mp \frac{1}{4}l^2)$ $s = Se^{-\frac{1}{2}lt}(\cos\sqrt{[k - \frac{1}{4}l^2]}\,t + (\frac{1}{2}l/\sqrt{[k - \frac{1}{4}l^2]})\sin\sqrt{[k - \frac{1}{4}l^2]}\,t)$, so determining

$$v = (k/\sqrt{[k - \tfrac{1}{4}l^2]})\,Se^{-\frac{1}{2}lt}\sin\sqrt{[k - \tfrac{1}{4}l^2]}\,t:$$

at once, since the cycloidal pendulum thus resisted returns instantaneously to rest after a period $2\pi/\sqrt{[k - \frac{1}{4}l^2]}$, its oscillations (decreasing with each swing in geometrical progression by a factor $e^{-\frac{1}{2}l\pi/\sqrt{[k - \frac{1}{4}l^2]}}$) will indeed be isochronous as Newton here—to our mind, not a little luckily!—in 1685 affirmed. Thirty-four years later, by then an old man stiff in his emotional attitudes and rigid in his patterns of thought, he was, it is clear, prepared to regard the query communicated by Stirling in 1719 as but one more captious and unfounded Bernoullian objection for objection's sake, and some six years afterwards permitted his initial 'demonstration' of Proposition XXVI to pass unaltered into the *Principia's editio ultima* ($_3$1726: 297), whence it is repeated without variation or critical comment in all subsequent editions.

(5) Since the proof of this theorem—rawly approximate at best—is founded on the

Designet BC arcum descensu descriptum, Ca arcum ascensu descriptum, & Aa differentiam arcuum: & stantibus quæ in Propositione XXV constructa & demonstrata sunt, erit vis qua corpus oscillans urgetur in loco quovis D ad vim resistentiæ ut arcus CD ad arcum CO, qui semissis est differentiæ illius Aa. Ideoq; vis qua corpus oscillans urgetur in Cycloidis principio seu puncto altissimo, id est vis gravitatis, erit ad resistentiam ut arcus Cycloidis inter punctum illud supremum & punctum infimum C ad arcum CO; id est (si arcus duplicentur) ut Cycloidis totius arcus seu dupla penduli longitudo ad arcum Aa. Q.E.D.[(6)]

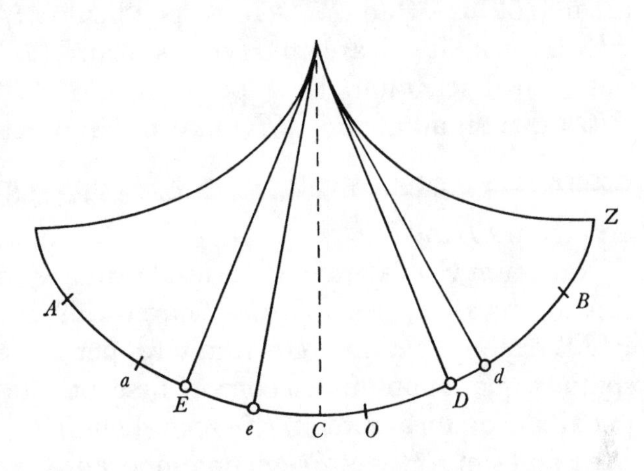

Prop. XXIX. Prob. VII.

Posito quod corpus in Cycloide oscillans resistitur in duplicata ratione velocitatis: invenire resistentiam in locis singulis.

Sit Ba arcus oscillatione integra descriptus, sitq; C infimum Cycloidis punctum & CZ semissis arcus Cycloidis totius longitudini Penduli æqualis; & quæratur resistentia corporis in loco quovis D. Secetur recta infinita OQ in punctis O, S, P, Q[(7)] ea lege ut (si erigantur perpendicula OK, ST, PI, QE centroq; O & Asymptotis OK, OQ describatur Hyperbola $TIGE$ secans perpendicula ST, PI, QE in T, I & E, & per punctum I [parallela Asymptoto OQ][(8)] agatur KF occurrens Asymptoto OK in K, & perpendiculis ST & QE in L & F)

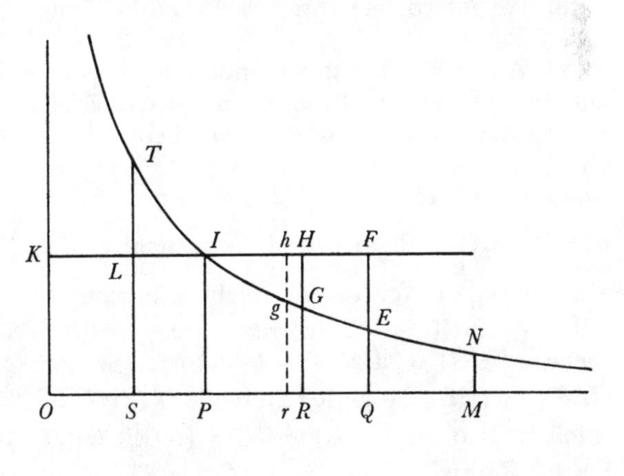

erroneously demonstrated preceding Proposition XXVI, we see no point in repeating it. The *princeps* text (*Principia*, $_1$1687: 307–8) was only verbally refined in succeeding editions.

(6) It follows equivalently from note (3) above that, since $\widehat{CO} = l/k$ where $l = \rho$ and $k = g/\widehat{CZ}$, therefore $g/\rho = \widehat{CZ}/\widehat{CO}$.

(7) To avoid possible confusion with the base-point of the cycloid, here and in sequel we follow both Humphrey Newton's press transcript (see note (2)) and the *Principia's editio*

fuerit area Hyperbolica *PIEQ* ad aream Hyperbolicam *PITS* ut arcus *BC* descensu corporis descriptus ad arcum *CA* ascensu descriptum, & area *IEF* ad aream *ILT* ut *OQ* ad *OS*. Dein perpendiculo *MN* abscindatur area Hyperbolica *PINM* quæ sit ad aream Hyperbolicam *PIEQ* ut arcus *CZ* ad arcum *BC* descensu descriptum. Et si perpendiculo *RG* abscindatur area Hyperbolica *PIGR* quæ sit ad aream *PIEQ* ut arcus quilibet *CD* ad arcum *BC* descensu toto descriptum: erit resistentia in loco *D* ad vim gravitatis ut area $\frac{OR}{OQ}IEF-IGH$ ad aream *PIENM*.(9)

Nam cum vires a gravitate oriundæ quibus corpus in locis *Z*, *B*, *D*, *a* urgetur sint ut arcus *CZ*, *CB*, *CD*, *Ca*, & arcus illi sint ut areæ *PINM*, *PIEQ*, *PIGR*, *PITS*, exponatur tum arcus tum vires per has areas respective. Sit insuper *Dd* spatium quam minimum a corpore descendente descriptum, & exponatur idem per aream quam minimam *RGgr* parallelis *RG*, *rg* comprehensam, & producatur *rg* ad *h* ut sint *GHhg* & *RGgr* contemporanea arearum *IGH*, *PIGR* decrementa. Et areæ $\frac{OR}{OQ}IEF-IGH$ incrementum $GHhg-\frac{Rr}{OQ}IEF$ seu $Rr\times HG-\frac{Rr}{OQ}IEF$ erit ad areæ *PIGR* decrementum *RGgr* seu $Rr\times RG$ ut $HG-\frac{IEF}{OQ}$ ad *RG*, adeoq;

---

ultima ($_3$1726: 300–1) in denoting by *S* the point in *OQ* which in the first and second published editions ($_1$1687: 309–10; $_2$1713: 277–8) is designated by *C*.

(8) We interpolate this necessary qualifying phrase from the *Principia's* second edition ($_2$1713: 277).

(9) Where (further to continue the terms of note (1)) the resistance is $\rho = lv^2$, the ensuing equation of motion in the given cycloid *ZCA* is $v.dv/ds = \frac{1}{2}d(v^2)/ds = -ks+lv^2$. On introducing the substitution $v^2 = u\sigma$ there results $u(d\sigma/ds-2l\sigma)+\sigma.du/ds = -2ks$, which by choosing $d\sigma/ds = 2l\sigma$ (that is, $\sigma = e^{2ls}$ in modern Eulerian terms) reduces to be $du/d\sigma = (-ks/l\sigma^2 \text{ or}) - (k/2l^2)\,\sigma^{-2}\log\sigma$, whence, if $\sigma = \Sigma$ when $s = S$, there comes $u = (k/2l^2).\left[\sigma^{-1}\log\sigma+\sigma^{-1}\right]_{\Sigma}^{\sigma}$ and finally $v^2 = (k/2l^2)\,(\log\sigma+1-\sigma\Sigma^{-1}(\log\Sigma+1))$. In his auxiliary figure Newton effectively constructs $OP = 1$, $PI = 1/2l$, $OR = \sigma$ (corresponding to $\widehat{CD} = s$) and $OQ = \Sigma$ (corresponding to $\widehat{CB} = S$), so that $(PRGI) = (1/2l)\log\sigma = s$ and thence $(IGH) = (1/2l)\,(\sigma-1-\log\sigma)$; similarly $(PQEI) = (1/2l)\log\Sigma = S$ and thence $(IEF) = (1/2l)\,(\Sigma-1-\log\Sigma)$, while $(PMNI) = \widehat{CZ}$; accordingly, since $\rho = lv^2$ and $k=g/\widehat{CZ}$, follows that $\rho/g = (l/k)\,v^2/\widehat{CZ} = ((OR/OQ)(IEF)-(IGH))/(PMNI)$ as Newton accurately asserts. (Likewise, if on putting $\widehat{Ca} = -S'$ we set $OS = e^{-2lS'}$, then also

$$\rho/g = ((OR/OS)(ITL)-(IGH))/(PMNI).)$$

This constructed proportion he justifies in sequel by demonstrating that the instantaneous decrement (with respect to $(PRGI) = \widehat{CD} = s$) of the area $\rho/k = ((OR/OQ)(IEF)-(IGH))$ satisfies the infinitesimal equality $-d(\rho/k) = 2l(s-\rho/k).ds$, equivalent to the primary equation of motion from which his preliminary analysis—much as our present reconstruction of it—doubtless evolved.

ut $OR\times HG-\frac{OR}{OQ}IEF$ ad $OR\times GR$ seu $OP\times PI$: hoc est (ob æqualia $OR\times HG$, $OR\times HR-OR\times GR$, $ORHK-OPIK$, $PIHR$ & $PIGR+IGH$) ut

$$PIGR+IGH-\frac{OR}{OQ}IEF$$

ad $OPIK$. Igitur si area $\frac{OR}{OQ}IEF-IGH$ dicatur Y, atq; areæ $PIGR$ decrementum $RGgr$ detur, erit incrementum areæ Y ut $PIGR-Y$.

Quod si V designet vim a gravitate oriundam arcui describendo CD proportionalem qua corpus urgetur in D, & R pro resistentia ponatur: erit $V-R$ vis tota qua corpus urgetur in D, adeoq; ut incrementum velocitatis in data temporis particula factum. Est autem resistentia R (per Hypothesin) ut quadratum velocitatis, & inde (per Lem. II)[10] incrementum resistentiæ ut velocitas & incrementum velocitatis conjunctim, id est ut spatium data temporis particula dscriptum & $V-R$ conjunctim, atq; adeo si momentum spatij detur, ut $V-R$; id est, si pro vi V scribatur ejus exponens $PIGR$ & resistentia R exponatur per aliam aliquam aream Z, ut $PIGR-Z$.

Igitur area $PIGR$ per datorum momentorum subductionem uniformiter descrescente, crescunt area Y in ratione $PIGR-Y$ & area Z in ratione $PIGR-Z$. Et propterea si areæ Y & Z simul incipiant & sub initio æquales sint, hæ per additionem æqualium momentorum pergent esse æquales, & æqualibus itidem momentis subinde decrescentes simul evanescent. Et vicissim, si simul incipiunt & simul evanescunt, æqualia habebunt momenta & semper erunt æquales: id adeo quia si resistentia Z augeatur, velocitas una cum arcu illo Ca qui in ascensu corporis describitur, diminuetur, & puncto in quo motus omnis una cum resistentia cessat propius accedente ad punctum C, resistentia citius evanescet quam area Y. Et contrarium eveniet ubi resistentia diminuitur.

Jam vero area Z incipit desinitq; ubi resistentia nulla est, hoc est in principio & fine motus ubi arcus CD, CD arcubus CB & Ca æquantur, adeoq; ubi recta RG incidit in rectas QE & ST.[11] Et area Y seu $\frac{OR}{OQ}IEF-IGH$ incipit desinitq;

(10) See IV: 521–5. In the *Principia's* second edition ($_2$1713: 279) Newton deleted the terminal clause 'adeoq; ut incrementum...factum' of the previous sentence and thereafter continued: 'Est itaque incrementum velocitatis ut $V-R$ & particula illa temporis in qua factum est conjunctim: Sed & velocitas ipsa est ut incrementum contemporaneum spatii descripti directe & particula eadem temporis inverse. Unde, cum resistentia (per Hypothesin) sit ut quadratum velocitatis, incrementum resistentiæ (per Lem. II) erit ut velocitas & incrementum velocitatis conjunctim, id est, ut momentum spatii & $V-R$ conjunctim...'.

(11) In his *editio ultima* of the *Principia* ($_3$1726: 302) Pemberton slightly improved the wording of this to read '...hoc est, in principio motus ubi arcus CD arcui CB æquatur & recta RG incidit in rectam QE, & in fine motus ubi arcus CD arcui $C[a]$ æquatur & RG incidit in rectam SY'.

ubi nulla est, adeoq; ubi $\frac{OR}{OQ}IEF$ & IGH æqualia sunt: hoc est (per constructionem) ubi recta RG incidit in rectam QE & ST. Proindeq; areæ illæ simul incipiunt & simul evanescunt, & propterea semper sunt æquales. Igitur area $\frac{OR}{OQ}IEF-IGH$ æqualis est areæ Z per quam resistentia exponitur, & propterea est ad aream $PINM$ per quam gravitas exponitur ut resistentia ad gravitatem. Q.E.D.(12)

Corol. 1. Est igitur resistentia in loco infimo C ad vim gravitatis ut area $\frac{OP}{OQ}IEF$ ad aream $PINM$.

Corol. 2. Fit autem maxima ubi area $PIHR$ est ad aream IEF ut OR ad OQ. Eo enim in casu momentum ejus (nimirum $PIGR-Y$) evadit nullum.

Corol. 3. Hinc etiam innotescit velocitas in locis singulis: quippe quæ est in dimidiata ratione resistentiæ, & ipso motus initio æquatur velocitati corporis in eadem Cycloide absq; omni resistentia oscillantis.

Cæterum ob difficilem calculum quo resistentia & velocitas per hanc Propositionem inveniendæ sunt, visum est Propositionem sequentem subjungere, quæ & generalior sit & ad usus Philosophicos abunde satis accurata.

Prop. XXX. Theor. XXIII.

Si recta aB æqualis sit Cycloidis arcui quem corpus oscillando describit, & ad singula ejus puncta D erigantur perpendicula DK quæ sint ad longitudinem Penduli ut resistentia corporis in arcus punctis correspondentibus ad vim gravitatis: dico quod differentia inter arcum descensu toto descriptum & arcum ascensu toto subsequente descriptum, ducta in arcuum eorund[e]m semisummam, æqualis erit areæ BKaB a perpendiculis omnibus DK occupatæ quamproxime.(13)

Exponatur enim tum Cycloidis arcus oscillatione integra descriptus per rectam illam sibi æqualem aB, tum arcus qui describeretur in vacuo per longitudinem AB. Bisecetur AB in C, & punctum C repræsentabit infimum Cycloidis punctum, & erit CD ut vis a gravitate oriunda, qua corpus in $[D]$ secundum Tangentem Cycloidis urgetur, eamq; habebit rationem ad longitudinem Penduli quam habet vis in D ad vim gravitatis. Exponatur igitur vis illa per

(12) Not surprisingly, Newton attempts no parallel construction of the time $t = \int_s^S 1/v.ds$ of swing over the arc $\widehat{BD}$ where (see note (9)) the instantaneous speed v at D is $\sqrt{[(k/2l^2)\,(2ls+1-(2lS+1)\,e^{2l(s-S)})]}$, but observe that he cautiously refuses to commit himself even to an opinion as to whether this species of resisted cycloidal motion is isochronous or (as seems highly likely) it is not.

(13) Newton's stated equality is, however, rigorously exact. This mistaken terminal adverb was justly suppressed in later editions of the *Principia*.

longitudinem CD & vis gravitatis per longitudinem penduli, & si in DE[14] capiatur DK in ea ratione ad longitudinem penduli quam habet resistentia ad gravitatem, erit DK exponens resistentiæ. Centro C & intervallo CA vel CB construatur semicirculus $BEeA$. Describ[a]t autem corpus tempore quam minimo spatium Dd, & erectis perpendiculis DE, de circumferentiæ occurrentibus in E & e, erunt hæc ut velocitates quas corpus in vacuo descendendo a puncto B acquireret in locis D & d. Patet hoc per Prop. LII. Lib. I.[15] Exponantur itaq; hæ velocitates per perpendicula illa DE, de sitq; DF velocitas quam acquirit in D cadendo de B in Medio resistente. Et si centro C & intervallo CF describatur circulus FfM occurrens rectis de & AB in f & M, erit M locus ad quem deinceps absq; ulteriore resistentia ascenderet & df velocitas quam acquireret in d. Unde etiam si Fg designet velocitatis momentum quod corpus D describendo spatium quam minimum Dd ex resistentia Medij amittit, & sumatur CN æqualis Cg: erit N locus ad quem corpus deinceps absq; ulteriore resistentia ascenderet, & MN erit decrementum ascensus ex velocitatis illius amissione oriundum. Ad df demittatur perpendiculum Fm, & velocitatis DF decrementum $[F]g$ a resistentia DK genitum erit ad velocitatis ejusdem incrementum fm a vi CD genitum ut vis generans DK ad vim generantem CD. Sed & ob similia triangula Fmf, Fhg, FDC est fm ad Fm seu Dd ut CD ad DF, & ex æquo Fg ad Dd ut DK ad DF. Item Fg ad Fh ut CF ad DF, & ex æquo perturbate Fh seu MN ad Dd ut DK ad CF.[16] Sumatur DR ad $\frac{1}{2}aB$ ut DK ad CF, & erit MN ad Dd ut DR ad $\frac{1}{2}aB$; ideoq; summa omnium $MN\times\frac{1}{2}aB$, id est $Aa\times\frac{1}{2}aB$, æqualis erit summæ omnium $Dd\times DR$, id est areæ $BRrSa$ quam rectangula omnia $Dd\times DR$ seu $DRrd$ componunt. Bisecentur Aa & aB in P & O, & erit $\frac{1}{2}aB$ seu OB æqualis CP, ideoq; DR est ad DK ut CP ad CF vel CM, & divisim KR ad

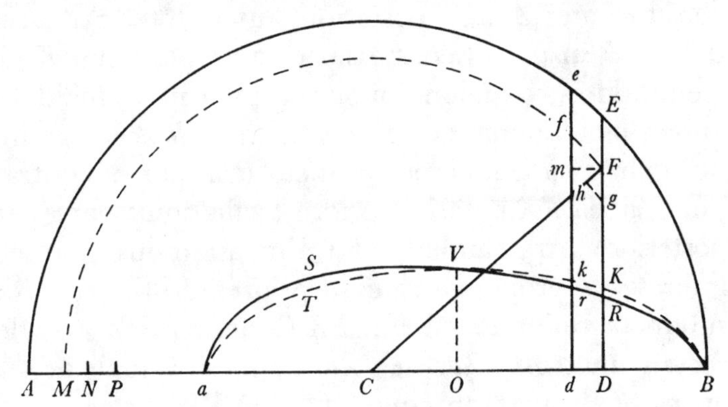

(14) Understand this to be perpendicularly ordinate to CD.

(15) And of the preceding 'De motu Corporum Liber primus' which (here) differs minimally from the *editio princeps*. Compare §3: note (296).

(16) In the *Principia's* second edition ($_2$1713: 281) Newton drastically curtailed the remainder of this demonstration to be 'seu CM; ideoque summa omnium $MN\times CM$ æqualis erit summæ omnium $Dd\times DK$. Ad punctum mobile M erigi semper intelligatur ordinata rectangula æqualis indeterminatæ CM, quæ motu continuo ducatur in totam longitudinem Aa; & trapezium ex illo motu descriptum sive huic æquale rectangulum $Aa\times\frac{1}{2}aB$ æquabitur summæ omnium $MN\times CM$, adeoque summæ omnium $Dd\times DK$, id est, areæ $BKkVTa$. Q.E.D.'

DR ut *PM* ad *CP*. Ideoq cum punctum *M* ubi corpus versatur in medio oscillationis loco *O*, incidat circiter in punctum *P* & priore oscillationis parte versetur inter *A* & *P*, posteriore autem inter *P* & *a*, utroq in casu æqualiter a puncto *P* in partes contrarias errans: punctum *K* circa medium oscillationis locum, id est e regione puncti *O* puta in *V*, incidet in punctum *R*, in priore autem oscillationis parte jacebit inter *R* & *E*, & in posteriore inter *R* & *D*, utroq in casu æqualiter a puncto *R* in partes contrarias errans. Proinde area quam linea *KR* describit priore oscillationis parte jacebit extra aream *BRSa*, posteriore intra eandem, idq dimensionibus hinc inde propemodum æquatis inter se, & propterea in casu priore addita areæ *BRSa*, in posteriore eidem subducta relinquet aream *BKTa* areæ *BRSa* æqualem quam proxime. Ergo rectangulum $Aa \times \frac{1}{2}aB$ seu *AaO*, cum sit æquale areæ *BRSa*, erit etiam æquale areæ *BKTa* quamproxime. Q.E.D.(17)

Corol. Hinc ex lege resistentiæ & arcuum *Ca*, *CB* differentia *Aa* colligi potest proportio resistentiæ ad gravitatem quamproxime.

Nam si uniformis sit resistentia *DK*, figura *aBKk*[*T*] rectangulum erit sub *Ba* & *DK*, & inde rectangulum sub $\frac{1}{2}Ba \times Aa$ æqual[e] erit rectangulo sub *Ba* & *DK*, & *DK* æqualis erit $\frac{1}{2}Aa$. Quare cum *DK* sit exponens resistentiæ & longitudo penduli exponens gravitatis, erit resistentia ad gravitatem ut $\frac{1}{2}Aa$ ad longitudinem Penduli, omnino ut in Propositione XXVIII demonstratum est.(18)

---

(17) Evidently, where the downwards *vis gravitatis* g is 'expressed' by the length $\widehat{CZ}$ of the given cycloidal pendulum, the arc $\widehat{CD} = s$ similarly expresses the component $(g/\widehat{CZ})\, s = ks$ of that uniform force of gravity acting instantaneously in the direction of motion at D; whence, if we again take v to be the speed attained at that point and ρ the deceleration there due to resistance, the defining equation of motion will once more (as in note (1)) prove to be $v \,.\, dv/ds = -ks+\rho$, that is, $\frac{1}{2}d(v^2+ks^2)/ds = \rho$. Accordingly, if the limits B and a of pendular oscillation from rest to rest (at each of which $v = 0$) are fixed by $\widehat{CB} = S$ and $\widehat{Ca} = -S'$, and $DK = \rho/k$ (constructed ordinate to CD at D) is taken to express the resistance ρ at D in the arc $\widehat{CB}$, then, because $\widehat{Aa} = S-S'$ and $\widehat{aB} = S+S'$, from $\frac{1}{2}\left[v^2+ks^2\right]_{s=-S'}^{s=S} = \int_{-S'}^{S} \rho \,.\, ds$ it follows that $\widehat{Aa} \times \frac{1}{2}\widehat{aB} = \frac{1}{2}(S^2-S'^2) = \int_{-S'}^{S} (\rho/k) \,.\, ds = (BKTa)$.

In sequel, assuming that (for small ρ at least) the speed attained at the corresponding point of resisted motion in the arc $\widehat{Ba}$ is very nearly proportional to that, $k^{\frac{1}{2}} \,.\, DE = k^{\frac{1}{2}} \sqrt{[S^2-s^2]}$ (see §3: note (286)), at D in the similar unresisted cycloidal motion from rest at B, and taking the resistance ρ to be some power lv^n of the resisted speed v at D, Newton deduces that, where $\widehat{aO} = \widehat{OB} = \frac{1}{2}(S+S') = \Sigma$ and $\widehat{OD} = s-\frac{1}{2}(S-S') = \sigma$, then $v \approx k^{\frac{1}{2}} \sqrt{[\Sigma^2-\sigma^2]}$, very nearly. In consequence, $\widehat{Aa} \approx \int_{-S'}^{S} (l/k)\, v^n \,.\, ds \Big/ \frac{1}{2}(S+S') = 2(R/k) \int_0^1 (1-x^2)^{\frac{1}{2}n} \,.\, dx$, where $R = l(k^{\frac{1}{2}}\Sigma)^n$ is the resistance at O (when $\sigma = 0$) and $x = \sigma/\Sigma$, so that $\left(\frac{1}{2} \Big/ \int_0^1 (1-x^2)^{\frac{1}{2}n} \,.\, dx\right) \widehat{Aa}/\widehat{CZ} \approx R/g$.

(18) Here (to continue the terms of the previous note) $n = 0$ and hence $\rho = l = R$, so that at once $\frac{1}{2}\widehat{Aa} = R/k = DK$, where (as ever) $k = g/\widehat{CZ}$.

Si resistentia sit ut velocitas, Figura $aBKk[T]$ Ellipsis erit quam proxime. Nam si corpus in Medio non resistente oscillatione integra describeret longitudinem BA, velocitas in loco quovis D foret ut circuli diametro AB descripti ordinatim applicata DE. Proinde cum Ba in Medio resistente & BA in Medio non resistente æqualibus circiter temporibus describantur, adeoq velocitates in singulis ipsius Ba punctis sint quam proxime ad velocitates in punctis correspondentibus longitudinis BA ut est Ba ad BA, erit velocitas DK(19) in Medio resistente ut circuli vel Ellipseos super diametro Ba descripti ordinatim applicata, adeoq figura $BKVTa$ Ellipsis quam proxime. Cum resistentia velocitati proportionalis supponatur, sit OV exponens resistentiæ in puncto medio O, & Ellipsis centro O semiaxibus OB, OV descripta figuram $aBKVT$ eiq æquale rectangulum $Aa \times BO$ æquabit quam proxime. Est igitur $Aa \times BO$ ad $OV \times BO$ ut area Ellipseos hujus ad $OV \times BO$, id est Aa ad OV ut area semicirculi ad quadratum radij sive ut 11 ad 7 circiter: Et propterea $\frac{7}{11}Aa$ ad longitudinem penduli ut corporis oscillantis resistentia in O ad ejusdem gravitatem.(20)

Quod si resistentia DK sit in duplicata ratione velocitatis, figura $BKVTa$ Parabola erit verticem habens V & axem OV, ideoq æqualis erit duabus tertijs partibus rectanguli sub Ba & OV quam proxime. Est igitur rectangulum sub $\frac{1}{2}Ba$ & Aa æquale rectangulo sub $\frac{2}{3}Ba$ & OV, adeoq OV æqualis $\frac{3}{4}Aa$, & propterea corporis oscillantis resistentia in O ad ipsius gravitatem ut $\frac{3}{4}Aa$ ad longitudinem Penduli.(21)

Atq has conclusiones in rebus practicis abunde satis accuratas esse censeo. Nam cum Ellipsis vel Parabola congruat cum figura $BKVTa$ in puncto medio V, hæc si ad partem alterutram BKV vel VTa excedit figuram illam, deficiet ab eadem ad partem alteram, & sic eidem æquabitur quam proxime.(22)

---

(19) That is, 'in puncto D' as Pemberton specified in the *Principia*'s *editio ultima* ($_3$1726: 305).

(20) Here, on putting $n = 1$ in the terminal result of note (17), there will be

$$\tfrac{1}{2}\Big/\int_0^1 (1-x^2)^{\frac{1}{2}}.dx = 2/\pi \approx \tfrac{7}{11}.$$

(21) Correspondingly, on setting $n = 2$ in note (17), there will come

$$\tfrac{1}{2}\Big/\int_0^1 (1-x^2).dx = \tfrac{3}{4}.$$

(22) We omit a following 'Prop. XXXI. Theor. XXIV'. (*Principia*, $_1$1687: 315–16) in which Newton emphasises the evident corollary that '*Si corporis oscillantis resistentia in singulis arcuum descriptorum partibus proportionalibus augeatur vel minuatur in data ratione, differentia inter arcum descensu descriptum & arcum subsequente ascensu descriptum augebitur vel diminuetur in eadem ratione quamproxime.* Oritur enim differentia illa [$\widehat{Aa}$] ex retardatione Penduli per resistentiam Medij, adeoq est ut retardatio tota eiq proportionalis resistentia retardans.' In an ensuing 'Scholium Generale', set after Proposition XL in the first edition ($_1$1687: 339–54) but, after considerable abridgment of its text, advanced to follow immediately upon Proposition XXXI

APPENDIX 6. DRAFTS OF AN INTENDED PROPOSITION SKETCHING THE GENERAL ALGEBRAIC 'QUADRATURE OF FIGURES' (SUMMER? 1685).[1]

From an autograph worksheet[2] in the University Library, Cambridge

[1][3]

Tandem ut compleatur solutio[4] superiorum Problematum adjicienda est quadratura Figurarum toties assumpta.

---

in the second ($_2$1713: 284–93), Newton made this the basis of a procedure for empirically measuring the relative variation of the resistance of air, water and other fluids with differing powers of the instantaneous speed of motion in small swings of a circular pendulum formed by hanging a heavy sphere (of wood, lead and other like materials) on a long, thin cord: since the length $\widehat{CB}$ of each semi-oscillation—virtually in a cycloidal arc—from rest at B is, to a high approximation, proportional to the maximum speed V of swing, by counting the number of vibrations in which a pendulum, after release from a measured point, comes to lose a given portion of its 'motion'—as measured by its arc of swing—the average difference in length $\widehat{Aa} = \widehat{CB} - \widehat{aC}$ between successive semi-oscillations may be determined for a suitable range of maximum speeds of swing, and postulated laws of resistance $\rho = \sum_i (l_i v^i)$ thereby tested. In his prime example (*Principia*, $_1$1687: 339–41) Newton states that 'Globum ligneum pondere unciarum *Romanarum* $57\frac{7}{22}$ diametro digitorum *Londinensium* $6\frac{7}{8}$ fabricatum, filo tenui ab unco satis firmo suspendi, ita ut inter uncum & centrum oscillationis Globi distantia esset pedum $10\frac{1}{2}$.... Deinde numeravi oscillationes quibus Globus quartam motus sui partem amitteret'; by thus comparing the average difference in length between successive oscillations in swings starting from rest at distances of 4″, 16″ and 64″ from the base-point he was able therefrom to determine that $\widehat{Aa} \propto AV + BV^{\frac{3}{2}} + CV^2$ where $A:B:C = 2097:8955:30298$, whence by Proposition XXX (see note (17)) $\rho \propto av + bv^{\frac{3}{2}} + cv^2$ on setting $a/A = \frac{1}{2} \Big/ \int_0^1 (1-x^2)^{\frac{1}{2}}.dx = 2/\pi \approx \frac{7}{11}$ (compare note (20)) and $c/C = \frac{1}{2} \Big/ \int_0^1 (1-x^2).dx = \frac{3}{4} = \frac{9}{12}$ (compare note (21)), and thence deducing $b/B = \frac{1}{2} \Big/ \int_0^1 (1-x^2)^{\frac{3}{4}}.dx \approx \frac{7+9}{11+12} = \frac{16}{23}$, rounded off in the *Principia*'s *editio ultima* ($_3$1726: 309) to be $\frac{7}{10}$. (Accurately, $b/B = \frac{1}{2}\Gamma(\frac{9}{4})/\Gamma(\frac{7}{4}).\Gamma(\frac{3}{2}) \approx 0{\cdot}69554$. We will return in the next volume to discuss the technique of numerical interpolation employed by Newton in his reduction.) A similar following test (*Principia*, $_1$1687: 348–9) of the simpler putative law $\rho \propto a'v + b'v^2$ was likewise effected by computing the relative values of A' and B' which satisfy $\widehat{Aa} \propto A'V + B'V^2$. (This is the only law of resistance considered in the preliminary version of the 'Scholium Generale' which is now found on ULC. Add. 3965.10: 104^r/105^r, reproduced in I. B. Cohen's *Introduction to Newton's 'Principia'* (Cambridge, 1971): 101–3. Newton there ascribes discrepancies between theory and observational test somewhat ingenuously to the fact 'quod oscillationes non fierent in Cycloide, adeoq; non essent satis isochronæ. Nam pendulis æqualibus tempora oscillationum in circulo paulo majora sunt quàm in Cycloide idq; excessu quodam qui est in duplicata circiter ratione arcûs a pendulo descripti'; compare III: 393, note (6).) A few years later, as part of his general scheme of revision of the *Principia*'s *editio princeps*, Newton briefly made trial of such further functional relationships as $\widehat{Aa} \propto A + BV + CV^2$ and $\widehat{Aa} \propto AV^{\frac{1}{2}} + BV + CV^2$ (see ULC. Add. 3965.18: 666^r) but found them wanting.

Prop. Prob.

Figuras curvilineas geometricè rationales, si fieri possit, quadrare.

Ad Abscissas positione datas *VB*, *vb* ad data puncta *V*, *v* terminatas ordinatim applicentur *BC*, *bc* in angulis rectis *VBC*, *vbc* sintꝗ *CE*, *ce* curvæ quas puncta *C*, *c*

---

(1) Having in his preceding 'De motu Corporum Liber primus', notably in Propositions XLI, LIII and LIV (and also in Propositions LV and LVI which follow these in the fuller text of the printed *Principia*), several times set out a general construction for the motion ensuing from a point under given conditions 'concessis figurarum curvilinearum quadraturis', Newton shortly afterwards roughed out a complementary 'Prop. [LVI *bis*?]' wherein he outlined the principles of 'the quadrature of (algebraic) curves' to which he thereby appealed. The extant drafts of this intended *addendum* to Section X of the 'Liber primus', briefly described by W. W. Rouse Ball in his *Essay on Newton's 'Principia'* (London, 1893): 86–7, are here reproduced (for the first time) in their entirety. In this preliminary sketch of the quadrature of 'geometrically rational figures', departing from the not unreasonable assumption that the squaring of curves defined by a Cartesian equation of two terms—the determination, that is, of $\int v.dx$ where (say) $bv^{\alpha}+cv^{\beta}x^{\epsilon}=0$ and so $v = {}^{\alpha-\beta}\sqrt{[-(c/b)\,x^{\epsilon}]}$—is 'well known' (see note (19) below), Newton borrows heavily from an earlier paper (III: 380–2) on 'The Quadrature of all curves whose æquations consist of but three terms' to reduce the trinomial case to an equivalent, immediately quadrable form. What further techniques of algebraic quadrature he had it in mind (see note (17)) here to subjoin we can only surmise. Presumably, however, he would there have summarised the 'Catalogus Curvarum aliquot ad Conicas Sectiones relatarum' which he had earlier set down in Problem 9 of his 1671 tract (see III: 244–54) and on particular tabulated integrals in which he founded both Corollary 3 to Proposition XLI, and Corollary 2 to Proposition XCI in the 'Liber primus' (compare §3: notes (209) and (213); and §2, Appendix 4: note (32) respectively). For all its mathematical elegance this scheme of reduction of trinomial equations to quadrable form finds no application in the context of central-force orbits, and when Newton came subsequently to omit it from the *Principia*'s press copy he had by then no doubt realised that it was, in the absence of an ancillary table of conic integrals at least, essentially irrelevant to the several foregoing appeals to a general theory of algebraic quadrature which it was tentatively introduced to clarify.

(2) Add. 3962.5: 68r/69r–69v.

(3) This first version is entered on f. 69v beneath a draft 'Schol[ium]', perhaps intended initially to conclude Section XI of the 'Liber primus' (in place of that printed in *Principia*, ₁1687: 191–2), which reads: 'In præcedentibus supponimus corpora mota aut non rotari circum propria gravitatis centra aut rotari uniformiter. Nam si motus vertiginis acceleratur ac retardatur (uti fieri solet in corporibus pendulis reciproco motu agitatis inꝗ globis qui rotando progrediuntur) minuetur acceleratio et retardatio motus progressivi per resistentiam a qua motus vertiginis oritur. Debet igitur vis resistentiæ inveniri deꝗ vi centripeta detrahi et motus progressivus ex parte reliqua determinari. Quanta autem pars in motum vertiginis accelerandū vel retardandū impenditur constabit per Propositionem sequentem et ejus corollaria'. Of that following 'Prop. Theor.' only the head was written, and Newton's accurate insight that the resistance to change of uniform motion—such as exists instantaneously in both a rolling body and an oscillating pendulum—distorts the 'progressive' motion of the centre of gravity of bodies found no place in the printed *Principia*.

(4) Changed from 'Cæterum ad complendam solutionem'. Initially Newton wrote 'Tandem cum Propositiones tam multæ depende[a]nt ex quadratura Figurarum, [paucis exponenda est]', and then altered this opening to be 'Tandem quadratura Figurarum quam toties assumpsi, ita se habet.'

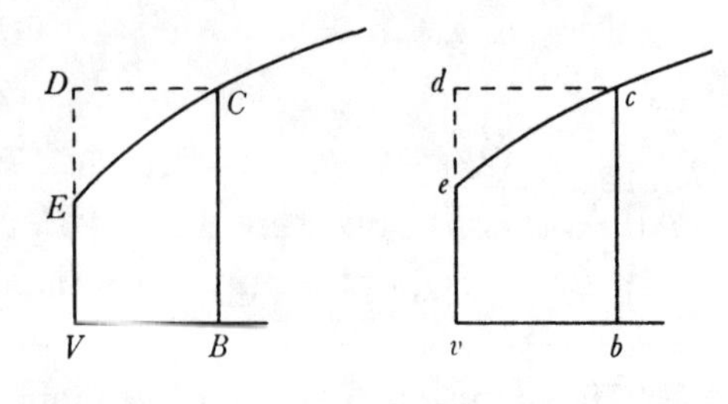

perpetuo tangunt: et completis rectangulis(5) $VBCD$, $vbcd$, occurrant VD, vd curvis in E et e, et si nominentur VB, BC, vb, bc, area $VECB$ et area $vecb$, v, x, y, z, A et B respectivè, sintꝗ f, g, h, k, l, m &c quantitates quævis datæ, et α, β, γ, δ &c indices numerales & curva $VBCE$ definiatur per æquationes in columna prima Tabulæ subjunctæ, habebitur quadratura ejus in columnis sequentibus.(6)

Figura quadranda.
$$fv^{\alpha}+gv^{\beta}x^{\epsilon}+hx^{\zeta}=0.$$

$\dfrac{\alpha-\beta+\epsilon}{\alpha-\beta}=\tau.$	$x^{\tau}=z.$	$fy^{\alpha}+gy^{\beta}+hz^{\frac{\alpha\epsilon+\beta\zeta-\alpha\zeta}{\beta-\alpha-\epsilon}}[=0].$	$A=\dfrac{B}{\tau}.$
$\dfrac{\alpha+\zeta}{\alpha}=\tau.$	$x^{\tau}=z.$	$fy^{\alpha-\beta}+[h]y^{-\beta}+[g]z^{\frac{\alpha\epsilon+\beta\zeta-\alpha\zeta}{\alpha+\zeta}}[=0].$	$A=\dfrac{B}{\tau}.$
$\dfrac{-\epsilon+\beta+\zeta}{\beta}=\tau.$	$x^{\tau}=z.$	$[g]y^{\beta-\alpha}+[h]y^{-\alpha}+[f]z^{\frac{\alpha\epsilon+\beta\zeta-\alpha\zeta}{\epsilon-\beta-\zeta}}[=0].$	$A=\dfrac{B}{\tau}.$

[2](7)

Casus primus curvarũ quæ per æquationes duorum terminorum(8) definiuntur.

$$\text{Æquatio } av^{\alpha}=x. \quad \text{Area } \frac{a}{\alpha[+1]}v^{\alpha+1}.$$

Casus secundus curvarum(9) quæ per æquationes trium terminorum definiuntur.

(5) Originally 'parallelogrammis'; correspondingly, the ordination angle at b in the right-hand figure was initially drawn obtuse by Newton. As he at once saw, to assume that one (or both) of $\widehat{VBC}$, $\widehat{vbc}$ differs from a right angle here introduces a needless complication.

(6) Much as in III: 380, note (3), Newton's stipulations in sequel that $x^{\tau}=z$ and (where $A\equiv\int v.dx$ and $B\equiv\int y.dz$) $\tau A=B$ readily determine that $v=(1/\tau)\,y.dz/dx=yz^{1-1/\tau}$; whence the equation defining the given 'curva quadranda' is converted into the equivalent trinomial $fy^{\alpha}z^{\alpha(1-1/\tau)}+gy^{\beta}z^{\beta(1-1/\tau)+\epsilon/\tau}+hz^{\zeta/\tau}=0$, which, according as there is set
$$\alpha(\tau-1)=\beta(\tau-1)+\epsilon \quad\text{or}\quad \alpha(\tau-1)=\zeta \quad\text{or}\quad \beta(\tau-1)+\epsilon=\zeta,$$
that is, as $\tau=(\alpha-\beta+\epsilon)/(\alpha-\beta)$ or $(\alpha+\zeta)/\alpha$ or $(\beta-\epsilon+\zeta)/\beta$, may be reduced straightforwardly to the forms listed below.

(7) A preliminary recasting (on f. 68r) of the terminal scheme of reduction in [1] preceding.

(8) Originally '...æquationum ex duobus terminis constantiũ'.

(9) Newton first began to write 'æquati[onum]'; compare the previous note.

Æquatio $dv^{\alpha}+ev^{\beta}x^{\epsilon}+fx^{\zeta}=0$.

	Longitudo vb.	Longitudo bc.(10)	Area $VBCE$.
Modus 1.	$x^{\frac{\alpha-\beta+\epsilon}{\alpha-\beta}}=z$.	$dy^{\alpha}+ey^{\beta}+fz^{\frac{\alpha\epsilon+\beta\zeta-\alpha\zeta}{\beta-\alpha-\epsilon}}[=0]$.	$\frac{\alpha-\beta}{\alpha-\beta+\epsilon}B=A$.
Modus 2.	$x^{\frac{\alpha+\zeta}{\alpha}}=z$.	$dy^{\alpha-\beta}+[f]y^{-\beta}+[e]z^{\frac{\alpha\epsilon+\beta\zeta-\alpha\zeta}{\alpha+\zeta}}[=0]$.	$\frac{\alpha}{\alpha+\zeta}B=A$.
Modus 3.	$x^{\frac{\epsilon-\beta-\zeta}{\beta}}=z$.	$[e]y^{\beta-\alpha}+[f]y^{-\alpha}+[d]z^{\frac{\alpha\epsilon+\beta\zeta-\alpha\zeta}{\epsilon-\beta-\zeta}}[=0]$.	$\frac{\beta}{-\epsilon+\beta+\zeta}B=A$.

Casus tertius Curvarum quæ per æquationes quatuor terminorum definiuntur.

Æquatio $dv^{\alpha}+ev^{\beta}x^{\epsilon}+v^{\gamma}x^{\zeta}+gx^{\eta}=0$.

Modus 1.(11)

---

[3](12)

$$bv^{\alpha}+cv^{\alpha+\beta}+dv^{\alpha+2\beta}+ev^{\alpha+3\beta}+fv^{\alpha+4\beta}+gx^{\eta}[=0].$$

[fit]
$$by^{\alpha}z^{\frac{\alpha\tau-\alpha}{\tau}}+cy^{\alpha+\beta}z^{\frac{\alpha\tau+\beta\tau-\alpha-\beta}{\tau}}+dy^{\alpha+2\beta}z^{\frac{\alpha\tau+2\beta\tau-\alpha-2\beta}{\tau}}$$
$$+ey^{\alpha+3\beta}z^{\frac{\alpha\tau+3\beta\tau-\alpha-3\beta}{\tau}}+fy^{\alpha+4\beta}z^{\frac{\alpha\tau+4\beta\tau-\alpha-4\beta}{\tau}}+gz^{\frac{\eta}{\tau}}[=0].$$(13)

$$\begin{matrix}\zeta ma\, x^{\zeta-1+\frac{1}{\lambda}} + \eta mb\, x^{\eta+\frac{1}{\lambda}-1} + \theta mc\, x^{\theta+\frac{1}{\lambda}-1} \\ +na \qquad +nb \qquad +nc\end{matrix} \quad \text{in} \quad \overline{ax^{\zeta+\frac{1}{\lambda}}+bx^{\eta+\frac{1}{\lambda}}+cx^{\theta+\frac{1}{\lambda}}}\Big|^{\lambda-1}.$$(14)

$m=\lambda$. $n=\pi=1=k$. $vx+A-A=B$. $x\times\overline{ax^{\zeta}+bx^{\eta}+cx^{\theta}}|^{\lambda}=$ areæ B.(15)

---

(10) Understand 'definitur per'.

(11) Newton would evidently here proceed, much as in his earlier paper on 'The Quadrature of many Curves whose æquations consist of more then three terms' (III: 383–5), by listing in succession the six 'Modes' of reduction which ensue on equating in pairs the indices of powers of z in the equivalent quadrinomial

$$dy^{\alpha}z^{\alpha(1-1/\tau)}+ey^{\beta}z^{\beta(1-1/\tau)+\epsilon/\tau}+fy^{\gamma}z^{\gamma(1-1/\tau)+\zeta/\tau}+gz^{\eta/\tau}=0.$$

(12) These intermediate computations follow [2] on f. 68r.

(13) Newton's attempt to simplify the area $\int v.dx$ under the curve defined by his given polynomial is manifestly abortive since the equal area $(1/\tau)\int y.dz$ under the polynomial which ensues on substituting $x=z^{1/\tau}$ and $v=yz^{1-1/\tau}$ is not notably simplified by equating $(\alpha+i\beta)(1-1/\tau)$ to η/τ and so determining $\tau=(\alpha+i\beta+\eta)/(\alpha+i\beta)$, $i=0, 1, 2, 3$ or 4.

(14) Where $v=(ax^{\zeta}+bx^{\eta}+cx^{\theta})^{\lambda}$, this is $(m/\lambda)\,x.dv/dx+nv$, the derivative of

$$(m/\lambda)\,(cx-A)+nA$$

on taking (as before) $A\equiv\int v.dx$.

(15) Where, that is, $B\equiv\int((\lambda\zeta+1)\,ax^{\zeta}+(\lambda\eta+1)\,bx^{\eta}+(\lambda\theta+1)\,cx^{\theta})\,(ax^{\zeta}+bx^{\eta}+cx^{\theta})^{\lambda-1}.dx$,

[4][16]

Tandem de quadratura figurarum quam toties assumpsi visum est Propositionem unam et alteram[17] subjungere.

Prop. Prob.

Quadrare figuras curvilineas quarum ordinatim applicatæ ex[18] *abscissarū longitudinibus per æquationes trium terminorū et areæ per æquationes numero terminorum finitas generaliter definiri possunt.*

Nota est quadratura curvarum quarum ordinatim applicatæ per æquationes duorum terminorum determinantur.[19] Quapropter hunc casum prætereo et curvarum quæ applicatas habent per æquationes trium terminorum definitas quadraturam hic exhibeo. Ad abscissas *GH*, *LM* a datis punctis *G*, *L* in data positione ductas ordinatim applicentur *HI*, *MN* in angulis rectis *GHI*, *LMN* tangantq͡3 earum termini *I* et *N* curvas *KI*, *PN*, et ad abscissarum principia *G* et *L* erigantur perpendicula *GK*, *LP* curvis illis occurrentia in *K* et *P*. Dicantur autem *GH*, x; *HI*, v; *LM*, z; *MN*, y; area *GHIK*, *A* et area *LMNP*, *B* sintq͡3 α, β, γ, δ, ϵ dignitatum indices numerales et b, c, d &c quantitates quævis datæ. Et si ordinatim applicata *HI* definiatur per hanc æquationem $bv^{\alpha}+cv^{\beta}x^{\epsilon}+dx^{\zeta}=0$, invenietur area *GHIK* per tabulam sequentem.[20]

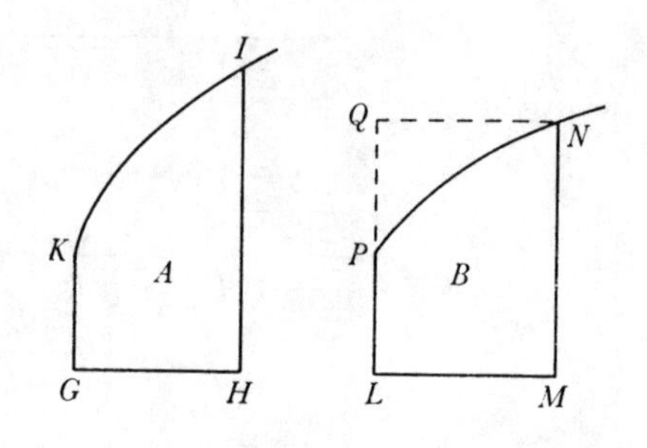

the particular value $k = \pi = 1$ of the expansion on III: 385 on setting $\lambda \rightarrow \lambda-1$. It will be clear that, as there, Newton is toying with the idea of introducing into his 'Liber primus' some discussion of his general technique (see III: 385, note (31)) for reducing algebraic integrals $hx^{\epsilon}(ax^{\zeta}+bx^{\eta}+cx^{\theta}\ldots)^{\lambda-1}.dx$ to simpler form.

(16) This revised version follows immediately after [1] on ff. 69ᵛ/69ʳ.

(17) We have previously suggested (see note (1)) that Newton would have been likely in sequel to have summarised the table of conic integrals which he had earlier presented in his 1671 tract, perhaps with some appended discussion of the techniques (see III: 256–9, notes (549)–(569)) employed in its construction.

(18) 'assignatis' is deleted.

(19) This case evidently reduces to the 'Casus primus' of quadrature of algebraic curves listed in [2] above.

(20) Effectively as in [1]: the earlier coefficients f, g, h of the defining equation of the 'curva quadranda' are here trivially replaced by b, c, d, while in the three following 'Modes' there is successively substituted for τ its explicit value (that ensuing from equating the pertinent pairs of indices of z in the unreduced transformed equation; compare note (6) above).

Modus primus.

Longitudo Abscissæ LM	$x^{\frac{\alpha-\beta+\epsilon}{\alpha-\beta}}=z.$
Longitudo Applicatæ MN	$by^{\alpha}+cy^{\beta}+dz^{\frac{\alpha\epsilon+\beta\zeta-\alpha\zeta}{\beta-\alpha-\epsilon}}[=0].$
Area $GHIK$	$\frac{\alpha-\beta}{\alpha-\beta+\epsilon}B=A.$

Modus secundus.

Longitudo Abscissæ LM	$x^{\frac{\alpha+\zeta}{\alpha}}=z.$
Longitudo Applicatæ MN	$by^{\alpha-\beta}+dy^{-\beta}+cz^{\frac{\alpha\epsilon+\beta\zeta-\alpha\zeta}{\alpha+\zeta}}[=0].$
Area $GHIK$	$\frac{\alpha}{\alpha+\zeta}B=A.$

Modus tertius.

Longitudo Abscissæ LM	$x^{\frac{\beta+\zeta-\epsilon}{\beta}}=z.$
Longitudo Applicatæ MN	$[c]y^{\beta-\alpha}+dy^{-\alpha}+bz^{\frac{\alpha\epsilon+\beta\zeta-\alpha\zeta}{\epsilon-\beta-\zeta}}[=0].$
Area $GHIK$	$\frac{\beta}{\beta+\zeta-\epsilon}B=A.$

In aliqua æquationum quibus longitudo ordinatim Applicatæ MN in hisce tribus modis definitur, index valoris quantitatis z erit unitas si modò Figura proposita quadrari possit.(21)

---

(21) This is not quite accurate. The most general rationally quadrable curve defined by a trinomial equation is ($z = ey^{rm-1}\,{}^{n}\sqrt{[b+cy^{m}]}$ or) $by^{\alpha}+cy^{\beta}+dz^{n} = 0$, where $\alpha = (rm-1)\,n$, $\beta = (rm-1)\,n+m$ and $d = -e^{-n}$; and this is the transform by Newton's Mode 1 of

$$bv^{\alpha}+cv^{\beta}x^{\epsilon}+dx^{\zeta} = 0,$$

where $\alpha\epsilon+(\beta-\alpha)\,\zeta = (\beta-\alpha-\epsilon)\,n$ or $(rm-1)\,n\epsilon+m\zeta = (m-\epsilon)\,n$, that is, $\zeta = (1-r\epsilon)\,n$. Accordingly, since then $\alpha+\zeta = (m-\epsilon)\,nr$ and $\epsilon-\beta-\zeta = -(m-\epsilon)\,(nr+1)$, this derived 'curva quadranda' may be transformed back by Mode 2 into the trinomial

$$by^{\alpha-\beta}+dy^{-\beta}+cz^{1/r} = 0$$

and by Mode 3 into $cy^{\beta-\alpha}+dy^{-\alpha}+bz^{-n/(nr+1)} = 0$. A like result holds if the primary equation be transmuted by Mode 2 or Mode 3, as may readily be proved. Newton's assertion that the resulting index of z be unity therefore requires that the initial equation be reducible (when $r=1$) to $z = ey^{m-1}\,{}^{n}\sqrt{[b+cy^{m}]}$ or (when $r = -1-1/n$) to $z = ey^{-m-1}\,{}^{n}\sqrt{[by^{-m}+c]}$, and doubtless derives from wrongly assuming one or other of these to be the most general quadrable equation possible which may be set in an equivalent, rational trinomial form.

2

APPROACHES TO THE MATHEMATICAL SOLUTION OF PARTICULAR PROBLEMS[1]

From autograph drafts in the University Library, Cambridge

§1. THE SOLID OF REVOLUTION OF LEAST RESISTANCE TO MOTION IN A UNIFORM FLUID.[2]

[*c.* late 1685]

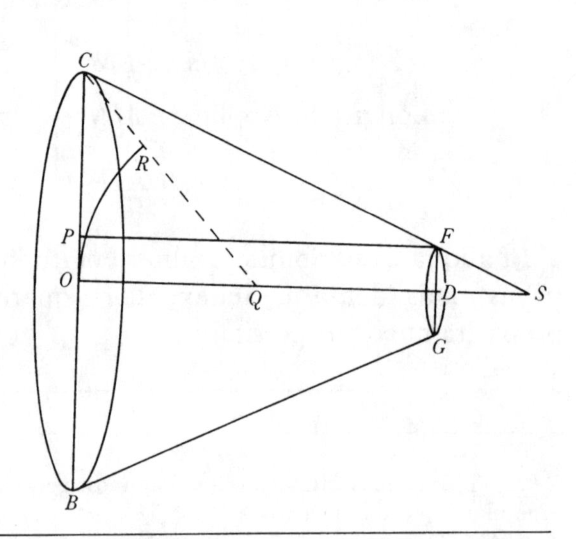

[1][3]

[Pone] $OC=a.$ $OD=b.$ $CP=e.$[4] [erit] $CF=\sqrt{ee+bb}.$ [adeoq] Resis[tentia]

$$=\frac{aaee}{ee+bb}^{(5)}+aa-2ae+ee.$$

(1) Though in the several tracts 'De motu Corporum' reproduced in previous sections it had, with one eye set firmly on the astronomical model of planetary motion round the Sun in which this dynamical structure is principally exemplified, been Newton's main purpose to analyse the motion of an individual body travelling without external perturbation (and so perforcedly in a plane) in a simple, central force-field, he afterwards in the fuller, more developed text he published to the world in 1687 as his proclaimed 'Mathematical principles of natural philosophy' came both, in its second Book, tentatively—and not a little erroneously —to broach topics in the mechanics of continuous fluids, and also, in its third, to attack the equally tricky problems of accurately determining the (near) parabolic paths of comets round the Sun from given terrestrial sightings and above all of embracing the observed inequalities of the Moon's wobbling oval orbit round the Earth within a rigorous dynamical theory of its continually changing deviation from 'true' elliptical form under the disturbance of solar gravity. In a number of instances, however, the printed *Principia* does scant justice to the

[hoc est]

$$\text{Resist} = \frac{+2aaee + aabb - 2ae^3 - 2bbae + e^4 + bbee - aaee + 2ae^3 - e^4}{ee + bb} = R$$

$$= \frac{aaee + aabb - 2aebb + bbee}{ee + bb}.^{(6)}$$

---

complexity and sophistication of Newton's preliminary analysis as we may know it from his preceding autograph worksheets and drafts. This is most notably true of his determination of the solid of revolution of least resistance to rectilinear motion, whose geometrically framed defining differential equation is merely enunciated without any explanation in the concluding paragraph of the scholium to Proposition XXXV in Book 2 of the published work; and of his statement in the parallel scholium to Proposition XXXV in its Book 3 that he had computed the mean secular advance of lunar perigee to be 'some 40°' per year but was reluctant to append his computation 'as being too intricate and encumbered with approximations and insufficiently accurate' (see §3: note (1)): Newton's reticence here caused his contemporaries and eighteenth-century successors a good deal of effort and frustration when they sought to replicate his calculations, and the question as to exactly how in each of these cases he achieved his stated end has remained to plague historians down to the present day. In sequel (in §§1 and 3 respectively) we reproduce, for the first time in full, the extant manuscript record of Newton's prior investigations of both these topics, further adjoining (in §2) the scarcely better known concluding paragraphs of his initial 'De motu Corporum Liber secundus' where he elaborated a first method—as cumbrous as it is impractical—for approximately constructing cometary paths in the (then novel) hypothesis that these are highly parabolic inverse-square force orbits traversed round the Sun at a focus.

(2) In an immediately preceding draft (reproduced in Appendix 1 below) of what, with minor rephrasing, was to become Proposition XXXV of Book 2 of his published *Principia* ($_1$1687: 324–6) wherein he compared the resistance to such motion of a sphere and a cylindrical drum of equal radius, Newton laid down his conception of how a solid, understood to be advancing at a uniform but 'rapid' speed in a 'rare, elastic fluid', would at each point of its exposed fore-surface be impeded: ignoring the physically critical distortions due to surface friction and the disturbance of wave-flow, he postulated namely that the resisting effect of such a medium on the solid's progress would (by his third Law of Motion) be that to which the solid would be subjected were it to remain stationary while the fluid moved uniformly in the contrary direction at the same speed. It follows that, where any element of the solid's surface is inclined at an angle θ to the direction of flow of the fluid stream, the pressure, p say, which the fluid exerts on that element is proportional to $\sin\theta$, and therefore its resistance on the element—due, in Newton's hypothesis, solely to the normal component $p\sin\theta$ of this pressure—will be proportional to $\sin^2\theta$. The restriction of the solid to be one of revolution is, of course, inessential and serves merely to simplify the ensuing mathematical discussion.

(3) Add. 3965.10: 134^v. Having (see Appendix 1) computed the resistance to motion of a sphere to be but half that of a cylinder of equal transverse section, Newton is here led straightforwardly to calculate the analogous resistance of (the exposed surface of) a frustum of a right circular cone of given height and base diameter, and thence to determine the inclination of the curved surface to the axis which makes the resistance minimal. From this it is an immediate step—which he at once takes—to compute the differential equation of the diametral cross-section of a general surface of revolution whose resistance to uniform rectilinear motion in the direction of its axis shall, at its every point, be locally a minimum.

(4) Understand in Newton's accompanying figure (from which an irrelevant broken line joining *CD* is here deleted) that *BCFG* is the frustum of a cone swept out when the trapezium

[id est $aaee+aabb-2aebb+bbee=eeR+bbR$. Unde][7]

$$2a\dot{a}e+2bbe-2abb=2eR=2e\dot{a}a+\frac{2e^3bb-4aeebb}{ee+bb}.$$

[sive] $$2b\dot{b}e^3-2abbee+2b^4e-2ab^4=2e^3\dot{b}b-4aeebb.$$

[vel] $-2abbee+2ab^4-2eb^4=0$. [id est] $aee=abb-ebb$.

[seu] $ee=-\frac{bb}{a}e+bb$. [Unde radicem extrahendo] $e=-\frac{bb}{2a}\pm\sqrt{\frac{b^4}{4aa}+bb}$. [hoc est] $a.b::\sqrt{\tfrac{1}{4}bb+aa}-\tfrac{1}{2}b.e$[8] [vel] $a\sqrt{\tfrac{1}{4}bb+aa}+\tfrac{1}{2}ab.b::aa.e$. [adeoq;]

$$\sqrt{\tfrac{1}{4}bb+aa}+\tfrac{1}{2}b.b::a.e.$$

[Fit igitur] $OC.OD::CQ-OQ$[9]$.CP=CO-FD.$

[et] $CQ+OQ.OD::OC.OC-FD=CP$. Fac [ergo] $QS=QC$[10] et duc CS secantem FD in F.

$OCFD$ is rotated round OD as axis. We omit an equivalent initial calculation in which he similarly began '$OC = a$. $OD = b$. $DF = c$. $PC = a-c = e$. $CF = \sqrt{bb+cc-2ac+aa} = d$. [adeoq;] $\frac{ee}{dd}\overline{aa-cc}+cc = \frac{ee}{dd}aa+\frac{dd-ee}{dd}cc = R$' and then eliminated '$d = \sqrt{bb+ee}$. $c = a-e$'. to achieve '$bbR+eeR = eeaa+bbaa-2bbae+bbee$'.

(5) The numerator should read '$aa-\overline{a-e}|^2\times ee$'. Newton at once corrects this momentary slip by appending the terms '$-aaee+2ae^3-e^4$' to the numerator of his following fraction, and the sequel is unaffected.

(6) That is, $(CO^2-FD^2).\sin^2\theta+FD^2$ where $\widehat{CSO} = \widehat{CFP} = \theta = \tan^{-1}(e/b)$ is the inclination of CF to the axis OD. Strictly, according to Newton's preferred hypothesis (see note (2)) the resistance on the surface formed by rotating CF round OD is to that on the circular ring formed by similarly revolving CP as $\sin^2\theta$ to 1, and hence proportional to $\pi.(CO^2-PO^2).\sin^2\theta$, while the resistance on the end-circle FG is likewise proportional to its area $\pi.FD^2$; but in the sequel only the relative magnitude of these component resistances is pertinent and no harm ensues from his here absorbing the constants π into the proportionality factor.

(7) To determine minimum R as the point P varies in the line CO, thus altering the inclination of the cone's curved suface, Newton differentiates the preceding equation with respect to $CP = e$, setting $dR/de = 0$. As elsewhere in his pre-1691 worksheets, the superscript dots in the two following equations cancel equal terms on their respective sides, and must not be confused with Newton's later equivalent use of the point to signify the (first) fluxion of the letter over which it is set.

(8) Newton chooses the root e which lies in the interval $[0, a]$: the second root is physically meaningless.

(9) That is, CR where (as in the figure) the circle arc $\widehat{OR}$ of centre Q is drawn to meet CQ in R.

(10) Whence at once $CO:CP = OS:(OD$ or$)\ PF$ and so the point F is in line with C and S.

[Pone $BA = a$. $AD = x$. $CD = o$. fit]

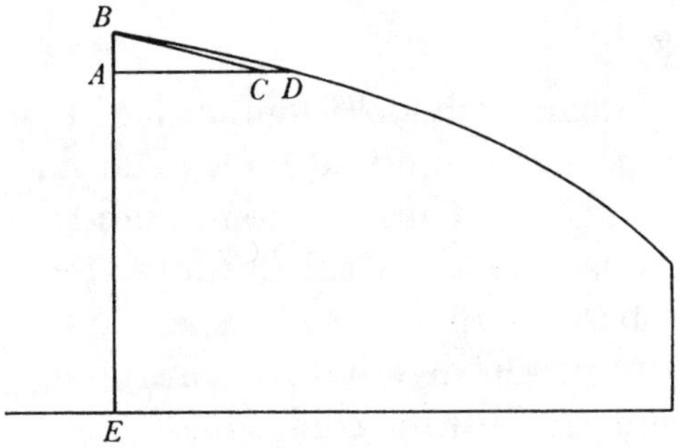

$$-\frac{1}{aa+xx}+\frac{1}{aa+xx-2ox+oo}[\times BE]$$

$$=\frac{2xo-oo}{a^4+2aaxx+x^4}\times B[E]$$

$$=\text{dato}=\frac{2,o}{pp}.^{(11)}$$

[Dic] $B[E] = y$. [Erit][12]

$$ppxy = a^4+2aaxx+x^4.$$

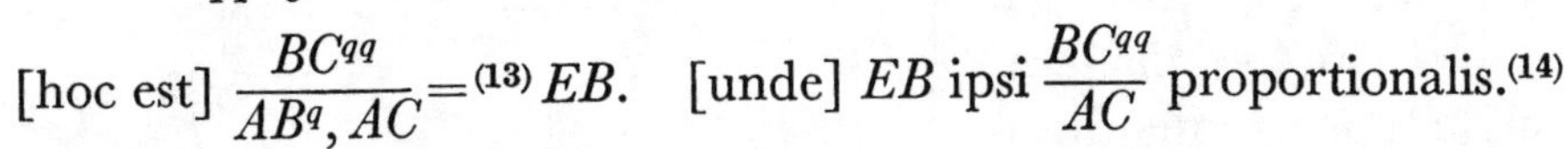

[hoc est] $\dfrac{BC^{qq}}{AB^q, AC} =^{(13)} EB$. [unde] EB ipsi $\dfrac{BC^{qq}}{AC}$ proportionalis.[14]

---

(11) Where understand that p is some constant.

(12) On ignoring the term $-oo$ in the previous numerator as being, in the limit as o vanishes, infinitesimal in regard to $2xo$.

(13) A necessary proportionality factor '$\dfrac{pp}{aa}\times$' is here omitted.

(14) In these brief recorded calculations Newton only hints at the complexity of his underlying argument. In amplification, where $EB = y$ and (say) $FE = z$ are Cartesian coordinates of the general point $B(z, y)$ of the curve $\widehat{BG}$ by whose rotation round EF the required surface of minimum resistance to the 'pressure' of uniform motion in the direction EF is engendered, it is clear that he seeks the relationship between these coordinates and their related instantaneous increments $BA\ (= dy) = a$ and

$$AD\ (= dz) = x$$

thus defined. Since, on taking

$$\widehat{ADB} = \tan^{-1}(a/x) = \theta$$

and $FG = Y$ (at $z = 0$), the resistance on the element $BD = \sqrt{[a^2+x^2]}$ rotated round EF at the distance $EB = y$ is (see note (2))

$$2\pi y \sin^2\theta . dy = 2\pi F$$

where $F = a^3y/(a^2+x^2)$, the problem reduces to determining the condition which makes $2\pi\int_Y^y F$ a minimum. For this it is clearly necessary that the resistance on each local neighbourhood of the surface shall be minimal: whence, if an equal increment $D\alpha\ (= Aa) = a$ of $eD = y-a$ be set off and the corresponding increment $\alpha d = x'$ of $Fe = z-x$ constructed, the total resistance $2\pi\left(\dfrac{a^3y}{a^2+x^2}+\dfrac{a^3(y-a)}{a^2+x'^2}\right)$ on the surface generated by the rotation of BDd must be a

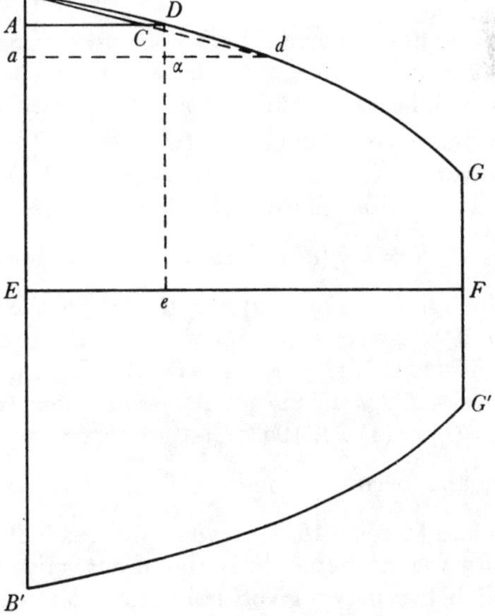

[2][15] *Schol.*[16]

Eadem methodo[17] figuræ aliæ inter se quoad resistentiam comparari possunt, eæq; inveniri quæ ad motus suos in Medijs resistentibus continuandos aptiores sunt.[18] Ut si altitudine *OD* et basi circulari *CEBH* quæ centro *O* radio *OC* describitur, construendum sit frustum coni *CBGF* quod omnium eadem basi et altitudine constructorum et secundum plagam axis sui versus *D* progredientium frustorum minimè resistatur: biseca altitudinē *OD* in *Q* et produc *OQ* ad *S* ut sit *QS* æqualis *QC*, et erit *S* vertex coni cujus frustum quæritur.[19]

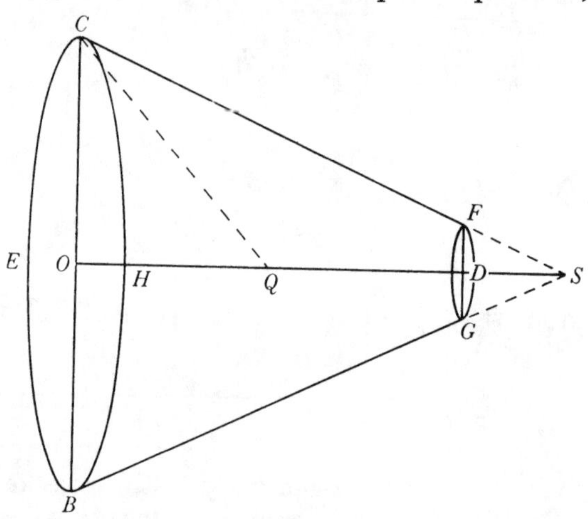

minimum, and so constant in magnitude for infinitesimal variations in the position of *D* as the end-points *B* and *d* (and hence their horizontal distance $ad = x+x'$) remain fixed. Accordingly, if *D* alters its position minimally to *C* where $CD = o$, there must, on suppressing the common factor $2\pi a^3$, be $\frac{y}{a^2+x^2}+\frac{y-a}{a^2+x'^2} = \frac{y}{a^2+(x-o)^2}+\frac{y-a}{a^2+(x'+o)^2}$ and therefore

$$\left(-\frac{1}{a^2+x^2}+\frac{1}{a^2+(x-o)^2}\right) y = \left(\frac{1}{a^2+x'^2}-\frac{1}{a^2+(x'+o)^2}\right) (y-a)$$

in the limit as *o* vanishes. Since on dividing through by *o*, the common value of each side is $-\partial(y/(a^2+x^2))/\partial x = 2xy/(a^2+x^2)^2$, Newton has merely to set this equal to some constant (which he chooses to be $2/p^2$) and is at once able to derive the defining differential 'symptom' of the curve $\widehat{BG}$ to be $y \propto (a^2+x^2)^2/a^3x$ $[= (dy^2+dz^2)^2/dy^3dz]$ on reinserting the omitted cubic power of the 'constant' increment *a* of the base variable *y*.

In a wider perspective Newton's necessary criterion $\partial F/\partial x =$ constant for the integral $\int_Y^y F$, $F = a^3y/(a^2+x^2)$, to be minimal may likewise be applied in the case of any continuous function $F \equiv F(y, x, a)$ which is free of the second coordinate *z*: indeed, as Leonard Euler was to show in a series of papers written in the decade after Newton's death (and now conveniently collected in his *Opera Omnia* (1) **25**, Zurich, 1952) which culminated in his magisterial *Methodus inveniendi Lineas Curvas Maximi Minimive Proprietate gaudentes* (Lausanne/Geneva, 1744 [= *Opera* (1) **24**, 1952]), an analogous argument establishes that $d(\partial F/\partial x)/dy = \partial(F/a)/\partial z$ is a general, necessary condition for $\int_Y^y F(z, y, x, a)$ to be extremal. When a dozen years afterwards, in late January 1697, Newton was (as will be explained in the eighth volume) abruptly faced with Johann Bernoulli's printed challenge to construct the *curva brachystochrona* of least time of fall between two given points in a simple gravitational field of intensity *g*, we will not be surprised that, having once reduced the problem to the formal requirement that $\int_0^y \sqrt{[a^2+x^2]}/v$, $v = \sqrt{[2gy]}$, be a minimum, he found no difficulty in meeting it by setting

$$[\partial(\sqrt{[a^2+x^2]}/v)/\partial x =]\ x/\sqrt{[a^2+x^2]}\ v = k, \text{ constant}$$

Scholium(16)

By the same method(17) other figures can be compared together in regard to their resistance, and those more suited to continuing their motion in resisting media ascertained.(18) If, for instance, we need with height OD and circular base $CEBH$ described round centre O with radius OC to construct a cone frustum $CBGF$ which of all frustums constructed with the same base and height and moving forward along the line of its axis towards D shall be least resisted, bisect the height OD in Q and extend OQ to S so that QS be equal to QC, and then S will be the vertex of the cone whose frustum is sought.(19)

---

and then readily identifying the ensuing differential equation

$$x/a \text{ (or } dz/dy) = \sqrt{[y/(1/2gk^2-y)]}$$

as that of a common cycloid traced by a point in the perimeter of a rolling circle of radius $1/4gk^2$. The further subtlety that the existence everywhere in the range of the integrand $F(y, x, a)$ of a local minimum defined by $\partial F/\partial x = k$ does not necessarily guarantee that the global integral $\int_Y^y F$ thus constrained is an absolute minimum—that is, that it can never by drastically varying the restriction on F attain some lower total value in the interval $[Y, y]$ at issue—did not in 1685 pass Newton by. For when (in the last paragraph of the draft reproduced in [2] following) he came soon after to incorporate its present differential defining symptom $y \propto (a^2+x^2)^2/a^3x$ into his *Principia*, he accurately and percipiently there restricted his equivalently enunciated geometrical proportion determining the diametral curve of the surface of revolution 'resisting least of all' to hold only for the range $x \geqslant a$; see note (24) below.

(15) Add. 3965.10: 107v. This augmented verbal presentation of the defining properties of the cone frustum and general surface of revolution of least resistance computed in [1] is little different from the revised version published by Newton in his *Principia* ($_1$1687: 326–7).

(16) Namely to 'Prop. XXXV. Theor. XXVIII' of the second book of the published *Principia* ($_1$1687: 324–6)—afterwards advanced to be Proposition XXXIV in the second edition ($_2$1713: 298–9)—whose initial draft (ULC. Add. 3965.10: 134r/134v, reproduced in Appendix 1) is here, having been restyled on the preceding manuscript folios 106r/106v to embrace the newly contrived text of his following Proposition XXXVI (*Principia*, $_1$1687: 327–9), once more scheduled by Newton to appear separately under the revised enunciation 'Si globus et cylindrus æqualibus diametris descriptis in Medio raro et elastico secundum plagam axium suorum æquali cum velocitate moveantur: erit resistentia globi duplo minor quàm resistentia cylindri' (compare *ibid.*: 324).

(17) In the hypothesis (see note (22)) that the resistance of a plane surface to uniform rectilinear motion is proportional to the square of the sine of its inclination to that direction.

(18) Newton initially wrote—explicitly as a 'Corol.' to the preceding Proposition XXXV—'Hinc eadem methodo comparari possunt resistentiæ diversarum figurarum' (Hence by the same method there can be compared the resistances of different figures), continuing thereafter 'Ut si sphærois [initially 'solidum Ellipticum'] secundum axis sui plagam progrediatur, erit hujus resistentia ad resistentiam Cylindri circumscripti' (Should, for instance, a spheroid advance in the direction of its [main] axis, then its resistance will be to the resistance of the circumscribed cylinder...) but breaking off in mid-sentence to draft the revision reproduced above.

(19) A first intended conclusion reads 'axeq; OS, vertice S ac basi $CEBH$ describe [conum]' (and with axis OS, vertex S and base $CEBH$ describe the cone). The construction is identical with that derived in [1]; compare note (10).

Unde obiter, cùm angulus *CSB* semper sit acutus,[20] consequens est quod si solidum *ADBE* convolutione figuræ Ellipticæ vel Ovalis *ABBE* circa axem *AB* generetur,[21] et tangatur figura generans a rectis tribus *FG, GH, HI* in punctis *F, B* et *I* ea lege ut *GH* sit perpendicularis ad axem in puncto contactus *B* et *FG, HI* cum *GH* contineant angulos *FGB*[,] *BHI* graduum 135: solidum quod convolutione figuræ *ADFGHIE* circa axem eundem generetur minus resistetur quam solidum prius si modo utrumqꝫ secundum plagam axis sui *AB* progrediatur, & utriusqꝫ terminus *B* præcedat.[22] Quam quidem Propositionem in construendis navibus non inutilem futuram esse censeo.[23]

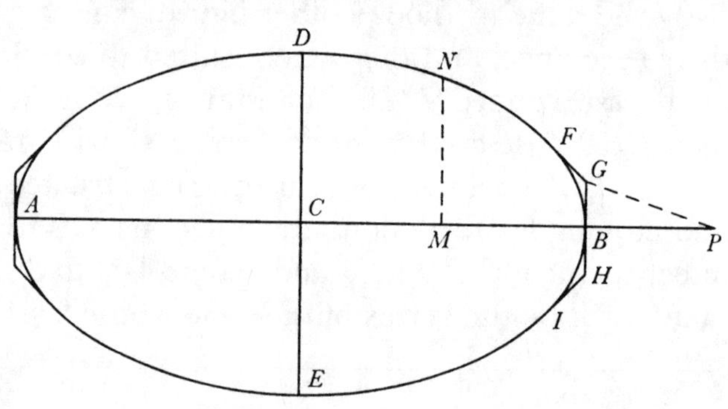

Quod si figura *DNFG* ejusmod[i] sit ut, si ab ejus puncto quovis *N* ad axem

(20) Originally—and equivalently—'cùm angulus *CSO* m[ajo]r esse nequeat quam angulus graduum 45' (since the angle $\widehat{CSO}$ cannot be greater than 45°).

(21) Initially 'genitum sit' (should be begotten), after which Newton wrote 'per axis terminum B erigatur perpendiculum *GBH*' (through the end-point *B* of the axis erect the perpendicular *GBH*).

(22) Where the height *OD* of the cone frustum *CFGB* in the previous paragraph is infinitesimally small, it will be clear that $(CQ \approx)$ *CO* and $(QS \approx)$ *OS* come to be equal, so that *CF* is in this limiting case inclined at 45° to the axis *OD*. Whence in the present instance, if *N, n* are two indefinitely close points in the arc $\widehat{BF}$ wholly contained by the tangent *FG* (inclined at 45° to *CB*) and the vertical *BG* at its respective end-points *F* and *B*, on constructing the (perpendicular) ordinates *MN, mn* at these points and drawing *no* parallel to *FG* till it meets *MN* in *o* it at once follows that the resistance on the surface formed by rotating $MN+Nn$ round *CB* is greater than that on the cone frustum generated by likewise rotating $Mo+on$, and therefore the resistance on *Nn* rotated round *CB* is greater than that on the frustum region formed by rotating $No+on$; that is, (because, where $N\nu$ and $o\omega$ are let fall perpendicularly to *mn*, there is

$$\omega n = o\omega = Mm = d(BM)$$

and $\nu n = mn - MN = d(MN)$) it is greater than

$$2\pi . MN . \nu n + 2\pi . (\omega n/on)^2 . Mo . \omega n$$
$$= \pi . (2MN . d(MN) - MN . d(BM)).$$

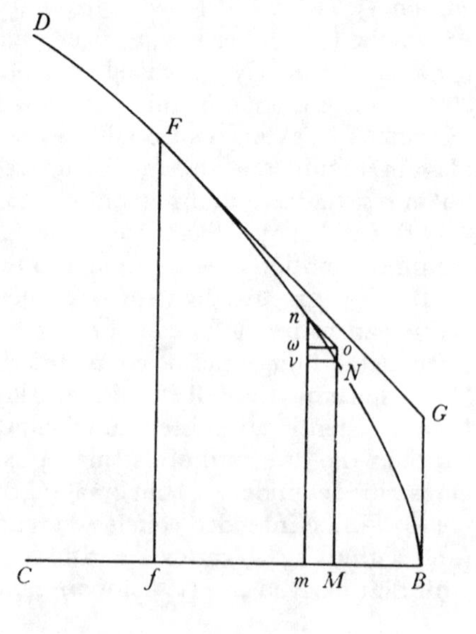

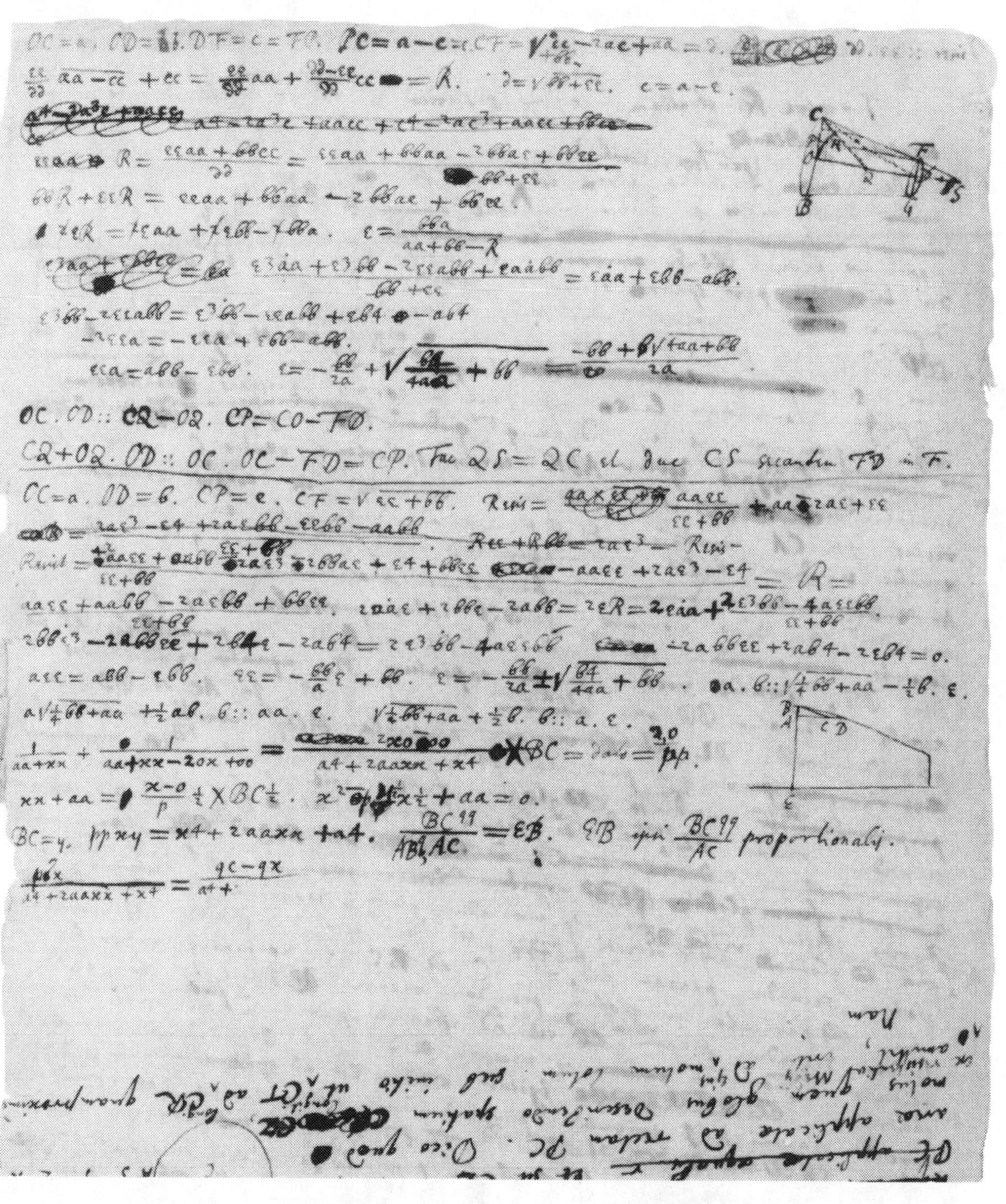

Plate III. Computation of the cone frustum and solid of revolution of least resistance to translation along its axis (2, §1.1).

Whence, by the way, since the angle $C\widehat{S}B$ is always acute,[20] the consequence is: if the solid *ADBE* be generated[21] by rotating the elliptical or oval figure *ADBE* round its axis *AB*, and the generating figure be touched at the points *F*, *B* and *I* by three straight lines *FG*, *GH*, *HI*, where *GH* is restricted to be perpendicular to the axis at the contact point *B*, while *FG*, *HI* contain with *GH* angles $F\widehat{G}B$, $B\widehat{H}I$ of 135°, then the solid generated by the rotation of the figure *ADFGHIE* round the axis will be resisted less than the previous solid, provided that each moves forward along the line of its axis *AB*, the end-point *B* leading in each case.[22] This proposition will, I reckon, be not without application in the building of ships.[23]

But should the figure *DNFG* be so shaped that, if from any point *N* in it the

---

Accordingly, the total resistance on the surface formed by rotating $\widehat{BF}$ is greater than

$$\pi\left(\int_0^{Ff} 2MN.d(MN) - \int_0^{Bf} MN.d(BM)\right),$$

that is, $\pi(Ff^2-(BFf))$. The comparable resistance on the cone frustum *FGHI* generated by rotating $BG+GF$, namely, $\pi.BG^2+\pi.(Ff-BG).\frac{1}{2}(Ff+BG) = \pi(Ff^2-\frac{1}{2}(Ff^2-BG^2))$, is less than this, since $\frac{1}{2}(Ff^2-BG^2) = (BGFf) > (BFf)$; *a fortiori*, therefore, it is less than the resistance on the surface of revolution *FBI*. While no record of Newton's own demonstration of this inequality has survived, we have no reason to think that it could have been greatly different in its structure.

(23) Originally 'Quæ quidem Propositio in construendis navibus usus sit non contemnendi' (Which proposition, indeed, should be of no trifling use in constructing ships). In the margin of his copy of the published *Principia* at this point ($_1$1687: 327) the Scots mathematician John Craige later wrote alongside: 'Hujus meditationis occasionem ipse præbui, dum Cantabridgiæ de Figura navium aptissimâ invenienda, problema celeberrimo Autori proponerem' (see I. B. Cohen, *Introduction to Newton's 'Principia'* (Cambridge, 1971): 204). While Craige certainly was in mid-1685 on an extended stay in Cambridge—'diutius commoratus' as Newton observed some half dozen years later in the prefatory paragraph (ULC. Add. 3962.2: 29^v) of the first (unpublished) version of his 'De quadratura Curvarum'—during which he came to know Newton well enough to be shown his private papers on algebraic quadrature, his assertion is scarcely supported by the physical appearance of the manuscript text (see Plate III), in which the present sentence, far from being a later addition squashed in between the lines, evidently continues the flow of the previous written passage and afterwards carries on uniformly (and initially without break of paragraph) into the sequel. It might be that Craige intended to refer to a somewhat similar sentence in the following 'Scholium Generale' (to Section VII of Book 2 of the *editio princeps*), which Newton concluded by offering the eminently practical suggestion that the 'aptest shape for ships' might be determined 'at little cost' by noting the resistance to motion in a water tank of model hulls, suitably scaled down: 'Eadem methodo qua invenimus resistentiam corporum Sphæricorum in Aqua..., inveniri potest resistentia corporum figurarum aliarum; & sic Navium figuræ variae in Typis exiguis constructæ inter se conferri, ut quænam ad navigandum aptissimæ sint, sumptibus parvis tentetur' (*Principia*, $_1$1687: 354, omitted in all later editions). If Newton had made such a controlled experiment for himself, he would rapidly have come to appreciate the artificiality of his supposition that resistance to fluid motion varies simply as the square of the sine of the angle of slope, even if he were still unable to distinguish such considerable ancillary distorting factors as skin friction and flow disturbance.

AB demittatur perpendiculum *NM*, et ducatur recta *GP* quæ parallela sit rectæ figuram tangenti in *N*, et axem productam secet in *P*, fuerit *MN* ad *GP* ut GP^{cub} ad $4BP \times GB^q$,(24) solidum quod figuræ hujus revolutione circa axem *AB* describitur resistetur minimè omnium ejusdem longitudinis & latitudinis.(25)

---

(24) That is, $MN/BG = \frac{1}{4}(BG^2+BP^2)^2/BG^3 \times BP$ where, since *GP* is drawn parallel to the tangent at *N*, $BG:BP = d(MN):d(BM)$. This is but a geometrical restatement of Newton's previously computed defining differential symptom $y \propto (a^2+x^2)^2/a^3x$ of the point $N(z, y)$ whose coordinates $MN = y$ and $BM = z$ have the instantaneous increments $dy = a$ and $dz = x$ respectively (see note (14) above), though now, however, it is implicitly restricted by the condition that at the point *G* where *MN* comes to coincide with $BG = Y$ the angle of slope of the curve to the axis *CB* shall be 45°, or $BG = BP$ and so $x = a$; whence Newton's analytical symptom becomes $y/Y = (\frac{1}{4}a^2+x^2)^2/a^3x = \frac{1}{4}(q^2+1)^2/q$ on setting $x/a\ (= dz/dy) = q$. From his original manuscript sheet (see Plate III) it is clear that Newton found great difficulty in thereafter constructing the curve $\widehat{GN}$ thus defined. Initially, presuming that *MN* does not greatly differ (in size and position) from *BG*, he deduced the consequence that

$$NP/NM \approx GP/GB \approx (GP^4/MN \times GB^3)^{\frac{1}{4}} \approx (4MP/BG)^{\frac{1}{4}}$$

and then went on to write '...fuerit *NP* and *NM* in quarta parte rationis 4*MP* ad *BG*, seu $NP^{qq} \times BG$ æquale $4NM^{qq} \times NP$', but immediately converted it into the accurate defining proportion which here replaces it. Subsequently, as two abortive figures roughed out on the sheet (and a third overleaf on f. 107r) indicate, Newton essayed an approximate 'step' construction of the curve, setting $q = x/a$ to advance from unity by small finite increments, Δq say, and then corresponding to each segment $BP_k = (1+k.\Delta q)\ Y \equiv q_k Y$, $k = 1, 2, 3, \ldots$, successively computing the related length $M_kN_k = \frac{1}{4}q_k^{-1}(q_k^2+1)^2 \approx Y+q_kY(k.\Delta q)$, that is, BP_k very nearly when q_k does not greatly differ from 1, and erecting it ordinate to *CB* so that the inclination of its end-point N_k to the preceding N_{k-1} is that $(\cot^{-1}q_{k-1})$ of GP_{k-1}. It did not apparently occur to him till several years afterwards (see Appendix 2.4) that from

$$MN = y = \tfrac{1}{4}(q^3+2q+q^{-1})\ Y$$

he might, since $dy/dq = \frac{1}{4}(3q^2+2-q^{-2})Y = \frac{1}{4}(3q^2-1)(1+q^{-2})Y$, at once derive

$$BM = z = \int_Y^y q.dy$$
$$= \int_1^q \tfrac{1}{4}(3q^3+2q-q^{-1})\ Y.dq$$
$$= \tfrac{1}{4}(\tfrac{3}{4}q^4+q^2-\log q-\tfrac{7}{4})\ Y$$

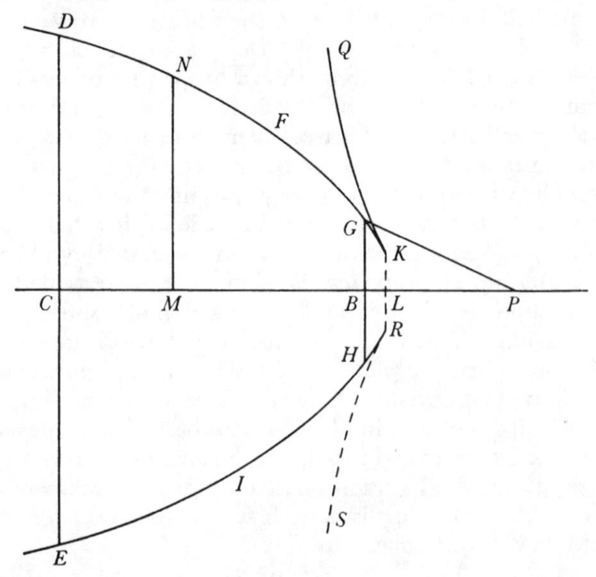

and then parametrically define $N(z, y)$ by allowing q to vary freely. Because both y and z are minimal for $dy/dq = 0$ and so $q = 1/\sqrt{3}$, it readily ensues that the full diametral curve *DNGKQ* (and its mirror-image *EIHRS* in *CB*) defined by Newton's symptom enters parabolically from infinity along $\widehat{DNF}$, continuing through *G* to be cusped at *K* (where its angle of inclination to *CB*, distant $(4/3\sqrt{3})\ Y$ away, is

$$\cot^{-1}(1/\sqrt{3}) = 60°)$$

perpendicular NM be let fall to the axis AB, and the straight line GP be drawn parallel to the tangent to the figure at N, cutting the axis (extended) in P, there shall be MN to GP as GP^3 to $4BP \times GB^2$,(24) then will the solid described by this figure's rotation round the axis AB be resisted less than all others of the same length and breadth.(25)

---

and thereafter passing in the reverse direction to infinity along the concave arc $\widehat{KQ}$. Since by the previous paragraph (see note (22)) the resistance to the surface generated by rotating $\widehat{GK}+KL$ round BL is greater than that to the circumscribing cone frustum formed by similarly rotating $GK'+K'L$, where K' is the meet with KL of the tangent at G (where the curve's slope is 45°), the (locally minimal) surface generated by rotating the portion $\widehat{GK}$ of the diametral curve does not define the solid of (global) least resistance; likewise its concave branch $\widehat{KQ}$, because it completely shields its companion arc $\widehat{DK}$ from any contact with the fluid resisting uniform passage of the solid along its axis CB, may be omitted as irrelevant to any discussion of resistance on the surface determined by the latter's revolution. Accordingly, as Newton accurately restricts it, only the portion $\widehat{GND}$ of the diametral curve can possibly, by its revolution round CB, define a surface of least resistance to uniform motion along its axis. Whether, of all possible diametral curves joining the fixed general point D to B, even $\widehat{DGB}$ defines by its rotation round CB a surface of absolutely minimal resistance is manifestly beyond Newton's present capacity to determine. Neither he nor his several younger contemporaries who a decade and a half later (see note (25)) presented their independent verifications of Newton's defining symptom, nor even Euler (who in his *Methodus inveniendi Lineas Curvas Maximi Minimive Proprietate gaudentes* (see note (14)): 51 [= *Opera Omnia* (1) **24**, 1952: 46–7] consigned Newton's problem to be an unidentified, briefly evaluated 'Exemplum V' to Proposition III of his 'Caput II. De Methodo Maximorum ac Minimorum ad Lineas Curvas inveniendas absoluta') foresaw a disturbing possibility which was adumbrated by Legendre in his ensuing 'Mémoire sur la Manière de distinguer les Maxima des Minima dans le Calcul des Variations' (*Mémoires de l'Académie Royale des Sciences. Année MDCCLXXXVI* (Paris, 1788): 7–37). Namely, if we allow the curve $\widehat{DB}$ to be made up of a number of zigzag lines

$DN_1, N_1N_2, N_2N_3, \ldots, N_{2p-1}N_{2p}, N_{2p}B$

each of the same slope $\pm\theta$ to CB, then the multiply re-entrant surface formed by rotating this round CB will suffer a Newtonian resistance, to uniform motion along that axis, of magnitude $\pi.CD^2\sin^2\theta$—which by decreasing θ (and suitably augmenting the number of zigzags joining D to B) may be made arbitrarily small.

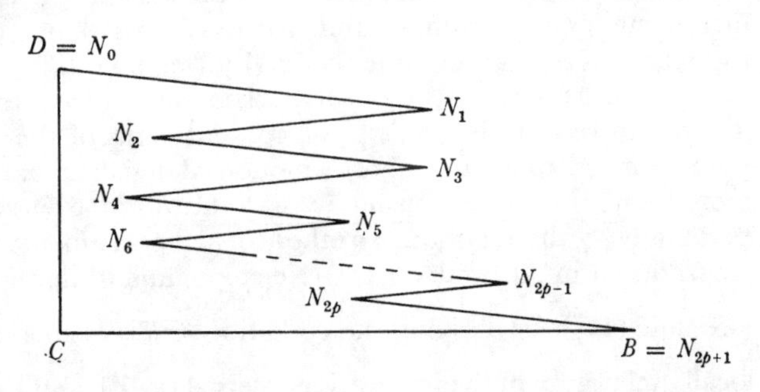

Here, of course, no account is taken of the 'pockets' of total resistance at the individual points N_{2i}, $i = 1, 2, \ldots, p$, which gainsay this anomaly. If, to repair the argument, we add the restriction that at every point N of $\widehat{DB}$ the angle of slope $\theta = \tan^{-1}(a/x)$ shall lie in the interval $[0, \frac{1}{2}\pi]$, then, as Weierstrass first demonstrated in unpublished lectures at Berlin in 1872, it is possible rigorously to prove that of all possible diametral curves $\widehat{DB}$

APPENDIX 1. THE RESISTANCE OF A SPHERE TO 'RAPID' RECTILINEAR MOTION.(1)

From a preliminary draft(2) in the University Library, Cambridge

Prop. [*XXXV.*] *Theor.* [*XXVIII.*]

Invenire Resistentiam corporis sphærici in fluido raro elastico velocissimè progredientis.

Quoniam resistentia eadem est quamproximè ac si partes fluidi viribus nullis se mutuò fugerent, supponamus partes Fluidi ejusmodi viribus destitutas per spatia omnia uniformiter dispergi. Et quoniam actio Medij in corpus eadem est

---

Newton's *figura* $\widehat{DGB}$ defines by its revolution a surface whose resistance is an absolute minimum. (Compare L. A. Pars, *An Introduction to the Calculus of Variations* (London, 1962): 285–92; see also A. R. Forsyth, 'Newton's Problem of the Solid of Least Resistance' [= (ed. W. J. Greenstreet) *Isaac Newton 1642–1727* (London, 1927): 75–86]: 75–6, 82–3.)

(25) The published *Principia* ($_1$1687: 327) concludes a little differently that 'Solidum quod figuræ hujus revolutione circa axem *AB* facta describitur, in Medio raro & Elastico ab *A* versus *B* velocissime movendo, minus resistetur quam aliud quodvis eadem longitudine & latitudine descriptum Solidum circulare' (...the solid which is described by the revolution of this figure made round the axis *AB* shall, on moving very rapidly in a rare and 'elastic' [frictionless] medium from *A* to *B*, be less resisted than any other circular solid described with the same length and breadth).

The immediate reaction of Newton's contemporaries to this scholium on its publication in the 1687 *Principia* was one of near-total incomprehension. Even Leibniz in his annotated copy (see 1, §2: note (49)) could only pen alongside its last paragraph an indecisive observation that 'investigandum est isoclinis facillimè progrediens'. The sole exception was, as we might have expected, Christiaan Huygens: between '22' and '25 Apr. [16]91' (as he dated his notes, now reproduced in his *Œuvres complètes*, **22** (The Hague, 1950): 335–41) he succeeded in accurately reproducing Newton's suppressed analysis both of the cone frustum and—despite an impercipient editorial *avertissement* (*ibid.*: 330) to the contrary—also of the general solid of revolution of least resistance (even down to assigning a and x to be the respective infinitesimal increments of the ordinate and abscissa of its diametral curve), though he was not able rigorously to substantiate the inequality (see note (22)) enunciated by Newton in his second paragraph. Much more typically, when some two years afterwards David Gregory sought likewise to repeat Newton's prior computations of the cone frustum and general surface of minimum resistance to uniform motion along their axis, he failed to begin to do so. Some months later, however, during a visit to Cambridge early in May 1694, he was able to appeal personally to the scholium's author for enlightenment: in his ensuing checking calculations (reproduced in Appendix 2.1/2) Newton came to develop a refinement of his initial method (see note (14)) for deriving the condition $\partial F/\partial x$ = constant for $\int_Y^y F(y, x, a)$ to be everywhere locally minimum in the integration interval $[y, Y]$, and this, after he passed it on to Gregory on the following 14 July (compare the draft reproduced in Appendix 2.3), was subsequently incorporated by the latter in his widely circulated manuscript compendium on 'Newtoni Methodus Fluxionum' and ultimately published by Andrew Motte in appendix to his 1729 English translation of the *Principia*. Meanwhile all was quiet for five years till in the late spring of 1699 Fatio de Duillier once more brought the general Newtonian problem to public notice in his *Investigatio Geometrica Solidi Rotundi, in quod Minima fiat Resistentia* (appended to his

(per Legum corol [5]) sive corpus in Medio quiescente moveatur sive Medij particulæ eadem cum velocitate impingant in corpus quiescens: consideremus corpus tanquam quiescens et videamus quo impetu urgebitur idem a Medio

---

Lineæ Brevissimi Descensus Investigatio Geometrica Duplex (London, 1699), wherein he constructed the defining differential equation of the cycloidal brachistochrone by an analogous curvature technique; see D. T. Whiteside, 'Patterns of Mathematical Thought in the later Seventeenth Century' (*Archive for History of Exact Sciences*, **1**, 1961: 179–388): 381). On recasting his argument into the terms of note (14), Fatio was there able to derive the second-order differential equation of the diametral curve to be $y.dx = ax(a^2+x^2)/(3x^2-a^2)$ (where, as before, $a = dy$ and $x = dz$) and then integrate it, in the form $a/y = (4x/(a^2+x^2)-1/x).dx$, to derive $\log y = 2\log(a^2+x^2)-\log x-\log a^3$, thereafter—probably in ignorance of Newton's earlier attainment of this parametric representation (see Appendix 2.4)—going on to deduce that $y = \frac{1}{4}(q^2+1)^2Y/q$ and thence $z = \frac{1}{4}(\frac{3}{4}q^4+q^2-\frac{7}{4})Y-u$, where $q = (x/a$ or$)\ dz/dy$ and 'Quantitas $u\left[\text{namely, }\int_1^q p.dq = \frac{1}{4}Y\log q\right]$ est Integralis Quantitas, orta ex Termino Analytico $[p = \frac{1}{4}Y/q]$'. Stung not least by his accompanying remark in his monograph—copies of which Fatio took care to send to the leading Continental mathematicians—that 'Newtonum...primum, ac pluribus Annis vetustissimum, ...Calculi Inventorem, ipsa rerum evidentia coactus, agnosco: a quo utrum quicquam mutuatus sit Leibnitius, secundus ejus Inventor, malo eorum...sit Judicium, quibus visæ fuerint Newtoni Litteræ, aliique ejusdem Manuscripti Codices. Neque modestioris Newtoni Silentium, aut prona Leibnitii Sedulitas, Inventionem hujus Calculi sibi passim tribuentis, ullis imponet, qui ea pertractarint, quæ ipse evolvi, Instrumenta', Leibniz at once swung into a counter-attack, first in a flurry of private correspondence with Johann Bernoulli, and then more publicly. As an interim measure L'Hospital, who had already on 20 June read to the Paris Académie an innocent paper detailing his 'Methode Facile pour trouver un Solide Rond, qui...rencontre moins de Résistance que tout antre Solide...' (subsequently published in its *Memoires...Année MDCXCIX* (Paris, 1702): 107–12), was persuaded to let its Latin version appear at once in the *Acta Eruditorum* (August 1699): 354–9. Leibniz himself, after further correspondence with Bernoulli, delivered a stinging review of Fatio's pamphlet in the *Acta* (November 1699): 510–13 in which, *inter alia*, he asserted that 'Dn. *Hugenium* id problema [solidi rotundi cui minimum resistatur] A[nnis] 1690, 1691 frustra tentasse'; and thereto appended (*ibid.*: 513–16) a long, edited excerpt from Bernoulli's letter to him of 7 August preceding (whose original was first published in *G. G. Leibnitii et Johan. Bernoulli Commercium Philosophicum et Mathematicum*, **1** (Lausanne/Geneva, 1745): 463–70) imparting a simplified solution of the problem which 'charta & calamo destitutus, & in lecto decumbens, solius imaginationis ope plenarie...solvi'. Six months later he published, this time under his name, a further long 'Responsio ed Dn. Nic. Fatii Duillerii Imputationes' (*Acta* (May 1700): 198–200), yet again with an *addendum* by Bernoulli presenting a lightly revised version of his method of solution. In England, though John Craige—without even mentioning Fatio's name—sent an open letter to Sloane on 21 December 1700 (printed in *Philosophical Transactions*, **22**, No. 268 [January 1700/1]: 746–51) including his own minimally variant derivation (*ibid.*: 747–9) of the defining equation of the solid's diametral curve, Fatio found no support: whatever his thoughts, Newton himself kept a tactful silence appropriate to his lordly status as Master of the Mint. On the Continent, to Leibniz' aspersions and Bernoulli's technical criticisms Fatio was permitted only a summary response (*Acta Eruditorum* (March 1701): 134–6), in which he yet managed neatly to turn the tables on Bernoulli—who in the *Acta* (May 1697): 208–9 had produced a construction of the brachistochrone problem on the same Fermatian principle of least time—in a brief 'Solidi rotundi minime resistentis investigatio ex Fermatii doctrina refractionum', taking $dt = a^3y/(a^2+x^2)$ to be the element of 'time'

movente. [(3)]Incidant igitur particulæ Medij data cum velocitate in corpus ABK secundum rectas ipsi CA parallelas: sitq; BF ejusmodi recta. In ea capiatur BL semidiametro æqualis et ducatur BD quæ sphæram tangat in B. In $[K]C$ et BD demittantur perpendiculares BE, DL, et vis qua particula Medij secundum rectam FB oblique incidendo globum feriet erit ad vim qua particula secundum rectam $[BC]$ per-

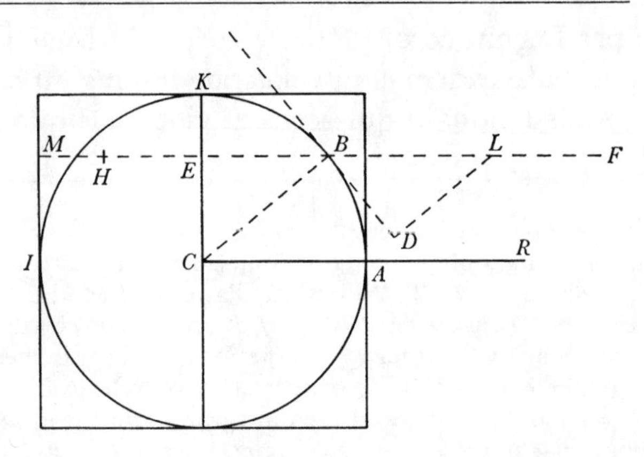

in which a light-corpuscle traverses the infinitesimal arc-length $\sqrt{[a^2+x^2]}$, and thence deriving as his criterion for the total time $t = \int_Y^y a^3y/(a^2+x^2)$ to be a minimum the condition that the instantaneous 'speed' $\sqrt{[a^2+x^2]}/dt$ shall everywhere be proportional to the 'sine' $x/\sqrt{[a^2+x^2]}$ of refraction. Even though many years later Nicolas felt impelled to publish a letter to his brother Jean Christophe 'qua vendicat Solutionem suam Problematis de Inveniendo Solido Rotundo... in quod Minima fiat Resistentia' (*Philosophical Transactions*, **28**, 1713: §XIX: 172–6), the controversy had ceased to spawn any fruitful offspring by the time Charles Hayes came to 'Investigate the Nature of the Solid of Least Resistance' in his *Treatise of Fluxions: or an Introduction to Mathematical Philosophy* (London, 1704): Section V, Proposition XVII: 146–50, there distilling 'for those that would know how to apply Mathematicks to Nature' the essence of the 'Methods of investigating the same' contrived by 'the Incomparable Analysts, the Noble Marquess *de l'Hospital*; M. *Jo. Bernoulli*; M. *Craig*, and M. *Fatio*'.

(1) This unpublished initial draft of what in the second book of the *Principia*'s *editio princeps* was, with but minor revisions, to become its Proposition XXXV is here adjoined to illuminate the context in which, departing from his preferred hypothesis of the resistance of an inclined plane surface to uniform rectilinear motion (see note (5) below), Newton came to evaluate the relative resistance to such motion along their central axis of given surfaces of revolution (here instanced as the vertical circular front surface of a cylinder and the comparable fore hemispherical surface of a globe of equal radius/cross-section, though his argument is readily generalisable), and thereafter (as we have seen in § 1) to seek to define surfaces of revolution—namely, a cone frustum and a general 'rotund solid'—which shall, of all of a given type, suffer a minimum resistance.

(2) Add. 3965.10: 134r; a revise on f. 106r (compare §1: note (2)) is virtually identical with the version published in the *Principia* ($_1$1687: 324–6). In our reproduction of the manuscript figure we omit a roughly sketched and unlettered auxiliary hyperbola which Newton afterwards incorporated in his original scheme, thereby (as in the published text) enlarging it to embrace the diagram which he initially set separately to illustrate his companion draft (on f. 134v overleaf) of 'Prop. [XXXVI.] Prob. [VIII.] Invenire resistentiam corporis sphærici in fluido elastico velocissimè progredientis'.

(3) Newton has deleted a preliminary passage in which he initially continued: 'Designet igitur ABK corpus sphæricum centro C intervallo CA descriptum et incidant particulæ ...'. He later changed his mind, replacing this omitted opening phrase in the published text (*Principia*, $_1$1687: 325).

pendiculariter incidendo globum feriret, ut *LD* ad *LB* vel *EB* ad *BC*. Rursus effectus[4] hujus vis ad movendum globum secundum plagam *AC* vel *FB* ut *BC* ad *BE*[:] et conjunctis rationibus, effectus particulæ in globum secundum rectam *FB* incidentis ad movendum eundem secundum plagam incidentiæ suæ est ad effectum particulæ in globum secundum rectam *RA* incidentis ad ipsum movendum in plagam eandem, ut *BE* quadratum ad *BC* quadratum.[5] Igitur ad planū circuli maximi *KCI* erigatur perpendiculum *EHM*, et sit *EM* æqualis radio *CA* et *MH* æqualis $\frac{CE^{\text{quad}}}{CB}$:[6] et erit *EH* ad *EM* ut effectus particulæ incidentis secundum rectam *FB* ad effectum particulæ incidentis secundum rectam *RA*, et propterea solidum quod a rectis omnibus *EH* occupatur est ad solidum quod a rectis omnibus *EM* occupatur ut effectus particularum omnium incidentium in sphæram totam ad effectum particularum totidem perpendiculariter incidentium in punctum *A*. Sed solidum prius est parabolois[7] et solidum posterius est cylindrus circumscriptus. Et notum est quod cylindrus sit Paraboloide inscripto duplo major.[8] Quare vis tota particularum incidentium in globum duplo minor est quam foret vis tota particularum totidem perpendiculariter incidentium in punctum *A*. Proindeq; si cylindrus constitueretur cujus diameter æqualis esset diametro globi & particulæ fluidi incidant perpendiculariter in basem cylindri quiescentis, vis tota fluidi in globum duplo minor foret quàm vis tota fluidi in cylindrum. Et propterea si particulæ fluidi quiescerent & cylindrus ac globus æquali cum velocitate moverentur[,] foret resistentia cylindri duplo major quàm resistentia globi. Hæc ita se habent in Medio raro.

---

(4) Altered to be 'efficacia' in the published text.

(5) That is, as $\sin^2\theta$ on taking $\widehat{LBD}$ $(= \widehat{BCE}) = \theta = \tan^{-1}[d(CE)/d(EB)]$ to be the angle at which $\widehat{KA}$ is at *B* inclined to the direction of motion along the axis. It follows that the resistance to the surface formed by rotating $\widehat{KA}$ round *CA* is proportional to $\int_0^{CK} 2\pi . CE \sin^2\theta . d(CE)$ and this, in the present instance in which $\widehat{KA}$ is a circle quadrant of centre *C* (and so $\tan\theta$ is $-EB/CE$), Newton proceeds straightforwardly to evaluate.

(6) Whence, since $EM = CB$, there is $EH = (CB^2 - CE^2)/CB = BE^2/CB$. Except for a venial slip set right in its concluding list of *errata*, this later equivalent expression was preferred in the printed *Principia*.

(7) Because the defining equation of the locus (H) is $MH = (CE^2/CB$ or$)$ IM^2/CK.

(8) Alternatively, since $BE^2 = (BC^2$ or$)$ $CK^2 - CE^2$ and therefore (by note (5)) the resistance on the hemisphere formed by rotating $\widehat{KBA}$ round the axis of motion *CA* varies as

$$\pi \int_0^{CK} 2CE(1 - CE^2/CK^2) . d(CE) = \pi \Big[CE^2 - \tfrac{1}{2} CE^4/CK^2 \Big]_{CE=0}^{CE=CK} = \tfrac{1}{2}\pi . CK^2,$$

this is manifestly half the comparable resistance on the transverse diametral cross-section generated by similarly rotating *CK*. Newton's elegant equivalent appeal to the standard result

APPENDIX 2. RECOMPUTATION OF SURFACES OF LEAST RESISTANCE (1694).(1)

From originals in the Royal Society, London and the University Library, Cambridge

[1](2) Vis in con[icam] superf[iciem(3) est] ad vim in circ[ulum](3) ut $OC^q - OK^q$ ad OK^q et CL ad CF seu CK^q ad CF^q(4) hoc est ut $OC^q - OK^q$ in CK^q ad OK^q in CF^q, seu $\dfrac{OC^q, CK^q - OK^q, CK^q}{CF^q}$ ad OK^q. Sed vis in circ[ulum] est ut OK^q[,] ergo vis in frustum totum est ut summa

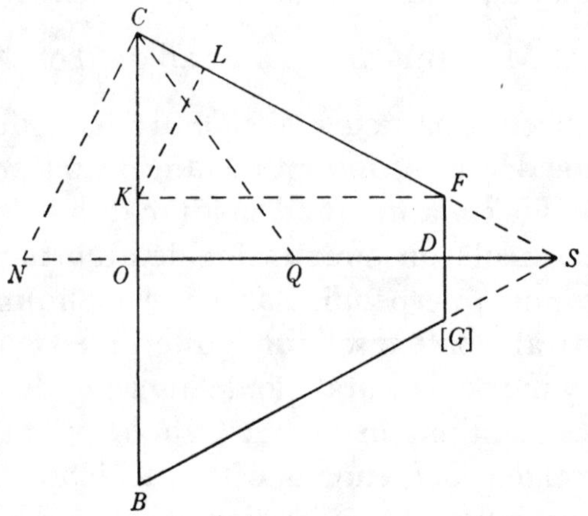

$$\frac{OC^q, CK^q - OK^q, CK^q + OK^q, CF^q}{CF^q}$$

hoc est ut $\dfrac{OC^q, CK^q + OK^q, KF^q}{CF^q}$ hoc est (si sit $2OQ = KF = a$. $OC = b$. $CK = x$. [adeoq3] $CF = \sqrt{aa+xx}$. $OK = b - x$[)] ut $\dfrac{aabb - 2aabx + aaxx + bbxx}{aa + xx} = y$. Unde $aabb - 2aabx + aaxx + bbxx = aay + xxy$ et [capiendo fluxiones](5)

$$-2aab\dot{x} + 2aax\dot{x} + 2bbx\dot{x} = 2x\dot{x}y + aa\dot{y} + xx\dot{y}.$$

(first demonstrated by Archimedes in his *On Conoids and Spheroids*, 20–2) that the volume of the frustum of a paraboloid of revolution is half that of the circumscribing cylinder somewhat obscures the basic simplicity of his deduction.

(1) The impulse which led Newton in the summer of 1694 to recompute the analytical basis of his published construction of the cone frustum and general solid of revolution whose exposed fore-surface offers minimum resistance was, it would strongly appear, an unrecorded plea for enlightenment thereon addressed to him by David Gregory, either in person during his stay in Cambridge in early May of that year or in a lost letter to which Newton replied on 14 July (see his *Correspondence*, **3**, 1961: 382). The paper 'manu propria' (reproduced in [1] following) in which Newton elaborates his prior analysis of the shape of the extremal cone frustum is seemingly the one which he subsequently passed back to Gregory, while his preliminary recomputation (reproduced in [2]) of the latter construction is now (see note (8)) preserved with the draft (partly given in [3]) of his July letter to Gregory wherein he expounded its essence, though slightly modifying its mode of solution: both reworked analyses were quickly incorporated by Gregory in a compendium of problems exemplifying Newton's 'Method of Fluxions' which he compiled in late autumn 1694 for the instruction of his Oxford students (see notes (7) and (29) below). We further adjoin (in [4]) the text of a considerably augmented revise of the parent *Principia* scholium (see §1.2 preceding) which Newton went on to prepare at about this time for future incorporation in its *editio secunda*, but which—like many another of his planned additions in the 1690's—has remained unprinted despite its polished refinements.

hoc est ([ponendo $\dot{y}=0$] ob minimam y) $=2x\dot{x}y$, et

$$-aab+aax+bbx=xy=\frac{aabbx-2aabxx+aax^3+bbx^3}{aa+xx},$$

et[(6)]

$$-a^4b+a^4x+aab\dot{b}x-aabxx+aa\dot{x}^3+bb\dot{x}^3=aab\dot{b}x-2aabxx+aa\dot{x}^3+bb\dot{x}^3,$$

et $-a^4b+a^4x+aabxx=0$. et $xx=-\frac{aa}{b}x+aa$. [sive]

$$x=-\frac{aa}{2b}\pm\sqrt{\frac{a^4+4a^2b^2}{4b^2}}=\frac{-aa+a\sqrt{aa+4bb}}{2b}$$

$$=CK=\frac{OQ}{OC}\times\overline{2CQ-2OQ}=\frac{OQ}{OC}\times 2DS=\frac{OD,DS}{OC}=\frac{OD,NO}{OC}=\frac{KF,NO}{OC},$$

posito scilicet quod sit $NQ=QS=QC$. Est ergo $CK=\frac{KF\times NO}{OC}$ et inde $CK.KF::NO.OC$. Nam triangula CKF, NOC similia sunt. Construetur igitur Problema capiendo $QS=CQ$.[(7)]

---

(2) Royal Society. Gregory MS: 165, first published in *The Correspondence of Isaac Newton*, **3**: 323. In David Gregory's 'Index Chartarum C in folio' it is listed as '[34] De Cono truncato cui minime resistitur: Neutoni manu propria'. An earlier draft (C 48) also exists.

(3) Understand the surfaces formed by rotating CF and FD respectively round the axis OD.

(4) That is, $\sin^2\theta$ where $(\widehat{CFK}=)\ \widehat{CSO}=\theta$ is the angle of slope of CF to OD; compare §1: note (3).

(5) The dotted letters in immediate sequel are Newtonian fluxions; that is, $\dot{x}=dx/dt$ and $\dot{y}=dy/dt$ where t is some independent variable of 'time'.

(6) Here Newton confusingly uses superscript points to cancel equal terms a^2b^2x, a^2x^3 and b^2x^3 on each side of the equation.

(7) It will be evident that this revised analysis of his published construction of the cone frustum of least resistance is effectively identical with Newton's original scheme of computation (see § 1.1) in 1685. We may add that a few months after receiving its text in or shortly after May 1694 (see note (1)) David Gregory inserted it in minimally recast form as Proposition 26 of his student compendium, 'Isaaci Newtoni Methodus Fluxionum; ubi Calculus Differentialis Leibnitij, et Methodus Tangentium Barrovij explicantur, et exemplis plurimis omnis generis illustrantur'. (Gregory's original draft of this unpublished compilation of contemporary calculus problems, now St Andrew's QA33 G8D12, was, as several entries beginning 'Ox[ford] 23/10/1694' reveal, composed over a period of several weeks in the late autumn of 1694; an autograph fair copy in Christ Church, Oxford is undated. Contemporary transcripts by John Keill and William Jones exist in the University Library, Cambridge (Lucasian Papers, Packet No. 13) and in private possession.) It was eventually printed therefrom by Andrew Motte in his *The Mathematical Principles of Natural Philosophy. By Sir Isaac Newton. Translated into English*, **2** (London, 1729): v–vii as the second of the terminal 'Explications, (given by a Friend [Jones?],) of some Propositions in this Book not demonstrated by the Author'.

[2][8]

[Pone] $AH = a$. $BH = FG = b$. $CD = DE = c$. $BD = x$.

$$CH = m = [\sqrt{bb+cc+2cx+xx}].\ HG = n = \sqrt{bb+cc-2cx+xx}.^{(9)}$$

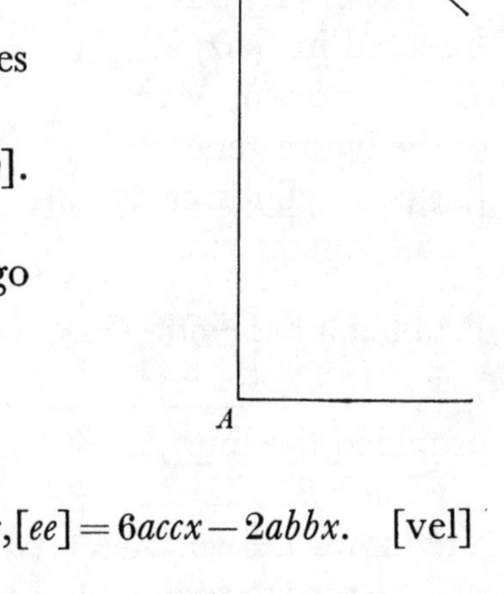

Vis in $CH = \frac{bbc}{mm} \times [\overline{2ab+bb}]$. Vis in $HG = \frac{bbc}{nn} \times \overline{2ab-bb}$.

[adeoq3] Vis in $CH + HG$

$$= \frac{b^3c}{mm} \times \overline{2a+b} + \frac{b^3c}{nn} \times [\overline{2a-b}] = b^3cv.$$

[ubi][10] $\frac{2a+b}{mm} + \frac{2a-b}{nn} = v$. [Igitur capiendo fluxiones prodit] $\frac{\overline{2a+b},\dot{m}}{m^3} + \frac{\overline{2a-b},\dot{n}}{n^3} [= \dot{v}] = 0$ [ob minimam v].

[hoc est][11] $\frac{\overline{2a+b},\overline{c+x}}{m^4} [-] \frac{\overline{2a-b},\overline{c-x}}{n^4} = 0$. [Ergo ponendo] $bb+cc = ee$. [fit][12]

$$\frac{2ac+bc+2ax+bx}{e^4+4eecx} [-] \frac{2ac-2ax-bc+bx}{e^4-4eecx} = 0,$$

[adeoq3 $2bce^4 - 16aeeccx + 4ae^4x = 0$. hoc est] $bc,[ee] = 6accx - 2abbx$. [vel] $x = \frac{\frac{1}{2}bcee}{3acc-abb}$ rectissime.

(8) ULC. Add. 3967.2: $12^r/12^v$, partially printed in *The Correspondence of Isaac Newton*, **3**, 1961: 375–7. These tentative calculations and ensuing verbal draft evidently represent Newton's preliminary reworking in early summer 1694 of his original 1685 analysis (see § 1: note (14)) of the diametral curve of the solid of revolution of least resistance, preparatory to his communicating it to Gregory on 14 July in somewhat modified form. In parallel with his draft of this letter (partly given in [3]) in which Newton afterwards enclosed it, a corner of the manuscript sheet has been completely burnt away: the portion of the text thus obliterated is here restored in square brackets.

(9) Except for a remaining scrap at its bottom left around 'A' Newton's accompanying manuscript figure has (see note (8)) perished: that here reproduced is our restoration.

(10) We omit an initial calculation (deleted by Newton) which proceeds from

$$\text{'}\overline{2a+b} \times nn + \overline{2a-b} \times mm = mmnnv\text{'}$$

similarly but at appreciably greater length to derive the equivalent result

$$\text{'}x = \frac{bc,\ bb+cc}{2a,\ 3cc-bb} \text{ rectissime'}.$$

(11) Since $\dot{m}/\dot{x}$ $(= dm/dx) = (c+x)/m$ and likewise $\dot{n}/\dot{x} = (-c+x)/n$, so that

$$\dot{m}:\dot{n} = (c+x)/m : -(c-x)/n.$$

(12) On neglecting powers of the (second-order) infinitesimal x higher than the first as being irrelevant to the sequel (in which b, c and so $e = \sqrt{[b^2+c^2]}$ are taken to be vanishingly small).

$$[\text{Aliter}]\ \frac{2a+b}{ee+2cx+x^2}+\frac{2a-b}{e^2-2cx+x^2}=v.\ [\text{seu}]\ \frac{4aee+4axx-4bcx}{e^4+2eexx-4ccxx}=v.\ [\text{id est}]^{(13)}$$

$$-4bc+8ax=4eexv-8ccxv=\frac{16ae^4x-16bceexx-32ccaeex}{e^4+2eexx-4ccxx}.$$

[sive] $-4bce^4+8axe^4=16axe^4-32acceex$. [adeoq͡z]

$$\frac{bcee}{8acc-2aee}=x=\frac{bcee}{6acc-2abb}\ \text{rectissime.}$$

(14)*Lem.* [*1.*] If the figure *AECB* rectanguled at *A* & *B* be given & *EC* be bisected in *G* & *GD* be parallel to *AB* & of any length & *ED DC* be drawn, & the figure *AEDC* be rolled about y^e^ axis *AB*, & the figure generated be moved in water(15) from *A* towards *B*: the two cylindrical superficies generated by the lines *ED*, *DC* [shall together feel y^e^ least] resistance when *GD* is to

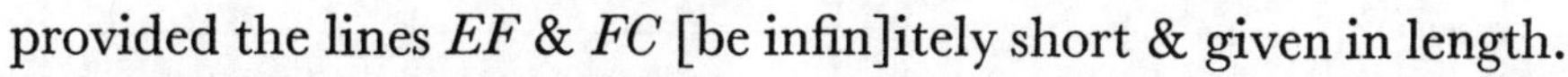

$$\left[\frac{EF\times FC}{8AH}\ \text{as}\ EC^q\ \text{to}\ 3F\right]C^q-EF^q,^{(16)}$$

provided the lines *EF* & *FC* [be infin]itely short & given in length.

[*Lem.*] *2.* If upon *A*[*B*] be erected perpendiculars *A*[*S*], [*B*]*T* [of] given lengths & between y^e^ given points [*S*] & [*T*] be drawn such a curve line [*ST*] that if the figure *A*[*STB*] be revolved about y^e^ axis *A*[*B*] & the figure generated be moved in water from *A* towards [*B*] the resistance of ye surface generated by the curve line [*ST*] shall be the least that can be[,] & if in *A*[*S*] you take any three points *P*, *Q*, *R* at infinitely little equal distances & erect y^e^ perp^s^(17) *PC*, *QD*, *RE* cutting y^e^ curve in *C*, *D* & *E* & draw y^e^ chord *CE* cutting *QD* in *G*, & upon *PC* let

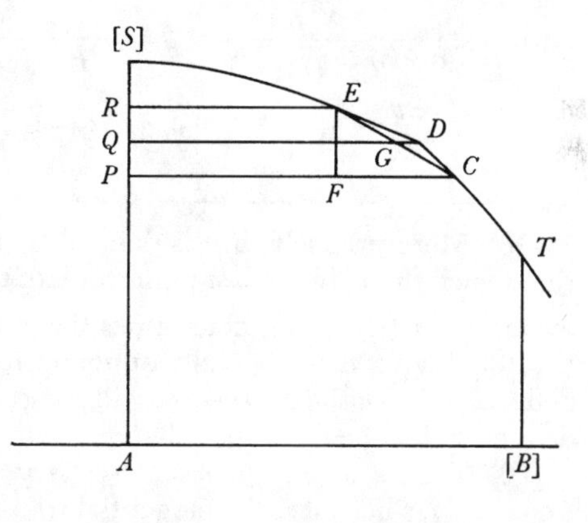

(13) On multiplying through by $e^4+2e^2x^2-4c^2x^2$, taking fluxions and then setting $\dot{v}/\dot{x}=0$.

(14) Newton now elaborates his preceding result in a first, tentative verbal draft.

(15) Or, of course, any similarly resisting fluid.

(16) On setting $GD=x$, $AH=a$, $EH=HF=b$, $HG=\frac{1}{2}FC=c$ and so

$$EG=GC=e=\sqrt{[b^2+c^2]},$$

this proportion at once reduces to the preceding defining symptom $x=\frac{1}{2}bce^2/a(3c^2-b^2)$ from which it is manifestly constructed.

(17) 'perpendiculars'.

fall the perpendicular EF: the sagitta GD shall be to $\frac{EF \times FC}{8AQ}$ as EC^q to $3FC^q - EF^q$. For its plain by the first Lemma that if the sagitta GD be made either longer or shorter without altering the points E & C[,] the resistance of y^e^ surface generated by y^e^ arch EC will become greater.

Lem. 3. If the curve [S]T be such as is described in the book(18) then the sagitta GD will have y^e^ property here described & by consequence the solid generated by the revolution of this curve will feel the least resistance.(19)

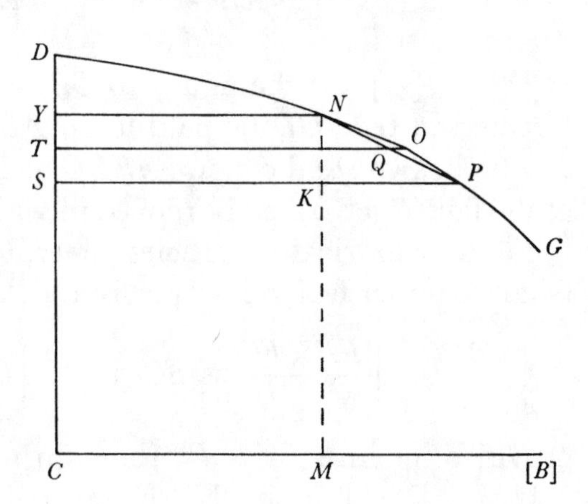

[Per Schol. Prop. XXXV prodit] $MN . NP :: NP^3 . 4KP, NK^2$.(20) et [per hujus Lem. 2]

$$OQ . \frac{NK, NP}{8MN} :: NP^q . \overline{3KP^q - KN^q}.$$

[Per priorem fit ponendo $MN = x$, $CM = y$](21) $\quad x \times 4\dot{y}\dot{x}^2 = \frac{\dot{x}^4 + 2\dot{x}\dot{x}\dot{y}\dot{y} + \dot{y}^4}{\dot{x}}$. [sive] $4x\dot{y}\dot{x}^3 = \dot{x}^4 + 2\dot{x}\dot{x}\dot{y}\dot{y} + \dot{y}^4$. [adeoq; capiendo fluxiones](22)

$$4\dot{x}\dot{y}\dot{x}^3 + 4x\ddot{y}\dot{x}^3 = 4\dot{x}\dot{x}\dot{y}\ddot{y} + 4\ddot{y}\dot{y}^3.$$

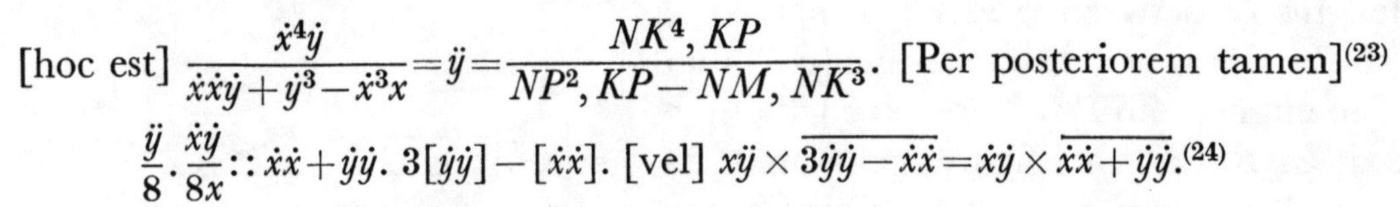

[hoc est] $\frac{\dot{x}^4\dot{y}}{\dot{x}\dot{x}\dot{y} + \dot{y}^3 - \dot{x}^3x} = \ddot{y} = \frac{NK^4, KP}{NP^2, KP - NM, NK^3}$. [Per posteriorem tamen](23)

$$\frac{\ddot{y}}{8} . \frac{\dot{x}\dot{y}}{8x} :: \dot{x}\dot{x} + \dot{y}\dot{y} . 3[\dot{y}\dot{y}] - [\dot{x}\dot{x}]. \text{ [vel] } x\ddot{y} \times \overline{3\dot{y}\dot{y} - \dot{x}\dot{x}} = \dot{x}\dot{y} \times \overline{\dot{x}\dot{x} + \dot{y}\dot{y}}.(24)$$

(18) More precisely, if it is defined by a differential property identical (structurally) with that which there, in the last paragraph of the scholium to Proposition XXXV of the *Principia*'s second book (see § 1.2), determines the generating curve $\widehat{DNG}$.

(19) Having thus hopefully announced the effective identity of these two formally distinct defining relationships, Newton will in sequel seek to substantiate this stated equivalence by deducing the one from the other.

(20) This is an erroneous consectary of the earlier defining symptom in the *Principia.* Accurately, since there the tangential triangle BGP is similar to the present differential triangle KNP, it is '$MN, NK . BG, NP :: NP^3 . 4PK, NK^2$'. Newton's ensuing fluxional 'translation' of this proportion silently incorporates the missing factor NK—or rather its finite proportional $\dot{x}$—but by continuing to omit BG unnecessarily restricts it to be unity.

(21) Because the vanishingly small increments $NK = d(MN)$ and $KP = d(CM)$ are to one another as the respective fluxions $\dot{x}$ and $\dot{y}$ of $CY = MN = x$ and $YN = CM = y$.

(22) And understanding x to be the base variable, so that $\ddot{x} = 0$.

(23) Because, on setting $NK = o$, the incremented value of $CM = y \equiv y_x$ is

$$SP = y_{x+o} = y + o\dot{y}/\dot{x} + \tfrac{1}{2}o^2\ddot{y}/\dot{x}^2 \ldots,$$

whence $QO = TO - (TQ \text{ or}) \tfrac{1}{2}(YN + SP) = y_{x+\frac{1}{2}o} - \tfrac{1}{2}(y + y_{x+o}) = -\tfrac{1}{8}o^2\ddot{y}^2/\dot{x}^2 \ldots$ and therefore $\lim_{0 \to \text{zero}} (QO/NK^2) = -\tfrac{1}{8}\ddot{y}/\dot{x}^2$.

[Finge] $ax\dot{x}\dot{y}+bx\dot{y}\dot{y}+c\dot{x}\dot{x}y+d\dot{x}\dot{y}y=0$. [Erit capiendo fluxiones]

$$a\dot{x}^3\dot{y}+ax\dot{x}\ddot{y}+b\dot{x}\dot{y}^3+3bx\dot{y}\dot{y}\ddot{y}+c\dot{x}^3\dot{y}+d\dot{x}\dot{y}^3+2d\dot{x}\dot{y}\ddot{y}y[=0].$$

[id est] $$\overline{a+c}\,\dot{x}^3\dot{y}+\overline{b+d}\,\dot{x}\dot{y}^3+ax\dot{x}\dot{x}\ddot{y}+3bx\dot{y}\dot{y}\ddot{y}+2d\dot{x}\dot{y}\ddot{y}y=0.$$

[Cape] $d=0$.(25) $b=a+c$. [ut et] $b=1$. $a=-1$. [adeoq;] $c=2$.(26) [fit]

$$x\ddot{y}^3-x\ddot{y}\dot{x}\dot{x}=2y\dot{x}^3.\ ^{(27)}$$

[hoc est] $x.y::2\dot{x}^3.\dot{y}^3-\dot{y}\dot{x}\dot{x}$. [sive] $CT.TO::2NK^3.KP^3-KP,NK^2$. [Sed iterum capiendo fluxiones] $\dot{x}\dot{y}^3+3x\dot{y}\dot{y}\ddot{y}-\dot{x}^3\dot{y}-x\ddot{y}\dot{x}\dot{x}=2\dot{y}\dot{x}^3$.(28)

---

[3](29)

The figure wch feels y^e^ least resistance in y^e^ Schol. of Prop. XXXV. Lib. II is demonstrable by these steps.(30)

---

(24) Which is but a variant on the preceding derivative form

$$\ddot{y}=\dot{x}\dot{y}/(\dot{y}(\dot{x}^2+\dot{y}^2)/\dot{x}^3-x)=\dot{x}\dot{y}(\dot{x}^2+\dot{y}^2)/x(3\dot{y}^2-\dot{x}^2)$$

of the prior defining equation $(\dot{x}^2+\dot{y}^2)^2=4x\dot{x}^3\dot{y}$. Failing to observe this equivalence, Newton proceeds to attempt a direct integration to a first-order fluxional equation of type

$$x\dot{y}(a\dot{x}^2+b\dot{y}^2)+\dot{x}y(c\dot{x}^2+d\dot{y}^2)=0.$$

(25) Whence the preceding canonical equation assumes the form

$$x\ddot{y}(3b\dot{y}^2+a\dot{x}^2)+\dot{x}\dot{y}((a+c)\dot{x}^2+b\dot{y}^2)=0.$$

(26) Unfortunately, these particular values of a, b and c define the fluxional equation $x\ddot{y}(3\dot{y}^2-\dot{x}^2)+\dot{x}\dot{y}(\dot{x}^2+\dot{y}^2)=0$ which, on correcting Newton's following slip in sign, is at once determined to have a first integral $x\dot{y}(-\dot{x}^2+\dot{y}^2)+2y\dot{x}^3=0$.

(27) This should read '$-2y\dot{x}^3$'.

(28) On checking back Newton noticed his preceding numerical slip (see note (27)) and then converted this to be '$-2\dot{y}\dot{x}^3$'. He was doubtless then quickly led to observe that, on reordering its elements, the corrected equation is of the form $x\ddot{y}(3\dot{y}^2)=-\dot{x}\dot{y}(\dot{x}^2+\dot{y}^2)$ and thereafter to confirm that the nearly identical equation $x\ddot{y}(3\dot{y}^2-\dot{x}^2)=+\dot{x}\dot{y}(\dot{x}^2+\dot{y}^2)$ has indeed the considerably different first integral $4x\dot{x}^3\dot{y}-(\dot{x}^2+\dot{y}^2)^2=0$ equivalent to the defining symptom of the solid of least resistance as he had published it in his *Principia*.

(29) Extracted from Newton's roughly written and much cancelled and interlineated draft (now ULC. Add. 3967.2: 10^r^–11^r^) of his letter of 14 July 1694 to David Gregory. At different times part of the second leaf of the original (folded) manuscript sheet has been ripped off and a considerable portion of its page-bottom burnt away, and much of the remaining paper surface is charred or otherwise soiled. (See the photocopy of f. 10^r^ reproduced as Plate VI [facing page 380] in *The Correspondence of Isaac Newton*, **3**, 1961.) The extensive lacunæ thus created in its text were first—and at one or two places none too satisfactorily—filled by John Couch Adams in his *editio princeps* of the draft in appendix to the editorial preface of *A Catalogue of the Portsmouth Collection of Books and Papers written by or belonging to Sir Isaac Newton* (Cambridge, 1888): xxi–xiii, and his scheme of reparation has been more or less exactly followed by those—W. W. Rouse Ball (*An Essay on Newton's 'Principia'*, London, 1893: 101–3), F. Cajori (*Sir Isaac Newton's Mathematical Principles of Natural Philosophy...*, Berkeley/Los Angeles, 1934: 657–9) and H. W. Turnbull (*Correspondence of Isaac Newton*, **3**: 380–2)—who have since reproduced it.

1. If upon BM be erected infinitely narrow parallelograms $BGhb$ & $MNom$[31] & their distance Mb & altitudes MN, BG be given & the semisum of their bases $\frac{Mm+Bb}{2}$ be also given & called s & their semidifference $\frac{Mm-Bb}{2}$ be called x[,] and if the lines BG, bh, MN, mo butt upon y^e^ curve $nNgG$ in y^e^ points

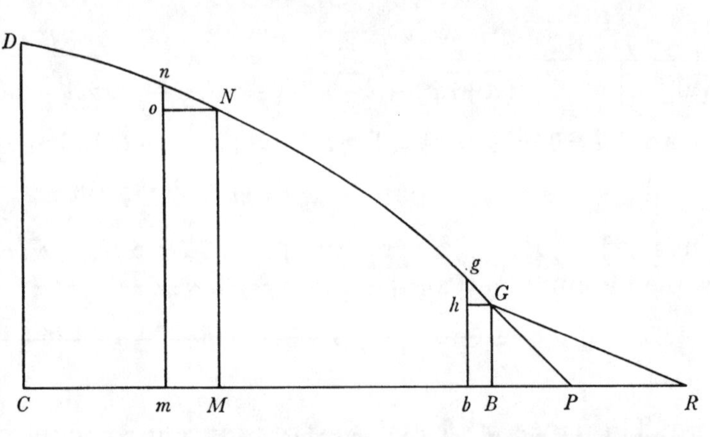

n, N, g & G, & the infinitely little lines on & hg be equal to one another & called c, & the figure $mnNgGB$ be turned about its axis BM to generate a solid, & this solid move uniformly in water[32] from M to B according to y^e^ direction of its axis BM: the summ of the resistances of the two surfaces generated by the infinitely little lines Gg Nn shall be least when gG^{qq} is to nN^{qq} as $BG \times Bb$ to $MN \times Mm$.[33]

(Cajori did, however, support the criticism made by Oskar Bolza in a footnote to his 'Bemerkungen zu Newtons Beweis seines Satzes über den Rotationskörper kleinsten Widerstandes' (*Bibliotheca Mathematica*, $_3$**13**, 1912–13: 146–9, especially 147) that Adams' restoration of a key phrase in Newton's argument was meaningless and needed to be replaced.) Our present departures from it are, we may hope, for the better. While the letter (now seemingly lost) which Newton ultimately sent to Gregory no doubt presented a verbally more finished form of the present draft argument, the divergences could not have been substantial; for the version of it which Gregory a few months later entered into his manuscript compendium 'Isaaci Newtoni Methodus Fluxionum...' (see note (7)) as its 'Prop. 42. Problema' (and which was long afterwards in 1729 published therefrom by Andrew Motte as the last of the editorial 'Explications' appended by him to his English translation of *The Mathematical Principles of Natural Philosophy. By Sir Isaac Newton*, **2**: vii–viii) does not essentially differ in its basis from that here reproduced.

(30) Orginally 'upon these principles'. As in [2] preceding, Newton initially listed his three following 'steps' under the heads '*Lem. 1*', *Lem. 2*' and—or so we presume since the text here is lacking—'*Lem. 3*'.

(31) Notice that these are now removed from one another in distance. (In an unlettered preliminary figure at this point Newton still drew them as joined, MN coinciding with bg, so that his decision to separate them was made at the last minute.) While there is no mathematical gain in introducing this refinement, it serves the heuristic purpose of allowing the unsophiscated reader—and Gregory in particular—more readily to accept the accuracy of Newton's infinitesimal approximations, according to which no distinction need be made between BG and bg or MN and mn, but the second-order difference $No-Gh$ is of crucial significance.

(32) Or some general fluid of uniform density; compare note (15).

(33) Newton first concluded '...when $MN \times Gg^{qq}$ is to $BG \times nN^{qq}$ as Bb to Mm. Or gG^{qq} is to nN^{qq} as $BGhb$ to $MNom$'.

For the resistances of the surfaces generated by the revolution of Gg & Nn are as $\frac{BG}{Gg^{\text{quad}}}$ & $\frac{MN}{Nn^{\text{quad}}}$, that is, if Gg^{quad} & Nn^{quad} be called p & q, as $\frac{BG}{p}$ & $\frac{MN}{q}$[,] & their summ $\frac{BG}{p}+\frac{MN}{q}$ is least when the fluxion thereof $-\frac{BG\times\dot{p}}{pp}-\frac{MN\times\dot{q}}{qq}$ is nothing, or $-\frac{BG\times\dot{p}}{pp}=+\frac{MN\times\dot{q}}{qq}$. Now

$$p = Gg^{\text{quad}} = Bb^{\text{quad}}+gh^{\text{quad}} = ss-2sx+xx+cc$$

& therefore $\dot{p} = -2s\dot{x}+2x\dot{x}$ & by ye same argument(34) $\dot{q}=2s\dot{x}+2x[\dot{x}]$ & therefore $\frac{BG\times\overline{2s\dot{x}-2x\dot{x}}}{pp}=\frac{MN\times\overline{2s\dot{x}+2x\dot{x}}}{qq}$, or $\frac{BG\times\overline{s-x}}{pp}=\frac{MN\times\overline{s+x}}{qq}$ & thence pp is to qq as $BG\times\overline{s-x}$ is to $MN\times\overline{s+x}$[,] that is gG^{qq} to nN^{qq} as $BG\times B[b$ to$]$ $MN\times Mm$.(35)

2. If the curve line $DnNgG$ be such that the surface of the solid generated by [its] revolution feels ye least resistance of any solid with the same top & bottom BG & CD,(36) then the resistance of the two narrow annular surfaces generated by ye revolution of the [arches Nn &] $G[g$ is] less then if the intermediate solid $bgNM$ be removed [along BM a little from or] to BG & by consequence it is the least that can be & therefore gG^{qq} is to nN^{qq} as $BG\times Bb$ [to $MN\times Mm$.]

[3. If GR be drawn parallel to the tangent at N &] gh be equal to hG so that ye angle [BGg is 135degr, & thus $gG^q = 2Bb^q$, t]hen will [$4B]b^{qq}$ be [to Nn^{qq} as $BG\times Bb$ to] $MN\times Mm$,(37) & by consequence $4BG^{qq}$ [to GR^{qq} as $BG\times BP$ to $MN\times BR$ & therefore] $4BG^q\times BR$ to GR^{cub} [as GR to MN.](38)

---

[4](39)

[... si solidum $ADBE$(40) convolutione figuræ Ellipticæ vel Ovalis $ADBE$ circa axem AB facta generetur, & tangatur figura generans a rectis tribus FG,

---

(34) Namely, since $q = s^2+2sx+x^2+c^2$.

(35) Originally 'as $BGhb$ [to] $MNom$'; compare note (33).

(36) Newton had first begun to write 'between ye given circles BG & [CD]'.

(37) A cancelled preliminary conclusion here continues: 'And by reduction $4Bb^{\text{cub}}\times Mm$ to nN^{qq} as BG to MN. Whence the Proposition to be demonstrated easily follows.'

(38) Some further indication of the considerable difficulty Newton experienced in seeking to repeat—or, more accurately perhaps, rigorously justify—for Gregory his initial derivation of the defining symptom of the solid of revolution of least resistance may be glimpsed in his apologetic concluding remark in sequel that 'I had answered yor letter [of *c.* late May, now lost] sooner but that I wanted time to examin this Theoreme'.

(39) ULC. Add. 3965.12: 207r, a lightly corrected but otherwise carefully written out intended addition 'Ad *Philos. Nat. Princip. Math.* [11687] p. 327 lin. 3', that is, to the scholium (to Proposition XXXV of its second book) whose preliminary draft is reproduced in § 1.2. Nothing is known of the circumstances in which this *addendum*—never in fact incorporated in

GH, *HI* in punctis *F*, *B* & *I* ea lege ut *GH* sit perpendicularis ad axem in puncto contactus *B*, & *FG*, *HI* cum eadem *GH* contineant angulos *FGB*, *BHI* graduum 135: solidum quod convolutione figuræ *ADFGHIE* circa axem eandem *CB* generatur, minus resistitur quam solidum prius; si modo utrumq; secundum plagam axis sui *AB* progrediatur, &] utriusq; terminus *B* præcedat.(41) Et si manentibus longitudine *AB* et latitudine *DE* curva *DNG* sit arcus Ellipseos centro *C* axe *DE* descriptæ & angulus *FGB* quem hic arcus cum recta *BG* continet sit 135gr.: solidum quod convolutione figuræ *ADNFGB* circa axem eundem *AB* facta generatur resistentiam adhuc minorem sentiet.(42) Quas quidem Propositiones in construendis navibus non inutiles futuras esse censeo.

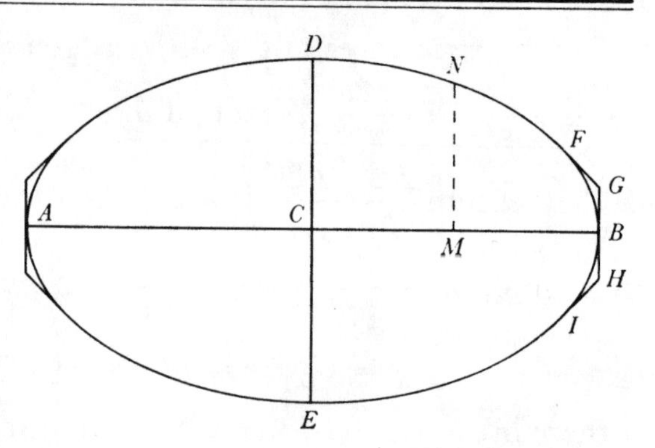

---

any subsequent edition of the *Principia*—was composed, but the internal evidence of its handwriting style suggests that it was penned in the middle 1690's. If this be so, we may conjecture that it was David Gregory—rather than, say, Fatio de Duillier (see note (45) below)—who provided the external stimulus which goaded Newton into preparing this sophisticated enunciation and construction of the defining properties of 'oval' surfaces of minimum resistance to uniform translation along their axis. It will be seen that Newton's text begins abruptly in mid-sentence ('—utriusq; terminus *B*...') at the point where it starts to diverge from the 1687 printed version: to render it independently intelligible we have here introduced (within square brackets) an appropriate opening passage therefrom.

(40) The manuscript lacks an accompanying figure: that reproduced is the variant of the 1687 published diagram which ensues on obeying Newton's present marginal instruction 'In the Book blot out the lines *GP*, *HP*, *GR* [the extension of *FG* to its meet with *BP*]'.

(41) See §1: note (22).

(42) This must surely be merely an educated guess on Newton's part, one founded on the 'evident' truth that the surface formed by rotating the (new) $\widehat{DNG}+GB$ round *CB* more nearly approximates the ideal surface of revolution of minimal resistance constructed in sequel. Accurate analytical justification of the assertion—and determination of the limits within which it is valid if it be not universally true—is a formidably complicated operation to define, let alone implement, and we may be certain that, if Newton made the attempt, he rapidly abandoned it. If in standard Cartesian terms we take the defining equation of the original ellipse quadrant $\widehat{DNB}$ of axes $AB = q$ and $DE = \sqrt{[qr]}$ to be $y^2 = rx-(r/q)\,x^2$ where $BM = x$ and $MN = y$ are perpendicular coordinates of its general point $N(x, y)$, then the condition that the slope $\tan^{-1}[dy/dx] = \tan^{-1}[(r/q)\,(\frac{1}{2}q-x)/y]$ at $F(X, Y)$ be 45° determines $X = \frac{1}{2}q(1-\sqrt{[q/(q+r)]})$ and $Y = \frac{1}{2}r\sqrt{[q/(q+r)]}$; whence the resistance on the arc $\widehat{DF}$ rotated round *CB* is

$$\int_{Y}^{\frac{1}{2}\sqrt{[qr]}} 2\pi y(dy/ds)^2.dy = \pi\int_{Y^2}^{\frac{1}{4}qr} (\tfrac{1}{4}qr^2-ry^2)/(\tfrac{1}{4}qr^2+(q-r)\,y^2).d(y^2)$$
$$= \pi(-\tfrac{1}{4}q^2r^2/(q^2-r^2)+\tfrac{1}{4}q^2r^2(q-r)^{-2}\log\tfrac{1}{2}(q/r+1)),\quad q \neq r,$$

Quod si arcus DNG ejusmodi sit ut ab ejus puncto quovis N ad rectam BG productam demittatur perpendiculum NS et ducatur etiam NT quæ curvam tangat in N rectæq; BS occurrat in T, sit BS ad BG ut NT^{qq} ad $4NS \times ST^{\text{cub}}$;(43) solidum quod figuræ hujus revolutione circa axem BC facta describitur, in Medio raro et elastico ab A versus B progrediendo minus resistetur quam aliud quodvis eadem longitudine BC & latitudine $2CD$ descriptum solidum circulare. Determinatur vero solidum illud in hunc modum.

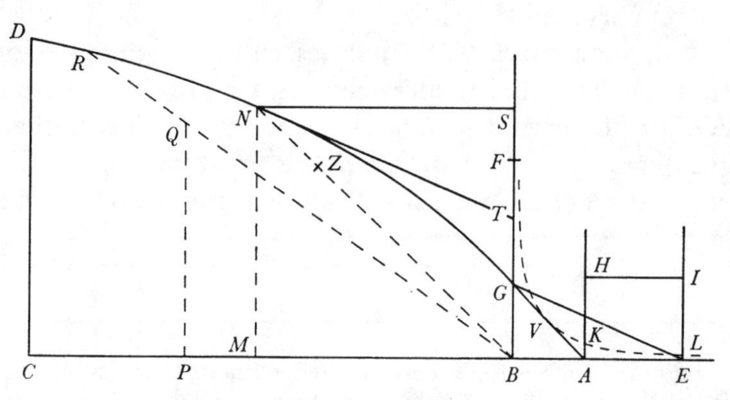

A puncto dato G ducatur recta quævis GE rectæ CB productæ occurrens in E, et in recta BG producta capiatur tum BS quæ sit ad GE ut GE^{cub} ad $4BE \times BG^q$, tum etiam GF quæ sit ad $BG+BE$ ut $5BG^q+BE^q$ ad $16BG^q$. In BE capiatur BA æqualis BG, erigatur perpendiculum AH æquale FS,(44) compleatur rectangulum $HAEI$ et centro B asymptotis BE BS describatur Hyperbola VKL quæ actam rectam AG tangat in medio ejus puncto V et rectis AH, EI occurrat in K et L. Deniq; ad BS erigatur perpendiculum SN æquale summæ longitudinis GS et longitudinis quam area $HKLI$ ad longitudinem BG applicata generat, et punctum N locabitur in curva GND cujus revolutione circa axem BC facta

---

while that on the cone frustum formed by rotating $FG+GB$ is

$$\pi.(\tfrac{1}{2}(Y^2-(Y-X)^2)+(Y-X)^2) = \pi.\tfrac{1}{2}(Y^2+(Y-X)^2)$$
$$= \pi.\tfrac{1}{2}(\tfrac{1}{4}qr^2/(q+r)+\tfrac{1}{2}q^2+\tfrac{1}{4}qr-\tfrac{1}{2}q\sqrt{[q(q+r)]}).$$

If the ellipse $\widehat{DN'G'}$ of major axis $q' = q+2X'$ and the same minor axis DE $(= \sqrt{[qr]}) = \sqrt{[q'r']}$ is to meet BG in the point $G'(X', Y')$ at which its slope to CB is 45°, then similarly

$$X' = \tfrac{1}{2}q'(1-\sqrt{[q'/(q'+r')]}) \quad \text{and} \quad Y' = \tfrac{1}{2}r'\sqrt{[q'/(q'+r')]};$$

from which it follows that $q = q'\sqrt{[q'/(q'+r')]}$ and $r = r'\sqrt{[(q'+r')/q']}$, and therefore $q'^2 = q(\sqrt{[\tfrac{1}{4}q^2+qr]}+\tfrac{1}{2}q)$, $r'^2 = r(\sqrt{[\tfrac{1}{4}q^2+qr]}-\tfrac{1}{2}q)$. We leave the reader to check whether the resistance $\pi\int_{Y'^2}^{\frac{1}{4}qr}(dy/ds)^2.d(y^2)+\pi Y'^2$ on the surface formed by rotating $\widehat{DG'}+G'B$ round CB is in fact less than the total preceding (on the surface generated by rotating $\widehat{DF}+FG+GB$).

(43) That is, since $BS = MN$ and GE is drawn parallel to NT in the figure (see note (46) below), so that the triangles STN and BGE are similar, 'MN ad BG ut GE^{qq} ad $4BE \times BG^{\text{cub}}$' effectively as in the printed scholium (compare §1: note (24)).

(44) This should be '$BG+FS$'; see note (45) following.

generabitur solidum quæsitum. Recta autem NT quæ rectæ GE parallela est tanget curvam illam in N.(45)

Postquam curva GN semel descripta est, potest solidum data quavis longitudine et latitudine sic determinari. Detur solidi longitudo BP et latitudo $2PQ$. Agatur BQ curvæ descriptæ occurrens in R. Ducatur quævis BN et in ea capiatur BZ quæ sit ad BN ut BQ ad BR et punctum Z incidet in curvam novam cujus revolutione circa axem BM(46) generabitur solidum quæsitum.(47)

---

(45) That is, Newton's construction determines the point N of the diametral curve at which it has the given slope $BE/BG = q$. On setting $BG = Y$, it follows at once from the curve's defining differential equation that $MN = BS = \frac{1}{4}Y(q^2+1)^2/q$, and from this (see §1: note (24)) there ensues $BM = SN = \frac{1}{4}Y(\frac{3}{4}q^4+q^2-\frac{7}{4}-\log q)$. Since $BE = Yq$ and hence

$$GE = Y\sqrt{[q^2+1]},$$

the first of these coordinates is readily constructible by making $BS = \frac{1}{4}GE^4/BE \times BG^2$ as Newton requires, but fabricating the latter requires a little more ingenuity. His auxiliary rectangular hyperbola VKL evidently defines the area

$$(AKLE) = \tfrac{1}{4}BG^2 \log(BE/BA) = \tfrac{1}{4}Y^2 \log q,$$

so that

$$SN+(AKLE)/BG+BG = \tfrac{1}{4}Y(\tfrac{3}{4}q^4+q^2+\tfrac{9}{4}) = \tfrac{1}{4}Y((q^2+1)^2-\tfrac{1}{4}(q^2-1)\,(q^2+5)),$$

that is, $q.BS-(q-1).GF = BS+(q-1).AH$ upon setting

$$GF = \tfrac{1}{16}Y(q+1)\,(q^2+5) = \tfrac{1}{16}(BE+BG)\,(BE^2+5BG^2)/BG^2$$

and $AH = BS-GF = BG+FS$; whence $SN = GS+(AE \times AH-(AKLE))/BF$. If we are accurate in assigning the data of the present paper to the middle 1690's, the general parametric definition of the curve here implicit was attained by Newton independently of—and several years earlier than—the equivalent parametrisation published by Fatio de Duillier in 1699 in his *Investigatio Geometrica Solidi Rotundi in quod Minima fiat Resistentia* (on which see §1: note (25)).

(46) 'facta' is deleted in sequel.

(47) For since BM/BG and MN/BG are functions solely of the slope $d(BM)/d(MN) = q$ at N, the family of curves $(\widehat{DNG})$ generated as BG varies in length will be in homothety from the pole B and hence similar one to another.

§2. THE APPROXIMATE DETERMINATION OF A PARABOLIC COMETARY PATH.(1)

[late summer? 1685]

(1) We have already remarked (see v: 524, note (1)) that Newton came only very slowly to agree with Flamsteed's notion that a comet, like the solar planets, is 'made to fetch a compass about the sun' under some power of attraction towards it, and in particular that the comets observed during the winter of 1680–1, the one in ingress between 11 and 27 November 1680, the other in regress between 12 December 1680 and 9 March 1681 (see *Principia*, $_1$1687: 490–8; and compare *The Correspondence of Isaac Newton*, **2**: 354, 357), were one and the same body circling the sun in some highly elongated curved orbit. After his researches in the late summer and autumn of 1684 into the various species of inverse-square orbit and his concomitant rejection of all resisting vortices and ethereal fluids which might be conceived to fill interplanetary space, it became clear to him that, if Flamsteed's hypothesis were correct, then —external perturbations apart—all cometary paths must be near-parabolic conics traced (in different planes) round the sun at their common focus. Since the orthogonal projection of such a conic orbit upon the ecliptic is an equivalent focal conic, to test the empirical accuracy of this hypothesis it remained to show that the longitudinal terrestrial sightings of the 1680–1 comet (and of other earlier ones less accurately observed) were in accord with its trajectory being, to high approximation, a parabola. His first notion, outlined in the scholium to Problem 4 of his initial tract 'De motu Corporum' (see 1, §1: note (79)), of seeking to adapt the simple Wrennian model of a uniformly traversed rectilinear path to determine the general focal conic cometary orbit quickly proved to be hopelessly impractical (compare 1, §1. Appendix 2) principally because every comet, in the portion of its orbit in the vicinity of the sun which is alone visible to a terrestrial observer, markedly alters its speed over even a small interval of time. Although by the late summer of 1685 he had, as he affirmed to John Flamsteed on 19 September, 'not yet computed ye orbit of a comet', he was 'now going about it: & taking that of 1680 into fresh consideration, it seems very probable that those of November & December were ye same' (*Correspondence*, **2**: 419). The argument by which Newton had thus hesitantly convinced himself that these two observed trajectories were indeed traversed in succession by one and the same comet was no doubt the 'rough computation' to this effect (see note (10) below) which he had some while before adduced in his 'De motu Corporum Liber secundus', but later suppressed when he came afterwards to subsume this preliminary text into the published *Principia*'s third book. From this same manuscript, initially transcribed by his amanuensis Humphrey Newton but much corrected by Newton himself, we here reproduce a following scheme of 'calculation of ye orbit' which, he went on to inform Flamsteed in his letter, is dependent for its success 'only on three observations & if I can get three at convenient distances exact to a minute or less I hope ye orbit [of the 1680 comet] will answer exactly enough not only to ye observations of December January February & March but also to those of November before ye Comet was conjoyned wth ye sun' (*ibid.*: 420). Though he wrote back a month later on 14 October sending Flamsteed his 'hearty thanks' for at once communicating a corrected 'tablet' (now lost, but see v: 525, note (3)) of his Greenwich sightings, which 'being so exact...will save me a great deale of pains' (*Correspondence*, **2**: 430), his optimism was to be short-lived: when Newton attempted to apply his present general method of construction to compute the 1680 comet's 'true' parabolic path, his efforts were unrewarded. Even the next summer, reporting to Halley on 20 June 1686 his current progress in completing the *Principia*'s press copy, he noted that its (revised) third book still 'wants ye Theory of Comets. In Autumn last I spent two months in calculations to no purpose for want of a good method, wch made me afterwards return to ye first Book & enlarge it wth divers

De motu Corporum
Liber secundus(2)

. . . .

Cometas moveri in Sectionibus Conicis umbilicum habentibus in centro Solis & radijs ad centrum illud ductis areas describere temporibus proportionales.

Versantur igitur Cometæ toto apparitionis tempore intra sphæram activitatis vis circumsolaris, adeoq; agitantur ipsius impulsu et propterea (per Coroll. 1 Prop. XII(3)) describunt Conicas Sectiones umbilicos habentes in centro Solis, et radijs ad Solem ductis conficiunt(4) areas proportionales temporibus. Nam vis illa in immensum propagata reget motus corporum longe ultra orbem Saturni.

Sectiones illas Conicas esse Parabolis finitimas. Id ex velocitate Cometarum colligitur.

Cæterum de Cometis Hypothesis est triplex: eos vel generari et interire quoties apparent et evanescunt, vel de fixarum regionibus(5) venientes præterire Systema nostrorum Planetarum, vel deniq; circa Solem in orbibus valde excentricis perpetuo revolvi. Casu primo Cometæ pro varia sua velocitate movebuntur in Sectionibus quibuscunq; Conicis, secundo movebuntur in Hyperbolis, utroq; indifferenter frequentabunt regiones omnes tam polorum quam eclipticæ, tertio motus peragentur in Ellipsibus valde excentricis et ad speciem Parabolarum quamproxime accedentibus[,] orbes autem (si lex Planetarum servetur) haud multum divaricabunt(6) a plano Eclipticæ. Et quantum hactenus animadvertere potui casus tertius obtinet. Nam Cometæ maximè frequentant Zodiacum(7) et vix

---

Propositions some relating to Comets others to other things found out last Winter' (*ibid.*: 437; compare 1, §3: note (1) above). It is, however, only fair that we should cap this long sequence of failure on Newton's part to contrive a viable method of fixing the orbit of a solar comet from given terrestrial sightings of it by reproducing in appendix the reasonably approximate mode of its construction (*Principia*, $_1$1687: 484–90) which he was then led in final triumph to devise and thereafter (*ibid.*: 490–4) successfully to employ in determining the elements of the comet of 1680/1.

(2) The following (terminal) portion of this unfinished 'second book'—of the two Newton initially planned to publish—'on the (dynamical) motion of bodies' is taken from ff. 48^r/51^r–56^r of the original manuscript, now ULC. Add. 3990, which was (see I: xx, note (12)) first printed, with minimal editorial ameliorations by John Conduitt, as *De Mundi Systemate Liber Isaaci Newtoni* (London, 1728): a slightly variant English version also came out the same year—and perhaps a little in advance of it (see note (10) below)—under the title *A Treatise of the System of the World. By Sir Isaac Newton.* On the background and general character of this 'Liber secundus' see I. B. Cohen's *Introduction to Newton's 'Principia'* (Cambridge, 1971): 109–15 and especially 327–35; compare also his 'Newton's *System of the World*: some textual and bibliographical notes', *Physis* **11**, 1969: 152–66. Since its citations of propositions relate not to those of its apparently complementary 'Liber primus' (1, §3) but to the preceding revised tract 'De motu Corporum' (1, §2)—a fact we there (*ibid.*: note (188)) used to advantage in describing its now missing latter part—our conjectured date of composition will not be greatly out. The accompanying figures, lacking in the surviving manuscript, are reproduced from 'Tab. II' of Conduitt's *editio princeps*.

ON THE MOTION OF BODIES.
BOOK TWO.[2]

. . . .

Comets move in conics having a focus at the sun's centre, and by radii drawn to that centre describe areas proportional to the times.

Comets, therefore, are located during the whole time of their appearance within the sphere of activity of the circumsolar force, and hence are acted upon by its impulsion, and consequently (by Corollary 1 to Proposition XII[3]) describe conics having foci at the sun's centre, and by rays drawn to the sun complete[4] areas proportional to the times. For that force, being propagated to an immense distance, will govern the motions of bodies far beyond the orbit of Saturn.

Those conics are, to a high approximation, parabolas. That is gathered from the speed of comets.

Regarding comets, however, existing hypothesis is three-fold: either they are generated and decay as often as they appear and vanish, or in their journey from the regions of the fixed stars[5] they pass through our planetary system, or, lastly, they revolve round the sun in extremely eccentric orbits. In the first case comets will, according to their varying speed, move in conics of all species, in the second they will move in hyperbolas, and in either case they will frequent all regions (of the sky) indifferently, both polar and ecliptic; in the third their motions will be performed in extremely eccentric ellipses approaching very nearly the species of parabolas, but their orbits will (if the law of the planets be observed) not much depart[6] from the plane of the ecliptic. And, as far as I have been able till now to discern, the thirdcase obtains. For comets mostly frequent the zodiac[7] and

---

(3) Namely, of the revised treatise 'De motu Corporum' (1, §2) [= Proposition XIII of the ensuing 'Liber primus' (1, §3)]; see note (2).

(4) The more vivid 'verrunt' (sweep out) was first written.

(5) Effectively from an infinite distance, that is, whither they will depart (continuing in the same near-rectilinear direction) after their brief, once-for-all passage through the solar system. As we have earlier remarked (see v: 524, note (1)) it had been an *idée fixe* with Newton in the early 1680's that 'ye Comets of November & December [1680]' were not, as Flamsteed supposed, 'one & ye same Comet' but apparitions of 'two different ones', and that 'to make ye Comets of November & December but one is to make that one paradoxical' (as he wrote to Flamsteed on 28 February and 16 April 1681; see his *Correspondence*, **2**: 342, 364).

(6) Newton initially wrote 'recedent' (will...recede).

(7) '...a Broad Circle [of the celestial sphere]...thro' the middle of [which] is drawn... the *Ecliptick*, or...way of the Sun....The Breadth of this Circle is 20 Degrees, for beyond 10 Degrees North, and 10 Degrees South, the Latitude of no Planet ever reaches' (John Harris, *Lexicon Technicum: Or, An Universal English Dictionary of Arts and Sciences*, London, 1704). Newton's remark is valid in regard to the portions of cometary paths (those in the vicinity of the sun) which are alone visible to a terrestrial observer, but is untrue of their orbits (in large part unseen) as a whole. His following assertion is particularly unfortunate when viewed in retrospect: of the three comets (those of 1680/1, 1682 and 1683) which had been observed in

unquam pertingunt ad latitudinem heliocentricam graduum quadraginta: moventur etiam in orbibus quamproxime Parabolicis, uti colligo ex eorum velocitate. Nam velocitas qua Parabola describetur est ubivis ad velocitatem qua Cometa vel Planeta in circulo ad eandem a Sole distantiam revolvi posset in dimidiatâ ratione numeri binarij ad unitatem (per Coroll. 7 Prop. XV.(8)) Et meo quidem calculo talis circiter(9) reperta est Cometarum velocitas. . . .

. . . .(10)

Proponitur inventio Trajectoriæ Cometarũ. Est igitur Cometarum velocitas, quatenus ea per hujusmodi computationes nimium rudes definiri queat, ea ipsa quâcum Parabolæ vel his affines Ellipses describi debeant: et propterea ex data inter

England in the recent past, Newton himself was within a few months to compute the inclination to the ecliptic of the first to be some 61° 20′ (*Principia*, ₁1687: 494), while ten years afterwards Edmond Halley similarly determined that of the last to be more than 83° (see his letter to Newton of 7 September 1695 in *Correspondence*, 4: 165, and his subsequent *Astronomiæ Cometicæ Synopsis* (Oxford, 1705): signature A1ᵛ, where the accurately computed elements of all three comets are listed in a table of twenty-two sighted between 1337 and 1698, ten of these being inclined at more than 50° to the ecliptic).

(8) Understand of the 'De motu Corporum' (1, §2) [= Proposition XVI of the expanded 'Liber primus' (1, §3)].

(9) 'fere' (almost) is replaced.

(10) In this omitted passage (f. 48ʳ–51ʳ [= *De Mundi Systemate:* 92–8]) Newton added in substantiation: 'Rem examinavi colligendo præterpropter velocitates ex distantijs ac distantias ex parallaxi et phænomenis caudæ conjunctim. Errores in velocitatis excessu vel defectu haud majores obvenere quam qui ex erroribus in distantijs eâ ratione collectis oriri potuerint' (I examined the matter by loosely determining the speeds from the distances and the distances from the parallax and visual appearance of their tail jointly. The errors in speed by excess or defect came out to be no greater than might have arisen from the errors in the distances determined in this manner). The apparent length and position of a comet's tail—taken by Newton to point directly away from the sun—can, however, give only the crudest knowledge of the true location and shape of its orbit, and he passed straightaway on to outline an improved 'ratiocinium' for the purpose, dependent on two tables relating the solar distance of a comet, its time of travel from perihelion and its apparent diurnal motion and solar elongation when viewed from the earth. On taking the mean distance of the earth from the sun as the astronomical unit, it follows that the solar gravitation at this distance is $4\pi^2$ units/year², whence the speed of a comet at a distance r from the sun is $v_r = 2\pi\sqrt{[2/r]}$ units/year, and therefore the time of its passage thither from its perihelion point, distant ϵ $(= Q/v_\epsilon)$ from the sun, is

$$t_{r,\epsilon} = \int_\epsilon^r 1/[v_r^2 - Q^2/r^2]\,.\,dr = (1/2\pi\sqrt{2})\int_\epsilon^r r/\sqrt{[r-\epsilon]}\,.\,dr = (1/3\pi\sqrt{2})\,(r+2\epsilon)\sqrt{[r-\epsilon]}$$

years; again, since the earth's diurnal angular motion in its annual orbit round the sun is $360°/365.25\ldots \approx 59'$, where the solar elongation of a comet at unit distance from the sun is θ (and so its instantaneous distance from the earth is $2\cos\theta$ when the earth is at its mean distance from the sun) its apparent diurnal motion is very nearly equal to

$$s_\theta = 59'.(\sqrt{2}/2\cos\theta \pm 1),$$

the ambiguous sign being taken positive for a direct comet and negative for a retrograde one. In his 'Tab. I' Newton lists his none too accurately computed values of $t_{r,\epsilon}$, $r = 1, 2, 3, 4$ and

scarcely ever attain a heliocentric latitude of 40°: they also move in orbits very nearly parabolic, as I gather from their speed. For (by Corollary 7 to Proposition XV(8)) the speed whereby a parabola shall be described is everywhere to the speed by which a comet or planet might revolve in a circle at the same distance from the sun as the square root of 2 to 1, and by my calculation, indeed, such or thereabouts(9) is the speed of comets found to be....

... ...(10)

The ascertaining of cometary trajectories is proposed.

The speed of comets, therefore, is—insofar as we may fix it by over-rough computations of this sort—exactly that with which parabolas, or ellipses akin to them, ought to be described; and in consequence, given the distance between a comet and the sun,

---

$\epsilon = 2^i \times 0.005$, $i = 0, 1, 2, ..., 9$, setting $t_{1,0} = 1/3\pi\sqrt{2}$ years to be 27 days 11 hours 12 minutes; and correspondingly in his 'Tab. II' he sets out in parallel columns s_θ and $2\cos\theta$, where $\theta = (60+\mu)°$, $\mu = 0, 5, 10(2)\ 30$. (The fact that in the first table the two final values $\epsilon = 1.28$ and $\epsilon = 2.56$ are thoughtlessly made by Newton to yield real corresponding values of $t_{r,\epsilon}$ seems to have passed his eighteenth-century editors by, even though its anonymous English translator—might it be Edmond Halley?—noted in the 1728 *Treatise of the System of the World* (note (2)): 134 that the numerical tabulations there printed were 'here corrected' from Newton's own: these amended values were silently incorporated by Conduitt in his Latin *editio princeps* the same year without any indication to the reader that the manuscript text was not faithfully reproduced.) From these Newton was able to satisfy himself by a 'rudis computus' that the 'comets' of 1680/1 were indeed one and the same; for when, at an interval of some 56 days, the ingressing and egressing comet was at unit distance from the sun, it was in each case by observation about 28 days from perihelion 'Atq; tot dies (per Tab. I) in trajectorijs Parabolicis consumi debebant' (and the spending of this many days was, by Table I, in parabolic trajectories called for), while a slightly more refined calculation deriving 'ex Tabula II' showed that, because by reason of the observed solar elongations of the comet at the points of its unit distance from the sun it was then about 0.36 units (in ingress) and 0.63 units (in egress) from the circling earth, these points were about 0.28 units apart, so that the perihelion distance of the comet was about $\frac{1}{4}(0.14)^2 \approx 0{\cdot}005$ units (about 20% short of true). Therefrom he was happy to conclude (ff. $49^v/50^r$) that 'ex coincidentia periheliorum & velocitatum consensu verisimile fit Cometas hosce, quos ut duos jam consideravi, non duos fuisse sed unum et eundem Cometam. Ea lege orbita hujus Cometa vel Parabola erit vel Conisectio a Parabola parum discrepans et superficiem Solis propemodum tanget in vertice' (from the coincidence of their perihelia and the agreement of their speeds it comes to be likely true that these comets, which I have previously considered as two, were not two but one and the same comet. And with that stipulation the orbit of this comet will be either a parabola or a conic little differing, and shall all but touch the sun's surface at its vertex). In sequel Newton attempted 'similibus computis' roughly to define the perihelion distances of other past comets (namely those of 1472, 1607, 1618_1, 1618_2, 1665 and 1682) but lack of accurate observations prevented him from there achieving a like accuracy: in particular, he failed to remark on the close parallel between the 1607 and 1682 apparitions of 'Halley's' comet, being able to compute 'their' perihelion distances only very crudely (as 0.39 and 0.35 units respectively, instead of the correct 0.58_+). Encouraged none the less by the over-optimistic agreement between parabolic theory and observed orbital reality at which he here contrived to arrive, he proceeded undismayed to draw from it the ensuing conclusion.

Cometam et Solem distantia datur quamproximè.(11) Et hinc oritur Problema tale.

Datur relatio inter velocitatem Cometæ et distantiam ipsius a centro Solis, requiritur Trajectoria.

Soluto autem hoc Problemate habebitur tandem methodus determinandi Trajectorias Cometarum quam exactissimè. Nam si relatio illa bis assumatur et inde trajectoria bis computetur, deinde ex observationibus inveniatur error Trajectoriæ utriusq; potest assumptio per regulam falsæ positionis(12) corrigi et sic tertia inveniri trajectoria quæ cum observationibus plane congruet. Et hac methodo determinando Cometarum trajectorias sciri tandem potest exactius quasnam regiones hæc corpora percurrunt, quantæ sint eorum velocitates, cujus generis trajectorias describunt.... Quinetiam sperandum est quod hac methodo sciri tandem possit utrum Cometæ statis temporibus in orbem redeant et quibus temporum periodis revolutiones singulorum complentur. Solvitur autem Problema colligendo primum ex tribus vel pluribus observationibus motum horarium Cometæ ad tempus datum, deinde ex hoc motu trajectoriam derivando. Sic pendet inventio trajectoriæ ab unica observatione motuq; horario tempore illius observationis atq; adeo seipsam vel probabit vel redarguet. Nam conclusio quæ ex motu horæ unius et alterius et hypothesi falsa deducitur nunquam congruet cum motibus Cometarum a principio ad finem. Methodus computationis totius ita se habet.

In Problematis solutionem præmittuntur Lemmata.

Lemma 1.

Datas positione duas rectas OR, TP tertia RP ad rectum angulum TRP secare ea lege ut si ad datum aliquod punctum S(13) *ducatur recta SP, datum sit contentum solidum sub ducta illa SP et quadrato rectæ OR terminatæ ad datum punctum O.*

Confit graphice sic. Sit contentum illud datum $M^{\text{quad}}\times N$. Ad rectæ [OR] punctum quodvis r erige perpendiculum rp occurrens ipsi [T]P in p. Et per puncta $S_{[,]}$ p age Sq æqualem $\dfrac{M^q\times N}{Or^{\text{quad}}}$. Simili methodo age tres vel plures re[c]tas S^2q, S^3q &c et per puncta omnia q, ${}^2q$, ${}^3q$ [&c] acta linea regularis q^2q^3q(14) secabit rectam TP in puncto P, a quo demittendum est perpendiculum PR. Q.E.F.

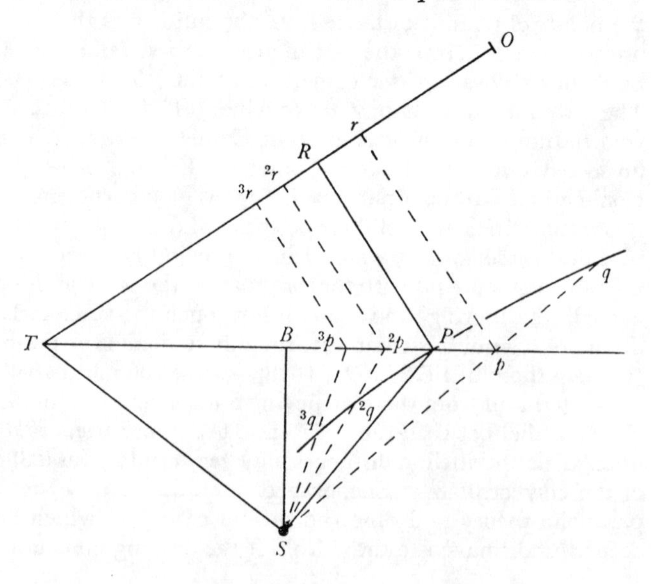

its speed is given therefrom to a close approximation.[11] And hence arises the following problem.

Given the proportion between a comet's speed and its distance from the centre of the sun, its trajectory is required.

Once this problem is solved, however, there will at length be obtained a method for determining the trajectories of comets with great precision. For if that proportion be twice assumed and the trajectory thence twice computed, and then the error in each trajectory be ascertained from observations, the assumption can be corrected by the rule of false position[12] and in this way a third trajectory found which shall be entirely in agreement with the observations. And from determining the trajectories of comets by this method we may ultimately know more exactly which regions these bodies pass through, the magnitude of their speeds, what kind of orbits they describe.... Indeed, it is to be hoped that we might by this method come at length to know whether comets return periodically at regular intervals and in what period of time the revolution of each individual one is completed. The problem itself is solved by first gathering the comet's hourly motion at a given time from three or more observations and then deriving its trajectory from this motion. In this way the determining of the trajectory depends upon a single observation and the hourly motion at the time of observation, and will hence either confirm or refute itself. For a conclusion which is drawn from the motion of an hour or two and from a false hypothesis will never be in agreement with the motions of comets from beginning to end. The method of the whole computation is thus characterized.

Lemma 1.

Lemmas premised to the problem's solution.

To cut two straight lines OR, TP given in position by a third RP at the right angle $\widehat{TRP}$ *in such a fashion that, if to some given point S*[13] *the straight line SP be drawn, there shall be given the 'solid' contained by the line SP so drawn and the square of the line-segment OR terminated at the given point O.*

It is accomplished by drawing as follows. Let the given 'solid' be M^2N. At any point r of the line OR erect the perpendicular rp meeting TP in p, and then through the points S and p draw Sq equal to M^2N/Or^2. In like manner draw three or more straight lines S^2q, S^3q, ..., and when a regular line q^2q^3q... is drawn[14] through all the points q, ${}^2q$, ${}^3q$, ... it will intersect the straight line TP in the point P from which the perpendicular PR has to be dropped. As was to be done.

(11) As Newton has earlier observed, the instantaneous orbital speed in a parabolic cometary path at distance r from the sun is 'per Coroll. 7 Prop. XV' (see note (8)) $\sqrt{2}$ times that ($2\pi/\sqrt{r}$ units/year) of a 'planet' circling the sun at the same distance.

(12) That is, by linear interpolation; compare Newton's 'trigonometrical' solution of Lemma 1 below (see note (16)).

Trigonometricè sic. Assumatur linea novissimè inventa $[q]P$, et inde[15] in triangulis $T[pr]$, $T[p]S$ dabuntur perpendicula $[pr]$, SB et in triangulo $SB[p]$ latus $S[p]$ et error $[pq=]\frac{M^qN}{Or^q}-S[p]$. Fac ut error iste puta D sit ad errorem novum puta E ut error ${}^2p^2q[\pm]pq$, vel ${}^2p^2q+\text{vel}-D$ ad errorem ${}^2p[p]$ et error ille novus additus vel subductus longitudini $T[p]$ dabit longitudinem correctam[16] $T[p]+\text{vel}-E$. Electio additionis vel subductionis inspectione schematis dirigitur. Siquando opus fuerit ulteriore correctione, operationem repete.

Nota $\pm$ significat $+$ vel $-$ ambiguè et $\mp$ ponitur pro signo contrario.

Arithmeticè sic. Puta factum sitqꝫ lineæ TP graphicè inventæ correcta longitudo $TP\pm e$: et correctæ longitudines linearum OR, BP et SP erunt $OR[\mp]\frac{TR}{TP}e$, $BP\pm e$ et $\sqrt{SP^q\pm 2BP\times e+ee}=\dfrac{M^qN}{OR^q[\mp]\frac{2OR\times TR}{TP}e+\frac{TR^q}{TP^q}ee}$. Unde per methodum serierum convergentium fit

$$SP\pm\frac{BP}{SP}e+\frac{SB^q}{2SP^c}ee\ \&\text{c}=\frac{M^qN}{OR^q}\pm\left[2\frac{TR}{TP}\times\right]\frac{M^qN}{OR^c}e+\left[3\frac{TR^q}{TP^q}\times\right]\frac{M^qN}{OR^{qq}}ee\ \&\text{c}.$$

Pro datis $\frac{M^qN}{OR^q}-SP$, $\pm\frac{BP}{SP}\left[\mp2\frac{TR}{TP}\times\right]\frac{M^qN}{OR^c}$, $\left[-3\frac{TR^q}{TP^q}\times\right]\frac{M^qN}{OR^{qq}}[+]\frac{SB^q}{2SP^c}$

scribe F, $\frac{F}{G}$, $\frac{F}{GH}$, et signis probe observatis fiet $GH-ee=He$[17] seu $G-\frac{ee}{H}=e$. Hinc,[18] neglecto termino perexiguo $\frac{ee}{H}$, prodit $e=G$. Si error $\frac{ee}{H}$ non est contemnendus, pone $G-\frac{GG}{H}=e$.

Et nota quod hic insinuatur methodus generalis[19] solvendi difficiliora

(13) It is patently unnecessary that S should lie in the same plane as the lines ORT and TP; indeed, this is not the case in the configuration of his following Problem 1 to which Newton will apply it.

(14) Understand 'æquo manus motu' (with an even motion of the hand) as Newton expressed this 'graphic' operation a few months earlier in Chapter 3 of his 'Matheseos Universalis Specimina' (see IV: 560, note (112)). In equivalent analytical terms, if $S(0, 0)$ is the origin of the perpendicular coordinate system in which the line PT through $T(a, b)$ has the defining equation $y = b$ and $OT = c$ (not necessarily in the plane STP) is inclined to it at the angle $\cos^{-1}(TR/TP) = \cos^{-1}k$, then the general transversal Spq through $q(x, y)$ meets TB in the point $p(bx/y, b)$ distant $pT = a+bx/y$ from T, whence $Or = c-k(a+bx/y)$; accordingly, the locus (q) defined by $Sq\ (=\sqrt{[x^2+y^2]}) = M^2N/Or^2$ has the defining equation

$$(x^2+y^2)\,(bx-(c/k-a)\,y)^4 = (M^4N^2/k^4)\,y^2,$$

a 'circular' sextic through S (when Or is infinitely great and so the vanishingly small tangential chord Sq is parallel to PT) and doubling back thereafter to approach a parabolic point in the

Trigonometrically, it is done this way. Assume the curve qP most recently found, and therefrom[15] in the triangles TPr, TpS there will be given the perpendiculars pr, SB and so in the triangle SBp the side Sp, and thence the error $(pq =)\ M^2N/Or^2 - Sp$. Make that error, D say, to a new error, E say, as the error ${}^2p^2q \pm pq$, or ${}^2p^2q \pm D$ to the error ${}^2pp$, and that new error added to or subtracted from the length Tp will yield the corrected length[16] $Tp \pm E$. The choice of adding or subtracting is governed by inspection of the figure. Should you ever have need of a further correction, repeat the procedure.

Arithmetically, this way. Suppose it done, and let the corrected length of the line TP found by drawing be $TP \pm e$: the corrected lengths of the lines OR, BP and SP will then be $OR \mp (TR/TP)\,e$, $BP \pm e$ and

The symbol $\pm$ denotes $+$ or $-$ ambiguously and $\mp$ is set for the contrary sign.

$$\surd[SP^2 \pm 2BP.e + e^2] = M^2N/(OR^2 \mp 2OR \times (TR/TP)\,e + (TR^2/TP^2)\,e^2).$$

Whence by the method of converging series there comes

$$SP \pm (BP/SP)\,e + \tfrac{1}{2}(SB^2/SP^3)\,e^2 \ldots$$
$$= M^2N/OR^2 \pm 2(TR/TP)\,(M^2N/OR^3)\,e + 3(TR^2/TP^2)\,(M^2N/OR^4)\,e^2 \ldots.$$

In place of the given quantities $M^2N/OR^2 - SP$, $\pm BP/SP \mp 2(TR/TP)\,M^2N/OR^3$ and $-3(TR^2/TP^2)\,M^2N/OR^4 + \tfrac{1}{2}SB^2/SP^3$ write F, F/G and F/GH, and then with the signs properly observed there will come to be $GH - e^2 = He$,[17] or $G - e^2/H = e$. From this,[18] upon neglecting the exceedingly slight term e^2/H, there results $e = G$. If the error e^2/H is not to be disregarded, set $e = G - G^2/H$.

And note that here the hint is given of a general method[19] for solving more

---

direction $S\alpha$, where α is the meet of TP with the parallel to RP through O. In our accompanying figure, founded on the ill-drawn equivalent printed in the 1728 *editio princeps* (see note (2)) for want of Newton's original, we have correctly adjusted the curvature and position of the 'linea irregularis' (q).

(15) Understand 'dato puncto r adeoq; puncto q' (given the point r and hence the point q). Here and in immediate sequel we have had considerably to amend Humphrey Newton's transcription in the manuscript text to make sense of it.

(16) Approximating TP, that is. When both p and ${}^2p$ are close to P, it will be clear that the 'errors' are virtually parallel and that the included arc $\widehat{q^2q}$ of the curve qP is very nearly straight, whence P is to a close approximation the meet of Tp with the chord q^2q; accordingly, $Pp \approx {}^2pp \times pq/(pq - {}^2p^2q)$ and therefore $TP(= Tp - Pp) \approx Tp + {}^2pp \times pq/({}^2p^2q - pq)$, where, as Newton observes in his following marginal note, due regard is to be paid to the signs ($+$ or $-$) of the individual directed line segments. The procedure is (compare I: 491, note (13)) equivalent to linear interpolation by first-order divided differences.

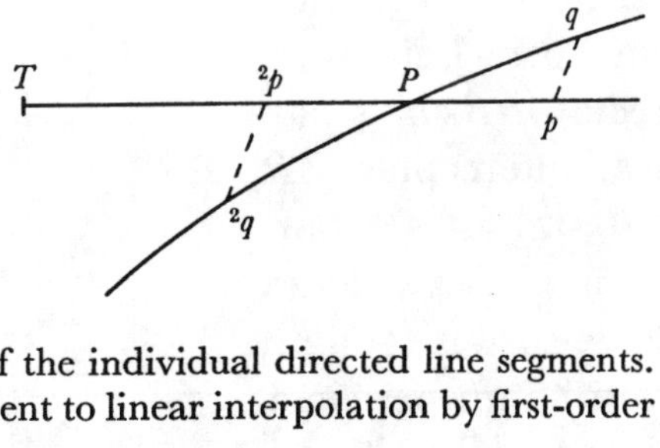

(17) On ignoring terms in the cube and higher powers of e, that is.

(18) 'Primum' (First) was initially written.

(19) That which proceeds from the assumption that any required value may adequately be approximated by a 'converging series' of powers of some base variable. Newton had

Problemata tam Trigonometricè quam Arithmeticè absq; perplexis illis computis et resolutionibus affectarum æquationum quæ hactenus in usu fuerunt.

Lemma 2.

Datas positione tres rectas quarta secare cujus partes interceptæ datam habeant proportionem ad invicem, quæq; transeat per punctum quod in una earum datur.(20)

Dentur positione AB AC BC et in AC detur punctum D. Ipsi AB agatur parallela DG occurrens BC in G: capiatur GF ad BG in data illa ratione et agatur FDE. Erit FD ad DE ut FG ad BG. Q.E.F.

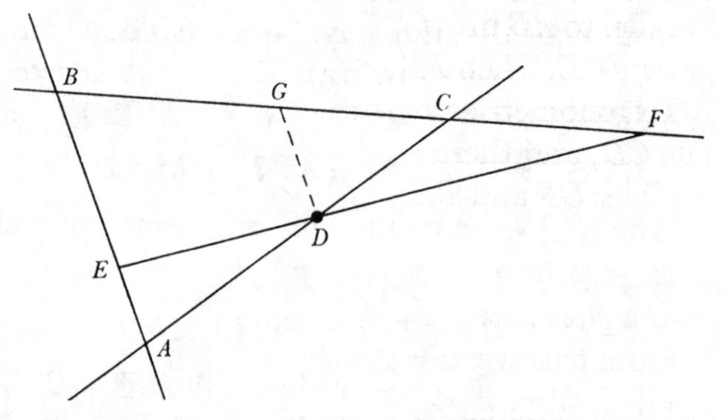

Trigonometricè sic. In triangulo CGD dantur anguli et latus CD, et inde(21) latera reliqua, et ex datis rationibus dantur lineæ GF et BE.

Lemma 3.

Ad datum tempus invenire et graphicè exponere motum horarium Cometæ.

Ex observationibus probæ fidei(22) dentur tres Longitudines Cometæ. Sunto harum differentiæ ATR, RTB, et requiratur motus horarius ad tempus observationis longitudinis intermediæ TR. Ducatur per Lemma 2 recta ARB cujus partes interceptæ AR, RB sint ut tempora inter observationes. Et si corpus tempore toto totam percurrat lineam AB uniformiter et interea spectetur de loco T[,] is erit motus ejus apparens circa punctum R qui fuit Cometæ tempore observationis TR quamproxime.

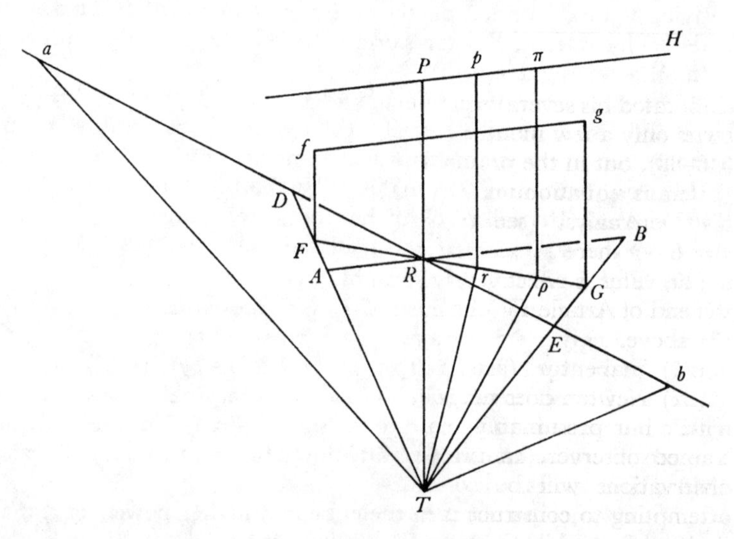

difficult kinds of problem both trigonometrically and arithmetically, and without those intricate computations and resolutions of affected equations which have till now been in use.

Lemma 2.

To cut three straight lines given in position by a fourth whose intercepted parts shall have a given ratio to one another and which shall pass through a point given in one of them.[20]

Let AB, AC, BC be given in position, and in AC let there be given the point D. Parallel to AB draw DG meeting BC in G: take GF to BG in the given ratio and draw FDE. Then FD will be to DE as FG to BG. As was to be done.

Trigonometrically thus. In the triangle CGD there are given the angles and side CD, and therefrom[21] the remaining sides, and then from the given ratios the lines GF and BE are given.

Lemma 3.

At a given time to ascertain and graphically represent the hourly motion of a comet.

From trustworthy observations[22] let three longitudes of the comet be given. Let their differences be $A\widehat{T}R$, $R\widehat{T}B$ and let the hourly motion at the time of the intermediate longitude TR be required. Draw, by Lemma 2, the straight line ARB whose intercepted parts AR, RB shall be as the times between the observations. Then, if a body in the whole time shall uniformly traverse the whole line AB and as it does so be viewed from the place T, its apparent motion about the point R will be that which the comet had at the time of the observation TR, very nearly.

---

elaborated his several equivalent techniques for successively deriving the terms of such power-series only a few months before in his unfinished 'Matheseos Universalis Specimina' (see IV: 540–88), but in the present case of the equation $G-e-(1/H)e^2\ [+ae^3+be^4...] = 0$ the simple 'Literalis æquationum affectarum resolutio' which he had expounded fifteen years earlier in his 'De Analysi' (see II: 222–6) is alone 'insinuated': for, since $e \approx G$ to $O(e^2)$, on setting $e = G+p$ there at once ensues $p = -G^2/H+O(e^3)$.

(20) This is effectively 'Cas. 1' of the second of the two lemmas rejected by Newton from the end of Article IV of his 'De motu Corporum Liber primus'; see 2, §3, Appendix 1: note (7) above.

(21) 'habentur' (are had) is deleted.

(22) Newton does not specify his criterion for distinguishing such observations 'of superior trust', but presumably intends that they be ones made in good sighting conditions by a trained observer. In practice this naïve notion of the empirical accuracy of individual observations will be considerably modified by the inevitable feed-back from similarly attempting to construct the comet's hourly motion from other trios of less favoured observations—a fact which he is well aware of and indeed in sequel uses to advantage in refining the present construction based on the 'simple' supposition that over a small arc of its trajectory a comet's motion may be taken to be uniform and rectilinear.

Idem accuratiùs.[(23)]

Dentur longitudines hinc inde magis distantes *Ta*, *Tb* et per Lemma 2 ducatur *aRb* cujus partes *aR*, *Rb* sint ut tempora inter observationes *aTR*, *RTb*. Secet hæc lineas *TA*, *TB* in *D* et *E* et quoniam error inclinationis [*A*]*Ra* crescit quasi in duplicata ratione temporis inter observationes[(24)] age *FRG* ea lege ut vel angulus *DRF* ad angulum *ARF* vel linea *DF* ad lineam *AF* sit in duplicata ratione temporis totius inter observationes *aTb* ad tempus totum inter observationes *ATB*, atq; loco lineæ superius inventæ usurpetur linea jam inventa *FG*.

Convenit angulos *ATR*, *RTB*, *aTA*, *BTb* haud minores adhiberi quam decem vel quindecim graduum ac tempora ipsis respondentia haud majora quam dierum octo vel duodecim, atq; longitudines capi ubi Cometa celerrimè movetur. Hoc enim pacto errores observationum minorem habebunt rationem ad differentias longitudinum.

Lemma 4.

Ad data tempora invenire longitudines Cometæ.

Fit capiendo in linea *FG* distantias [*R*]*r*, *rρ* temporibus proportionales, et ducendo lineas *Tr*, *Tρ*. Operatio trigonometrica palam est.

Lemma 5.

Invenire latitudines.

Ad radios *TF*, *TR*, *TG* erigantur normaliter tangentes observatarum latitudinum *Ff*, *RP*, *Gg*. Ipsi *fg* parallela ducatur *PH*. Huic occurrentia perpendicula *rp*, *ρπ* tangentes erunt latitudinum quæsitarum ad radios *Tr*, *Tρ*.[(25)]

Problema 1.

Ex assumpta ratione velocitatis determinare Trajectoriam Cometæ.

Solvitur Problema.

Designent *S* Solem; *t*, *T*, *τ* loca tria æquidistantia telluris in ipsius orbita; *p*, *P*, *π* loca totidem respondentia Cometæ in ipsius Trajectoria, interpositis inter singula intervallis horæ unius; *pr*, *PR*, *πρ* perpendicula demissa ad planum Eclipticæ, et *rRρ* vestigium Trajectoriæ in hoc plano. Jungantur *Sp*, *SP*, *Sπ*,

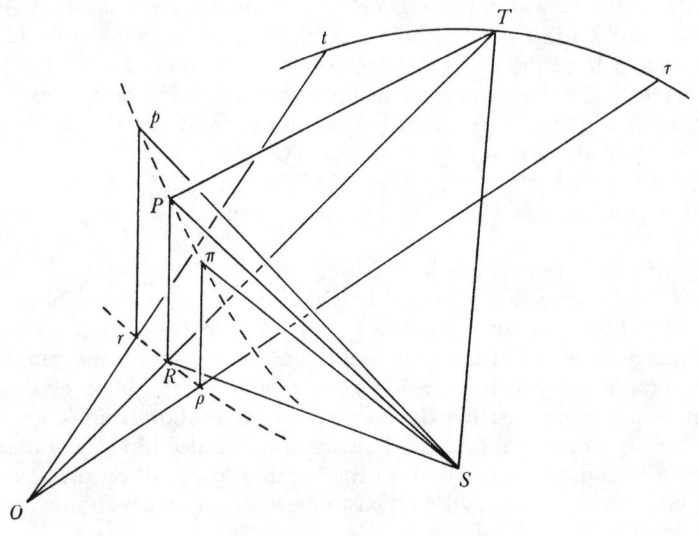

The same more accurately.(23)

Let longitudes Ta, Tb more distant on either side be given and, by Lemma 2, draw aRb whose parts aR, Rb shall be as the times between the observations Ta, TR, Tb. Let this cut the lines TA, TB in D and E, and then, because the error $\widehat{ARa}$ in inclination increases virtually in the doubled ratio of the time between the observations,(24) draw FRG in such a way that either the angle $\widehat{DRF}$ to the angle $\widehat{ARF}$ or the line DF to the line AF shall be in the doubled ratio of the whole time between the observations Ta, Tb to the whole time between the observations TA, TB, and in place of the line previously ascertained employ the line FG now found.

It is convenient to make use of angles $A\hat{T}R$, $R\hat{T}B$, $a\hat{T}A$, $b\hat{T}B$ of not less than ten or fifteen degrees, and times corresponding to them of not more than eight or twelve days, taking the longitudes at which the comet is moving most rapidly. For by this means the errors in observation will bear a lesser ratio to the differences in longitude.

Lemma 4.

At given times to ascertain the longitudes of a comet.

This is done by taking in the line FG distances Rr, $r\rho$ proportional to the times, and drawing the lines Tr, $T\rho$. The trigonometrical working is obvious.

Lemma 5.

To find the latitudes.

Normally to (the plane of) the radii TF, TR, TG erect the tangents Ff, RP, Gg of the observed latitudes. Parallel to fg draw PH. The perpendiculars rp, $\rho\pi$ meeting this will then be the tangents of the required latitudes with respect to the radii Tr, $T\rho$.(25)

Problem 1.

Assuming the ratio of the speed, to determine therefrom a comet's trajectory.

The problem is solved.

Let S denote the sun; t, T, τ three equidistant places of the earth in its orbit; p, P, π an equal number of corresponding positions of the comet in its trajectory, with intervals of one hour interposing between each; pr, PR, $\pi\rho$ perpendiculars let fall to the plane of the ecliptic; and $rR\rho$ the track of the trajectory in this plane. Join Sp, SP, $S\pi$, SR, ST, tr, TR, $\tau\rho$, TP and let tr, $\tau\rho$ meet in O: TR will then

(23) Originally 'exactiùs' (more exactly).

(24) Assuming that the true cometary orbit is parabolic, it would appear; even so, the supposition is only loosely valid.

(25) Understand 'quamproximè' (very nearly) as before.

SR, ST, tr, TR, $\tau\rho$, TP et coeant tr, $\tau\rho$ in O. Converget TR ad idem O quamproximè. Error contemnendus est.[26] Per Lemmata præcedentia dantur anguli rOR, $RO\rho$ et proportiones pr ad tr, PR ad TR, $\pi\rho$ ad $\tau\rho$. Datur etiam figura $tT\tau O$ magnitudine et positione una cum distantia TS et angulis STR, PTR, STP. Assumamus velocitatem Cometæ in loco P esse ad velocitatem Planetæ gyrantis in circulo ad eandem a Sole distantiam SP, ut V ad 1[27] et determinanda erit linea $pP\pi$ hac lege ut sit spatium a Cometa duabus horis descriptum $p\pi$ ad spatium $V\times t\tau$ (hoc est ad spatium qu[od] tellus eodem tempore describit multiplicatum per numerum V) in dimidiata ratione distantiæ telluris a sole ST ad distantiam Cometæ a Sole SP;[28] utq sit spatium pP hora prima a Cometa descriptum ad spatium $P\pi$ hora secunda a Cometa descriptum ut velocitas in p ad velocitatem in P[,] hoc est in dimidiata ratione distantiæ SP ad distantiam Sp, sive in ratione $2SP$ ad $SP+Sp$.[29] Minutias enim toto hoc opere negligo quæ errorem sensibilem creare nequeunt.

Imprimis igitur ut in resolutione æquationum affectarum Mathematici prima vice radicem conjecturâ[30] colligunt, sic in hoc opere analytico conjecturam faciendo assequor ut possim distantiam quæsitam TR et per Lemma 2 duco $r\rho$ primùm ita ut rR et $R\rho$ æquentur, deinde (ubi proportio SP ad Sp hinc innotuerit) ita ut sit rR ad $R\rho$ uti $2SP$ ad $Sp+SP$, et invenio rationes linearum $p\pi$, $r\rho$, et OR ad invicem. Ponatur esse M ad $V\times t\tau$ ut OR ad $p\pi$, et ob proportionalia $p\pi$ quadratum ad quadratum ex $V\times t\tau$ et ST ad SP erit ex æquo OR^q ad M^q ut ST ad SP, adeoq contentum solidum $OR^q\times SP$ æquale dato $M^q\times ST$. Unde (si triangula STP PTR jam locentur in eodem plano)[31] dabuntur TR, TP, SP, PR per Lemma 1. Hæc omnia perago primum graphicè opere celeri et rudi, dein graphicè majori cum diligentia, ultimò per computationem numeralem. Tum denuò situm linearum $r\rho$ $p\pi$ determino accuratissimè una cum nodis[32] et inclinatione plani $Sp\pi$ ad planum Eclipticum, inq plano illo $Sp\pi$, per Prop. XVI[33] describo Trajectoriam in qua corpus movebitur emissū de dato loco P secundum datam rectam $p\pi$, ea cum velocitate quæ sit ad velocitatem telluris ut $p\pi$ ad $V\times t\tau$. Q.E.F.[34]

(26) This invalid assumption—one which Newton dared not make in his parallel computations shortly afterwards (see v: 524–9) of the regressing trajectory of the 1680/1 comet in the Wrennian hypothesis that it is (near enough) a uniformly traversed straight line—ineluctably, as he came soon to see, imposes a crippling bound on any attempt to apply the present mode of construction to determine the orbit of any comet observed from the circling earth.

(27) For consistency with the immediate sequel this should read 'ut 1 ad V' (as 1 to V), where 'per Coroll. 7 Prop. XV' (see note (8)) $V = \sqrt{2}$ accurately.

(28) That is, by Proposition IV of the revised 'De motu Corporum' (1, §2), as the ratio—$p\pi/V$ to $t\tau$ *quamproximè*—of the speeds at P and T in concentric circle orbits round the centre S of the inverse-square solar gravitational field in which both the comet and earth move.

(29) For, because the difference $Sp-SP = \epsilon$, say, is presumed to be very small, there is in consequence $(SP/Sp)^{\frac{1}{2}} = 1/(1+\epsilon/SP)^{\frac{1}{2}} \approx 1/(1+\frac{1}{2}\epsilon/SP)$.

(30) Thus minimally changed in the manuscript from 'conjectando' (by making a guess).

converge on the same O very nearly—the error is negligible.(26) By the preceding lemmas there are given the angles $r\widehat{O}R$, $R\widehat{O}\rho$ and the ratios of pr to tr, PR to TR, $\pi\rho$ to $\tau\rho$. Also the figure $tT\tau O$ is given in magnitude and position, together with the distance TS and the angles $S\widehat{T}R$, $P\widehat{T}R$, $S\widehat{T}P$. Let us assume that the speed of the comet at the place P is to the speed of a planet orbiting in a circle at the same distance SP from the sun as V to 1,(27) and then the curve $pP\pi$ will need to be determined subject to the restriction that the space $p\pi$ described by the comet in two hours be to the space $V\times t\tau$ (that is, the space which the earth describes in the same time, multiplied by the number V) in the halved ratio of the earth's distance ST from the sun to the comet's distance SP from the sun;(28) and that the space pP described by the comet in the first hour be to the space $P\pi$ described by the comet in the second hour as the speed at p to the speed at P, that is, in the halved ratio of the distance SP to the distance Sp, or in the ratio of $2SP$ to $SP+Sp$:(29) for throughout this operation I neglect minute quantities which can lead to no perceptible error.

In the first instance, then, just as arithmeticians at a first stage in the solution of affected equations gather a root by guess,(30) so in the present analytical operation I attain the required distance TR as best I may be hazarding a guess, and then by Lemma 2 I draw $r\rho$, first so that rR and $R\rho$ may be equal, and subsequently (when the ratio of SP to Sp has been discovered therefrom) so that rR shall be to $R\rho$ as $2SP$ to $SP+Sp$, and I ascertain the ratios of the lines $p\pi$, $r\rho$ and OR to one another. Set M to be to $V\times tr$ as OR to $p\pi$, and, because $p\pi^2$ to $(V\times t\tau)^2$ is in the same ratio as ST to SP, there will *ex æquo* be OR^2 to M^2 as ST to SP, and hence the 'solid' product $OR^2\times SP$ is equal to the given $M^2\times ST$. Whence by Lemma 1 (if the triangles STP, PTR be now located in the same plane)(31) there will be given TR, TP, SP and PR. All these things I accomplish first graphically by a rough, swift operation, then graphically still but with greater care, and lastly by a numerical computation. Then I determine the position of the lines $r\rho$, $p\pi$ afresh with very great accuracy, along with the nodes(32) and the inclination of the plane $Sp\pi$ to that of the ecliptic, and in that plane $Sp\pi$, by Proposition XVI,(33) I describe the trajectory in which a body sent off from the given place P shall move away along the given straight line $p\pi$ with a speed that is to the speed of the earth as $p\pi$ to $V\times t\tau$. As was to be done.(34)

(31) Most simply, by rotating the triangle SPT round PT till it comes to lie in the plane. As we have already observed (see note (13)), this heuristic prop to the reader's comprehension of the construction is mathematically superfluous.

(32) The intersection of the parabolic orbit $pP\pi$ with the ecliptic plane $ORST$.

(33) Once more understand of the 'De motu Corporum' (1, §2) [= Proposition XVII of the expanded 'Liber primus'].

(34) It is significant that at this point in the manuscript no space is left in sequel for any résumé of the abortive 'calculations to no purpose' by which (see note (1)) Newton sought

Prob. 2.

Assumptam velocitatis rationem et inventam Trajectoriam corrigere.

Adhibeatur observatio Cometæ sub finem motus[35] aliave aliqua quam longissimè distans ab observationibus prius adhibitis[,] et radij qui in illa observatione ad Cometam ducitur planiꝗ $Sp\pi$ quæratur intersectio, ut et locus Cometæ in trajectoria ad tempus illius observationis. Si intersectio ista incidit in hunc locum, argumento est Trajectoriam rectè inventam esse. Sin minus, sumendus erit novus numerus V[36] et trajectoria nova invenienda, dein locus Cometæ in hac Trajectoria tempore observationis illius probatoriæ et intersectio radij cum plano Trajectoriæ determinandi ut prius. Et ex variatione erroris collata cum variatione aliarum quantitatum colligetur per regulam auream[37] quantæ debeant esse variationes seu correctiones illarū aliarum quantitatum ut error evadat quam minimus. Quibus adhibitis correctionibus habebitur Trajectoria exactè satis, posito quod computatio innixa fuit observationibus exactis quodꝗ non multum erratum fuit in assumptione quantitatis V. Nam si multum erratum fuit iterandum est opus eousꝗ dum Trajectoria inveniatur exactè satis. Q.E.F.[38]

---

over several following weeks to apply this defective construction to determining the trajectory of the 1680/1 comet. Had these computations thus (or indeed otherwise) survived, it would have been instructive to find out from them whether his present over-crude assumption (see note (26)) that a close trio of longitudinal sightings tr, TR, $\tau\rho$ of the comet may be taken as effectively passing through a common point O in the ecliptic was the sole reason for their failure.

(35) More accurately 'apparitionis suæ' (of its apparition) as the manuscript initially read.

(36) Though in the case of an exactly parabolic cometary trajectory V has (compare note (27)) the constant value $\sqrt{2}$, the minimally elliptical/hyperbolic orbits in which $V = \sqrt{2} \mp e$, e small, do not significantly diverge from being parabolic in the neighbourhood of the sun where alone they are visible, and in consequence this construction permits a small variation in the magnitude of V around its exact parabolic value $\sqrt{2}$.

(37) Understand 'proportionis' (of proportion). While the 'golden rule' usually signifies the 'Rule of Three' (see IV: 135, note (11)), it will be evident that here, much as in his preceding 'trigonometrical' solution to Lemma 1 above, Newton invokes the simple inter-

Problem 2.

To correct the assumed ratio of the speed and the trajectory found.

Apply an observation of the comet at the end of its motion(35) or some other very far distant from the observations before employed, and seek the intersection of the radius drawn to the comet in that observation with the plane $Sp\pi$, and also the place of the comet in its trajectory at the time of that observation. If that intersection falls at this place, it is a proof that the trajectory has been correctly ascertained. But if not, you will need to take a new number V(36) and find out a new trajectory, and then determine the place of the comet in this trajectory at the time of that test observation and the intersection of its radius with the plane of the trajectory, as before. And from a comparison of the variation of the error with the variation of the other quantities you will infer by the golden rule(37) the size of the variations or corrections of those other quantities necessary to make the error become minimal. Once these corrections are introduced, the trajectory will be obtained exactly enough, presuming that the computation is based on exact observations and that you were not too much out in your assumption of the quantity V. For if you were much out, the operation needs to be repeated until the trajectory proves to be exactly enough determined. As was to be done.(38)

---

polation by constant first differences which derives from presupposing the 'error' from true to vary linearly with some base argument (compare note (16)).

(38) The manuscript here breaks abruptly off at the top of f. 56r, again failing to adduce any numerical example in application of the preceding generalized argument, doubtless because Newton was unable to frame an adequate one in the primary instance of the comet of 1680/1. It is our conjecture (see note (1) above) that he then went briefly back to the Wrennian hypothesis of a uniformly traversed rectilinear cometary orbit in one last effort (about late September/early October 1685) to make it work as a viable approximation, but there too had his hopes dashed; and that only then did he return to attack the general problem of the parabolic orbit, so succeeding at length in contriving the complicated approach which he ultimately presented in Proposition XLI of the third book of his published *Principia* (and which we reproduce in the opening paragraphs of the following Appendix).

APPENDIX. REVISED COMPUTATION OF THE ELEMENTS OF A PARABOLIC COMETARY ORBIT(1)

[late 1685?/mid-1692]

From the first edition of Newton's *Principia* and an autograph fragment in the University Library, Cambridge

[1](2) *Lemma VII.*

Per datum punctum P ducere rectam lineam BC cujus partes PB, PC rectis duabus positione datis AB, AC abscissæ datam habeant rationem ad invicem.

A puncto illo P ad rectarum alterutram AB ducatur recta quævis PD, & producatur eadem versus rectam alteram AC usq; ad E ut sit PE ad PD in data illa ratione. Ipsi AD parallela sit EC, & si agatur CPB erit PC ad PB ut PE ad PD. Q.E.F.(3)

Lemma VIII

Sit ABC Parabola umbilicum habens S. Chordâ AC bisectâ in I abscindatur segmentum ABCI cujus diameter sit Iμ & vertex μ. In Iμ producta capiatur μO æqualis dimidio ipsius Iμ. Jungatur OS & producatur ea ad ξ ut sit Sξ æqualis 2SO. Et si Cometa B moveatur in arcu CBA & agatur ξB secans AC in E: dico quod punctum E abscindet de chorda AC segmentum AE tempori proportionale quamproximè.

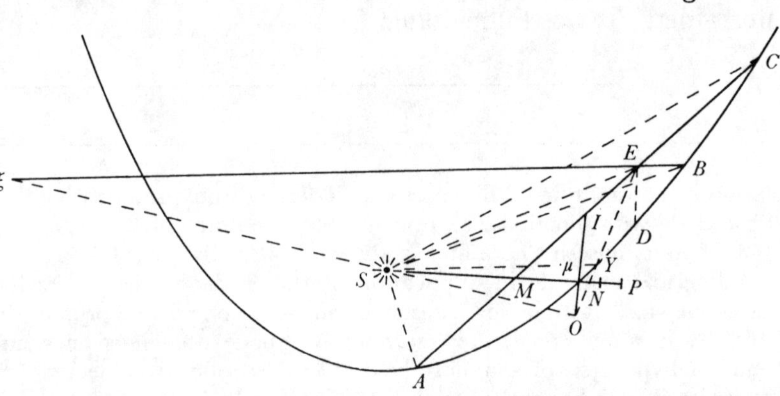

(1) This first successful approximate construction of a parabolic cometary orbit from given terrestrial sightings of its longitudes—one carefully applied by Newton himself in a following *Exemplum* in the published *Principia* to determine the elements of the 1680/1 comet on which he had earlier spent so much frustrated effort (see note (26) below) and afterwards analogously extended by Edmond Halley to fix the orbits of some twenty more in his *Astronomiæ Cometicæ Synopsis*—is here reproduced as a fitting climax to his previously unrewarded labours directed to achieving the same end (compare §2: note (1) preceding). Since Newton's inspiration for its initial formulation came shortly after he had given over the 'calculations to no purpose for want of a good method' on which, so he informed Halley on 20 June 1686, he had 'In Autumn last...spent two months' (*Correspondence*, 2: 437), a date of early winter 1685/6 for his discovery

Jungatur enim *EO* secans arcum Parabolicum *ABC* in *Y*,(4) & erit area curvilinea *AEY* ad aream curvilineam *ACY* ut *AE* ad *AC* quamproximè. Ideoqȝ cum triangulum *ASE* sit ad triangulum *ASC* in eadem ratione, erit area tota *ASEY* ad aream totam *ASCY* ut *AE* ad *AC* quamproximè. Cum autem ξO sit ad *SO* ut 3 ad 1 & *EO* ad *YO* prope in eadem ratione, erit *SY* ipsi *EB* parallela quamproximè, & propterea triangulum *SEB* triangulo *YEB* quamproximè æquale. Unde si ad aream *ASEY* addatur triangulum *EYB* & de summa auferatur triangulum *SEB*, manebit area *ASBY* areæ *ASEY* æqualis quamproximè, atqȝ adeo ad aream *ASCY* ut *AE* ad *AC*. Sed area *ASBY* est ad aream *ASCY* ut tempus descripti arcus *AB* ad tempus descripti arcus totius. Ideoqȝ *AE* est ad *AC* in ratione temporum quamproximè. Q.E.D.(5)

Lemma IX.

Rectæ $I\mu$ *&* μM & *longitudo* $\frac{AIC}{4S\mu}$ *æquantur inter se.*

Nam $4S\mu$ est latus rectum Parabolæ pertinens ad verticem $[\mu]$.(6)

---

of the essence of this renewed approach to the problem cannot be too far out, although the detail of its text (here, for want of any prior autograph manuscript, reprinted from the *editio princeps* in [1]) was doubtless considerably refined before being entered into the finished secretary transcript of the *Principia*'s third book some time before late March 1687 when it was sent off to Halley in London. For the minor complement which we give in [2] following a firm dating of *c.* early August 1692 is assured (see note (27)).

(2) *Principia*, $_1$1687: 484–90. In following footnotes we have made good use of A. N. Kriloff's excellent exegesis of this passage in his article 'On Sir Isaac Newton's Method of Determining the Parabolic Orbit of a Comet' (*Monthly Notices of the Royal Astronomical Society*, **85**, 1925: 640–56, especially 648–51, 655–6).

(3) This light remodelling of Lemma 2 in §2 preceding yields a trivially variant construction of the required intercept *BC*.

(4) In a slight emendation of his ensuing argument in the *Principia*'s second edition ($_2$1713: 449) Newton in his figure drew the tangent at μ as far as its intersection, at *X*, with *OE* and then here continued analogously: '& agatur μX quæ tangat eundem arcum in vertice μ & actæ *EO* occurrat in *X*; & erit area curvilinea $AEX\mu A$ ad aream curvilineam $ACY\mu A$ ut *AE* ad *AC*. Ideoqȝ...erit area tota $ASEX\mu A$ ad aream totam $ASCY\mu A$ ut *AE* ad *AC*. Cum autem ξO sit ad *SO* ut 3 ad 1, & *EO* ad *XO* in eadem ratione, erit *SX* ipsi *EB* parallela: & propterea si jungatur *BX*, erit triangulum *SEB* triangulo *XEB* æquale. Unde si ad aream $ASEX\mu A$ addatur triangulum *EXB*, & de summa auferatur triangulum *SEB*, manebit area $ASBX\mu A$ areæ $ASEX\mu A$ æqualis, atque adeo ad aream $ASCY\mu A$ ut *AE* ad *AC*. Sed areæ $ASBX\mu A$ æqualis est area $ASBY\mu A$ quamproxime, & hæc area $ASBY\mu A$ est ad aream $ASCY\mu A$, ut tempus descripti arcus *AB* ad tempus descripti arcus totius *AC*. Ideoque...'.

(5) In revise (*Principia*, $_2$1713: 449) Newton afterwards appended a '*Corol.* Ubi punctum *B* incidit in Parabolæ verticem μ, est *AE* ad *AC* in ratione temporum accurate': in this case, of course, when (it is understood) *A* and *C* are roughly equidistant from *B*, the diameter $I\mu$ will effectively coincide with the axis *SB* and so the area (μXY) will be wholly negligible. The autograph draft of a following '*Scholium.* Si jungatur $\mu\xi$ secans *AC* in δ, & in ea capiatur ξn quæ sit ad μB ut $27MI$ ad $16M\mu$: acta *Bn* secabit chordam *AC* in ratione temporum magis accurate quam prius...' is reproduced in [2] below.

Lemma X.

Si producatur $S\mu$ ad N & P ut μN sit pars tertia ipsius μI, & SP sit ad SN ut SN ad $S\mu$: Cometa quo tempore describit $A\mu C$, si progrederetur ea semper cum velocitate quam habet in altitudine ipsi SP æquali, describeret longitudinem æqualem chordæ AC.

Nam si(7) velocitate quam habet in μ eodem tempore progrediatur uniformiter in recta quæ Parabolam tangit in μ, area quam Radio ad punctum S ducto describeret æqualis esset areæ Parabolicæ $ASC\mu$. Ideoq contentum sub longitudine in Tangente descripta & longitudine $S\mu$ esset ad contentum sub longitudinibus AC & SM ut area $ASC\mu$ ad triangulum $ASCM$, id est ut SN ad SM. Quare AC est ad longitudinem in tangente descriptam ut $S\mu$ ad SN. Cum autem velocitas Cometæ in altitudine SP sit ad velocitatem in altitudine $S\mu$ in dimidiata ratione SP ad $S\mu$ inversè,(8) id est in ratione $S\mu$ ad SN, longitudo hac velocitate eodem tempore descripta erit ad longitudinem in Tangente descriptam ut $S\mu$ ad SN. Igitur AC & longitudo hac nova velocitate descripta, cum sint ad longitudinem in Tangente descriptam in eadem ratione, æquantur inter se. Q.E.D.(9)

Corol. Cometa igitur ea cum velocitate quam habet in altitudine $S\mu+\frac{2}{3}I\mu$, eodem tempore describeret chordam AC quamproximè.(10)

(6) By Lemma XIII of the *Principia*'s 'Liber primus' (see 1, §2, Appendix 2.9: note (38)). The equality of $I\mu$ and $M\mu$ here assumed is, we scarcely need to say, an immediate consequence of the parabola's property that the focal radius $S\mu$ and the diameter $I\mu$ are equally inclined to the tangent at μ, and hence to the chord AC parallel to it.

(7) Understand 'Cometa', a clarification afterwards inserted into the *Principia*'s revised text ($_2$1713: 450).

(8) Namely 'per Corol. 6. Prop. XVI. Lib. I' to quote Newton's later justification of this premiss (see *ibid.*).

(9) Since (as we showed in §2: note (10) preceding) the instantaneous speed in a parabola at a distance r from the sun is $2\pi\sqrt{[2/r]}$ astronomical units/year, it follows that the time, say $t_{A\to C}$, of orbit over the parabolic arc $\widehat{ABC}$ is $AC/2\pi\sqrt{2}.SP^{-\frac{1}{2}}$ years, where by Newton's present construction $SP = SN^2/S\mu = (S\mu+\frac{1}{3}I\mu)^2/S\mu$. As Lagrange long afterwards pointed out (*Mécanique Analytique. Nouvelle édition, revue et augmentée*, **2** (Paris, 1815): Partie II, Section VII, §26: 31), this is equivalent to an analytical result independently derived by Euler in his 'Determinatio Orbitæ Cometæ qui Mense Martio...anni 1742...fuit observatus' (*Miscellanea Berolinensia*, **7**, 1743: 1–90 [= *Leonhardi Euleri Opera Omnia* (2) **28** (Zurich, 1959): 28–104]): 15–17 [= 37–9]. In proof, because the distance of the general Cartesian point (x, y) of the parabola $y^2 = px$ from its focus $(\frac{1}{4}p, 0)$ is $x+\frac{1}{4}p$, at once $S\mu-\frac{1}{2}(SA+SC) = I\mu$, that is, $AC^2/16S\mu$; whence on putting $SA+SC+AC = \alpha$ and $SA+SC-AC = \beta$ there ensues

$$S\mu^2-\tfrac{1}{4}(\alpha+\beta)\,S\mu+\tfrac{1}{64}(\alpha-\beta)^2 = 0 \text{ and thence } S\mu = \tfrac{1}{8}(\alpha^{\frac{1}{2}}+\beta^{\frac{1}{2}})^2,$$

so that

$$t_{A\to C} = AC(S\mu+\tfrac{1}{48}AC^2/S\mu)/2\pi\sqrt{2}.S\mu^{\frac{1}{2}} = (\alpha^{\frac{3}{2}}-\beta^{\frac{3}{2}})/12\pi \text{ years.}$$

(Compare A. N. Kriloff, 'On a Theorem of Sir Isaac Newton', *Monthly Notices of the Royal Astronomical Society*, **84**, 1924: 392–5.)

(10) For $SP(= S\mu+\frac{2}{3}I\mu+\frac{1}{9}I\mu^2/S\mu) \approx S\mu+\frac{2}{3}I\mu$.

Lemma XI.

Si Cometa motu omni privatus de altitudine SN seu $S\mu+\frac{1}{3}I\mu$ demitteretur ut caderet in Solem, & ea semper vi uniformiter continuata urgeretur in Solem qua urgetur sub initio, idem quo tempore in orbe suo describat arcum AC descensu suo describeret spatium longitudini $I\mu$ æquale.

Nam Cometa quo tempore describat arcum Parabolicum AC, eodem tempore ea cum velocitate quam habet in altitudine SP (per Lemma novissimum) describet chordam AC, adeoq;[(11)] eodem tempore in circulo cujus semidiameter esset SP revolvendo,[(12)] describeret arcum cujus longitudo esset ad arcus Parabolici chordam AC in dimidiata ratione unius ad duo. Et propterea eo cum pondere quod habet in Solem in altitudine SP cadendo de altitudine illa in Solem, describeret eodem tempore (per Scholium[(13)] Prop. IV. Lib. I.) spatium æquale quadrato semissis chordæ illius applicato ad quadruplum altitudinis SP, id est spatium $\frac{AI^q}{4SP}$. Unde cum pondus Cometæ in Solem in altitudine SN sit ad ipsius pondus in Solem in altitudine SP ut SP ad $S\mu$: Cometa pondere quod habet in altitudine SN eodem tempore in Solem cadendo describet spatium $\frac{AI^q}{4S\mu}$, id est spatium longitudini $I\mu$ vel $M\mu$ æquale. Q.E.D.

Prop. XLI. Prob. XX.

Cometæ in Parabola moventis Trajectoriam ex datis tribus observationibus determinare.

Problema hocce longe difficillimum multimodè aggressus, composui Problemata quædam in Libro primo quæ ad ejus solutionem spectant.[(14)] Postea solutionem sequentem paulò simpliciorem excogitavi.

Seligantur tres observationes æqualibus temporum intervallis ab invicem quamproximè distantes. Sit autem temporis intervallum illud ubi Cometa tardius movetur paulo majus altero, ita videlicet ut temporum differentia sit

(11) Understand 'per Corol. 7. Prop. XVI. Lib. I' as Newton interjected in the *Principia*'s second edition ($_2$1713: 451).

(12) That is, 'vi gravitatis suæ' (compare *ibid.*).

(13) Changed to 'Corol. 9.' in the second edition (see *ibid.*).

(14) As we have earlier remarked (see 1, §3: note (1)) there are no traces in the surviving manuscript of the 'Liber primus' of any added problems relating to the construction of parabolic cometary paths from given sighting lines, even though Newton independently observed to Halley on 20 June 1686 that 'two months' of 'calculations to no purpose for want of a good method' the previous autumn had led him 'afterwards' to 'return to ye first Book & enlarge it wth divers Propositions some relating to Comets' (*Correspondence*, 2: 437); it may be that he here refers obliquely to his elegant construction, in the scholium to Lemma XXVII of his first book (see 1, §3: note (110)), of the 'ideal' uniformly traversed rectilinear path which, after further unsuccessful attempts to apply it to the 1680/1 comet (see v: 524–9), he had finally abandoned as a viable approximation to celestial reality by the late autumn of 1685.

ad summam temporum ut summa temporum ad dies plus minus sexcentos.[15] Si tales observationes non præsto sint, inveniendus est novus Cometæ locus per Lemma sextum.[16]

Designent S Solem, T, t, τ tria loca Terræ in orbe magno, TA, tB, τC observatas tres longitudines Cometæ, V tempus inter observationem primam & secundam, W tempus inter secundam ac tertiam, X longitudinem quam Cometa

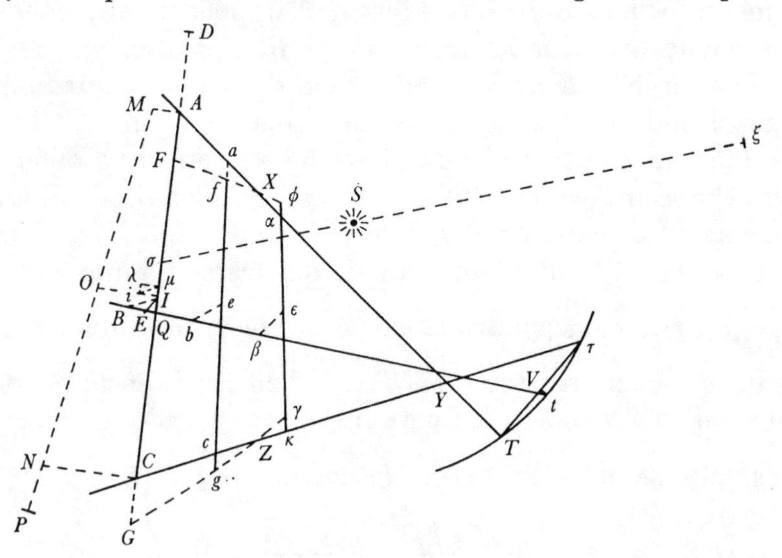

toto illo tempore ea cum velocitate quam habet in mediocri Telluris à Sole distantia describere posset, & tV perpendiculum in chordam $T\tau$. In longitudine media tB sumatur utcunq; punctum B, & inde versus Solem S ducatur linea BE quæ sit ad sagittam tV ut contentum sub SB & St quadrato ad cubum hypotenusæ[17] trianguli rectanguli cujus latera sunt SB & tangens latitudinis

(15) A loose empirical rule which Newton elected not to follow in his ensuing *Exemplum* of the 1680/1 comet, there choosing as his primary observations 'tres, quas *Flamstedius* habuit *Dec.* 21, *Jan* 5. & *Jan.* 25' (see note (26) below), differing respectively by 15 and 20 days. In an inserted clarification at this point in the second edition (*Principia*, ${}_{2}1713$: 452) he here went on: 'vel ut punctum E incidat in punctum M quamproxime, & inde aberret versus I potius quam versus A'.

(16) Namely, by applying the general formula for interpolation by finite differences which is expounded in the preceding Lemma V (*Principia*, ${}_{1}1687$: 481–3, reproduced on IV: 70–3) so as 'Cometæ in Parabola moventis Trajectoriam ex datis tribus observationibus determinare' (see *ibid.*: 483).

(17) Namely, the solar *radius vector* SB' at the point B' of the true cometary parabola $\widehat{A'B'C'}$ (in a plane inclined through S to the ecliptic on which its orthogonal projection is the parabolic arc $\widehat{ABC}$ here understood). Since the corresponding cometary *sagitta*

$$B'E'(=BE\times SB'/SB)$$

is to the terrestrial *sagitta* tV (generated in the same time) very nearly as the solar *vis gravitatis* acting at B' to that acting at t, that is, as SB'^{-2} to St^{-2}, Newton's stated proportion is an immediate consequence.

Cometæ in observatione secunda ad radium tB. Et per punctum E agatur recta AEC,[18] cujus partes AE, EC ad rectas TA & τC terminatæ sint ad invicem ut tempora V & W: Tum per puncta A, B, C duc circumferentiam circuli[19] eamq; biseca in i, ut & chordam AC in I. Age occultam Si secantem AC in λ & comple parallelogrammum $iI\lambda\mu$. Cape $I\sigma$ æqualem $3I\lambda$ & per Solem S age occultam[20] $\sigma\xi$ æqualem $3S\sigma + 3i\lambda$. Et deletis jam literis A, E, C, I à puncto B versus punctum ξ duc occultam novam BE quæ sit ad priorem BE in duplicata ratione distantiæ BS ad quantitatem $S\mu + \frac{1}{3}i\lambda$. Et per punctum E iterum duc rectam AEC eadem lege ac prius, id est ita ut ejus partes AE & EC sint ad invicem ut tempora inter observationes V & W.[21]

Ad AC bisectam in I erigantur perpendicula AM, CN, IO, qu[o]rum AM & CN sint tangentes latitudinum in observatione prima ac tertia ad radios TA & $\tau[C]$. Jungatur MN secans IO in O. Constituatur rectangulum $iI\lambda\mu$ ut prius. In IA producta capiatur ID æqualis $S\mu + \frac{2}{3}i\lambda$, & agatur occulta OD. Deinde in MN versus N capiatur MP quæ sit ad longitudinem supra inventam X in dimidiata ratione mediocris distantiæ Telluris à Sole (seu semidiametri orbis magni) ad distantiam OD.[22] Et in AC capiatur CG ipsi NP æqualis ita ut puncta G & P ad easdem partes rectæ NC jaceant.

Eadem methodo qua puncta E, A, C, G ex assumpto puncto B inventa sunt, inveniantur ex assumptis utcunq; punctis alijs b & β puncta nova e, a, c, g & ϵ, α, κ, γ. Deinde si per G, g, γ ducatur circumferentia circuli[23] $Gg\gamma$ secans rectam τC in Z, erit Z locus Cometæ in plano Eclipticæ. Et si in AC, ac, $\alpha\kappa$ capiantur AF, af, $\alpha\phi$ ipsis CG, cg, $\kappa\gamma$ respectivè æquales, & per puncta F, f, ϕ ducatur circumferentia circuli[23] $Ff\phi$ secans rectam AT in X, erit punctum X alius Cometæ locus in plano Eclipticæ. Ad puncta X & Z erigantur tangentes latitudinum Cometæ ad radios TX & τZ, & habebuntur loca duo Cometæ in orbe proprio. Deniq; (per Prop. XIX Lib. I.) umbilico S per loca illa duo describatur Parabola[24] & hæc erit Trajectoria Cometæ. Q.E.I.

(18) That is, 'per Lem. VII' as Newton afterwards specified in the second edition (*Principia*, ${}_2$1713: 452).

(19) This will evidently roughly approximate the true parabolic arc of focus S which Newton requires.

(20) In strict truth the line $\sigma\xi$ is not (partly) 'hidden' from our sight in Newton's figure, but rather suitably foreshortened so that the diagram may conveniently be contained within the confines of the printed page.

(21) 'Et erunt A & C loca Cometæ magis accurate' (*Principia*, ${}_2$1713: 453).

(22) Newton afterwards augmented the sequel to read: 'Si punctum P incidat in punctum N; erunt A, B, C tria loca Cometæ, per quæ Orbis ejus in plano Eclipticæ describi debet. Sin punctum P non incidat in punctum N; in recta AC capiatur CG...' (*ibid.*).

(23) Or any other 'smooth' freehand curve drawn through the points would, of course, serve equally well. It is not necessary, however, that the points F, f, ϕ and G, g, γ should straddle the sighting lines TA and τC, as Newton's figure has it.

(24) Namely, by first constructing its directrix (at distances from X and Z equal to SX and

Constructionis hujus demonstratio ex Lemmatibus consequitur: quippe cum recta AC secetur in E in ratione temporum per Lemma VIII; & BE per Lem. XI sit pars rectæ BS in plano Eclipticæ arcui ABC & chordæ $AE[C]$ interjecta; & MP (per Lem. VIII) longitudo sit chordæ arcus quem Cometa in orbe proprio inter observationem primam ac tertiam describere debet, ideoq; ipsi MN æqualis fuerit si modò B sit verus Cometæ locus in plano Eclipticæ.

Cæterum puncta B, b, β non quælibet sed vero proxima eligere convenit. Si angulus AQt in quo vestigium orbis in plano Eclipticæ descriptum secabit rectam tB præterpropter innotescat, in angulo illo ducenda erit recta occulta AC quæ sit ad $\frac{4}{3}T[\tau]$ in dimidiata ratione SQ ad St.(25) Et agendo rectam SEB cujus pars EB æquetur longitudini Vt, determinabitur punctum B quod prima vice usurpare licet. Tum rectâ AC deletâ & secundum præcedentem constructionem iterum ductâ, & inventâ insuper longitudine MP, in tB capiatur punctum b ea lege ut si TA, $[\tau]C$ se mutuò secuerint in Y, sit distantia Yb ad distantiam YB in ratione composita ex ratione MN ad MP & ratione dimidiata SB ad Sb. Et eadem methodo inveniendum erit punctum tertium β si modò operationem tertiò repetere lubet. Sed hac methodo operationes duæ ut plurimum suffecerint. Nam si distantia Bb perexigua obvenerit, postquam inventa sunt puncta F, f & G, g, actæ rectæ Ff & Gg secabunt TA & τC in punctis quæsitis X & Z.(26)

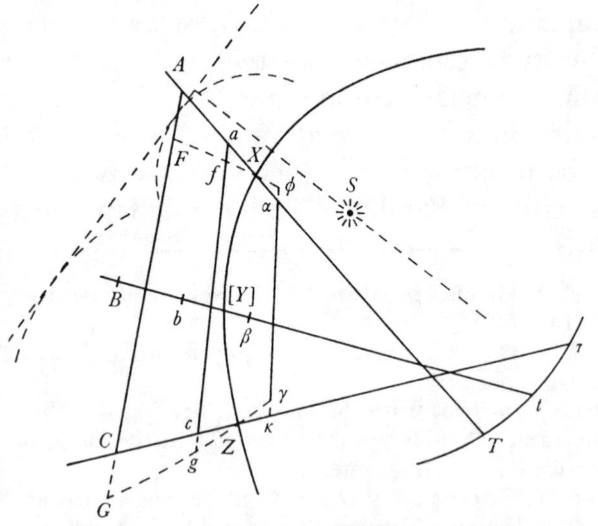

SZ respectively). Prior knowledge of the approximate position of the parabola's perihelion point will readily determine which of the two possible solutions is appropriate to the physical problem.

(25) The printed text (*Principia*, ₁1687: 490, corrected in ₂1713: 454) reads '*St* ad *SQ*' by a trivial transpositional error.

(26) Notice that Newton here implicitly assumes—whether forgetfully or for simplicity—that under orthogonal projection onto the ecliptic plane the focus of the cometary parabola will remain exactly centred in the sun and that its perihelion point passes into the vertex of its ecliptical projection. Neither supposition is rigorously true, but the error in the case of a 'normal' comet which grazes the sun's surface near perihelion will be inconsiderable and, where necessary, correction factors are easily applied. (Compare A. N. Kriloff, 'Newton's Method of Determining the Parabolic Orbit of a Comet' (note (2)): 648–51.)

In an immediately following '*Exemplum*' of this construction Newton applied it (*Principia*, ₁1687: 490–4) to determine the elements of the comet, that of 1680/1, upon which he had

[2][(27)]

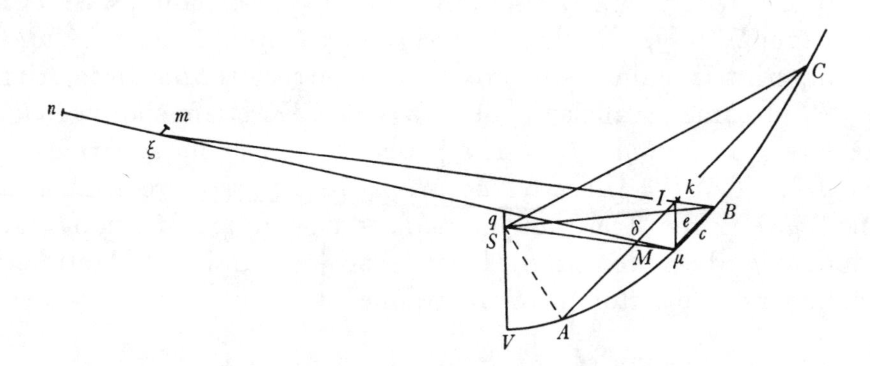

previously spent so much fruitless effort, taking as his empirical basis the set of observations of it, refined and corrected for refraction, which Flamsteed had, it would appear, sent him on 26 September 1685 in a now lost 'Tablet' (see v: 525, note (3)). As he went on to write (*Principia*: 493–4), 'selegi ex observationibus hactenus descriptis tres, quas *Flamstedius* habuit *Dec.* 21 [.6.36.59], *Jan.* 5.[6.1.38] & *Jan.* 25 [.7.58.42]. Ex his inveni *St* partium 9842,1 & *Vt* partium 455, quales 10000 sunt semidiameter orbis magni. Tum ad operationem primam assumendo *tB* partium 5657, inveni *SB* 9747, *BE* prima vice 412, $S\mu$ 9503, $i\lambda = 413$: *BE* secunda vice 421, *OD* 10186, *X* 8528,4, *MP* 8450, *MN* 8475, *NP* 25. Unde ad operationem secundam collegi distantiam *tb* 5640. Et per hanc operationem inveni tandem distantias *TX* 4775 & τZ 11322. Ex quibus orbem definiendo inveni Nodos ejus in ♋ & ♑ $0^{gr}. 53'$; Inclinationem plani ejus ad planum Eclipticæ $61^{gr}. 21\frac{1}{3}$[?]; verticem ejus (seu perihelium Cometæ) in [♐] $27^{gr}. 43'$ cum latitudine australi $7^{gr}. 34'$; & ejus latus rectum 236,8, areamq; radio ad Solem ducto singulis diebus descriptam 93585; Cometam verò *Decemb.* $8^{d}. 0^{h}. 4'$. P.M. in vertice orbis seu perihelio fuisse. Hæc omnia per scalam partium æqualium & chordas angulorum ex Tabula Sinuum naturalium collectas determinavi graphicè; construendo Schema satis amplum, in quo videlicet semidiameter orbis magni (partium 10000) æqualis esset digitis $16\frac{1}{3}$ pedis Anglicani'. These elements—accurately calculated by Newton (within the limits of empirical error of his three chosen observations) to within a unit or two, as an independent computation will confirm—do not differ significantly from the refined ones obtained more than a century later by J. F. Encke (in his 'Versuch einer Bestimmung der wahrscheinlichsten Bahn des Cometen von 1680', *Zeitschrift für Astronomie und verwandte Wissenschaften*, **6** (Tübingen, 1818): 27–120/129–208) from the totality of the thirty observations, corrected not only for atmospheric refraction but for optical aberration and further refined by least-squares analysis, which are recorded of the comet: namely, perihelion (in the true orbit) at December 17. $23^{h}\ 46'\ 9''$ at longitude ♐ 22° 49' at a distance from the sun (one-quarter its *latus rectum*) of 62.2, with its plane inclined at 60° 40' to the ecliptic and meeting it, in its ascending node, at a longitude ♑ 2° 9'. (The only serious discrepancy is in its computed period. Encke, relying heavily on a first, otherwise unchecked observation by Kirch on 3 November (O.S.), reckoned this to be 'most probably' between 6179.3 and 14030.6 years—usually rounded off in recent texts as 8813 years in a spurious show of precision—by determining the elliptical orbit which best fitted his 'good' observations. Borrowing heavily from Halley's privately circulated pamphlet 'De Motu Cometarum in Orbibus Ellipticis' (1719?), Newton himself in the third edition of his *Principia* ($_3$1726: 501–2) later came to accept the period of some 575 years which ensued from identifying 'previous apparitions' of the comet in 1106, 531 and 44 B.C. J. R. Hind has adequately disposed of this mythical cometary period—which Whiston afterwards made basic to his theory of the deluge and wherein too many

[In Parabola ABC vertice V, foco S descripta finge sectorem SAB æqualem esse sectori SBC; chordam AC biseca in I, duc $I\mu$ parallelam ipsi SV et in VS producta cape] $Sq = \frac{1}{3}I\mu$. [Incidat BI ipsi μq in ξ. Jam ubi puncta A, B, C sibi vicinia sunt, erit triangulum $SB\mu$ proximè æquale sectori $SB\mu$ adeoq; triangulo $SI\mu$, itaq; BI proximè parallela $S\mu$, ut et $\xi q = 2q\mu$. Est etiam chorda $B\mu$ quamproximè parallela CA, unde] $B\mu = IM$. [Biseca $B\mu$ in e et duc ec parallelam SV; fit] $AC^3 . MI^3 :: ACA . \mu B\mu$. [ut et] $ce . I\mu :: MI^q . AC^q$. [Pone sectorem $SB\mu$ æqualem triangulo $Sk\mu$ verissimè. Erit triangulum $SI\mu$ ad triangulum $MI\mu$ ut $S\mu$ ad $M\mu$ et triangulum $MI\mu$ ad sectorem $\mu B\mu$ ut $\frac{1}{2}I\mu \times MI$ ad $\frac{2}{3}ce \times \mu B$ sive ut $\frac{1}{2}I\mu$ ad $\frac{2}{3}ce$, unde triangulum $SI\mu$ est ad sectorem $\mu B\mu$ in ratione]

$$\tfrac{1}{2}I\mu \times S\mu . \tfrac{3}{2}^{(28)}ce \left(= \frac{I\mu,\, MI^q}{AC^q}\right) \times M\mu [\text{seu}]\ IB^{(29)} :: \delta I . Ik.$$

Newtonian historians still naïvely place their trust—in his excellent survey 'On the supposed Period of Revolution of the Great Comet of 1680', *Monthly Notices of the Royal Astronomical Society*, **12**, 1852: 142–50.) As to how Newton consistently achieved this incredible accuracy of to within 1 unit or some $16\frac{1}{3} . 10^{-4} \approx \frac{1}{600}$ inches in his scale drawing, Kriloff has observed that 'such a precision can be obtained if at the same time two drawings be made, the principal one which is described...and the other only an auxiliary one enlarged 50 to 100 times, containing only the points B, E and the parallelogram $I\lambda\mu i$....On the first drawing the directions of the chord and of the other lines are determined graphically, and their lengths computed; these directions are to be transferred to the second drawing on which the "corrections" of the lengths are then determined graphically' ('Newton's Method...': 655–6). His further query (*ibid.*: 656) as to whether Newton, who cites only the single value $St = 9842.1$, computed the lengths of the three basic solar *radii vectores* ST, St and $S\tau$ of the earth's orbital arc $\widehat{Tt\tau}$ by taking it to be a true ellipse, or—as on the face of it seems more likely—merely by applying a suitable correction factor to the corresponding vectors of the approximating circle drawn round the sun with the mean radius 10000, can only be answered when (and if) the worksheet calculations whose results Newton here summarises turn up. Though he normally elsewhere employed Flamsteed's solar longitudes in his astronomical calculations, there are rare manuscripts—notably ULC. Add. 3965.11: 15r (on which see 1,§ 3: note (160) above)—where he made use of a full elliptical theory with a simple 'Boulliau' upper-focus equant. (Compare J. A. Ruffner, 'The Background and Early Development of Newton's Theory of Comets' [Ph.D. thesis, Indiana University, 1966]: 346, note 63.)

(27) ULC. Add. 3965.13: 469r. These calculations—marred at a crucial point by a numerical slip which is carried into Newton's final proportion—and the concluding verbal enunciation of their result are the basis for the minor refinement of Lemma VIII in [1] preceding (see note (5) above) which was eventually introduced as a scholium thereto in the *Principia*'s second edition ($_2$1713: 449; compare note (32) below). Following our now established practice we have here fleshed out their bare skeleton with a number of editorial insertions which, we may hope, serve to render their argument more readily intelligible. A date for their penning in mid/late summer 1692 is all but irrefutably guaranteed by the circumstance that elsewhere on the same manuscript sheet Newton has drafted an early (and still unpublished) version of the letter which he subsequently sent to Locke on 2 August 1692 (see his *Correspondence*, **3**, 1961: 217–19, especially 218).

(28) Read '$\frac{2}{3}$'! This momentary inversion of numerator and denominator, here ostenta-

[id est] $S\mu, AC^q.3IB, MI^q :: \delta I.Ik$. [Duc ξm parallelam MI et occurrat Bk ipsis ξm, $\mu\xi$ in m et n. Erit]

$$BI.Ik\left(=\frac{3IB, MI^q, \delta I}{S\mu, AC^q}\right) :: B\xi.\xi m = \frac{3B\xi, MI^q, \delta I}{S\mu, AC^q}\text{ [adeoq}_3\text{]}$$

$$B\mu\text{ [seu } MI\text{]}.\xi[m] :: \mu n.\xi n = \frac{3B\xi, MI, \mu n, \delta I}{S\mu, AC^q} = \frac{3B\xi, \mu n, MI}{16S\mu^q, M\mu}, I\delta.^{(30)}$$

[hoc est, quia $B\xi = \mu n = 3S\mu$ proxime,] $16M\mu.27MI :: I\delta.\xi n$.[31]

Scholium. Si jungatur $\mu\xi$ secans AC in δ et in ea capiatur ξn quæ sit ad μB ut MI ad $\frac{16}{27}M\mu$: acta Bn secabit chordam AC in ratione temporum magis accuratè.[32]

---

tiously evident in sequel to our preceding interpolation, but hard to catch in the terse bleakness of the original manuscript, badly distorts the remainder of the computation. 'IB'

(29) Here and also in the next two lines the orginal reads $M\mu$, but in sequel we have silently made Newton's intended substitution.

(30) Since by Lemma XIII of the preceding 'Liber primus' (see 1, §2, Appendix 2.9) $\frac{1}{4}AC^2 = AI^2 = 4S\mu \times I\mu$.

(31) On correcting Newton's earlier numerical slip (see note (28)) this final proportion should read '$4M\mu.3MI :: I\delta.\xi n$'.

(32) Except that (to suit the printer's convenience) the proportion has been minimally altered to read equivalently 'ξn...ad μB ut $27MI$ ad $16M\mu$', this scholium duly passed into the *editio secunda* of the *Principia* ($_2$1713: 449) along with the cautionary remark 'Jaceat autem punctum n ultra punctum ξ, si punctum B magis distat a vertice principali Parabolæ quam punctum μ; & citra, si minus distat ab eodem vertice'. And so—the incorrectness (see notes (28) and (31)) of the proportion notwithstanding—it is reproduced therefrom in all subsequent editions down to the present day.

§3. COMPUTATION OF THE RATE OF MOTION OF THE MOON'S APOGEE AND ITS MEAN SECULAR ADVANCE.(1)

[*c.* late 1686]

[1](2) *Lemma* [α]

Si Luna P in orbe Elliptico QPR axem QR umbilicos S, F habente, revolvatur circa Terram S et interea vi aliqua V a pondere suo in Terram diversa continuò impellatur versus Terram; sit autem umbilicorum distantia SF infinitè parva: erit motus Apogæi(3) *ab impulsibus illis oriundus ad motum medium Lunæ circa Terram in ratione composita ex ratione duplæ vis V ad Lunæ pondus mediocre P, et ratione lineæ SE quæ centro Terræ et perpendiculo PE interjacet*(4) *ad umbilicorum distantiam SF.*

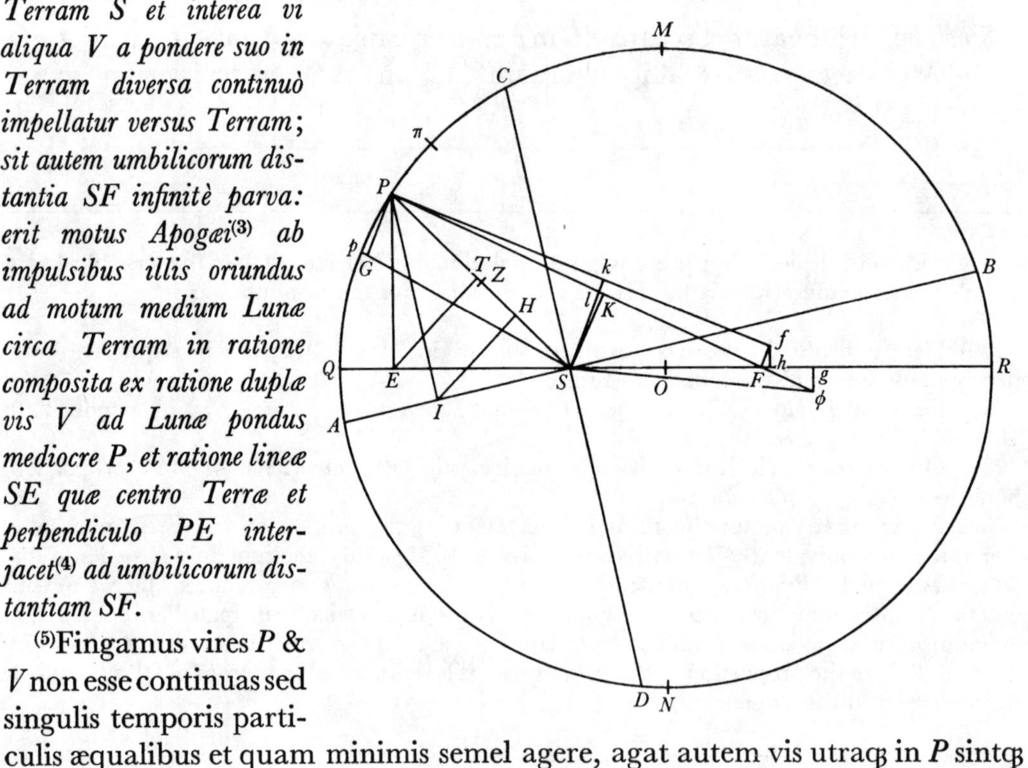

(5)Fingamus vires *P* & *V* non esse continuas sed singulis temporis particulis æqualibus et quam minimis semel agere, agat autem vis utraqꝫ in *P* sintqꝫ

(1) As our final illustration of the ways in which Newton's manuscript drafts and worksheets are, by preserving a record of the general structure (if not every fine detail) of his preliminary analyses, able to shed considerable light upon his treatment of a number of essentially mathematical topics which in the published *Principia* are given a minimally informative synopsis, we here take up his attempted solution of this most difficult problem of basic lunar theory—one which continued to tax the efforts and ingenuity of such skilled eighteenth-century geometers as Euler, D'Alembert and Clairaut till it was in 1748 resolved by the last on carrying the then standard analytical technique of approximation through to a higher order of accuracy. (Compare Robert Grant's *History of Physical Astronomy from the Earliest Ages to the Middle of the Nineteenth Century* (London, 1852): 44–6; see also H. Godfray, *An Elementary Treatise on the Lunar Theory* (London, $_3$1871): 67–8.) Newton's approach to the problem some sixty years earlier was far more geometrical in character and heavily founded on the brief survey of the principal lunar motions—which it was, indeed, initially framed to climax—given by him in

Translation

[1][2] *Lemma* [α].

If the moon P should revolve round the earth S in the elliptical orbit QPR having axis QR and foci S, F, and all the while be continuously impelled towards the earth by some force V different from its weight to the earth, but let the distance SF of its foci be indefinitely small: the motion of the apogee[3] *ensuing from that impulse will be to the mean motion of the moon round the earth in a ratio compounded of the ratio of twice the force V to the moon's average weight P and the ratio of the line SE lying between the earth's centre and the perpendicular PE*[4] *to the distance SF of the foci.*

[5]Let us imagine that the forces *P* and *V* are not continuous but that they act once in each individual equal and minimally small particle of time; let each

---

Propositions XXV–XXXV of the third book of his published *Principia* ($_1$1687: 434–62). There, taking as his basis the simple model in which the moon's orbit is, in the absence of any solar perturbation, assumed to be effectively a circle uniformly traversed round the earth at its centre, he was able hereby, on introducing (in Proposition XXV) a small disturbing force instantaneously directed towards the sun and varying as the square of the distance from its centre, to account with considerable success for the observed magnitudes of such major periodic and secular inequalities in the moon's motion as (in Propositions XXVI–XXIX) its variation in angular distance from uniform rotation in the orbital plane, and (in Propositions XXX—with a simple correction for a modified elliptical orbit in XXXI—and XXXII/XXXIII) the 'hourly motion' and mean regress of the nodes in which the moon's orbit intersects the ecliptic, together with (in Propositions XXXIV/XXXV) the latitudinal wobble of the plane of that orbit around its mean inclination to the ecliptic. In the concluding Corollary 2 to Proposition XLV of his 'De motu Corporum Liber primus', as we have seen (1, §3: note (260)), Newton had already broached the related problem of determining (within the confines of the simple circular model of the undisturbed orbit) the mean angular progress of the moon's apsides: having computed the component of the sun's perturbing force which is instantaneously directed radially through the earth's centre to be about $\frac{1}{357}$th part of the similar pull of terrestrial gravity upon the moon, he there deduced that this would of itself—implicitly ignoring, that is, the complementary transverse component of solar perturbation, which likewise acts (on average) to rotate the lunar apsidal line—be able to produce only some half of the mean annual advance of apogee which is in fact observed. In the manuscript we now reproduce Newton sought to refine this inconsequential calculation not only by taking into consideration both the radial and transverse components of the sun's disturbing force acting to deviate the moon from its simple terrestrial orbit, but also by introducing a sophisticated Horrocksian model of the orbit wherein (see note (28) below) it is assumed that the moon's pristine path is a Keplerian ellipse traversed round the earth at a focus and that the effect of solar perturbation is slightly to alter its eccentricity while maintaining the length (but not direction) of its major axis. Although by suitably adjusting a crucial ratio (see note (26))—'having', in his ambiguous accompanying phrase, 'undertaken the calculations'—he here contrives to arrive ultimately at a close approximation to the observed annual progress of lunar apogee, his computations seemed afterwards to him 'too complicated and cluttered with approximations, and not accurate enough' to be set in print after his other lunar propositions. Or in these apologetic words at least he contented himself with publicly appending thereto a short scholium (*Principia*, $_1$1687: 462–3, quoted in full in note (63) below) where he gave

πP particula Ellipseos quam Luna præcedente temporis particula descripsit, Pp particula ejusdem Ellipseos [quam] per impulsum vis solius P absq; impulsu vis V posteriore temporis particula describere deberet et PG particula orbis novi quem Luna per impulsum vis utriusq; V & P in loco P factum eadem posteriore temporis particula describeret. Et erit angulus pPG ad angulũ quem lineola pP cum lineola proximè ante appulsum Lunæ ad locum P descripta et producta contineat, id est ad angulum PSG(6) seu motum angularem Lunæ ut vis V qua angulus prior genitus est ad vim ponderis P qua angulus posterior genitus est. Agatur Pf ea lege ut angulus fPG complementum sit anguli SPG ad duos rectos et Pf transibit per umbilicum superiorem Ellipseos novæ, et quoniam angulus FPp (ex natura Ellipseos) complementum sit anguli SPp ad duos rectos, angulus FPf duplo major erit angulo $[p]PG$,(7) adeoq; eam habebit rationem ad angulum PSG quam habet vis $2V$ ad vim P. Sit f umbilicus iste superior, et in PF ac Pf demittantur perpendicula SK et Sk, quorum Sk secet PF in $[l]$. Et per ea quæ in Prop. [XVII] Lib. I(8) ostensa sunt, erit PF ad $SP+PF$ ut Ellipseos latus rectum quod nominabimus L ad $2SP+2PK$, et divisim PF [ad] SP ut L ad $2SP+2PK-L$ seu PF æqualis $\frac{L\times SP}{2SP+2PK-L}$, et [eo]dem argumento Pf æqualis $\frac{L\times SP}{2SP+2Pk-L}$. Nam latus rectum $[L]$ quod est (per [Prop. XIV] Lib I Princip.(9)) in duplicata

---

a bare announcement of his present numerical results, slightly scaled up in size for the occasion (or so it would appear). Even that minimal printed record of his lengthy efforts to calculate the rate of the moon's apsidal advance failed to survive into the *Principia*'s second edition, and was quickly forgotten till J. C. Adams called attention to it in 1888 (see the next note).

(2) ULC. Add. 3966.12: 105^r–$107^r/102^r[+107^v]$–$104^r/110^r$–111^r. The manuscript, initially transcribed (with minor additions) by his amanuensis Humphrey Newton from a much cancelled and reworked preliminary holograph draft (Add. 3966.12: $112^r[\rightarrow 108^r/108^v]$–$112^v/108^v$–$109^r$) in which the introductory lemmas (unnumbered in the original text but here distinguished as α and β respectively) formed the opening paragraphs of the first of the two following propositions (here named A and B), has been subsequently much revised and augmented by Newton himself. (Compare I. B. Cohen, *Introduction to Newton's 'Principia'* (Cambridge, 1971): 116–19, 350–1.) The variants between these states are largely verbal, and only a few of present significance are here noticed in sequel; because of its interest and importance, however, we reproduce Newton's unfinished terminal revise of the text of Proposition A separately in [2] following. Lemmas α and β have previously been published by J. C. Adams in preface to *A Catalogue of the Portsmouth Collection of Books and Papers written by or belonging to Sir Isaac Newton* (Cambridge, 1888): xxvi–xxx, along with a short introductory summary (*ibid.*: xii–xiii) of how they are further applied to 'deduce the horary motion of the moon's apogee...and [its] mean hourly motion...[although] the investigation on this point is not entirely satisfactory'. For want of any surviving record of the chronological sequence of Newton's researches during 1686–7 the date of composition can only roughly and tentatively be assigned, but they were evidently given a final reworking some little time before late March 1687 when Newton sent off to Halley in London the printer's script of the *Principia*'s third book containing the short scholium (see note (63) below) whose meagre résumé of results ultimately replaced it.

(3) '*Aphelij*' (*aphelion*) was carelessly first written; compare note (12).

force, however, act at P, and let πP be the particle of the ellipse which the moon has described in the preceding particle of time, Pp the particle of the same ellipse which it were to describe in the next particle of time after through the impulse of force P alone without the impulse of force V, and PG the particle of the new orbit which the moon would describe in the same next particle of time through the joint impulse of the forces V and P effected at the place P. The angle $p\widehat{P}G$ will then be to the angle contained by the arc-element pP with the extension of the element described immediately before the moon's arrival at the place P, that is, the angle $P\widehat{S}G$[6] or the angular motion of the moon, as the force V by which the former angle was generated to the force of the weight P whereby the latter angle was generated. Draw Pf in such a way that the angle $f\widehat{P}G$ shall be the supplement of $S\widehat{P}G$, and Pf will pass through the upper focus of the new ellipse; and, because the angle $F\widehat{P}p$ is (from the nature of an ellipse) the supplement of $S\widehat{P}p$, the angle $F\widehat{P}f$ will be twice as great as $p\widehat{P}G$,[7] and hence will have to $P\widehat{S}G$ the ratio which the force $2V$ bears to the force P. Let f be that upper focus, and onto PF and Pf let fall the perpendiculars SK and Sk, of which Sk shall intersect PF in l. Then, by what has been shown in Proposition XVII of Book 1,[8] PF will be to $SP+PF$ as the ellipse's *latus rectum*—we shall name it L—to $2SP+2PK$, and *dividendo* PF to SP as L to $2SP+2KP-L$, that is, PF will be equal to $L\times SP/(2SP+2PK-L)$, and by the same argument Pf is equal to

$$L\times SP/(2SP+2Pk-L);$$

for the *latus rectum* L, which is (by *Principia*,[9] Book 1, Proposition XIV) in the

(4) Originally just 'perpendiculo PE subtensæ' (subtended by the perpendicular PE).

(5) A superfluous introductory '*Cas. 1*', matched by no corresponding second case of the lemma, is here omitted.

(6) Newton assumes that the angles $\pi\widehat{P}S$ and $P\widehat{p}S$ differ only negligibly in size.

(7) The manuscript mistakenly reads 'QPG'. From Newton's ubiquitous assumption (see note (1)) that the disturbed lunar orbit is again instantaneously a Keplerian ellipse continuing to be traversed round the earth at its primary focus S but now with the new second focus f, it readily follows that

$$F\widehat{P}f+F\widehat{P}G \text{ (or } f\widehat{P}G) = 180°-S\widehat{P}G \quad\text{and also}\quad F\widehat{P}G+p\widehat{P}G \text{ (or } F\widehat{P}p) = 180°-S\widehat{P}p,$$

whence at once $F\widehat{P}f = 2p\widehat{P}G$.

(8) That is, of the 'De motu Corporum Liber primus', the published *Principia*'s first book. In the manuscript Newton has here omitted to specify his reference thereto, but we have filled the lacuna by citing his intended Proposition XVII [= Proposition XVI of the preliminary 'De motu Corporum' reproduced in 1, §2 above]. A number of similarly incomplete references in Newton's text are likewise mended in sequel by our editorial hand.

(9) Compare the previous note. We may observe *en passant* that Newton had already settled on this title for his work by late April 1686 when (see *Correspondence*, 2: 432, note 4) its 'Liber primus' was formally presented by Vincent to the Royal Society. The badly fragmented

ratione [areæ quam] Luna radio ad Terram ducto singulis temporis par[ticulis describit, quoniam] quantitas areæ illius per impulsum vis V nil mutetur, idem manet in Ellipsi utraq;. Cum autem $2SP$ et L ob infinitè parvam distantiam SF æquentur, deleatur $2SP-L$ et erit PF æqualis $\frac{SP^q}{PK}$ et Pf æqualis $\frac{SP^q}{Pk}$ quarum differentia est $\frac{SP^q\times lK}{PK^q}$ seu lK. Est autem lK ad lk ut SK ad Pk, ideoq; (ob infinitè parvam SF) est lK infinitè minor quàm lk seu Ff et propterea Ff perpendicularis est ad PK.(10) Quare si jungatur Ef, anguli FEf & FPf, in segmento circuli per puncta P, E, F, f transeuntis consistentes, æquales erunt inter se. Ideoq; cùm angulus FSf sit ad angulum FEf ut FE vel SE ad FS seu $2OS$, et angulus FPf supra fuerit ad angulum PSG ut vis $2V$ ad vim P: erit ex æquo angulus FSf ad angulum PSG, id est motus Apogæi ad motum medium Lunæ ut $2V\times SE$ ad $P\times SP$ seu $V\times SE$ ad $P\times OS$. Concipe jam numerum impulsuum augeri et intervalla diminui in infinitum ut actiones virium V et P reddantur continuæ et constabit Propositio. Q.E.D.

Corol. Valet Propositio quamproximè ubi excentricitas finitæ est magnitudinis, si modo parva sit.

Lemma [β].

Si Luna P, in orbe Elliptico QPR axem QR et umbilicos S, F habente revolvatur circa Terram, et interea a vi aliqua W a pondere suo(11) *diversa secundum lineam distantiæ SP perpendicularem impellatur; sit autem excentricitas OS infinite parva: erit motus* [*Apogæi*](12) *ab impulsu illo oriundus ad motum medium Lunæ in ratione composita ex ratione* [*duplæ*](13) *vis W ad pondus P et ratione perpendiculi PE ad excentricitatem OS.*

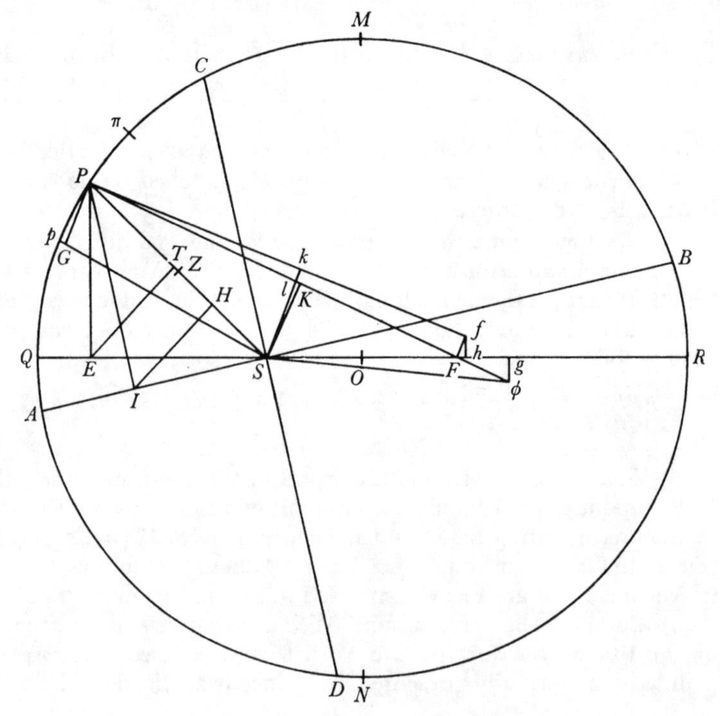

(14)Distinguatur enim tempus in particulas æquales et quàm minimas, et agat vis W non continuò sed singulis temporis particulis semel. Sit autem T velocitas Lunæ [in] P ante

doubled ratio of the area which the moon by a radius drawn to the earth describes in individual particles of time, will, because the quantity of that area is totally unchanged by the impulse of force V, remain the same in either ellipse. But since, in consequence of the infinitely small distance SF, $2SP$ and L are equal, delete $2SP-L$, and PF will then be equal to SP^2/PK and Pf equal to SP^2/Pk, the difference of which is $(SP^2/PK^2)\times lK$ or lK. However, lK is to lk as SK to Pk, and in consequence (because SF is infinitely small) lK is infinitely less than lk, that is, Ff, and accordingly Ff is perpendicular to PK.(10) Wherefore, if Ef be joined, the angles $\widehat{FEf}$ and $\widehat{FPf}$ standing in a segment of the circle passing through the points P, E, F, f will be equal to one another. Consequently, since $\widehat{FSf}$ is to $\widehat{FEf}$ as FE or SE to FS or $2OS$, while $\widehat{FPf}$ was above to $\widehat{PSG}$ as the force $2V$ to the force P, then *ex æquo* will $\widehat{FSf}$ to $\widehat{PSG}$, that is, the motion of the apogee to the mean motion of the moon, be as $2V\times SE$ to $P\times SP$, or as $V\times SE$ to $P\times OS$. Conceive now that the impulses are increased in number, and their intervals diminshed, indefinitely so as to render the actions of the forces V and P continuous, and the proposition will be established. As was to be proved.

Corollary. The proposition is valid to a close approximation when the eccentricity is of finite size, provided it be small.

Lemma [β].

If the moon P should revolve round the earth in the elliptical orbit QPR having axis QR and foci S, F, and all the while be impelled by some force W different from its weight(11) *along a line perpendicular to the distance SP, but let the eccentricity OS be indefinitely small: the motion of the apogee*(12) *ensuing from that impulse will be to the mean motion of the moon in a ratio compounded of the ratio of twice*(13) *the force W to the weight P and the ratio of the perpendicular PE to the eccentricity OS.*

(14)For let the time be separated into equal, minimally small particles, and the force W act not continuously but once in each individual particle of time. Let,

bottom left-hand corner of the manuscript sheet has at this point—and in patches at corresponding intervals below—necessitated a certain amount of editorial restoration (within square brackets) of its original text.

(10) And consequently, since the angle $\widehat{FPf}$ of their mutual inclination is conceived to be vanishingly small, to Pf also; accordingly, the points E and f will (to within an infinitesimal error) lie on the circle whose diameter is PF, as Newton goes on to assume.

(11) Understand '*in terram*' (*towards the earth*), as the manuscript originally continued. Why Newton cancelled this qualification is not clear (perhaps he took its point to be self-evident?) Initially, in the previous line, he had correspondingly described the moon as being continually deviated into its changing elliptical orbit 'vi ponderis sui qui Terram petit' (by the force of its weight directed to the earth).

(12) In a repetition, here uncorrected, of its earlier slip in naming the moon's upper apsis (see note (3)) the manuscript has '*Aphelij*' (*aphelion*).

(13) Inconsistently with the lemma's result, the manuscript reads '*quadruplæ*' (*four times*) in error.

impulsum vis W ibi factum et t incrementum [velo]citatis ex impulsu et L latus rectum orbis Lunaris ante [impulsu]m. Et quoniam area, quam Luna radio ad Terram [ducto singulis tem]poris particulis æqualibus describit, sit ante impulsum ad eandem aream post impulsum ut T ad $T+t$, et latus rectum (per Prop. XIV Lib. I Princip.) sit in duplicata ratione areæ, erit (per Lem. [II] Lib. II Princip.)[15] $\frac{T+2t}{T}L$ seu $L+\frac{2t}{T}L$ latus rectum post impulsum. Est autem (ut in Lemmate superiore) $\frac{SP\times L}{2SP+2PK-L}$ longitudo PF qua Luna distabat ab umbilico superiore ante impulsum; et propterea cum situs lineæ PF, si modò excentricitas $S[O]$ infinitè parva sit, ex impulsu illo nil mutetur,[16] ideoq; PK maneat eadem quæ prius[17] et solum L mutetur, si producatur PF ad ϕ ut sit ϕ umbilicus superior post impulsum, erit $P\phi$ æqualis

$$\frac{SP\times\frac{T+2t}{T}L}{2SP+2PK-L-\frac{2t}{T}L}.$$

De hac longitudine subducatur longitudo ipsius PF superius inventa, nempe $\frac{SP\times L}{2SP+2PK-L}$, et interea in utraq; pro $2SP+2PK$ scribatur $2L$ & manebit differentia $F\phi$ æqualis $\frac{4t}{T-2t}SP$ seu[18] $\frac{4t}{T}SP$. Unde longitudo perpendiculi ϕg quod in diametrum QR ab umbilico ϕ demittitur, erit $\frac{4t}{T}PE$.[19] Jam vero in Lemmate superiore, velocitas quam vis V impulsu unico generare potest, est ad velocitatem[20] Lunæ ut lineola pG quam Luna vi impulsus illius dato tempore describere posset ad lineolam Pp quam Luna velocitate sua data T eodem tempore describat, id est ut $[\frac{1}{2}]\,Ff$ ad PF ob angulum FPf anguli GPp duplum,[21] ideoq; si velocitas prior nominetur S erit Ff æqualis $\frac{2S\times PF}{T}$, et perpendiculum fh quod ab umbilico f in Ellipseos axem QR demittitur æquale $\frac{2S}{T}EF$. Proinde cum angulus ϕSF sit ad angulum FSf ut ϕg ad fh, et angulus FSf ad angulum PSp ut $V\times SE$ ad $P\times OS$,[22] erit angulus ϕSF ad angulum PSp, hoc est motus Apogæi a vi W genitus ad motum medium

(14) A superfluous subhead '*Cas. 1*', again without any mate in the ensuing text of the lemma, is once more (compare note (5)) here omitted.

(15) See IV: 521–5. In the terminology of that lemma, of course, t is the infinitesimal 'moment' of the speed T, whence in the limit as it vanishes $((T+t)/T)^2 = (T+2t)/T$. Observe that Newton again implicitly assumes that the disturbed lunar orbit will at all times be instantaneously a Keplerian ellipse.

(16) More accurately, such a transverse impulse will induce a slight rotary movement of SP round the focus S which is wholly negligible in comparison with the increment $\widehat{Pp}$ traversed

again, T be the moon's speed at P before the impulse of the force W is effected upon it, t the increment in the speed resulting from the impulse, and L the *latus rectum* of the lunar orbit before the impulse. Then, because the area which the moon describes by a radius drawn to the earth in each equal, individual particle of time before the impulse is to the same area after the impulse as T to $T+t$, while the *latus rectum* is (by *Principia*, Book 1, Proposition XIV) in the doubled ratio of the area, $((T+2t)/T)\,L$ or $L+(2t/T)\,L$ will (by *Principia*, Book 2, Lemma II)(15) be the *latus rectum* after impulse. However, (as in the previous Lemma) $SP\times L/(2SP+2PK-L)$ is the length PF the moon was distant from the upper focus before impulse; and accordingly, since the position of the line PF is, provided the eccentricity SO be infinitely small, changed not at all as a result of the impulse(16) and consequently PK remains the same as before,(17) L alone being changed, if PF be produced to ϕ such that ϕ be the upper focus after impulse, then $P\phi$ will be equal to $SP\times(L+(2t/T)\,L)/(2SP+2PK-L-(2t/T)\,L)$. From this length take the length of PF found above, namely $SP\times L/(2SP+2PK-L)$, and in so doing in place of $2SP+2PK$ in each write $2L$, and there will remain the difference $F\phi$ equal to $(4t/(T-2t))\,SP$, that is,(18) $(4t/T)\,SP$. Whence the length of the perpendicular ϕg let fall to the diameter QR from the focus ϕ will be $(4t/T)\,PE$.(19) Now it is in the previous Lemma that the speed which the force V can generate by a single impulse is to the speed(20) of the moon as the line-element pG which the moon were able to describe by the force of that impulse in given time to the element Pp which the moon shall describe with its given speed T in the same time, that is, as $\frac{1}{2}Ff$ to PF because the angle $F\widehat{P}f$ is twice $G\widehat{P}p$;(21) consequently, if the former speed be named S, then Ff will be equal to $(2S/T)\,PF$, and the perpendicular fh let fall from the focus f onto the ellipse's axis QR equal to $(2S/T)\,EF$. As a result, since the angle $\phi\widehat{S}F$ is to $F\widehat{S}f$ as ϕg to fh, and the angle $F\widehat{S}f$ to $P\widehat{S}p$ as $V\times SE$ to $P\times OS$,(22) the angle $\phi\widehat{S}F$ to $P\widehat{S}p$, that is, the motion of the apogee generated by the force W to the mean motion of the moon, will be as

(in the disturbed orbit) in the same infinitesimal time and (since the orbital eccentricity is taken to be indefinitely small) in the same direction; by itself, therefore, it will fail appreciably to alter the position either of SP or of the infinitely close line PF drawn to the second focus F. It readily follows, as Newton goes on to conclude, that the impulse likewise leaves the segment PK (cut off in PF by the perpendicular SK) effectively unchanged in magnitude.

(17) As an explicit premiss to the following assertion Newton needs to insert some such phrase as 'ut et *SP*' (as also does *SP*), evident though this is from his assumption (see the previous note) that the transverse impulse leaves the focal line unaltered in position.

(18) Since the increment t is infinitesimally small in comparison with T.

(19) For, because the triangles $F\phi g$ and FPE are similar, and since the focal distance SF is assumed to be vanishingly small, there is $F\phi:\phi g = FP(\approx SP):PE$.

(20) Namely T.

(21) See note (7) above. We have advanced this justificatory phrase—a late, hasty insertion by Newton in the manuscript—to its more logical position in the text.

(22) By Lemma α, of course.

Lunæ ut $\frac{4t}{T}PE$ ad $\frac{2S}{T}EF$ et $V \times SE$ ad $P \times OS$ conjunctim, id est (ob æquales[23] EF [&] SE et proportionales t & S, W & V) ut $2W \times PE$ ad $P \times OS$. Q.E.D.

Corol. Obtinet etiam Propositio quamproxime ubi [exigua est] excentricitas, etiamsi non sit infinitè parva.

Prop. [*A*]. *Prob.*

Invenire motum horarium Apogæi Lunæ.

Designet $ABCD$ orbem Lunæ, A & B syzygias, C & D quadraturas, S Terram, P Lunam & PI & $2IS$ vires Solis ad perturbandos motus Lunares ut supra;[24] quarum virium altera PI trahat Lunam versus I, altera $2IS$ distrahat Lunam a linea CD. Jungatur PS & in eam demittatur normalis IH et vis PI resolvetur in vires PH & IH, atq; vis $2IS$ in vires $2IH$ & $2HS$. Hæ vires ubi Luna in orbe illo versatur in quo per easdem sine excentricitate revolvi posset, ad motum Apogæi nil conducunt. Oritur motus Apogæi a differentijs inter has vires & vires quæ in recessu Lunæ ab orbe illo concentrico si centripetæ sunt vel centrifugæ[25] decrescunt in duplicata ratione qua distantia inter Lunam & centrum Terræ augetur[,] sin agant

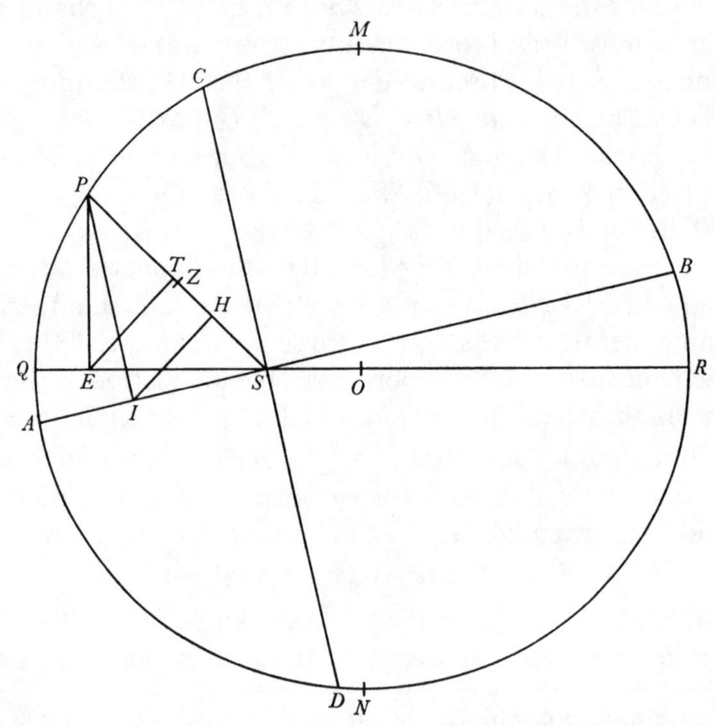

(23) On ignoring, that is, the indefinitely small focal distance SF by which they differ.

(24) Understand, *mutatis mutandis*, Proposition XXVI of the published *Principia* third book ($_1$1687: 435): in this limit form of the preceding Proposition XXV (*ibid.*: 434–5) the instantaneous perturbing effect of the sun (assumed to be effectively at infinity) on the moon P (taken to orbit the earth in a circle, with the present elliptical eccentricity SO vanishing) is, on taking the radius of the earth's solar orbit to represent its gravity towards the sun, there shown to be compounded of a force $MS = 3KP$ directed through S along the line of syzygies BA and of a force LM equal and parallel to the lunar radius PS; that is, equivalently, of a force

$(4t/T)\,PE$ to $(2S/T)\,EF$ and $V \times SE$ to $P \times OS$ jointly, that is, (because EF and SE are[23] equal, and t to S is the same ratio as W to V) as $2W \times PE$ to $P \times OS$. As was to be proved.

Corollary. The proposition also holds very nearly true when the eccentricity is slight, even though it be not indefinitely small.

Proposition [*A*], *Problem.*

To ascertain the hourly motion of the moon's apogee.

Let $ABCD$ denote the orbit of the moon, A and B its syzygies, C and D its quadratures, S the earth, P the moon, and PI and $2IS$ the forces of the sun to perturb the lunar motions, as above;[24] of these forces one, PI, shall draw the moon towards I and the other, $2IS$, shall draw the moon away from the line CD. Join PS and to it let fall the normal IH, and the force PI will then be resolved into the forces PH and IH, and the force $2IS$ into the forces $2IH$ and $2HS$. These forces do not, when the moon is located in the orbit in which by their means it might revolve without eccentricity, contribute at all to the motion of the apogee. The motion of the apogee arises from the differences between these forces and forces which in the moon's recession from that concentric orbit decrease, if they are centripetal or centrifugal,[25] in the doubled ratio of the (increasing) distance between the moon and the earth's centre, but, should they act laterally, in the

$LN = PI$ perpendicular to BA and of a force $NS = 2SI$ acting along it. (For where σ is the place of the sun—in loose terms, the 'meet at infinity' of the parallels LKP and BSA—and $\sigma S\,(=\sigma K)$ denotes the earth's gravity to the sun, then

$$\sigma L = \sigma S.(\sigma P^{-2}/\sigma S^{-2})$$

represents the moon's inverse-square gravitation to the sun along $P\sigma$. Accordingly, the vectorial difference $\overline{\sigma L}-\overline{\sigma S} = \overline{LM}+\overline{MS}$, where

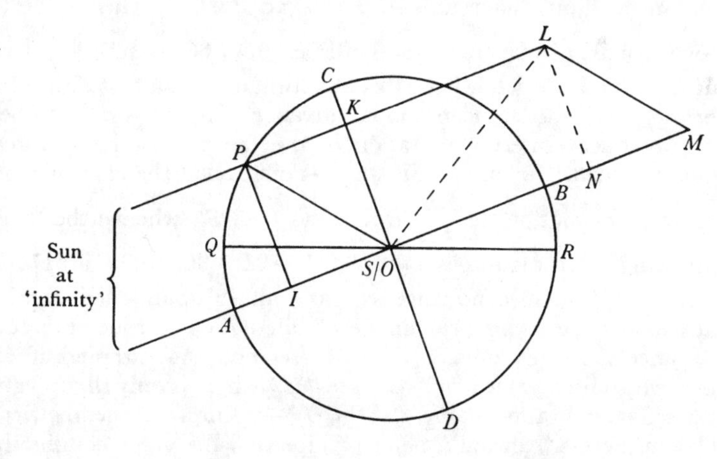

$$MS = LP = \sigma K^3/\sigma P^2 - \sigma P \approx \sigma K(1+2KP/\sigma P) - \sigma P \approx 3KP,$$

evidently represents in this way the perturbing action of solar gravity in deflecting the moon from its (virtually) uniform circular orbit round the earth.)

(25) This inserted proviso was unnecessary in the original version of the manuscript sentence, which concluded *tout court* '...inter has vires & vires quæ in recessu Lunæ ab Orbe illo concentrico decrescunt in duplicata ratione qua distantia inter ipsam & centrum Terræ augetur' (...between these forces and forces which in the moon's recession from that orbit decrease in the doubled ratio of the distance between it and the earth's centre). See the next note.

in latus decrescunt in eadem ratione triplicata uti calculis initis inveni.[(26)] Hunc orbem Ovalem esse supra[(27)] ostensum est. Et inde fit ut orbis excentricus, in quo Luna verè revolvitur, non sit Ellipsis sed alterius cujusdam generis Ovalis. At si ad minuendam difficultatem calculi supponamus orbem illum concentricum esse circulū, orbis excentricus evadet Ellipsis, cujus axis major $Q[R]$ æqualis erit diametro circuli.[(28)] Et vires prædictæ PH, IH, $2IH$ & $2HS$, si Luna de orbe Elliptico in circularem transferretur, evaderent $\frac{QO}{SP}PH$, $\frac{QO}{SP}IH$, $\frac{2QO}{SP}IH$ & $\frac{2QO}{SP}HS$, ac deinde si Luna hinc in orbem ellipticum redeunte vires centripetæ $\frac{QO}{SP}PH$ et $\frac{2QO}{SP}HS$ diminuerentur in duplicata ratione distantiæ SP Lunæ a centro Terræ [ad QO] et vires laterales

---

(26) By which Newton signifies, however ambiguously, not that he has some sophisticated dynamical reason for adducing so strange a ratio for the lateral disturbing force, but simply that the cube of the moon's varying distance from the earth proves, on making the requisite calculations, best geared to yielding a physically realistic value for the mean annual motion of its apogee. In fact, in the first version of this sentence quoted in the previous note (though essentially equivalent, Newton's autograph draft—unfinished—of the present proposition on ff. $108^r/108^v$ is somewhat different in its detail) Newton had supposed the lateral ratio to be as $(SP/QO)^2$, and then went on to deduce that the 'motus Apogæi a vi illa oriundus est ad motum medium Lunæ ut $\frac{18x}{SP}IH\times PE$ ad $P\times OS$', whence the total 'motus Apogæi' to the corresponding lunar mean motion is as '$6D+6C$, $-3D+3C$, $+18C$ ad $178\llcorner 725\times 4D$, hoc est ut $D+9C$ ad $238\frac{3}{10}D$'; from this he ultimately, in a first draft of Proposition B (see note (50) below), derived the mean annual advance of lunar perigee to be $37°\ 2'$, some $3\frac{1}{2}°$ too small. A subsequent interim alteration of the lateral disturbing force to be 'in eadem ratione quadruplicata'—namely, $(SP/QO)^4$—determined the resulting 'motus Apogæi' to be to the mean lunar motion 'ut $\frac{30x}{SP}IH\times PE$ ad $P\times OS$', whence the total such 'motus' to the corresponding mean motion is as '$6D+6C$, $-3D+3C$, $+30C$ ad $178\llcorner 725\times 4D$, hoc est ut $D+13C$ ad $238\frac{3}{10}D$'; this second tentative was similarly abandoned after Newton, forgetfully returning at a late stage to his prior mode of calculation, erroneously computed the ensuing annual advance of apogee to be only $35°\ 51'$ (see note (54)), though if we correct his terminal deviation we attain a value of $42°\ 49'$ (see *ibid.*) which is only slightly greater than true. The present revised lateral ratio—namely, $(SP/QO)^3$—is simply the geometrical mean of his two previously chosen powers. Newton's manifest hope that the value computable therefrom for the annual advance of apogee will lie close to that, $40°\ 41'$, which Flamsteed had recently obtained by observation (see note (63) below) is here straightforwardly realised in sequel. (It is not difficult, in fact, to show that where the lateral ratio is taken to be $(SP/QO)^i$, so yielding a 'motus Apogæi' of $(D+(2i+5)\,C)/238.3D$, the resulting mean annual advance is roughly

$$\left(20+\tfrac{2}{3}\left(i+\tfrac{5}{2}\right)^2\right)°;$$

compare note (58) below.)

Whatever we may think of so accurately and adeptly fudged a result, Newton himself for one reason and another remained considerably dissatisfied with his amended value of almost 39°

same ratio tripled—as I found once I undertook the calculations.[26] This orbit has above[27] been shown to be oval. And thence it results that the eccentric orbit in which the moon really revolves is not an ellipse but an oval of another kind. But if, to lessen the difficulty of the computation, we should suppose that the concentric orbit is a circle, the eccentric orbit will turn out to be an ellipse whose major axis QR will be equal to the circle's diameter.[28] And the above-mentioned forces PH, IH, $2IH$ and $2HS$ would, if the moon were transferred from the elliptical orbit into the circular one, come to be $(QO/SP)\,PH$, $(QO/SP)\,IH$, $2(QO/SP)\,IH$ and $2(QO/SP)\,HS$; and thereafter if, with the moon returning from this orbit to the elliptical one, the centripetal forces

$$(QO/SP)\,PH \quad \text{and} \quad 2(QO/SP)\,HS$$

were to diminish in the doubled ratio of the distance SP of the moon from the earth's centre [to QO], and the lateral forces $(QO/SP)\,IH$ and $2(QO/SP)\,IH$ in

for the mean annual advance, and in a sequence of interlineations in the manuscript (reproduced separately in [2] following for convenience) mapped out the changes consequent on further modifying this ratio of lateral perturbation to be $(SQ/OQ)^2$. The end effect is disastrous: the resulting mean advance of lunar apogee—some 24° per year ensues (see note (63)) if we complete Newton's calculation—falls hopelessly short of empirical truth, and the abortive refinement was abandoned unfinished.

(27) Understand Proposition XXVIII of the published *Principia*'s third book ($_1$1687: 439–41) where, on neglecting the earth's eccentric displacement within it as being not appreciable, Newton deduces that the disturbance of solar gravity will effectively distort the moon's pristine circular orbit into an ellipse, still centred on the earth, whose minor axis, lying along the line of syzygies, is in proportion to its major axis, marking the quadratures, very nearly as 69 to 70.

(28) In 1888 J. C. Adams paraphrased this simplifying assumption of Newton's to state that 'the form of the orbit in which the moon really moves will be related to the form of the oval orbit before mentioned, nearly as an elliptic orbit of small eccentricity with the earth in its focus is related to a circular orbit about the earth in the centre' (*A Catalogue of the Portsmouth Collection*...(note (2)): xii). To be strict, however, while Newton may or may not have afterwards been tempted to justify it in this way, he here introduces only the minimal Horrocksian assumption that—much as in his 'preceding' lunar propositions (see note (1)) where he had taken the moon's pristine orbit to be a circle uniformly traversed round the earth—, when it is no longer permitted to ignore the eccentric position of the earth within the lunar orbit, this shall be considered instantaneously to be a Keplerian ellipse (ever with the earth at its primary focus) whose major axis remains unchanged in length during its periodic distortions. (Jeremiah Horrocks briefly outlines his basic kinematical model for what he termed the 'Equation of lunar Apogee' in his letter to William Crabtree of 20 December 1638, published in his *Opera Posthuma* (London, $_1$1672): 465–70, but the extension which takes account of the varying eccentricity of the lunar orbit was contrived by Flamsteed on the basis of an explanatory letter by Crabtree to William Gascoigne on 21 July 1642 and first published as an epilogue in the second edition of the *Opera* ($_2$1673): 491; compare Francis Baily's *Supplement to the Account of the Revd John Flamsteed* (London, 1837): 683. We may pertinently remark that Newton's well-read library copy of the 1678 edition of Horrocks' *Opera* is now Trinity College, Cambridge. NQ.8.19.)

$\frac{QO}{SP}IH$ et $\frac{2QO}{SP}IH$ in ratione triplicata[,] fierent $\frac{QO^{cub}}{SP^{cub}}PH$, $\frac{QO^{qq}}{SP^{qq}}IH$,(29) $\frac{2QO^{qq}}{SP^{qq}}IH$(29) et $\frac{2QO^{cub}}{SP^{cub}}HS$, quæ (cùm distantiarum SP & QO differentia $SP-QO$ (ob infinite parvam excentricitatem) sit infinite parva) si distantia illa dicatur x, fiunt (per Lem. 2 Lib. II Princip.)(30) $\frac{SP-3x}{SP}PH$, $\frac{SP-4x}{SP}IH$,(31) $\frac{2SP-8x}{SP}IH$(31) & $\frac{2[SP]-6x}{SP}HS$. Et his viribus Luna absq motu Apogæi revolveretur. Proindeq [s]i hæ vires de viribus PH, IH, $2IH$ & $2HS$ subducantur, vires [re]liquæ $\frac{3x}{SP}PH$, $\frac{4x}{SP}IH$, $\frac{8x}{SP}IH$ & $\frac{6x}{SP}HS$ eæ ipsæ erunt a [quibu]s Apogæum movetur. Et vi quidem $\frac{6x}{SP}HS-\frac{3x}{SP}PH$ Luna, si [extra] circulum versatur, distrahitur a Terra, & ubi [intra versatur at]trahitur in terram. Unde vis illa eadem est cum vi V in Lemmate [α] & propterea (per Lemma illud) motus Apogæi a vi illa oriundus est ad motum medium Lunæ ut $\frac{6x}{SP}HS\times SE-\frac{3x}{SP}PH\times SE$ ad $P\times OS$. Vi autem $\frac{4x}{SP}IH+\frac{8x}{SP}IH$ seu $\frac{12x}{SP}IH$(32) Luna, ubi extra circulum versatur, trahitur versus syzygias, et ubi intra circulum est, trahitur in partes contrarias, ideoq hæc est W in Lemmate [β] et propterea (per Lemma illud) motus Apogæi a vi illa oriundus est ad motum medium Lunæ ut $\frac{24x}{SP}IH\times PE$(33) ad $P\times OS$. Et compositè, motus totus Apogæi a vi utraq oriundus est ad motum mediũ Lunæ ut

$$\frac{6x}{SP}HS\times OE-\frac{3x}{SP}PH\times OE+\frac{24x}{SP}IH\times PE^{(33)}$$

ad $P\times OS$, seu $6HS\times SE^{q}-3PH\times SE^{q}+24IH\times PE\times SE$ ad $178\llcorner725SP^{cub}$. Nam x est semidifferentia rectarum PF & PS, adeoq est ad OS ut SE ad SP. Et Lunæ pondus P supra (in Prop. [XXV])(34) erat ad vim SP (quæ in vires PH, IH, HS hic resolvebatur) ut $178\llcorner725$ ad 1, ideoq P valet $178\llcorner725SP$. Jam verò si in PS

(29) The indices of QO and SP were originally 'cub' (3) and thereafter '*qc*' (5) in line with Newton's preliminary suppositions (see note (26) above) that the lateral solar perturbation was to be diminished in the ratio $(SP/QO)^2$ and then $(SP/QO)^4$.

(30) Compare note (15), and again see IV: 521–5. In the terms of his preceding lemma Newton here takes $SP-QO = x$ to be the vanishingly small moment of QO, whence

$$(QO/SP)^{i} = ((QO+x)/QO)^{-i} = (QO-ix)/QO = (SP-ix)/SP$$

in the limit as SP comes to coincide with QO.

that ratio tripled, then they would become $(QO^3/SP^3)\,PH$, $(QO^4/SP^4)\,IH$,(29) $2(QO^4/SP^4)\,IH$(29) and $2(QO^3/SP^3)\,HS$: since the difference $SP-QO$ of the distances SP and QO is (because of the infinitely small eccentricity) indefinitely small, if that distance be called x, these become (by *Principia*, Book II, Lemma 2)(30)

$$(1-3x/SP)\,PH,\quad (1-4x/SP)\,IH,^{(31)}\quad (2-8x/SP)\,IH^{(31)}\quad \text{and}\quad (2-6x/SP)\,HS.$$

And so under these forces the moon would revolve without any motion of its apogee. Consequently, if these forces be taken away from the forces PH, IH, $2IH$ and $2HS$, the forces $(3x/SP)\,PH$, $(4x/SP)\,IH$, $(8x/SP)\,IH$ and $(6x/SP)\,HS$ remaining will be the very ones whereby the apogee moves. By the force $(6x/SP)\,HS-(3x/SP)\,PH$, indeed, the moon is, if it is located outside the circle, drawn away from the earth, but, when inside, attracted to the earth; whence that force is the same as the force V in Lemma [α], and accordingly (by that Lemma) the motion of the apogee arising from that force is to the moon's mean motion as $(6x/SP)\,HS\times SE-(3x/SP)\,PH\times SE$ to $P\times OS$. But by the force $(4x/SP)\,IH+(8x/SP)\,IH$, that is, $(12x/SP)\,IH$,(32) the moon is, when located outside the circle, drawn towards the syzygies, and, when it is inside the circle, in the opposite direction; in consequence, this is the force W in Lemma [β], and accordingly (by that Lemma) the motion of the apogee arising from that force is to the moon's mean motion as $(24x/SP)\,IH\times PE$(33) to $P\times OS$. And, on compounding, the total motion of the apogee arising from both forces is to the moon's mean motion as $(6x/SP)\,HS\times OE-(3x/SP)\,PH\times OE+(24x/SP)\,IH\times PE$(33) to $P\times OS$, that is, $6HS\times SE^2-3PH\times SE^2+24IH\times PE\times SE$ to $178.725SP^3$. For x is half the difference of the straight lines PF and PS, and hence is to OS as SE to SP; and the moon's weight P was above (in Proposition XXV)(34) to the force SP (here resolved into the forces PH, IH, HS) as 178.725 to 1, and P's value is in

(31) Originally (compare notes (29) and (30) preceding) the numerators were '$SP-3x$' and '$2SP-6x$', and thereafter '$SP-5x$' and '$2SP$—$10x$' respectively, so producing in sequel corresponding forces '$\frac{3x}{SP}IH, \frac{6x}{SP}IH$', and then '$\frac{5x}{SP}IH, \frac{10x}{SP}IH$' acting to move the apogee.

(32) Initially, in line with preceding original tentatives (see notes (26) and (29)–(31)), '$\frac{3x}{SP}IH+\frac{6x}{SP}IH$ seu $\frac{9x}{SP}IH$' and then '$\frac{5x}{SP}IH+\frac{10x}{SP}IH$ seu $\frac{15x}{SP}IH$'.

(33) Originally '$\frac{18x}{SP}IH\times PE$' and then '$\frac{30x}{SP}IH\times PE$' in sequel to the earlier variants recorded above. At the next stage, correspondingly, these produced '$18IH\times PE\times SE$' and then '$30IH\times PE\times SE$'.

(34) Namely, of the published *Principia*'s third book ($_1$1687: 434–5). The following ratio is there, 'per Corol. 17. Prop. LXVI. [and thence Proposition IV] Lib I', straightforwardly computed as $(T_E/T_M)^2$ where T_E = 365 days, 6 hours, 9' and T_M = 27 days, 7 hours, 43' are the periodic times of orbit of the earth round the sun and of the moon round the earth respectively.

demittatur perpendiculum ET et pro ES^q et $PE\times ES$ scribantur PST & $PS\times ET$: fiet motus horarius Apogæi ad motum medium horarium Lunæ id est ad $32'.56''.27'''.12^{iv}$, ut $6HST-3PH\times ST+24IH\times ET$(35) ad $178{,}725SP^q$. Ubi terminus $6HST$ semper affirmativus est et terminus $3PH\times ST$ semper negativus, at terminus $24IH\times ET$(35) proinde ut rectæ IH & ET jaceant ad easdem vel ad contrarias partes rectæ PS, affirmativus est vel negativus. Apogæum autem, ubi quantitas $6HST-3PH\times ST\pm 24IH\times ET$(35) negativa est, regreditur; ubi verò affirmativa progreditur. Inventus est igitur motus horarius Apogæi. Q.E.I.

Corol. Hinc innotescit etiam motus mediocris horarius Apogæi in situ quocunq; dato. Concipe Apogæum dum Luna revolvendo cursum periodicum complet, horis singulis a loco [no]vissimè occupato retrahi & reduci semper in locum priorem [eò u]t ejus situs toto revolutionis tempore datus maneat, et [quia mo]tus horarius Apogæi est ad motum medium hora[rium Lunæ ut

$$6HST-]\,3PH\times ST\pm 24IH\times ET^{(35)}$$

ad $178{,}725SP^q$[,] summa motuum Apogæi erit ad summam motuum horariorum Lunæ, id est ad $360^{gr.}$ ut summa omnium $6HST-3PH\times ST\pm 24IH\times ET$(35) ad summam totidem $178{,}725SP^q$ in revolutione integra. Sed si bisecetur PS in Z, sunt omnia $IH\times ET$ ad totidem SZ^q ut cosinus duplicati anguli QSA ad duos Radios per Corol. Prop. (36) et omnia $PH\times ST$ id est $SZ+ZH$ in $SZ-ZT$, seu $SZ^q-SZT+SZH-HZT$, id est (cum SZT & SZH æquales summas(37) generando se mutuo eluant) omnia SZ^q-HZT sunt ad totidem SZ^q ut excessus diametri supra cosinum anguli QSA duplicati ad diametrum per idem Corollarium.(38) Ac deniq; omnia HST seu $SZ-ZH$ in $SZ-ZT$ id est

$$SZ^q-SZH-SZT+HZT,$$

hoc est (cum omnia SZH & SZT toties affirmativè sumpta quoties negativè sese eluant)(39) omnia SZ^q+HZT sunt ad totidem SZ^q ut summa diametri & cosinus

(35) Originally '$18IH\times ET$' and then '$30IH\times ET$' much as in note (33).

(36) While Corollary 1 to Proposition XXX of its third book ($_1$1687: 446–7) gives, *mutatis mutandis*, the comparable calculation of 'omnia $PE\times PI$ ad totidem SZ^q', Newton set in his published *Principia* no proposition to which this unfilled citation refers. The integral to which he here appeals is, however, readily effected: in straightforward analytical equivalent (to which a corresponding geometrical form is easily furnishable), on setting the sun's elongation from perigee $\widehat{QAS}=a$ (taken to be effectively constant ever the range of integration) and the moon's elongation $\widehat{QSP}=x$, there ensues $IS/SP=\cos(a+x)$ and $PE/SP=\sin x$, so that $IH/SZ=\sin 2(a+x)$ and $ET/SZ=\sin 2x$, and therefore

$$\textit{omnia } IH\times ET/\textit{totidem } SZ^2 = \int_0^{2\pi}\sin 2(a+x)\sin 2x\,.\,dx\Big/2\pi$$
$$= \int_0^{2\pi}\tfrac{1}{2}(\cos 2a-\cos(2a+4x))\,.\,dx\Big/2\pi = \tfrac{1}{2}\cos 2a.$$

consequence $178.725SP$. Now indeed, if the perpendicular ET be let fall to PS, and in place of ES^2 and $PE \times ES$ there be written $PS \times ST$ and $PS \times ET$, the hourly motion of the apogee will come to be to the moon's mean hourly motion, that is, to 32′ 56″ 27‴ 12^iv, as $6HS \times ST - 3PH \times ST + 24IH \times ET$(35) to $178.725SP^2$. Here the term $6HS \times ST$ is always positive, the term $3PH \times ST$ always negative, but the term $24IH \times ET$(35) is, accordingly as the lines IH and ET lie on the same or opposite sides of the line PS, positive or negative. The apogee, however, when the quantity $6HS \times ST - 3PH \times ST \pm 24IH \times ET$(35) is negative regresses; when positive, indeed, it advances. The hourly motion of the apogee has therefore been ascertained. As was to be found.

Corollary. Hence is known also the average hourly motion of the apogee at any given position whatever. Conceive that the apogee is, while the moon in its revolution completes its periodic course, drawn back each separate hour from its most recently occupied place and ever returned to its former one, to the end that throughout the time of revolution its position remains fixed. Then, seeing that the hourly motion of the apogee is to the moon's mean hourly motion as $6HS \times ST - 3PH \times ST \pm 24IH \times ET$(35) to $178.725SP^2$, the sum of the motions of the apogee will, in an entire revolution, be to the sum of the moon's hourly motions, that is, 360°, as the sum of all $6HS \times ST - 3PH \times ST \pm 24IH \times ET$(35) to the sum of an equal number of $178.725SP^2$. But, if PS be bisected at Z, all the $IH \times ET$ are (by Corollary of Proposition)(36) to as many SZ^2 as the cosine of double the angle $\widehat{QSA}$ to twice the radius; and all the $PH \times ST$, that is, $(SZ+ZH) \times (SZ-ZT)$ or $SZ^2 - SZ \times ZT + SZ \times ZH - HZ \times ZT$, or (since $SZ \times ZT$ and $SZ \times ZH$ by generating equal sums(37) mutually destroy one another) all the $SZ^2 - HZ \times ZT$, are, by the same Corollary, to as many SZ^2 as the excess of the diameter over the cosine of twice $\widehat{QSA}$ to the diameter;(38) while, finally, all the $HS \times ST$ or $(SZ-ZH) \times (SZ-ZT)$, that is,

$$SZ^2 - SZ \times ZH - SZ \times ZT + HZ \times ZT,$$

or (since all the $SZ \times ZH$ and $SZ \times ZT$, being taken as many times positive as negative, destroy themselves)(39) all the $SZ^2 + HZ \times ZT$ are to as many SZ^2 as

(37) Namely zero, since (in the analytical terms of the previous note) $ZT/SZ = \cos 2x$ and $ZH/SZ = \cos 2(a+x)$, whence

$$\textit{omnia } ZT/\textit{totidem } SZ \left(= \int_0^{2\pi} \cos 2x \, . \, dx \Big/ 2\pi \right)$$
$$= \textit{omnia } ZH/\textit{totidem } SZ \left(= \int_0^{2\pi} \cos 2(a+x) \, . \, dx \Big/ 2\pi \right) = 0.$$

(38) See note (36). In the analytical equivalent which we there introduced it straightforwardly follows that

$$\textit{omnia } (SZ^2 \mp HZ \times ZT)/\textit{totidem } SZ^2 = \int_0^{2\pi} (1 \mp \cos 2(a+x) \cos 2x \, . \, dx) \Big/ 2\pi$$
$$= \int_0^{2\pi} (1 \mp \tfrac{1}{2}(\cos 2a - \cos(2a+4x))) \, . \, dx \Big/ 2\pi = 1 \mp \tfrac{1}{2} \cos 2a.$$

(39) See note (37) above in justification.

anguli $2QSA$ ad diametrum.[38] Igitur cum quatuor SZ^q æquentur SP^q, si D pro diametro et C pro cosinu prædicto scribatur, summa omnium

$$6HST - 3PH \times ST + 24IH \times ET$$

erit ad summam totidem $178\llcorner 725SP^q$ ut $6D+6C$, $-3D+3C$, $+24C$[40] ad $178\llcorner 725 \times 4D$[,] hoc est ut $D+11C$[41] ad $238\frac{3}{10}D$. Ubi $11C$,[41] quoties Apogæum inter octantes et syzygias versatur affirmativè sumitur, & quoties inter octantes et quadraturas, negativè. Nam cum C sinus sit duplicatæ distantiæ Apogæi ab octante,[42] is in transitu Apogæi per octantes de affirmativo in negativum deqꝫ negativo in affirmativum semper mutatur. Est igitur summa motuum horariorum Apogæi ad summam motuum mediorum horariorum Lunæ in revolutione integra ut $D \pm 11C$[41] ad $238\frac{3}{10}D$, et propterea etiam motus mediocris horarius Apogæi est ad motum medium horarium Lunæ, id est ad $32'.\ 56''.\ 27'''$, ut $D \pm 11C$[41] ad $238\frac{3}{10}D$.[43]

Prop. [*B*] *Prob.*

[*Posi*]*to quod excentricitas orbis Lunaris sit infinite parva,* [*invenire motum*] *medium Apogæi.*

[Designet S Terram], $AEBD$ circulum centro S intervallo quovis [S]A descriptum diametrisqꝫ perpendicularibus AB, DE in quadrantes divisum[,] AG duplicatam distantiam Apogæi a Sole[,] Gg datum incrementum[44] arcus AG[,] & GH ac gh cosinus arcuum AG et Ag. Et cum motus Apogæi in G sit ad motum medium Lunæ ut $D+11C$[41] ad $238\frac{3}{10}D$ id est ut $AB+11GH$[45] ad $238\frac{3}{10}AB$, adeoqꝫ ut $\frac{2}{11}AS+GH$ ad $43\frac{18}{55}AS$, hoc est ut $\frac{2}{11}Gg+Hh$ ad $43\frac{18}{55}Gg$. Proinde si tempus quo distantia Apogæi & Solis datum accipit incrementum Gg exponatur per t, et motus medius Lunæ eo tempore exponatur per $43\frac{18}{55}Gg \times t$, motus Apogæi eodem tempore exponetur per $\frac{2}{11}Gg \times t + Hh \times t$. Concipe autem Apogæum singulis horis a loco novissime occupato retrahi & in locum priorem reduci eò ut anno toto sidereo datum servet situm ad fixas et summa omnium motuum mediorum Lunæ in integra Solis ab Apogæo ad Apogæum revolutione seu diebus $365.\ 6^h.\ 7'$, id est summa

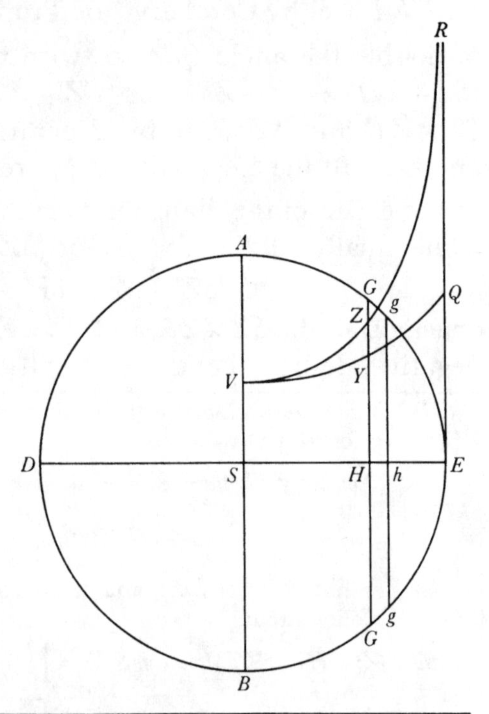

(40) In line with the preliminary variants recorded in note (35), the manuscript originally read '$+18C$' and then '$+30C$', thence determining the first member of the succeeding ratio

the sum of the diameter and the cosine of $2\widehat{QSA}$ to the diameter.(38) Therefore, since $4SZ^2$ are equal to SP^2, if D be written for the diameter and C for the above-named cosine, the sum of all the $6HS \times ST - 3PH \times ST + 24IH \times ET$ will be to the sum of as many $178.725SP^2$ as $6(D+C)-3(D-C)+24C$(40) to $178.725 \times 4D$, that is, as $D+11C$(41) to $238.3D$. Here $11C$(41) is, whenever the apogee lies between the octants and syzygies, taken positive and, whenever between the octants and quadratures, negative; for, since C is the sine of double the distance of the apogee from octant,(42) in the passage of the apogee through the octants it is always changed from positive to negative and from negative to positive. Therefore the sum of the hourly motions of the apogee is to that of the moon's mean hourly motions in an entire revolution as $D \pm 11C$(41) to $238.3D$, and accordingly also the average hourly motion of the apogee is to the moon's mean hourly motion, that is, 32′ 56″ 27‴, as $D \pm 11C$(41) to $238.3D$.(43)

Proposition [*B*], *Problem.*

Supposing that the eccentricity of the lunar orbit be indefinitely small, to ascertain the mean motion of the apogee.

Let S denote the earth, and $AEBD$ a circle described with centre S and any radius SA and divided by the perpendicular diameters AB, DE; let $\widehat{AG}$ be double the distance of the apogee from the sun, Gg a given increment(44) of the arc $\widehat{AG}$, and GH and gh the cosines of the arcs $\widehat{AG}$ and $\widehat{Ag}$. Now since the motion of the apogee at G is to the moon's mean motion as $D+11C$(41) to $238.3D$, that is, as $AB+11GH$(45) to $238.3AB$ and hence as $\frac{2}{11}AS+GH$ to $43\frac{18}{55}AS$, that is, as $\frac{2}{11}Gg+Hh$ to $43\frac{18}{55}Gg$, it follows that, if the time in which the distance between the apogee and the sun receives the given increment Gg be expressed by t, and the mean motion of the moon in that time be expressed by $43\frac{18}{55}Gg \times t$,(46) then the motion of the apogee in the same time will be expressed by $\frac{2}{11}Gg \times t + Hh \times t$. Conceive, however, that the apogee is drawn back each single hour from the place it has most recently occupied and returned to its former place, to the end that in a whole sidereal year it preserves a given position with regard to the fixed stars, and the sum of all the mean motions of the moon in an entire revolution of the sun from apogee to apogee, or 365 days, 6 hours, 7 minutes—that is,

---

to be $\frac{1}{3}(6(D+C)-3(D-C)+18C) = D+9C$ and thereafter

$$\tfrac{1}{3}(6(D+C)-3(D-C)+30C) = D+13C$$

correspondingly.

(41) Initially '$9C$' and then '$13C$'; see previous note.

(42) For $C\,(=\frac{1}{2}D\cos 2\widehat{QSP}) = \frac{1}{2}D\sin 2\widehat{Q'SP}$ where $\widehat{QSQ'} = \pm\frac{1}{4}\pi$ radians.

(43) That is, as $1 \pm \frac{11}{2}\cos 2\widehat{QSP}$ to 238.3. It follows from this that in syzygy ($\widehat{QSP} = 0$ or π) the mean daily advance of apogee is about $24 \times (\frac{13}{2}/238.3) \times 33' \approx 21\frac{1}{2}'$, while at quadratures ($\widehat{QSP} = \pm\frac{1}{2}\pi$) there will be a mean daily regress of some $24 \times (\frac{9}{2}/238.3) \times 33' \approx 15'$. Newton's

omnium $43\frac{18}{55}Gg\times t$(46) eo tempore erit 4812^{gr}. 45′. 38″. atq summa omnium motuum Apogæi, id est summa omnium $\frac{2}{11}Gg\times t+Hh\times t$ sic invenietur.

Sunt omnia $\frac{2}{11}Gg\times t$(47) ad omnia $43\frac{18}{55}Gg\times t$(47) (id est ad 4812^{gr}. 45′. 38″) ut $\frac{2}{11}$ ad $43\frac{18}{55}$, seu 1 ad $238\frac{3}{10}$, ideoq valent 20^{gr}. 11′. 46″. 24‴. Tempus autem t est reciprocè ut differentia inter motum horarium Solis et motum progressivum horarium Apogæi si modò Apogæum(48) progreditur, sin Apogæum(48) regreditur tempus est reciprocè ut summa eorundem motuum; id est (cum motus medius horarius Solis sit ad motum medium Lunæ ut 360^{gr} ad 4812^{gr}. 45′. 38″, seu 100000 ad 1336878, id est ut $3{\llcorner}24093AS$ ad $43\frac{18}{55}AS$,(49) et motus medius horarius Lunæ ad motum horarium Apogæi ut $[43\frac{18}{55}]AS$ ad $\frac{2}{11}AS\pm GH$[,] et idcirco motus medius Solis ad motum[horarium] Apogæi ut $3{\llcorner}24093AS$ ad $\frac{2}{11}AS\pm GH$ id est ad $0{\llcorner}18181818AS\pm GH$(50)) [te]mpus t est reciprocè ut $3{\llcorner}0591113AD\mp GH$. Et propterea omnia $[Hh]\times t$ sunt ad totidem $Gg\times t$ ut omnia $\dfrac{Hh}{3{\llcorner}0591113AS\mp GH}$(51) ad [totidem] $\dfrac{Gg}{3{\llcorner}0591113AS\mp GH}$, adeoq (si jam detur Hh & recta SH [per additionem] particularum æqualium Hh uniformiter crescat,) ut [omnia $\dfrac{AS}{3{\llcorner}0591113AS\mp GH}$ a]d omnia $\dfrac{AS^q}{3{\llcorner}0591113AS,\ GH\mp GH^q}$, id est cum GH toties negative sumatur quoties affirmative[,] ut omnia

$$\frac{AS}{3{\llcorner}0591113AS-GH}\ \frac{AS}{3{\llcorner}0591113AS+GH}$$

---

prior ratios '$D\pm 9C$ ad $238\frac{3}{10}D$' and '$D\pm 13C$ ad $238\frac{3}{10}$' would, it will be clear, respectively decrease these mean motions in the ratios $\frac{11}{13}$ and $\frac{7}{9}$, and increase them in the ratios $\frac{15}{13}$ and $\frac{11}{9}$.

(44) Understand in an indefinitely small unit of time.

(45) Originally '$9GH$' and then '$13GH$' in line with the preceding variants listed in note (41). In sequel this led to 'adeoq ut $\frac{2}{9}AS+GH$ ad $52\frac{43}{45}AS$, hoc est ut $\frac{2}{9}Gg+Hh$ ad $52\frac{43}{45}Gg$', and then correspondingly to '...ut $\frac{2}{13}AS+GH$ ad $36\frac{43}{65}AS$, hoc est ut $\frac{2}{13}Gg+Hh$ ad $36\frac{43}{65}Gg$'.

(46) Initially (compare previous note) '$52\frac{43}{45}Gg\times t$' and then '$36\frac{43}{45}Gg\times t$', with

'$\frac{2}{9}Gg\times t+Hh\times t$' and then '$\frac{2}{13}Gg\times t+Hh\times t$'

in sequel.

(47) The coefficients were again (compare notes (46) and (47)) originally '$\frac{2}{9}$' and '$52\frac{43}{45}$', and then '$\frac{2}{13}$' and '$36\frac{43}{65}$' respectively.

(48) Initially specified in the manuscript as 'inter Octantes & Syzygias vers[ans]' (where it roams between the octants and syzygies) and 'inter Octantes & Quadraturas consist[ens]' (where it is positioned between octants and quadratures) respectively.

(49) Following on his previous variants (see notes (45) and (47)) Newton here orginally computed the equivalent ratios '$3{\llcorner}961136AS$ ad $52\frac{43}{45}AS$' and then '$2{\llcorner}742325AS$ ad $36\frac{43}{65}AS$', and then continued with a repetition from above of the mean horary motion of apogee to the moon's mean motion 'ut $52\frac{43}{45}AS$ ad $\frac{2}{9}AS+GH$' and then 'ut $36\frac{43}{65}AS$ ad $\frac{2}{13}AS+GH$'.

(50) Originally, in direct consequence of the variants recorded in the previous note, 'ut

the sum of all $43\frac{18}{55}Gg\times t$(46) in that time—will be 4812° 45′ 38″, while the sum of all the motions of the apogee—the sum of all $\frac{2}{11}Gg\times t+Hh\times t$, that is—will be found as follows.

All $\frac{2}{11}Gg\times t$(47) are to all $43\frac{18}{55}Gg\times t$(47) (that is, to 4812° 45′ 38″) as $\frac{2}{11}$ to $43\frac{18}{55}$, that is, as 1 to 238.3, and their value is consequently 20° 11′ 46″ 24‴. The time t, however, is reciprocally as the difference between the hourly motion of the sun and the progressive hourly motion of the apogee should the apogee(48) advance, but if the apogee(48) should regress, the time is reciprocally as the sum of those motions; that is, (since the mean hourly motion of the sun is to the mean motion of the moon as 360° to 4812° 45′ 38″, or 100000 to 1336878, that is, as $3.24093AS$ to $43\frac{18}{55}AS$,(49) and the mean hourly motion of the moon to the hourly motion of the apogee as $43\frac{18}{55}AS$ to $\frac{2}{11}AS\pm GH$, and in consequence the mean motion of the sun to the hourly motion of the apogee as $3.24093AS$ to $\frac{2}{11}AS\pm GH$, that is, $0.18181818AS\pm GH$(50)) the time t is reciprocally as $3.0591113AS\mp GH$. Accordingly, all $Hh\times t$ are to as many $Gg\times t$ as all $Hh/(3.0591113AS\mp GH)$(51) to as many $Gg/(3.0591113AS\mp GH)$, and hence (if Hh now be given and the straight line SH shall uniformly increase by the addition of equal particles Hh) as all $\dfrac{AS}{3.0591113AS\mp GH}$ to all $\dfrac{AS^2}{(3.0591113AS\mp GH)\,GH}$, that is, since GH is taken as may times negatively as positively, as all

$$\frac{AS}{3.0591113AS-GH}-\frac{AS}{3.0591113AS+GH}$$

$3\llcorner961136AS$ ad $\frac{2}{9}AS+GH$ id est ad $0\llcorner22222AS+GH$' and then 'ut $2\llcorner742325AS$ ad $\frac{2}{13}AS+GH$ id est ad $0\llcorner153846AS+GH$', whence in immediate sequel

'tempus t est reciprocè ut $3\llcorner738914AS+GH$'

and thereafter '...reciprocè ut $2\llcorner588479AS+GH$'. In the final manuscript Newton at this point (at the bottom of f. 110r) ceases to concern himself further with the first of these tentative schemes of calculation, but his further computation from the supposition that t is reciprocally proportional to $3.738914AS+GH$ is preserved in preliminary draft on f. 109r: there, on reducing his integrands into 'series convergentes' (by Mercator division; compare note (54) below) and then integrating these term by term 'per notam quadraturarum methodum' as he had taught nearly twenty years before in his 'De Analysi' (see II: 212), he determined ultimately that 'omnia $Hh\times t$ s[u]nt ad omnia $Gg\times t$ ut 9811 ad 53910' and so 'omnia $Hh\times t$ ad omnia $\frac{2}{9}Gg\times t$ id est ad 20gr. 11′. 46″ ut 9811 ad 11980 ideoqȝ valebunt 16gr. 54′', whence 'summa omnium $\frac{2}{9}Gg\times t+Hh\times t$ est 37gr. 2′'. This computed mean annual advance of lunar apogee is nearly 4° less than the value (derived from observation) which Flamsteed had published in 1680, as Newton was well aware (see note (63) below): rather than meekly admit the possibility that such a discrepancy might be contained within the limits of error of his general dynamical approach, he was led in consequence to modify his initial square power of SP/QO defining the lateral solar disturbance deviating the moon from a 'concentric' terrestrial orbit (see note (26)), first changing it, as we have seen, to be the fourth power, and then, having somewhat bungled the final stage of his calculations departing therefrom and so

ad omnia $\dfrac{AS^q}{3\llcorner 0591113AS\times GH-GH^q}+\dfrac{AS^q}{3\llcorner 0591113AS\times GH+GH^q}$, hoc est, si $3\llcorner 0591113$ dicatur n,(52) ut omnia $\dfrac{AS^q\times GH}{nnAS^q-GH^q}$ ad omnia $\dfrac{n\times AS^{qq}}{nnAS^q\times GH-GH^{cub}}$ ac proinde (si in GH capiatur HY æqualis $\dfrac{AS^q\times GH}{nn\times AS^q-GH^q}$ et HZ æqualis

$$\frac{n\times AS^{qq}}{nn\times AS^q\times GH-GH^{cub}},$$

sintq; VYQ(53) et VZR lineæ curvæ quas puncta Y et Z perpetuo tangunt,) ut area $SVYQE$ ad aream $SVZRE$.(54) Est autem area prior ad aream circuli cujus

---

attained a falsely low value for the advance of lunar apogee (see note (54) below), further amending it to be the cube of SP/QO and thereby ultimately gaining in the revised text reproduced a 'corrected' value whose difference from true is permissibly 'very slight'.

(51) Here and throughout the remainder of this sentence the coefficient of AS was originally '$2\llcorner 588479$' (the preliminary value '$3\llcorner 738914$' being henceforth suppressed; see the previous note).

(52) Notice that Newton also here multiplies through by the constant AS.

(53) In Newton's preceding figure (lacking in the revised manuscript, but here reproduced from his preliminary autograph draft on f. 108v) this locus of Y is erroneously drawn as convex downwards, departing from the point V where the locus of Z cuts AS to intersect ER in a point Q distinct from E. In fact, since $\widehat{VZR}$ meets AS such that $SV = AS.n/(n^2-1)$, the locus (Y) meets it correspondingly in V' such that $SV'(= AS/(n^2-1)) = SV/n$, curving downwards therefrom to intersect ER (where $GH = 0$) in E. The geometrical deficiencies in Newton's visual illustration do not, of course, in any way impair his following analytical argument.

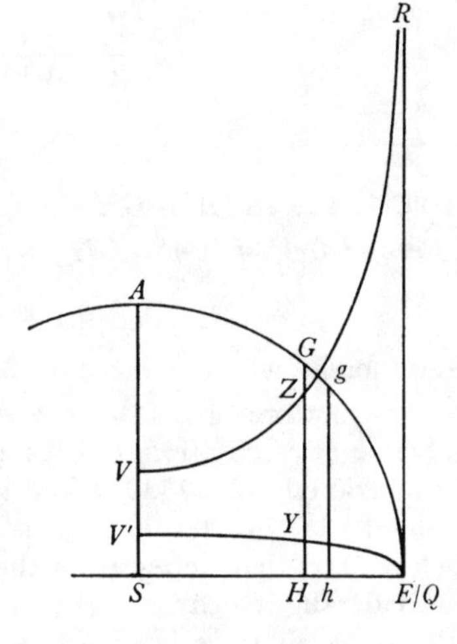

(54) In a cancelled continuation at this point Newton first went on to compute, departing from the preliminary value $n = 2.588479$ (see note (51)), that 'Est autem area prior [$SV'YE/Q$] per methodum serierum convergentium $0.075153AS^q$ atq; area posterior [$\widehat{SVZRE}$] per eandem methodum $0\llcorner 4362AS^q$. Nam si pro $\overline{nn-1}[\times SA^q]$ scribatur pp, area illa [*sc.* posterior] fiet ad circulum cujus diameter est [AS] ut series convergens

$$[SA^q \text{ in}]\ \frac{2n}{pp}-\frac{n}{p^4}SA^q+\frac{3n}{4p^6}SA^{qq}-\frac{5n}{8p^8}SA^{cc}+\frac{35n}{64p^{10}}SA^{qcc}\ \&c$$

ad unitatem. Quæ series per debitam reductionem convertitur in quantitatẽ finitam $\dfrac{2n[\times SA^q]}{p\sqrt{pp+SA^q}}$' (The former area ($SV'YE$) is, however, by the method of converging series $0.075153SA^2$, while the latter area ($\widehat{SVZRE}$) is by the same method $0.4362AS^2$. For if in place of $(n^2-1)\ SA^2$ there be written p^2, the (latter) area will come to be to a circle whose diameter is AS as the con-

to all $\dfrac{AS^2}{(3.0591113AS-GH)\,GH}+\dfrac{AS^2}{(3.0591113AS+GH)\,GH}$; in other terms, if 3.0591113 be called n,(52) as all $AS^2\times GH/(n^2.AS^2-GH^2)$ to all

$$n.AS^4/(n^2.AS^2-GH^2)\,GH,$$

and consequently (if in GH there be taken HY equal to

$$AS^2\times GH/(n^2.AS^2-GH^2)$$

and HZ equal to $n.AS^4/(n^2.AS^2-GH^2)\,GH$, and VYQ(53) and VZR be the curves which the points Y and Z perpetually trace) as the area $(SVYQE)$ to the area $(SVZRE)$.(54) But the first area is to the area of a circle whose diameter is unity(55)

---

verging series $SA^2(2n/p^2-(n/p^4)\,SA^2+\frac{3}{4}(n/p^6)\,SA^4-\frac{5}{8}(n/p^8)\,SA^6+\frac{35}{64}(n/p^{10})\,SA^8\ldots)$ to 1. This series is by due reduction converted into the finite quantity $2n.AS^2/p\sqrt{[p^2+AS^2]}$). In proof, since

$$omnia\,\frac{n.AS^4}{GH(n^2.AS^2-GH^2)}\,[\times d(SH)] = \int_0^{SA}\frac{n.AS^4}{GH(p^2+SH^2)}.d(SH),$$

on setting $\widehat{ASG}=\theta$ (whence $SH=AS.\sin\theta$ and so $d(SH)/d\theta=GH$) the latter area $(\widetilde{SVZRE})$ proves to be equal to

$$(n/p^2)AS^4\int_0^{\frac{1}{2}\pi}(1+(AS^2/p^2)\sin^2\theta)^{-1}.d\theta = (n/p^2)AS^4\int_0^{\frac{1}{2}\pi}\sum_{0\leqslant i\leqslant\infty}(-(AS^2/p^2)\sin^2\theta)^i.d\theta$$

$$= (n/p^2)AS^4.\sum_{0\leqslant i\leqslant\infty}\tfrac{1}{2}\pi\binom{\frac{1}{2}}{i}(AS^2/p^2)^i = (n/p^2)AS^4.\tfrac{1}{2}\pi(1+AS^2/p^2)^{-\frac{1}{2}},$$

that is, $\pi(\frac{1}{2}AS)^2.2n.AS^2/p\sqrt{[p^2+AS^2]}$ as Newton here affirms. At once, on replacing p by $AS\sqrt{[n^2-1]}$, this reduces to be $\pi(\frac{1}{2}AS)^2.2/\sqrt{[n^2-1]}$; whence since, in these terms, the former area $(SV'YE) = omnia\,\dfrac{AS^2\times GH}{n^2.AS^2-GH^2}[\times d(SH)]$ is evidently equal to

$$AS^2\int_0^{\frac{1}{2}\pi}\frac{\cos^2\theta}{n^2-\cos^2\theta}.d\theta = AS^2\int_0^{\frac{1}{2}\pi}\left(-1+\frac{(n^2/p^2)\,AS^2}{1+(AS^2/p^2)\sin^2\theta}\right).d\theta,$$

that is, $-\frac{1}{2}\pi.AS^2+n.(\widetilde{SVZRE}) = \pi(\frac{1}{2}AS)^2.(-2+2n/\sqrt{[n^2-1]})$, it will be in proportion to $(\widetilde{SVZRE})$ as $n-\sqrt{[n^2-1]}$ to 1, as Newton will state in immediate sequel in his revised text.

In a further cancelled passage following, we may add, he originally went on to draw from this draft evaluation of the two preceding areal integrals the straightforward conclusion that 'omnia $Hh\times t$ sunt ad totidem $Gg\times t$ ut 0⌊075153 ad 0⌊4362', but thereafter—momentarily forgetting that his present choice of $n=2.588479$ requires him to determine 'summa omnium $Hh\times t+\frac{2}{13}Gg\times t$', given that 'summa omnium $\frac{2}{13}Gg\times t$' = 20° 11′ 46″—he proceeded inconsistently to compute by his now superseded initial approach (see note (50)) that 'omnia $Hh\times t$ [sunt] ad omnia $\frac{2}{9}Gg\times t$ ut 0⌊07515 ad 0⌊09693, & composite omnia $Hh\times t+\frac{2}{9}Gg\times t$ ad omnia $\frac{2}{9}Gg\times t$ ut 0⌊17208 ad 0⌊09693. Unde cùm omnia $Hh\times t+\frac{2}{9}Gg\times t$ sint motus medius Apogæi ...et...summa omnium $\frac{2}{9}Gg\times t$ sit tempori proportionalis: erit motus medius in anno ad summam omnium $\frac{2}{9}Gg\times t$ in eodem anno, id est ad 20gr. 11′. 46″ ut 0⌊17208 ad 0⌊09693, ideoqȝ valebit 35gr. 51″. Correctly, since 'omnia $Hh\times t$ sunt ad totidem $\frac{2}{13}Gg\times t$ ut 0⌊07515 ad [$\frac{2}{13}\times 0.4362=$] 0⌊0671', Newton's conclusion should rather have been that the value of the apogee's mean motion 'summa omnium $Hh\times t+\frac{2}{13}Gg\times t$' is some 42° 49′, a little greater than true, but not alarmingly so. Spurred on, however, by the gross deficiency of his present falsely calculated annual advance of 35° 51′, he forthwith rejected this second prior approach and returned once more to cook the ratio of the solar lateral force into the form $(SP/QO)^3$ as it is reproduced above in the revised text here given (see note (26)).

diameter est unitas(55) ut $2n-2\sqrt{nn-1}$ ad $\sqrt{nn-1}$, et area posterior ad aream circuli cujus diameter est unitas(55) ut 2 ad $\sqrt{nn-1}$ et propterea area prior ad aream posteriorem ut $2n-2\sqrt{nn-1}$ ad 2, seu $n-\sqrt{nn-1}$ ad 1, id est ut 0,1680627 ad 1. Eadem arearum proportio per methodum serierum convergentium inveniri potest.(56) Sunt igitur omnia $Hh\times t$ ad omnia $Gg\times t$ ut 0,1680627 ad 1, et idcirco omnia $Hh\times t$ ad omnia $\frac{2}{11}Gg\times t$ ut 0,1680627 ad $\frac{2}{11}$ sive ut 1,848689 ad 2[,] et compositè omnia $Hh\times t+\frac{2}{11}Gg\times t$ ad omnia $\frac{2}{11}Gg\times t$ ut 3,848689 ad 2. Unde cùm summa omnium $Hh\times t+\frac{2}{11}Gg\times t$ sit motus medius Apogæi quo tempore Sol cursum suum ab Apogæo ad Apogæum complet, et tam motus ille medius quam summa omnium $\frac{2}{11}Gg\times t$ sit tempori proportionalis: erit motus idem medius in anno sidereo ad summam omnium $\frac{2}{11}Gg\times t$ in eodem anno ut 3,848689 ad 2. Ideoq cùm summa omnium $\frac{2}{11}Gg\times t$ in anno sidereo sit 20gr. 11′. 46″, motus ille medius Apogæi in eodem anno erit 38gr. 51′. 51″. Hic motus per Tabulas Astronomicas(57) est 40gr. 41½′. Differentia perexigua est et ab differentia excentricitatis oriri potest. Nam in calculi initio excentricitatem infinitè parvam esse supposuimus quæ tamen in orbe vero Lunari finitæ est magnitudinis.(58)

Corol.(59)

---

(55) Read '*AS*'. An equivalent slip by Newton has already been thus corrected in the preliminary version of this passage reproduced in the previous note.

(56) Such indeed (see note (54)) was Newton's own preferred way of evaluating the latter area ($\widehat{SVZRE}$), though, of course, the related integral

$$\int_0^{SA} \frac{n.AS^4}{GH(n^2.AS^2-GH^2)}.d(SH) = AS^2.\int_0^{\frac{1}{2}\pi} \frac{n}{n^2-\cos^2\theta}.d\theta$$

is more directly resolvable as $\frac{1}{2}AS^2.\int_0^{\frac{1}{2}\pi}\left(\frac{1}{n+\cos\theta}+\frac{1}{n-\cos\theta}\right).d\theta,$

where $$\int\frac{1}{n\pm\cos\theta}.d\theta = \frac{1}{\sqrt{[n^2-1]}}\cos^{-1}\left(\frac{n\cos\theta\pm 1}{n\pm\cos\theta}\right).$$

(57) Understand those appended by John Flamsteed to his *Doctrine of the Sphere* (London, 1680): 95–104; compare note (63) below.

(58) To abstract the general structure of Newton's argument, if A and S be the geocentric longitudes at time T of the moon's apogee and the sun respectively, so that (where dT is Newton's 'instantaneous' moment t, conventionally taken in contemporary astronomical practice to be a unit hour) the sun's 'horary motion'—assumed to be effectively uniform—is $dS/dT = 360°$ per year, and if furthermore (in consequence of some suitable choice of lateral disturbing ratio; see note (26)) the corresponding 'horary motion' of the moon's apogee shall have been determined in Proposition A to be $dA/dT = k(1+\lambda\cos 2(S-A))$, in which $k = (32'\ 56''\ 27'''/238.3 \approx)\ 8''\ 9'''$ per hour $= 20°\ 11'$ per year and $\lambda = i+\frac{5}{2}$, then

$$d(S-A)/dT = k(\mu-1-\lambda\cos 2(S-A))$$

as $2n-2\sqrt{[n^2-1]}$ to $\sqrt{[n^2-1]}$, and the latter area to the area of a circle whose diameter is unity(55) as 2 to $\sqrt{[n^2-1]}$, and accordingly the first area is to the latter one as $2n-2\sqrt{[n^2-1]}$ to 2, or $n-\sqrt{[n^2-1]}$ to 1, that is, as 0.1680627 to 1. The same ratio of the areas can be found by the method of converging series.(56) Therefore, all $Hh\times t$ are to all $Gg\times t$ as 0.1680627 to 1, and as a consequence all $\frac{2}{11}Gg\times t$ as 0.1680627 to $\frac{2}{11}$, or as 1.848689 to 2, and, on compounding, all $Hh\times t+\frac{2}{11}Gg\times t$ to all $\frac{2}{11}Gg\times t$ as 3.848689 to 2. Whence, since the sum of all the $Hh\times t+\frac{2}{11}Gg\times t$ is the mean motion of the apogee during the time in which the sun completes its course from apogee to apogee, while both that mean motion and the sum of all the $\frac{2}{11}Gg\times t$ is proportional to the time, on that basis, since the sum of all the $\frac{2}{11}Gg\times t$ in a sidereal year is 20° 11′ 46″, the mean motion of the apogee in the same year will be 38° 51′ 51″. By astronomical tables(57) this motion is 40° 41½′. The difference is very slight and may arise from the difference in eccentricity. For at the start of the calculation we supposed the eccentricity to be infinitely small, whereas in the true lunar orbit it is of finite size.(58)

Corollary.(59)

---

where $\mu = 360°/20°11' \approx 17.83$, and hence

$$dA/d(S-A) = (1+\lambda\cos 2(S-A))/(\mu-1-\lambda\cos 2(S-A)).$$

Accordingly, the mean advance of apogee 'quo tempore Sol cursum suum ab Apogæo ad Apogæum complet' is (in radians)

$$\Big[A\Big]_{S-A=0}^{S-A=2\pi} = \int_0^{2\pi}\frac{1+\lambda\cos 2x}{\mu-1-\lambda\cos 2x}.dx = 2\int_0^{\pi}\frac{1/\lambda+\cos y}{n-\cos y}.dy$$
$$= 4\int_0^{\frac{1}{2}\pi}\frac{n/\lambda+\cos^2 y}{n^2-\cos^2 y}.dy = 2\pi\left(\frac{1/\lambda+n}{\sqrt{[n^2-1]}}-1\right) = 2\pi\left(\frac{\mu}{\sqrt{[(\mu-1)^2-\lambda^2]}}-1\right),$$

that is, very nearly $2\pi(1/(\mu-1)+\frac{1}{2}\mu\lambda^2/(\mu-1)^2)$, where (as Newton so constructs it to be) $n = (\mu-1)/\lambda$.

(59) Newton evidently here intended to append, in copy of his preliminary draft on f. 109r, a remark that 'Eodem computo liquet æquatio motus Apogæi in octantibus' (By the same computation the equation of the motion of the apogee at the octants is clear): in the terms of the previous note the mean annual advance of apogee 'quo tempore Sol cursum suum ab Octante ad Octantem complet' is evidently

$$\Big[A\Big]_{S-A=\frac{1}{4}\pi}^{S-A=2\frac{1}{4}\pi} = 4\int_{-\frac{1}{4}\pi}^{\frac{1}{4}\pi}\frac{n/\lambda+\cos^2 y}{n^2-\cos^2 y}.dy = \frac{4(1/\lambda+n)}{\sqrt{[n^2-1]}}\tan^{-1}\frac{\sqrt{[8(n^2-1)]}}{n}-2\pi.$$

This stray '*Corol.*' was subsequently engulfed by the final sentence of the preceding paragraph (which is a late addition in the manuscript) and Newton afterwards struck it out, along with an abortive following '*Schol[ium]*' in which he no doubt intended to pass a general comment on the possibly damaging cumulative effect of the several approximations he has adduced in attaining his final 'accurate' evaluation of the mean advance of lunar apogee.

[2][(60)]

Designet $ABCD$ orbem Lunæ, A & B syzygias, C & D quadraturas, S Terram, P Lunam & PI & $2IS$ vires Solis ad perturbandos motus Lunares........vis PI resolvetur in vires PH & IH, atq; vis $2IS$ in vires $2IH$ & $2HS$. Hæ vires ubi Luna in orbe illo versatur in quo per easdem sine excentricitate revolvi posset, ad motum Apogæi nil conducunt. Oritur motus Apogæi a differentijs inter has vires & vires quæ in recessu Lunæ ab orbe illo concentrico si centripetæ sunt vel centrifugæ decrescunt in duplicata ratione qua distantia inter Lunam & centrum Terræ augetur[,] sin agant in latus decrescunt in ratione duplicata SQ ad OQ[(61)] uti calculis initis inveni. Hunc orbem Ovalem esse supra ostensum est. Et inde fit ut orbis excentricus, in quo Luna verè revolvitur, non sit Ellipsis sed alterius cujusdam generis Ovalis. At si ad minuendam difficultatem calculi supponamus orbem illum concentricum esse circulū, orbis excentricus evadet Ellipsis, cujus axis major PQ æqualis erit diametro circuli. Et vires prædictæ PH, IH, $2IH$ & $2HS$, si Luna de orbe Elliptico in circularem transferretur, evaderent $\frac{QO}{SP}PH$, $\frac{QO}{SP}IH$, $\frac{2QO}{SP}IH$ & $\frac{2QO}{SP}HS$, ac deinde si Luna hinc in orbem ellipticum redeunte vires centripetæ $\frac{QO}{SP}PH$ et $\frac{2QO}{SP}HS$ diminuerentur in duplicata ratione distantiæ SP Lunæ a centro Terræ et vires laterales $\frac{QO}{SP}IH$ et $\frac{2QO}{SP}IH$ in ratione duplicata SQ ad QO, fierent $\frac{QO^{\text{cub}}}{SP^{\text{cub}}}PH$, $\frac{QO^{\text{cub}}}{SP, SQ^{q}}IH$, $\frac{2QO^{\text{cub}}}{SP, SQ^{q}}IH$ et $\frac{2QO^{\text{cub}}}{SP^{\text{cub}}}HS$, quæ (cùm distantiarum SP & QO differentia $SP-QO$ (ob infinite

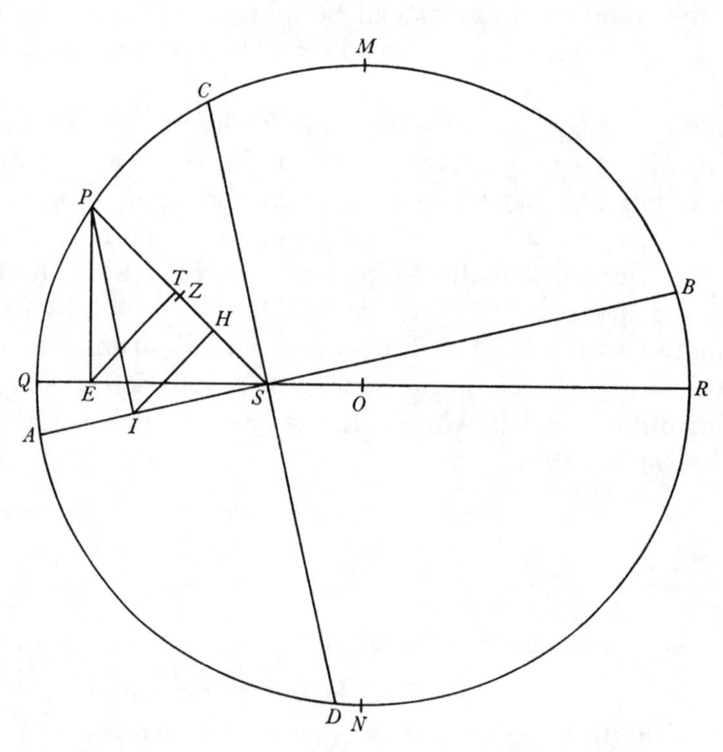

[2][60]

Let $ABCD$ denote the orbit of the moon, A and B its syzygies, C and D its quadratures, S the earth, P the moon, and PI and $2IS$ the forces of the sun to perturb the lunar motions.... ...the force PI will then be resolved into the forces PH and IH, and the force $2IS$ into the forces $2IH$ and $2HS$. These forces do not, when the moon is located in the orbit in which by their means it might revolve without eccentricity, contribute at all to the motion of the apogee. The motion of the apogee arises from the differences between these forces and forces which in the moon's recession from that concentric orbit decrease, if they are centripetal or centrifugal, in the doubled ratio of the (increasing) distance between the moon and the earth's centre, but, should they act laterally, in the doubled ratio of SQ to OQ[61]—as I found once I undertook the calculations. This orbit has above been shown to be oval. And thence it results that the eccentric orbit in which the moon really revolves is not an ellipse but an oval of another kind. But if, to lessen the difficulty of the computation, we should suppose that the concentric orbit is a circle, the eccentric orbit will turn out to be an ellipse whose major axis PQ will be equal to the circle's diameter. And the above-mentioned forces PH, IH, $2IH$ and $2HS$ would, if the moon were transferred from the elliptical orbit into the circular one, come to be $(QO/SP)\,PH$, $(QO/SP)\,IH$, $2(QO/SP)\,IH$ and $2(QO/SP)\,HS$; and thereafter if, with the moon returning from this orbit to the elliptical one, the centripetal forces

$$(QO/SP)\,PH \quad \text{and} \quad 2(QO/SP)\,HS$$

were to diminish in the doubled ratio of the distance SP of the moon from the earth's centre, and the lateral forces $(QO/SP)\,IH$ and $2(QO/SP)\,IH$ in the doubled ratio of SQ to OQ, then they would become

$$(QO^3/SP^3)\,PH, \quad (QO^3/SP\times SQ^2)\,IH, \quad 2(QO^3/SP\times SQ^2)\,IH$$

and

$$2(QO^3/SP^3)\,HS:$$

since the difference $SP-QO$ of the distances SP and QO is (because of the

(60) As earlier announced (see notes (2) and (26)) we here separately reproduce the much changed text of Proposition A in [1] which ensues on introducing therein the extensive emendations—uniquely consequent on his alteration of the ratio of lateral perturbation to be $(SQ/OQ)^2$ (see next note)—which Newton afterwards made in the manuscript (ff. $102^r/103^r$) but abandoned without likewise 'mending' its scholium or the following Proposition B.

(61) The ultimate effect of this crucial change in the lateral ratio (from $(SP/QO)^i$, $i = 2$, then 4, and lastly 3; compare note (26) above) is assessed in note (63) below.

parvam excentricitatem) sit infinite parva) si distantia illa $[SP-QO]$ dicatur x et SO dicatur y, fiunt (per Lem. 2 Lib. II Princip.) $\frac{SP-3x}{SP}PH$, $\frac{SP-x-2y}{S[P]}IH$,(62) $\frac{2SP-2x-4y}{S[P]}IH$(62) & $\frac{2[SP]-6x}{SP}HS$. Itaq; si Luna his viribus revolveretur, Apogæum quiesceret. Proindeq; [s]i hæ vires de viribus PH, IH, $2IH$ & $2HS$ subducantur, vires [re]liquæ $\frac{3x}{SP}PH$, $\frac{x+2y}{SP}IH$, $\frac{2x+4y}{SP}IH$ & $\frac{6x}{SP}HS$ eæ ipsæ erunt a [quibu]s Apogæum movetur. Et vi quidem $\frac{6x}{SP}HS-\frac{3x}{SP}PH$ Luna, si [extra] circulum versatur, distrahitur a Terra, & ubi [intra versatur at]trahitur in terram. Unde vis illa eadem est cum vi V in Lemmate [α] & propterea (per Lemma illud) motus Apogæi a vi illa oriundus est ad motum medium Lunæ ut $\frac{6x}{SP}HS\times SE-\frac{3x}{SP}PH\times SE$ ad $P\times OS$. Vi autem $\frac{x+2[y]}{SP}IH+\frac{2x+4y}{SP}IH$ seu $\frac{3x+6y}{SP}IH$ Luna, ubi extra circulum versatur, trahitur versus syzygias, et ubi intra circulum est, trahitur in partes contrarias, ideoq; hæc est vis W in Lemmate [β] et propterea (per Lemma illud) motus Apogæi a vi illa oriundus est ad motum medium Lunæ ut $\frac{6x+12y}{SP}IH\times PE$ ad $P\times OS$. Et compositè, motus totus Apogæi a vi utraq; oriundus est ad motum medium Lunæ ut

$$\frac{6x}{SP}HS\times OE-\frac{3x}{SP}PH\times OE+\frac{6x+12y}{SP}IH\times PE$$

ad $P\times OS$ seu $6HS\times SE^q-3PH\times SE^q+6IH\times PE\times SE+12IH\times PE\times SP$ ad $178{\llcorner}725SP^{cub}$. Nam x est semidifferentia rectarum PF & PS, adeoq; est ad OS seu y ut SE ad SP. Et Lunæ pondus P supra (in Prop. [XXV]) erat ad vim SP (quæ in vires PH, IH, HS hic resolvebatur) ut $178{\llcorner}725$ ad 1, ideoq; P valet $178{\llcorner}725SP$. Jam verò si in PS demittatur perpendiculum ET et pro ES^q et $PE\times ES$ scribantur PST & $PS\times ET$: fiet motus horarius Apogæi ad motum medium horarium Lunæ id est ad 32′. 56″. 27‴, 12^{iv}, ut

$$6HST-3PH\times ST+6IH\times ET+12IH\times PE$$

ad $178{\llcorner}725SP^q$. Ubi terminus $6HST$ semper affirmativus est et terminus $3PH\times ST$ semper negativus, at termin[i] $6IH\times ET+12IH\times PE$ proinde ut rectæ IH & ET jaceant ad easdem vel ad contrarias partes rectæ PS, affirmativ[i sunt] vel negativ[i]. Apogæum autem, ubi quantitas

$$6HST-3PH\times ST\pm 6IH\times ET\pm 12IH\times PE$$

(62) For, much as in note (30) above, on taking $SP-QO=x$ and $SO\ (=SQ-QO)=y$

infinitely small eccentricity) indefinitely small, if that distance $[SP-QO]$ be called x and SO called y, these become (by *Principia*, Book II, Lemma 2)

$$(1-3x/SP)\,PH, \quad (1-(x+2y)/SP)\,IH,^{(62)} \quad (2-(2x+4y)/SP)\,IH^{(62)}$$

and
$$(2-6x/SP)\,HS.$$

And so, if the moon were to revolve under these forces, the apogee would stay at rest. Consequently, if these forces be taken away from the forces PH, IH, $2IH$ and $2HS$, the forces $(3x/SP)\,PH$, $((x+2y)/SP)\,IH$, $((2x+4y)/SP)\,IH$ and $(6x/SP)\,HS$ remaining will be the very ones whereby the apogee moves. By the force $(6x/SP)\,HS-(3x/SP)\,PH$, indeed, the moon is, if located outside the circle, drawn away from the earth, but, when inside, attracted to the earth; whence that force is the same as the force V in Lemma [α], and accordingly (by that Lemma) the motion of the apogee arising from that force is to the moon's mean motion as $(6x/SP)\,HS\times SE-(3x/SP)\,PH\times SE$ to $P\times OS$. But by the force $((x+2y)/SP)\,IH+((2x+4y)/SP)\,IH$, or $((3x+6y)/SP)\,IH$, the moon is, when located outside the circle, drawn towards the syzygies, and, when inside the circle, in the opposite direction; in consequence, this is the force W in Lemma [β], and accordingly (by that Lemma) the motion of the apogee arising from that force is to the moon's mean motion as $((6x+12y)/SP)\,IH\times PE$ to $P\times OS$. And, on compounding, the total motion of the apogee arising from both forces is to the moon's mean motion as

$$(6x/SP)\,HS\times OE-(3x/SP)\,PH\times OE+((6x+12y)/SP)\,IH\times PE$$

to $P\times OS$, that is, $6HS\times SE^2-3PH\times SE^2+6IH\times PE\times SE+12IH\times PE\times SP$ to $178.725SP^3$. For x is half the difference of the straight lines PF and PS, and hence is to OS or y as SE to SP; and the moon's weight P was above (in Proposition XXV) to the force SP (here resolved into the forces PH, IH, HS) as 178.725 to 1, and P's value is in consequence $178.725SP$. Now indeed, if the perpendicular ET be let fall to PS, and in place of ES^2 and $PE\times ES$ there be written $PS\times ST$ and $PS\times ET$, the hourly motion of the apogee will be to the moon's mean hourly motion, that is, $32'\ 56''\ 27'''\ 12^{\text{iv}}$, as

$$6HS\times ST-3PH\times ST+6IH\times ET+12IH\times PE$$

to $178.725SP^2$. Here the term $6HS\times ST$ is always positive, the term $3PH\times ST$ always negative, but the terms $6IH\times ET+12IH\times PE$ are, accordingly as the lines IH and ET lie on the same or opposite sides of the line PS, positive or negative. The apogee, however, when the quantity

$$6HS\times ST-3PH\times ST\pm 6IH\times ET\pm 12IH\times PE$$

---

to be contemporaneous moments of QO there results

$$QO^3/SP\times SQ^2 = ((QO+x)/QO)^{-1}\times((QO+y)/QO)^{-2} = (QO-x-2y)/QO,$$

that is, $(SP-x-2y)/SP$, in the limit as SP and SQ come to coincide with QO.

negativa est, regreditur; ubi vero affirmativa progreditur. Inventus est igitur motus horarius Apogæi. Q.E.I.[63]

---

(63) Newton proceeded no further with his recasting of the text reproduced in [1], failing to continue his emendations even into the following Corollary to Proposition A. If we pursue their effect in the sequel we will readily appreciate his reason for so summarily abandoning an approach which he had initially found attractive. For since, in the terms of note (36) above, there is

$$\textit{omnia } IH \times PE/\textit{totidem } SZ^2 = \int_0^{2\pi} \sin 2(a+x)\sin x\,.\,dx \Big/ 2\pi$$
$$= \int_0^{2\pi} \tfrac{1}{2}(\cos(2a+x) - \cos(2a+3x))\,.\,dx \Big/ 2\pi = 0,$$

at once 'summa omnium $6HST - 3PH \times ST + 6IH \times ET + 12IH \times PE$ erit ad summam totidem $178\llcorner 725SP^q$ ut $6D+6C$, $-3D+3C$, $+6C$ ad $178\llcorner 725 \times 4D$, hoc est ut $D+5C$ ad $238\frac{3}{10}D$'. By suitably modifying Proposition B it then follows that the 'motus Apogæi in G' comes to be to the moon's mean motion 'ut $\frac{2}{5}AS+GH$ ad [$\frac{2}{5} \times 238.3 =$] $95\frac{8}{25}AS$, hoc est ut $\frac{2}{5}Gg+Hh$ ad $95\frac{8}{25}Gg$'; whence, since the ratio of the latter motion to the mean 'horary' motion of the sun is ($4812°\,45'\,38''/360° =$) 13.36878 to 1 'id est ut $95\frac{8}{25}AS$ ad $7\llcorner 13004AS$', the sun's motion will thereby be to the mean motion of lunar apogee 'ut $7\llcorner 13004AS$ ad $\frac{2}{5}AS+GH$ id est ad $0\llcorner 4AS+GH$'; so that by taking $n = (7.13004-0.4) = 6.73004$ in the ensuing integration there is 'omnia $Hh \times t$ ad omnia $Gg \times t$ ut $0\llcorner 0747$ as 1', and in consequence

$$\text{'omnia } Hh \times t + \tfrac{2}{5}Gg \times t \text{ ad omnia } \tfrac{2}{5}Gg \times t\,[= 20°\,11'\,46''] \text{ ut } 2\llcorner 373 \text{ ad } 2\text{'},$$

from which the 'motus medius Apogæi quo tempore Sol cursum suum ab Apogæo ad Apogæum complet' proves to be about 23° 58′, some 17° less than the 'motus per Tabulas Astronomicas' —a sad disappointment indeed.

In print Newton afterwards contented himself with appending to his more elementary propositions on the motion of the moon (assumed to orbit the earth in a 'concentric' circle; see note (1)) a short scholium (*Principia*, ${}_{1}1687$: 462–3; suppressed in later editions) where he

is negative regresses; when positive, indeed, it advances. The hourly motion of the apogee has therefore been ascertained. As was to be found.[63]

---

announced: 'Hactenus de motibus Lunæ quatenus Excentricitas Orbis non consideratur. Similibus computationibus inveni quod Apogæum, ubi in Conjunctione vel Oppositione Solis versatur, progreditur singulis diebus 23′ respectu [stellarum] Fixarum; ubi verò in Quadraturis est, regreditur singulis diebus $16\frac{1}{3}$['] circiter: quodq; ipsius motus medius annuus sit quasi 40^gr^. Per Tabulas Astronomicas [*sc.* in his *Doctrine of the Sphere* (London, 1680)] à Cl. *Flamstedio* ad Hypothesin *Horroxii* [see note (28)] accommodatas, Apogæum in ipsius Syzygijs progreditur cum motu diurno 24′. 28″, in Quadraturis autem regreditur cum motu diurno 20′. 12″, & motu medio annuo 40^gr^. 41′ fertur in consequentia. Quod differentia inter motum diurnum progressivum Apogæi in ipsius Syzygijs & motum diurnum regressivum in ipsius Quadraturis per Tabulas fit 4′. 16″, per computationem verò nostram $6\frac{2}{3}$′, vitio Tabularum tribuendum esse suspicamur. Sed neque computationem nostram satis accuratam esse putamus. Nam rationem quandam ineundo prodiere Apogæi motus diurnus progressivus in ipsius Syzygijs & motus diurnus regressivus in ipsius Quadraturis paulo majores. Computationes autem, ut nimis perplexas & approximationibus impeditas neque satis accuratas, apponere non lubet'. There is here no necessity to adopt J. C. Adams' unsupported conjecture (*A Catalogue of the Portsmouth Collection* (note (2)): xiii) that Newton derived these variant published values for the moon's extremes of horary motion (see note (43) above) and the mean annual advance of its apogee 'by a more complete and probably a much more complicated investigation than that contained in the extant MSS'. Rather, these values represent a minimal modification of Newton's manuscript scheme: for in much the same way as he had before, in [1] above (compare note (58)), departed from a 'computed' hourly apogee motion of $(1+\lambda\cos 2\widehat{QSA})\times 8''9'''$, $\lambda = \frac{11}{2}$, to deduce an annual mean advance of some 38° 52′, the upgraded parameter $\lambda = 6$ yielding a virtually exact annual advance (about 40° 43′ to be precise) determines in syzygies ($\widehat{QSA} = 0°$ or 180°) a daily progress of about 22′ 49″ and in quadratures ($\widehat{QSA} = 90°$ or 270°) a daily regress of about 16′ 18″. Further slight increase in λ 'rationem quandam ineundo' will, it is evident, marginally augment the values of the apogee's daily motion in its extremes, but at the expense of producing an annual mean advance which is a little greater than true.

3

THE 'DE MOTU CORPORUM LIBER PRIMUS' REMODELLED[1]

[early 1690's]

Extracts from original drafts in the University Library, Cambridge

§1. PRELIMINARY AMELIORATIONS OF THE PUBLISHED TEXT.[2]

IN PHILOSOPHIÆ NATURALIS PRINCIPIJS MATHEMATICIS CORRIGENDA ET ADDENDA.[3]

[1][4] *pag. 12.* *Lex. II.*

Motum omnem novum quo status corporis mutatur vi motrici impressæ proportionalem esse, & fieri a loco quem corpus alias occuparet, in metam[5] *quam vis impressa petit.*[6]

Si vis aliqua motum quemvis generet, dupla duplum, tripla triplum generabit, sive simul et semel[7] sive gradatim et successive impressa fuerit. Et

---

(1) In addition to the many minor ameliorations of its textual detail which Newton started to record soon after publication in his annotated working and library copies (now ULC. Adv. b. 39.1 and Trinity College, Cambridge. NQ. 16.200 respectively) of his 1687 *Principia* and ultimately, with still other additions and certain more substantial changes in its second and third books, to incorporate in its revised 'Editio secunda, auctior et emendatior' in 1713, he also concurrently drafted a number of tentative further improvements which never found their way into print, commencing with the remoulding of individual axioms, lemmas and propositions, but soon—no later than May 1694 when (see §2: note (1) following) he allowed David Gregory a glimpse of this more radical scheme of revision—passing on to begin a major rebuilding of the *Principia*'s structure which he never finished, yet of which extensive fragments survive in his unpublished papers. From the various remodellings of the *Principia*'s opening propositions 'De motu Corporum' which now exist in ULC. Add. 3965 we here, in a not unfitting epilogue to our earlier reproduction of their preliminary versions (in 1, §§1/2 above), select the mathematically more important of those which relate to Newton's definition of the measure of a force by its accelerative deviation from a rectilinear inertial path and his application of it to determine the free and disturbed motion of a body orbiting in a general conic round a force-centre in its plane.

(2) Restylings and augmentations of Lemma II and Propositions I, VI, VII, X, XIII and XVII of the 1687 *Principia*'s first book, to whose printed pages they are individually keyed. The corresponding portions of the preliminary tract 'De motu Corporum' (1, §2 above) will readily be identified.

Translation

CORRECTIONS AND ADDITIONS TO BE MADE IN THE 'PHILOSOPHIÆ NATURALIS PRINCIPIA MATHEMATICA'.(3)

[1](4) *page 12.* *Law II.*

All new motion by which the state of a body is changed is proportional to the motive force impressed, and occurs from the place which the body would otherwise occupy towards the goal(5) *at which the impressed force aims.*(6)

Should some force generate any motion, then twice the force will generate its double, and three times it its triple, whether it be impressed once and instantaneously(7) or successively and by degrees. And this motion will transfer the

(3) This generic title is taken from Add. 3965.12: 182r, where it heads the manuscript sheet from which we have extracted [2] below.

(4) Add. 3965.19: 731v/731r. The lower half of the manuscript is badly charred and a considerable section of it is completely burnt away: the ensuing gaps in the text are here restored (within square brackets) on the pattern of Newton's preliminary rough drafts on Add. 3965.12: 274r/274v. An extended discussion of this revision and its related drafts (notably Add. 3965.6: 86r, reproduced in Appendix 1 below) is given by I. B. Cohen in his 'Newton's Second Law and the Concept of Force in the *Principia*' (= (ed. R. Palter) *The 'Annus Mirabilis' of Sir Isaac Newton, 1666–1966* (Cambridge, Massachusetts, 1970): 143–85): 160–71 (+178–85): 'Newton's revision of the second law of motion'. (A rawer earlier version of the article appeared in *The Texas Quarterly*, **10**, 3 (Autumn, 1967): 127–57; see especially 143–52.)

(5) That is, the instantaneous centre of force. Newton first wrote 'plagam' (direction) and then 'regionem' (region).

(6) In preliminary drafts of this revised enunciation on f. 274r (see note (4)) Newton initially asserted '*Motum in spatio vel immobili vel mobili genitum proportionalem esse vi motrici impressæ & fieri secundum lineam rectam qua vis illa imprimitur*' (*The motion begotten in either a stationary or mobile space is proportional to the motive force impressed and takes place following the straight line along which that force is impressed*) and subsequently '*Motum a loco quem corpus alias occuparet, vi motrici impressæ proportionalem esse et in plagam ejus dirigi*' (*The motion from the place which a body would otherwise occupy is proportional to the motive force impressed and is in the line of its direction*) before attaining the present finally preferred statement. Observe that his earlier ambiguous appeal to a 'mutatio motus' in the corresponding enunciation of his revised 'De motu Corporum' (1, §2 preceding)—one which has misled many recent historians and philosophers of science needlessly to make therein a rigid distinction between an incremental 'change in motion' and the rate of that change when in Newton's context 'dato tempore' none such is possible (or indeed fruitful)—is here clarified by the implicit insistence that this 'new' motion is qualitatively comparable as well as quantitatively compoundable with the 'old' one which it alters in course and size. The better to illustrate the meaning of this key principle that the motion so induced by an impressed force—or rather the (vanishingly small) linear distance which it traverses, in deviation from the path of the 'old' motion, in given (infinitesimal) time—is in the direction of its action and quantitatively proportional to its magnitude, Newton will in sequel advance his previous prime corollary, now (see note (8)) converted to instance the case of a continuous *vis impressa*, to be with the main explanatory text.

(7) See 1, §2: note (15); compare also the next note.

hic motus transferet corpus de loco in quo alias reperiri deberet, versus mundi regi[onem] in quam vis impressa dirigitur, ideoqꝫ si cor[pus antea move]batur, motui ejus vel conspiranti additur [vel contrari]o subducitur, vel obliquo oblique adjicitur, [et cum eo secundum] utriusqꝫ determinationem componi[tur. Si corpus *A*] in loco suo *A* [ubi vis in illud imprimitur motum [habeat quo uniformiter c]ontinuato [describeret lineam rectam *Aa*, s]ed per vim impressam deflectatur de [hac linea in] aliam(8) *Ab*, et ubi reperiri deberet in [loco *a* invenia]tur in loco *b*: quoniam corpus absqꝫ vi [impressa occuparet] locum *a* & de hoc loco per vim illam [deturbatur et in]de transfertur ad locum *b*, transla[tio corporis de] loco *a* ad locum *b* ex mente legis hujus erit [huic v]i proportionalis & in metam(9) dirigetur in quam [hæc vis] imprimitur. Unde si corpus idem motu omni privatū [a vi eadem] in eandem metam impressa, eodem tempore de loco *A* [in locum *B*] transferri posset: rectæ duæ *AB* et *ab* parallelæ erunt et æquales.(10)

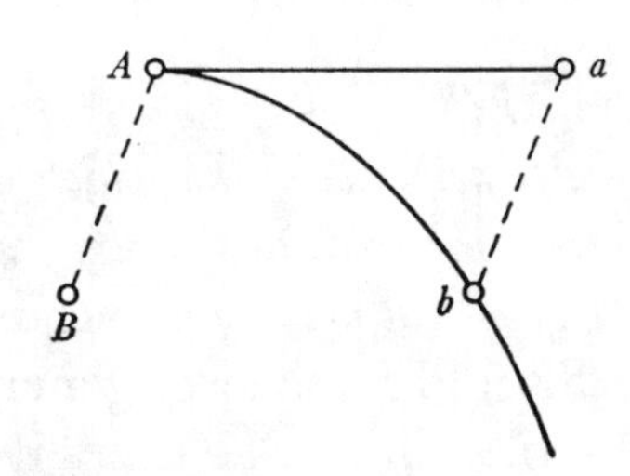

(8) Understand 'curvam' (curved) as it appears in Newton's accompanying figure, drawn to illustrate his 'Cas. 2' (compare Appendix 1 below) of a continuous impressed force whose action in deflecting the 'old' uniform motion over *Aa* is broken down into the step-by-step application of an infinity of component impulses, each infinitesimally small and acting 'instantaneously' at successive infinitesimal instants of time to produce a total effect of continuous curvilinear motion in the resulting *linea Ab* (a parabola of diameter *AB* under Newton's implicit further assumption—see note (10)—that the time in which it is traversed is itself infinitesimal, so determining the impressed force to be locally constant in magnitude and direction). In [2] following (see note (13)) Newton will assume the action of a continuous *vis impressa*—f, say, in infinitesimal time dt—in producing 'gradatim' the deviation $ab = \frac{1}{2}f.dt^2$ (see 1, §1: note (19)) to be equivalently determinable as that of the same total force f when applied as a single impulse 'simul et semel' at A to produce over time dt the deflection $\beta b = f.dt^2$ of a previous motion along the chord αA into one along the contiguous chord Ab, with Aa interpreted as the 'mean' direction at A (so that $\beta a = ab = \frac{1}{2}\beta b$). This confused notion was to remain to plague such competent dynamical theorists as Varignon and Leibniz during 1704–6 (see E. J. Aiton, 'The Celestial Mechanics of Leibniz', *Annals of Science*, **16**, 1960: 65–82, especially 75–82) and, forty years later, D'Alembert (compare T. L. Hankins, *Jean D'Alembert: Science and the Enlightenment* (Oxford, 1970): 228–31). The related mathematical problem of defining the 'tangential' direction at a vertex of an infinitely-sided limit-polygon similarly taxed the comprehension of most early eighteenth-century geometers, and led frequently to argument. One such disagreement between Abraham de Moivre and Niklaus

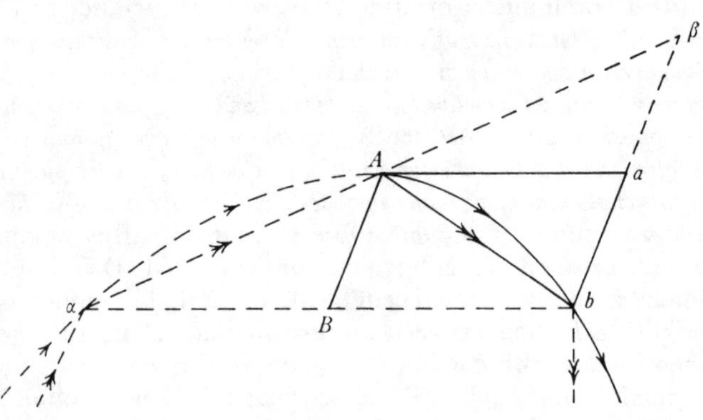

body from the place in which it ought otherwise to be found towards the region of the world at which the impressed force is directed; consequently, if the body was beforehand in motion, it is either added to it when in unison, or taken from it when contrary, or adjoined obliquely to it when oblique and combined with it according to the determined direction of each. If the body *A* should, at its place *A* where a force is impressed upon it, have a motion by which, when uniformly continued, it would describe the straight line *Aa*, but shall by the impressed force be deflected from this line into another[(8)] one *Ab* and, when it ought to be located at the place *a*, be found at the place *b*, then, because the body, free of the impressed force, would have occupied the place *a* and is thrust out from this place by that force and transferred therefrom to the place *b*, the translation of the body from the place *a* to the place *b* will, in the meaning of this Law, be proportional to this force and directed to the same goal[(9)] towards which this force is impressed. Whence, if the same body deprived of all motion and impressed by the same force with the same direction, could in the same time be transported from the place *A* to the place *B*, the two straight lines *AB* and *ab* will be parallel

I Bernoulli in late 1712 was sagely concluded by the latter's uncle Johann in a letter of 18 February 1713 (N.S.) to de Moivre with the judgement: 'La dispute qui a été entre vous... et mon neveu sur les tangentes des courbes n'est effectivement qu'une logomachie, car on peut considérer une ligne courbe en deux différentes manieres, à sçavoir comme un polygone d'une infinité...de côtés droites, et aussi comme l'assemblage d'une infinité d'arcs de cercle qui se décrivent successivement...; une courbe étant donc prise dans le premier sens, il est évident que la ligne infiniment proche de l'ordonnée, interceptée entre la tangente et la courbe, doit être regardée comme la seconde fluxion de l'ordonnée, en prenant la fluxion de l'abscisse comme invariable, et le côté prolongée pour la tangente; mais la courbe étant considérée dans le second sens, les arcs de cercle dont la courbe est censée être composée, quoiqu'ils soient eux-mêmes infiniment petits par rapport à la courbe, ils sont pourtant infiniment grands par rapport aux petites lignes droites, dont chacun de ces petits arcs est composé comme un polygone, et alors la tangente purement géométrique ne differe point du prolongement d'un des côtés dont un nombre infini fait un arc...infiniment petit....Cependant ces deux idées peuvent mener...à une même conclusion, pourvu qu'on raisonne toujours conséquemment, et qu'on ne quitte jamais son hypothèse, comme il arriva il y a quelques années à M. Varignon...' (see K. Wollenschläger, 'Der mathematische Briefwechsel zwischen Johann I Bernoulli und Abraham de Moivre' [= *Verhandlungen der Naturforschenden Gesellschaft in Basel*, **43**, 1933: 151–317]: 281). With Bernoulli's general point Newton would wholeheartedly have agreed, though he would rightly have rejected his added surmise (*ibid.*: 282) that his own earlier numerical 'méprise [in Proposition X of the *Principia*'s second book as set out in the 1687 *editio princeps*] sur la raison de la résistance à la force centripète, semble tirer son origine du même principe des deux idées des tangentes mal conciliées' since—as admittedly he came to understand only in late September 1712 after Bernoulli had, by a variant analytical derivation, faulted its primary *Exemplum* of resisted motion in an upright semicircle—his mistake there was far different. (See D. T. Whiteside, 'The Mathematical Principles underlying Newton's *Principia Mathematica*' (*Journal for the History of Astronomy*, **1**, 1970: 116–38): 126–30, especially 128–9. We will return to treat this point in detail in the eighth volume.)

(9) See note (5).

Nam vis eadem cum eadem directione eodem tempore in corpus idem sive quiescens sive motu quocunq; latum agendo, eandem translationem versus eandem metam[9] ex mente Legis hujus efficiet. Et translatio in hoc casu est *AB* ubi corpus ante vim impressam quiescebat & *ab* u[bi in statu] illo [mov]ebatur. At corporis majoris translatio per vim [impressam] minor est et minoris major ut motus[11] sit idem. [Principij] autem hujus veritas abunde satis ex [experientia] confirmatur cum potestates quinq; mechanicæ[12] [ex illo] demonstrentur ut in sequentibus docebitur.

---

[2][13] *p. 38 l. 15.* [*Prop. I, Theor. I.*]

Dele Corollaria duo[14] *et eorum vice scribe hæcce.*

Corol. 1. Velocitas corporis in centrum immobile attracti est in spatijs non resistentibus reciproce ut perpendiculum a centro illo in orbis tangentem rectilineam demissum. Est enim velocitas in *A*, *B*, *C*, *D*, *E* ut sunt bases æqualium triangulorum *AB*, *BC*, *CD*, *DE*, *EF* et hæ bases sunt reciproce ut perpendicula in ipsas demissa.

Corol. 2. Si arcuum duorum æqualibus temporibus in spatijs non resistentibus descriptorum chordæ *AB*, *BC* compleantur in parallelogrammum *ABCV*: hujus diagonalis *BV* in ea positione quam ultimò habet ubi arcus illi in infinitum diminuuntur, si producatur, transibit per centrum virium.

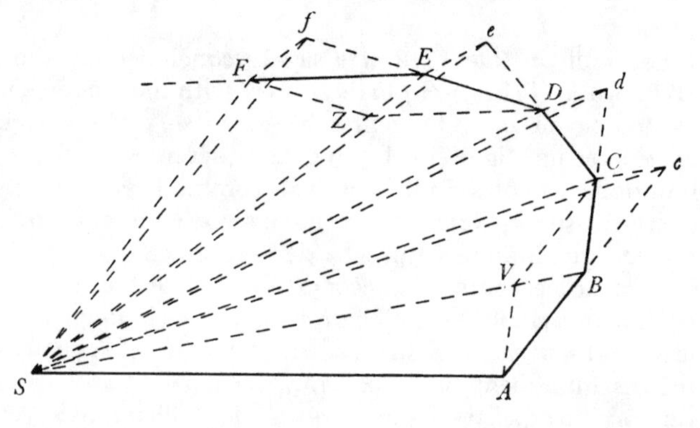

---

(10) Much as before (see 1, §1: note (19)), Newton's present assumption that the continuously induced deviation *ab* from the pristine rectilinear path *Aa* is both (effectively) straight and parallel to *AB* again requires, to be valid, that the total time of passage over $\widehat{Ab}$ be infinitesimal unless the 'goal' at which the *vis impressa* is directed shall be at infinity.

(11) That is, momentum (since now the mass of the moving body enters into the reckoning).

(12) See 1, §2: note (24).

(13) Add. 3965.12: 182ᵛ. The purpose of these new opening Corollaries 1–3 is to clarify the way in which the basic parallelogram of 'forces' (that is, velocities generated 'dato tempore' by given forces, one of which is here taken to be the ideal *vis insita* which Newton introduces to sustain uniform inertial motion onwards in a straight line at any instant, and the other is a general *vis centripeta*; compare 1, §2: note (19)) serves to yield a measure of an impressed force in terms of the velocity increment $BV(=cC)$ which its instantaneous impulse at *B* towards the centre *S* produces in the time that the orbiting body so impelled would otherwise move uniformly from *B* to *c* at a pristine inertial speed represented by $AB = Bc$, thereby producing a new resultant uniform motion at *B*, now towards *C*, which is correspondingly represented by

and equal.(10) For the same force, by acting with the same direction and in the same time on the same body whether at rest or carried on with any motion whatever, will in the meaning of this Law achieve an identical translation towards the same goal;(9) and in this present case the translation is AB where the body was at rest before the force was impressed, and ab where it was there in a state of motion. But, to achieve the same motion,(11) the translation of a greater body by an impressed force is less, and of a lesser one greater. The truth of this principle, I may add, is amply enough confirmed from experience since, as will be explained in the sequel, the five mechanical powers(12) may be demonstrated from it.

---

[2](13) *page 38, line 15.* [*Proposition I, Theorem I.*]

Delete the two corollaries(14) *and in their stead write these.*

Corollary 1. The speed of a body attracted to a stationary centre is in non-resisting spaces reciprocally as the perpendicular let fall from that centre onto the orbit's rectilinear tangent. For the speeds at $A, B, C, D, E, \ldots$ are as the bases $AB, BC, CD, DE, EF, \ldots$, and these bases are reciprocally as the perpendiculars let fall to them.

Corollary 2. If the chords AB, BC of two arcs described in equal times in non-resisting spaces should be completed into the parallelogram $ABCV$, then the diagonal BV of this shall, in the position which it ultimately has when those arcs are indefinitely diminished, pass (if produced) through the centre of force.

$AV = BC$. Corollaries 4 and 5 following, which in an earlier draft on Add. 3965.6: 36r appear as a separate 'Prop. V. Theor. IV' and its prime corollary, pass confusingly and without prior warning to the equivalent (see note (8) above) in which the impulse deviation $BV = cC$ generated in given time by the force impressed 'simul et semel' at B towards S is taken to match the effect of an equal force applied in infinitesimal component impulses 'gradatim et successive' at an infinity of separate instants in the same interval of time to produce (in the limit) a curved orbital arc $\widehat{BC}$ whose *sagitta* $B\beta$ $(= \frac{1}{2}BV)$ is equal in magnitude and direction to the corresponding 'perpetual' deviation γC from the mean direction of motion (along $B\gamma$ parallel to the chord AC) at B in the continuous—and effectively parabolic—arc $\widehat{ABC}$. The final Corollary 6 is an immediate consequence of the previous 'Leges Motûs' (and is so justified by Newton). All six corollaries went afterwards into the *Principia*'s second edition ($_2$1713: 35–6) without further significant change.

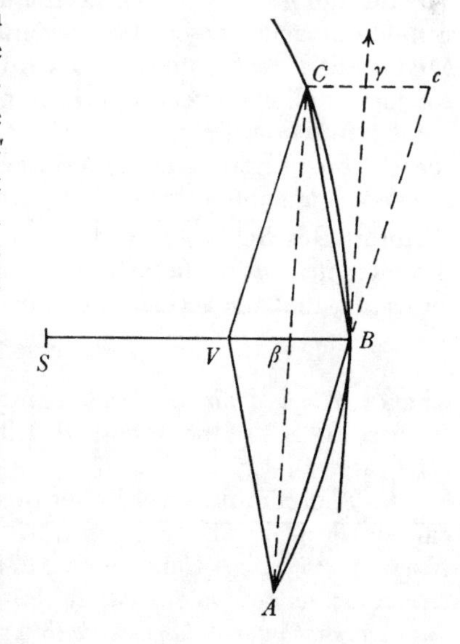

(14) Those reproduced in 1, §2, Appendix 2.3 preceding, that is. These were, in fact, retained in the second edition but resited to follow 'Propositio II. Theorema II' (*Principia*,

Corol. 3. Si arcuum æqualibus temporibus in spatijs non resistentibus descriptorum chordæ AB, BC ac DE, EF compleantur in parallelogramma $ABCV$, $DEFZ$: vires in B et E sunt ad invicem in ultima ratione diagonalium BV, EZ ubi arcus isti in infinitum diminuuntur. Nam corporis motus BC et EF componitur (per Legum Corol 1) ex motibus Bc, BV et Ef, EZ quorum BV et EZ ipsis Cc et Ff æquales in Demonstratione Propositionis hujus generabantur ab impulsibus vis centripetæ in B et E, ideoq; sunt his impulsibus proportionales.(15)

Corol. 4.(16) Vires quibus corpora quælibet in spatijs non resistentibus a motibus rectilineis retrahuntur ac detorquentur in orbes curvos sunt inter se ut arcuum quam minimorum æqualibus temporibus descriptorum sagittæ(17) illæ quæ convergunt ad centrum virium & chordas bisecant. Nam hæ sagittæ sunt semisses diagonalium de quibus egimus in Corollario tertio.

Corol. 5. Ideoq; vires eædem sunt ad vim gravitatis ut hæ sagittæ ad sagittas horizonti perpendiculares arcuum Parabolicorum quos projectilia eodem temporis spatio describunt.(18)

Corol. 6. Eadem omnia obtinent (per Legum Corol. 4) ubi plana in quibus corpora moventur, una cum centris virium quæ in ipsis sita sunt, non quiescunt sed moventur uniformiter in directum.

$_2$1713: 36) where in slightly modified form they assert respectively that 'In Spatiis vel Mediis non resistentibus, si areæ non sunt temporibus proportionales, vires non tendunt ad concursum radiorum; sed inde declinant in consequentia seu versus plagam in quam fit motus, si modo arearum descriptio acceleratur: sin retardatur, declinant in antecedentia' and 'In Mediis etiam resistentibus, si arearum descriptio acceleratur, virium directiones declinant a concursu radiorum versus plagam in quam fit motus'.

(15) In modern equivalent, if the orbital speed at B (along ABc) is $AB = Bc = v$ and the central force impressed 'simul et semel' at B towards S generates in the (infinitesimal) time dt of inertial motion from B to c the impulse $cC = BV = f.dt$, then the resultant velocity vector at B is $\overline{BC} = \overline{AV} = \overline{AB}+\overline{BV}$. It straightforwardly follows, where $AB\,(=Bc) = v.dt$ is the increment ds of the orbital arc s terminating at B, distant $SB = r$ from the force-centre let us say, that the accelerative increment dv in the orbital speed at B is

$$BV.\cos\widehat{ABS} = (f.dt).(dr/ds) = f.dr/v,$$

whence $f = v.dv/dr$: a fundamental result earlier equivalently obtained by Newton in Proposition XL of the published 'Liber primus' (*Principia*, $_1$1687: 125–7; compare 1, §3: note (198) above).

(16) A preliminary version of this corollary on Add. 3965.6: 36r (itself a revision of a still earlier 'Prop. I. Theor. I' on f. 37v which is reproduced by I. B. Cohen in his *Introduction to Newton's 'Principia'* (Cambridge, 1971): 167) is there set to be a separate 'Prop. V. Theor. IV' which asserts equivalently that '*Vires quibus corpora in medijs non resistentibus a motibus rectilineis perpetuo retrahuntur & in vias curvilineas detorquentur sunt inter se ut sagittæ arcuum æqualibus temporibus descriptorum si modo eæ sumantur sagittarum rationes quæ ultimo fiunt ubi arcus illi simul descripti in infinitum diminuuntur.* Nam hæ sagittæ sunt semisses diagonalium quibus vires proportionales

Corollary 3. If the chords AB, BC and DE, EF of arcs described in equal times in non-resisting spaces be completed into the parallelograms $ABCV$, $DEFZ$, then the forces at B and E are to one another in the ultimate ratio of the diagonals BV, EZ when those arcs are indefinitely diminished. For the motions BC and EF of the body are (by Corollary 1 of the Laws) compounded of the motions Bc, BV and Ef, EZ, and of these BV and EZ, equal to Cc and Ff, were in the proof of the present proposition generated by impulses of the centripetal force at B and E, and are in consequence proportional to these impulses.(15)

Corollary 4.(16) The forces by which any bodies are in non-resisting spaces drawn and twisted away from their rectilinear motions into curved orbits are to one another as those *sagittæ*(17) of the minimal arcs described in equal times which are directed through the centre of force and bisect their chords. For these *sagittæ* are halves of the diagonals with which we had to do in Corollary 3.

Corollary 5. In consequence, the same forces are to the force of gravity as these *sagittæ* to the 'arrows', perpendicular to the horizon, of the parabolic arcs which projectiles describe in the same space of time.(18)

Corollary 6. The same inferences all (by Corollary 4 of the Laws) hold true when the planes in which the bodies move, together with the force-centre sited in them, are not at rest but in uniform motion straight on.

---

esse in Corol. [3] Prop. I ostendimus'. To this the old Propositio IV (*Principia*, $_1$1687: 41–2; see pages 128–30 above) was appended in summary form as '*Corol. 2.* Si corpus uniformi cum motu in circulo revolvitur erit vis qua retrahitur a motu rectilineo ad ipsius pondus ut quadratum arcus dato quocunq; tempore descripti ad rectang[ulum] sub diametro circuli et spatium quod grave eodem tempore cadendo describere posset. Nam sagitta arcûs minimo tempore descripti æqualis est quadrato arcûs ejusdem applicato ad diametrum circuli, hoc est spatio quod grave eodem tempore describit. Et aucto tempore in ratione quacunq; augentur æqualia illa in eadem ratione duplicata et propterea semper sunt æqualia' (compare Cohen, *Introduction*: 168–9).

(17) Namely, the 'arrows' ($B\beta$ in the figure accompanying note (13) above) set between the 'bows' (*arcūs*) $\widehat{ABC}$ of the orbital curve here understood to pass through the vertices $A, B, C, D, E, \ldots$ and the (unstretched) 'strings' (*chordæ*) AC joining their end-points, such that when extended they pass through the centre S: because they lie along the diagonals BV of the parallelograms $ABCV$ they evidently bisect the 'bow-cords' AC and are in length equal to $\frac{1}{2}BV$. Newton here introduces a happy generalisation of what, in the elementary trigonometry of the circle, was in his day a familiar *terminus technicus* for the versine—one which in Arabic (*saḥem*) is at least as old as al-Ḫwārizmī in the early ninth century, and in the medieval Latin West was already used by Levi ben Gerson in his *De sinibus, chordis et arcubus* of about 1330. (See A. von Braunmühl, *Vorlesungen über die Geschichte der Trigonometrie*, **1** (Leipzig, 1900): 84, note 1; and compare D. E. Smith, *History of Mathematics*, **2** (Boston, 1925): 618–19.)

(18) Understand at or near the surface of the earth where the terrestrial *vis gravitatis* may effectively be assumed to be locally constant. Newton's preliminary version on Add. 3965.6: 36r—where it is set as 'Corol. 1' to his intended 'Prop. V. Theor. IV' (see note (13))—

[3][19] *pag. 44. lin 20.*

Prop. VI. Theor. V.

Si corpus in spatio non resistente circa centrum immobile in orbe quocunꝗ revolvatur et arcum quemvis jamjam nascentem tempore quam minimo describat & sagitta arcus ducatur quæ chordam bisecet & producta transeat per centrum virium: erit vis centripeta in medio arcus ut sagitta directe & quadratum temporis inversè.[20]

Nam sagitta dato tempore[21] est ut vis per Corol. 4 Prop. I,[22] & augendo tempus in ratione quavis, ob auctum arcum in eadem ratione augetur in ratione illa duplicata per Corol. 2 & 3 Lem. XI, adeoꝗ est ut vis et quadratum temporis conjunctim. Subducatur duplicata ratio temporis utrinꝗ[23] & fiet vis ut sagitta directe & quadratum temporis inverse. Q.E.D.

Idem quoꝗ facile demonstrari potest per Corol. 4 Lem. X.[24]

Corol. 1. Si corpus P revolvendo circa centrum S, describat lineam curvam APQ, tangat verò recta ZPR curvam illam in puncto quovis P, et ad tangentem ab alio quovis Curvæ puncto Q agatur QR distantiæ SP parallela, ac demittatur QT perpendicularis ad distantiam SP: erit vis centripeta reciproce ut solidum $\dfrac{SP^{\text{quad}} \times QT^{\text{quad}}}{QR}$, si modò solidi illius ea semper sumatur quantitas quæ ultimò fit ubi coeunt puncta P et Q. Nam QR æqualis est sagittæ arcus QP, et triangulum SQP ejusꝗ duplum $SP \times QT$ tempori quo arcus

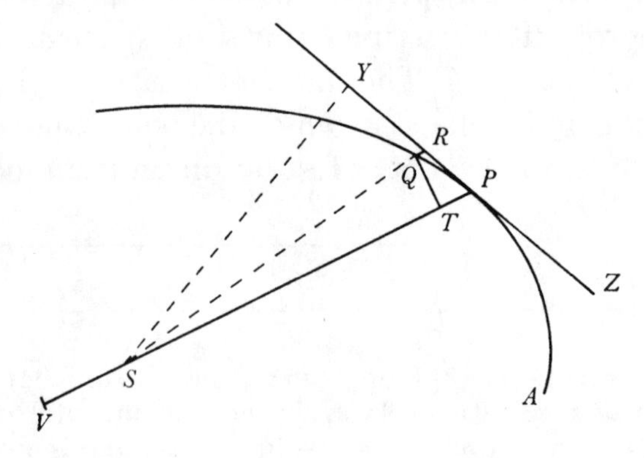

affirms equivalently, if at slightly greater length, that 'Vis qua corpus a motu rectilineo trahitur et in viam curvam detorquetur est ad ipsius vim gravitatis seu pondus ut sagitta arcûs cujusvis minimo tempore descripti ad spatium quod grave cadendo eodem tempore describere posset. Nam spatium illud est sagitta Parabolæ quam corpus projectum describeret'.

(19) Add. 3965.12: 183r/183v. Newton here augments and generalises the content of Propositions VI and VII/VIII respectively of the published 'Liber primus' (*Principia*, ₁1687: 44–6; compare pages 132–6 above).

(20) In his otherwise little variant preliminary draft of this Proposition and its five Corollaries on Add. 3965.12: 181r/181v Newton originally framed (f.181r) the equivalent enunciation: '*Corporis in orbe quocunꝗ in Medio non resistente circa centrum immobile revolventis vis centripeta est reciproce ut quadratum lineæ tempori proportionalis applicatum ad sagittam arcus*' (*Where a body revolves in any orbit whatever in a non-resisting medium round a stationary centre, the centripetal force is reciprocally as the square of a line proportional to the time, divided by the sagitta of the arc*).

(21) In his draft on f. 181r (see the previous note) Newton had again been careful to specify 'quam minimo' (minimal).

(22) In the revised version reproduced in [2] preceding.

(23) Of the proportion 'sagitta est ut vis et quadratum temporis conjunctim' (the *sagitta* is

[3][19] *page 44, line 20.*

Proposition VI, Theorem V.

If a body should, in a non-resisting space, revolve round a stationary centre in any orbit whatever and decribe any just barely nascent arc in a minimal time, and an 'arrow' of the arc be drawn to bisect its chord and pass, when produced, through the centre of force, then the centripetal force at the mid-point of the arc will be as that sagitta directly and the square of the time inversely.[20]

For the *sagitta* in a given[21] time is (by Proposition I, Corollary 4[22]) as the force, while on increasing the time in any ratio, because the arc is increased in the same ratio, it is (by Corollaries 2 and 3 of Lemma XI) increased in that ratio doubled, and hence is as the force and the square of the time jointly. Take away the doubled ratio of the time from each side[23] and the force will come to be as the *sagitta* directly and the square of the time inversely. As was to be proved.

The same can also easily be demonstrated by means of Corollary 4 of Lemma X.[24]

Corollary 1. If a body P in revolving round the centre S should describe the curved line APQ, while however the straight line ZPR touches that curve in any point P, to the tangent from any other point Q of the curve draw QR parallel to the distance SP, and perpendicular to it let fall QT: the centripetal force will then be reciprocally as the 'solid' $SP^2 \times QT^2/QR$, provided that the ultimate quantity of that solid occurring when the points P and Q come to coincide is always taken. For QR is equal to the *sagitta* of the arc $\widehat{QP}$, and the triangle SQP, and so its double $SP \times QT$, is

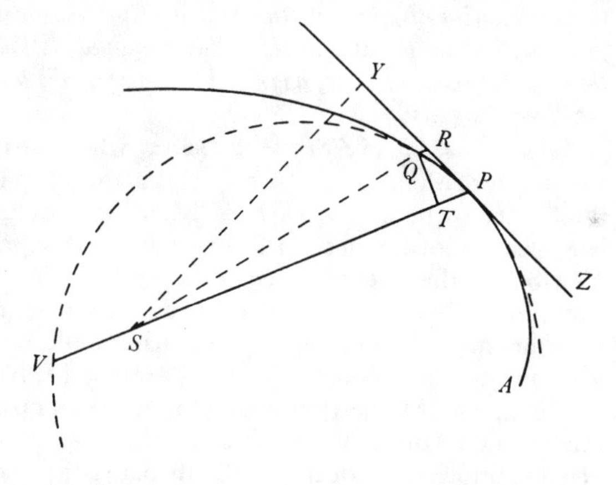

as the force and the square of the time jointly), that is—in other words, we are to divide through by the square of the time.

(24) This variant demonstration is preserved in the draft on f. 181r (see note (20)) where Newton argues similarly in proof: 'Nam vis centripeta est ut sagitta arcus dato tempore quam minimo descripti per Corol. 4 Prop. I & hæc sagitta est ad sagittam arcus alio quovis tempore quam minimo descripti in duplicata ratione temporis dati ad tempus alterum per Lem. [X], et propterea sagitta dato tempore est reciproce ut quadratū exponentis temporis alterius applicat[um] ad sagittam arcus illo tempore geniti' (For the centripetal force is as the *sagitta* of the arc described in the given minimal time (by Corollary 4 of Proposition I) and this *sagitta* is to the *sagitta* of an arc described in any other minimal time in the doubled ratio of the given time to the latter time (by Lemma X), and consequently the *sagitta* in the given time is reciprocally as the square of the exponent of the latter time divided by the *sagitta* of the arc engendered in that time).

iste describitur proportionale est ideoq; pro temporis exponente scribi potest.

Corol. 2. Eodem argumento vis centripeta est reciproce ut solidum $\frac{SY^q \times QP^q}{QR}$, si modo SY perpendiculum sit a centro virium in Orbis tangentem PR demissum. Nam rectangula $SY \times QP$ et $SP \times QT$ æquantur.

Corol. 3. Si Orbis vel circulus est, vel angulum contactus cum circulo quam minimum continet, eandem habens curvaturam eundemq; radium curvaturæ ad punctum contactus P; et si PV Chorda sit circuli hujus a corpore per centrum virium acta: erit vis centripeta reciprocè ut solidum $SY^q \times PV$.(25) Nam PV est $\frac{QP^q}{QR}$.

---

(25) In a preliminary recasting on Add. 3965.6: 37r Newton toyed with the notion of making this derived property basic in a variant 'Prop. VI. Theor. V' which announced: '*Si circuli tangant orbes concavos ad corpora et sint ejusdem curvaturæ cum orbibus in punctis contactuum: vires erunt reciproce ut solida contenta sub chordis arcuum circulorum ductis a corporibus per centra virium et quadratis perpendiculorum demissorum a centris ijsdem in tangentes rectilineas*' (*If circles should be tangent in the concavity of orbits at bodies, and of the same curvature as the orbits at the points of contact: the forces will be reciprocally as the solids contained by the chords of the circle-arcs drawn from the bodies through the centres of force, and by the squares of the perpendiculars let fall from the same centres onto the rectilinear tangents*).

Since $PV = 2\rho.(SY/SP) = 2\rho.\sin\alpha$, where ρ is the radius of curvature at P and $\widehat{SPY} = \alpha$ the angle of slope of the orbit to the radius vector SP, Newton's result straightforwardly implies that, where v ($\propto 1/SY$) is the orbital speed, then $(2/SY^2 \times PV =)\ SP/\rho.SY^3 \propto v^2/\rho.\sin\alpha$ is a measure of the central force at P directed to S. This fundamental formula, implicit in Newton's earlier Corollary 5 to Proposition IV of the 'Liber primus' (see 1, §2: note (86)), was independently discovered some dozen years afterwards by Abraham de Moivre who, when he announced the finding to Johann Bernoulli on 27 July 1705 (N.S.), observed that 'Après avoir trouvé ce théoreme je le montrai à M. Newton et je me flattois qu'il lui paroîtroit nouveau, mais M. Newton m'avoit prévenu; il me le fit voir dans les papiers qu'il prépare pour une seconde édition de ses *Principia Mathematica:* toute la différence qu'il y avoit, c'est qu'au lieu d'exprimer la loi de la force centripete par le moyen du rayon de la concavité, il l'exprimoit par le moyen d'une corde inscrite dans le cercle de la concavité: mais il me dit qu'il valoit mieux l'exprimer par le rayon, comme j'avois fait' ('Der mathematische Briefwechsel...' (note (8)): 214). As we would expect, Johann lost no time in concocting his own *ad hoc* demonstration of the 'théoreme', communicating it to de Moivre in his reply on 16 February 1706 (*ibid.*: 224–5) and later publishing it in the *Mémoires de l'Académie Royale des Sciences. Année M.DCC.X* (Paris, $[_1 1713 =]_2$ 1732: 529–30) as 'un assez beau Theorême [pour] trouver les forces centripetes, les Courbes étant données...: je la communiquai à M. Moivre dans une Lettre du 16. Fevrier 1706' without mention of having received its enunciation from de Moivre in the first place or of Newton's earlier equivalent discovery. Though de Moivre did not print his own variant proof of his 'théoreme'—sent privately to Bernoulli on 25 April 1706 ('Briefwechsel': 228)—till more than ten years later in his 'Proprietates quædam Sectionum Conicarum ex natura Focorum deducta; cum Theoremate generali de Viribus Centripetis; quorum ope Lex Virium Centripetarum ad Focos Sectionum tendentium, Velocitates Corporum in illis revolventium, & Descriptio Orbium facillime determinantur' (*Philosophical Transactions*, **30**, No. 352 [April–June 1717], §5: 622–8; the 'Theorema Generale' itself is

proportional to the time in which that arc is described, and consequently can be written as an expression of the time.

Corollary 2. By the same argument the centripetal force is reciprocally as the solid $SY^2 \times QP^2/QR$, if SY but be a perpendicular let fall from the centre of force onto the orbit's tangent PR. For the rectangles $SY \times QP$ and $SP \times QT$ are equal.

Corollary 3. If the orbit either is a circle or contains with a circle a minimal angle of contact, having the same curvature and same radius of curvature at the point P of contact, and if PV be the chord of this circle drawn from the body through the centre of force, then the centripetal force will be reciprocally as the solid $SY^2 \times PV$:(25) for PV is QP^2/QR.

---

presented on p. 624), he was not shy in the meantime of communicating its statement to his 'fellow' British mathematicians, and notably to his friend Edmond Halley who, in turn, passed it on to his Oxford colleague David Gregory and then to his successor, John Keill, in the Savilian chair of astronomy. Neither of the latter found any difficulty in deriving the result as an immediate consequence of Proposition IV of the 1687 *Principia*'s 'Liber primus': Gregory's proof (of about April 1707) is set down in his 'Codex E' (now Christ Church, Oxford. MS 346): 78 without added remark; but Keill enlarged his 'demonstratio perfacilis' into an open 'Epistola ad Clarissimum Virum Edmundum Halleium...de Legibus Virium Centripetarum' (*Philosophical Transactions*, **26**, No. 317 [for September/October 1708], §II: 174–88, especially 174–6) where he was careful to remind its ostensible recipient that 'Theorema tibi monstravit egregius Mathematicus D. Abrahamus De Moivre, dixitque Dominum Isaacum Newtonum theorema huic simile prius invenisse' (*ibid.*: 174). Though Bernoulli in a further article 'De Motu Corporum Gravium..., supposita Gravitate...ad quodvis datum punctum tendente...' (*Acta Eruditorum* (February/March 1713): 77–95/115–32) was led more accurately and charitably to refer to the 'Theorema' as 'mihi olim sed sine demonstratione transmissum a perexim[i]o Abrahamo Moyvræo, a me postea cum demonstratione quamvis ex alio fundamento petita remissum' (*ibid.*: 127), Keill riposted in a general 'Defense du Chevalier Newton. Dans laquelle on répond aux Remarques de Messieurs Jean et Nicolas Bernoulli...' (*Journal Literaire*, **8**, 1716: 418–33) not only to insist that de Moivre should be acknowledged as the 'Auteur du Théoreme', but to stress the ease with which it could be demonstrated, 'car le théoreme me fut communiqué à peu prés dans le même tems qu'il l'avoit été à M. *Bernoully*, & je puis dire que la Démonstration ne me couta pas plus d'un quart d'heure à trouver' (p. 421). Bernoulli could only answer this—from behind the none too efficient disguise of an outsider's borrowed name—with a counter-punch: 'Vix operæ pretium videtur ad hæc responderi: sed ut mysterium totum detegatur, ...monemus, *Bernoullium* inventionem theorematis non magis difficilem quam demonstrationem judicasse...: data vero occasione a nemine coactus...*Act. Erudit.* [1713 p. 127] *Moivreum* inventorem publice pronunciavit.... Cumque [*Keilius*] sibi gloriosum reputet, quod demonstrationem intra unius horæ quadrantem repererit, notet velim, *Bernoullium* vix quatuor horæ minutæ eidem impendisse...' ('M. Jo. Henr. Crusij Responsio ad Cl. Viri Johannis Keil...Defensionem pro...Is. Newtono...' *Acta Eruditorum* (October 1718): 454-66, especially 455–6). Such sterile squabbling carefully skirted the fact—well known to Bernoulli—that Pierre Varignon had long before published his similarly derived equivalent analytical measure $v^2/\rho.(rd\phi/ds)$ of the force acting on the point $P(r, \phi)$ instantaneously in the direction of the centre $S(0, 0)$, where v is the orbital speed over the vanishingly small arc $\widehat{PQ} = ds$. (See Varignon's 'Autre Regle Generale des Forces Centrales. Avec une maniere d'en déduire & d'en trouver une infinité d'autres à la fois, dépendemment & indépendemment des Rayons osculateurs...' (*Mémoires de l'Académie Royale des Sciences de Paris de l'Année M.DCCI*, Paris, 1704: 20–38): 21; and also

Corol. 4. Iisdem positis, est vis centripeta ut quadratum velocitatis directè & chorda illa inversè. Nam velocitas est reciproce ut perpendiculum SY per Corol. 1, Prop. I.

Corol. 5. Hinc si detur figura quævis curvilinea APQ, et in ea detur etiam punctum S ad quod vis centripeta perpetuo dirigitur: inveniri potest lex vis centripetæ qua corpus P a cursu rectilineo perpetuo retractum in figuræ illius perimetro detinebitur eamq revolvendo describet. Nimirum computandum est vel solidum $\frac{SP^q \times QT^q}{QR}$ vel solidum $SY^q \times PV$ huic vi reciproce proportionale. Ejus rei dabimus exempla in Problematis sequentibus.(26)

Prop. VII. Prob. II.(27)

Gyretur corpus in circumferentia circuli, requiritur Lex vis centripetæ tendentis ad punctum quodcumq datum.

Esto circuli circumferentia $VQPA$, punctum datum ad quod vis ceu centrum suum tendit S, corpus in circumferentia latum P, locus proximus in quem movebitur Q et circuli tangens ad locum priorem PRZ. Per punctum S ducatur chorda PV & acta circuli diametro VA jungatur AP, et ad SP demittatur perpendiculum QT quod productum occurrat tangenti PR in Z, ac deniq per punctum Q agatur LR quæ ipsi SP parallela sit et occurrat tum circulo in L tum tangenti PR in R. Et ob similia triangula ZQR, ZTP, VPA erit RP^{quad} (hoc est QRL) ad QT^{quad} ut AV^{quad} ad PV^{quad}. Ideoq $\frac{QRL \times PV^{\text{quad}}}{AV^{\text{quad}}}$ æquatur QT^{quad}. Ducantur hæc æqualia in $\frac{SP^{\text{quad}}}{QR}$, et punctis P et Q coeuntibus scribatur PV pro RL. Sic fiet $\frac{SP^{\text{quad}} \times PV^{\text{cub}}}{AV^{\text{quad}}}$ æquale $\frac{QT^q \times SP^{\text{quad}}}{QR}$. Ergo (per Corol. 1 Prop. VI) vis centripeta est reciprocè ut $\frac{SP^{\text{quad}} \times PV^{\text{cub}}}{AV^{\text{quad}}}$, id est (ob datum $AV^{\text{quad.}}$) reciproce ut quadratum distantiæ seu altitudinis SP et cubus chordæ PV conjunctim. Quod erat inveniendum.(28)

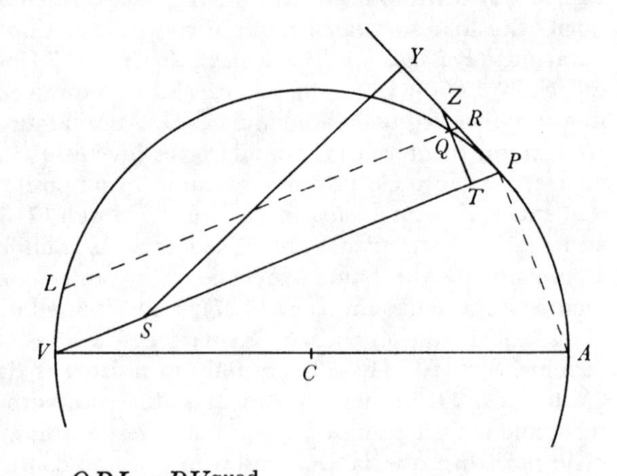

compare E. J. Aiton, 'The Inverse Problem of Central Forces', *Annals of Science*, **20**, 1964: 81–99, especially 89–90.)

(26) In a trivially amended autograph transcript (now Trinity College, Cambridge.

Corollary 4. With the same suppositions, the centripetal force is as the square of the speed directly and that chord inversely. For the speed is, by Corollary 1 of Proposition I, reciprocally as the perpendicular SY.

Corollary 5. Hence if there be given any curvilinear figure APQ and within it be also given the point S to which the centripetal force is perpetually directed, there can be ascertained the law of centripetal force whereby the body P, perpetually drawn aside from a rectilinear course, will be confined in the perimeter of that figure and describe it in its revolution. Specifically, there needs to be computed either the 'solid' $SP^2 \times QT^2/QR$ or the 'solid' $SY^2 \times PV$ reciprocally proportional to this force. Of this technique we shall give illustrations in following problems.(26)

Proposition VII, Problem II.(27)

Let a body orbit in the circumference of a circle; there is required the law of centripetal force tending to any given point whatever.

Let $VQPA$ be the circle's circumference, S the given point to which the force tends (its centre, as it were), P the body carried in the circumference, Q the very next place to which it shall move, and PRZ the circle's tangent at the former place. Through the point S draw the chord PV and, with the circle's diameter VA traced, join AP and to SP let fall the perpendicular QT, which shall, when produced, meet the tangent PR in Z; finally, through the point Q draw LR to be parallel to SP and meet both the circle in L and the tangent PR in R. Then, because the triangles ZQR, ZTP, VPA are similar, there will be RP^2, that is, $QR \times RL$, to QT^2 as AV^2 to PV^2; and consequently $QR \times RL \times PV^2/AV^2$ is equal to QT^2. Multiply these equals by SP^2/QR and then, with the points P and Q coalescing, write PV in place of RL. In this way there will come to be

$$SP^2 \times PV^3/AV^2$$

equal to $QT^2 \times SP^2/QR$. Therefore (by Corollary 1 of Proposition VI) the centripetal force is reciprocally as $SP^2 \times PV^3/AV^2$, that is, (because AV^2 is given) reciprocally as the square of the distance or height SP and the cube of the chord PV jointly. As was to be found.(28)

R. 16. 38: 4–5) made specially for the purpose, this theorem was afterwards passed on to Roger Cotes in the early autumn of 1709 and duly appeared without further alteration in the *Principia*'s second edition ($_2$1713: 41–2).

(27) Newton's little variant rough draft (omitting the *idem aliter*) of this generalised problem and its first corollary—in which (see note (30)) the earlier Proposition VII is now comprehended as but a special case—exists on Add. 3965.12: 186r, there keyed to be inserted on [*Principia*, $_1$1687:] 'pag. 45 post l. 13'. A preliminary intention of dividing its content between three concluding Corollaries 10–12 to a new 'Prop. V. Theor. IV' on Add. 3965.6: 36v (reproduced in Cohen's *Introduction to Newton's 'Principia'*: 169; compare note (16) above) was quickly cancelled.

(28) Compare our earlier equivalent analytic derivation of this result in 1, §2: note (94).

Idem aliter.

Ad tangentem PR productam demittatur perpendiculum SY et ob similia triangula SYP, VPA erit AV ad PV ut SP ad SY ideoq $\frac{SP\times PV}{AV}$ æquale SY & $\frac{SP^q\times PV^{\text{cub}}}{AV^q}=SY^q\times PV$ et propterea (per Corol 5 Prop VI) vis centripeta reciproce ut $\frac{SP^q\times PV^{\text{cub}}}{AV^q}$, hoc est (ob datum AV) reciproce ut $SP^q\times PV^{\text{cub}}$. Q.E.I.(29)

Corol. 1. Igitur si punctum datum S ad quod vis centripeta semper tendit, locetur in circumferentia hujus circuli puta ad [V], erit vis centripeta reciproce ut quadrato-cubus altitudinis SP.(30)

Corol. 2. Et si punctum S locetur ad infinitam distantiam, erit vis reciproce ut cubus chordæ PV.(31)

---

[4](32) *p 51 l 17.* [*Prop. XI. Prob. VI.*]

Idem aliter.(33)

Actis $SCHA$, SP, $KCED$, RPZ, RQ, QT, QX, PF ut supra, producatur PS donec Ellipsi occurrat in L et eidem PS agantur parallelæ HO, CM Ellipsi occurrentes in O et M. Jungatur PO secans CM in N et producatur CM donec tangenti RP occurrat in Z. Et ex natura Ellipseos RP^q est ad $RQ\times PL$ vel quod perinde est(34) QX^q ad PXL ut DC^{quad} ad CM^{quad}[,] id est (ex natura Ellipseos)

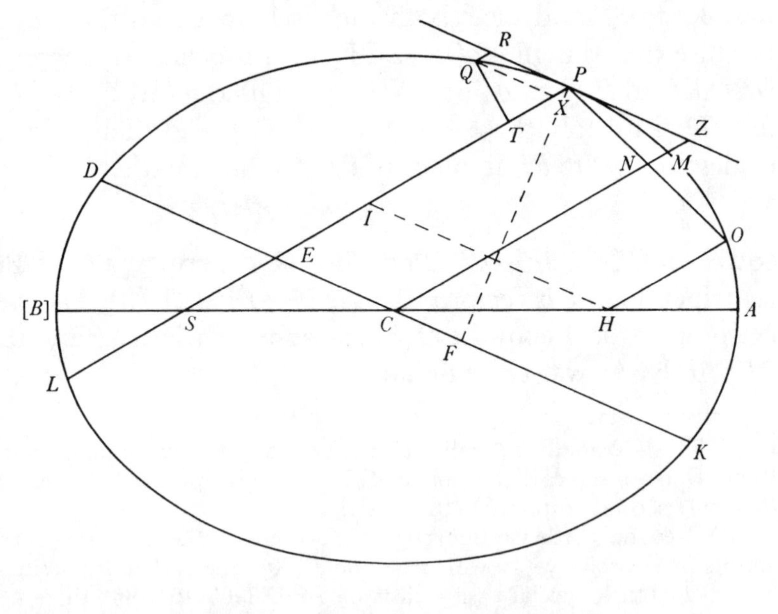

(29) Conversely, of course, the present *idem aliter* may be transposed to yield a proof of the general measure of centripetal force exhibited in Proposition VI, Corollary 5.

(30) This is the result achieved in the published Proposition VII (*Principia*, $_1$1687: 45–6) which is generalised in the above revised 'Prob. II'.

The same another way.

To the tangent PR produced let fall the perpendicular SY, and then, because the triangles SYP, VPA are similar, AV will be to PV as SP to SY, and in consequence $SP\times PV/AV$ will be equal to SY and so $SP^2\times PV^3/AV^2 = SY^2\times PV$; accordingly, (by Corollary 5 of Proposition VI) the centripetal force is reciprocally as $SP^2\times PV^3/AV^2$, that is, (because AV is given) reciprocally as $SP^2\times PV^3$. As was to be found.(29)

Corollary 1. Therefore if the given point S to which the centripetal force ever tends be located in the circumference of this circle, say at V, the centripetal force will be reciprocally as the fifth power of the height SP.(30)

Corollary 2. While if the point S be located at an infinite distance, the force will be reciprocally as the cube of the chord PV.(31)

---

[4](32) *page 51, line 17.* [*Proposition XI, Problem VI.*]

The same another way.(33)

Having drawn $SCHA$, SP, $KCED$, RPZ, RQ, QT, QX, PF as above, extend PS till it meets the ellipse in L, and parallel to the same PS draw HO, CM meeting the ellipse in O and M. Join PO cutting CM in N and extend CM till it meets the tangent RP in Z. Then RP^2 to $RQ\times PL$ or, what is essentially the same,(34) QX^2 to $PX\times XL$ is, from the nature of an ellipse, as DC^2 to CM^2, that is, (from

---

(31) This similarly comprehends the published following 'Prop. VIII. Prob. III' (*Principia*, ${}_1$1687: 46). In the *Principia*'s second edition (${}_2$1713: 42–4), where—with the addition of the new Corollaries 2–3 reproduced in Appendix 2 below—the present text is printed with only trivial differences, this second instance of the preceding generalised 'Prop. VII' is deleted, the original Proposition VIII there (*ibid.*: 44) being allowed to stand with an appended final sentence 'Idem facile colligitur etiam ex Propositione præcedente' (The same is easily gathered also from the preceding proposition).

(32) Add. 3965.12: 186r/187r. In this pair of intended revisions of his discussion of orbital motion in a conic in Propositions XI–XIII and their corollaries (*Principia*, ${}_1$1687: 50–5) Newton introduces in succession an alternative proof (see next note) that the central force thereby induced towards a focus varies instantaneously as the inverse square of the distance away from it, and then two revamped following corollaries in which (compare note (38)) he seeks to substantiate the converse truth that only focal conics may be traversed under the deviating action of an inverse-square force directed to a given point-centre.

(33) This variant reproduces the essence of 'A Demonstration That the Planets by their gravity towards the Sun may move in Ellipses' which, according to an inscription by his secretary Brownover on the cover of his contemporary transcript of the piece (now Bodleian. MS Locke. c. 31: 101r/104r/102r [+104v] − 103r, first published by Lord King in his *Life of John Locke, with Extracts from his Correspondence, Journals and Common-place Books*, **1** (London, ${}_2$1830): 389–400), was received by Locke from 'Mr Newton Mar $\frac{89}{90}$'. Newton's original autograph, here copied, does not survive but his later version—augmented with a new

ad $CZ \times CN$ hoc est ad EP in $\frac{PS+HO}{2}$ hoc est ad CA in $\frac{1}{2}PL$.(35) Est ergo $\frac{RP^q}{RQ}$ æquale $\frac{DC^q}{[\frac{1}{2}]CA}$ et propterea ut DC^q[,] hoc est(36) reciprocè ut PF^q, ideoq $\frac{RP^q \times PF^q}{RQ}$ est ut datum. Sed RP est ad QT ut PE seu CA ad PF. Ergo $\frac{QT^q \times CA^q}{RQ}$ æquale est $\frac{RP^q \times PF^q}{RQ}$ et propterea est ut datum, et ob datum CA^q est $\frac{QT^q}{RQ}$ ut datum et $\frac{QT^q \times SP^q}{RQ}$ ut SP^q. Ideoq per Corol. Prop. VI. vis centripeta est reciprocè ut quadratum distantiæ SP. Q.E.I.

Eadem brevitate &c(37).

pag 55 lin 1. [*Prop. XIII. Prob. VIII.*]

Corol. 1. Ex tribus novissimis Propositionibus consequens est quod si velocitas quacum corpus quodvis P secundum lineam quamvis rectam PR exit de loco suo quovis P, ea sit qua lineola PR in minima aliqua temporis particula describi possit, et vis centripeta potis sit eodem tempore corpus idem movere per spatium QR sitq etiam reciproce proportionalis quadrato distantiæ locorum a centro: movebitur hoc corpus in Conica aliqua sectione cujus latus rectum est quantitas illa $\frac{QT^q}{QR}$ quæ ultimò fit ubi lineolæ PR, QR in infinitum diminuuntur. Nam si corpus idem eadem cum velocitate exiret de eodem loco P et movere pergeret in hac Conica Sectione: vis centripeta qua retineretur in hoc orbe ea esset quam diximus (per hasce tres novissimas Propositiones), et ab eodem corpore eadem cum velocitate eodem de loco exeunte, non possunt duæ lineæ ab invicem diversæ impellentibus viribus ijsdem centripetis describi, ne motus inæquales ab

'Prop. 2' and otherwise considerably amended in its textual detail—is preserved in Cambridge University Library. Add. 3965.1: 1r–3v, incompletely reproduced by W. W. Rouse Ball in his *Essay on Newton's "Principia"* (London, 1893): 116–20 and first accurately published by A. R. and M. B. Hall in their *Unpublished Scientific Papers of Isaac Newton* (Cambridge, 1962): 293–301. (We shall not here enter into the question—ultimately not very rewarding, in our opinion—as to whether Newton's lost original belongs to a more primitive earlier phase in the growth of his dynamical understanding. On this essentially undocumentable hypothesis, first put forward by J. W. Herivel in 'The Originals of the two Propositions discovered by Newton in December 1679?', *Archives Internationales d'Histoire des Sciences*, **14**, 1961: 23–33 and more recently supported by R. S. Westfall in 'A note on Newton's demonstration of motion in ellipses', *ibid.* **22**, 1969: 111–17, see our critiques in *History of Science*, **5**, 1966: 107–8/115, note 4 and *Zentralblatt für Mathematik und ihre Grenzgebiete*, **194** (1970): 2–3.) It will be evident that a similar *idem aliter*, dealing *mutatis mutandis* with motion in a hyperbola, may be attached to the following published Proposition 12 'ad pag. 52 lin. ult.'

(34) Since $RP = QX$ by construction, and the difference $RQ = PX$ between PL and XL is taken to be vanishingly small.

the nature of an ellipse) $CZ \times CN$ or $EP \times \frac{1}{2}(PS+HO)$, that is, $CA \times \frac{1}{2}PL$.(35) Therefore RP^2/RQ is equal to $DC^2/\frac{1}{2}CA$ and accordingly as DC^2, that is,(36) reciprocally as PF^2, and in consequence $RP^2 \times PF^2/RQ$ is as a given magnitude. But RP is to QT as PE or CA to PF. Therefore $QT^2 \times CA^2/RQ$ is equal to $RP^2 \times PF^2/RQ$ and accordingly as a given magnitude, and so, because CA^2 is given, QT^2/RQ is as a given magnitude and $QT^2 \times SP^2/RQ$ as SP^2. Consequently by the Corollary to Proposition VI, the centripetal force is reciprocally as the square of the distance SP. As was to be found.

With the same conciseness....(37)

page 55, line 1. [*Proposition XIII, Problem VIII.*]

Corollary 1. It is a consequence of the three most recent propositions that, if the speed with which any body P departs from any of its places P following any straight line PR be that whereby the line-element PR might be described in some minimum particle of time, and the centripetal force be capable in the same time of moving the same body through the space QR and be also reciprocally proportional to the square of the distance of the places from the centre, then this body shall move in some conic whose *latus rectum* is the ultimate value of the quantity QT^2/QR when the elements PR, QR are infinitely diminished. For, should the same body depart from the same place P with the same velocity and proceed to move in this conic, the centripetal force by which it would be kept in this orbit would (by these three most recent propositions) be what we said, and by the same body departing from the same place with the same speed two curves differing from one another cannot be described under the impulsion of the same centripetal forces—otherwise unequal motions would (contrary to

(35) Whence $PL \times AB = (2CM)^2$. In his earlier 'Demonstration' for Locke (see note (33)) this—in sequel to a 'Lemma 1' demonstrating, much as in the published proof of Proposition XI (*Principia*, ₁1687: 50; see page 140 above), that 'If a right line [RZ] touch an Ellipsis in any point [P] thereof & parallel to that tangent be drawn another right line [DK] from the center [C]...wch shall intersect a third right line [PL] drawn from ye touch point through either focus [S]...: the segment [EP] lying between ye point of intersection & ye point of contact shall be equal to half y^e long axis [AB]'—is made a separate 'Lemma 2. Every line drawn through either Focus of any Ellipsis & terminated at both ends by the Ellipsis is to that diameter...wch is parallel to this line as the same Diameter is to the long Axis of the Ellipsis' (ULC. Add. 3965.1: 2^r [= Bodleian. MS Locke, c. 32: 104^r]).

(36) Namely, by the preceding Lemma XII (on *Principia*, $_1$1687: 47; compare page 138 above).

(37) Understand that the printed text is to be picked up at the concluding paragraph of Proposition XI (*Principia*, $_1$1687: 51, line 14 ff): 'Eadem brevitate qua traduximus Problema quintum ad Parabolam & Hyperbolam, liceret idem hic facere...' (With the same conciseness with which we transposed Problem V to treat of the parabola and hyperbola, we would be at liberty to do the same here...); compare page 142 above.

æqualibus viribus generentur contra motus Legem II. Circulum in hoc Corollario refero ad Ellipsin & casum excipio ubi corpus recta descendit ad centrum.

Corol. 2 Igitur si corpus quodvis *P* secundum lineam quamvis rectam *PR* quacunq cum velocitate exeat de loco quovis *P*, & vi quacunq centripeta quæ sit reciprocè proportionalis quadrato distantiæ locorum a centro simul agitetur: movebitur hoc corpus in aliqua Sectionum Conicarum umbilicum habente in centro virium.(38)

---

[5](39) *Pag. 58 ante Prop. XVII.* [*Prob. IX.*]

Lemma 1.

Si umbilicis S, H et filis SP, HP datam summam(40) habentibus describatur Ellipsis(41) cujus axis principalis sit AD et centrum C, et si per centrum C tangenti PR parallela agatur CE filo alterutro PS occurrens in E: erit PE dimidium axis principalis AD.

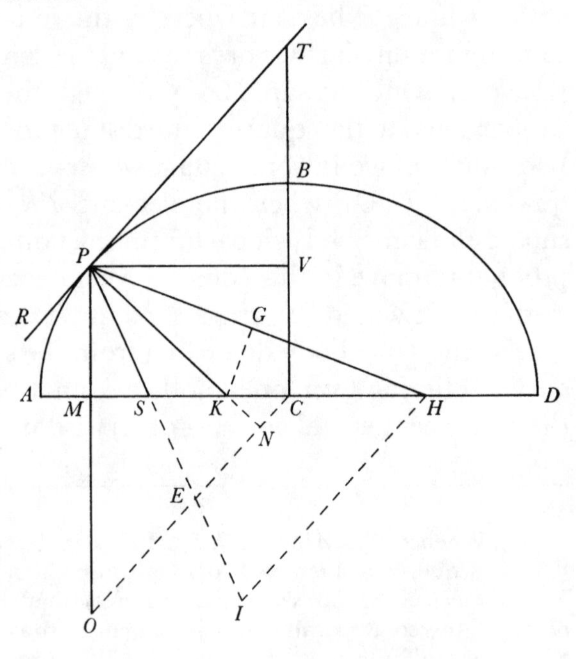

Nam si tangenti eidem per fili alterius umbilicum *H* agatur parallela *HI*, ob æquales *SC*, *CH* æquales erunt *SE*, *EI* ideoq *PE* semisumma erit longitudinum *PS*, *PI* id est longitudinum *PS*, *PH* quæ conjunctim axem principalem *AD* adæquant. Est ergo *PE* dimidium hujus Axis. Q.E.D.(42)

---

(38) These minimal transpositions of the texts of the published Corollaries 1 and 2 (*Principia*, $_1$1687: 55) which they are set to replace do little to illuminate Newton's obscure published argument that focal conics—but, by implication, no other kinds of curvilinear orbit—may be constructed to satisfy all possibilities of motion from a point under the urge of an inverse-square central force. As we have earlier remarked (see 1, §2: note (124)) it was not till long afterwards in late October 1709 that Newton came to implement the bare sentences of his published Corollary 1 with the pithy but cogent *addendum* 'Nam datis umbilico & puncto contactus & positione tangentis, describi potest sectio Conica quæ curvaturam datam ad punctum illud habebit. Datur autem curvatura ex data vi centripeta: & Orbes duo se mutuo tangentes, eadem vi centripeta describi non possunt' (*Principia*, $_2$1713: 53).

Law II of motion) be generated by equal forces. The circle in this corollary I list with the ellipse and I exclude the case where the body descends straight downwards towards the centre.

Corollary 2. Therefore if any body *P* should depart from any place *P* following any straight line *PR* with any speed whatever and simultaneously be acted on by any force whatever which shall be reciprocally proportional to the square of the distance of the places from the centre, this body shall then move in some one of the concis having a focus at the centre of force.(38)

---

[5](39) *Page 58, before Proposition XVII,* [*Problem IX*].

Lemma 1.

If with foci S, H and threads SP, HP having a given sum(40) *there be described an ellipse*(41) *whose main axis shall be AD and centre C, and if through the centre C parallel to the tangent PR there be drawn CE meeting one or other of the threads PS, then PE will be half the main axis AD.*

For if parallel to the same tangent through the focus *H* on the other thread there be drawn *HI*, because *SC*, *CH* are equal *SE*, *EI* will be equal, and in consequence *PE* will be the half-sum of the lengths *PS*, *PI*, that is, of the lengths *PS*, *PH* which taken together equal the main axis *AD*. Hence *PE* is half this axis. As was to be proved.(42)

---

(39) Add. 3966.2: 18r/18v. Having minimally refined the text of his published Proposition XVII and its first three corollaries, Newton here elaborates its concluding 'Corol. 4' (the old Corollary 2 is here suppressed as being superfluous) into four new ones in which—with a weather eye on their application to computing the hourly motion of the lunar apsides, much as in Lemmas α and β of his earlier abortive scheme in 1686 (see 2, §3 above)—he seeks approximately to determine the rotatory effects round the force-centre (and focus) *S* of small radial and transverse disturbing impulses *V* and *W* on a body orbiting in a Keplerian ellipse. To facilitate the mathematical argument he sets 'ante Prop. XVII' two new Lemmas '1' and '2' to serve as riders. In an unfinished preliminary draft (on Add. 3966.2: 17r/17v, reproduced below in Appendix 3) he had originally intended to develop this *approximatio* as a separate new 'Prop. XVIII. Theor.' following.

(40) 'vel differentiam' (or difference) is cancelled. Since Newton is in sequel concerned only with disturbed periodic conic orbits, he has no need here to mention this complementary hyperbolic case.

(41) Originally 'Conica sectio' ([central] conic); compare the previous note.

(42) Since Newton has earlier made use of this lemma—in its generalised form—in the proof of his preceding Proposition XI (*Principia*, $_1$1687: 50), it is curious that he should only now, 'post Prop. XVI', wish to give it independent status. At a later date (see §3: note (5) below) it was his more logical intention to set its generalisation as one of a group of initial 'Lemmata Generalia'.

Lemma 2.

Iisdem positis si a perpendiculo PK Axi AD in K et rectæ CE in N occurrente bisecetur angulus SPH et ad filorum alterutrum PH demittatur perpendiculum KG, erit PG dimidium lateris recti principalis.

Etenim propter sim[ilia] tri[angula] *PKG*, *PEN* est *PG* ad *PK* ut *PN* ad *PE*[,] et erecto ad *AD* perpendiculo *CVBT* occurrente curvæ in *B* et ejus tangenti in *T* completisꝗ parallelogrammis *CTPO CVPM*, ob similia triangula *PMK PN*[*O*] est *PK* ad *PM* vel *CV* ut *PO* vel *CT* ad *PN*: ergo ex æquo est *PG* ad *CV* ut *CT* ad *PE*. Est autem ex natura sectionum Conicarum axis conjugatus *CB* medium proportionale tam inter *CT* et *CV*, quàm inter semiaxem principalem et semissem Lateris recti principalis[,] et propterea ex æquo est *PG* ad semissem lateris recti principalis ut semiaxis principalis ad *PE*. Sed sem[iaxi]s principalis et *PE* æquantur. Ergo et *PG* æqualis est semissi l[ater]is recti. Q.E.D.(43)

pag 59. lin. 11. [*Prop XVII. Prob. IX*]

[...latus rectum Conisectionis...datur. Sit istud *L*. Datur præterea Conisectionis umbilicus *S*. Anguli *RPS* complementum ad duos rectos fiat angulus *RPH*, & dabitur positione linea *PH*] in qua umbilicus alter locatur. In *PH* cape *PG* æqualem dimidio lateris recti *L*. Et erige perpendiculum *GK* quod rectæ *PK* angulum datum *SPH* bisecanti occurrat in *K* et acta *SK* (per Lemma novissimum) erit axis fi[guræ] occurrens rectæ *PG* in umbilico altero *H*. Fig[ura] Ellipsis erit ubi umbilici *S*, *H* jacent ad easdem partes tangenti[s, Hy]perbola ubi ad contrarias et Parabola ubi rect[is *SK*, *PG*] parallelis existentibus umbilicus [*H* in infin]it[um] ev[anuit]. Quæ erant determinanda.(44)

Corol. 1. Hinc si datur [umbilicus *S*, vertex principalis *A* et latus rectum *L*,] expedite determinabitur orbis capiendo [*AH* ad *AS* ut *L* ad $4AS-L$. Nam] ob æquales angulos *SPK*, *KPH*, est *PH* ad *KH* ut *SP* ad *SK*, id est (ob æquales et coincidentes *PG*, *AK*) *AH* ad *KH* ut *AS* ad *SK*[,] et divisim *AH* ad *AK* seu $\frac{1}{2}L$ ut *AS* ad $AS-SK$ id est ad $2AS-AK$ seu $2AS-\frac{1}{2}L$. Et duplicatis consequentibus *AH* ad *L* ut *AS* ad $4AS-L$.

(43) This property of the ellipse, evidently Newton's present discovery, was first published a decade afterwards by James Milnes in his popular undergraduate textbook *Sectionum Conicarum Elementa Nova Methodo Demonstrata* (Oxford, $_1$1702): Pars IV, Prop. VI, Theor. IV: 86, and afterwards frequently cited therefrom by eighteenth-century geometrical writers such as John Keill and Robert Simson. Newton had earlier shown the theorem to David Gregory during the latter's visit to Cambridge in early May 1694, and it would seem not unlikely that Gregory passed its enunciation on to his Oxford colleague. Gregory himself found no difficulty in afterwards repeating Newton's proof of the property in the present elliptical case, but in attempting to extend it to the hyperbola—where it holds without modification—went badly wrong in claiming that there 'loco *P*[*K*] normalis ad Curvam sumenda est *P*[*K*] tangens Curvam' (Gregory Codex C33 [now Royal Society. Gregory Volume: 65], reproduced in *The*

Lemma 2.

With the same suppositions, if the angle SPH be bisected by the perpendicular PK meeting the axis AD in K and the straight line CE in N, and to one or other of the threads PH there be let fall the perpendicular KG, then PG will be half the main latus rectum.

For indeed, since the triangles *PKG*, *PEN* are similar, *PG* is to *PK* as *PN* to *PE*, and, when the perpendicular *CVBT* is erected to *AD*, meeting the curve in *B* and its tangent in *T*, and the parallelograms *CTPO*, *CVPM* are completed, since the triangles *PMK*, *PNO* are similar *PK* is to *PM* or *CV* as *PO* or *CT* to *PN*; therefore *ex æquo PG* is to *CV* as *CT* to *PE*. However, from the nature of conics the conjugate axis *CB* is a mean proportional both between *CT* and *CV* and between the main semi-axis and half the main *latus rectum*; *ex æquo*, accordingly, *PG* is to half the main *latus rectum* as the main semi-axis to *PE*. But the main semi-axis and *PE* are equal, and so too therefore *PG* is equal to half the *latus rectum*.(43)

page 59, line 11. [*Proposition XVII, Problem IX.*]

[...the conic's *latus rectum*...is given. Let it be *L*. There is given, moreover, the conic's focus *S*. Make the angle $\widehat{RPH}$ the supplement of the angle $\widehat{RPS}$, and there will be given in position the line *PH*] in which the other focus is located. In *PH* take *PG* equal to half the *latus rectum L*, and erect the perpendicular *GK* to meet in *K* the straight line *PK* bisecting the given angle $\widehat{SPH}$; then, once *SK* is drawn, it will (by the most recent Lemma) be the figure's axis, meeting the line *PG* in the second focus *H*. The figure will be an ellipse when the foci *S*, *H* lie on the same side of the tangent, a hyperbola when they are on opposite sides, and a parabola when, with the lines *SK*, *PG* proving to be parallel, the focus *H* disappears to infinity. As were to be determined.(44)

Corollary 1. Hence, if there is given the focus *S*, main vertex *A* and *latus rectum L*, the orbit will speedily be determined by taking *AH* to *AS* as *L* to $4AS-L$. For, because the angles $\widehat{SPK}$, $\widehat{KPH}$ are equal, *PH* is to *KH* as *SP* to *SK*, that is, (because *PG* and *AK* are equal and coincident) *AH* is to *KH* as *AS* is to *SK*, and *dividendo AH* to *AK* or $\frac{1}{2}L$ as *AS* to $AS-SK$, that is, $2AS-AK$ or $2AS-\frac{1}{2}L$. And so, with the latter members doubled, *AH* is to *L* as *AS* to $4AS-L$.

Correspondence of Isaac Newton, **3** (Cambridge, 1961): 311–15; see especially 314–15). The property is, of course, effectively the polar defining equation of a general conic in a lightly disguised geometrical form: for, where $AC = CD = a$, $SC = CH = ae$, $PH = r$ and $\widehat{AHP} = \phi$, at once $KH = PH \times SH/(SP+PH) = re$, so that $PG = r(1-e\cos\phi) = a(1-e^2)$, half the *latus rectum*.

(44) A minimal recasting of the conclusion of the published demonstration (*Principia*, $_1$1687: 58–9). Corollary 1 following is likewise superficially reshaped to accord with the emendations.

Corol. 2.[45] Hinc etiam si corpus moveatur in sectione quacunq͡ȝ conica et ex orbe suo impulsu quocunq͡ȝ [exturbetur, cognosci potest orbis in quo postea cursum suum peraget. Nam componendo proprium corporis motum cum motu illo quem impulsus solus generaret, habebitur motus quocum corpus de dato impulsus loco secundum rectam positione datam] exibit.

Corol. 3. Et si corpus illud [vi aliqua extrinsecus impressa continuo perturbetur, innotescet cursus quamproxime colligendo mutationes quas vis illa in punctis quibusdam inducit et ex seriei analogia mutationes continuas in locis intermedijs] æstimando.[46] Id quod expedite fit per approximationem sequentem. Revolvatur Corpus P in orbe DPA circa centrum virium S et interea ab impulsu aliquo in ipsum secundum rectam IP incidente de hoc orbe deturbetur. In recta PH quæ ad orbis umbilicum mobilem[47] H ducitur, capiatur PG dimidium lateris recti figuræ. Fac angulum PGg angulo IPR æqualem quem linea incidentiæ IP continet cum orbis tangente PR in plagam motus ducto, et sit Gg ad latus rectum $2PG$ ut corporis velocitas quam impulsus ille generare posset ad velocitatem ante impulsum. Jaceat autem Gg ad partes rectæ PG versus centrum virium S si modo linea incidentiæ IP jaceat ad partes tangentis versus centrum illud: sin minus jaceat Gg ad partes contrarias[,] et acta Pg erit dimidium lateris recti in orbe novo et producta transibit per ejus umbilicum mobilem h. Ideoq͡ȝ si ad Pg erigatur perpendiculū gk quæ rectæ Pk angulum SPg bisecanti occurrat in k & rectæ Sk & Pg productæ concurrant in h, erit h umbilicus ille mobilis, & $[\frac{1}{2}]Sh$ excentricitas orbis novi. Si lineola Gg infinite parva est, accurata est hæc constructio.

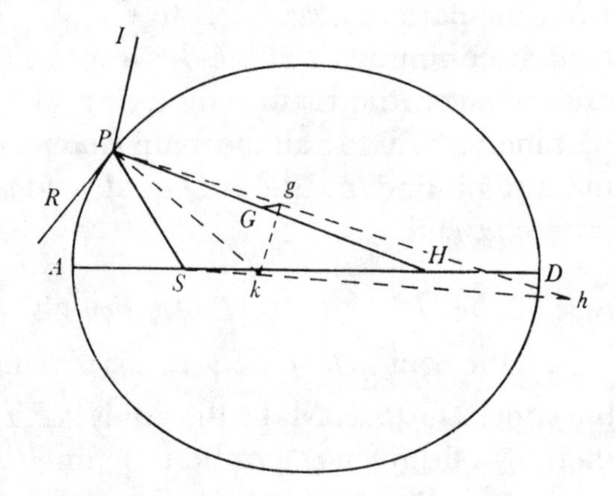

Vel sic. Sit PG latus rectum orbis primi sitq͡ȝ Pg ad PG in duplicata ratione velocitatis post impulsum ad velocitatem ante impulsū & ad lineam Pg & lineam illam in qu[a] corpus post impulsum incipit moveri erigantur perpendicula gk, Pk concurrentia in k, et acta Sk transibit per umbilicum h ut ante.

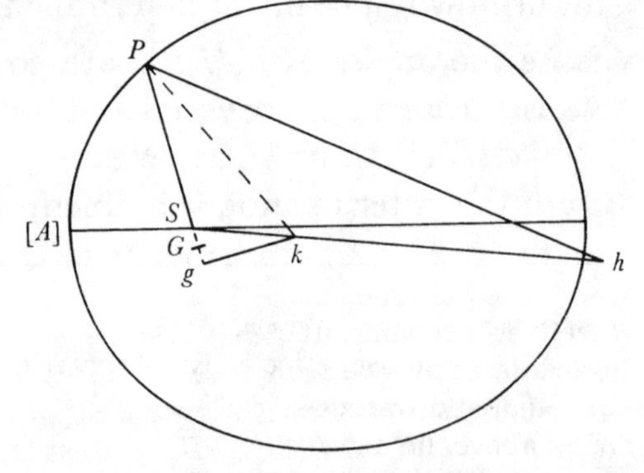

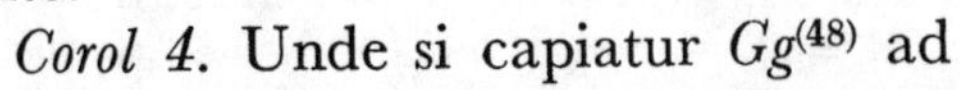

Corol 4. Unde si capiatur Gg[48] ad

(45) The original Corollary 2 (*Principia*, ₁1687: 60)—which merely serves trivially to restate Corollary 1—is now abandoned.

Corollary 2.(45) Hence also, if a body should move in any conic whatever and be thrust out from its orbit by any impulse, there can be found out the orbit in which thereafter it pursues its course. For by compounding the individual motion of the body with the motion which the impulse by itself were to generate there will be had the motion with which the body shall depart from the given place of impulse along a straight line given in position.

Corollary 3. And if that body be continuously perturbed by some externally impressed force, its course will come to be known to a close approach by assembling the changes which that force induces at certain points and then from the pattern of their sequence estimating the continuous changes at intermediate points.(46) That is speedily achieved by the following approximation. Let the body P revolve in the orbit DPA round the force-centre S and during its passage be driven aside from this orbit by some impulse incident upon it along the straight line IP. In the straight line PH which is drawn to the mobile(47) focus H of the orbit take PG half the figure's *latus rectum*. Make the angle $\widehat{PGg}$ equal to the angle $\widehat{IPR}$ contained by the line IP of incidence with the orbit's tangent PR (drawn in the direction of motion), and let Gg be to the *latus rectum* $2PG$ as the body's speed which the impulse were able to generate to its speed before impulse—let, however, Gg lie on the side of the line PG towards the force-centre S provided the line IP of incidence lies on the side of the tangent towards that centre; but if it does not, let Gg lie the opposite way. Then, when Pg is drawn, it will be half the *latus rectum* in the new orbit and, when extended, will pass through its mobile focus h. Consequently, if to Pg there be erected the perpendicular gk to meet the line Pk bisecting the angle $\widehat{SPg}$ in k, and the lines Sk and Pg produced concur in h, then h will be that mobile focus, and $\frac{1}{2}Sh$ the eccentricity of the new orbit. If the line-element Gg be infinitely small, this construction is accurate.

Or thus. Let PG be the *latus rectum* of the first orbit, and Pg to PG in the doubled ratio of the speed after impulse to the speed before impulse; to the line Pg and the line in which the body after impulse begins to move erect perpendiculars gk, Pk concurrent in k, and then, once drawn, Sk will pass through the focus h as before.

Corollary 4. Whence, if Gg be taken(48) to PG as the increment in the speed to

(46) The published Corollary 4 (*ibid.*), repeated without verbal change. In sequel Newton proceeds to exemplify the 'continuous' action of such disturbing 'mutations' of a pristine elliptical orbit by evaluating the changes in 'hourly' motion round the principal focus which their radial and transverse components induce; as in his earlier equivalent Lemmas α and β in 3, §2 above, he makes the basic Horrocksian hypothesis that the disturbed orbit is likewise instantaneously a Keplerian ellipse.

(47) Newton first wrote 'alterum' (second).

(48) A superfluous 'in ratione' (in ratio) is cancelled.

PG ut incrementum velocitatis ad velocitatem ante impulsum, [sta]ntibus *gk*, *Pk* et *Skh* ut supra, erit *h* umbilicus orbis novi quamproxime.

Corol. 5. Et hinc si impulsus agat secundum tangentem orbis ac dicatur *V*; [et im]pulsus quo totus corporis motus generari posset dicatur *I* & sinus anguli *PKS* dicatur *S* & Radius sit *R*: erit [mo]tus horarius apsidis ut $\frac{2S, PK}{R, SK}, V$ quamproxime, et hic motus ad motū [hora]rium corporis *P* circum *S* ut $\frac{2S, PK}{R, SK}V$ ad $\frac{PG^2}{PK, PS}I$, seu $2V$ ad $\frac{R, SK, PG^2}{S, PS, PK^2}I$.

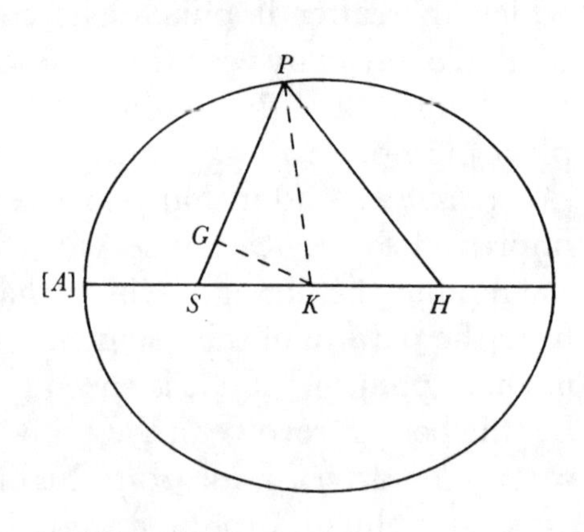

Corol. 6. Et si impulsus agat secundum perpendiculum ac dicatur *W*, erit motus [horarius aps]idis ut $\frac{GS, PK^2}{PG, SK^2}W$, & hic motus ad motum horarium Lunæ[49] ut $\frac{GS, PK^2}{PG, SK^2}W$ [ad $\frac{PG^2}{PK, PS}I$. Ubi $\frac{SK^2}{PK^2}$ quadra]tum excentricitatis est.[50]

APPENDIX 1. INITIAL REVISION OF THE PRIME COROLLARY TO THE PUBLISHED 'LEX II'.[1]

From the original draft[2] in the University Library, Cambridge

[*Ad pag. 13, lin. 20*)

[*Corol. 1.*

Corpus viribus conjunctis reperiri in concursu linearum quas eodem tempore separatis describit.][3]

Cas. 1. Si corpus[4] [dato tempore vi sola *M* ferretur ab *A* ad *B* & vi sola *N* ab *A* ad *C*, compleatur parallelogrammum *ABDC* & vi utraq; feretur id eodem tempore ab *A* ad *D*. Nam quoniam vis *N* agit secundum lineam *AC* ipsi *BD* parallelam, hæc vis nihil mutabit velocitatem accedendi ad lineam illam *BD* a vi altera genitam. Accedet igitur corpus eodem tempore ad lineam *BD* sive vis *N* imprimatur sive non, atq; adeo in fine illius temporis

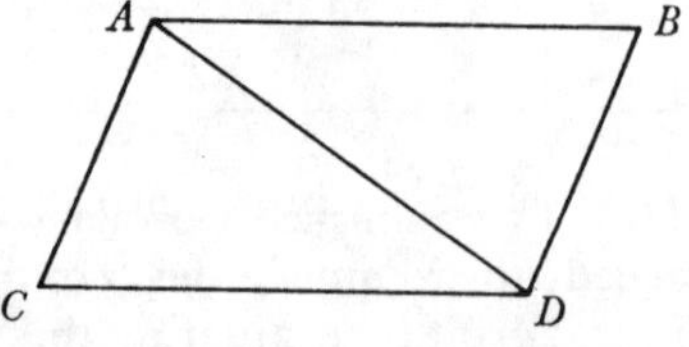

---

(49) Read 'corporis *P* circum *S*' (of the body *P* round *S*)! Newton unconsciously specifies the particular celestial orbiting body which is in the forefront of his mind as he writes.

the speed before impulse, then, with *gk*, *Pk* and *Skh* standing as above, to a close approximation *h* will be the focus of the new orbit.

Corollary 5. And hence, if the impulse should act along the tangent to the orbit, call it *V* and then, if the impulse whereby the whole motion of the body were able to be generated be called *I*, the sine of the angle $P\widehat{K}S$ be called *S* and its radius be *R*, the hourly motion of the apsis will be as $2(S \times PK/R \times SK)\ V$ very nearly, and this motion to the hourly motion of the body *P* round *S* as $2(S \times PK/R \times SK)\ V$ to $(PG^2/PK \times PS)\ I$, or $2V$ to $(R \times SK \times PG^2/S \times PS \times PK^2)\ I$.

Corollary 6. While if the impulse should act along the perpendicular, call it *W* and then the hourly motion of the apsis will be as $(GS \times PK^2/PG \times SK^2)\ W$, and this motion to the hourly motion of the moon[49] as $(GS \times PK^2/PG \times SK^2)\ W$ to $(PG^2/PK \times PS)\ I$. Here SK^2/PK^2 is the square of the eccentricity.[50]

---

reperietur alicubi in linea illa *BD*. Eodem argumento in fine temporis ejusdem reperietur alicubi in linea *CD*, et idcirco in utriusꝗ lineæ concursu *D* reperiri necesse est.]

Cas. 2. Eodem argumento si corpus dato tempore vi sola *M* in loco *A* impressa ferretur uniformi cum motu ab *A* ad *B* & vi sola *N* non simul & semel sed[5] perpetuo impressa ferretur accelerato cum motu in recta *AC* ab *A* ad *C*[,] compleatur parallelogrammum *ABDC* & corpus eodem temp[ore] vi utraꝗ feretur ab *A* ad *D*. Nam reperietur in fine temporis tam in linea *CD* quam in linea *BD* et propterea[6] [in utriusꝗ lineæ concursu *D* inveniri debet].[7]

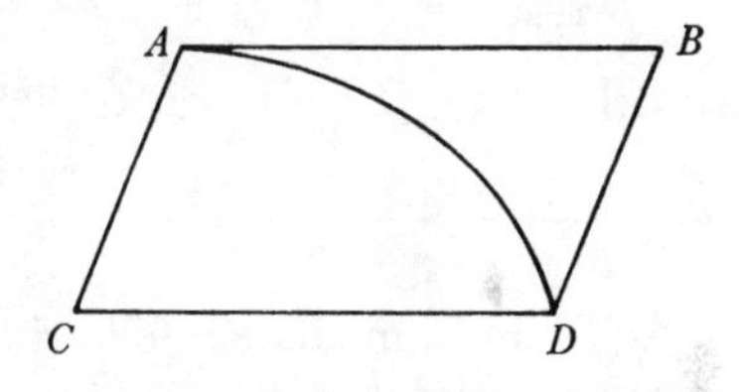

---

(50) The latter part of this concluding paragraph, whose manuscript text has in the course of time almost wholly flaked away, is here freely—but nonetheless accurately, we hope—restored by us. As with his equivalent earlier Lemmas α and β in 3, §2 above it was doubtless Newton's intention (compare the previous note) to apply these new Corollaries 3–6 to evaluate the moon's hourly motion, and thence compute the mean annual advance of its perigee.

(1) As a preliminary to combining these in a single collective scheme (see note (7) below) Newton here, in a first 1690's revise of Corollary 1 to 'Lex II' in his published *Principia* ($_1$1687: 13–14; compare pages 98–100 preceding), adjoins to the case there treated—that of a single impulse of impressed force acting 'simul et semel' to alter the size and direction of a given uniform motion—a complementary 'Cas. 2' where the force is presumed to act 'continuously' over an equal time-interval as an infinite sequence of infinitesimal component impulses applied 'gradatim et successive'.

(2) Add. 3965.6: 86^r. The diagrams (lacking in the manuscript) are here restored.

(3) In this inserted heading we appropriately amend the enunciation of the published 'Corol. 1' to embrace both cases following.

(4) The remainder of 'Cas. 1' is not found in the manuscript but, as Newton intends, we complete its text from the published *Principia.*

APPENDIX 2. REVISED COROLLARIES TO THE NEW 'PROPOSITIO VII'[1]

From the autograph insert[2] in Newton's interleaved copy of the *Principia*'s first edition in the University Library, Cambridge

Corol. 2. Vis qua corpus P in circulo $APTV$ circum virium centrum S revolvitur, est ad vim qua corpus idem P in eodem circulo et eodem tempore periodico circum aliud quodvis virium centrum R revolvi potest ut $RP^{\text{quad}}\times SP$ ad cubum rectæ SG quæ a primo virium centro S ad orbis tangentem PG ducitur et distantiæ corporis a secundo virium centro parallela est. Nam per constructionem hujus Propositionis[3] vis prior est ad vim posteriorem ut $RP^q\times PT^{\text{cub}}$ ad $SP^q\times PV^{\text{cub}}$, id est ut $SP\times RP^q$ ad

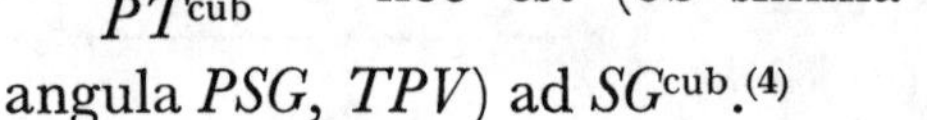

$\dfrac{SP^{\text{cub}}\times PV^{\text{cub}}}{PT^{\text{cub}}}$ hoc est (ob similia triangula PSG, TPV) ad SG^{cub}.[4]

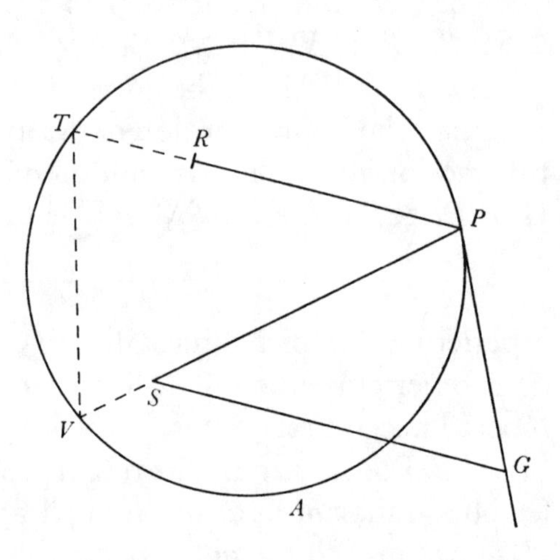

(5) In an immediately following redraft (compare note (7)) Newton here inserted a very necessary 'paulatim et'.

(6) The draft terminates at this point, but we suitably complete the argument in line with that of the preceding 'Cas. 1'.

(7) On the same manuscript page Newton proceeded straightaway to blend these complementary cases into a single paragraph reading: 'Ideoq; si vires M et N secundum lineas AB et AC simul et semel imprimantur, sic ut motus seorsim generarent in lineis istis AB et AC uniformes: corpus dato illo tempore perget ab A ad D in diagonali rectilinea AD uniformi cum motu ex vi utraq; oriundo. Sin vires istæ M et N secundum lineas easdem AB et AC imprimantur paulatim & perpetuò sic ut motus seorsim generarent in lineis istis AB et AC vel in earum alterutra acceleratos: corpus perget eodem tempore ab A ad D in diagonali curvilinea AD cum motu accelerato ex vi utraq; oriundo.'

(1) See §1: note (31) preceding. From the measure there derived of the force, instantaneously directed to a given point in its plane, which is induced by motion constrained to be in a given circle—and hence, more generally, in a given curve of identical curvature at the point—Newton here deduces the corresponding relative magnitude of the force thereby induced to any other given point in the plane.

(2) Adv. b. 39. 1: facing page 46. The handwriting shows that the insertion was penned in the early 1690's.

(3) See page 550 above.

(4) This elegant variant deduction of a theorem which he demonstrates from first principles in Proposition VIII in §2 following evidently came to Newton in sequel thereto as an afterthought. Here too, some dozen years later, de Moivre learnt that Newton had anticipated

Corol. 3. Vis qua corpus P in orbe quocunꝗ circum virium centrum S revolvitur, est ad vim qua corpus idem P in eodem orbe eodemꝗ tempore periodico circum aliud quodvis virium centrum R revolvi potest ut contentum sub distantia corporis a primo virium centro et quadrato distantiæ ejus a secundo virium centro $SP \times RP^q$ ad cubum rectæ SG quæ a primo virium centro S ad orbis tangentem PG ducatur et distantiæ corporis a secundo virium centro parallela est. Nam vires in hoc orbe, ad ejus punctum quodvis P, eædem sunt ac in circulo ejusdem curvaturæ.(5)

APPENDIX 3. AN INTENDED INSERTION ON DISTURBED ELLIPTICAL ORBITS.(1)

From the original draft(2) in the University Library, Cambridge

In Philos. Natural. [P]rincip. Math. pag 59
[Prop. XVII. Prob. IX]

. . . .(3)

Corol. 4. Et si corpus - - - - æstimando. Sed ut hoc promptius fiat, visum est approximationem sequentem subjungere.

---

him in drawing this consequence of the measure of force which the latter set down in Corollary 5 to his revised Proposition VI of the 'Liber primus' (see §1: note (25) preceding); for, he continued in his letter to Bernoulli on 27 July 1705, after being shown the 'théoreme' in Newton's private 'papiers qu'il prépare pour une seconde édition de ses *Principia*...' he had, 'en riant', expressed to him the hope 'que vous ne m'aurez pas ravi le plaisir d'avoir tiré le premier un très beau corollaire de ce théoreme. Quel est ce corollaire? dit-il. C'est, répondis-je, qu'il s'ensuit de là que dans une courbe la loi de la force centripete qui tend à un point quelconque étant donnée, on en peut dériver facilement la loi de la force centripete qui tend à un autre point. Cela, dit-il, ne m'est point échappé; et il me le montra encore' (Wollenschläger, 'Briefwechsel': 214).

(5) With these replacements for its original Corollary 2 (see §1: note (31)) Newton passed his revised transcript (now Trinity College, Cambridge. R. 16. 38: 5–6) of 'Propositio VII' on to Roger Cotes in September 1709, and it was duly printed without further variant in the *Principia*'s second edition ($_2$1713: 42–4).

(1) In this preliminary version of §1.5 preceding Newton comprehends his approximate determination of the action of 'continuous' disturbing radial and transverse 'mutations' in an intended separate following 'Prop. XVIII. Theor.'

(2) Add. 3966.2: 17r/17v. The bottom right-hand corner of the manuscript is badly eroded; we have made appropriate restoration of the text within square brackets.

(3) We omit to reproduce a trivial rearrangement of the published text which is outlined in the manuscript (f. 17r), and also Newton's indication there that the first three of the corollaries following are to be repeated without change.

Prop. XVIII. Theor. [*IX.*][4]

Si corpus quodvis P in orbe quocunq; propemodum circulari DPA[5] *descripto circa virium centrum S revolvens impulsu parvo P*[*N sec*]*und*[*um*] *rectam NP incidente de orbe suo* [*deturbetur*] *et ad t*[*angentem PM*] *demisso perpendiculo NM, distinguatur impulsus NP in impulsus duos NM et MP quorum MP agat secundum orbis tangentem & NM secundum orbis perpendiculum, et si a recta* [*P*]*Q ipsi SH occurrente in Q bisecetur angulus SPH et in hac recta capiatur QR quæ sit ad* $\frac{1}{2}$*PQ*[6] *ut velocitas corporis P quem impulsus MP generare posset ad velocitatem ejus ante impulsum, jaceat autem punctum R inter puncta P et Q si velocitas corporis impulsu MP diminuitur, vel ad contrarias partes puncti Q si ejus velocitas augeatur; deinde ad PR erigatur perpendiculum Rq quæ sit ad PR ut corporis velocitas quam impulsus NM generare posset ad ejus velocitatem ante impulsum, jaceat autem Rq ad partes ipsius PR versus umbilicum H si corpus impu*[*l*]*su NM urgeatur versus centrum virium S, secus ad partes contrarias, ac deniq; si angulo SPq æqualis fiat angulus qPh & producatur Sq donec occurrat Ph in h et umbilicis S, h per punctum P describatur Ellipsis; hæc Ellipsis erit Orbis in quo corpus post impulsum revolvetur quamproxime.*

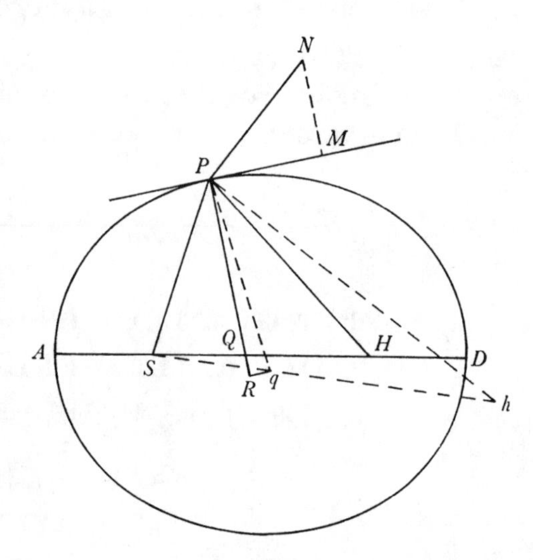

Nam cum *Rq* sit ad *PR* ut motus corporis quem impulsus orbi perpendicularis *NM* generet ad ejus motum totum, erit angulus *RPq* æqualis angulo quem orbis novus cum orbe veteri continebit ad punctum *P*, ideoq; recta *Pq* perpendicularis est ad orbem novum et intersectio perpendiculi illius & intervalli umbilicorum *SH* quod ante impulsum reperiebatur in *Q*, post impulsum reperietur alicubi in recta *Pq*. Longitudo autem *PQ* æqualis est semissi lateris recti figuræ quamproximè et ejus incrementa ac decrementa per impulsus facta sunt quamproximè ut incrementa ac decrementa [7]Lateris recti, hoc est eam habent rationem ad semissem totius quamproxime quam habet motus corporis quem impulsus *MP* secundum tangentem impressus generare posset ad ejus motum

(4) Presuming, namely, that Newton here meant to retain his previous published Theorems I–VIII unaltered. This head was afterwards cancelled by him.

(5) Understand 'elliptico', of course. For some reason Newton has deleted in sequel 'centro *C* umbilicis *S, H*'.

(6) Read '*PQ*' simply.

(7) Understand 'semissis' to be consistent.

totum ante impulsum, ideoq; longitudo illa PQ per impul[s]um convertitur in longitudinem Pq quamproxime et recta Sq ducta et utrinq; producta transibit per Orbis umbilicum ulteriorem h.[8].

---

(8) A terminal 'Ellipseos' is cancelled.

§2. MORE RADICAL RESTRUCTURINGS OF THE 'LIBER PRIMUS'.(1)

[1](2) *pag. 41.*

Prop. IV. Prob. I.(3)

Data quibuscunq; in locis velocitate qua corpus figuram datam viribus ad commune aliquod centrum tendentibus describit, centrum illud invenire.

Constructio.

Figuram descriptam tangant rectæ tres PT, TQV, VR in punctis totidem P, Q, R concurrentes in T et V. Ad tangentes erigantur perpendicula PA, QB, RC velocitatibus corporis in punctis illis P, Q, R a quibus eriguntur reciprocè proportionalia; id est ita ut sit PA ad QB ut velocitas in Q ad velocitatem in P, & QB ad RC ut velocitas in R ad velocitatem in Q. Per [per]pendiculorum terminos A, B, C ad angulos rectos ducantur AD, DBE, EC concurrentes in D et E & actæ TD, VE concurrent in centro quæsito S.

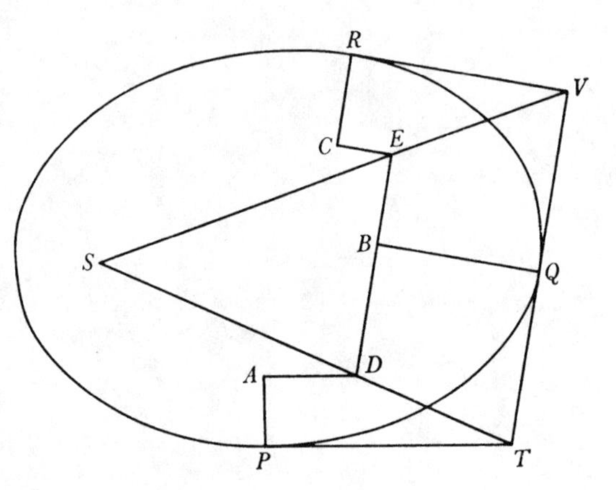

Demonstratio.

Nam perpendicula a centro S in tangentes PT, QT demissa, sunt (per Corol. 1. Prop. I(4)) ut velocitates reciprocè, adeoq; (per Constructionem) ut perpendicula

(1) As we have already remarked (see §1: note (1)) Newton came in the early 1690's to conceive a grand scheme of revision of the published *Principia*, in which not only its particular verbal and mathematical errors were to be corrected but, much more radically, the redundant in its logical and expository framework was to be cut out and the flimsier portions of the remaining structure were to be strengthened and supported and (where necessary) completely rebuilt. To David Gregory, during a visit by him to Cambridge in early May 1694 he proved unprecedentedly expansive regarding his intentions, elaborating for him a detailed overview of his plans for revision, and also showing him the manuscript papers in which he had already himself gone some way towards implementing them. In the retrospective summary Gregory afterwards compiled of what 'In Editione nova Philosophiæ Newtonianæ...ab Auctore fient' (Codex C42 [now in the University Library, Edinburgh]; reproduced in H. W. Turnbull's not always accurate English translation in *The Correspondence of Isaac Newton*, **3**, 1961: 384–6) he subsequently—building on more fragmentary earlier memoranda of his conversations with Newton (compare notes (21), (23) and (34) below)—wrote that 'Plurima circa initium corriguntur. Corrollaria quædam adduntur, ordo propositionum immutatur et quædam etiam omittuntur et delentur. Rationem vis Centripetæ corporis ad focum Conisectionis tendentis deducit ex ratione vis centripetæ tendentis ad centrum, et hanc rursus de ratione vis centripetæ æquabilis

Translation

[1][2] *page 41.*

Proposition IV, Problem I.[3]

Given in any places whatever the speed with which a body describes a given figure under forces tending to some common centre, to find that centre.

Construction.

Let the figure described be touched at an equal number of points P, Q, R by three straight lines PT, TQV, VR meeting in T and V. To the tangents erect perpendiculars PA, QB, RC reciprocally proportional to the speeds of the body at the points P, Q, R from which they are erected; that is, so that PA be to QB as the speed at Q to the speed at P, and QB to RC as the speed at R to the speed at Q. At right angles through the end-points A, B, C of the perpendiculars draw AD, DBE, EC meeting in D and E, and when TD, VE are drawn they will meet at the required centre S.

Demonstration.

For the perpendiculars let fall from the centre S to the tangents PT, QT are (by Corollary 1 of Proposition I[4]) as the speeds reciprocally, and hence (by

---

tendentis ad circuli centrum; adeoq; quæ a propositione VII ad XIII inclusive demonstrantur exhinc tanquam corrollaria fluunt'. In [1] following we reproduce the manuscript sequence on which Gregory based this observation, adjoining in [2] an allied scholium (presumably intended to be set 'post Prop. XVII', as its first paragraph was in the *editio secunda*) in which Newton went on skilfully to develop a unique general fluxional measure of the *vis centripeta* inducing orbital motion in a given curve, but twice failed accurately to apply it in the particular case of a focal conic (see note (58) below and Appendix 2.3: note (12)). Only fragments of each of these were to appear in print, and the remainder was quickly to pass into oblivion after Gregory's death in 1708 till its resurrection at the hands of modern historical scholars.

(2) Add. 3965.6/12: 27^r–28^v/188^r–189^v/183^r–183^v/31^r–32^r. Newton's preliminary partial drafts of this final textual sequence will be cited at appropriate places in following footnotes. At one point he subsequently began to enter its content into his interleaved 1687 *Principia* (ULC. Adv. b. 39. 1: facing pages 43–5) but broke off at the beginning of the demonstration of Proposition VII after having transcribed only its enunciation and the following initial subhead '*Cas. 1*'. The new set of corollaries to the published 'Propositio I' (see §1.2) to which repeated appeal is made in the present Propositions IV–VI are here to be understood in preliminary.

(3) A lightly remodelled version of the published 'Prop. V. Prob. I' (*Principia*, $_1$1687: 43–4, there printed without change from the original autograph insert in the manuscript 'Liber primus' which is reproduced in 1, §2, Appendix 2.7 above), now interchanged in sequence with the immediately following 'Prop. V. Theor. IV' (*ibid.*: 41–3); in his interleaved copy of the *editio princeps* (see the previous note) Newton later inserted at this point a corresponding editorial instruction to 'Invert the order of this & the following Proposition', but the original sequence is retained in later editions of the *Principia*.

(4) Understand the revised version reproduced in §1.2 above; see note (2). At the corre-

AP, *BQ* directè, id est ut perpendicula a puncto *D* in tangentes demissa. Unde facile colligitur quod puncta *S*, *D*, *T* sunt in una recta. Et simili argumento puncta *S*, *E*, *V* sunt etiam in una recta; & propterea centrum *S* in concursu rectarum *TD*, *VE* versatur. Q.E.D.

Prop. V. Theor. IV.(5)

Corporum quæ diversos circulos æquabili motu describunt vires centripetas ad centra circulorum tendere & esse inter se ut sunt arcuum simul descriptorum quadrata applicata ad altitudines corporum supra centrum virium.

Demonstratio.

(6)Tendunt hæ vires ad centra circulorum per Prop. II & Corol. 2 Prop. I & sunt inter se ut arcuum æqualibus temporibus quamminimis descriptorum sagittæ vel(7) sinus versi (per Corol. 4 & 5 Prop. I) hoc est ut quadrata arcuum illorum ad diametros circulorum applicata, & propterea (cum semidiametri diametri & arcus majori simul tempori descripti arcubus minori tempore descriptis sint proportionales) vires illæ sunt ut quadrata arcuum æqualibus quibuscunq temporibus(8) descriptorum applicata ad circulorum semidiametros. Q.E.D.

Corol. 1. Unde cùm arcus simul descripti sint ut velocitates corporum[,] hæ vires sunt ut quadrata velocitatum applicata ad altitudines corporum. Vel ut cum Geometris loquar, hæ vires sunt in ratione quæ componitur ex duplicata ratione velocitatum directe et ratione simplici altitudinum inverse.

Corol. [2]. Et cùm tempora periodica sint in ratione quæ componitur ex ratione altitudinum directe & ratione velocitatum inverse: Vires centripetæ sunt reciproce ut quadrata temporum periodicorum applicata ad corporum

sponding place in the published 'Prop. V', as we have observed (see 1, §2, Appendix 2.7: note (30)), demonstration that at any instant the orbital speed is inversely proportional to the corresponding tangential polar had rather inconveniently—for want of an appropriate preceding corollary—there to be made from first principles.

(5) This revised version of the published 'Prop. IV' (*Principia*, $_1$1687: 41–2) is here given concise proof in terms of the new set of corollaries understood (see note (2)) to be appended to Proposition I preceding; for which see §1.2 above.

(6) A cancelled first version of the following proof (on f. 27^r) began 'Quoniam circuli æquabili motu describuntur, ideoq areæ radijs circulorum descriptæ sunt temporibus proportionales, vires centripetæ per Prop. II tendunt ad centra circulorum. Quod erat primò demonstrandum' (Seeing that the circles are described with a uniform motion, and consequently the areas described by the circles' radii are proportional to the times, the centripetal forces (by Proposition II) tend to the centres of the circles. As was first to be demonstrated), continuing thereafter with a slightly reshaped form of the published demonstration (*Principia*, $_1$1687: 41): 'Corpora autem *B*, *b* in circumferentijs circulorum *BD*, *bd* gyrantia, ... Et quoniam... tangentes *BC*, *bc* his arcubus æquales et viribus centripetis perpetuo retrahuntur ab his tangentibus ad circumferentias circulorum & ea retractione singulis momentis transferuntur per

construction) as the perpendiculars AP, BQ directly, that is, as the perpendiculars let fall from the point D to the tangents. Whence it is easily gathered that the points S, D, T are in a single straight line. And by a similar argument the points S, E, V are also in one straight line; and accordingly the centre S is located at the meet of the lines TD, VE. As was to be proved.

Proposition V, Theorem IV.(5)

Where bodies describe differing circles with a uniform motion, their centripetal forces tend to the centres of the circles and are to one another as the squares of the arcs simultaneously described, divided by the heights of the bodies above the centre of force.

Demonstration.

(6)These forces tend (by Proposition II and Corollary 2 of Proposition I) to the centres of the circles, and (by Corollaries 4 and 5 of Proposition I) are as the *sagittæ* or(7) the versed sines of the arcs described in equal minimal times, that is, as the squares of those arcs divided by the diameters of their circles, and accordingly (since the radii, diameters and arcs simultaneously described in a greater time are proportional to the arcs described in a lesser time) those forces are as the squares of the arcs described in any equal(8) times, divided by the radii of their circles. As was to be proved.

Corollary 1. Whence, since the arcs simultaneously described are as the speeds of the bodies, these forces are as the squares of the speeds divided by the heights of the bodies. Or (to speak in geometrical parlance) these forces are in a ratio compounded of the doubled ratio of the speeds directly and the simple ratio of the heights inversely.

Corollary 2. And since the periodic times are in a ratio compounded of the ratio of the heights directly and the ratio of the speeds inversely, the centripetal forces are reciprocally as the squares of the periodic times divided by the heights of the

---

intervalla semper nascentia CD, cd, vires autem (per Leg. II) sunt ut motus simul geniti atq; adeo ut intervalla motibus istis simul descripta: manifestum est quod vires hæ centripetæ sunt in ratione prima intervallorum nascentium CD, cd. Nam et per Corol. 4, Prop. I, hæ vires sunt ut arcuum minimorum simul descriptorum $2BD$, $2bd$ sagittæ CD, cd. Fiat jam figura tkb figuræ DCB similis, &...ut $\frac{BD^{\text{quad.}}}{SB}$ ad $\frac{bd^{\text{quad.}}}{Sb}$. Et propterea cùm vires centripetæ supra essent ad invicem ut CD ad cd, hæ vires sunt ad invicem ut $\frac{BD^{\text{quad.}}}{SB}$ ad $\frac{bd^{\text{quad.}}}{Sb}$, id est ut arcuũ simul descriptorum quadrata applicata ad radios circulorum. Q.E.D.' The preliminary draft on Add. 3965.6: 86v differs only trivially from this.

(7) Since in the present instance the orbits are circles set concentrically round the force-centre, whence Newton's generalised 'arrows' here reduce to their standard signification of circular 'turned sines'; compare §1: note (17).

(8) Understand 'quamminimorum' (minimal) once more.

altitudines, id est in ratione quæ componitur ex ratione altitudinum directe & duplicata ratione temporum periodicorum inverse.

Corol. 3. Unde si tempora periodica æquantur erunt tum vires centripetæ tum velocitates ut altitudines: & contra.

Corol. 4. Si quadrata temporum periodicorum sunt ut radij vires centripetæ sunt æquales & velocitates in subduplicata ratione altitudinum: & contra.

Corol. 5. Si quadrata temporum periodicorum sunt ut quadrata radiorum id est tempora ut radij, vires centripetæ sunt reciproce ut altitudines et velocitates æquales: et contra.

Corol. 6. Si quadrata temporum periodicorum sunt ut cubi radiorum vires centripetæ sunt reciprocè ut quadrata altitudinum, velocitates autem reciproce in altitudinum subduplicata ratione: et contra.

Corol. 7. Et universaliter si tempora periodica sint ut altitudinum R, S dignitates R^n, S^n, vires centripetæ sunt reciproce ut dignitates R^{2n-1}, S^{2n-1} & velocitates reciproce ut dignitates R^{n-1}, S^{n-1}: & contra.(9)

Schol.

Hic et in sequentibus considero centrum virium ut punctum infimum in systemate revolventium corporum & per altitudines corporum revolventium intelligo eorum distantias ab hoc centro.(10) Casus autem Corollarij sexti obtinet in corporibus cœlestibus (ut seorsim collegerunt etiam Wrennus, Hookius[,] Halleius) & propterea ------ centrum versus.

Prop. VI. Theor. V.

Corporum quæ similes omnes similium figurarum partes temporibus proportionalibus describunt, vires centripetas ad centra tendere in figuris istis similiter posita, & esse inter se in ratione quæ componitur ex ratione altitudinum directè & duplicata ratione temporum inverse.

Demonstratio.

Nam altitudines corporum ob orbium similitudinem sunt ut sagittæ arcuum similium temporibus proportionalibus descriptorum & vires per Corol. 7 Prop. I & Corol. Lem. [X] sunt ut hæ sagittæ directe et quadrata temporum inverse.(11)

(9) An unwanted concluding sentence 'Hic nota quod n, $n-1$, $2n-1$ indices significant' (Here note that n, $n-1$, $2n-1$ denote indices) is cancelled. The final published '*Corol. 7.* Eadem omnia de temporibus, velocitatibus & viribus, quibus corpora similes figurarum quarumcunq; similium, centraq; similiter posita habentium, partes describunt, consequuntur ex Demonstratione præcedentium ad hosce casus applicata' (*Principia*, $_1$1687: 42; see also page 198 above) is now elaborated to be a separate 'Prop. VI. Theor. V' below, and so here suppressed. (In the second edition (*Principia*, $_2$1713: 39) it was to be reintroduced, along with an added sentence 'Applicatur autem substituendo æquabilem arearum descriptionem pro æquabili motu, & distantias corporum a centris pro radiis usurpando' and a new concluding

bodies, that is, in a ratio compounded of the ratio of the heights directly and the doubled ratio of the periodic times inversely.

Corollary 3. Whence, if the periodic times are equal, then both the centripetal forces and the speeds will be as the heights; and conversely.

Corollary 4. If the squares of the periodic times are as the radii, the centripetal forces are equal and the speeds in the halved ratio of the heights; and conversely.

Corollary 5. If the squares of the periodic times are as the squares of the radii, that is, the times as the radii, the centripetal forces are reciprocally as the heights, and the speeds are equal; and conversely.

Corollary 6. If the squares of the periodic times are as the cubes of the radii, the centripetal forces are reciprocally as the squares of the heights, but the speeds reciprocally in the halved ratio of the heights; and conversely.

Corollary 7. And universally, if the periodic times be as the powers R^n, S^n of the heights R, S, then the centripetal forces are reciprocally as the powers R^{2n-1}, S^{2n-1} and the speeds reciprocally as the powers R^{n-1}, S^{n-1}; and conversely.(9)

Scholium.

Here and in the sequel I consider the centre of force as the lowest point in the system of revolving bodies, and so by the 'heights' of revolving bodies I understand their distances from this centre.(10) The case of Corollary 6, however, holds true in the heavenly bodies (as separately Wren, Hooke and Halley have also gathered) and accordingly...... towards the centre.

Proposition VI, Theorem V.

Where bodies describe all similar parts of similar figures in proportional times, their centripetal forces tend to centres similarly positioned in those figures and are to one another in a ratio compounded of the ratio of the heights directly and the doubled ratio of the times inversely.

Demonstration.

For because of the similarity of the orbits the heights of the bodies are as the *sagittæ* of similar arcs described in proportional times, while (by Corollary 7 of Proposition I and the Corollary to Lemma X) the forces are as these *sagittæ* directly and the squares of the times inversely.(11)

---

corollary which affirmed that 'Ex eadem demonstratione consequitur etiam, quod arcus quem corpus in circulo data vi centripeta uniformiter revolvendo tempore quovis describit, medius est proportionalis inter diametrum circuli & descensum corporis eadem data vi eodemque tempore cadendo confectum'.)

(10) A minor explanatory insertion. Newton indicates in sequel that the original published scholium (*Principia*, $_1$1687: 42–3; compare pages 199–200 above) is here to follow without variant.

(11) The otherwise insignificantly variant preliminary draft on f. 188r here adds: 'Hæc ita

Prop. VII. Theor VI.(12)

Si orbium duorum proportionales ordinatæ abscissis proportionalibus in angulis quibusvis datis insistunt, & virium centra in abscissis(13) similiter locantur, corpora correspondentes orbium partes temporibus proportionalibus describent et vires centripetæ erunt ut altitudines corporum directe et quadrata temporum inverse.

Cas. 1.(14) Si orbium *APpB*. *AQqB* ordinatæ omnes *PR* & *QR*, eidem abscissæ *AR* in angulis datis *ARP*, *ARQ* insistentes datam habent rationem ad invicem & corpora *P* et *Q* in his orbibus revolventia ad commune virium centrum *S* in abscissa illa locatum trahantur: hæc corpora correspondentes orbium partes *Pp*, *Qq* simul percurrendo, radijs *PS*, *QS* ad centrum illud ductis describent areas sibi invicem proportionales *PSp*, *QSq* & vires illæ quibus in orbibus retinentur erunt ad invicem ut altitudines corporum.

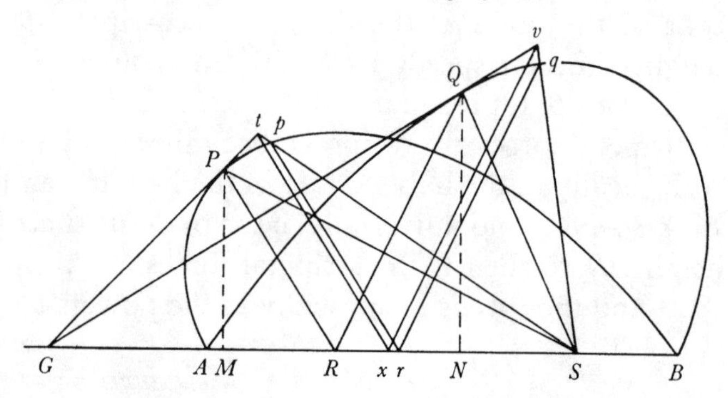

Nam si a corporibus (15) ad abscissam demittantur perpendicula *PM*, *QN*, et ordinatæ *PR* & *pr*, *QR* & *qr* intervallo quàm minimo ab invicem distent, erunt tam areæ quadrilateræ *PRrp*, *QRrq* quam triangula *PRS* & *QRS*, *prS* et *qrS* ad invicem ut eorum altitudines *PM* et *QN*. Et propterea si areis illis quadrilateris addantur triangula *prS*, *qrS* et auferantur triangula *PRS*, *QRS* manebunt triangula *PSp*, *QSq* in eadem perpendiculorum ratione ad invicem[,] et hæc ratio ob datos angulos *ARP*, *ARQ* & datam ordinatarum *PR QR* rationem data est. Q.E.D. Porrò cum ordinatæ *PR*, *p*[*r*] sunt ad invicem ut ordinatæ [*Q*]*R*, *q*[*r*], si ad puncta *P*, *Q* ducantur orbium tangentes *PG*, *QG*, incidet earum concursus *G* in abscissam *AB*.(16) Producantur *GP*, *Sp* donec concurrant in *t*. Ducantur

se habent ubi centra viriū in figuris similibus similiter posita sunt' (These conditions hold when the centres of forces are similarly positioned in similar figures)—presuming, of course, that the several force-fields do not mutually disturb one another's action.

(12) In a preliminary draft of this on Add. 3966.2: 19r/19v, where it was initially sketched out as a complementary pair of unnumbered 'Prop.'s, the present 'Theor. VII' was originally set (on f. 19v) as a subsidiary 'Corol. M' to the latter (see note (22) below), there followed by a 'Corol. N' demonstrating essentially the main result here proved, and also by a 'Corol. [O]' stating its particular application in the case where the orbit is a circle or ellipse. Subsequently in a preparatory revise (*ibid.*: 20r) Newton refashioned 'Corol. N' to be a 'Prop. VI' enunciating the special case of the present 'Theor VI' in which the ordinates *PR* and *QR* coincide, and set 'Corol. [O]' (as here) in consectary to it; while in sequel 'Corol. M' was similarly reshaped to be a new 'Pro[p] VII', supported by four corollaries and a scholium which differ

Proposition VII, Theorem VI.(12)

If in two orbits proportional ordinates stand at any given angles on proportional abscissas, and the centres of force are similarly located in the abscissas,(13) *bodies shall describe corresponding parts of the orbits in proportional times and the centripetal forces will be as the heights of the bodies directly and the squares of the times inversely.*

Case 1.(14) If in the orbits *APpB*, *AQqB* all ordinates *PR* and *QR* standing on the same abscissa *AR* at given angles $\widehat{ARP}$, $\widehat{ARQ}$ have a given ratio to one another, and the bodies *P* and *Q* revolving in these orbits be drawn to a common centre *S* of force located in that abscissa, then these bodies, by simultaneously traversing corresponding parts $\widehat{Pp}$, $\widehat{Qq}$ of the orbits, will by the radii *PS*, *QS* drawn to that centre describe areas (*PSp*), (*QSq*) proportional to one another, and the forces whereby they are kept in the orbits will be to one another as the heights of the bodies.

For, if perpendiculars *PM*, *QN* be let fall from the bodies(15) to the abscissa, and the ordinates *PR* and *pr*, *QR* and *qr* be a minimal distance apart from each other, then both the quadrilateral areas (*PRrp*), (*QRrq*) and the triangles *PRS* and *QRS*, *prS* and *qrS* will be to each other as their heights *PM* and *QN*. Accordingly, if to those quadrilateral areas there be added the triangles *prS*, *qrS* and from them subtracted the triangles *PRS*, *QRS*, there will remain the triangles *PSp*, *QSq* in the same ratio to each of the perpendiculars, and this ratio, because of the given angles $\widehat{ARP}$, $\widehat{ARQ}$ and given ratio of the ordinates *PR*, *QR*, is given. As was to be proved. Further, since the ordinates *PR*, *pr* are to one another as the ordinates *QR*, *qr*, if at the points *P*, *Q* tangents *PG*, *QC* to the orbits be drawn, their meet *G* will fall in the abscissa *AB*.(16) Extend *GP*, *Sp* till they meet in *t*, then

---

only minimally from those appended to the equivalent revised 'Theor VII' here reproduced below.

(13) This replaces '*in ordinatis quibusvis correspondentibus*' (*in any corresponding ordinates whatever*). In his final version Newton again (compare note (11) above) evades having unrealistically to presuppose the simultaneous existence of two separate neighbouring centres of force which do not in any way disturb each other's action.

(14) In an immediately preceding version on f. 188r—evidently cancelled as a superfluous distinction—this first case was tentatively split into a component pair, '*Cas. 1*' and '*Cas. 2*', according as the ordinates *PR* and *QR* are inclined to their common base *AR* 'in eodem angulo *ARPQ*' (in the same angle $\widehat{ARP}/\widehat{ARQ}$) or 'in angulis diversis *ARP ARQ*' (in differing angles $\widehat{ARP}$, $\widehat{ARQ}$) respectively.

(15) Without any significant gain this was so changed from 'a punctis [*P*, *Q*]' (from the points *P*, *Q*).

(16) For, as *p*, *t* and *q*, *v* come to coincide with *P* and *Q* respectively, where the tangent at *P* meets the base *AR* in *G* there is $GR : Gx = PR : (tx \approx)\, pr = QR : (qr \approx)\, vx$, so that the tangent *Qv* at *Q* also meets *AR* in *G*.

ordinatis *PR*, *QR* parallelæ *tx*, *vx* quarum *xv* concurrat cum tangente *GQ* producta in *v* et erit *tx* ad *PR* ut *Gx* ad *GR* adeoq; ut *vx* ad *QR*. Et vicissim *tx* ad *vx* ut *PR* ad *QR* atq; adeo ut *pr* ad *qr*[,] et vicissim *tx* ad *pr* ut *vx* ad *qr*. Sed ut *tx* est ad *pr* ita *Sx* est ad *Sr* et propterea *Sx* est ad *Sr* ut *v*[*x*] ad *q*[*r*]. Jacent ergo in directum puncta *S*, *q* et *v*. estq; *tp* ad *pS* ut *xr* ad *rS* et *vq* ad *qS*[,] et vicissim *tp* ad *vq* ut *pS* ad *qS*. Sed *tp* et *vq* sunt spatia per quæ corpora *P*, *Q* de tangentibus ad Orbes a viribus suis centripetis simul trahuntur & propterea vires istæ sunt ad invicem ut hæc spatia. Quare cum hæc spatia sunt ut altitudines *pS*, *qS*, atq; adeo[17] si jam coincidant ordinatæ *PR* et *pr*, *Q*[*R*] et *qr*, ut altitudines *PS* et *QS*, erunt vires ut altitudines *PS* et *QS*.[18] Q.E.D.

Cas. 2. Ponatur jam orbis tertius orbi *APB* similis et centrum habens similiter positum; et quod corpora duo simul percurrunt similes horum orbium partes[,] & cum vires in his orbibus sint ad invicem ut altitudines per Prop VI, corpora tria simul percurrent correspondentes horum trium orbium partes & vires in his omnibus erunt ut altitudines. Q.E.D.

Cas. 3. Augeatur jam velocitas corporis in orbe quovis et vires augendæ erunt in duplicata ratione velocitatum (per Corol. [1. Prop. V]) adeoq; cùm tempus diminuatur in ratione augmenti velocitatis vires erunt ut altitudines directe et quadrata temporum inversè. Q.E.D.[19]

Corol.[20] Igitur si Orbis alteruter *APB* circulus et Orbis alter *AQB* Ellipsis quælibet & punctum *S* utriusq; centrum; cum vis qua corpus uniformi cum motu in circulo revolvitur data sit et propterea datæ corporis altitudini *SP* proportionalis, erit vis qua corpus aliud in Ellipsi simul revolvi possit ut corporis illius altitudo *SQ*.[21]

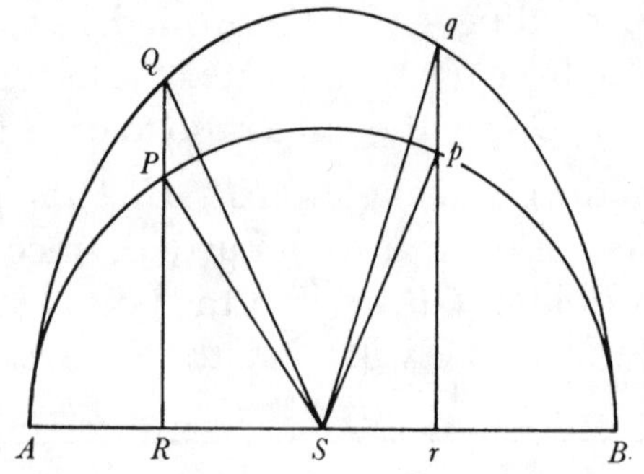

(17) Newton began in immediate sequel to write 'ob infinitam parvitatem [ipsius *Rr*]' (because of *Rr*'s infinite smallness).

(18) A somewhat over-long elaboration of Newton's draft proof (Add. 3966.2: 19r) that 'vires quibus corpora ad centrum illud [*S*] trahuntur...sunt ut arcuum synchronorum *Pp*, *Qq* sagittæ [*tp*, *vq*]...& hæ sagittæ sunt ut altitudines [*S*]*P*, [*S*]*Q*' (the forces whereby the bodies are drawn to the centre *S* are as the *sagittæ tp*, *vq* of the contemporaneous arcs $\widehat{Pp}$ and $\widehat{Qq}$, and these *sagittæ* are as the heights *SP*, *SQ*).

(19) As we shall see in Appendix 2. 1/2 below, Newton not long afterwards conceived the alternative notion of adding the gist of this theorem as 'Prop. 1' of a scholium (on Add 3965.2: 6v) scheduled, it would seem, to be introduced after Proposition VI in the published 'Liber primus' in the first instance, and then, in a further change of mind, subsequently determined to annex its essence to the following published Proposition X.

(20) A cancelled first version of this corollary (immediately preceding it on f. 189r) reads: 'Hinc si corpus in Ellipsi revolvatur & vires ad centrum Ellipseos dirigantur, vires illæ erunt ut altitudo corporis. Nam vires in Ellipsi sunt ad vires in circulo ut altitudo in Ellipsi ad altitudinem in circulo et in circulo tam vires quam altitudines æquantur. Sit *APpB* circulus & *AQqB*

parallel to the ordinates *PR*, *QR* draw *tx*, *xv*, of which *xv* shall meet the tangent *GQ* produced in *v*, and *tx* will be to *PR* as *Gx* to *GR*, and hence as *vx* to *QR*; and so *alternando tx* to *vx* as *PR* to *QR*, and hence as *pr* to *qr*, and *alternando tx* to *pr* as *vx* to *qr*. But as *tx* to *pr*, so is *Sx* to *Sr*, and accordingly *Sx* is to *Sr* as *vx* to *qr*. The points *S*, *q* and *v* therefore lie in a straight line, and *tp* is to *pS* as *xr* to *rS* and so *vq* to *qS*, and *alternando tp* to *vq* as *pS* to *qS*. But *tp* and *vq* are the spaces through which the bodies *P*, *Q* are simultaneously drawn away from the tangents to the orbits by their centripetal forces, and accordingly those forces are to one another as these spaces. Consequently, since these spaces are as the heights *pS*, *qS* and therefore,(17) if the ordinates *PR* and *pr*, *QR* and *qr* should now coincide, as the heights *PS* and *QS*, the forces will be as the heights *PS* and *QS*.(18) As was to be proved.

Case 2. Now suppose a third orbit similar to the orbit *APB* and having its centre similarly positioned, and that two bodies simultaneously traverse similar parts of these orbits; then, since (by Proposition VI) the forces in these orbits are to one another as the heights, the three bodies will simultaneously traverse corresponding parts of these three orbits and the forces in all of them will be as the heights. As was to be proved.

Case 3. Let now the speed of a body in any orbit be increased and the forces will (by Corollary 1 of Proposition V) need to be increased in the doubled ratio of the speeds; hence, since the time diminishes in the ratio of the increase in speed, the forces will be as the heights directly and the squares of the times inversely. As was to be proved.(19)

Corollary.(20) Therefore if one of the orbits *APB* be a circle and the other orbit *AQB* any ellipse, and the point *S* be the centre of both, since the force whereby a body revolves with uniform motion in the circle is given and accordingly proportional to the body's given height *SP*, the force whereby another body might simultaneously revolve in the ellipse will be as the height *SQ* of that body.(21)

Ellipsis communem habentes diametrum *AB* et commune centrum *S* et communes ordinatas *QPR*, *qpr*. Et vis in *Q* erit ad vim in *P* ut *QS* ad *PS* et vis in *P* ad vim in *p* ut *PS* ad *pS*[,] hoc est in ratione æqualitatis et vis in *p* ad vim in *q* ut *pS* ad *qS*[,] et ex æquo vis in *Q* ad vim [in] *q* ut *QS* ad *qS*' (Hence, if the body revolve in an ellipse and the forces be directed to the ellipse's centre, those forces will be as the height of the body. For the forces in an ellipse are to the forces in a circle as the height in the ellipse to the height in the circle, while in a circle both the forces and heights are equal. Let $\widehat{APpB}$ be a circle and $\widehat{AQqB}$ an ellipse having a common diameter *AB*, a common centre *S* and common ordinates *QPR*, *qpr*. The force at *Q* will then be to the force at *P* as *QS* to *PS*, and the force at *P* to the force at *p* as *PS* to *pS*—that is, in a ratio of equality, and the force at *p* to the force at *q* as *pS* to *qS*; whence *ex æquo* the force at *Q* will be to the force at *q* as *QS* to *qS*).

(21) An unfinished rough note by Newton on Add. 3965.17: 634r, drafted some fifteen years later if our judgement of its handwriting style is correct, suggests that he then had it in mind to append a suitably enlarged version of this corollary as an *idem aliter* to his revise of the published Proposition X (*Principia*, $_1$1687: 48–9) which establishes an equivalent result by

Prop. VIII. Theor VII.

Vis qua corpus quodvis P in orbe quocunq; APB circa virium centrum S revolvi potest est ad vim qua corpus aliud P in eodem orbe eodemq; tempore periodico circa aliud quodvis virium centrum R revolvi potest ut contentum sub altitudine primi corporis et quadrato altitudinis secundi corporis $SP \times RP^{\text{quad}}$ *ad cubum rectæ PT quam recta ST tangenti Orbis parallela de altitudine secundi corporis versus corpus illud abscindit.*(22)

Sit enim *PG* tangens illa et huic parallela ducatur arcus quam minimi chorda *MN* secans altitudinem *PR* in *e* et *PS* in *f*. Et vires quibus corpora petunt centra *S* et *R* erunt directè ut arcus hujus sagittæ *Pf*, *Pe* ad centra illa convergentes hoc est ut *PT*, *PS* & inverse ut quadrata temporum quibus arcus ille a corporibus describitur[,] hoc est in duplicata ratione arearum *NSM*, *NRM* quas corpora radijs ad centra illa ductis describunt, id est inverse in duplicata ratione rectarum *PT*, *PR* areis illis proportionalium. Et hæ rationes, nempe *PS* ad *PT* et PR^{quad} ad PT^{quad}, conjunctæ faciunt rationem $PS \times PR^{\text{quad}}$ ad PT^{cub}. Q.E.D.(23)

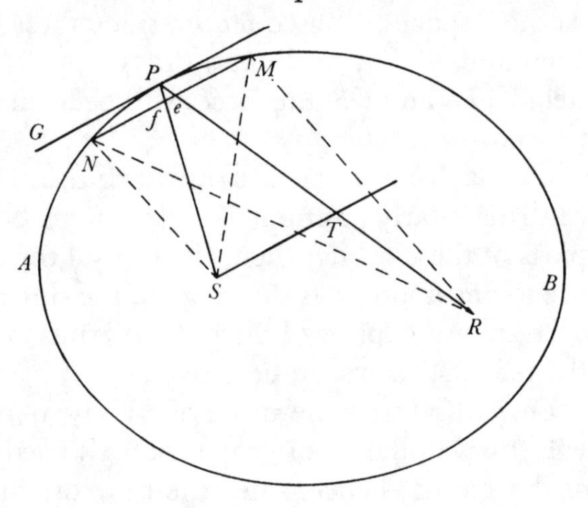

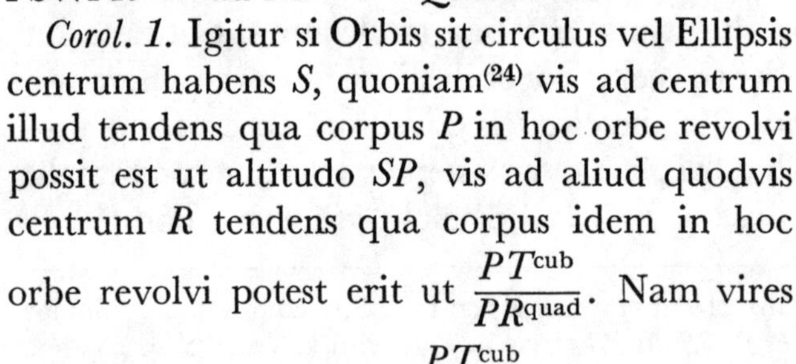

Corol. 1. Igitur si Orbis sit circulus vel Ellipsis centrum habens *S*, quoniam(24) vis ad centrum illud tendens qua corpus *P* in hoc orbe revolvi possit est ut altitudo *SP*, vis ad aliud quodvis centrum *R* tendens qua corpus idem in hoc orbe revolvi potest erit ut $\frac{PT^{\text{cub}}}{PR^{\text{quad}}}$. Nam vires sunt ad invicem ut *SP* et $\frac{PT^{\text{cub}}}{PR^{\text{quad}}}$.

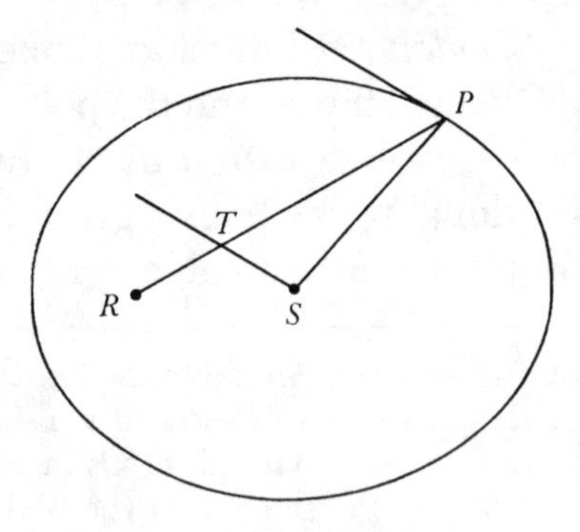

applying to this particular case of elliptical motion round a force-centre at its geometrical centre the general measure of *vis centripeta* derived in the preceding Proposition VI (*ibid.*: 44–5; compare pages 132 and 138–40 above). David Gregory had earlier jotted down a précis of the present theorem among other 'Adnotata Phys: [et] Math:' which he took down 'ex Neutono 5.6.7 Maij 1694' during a visit to Cambridge, there noting that 'Si Ordinatæ [*RP*] proportionaliter augeantur ubiq; fiantq; [*RQ*], ex data vi ad *S* in orbita [*APB*], dabitur vis ad idem *S* in orbita [*AQB*], etiam dato quocunq; angulo applicationis [*ARQ*]. Hinc a[b] æquabili vi ad circuli Centrum transibitur ad vim ad Ellipseos centrum' and commenting on the latter corollary that 'transferendo vim a Centro ellipsis ad focum [per Theor VII]...hic est facillimus modus pro Motu planetarum in Sectionibus Conicis'. (See his Codex C44 [now in

Proposition VIII, Theorem VII.

The force whereby any body P can revolve in any orbit APB whatever round the centre S of force is to the force whereby another body P can revolve in the same orbit and in the same periodic time round any other centre R of force as the product of the height of the first body and the square of the height of the second body, $SP \times RP^2$, to the cube of the straight line PT which the straight line ST parallel to the orbit's tangent cuts off from the height of the second body in the direction of that body.(22)

For let *PG* be that tangent and parallel to this draw the chord *MN* of a minimal arc, intersecting the height *PR* in *e* and *PS* in *f*. Then the forces whereby the bodies seek the centres *S* and *R* will be directly as the *sagittæ Pf*, *Pe* of this arc converging on those centres—that is, as *PT* and *PS*—and inversely as the squares of the times in which that arc is described by the bodies, that is, in the doubled ratio of the areas (*NSM*), (*NRM*) which the bodies describe by radii drawn to those centres—that is, inversely in the doubled ratio of the straight lines *PT*, *PR* proportional to those areas. And these ratios—namely *PS* to *PT*, and PR^2 to PT^2—jointly form the ratio $PS \times PR^2$ to PT^3. As was to be proved.(23)

Corollary 1. Therefore if the orbit be a circle or ellipse having its centre at *S*, because(24) the force tending to that centre whereby the body *P* may revolve in this orbit is as the height *SP*, the force tending to any other centre *R* whereby the same body may revolve in this orbit will be as PT^3/PR^2. For the forces are to each other as *SP* and PT^3/PR^2.

---

the Royal Society], reproduced in *The Correspondence of Isaac Newton*, **3**, 1961: 334–5, especially 334.)

(22) Newton's original 'Corol. M' on Add. 3966.2: 19v (see note (12)) announced little differently that 'Vis qua corpus [*P*] in orbe quocunꝗ [*PAB*] circa centrum [*S*] revolvi potest est ad vim qu[a] in eodem orbe eodemꝗ tempore periodico circa centrum quodvis aliud [*R*] revolvi potest, ut [*SP*] ad $\frac{[TP]^{\text{cub}}}{[RP]^{\text{quad}}}$, si modo [*ST*] ducatur orbis tangenti [*PG*] parallela'. He initially concluded the present revised enunciation in much the same fashion by writing '...*ad cubum rectæ PT quæ altitudo est secundi corporis supra rectam ST quæ per centrum virium primi corporis ducitur & tangenti orbis parallela est*' (...*to the cube of the straight line PT which is the* [*oblique*] *height of the second body above the line ST which is drawn through the force-centre of the first body parallel to the tangent to the orbit*).

(23) Along with a miscellany of other 'adnotata Phys: et Math:' jotted down 'Cantabrigiæ. 4 Maij 1694' (compare note (21) above) David Gregory later duly recorded that Newton 'Theorema condidit quo ex data vi qua dato tempore in data Orbita circa datum centrum revolvitur vim inveniat qua in eadem Orbita circa aliud datum Centrum revolvatur. V: g: est vis circa [*S*] ad vim circa [*R*] ut [*PS*] × $[PR]^q$ ad $[PT]^{\text{cub}}$' (Codex C33 [Royal Society], reproduced in *Correspondence*, **3**: 312), afterwards adjoining that 'Nos hoc Theorema clare enunciatum vel potius de novo inventum [!] (arrepta solum hinc occasione) de novo plene demonstravimus' (*ibid.*). Gregory's retrospective proof of Newton's theorem does not appear to have survived, but it was doubtless little different in its essentials; it is hardly likely that he would, unaided, have thought of the elegant manner in which (see §1, Appendix 2: note (4) above)

Corol. 2. Et si punctum R sit Ellipseos umbilicus alteruter, ob longitudinem PT[25] in hoc casu datam, erit vis centripeta reciprocè ut quadratum altitudinis PR.

Corol. 3. Et si manente Ellipseos umbilico centrum ejus in infinitum abeat, vis quæ semper est reciprocè ut quadratum altitudinis, erit etiam in hoc casu reciproce ut quadratum altitudinis, hoc est ubi Ellipsis in Parabolam migraverit.

Corol. 4. Et cum proprietates omnes quæ Ellipsi et Parabolæ communes sunt congruunt etiam Hyperbolæ,[26] erit vis qua corpus in Hyperbola, circa ejus umbilicum ceu centrum moveri potest, reciprocè ut quadratum altitudinis.[27]

Schol.

In hac Propositione & Corollarijs ejus, centrum viriū S ad concavas orbis partes locatur. Si centrum illud ad convexas orbis partes migret, vis centripeta, ob mutatum altitudinis RP signum affirmativum in signum negativum, vertetur in vim centrifugam, cæteris manentibus.

Prop. IX. Theor. VIII.[28]

Si corpus in spatio non resistente circa centrum immobile in orbe quocunꝗ revolvatur et arcum quemvis jamjam nascentem tempore quam minimo describat & sagitta arcus ducatur quæ chordam bisecet & producta transeat per centrum virium: erit vis centripeta in medio arcus ut sagitta directe & quadratum temporis inversè.

Nam sagitta...inverse. Q.E.D.

Corol 1.... Corol. 2.... Corol. 3.... Corol. 4.... Corol. 5.... vel solidum $SY^q \times PV$ huic vi reciproce proportionale.[29]

Newton subsequently presented it as Corollary 3 of the revised Proposition VII of the published *Principia* ($_2$1713: 44).

(24) Understand 'per Corol. Prop. VII' (by the Corollary to Proposition VII); see also note (23) above.

(25) Equal to half the main axis, namely, as Newton had before proved incidentally in Proposition XI of his published 'Liber primus' (*Principia*, $_1$1687: 50) and subsequently demonstrated as a separate introductory 'Lemma 1' in the preliminary revise 'ante Prop. XVII' reproduced in §1.5 preceding.

(26) A curious appeal in a dynamical context to a Keplerian principle of continuity between the various species of conic; while (as Newton well knows) the extension is valid for the present focal conic traversed in an inverse-square force field, in the comparable instance of a direct-distance field (where a hyperbolic orbit—unlike an elliptical one—cannot be traversed under the action of a *vis centripeta*; and conversely so where the force is centrifugal) the 'congruence' is inexact.

(27) In his jotted 'adnotata...Cantabrigiæ 4 Maij 1694' (see note (23)) David Gregory shortly afterwards correspondingly recorded that 'Facillime...proprietates Coni sectionum

Corollary 2. And if the point R be one or other focus of the ellipse, because the length of PT[25] is in this case given, the centripetal force will be reciprocally as the square of the height PR.

Corollary 3. And if, with the ellipse's focus remaining fixed, its centre should go off to infinity, the force, which is ever reciprocally as the square of the height, will also in this case—where, that is, the ellipse has passed into a parabola—be reciprocally as the square of the height.

Corollary 4. And since all properties which are common to the ellipse and parabola accrue also to the hyperbola,[26] the force whereby a body can revolve in a hyperbola about its focus as centre will be reciprocally as the square of the height.[27]

Scholium.

In this proposition and its corollaries the force-centre S is located on the concave side of the orbit. If that centre shall pass over to the orbit's convex side, the centripetal force will, because the positive sign of the height RP is changed to negative, be turned into a centrifugal force, with the rest remaining the same.

Proposition IX, Theorem VIII.[28]

If a body should, in a non-resisting space, revolve round a stationary centre in any orbit whatever and describe any just barely nascent arc in a minimal time, and an 'arrow' of the arc be drawn to bisect its chord and pass, when produced, through the centre of force, then the centripetal force at the mid-point of the arc will be as that sagitta directly and the square of the time inversely.

For the *sagitta*...inversely. As was to be proved.

*Corollary 1....Corollary 2....Corollary 3....Corollary 4....Corollary 5....*or the 'solid' $SY^2 \times PV$ reciprocally proportional to this force.[29]

---

prosequitur quoad vim centripetam. Nam quæ et Ellipsi et parabolæ conveniunt[,] etiam et hyperbolæ' (*Correspondence*, **3**: 123).

(28) Except for this revised head, this is the 'Prop. VI. Theor. V' (on *f.* 188r) which Newton earlier planned to introduce on 'pag. 44. lin. 20' of his 1687 *Principia*; since we have already reproduced its full text in §1.3 preceding, it will be enough here to repeat its enunciation with a mere indication of the following demonstration and ensuing corollaries.

(29) Since the earlier final sentence (page 550 above) promising 'Ejus rei...exempla in Problematis sequentibus' is now an inappropriate harbinger of what is here immediately to come, we choose to omit it. If it still, on the model of Propositions IX, X and XI–XIII of the 1687 'Liber primus', remained Newton's intention to apply his original measure $QR/SP^2 \times QT^2$ of impressed *vis centripeta* (see 1, §1: note (30)) ultimately to compute the forces (inverse-cube, direct-distance and inverse-square respectively) induced by orbital motion in a logarithmic spiral (round its pole), in an ellipse (round its centre) or in a general conic (round a focus), doubtless these worked examples would be presented in a more summary form.

Lemma XII.

Si in Sectionis conicæ diametro quavis PG ad partes concavas sumatur PM æqualis lateri recto ad diametrum illam pertinenti, & per puncta P[,] *M describatur circulus qui sectionem conicam tangat ad P: hic circulus eandem habebit curvaturam cum sectione illa Conica ad P.*(30)

Nam si circulus ille secet sectionem conicam in Q et si ad diametrū PG agatur ordinatim applicata QV, et compleatur pgrū(31) $QVPR$ ac producatur RQ donec circulo occurrat in N sitq; conicæ sectionis latus rectum R et latus transversum(32) T: erit QRN æquale RP^q seu QV^q[,] hoc est (ex natura sectionum conicarum) rectangulo sub PV et $R-\frac{R}{T}PV$;(33) et applicandis his æqualibus ad æquales QR et PV erit RN æqualis $R-\frac{R}{T}PV$. Coeant jam puncta P et Q eò ut circulus evadat ejusdem curvaturæ cum sectione conica in P et quantitate PV evanescente quantitates RN et $R-\frac{R}{T}PV$ evadent PM et R et propterea PM æqualis est lateri recto.(34) Q.E.D.

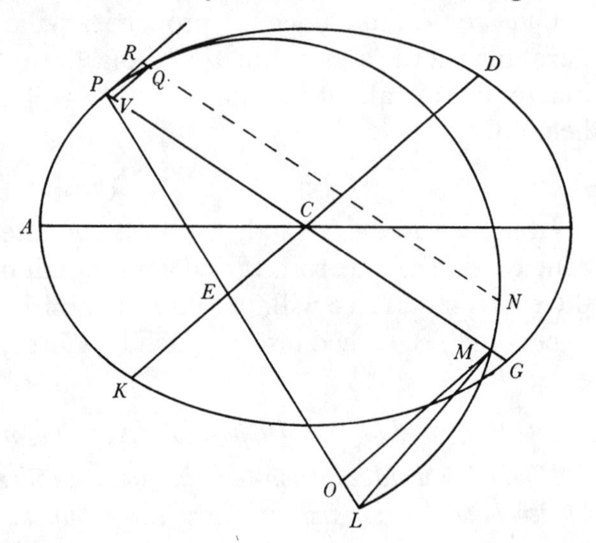

Corol. 1. Si in sectionis conicæ diametro qu[a]vis PG sumatur PM lateri recto æqualis et per punctum M agatur recta MO tangenti PR parallela, deinde a puncto P agatur alia quævis recta PO priori MO occurrens in O et in ea sumatur

(30) A cancelled first version (on f. 31r) of this enunciation asserts the result obtained in Corollary 2: '*Si centro C diametris conjugatis PG, DK describatur sectio Conica DQPAKG, et a vertice in concavas partes figuræ agatur recta quævis P[L] quæ secet Diametrum DK in E & sit ad DK ut DK ad 2PE; locabitur punctum [L] in circulo qui tanget sectionem conicam in P et erit ejusdem curvaturæ cum sectione Conica in puncto contactûs.* (*If with centre C and conjugate diameters PG, DK there be described a [central] conic DQPAKG, and from a vertex to the concave side of the figure there be drawn any straight line PL which shall cut the diameter DK in E and be to DK as DK to 2PE, then the point L will be located in the circle tangent to the conic at P and of the same curvature as the conic at the point of contact*). In sequel Newton penned the unfinished proof 'Nam si recta quam minima PR sectionem conicam tangat in P et agatur RQL ipsi $P[M]$ parallela quæ sectioni conicæ occurrat in Q [et L] ac describatur circulus qui rectam QR secet in Q et [N]; erit RL ad $R[N]$ ut QRL ad $QR[N]$ hoc est ad RP^q adeoq; ut PG^q ad DK^q. Coeant jam puncta R et P ut fiat RL æqualis [PG] et RV æqualis $P[M]$ et erit PG ad $P[M]$ ut PG^q ad DK^q [adeoq; PM ad DK ut DK ad PG id est $2PC$]'. (For, if the minimal straight line PR should touch the conic at P, and parallel

Lemma XII.

If in any diameter PG of a conic there be taken, on its concave side, PM equal to the latus rectum pertaining to that diameter, and through the points P, M a circle be described to touch the conic at P, then this circle will have the same curvature as that conic at P.(30)

For, if that circle should intersect the conic in Q, to the diameter PG draw the ordinate QV, complete the parallelogram $QVPR$ and extend RQ till it meets the circle in N, and let the conic's *latus rectum* be R and the transverse diameter(32) T: then $RQ \times RN$ will be equal to RP^2 or QV^2, that is, (from the nature of conics) to the rectangle contained by PV and $R-(R/T)\,PV$;(33) and, on dividing these equals by the equals RQ and PV, there will be RN equal to $R-(R/T)\,PV$. Now let the points P and Q coincide, so that the circle may come to be of the same curvature as the conic at P, and with the quantity PV vanishing the quantities RN and $R-(R/T)\,PV$ come to be PM and R; accordingly, PM is equal to the *latus rectum*.(34) As was to be proved.

Corollary 1. If in any diameter PG of a conic there be taken PM equal to the *latus rectum* and through the point M be drawn the straight line MO parallel to the tangent PR, and if then from the point P there be drawn any other straight line PO meeting the former, MO, in O and in it be taken PL which shall be to the

---

to PM there be drawn RQL to meet the conic in Q and L, and a circle described to cut the line QR in Q and N: there will then be RL to RN as $RQ \times RL$ to ($RQ \times RN$, that is) RP^2, and so as PG^2 to DK^2. Let the points R and P coalesce, so that RL becomes equal to PG and RV to PM, and then PG will be to PM as PG^2 to DK^2, [and hence PM to DK as DK to PG, that is, $2PC$].) As below, adjoining the proportion $PC : PE = PL : PM$ completes the demonstration.

(31) 'parallelogrammum'.

(32) Namely, PG (to which the related *latus rectum* $R\,[= DK^2/PG]$ is as DK^2 to PG^2 by Apollonian definition).

(33) That is, $(R/T).(T-PV) = (R/T).VG = QV^2/PV$ 'per Apollonij *Conica* I, 11–13'.

(34) This result, too, was recorded by David Gregory in his 'adnotata Phys: et Math: Cantabrigiæ 4 Maij 1694' (see note (23) above): 'Sit quævis sectio Conica PDG[:] a quovis puncto [P] ducta ad partes concavas diametro [PG] inqꝫ ea sumpto [PM] æquali lateri recto ad prædictam [PG] pertinente, ducatur Circulus cujus diameter sit recta positione data [PL] ad tangentem [PR] normalis et qui transeat per punctum superius determinatum [M], erit Curvæ datæ Curvitas in [P] eadem cum Curvitate circuli [PML]'. The extended proof of this construction of the 'Problema. Invenire circulum equicurvum cum Sectione Conica ad datum punctum' which Gregory subsequently wrote out on a separate sheet (Codex C45 [Royal Society], none too adequately reproduced in *Correspondence*, **3**: 340–2) is little different in its structure from the present demonstration on which it is based. Newton himself afterwards unobtrusively introduced the result into the *Principia*'s second edition as the basis of an elegant *idem aliter*, styled on Corollary 1 of Proposition X here following, to the proof of his earlier Proposition X (in the published 'Liber primus') that a direct-distance *vis centripeta* induces elliptic motion round its geometrical (and force-) centre; see Appendix 1 below, where (from Add. 3965.19: 744r) we reproduce his autograph drafts of the insertion, and also that of a related *idem aliter* to the following published Proposition XII dealing with the motion in a hyperbola induced by an inverse-square force 'attraction' to a focus.

PL quæ sit ad latus rectum *PM* ut *PM* ad *PO*[,] incidet punctum *L* in circulum qui tangit sectionem conicam in *P* et est ejusdem curvaturæ cum sectione illa in puncto contactus. Patet per Loca plana Apollonij Prop. [6 Lib. II].(35)

Corol. 2. Idem fiet si recta *PL* secet Sectionis diametrum conjugatam *DK* in *E*, sitq3 ad latus rectum ut latus transversum ad 2*PE*: vel (quod perinde est) ad diametrum [conjugat]am *DK* ut *DK* ad 2*PE*. Nam latus transversum 2*PC* est ad 2*PE* ut *PM* ad *PO*.

Corol. 3. Si recta *PE* transit per Sectionis umbilicum *S* erit *PL* ad latus rectum(36) ut latus transversum(36) ad axem principalem sectionis. Nam 2*PE* in hoc casu æqualis est axi principali 2*AC* propterea quod acta ab altero sectionis umbilico *H* linea *HI* ipsi *EC* parallela, rectæ *ES*, *EI* (ob æquales *CS*, *CH*) æquentur adeo ut 2*PE* in Ellipsi summa sit ipsarum *PS*, *PI*, id est (ob parallelas *HI*, *PR* & angulos æquales *IPR*, *HPZ*) ipsarum *PS*, *PH*, in Hyperbola verò earundem differentia[,] et hæc summa vel differentia axi principali æquetur.(37) Axem principalem voco qui in Ellipsi major est axis, quive in Hyperbola interjacet verticibus figuræ.

Corol. 4. Ijsdem positis erit *PL* ad diametrum conjugatam *DK* ut hæc diameter ad axem principalem. Nam DK^q est rectangulum sub latere recto et latere transverso.(38)

Prop. X. Prob. [*II*].(39)

Moveatur corpus in perimetro sectionis Conicæ PQ: requiritur vis centripeta tendens ad punctum quodvis datum S.

(35) Understand the sixth class of 'plane' (rectilinear/circular) loci described by Pappus in his summary of the second book of Apollonius' lost treatise *De locis planis* which is inserted in the preamble to the seventh book of his own later *Mathematical Collection*: 'If a point is given in a circle given in position, and a straight line be drawn through this point, then, if an external point be taken in this line and the square constructed on the straight line going from the given point to a given interior point...be equal to the rectangle comprised by the entire line and that cut off inside [the circle], the external point will lie in a straight line given in position'. (Compare P. ver Eecke, *Pappus d'Alexandrie: La Collection Mathématique...traduite...en français avec une introduction et des notes*, **2** (Paris/Brussels): 500.) The Latin version given by Federigo Commandino in his *Pappi Alexandrini Mathematicæ Collectiones...in Latinum conuersæ, & Commentarijs illustratæ* ([Pesaro, $_1$1588 →] Bologna, $_2$1660): 248–9 is hopelessly confused in its intention meaningfully to render Pappus' obscure sentence. By the very fact of his own present citation of the Apollonian locus—where an equivalent argument from Euclid's *Elements* would have done as well—it will be evident that Newton has anticipated Robert Simson in the correct enodation of the text which the latter set out in his *Apollonii Pergæi Locorum Planorum Libri II restituti* (Glasgow, 1749): 194–201. (We will return to this point in the next volume when we come to reproduce the extant manuscript record of Newton's extensive contemporary researches into—and in partial restoration of—the higher geometrical analysis of the 'ancients'.) In equivalent proof of Newton's corollary—strictly, the converse (also obscurely stated by Pappus) of the Apollonian circle locus—we need only show that, since (by Euclid, *Elements* III, 32 and I, 29 respectively) $\widehat{PLM} = \widehat{MPR} = \widehat{PMO}$, the triangles *PLM* and *PMO* are equiangular

latus rectum PM as PM to PO, the point L will fall on the circle which touches the conic at P and is of the same curvature as that section at the point of contact. This is evident by Apollonius' *Plane Loci*, Book II, Proposition 6.(35)

Corollary 2. The same will happen if the straight line PL shall cut the section's conjugate diameter DK in E, and be to the *latus rectum* as the transverse diameter to $2PE$, or (what is exactly the same) be to the conjugate diameter DK as DK to $2PE$. For the transverse diameter $2PC$ is to $2PE$ as PM to PO.

Corollary 3. If the straight line PE passes through the focus S of the section, there will be PL to the *latus rectum*(36) as the transverse diameter(36) to the section's main axis. For $2PE$ in this case is equal to the main axis $2AC$, seeing that, when the line HI is drawn from the conic's other focus H parallel to EC, the straight lines ES, EI are (because of the equals CS, CH) equal, and hence $2PE$ is in the ellipse the sum of PS and PI, that is, (because HI, PR are parallel and the angles $I\widehat{P}R$, $H\widehat{P}Z$ equal) of PS and PH, but in the hyperbola it is the difference of these, and this sum or difference is equal to the main axis.(37) I call the main axis that which in the ellipse is the major axis, or which in the hyperbola lies between the vertices of the figure.

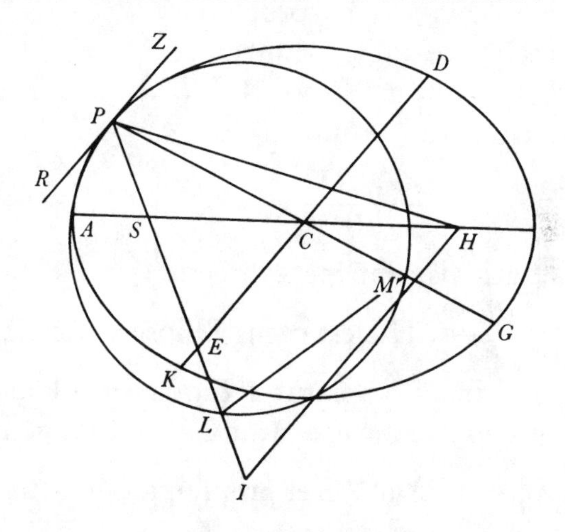

Corollary 4. With the same suppositions, PL will be to the conjugate diameter as this diameter to the main axis. For DK^2 is the product of the *latus rectum* and the transverse diameter.(38)

Proposition X, Problem II.(39)

Let a body move in the perimeter of the conic PQ: there is required the centripetal force tending to any given point S.

---

and therefore similar, whence at once $PL:PM = PM:PO$. The direct consectary that $PL \times PO = PM^2$, constant, underpins the more modern view according to which the straight line $(O) \equiv OM$ is the inverse of the circle $(L) \equiv LMP$ with respect to centre P (on the circle) and 'power' PM^2.

(36) That is, PM and PG $(= 2PC)$ respectively.

(37) Accordingly $PL:PM = (PM:PO$ or) $PC:PE = PG:2AC$.

(38) Namely, $PM \times PG =$ (by Corollary 3) $PL \times 2AC$, from which Newton's asserted proportion $PL:DK = DK:2AC$ derives in immediate consequence.

(39) The unadjusted manuscript reads 'IV' (indicating that Newton originally destined this proposition to follow after 'Prop. IX. Prob. III' in the published 'Liber primus'?).

Stante Lemmatis præcedentis constructione [atq; demissis perpendiculis PF in CE et SY in sectionis tangentem PR](40) est PL (per Cor. 2 Lem. XII) æqu[alis] $\frac{DK^q}{2PE}$. Est et(41) SY æqu[alis] $\frac{PS,PF}{PE}$ ideoq;

$$\frac{PS^q \times PF^q}{PE^q} \times \frac{DK^q}{2PE}$$

est $SY^q \times PL$ et propterea reciproce ut vis centripeta per Cor. 5 Prop. [IX]. Negligatur datum(42)

$$\frac{PF^q, DK^q}{2}$$

et erit vis centripeta directè ut $\frac{PE^{\text{cub}}}{PS^q}$. Hic est casus Ellipseos atq; Hyperbolæ.

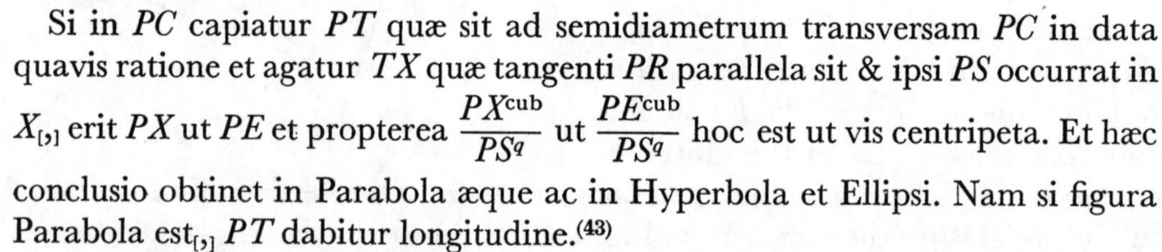

Si in PC capiatur PT quæ sit ad semidiametrum transversam PC in data quavis ratione et agatur TX quæ tangenti PR parallela sit & ipsi PS occurrat in X[,] erit PX ut PE et propterea $\frac{PX^{\text{cub}}}{PS^q}$ ut $\frac{PE^{\text{cub}}}{PS^q}$ hoc est ut vis centripeta. Et hæc conclusio obtinet in Parabola æque ac in Hyperbola et Ellipsi. Nam si figura Parabola est[,] PT dabitur longitudine.(43)

Corol. 1. Si vis centripeta tendit ad centrum sectionis Conicæ, erit vis illa ut distantia corporis a centro. Nam PS et PE in hoc casu evadunt æquales ipsi PC ideoq; $\frac{PE^{\text{cub}}}{PS^{\text{quad}}}$ fit PC.(44)

Corol. 2. Si vis centripeta tendit ad umbilicum sectionis Conicæ ceu centrum erit vis illa reciproce ut quadratum distantiæ corporis ab eodem centro. Nam PE

(40) We insert this necessary preliminary in line with a preceding cancelled opening: 'Sit sectio conica PQ, ejus centrum C, centrum vis centripetæ S, corpus P, tangens Orbis PY, & perpendiculum in tangentem demissum SY. Agantur CE quæ tangenti parallela sit & ipsi SP productæ occurrat in E' (Let the conic be PQ, its centre C, the centre of centripetal force S, the body P, the orbital tangent PY, and the perpendicular let fall to the tangent SY. Draw CE parallel to the tangent and to meet SP produced in E). The accompanying figure now rather confusingly alters its species to be a hyperbola: in our English version we introduce an equivalent ellipse which should render the transition from the preceding Lemma XII visually smoother.

(41) Understand 'ob proportionales $PE.PF :: PS.SY$' (because of the proportionals $PE: PF = PS: SY$), as Newton first added parenthetically in an otherwise insignificantly variant preceding draft.

With the construction of the preceding Lemma standing [and letting fall the perpendiculars PF to CE and SY to the conic's tangent PR],[(40)] PL is (by Corollary 2 of Lemma XII) equal to $DK^2/2PE$. Also[(41)] SY is equal to

$$PS \times PF/PE,$$

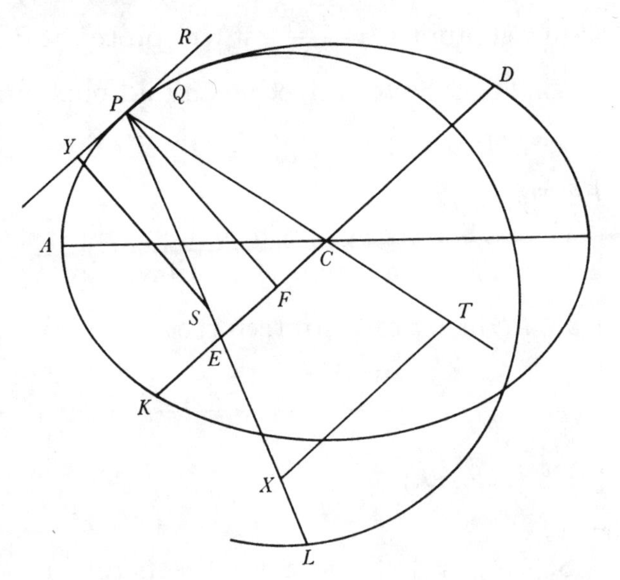

and in consequence

$$(PS^2 \times PF^2/PE^2) \times (DK^2/2PE)$$

is $SY^2 \times PL$ and accordingly (by Corollary 5 of Proposition IX) reciprocally as the centripetal force. Neglect the given[(42)] $PF^2 \times DK^2/2$, and the centripetal force will be directly as PE^3/PS^2. This is the case of the ellipse and hyperbola.

If in PC there be taken PT to be to the transverse semi-diameter PC in any given ratio, and TX be drawn parallel to the tangent PR to meet PS in X, then PX will be as PE and accordingly PX^3/PS^2 as PE^3/PS^2, that is, as the centripetal force. And this conclusion holds true in the parabola equally as much as in the hyperbola and ellipse. For if the figure is a parabola, PT will be given in length.[(43)]

Corollary 1. If the centripetal force tends to the centre of the conic, that force will be as the distance of the body from the centre. For PS and PE in this case prove to be equal to PC, and consequently PE^3/PS^2 comes to be PC.[(44)]

Corollary 2. If the centripetal force tends to the conic's focus as centre, that force will be reciprocally as the square of the distance of the body from the same

(42) For the 'rectangle' $PF \times DK$ is half the area of the conic's circumscribing (or, in the hyperbolic case, inscribed) parallelogram, which is, by [Apollonius, *Conics* VII, 31 =] Lemma XII of the published 'Liber primus', constant in magnitude; compare 1, §1: note (14) preceding.

(43) In a first version of this limiting case of motion in an ellipse or hyperbola where the centre C passes to infinity, Newton did not introduce the auxiliary parallel TX but concluded more directly—if less persuasively—that '...PE dabitur longitudine sed quidem infinita est' (PE will be given in length, though to be sure it is infinite). In revise, the reader's attention is here tactfully diverted from the uncomfortable equivalent consequence that the ratio of PC to PT comes to be infinitely great.

(44) This corollary—ultimately enlarged by a summary proof of the preceding Lemma XII on which its above generalisation to an arbitrarily placed force-centre S is founded—was afterwards elaborated to be a separate *idem aliter* to the published Proposition X which directly derives its equivalent result; see Appendix 1. 1/2 following.

in hoc casu æqualis est semiaxi AC (ut in Corol 3 Lem [XII] ostensum est) ideoq; datur, et propterea $\frac{PE^{cub}}{PS^q}$ est reci[proce] ut PS^q.(45)

Corol 3. Si vis centripeta tendit ad punctum infinite distans, erit vis illa ut PE^{cub}. Nam distantia infinita PS erit ut quantitas data et propterea $\frac{PE^{cub}}{PS^q}$ ut PE^{cub}.(46)

Corol 4. Si conica sectio in circulum abeat et vis centripeta tendat ad punctum in circumferentia datum, erit vis illa recipr[oce] ut quadratocubus distantiæ PS. Nam PE in hoc casu erit reciproce ut PS & propterea $\frac{PE^{cub}}{PS^q}$ reciproce ut PS^{qc}.(47)

---

[2](48) [*pag. 60.*]

Schol.

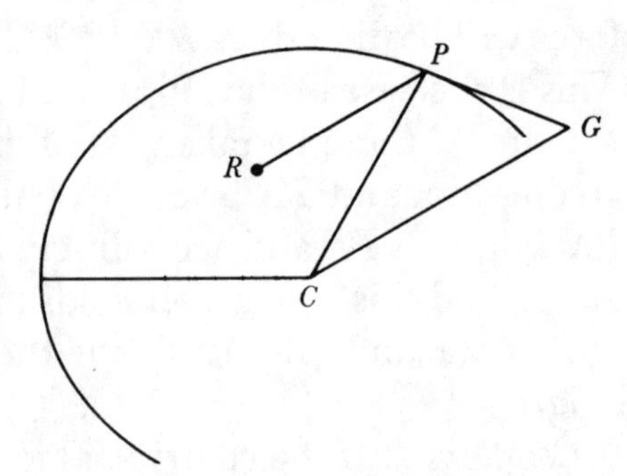

Si corpus [P] moveatur in perimetro datæ cujuscunq; Sectionis Conicæ cujus centrum sit C & requiratur vis centripeta tendens ad punctum quodcunq; datum R: Radio RP parallela ducatur CG Orbis tangenti PG occurrens in G et vis centripeta quæsita (per Prop. [X] et Cor 3 Prop. VII(49)) erit ut $\frac{CG^{cub.}}{RP^{quad.}}$. Hæc est Lex vis centripetæ qua corpus in sectione quacunq; Conica circa centrum quodcunq; movebitur.(50)

---

(45) This corollary, too, was later (see Appendix 1.3 below) similarly adapted to be an *idem aliter* to the published Proposition XII where its result is more directly derived from Newton's prior general measure of centripetal force; subsequently, likewise to embrace the case of elliptical orbit a parallel *idem aliter* was appended by him to the preceding Proposition XI in the *Principia*'s second edition ($_2$1713: 49; compare Appendix 1.3: note (11)).

(46) Though Newton came afterwards (see note (50) below) to introduce the gist of the preceding main theorem into the *Principia*'s second edition as a general scholium to Section III of its first book, for some reason he omitted ever to oust the earlier scholium to Proposition VIII of the published 'Liber primus' where he had in 1685 (see 1, §2: note (99)) erroneously attempted its generalisation, blatantly inconsistent though it henceforward was with the present correct extension.

(47) A result directly deduced from Newton's general measure of centripetal force in Proposition VII of the 1687 *Principia*'s first book (compare page 134 above) and afterwards equivalently derived as the prime corollary to its subsequent generalisation (see § 1.3 preceding) in later editions.

The remainder of the sheet (f. 32) on which the manuscript text here terminates is blank.

centre. For PE in this case is equal to the semi-axis AC (as was shown in Corollary 3 of Lemma XII) and is consequently given, and accordingly PE^3/PS^2 is reciprocally as PS^2.(45)

Corollary 3. If the centripetal force tends to an infinitely distant point, then that force will be as PE^3. For the infinite distance PS will be as a given quantity, and accordingly PE^3/PS^2 will be as PE^3.(46)

Corollary 4. If the conic passes into a circle and the centripetal force should tend to a point given in its circumference, then that force will be reciprocally as the fifth power of the distance PS. For PE in this case will be reciprocally as PS, and accordingly PE^3/PS^2 reciprocally as PS^5.(47)

---

[2](48) [*page 60.*]

Scholium.

If a body P should move in the perimeter of any given conic whose centre be C, and there be required the centripetal force tending to any given point R whatever, parallel to the radius RP draw CG, meeting the orbit's tangent PG in G, and the required centripetal force will then (by Proposition X and Corollary 3 of Proposition VII(49)) be as CG^3/RP^2. This is the law of centripetal force whereby a body shall move in any conic whatever around any centre whatever.(50)

What further novelties, if any, Newton there initially planned to adjoin, and how he proposed to dovetail them into his published text (beginning with Propositions XIV–XVII?), we may only conjecture.

(48) Extracted from Add. 3965.2: 5r/6v. (Other related drafts on these two outer pages of the same 1690's worksheet are reproduced in Appendix 2 below.) This manuscript scholium was first described—and partly printed—by W. W. Rouse Ball in 'A Newtonian Fragment relating to Centripetal Forces', *Proceedings of the London Mathematical Society*, **23**, 1892: 226–31, especially 228–31; and more recently (again not adequately) by A. R. and M. B. Hall in their *Unpublished Scientific Papers of Isaac Newton* (Cambridge, 1962): 13–14, 65–8 (see also our essay review in *History of Science*, **2**, 1963: 125–30, especially 129, note 4). The prospective location in the 1687 *Principia* which is interpolated by us in sequel is that—following on Proposition XVII of the 'Liber primus'—where an abridged version of the opening paragraph was afterwards added in scholium in the second edition; see note (50).

(49) Understand the comparison theorem, initially derived by Newton *ab initio* in Proposition VIII in [1] preceding (see note (23) above) but subsequently deduced as a consectary to the generalised Proposition VII in the revised 'Liber primus' (*Principia*, $_2$1713: 44; compare §1, Appendix 2: note (4) above), according to which in the arbitrary orbit (P) the centripetal force tending instantaneously to C is to that directed at the same moment to R as $CG \times RP^2$ to CG^3, where CG is drawn parallel to RP to meet the tangent at P in G.

(50) With the omission of this final sentence and some slight abridgment of the preceding text, this reshaped summary of Proposition X in [1] preceding was subsequently published 'post Prop. XVII' as a general scholium to Section III of the revised *Principia*'s first book

Si Lex vis centripetæ investiganda sit qua corpus P in orbe quocunq; APE circa centrum quodcunq; datum C movebitur: institui[51] potest calculus per methodum sequentem. Sit AB Abscissa Curvæ propositæ per centrum[52] C transiens sitq; BP ejus Ordinata in dato quovis angulo[53] Abscissæ insistens. Ducatur TPG Orbem tangens in P & Abscissæ occurrens in T et Ordinatæ BP parallela agatur CG tangenti occurrens in G. Fluat Abscissa uniformiter & exponatur ejus fluxio per unitatem. Et si Ordinata $B[P]$ dicatur $v_{[,]}$ vis centripeta

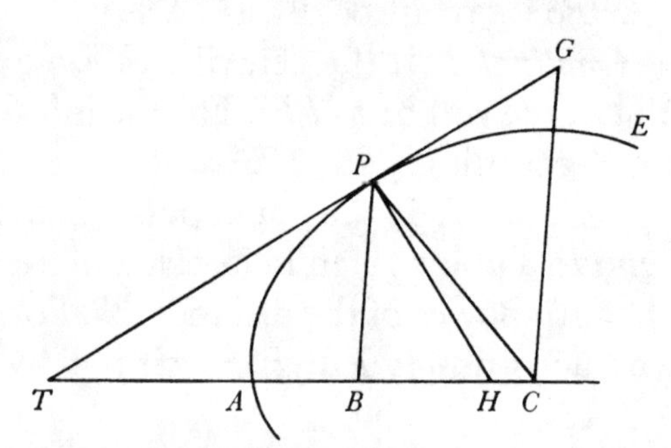

qua corpus P in Orbe APE circa centrum C movebitur erit ut $-\frac{CP \times \ddot{v}}{CG^{\text{cub}}}$.[54] Ut[55] si æquatio ad Curvam sit $ax = v^n$ (ubi a quantitatem quamvis datam et x Abscissam denotat, & n index est dignitatis Ordinatæ v,) Methodus fluxionum

---

($_2$1713: 58). Earlier, in his interleaved copy of the 1687 *editio princeps* (ULC. Adv. b. 39. 1: facing page 60) he had thought of appending thereto the additional observation that 'hæc quantitas $\frac{CG^{\text{cub.}}}{RP^{\text{quad.}}}$ ubi virium centrum R ad convexas orbis partes locatur negativa evadit ac denotat vim centrifugam. Hoc uno Theoremate exhibetur vis qua corpus in Sectione quacunq; conica circum quodcunq; virium centrum moveri potest' (this quantity CG^3/RP^2, when the centre R of force is located on the convex side of the orbit, comes out to be negative and denotes a centrifugal force. In this one theorem there is exhibited the force whereby a body can move in any conic whatever round any centre of force whatever). We may add that two preliminary sketches of the accompanying figure—one having C impossibly far distant from the orbit's geometrical centre (and positioned, along with CG, above RP), the other unnecessarily restricting the conic (P) to be a closed ellipse—are preserved on Add. 3965.17: 637^v.

(51) 'etiam' (also) is deleted.

(52) Once again understand 'virium' (of force).

(53) This replaces 'ad rectos angulos' (at right angles). Although Newton initially (compare Appendix 2.3 below) derived his variant formula for the *vis centripeta* at P tending to C from the Moivrean measure $CP/\rho . CG^3$ (where ρ is the radius of curvature of the orbit at P) which holds only where the ordination angle $\widehat{ABP}$ is right, it is itself, as he now accurately observes, valid where this (constant) angle is arbitrary in size. Rouse Ball's tentative contrary assertion that this 'subsequent alteration to oblique axes was an error' ('A Newtonian Fragment...' (note (48)): 230) is mistaken. In his terms—in which $\widehat{ABP} = \omega$ (fixed), $\widehat{BTP} = \psi$ and $\widehat{AP} = s$, so that $\rho = ds/d\psi$ and $ds/dx = \sin\omega/\sin(\omega-\psi)$—it follows that $\dot{v} = dv/dx = \sin\psi/\sin(\omega-\psi) = \sin\omega\cot(\omega-\psi)-\cos\omega$ and so

$$\ddot{v} = d\dot{v}/dx = \sin\omega . (d\psi/dx) . d(\cot(\omega-\psi))/d\psi = \sin^2\omega/\rho . \sin^3(\omega-\psi),$$

whence the generalised Moivrean measure $CP/\rho . (CG\sin(\omega-\psi))^3$ of the force at P tending to C is $CP/CG^3 . (\sin^2\omega/\ddot{v}) \propto CP . \ddot{v}/CG^3$, as Newton now asserts 'dato quovis angulo [$\widehat{ABP} = \omega$]'. (Rouse Ball's confusion is founded on his incautious omission of the necessary constant, $-\cot\omega$, of integration in 'concluding' from $\ddot{v} \propto d(\cot(\omega-\psi))/dx$ that $\dot{v} \propto \cot(\omega-\psi)$, whence he

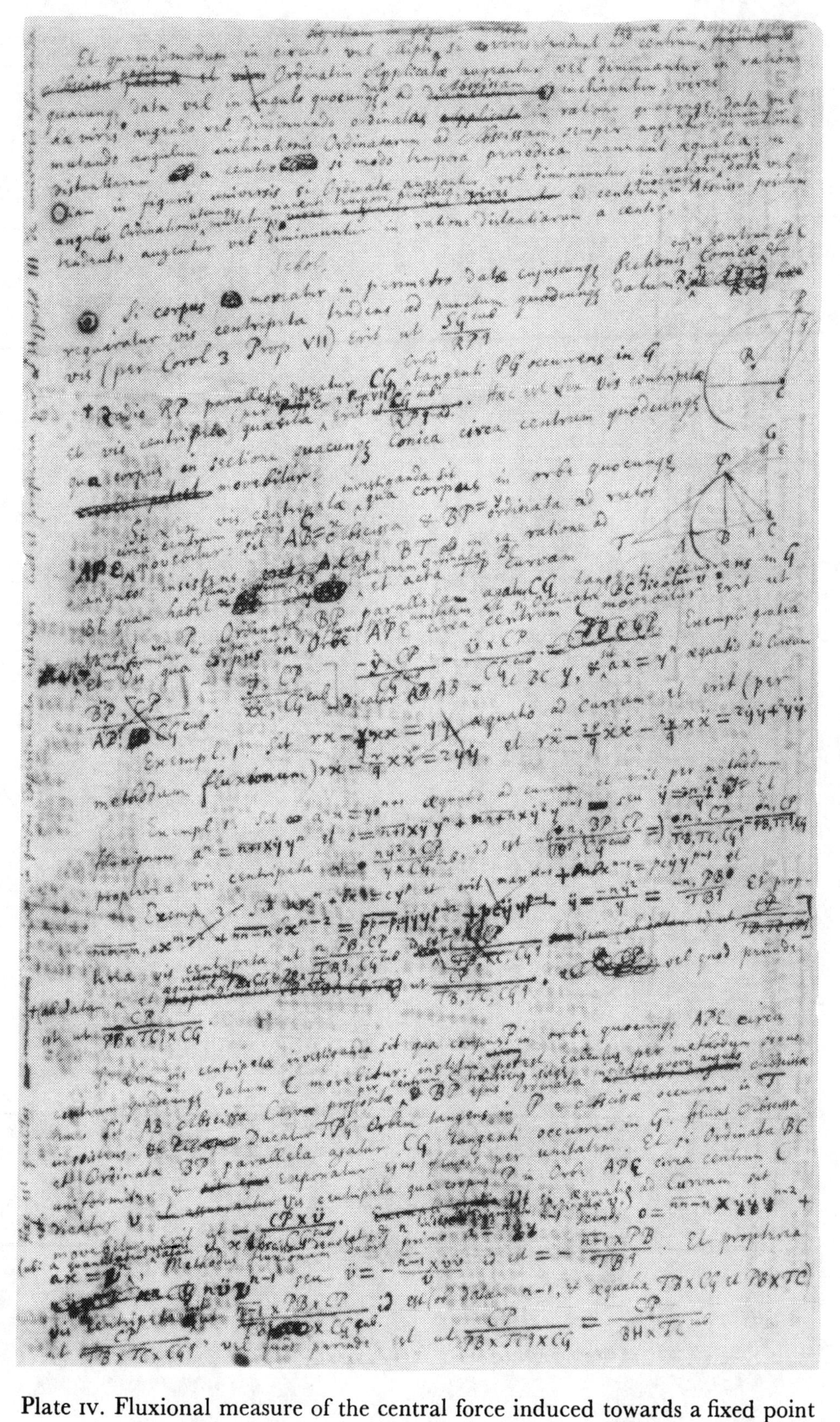

Plate IV. Fluxional measure of the central force induced towards a fixed point in its plane by motion in a given orbit (3, §2.2 and Appendix 2.3).

If investigation needs to be made into the law of centripetal force whereby a body P shall move in any orbit whatever around any given centre C whatever, the computation can(51) be set up by the following method. Let AB be the abscissa of the curves proposed, passing through the centre(52) C, and BP its ordinate, standing at any given angle(53) to the abscissa. Draw TPG tangent to the orbit at P and meeting the abscissa in T, and parallel to the ordinate BP draw CG meeting the tangent in G. Let the abscissa flow uniformly and represent its fluxion by unity. Then, if the ordinate BP be called v, the centripetal force whereby the body P shall move in the orbit APE around the centre C will be as $-CP.\ddot{v}/CG^3$.(54) If, for instance,(55) the equation to the curve be $ax = v^n$ (where a denotes any given quantity and x the abscissa, and n is the index of the

---

falsely infers by identifying the latter erroneous proportion with $\dot{v} = \sin\psi/\sin(\omega-\psi)$ that $\omega = \frac{1}{2}\pi$.)

(54) While Newton could feasibly have justified this extended fluxional formula for the *vis centripeta* from the equivalent generalised Moivrean measure $CP/\rho.CG^3.\sin^3(\omega-\psi)$ (see the previous note), a more direct justification *ab initio* would certainly have been more to his taste. Most simply, perhaps, if we conceive that in some infinitesimal time the orbiting body would, unforced, pass inertially from P along the tangent PG to π but is in the same time 'pulled' along πQ (effectively parallel to PC) to the neighbouring point Q of orbit, then, since—on supposing (with Newton) that $AB = x$ is the base variable (of uniform fluxion $\dot{x} = 1$) and $BP = v \equiv v_x$ is the related ordinate defining the point P—the abscissal increment $Bb = o$ determines the corresponding incremented ordinate $bQ = v_{x+o}$ and thence the oblique tangential distance

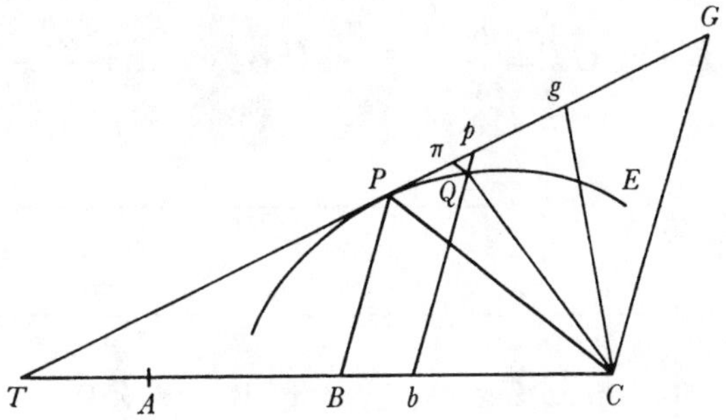

$$Qp = v_{x+o} - v - o\dot{v} = \tfrac{1}{2}o^2\ddot{v} + O(o^3),$$

it follows that the deviation from the tangent in the direction PC is $\pi Q = -\frac{1}{2}\ddot{v}.Bb^2.CP/CG - O(Bb^3)$; accordingly, since on drawing Cg to meet PG at the angle $\widehat{CgG} = \widehat{ABP}$ (so cutting off the triangle CgG similar to the triangle TBP) the time in which this deviation is induced is proportional to

$$2(PCQ) = (P\pi \approx)\, Pp \times Cg.\sin\widehat{ABP} = Bb \times CG.\sin\widehat{ABP},$$

the force acting at P towards C is consequently measured by

$$\lim_{Bb\to 0} (\pi Q/Bb^2 \times CG^2.\sin^2\widehat{ABP}) = -\tfrac{1}{2}\operatorname{cosec}^2\widehat{ABP}.(CP \times \ddot{v}/CG^3),$$

proportional to Newton's stated formula.

(55) Originally 'Exempli gratia' (for instance's sake). The ensuing application of the preceding formula to compute the force inducing motion in the general parabola $ax = v^n$—where, Newton now similarly assumes (see note (57) below), the angle between the coordinates $AB = x$ and $BP = v$ is right—is narrowly based on the parallel computation for the analogous curve $a^n x = [v]^{n+1}$ which is set as 'Exempl. 1' in the preliminary draft earlier on the same manuscript page (f. 5r, reproduced in Appendix 2.3 following).

dabit primò[56] $a = n\dot{v}v^{n-1}$[,] deinde $0 = \overline{nn-n} \times \dot{v}\dot{v}v^{n-2} + n\ddot{v}v^{n-1}$ seu $\ddot{v} = -\dfrac{\overline{n-1} \times \dot{v}\dot{v}}{v}$ id est $= -\dfrac{\overline{n-1} \times PB}{TB^q}$. Et propterea vis centripeta erit ut $\dfrac{\overline{n-1} \times PB \times CP}{TB^q \times CG^{\text{cub.}}}$ id est (ob datam $n-1$, & æqualia $TB \times CG$ et $PB \times TC$) ut $\dfrac{CP}{TB \times TC \times CG^q}$, vel quod perinde est ut $\dfrac{CP}{PB \times TC^q \times CG}$ [hoc est][57] $= \dfrac{CP}{BH \times TC^{\text{cub}}}$.

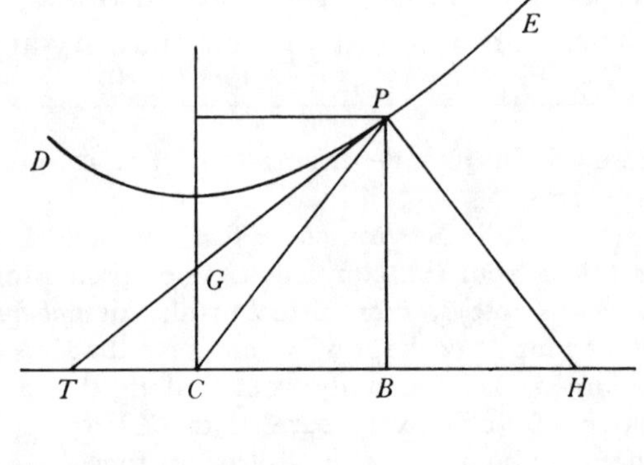

[58][Posito $CB = x$ et $BP = y$ sit] $1 + xx = yy$ [æquatio ad Hyperbolam centro C descriptam, erit per methodum fluxionum] $x = y\dot{y}$.[59] [adeoque]

$$1 = \ddot{y}y + \dot{y}\dot{y} = \ddot{y}y + \frac{xx}{yy}.$$

[sive] $\ddot{y} = \dfrac{yy - xx}{y^3} = \dfrac{1}{y^3}$. [Vis centripeta ad C est ut] $\dfrac{CP}{y^3, BH, CT^3}$.[60] [ubi] $BH = x$. $CT = \dfrac{yy}{x} - x = \dfrac{1}{x}$. [hoc est ut] $\dfrac{CP, xx}{y^3}$.[61]

---

(56) On setting $\dot{x} = 1$; compare note (54).

(57) Understand 'ubi *PH* perpendiculum est ad curvam' (where *PH* is the normal to the curve). From the implied equality $PB \times CG = BH \times TC$, that is, $PB : BH = TC : CG$, it further ensues that the triangles *PBH* and *TCG*, and hence *TBP*, are similar, so necessitating—as in the equivalent 'Exempl. 1' of the preliminary draft on which (see note (55)) the present computation is founded—that the angle $\widehat{ABP}$ at which the coordinates are inclined be right. As before, also, in the special case of the Apollonian parabola $ax = v^2$ when the force-centre is at the focus $(\frac{1}{4}a, 0)$, it may readily be shown (compare Appendix 2.3: note (15)) that $CP = CT = CG \times PB / \frac{1}{2}a$, so that the force at *P* directed to *C* does indeed vary as CP^{-2} in agreement with Proposition XIII of the published 'Liber primus'; no other instance is seemingly of other than purely mathematical interest.

power of the ordinate v), the method of fluxions will yield, first,[56] $a = n\dot{v}v^{n-1}$, then $0 = n(n-1)\,\dot{v}^2v^{n-2} + n\ddot{v}v^{n-1}$ or $\ddot{v} = -(n-1)\,\dot{v}^2/v$, that is,

$$\ddot{v} = -(n-1).PB/TB^2.$$

And accordingly the centripetal force will be as $(n-1).PB \times CP/TB^2 \times CG^3$, that is, (because $n-1$ is given and $TB \times CG$, $PB \times TC$ are equal) as

$$CP/TB \times TC \times CG^2,$$

or, what is the same, as $CP/PB \times TC^2 \times CG$, that is,[57] $CP/BH \times TC^3$.

[58][On setting $CB = x$ and $BP = y$ let] $1 + x^2 = y^2$ [be the equation to a hyperbola described with centre C, and then by the method of fluxions there will be] $x = y\dot{y}$[59] [and therefore] $1 = \ddot{y}y + \dot{y}^2 = \ddot{y}y + x^2/y^2$ [or] $\ddot{y} = (y^2 - x^2)/y^3 = 1/y^3$. [The centripetal force to C is as] $CP/y^3.BH \times CT^3$[60] [where] $BH = x$,

$$CT = y^2/x - x = 1/x;$$

[that is, as] $CP.x^2/y^3$.[61]

---

(58) We append this second illustrative example, plumped out from the bare bones of an isolated calculation on f. 6ᵛ with appropriate editorial interpolations, to document Newton's curious inability (see also Appendix 2.3: note (12) below) accurately and conclusively to apply his present fluxional measure of *vis centripeta* to compute the force induced towards a given point by motion in any species of conic—here a rectangular hyperbola defined by its simplest possible Cartesian equation, with the force-centre taken to be at the origin (its geometrical centre).

(59) That is, $CB = BH$ in the figure (whose manuscript original Newton has by an oversight omitted to letter, but whose points are here designated in the manner of the preceding diagram illustrating the general case of an arbitrary given curve).

(60) Read '$-\dfrac{CP \times \ddot{y}}{CG^{\text{cub}}}$' $(-CP.\ddot{y}/CG^3)$! In a momentary slip which is continued into the sequel Newton forgetfully transposes, not the preceding general fluxional formula, but the computed result of applying it to the parabolic curve $ax = y^n$ which follows.

(61) In correct application of the formula $-CP.\ddot{y}/CG^3$, since here $CG = CT.\dot{y} = 1/y$, this should be $CP.y^{-3}/(y^{-1})^3$, that is, CP simply.

APPENDIX 1. PROPOSITIONS X AND XII OF THE ‘LIBER PRIMUS’ DONE ‘ANOTHER WAY’.(1)

From the original drafts(2) in the University Library, Cambridge

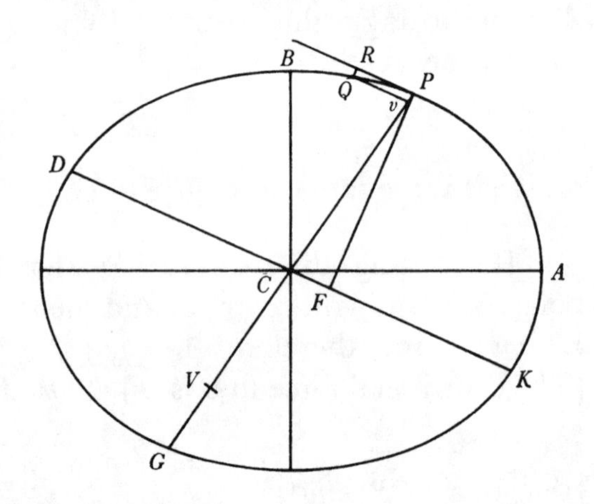

[1] Est $\frac{2CK^q}{CP}=L$ (ex Conicis(3)) ideoq; PF^q, L est ut $\frac{PF^q, CK^q}{C[P]}$. id est (ob datum rectangulum PF, CK(4)) reciproce ut CP. Est $\frac{2CK^q}{CP}=PV$ (ex Conicis(5)) ideoq; PF^q, PV est reciproce ut CP.

Idem aliter.(6)

Si ad Ellipseos diametrum quamvis $PC[G]$ ducatur conjugata diameter DK tangenti PR parallela, & ad hanc demittatur perpendiculum PF & in PG capiatur PV æqualis lateri recto Ellipseos ad ejus latus transversum PG pertinenti: circulus qui Ellipsin tangit in P et diametrum ejus PG secat in V, ejusdem erit curvaturæ cum Ellipsi in puncto P ut ex conicis facile colligitur, ideoq; vis qua corpus in Ellipsi revolvi potest erit in puncto quovis P reciproce ut $PF^q \times PV$

(1) Namely, by way of equivalents to (Corollary 3 of) Proposition IX, Lemma X—here initially (in [1]) assumed to be easily derivable ‘ex conicis’, and then (in [2]) given explicit proof *en passant*—and also Proposition VIII in §2 preceding. (Compare notes (7), (5) and (13) below.) With minimal further verbal rephrasing these variant demonstrations were afterwards inserted by Newton in the *Principia*'s second edition ($_2$1713: 46–7 and 51 respectively).

(2) Add. 3965.19: 744r. The handwriting of this roughly penned sheet suggests a date of composition some time about 1708.

(3) Since by Apollonian definition (*Conics* I, 11–13) the *latus rectum* $L = DK^2/GP$.

(4) A corollary to [Apollonius, *Conics* VII, 31 =] Lemma XII of the published ‘Liber primus’ (*Principia*, $_1$1687: 47; compare 1, §1: note (14) preceding).

(5) Understand ‘ut notum est’, or rather ‘ut facile colligitur’ as Newton will add below in clarification; in fact (as initially derived by him in Lemma X of §2 preceding) this elegant property of a conic that the circle of curvature at a point cuts off from the diameter through the point a length equal to the related *latus rectum* (here $PV = L$) is seemingly his own first discovery.

(6) That is, to Proposition X of the published ‘Liber primus’ (*Principia*, $_1$1687: 48–9). The following demonstration merely elaborates in independent form the particular case of Proposition X in §2 preceding which is outlined in its Corollary 1 (see page 586 above).

per Cor. 3. Prop. VI.(7) hoc est directe ut $\frac{DK^q}{PV}$ per Lem XII,(8) id est ut diameter *PG* vel ut ejus dimidium *PC*. Nam *PV*, *DK* et *PG* continue proportionales esse notum est in conicis. Est igitur vis illa ut distantia *PC*. Q.E.I.

---

[2] *Idem aliter*.(9)

In *PG* cape *Vv* quæ sit ad *vG* ut DC^q ad $PCG_{[,]}$ et quoniam (ex Conicis) est *PvG* ad Qv^q ut *PCG* ad DC^q erit *PvV* æquale Qv^q, adeoq; circulus qui tangit sectionem conicam(10) in *P* et transit per punctum *Q* transibit etiam per punctum *V*. Coeant puncta *P* et [*Q*] et hic circulus ejusdem erit curvaturæ cum sectione conica in *P*, et [*PV*] æqualis erit $\frac{2DC^q}{PC}$. Proinde vis qua corpus *P* in [Ellipsi] revolvi potest, erit reciproce ut $\frac{2DC^q}{PC}$ in PF^q (per Cor 3. Prop VI(7)) hoc [est directe] ut *PC*. Nam $2DC^q$ in PF^q datur per Lem [XII.(8) Q.E.I.]

---

[3] *Idem aliter*.(11)

Inveniatur vis quæ tendit ab Hyperbolæ centro *C*. Prodib[it hæc distantiæ](12) *PC* proportionalis. Inde vero (per Cor. 3(13) Prop. VII) vis [ad umbilicum] *S* tendens erit ut $\frac{PE^{cub}}{SP^q}$, hoc est (ob datam *PE*) recipr[oce ut SP^q. Q.E.I.]

---

(7) That is, Proposition IX in §2 preceding.

(8) Namely, of the published 'Liber primus'; see note (4) above.

(9) A lightly reshaped revise of the previous paragraph in which the conic property $PV = (L \text{ or}) \ 2DC^2/PC$ is now given explicit proof (effectively identical with that of Lemma X in §2 preceding).

(10) This replaces an unnecessarily specific 'Ellipsin'.

(11) Understand to Proposition XII of the published 'Liber primus' (*Principia*, $_1$1687: 51–2) 'ad pag. 52 lin. ult.' Using the result (originally derived by him in Proposition VIII of §2 preceding) that 'vis ad *C*': 'vis ad *S*' $= PC \times PS^2 : PE^3$, where $PE = AC$ is constant, Newton now extends the result of his previous *idem aliter* to determine the *vis centripeta* which sustains orbit in a hyperbola by continuously deviating its tangential motion towards a focus. A similar argument evidently holds *mutatis mutandis* for focally deviated orbit in an ellipse, and to be sure Newton subsequently inserted an analogous *idem aliter* to the preceding published Proposition XI, there (*Principia*, $_2$1713: 49) affirming: 'Cum vis ad centrum Ellipseos tendens, qua corpus *P* in Ellipsi illa revolvi potest, sit (per Corol. 1 Prop. X) ut *CP* distantia corporis ab Ellipseos centro *C*; ducatur *CE* parallela Ellipseos tangenti *PR*: & vis qua corpus idem *P*, circum aliud quodvis Ellipseos punctum *S* revolvi potest, si *CE* & *PS* concurrant in *E*, erit ut $\frac{PE^{cub}}{SP^q}$ (per Corol. 3 Prop. VII,) hoc est, si punctum *S* sit umbilicus Ellipseos, adeoq; *PE* detur, ut SP^q reciproce. Q.E.I.' See the figures on pages 138 and 142 above.

APPENDIX 2. RELATED DRAFT SCHOLIA 'INVESTIGATING THE FORCES BY WHICH BODIES SHALL BE REVOLVED IN PROPOSED ORBITS'.(1)

From an autograph worksheet in the University Library, Cambridge

[1](2) [*pag. 45*?]

Scolium.

Methodus investigandi vires quibus corpora in Orbibus propositis revolventur amplior reddi potest per Propositiones sequentes.

Prop. 1. Si Orbium duorum APD, apd proportionales ordinatæ BP, bp Abscissis proportionalibus AB, ab in datis angulis ABP, abp insistunt; et virium centra S, s in Abscissis similiter locantur: corpora correspondentes orbium partes AP, ap temporibus proportion-

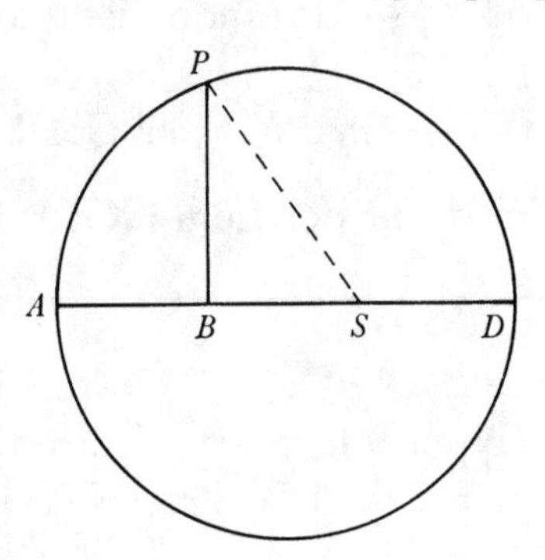

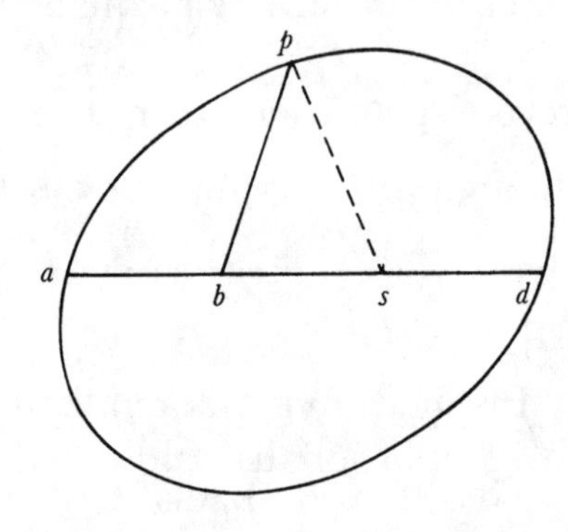

(12) 'corporis a cent[ro]' is cancelled.

(13) See §1, Appendix 2 above. This corollary to the generalised Proposition VII of the published 'Liber primus' is of course (compare *ibid.*: note (4)) a variant later deviation of the comparison theorem deduced from basic principles in Proposition VIII of §2 preceding.

(1) These tentative preliminary versions of scholia which Newton evidently intended to insert at one place or another in Sections II and III of the published 'Liber primus' are roughly written out (with frequent cancellations, interlineations and appended afterthoughts) on the two outside pages (Add. 3965.2: $5^r/6^v$) of a folded sheet whose inner portion contains a variety of calculations and draft verbal revisions destined ultimately for the *Principia*'s 'editio secunda, auctior et emendatior' as it appeared in 1713. (See W. W. Rouse Ball, 'A Newtonian Fragment relating to Centripetal Forces', *Proceedings of the London Mathematical Society*, **23**, 1892: 226–31, especially 227 where the contents of the sheet—mainly relating to the computation of the speed of sound (Book II, Proposition L, Scholium) and of the moon's mean distance from the earth (Book III, Proposition IV), but also hazarding (in the terms of 1, §3: note (160) preceding) the equally faulty new 'ang. med. mot. æquat.' $\psi = T + Y\sin 2(T + \sin^{-1}[\frac{4}{3}e\sin T])$, that is, $T + \frac{1}{4}e^2\sin 2T + \frac{2}{3}e^3\sin T\cos 2T + O(e^4)$, in replacement of the incorrect upper-focus equant mechanism proposed in the (last paragraph of the) scholium to Proposition XXXI in the 1687 'Liber primus'—are briefly reviewed *en passant* as a preliminary to discussing the revised version of [3] following which is reproduced in §2.2 above.) Since the three propositions which are summarily enunciated in [1] (and the further revise of its 'Prop. 1' which is given in [2]) reshape Propositions VII, VIII and Corollary 3 to Proposition IX in §2.1 preceding, and more particularly because standard 'dotted' fluxions (first introduced by Newton in the first version of his 'De Quadratura Curvarum' in about December 1691, as we shall see in the next

alibus describent, et vires centripetæ erunt inter se ut sunt radij seu corporum altitudines *SP*, *sp* directe et quadrata temporum inverse.(3)

Prop. 2. Vis qua corpus quodvis *P* in orbe quocunq; *APB* circa virium centrum *S* revolvi potest, est ad vim qua corpus aliud *P* in eodem orbe eodemq; tempore periodico circa aliud quodvis virium centrum *R* revolvi potest ut contentum sub altitudine primi corporis et quadrato altitudinis secundi corporis $SP \times RP^{\text{quad}}$ ad cubum rectæ *SG* quæ a primo virium centro *S* ad Orbis tangentem *PG* ducitur et secundi corporis altitudini *RP* parallela est.(4)

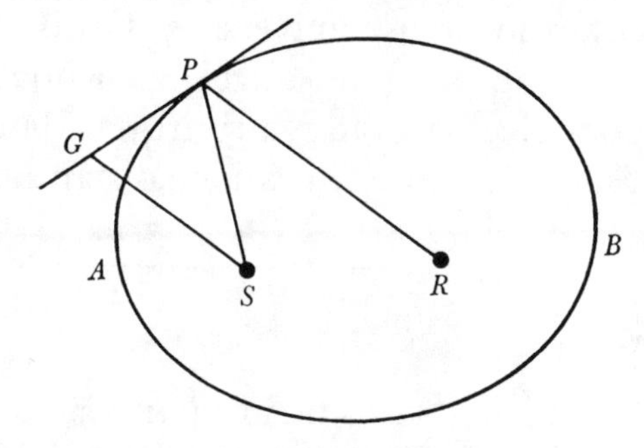

Prop. 3. Vis qua corpus quodvis *P* in Orbe quocunq; *APB* circa quodvis virium centrum *S* revolvi potest est [reciproce] ut cubus rectæ *SG* quæ a centro illo ad Orbis tangentem perpendiculariter demittitur ductus in radium curvaturæ quam Orbis habet ad corpus revolvens & applicatus ad corporis altitudinem *SP*.(5)

---

[2](6) [*pag. 49, lin. ult.*]

[...vi centripeta in centrifugam versa.] Et quemadmodum in circulo vel Ellipsi si vires tendunt ad centrum figuræ in Abscissa positum(7) hæ vires augendo

---

volume) are employed in [3], the date of their composition cannot be earlier than 1692; our assessment of the handwriting style places it as not much later than this *terminus ante quem non*.

(2) Add. 3965.2: 6v. While our conjecture in sequel that this scholium—or rather its elaboration with proofs of its component propositions attached—was scheduled to be introduced into the published 'Liber primus' after Proposition VI (*Principia*, $_1$1687: 44–5) is open to query, it is difficult to see where else it might more appropriately have been inserted.

(3) This enunciation effectively repeats that of 'Cas. 3' of the new Proposition VII in § 2.1 above. By way of the preliminary draft reproduced in [2] following, its content was eventually subsumed into the *Principia's* second edition as an *addendum* to the published scholium to Proposition X of the 'Liber primus'; see note (8) below.

(4) As in the equivalent preceding Proposition VIII in §2.1, Newton in his accompanying figure originally drew *ST* parallel to the tangent at *P* to meet *PR* in *T*, and then here equivalently concluded with the words '...ad cubum rectæ *PT* quam recta *ST* tangenti *GP* Orbis quæ per corpus revolvens *P* ducitur parallela, de altitudine [*PR*] secundi corporis tangentem versus abscindit'. As we have earlier seen (§1, Appendix 2 above), with minimal further alteration this revised 'Prop. 2' was afterwards set by Newton as Corollary 3 to the generalised Proposition which was subsequently published in the *Principia*'s second edition ($_2$1713: 44), there elegantly derived *modo novo* as an immediate consectary to the preceding Corollary 2.

(5) A first explicit enunciation of the 'Moivrean' measure of central force which, as we have previously remarked (see §1.3: note (25) above) is effectively contained in the new Corollary 3 which Newton came, in about 1692, to add to his augmented Proposition VI [→ IX in §2 preceding] in his early revisions of the 1687 *Principia*'s first book. In [3] below he will at once further modify its mathematical form; see note (10).

vel diminuendo ordinatas in ratione quacunq; data vel mutando angulum inclinationis Ordinatarum ad Abscissam, semper augentur vel diminuuntur in ratione distantiarum a centro si modo tempora periodica maneant æqualia: sic etiam in figuris universis si Ordinatæ augeantur vel diminuantur in ratione quacunq; data vel angulus Ordinationis utcunq; mutetur manente tempore periodico, vires ad centrum quo[d]cunq; in Abscissa positum tendentes augentur vel diminuuntur in ratione distantiarum a centro.(8)

---

[3](9)

Si Lex vis centripetæ investiganda sit qua corpus in orbe quocunq; APE circa centrum quodvis C movebitur: sit $AB = x$ Abscissa & $BP = y$ ordinata ad rectos angulos insistens. Cape BT in ea ratione ad BP quam habet $\dot{x}$ ad $\dot{y}$ et acta TP curvam tanget in P. Ordinatæ BP parallelam age CG tangenti occurrens in G et vis qua corpus in Orbe APE circa centrum C movebitur erit ut $\dfrac{\ddot{y}, CP}{\dot{x}\dot{x}, CG^{cub}}$.(10)

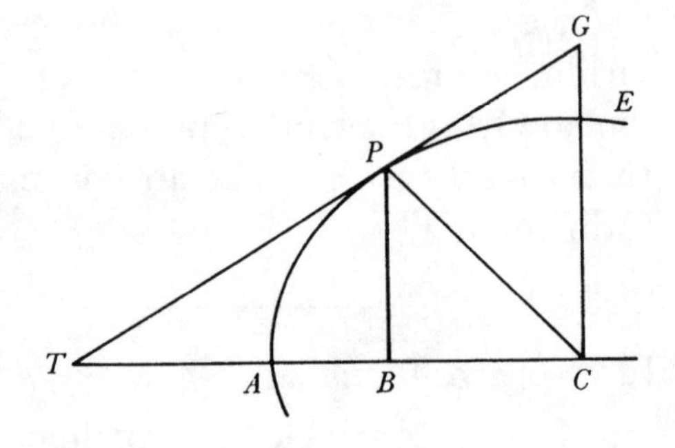

---

(6) Add. 3965.2: 5r (top). The following interpolated location—marking the last line of the scholium to Proposition X in the 1687 'Liber primus'—is the place where this *addendum* was afterwards equivalently introduced in the *Principia*'s second edition; see note (8).

(7) Originally 'jacente', after which Newton first continued: 'et Ordinatim Applicatæ augeantur vel diminuantur in ratione quacunq; data vel in angulo quocunq; dato ad Abscissam inclinentur, vires [semper augentur vel diminuuntur in ratione distantiarum a centro...]'.

(8) This light summary rephrasing of Proposition VIII in §2.1 preceding was, with a few trivial verbal changes, afterwards published at its appointed place in the *Principia*'s second edition ($_2$1713: 47).

(9) Add. 3965.2: 5r (in sequel to the preceding text). We reproduce this preliminary draft of the latter portion of §2.2 preceding both because it clarifies the way in which the Moivrean measure of central force enunciated in 'Prop. 3' in [1] above is here modified (and the result thereafter generalised) and because (see note (12) below) it revealingly documents the failure —when to us, in retrospect, the few manipulations needed to achieve a successful outcome appear so very obvious—of Newton's ensuing attempt to apply his derived fluxional formula to the primary example of motion in an ellipse.

(10) For in direct application of the Moivrean measure announced in 'Prop. 3' in [1] preceding (compare note (5) above) the central force directed to the centre C is proportional to $CP/\rho.Cg^3$, where ρ is the radius of curvature of the orbit at P and Cg is the perpendicular let fall from C to TPG; at once (since the right triangles gGC and BPT are similar) there is $Gg/Cg = BP/BT = \dot{y}/\dot{x}$ and therefore $\rho = (CG/Cg)^3/(\ddot{y}/\dot{x}^2)$, whence Newton's formula ensues on replacing $\rho.Cg^3$ by its equal $CG^3.\dot{x}^2/\ddot{y}$.

Exempl. 1. Sit $rx-\frac{r}{q}xx=yy$ æquatio ad Curvam(11) et erit (per methodum fluxionum) $r\dot{x}-2\frac{r}{q}x\dot{x}=2y\dot{y}$ et $-2\frac{r}{q}\dot{x}\dot{x}=2\dot{y}\dot{y}+2y\ddot{y}$.(12)

Exempl. 1. Sit $a^n x=y^{n+1}$ æquatio ad curvam(13) et erit per methodum fluxionum(14) $a^n=\overline{n+1}\times\dot{y}y^n$ et

$$0=\overline{n+1}\times\ddot{y}y^n+\overline{nn+n}\times\dot{y}^2y^{n-1} \text{ seu } \ddot{y}=\frac{-n\dot{y}^2}{y}=\frac{-n, PB}{TB^q}.$$

Et propterea vis centripeta ut $\frac{n\,PB, CP}{TB^q, CG^{\text{cub}[\text{?}]}}$ id est (ob datum n et æqualia $TB\times CG$ & $PB\times TC$) ut $\frac{CP}{TB, TC, CG^q}$, vel quod perinde est ut $\frac{CP}{PB\times TC^q\times CG}$.(15)

Exemp. 2. Sit $ax^m+bx^n=cy^p$ et erit(14) $max^{m-1}+nbx^{n-1}=pc\dot{y}y^{p-1}$ et

$$\overline{mm-m}, ax^{m-2}+\overline{nn-n}\,bx^{n-2}=\overline{pp-p}, c\dot{y}\dot{y}y^{p-2}+pc\ddot{y}y^{p-1}.\text{(16)}$$

---

(11) An ellipse (P) of *latus rectum* r and main axis q, here referred in standard Cartesian analytical form (compare I: 44) to a principal vertex A as origin of the perpendicular coordinates $AB = x$, $BP = y$.

(12) Doubtless put off by the complexity of these derived fluxional equations, Newton breaks off this example to take up the simpler one which follows. The complications are, however, more apparent than real; for, since at once

$$\dot{y}/\dot{x} = r(1-2x/q)/2y \quad\text{and so}\quad \ddot{y}/\dot{x}^2 = -(r/q+\dot{y}^2/\dot{x}^2)/y = -r^2/4y^3,$$

the force to C is proportional to $-\frac{1}{4}r^2.CP/y^3.CG^3$. As the first of the two chief corollaries, if C is the ellipse's centre $(\frac{1}{2}q, 0)$, then $CT = CA^2/CB = \frac{1}{2}q/(1-2x/q)$ and so $CG = CT.\dot{y}/\dot{x} = \frac{1}{4}qr/y$, whence the central force at P varies as $-(16/q^3r).CP \propto CP$; secondly, if C is the focus $(\frac{1}{2}q(1-e), 0)$ where $e = \sqrt{[1-r/q]}$ is the ellipse's eccentricity, it readily follows that

$$CT = CP/(1-2x/q) \quad\text{and so}\quad CP/CG = (CP/CT).\dot{x}/\dot{y} = y/\tfrac{1}{2}r,$$

from which the central force at P varies as $-(2/r)/CP^2 \propto CP^{-2}$. For all their elegance these variant analytical derivations manifestly lack the direct, visual simplicity of their published equivalents in Propositions X and XI of the 'Liber primus'—a factor which would have had its weight with Newton if he had persevered with this initial 'Exempl. 1'.

(13) A general parabola, symmetrical round the vertex A for odd integers n, but mirror-symmetric round the diameter AB for even ones.

(14) Assuming that $\dot{x} = 1$ for simplicity.

(15) When $n = 1$ and the curve is the Apollonian parabola defined by $ax = y^2$, it follows, where (as shown in Newton's figure) C is a point in the diameter AB, that

$$CG/CT = (\dot{y}/\dot{x} \text{ or}) \tfrac{1}{2}a/y,$$

and hence that the central force at P instantaneously directed to C varies as CP/CT^3; in particular, if C is the focus $(\frac{1}{4}a, 0)$, then $CP = \sqrt{[y^2+(\frac{1}{4}a-x)^2]} = \frac{1}{4}a+x = CT$ and therefore the force to C varies as CP^{-2}. This variant derivation of the result more directly established in Proposition XIII of the published 'Liber primus' again lacks its heuristic simplicity.

(16) The complexities of this more general algebraic curve prove quickly formidable, and Newton again breaks off his calculation after computing its related first- and second-order fluxional equations.

§3. THE PRINCIPAL LEMMATICAL RIDERS TO THE REVISED *PRINCIPIA* RESET AS A SEPARATE INTRODUCTORY GROUP.(1)

[1](2) LEMMATA GENERALIA

Lem. I. Lem. II. &c [*Lem. XI.*](3)

Lem. XII.

Parallelogramma omnia circa datam Ellipsin descripta ad quorū latera diametri conjugatæ terminantur sunt inter se æqualia. Idem intellige de Parallelogrammis in Hyperbola circa diametros ejus descriptis.(4)

Lem. XIII.

Recta quæ per centrum sectionis conicæ ducitur & cuilibet ejus tangenti parallela est, de recta quæ ducitur a puncto contactus ad umbilicum sectionis, abscindit versus punctum contactûs longitudinem semissi axis principalis æqualem.(5)

Lem. XIV.

In Ellipsi et Hyperbola(6) *axis principalis est ad diametrū quamvis ut diameter illa ad rectam quæ per umbilicum ad Curvam*(7) *utrinq͡ ducitur ac diametro illi parallela est.*(8)

---

(1) Because of their generality and repeated application it is understandable that Newton should, in revising its 1687 text, consider gathering together the various preparatory lemmas in his *Principia* as a single collective preliminary to its main text. In [1] and [2] following we reproduce two such related schemes, each with their own idiosyncracies and novelties, and in [3] thereto adjoin the rough implementation on the same manuscript sheet of a new 'Lem. [XX →] XIII' which (in generalisation of Lemma XII in §2.1 preceding) constructs the circle of curvature at an arbitrary point on a general algebraic curve. In the outcome nothing of this appeared in print. Though Abraham de Moivre in a letter to Johann Bernoulli on 6 July 1708 in which he announced that 'Le livre de M. Newton *Philosophiæ naturalis principia mathematica* se réimprime à Cambridge' went on to remark that 'Il y aura beaucoup de changemens: les lemmes du commencement sont changés pour le mieux; je parle seulement des lemmes parce qu'il n'y a que cela que j'aye vu, M. Newton m'ayant seulement montré le commencement des corrections qu'il a faites' (Karl Wollenschläger, 'Der mathematische Briefwechsel zwischen Johann I Bernoulli und Abraham de Moivre' (*Verhandlungen der Naturforschenden Gesellschaft in Basel*, **43**, 1933: 151–317): 250), inertia prevailed and in the revised 'editio secunda, auctior et emendatior' of the *Principia* (and in all subsequent editions thereafter) the lemmas of the 1687 *editio princeps* all re-appear unchanged in location and with only minimal alterations in their verbal text.

(2) Add. 3965.17: 635r/635v, briefly described by I. B. Cohen in his *Introduction to Newton's 'Principia'* (Cambridge, 1971): 171–2. This first tentative scheme collects together the introductory Lemmas I–XI 'De methodo Rationum primarum & ultimarum' and the remainder of those in the published 'Liber primus' with the exception of its geometrical Sections IV/V, augmenting these with new Lemmas XIII/XIV (intended, we conjecture, to preface a variant proof of Proposition XI on the pattern of the intended *idem aliter* thereof reproduced in §1.4 preceding) and XIX/XX (a lightly revised repeat of Lemma XII in §2.1 on the circle of curvature at a point of a conic, and its extension to a general algebraic curve), and further

Translation

[1][2] GENERAL LEMMAS

Lemma I. Lemma II. ... [Lemma XI.][3]

Lemma XII.

All parallelograms described about a given ellipse with conjugate diameters terminating at their sides are equal to one another. Understand the same for parallelograms described in a hyperbola about diameters of it.[4]

Lemma XIII.

A straight line which is drawn through the centre of a conic, parallel to any tangent of it, cuts off from the straight line drawn from the point of contact to a focus of the section a length, in the direction of the point of contact, equal to the main semi-axis.[5]

Lemma XIV.

In the ellipse and hyperbola[6] *the main axis is to any diameter as that diameter to the straight line drawn through a focus either way to the curve,*[7] *parallel to that diameter.*[8]

---

adjoining thereto Lemma II (on fluxional 'moments') from the second book and Lemma V (on the theory of interpolation by advancing finite differences) from the third. As Newton told David Gregory during the latter's visit to Cambridge early in May 1694, 'In Editione nova *Philosophiæ*...Sectiones IV et V (Lib. I] eximuntur et tractatûs separati [de Veterum Geometria] partes fiunt....In hoc tractatu genuinum Veterum institutum explicatur, ...Euclid[is] Porismatum liber, Appollonij libri deperditi reliquique Veterum explicabuntur ex illis quæ a Pappo alijsque de ijs dicuntur quæque a nemine hactenus intellecta [et] pars erit de inventione orbium ex dato umbilico vel etiam neutro umbilico dato, quæ Sect: IV et V constituunt' (Gregory Codex C42 [now in the University Library, Edinburgh]; compare H. W. Turnbull's English translation in *The Correspondence of Isaac Newton*, **3**, 1961: 384, 385). Elsewhere (Codex C44 [now in the Royal Society], reproduced in *Correspondence*, **3**: 335) Gregory noted that Newton then had 'duo aut 3 folia quibus proprietates coni sectionum (quatenus sint Orbitæ planetarum vel Cometarum) facillime ex facilibus prop: Generalibus deducuntur'. We will return to examine the detail of this unpublished 'Geometriæ Liber primus' of Newton's when we reproduce its several variant manuscript drafts in the next volume.

(3) These are evidently (compare [2] following) the eleven introductory lemmas of Section I of the published 'Liber primus' (*Principia*, $_1$1687: 26–34; see also pages 106–18 above) 'cujus ope sequentia demonstrantur' (*ibid.*: 26).

(4) An unchanged repeat of the equivalent Lemma XII in the 'Liber primus' (*Principia*, $_1$1687: 47; compare page 138 above).

(5) Newton explicitly enunciates a property of the central conic on which the proofs of Propositions XI and XII in the published 'Liber primus' (*Principia*, $_1$1687: 50–2; compare pages 140–4 preceding) are founded. Subsequently, he had contemplated introducing its elliptical case as a preparatory 'Lemma 1' to a recasting of the latter portion of the following Proposition XVII; see §1.5: note (42).

(6) Newton began forgetfully to write '*In omni secti[one Conica]*' (*In every conic*) but hastily amended his enunciation to exclude the parabola (which lacks a finite centre).

(7) Originally '*ad chordam quæ per nodum ad figuram*' (*to the chord...through a 'node'...to the figure*). Newton also tentatively introduced this novel *terminus technicus* for a (conic) focus into the two following lemmas; see note (9).

Lem XV.

Latus rectum Parabolæ ad verticem quemvis pertinens est quadruplum distantiæ verticis illius ab umbilico figuræ.(9)

Lem. XVI.

Perpendiculum quod ab umbilico Parabolæ ad tangentem ejus dimittitur, medium est proportionale inter distantias umbilici a puncto contactus & a vertice principali figuræ.(9)

Lem. XVII.

Nulla extat Figura Ovalis cujus area rectis pro lubitu abscissa possit per æquationes numero terminorū ac dimensionum finitas generaliter inveniri.(14)

Lem. XVIII. vide pag. 203.(11)

[*Si describantur centro S circulus quilibet AEB...ad lineam PS.*]

Lem. XIX.

Si in Sectionis conicæ diametro quavis PG ad partes concavas sumatur longitudo PM æqualis lateri recto ad diametrū illam pertinenti & per puncta P M describatur circulus qui

(8) As we have already observed (see §1.4: note (35) above) Newton had previously set this focal property of a central conic as 'Lemma 2' of 'A Demonstration that the Planets by their Gravity towards the Sun may move in Ellipses' which he passed on to John Locke about March 1690 (when the secretary transcript now preserved as Bodleian MS. Locke. c. 31: 101^r–104^r was made). Subsequently (compare §1.4: note (33)) he had equivalently employed it in a summary of the 'Demonstration' which he had it in mind to append as an *idem aliter* to the published Proposition XI 'ad [*Principia*, $_1$1687:] p. 51 l. 11' where its result is more directly derived, and no doubt a similar ultimate purpose affords the present Lemma its *raison d'être*. We may here fitly adjoin its 'Lockean' proof, adapted to the latter context: 'Let *APB*[*L*] be y^e^ Ellipsis [or Hyperbola], *AB* its long Axis, [*S*, *H*] its foci, *C* its centre, *P*[*L*] ye line drawn through its focus [*S*] & *C*[*M*] its [semi] diameter parallel to *P*[*L*], & ... draw [*HO*] parallel to [*LS*]*P* & cutting the Ellipsis in [*O*]. Joyn *P*[*O*] cutting *C*[*M*] in [*N*] & draw *P*[*Z*] wch shall touch the Ellipsis in *P* & cut *C*[*M*] produced in [*Z*], & *C*[*N*] will be to *C*[*M*] as *C*[*M*] to *C*[*Z*], as has been shewed by [Apollonius, *Conics* I, 37]. But *C*[*N*] is y^e^ semisumm of

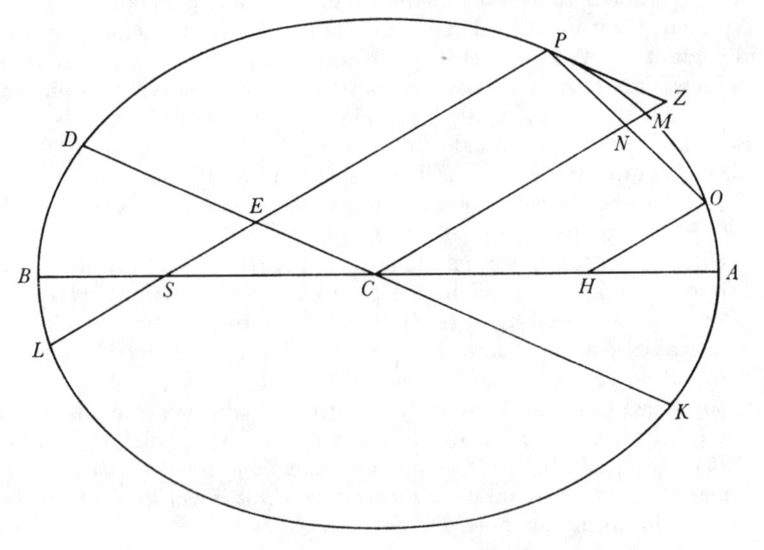

Lemma XV.

In a parabola the latus rectum pertaining to any vertex is four times the distance of that vertex from the figure's focus.(9)

Lemma XVI.

The perpendicular let fall from the focus of a parabola to its tangent is a mean proportional between the distances of the focus from the point of contact and from the principal vertex of the figure.(9)

Lemma XVII.

There exists no oval figure whose area cut off at will by straight lines might generally be found by equations finite in the number of their terms and dimensions.(10)

Lemma XVIII. see page 203.(11)

[*If there be described with centre S any circle AEB...to the line PS.*]

Lemma XIX.

If in any diameter PG of a conic there be taken, on its concave side, a length PM equal to the latus rectum pertaining to that diameter, and through the points P and M be

---

$[S]P$ & $[HO]$ that is of $[S]P$ & $[LS]$, & therefore $2C[N]$ is equal to $P[L]$. Also... (by y^e^ foregoing Lemma [1]) $2C[Z]$ is equal to $[2PE=]AB$. Wherefore $P[L]$ is to $[2CM]$ as $[2CM]$ is to AB. W. w. to be Dem.' ([Bodleian MS. Locke. c. 31: 104^r^ =] ULC. Add. 3965.1: 2^r^/2^v^; compare A. R. and M. B. Hall, *Unpublished Scientific Papers of Isaac Newton* (Cambridge, 1962): 297). Alternatively, where $\widehat{ACM} = \widehat{ASP} = \phi$ and $SC/BC = e$, it follows more shortly from the central conic's defining polar equation that

$$SP/BC = (1-e^2)/(1-e\cos\phi) \quad \text{and} \quad LS/BC = (1-e^2)/(1+e\cos\phi),$$

while $(CM.\cos\phi)^2/BC^2+(CM.\sin\phi)^2/BC^2(1-e^2) = 1$, so that

$$\tfrac{1}{2}LP/BC = CM^2/BC^2 = (1-e^2)/(1-e^2\cos^2\phi).$$

(9) Unchanged repeats of Lemmas XIII and XIV respectively of the published 'Liber primus' (*Principia*, $_1$1687: 53; compare page 204 above). In both (compare note (7) preceding) Newton throughout replaced the technical term 'umbilicus' by 'nodus' before returning to the prior, printed form.

(10) Lemma XXVIII of the published 'Liber primus' (*Principia*, $_1$1687: 105 ff; compare pages 302–8 above), here enunciated without variant.

(11) This rider (*Principia*, $_1$1687: 203–4) to Propositions LXXIX–LXXXI of the published first book is reproduced in 1, §2, Appendix 4.1 above (see page 213).

sectionem conicam tangat ad punctum P: hic circulus eandem habebit curvaturam cum sectione conica ad punctū illud P.(12)

Lem XX.

Invenire Curvarum curvaturas ad puncta data.(13)

Lem. XXI.

Lineam curvam generis Parabolici per data quotcunq; puncta describere.(14)

Lem. XXII.

Momentum Genitæ &c(15)

---

[2](16) LEMMATA GENERALIA.

Lemmata undecim sequentia spectant ad methodum rationum primarum et ultimarum: reliqua quatuor ad alias methodos quæ usui sunt in Libris sequentibus.

[*Lem. I.* *Lem II.* &c *Lem. XI*]

Lem. XII.

Si in Sectionis conicæ diametro quavis PG... ad punctum illud P.

Lem XIII.

Invenire Curvarum curvaturas ad puncta data.

Lem. XIV.

Lineam curvam generis Parabolici per data quotcunq; puncta describere.

Lem. XV.

Momentum Genitæ &c

---

(12) The new Lemma XII in §2.1 preceding (see page 582), here faithfully transcribed with two small verbal clarifications.

(13) This unprecedented extension of the previous Lemma XIX is elaborated in [3] below.

(14) A lightly amended rephrasing of the enunciation of Lemma V of the published third book (*Principia*, $_1$1687: 481–3, reproduced in IV: 70–3).

(15) Lemma II of the published 'Liber secundus' (*Principia*, $_1$1687: 250–3, reproduced in

described a circle which shall touch the conic at the point P, this circle will have the same curvature as the conic at that point P.(12)

Lemma XX.

To find the curvatures of curves at given points.(13)

Lemma XXI.

To describe a curve of parabolic kind through any number of given points.(14)

Lemma XXII.

The moment of a begotten [quantity]....(15)

---

[2](16) GENERAL LEMMAS.

The eleven following lemmas have regard to the method of first and last ratios, the remaining four to other methods which are of use in the following books.

Lemma I. Lemma II. ... [Lemma XI.]

Lemma XII.

If in any diameter PG of a conic...to that point P.

Lemma XIII.

To find the curvatures of curves at given points.

Lemma XIV.

To describe a curve of parabolic kind through any number of given points.

Lemma XV.

The moment of a begotten [quantity]....

IV: 521–5). The following scholium (*ibid.*: 253–4; see IV: 524, note (11)) would presumably be appended here also.

(16) Add. 3965.17: 636v/635v. This conservative revision of the previous scheme retains only its opening Lemmas I–XI 'de methodo rationum primarum et ultimarum'—here cited once more without change from their equivalents in Section I of the published 'Liber primus'—and its (renumbered) terminal Lemmas XIX–XXII. Into these (see notes (3) and (12)–(15) above respectively) we do not here need to go into detail.

[3][17] *Lem XIII.*

Invenire curvarum curvaturas.

Cas. 1. Sit MPN curva proposita, PT tangens ejus, P punctum contactus, MN recta quævis tangenti parallela, et PG recta quævis alia priores secans in P ac G: et si curva MPN sit sectio conica secans rectam PG in P et D et rectam MN in M et N cape PX ad PD ut NGM ad PGD.[18]

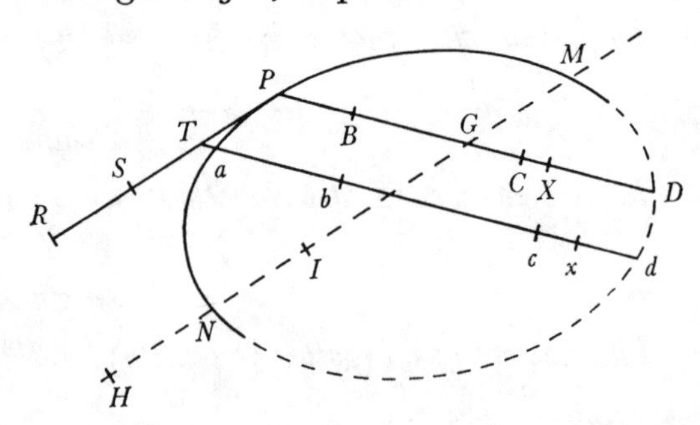

Cas. 2. Si curva MPN sit trium dimensionum secans rectam PG in P, B, D et rectam MN in M, N, H, et tangentem PT in R: cape PX ad $\dfrac{PB\times PD}{PR}$ ut $GM\times GN\times GH$ ad $GP\times GB\times GD$.

Cas. 3. Si curva sit quatuor dimensionum secans rectā PG in P, B, C, D et rectam MN in M, N, H, I et tangentem PT in R et S: cape PX ad

$$\frac{PB\times PC\times PD}{PR\times PS}$$

ut $GM\times GN\times GH\times GI$ ad $GP\times GB\times GC\times GD$. Et sic deinceps in infinitum.

Deniq; per puncta P et X describatur circulus qui tangat Curvam in P et hic circulus eandem habebit curvaturam cum figura MPN ad punctum P.

[Nam in casu generali] sec[et] recta PG curvam MPN in pleno punctorum numero P, B, C, D, &c. Deinde per hujus punctum quodvis G ducatur recta quævis alia MN quæ tangenti PT parallela sit & secet curvam in eodem punctorum numero M, N, H, I &c[,] & secet etiam Tangens PT Curvā in eodem punctorum numero R, S &c demptis duobus. Nam punctum contactus pro duobus habendum est. [Ducatur etiam recta Td ipsi PG parallela, quæ curvam secet in pleno punctorum numero a, b, c, d &c ut et circulum aPX, ejusdem curvaturæ cum curva ad P, in x. Fac

$$GM\times GN\times GH\times GI\,\&c.\,GP\times GB\times GC\times GD\,\&c :: d.e$$

(17) Add. 3965.17: 636r/636v. The following text of Newton's elaboration of his previously enunciated Lemma [XX →] XIII—in which (see note (18)) the preceding Lemma [XIX →] XII is now comprehended as its primary particular case—has been fashioned by us, with some silent editorial interchange of upper- and lower-case letters to render its citation of points in the accompanying figure consistent, by blending together in inverse sequence a preliminary sketch (on f. 636r) of its general proof and a more finished following induction (on ff. 636r–636v) of its pattern of construction from specified individual cases.

[3][17]

Lemma XIII.

To find the curvatures of curves.

Case 1. Let MPN be the curve proposed, PT its tangent, P the point of contact, MN any straight line parallel to the tangent, and PG any other straight line cutting the previous ones in P and G: then, if the curve MPN be a conic intersecting the straight line PG in P and D and the straight line MN in M and N, take PX to PD as $GM \times GN$ to $GP \times GD$.[18]

Case 2. If the curve MPN be of three dimensions, intersecting the straight line PG in P, B, D, the straight line MN in M, N, H and the tangent PT in R, take PX to $PB \times PD/PR$ as $GM \times GN \times GH$ to $GP \times GB \times GD$.

Case 3. If the curve be of four dimensions, intersecting the straight line PG in P, B, C, D, the straight line MN in M, N, H, I and the tangent PT in R and S, take PX to $PB \times PC \times PD/PR \times PS$ as $GM \times GN \times GH \times GI$ to

$$GP \times GB \times GC \times GD.$$

And so on in turn indefinitely.

Finally, through the points P and X describe a circle to touch the curve at P and this circle will have the same curvature as the figure MPN at the point P.

[For in the general case] let the straight line PG cut the curve MPN in the full number of points P, B, C, D, ... Then through any point G in this draw any other straight line MN to be parallel to the tangent PT and intersect the curve in the same number of points M, N, H, I, ..., and let also the tangent PT cut the curve in the same number of points R, S, ... less two (for the point of contact is to be considered as two). [Draw also the straight line Td parallel to PG to cut the curve in the full number of points a, b, c, d, ... and also the circle aPX, of the same curvature as the curve at P, in x. Make

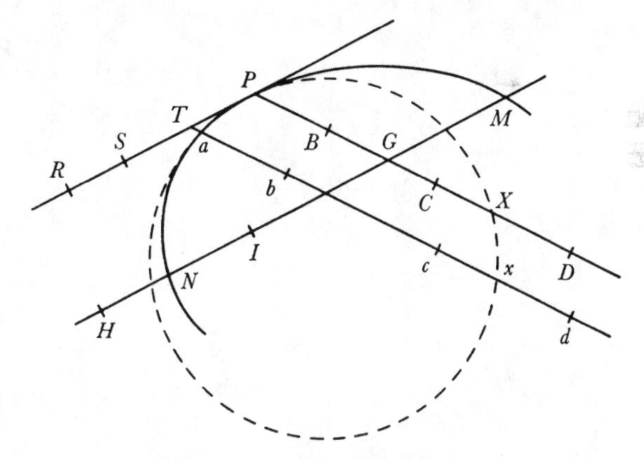

$$GM \times GN \times GH \times GI \ldots : GP \times GB \times GC \times GD \ldots = d : e$$

(18) Whence, as Newton had first announced in Lemma XII of §2.1 preceding, it follows that PX is the conic's *latus rectum* pertaining to the diametral *latus transversum* through P. Whether he still had it in mind to preface its present generalisation by a separate Lemma [XIX →] XII likewise repeating its enunciation (see note (12) above) is not clear.

et erit] $d.e :: TP^q, TR, TS$ [&c]. Ta, Tb, Tc, Td [&c sive]

$$\frac{TP^q}{Ta} = \frac{d, Tb, Tc, Td\,[\&c]}{e, TR, TS\,[\&c]} = Tx.^{(19)}$$

[Coincidat jam recta Td cum recta PD et fit] $\dfrac{d, PB, PC, PD\,[\&c]}{e, PR, PS\,[\&c]} = PX.$

---

(19) For, by Newton's generalised intercept theorem (see IV: 358, note (18)) or by Apollonius, *Conics* III, 17/18 in the primary conic case, the parallels PT, MN and PG, Tx cut the given algebraic curve—'in pleno numero punctorum', it is understood—such that

$$TP^2 \times TR \times TS \times \ldots : Ta \times Tb \times Tc \times Td \times \ldots$$
$$= GM \times GN \times GH \times GI \times \ldots : GP \times GB \times GC \times GD \times \ldots,$$

and there will be] $d:e = TP^2 \times TR \times TS\ldots : Ta \times Tb \times Tc \times Td\ldots$, that is, $TP^2/Ta = (d/e).\ Tb \times Tc \times Td\ldots/TR \times TS\ldots = Tx$.(19) [Now let the straight line Td coincide with the straight line PD and there comes]

$$(d/e).PB \times PC \times PD\ldots/PR \times PS\ldots = PX.$$

---

while in the circle of curvature PXx at P (by the particular case which is Euclid, *Elements* III, 36) the corresponding ratio $TP^2: Ta \times Tx$ is unity.

INDEX OF NAMES